QUANTUM MECHANICS AND MIND

2

QUANTUM MECHANICS AND MIND

The Alternative Realist Thesis on Physics, Quantum Mechanics, and the Critique of the Philosophy of Mind

Emre Asena

DEDICATION

To Martin Hayward, without whose life-saving surgical intervention in 2003 this work would never have reached the light of day.

Eternal gratitude.

CONTENTS

INTRODUCTION TO QUANTUM MECHANICS AND MIND

Subject & Object

We begin with the perennial **object-subject relation**. The relation is directly pertinent to the mind-world conundrum. Two principal approaches bestride the relation: On the one hand, **materialism** seeks to collapse subject to object, or *nous* to 'matter in space'[1]… to the material brain. Therein, matter is asserted the primary existent. In **eliminative materialism**, the subject does not exist: 'illusory' nous is eliminated from brain, world, and ontology. Alternatively, other materialist approaches posit nous as epiphenomenal to matter and brains, in the sense that magnetism is epiphenomenal to electricity. Alternatively, emergence asserts nous as supervenient to brains; arising from the combined sum of constituents comprising brains; perhaps as a state greater than the sum of the constituents.

While **epiphenomenalism** and **emergence** theory appear to admit to its onticity, neither permit to nous causal power vis-à-vis the brain and world. This is in order to save the much-abused conservation principle: a demand central to materialist philosophy of mind[2]. Yet, mind neutered of causal power vis-à-vis brain and world is as good as eliminated. Indeed, all materialist approaches to nous constitute implicit eliminative theories. Thus, in epiphenomenalism and emergence, core eliminative materialism is merely obscured.

Hence, in materialist philosophy of mind, the subject supposedly collapses to object… to brains… to matter. Effectively, the subject, or nous, constitutes an *illusion*: *not* part of real existence; not part of the causality-chain, given the causal neutering of nous.

On the other hand, the popular opposition to materialism is **idealism**, exemplified by Bishop Berkeley's approach and by certain quantum-mind theories inspired by Wigner's 'consciousness causes collapse'. Therein, object collapses to subject and nous constitutes the primary ontology: The world or 'matter' is reduced to the mere phantomatic contrivance of nous… wherein there exist no independent stand-alone world or object distinct from subject, save as the 'illusion' of nous; perhaps as Eddington's 'mind-stuff'. Effectively, idealism constitutes the inversion of the eliminative materialist contention.

There is an approach to object-subject relations that accords ontic reality to nous *and* world. As such, it is no more implausible and, contrary to prejudgement, much less problem-ridden than the three-hundred year-long materialist failure; a failure signalled by the profusion of materialist theories within academic philosophy of mind, none of which resolve how consciousness and intentionality is generated (or eliminated) by matter and brains, and per the fact that all hard problems of consciousness remain hitherto unsolved. This third approach is, of course, **mind-brain dualism** from Plato to Descartes. Therein, the object-subject relation is typically treated as a given, typically without explanation[3], as opposed to endless futile attempts to collapse one to the other term in materialism or idealism.

If philosophy since Plato constitutes the two-millennia long compilation of footnotes *against* Plato… wherein philosophy is ironically constituted as *philosophy against philosophy*…in its turn, academic philosophy of mind constitutes the three-century long materialist failure and apologetics arrayed against Cartesian dualism; constituted as ironic apagogic *philosophy of mind, against mind*.

Through the critique of the philosophy of physics and of quantum mechanics, our Alternative Realist thesis and interpretation seeks the repudiation of materialist academic philosophy of mind. Through physics, our thesis seeks the recovery of philosophy *and* mind.

The Relativity of Simultaneity Rescues Object-Subject Relations

In Einstein's Special Relativity (SR), **the relativity of simultaneity** emerged out of the **demise of absolute time:** the latter constituting the notional 'common present moment', or the general 'same-time' state in which all things and events supposedly arise. One of our contentions is that the relativity of simultaneity rescues the onticity of object-subject relations and prohibits the want to collapse the relation to one or the other of its terms, with implications to both materialism and idealism.

To clarify the meaning of the relativity of simultaneity, we distribute observers to the Moon, to the Earth, and a third to the middle point between the Moon-Earth. We attach powerful flashbulbs to the observers on the Moon and the Earth: a red versus a blue flashbulb, respectively. The flashbulbs are pre-synchronised to flash at a given moment in the future. The arrangement assumes the ontic reality of space in terms of the Moon-Earth distance. The arrangement also presupposes the putative 'speed of light': (more will be said about both space and the speed of light in Book-IV and elsewhere).

Per the speed of light, it takes light in the form of the pertinent flashbulb emissions some 1.5 seconds to cross the Moon-Earth distance. Hence, observer-A on the Moon will observe the light from his red-flashbulb first. Then, 1.5 seconds later, he will observe the blue flash from Earth. Thus, a **delayed choice time-interval relation** equal to 1.5 seconds abides throughout, in the sense that observer 'A' must wait a minimum of 1.5 seconds to observe the blue flash from the Earth. He cannot make that observation before the 1.5 second delimit. The delimit is not a matter of choice or subjective whim.

---

[1] *Nous* encapsulates both putative active causal sub-conscious mind *and* passive retrospective witness or consciousness, as we shall discover.

[2] In its modern form, the principle of conservation states that energy and matter are interconvertible ($E = mc^2$) and the sum of energy-matter remains constant, with the sum *after* state-change the same as the sum *before* it. Materialism claims that the conservation principle would be violated by any putative causal mind extant energy-matter or material brains. The pervasiveness of this falsifiable view has hampered the development of noetics and, with the equally forlorn notion of 'space', resides at the centre of dismal academic philosophy of mind.

[3] Our contention is for a neutral monism or neutral ontology: one that subsumes non-illusory object-world *and* non-illusory subject-nous, but supersedes both object and subject *without* reduction to or identification with either. Ultimate ontology is neither object nor subject…neither 'matter' nor nous: It is ineffable existence or *Being*.

i

On the other hand, the order of events is reversed for observer-B on Earth: For observer-B, it will be blue flash first and red flash from the Moon 1.5 seconds later. But observer-C at the Earth-Moon mid-point will be subject to a delayed choice time-interval of 0.75 seconds and will observe both the red-flash from the Moon *and* the blue-flash from the Earth simultaneously.

The term 'choice' in 'delayed choice' is bereft of any idealist connotations: The issue is not about subjectivity over reality. On the contrary, the time-interval delays and restrictions with which all observers must abide remain inherent and *objective* to the system. Also, the composition of facts remains the same for all observers: The constitution of facts are conserved, in that all will observe the same composition of events and terms, even if as relativised different order of events: i.e., all will encounter the red flash and the blue flash; the pertinent flashbulbs; observers A, B and C; the Earth and the Moon. Only the order and delayed choice time-interval distributions of observers and events will be different for different observers or 'frames of reference'.

Now, if absolute time held true... if all the said events transpired at an all-encompassing same-time or 'common present moment' *without* any delayed choice time-intervals, and *without* the relativisation of simultaneity... observer-A would see his red flash and the blue flash simultaneously; as would observer-B on Earth. Thus, observers-A, B, and the mid-point observer-C would see the same events in the same order, simultaneously... without *any* time-delays, without any differentiation in the order of events. Indeed, given that all observers would experience the same events, in the same order, and as a same one-experience, all observers would become reduced to a one-observer... to a singular reference frame: i.e., no unique or differentiated observers or frames of reference could arise.

Moreover, under absolute time, the intensity and luminosity of the emitted flash-light could not diminish per the Earth-Moon distance or per the equivalent delayed choice time-interval delay. The intensity and luminosity of all flashes of light at their points of arrival would be the same as at their respective points of emissions, with no reduction in either luminosity or intensity. It will be as if the light did not travers any distance and never diminished. Consequently, the notion of 'distance' would unravel, as would the attendant time delays in the distribution of events. Indeed, the total arrangement of flashbulbs, flashes, observers and events would collapse to a same place point-singularity, and the object-subject relation would be rendered null, both before and upon the said collapse of the object and subject into the point-singularity.

The implausibility of absolute time, exemplified in its self-collapse to singularity, highlights how the relativity of simultaneity and attendant delayed choice time-intervals distributive of events and observers must remain essential to the viability and actuality of object subject relations, to the viability of witnessing observers, as much as to the viability of non-conscious measuring devices. In a universe based on absolute time, distinct conscious observers and a-noetic devices could never exist. At best, all subject-observers and devices, and all object-events, would collapse to a one-and-same singularity, without differentiation into *any* relative order of events and subjects, and without any time delays or distributions.

Only if events are distributed per function of delayed choice time-intervals into inequivalent unique event-order arrangements of otherwise conserved events could unique observers and frames be rendered viable, insofar as the order of conserved events and their time distributions would be different vis-à-vis different observers, so rendering the events and observers unique, and rendering witnessing observers into unique first-person frames and first-person versus third-person relations.

In short, the relativity of simultaneity supports both the subject and the object, and renders viable the object-subject relation and differentiation, now recuperated into a system of time-distributed and time-differentiated events and frames.

*

In Einstein's Special Relativity (SR), the observers need not be comprised of conscious or witnessing beings. These could equally be non-conscious a-noetic detectors and machines. However, SR does not preclude conscious beings in place of or in concert with expected a-noetic detectors. Therefore, our previous examination of the object-subject relation in terms of delayed choice time-intervals and the relativity of simultaneity vis-à-vis conscious witnessing subjects or observers, remains valid. Indeed, insofar as object-events transpire vis-à-vis subjects, and do so in the form of pre-constituted past events... and this is per function of the inherent delayed choice time-interval relations that reside between events and their observers, or due to the 'speed of light', the object-events are *always* observed by the subjects *retrospectively*.

Note that the past-in-retrospect does not arise from the subsequent witnessing subject or from consciousness: This is guaranteed by the fact that the subject-observers have no choice about the delayed choice time-interval restrictions inherent to their relations to the object-events that are constituted as the past, grasped retrospectively. That is, the observers cannot subjectively wish away the delayed choice time structure with respect to which they must abide. Indeed, the delayed-choice time-structure is key to the differentiation of the object in relation to the subject. Hence, the delayed choice time-structure, and the relativity of simultaneity mutually attending it, saves the perennial object-subject relation.

*

What are the implications of the relativity of simultaneity to the attempt to collapse object to subject under the idealist project? The idealist would be hard pressed to explain why, if the subject creates the object, or if nous creates the world as its illusory fantasy... it deems to constitute that world as a distribution of past events *not* in absolute time, and *not* constituted into a common present 'now' *with* the moment of consciousness. The idealist would be hard pressed to explain why the world does not arise in absolute synchronicity with the noetic moment at which it is purportedly created... or why there exists a delayed-choice time-relation vis-à-vis the retrospective apprehension of the 'object as the past' by the subject that supposedly created the object per function of its ontic primacy. Indeed, the object-world must resist collapse to the subject-mind precisely because the object, as a distribution of past events in time, was not constituted at the same time as the subsequent apprehending witnessing subject: The object-world was temporally constituted as an admixture of events from different moments of the past *before* the moment of its apprehension by subject-nous: The object-world arose to the subsequent noetic moment as a pre-constituted corpus of time-distributed past events… furnished to nous, but *not from* nous.

Hence, the object-subject relation is recapitulated as a time-distributed relation, wherein the object is *not* temporally simultaneous with the subject, and vice versa; and wherein mind and world belong to different moments in time, *not* to a same-time moment; *not* to a state of absolute time. Hence, the object cannot temporally collapse to the subject as the latter's creation or illusion: Hence, nous cannot constitute the primary ontology from which world or object supposedly arises as Schopenhauer's 'will and idea' or as Eddington's 'mind-stuff'.

As a lesser consideration, Heidegger's anti-Cartesian *Dasein*, or 'being in the world', must also unravel, given that, if we take for granted that nous is the being through which the question of Being is posed, it turns out that being is *not* in the world (*not Dasein*). Instead, the world, as the pre-constituted past subsequently apprehended in retrospect by being, is furnished *to* being, but *not from* being… given the a-synchronicity between nous versus world by function of the relativity of simultaneity *and* the implausibility of absolute time. A lesser *Dasein* could be asserted, but as *enmeshment*: Therein, despite the time distribution and differentiation of subject from object, the separation is obscured by a psychological gross over-involvement of the subject with a now-fetishised object-world.

*

If the object cannot collapse to or arise from the subject, it also follows that the subject cannot collapse to or arise from the object: The world is not merely comprised of observables distributed in time: The world of objects also subsumes the brain itself. The succinct assertion of nihilistic materialism consists in the narrow claim that nous (should it be granted by materialism *any* existence) is in identity to, or emergent from, or epiphenomenal to the world, to matter', to brains. But this assertion requires that, in order to generate the 'illusion' of nous, the brain must constitute as a state or frame in absolute time.

In general, mind and consciousness cannot emerge from the world or its subsumed brain simply because the world and brain temporally transpired first, albeit as an admixture of different events from different moments of the past, all subject to the relativity of simultaneity: (a further elaboration of this insight is given below). Only later in time did nous transpire… *not* at the same time as the world and brain, but temporally *after* the world and brain, with the brain itself furnished to witnessing consciousness as a pre-constituted past apprehended by nous, but *only and always* in retrospect. That is, since world and brains versus nous did not happen at the same time and did not arise in a common state of absolute time… and since the consciousness of the world happened temporally *after* the world and brain had transpired, with these latter constituted as the complex past, the temporally subsequent conscious witness of the past-world and past-brain could not have arisen from that world and brain.

In short, given the temporal relation that must exist between subject and object, it necessarily follows that object and subject do not arise at the same time, and the subject is no more capable of being temporally generated by the object (or the brain) than is object temporally generated by the subject. We thus find that both idealism and materialism are hard pressed to explain how things that transpire at different times (namely, object-world versus subject-nous, or brain versus nous) could ever collapse one to the other, much less arise from one or the other.

*

We must unravel the above astonishing-seeming claim into greater definition. To this end, internally, the events that constitute the brain are necessarily events distributed in time per function of an internal relativity of simultaneity. That is, the various events comprising the brain belong to different times and do not arise in a 'same-time' or in absolute time, as is also true across the universe. Thus, the events of the brain are distributed and relativised in time.

The relativity of simultaneity abides in the brain, whether we chose to model the internal relativity of simultaneity of the brain according to the 'speed of light' or on the basis of biochemical 'camel speed' impulses that supposedly transpire across the brain, or as an admixture of both.

Hence, the brain is *not* and *cannot* constitute a unity in absolute time: It cannot constitute a 'common present moment' or instantaneous whole. Indeed, we can transplant the previous flashbulb Moon-Earth relation to the brain to illustrate this: Therein, the red-flash is placed in the left hemisphere of the brain and the blue-flash is placed in the right hemisphere. Where is the moment of consciousness of the flash-events located? If the moment of 'illusory' conscious rapport of the events originates and attends the moment associated with the light-flash in the left hemisphere, the order of events apprehended will be "red-flash / time-delay / blue flash". If the conscious apprehension of events is attached to the flash in the right hemisphere, the order of apprehended events will be "blue flash / time-delay / red-flash". But if the moment of consciousness is located in the middle, synchronic apprehension of the events will be achieved, albeit subject to a half-time delay. And if the moment of consciousness occurs at every one of these locations in the brain, assuming this could be permitted by the relativity of simultaneity (and it cannot be so permitted) nous would apprehend a manifold schizophrenic disconsolate order-of-events, making for a non-unitary disconsolate fragmented consciousness.

However, if the brain is surgically severed into a split-brain, as was the practice in the 1940s to treat epilepsy, wherein the mediation of the events cannot transpire from one hemisphere to the other, much less attain synchronicity in the middle, then *where* in the brain is the common unitary experience of events produced? Where in the brain is the visceral sense of the common present moment furnished, such that the myriad temporally disparate events of the brain arise to a singular 'illusory' consciousness? Indeed, in split-brain patients, unitary nous and identity prevail, and patients do not experience a split identity or consciousness, unless specifically tested for. But recent tests by Yair Pinto have demonstrated that unity of consciousness prevails even when specifically tested against. (More will be said about the conundrum of the split-brain and Pinto's findings in the main work).

iii

Insofar as the relativity of simultaneity and the unequal temporal differentiation of conserved-events across the brain must abide, as it does across the whole universe, it ought to be impossible for the brain to furnish even an 'illusory' moment of unitary consciousness, and even less so under split-brain conditions.

Faced with such challenges, materialism cannot address 'where' mind and consciousness (or the supposed 'illusion' of it) resides in the brain, or the 'where-in-brains' at which illusory consciousness is produced. In the face of the relativity of simultaneity, identity theory, or epiphenomenalism, or even emergence theory, and the materialist cause cannot avail. This is simply because the complex of events constitutive of the brain relative to its outside observers (i.e., to the observing materialists in forlorn search of 'correlates of consciousness') or even relative to any frame of reference within the brain itself… or even relative to and in temporal relation to the brain's own moment of 'illusory' consciousness… remains ultimately composed of a complex set of past events constituted as relative *whens*. Hence, any putative 'where' that one might want to specify as the candidate for 'where the consciousness is' in the brain, would constitute nothing more than a mere past-event related to the specifier, whether this be the experimenter, or some another part of the brain, or even its own witnessing consciousness. The events are *always* related to the specifier as a complex of past events distributed per function of the relativity of simultaneity and the inherent delayed choice time-interval pattern that abides across the brain and across the universe. This fact is not a trivial or tautological one: To find mind and consciousness (or the 'illusion' of either) in the brain, one must specify for it a spatial loci at which simultaneity is achieved vis-à-vis all other events and frames constituting that brain: a 'midpoint' or Lagrange at which the events of the brain converge in time into simultaneity (the equivalent of observer-C in our previous depiction of the relativity of simultaneity) and at which consciousness, or its 'illusion', is supposedly 'produced'… at which the brain is apprehended by consciousness in unitary concerted form. However, per the relativity of simultaneity, such unity cannot be furnished from the brain or by function of it.

The alternative would require that the brain somehow furnish an absolute time-state to all its events before it presents these to an all-brain synchronous nous, in contradiction to the physics that abides across nature. Such a thing must also remain impossible if we must abide to a materialist philosophy of nature: a philosophy that must necessarily conform to the relativity of simultaneity, as must any other philosophy of physics and nous.

Moreover, a locus for the 'production' of nous in the brain could not be constituted as an in-brain spatial functional area or corpus[4]: Such a functional area must also be internally composed of past-events, and must *also* be rendered into a time-distribution; subject to the relativity of simultaneity. The putative corpus could not enjoy or furnish any form of common simultaneity or any absolute time state, such as to 'produce' or 'excrete' a unitary moment of apprehending nous, 'illusory' or otherwise.

Nor could consciousness be generated as a clouded process across the whole brain, or generated 'everywhere in brains', as was previously noted. (Nor can quantum non-locality avail this view, as we shall discover below). Given the relativity of simultaneity, all that this could obtain would constitute a schizophrenic admixture of various in-brain frames of reference with respect to which the order of brain-related events and their delayed choice time-relations would be arranged differently and *unequally* from one in-brain frame to every other in-brain frame… from which a single unitary consciousness could never arise, save as a schizophrenic, much less as an 'illusion'… unless the brain *somehow* constituted a state in absolute time *without* collapsing to the previously stated singularity.

Indeed, given the ineliminability of the relativity of simultaneity and the delayed choice organisation of all events to every other such across the brain, and given the impossibility of absolute time owing to its tendency to collapse to singularity, it follows that the 'illusory' noetic moment, and the experiential sense and moment in which all things arise (i.e., consciousness), must *somehow* attend an in-brain Lagrange constituted as a dimensionless point or a 'noetic point in space': a noetic monad. This is a bizarre requirement necessitated by the materialist demand to collapse subject to object or to the brain: a result wholly apagogic to materialism itself, as much as it is contrary to the stated physics.

*

One might hope in quantum non-locality to rescue the materialist contention. In the physics-example of non-locality, a quantum spin for a particle resolved 'over here' will *instantaneously* resolve the correlate quantum spin for the entangled partner-particle 'over there', even if the latter is a hundred lightyears away. Could this sort of thing furnish an absolute time-like unity for and between the time-disparate events of the brain and, in the process, 'excrete' the illusion of the singular unitary moment of consciousness?

Ignoring the impossibility of the communication of meaningful information via non-locality, to confirm said non-locality requires that attendant observers attached to each entangled outcome subsequently inter-communicate their findings via normal means, at the speed of light, requiring a hundred year long delay, or fifty years for an observer located at the mind-point between the entangled particles. The same problem would arise in the non-local entanglement of one event in the brain vis-à-vis every other brain-event therein, albeit with attendant time-delays less accentuated, but *not* objectively obviated. That is, even with hypothetical non-locality across the brain, the relativity of simultaneity must yet abide in brains thereafter between non-locally entangled in-brain events. Why? A noetic moment attending one entangled frame would need to confirm such non-locality vis-à-vis every other frame and noetic moment in the brain via subsequent confirmatory signals necessarily subject to delayed choice time-intervals and to the relativity of simultaneity, with the subsequent confirmatory signals affected via 'camel speed' transactions, but with no difference in the outcome even if affected at the 'speed of light'… in each case *without* the possibility of absolute time, and *without* the generation of the sought unitary nous.

---

[4] The middle prefrontal cortex would likely constitute the best candidate for any putative locus for the 'production' of consciousness (the latter supposedly 'illusory').

QUANTUM MECHANICS AND MIND

The fact that the brain cannot constitute a state in absolute time even with putative quantum non-locality, and that, in the confirmatory phase, it must yet again be subject to an internal relativity of simultaneity of events and frames *even with the assumed prior non-locality*, and that, consequently, the materialist cannot find the locus or 'cloud' at which the 'illusion' of unitary consciousness emerges… or to which it is epiphenomenal… or from which it is supposedly 'excreted'… must again segue into the problem of how unitary coherent brain *and* unitary-singular 'illusory' nous could ever be obtained in, by, or from the brain.

Indeed, the neurosurgeon Wilfred Penfield spent a lifetime in search of an in-brain basis for unitary brain and nous. He could not find it, and rightly subscribed to mind-brain dualism.

*

One last problem besets the materialist want to collapse subject to object or to brains: While the rest of the universe including brains cannot furnish a state in absolute time, it turns out that 'illusory' nous *does* constitute *a* visceral sense of the 'common present moment', even if the pre-constituted past arising to nous is subject to the relativity of simultaneity and internally temporally disparate. In short, consciousness, or the 'illusory' noetic moment, possesses absolute time-like characteristics. That is, the object-brain and world, internally subject to the relativity of simultaneity and without absolute time-like characteristics, interrelates to, arises to, and is retrospectively apprehended by a witnessing nous that *is* constituted as an undivided noetic temporal monad… *with* absolute time-like characteristics.

Such a state defies any materialist ontology and possibility, even when dismissed as 'illusory' or non-existent, as is the materialist tendency to do so *ad nauseum*: It cannot be furnished, produced, generated, or emerge from, or be 'excreted' by brains, or by any functional area or corpus therein; not even as a causally neutered 'illusion'.

All of this culminates into the revival of Cartesian dualism, recapitulated as a dualism between subject or nous constituted as a temporal monad in absolute time, versus object or world (and brains), with the latter always subject to the relativity of simultaneity and into a temporally distributed complex of time-disparate events.

De-Spatialisation and the Temporisation of Physics

In the preceding account of the relativity of simultaneity, in the object-subject relations explored therein, and in the critique of attempts to collapse object vis-à-vis subject in the dismal philosophies of materialism and idealism, we utilised an almost pure temporal account of the relations attendant objects and subjects. This was not an accident. It turns out that **space is pleonastic**[5], **and only time is real**. Hence, 'space' has no ontic reality. Indeed, its presumption has obstructed the further advancement of physics, as much as it has undermined noetics.

But is it not the case that light radiation, the bearer of events, has a constant 'speed', and that it displaces across really-existing space? Is it not *this* that furnishes the temporal distribution of events and the attendant relativity of simultaneity? Does 'space' yet abide, despite our doubts? We shall return to this conundrum shortly.

Where did the notion of space come from? To address this, we again presume the 'speed of light' and recapitulate the relativity of simultaneity. But, instead of the Earth-Moon relation, we place our observers and their less-powerful flashbulbs in a large schoolroom, with the two observers and flashbulb each placed at the two ends of the room, with one observer placed in the middle. Therein, we do not obtain the extreme delayed choice time-relations found in the Earth-Moon relation or in the cosmic environment: The putative 'speed of light' is so swift that we do not notice the delayed choice time-interval relation inherent to the schoolroom arrangement. Therein, per the limits of human perception, the observers will not notice the objective time-differences of the events, nor discern the objective differences in the time-order of the same. None will notice that simultaneity only applies to the middle observer. All will assume that *all* events are transpiring in a common present same-time moment, in absolute time, such as to form an equivalent 'common unifying surface'; a collection of *wheres*: i.e. the floor: a *presumed* contiguous two-dimensional surface, further elaborated into a contiguous three-dimensional block. Thus, 'space', 'locations', *wheres*, and the assumption that all things are same-time distributions, hence 'distributions in space', arises from the inability to discern very short time-interval relations between otherwise time-disparate non-synchronous purely temporal *whens*, *and* the erroneous notion that all things arise in absolute time, or on a unifying 'surface' or 'space' that attends absolute time. But, in truth, all we are permitted to empirically attest are non-spatial time-distribution of *whens*. Events have no 'place': the notion is pleonastic. Hence, we drop all reference to 'place' and keep all relative time-markers for the events (i.e., the *whens*, and the *frequencies*), *and* recuperate all of viable physics, *without* resort to space. The only way to bring back space and 'place' is to presume absolute time. Only then do the time markers disappear into a one same-time marker or 'common present moment', such as to restore or permit the assumptional foisting of 'surface-places and locations', or the foisting of *wheres* to the *whens*… just before it all collapses into a point-singularity, and the very attempt to rescue 'place and space' via absolute time self-voids. This assertion will be fully elaborated when we posit *the impossibility of contiguity and extensionality through simultaneity* in Book-IV.

The belief in absolute time… hence, the belief in 'space'… emerged per function of the very short delayed choice time-interval relations furnished by the supposedly swift *putative* 'speed of light', further aided by the limitations of human perception. However,

---

[5] *Pleonastic* space: space grasped as unnecessary and superfluous to physics, succeeded by a purely temporal distribution of physical information. De-spatialisation and temporisation usurp contact or spatially contiguous causality and unravel the notion of 'matter in space'; forcing the recuperation of world and brain as a distribution in time, with the brain transformed into a perduring Bergsonian 'worm' within a purely temporal and de-spatialised growing block system divided into the domains of past-memory and future-possibility.

this false presumption of 'absolute time' or 'space' at the schoolroom level disappears when we scale-up to the Earth-Moon, wherein the reality of the delayed choice time-interval distribution of events and frames, but *not* in space, is rendered fully salient, and absolute time is seen for the artefact that it is, while 'space' is finally rendered suspect. But we cannot presume that, when we return to the schoolroom scale, absolute time is in any way restored, or that synonymous 'space' is restored with it. The limits of perception cannot obviate the reality of the relativity of simultaneity of pure temporal event-order inequalities and unique viewpoints, regardless of scale.

Indeed, in the previous essay, we argued that, if absolute time were at all possible, all things would not only transpire at the same time or in a common moment, but that all would collapse to a same-place singularity-point, and that both object-subject relations *and* the presumption of places interspersed by 'distance' would unravel. Yet, the observers and events recast to the schoolroom framework do not suffer a collapse to singularity. At the schoolroom scale, we get to keep the false perception of 'absolute time', and, with it, the false presumption of synonymous space, thanks to the limits and errors of human perception.

If the notion of absolute time emerged as folly born of the limitations of human perception, so had 'space', now synonymous with forlorn absolute time. Thus, 'space' is no more viable than is absolute time, at *any* scale. Indeed, space should have gone the way of the eather *and* space-synonymous absolute time with the advent of Special Relativity in 1905. We will fully argue this point in Book IV.

*

But, again, is it not the case that light radiation, the bearer of the above events, has a 'speed' and displaces across space? Is it not *this* that furnishes the said temporal distribution of events and the relativity of simultaneity? Can 'space' yet abide, despite our doubts? The assertion that light radiation displaces across space must be proven and cannot be merely assumed. We must prove it through direct observations of it undergoing its purported 'motion in space'. Yet, this is not possible on many terms. Indeed, the light radiation is ultimately quantum mechanical in nature, and there is a stricture in quantum mechanics to the effect that, if we want to observe the quantum mechanical wave of radiation undergoing its putative motion at the 'speed of light', we need to disturb it via continuous physical interaction-inputs and outputs of observation. Putting aside the fact that continuous observation is impossible despite the Quantum Zeno effect[6], such observation would have the effect of wiping out the very quantum mechanical wave and the purported 'motion' we seek to observe. In other words, we *cannot* observe putative 'spatially displacing light radiation': we can only entertain it and its 'motion'… *and* its 'space'… as *assumption*. Indeed, it is *nothing but* an assumption; ultimately superfluous to physics.

Furthermore, when we introduce the growing block universe in the essay below, it will turn out that the quantum mechanical radiation (the putative light radiation) constitutes the ontic future. As such, it is at best spatially static; *not* moving. If so, what need of the 'space' across which it is *not* moving? Indeed, to repeat, it is possible to purge *all* spatial notions from physics, including wavelength, and apply only temporal markers (e.g., frequency and delayed-choice time-interval relations and distributions) *and* recuperate almost all of the physics we know into a pure temporised daz, *without* 'space'. Again, it is our contention that absolute time and space are synonymous. Again, we contend that, in 1905, in SR, space should have gone the way of absolute time and the eather. If nothing else, 'space' should have met its demise in 1927 with the advent of quantum mechanics and *its* implications to the limits of empirical observation vis-à-vis 'moving' quantum mechanical light radiation and de Broglie matter.

The full case for de-spatialisation and the temporisation of physics will emerge out of five arguments furnished in Book-IV. These will include two variations on John Wheeler's delayed choice experiment; first in an attack against the notion of the motion of the quantum mechanical wave, and, finally, against the very notion of the 'space' presumed to that motion. Thus, we shall discover in full that there is no space; that space is pleonastic, and that the physical order is purely temporal.

Through temporisation, the object-subject relation must yet again transform into a pure time relation and time-distribution function, and, per the relativity of simultaneity, the want to collapse object to the subject in idealism must again be rendered as forlorn as the want to collapse subject to object in materialism: both impossible in a purely temporised universe. Hence, inexorable Cartesian revival.

Quantum Mechanics and the Growing Block Universe

The Alternative Realist thesis (Alt-R) is primarily concerned with quantum mechanics in three areas: i.e., the meaning of quantum mechanics; its direct pertinence to the growing block universe; and the pertinence of both to prospective Cartesian revival.

While the classical 'common sense view' posits that any state in physics must be comprised as a set of resolved attributes of position, spin, etc. quantum mechanics asserts that, before measurement, the system must be described as a 'smear' of possible and potential positions, spin, etc. The former classical-relativistic description of nature fashions for the OR-form 'particle' description, wherein position can only be **a** OR **b**, *not* both. The latter quantum mechanical description is the AND-form 'wave' description: wherein position is cast as *both* **a** AND **b**, at least before 'measurement'. Both OR-form *and* AND-form logic, when combined, describe wave-particle duality, or what is the same expressed as AND-OR duality. There is a mysterious process through which AND-form possibility 'collapses' and converts into OR-form actua. This is modelled as 'wavefunction collapse'.

Convention asserts that the classical-relativistic and quantum mechanical descriptions are incompatible. But the growing block approach incorporates both into consistent mutuality *and* primitive unification. It turns out that 'spacetime' is not a fully resolved **block universe**, wherein the future is fully pre-constituted. Instead, we live in a **growing block universe**, wherein the future is objectively

---

[6] The Quantum Zeno permits a series of very high frequency observations of a physical state such a to slow its time-evolution. Yet, the high frequency observations do not constitute seamless-continuous observation. This would require infinite-frequency and infinite energy observational inputs and outputs that are physically impossible to obtain. In short, continuous observation is inherently impossible.

real and not-yet resolved: *not* a bloss subjective state, but one configured as an objective AND-form potentiality that elaborates to the infinitely removed future. Thus, quantum phenomena, or AND-form states and waves, constitute the objective future phenomenalised as the quantum mechanical wave of future-potentiality.

On the other hand, the OR-form description of nature constitutes the most recent set of resolved states and outcomes of the universe. Such OR-form events, or their abstract basis, cumulate and contribute to the growing block universe as accumulating history or *memory*, generated from the perpetual AND-to-OR 'wavefunction collapse' of the objective ontic future. One surprising outcome from this is that time is rendered synonymous with perpetual AND-to-OR wavefunction collapse. That is, time is *not* a geometric 'fourth dimension'. Instead, time is perpetual AND-to-OR wavefunction collapse itself.

But is any of this true? A key argument that justifies the growing block approach is per the fact that a bock universe could not produce the interference pattern we observe in the two-slit experiment: This can only arise if the future state of the self-interfering particles within that experiment constitute as-yet not-happened AND-form future potentialities, and are *not* pre-resolved into definite putative OR-form paths, positions, or attributes. Hence, in a block universe, wherein the future is fully resolved, AND-form states, quantum mechanical phenomena, and quantum mechanics itself, could not exist. Therefore, since both the two-slit experiment *and* quantum mechanics *are* tenable, it follows that we inhabit a growing block universe: Thus, *the growing block universe really is true*.

In the growing block universe, the decay of the AND-form futures that elaborate to the infinite future resolve into successions of pertinent OR-form events from which we infer (or for which we *assume… or* upon which we foist) the notion of the motion of objects and, more critically, the notion of the motion of the quantum mechanical light radiation across attendant space. Yet the future, and the total quantum wave that constitutes it, hence the quantum mechanical light radiation embedded to it, does not 'move'. Instead, light radiation constitutes a remarkable static non-moving and non-undulating future potentiality state, as does the total future. Indeed, the notion that the future can move is absurd: and so is the notion of the motion of future-embedded quantum mechanical waves and of quantum mechanical light radiation: the very thing that fashions the temporal distribution of events *and* attendant relativity of simultaneity through its alleged 'motion in space'. The succession of OR-form events resolved out of future-potentiality requires that we clock each event purely as a temporal succession, organised into delayed choice time-interval relata and pure time-distributions, *without* resort to 'motion' or 'space'. Hence, the growing block furnishes proof of de-spatialisation: Thus, *space really is pleonastic*.

In its first iteration, the growing block universe approach will be developed in Book-II as the *intermediate model spacetime*. The model will assume 'space'. But, by Book-IV, intermediate spacetime will evolve into a pure temporised Bergson-Whitehead amalgam…into spacetime *without* 'space'. It will emerge that, in its de-spatialised form, the growing block universe retains the perennial object-subject relation as a temporal relation, and, as such, designs for Cartesian revival. Decisively, by Book-V, the purely temporised growing block universe and its processes of memory, possibility and time will comprise the framework for mind-brain dualism and Cartesian causality, wherein mind exploits the process of AND-to-OR collapse and time to intercede in how the growing block future decoheres and subsequently resolves into the sought brain activity, and into subsequent intended world-consequences.

The Nature of Time and Mind: The Collapse of Materialism and the Demise of Philosophy of Mind

Time is AND-to-OR wavefunction collapse, now recapitulated as the partial-perpetual decay of AND-form potential futures into OR-form realised events, with the basis of the latter retained within growing block memory, albeit in pure temporised abstract form. Nous and consciousness must relate to the system of growing block memory, futurity *and* AND-to-OR collapse and time in a purely temporal way, without Cartesian *res extensa… without* 'space'… *without* contiguous contact-causality. Therein, the brain is no longer a spatial state. Instead, it is a pure temporised distribution that requires a growing block description in its own right: wherein the brain is partially recuperated as an abstract temporised mnemonic 'worm' within a de-spatialised temporally accumulated Bergsonian history or memory. Simultaneously, within the same growing block system, the brain must also be described as a projection into the future; an AND-form state of future-potential configurations into which the brain *might* resolve in future moments. This quantum mechanical addendum to the brain otherwise recuperated as a pure temporal Bergsonian 'worm-in-memory' constitutes an indispensable quantum mechanical future-form description of brains; one supplementary to the temporal mnemonic worm. Thus, against the grain, the brain *is* quantum mechanical per function of its Whiteheadian futurity. And it is through spontaneous perpetual AND-to-OR collapse of the future-potential of the brain that mind comes into the fray from a sidereal domain extant the growing block system of memory, possibility, and time… to decohere the futures pertinent to its brain into consistent potentialities in line with its intent and will.

None of this requires that nous collapse the wavefunction or drive time itself. Nor does it require nous to orchestrate pertinent quantum indeterminate terms into determinate global brain-outcomes. Time, or AND-to-OR collapse, happens spontaneously, and there is a solution to the problem of quantum indeterminacy that requires neither orchestrating a-noetic hidden gears nor any resort to a secret quantum orchestrating nous. In short, we do not espouse quantum mind theory, and Cartesian mind is *not* a quantum mind.

The account of the requisite gear-less 'drive of time' will be furnished in Book-III and V, as will solutions to quantum indeterminism that obviate types of quantum mind theory that espouse consciousness as the 'collapsor' of future-possibilities, or as the noetic orchestrator of quantum indeterminism, or as both.

Insofar as nous constitutes the visceral sense of the common present moment to and within which memory (the past) arises, and within which possibility transforms into actuality, neither consciousness nor mind can be internally subject to the relativity of simultaneity. Nous constitutes the closest we might come to a 'spatial point' in a de-spatialised universe. Indeed, nous is constituted as a temporal monad to which possibility and memory arise, in full accord with the subject-object relation. Hence, nous is *extant* the world and its brain: *extant* the growing block system of possibility, memory, and time. Cartesian dualism necessarily follows suit.

How extant nous decoheres future potentialities into sought outcomes in line with its will is to be clarified in Book-V. Also, a solution to how mind circumvents the principle of conservation often weaponised against Cartesian dualism will emerge from several findings: First, AND-to-OR collapse or time presupposed to the succession of events upon which we impose the generic form of the principle of conservation, is not itself an energy or work-driven process: The ontologically primary AND-to-OR time that furnishes the events and state-changes which we then relate to conservation principles and to energy-work relations constitutes a work function zero process. Nous exploits this remarkable work-function zero profile of AND-to-OR collapse and time to intercede into the growing block system of memory and possibility, and to decohere future potentials of its brain to those complex wavefunctions that suit its intents. In effect, mind-brain causality is itself a work function zero process: it does not require energy or work to induce state-change to brains.

Second, the said noetic decoherence of futures is also achieved on a work function zero basis, *without* violation of the principle of conservation. Indeed, this contention is augmented by de-spatialisation itself: De-spatialisation undercuts the requirement for contiguous or contact causality between physical states and 'particles'. Thus, putative force, work, or causality is *not* mediated via contact causality between particles or matter: There are no spatial points-of-contact at which contiguity or contact could transpire in a de-spatialised and temporised universe. There is no space. Hence, there are no spatial points-of-contact or 'impact' for force-mediation. Thus, the expectation that nous must engage in contact causality vis-à-vis its brain or future potentials via 'particle interactions', work or energy, and consequently inflict attendant violations of the principle of conservation, completely unravels, if only because the expectation does not abide across nature, given implications from de-spatialisation and temporisation. So why should it abide in mind-brain causality?

Third, in a de-spatialised temporised growing block universe, the principle of conservation itself transforms into an alternative abstract form: wherein growing block memory is conserved and cannot be destroyed or altered, while future-potentiality must perpetually decohere into consistent futures vis-à-vis conserved events retained in abstract memory: a form of the conservation principle that cannot be used or abused against Cartesian causality, given that Cartesian causality does not entail the erasure or alteration of memory or of the conserved past, much less impose on AND-to-OR collapse any inconsistent futures contrary to that past. Thus the utility of the principle of conservation as a materialist trope against Cartesian dualism withers, as must the whole of materialist academic philosophy of mind founded and totally dependent on the anti-Cartesian premise and abuse of the conservation principle.

The notion that 'all is matter in space', and that mind emerges from the contiguous interactions of matter in space, wherein causality is modelled as a contiguity-based input-operation-output schema... wherein the universe is presented as causally closed...and wherein memory exists only as a 'matter-trace in space' as opposed to Bergsonian non-contiguous time-distributed abstract memory *without* 'space'... is no longer tenable. Hence, the three-century long materialist reaction against Cartesian dualism *must* unravel.

Genesis of the Book

This endeavour did not begin with de-spatialisation and temporisation, nor with the growing block approach, nor with Cartesian revival fully formed. It began with a puzzle about how physical information might relate to radiation, matter and space: a puzzle that emerged out of readings made between 1998 to 2001. In his mind-brain interaction theory, John Eccles proposed that the quantum wave pertinent to the operation of the bouton structure in brain cells contained information, and that its probability structure could be modified by mind to control vesicular discharges in boutons. In *The Emperor's New Mind*, Roger Penrose criticised Eccles to the effect that the quantum wave does not contain information: i.e., in our terms, OR-form information is *not* contained in AND-form quantum mechanical waves, which constitute only potentialities and futures, *not* concrete resolutions of the relative present or of the past-in-memory.

Eccles aside, how is nature able to retain OR-form information in an AND-form quantum mechanical universe? How could such OR-form information be transported by AND-form quantum mechanical radiation and matter, *and* survive the scramble from quantum indeterminism at the destination-point? In photography, how could information pertaining to imagery... the vase on the table... get transported by AND-form radiation inimical to the OR-form image, *and* get to the film-surface, *and* survive the AND-to-OR quantum indeterminate collapse of the radiation at the film-surface... *and* get to form a coherent image true to the originary? Would not the same sort of problem beset the moment-to-moment resolution of the brain, its purported memory trace-states, *and* its 'illusory' consciousness? Hence was born the problem of **information-matter dualism**.

The Preliminary and Book-I will adumbrate on information-matter dualism vis-à-vis radiation, matter and space. Yet, by Book-II, in the first iteration of our growing block model universe, we will discover that information-matter dualism is a facile expression of the tripartite structure of the growing block system. This will segue into a surprise series of presage assertions about the basis of inertia, gravitation, and other things: all consequent upon the growing block universe, culminating into the revival of mind-brain dualism. For their part, Book-III and V will resolve critical problems pertaining to causality behind wavefunction collapse or time: a 'no hidden gears' solution to the drive of time *and* to the problem of quantum indeterminism. Book-IV will make the complete case for superlative de-spatialisation and temporisation, transforming the intermediate spacetime model from Book-II into the superlative temporised Bergson-Whitehead amalgam. We will find that OR-form information... indeed, the manifest brain we claim to observe... is restricted and confined only to the 'cut' or 'skin' that separates the past from the future, and that the past is not wholly configured in OR-form terms but largely assumes a de-spatialised temporised abstract-immaterial form: a Bergsonian temporal distribution of information and cosmic history. Hence, we will supersede the simple AND-OR dualism in Book-I and arrive at the retention of information as memory in pure temporised abstract-immaterial form, all within the mnemonic Bergsonian domain of the growing block universe.

Of course, by Book-V, all of the above will culminate, in addition to the demise of 'consciousness causes collapse' and the usurping of quantum idealism, into the general, intermediate, and fine-structure mind-world theories... into shameless Cartesian revival.

PRELIMINARY TO QUANTUM MECHANICS AND MIND

CONTENTS:

1

PRELIMINARY TO
QUANTUM MECHANICS AND THE MIND

GENERAL OVERVIEW

This work is divided into several books, including this Preliminary. The Preliminary consists of four parts. Part-I will explore the relationship between physical information and matter. It will accomplish this with some resort to quantum mechanics. The clarification of the relation between information versus matter, radiation and space is essential to a more complete understanding of nature: one that will include de-spatialisation and abstract memory as part of the **Alternative Realist thesis**: abbreviated Alt-R. Our exploration of the relation will also shed light on the hitherto unresolved mind-matter or mind-brain conundrum.

According to the naïve notion of seamless-continuous motion, particles and objects that undergo purported motion are expected to occupy *all* of the infinite number of positions along their trajectories. But **basic quantum mechanics forces the view that, at best, motion must be broken up into a finite number of realised putative positions, each interspersed by intervals of quantum potentials** that are devoid of any realised positional states. Hence, broken-discontinuous motion (i.e., grainy, 'broken up' motion) and the physical impossibility of seamless-continuous motion. This must have direct consequences to the character of the transportation of information in and by 'carrying' radiation and matter: **If the motion of the 'carrying' radiation or matter assumes the broken and discontinuous form, this must culminate into the broken-discontinuous relation of information to space, and to the 'carrying' matter or radiation: Hence, information-matter dualism**.

Of course, **the brain could not be exempt from such information-matter dualism,** and mind-brain and memory-brain dualism ought to follow from it.

However, information-matter dualism will not be the core culmination of Alt-R. Instead, it is the first salvo leading to conclusions that supersede even information-matter dualism, radical as it is.

*

Part-II and III of the Preliminary will constitute the core work that will anticipate the key claims of Alt-R. Part-II will constitute an introduction to quantum mechanics and mystery, leading to anticipations of key solutions to the mystery of wave-particle dualism, the superposition principle, the problem of wavefunction collapse, and quantum indeterminism: It will culminate into the measurement problem or, succinctly, the causality problem central to quantum mystery. Part-II will also unravel the ontology of the superposition principle, central to the quantum mechanical wavefunction. Hence, Part-II will assert that the quantum mechanical description constitutes the future-form description of nature: **The quantum wave constitutes the future phenomenalised into an objective physical state**. This ontology of the quantum wave will be **justified by the intertwine of the principle of causality with the principle of conservation of energy-matter**: both part of **the firewall principle**. This insight into the ontology of the quantum wave will revamp generic models of spacetime, integrating to these a quantum mechanical domain of superposed futures, further contextualising and clarifying the relation of information to radiation, matter, and space as part of **the growing block universe**. The intermediate model of spacetime will itself be superseded by inexorable implications from de-spatialisation: Its successor will be presented in Book-IV.

Part-II will also glimpse **the solution to the measurement problem**, which **constitutes the implicit crisis of the materialist contention about causality**: We will undermine the contiguity-driven input-operation-output schema (IOO-schema) implicit to classical contiguous or impact causality, both definitive of materialism.

Part-II will also introduce a radical take on the ontology of physical law, obtained through an application of the Einstein Podolski Rosen experiment (EPR). Beyond mere 'action at a distance', EPR exposes the relationship of physical law (specifically, the principle of conservation of spin) to matter, radiation, and space: culminating into an **overhaul of our assumptions about the ontology of physical law and of what constitutes legitimate naturalism and physicalism**.

Throughout Part-II and Part-III, we will anticipate **de-spatialisation as a central contention of Alt-R: namely, the falsification of space as a pleonastic term in physics**. This will engender extraordinary consequences to how we must model the distribution of information in the natural order; to causality; to the nature of time, and to the nature of spacetime. It will undermine the notion of 'matter in space' and the central notion of spatially mediated impact and contact-causality central to materialism.

Part-III will rely primarily on classical physics facts to augment the case for abstract physical law, but in a supplementary way to implications from EPR. It will also elucidate rarely acknowledged **'work function zero' processes in nature,** especially in gravitation, which sheds unexpected light on the causal process at the core of the physical order. This will also clarify how mind might relate to matter and brains on a similar work function zero process-basis. Part-III will also incorporate de-spatialisation and its implications to physics.

*

Galvanised by doubts raised about materialism from Part-I through to Part-III, Part-IV will challenge generic claims made on the basis of experimental investigations into memory and consciousness in brain science and neuroscience. Experiments from Karl Lashley through to those of Benjamin Libet have sought to clarify the relationship between memory, consciousness and the brain. The **experimental outcomes in brain science and neuroscience are often portrayed** *as if* **in support of the materialist view on the nature of memory, mind, and consciousness**: all posited as brain-writ. Thus, **the assumption of materialism is loaded into facts**

3

garnered from experiments. **The facts are then presented** *as if* **in proof of the loaded assumption. i.e., of materialism**. This is licensed per background culturally sanctioned *a priori* materialism. In its turn, materialism successfully contrives a false mandate from naturalism and physicalism to the point of synonymity with the latter. Part-IV will expose the often-unintentional assumption-loading just described, and it will break the false equivocation of naturalism and physicalism with materialism.

No explicit crucial experiment has been proposed to separate and test for the alternative assumptions (materialist, dualist, etc) about how memory and mind might relate to the brain, in the same sense that Isaac Newton used experiments with light to test for the-then two contending assumptions about light: or in the way that the two-slit experiment tests for the wave and particle assumptions. Such experimentation would require admission to contending assumptions, *not* the ready dismissal of alternatives in favour of culturally sanctioned materialism.

Part-IV will show that **critical experiments carried out by Lashley, Rose and Harding, Libet,** *et al*, **when interrelated in a specific way, and when integrated together, constitute implicit** *default* **crucial experiments** with conclusions that are contrary to materialism. Even if the materialist view could somehow survive such experiment-based critiques, the incontrovertible implications, even from facile information-matter dualism (Part-I), combined to the anticipated findings of Alt-R intimated in Part-II and Part-III, (especially de-spatialisation, if nothing else) independently undermine the background materialist assumption foisted on brain-studies, **unexpectedly reviving mind-brain dualism and Cartesian dualism.**

PRELIMINARY PART-I:
INFORMATION & MATTER

INFORMATION-MATTER RELATIONS

0.01: Information-Matter Relations

Our foray into the Alternative Realist thesis (Alt-R) begins with the analysis of the motion of a single particle traversing spatial interval-AB. We assume its motion is of the seamless-continuous form, wherein the particle gets to occupy each of the infinitely many positions along AB, with no gaps or discontinuities.

The transportation of information in or across putative space (of a photographic image by its light radiation, or the pattern of a snowflake carried by its material corpus) at first appears to be a function of the arrangement and displacement of the constituent group of particles belonging to the total radiation or matter. **If the photons of light involved in photography (or the electrons carrying the image of an electron micrograph) should displace across space in seamless-continuous form, the physical relationship of the pertinent information to the spatial interval and to the carrying matter or radiation will also assume a seamless-continuous form: with no physical separation of information from the pertinent radiation, or from the matter, or from the spatial interval of displacement**. Hence, our analysis of the motion of a particle along interval AB, insofar as it can relate in a fundamental way to the transportation of information across the same, is crucial to the clarification of information and matter relations.

Alternatively, we could characterise motion in the broken-discontinuous form, wherein the particle occupies a finite number of positions in discontinuous form along AB. Indeed, this is the inexorable form we will be driven to by sundry physics facts. It will garner dualistic consequences to the transportation of information and to the relation of information to group radiation, matter, and space.

*

Time evolution is indispensable to any corpus of matter and 'carried' information, even when a given frame of reference appears to be at rest. An information-carrying corpus at relative rest, from within its own frame, will appear to be subject only to time evolution. Thus, **we must also analyse information-matter relations in terms of pure time evolution, with or without motion.**

In time-evolution, 'common sense' might suggest that information must assume a seamless-continuous form to each moment of time that comprises the associated time-interval t_1 to t_2: a seamless-continuous relation of information to each infinitesimal moment comprising t_1-t_2… even when motion is occurring.

On the other hand, in the broken-discontinuous form of putative motion, a broken-discontinuous form of time-evolution must apply. Therein, the carried information will assume a finite number of instantiations along spatial interval-AB and associated finite number of moments along time interval t_1-t_2: i.e., a frame-by-frame manifestation of information in space and in time, analogised by a movie-film that runs as a finite number of frames per second.

In the following, we will show that, **in nature, it is the broken-discontinuous form of motion that holds true. The seamless and continuous form of motion is at best an idealisation, if not outright fiction. This will culminate into the dualism of information versus radiation, matter and space… into information-matter dualism.**

0.02: Practical Difficulties Imposed by Seamless-Continuous Motion

Any self-consistent perspective in favour of seamless-continuous motion must entail that the motion of our particle consist in the occupation by that particle of *all* of the infinitely many spatial positions available to it along interval-AB. Suppose we sought proof of this in physical-empirical terms, by taking photographs of the particle at its infinite number of positions and moments that it must occupy along AB and t_1-t_2. We confront an impossible problem: one that requires an infinite number of photographs of our particle at the said number of positions and moments.

Even if we had enough matter in the universe to constitute the infinite number of photographic exhibits, or space enough in which to store these, we would confront the daunting prospect of examining each photographic exhibit in its appropriate sequence in order to corroborate that the particle occupied an infinite number of positions, and so prove seamless-continuous motion. We could not have enough time in the universe to accomplish this. Indeed, **given the insurmountable infinities involved in any attempt at a practical proof of seamless-continuous motion, we can never evidentially prove the reality of it in practical experimental-observational terms, even if it turned out to be physically true**.

0.03: On Broken-Discontinuous Information-Matter Relations

The notion of seamless-continuous motion made its first philosophic appearance in the notorious Zeno's paradox. Therein, for an arrow to travel from A to B, it must first travel half of AB. Before this, it must first cover one-quarter of AB. Before that, one-eight of AB. Then one-sixteenth of AB… and so on, *ad infinitum*. In other words, there are infinitely many positions along spatial interval-AB. The leading point of an arrow would need to occupy an infinite number of spatial positions along its journey across AB, and an eternity of time to complete that journey. Hence, the arrow will never complete its journey. On this account, generalising it to *any* motion of a particle or object, motion ought to be illusory.

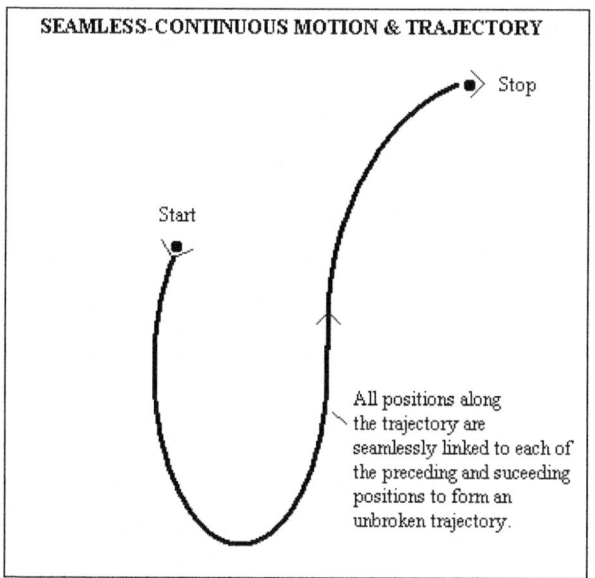
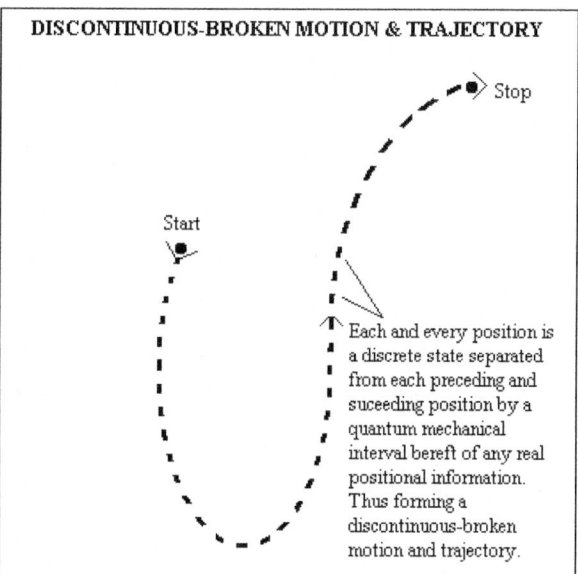

Fig. P1.00: Illustrating seamless-continuous versus discontinuous-broken motion and trajectory
Notes: The didactic illustration outlines the basic difference between seamless-continuous motion and trajectory versus discontinuous and broken motion and trajectory. The seamless-continuous form supposes that an object or particle will occupy a series of infinitely many positions from point of origin to its point of destination, with each seamlessly linked to the preceding and subsequent position (spatial contiguity), with no gaps or intervals, forming a continuous unbroken trajectory and motion: (See above-left). On the other hand, the discontinuous-broken form holds that the object or particle will *not* occupy infinitely many positions from origin to destination. The resulting motion and trajectory will be broken into a finite series of resolved positions, with intervals or gaps between each: gaps in which the object or particle will not assume *any* form of positional attribute: thus describing a discontinuous-broken trajectory: (See above-right). If one should, as it were, "zoom out" far enough away from an object undergoing discontinuous-broken motion, its trajectory will only appear to form a continuous line or path.

Zeno was in partial error, and not only because his view fails experience. Quantum mechanics shows us that seamless-continuous motion is neither physically true nor necessary. **Implicitly, generic quantum mechanics demands that a trajectory cannot observationally, experimentally, or even intrinsically consist of infinitely many seamlessly integrated realised positions, much less allow for the seamless-continuous observation of a particle in its motion.** Implicitly, quantum mechanics asserts that motion and trajectory is *always* at best comprised by a finite number of discrete resolved positions, both observationally and intrinsically. In short, **quantum mechanics permits only broken-discontinuous motion. This also implies tandem broken and discontinuous time evolution** of the arrow, with or without putative motion. We must now explore the reasons why this is so.

When we attempt to observe an arrow or particle in motion in the detail, it will be initially located at a given position along AB; followed by a succeeding position further along AB: then at a subsequent position closer to B… until, after a *finite* number of such observations and realised positions, the particle will attain B. Indeed, within the intervals that lie between each realised or observed position, our particle will *not* possess any physically realised positional attribute. The intervals will be quantum mechanical states comprised only of *potential positions, not* realised ones.

Let us augment: In order to make any observation, we must 'shine' or input some sort of radiation or de Broglie matter-wave to the object observed. The input must then 'bounce back' from the object to some detection apparatus: a photographic camera, or the human eye. With this in mind, consider the motion of, say, an electron, along AB: There is no physical observational input in nature that could accomplish an infinite series of positional determinations or observations of our electron such as to establish the reality of its purported seamless-continuous motion.

This realisation emerged from the historical facts about wavelength, frequency cut-off and energy considerations that arose from the investigation of black body radiation and the notorious ultraviolet catastrophe, with consequent constraints imposed on the resolution-power of physical-observational inputs that are always a function of wavelength or frequency. Thus, let us first restrict our analysis to the wavelength. **The detail and scale of observation afforded by the physical input vis-à-vis an object in motion will only be as good as the wavelength of that input**. Thus, the sorts of resolution-limits that attend light microscopy will apply to *any* observational physical input, or a series of such, to establish the putative seamless-continuous motion of a particle along AB.

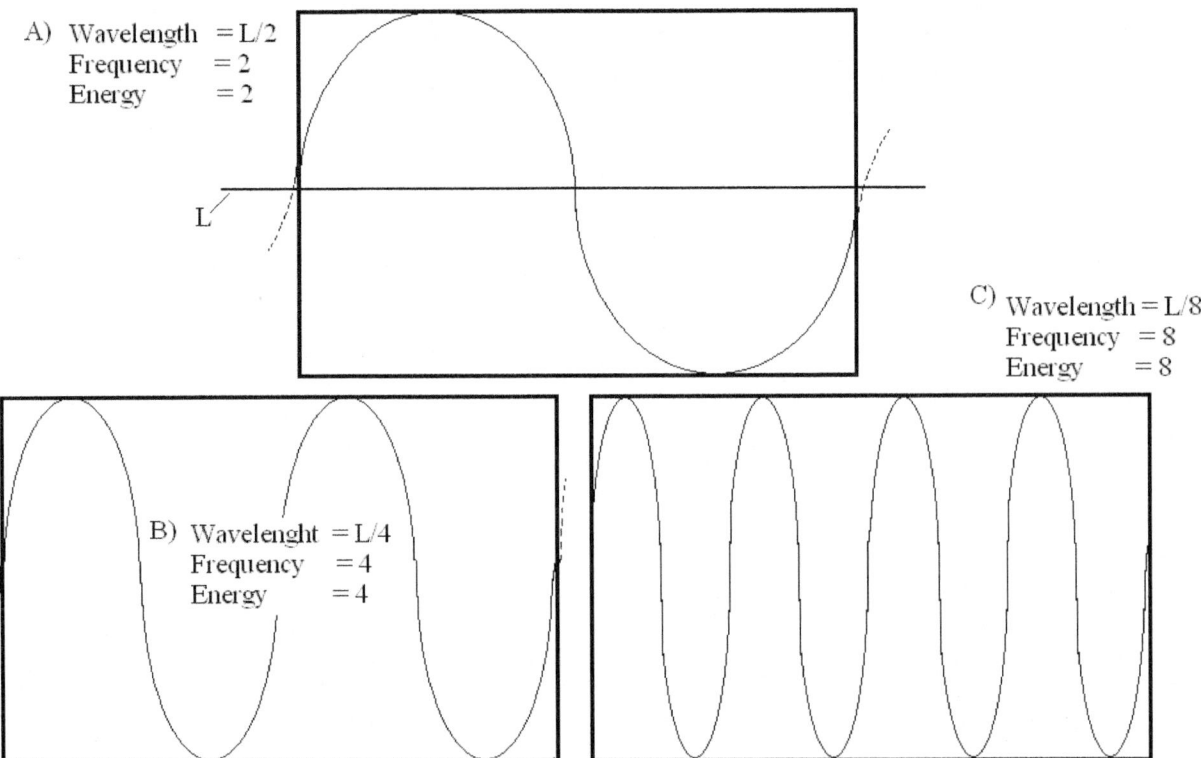

Fig. P1.01: The relationship between wavelength, frequency and energy
Notes: Imagine a box of length L. If the wavelength of a given radiation or de Broglie matter is the same length as that of the box, its frequency is 1, and its energy content is taken to be 1. If the wavelength is reduced by ½ to ½ of L, we will then fit two undulations (a peak and a trough) into the box. The frequency will then increase to 2, and energy content will also increase to 2, as depicted in **(A)**. If the wavelength is reduced to one-quarter of L, frequency will then become 4: energy will also increase to 4, as depicted in **(B)**. If we further reduce wavelength to 1/8 of L, we will then get a frequency of 8 and an energy content of 8, as depicted in **(C)**. Thus, the smaller the wavelength, the greater the number of undulations we can fit along L, and the greater the frequency and proportionate energy-content of the wave. If we could reduce wavelength to zero, frequency would shoot to infinity, and so would the energy content of the radiation or matter-wave. Observationally provable (or else intrinsic) seamless-continuous motion of radiation and matter would require observational inputs or intrinsic waves of infinite frequency and infinite energy. This is impossible.

QUANTUM MECHANICS AND MIND

If we divide the distance AB by the wavelength of our observational input, we obtain the maximum *finite* number of positional 'snapshots' we are eligible to make of our object on its journey along AB. The number of such 'snapshots' must always be finite because we can only ever fit a finite number of any given wavelengths along AB. If we want to take a greater number of positional 'snapshots' of the object on its journey along AB, or obtain a higher resolution of it, we need much smaller wavelength inputs: ultimately reducible to or approaching Plank's length.

Of course, any proof of the seamless-continuous trajectory of our particle will require an infinite number of 'snapshots' along AB, given the putative infinite number of positional states that the object traversing AB is expected to assume. To obtain such proof, we cannot simply rely on observational inputs of smaller and smaller wavelengths: **To resolve every infinite positional state of our object along AB, we would need zero-wavelength inputs**.

To reiterate, only zero-wavelength observational inputs could furnish the infinite number of 'snapshots' of an object in motion along spatial interval AB *and* establish the reality of its seamless and continuous motion. However, as the wavelength of a physical observational input is made smaller and smaller, its putative energy increases in a calculable way. **If it were possible to reduce the wavelength of the physical input to zero, the energy of that input would shoot to infinity. This is impossible.**

*

What about frequency? The smaller the wavelength of an observational input the greater the frequency of that input. To illustrate, imagine a box within which radiation or a matter-wave is suspended so as to form a standing wave: (See **Fig. P1.01**: A through to C). If the wavelength is the same length as the length of the box, we assign it a frequency of 1. To simplify, we treat a frequency of 1 as equal to energy-content equal to 1. If the wavelength is reduced to a wavelength equal to half the length of the box, we can then fit two undulations along that length. This will yield a frequency equal to 2, and energy equal to 2. If we further reduce the wavelength to one-fourth the length of the box, we will then get frequency that equals 4 and an energy equal to 4.

The more we reduce the wavelength the greater the number of waves we can fit along the length of our box, hence the greater the frequency obtained, and the greater the energy-content obtained. Now, **if we could reduce the wavelength to zero, we could then obtain a frequency for the radiation or matter equal to infinity, with an energy content also equal to infinity**.

There is no difference in principle between, on the one hand, fitting zero-wavelength radiation or matter waves into a box of a given length versus fitting an infinite number of observational inputs (each of zero wavelength) into or along interval AB, in the hope of obtaining an infinite number of 'snapshots' of our particle in its purported seamless-continuous motion along AB, requiring observational inputs of infinite frequency and infinite energy. **If this were at all possible, we would end with a result essentially identical to the notorious *ultraviolet catastrophe*;** the very problem that inspired pioneering work that led to the development of the 'quantum of energy', and, with it, the founding of quantum mechanics by Max Planck.

Not only does nature *not* allow zero-wavelength observational inputs per generic quantum mechanics solutions to the problem of ultraviolet catastrophe and frequency cut-off, but **the power source required to 'squeeze' the wavelength of our observational input to zero-wavelength, hence infinite frequency and energy, would itself need to be infinite. This is also impossible:** There can be no such thing as an infinite power-source.

Applying Planck's constant, the shortest possible wavelength for any observational input might approach, but possibly cannot be smaller than, Planck's length and related Planck-frequency and Plank-time. **Hence, the maximum number of 'snapshots' we might hope to get of a particle in motion will always be finite, approaching Plank length and time. Hence, seamless-continuous motion and trajectory is physically impossible**, at least as judged from an observational framework. **Only broken-discontinuous form of motion and time-evolution is possible**.

*

Now, most observations of motion involve objects exhibiting apparent spatial extension, such as an arrow, or a bullet. There are two ways we treat the motion of an extended object. To start with the first, we apply observational inputs of wavelength equal to the length of the object undergoing the motion. Yet, this cannot be useful to any proof of seamless-continuous motion. In the second approach, for brevity's sake, we focus on the leading point of the arrow. Therein, we would need to shift the second observational input slightly ahead of the first in order to obtain a second 'snapshot' of the leading point at the second position… followed by many similar shifts thereafter, and as many as are required to cover the interval-AB.

How far ahead would we need to shift the second and all subsequent observational inputs? An infinitesimally small distance. The problem is that there are an infinitely many such infinitesimally small shifts required to establish the seamless-continuous motion of the arrowhead along AB. While this will not entail instances of zero-wavelength and infinite frequency infinite-energy inputs, it will yet require an infinite number of such finite wavelength inputs vis-à-vis the infinite number of infinitesimally small observational shifts required.

Whether we observe a point-like object or an extended object, both must traverse space in a broken-discontinuous fashion: This is the only form of putative motion and trajectory possible for all objects, unless we want to entertain zero-wavelength observational inputs, ultraviolet catastrophes, and infinite power sources. It also follows that the informational content of putative radiation and matter must associated with the attendant quantum mechanical wave and with spatial interval-AB, and with the time interval t_1-t_2, in a wholly broken-discontinuous way… with consequent information-matter dualism.

*

Regardless of the limits that hold for physical observational inputs, it might yet be possible to entertain the reality of *intrinsic* seamless-continuous motion and trajectories for both point-like and extended objects. **An object in motion along AB could be imagined to *implicitly* cover an infinite number of positions, even if this cannot be proven observationally, for reasons previously given**. However, the particle of radiation, matter, or an extended body made out of such, will invariably possesses *intrinsic* wavelike characteristics. **The particle or object itself, even in the total absence of observational inputs, turns out to be wavelike,** as had been firmly established by the discovery of quantum mechanical radiation and de Broglie matter-waves. It follows that **the *intrinsic* positional resolution of our point-like particle in motion, of the sort necessary for the assumption of intrinsic seamless-continuous motion, will only be as good as the intrinsic wavelength of that particle**. Consequently, the number of intrinsic 'snapshots' of motion that we might indirectly infer for the particle, but *not* directly observe, will be equal to as many intrinsic wavelengths of that particle we can fit along AB. Any finer intrinsic positional resolution will require that we further shorten the intrinsic wavelength of the particle to increase the number of intrinsic 'snapshots' of it along AB. But **a full intrinsic seamless-continuous trajectory for our particle will require that it possesses an intrinsic zero-wavelength, and with it intrinsic infinite frequency and infinite energy**. The object would have an energy content obtaining infinity: The physical universe could not survive an encounter with such a particle.

This speaks against intrinsic zero-wavelength radiation or matter, and against intrinsic seamless-continuous motion generally. Again, the smallest non-observed intrinsic wavelength we might infer for a non-observed object in motion will likely just fall short of the Planck's length.

*

What if we again treated the non-observed object travelling along AB, not as a point-like state, but as a spatially extended object? Again, we need only concern ourselves with the *intrinsic* wavelength of our extended object, equal to the length of the object, and perhaps describable as group of non-observed de Broglie matter-wave complexes.

The motion of this object could be described in terms of a series of infinitesimally small shifts of the group wave toward B along AB: infinitely many such. Even in the strict absence of observational inputs, who is to say that the single de Broglie matter-wave complex cannot undergo such infinitesimally small shifts along AB, and hence undergo an infinite number of such shifts, even in a finite amount of time? Who is to say that the implicit seamless-continuous motion of the object is *not* real?

According to generic quantum mechanics, in the absence of observation, the system under consideration would constitute a system in *isolation*. A system in isolation is wholly a quantum mechanical state of possibilities or potentialities: It is *not* a state comprised of resolved or realised states or positions. A system in isolation is a system in superimposition of possibilities, at least until such time as an observation or 'measurement' *is* made upon it by means of applied physical inputs. The system will then 'collapse' into a set of realised outcomes and events.

The implication is as follows: **In terms of quantum mechanics, an arrowhead in presumed motion along AB, *totally* free of observational inputs, and necessarily in absolute or idealised physical isolation, is reduced to a group quantum mechanical state: a state *not* comprised of resolved constituents in real motion, given that the arrowhead constituent particles cannot possess resolved positional attributes under isolation. Instead, the arrowhead constitutes a *group potential* for potential positional attributes and potential paths. This potentiality is comprised, *not* of real positions or successions of such, but of non-realised *potential* positions and paths. These potentials are not in or along any realised space, but in *potential* space, or Hilbert's space.** Under absolute or perfect isolation, we cannot assert anything concrete about our extended object in terms of any real position or sequences of positions… until or unless the isolation is ended by the use of observational inputs comprised *only* of finite non-zero wavelength radiation or de Broglie matter.

Moreover, as we shall find in the main work beyond the Preliminary, any attempt to presume any realised motion and positional attribute to a particle, or even to an extended object subject to isolation in its quantum mechanical (future) state, would not only run into collision with causality in terms of requisite 'signals from the future', but would also imply the violation of the principle of conservation: i.e. the law that states that the putative sum of energy-matter must remain constant. Indeed, when we place any physical system of *any* scale in isolation, we defer the observation of that system to the future. In doing so, we cannot treat it as anything but a set of pure not yet realised possibilities that the system *might* assume *in the future*. Hence, we cannot treat it as an 'objective' or resolved state of events, whatever the scale. To treat it as such, and so assume the veracity of the block universe approach, would be tantamount to treating the future as a set of resolved events *before* it had transpired into such, which requires 'signals from that future' in evidence of such a presumption. This may be permissible in science fiction, but it is impossible in practice. Physical 'signals from that future' will entail attendant violations of principles of causality and principles of conservation of energy-matter, insofar as such signals would be registered as bits of energy or matter above the given unity of energy-matter. This insight is central to the resolution of the *raison d'etre* of the quantum superposition principle, the quantum mechanical wave, and its attendant wavefunction: It explains why we must live in a quantum mechanical universe: It validates the core of quantum mechanics in a way counter to the oft characterisation of the science as a-causal, 'crazy', or even 'beyond understanding'.

0.04: Broken-Discontinuous Motion & Information-Matter Relations

Insofar as there can be no seamless-continuous trajectory, the motion of objects in putative space must necessarily be structured as a succession of finite realised 'snapshots' or 'frames' along spatial intervals: analogous to the finite frames that make up a motion picture. Thus, the motion and transportation of information by radiation and matter in space must also assume a tandem broken discontinuous form.

At the proper observation level, a particle, or the point-like leading tip of a body in motion along AB, will first appear at one position. It will then be removed from space to be replaced by a quantum mechanical state of the pertinent radiation or matter: one dispossessed of any sort of realised or 'objective' positional information. The particle will then re-appear at a succeeding point further along AB. It will again be removed from space and be replaced by another quantum mechanical interval; again, bereft of any realised positional state.

The process will repeat a *finite* number of times along AB, until the particle is finally resolved at position-B.

The quantum mechanical intervals interspersed between the 'snapshots' composing the broken-discontinuous unfolding of an object's motion will not and *cannot* contain any realised positional information or *any* concrete information: For reasons given, such intervals can only be understood as pure possibilities or potentialities; *not* actualities. If this were otherwise, seamless-continuous motion and trajectories would become possible; the impossible physics of zero-wavelength states, ultraviolet catastrophes and infinite power sources would be at hand; and there would be no need or basis for quantum mechanics.

We now replace our original concern with position along AB with a related concern over the transportation (presumably by the pertinent radiation or by matter-waves) of complex states of organisation and pattern: i.e., *information*... such as a photographic image carried by light radiation. We may conceive the physical transportation of information as a literal place-to-place displacement of the information by the 'carrying' radiation or de Broglie matter-waves, or as information transported 'across' time, from moment to succeeding moment, with or without motion.

In the transportation of a photographic image by light radiation from a source-object at position-A to a developing film at succeeding position-B, the information is expected to be deposited as a coherent photographic image at B. In the case of information carried by matter or by de Broglie matter-waves, we might consider the transport or time-evolution of a snowflake pattern, embodied and 'carried' by the corpus of matter. In either case, to entertain the unity of information with carrying matter and radiation, we must obtain impossible seamless-continuous motion and time-evolution and trajectories. The impossibility of this leads to **radical information-matter dualism, and to the broken-discontinuous transportation and relation of information to space and to the 'carrying' radiation and matter**.

In short, in photography, the information appears to 'pulls out' of radiation and space, until it is restored to radiation and space at the destination point. The same applies to the snowflake pattern 'carried' by quantum mechanical de Broglie matter.

*

It is easy to admit the need for a more refined definition of information: This must await Book-I and beyond: The goal will not be completed until Book-IV. At this juncture, the definition of information is implicit to the photographic image, to the electron micrograph image carried by electron de Broglie matter; to the 'snowflake pattern' carried by and exhibited in matter. Such informational states must be subject to broken-discontinuous relations vis-à-vis their 'carrying' radiation, matter and space.

The complex information 'carried' by the corpus of radiation or matter (i.e. the photographic image; the snowflake pattern) will manifest at one position along AB; only to be replaced by a quantum mechanical superimposed group-wave entirely bereft of that resolved informational state: i.e. a superimposed state of *potential* information dispossessed of any realised form, in the same way that the particles that make up that group are also dispossessed of any realised positional attributes in such intervals. The complex information will then make a re-appearance in space amidst radiation and matter at a succeeding position along AB, only to be replaced by another quantum mechanical interval thereafter. This process will be repeated until the transportation of that information in its broken discontinuous form is concluded in its final re-manifestation at B. Throughout its transportation across AB, **the complex information will manifest in and along space and in its 'carrying' radiation or matter, at a finite number of points or 'frames'. Within the quantum mechanical intervals between each such 'frame', the concrete form of information will** *apparently* **withdraw from both space and from its 'carrying' radiation or matter. Yet, at each re-formed subsequent 'frame' further along AB, this information will return to and re-merge with space and the 'carrying' radiation or matter**, and finally so at B.

The central feature of what we have just described consists of two facile facts: The first fact consists in the *apparent* **physical withdrawal of information**: i.e., the *withdrawal* of the resolved photographic image or the snowflake pattern, from both space and its 'carrying' radiation or matter. The second fact consist in the **apparent implicit** *retention* **of the withdrawn information, retained independently and at a** *remove from* **the space, from radiation or from matter, across the quantum mechanical intervals**. That is, the information appears somehow physically retained in an abstract form at a remove from space, from radiation and from matter. Yet, information returns to space, radiation, or matter at moments of resolution along AB, and at the destination B.

Upon its withdrawal from radiation, matter, and space, where does the information go? Where is the information retained? This mystery will eventually be clarified in the various books of this work into the argument for abstract memory in a growing block universe: Simply, **nature has memory**, independent of what is manifested in putative radiation, matter, and space.

*

Radical information-matter dualism appears to follow from the above arguments, at least as a first approximation: one to be superseded by superlative conclusions based, amongst other things, on de-spatialisation, temporisation, the reformation of spacetime itself, and from the reformation of the relation of information to time in Books I through to V.

INFORMATION & MIND

0.05: Information-Matter Relations & Mind

With facile information-matter dualism in hand, we now come to a most radical implication: The dissociation of information from radiation, matter and space is not restricted to photography or to the drift of a snowflake. It holds true in *all* cases where there exists a manifest physical relation between information on the one hand and ostensive radiation, matter, and space on the other. As such, facile information-matter dualism ought to have direct implication to memory and mind vis-à-vis the brain. This is on the assumption that, illusion or not, mind and conscious memory are assumed to be identical or reducible to the structures, patterns, and processes of the brain. This is per *a priori* materialism: Therefore, **the brain cannot be exempted from the dualism of information versus matter and space, given the generality of information-matter dualism, especially when it is asserted that mind and memory must be one and the same with, or produced by, the material organisational pattern of the brain. Information-matter dualism will impose the physical dissociation of both memory and mind from the group corpus of matter that constitutes the brain, even when we grant a false veracity to materialism.** Nor could this conclusion become obviated by dubious scale-arguments about quantum mechanical phenomena. Even if we reduce the quantum effects to the imperceptibly small, it and the dissociation of information attending it abides i.e., Just because one cannot perceive the effect, it does not imply that it is not real or absent. Hence, insist in materialism, and one cannot circumvent information-matter dualism and consequent memory-brain and mind-brain dualism: In its dissociated form from its supposed carrying material brain, memory and mind will be retained in abstract-immaterial form at a remove from that brain. Of course, all of this assumes that memory and mind is in the first place reducible to brains. Is this last assumption true?

0.06: Towards a Definition of Mind

The question of what mind and consciousness constitutes has engendered wide dispute, ranging from outright *eliminative materialism* and the rejection of consciousness as a pseudo-term of folk psychology, through to *idealism* and *dualism*. Idealism is almost comprehensively rejected. Dualism has been relegated to the periphery; almost totally dismissed. The materialist contention dominates. Yet, dualism turns out to be the correct view, *if* our claim for facile information-matter dualism abides. Although facile information-matter dualism will be totally superseded by Book-IV, especially based on de-spatialisation, this will culminate into the same mind brain dualism and Cartesian revival obtained from facile information-matter dualism.

Let us postulate two preliminary working definitions of nous: The *phenomenal definition* and the *operation definition*. Both definitions are complementary and come prior to any assumption one might have about what mind is or what the world is like: Therefore, both definitions might elicit general starting agreement about the subject. **The phenomenal definition defines nous as a specific form or aspect of reality, but related only or solely to structures like the brain, and to no other structures in the universe.** In short, mind is the phenomenon that applies to brains; *not* to rocks, cups, pumpkins, etc.

We might all agree with this tautological phenomenal definition, save that materialists deny any causal power to mind. Indeed, the principal disagreement is on the causal locus: Is causality reducible and identifiable with the variables and structural process of the brain, such as to be wholly located in and generated by the brain? This is what forlorn materialism asserts. Or is the brain and the world the conjurance of an overarching mental causality; one that can conjure the brain and world as its illusion? This is what idealists from Bishop Berkley to the more haphazard New Age movement and *some* quantum consciousness advocates would have us consider.

Alternatively, is mind extant from and irreducible to the brain, belonging to an ontology coeval with putative material-spatial and temporal ontology, as Plato and Rene Descartes assumed, and as we assert? Whatever our disagreement, we can all agree that 'mind' refers to that specific *something* that relates only to brains.

In **the operational definition of mind, the *mind is defined in terms of what it does*.** While the list is quite extensive, we need only choose specific items: namely, **hard problems that engender serious challenges that disputants must confront and overcome, if their respective views are to prevail.**

Let us pose *renormalisation* as one example of a hard problem. Renormalisation predates quantum mechanics, although it emerged in a pronounced way in quantum field theory in the 1950s. Therein, when two particles approach, they purportedly interact by exchanging mediating particles. Usually, the main particles fly apart because of such exchanges. But if we 'zoom in' into the points or region of these interactions, we find not just one mediating particle but a complex network of possible particle exchanges. As we zoom in further, this explodes into even more complex networks of particle exchanges; and so on, *ad infinitum*. Indeed, the system 'blows up to infinity', and calculation becomes impossible. In order to overcome this problem, physicists turn to renormalisation: Simply put, they subtract the infinities from the infinities.

Whether one deems renormalisation a contrivance or a perfectly justifiable technique, it raises a problem that a valid theory of mind and consciousness *must* successfully account for: **What sort of universe can render possible the renormalising quantum mechanist? What sort of causality could operate within the brain (assuming materialism) such as to allow renormalisation?**

In order to renormalise, the physicist obviously needs to apprehend infinity. **From where does the physicist get the idea of 'infinity'? It cannot come from sense-experience:** There is nothing in the sensate-empirical data that can analogue infinity. Sensate experience is always experience of the finite. The fact that we can obtain insights into infinity without relying on sense-data forces the conclusion that knowledge cannot be wholly sourced in the empirical.

QUANTUM MECHANICS AND MIND

Does the brain of the physicist generate the idea of infinity to permit the possibility of renormalisation? If we assume materialism, we immediately confront two problems: The first is ***the problem of memory***: **How do we store 'infinity' in the finite structure and capacity of the brain?** Indeed, what would the 'memory of infinity' look like in biochemical and neural terms, or as brain 'code'? The second problem is ***the problem of process***: **How could the finite iterational clockable processes of the brain generate the apprehension of 'infinity' without crashing into insurmountable non-terminating or halting problems?** The materialist usually dismisses these problems by a sort of hand-waving: The materialist might claim that, once brains attain a level of sufficient complexity, both the storage and apprehension of infinity is solved: The complexity fallacy is a dubious and obvious evasion.

Nor can we avoid the problem by dismissing infinity as a *bloss* subjective idea, or even a false idea: In order for generic quantum field theory to work, and for calculations to be made, renormalisation is indispensable. Therefore, the intuition about 'false' infinities requisite to renormalisation are also indispensable. Moreover, quantum filed theory underpins quantum electronics. Quantum electronics is workable because underpinning quantum field theory permits calculations. Quantum field theory works because renormalisation makes possible such calculations, critically dependent on the physicist's grasp of infinities.

What about the pre-quantum infinitesimals and the succeeding limits-based approach to calculus, central to modern physics and to all manner of engineering? All of this presupposes intuition and apprehension of infinity. In short, without infinities, there would be no quantum electronics, no modern technology, no modern society.

Moreover, even basic dictionary definitions entail infinities: The infinitive, "to run" (for how long?); "to be", etc. No fallacy of equivocation here: "to run" specifies no time-limit and implies unboundedness. Hence, without the 'falsehood' of infinity, we could not engage even in basic dictionary definitions, meaning-apprehension, or language transactions.

Finally, **even if we sought a materialist account of our intuition of infinity in terms of the hands-on mechanical processes of the brain, information-matter dualism would have its way with it**. To grasp this, consider that, per materialism, renormalisation and our ability to handle infinities must resolve as a series of clockable iterations and outcomes in the brain, exhibited in and through successive moments of that brain: one necessarily structured into the broken-discontinuous form of spatial motion or time-evolution. Therefore, our grasp of infinity and the renormalisation process would be subject to information-matter dualism… *and* dissociation. Thus, even with materialism assumed, **we *must* look for the causal basis of the renormalising quantum mechanist, and for the intuition of infinities itself, in the causal process that resides *between* the succession of moments that engender the output-events of the brain**.

Both renormalisation and infinities, as consummate hard problems, smack of mind-brain dualism and intimate inescapable radical conclusions about causality pertinent to nous. Indeed, even discounting implications from information-matter dualism, **the intuition into infinities must involve a non-clockable process that cannot be realised through brain-writ material structure, since such an in-brain clockable process and structure would invariably run into an insurmountable non-terminating problem**, aside the question of what the in-brain 'memory of infinity' would look like. Thus, while generic hard problems, such as qualitative perceptions, emotions, and intentionality, can be unjustifiably dismissed as ambiguous, renormalisation and infinities suffer no such ambiguity. As such, both constitute a challenge to materialism: a challenge that it must ultimately lose.

0.07: Summary of Preliminary Part-I

- Two types of motion are conceivable: **seamless-continuous motion,** wherein a putative particle can occupy all infinite positions along AB… versus **broken-discontinuous motion**: wherein a particle jumps from one resolved position to a succeeding one, with gaps between each, within which positional attributes do not exist.

- **Putative seamless-continuous motion would crash into practical infinities**. It would also require observational inputs **of infinite frequency and energy** *and* even culminate into 'signals from the future', with violations of both causality and the principles of conservation. (See **0.02**)

- **Broken-discontinuous motion necessarily implies the broken-discontinuous association of the information with its 'carrying' radiation, matter, and space**. The photographic image carried by light radiation (or the electron micrograph image by de Broglie matter) appear to be withdrawn from space, radiation, and matter… and are somehow restored to space, radiation, and matter at a succeeding moment. Hence, **information-matter dualism**. (See **0.03** and **0.04**).

- **If, per materialism, we assume mind, consciousness and memory to be reducible to the brain, information-matter dualism will force the dissociation of such mind, consciousness and memory from brains** into an independent retention of mind and memory, extant the radiation, matter and space constituting the brain. Hence, at least *preliminary* **memory-brain and mind-brain dualism follows**. (See **0.05**).

- Our grasp and intuition about infinities cannot derive from empirical sense-data or from the world: **infinity cannot be computed by any clockable material iterational process in brains, given that this would crash into halting problems, aside the question of how the infinity would be retained in and as brain structure or as 'material memory'**. (See **0.06**).

- The intuition of infinity is made manifest in renormalisation in quantum field theory and in the very possibility of calculus, without which science and engineering would be rendered impossible. It is also inherent in basic dictionary definitions. **The intuition of infinity constitutes the clearest *hard problem of consciousness*: It short-circuits materialism and appears**

to force conclusions in favour of non-clockable processes extant the brain. Hence, mind-brain dualism, consistent with information-matter dualism. (See **0.06**).

Information-matter dualism appears to be true: its implications to memory and mind vis-à-vis the brain are also seemingly true. But it belies an even more complex superlative daz. It cannot be the last word, although it constitutes the first salvo of our Alternative Realist thesis.

Facile information-matter dualism presupposes space, with tandem notions of instances of 'matter in space', with instances of physical information distributed in space vis-à-vis radiation and matter. The dissociation of information from radiation, matter and space poses an immediate problem about 'where the information goes' upon dissociation. We need only pull on the remaining loose threads and all will unravel before our eyes, to reveal a more profound de-spatialised and purely temporised universe; an ephemeralised order of information as abstract memory and potentiality; and causality without contiguity or contact; and time, or 'wavefunction collapse', without gears.

The question of where the information goes is a problem solved by de-spatialisation and, with it, the emergence of a superlative form of information-matter dualism beyond the facile form so far explored. Therein, information must be reframed as a distribution in pure time. Therein, putative 'matter' becomes nothing more than the most recent resolution of events… wherein spatial *atomos* transforms into and is superseded by *durata*. Therein, the resonation of the past vis-à-vis the most recent events must transpire across time, *without* mediation across space; without being 'carried' by radiation or matter; and without contact or impact… simply because *there is no space*. Realised information transforms into the abstract de-spatialised past belonging to a growing block universe: a culminating discovery, the basis of which will be fully disclosed by Book-IV.

PRELIMINARY PART-II:
INTRODUCTION TO QUANTUM MECHANICS:
MYSTERIES AND SOLUTIONS

0.08: Introduction

The Copenhagen interpretation is the preeminent interpretation of quantum mechanics. Yet, all interpretations fall short of providing comprehensive solutions to the various quantum paradoxes, including the Copenhagen interpretation: None present decisive reasons for why nature exhibits the superposition principle. All interpretations tend to fail to offer any basis for novel phenomena, including even facile information-matter dualism, or furnish decisive insights into known physical mysteries, such as inertia and gravitation. Consequently, generic quantum interpretations tend to remain beyond decisive testability, despite the sharpening of matters since the Paris experiments in 1981, Bell's Inequality, and Wheeler's delayed choice.

Our Alternative Realist Interpretation (Alt-R) constitutes the groundwork for a future quantum mechanics-ii. It promises to resolve the ontology or *raison d'etre* of the superposition principle, opening the way to novel phenomena and the clarification of old physical mysteries; culminating into solutions to the conundrum of mind and consciousness.

Moreover, Alt-R will not constitute a naïve realist or implicate order critique of quantum mechanics: Succinctly, Alt-R will accept the reality of the superposition principle and objective quantum indeterminism. Finally, Alt-R will not constitute an attempt to invalidate quantum mechanics itself.

Finally, the solution to the problem of nous made possible by Alt-R will have no semblance to Wigner's 'consciousness causes collapse' approach. Indeed, critical facets of Alt-R will be developed out of the critique of Wigner's inadvertent quantum idealism (Book-V): Alt-R will conclude that consciousness, while ontically real, *does not* cause the perennial 'collapse' of the quantum wavefunction, and that consciousness is *not* the primary causality or ontology: The universe is not made up of 'mind stuff', as the 20th Century cosmologist Arthur Eddington assumed. Indeed, Alt-R resolves the relationship between nous versus the quantum mechanical order by unravelling *all* the extremisms: namely, idealism *and* materialism.

With these disclaimers, what follows is a summary appraisal of quantum mechanics: its core mysteries and intimated solutions, beginning with the general problem of wave-particle duality.

WAVE-PARTICLE DUALISM & MYSTERY:
THE DUALISM OF OR-FORM VERSUS AND-FORM LOGIC

0.09: OR-Form and AND-Form Logic

There is an entirely accessible way to apprehend both the nature of quantum mechanics and its central mystery, furnished by **the simple distinction between AND-form logic versus OR-form logic: synonymous with wave-particle duality**.

The 'common-sense' naïve realist or classical physics-view of the world asserts that we grasp nature in accordance with OR-form logic. Therein, **a particle or object is expected to be located at either one position OR at another: never at both simultaneously:**

Either here OR there. *Never* **here AND there**. This view appears validated in experience and, as the core of the naïve realist view, asserts that the universe *ought* to be constructed exclusively according to OR-form logic. Therein, what holds true for the position of a particle must also hold true with respect to its momentum. Thus, the particle's velocity multiplied by its mass *must* have just one value, not a range of different values, much less a *unitary superposition* comprised of many momentum-values.

Position and momentum are not the only attributes expected to be subject to OR-form logic. Using the somewhat misleading analogy of a spinning top viewed from above, according to OR-form logic, a given particle is expected to either spin 'clockwise' OR 'anticlockwise' (actually, "spin-up" OR "spin-down") … but *never* 'clockwise' AND 'anticlockwise' simultaneously.

Self-consistent quantum mechanics does not claim in any incontrovertible sense that the world based on OR-form logic cannot exist, even though it appears unable to accommodate the OR-form universe. Instead, quantum mechanics adds another contradictory seeming counter-logic to the undeniable experimental and experiential reality of the OR-form reality: one that appears incompatible with, and is yet complementary to, the OR-form order. Thus, **quantum mechanics demands the equal reality of AND-form logic in nature**… although it does not explain or account for the OR-form universe, in that the latter does not appear to arise from it by necessity or seamlessly. Yet, the reality of AND-form logic is as thoroughly affirmed in experiential and experimental terms as is OR-form logic, despite their seeming incompatibility.

Consequently, according to quantum mechanical AND-form descriptions, a **particle *can and does* simultaneously occupy one position AND another; and does so in a physically unitary form: as a *superposition*.** The same AND-form logic asserts that the particle need not be confined to just one momentum-value but a range of momentum-values in AND-form. Per the same AND-form logic, a particle is fully permitted to spin both 'clockwise' AND 'anticlockwise' and can assume the unitary superposition comprised of 'clockwise' AND 'anticlockwise' spin.

This **AND-form reality stands fully experimentally confirmed in the staple double-slit or two-slit experiment**. Consequently, the reality of AND-form logic in nature, **complemented by an equally undeniable OR-form logic**, cannot be denied, no matter how 'impossible' and contradictory it appears. Of course, the central paradox is how is it possible that the physical world can be informed and structured by both OR-form *and* AND-form logic: by logics that are seemingly mutually incompatible? How could such AND-OR duality ever arise without incurring the breakdown of order and sense? This problem is often approached as if incapable of any sensible solution: Richard Feynman said… "I think I can safely say that nobody understands quantum mechanics".

The paradox entailed in **AND-OR duality just outlined is typically known as *wave-particle duality***: Restated in our terms, wave versus particle duality is simply the dualism of apparently incompatible AND-form versus OR-form logics: Wave-particle duality is interchangeable with AND-OR duality. Hence, in quantum mechanics and physics, matter and radiation is both wave-like *and* particle-like. The wave-like description of matter and radiation pertains to the AND-form description, wherein a particle can assume a range of possible positions, momenta, and spin states, in unitary superposition AND-form: with some of the possibilities more probable than others; with the most probable possibility constituting the 'peak' or highest probability within the AND-form wavefunction (the highest amplitude). Projected in time, these probabilities are generically described as undulating and time-evolving, oft resembling a sinusoidal wave. Hence the waveform inherent to AND-form nature.

On the other hand, the particle-like description of matter and radiation pertain to the OR-form description of nature, akin to our 'common sense' naïve realist experience; wherein a particle can only have one concrete exclusive position; only one resolved momentum-value; and must spin in just one of two ways, *not* both.

0.10: Wave-Particle Duality: The Physical Meaning of the Quantum Mechanical Wave

With wave-particle duality comes the claim that the wave and particle descriptions of nature, the AND-form *and* OR-form description of nature, are 'complementary'. A mystery arises as to how such a complementarity can be possible between apparently mutually contradictory logics, or how nature could exhibit both 'particle-like' *and* 'wave-like' facets: This is a mystery that Alt-R promises to solve in a definite way.

As stated, the wave-like description of a particle is the AND-form description of that particle: a unitary superposition of *all* possible positions-states that the particle might get to assume in a mutually exclusive way upon subsequent 'measurement'. The same holds for particle momentum-values, for particle spin, etc.

In contrast, the particle-description, or OR-form description, asserts that the position, momentum and the spin of the same particle *must* assume a definite OR-form state, even in the absence of observation and measurement. The same ought to hold for momentum and spin.

Which of these descriptions is true? Is it the wave-description or the particle-description? Is it AND-form logic or OR-form logic? The paradoxical answer is that *both* wave and particle descriptions are correct: Thus, *both* AND-form logic *and* OR-form logic abide. The staple experiment which demonstrates the reality of AND-OR dualism is the famous double-slit or two-slit experiment: We will examine it in due course.

*

For understandable reasons, some have claimed that AND-form logic (the central, but by no means the only basis of quantum paradox) must be wrong. At some deeper level, reality must be purely OR-form. The 'common sense' OR-form world must be the only world and physics possible. This view is at the core of the naïve realist and implicate order approaches espoused by Albert Einstein, David Bohm, *et al*.

EMRE ASENA

Let us imagine a scenario involving a ship on a journey across the Atlantic from, say, New York to Southampton. This analogy was first used by James Jeans in his attempt to explain away the quantum mechanical wave (i.e. AND-form logic) as illusory[7]. The scenario is deliberately erroneous and, as such, useful as the profile of the false. We can use it as a springboard for the correct account of why nature exhibits AND-form logic.

In Jean's scenario, the navigator onboard the ship must make a series of accurate-as-possible determination of ships' position using old methods of dead reckoning at set moments throughout ship's journey. This is subject to an accumulation of uncertainties pertaining to exact position, which will increase over time. On the map, the navigator will describe the accumulating and increasing uncertainty in ship's position as an ever-expanding navigation circle, which only begins to shrink toward the end of the journey and finally reduces to zero on arrival at Southampton.

The uncertainty, drawn as the expanding navigation or dead-reckoning circle on the map, represents the navigator's informational ignorance about ship-position: It does not represent any literal or objective uncertainty: Thus, the ship's position is real and certain, and only the navigator's ignorance of an exact determination of its on-map position engenders uncertainty; an uncertainty purely in the navigator's mind. From the point of view of the navigator, the ship could be anywhere within the map-drawn navigation circle. Indeed, for purely abstract purposes, we might assert that the ship occupies *all* of the possible positions within the said navigation circle: i.e., "position-1" AND "position-2" AND... "position-10". As such, the navigation circle can be said to represent a *unitary superposition* of all positions that the ship assumes within the map-drawn navigation circle, by function of the navigator's accumulated ignorance and subjective limitations.

Thus, assuming the oversimplification of a total of ten possible positions within the navigation circle, by journey time t_n, the unitary superposition for the ship may be described by means of the following simple layperson's formalism or 'lay formalism':

Ship (position at t_n): {position-1 AND position-2 AND...., AND position-10:
Such that the ship assumes all positions}

Or, what is the same...

Ship (t_n): {[AND] position-1, position-2...., position-10 | Ship assumes all positions}

However, this is not objectively true of the ship itself: All the navigator needs to do is peek out of one of the portholes and grasp the fact that the ship has only one real position-set, *not* a unitary "smear" of ten distinct ones. Even so, the navigation circle will not go away, per inevitable informational ignorance. Thus, our navigator is right to assume that, despite what the navigation circle implies, the positional situation of the ship is not intrinsically an AND-form superposition of ten possibilities, but a collection of subjectively determined positions, with only one *real* position: all structured in accordance with OR-form logic.

Thus, the navigation circle *corrected* according to sensible OR-form logic and common-sense reality, will assert...

Ship (position t_n): {position-1 OR position-2 OR..., OR position-10: Such that, only one position is real}

...shortened further to...

Ship (t_n): {[OR] position-1, position-2...., position-10 | Only one position is **real**}

Is it possible that quantum mechanical AND-form logic is merely a product of our ignorance about a particle's position or its other attributes? Does AND-logic arise from similar factors as holds in ship navigation and dead reckoning? James Jeans speculated so and attempted to explain away quantum AND-form logic and the quantum wave as an illusion. Yet, his view simply cannot work. In the staple double-slit experiment, when single particles are allowed to pass across a barrier with two slits open, they are experimentally determined to pass through both slits: apparently simultaneously. This is ascertained from the interference pattern formed at a detector placed in front of the said barrier. The interference pattern is formed by way of the subsequent overlap of two quantum wavelets produced by the passage of a single particle through *both* open slits. Such a feat is beyond the purview of OR-form logic.

A ship that must make a choice between, say, two fiords leading to a common inland lake, will not literally sail through both fords, no matter what the navigator's on-map dead reckoning circle might suggest. Just because the dead reckoning circle happens to describe a region that overlaps across both fiords on the map does not imply the ship is travelling through both fords at the same time, or that it could form two separate positional wavelets upon exit from the fiords.

Moreover, the ship will not subsequently overlap and destructively or constructively interfere with its path from one fiord by its other path from the other fiord: It will not produce 'self-interference', as do wave-like AND-form particles exhibited in the famous double-slit experiment. In the double-slit experiment, a particle appears to occupy putative positions at "slit-1" AND at "slit-2". If the particle were a ship, and if it were able to pass through both slits at the same time, the ship would be described as something objectively occupying *both* positions simultaneously: a unitary smear of both positions. Unlike ships, particles are indeed experimentally determined to form such unitary "smear" of manifold putative positions, and it is no wonder we have ultimately falsifiable claim to the effect that quantum mechanics applies only to the very small.

---

[7] Jeans, J., Physics, and philosophy (Cambridge: Cambridge university Press, 1942

14

To reiterate, in the context of the double-slit experiment, the particle apparently gets to occupy positions at both slits *and* passes through both "slit-1" AND "slit-2", presumably at the same time. Thus, unlike ships, this unitary "smear" for possible positions appears to be an intrinsic-objective fact about particles: it cannot be dismissed as our subjective informational ignorance about its 'real position'.

How do we explain this? Why does nature assume this intrinsic non-illusory AND-form logic? Why is nature not restricted to only OR-form logic? How does nature reconcile the apparent contradiction between OR-form and AND-form logic, both of which she clearly manifests?

*

To glimpse a rational and sensible solution, we need to imagine a non-standard navigational description for a ship that has not yet left port.

This is absurd. Why would anyone want to perform a navigational determination for a ship that has not left port? Is not its position already obvious?

Imagine a U-boat captain during WWII: He needs to forecast the possible positions that an enemy ship presently anchored at New York *might* assume... *in the future*: In other words, he needs to make a *forecast of future possibilities* for the target ship currently at port. This forecast is critically necessary to his upcoming mission. The captain has access to intelligence to the effect that the ship will most likely take a direct route from New York to Southampton (i.e. Possibility-1). His navigator draws this path on a map of the Atlantic Ocean, stating it as the most probable position-set that the ship might assume over the course of its *future* journey. But a more comprehensive forecast must also assume other possible future-sets for ship's journey. Intelligence suggests that the ship might not depart from port at all (i.e., Possibility-2): The navigator treats this as the second most probable future possibility and makes note of it. The third most probable case is the case of the drunken navigator: Intelligence suggests that the enemy ship's captain, its first officer and its navigator are recovering alcoholics, likely to revert to their addiction due to the vagaries of war. Thus, the future prospect of a meandering ship's journey, one that almost never arrives at Southampton, designs for Possibility-3. Moreover, there is an added future possibility that the ship might never arrive at its set destination due to a sharp increase in the number of icebergs: Possibility-4: the iceberg-struck ship.

There are other future-possible paths: a gamut of least probable paths that the ship might take on its way to Southampton. Let us label these Possibility-5, 6..., up to possibility-n). The forecast for the ship that has not yet left port must also include these. In descending order of probability, our lay formalism describes the set of future possibilities for the target ship as....

Ship at New York (future):

{Possibility-1 **AND** Possibility-2 **AND** 3...**AND**..., **AND** Possibility-n: | Ship assumes all; most to least probable}

This can be shortened to...

Ship at New York (future):

{[**AND**] Possibility-1, Possibility-2,3,4,5,6, 7..., Possibility-**n**: | Ship assumes all; most to least probable}

The formalism amounts to an AND-form unitary superposition for navigational possibilities that the target-ship might get to assume *in the future*: This is essentially identical to AND-form unitary superposition found to hold for particles in quantum mechanical waves.

Note that the possible ship-positions in the future-forecast cannot obviously be rendered in OR-form 'common sense' terms. Why? This is because the possibilities have not yet transpired into concrete resolved events. Future events can only be described in purely AND-form abstract terms. In short, the future has not yet happened. Therefore, one cannot describe it as a set of resolved OR-form events. The future can only be described as an as-yet-to-happen AND-form *potentiality* for events, structured per their probability of occurrences.

On the other hand, resolved events that *have* come to pass may be described in OR-form terms: Insofar as the ship is, or has become, a resolved set of events in retrospect, it can only be 'here' OR 'there', *not* 'here' AND 'there'. On the other hand, a ship in its future-possible configuration must be *objectively* potentially *here* AND *there*, and its quantum mechanical characteristics will not be dependent on or be annulled even by considerations of scale: Whether we consider a single particle or a whole ship, **the quantum mechanical AND-form logic and consequent superposition principle, insofar as it constitutes an objective future pertinent at all scales (for particles *and* for ships) will apply**, though *if and only if* we are attempting to describe *the future-form* of that particle, ship, or **anything else... including brains**. Hence, the pertinence of AND-form logic is not exclusive to the very small scales or to very short time-intervals, as is often dubiously claimed. Indeed, we shall soon augment our radical scale-invariant view of quantum mechanics based on the intertwine between the principle of causality and the principle of conservation, and, later, in the context of the growing block universe.

*

Note that, within our lay formalism, with respect to the future-form of the ship, we can attach to each AND-form possibility a probability-value: from the most to the least probable. All these must sum to 1. Our lay formalism then assumes a similar but a stripped down variant of hypertrophic quantum formalism: Our lay formalism economises and omits the hypertrophy but conveys essentially the same quantum mechanical form and logic, albeit in accessible form.

Thus, assigning some meaningful probability values...

15

Ship at New York (future):
{[**AND**] Possibility-1(0.01), Possibility-2 (0.02), Possibility-3 (0.03), Possibility-4 (0.04) …, Possibility-**n** (0.nn) | Ship will assume all from most to least probable outcome, and all sum to 1}

Our lay formalism will be used throughout this work. It is highly recommended: It simplifies and renders accessible the central logic of quantum science, and it might prove useful as a shorthand method for communicating about the subject to laypersons, to new students, and to professionals in circumstances where precise calculation is not needed, and wherein communication of concept and logic is the goal.

Returning to our theme, in the case of quantum reality, the AND-form quantum mechanical superposed possibilities for a particle are *not* in the "mind's eye": The superpose is literally and physically "out there". The superposition AND-form logic is an objective fact about the particle *in its future-form state*. This also applies fully to the future state of macro-scale objects, like people, ships, planets, and even brains… *before* these are resolved into OR-form event-complexes. Thus, the answer to the question as to what the quantum mechanical wave physically means, hence the *raison d'etre* for why nature assumes AND-form logic, follows in an obvious way, resolving and harmonising the long-alleged 'incompatibility' between AND-form versus OR-form logics and physics. In short, **the AND-form quantum wave of a particle constitutes the physical manifestation of the future-form of that particle**. Indeed, it constitutes the future-form of spacetime pertaining to that particle. Thus:

- The future is objectively real, albeit as potentialities 'out there': *not* bloss subjective.
- The quantum wave is the physical manifestation of the future: the future is objectively real, even if not-yet resolved.

It is precisely because the AND-form state is physically real and intrinsic, and *not* some *bloss* subjective imposition of our ignorance or informational inadequacies, that the future identified with it turns out to be physically intrinsic and objective: a futurity independent of the existence of minds, their opinions, or their ignorance. The future is objectively real, and it physically manifests as the quantum mechanical wave.

For its part, the OR-form domain of affairs (i.e., the purview of classical-relativistic physics) constitutes, as a first approximation, the present-continuous forms of the physical order: comprised of AND-form future possibility collapsed into the most recent OR-form realised outcomes. Once grasped in this way, the supposed 'incompatibility' between quantum mechanical AND-form descriptions of nature versus classical-relativistic OR-form descriptions of the same, entirely disappear, if we relate and integrate these to each other in the correct *temporal* way. Indeed, the long-held AND-OR 'contradiction' is a product of our failure to grasp the real nature and significance of the quantum wave and of AND-form logic: i.e., as the objective-physical future-form state of possibilities… as the physical manifestation of the ontic future: the future grasped as objectively real.

The certain reasons why the quantum wave constitutes the objective future will be reiterated based on the **firewall principle of quantum mechanics: based on the intertwine of causality with the principle of conservation.** Therein, any attempt to treat the attributes of a futurised quantum object, or any futurised object at whatever scale, as if it were comprised purely and only of OR-form resolved events before the conversion of its AND-form futurity into such (hence, any attempt to treat of quantum mechanical waves as if these were implicitly OR-form states, and treat the future as if it had already transpired into resolved events) would entail the reception of 'signals from the future', with inescapable violations of the principle of causality, and, with it, the violation of the principle of conservation of energy-matter. Indeed, both the principle of conservation *and* causality turn out to be mutual and intertwined. Because of this, the future *must* remain non-resolved and in pure AND-form, until otherwise resolved into concrete OR-form events. Hence, the quantum mechanical wave *is* the future-form of spacetime: the future grasped as an objectively real physical existent; and as something 'out there': not a *bloss* subjective term.

*

There is one obvious strand that seems to have been missed: It immediately concludes in the objective reality and ontically real existence of the future as a set of as-yet not-happened possibilities. It emerges from the distinction between the block universe (without futures) versus the growing block universe *with* perpetually decaying futures… and how both relate to the two-slit experiment: (**see 0.13**). In the block universe, the future is pre-resolved into definite 'paths' and 'which way' states. In effect, there is no future. **If the block universe held true, the two-slit experiment could never yield self-interference or subsequent interference patterns, given that these could not emerge out of pre-resolved definite OR-form states and 'paths'. In real two-slit experiments, we *do* obtain self-interference and interference patterns. This can only happen if the future-state of the experiment exists, not as resolved 'paths', but as as-yet-to-happen possibilities rendered into probabilities. In short, AND-form logic and the quantum wave and wavefunction really do pertain to an ontic future**, and the quantum wave *is* the future-form of nature: one phenomenalised as an objective existent; *not* in our minds, and certainly 'out there'.

Insofar as AND-form and OR-form descriptions of nature are in fact mutually compatible and harmonious, we establish an unexpected harmonisation and **primitive unification of quantum mechanical descriptions of nature with those of classical-relativistic descriptions** of the same: key to the hitherto elusive unification of physics. We can make a preliminary presentation of this primitive unification as an adjoining of future spacetime (AND-form, quantum mechanical spacetime) with past spacetime (partially OR-form): Thus, the first iteration of our growing block universe approach. Indeed, the integration and harmonisation of AND-form logic with OR-form logic is essentially the integration and harmonisation of the future with the past, constituting a seamlessly unified mutually compatible and mutually necessary amalgam. This primitive unification is to be encapsulated in our intermediate model of spacetime in detail in Book-II and by its superlative de-spatialised successor in Book-IV. The intermediate model will be introduced in Part-II **0.29** and **0.30** of this Preliminary.

0.11: AND-OR Dualism, Future-Forms & Implications to Information-Matter Relations

What is the significance of AND-OR dualism to information-matter dualism? As was discussed in Part-I **0.03** to **0.04**, foundational facts of quantum science seem to compel conclusions in favour of the discontinuous-broken form of the motion and the same in the transportation of complex information by radiation and matter across putative space. We can now integrate quantum mechanical AND logic directly to the information-matter dualism from Part-I.

Using the analogy of succeeding frames evinced in a running old-school movie-reel, it appears that information dissociates from radiation, matter and space in one frame only to re-associate with that radiation, matter and space in a next frame: with each OR-form instance or frame separated from others by a quantum mechanical AND-form interval devoid of concrete information. The said 'frames' constitute OR-form states. The intervals between the frames constitute AND-form quantum mechanical intervals without OR-form attributes or resolved information. In short, **facile information-matter dualism arises from wave-particle AND-OR dualism in a seamless way: Wave-particle AND-OR dualism *and* information-matter dualism are almost synonymous.**

Radiation and matter propagating across space will always consist of an admixture of the quantum mechanical AND-form (the intervals between the 'frames') and the OR-form classical (the 'frames'). No physical system can ever be purely OR-form in the way naïve realism and 'common sense' presumes, no more than it could be rendered into pure AND-logic quantum mechanical form: The quantum mechanical intervals that reside between the OR-form frames constitute the not-yet realised future possibilities for position or information per the 'carrying' potential radiation, potential matter, and potential space, or "Hilbert space".

Once we grasp that AND-form logic is per function of the future-form of nature, this makes for an intermediate but superlative understanding of spacetime (to be fully detailed in Book-II) and a more radical de-spatialised model of the same, *without* 'space' in Book-IV. The intermediate model spacetime will design for deeper insights into information-matter dualism; succeeding the facile variant from Part-I; i.e., information-matter dualism as a facet of the processes of spacetime. Of course, for the full appreciation of this insight, the reader must await Book-II.

Obviously, mind-brain dualism wrested from facile information-matter dualism will attain an even deeper form from the deeper understanding of information-matter dualism that attends our anticipated physics-unifying intermediate spacetime model in Book-II and its superlative de-spatialised version furnished in Book-IV. This achievement will attend an even more profound conclusion about the mind-brain relation in Book-V.

0.12: Mind, Quantum Consciousness & the Quantum Mechanical "AND"

Does mind and consciousness belong to and arise from the AND-form quantum mechanical facet of nature? Is nous quantum mechanical? Is nous an AND-form superposition of all possible future thoughts and their correlate states of consciousness? If so, one would have to confine one's description of mind *only* to the future-form of nature, wherein nous would need to be conceived purely as a future-form and AND-form state of *potential* thought and consciousness. The problem therein is that the future-form of possible though, action and consciousness constitute an as-yet not-realised thought, action and consciousness. It is merely a set of alternative AND-form potentialities that mind might get to realise. Nous must surely be able to exploit such a menu of future possibilities. If this is so, nous cannot be confined solely to its future AND-form potentiality for thoughts, actions, and consciousness.

We shall discover that mind, as causal agency, is that which exploits the otherwise spontaneous decoherence and 'collapse' of the future-form wavefunction for potential brain-correlates of thoughts, actions, and consciousness in order to bring these into realisation. Thus, mind is *not* a quantum wave or a quantum mechanical phenomenon. Mind must be something independent of the quantum mechanical future-form of possibilities, consonant with Platonic and Cartesian theories.

There *is* an inferred need for a future-form potentiality for thought, action and consciousness, or for their brain-correlates: i.e. an AND-form superposition state for these from which mind can draw out what it seeks in order to bring these into realisation. Yet, mind must supersede and transcend its future-potential AND-form possibilities as thoroughly as it does its past and present expressions. Mind must be dualistic and irreducible to, and ontologically transcend, its AND-form futurity as much as its past. Book-V is destined to conclude in the same.

*

Is mind causally responsible for the wavefunction collapse and the conversion of AND-form future-possible thoughts, actions and consciousness into realised brain-correlate states? Does mind have the causal power to bring about such wavefunction collapse, especially per the brain? Only in idealism and quantum idealism is such a feat entertained; wherein mind purportedly collapses the wavefunction, not merely as it pertains to the brain, but for all observables, if not for the whole universe.

The answer that will emerge out of Alt-R will assert that mind is that which exploits otherwise spontaneous AND-to-OR collapse of the future-form wavefunction to draw out the correlates of potential thought, action, and consciousness. **Mind does not bring about wavefunction collapse: Co-requisite 'collapse' comes about on its own accord, per general causality that subsumes mind, yet transcends mind, as much as it transcends 'matter'.**

The process of wavefunction collapse of the brain and the physical world in general is not a one-time process, but a partial, perpetual one. This partial-perpetual wavefunction collapse can be likened to a perpetually moving water-wave that a 'surfing mind' does not itself generate but 'rides out'… albeit in any 'direction' it wants, but within permissible limits of that wave. Mind is a 'rider' of wavefunction collapse; not the instigator or its causal driver.

In short, the relation of mind to the quantum mechanical-AND is unlike the form entertained in various quantum mind theories: From our Alt-R perspective and its revived Cartesianism, mind is not a 'collapsor' of the quantum mechanical wave, no more than it is itself a quantum mechanical wave.

Nor does mind direct the outcome of perpetual wavefunction collapse. **Mind is not the secret orchestrator of quantum indeterminate incidences; in that it does not alter the probability or 'amplitude' of quantum indeterminate outcomes in the brain**. Instead, per the peculiar structural feature unique to brains, and from the overall consequences to the quantum mechanical future-form of the brain itself, mind is that which selects a specific global wavefunction from the future of the brain into which the brain subsequently collapses via non-directed *en masse* quantum indeterminate incidences. From the future, mind selects the sought whole-brain wavefunction from among other equally possible global wavefunctions, and then imposes its selection on subsequent spontaneous AND-to-OR wavefunction collapse processes, controlling the future development of the brain without having to secretly orchestrate otherwise quantum indeterminate discrete terms, and without needing to cause otherwise spontaneous wavefunction collapse or time.

*

A typical retort utilised against the attempts to link mind-phenomena with the quantum mechanical holds that quantum phenomena have no relevance to the brain. It is asserted that quantum phenomena, from the superposition principle to the reality of quantum indeterminacy, apply only to the micro-world, and has no relevance to the macro-scale brain. As such, the brain ought to be treated purely as a classical object, without resort to any 'quantum flapdoodle'. We ought to abstain from the want to integrate mind phenomena with quantum phenomena, even in terms that circumvent typical quantum mind theories.

But is this view sound?

Clearly, macro-scale billiard balls do not appear to assume any AND-form superposition states for alternate paths. Instead billiard balls seem only ever to assume one path. However, the particle constituents of billiard balls, within the confines of the boundary states that apply to these, may well assume AND-form superpositions of paths. These are not obvious or manifest at the usual scale. But the devil is in the details.

Consider the double-slit experiment, to be fully described in subsequent essays. The experiment manifests both the superposition principle through wave-interference as well as the reality of discrete quantum indeterminate outcomes obtained from AND-to-OR collapse. The wave-interference, and the quantum discrete reality that renders it evidentially manifest, are observationally simultaneously evident upon the detector screen. Yet, as we shall discover, the experiment produces a macro-scale outcome at the detector, from which quantum mechanical realities are clearly inferred, including the superposition principle and AND-form logic.

If the double-slit experiment is a macro-scale situation, as indeed it is, the quantum phenomena exhibited in the experiment must also necessarily be macro-scale phenomena, or scale-amplified into such.

Let us augment this dubious-seeming claim. As we shall find in our forthcoming description of the double-slit experiment, the behaviour of each particle released into the experimental system, and the fact that these abide by the superposition principle, is something we infer from the global macro-scale outcome resolved by the *en masse* deposition of many thousands of such particles on the detector. To infer that these particles are subject to self-interference and the superposition principle, it is not enough to resolve and deposit just one particle at the detector. Indeed, we need a threshold number of such depositions, such as to form a *specific* macro-scale distribution: i.e., the requisite macro-scale global outcome called 'the interference pattern': It is from *this* that we then infer the reality of self-interference and the operation of the superposition principle.

With depositions below the said threshold, we could never infer the reality of self-interference, nor the superposition principle. The point? The said requisite critical threshold is obtained at the macro-scale, not at the micro-scale. That is, we infer the reality of the superposition principle from the evidence generated at the macro-scale. In other words, we obtain quantum phenomena at the macro-scale, and only every infer these to the micro-scale, *subsequently* or in hindsight.

Thus, **quantum phenomena *are* manifest at the macro-scale: The very structure of the evidence for the superposition principle in the double-slit experiment demonstrates this in an obvious way.** As previously stated, in the double-slit experiment, one particle deposition is not enough to infer anything about that particle, let alone the operation of the superposition principle. **It is only per *en masse* distribution of many particles *at the macro-scale*, and per the macro-scale evidence of the superposition principle and self-interference furnished per the macro-scale interference pattern, that permits inference of the superposition principle *and* self-interference**.

The above is not the only proof that quantum mechanical phenomena are not restricted to the micro-scale. However, at this juncture, the example is more than sufficient to break the game: The point is that, at whatever scale one performs the double-slit experiment, one will obtain evidence of self-interference and the superposition principle, as well as of quantum indeterminacy… *at the macro-scale*. This is only possible if quantum mechanical phenomena can be amplified to the macro-scale, and are then rightly judged to be macro-scale phenomena.

Hence, **quantum phenomena do manifest at the macro-scale, and are *not* confined to the micro-scale. They are therefore pertinent to all macro-scale corpuses, including billiard balls and the brain itself**. In other words, what holds true across nature in general holds true for brains: One cannot abstract quantum phenomena from the study of the brain or from mind-brain research, or treat the brain as if it were purely an OR-form classical object…save after the fact, or at post-AND-to-OR collapse and resolution.

*

The other critical reason why we must assert for **the pertinence of quantum mechanics at the macro-scale is per the demands of our anticipated** *firewall principle*, **based on the intertwine between the principle of causality and the principle of conservation**. As previously asserted, both reside at the root of the ontology and *raison d'etre* of the quantum mechanical 'AND'. Recall that the AND-form description of any physical system is its future-form description: an objective state of that system before its observation, measurement or AND-to-OR transformation into realised events, regardless of that system's size or scale; billiard ball or particle.

All systems, at all scales, have a future-form state, insofar as there is a facet to them that has not yet resolved into concretised events and must remain suspended in a state of future potentiality, including macro-scale billiard balls and brains. This is part and parcel of our growing block approach to nature. To insist otherwise, and hence to restrict AND-form logic only to the very small, or to very short time-intervals, is to invite the violation at the macro-scale of both causality and the conservation of energy-matter.

Thus, **causality and the principle of conservation demands that AND-form quantum phenomena (i.e. the future-form descriptions of systems) abide at all scales and hold true at the macro-scale as much as they do at the micro-scale. Billiard balls and brain are not going to be exempt from this**. Therefore, one cannot use the perennial micro-scale restriction as an argument against quantum mind theories, or against our own quantum mechanics incorporative revived Cartesian dualism from Alt-R.

0.13: The Double-Slit Experiment: Particle-Wave Duality & de Broglie Matter-Waves

We now turn to the defining experiment of quantum mechanics: the one that established AND-OR (wave-particle) duality and the superposition principle: namely, the double-slit experiment. There will be two distinct appraisals of the experiment. The standard appraisal is encapsulated in this essay, and the non-standard appraisal furnished later.

The illustration in **Fig. P1.02** depicts the arrangement of the double-slit experiment with three basic components: An emitter (a photon or electron gun) emits a single quantum or 'particle' (a photon or electron) toward an intervening barrier with two slits, which may be set to the open or closed configurations in four possible combinations. Each combination will generate a unique outcome at the barrier from which key conclusions can be drawn. The combinations are…

- Both slits open
- Slit-A open and Slit-B closed.
- Slit-A closed and Slit-B open.
- Both slits closed.

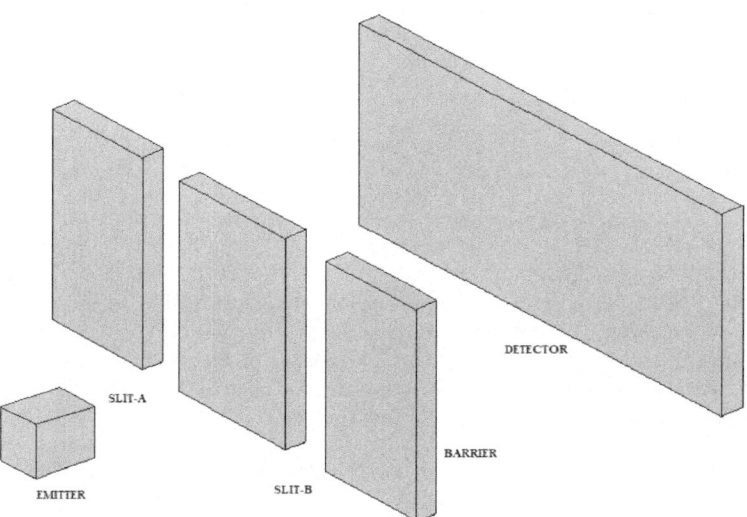

Fig. P1.02 Arrangement of the double-slit experiment
Notes: The arrangement of the double-slit experiment with its three main components: The emitter or electron gun from which particles of light or electrons are emitted. Released one at a time, these must negotiate the barrier on their way to the detector. The barrier has two slits which can be opened and closed in any combination. Beyond the barrier lies the detector on which particles must deposit.

The first and second combinations are the more important ones. The third setting is interchangeable with the second. The fourth produces a null and is typically ignored.

When photon 'particles' are emitted, a photographic film will suffice as the detector. If electron de Broglie matter is used, an electron-micrograph may suffice as the detector. Released one particle at a time, these get to deposit upon and manifest at the detector. By adjusting the barrier-settings, consequent patterns that the particles collectively form on the detector will inform us about the nature and physics of the radiation or matter. Thus, when photons of light are used, we obtain a double-slit experiment that tests for the physics of electromagnetic light radiation. Light radiation, radio waves, microwaves, all belong to electromagnetic radiation: The attendant radiation does not have any mass. When electrons are used, the experiment tests for the physics of matter. Electrons are 'material' by virtue of their mass and the related fact that they do not displace at the speed of light. Yet, like photons, electrons also exhibit wavelike properties. Thus, the double-slit experiment can test for the wave-like nature of matter. When we emit matter in the form of electrons, or neutrons, or protons, or some other exotic particle with mass, we discover the reality of what are called de Broglie matter-waves.

The experiment with just one slit open is illustrated in **Fig. P1.03**: We open slit-A, but close slit-B. We emit pertinent particles one at a time. Some of these will get across the barrier through the open slit. Over time, most of the particles will deposit on the surface of the detector in front of the open slit.

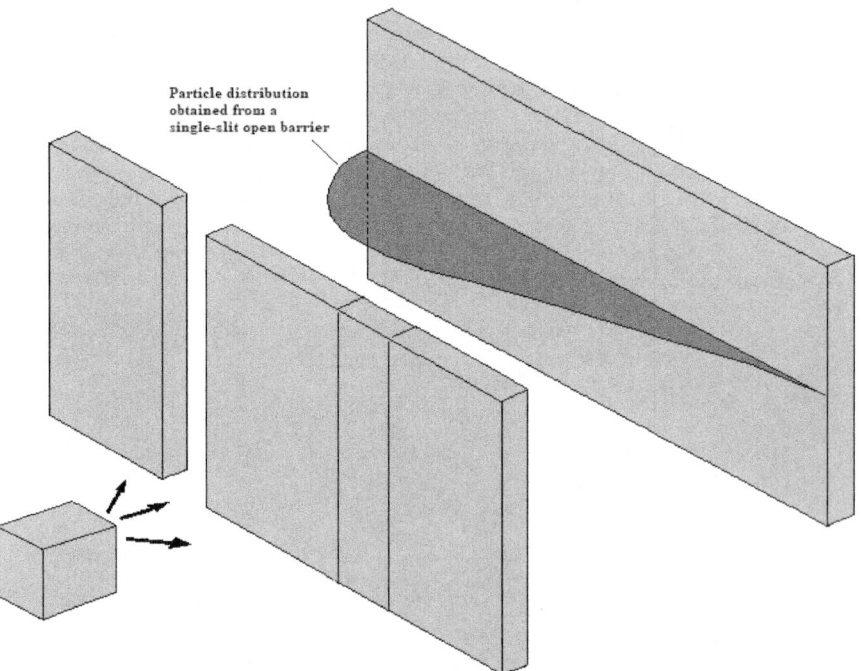

Particle distribution
obtained from a
single-slit open barrier

Fig. P1.03 The experiment with only one slit open
Notes: A didactic depiction of the distribution of particles obtained when only one slit is left open. The height of the curve of distribution along the detector-surface represents the number of particles deposited: Thus, the higher the curve above the detector-surface the greater the number of particles to be found on that detector-surface. The greatest number of particles are deposited just in front of the open slit. The number declines as we move away from the open slit. This distribution is presumed on the assumption that the emitted photons or electrons travel through the open slit as OR-form states set at definite 'which way' paths and trajectories and arrive at the detector-surface as OR-form putative particles consequent upon the said classical paths. However, the same distribution can also be obtained when we assume the quantum mechanical wave travelled as a pure AND-form wave, without resolved 'which way' paths, without *any* OR-form attributes throughout its journey, while the distribution of OR-form particle-depositions consequently obtained at the detector was a matter of quantum probability that culminated into the final total distribution (as shown). If true, the single slit condition does not necessarily imply any special emphasis of the particle-like OR-form character for the generic wave-particle duality.

Assuming a classical billiard-ball picture, one expects that some particles will ricochet off the open slit-sides and deposit on the detector, but in progressively diminishing numbers as we move away from the portion of the detector directly in front the open slit. If we carry out the same experiment with only slit-B open, the same result will be obtained, but in inverse.

QUANTUM MECHANICS AND MIND

It is often claimed that the single-slit experiment emphasises the particle-like aspect of the wave-particle duality: that it highlights the OR-form characteristics of radiation and matter, wherein we can specify a purported 'which way' path for our particle. Convention posits a description of the evolution of the emitted particle in the single-slit experiment in purely OR-form terms and as an unbroken trajectory that defines a 'which way' path from the emitter, through the barrier, to the detector. Thus, the single-slit experiment alleges to furnish evidence of definite 'which way' information about the particle's path. Hence, the standard appraisal of the single-slit experiment claims to amplify the spatial 'particle' aspect of the wave-particle duality.

However, our forthcoming non-standard appraisal of the same will develop a critically different version of affairs: While it is true that we can specify a particle-deposition on the detector, and, from it infer a *hypothetical* 'which way path' from the emitter, through the barrier slits and to the deposition, the supposition of any sort of OR-form 'which way' state prior to the final resolved event at the detector is not necessarily furnished by the outcome. Prior to any measurement, a system is purely an AND-form superposition state for future-possible putative 'which way' states. This is assuming the ontic reality of 'which way' states by disregarding the case for anticipated de-spatialisation, which renders the supposition for both 'positions' and 'paths' void.

However, even with the assumption of space, possible OR-form 'which way' states can never be observed in a direct real-time empirical fashion, and certainly not from the final outcome. Even with the forlorn attempt at a continuous observation of the particle on its journey, at best, we obtain a broken-discontinuous series of observations per facile information-matter dualism from Part-I. The ineliminable AND-form intervals therein cast doubt on the ontic reality of 'which way' paths, insofar as these require impossible seamless-continuous motion. In the absence of direct observation, and per the ineliminable AND-form intervals, and even with direct near-continuous observation, one can only *assume* an OR-form 'which way' particle path within the ineliminable AND-form intervals.

Even so, it is undeniably true that, after measurement we *do* obtain an OR-form quantum deposition at the detector. Generic conclusions that infer 'which way' claims upon this, and do so *as if* these are 'obvious', are to be doubted, even if we reduce the open slits to single particle-sized holes, as is done in Afshar's Experiment (see **0.17** to **0.18**).

In anticipation of the forthcoming non-standard appraisal, we assert that there are no resolved concrete or OR-form 'which way' states before or after the final OR-form outcome at the detector. The purported 'paths', 'trajectories' and 'which way' states turn out to have no real physical ontology per anticipatory de-spatialisation. Indeed, we shall discover that physics can furnish the outcomes of nature without resort to 'paths', 'trajectories', 'which way' states... *without* 'space' itself: (see Book-IV for the various explanations of de-spatialisation).

To augment, consider that, even the OR-form deposition formed on the detector upon which we impose a putative preceding 'which way' path is, like all events, apprehended in retrospect: apprehended as an event from the past. It is *not* a 'location' in 'space'. The 'location' is a mere assumption we impose on the observed event. There is no physical-empirical way for us to get to the putative 'location' of any past quantum deposition event at the detector, much less to the series of 'locations' that precede that event and supposedly formed up into a 'which way' path, even if we choose to ignore the ineliminable quantum mechanical AND-form intervals from that purported 'path'. Hence, the notion of 'position', 'location' and 'path' can be entirely dispensed with, wherein 'space' intervals give way to pure delayed choice time-interval relations, culminating into a temporised physics *without* space: (Book-IV).

What exactly is a delayed choice time-interval relation? When we observe the Moon, we see it as it was 1.5 seconds ago in the *past*. Instead of distance, we state that the Moon is separated from us by a pure delayed choice time-interval minimum: one approximately equal to 1.5 seconds. That is, if one wants to see the Moon 'now', one must wait a minimum of 1.5 seconds, delaying one's observation thus. Consequently, one can never get to see the Moon 'now'.

The same holds in all observations: All observations are of past events separated from the observer and from each other by pure delayed choice time-intervals. Thus, a sequence of resolved particle 'positions', such as to constitute a 'which way' pat,, in truth constitute a sequence of past events separated from the observer and from each other *purely* in terms of delayed choice time-intervals, purely in terms of time, ultimately *without* space... without 'position' or 'path'. When this insight is generalised, we remove the notion of space, position, and path from the lexicon of physics and from the interpretation of the double-slit experiment itself: a key achievement of Alt-R, and central to prospective quantum mechanics-ii.

*

What about the experiment with both slits open? **It is often posited that the experiment with two slits open emphasises the wave aspect of the wave-particle duality: the AND-form aspect.** The results are described in **Fig. P1.04**. Again, per the standard appraisal, and assuming the hypothetical ontic reality of 'which way' paths, and of 'space' itself, each emitted particle ought to pass through either slit-A OR slit-B, describing a definite 'which way' path through one slit, but *not* through both slits. But, in the absence of direct and near-continuous observation, AND-form logic demands that the single particle must pass through both slit-A AND slit-B.

If the particle's putative 'which way' states could be comprised of a series of seamlessly linked OR-form positions, the particle could only pass through either slit-A OR through slit-B. Instead, our single particle comprises a displacing AND-form wave that originates at the emitter; displaces toward both open slits; passes through both slits at the same time; and splits into two separate wavelets at slit-A and at slit-B. These AND-form wavelets of the same one-particle then head toward the detector, albeit from two different origins at slit-A and slit-B, respectively.

On their way to the detector, these AND-form wavelets interfere with each other: constructively in some places and destructively in others. This is called *self-interference*: The single particle interferes with itself, but can only do so if it assumes AND-form wave-like characteristics *throughout* its interim from emitter, through the barrier, to the detector. But none of this is directly observed: Given

21

experimental isolation per the withdrawal of all observational inputs, we cannot directly observe the putative displacement of the quantum mechanical AND-form wave in any continuous real-time way, or even witness the various moments of its self-interference during the putative approach of the wavelets from the two slits toward the detector: We can only infer... or *assume...* the supposed displacement of the AND-form quantum mechanical waves, i.e. at the end of isolation and per the global results generated at the detector.

Of course, prospective de-spatialisation undermines the notion of the motion of the AND-form quantum mechanical wave, as much as it undermines 'which way' states. De-spatialisation does not undermine AND-form logic or waves, but *does* force a stopgap spatially static version of the AND-form wave, with space totally obviated by the static character of the total future per the growing block universe approach: The future cannot obviously move in or across space, simply because the future does not 'move'. Indeed, if we recapitulate the *raison d'etre* for AND-form logic as the ontic reality of the future, *and* grasp the quantum mechanical wave as the physical manifestation of that future, the future cannot displace across 'space', and neither can the quantum wave that physically phenomenalised that future: At best, the future, as a static state, AND-to-OR decays into a succession of events (quantum depositions) upon which we then foist 'the notion of motion' of the quantum wave across its 'space'. This is indeed the case, and the growing block universe, combined to de-spatialisation will seal this conclusion on a permanent basis.

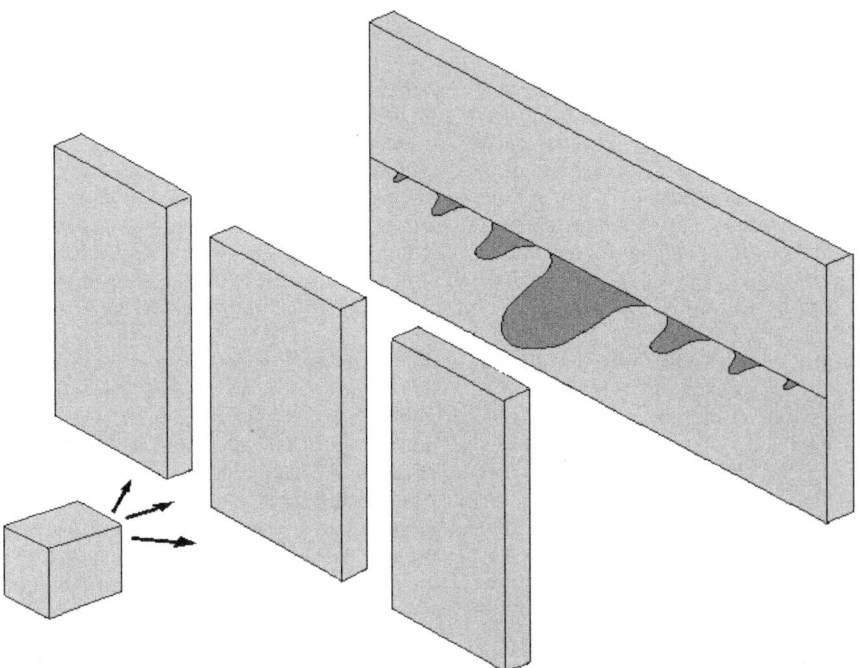

Fig. P1.04 Experiment with both slits open
Notes: A didactic depiction of results obtained when both slits are open. Again, the number of deposited particles is per function of the height of the curve vis-à-vis the surface of the detector. The areas under the curve shaded in dark grey represents where the particles deposit. The areas between the peaks, where there is no curve above the detector-surface, are areas at which no particles will be found. Thus, the peaks constitute the bands at which particles will be deposited, forming 'light bands', while the intervals between these will constitute 'dark bands' wherein no particle depositions will occur. Thus, an interference pattern is formed.

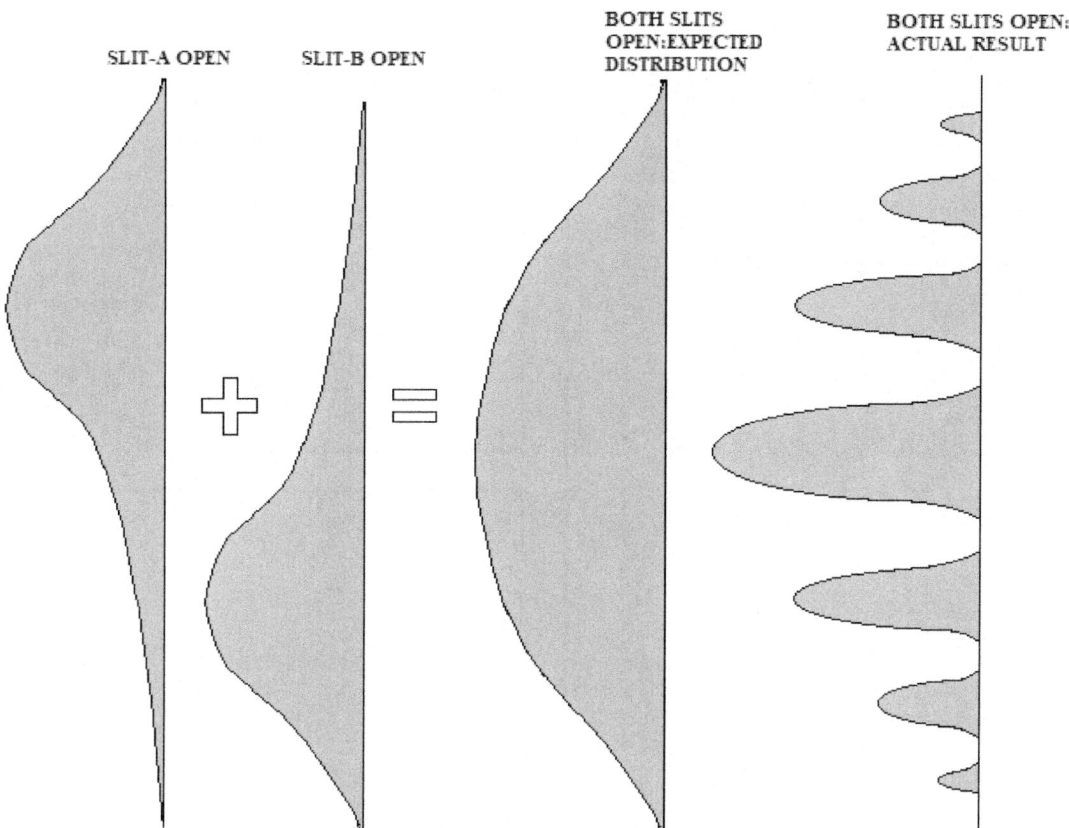

Fig. P1.05 Expected and unexpected experimental outcomes
Notes: The didactic illustration above contrasts the expected result when both slits are left open versus the actual result. If OR-form logic held true throughout, and the particles travelled as particles and arrived as such via definite OR-form 'which way' paths, the sum of the distribution patterns from slit-A open and from slit-B open would sum to something like the third distribution pattern (third from the left). However, when the two-slit open experiment is carried out, the pattern obtained is the fourth distribution pattern: the actual result: This is the notorious interference pattern; formed of alternating light bands comprised of large numbers of quantum depositions and dark bands with no depositions at all… possible only if entities released as putative single particles at the emitter pass through both slits at the same time *and* self-interfere on their way to the detector.

While the generic depiction of what is imagined to happen in the interval of experimental isolation is seemingly rational, the description of it in pseudo-realtime terms (*as if* we had any sort of direct continuous real-time observation of the isolated non-observed process) must be considered suspect. We *must* question the pseudo-realtime depiction of the 'displacing and self-interfering wave' as an unverifiable *assumption*. Nevertheless, the two-slit experimental outcome garnered at the end of isolation is clear: It exhibits expected AND-form processing of self-interference into mutual reinforcing and cancelling quantum mechanical waves, evinced as the globally materialised interference pattern.

*

How do we clarify the meaning of self-interference in generic terms? A depiction of wave-like or AND-form processes and self-interference entailed in the experiment with both slits open is given in **Fig. P1.05** and in **Fig. P1.06**: a didactic illustration of both destructive and constructive interference. In the two-slit experiment with both slits open, two wavelets get to form up: one at each of the two open slits. In the region between the barrier and the detector, where the wave-peaks of the overlapping wavelets coincide and superpose, they will reinforce: This is *constructive interference* of AND-form 'which way' potentialities. The result will be the concentration of quantum depositions on the detector immediately in front of the region of constructive interference. The quantum depositions will peak in terms of the numbers of OR-form particle-like incidences that form up on the detector-surface.

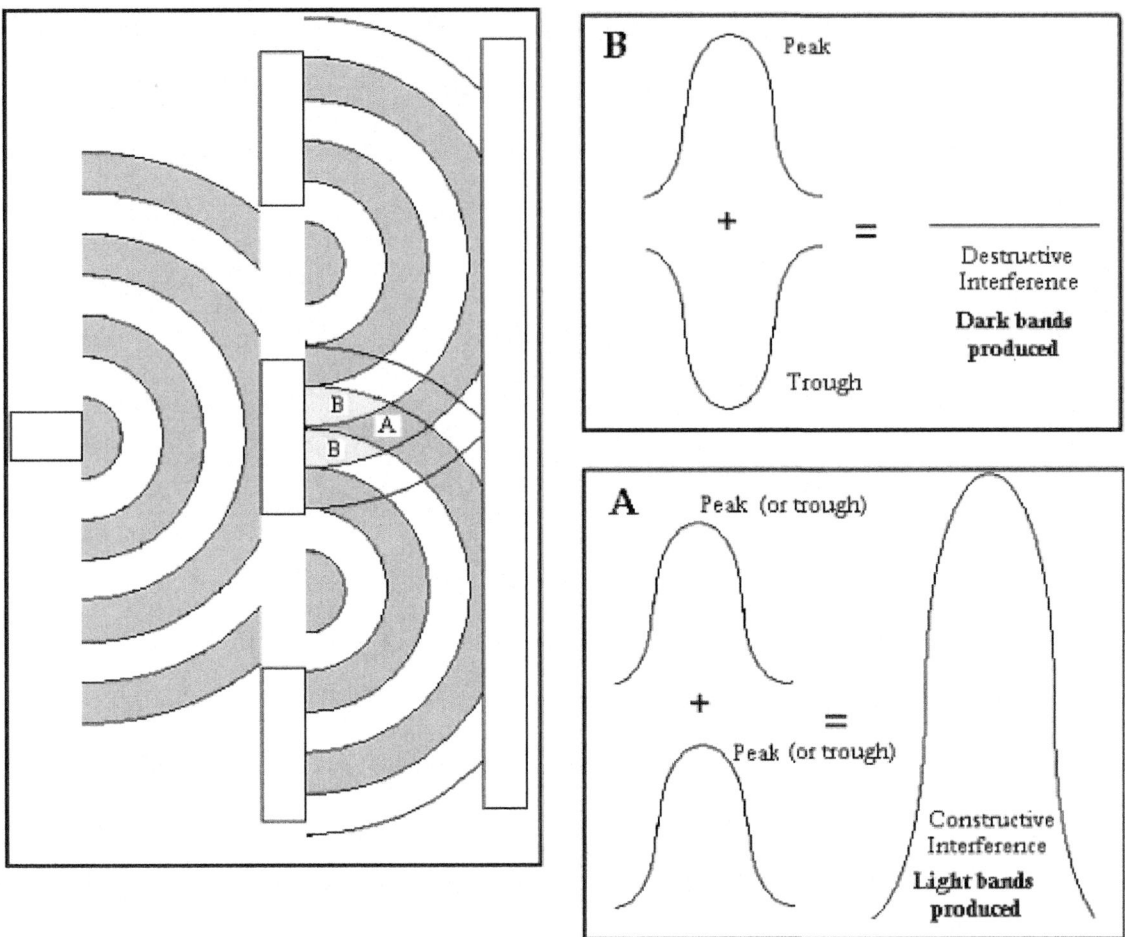

Fig. P1.06 Detailed look at the process of self-interference

Notes: In the left diagram, grey-shaded bands represent the peaks of a moving wave for a single photon of light or electron, while the light-shaded bands represent the troughs of that moving wave. Note that the wave splits into wavelets as it passes through both open slits in the double slit experiment. The interference pattern formed out of light bands and dark bands form on the detector per the interference of the wavelets. When a trough coincides with trough (or a peak coincides with peak), we obtain constructive interference (regions A on the left diagram and box-A on the right): This correlates on the detector as areas of maximum particle OR-form depositions and resolutions. When peak coincides with trough, we get destructive interference (regions-B and box-B): This correlates with dark bands on the detector when few or no particle OR-form depositions occur.

However, where the wavelets coincide in such a way that their peaks versus their troughs superpose, their putative 'which way' possibilities will cancel each other out. This is *destructive interference*: few if any particles will deposit on the parts of the detector that coincide with the area of destructive interference, so forming dark bands on the detector.

If many such particles are emitted and allowed to deposit on the detector, the global effect of attendant constructive and destructive self-interference will generate a distinctive interference pattern on that detector. This will be comprised of dark and light bands of *en masse* OR-form depositions. The greatest concentration of particles will coincide with areas associated with constructive self-interference. The darker bands will coincide with destructive self-interference and the absence of particle depositions. All of this indicates the operation of AND-form logic, wherein a single particle apparently *does* assume a superposed sum of all *possible* putative positions and 'which way' states, again assuming the ontic reality of space and paths. That is, in isolation, in its wave form, the particle will appear to be a superposition of AND-form putative alternative positional states.

Of course, throughout the experiment, experimental isolation is maintained. If we were to prematurely end this enforced isolation and intrude into the system using observational inputs on the particles on their alleged journeys from emitter to detector, or insert active detectors at the slits or at the points of constructive and destructive interference, the wave-like AND-form characteristic of the system would largely 'wash out'. That is, under almost continuous observation per, say, the Quantum Zeno effect, the particle traveling on its way to the barrier, and from barrier to the detector, would lose most of its AND-form characteristics. In doing so, it would approximate to a 'common sense' OR-form seamless-continuous trajectory (just as is observed at the macro-scale, or seemingly), limited only by the fact that, even with the Quantum Zeno effect, and per the incontrovertible implications of the logic behind facile information-matter dualism, we cannot obtain any absolute seamless-continuous real-time observation of the 'particle'. Consequently, the formed 'which way' trajectory must remain ultimately a discontinuous-broken one.

Moreover, with observational intrusion in the double-slit experiment, each emitted particle would be observed to pass through one slit or the other: *not* through both slits. The intrusion destroys the possibility of self-interference: No interference pattern would form at the detector... unless Afshar is correct: but more on this later.

However, we must again remind the reader that the notion of definitive 'which way' states is a pure assumption about the behaviour of 'particles' in both the single and double slit experiments. We reiterate that we cannot empirically secure the ontology of putative assumptional 'which way' states, especially before or even after the end of isolation and experimental conclusion. We reiterate that, ultimately, physics can abide *without* notions of 'position', 'which way' paths or 'trajectories', and 'space'. Physics can recuperate almost everything in terms of events distributed in pure delayed choice time-intervals, *without* 'space'. If so, if path and position have no ontic reality, what exactly is it that is being processed in destructive and constructive self-interference? What exactly does the quantum mechanical wave superpose in AND-form terms, if not 'position' and 'path'?

The answer is simple: We can discard 'position' and 'space' from the lexicon of physics, but we cannot discard the description of physical systems in terms of 'events organised in and by time'. It turns out that **the quantum mechanical AND-form wave superposes, *not* potential 'positions and paths' or 'which way' states, but only potential 'moments in time', or potential 'which moments'**: an AND-form superposition state for as-yet not-realised possible and probable future events structured in terms of probable delayed choice time-interval relations. Hence, **the self-interference is *not* an interference of potential 'positions' and 'paths', but of 'which moments' structured by potential delayed choice time-intervals**: The self-interference does not occur at 'locations in space' but at potential delayed choice time-interval *moments* in future-potential time.

0.14: The Non-Standard Appraisal of the Double-Slit Experiment

The generic appraisal of the double-slit experiment, partially criticised in the previous essay, asserts that a physical state is wholly particle-like under the single-slit condition and wave-like under two-slit condition, per function of the *observer effect*: i.e. per whether we choose to observe the particle under the one-slit or the two-slit condition: That is, how we measure or observe particles per the apparatus setup determines whether we get an OR-form particle outcome or an AND-form wave-like outcome. Outside the physics community, the observer effect is often misconstrued with what might be called the *consciousness effect*: the notion that consciousness decides whether we get a particle or a wave. This dubious notion is unfounded: the observer effect is purely about how we set up the barrier-slits or the conditions of the apparatus.

Conventionally, the particle versus wave aspects of the system do *not* manifest simultaneously, and the system is *not* AND-form and OR-form at the same time. Instead, the system starts as AND-form and only later ends in OR-form; the latter per function of measurement or the end of isolation. This is the famous *complementarity principle* of quantum mechanics. Hence, the standard appraisal of the double slit experiment can be summarised as...

- The single-slit experiment demonstrates the OR-form particle facet of nature.
- The double slit experiment demonstrates the AND-form wave aspect of nature.
- The OR-form particle and AND-form wave do not occur at the same time: i.e., the *principle of complementarity*.

However, our Alternative Realist non-standard appraisal asserts a more subtle view, even when we assume the false ontology of 'which way' states and the ontic reality of 'space'. Thus, **while the AND-form aspect is accentuated when both slits are open, in truth, both single-slit *and* double-slit conditions exhibit both the particle OR-form aspect *and* the wave-like AND-form aspects *observationally simultaneously*... at least at the detector**, at the end of isolation.

Note that we can only rely on observational-empirical evidence and on what we observe at the detector. We have no direct access to what happened in the interim of isolation. We have no observational-empirical mandate to make unwarranted assumptions about unobservable 'which way' paths regarding the interim. If we restrict ourselves to what we find on the detector at the end of the experiment, we discover OR-form particle like-depositions that collectively manifest interference patterns and AND-form wavelike outcomes: In other words, at the detector, **we require a collection of OR-form particles-like states to reveal the temporally prior AND-form wavelike aspect. Hence, at the detector, OR-form *and* AND-form realities manifest *at the same time*, in the sense that they are observationally simultaneous**. This is no small matter and applies even to the single-slit condition. This tautological seeming finding, combined to the limit of what we can and cannot justifiably infer about the state of the system in *any* preceding interim of isolation, has critical bearing on the Afshar experiment and its standing interpretations. (More on the Afshar experiment later).

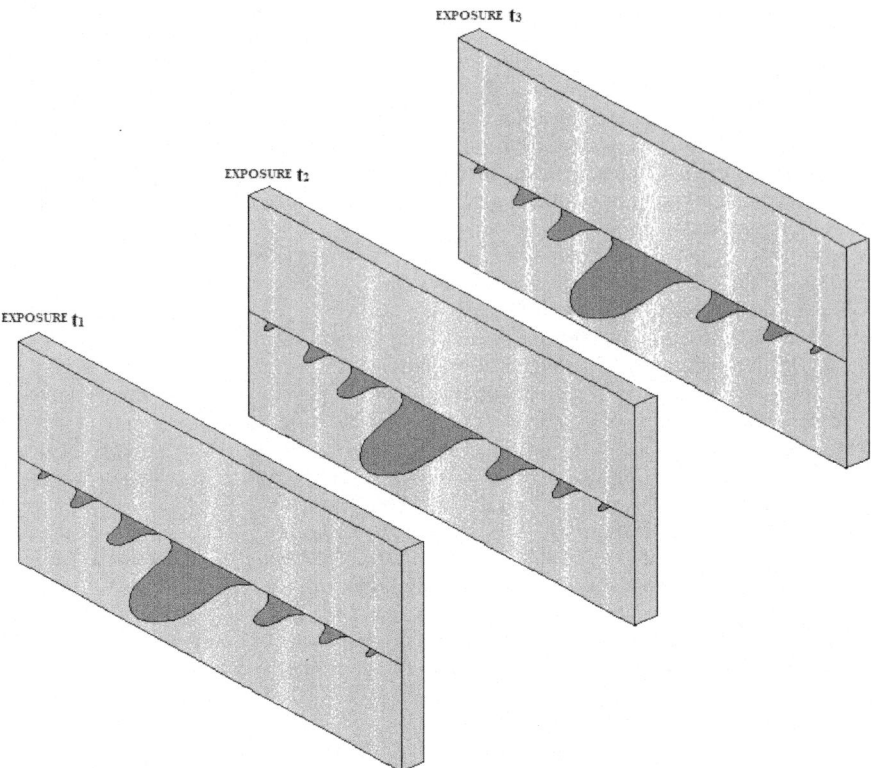

Fig. P1.07: A closer look at interference patterns
Notes: A didactic illustration showing how the interference pattern is built over time. Each time-exposure depicts the reality of the wave-like AND-form state per the self-interference of each particle emitted one at a time, and independently deposited on the detector. Each particle deposits on the film as a discrete OR-form incident, 'particle' or 'pixel': Collectively, these render salient the interference pattern, and evince the reality of self-interference and the AND-form superposition principle. In other words, the evidence for AND-form wave-like quantum phenomena are realised through *en masse* compounds of OR-form particle depositions. It follows that both AND-form *and* OR-form particle-wave facets are observationally simultaneous.

The only difference between the single-slit and double-slit experiment is relative accentuation and attenuation: The two-slit open experiment accentuates the AND-form wave-like aspect by structurally permitting for wave-interference. On the other hand, the one slit experiment structurally disallows wave-interference of the AND-form wave. Even so, in both experiments, the AND-form quantum wave-like facet is observationally simultaneous with the OR-form particle facet, at least at the end of isolation and end of experiment, given that **we need OR-form states to manifest evidence for prior AND-form realities in an observationally simultaneous way**.

To clarify, let us re-appraise the double-slit experiment with both slits open. **Fig. P1.07** clarifies that the evidence for AND-form logic and the superposition principle in the two-slit experiment is rendered evident by *en masse* particle-like OR-form depositions at the detector. Clearly, **OR-form particles are indispensable to and observationally simultaneous with the evidence for AND-form waves, constituted as the interference pattern rendered salient by the total collection of particle-depositions**.

Given that self-interference becomes evident *only* and *always* through compounds of OR-form particle-like incidences, we must always end in dual evidence: i.e., evidence *simultaneously obtained* for both OR-form particles *and* AND-form waves; for OR-form *and* AND-form logic. This happens at the end of isolation and at the end of experiment. Therefore, the two-slit outcome is *not* evidence exclusively or only for wave-like aspects of the duality, and this is *not* going to be any different for the single-slit experiment.

There now emerges a clear need to reframe the principle of complementarity in such a way that it is consistent with inevitable and incontrovertible observationally simultaneous manifestation of AND-form and OR-form logics, at least at the detector and at the end of isolation. That is, the claims of generic complementarity to the effect that AND-form and OR-form realities do not occur at the same time appears to come into conflict with the fact that they *do* occur at the same time in observational simultaneity, post-isolation. Again, we reiterate that one needs the OR-form depositions to render evident the AND-form reality *retrospectively*, and one needs the AND-form logic to structure and distribute the OR-form depositions into an evidence-state for both OR-form *and* AND-form realities, *simultaneously*.

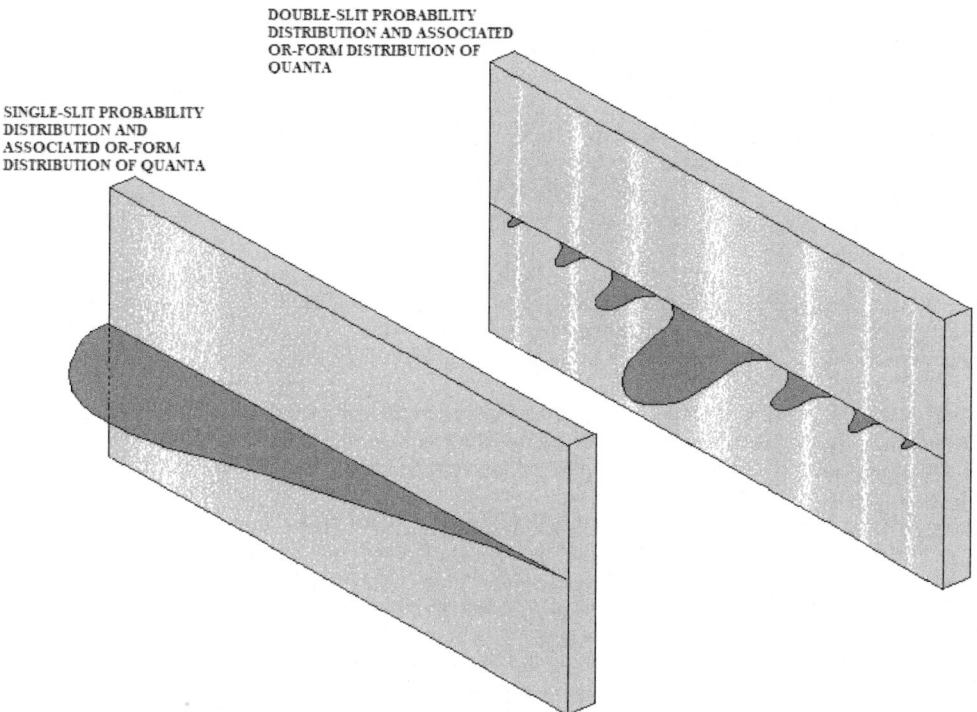

DOUBLE-SLIT PROBABILITY
DISTRIBUTION AND ASSOCIATED
OR-FORM DISTRIBUTION OF
QUANTA

SINGLE-SLIT PROBABILITY
DISTRIBUTION AND
ASSOCIATED OR-FORM
DISTRIBUTION OF QUANTA

Fig. P1.08 Comparing distribution patterns from single-slit and two-slit experiments
Notes: A comparison of the probability distributions and OR-form particle deposition-distributions that attend the single-slit and double-slit experiments, respectively. In the case of the double-slit, wherein wave-interference is made undeniably manifest, and the quantum mechanical AND-form of that wave is explicitly evinced, this is accomplished per *en masse* OR-form particle depositions. The wave-particle duality, or AND-OR duality, is therefore observationally simultaneous, given that we need particle depositions to render evident AND-form waves, in a simultaneous way, at the end of the experiment. In the single-slit experiment, there is no wave interference, given that the possibility for it is structurally obviated. This is often erroneously posed as if the single-slit exclusively evinces the OR-form particle facet. Yet, the same single-slit probability distribution is possible upon a pure AND-form wave passing through a single slit. The particle deposition-distribution attendant the single-slit probability distribution is also observationally simultaneous with the AND-form reality of its quantum wave: i.e., in the single slit, wave-particle AND-OR dualism is simultaneously evident at the detector, even if there is no interference pattern. The observational simultaneity of OR-form particle and AND-form wave realities is *not* exclusive to the double-slit condition. Both experiments involve wave-particle AND-OR duality.

The single-slit experiment also constitutes dual evidence, contrary to the usual claims. The evidence from the single-slit experiment does not manifest the OR-form particle facet exclusively. It also obtains evidence for the AND-form wave facet, again in an observationally simultaneous way. Unfortunately, the absence of an interference pattern tends to obscure this fact.

A typical distribution-pattern formed when only one slit is left open is given in **Fig. P1.08**: (left or front depiction). Notice therein that the quantum OR-form depositions are more concentrated in the detector-region directly in front of the single open slit. These diminish in number as we progress away from the region directly in front of the open slit. This distribution pattern is oft presented *as if* unique and peculiar to particle-like OR-form objects, given that the same pattern could be generated by firing bullets instead of fundamental particles toward the single open slit, or on the notion that the particle must be confined into an OR-form real position and path at the open slit, even though we cannot observationally corroborate this claim under conditions of isolation: The view is false.

However, the same pattern can also be obtained with the use of both mechanical and purely AND-form waves.

If the AND-form waveform can control for the deposition-density and distribution of OR-form outcomes in the two-slit experiment, the waveform in the single-slit experiment can do the same, albeit *without* forming an interference pattern. Thus, the OR-form depositions in the single-slit experiment comprises as much simultaneous evidence for AND-form waves of possibility as they do for OR-form states, as is also true for the two-slit experiment, albeit without forming an interference pattern in the single-slit condition.

To put the matter in another way, **the absence of interference patterns in the single-slit condition does not imply the absence of AND-form wave-like phenomena. (This is also true about the macro-scale 'classical' seeming world: billiard balls and brains). Moreover, the sort of particle distribution obtained in the single-slit experiment** *can* **be produced by an AND-form quantum mechanical wave. The only thing absent is the structural possibility for self-interference**. Hence, the single-slit experiment does not make for an absence of the AND-form wave-like realities. The single-slit condition is not the condition under which the 'wave behaves like a particle', as is commonly misconstrued: The system *always* behaves as an AND-form wave. And the outcome must *always* observationally simultaneously evince *both* OR-form *and* temporally prior AND-form realities.

If we were to perform the single-slit experiment, but instead used mechanical water waves vis-à-vis a detector array comprised of water buoys, these would undulate in a proportionate way to the energy of the incoming water waves. Therein, we would find that the energy of the incident-wave will be highest on the buoy detector immediately in front of the open slit but diminish in energy at buoys farthest from the open slit; all because it takes longer for parts of the water wavelet formed at the single slit to reach the more distant parts of the buoy detector array. Consequently, the water waves will arrive at those more distant parts with less energy and generate smaller undulations at the pertinent detector buoys.

The energy-distribution formed by water waves in single-slit experiments is of the same essential form as the particle probability of distribution pattern formed in single-slit experiments carried out with single-release quantum mechanical waves for photons or electrons, but with one key difference: Energy or intensity-measures apply to water waves, but quantum probability for deposition applies to quantum waves. Quantum waves are probability waves for putative 'which way' possibilities and potential 'positions' of the particles involved.

The point is simply this: The probability distribution pattern for particle deposition obtained from AND-form quantum probability waves in single-slit conditions has the same form as the energy distribution pattern obtained for single-slit experiments involving water waves. Thus, the distribution pattern obtained from experiments in which only one slit is open can indeed be produced by purely AND-form wave-based processes, grasped in terms of the probability of distribution. Therefore, we *cannot* assert the pattern unique to the single-slit experiment to be exclusive to particulate OR-form realities: We cannot assert single-slit experiments to be absent of wave-like AND-form processes, or infer definite 'which way' paths in 'space' to the interim of isolation, especially given that there is simply no empirical way to justify this later notion, given that the inference of 'which way' paths is always an assumption we impose on the system, *not* an observation of fact: given that the same sorts of outcomes can be obtained without the assumption of putative 'which way' information, both under single and double-slit conditions.

To reiterate, **the distribution pattern characteristic to single-slit experiments is as much evidence of wave-like AND-form phenomena as it is of particulate OR-form phenomena, through which the former becomes evident. Thus, the single-slit experiment comprises as much evidence for interim wave-like AND-form radiation and matter controlling for the probability of distribution of 'particles' as it does for those OR-form particles. Hence, we cannot assume or insist that the single-slit experiment only evinces OR-form particle-like realities, or that wave-like AND-form realities are absent from these. Hence, both OR-form and AND-form facets and logics are simultaneously evinced in single-slit experiments, but** *without* **interference... as they are in double-slit experiments, but** *with* **interference**.

*

Let us augment further: Utilising a dubious pseudo-realtime description previously criticised, or pretending that we could observe the process as it happens in realtime, we will assume that, when both slits are open, the quantum wave of a single particle will displace towards the barrier. At each of the barriers' open slits, the original single AND-form quantum wave will split into two independent distinct AND-form quantum wavelets. On the way to the detector, the AND-form wavelets will constructively and destructively interfere with each other. On arrival at the detector, the AND-form quantum waves will convert into single OR-form depositions, each exhibiting a clear and single OR-form 'particle' deposition on the detector-surface.

Thus, at the emitter, the particle will be emitted as a single OR-form state. It will immediately transform into an AND-form wave state and remain in that state throughout the rest of its putative realtime and directly unobservable journey to the detector... only to reconvert into an OR-form state upon its arrival at the detector. After many such processes, we will obtain *en masse* OR-form incidences that manifest the now-familiar interference pattern in proof of wave-like AND-form logic.

In the case of the single-slit experiment, we assert that the same sort of process occurs: Applying yet again the same dubious pseudo realtime description previously criticised, the particle is released into the system as an OR-form state. It immediately transforms into an AND-form superposition quantum wave. This wave then propagates to the barrier and passes through the single open slit, remaining in its pure AND-form wave-like state. That is, the particle *does not* travel to and across the single-slit barrier as an OR-form state comprised of any definite 'which way' path: So long as experimental isolation remains enforced, even in the single-slit experiment, and even when we cannot confirm this reality via direct realtime observation, the 'particle' remains wholly wave-like and AND-form quantum mechanical throughout its journey from emitter, through the single open slit, to the detector. But there is no possibility for its self-interference, given that only one slit is left open. As such, the single AND-form wavelet will get to and contact the detector-surface as a single non-self-interfering AND-form quantum mechanical wave. Its first point of contact with the detector will lie immediately in front of the open slit. As such, this constitutes the point or region of highest probability at which the AND-form wave will convert into an OR-form 'particle' deposition-outcome. Parts of the detector that are farther from this highest probability region will enjoy a lower probability for OR-form deposition-outcomes.

QUANTUM MECHANICS AND MIND

The point of the above description is that, so long as experimental physical isolation remains in force, from emitter to any point on the detector, the system evolves purely as an AND-form quantum mechanical wave, *even when only one slit is left open*. Only at some probable point at the detector does this AND-form wave 'collapse' into an OR-form deposition outcome. When we let large numbers of such waves successively get through to the detector, we obtain enough OR-form incidences at the detector to bring into saliency *both* the compounded OR-form 'particle' incidences themselves as well as the preceding AND-form reality evinced by the distribution pattern displayed by the *en masse* OR-form states. The OR-form incidences, the 'particles', will be distributed according to the probability structure implied by the AND-form wave that had passed through the single open slit. The reality of the AND-form wave will itself be made evident by the mass of OR-form outcomes that reflect the probability structure of the AND-form wave, except that no interference pattern will form under the single slit condition.

It is obvious that **the absence of the interference pattern in the single-slit conditional cannot imply the absence of AND-form characteristics from the system. We cannot use the single-slit experiment to assert a purely particle-like behaviour throughout the interim of isolation, or assert for any definite 'which way' path to a quantum mechanical wave dispossessed of any resolved OR-form attributes within that state of isolation, despite the expected confinement of the quantum wave to a single slit.**

The presence of AND-form logic throughout the interim of the single-slit experiment must necessarily be asserted, at least in the interim of experimental isolation. It follows that the observational simultaneity of OR-form particulate *and* AND-form wave applies equally to the single-slit variant of the experiment, as much as it does to the two-slit variant. Generic complementarity thesis must accommodate this fact: By doing so, it can avoid the pitfalls of the Afshar controversy examined later.

*

The crucial point of the non-standard appraisal is this: In the two-slit variant of the famous experiment, when we resolve in favour of self-interference and accentuate pertinent AND-form realities through *en masse* OR-form particle depositions, it is not possible to infer a definite putative 'which way' information for any of the deposited particles, even assuming the false ontology of 'space' and 'paths': From emitter to detector, the 'particle' was purely an AND-form state, with *no* resolved putative 'which way' path. Here is the clincher: In the single slit variant of the experiment, when we resolve OR-form particles at the detector *en masse*, it is still *not* possible to infer a definite putative *preceding* 'which way' information for any of the deposited particles, given that, from emitter, through the single slit, and to the detector thereafter, the 'particle' is in a pure AND-form state, with *no* OR-form putative 'which way' paths, even when the AND-form state is obviously confined to and restricted by a single open slit, or even a single open small hole. When so confined, this does not constitute a definite OR-form path. The confinement constitutes a state of AND-form possibility, not OR-form actuality: *not* OR-form position or 'path'.

In both single-slit and double-slit variants of the experiment, the OR-form reality may only be confined to the post-isolation resolution at the detector. We cannot allow for the reality or the assumption of any prior OR-form 'which way' information for the isolation-interim: The interim time-period up to the end of isolation cannot be legitimately described as a real or hypothetical OR-form 'which way' state, or as exhibiting any other resolved OR-form attribute, but *only* as a pure AND-form state *without* realised paths, even on the dubious assumption of the onticity of 'paths', 'positions', and 'space'.

Indeed, once we discard space itself from the ontology of physics, the whole notion of 'paths' falls apart, and the AND-form state is transformed into potential de-spatialised, purely temporised future-potential 'which moments' state, even in single-slit experiments. This last assertion contains the reformulation of the principle of complementarity, albeit in its now non-standard form. The non-standard or **new complementarity thesis asserts that…**

- **OR-form 'particle' realities are required to bring into evidence AND-form wave-like superposition-based distribution patterns at the detector**… in both single and double-slit experiments. One cannot obtain the latter without the former.

- **OR-form and AND-form realities are *always* observationally simultaneous at the detector**, and *always* occur at the same time at the end of isolation and at the post-experimental phase: This is true in both single and double-slit experiments, even if the former conveys no interference pattern: Its distribution is per function of the probability structure conveyed by the single slit AND-form wave.

- While OR-form and AND-form realities are always obtained in an observationally simultaneous way post-experiment, **in the interim of experimental isolation, only the AND-form reality abides: This applies equally in both single and double slit experiments: It is not possible to assign any sort of OR-form 'which way' state to the period of isolation, even in single slit experiments**. The 'which way' states constitute pure potentials: They are *not* OR-form resolved 'paths', even in single slit experiments.

The resulting *new complementarity thesis* thus asserts that…

- **At the end of isolation, OR-form and AND-form realities are observationally simultaneous and occur together, but this cannot permit any inference of prior definite OR-form 'which way' paths to and in the interim of isolation**, even under single slit conditions…and…

- **In the interval of isolation, the emitted 'particle' is purely an AND-form superposition state of possible, *not* realised, 'which way' paths…** or, per de-spatialisation, it is an AND-form 'which moment' configuration. The AND-form state in the interim of isolation is bereft of any OR-form information, in both single-slit *and* in double-slit conditions.

Thus, in this way, we save the complementarity principle, albeit in an entirely new non-standard form. The new complementarity principle will allow us to overcome the anticipated Afshar controversy.

0.15: The Single-Slit Experiment in the Light of the Firewall Principle
And the Ontology of the Quantum Mechanical 'AND'

How can we be certain that, so long as isolation holds, and the AND-form state of the system remains in force, we cannot assert *any* resolved OR-form 'which way' state to a putative particle, even when its AND-form possibilities are structurally confined to a single open slit, or even to a very small open hole? We can rely on the facts stated previously, inferred from the observationally simultaneous evidence of OR-form and AND-form logics, and from the fact that both double and single-slit experiments evince AND-form logic, and that the single-slit experiment is not exclusively particulate or OR-form... and that there is no empirical-observational or physical way of assuming that, in the interim of isolation, the system can assume *any* OR-form series of positions... even when we reduce the slit to a tiny hole. Indeed, when we interrupt isolation and bombard the system with observational inputs, these certainly destroy most of the AND-form characteristic of the system but cannot eliminate it. This is true even with a Quantum Zeno Effect of intense bombardment of the system and its purported particle 'path': The broken-discontinuous character of that path cannot be wiped out of those pesky AND-form intervals that intersperse and break that path up into a broken-discontinuous form.

We might further add, in anticipation of de-spatialisation, that *there is no space*. Therefore, ontically, there are no 'locations', series of positions, or 'which way paths' to begin with. At best, in a de-spatialised purely temporal physics, we can only hope for 'which moments' or 'which whens'. Even these will be *potentials* rendered into not-yet realised AND-form alternatives so long as isolation is enforced, or even when a Quantum Zeno effect is in force, given the ineliminability of AND-form intervals therein, no matter how much such AND-form interims are reduced or obscured.

However, there is a third key reason why we cannot infer or assume 'which way' paths: This reason derives from the *firewall principle*, which emerges inexorably from the *raison d'etre* for AND-form logic in nature.

*

When we place any system into isolation, we defer its resolution to the future. So deferred, the system assumes the logical structure of the future: i.e., a set of alternate possibilities. Hence, the system *must* assume a pure as-yet not-realised future-form of alternatives, rendered into an AND-form superposition state. The AND-form quantum mechanical state of a system is therefore the objective future-form of that system. Its future is 'out there', and *not* a bloss subjective state. As such we cannot treat that future as an already resolved OR-form state before subsequent AND-to-OR wavefunction collapse has transpired and ends isolation. **To treat the future *as if* it constituted an OR-form realisation is to treat it as if it was resolved into concrete events before it transpired into such. In such a case, we ought to be able to prove this in empirical-physical terms... by acquiring 'signals from the future',** from a system in its supposed already resolved state *before* it could become such.

However, such 'signals from the future' would be tantamount to the violation of principles of causality, and would also entail energy or matter above unity, in violation of the principle of conservation.

Both causality and the principle of conservation are intertwined: They enforce a firewall that asserts that one cannot gain concrete OR-form signals or physical effects from the future: Hence, the *firewall principle*. Consequently, the future must remain in pure AND-form until AND-to-OR collapse (the process of time itself) renders it otherwise. This explains basic quantum logic and the 'reason for being' for why AND-form logic and attendant quantum phenomena exist. i.e., because nature is sensible and rational, *not* 'crazy', as is often spuriously claimed.

It follows that, in both double-slit and single-slit experiments, isolation is typically enforced. Consequently, the system is relegated to the future. It is necessarily rendered AND-form, even if a single-slit or a single tiny hole confines the AND-form possibilities to a potential 'single path' or 'point'. This confinement does not constitute any realised path or point, given that it remains AND-form per its deference to the future, even assuming the veracity of 'space' and of 'paths'. **That is, the confined single AND-form 'path' is *not* an OR-form event... unless one can claim to receive 'signals from the future' and can treat it as such.**

The inference of 'which way' information to the experimental interim of isolation, even in single-slit or hole condition, is tantamount to the reception of 'signals from the future': an impossibility, for reasons given. Therefore, we cannot in principle gain any 'which way' information about a system deferred to the AND-form future, even in a single-slit or single-hole condition.

Moreover, even when we interrupt isolation and intensely observe the contents of the system via a Quantum Zeno effect, given the ineliminable broken-discontinuous evolution of what we observe, and the fact that we cannot eliminate the AND-form intervals from it, the AND-form future must abide. The facile 'which way path', grasped as a series of OR-form positions, will be invariably broken up by intervals of AND-form isolation. We cannot infer any realised 'path' for such AND-form intervals, while the OR-form series cannot ultimately be described as positions or locales in space, even when we assume the reality of space, given the retrospective nature of *any* empirical observation, and the fact that the events so observed are 'events of the past', and we have no retrospective physical access to their purported 'locations', to which we cannot now 'go'. Thus, the notion of 'which way paths' remain dubious, even when we assume the realty of space and ignore the ineliminable AND-form intervals of isolation and futures.

0.16: The Afshar Controversy

In summary, **the Afshar controversy is premised on the claim that we can obtain 'which way' information about a quantum mechanical AND-form system by generating a single-slit outcome from a two-slit arrangement.** This apparently calls into question the principle of complementarity central to generic quantum mechanics. That is, it appears to subvert the premise that the AND-form wave-like radiation and matter precedes the subsequent OR-form particle-like resolution, and that one can obtain OR-form paths from an *a priori* AND-form state.

Afshar's inference seems reasonable *only* if we assume that the single-slit variant of the two-slit experiment exemplifies exclusively particle-like OR-form behaviour *and* furnishes 'which way' information. All of this assumes the ontic reality of space, position and paths, despite counter-claims from our non-standard evaluation; despite implications from anticipated de-spatialisation against the ontics of space, position and path; despite ineliminable broken-discontinuous evolutions of systems per facile information-matter dualism; and despite the *firewall principle* previously clarified, which asserts that, to obtain putative 'which way' information from a system in isolation or deferred to the future would be tantamount to the reception of 'signals from the future'. In fairness, Afshar, was not and is not aware of such distinctions. This is also true for the plurality of the physics community.

*

Suppose we were to reduce the slit in the single slit experiment to a small hole; of a size that will allow the passage of one particle at a time. In having confined the particle's putative 'position' thus, we attribute to it a definite 'position' in its putative trajectory in and through that hole. We might further infer from this apparent 'certainty' a definite 'which way' path for our particle.

Even so, so long as isolation holds, only a purely AND-form description of the particle is permitted. By implication, we cannot assign any OR-form putative 'which way' position or path to our 'particle' at that hole, despite the implications from the single hole, even with the assumption of space. Despite the confinement imposed by the hole, we can only attribute to our particle at that hole an AND-form 'potential' pathway that must converge at the single hole and branch out thereafter into succeeding potential AND-form smear of putative 'which way' paths. But there is no realised or real path before the hole, at the hole, or thereafter, per all the stated reasons, including the non-standard evaluation, the new complementarity thesis, and the firewall principle... even when we exclude implications from prospective de-spatialisation.

The same must also apply to an experiment with two small holes: We will obtain the same sort of potential convergence and divergence of putative AND-form 'which way' paths, albeit per two open holes *and* with prospective interference, but always *in* superposition AND-form; again, even assuming the ontic reality of space and of 'paths'.

A single open hole, small enough for a single particle to pass through it, may afford us an *inferred* or *hypothetical* certainty about the particle at the hole, but it cannot furnish any directly observable OR-form state about any 'position' at that hole: For as long as isolation remains in force, this *must* remain purely an inferred and hypothetical notion. Such an inferred hypothetical non-empirical judgement about the particle, and its treatment *as if* it constituted an objectified position in 'space' at that hole before the end of isolation, can never be the same as an actual OR-form certainty wrested through direct empirical observation.

Thus, even when we reduce the potential AND-form paths to a single hole and might then entertain an inferred putative non-directly observable *assumption* of OR-form 'position' for the particle in the interim of isolation, this can never be equivalent to an OR-form information of that particle at that hole, which must yet remain purely AND-form per the enforced isolation. **The non-standard appraisal of the double-slit experiment and the new complementarity thesis, combined to the firewall principle, both of which retain the old thesis to the effect that the AND-form wave-like aspect of a system precedes its final OR-form resolution, and remains in that AND-form configuration throughout the interim of its isolation, abides... even with particle-sized holes in place of generic slits.**

*

We are now in a position to resolve the Afshar controversy[8]. In 2004, an experiment attacked the generic principle of complementarity by obtaining a one-slit outcome from a two-slit arrangement. Shariar Afshar showed that experimental conditions that exhibit OR-form characteristics also exhibit AND-form ones, *simultaneously*. Per the standard rendition of the double-slit experiment, Afshar's evidence appears to constitute a challenge to the viability of generic complementarity. However, as previously argued, the non-standard appraisal of the double-slit experiment, and the new complementarity thesis derived from it, can survive the facts from Afshar's experiment.

In the initial setup, Afshar passed photons one at a time through two open pinholes at the barrier. Each photon was expected to generate self-interference in the usual way. But the otherwise self-interfering photon was made to pass through a lens placed between the barrier and two lateral photon-detectors. The lens served to focus the photon to either one detector or the other: with the photon from pinhole-1 lensed to detector-1, while the photon from pinhole-2 ended up at detector-2 (see **Fig P1.09, Diagram-A** for the Afshar experiment setup).

Note the typical pseudo-realtime description of a 'displacing', self-interfering and lensed quantum mechanical wave. This pseudo realtime description of the evolution of the system, *as if* it was being observed in realtime... despite isolation... is presumed throughout the Afshar's experiment as it is in all generic single and double-slit experiments. But the pseudo-realtime description is invalid: So long as isolation holds, there can be no empirical continuous observation of a quantum mechanical wave displacing to the two holes, passing through these, self-interfering thereafter, then lensed to the two detectors. The attempt to accomplish such a thing would involve direct observation, the end of isolation, and the destruction of the very wave that supposedly undergoes the unwarranted pseudo-realtime process.

---

[8]S. S. Afshar (2004)."Waving Copenhagen Good-bye: Were the founders of Quantum Mechanics wrong?". *Harvard Seminar Announcement.*

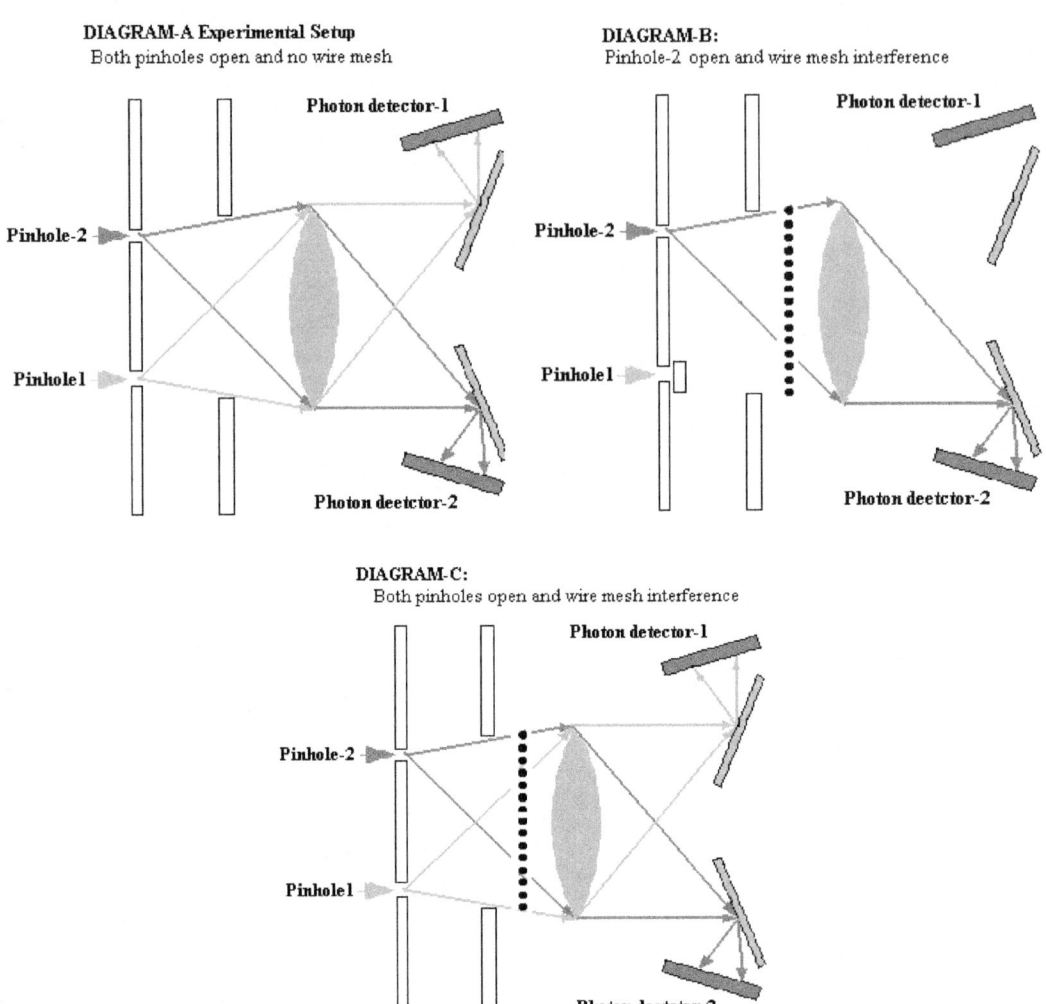

Fig P1.09: The Afshar experiment setup and results

Notes: **In Diagram-A**, a photon passes through two barrier pinholes; generates self-interference, and passes through a lens placed thereafter. The lens focuses the self-interfering photon to either one or the other lateral detectors. A photon from pinhole-1 will deposit at detector-1. The one from pinhole-2 will deposit at detector-2. In **Diagram-B**, only pinhole-2 is open. The lens focuses the photon to detector-2. The intervening wire mesh coincides with areas of destructive self-interference. Conversely, if only pinhole-1 is left open, the photon will end up at detector-1. The set up is expected to exhibit a definite 'which way' path for the particle per usual expectations from one slit experiments. The wires coincide with areas of destructive self-interference through which the photons are not expected to pass. This will not stop the expected outcome, but the quality of the image at detector-2 will be reduced. In **Diagram-C**, both holes are open: a self-interfering photon interacts with vertical wires coinciding with areas of destructive self-interference, but the quality of the image at the detectors is almost equivalent to the setup in which no wire was used, given that the wires only coincide with areas of self-interference through which the photons do not pass: And it is this set up that, according to Afshar, that exhibits both AND-form *and* OR-form realities at the same time, subverting the generic variant of complementarity, wherein the particles pass the wire as AND-form waves, but yet end up as particles at pinhole-detector correlations exhibiting certain 'which way' information equivalent to those obtained in Diagram-B. That is, we obtain definite which way information from wave-like processes... or single-slit outcomes from double-slit conditions.

All we can be certain of about the interim of isolation is the delayed choice time-interval between the emission of the particle at t_1 and the outcome at the detector at t_2: an interim of isolation measurable as a time-interval. Of course, anticipated de-spatialisation kills the notion of spatially displacing waves as much as it kills the notion of OR-form paths, and, with it, the whole notion of 'positions' and 'which way' spatial paths.

In the single-hole variant of Afshar's experiment, the intervening lens remains, but a wire mesh that coincides with areas that would otherwise involve the destructive self-interference of the photon is placed between the barrier with the single-hole and the lens. The said photon from open pinhole-2 will end up only at photon detector-2. Conversely, if only pinhole-1 is left open, the photon will end up at detector-1 (see **Fig P1.09, Diagram-B**). This setup should supposedly exhibit a single putative 'which way' path, or the inference of such on the assumption that, even in the absence of direct realtime observation, under conditions of isolation, the photon assumes an OR-form realised path, as is also claimed to be true in generic single-slit experiments: an erroneous assumption, as we have repeatedly avowed. The wires, insofar as these only coincide with areas of photon's possible destructive self-interference, and through which the photons would not be expected to pass, will not prohibit the usual expected outcome obtained from a single-hole experiment, although the quality of the image at detector-2 will be reduced due to diffraction effects from the wire.

The single-hole variant of Afshar's experiment needs to be re-evaluated per our non-standard appraisal and the new complementarity thesis. Therein, so long as isolation holds, we can only entertain a converging set of manifold *possible* (*not* OR-form realised) quantum mechanical AND-form 'which way' paths to and at the single hole; followed by a diverging set of AND-form (*not* OR-form) 'which way' paths after the hole, even when the lens and the wire system is factored in. These structures do not alter the logic, given that their presence do not end isolation, no more than barriers, slits and holes can end isolation. The wires affect the *potential* points of photon deposition, later resolved as OR-form outcomes at the pertinent lateral detector. The point again is that, throughout this single hole experiment, we *cannot* claim any real OR-form 'which way' path for the photon in the interim of isolation, so long as isolation holds: i.e. there *is no* realised 'which way' path to confine or decohere the possibilities inherent to the AND-form superposition of the photon… *unless we 'look' at the hole by means of directed observational inputs and end the enforced isolation at that moment.*

Moreover, no matter how convenient it might be, the pseudo-realtime description of a 'displacing and evolving AND-form wave' is not empirically and evidentially justifiable, for reasons already given. Again, all we can be certain of is the delayed choice time interval between the emission of the particle and the final outcome at the detector: an interim of isolation bounded in time.

In the two-hole variant of Afshar's experiment, both holes are left open. This time the wire is also used: The self-interfering photon is again made to interact with vertical wires placed to coincide with areas of destructive self-interference: The diffraction effects are much reduced, and the quality of the image formed at the detectors is better than that obtained in the single-hole experiment. The quality is almost equivalent to the initial two-hole setup in which no wire was used, given that the wires only coincide with areas of destructive self-interference through which the photons do not pass, given self-cancellation at those points: (See **Fig P1.09, Diagram-C**). The result is that, while the photon is subject to self-interference even with the wires placed, and is equivalent to the setup in which both holes are left open with no wire placed, so emulating and attenuating the AND-form wavelike character of the photon, the outcome on the detector is equivalent to a result obtained from a single-hole setup; with each photon ending up on just one pertinent detector; *supposedly* exhibiting evidence of a definite OR-form 'which way' path despite enforced wavelike form.

In short, *in the Afshar experiment, a two-hole arrangement is used to produce a one-hole outcome: the equivalent of using a two slit experiment to generate a one-slit outcome.* **If we subscribe to the error that one-hole or one-slit experiments exhibit particle-like behaviours and convey certain 'which way' information about the state of the system in the interim of its isolation, we will then commit the error of assuming that Afshar's two-hole arrangement, having produced a one-hole outcome, can yet furnish a definite and certain 'which way' information about the particle in the said interim, even when under isolation, consequently undermining the complementarity principle** which claims that one cannot have AND-form *and* OR-form realities furnished at the same time, or that, in the interim of isolation, the system must remain purely in AND-form.

However, from our own new complementarity thesis and the non-standard appraisal of the two-slit experiment on which it is based, augmented by the firewall principle, the evidence for both wave and particle realities are *always observationally simultaneous*, even under single-slit conditions. Thus, **Afshar's experiment, in which a two-hole arrangement leads to a one-hole outcome, does not imply the absence of AND-form reality per the interim of isolation, no more than it does in generic single slit experiments… and the result cannot be used to assert information about any definite OR-form 'which way' information from a two-hole arrangement**. Thus….

- At end of isolation, at the detectors in both generic and in Afshar's experiment, OR-form realities and AND-form realities are *always* observationally simultaneous and occur together. However, this outcome cannot allow for any inference of a temporally preceding OR-form 'which way' path in either two-hole or one-hole experiments. This remains true even with Afshar's pinhole, wire, lens, and dual detector setup... and even when Afshar obtains a one-hole outcome from a two-hole arrangement.

- In the interval of isolation, in both the one-hole and two-hole experiments, the emitted 'particle' is purely an AND-form superposition state of possible (*not* realised) putative 'which way' paths, albeit on the assumption of 'space'. As such, the interim is bereft of any OR-form putative 'which way' information, even with Afshar's pinhole, wire, lens and dual detector setup, and even when Afshar obtains a one-hole outcome from a two-hole arrangement.

- The non-standard and new complementary thesis states that the physical state of the photon or electron from emitter, through the pinholes, through the wire grid, through the lens, and to the detectors, is purely an AND-form state, *so long as experimental isolation is enforced throughout the experiment.* From emitter to the two lateral detectors, the electron or photon cannot in any way be characterised as a putative OR-form 'which way' state, nor as a resolved position-set, even under Afshar's two-hole, wire and lens system... at least *so long as experimental isolation holds.* Thus, **the AND-form and OR-form descriptions of a 'particle' cannot occur together before isolation: The AND-form description applies exclusively up until the end of isolation, although both OR-form *and* AND-form states are observationally simultaneously at the post-isolation and post-measurement stage,** e*ven in the absence of interference patterns,* in both single-hole and two-hole experiments, with or without a lens, wire-grids and lateral detectors. Thus, the principle of complementarity is saved, albeit in its new form.

0.17: Afshar's Experiment, the Non-Standard Account & the Ontology of the Quantum Mechanical "AND"

Our conclusion can be further justified from the ontology of quantum mechanical wave. The AND-form quantum wave constitutes the future phenomenalised as an objective reality. Indeed, the non-standard complementarity thesis is demanded by this truism and by the firewall principle that emerges from the intertwine of the principle of conservation of energy-matter with the principle of causality. Thus, the AND-form wave description of a system is the future-form of that system: its state *prior* to its resolution or conversion into realised OR-form events. Possibilities or potentialities in AND-form must ontologically precede OR-form realisations: That is, the future-form possibilities must ontologically precede the resolved present-continuous and past OR-form outcomes that decay out of that preceding future via AND-to-OR wavefunction collapse.

To assume a definite putative 'which way' path for a particle in generic two-slit experiments, or even in Afshar's experiment prior to the end of isolation and observation, is to assign to that particle the status of a resolved event-set... an OR-form putative 'which way' state... *before* it was resolved into such. Again, this would be tantamount to treating the future before it happened as a state that had already transpired... the empirical proof of which would be tantamount to 'signals from the future' and lead to the violation of the principle of causality, and usurp the principle of conservation.

By way of illustration, if one arrived at New York before having departed from London, and treated one's arrival at New York as a resolved set of events in OR-form terms, then, as a minimum, one would have doubled oneself into a London version *and* a New York version *before* having arrived there: Where will the extra energy-matter for one's New York version come from? If we could treat the future, even of particles, in the same way, as is implicitly suggested in the want to assign a definite putative 'which way' path to a particle in Afshar's experiment and in generic single-slit experiments, we would commit similar violations of the principle of conservation of energy-matter. Hence. the firewall principle and why AND-form logic exist in nature.

One might counter the above by claiming that the 'which way' information presumed to be obtained in single-slit conditions or in Afshar's experiment must pertain to the past state of the particle, before its leading OR-form resolution at the detector. This is not true, and a similar error is made in quantum eraser experiments carried out by Yoon-Ho Kim *et al* that purport to erase the past[9]. We reiterate that, when we enforce isolation on a system, we relegate its resolution *to the future*. The system is immediately rendered into a superpose of AND-form future possibilities. This abides until the end of isolation through AND-to-OR collapse. Thus, whenever we refer to a system under isolation, we are always referring to an interim state of future possibilities *before* an event gets to happen and the later OR-form outcome is obtained.

But when we make OR-form-like attributes, such as 'which way' paths, to a system in isolation *and* deferred to the future, hence attribute to that future before it has transpired the status of realised events, we are inadvertently treating the future *as if* it constituted OR-form events; tantamount to 'signals from the future'.

Moreover, the OR-form outcomes obtained at the end of isolation cannot allow any retrospective attribution of OR-form characteristics to the preceding period of isolation: this interim of isolation constituted the future, about which OR-form attributions could not be made. Therefore, no retrospective OR-form 'which way' paths can be attributed to that interim, even when two-hole arrangements are rigged to produce one-hole outcomes in Afshar's experiment.

In Afshar's experiment, isolation comes into play at the moment of the emission of the particle into the system and remains in force until the one-hole-like outcome is obtained at the end of isolation. Throughout, the system is suspended into a pure set of unresolved future possibilities, rendered purely in AND-form. Even a single-hole outcome at the end of isolation cannot secure OR-form information about the state of the system in its preceding interim of isolation, even when tricked to generate a single-slit outcome.

The holes certainly restrict and decohere the AND-form future scope of the system in isolation. The inclusion of the grid and the lens are additional decohering factors: All of these reduce or augment the potentialities available to the system to a collection of potential AND-form future states consistent with the holes, the mesh, and the lenses, such as to generate a single-hole outcome from a two-hole arrangement. But these do not and cannot eliminate the AND-form character of the system in its preceding isolation-interim, per function of the firewall principle and the demand to preserve both principles of causality that prohibit 'signals from the future' and obviate violations of principles of conservation that would attend such impossible signals.

---

[9]Kim, Yoon-Ho; R. Yu; S. P. Kulik; Y. H. Shih; Marlan Scully (2000). "A Delayed "Choice" Quantum Eraser". *Physical Review Letters.* 84 (1): 1–5

QUANTUM MECHANICS AND MIND

0.18: Afshar's Experiment, the Non-Standard Account & the Problem of 'Space'

We have already called into question the ontology of 'position', 'path', hence 'space'. These could be completely discarded and succeeded in a model of nature comprised purely of potential and realised events, organised and structured purely in and by time, *without* space. With de-spatialisation, in the interim of experimental isolation featured in the double slit experiment and in Afshar's experiment, even AND-form 'which way' paths cease to apply... simply because there is no space. Therefore, we cannot obtain 'paths': potential or otherwise. Instead of paths, we end in AND-form superposes of 'which moments', structured per potential delayed choice time-interval relations.

Thus, with the end of isolation, we will obtain the OR-form 'which moment': *not* 'position'... *not* 'which way'. The OR-form 'which moment' will not be distributed in space, but it will be integrated by way of delayed choice time-interval relations to the moment of particle-emission at the start of the experiment *and* to the moment of its later-observation. Per the assertions of the preceding essay, we cannot attribute any OR-form characteristics to the pure AND-form interim of isolation, given the firewall principle, which walls off the future from observation.

But why de-spatialisation? Why delayed choice time-intervals in lieu of 'space'?

When we observe the Moon at night, we see it as it was around 1.5 seconds ago in the past. Even assuming the Moon had a 'place' or 'position' 1.5 seconds ago, we cannot now go to that position or confirm its purported positional reality: The presumed 'position' or 'place' is now *passé:* It has been succeeded by new events. This is a problem for *all* position-claims and attendant to claims about 'which way' paths in double-slit and in Afshar experiments, and for general suppositions about 'spatial trajectories', supposedly made up of serial contiguous 'places': We can never go to purported places or 'positions in 'space': The moment we apparently 'get there', the signature-event has been succeeded by new events, and the presumed original 'place' is at best relegated to the past and to *time*, even assuming that space and 'place' could enjoy *any* ontic status to begin with.

Moreover, the moment of our conscious witness of events itself must defy locatability. The conscious observer can only assert itself as a 'moment of conscious observation': a 'which moment' related to other moments of the past under its observation: a relation that constitutes a pure time-relation to past moments, not a spatial relation to 'places'.

In Book-II, using John Wheeler's delayed choice approach, we will show that even the typical pseudo-realtime notion of a displacing and self-interfering quantum mechanical wave must be replaced. We will show that, at best, the quantum mechanical wave constitutes a static non-displacing state: hardly remarkable, given that the future (which the quantum mechanical state constitutes) *obviously* cannot move or displace. Indeed, in isolation, even a purportedly moving and evolving quantum wave can only do so presumably through Hilbert space: an AND-form future potential for 'space' that has yet to form up into realisation. If so, how can anything, even a quantum wave, displace through a 'space' that has not yet come into realisation?

Clearly, we must ruthlessly discard *bloss* assumptions about 'position', 'path' and 'space' from the rubric and lexicon of physics and quantum mechanics; from the double-slit experiment and from Afshar's experiment. There is no space. Therefore, the interim of isolation in the double-slit experiment or in Afshar's experiment cannot exhibit an AND-form superpose composed of 'which way' possibilities, let alone OR-form versions of the same. And a modified version of John Wheeler's delayed choice experiment will furnish us with the incontrovertible reasons for why we must discard the notion of 'space' from physical ontology, as we shall discover in Book-II through to Book-VI, notwithstanding direct and obvious de-spatialisation and temporisation from the growing block model.

0.19: The Double-Slit Experiment, AND-OR Duality & Information-Matter Relations

What is the relation between findings from the double-slit experiment and the case for facile information-matter dualism from **0.03** and **0.04**? Information-matter dualism asserts the broken-discontinuous form of information vis-à-vis radiation, matter, and space. Therein, the information purportedly 'carried' by radiation or de Broglie matter apparently dissociates from both the radiation and matter and from space. The information will re-associate with the radiation or the de Broglie matter and with putative space at a succeeding OR-form moment of resolution throughout the broken-discontinuous motion of radiation and matter, but with AND-form quantum mechanical gaps, throughout which the said dissociation will abide. Hence, information-matter dualism.

In the most extreme variant, the initial disassociation of information will occur at the beginning-point of the transmission: The information will re-associate with its radiation only at the endpoint or reception-point, or at the end of isolation. Therein, the signal or information dissociated from the radiation and from 'space' will be somehow retained by the physical order across the quantum mechanical AND-form intervals of isolation, but *not in* the AND-form interval. The resolved moments will constitute the OR-form 'frames', analogous to frames in a motion picture, while the gaps between these will constitute the AND-form intervals comprised of possible or potential, but never realised, future-potential states: all rendered in AND-form, and from which the 'carried' concrete information must dissociated.

Per the observational simultaneity of wave-particle duality at the end of isolation, and, with it, the new complementary thesis, the non-standard appraisal of the double-slit experiment serves to augment the assertion for the objective reality of AND-form states and attendant information-matter dualism necessitated by the ineliminability of AND-form intervals that break up the OR-form 'frames'. In short, the non-standard appraisal of the two-slit experiment asserts that AND-form reality is ineliminable from *all* conditions and systems, even in those where self-interference is structurally prohibited, even ignoring de-spatialisation. Insofar as this is true under *all* conditions, its implication to information-matter dualism is straight forward: The AND-form intervals that reside between successive OR-form moments exhibited in the broken-discontinuous relation of information vis-à-vis radiation, matter, and 'space', constitutes an objective fact, ineliminable from *all* information-matter relations. Therefore, information-matter dualism is also objectively real: i.e., information *must* dissociate from radiation, matter, and 'space': Its transportation across 'space' must be subject to the broken and discontinuous form attendant such dissociation.

What happens to information-matter dualism when we discard 'space' altogether, as will be done by Book-IV? What happens when we must reframe information pertaining to, say, a photographic image, in terms of events and 'which moments' organised into pure delayed choice time-interval relations, *without* 'space'? Even with de-spatialisation, the broken-discontinuous form of the relation between information versus radiation or matter must abide as a broken-discontinuous form of time-evolution of information, *without* 'space'. Hence, information-matter dualism abides even under conditions of de-spatialisation and temporisation.

Of course, information-matter dualism engenders implications to the nature of nous: Even if treated as an informational state supposedly carried by and initially reduced to 'matter in space', or purely in brains, such mind and memory must dissociated from 'matter and space' and from brains. With de-spatialisation, the initiatory Cartesian dualism inferred from facile information-matter dualism will transform into an inexorable conclusion: a dualism grasped as a function of the time-relation of memory and nous to a superlative de-spatialised purely temporised growing block model of the physical world and of brains.

EXPANDING THE LIST OF MYSTERIES: ANTICIPATING SOLUTIONS

0.20: More Mysteries:
The Measurement Problem, AND-to-OR Collapse, Time & Causality

Wave-particle duality, or AND-OR duality, constitutes the most accessible riddle of quantum mechanics. Its mystery unravels the moment we grasp the AND-form state as the objective phenomenalisation of the future. The greater mystery is causality: At some point, the quantum mechanical AND-form future must, by *some* act or process, decay into OR-form events. What is it that brings about this AND-to-OR wavefunction collapse? This is oft subsumed to the 'measurement problem' *and* pertains to the problem of causality, whose locus *is* AND-to-OR collapse.

There are two aspects to the measurement problem: The first pertains to **what causes AND-to-OR wavefunction collapse: Are there hidden gears behind this?** (i.e., causality). Is it observational inputs of measurement that bring about AND-to-OR collapse?

The second aspect of the measurement problem relates to **the question of whether there is a secret deterministic orchestrator of quantum indeterminate outcomes obtained from AND-to-OR collapse**. Both aspects constitute distinct facets of the same mystery: the mystery of causality.

In classical physics, all the variables pertaining to initial conditions, and the operations of physics wrought on those conditions, are *in principle* perfectly definable. The output is expected to follow on from the inputs and operations in a complete and closed way. In quantum mechanics, even if we could be sure of all the variables pertinent to the initial conditions, inputs, and operations, this will not causally account for AND-to-OR collapse, much less overcome the indeterminate resolution of outcomes.

Note that the input-operation-output schema (i.e., IOO schema) is central to the materialist conception of causality, and it is relevant to both AND-to-OR wavefunction collapse *and* quantum indeterminacy. 'Missing gears' remain synonymous with the missing middle term: the 'operation' in IOO. **The implicit or never openly stated crisis of causality in physics is necessarily the crisis of materialist IOO-schema causality or 'missing gears'.** The problem of measurement and AND-to-OR wavefunction collapse also undermines the presumption of closed causality, also central to the materialist notion of causality. **Putative 'missing gears' renders plausible the possibility of an extra causality outside of the presumed closed-chain causality of manifest inputs, operations, and outputs. This also opens the way to Cartesian mind-brain dualism.**

*

The matter of hidden variables has been sharpened by Bell's Inequality.[10] This described the conditions for hidden variables in Einstein Podolsky Rosen relations (EPR). Bell's inequalities undermine hidden variables in quantum indeterminacy. We will have more to say about Bell's inequality in Book-V. However, by Book-III we will discover an independent basis for concluding that there are no hidden variables behind either AND-to-OR wavefunction collapse or in the orchestration of quantum indeterminism. Book-II to IV will show that AND-to-OR collapse is mediated by a wholly structure-less ineffable causality.

Recall a central assertion of Alt-R: **AND-to-OR wavefunction collapse *is* time**: the coming into being of OR-form events out of AND-form futures. The mystery of causality is *essentially* the mystery of what causes time. **Alt-R will assert that any hidden gears behind the process of AND-to-OR collapse and time would crash into insurmountable infinities and never obtain the outcome: i.e., Time would 'freeze up'. Hence, the collapse-process cannot be mediated by any 'gears'.** The first infinity is inherent to the intermediate spacetime model in Book-II (itself to be superseded in Book-IV) and its attendant nested futures structure. Therein, a wavefunction does not constitute only the immanent future-potentials, but must also accommodate subsequent futures niched to each immanent possibility. The nested futures thus stretch to and totally converge at the indefinite or infinitely removed future. Consequently, an event 'here' will instantaneously decohere future possibilities elsewhere in a 'non-local' instantaneous way up to the indefinite future, and in a way that will render all futures for all reference frames consistent with what happened here. This is **counterfactual causality** and **grand decoherence**. And therein reside the critical infinities that usurp any want of causality as 'definable gears'.

---

[10]Bell J.S. (1964) "On the Einstein Podolsky Rosen Paradox" 195-200 Physics Physique Fizika 195-200

The AND-to-OR collapse of the immanent future must affect the grand decoherence of the system of attached nested futures to the infinitely removed future. In turn, this must restructure the system of nested futures, of which there are infinitely many, given that all futures converge at the indefinite future. Hence, any putative hidden 'algorithm', mechanism, or 'gears' requisite to the counterfactual elimination of the infinitely many non-consistent futures, in order to generate grand decoherence and culmination into AND-to-OR collapse of immanent futures, would crash into insurmountable infinities and halting problems: Time would 'freeze up'.

Hence, **whatever brings about AND-to-OR collapse and time is *not* a 'hidden gears' but must be a structure-less ineffable process that can sort out and eliminate infinite non-consistent futures to infinity, without crashing into halting problems.**

*

The fact that AND-to-OR collapse must be driven by a structure-less bizarre *something* was always implicit in renormalisation in quantum field theory. Therein, putative particle interactions, mediated by very short-lived particles, 'blow up to infinity'. The calculation problem that attends this is solved by subtracting the infinities: hence, *renormalisation*. Yet, the solution obscures the core problem of causality and what drives ontologically primary AND-to-OR collapse that mediates inputted particles to the outputted outcomes.

If the interval between particle-input and particle-output is comprised of infinitely many interaction potentials, and if this is to be mediated by some 'gears', the processing of that interval into the final outputs would crash into an insurmountable halting problem, and never obtain the final particle outputs. The solution to this requires that we grasp that **the process of AND-to-OR collapse or time simply 'leaps over' the infinities, if only because AND-to-OR collapse and time is driven by a an ineffable structure-less non-gears. Hence the *quantised time solution*,** which completely circumvents IOO-schema causality at the middle 'operation' term. Indeed, it is obvious that the very assumption of IOO causality in particle interactions is what engenders the 'blow up' to infinities: The process of the 'quantum leap in time' over the infinities obviates the need for the middle 'operation' in IOO. In short, the 'quantum leap in time' subverts materialism vested in IOO causality or 'hidden gears' per the middle 'operation' term.

*

Anticipated de-spatialisation also has a contribution to make: De-spatialisation implies that **there is no 'point of contact' for 'interacting' particles.** De-spatialisation totally falsifies the notion of contiguity, contact and impact-causality. **There is no 'space'. Therefore, there can be no 'locations' at which purported particle-contact can take place, much less some spatial outlay for an 'operation' that supposedly gears inputs to outputs, or brings about AND-to-OR collapse and time.**

0.21: The AND-to-OR Wavefunction Collapse, Information-Matter Relations & the Mind

What are the implications of AND-to-OR collapse to information-matter relations, if not to nous? The answers depend on the causal basis of wavefunction collapse and the way the crisis of materialist causality is solved. **In order to rescue materialism, one would need to establish that AND-to-OR wavefunction collapse is brought about by a discoverable 'hidden gears'.** Re-stated in terms of the broken-discontinuous format from information-matter dualism, **putative 'hidden gears' would have to be inherent to the quantum mechanical AND-form intervals that reside between successive OR-form 'frames', across nature *and* in brains.**

The requisite hidden gears would need to be causally responsible for the collapse of each AND-form interval into its OR-form moment in the time-evolution of nature and brains, vindicating IOO-schema causality per the middle 'operation' term. Even so, the causal factor that could resolve for the time-evolution of brains could not be identifiable with *any* materialised OR-form moment of that brain, but *only* with the hidden causality that operates from *between* the OR-form frames and resolutions of that brain.

However, if materialism could posit that the pertinent causality is constituted as a definable set of 'hidden gears'… *and* demonstrate the existence of such gears as the middle term in IOO… it would achieve the paradoxical outcome of subordinating apparently extant causality vis-à-vis brains to a closed-causality schema, one as good as the generic materialist conception, yet also different, given its extant relation vis-à-vis the brain and world, confined as it would be to the AND-form intervals residing between the OR-form moments constitutive of the 'materialised' brain.

However, **if the process of causality in AND-to-OR collapse is both extant *and* intrinsically structure-less and undefinable, conclusions inimical to materialist closed causality must follow. Cartesian revival must also follow.**

By Books-IV and V, critical insights into the measurement problem and to the nature of causality attendant AND-to-OR wavefunction collapse will make the case for ineffable causality *and* revived Cartesian mind-brain dualism.

0.22: Quantum Indeterminism: The Stochastic-Coherence Paradox & the Alternative Realist Thesis

What is quantum indeterminism? In the context of quantum spin, upon measurement, a particle will either convert into 'spin-up' OR to 'spin-down'. It will do so in a purely probabilistic and random way. We can only give a probability statement for the likelihood of obtaining one or the other spin: a 50:50 chance of one or the other becoming the OR-form outcome.

Previously, in **0.20**, we anticipated that wavefunction collapse is *not* brought about by any definable hidden gears. The basis for this will be fully explored in Book-II to III, and others. At this juncture, it is sufficient to state that what applies to causality behind wavefunction collapse also applies to the quantum indeterminacy obtained from that collapse. Thus, **the claim that there are no hidden gears behind AND-to-OR collapse must also imply no hidden gears behind the generation of quantum indeterministic outcomes.**

However, there is a way of arguing against hidden gears in quantum indeterminism: wholly independent from the AND-to-OR collapse process. The argument will be presented in Book-III, centred on the question of whether randomness and quantum indeterminacy is provably objectively real, or an artefact of our subjective ignorance and limits.

In Book-III, we will show that randomness *is* objectively real. By transference, quantum indeterminism is also objectively real. **In Book-III, the case for the objectivity of randomness will take some key process in which quantum mechanised randomness or straight quantum indeterminacy operates. It will systematically generate for that random process wholly deterministically orchestrating scripts**. The operation of the scripts, when hidden from view, will appear to engender randomness, or the illusion of it, even though the scripts operate in a wholly deterministic way: constituting putative 'hidden gears' behind facile randomness and quantum indeterminacy.

However, a very large number of such scripts will be generated. It will be found that **each script is equally adept at emulating facile randomness and indeterminacy. When we search amongst the very large number of such scripts for a *special* or exclusive script**, two things will happen: First, **we will not be able to find any such special script.** Second, since any script will work as well as any other, **we will discover that there exist no qualitative or structural basis for selecting the 'right' script from among the others.** In the absence of any qualitative or structural basis for doing so, **the only recourse left will be random selection from among the very large number of equally apt scripts.** In this way, we will show that **there is no way of eliminating some final resort to randomness and indeterminacy, even when we presume a hidden deterministically orchestrating underworld** that supposedly belies surface randomness and indeterminism. **In other words, we shall discover that there are no hidden gears behind quantum indeterminism, and that quantum indeterminism is objectively real; *not* a bloss subjective artefact of ignorance or human limits.**

*

The deeper consequence of the objective reality of randomness and quantum indeterminism is the stochastic-coherence paradox: This is the paradoxical fact of coherence, consistency, and pattern-persistence (or even perdurance) evinced throughout nature in the *endurance* of identities, or in the distribution and accumulation of identity from and as the sum of the past... despite quantum indeterminacy inherent to the process of the perpetuation of pattern and identity. Simply put, **the stochastic-coherence problem is the question of 'impossible' order from apparent quantum mechanical chaos. It is the problem of order in the context of the moment-to-moment quantum indeterminate 'scramble' inherent to the time-evolution and perdurance of pattern**.

If quantum indeterminacy is objectively real, as indeed it is, **any sort of coherence, consistency, and pattern-persistence, or even a self-determining consciousness vis-à-vis the brain and world, ought to remain wholly untenable**. Yet, we live in a universe in which the impossible is both possible *and* manifest.

How is this possible?

The terms 'order' or 'ordered' do not necessarily refer to a 'planned' or consciously designed state-of-affairs, although it could do so. To illustrate this, consider a pack of cards from the shop, arranged in its conventional card-sequence. We now shuffle this to produce a new 'scrambled order'. Our pack is unlikely to retain this new 'scrambled order' if it is subject to subsequent re-shuffling. If this new 'scrambled order' were to persist through a succession of re-shuffles, it would constitute a remarkable outcome: a miracle. Quantum indeterminacy is presumably equivalent to this sort of card re-shuffling: a first quantum indeterminate re-shuffling of the world ought to produce a 'scrambled order' equivalent to a first-time shuffled pack of cards. Yet, despite subsequent quantum indeterminate re-shuffling of our world, its founded 'scrambled order' abides and endures, *ceteris paribus*: The world retains its identity and pattern over great spans of time, and renders experience and the physics built upon it viable and possible.

For example, if we ran the notorious Schrödinger's cat experiment involving a box with a cat, with a poison vessel rigged to quantum spin outcomes in the box, there will be a fifty-fifty chance of the release of the poison into the box. Hence, the cat will die...or live. But the identity of the cat, and the identity of the apparatus itself, will not be scrambled by general quantum indeterminate re-shuffling. Both will persist over time, whether through endurance or perdurance: **The cat will *not* be a cat one moment, only to be re-shuffled into a pumpkin the next. Dead or alive, its identity will endure or perdure over a useful interval of time: The same will apply to the box and its arrangement. Such continuity of identity ought to be impossible in a universe governed by objective quantum indeterminacy. Hence, the stochastic coherence paradox.**

*

There *is* an Alternative Realist (Alt-R) solution to this great paradox. It will be presented in Book-III, and partly derive from an analysis of randomness in 'chaos games', such as the Sierpinski triangle. The anticipated Alt-R solution will frame these insights into a three-phase developmental approach, integrating this process to the facts garnered from facile information-matter dualism and the intermediate model spacetime developed in Book-II.

To glimpse the solution, consider a simple random process generated by the roll of three unbiased quantum mechanised dice, such as to generate a generic bell curve distribution. Over time, the dice will accumulate *en masse* outcomes structured by a **three-phase development process** comprised of the **below threshold phase**, in which indeterminate samples will not collectively evince any sort of pattern-structure... followed by a **transitional or chaotic phase**, which will exhibiting unstable and volatile pattern-structure... finally culminating into the **noise-reduction pattern-saliency phase**, wherein *en masse* random outcomes will culminate into the expected bell-curve distribution pattern, now fully materialised. The same sort of thing is observed in the development of the interference pattern in the two-slit experiment and in the development of a photographic image.

Hence, regardless of objective quantum indeterminacy, the bell-curve pattern turns out to be ordained and fated. Indeed, the staple bell-curve distribution *is* the order, as is the complex wavefunction that attends the formation of the interference pattern by *en masse* quantum indeterminate depositions in the double-slit experiment: as is the convoluted complex wavefunction that attends the formation of a photographic image. The same applies to the universal wavefunction, which constitutes the order that attends our physical universe as a whole.

The development and materialisation of all of these are pre-ordained, *not* by quantum indeterminacy itself, but by the bounded probability-context in which objective quantum indeterminate resolutions must always unfold through, the three-phase developmental process.

In summary, no matter how random and quantum indeterminate each quantum mechanised dice-outcome or quantum deposition remains, and no matter how causally non-connected these are to each other, whether consecutively, in parallel, or in succession, **'order' made evident as the materialised 'bell curve' distribution itself, or as the interference pattern, or as the photographic image, or as the universal wavefunction for the whole physical universe... will inexorably emerge out of the attendant *en masse* quantum indeterminate outcomes...** *so long as we identify 'order' with the abstract complex wavefunction itself, and do not seek it in some secret deterministic orchestration of the quantum indeterminate terms.*

We shall find that 'distribution curves' and abstract complex wavefunctions all derive from strict necessity: These are not typically randomly generated but enjoy a 'read only' status in nature: For example, the shape and configuration of the probability clouds that govern the electronic configuration of the atom (the s-p-d-f shells), and the forms these can assume, emerge from strict necessity. Order in nature is not typically subject to scrambling by quantum indeterminate processes, notwithstanding some exceptions, even when the order is made manifest in and through quantum indeterminate terms.

Hence, **'order' *must* emerge despite objectively quantum indeterminate processes because quantum indeterminate resolutions are bounded to implicit complex wavefunctions that frame the natural odds in favour of their own inexorable materialisations: This being the central insight entailed in the resolution of the stochastic coherence paradox in Alt-R.**

Of course, this conclusion contradicts the central assertion of facile information-matter dualism from **0.04**, which presumed the dissociation of information from the AND-form wavefunction. Yet, essential information-matter dualism will abide when we obtain the growing block universe *and* de-spatialisation: Both force the integration of information to a purely temporal physics, *without* space. Indeed, therein, the 'dissociation' of information from matter and radiation is in truth the separation of the resolved past from the future, or of memory from future-potentiality, with the future evinced as the generic *and* nested futures quantum mechanical wave.

*

Hence, the other key to the solution to the stochastic coherence paradox will be abstract memory: i.e., the *retention* of past wavefunctions, even after these have AND-to-OR collapse into outcomes and events. This implies radical abstract memory: a central finding of Alt-R that emerges seamlessly from de-spatialisation obtained in Book-IV. Abstract memory will force a reframe of facile information-matter dualism in the formers' terms, harmonising it with our general solution to the stochastic coherence paradox.

The retention of past abstract states by the natural order, which essentially involve the retention of past wavefunctions, implies abstract memory of a form that abides the vagaries of time, and wherein the past participates in the further evolution of events: Thus, **past wavefunctions also funnel *en masse* quantum indeterminate terms into global outcomes, manifesting pattern-persistence, and facile information-matter dualism**. The dissociation and extant retention of information outside of putative radiation, matter and 'space' is equivalent to the retention of the past in abstract form as memory: one that must temporally resonate to the relative present and decohere the as-yet not realised future into consistency with memory.

Is not such memory already obvious from anticipated de-spatialisation? There is no space. Therefore, there are no 'locations' or spatial distributions at which memory or future-possibility could be 'stored' or staged. Indeed, **anticipated de-spatialisation forces a pure temporal organisation of physical information, whether this is comprised of future-potentiality or of the retained past. What appear as 'objects in space' are in truth resonations of the past that get to us... not across 'space', but as resonations of the past realised in and across time.** That is, de-spatialisation seamlessly and perfectly implies abstract memory.

Again, such insights must incorporate and surpass facile information-matter dualism and entail obviously implications to memory-brain and mind-brain relations.

0.23: Quantum Indeterminacy, the Stochastic Coherence Problem & Mind

The greatest problem faced by mind-brain dualism does not derive from *a priori* materialism, nor from the attendant assumption of closed causality input-operation-output schema. Indeed, materialist notions of contiguous or impact causality suffer from rarely noticed precarity per the oft ignored question of missing variables or 'hidden gears' in quantum mechanics, central to the problem of measurement, and crucial to causality behind wavefunction collapse. Nor does the greatest problem posed to mind emerge out of the much abused principle of conservation of energy-matter, as we shall later discover. **The most obvious problem posed to mind-brain causality and dualism derives from quantum indeterminacy and the stochastic-coherence paradox.**

The anticipated solution to the stochastic-coherence paradox and quantum indeterminism (see **0.22**) is applicable to the problem of quantum indeterminacy in brains. **If quantum indeterminacy is objectively real (as indeed it is), and if it operates at the level of brains, then it ought to scramble the possibility of mind-to-brain control, or even internal materialist brain-based control.** Hence the problem posed by quantum indeterminacy to mind-brain control, and even usurp control in materialist theory.

*

Before we glimpse the solution, **we must first confront the claim that quantum mechanics, with both AND-form logic and quantum indeterminacy, has no relevance to the macro-scale world, including the brain... and that quantum mechanics is restricted only to the micro-scale.**

Recall from **0.12** that the double-slit experiment evinces the reality of both AND-form superposition (the wave) as well as OR-form particles (the *en masse* quantum indeterminate depositions on the detector) in an observationally simultaneous way. Hence the *new complementary thesis.* Yet, the double-slit experiment is a macro-scale frame. In principle, **one could carry out the double-slit experiment with very short wavelength high-frequency radiation and matter, confining it to the micro-scale. One could also use very long-wavelength low-frequency radiation and matter and perform the experiment at the large scale. Whatever the scale, quantum effects, including quantum interference and quantum indeterminacy, will evince. It follows that quantum phenomena are not restricted only to the very small. Indeed, science depends on the macro-scale manifestations and amplification of such phenomena in the double-slit experiment in order to infer their operation at the micro-scale.**

Photography, another macro-scale frame, also evinces quantum indeterminacy: The development of a photographic film proceeds according to the anticipated three-phase development process garnered in the solution to the stochastic coherence problem in Book-III: At the *below threshold phase*, indeterminate quantum photon depositions on the film-surface do not manifest any obvious pattern. This is followed by a *transitional or chaotic phase*, in which sufficient quantum indeterminate photon-depositions form developmental clusters or sites, but no definite coherent structure is yet obvious. As was intimated in **0.22**, this then transits into the *noise-reduction pattern-saliency phase*, in which a large number of indeterminate quantum photon-depositions fully manifest the ordained image in high detail and resolution. Thus, randomness in the form of quantum indeterminacy renders salient coherent and complex pattern. Indeed, the manifestation of the macro-scale interference pattern in the double-slit experiment follows essentially the same three-phase developmental stages as found in macro-scale photography.

The point is that the three-phase quantum mechanical account of macro-scale photographic development demonstrates that quantum indeterminacy cannot be abstracted out of the macro-scale or from long time-interval developmental frameworks. Herein resides the solution to how mind might overcome quantum indeterminacy in brains: **As is true of the photographic film, the brain, as a hot-body macro-scale, and as a long-time-interval framework for the said three-phase development process, is no more absent of quantum indeterminacy than the process of photographic development. Yet, coherent brain activity will form out of inherent quantum indeterminacy as it demonstrably does in photography.**

*

In our non-standard appraisal of the double-slit experiment (see **0.14 to 0.15**), we discovered that **AND-form states (the future form state of any physical systems, including that of the brain) must remain in AND-form, so long as isolation holds. Consequently, AND-form realities apply to the macro-scale to whatever degree isolation applies at that scale... as it must apply to the brain.** Indeed, from both the non-standard appraisal *and* from facile information-matter dualism, intervals of isolation, hence sought AND-form intervals, *cannot* be eliminated from any system, at *any* scale, including from the brain... even when subject to a Quantum Zeno effect and to almost continuous observation, given that we cannot have infinite frequency inputs or outputs.

Contrary to the implicit growing block universe thesis, the prospective elimination of AND-form states from *any* scale, including the brain, would be tantamount to the elimination of future possibilities and of the future as a whole, given that AND-form realities constitute the physical phenomenalisation of the future, as demanded by causality and principles of conservation constitutive of the *firewall principle.*

Indeed, recall that the critical reason why AND-form logic must apply at the macro-scale derives precisely from the fact that the principle of causality *and* the principle of conservation cannot be excluded from the macro-scale, any more than from the micro-scale. Hence, according to the *firewall principle,* causality and principles of conservation intertwined demand the compartmentalisation of the future from the relative present and from the past... at *all* scales. Thus, **to assert that AND-form logic and the superposition principle (and, with these, quantum indeterminacy) must be confined *only* to the very small is tantamount to the restriction of causality and the principle of conservation only to the very small**: a dubious claim.

Consistent with the growing block universe approach, **the firewall principle demands that that future at the macro-scale configure as an AND-form state, lest we treat the macro-scale future as a fully resolved state, with consequent violation of causality and the usurping of the principle of conservation: The brain is not going to be exempt from this truism. Therefore, the brain must also have a future possibility-set: one configured as an AND-form brain at the macro-scale. In other words, the brain is *indeed* quantum mechanical.**

*

By Book-V, Alt-R will show that, by exploiting both facile information-matter dualism and the processes that attended the successor to spacetime...

- Mind selects from the future AND-form potentiality of its brain those complex wavefunction for subsequent three-phase development through AND-to-OR collapse. This funnels *en mass* quantum indeterminate terms into culminating fated brain activity in accord with the intents of mind. Thus, circumventing the problem of quantum indeterminacy.
- **The mental orchestration of quantum indeterminate incidents in brains is no more necessary than it is in the development of a photographic image out of quantum indeterminate terms: Hence, mind *can* control the development and expression of its brain despite ineliminable quantum indeterminacy therein.**

40

0.24: The AND-to-OR Collapse Problem: Irreversibility, the Time-Symmetry of Physical Laws, & the Problem of Mind

With 'measurement', something *irreversible* has occurred: namely, the conversion of AND-form future possibilities into OR-form actualities. A permanent 'cut' has taken place in the cross-over from the quantum mechanical "AND" into the present-continuous *irreversible* OR-form event. **The irreversibility, or time-asymmetry, manifesting in and through the process of AND-to-OR wavefunction collapse, contrast with the oft-claimed time-symmetric character of physical laws: laws brought into evidence through time-irreversible and time-asymmetric AND-to-OR collapse.**

The irreversibility and time-asymmetry that comes with AND-to-OR wavefunction collapse constitutes another paradoxical element of wavefunction collapse: a matter to be explored in pertinent parts of this introduction and throughout the main work. However, the first full solution to the problem of irreversibility and the question of the time-symmetric physical law in the context of AND-to-OR collapse will be solved in the context of the *quantum theory of entropy* (i.e. the theory of potential entropy) in Book-I.

Both the irreversibility question and the time-symmetry question of physical law is fated to the explication of what brings about AND-to-OR collapse attendant prospective quantum theory of entropy. How? **Insofar as AND-to-OR collapse *is* the conversion of future-form potentiality for events into actual events, it follows that AND-to-OR collapse is synonymous with the process of 'coming into being'... with *time* itself...** grasped, *not* as a 'fourth dimension', but as the partial-perpetual decay of future-potentiality into new events out of the ontically real future. **Hence, the question of the causality responsible for AND-to-OR collapse is identical to the question of the causality responsible for *time* itself. Hence, the question of what drives AND-to-OR collapse is synonymous with the question of what drives time.**

These insights must be incorporated into our prospective growing block intermediate model of spacetime and its de-spatialised and purely temporised successor in Book-IV.

*

Further insights into the AND-to-OR collapse problem will also partially arise from the relation of mind and consciousness to the process of AND-to-OR collapse and time. The issue of the role of consciousness in 'measurement' leads to the question of whether, as Eugene Wigner originally held but later abandoned, consciousness is causally responsible for the process of AND-to-OR wavefunction collapse or time. If the answer is yes, mind and consciousness would constitute the 'drive of time': the ultimate causality and primary ontology.

Is there any possibility that quantum idealism might be true? Do we live in a consciousness-created universe? By Book-V, Alt-R will permanently unravel forlorn quantum idealism. **The notion that mind is the drive of time and of wavefunction collapse will be shown to be as dubious as the materialist contention for non-mental 'hidden gears' in the same.**

0.25: The Apparatus Problem, Radical Memory & Alt-R

Generic quantum mechanics tends to model the physical order as a pure AND-form system. **There is no intrinsic compelling logic within generic quantum mechanics to explain the fact of OR-form realities, or even delineate a 'cut' between the AND-form quantum mechanical versus the OR-form classical; or offer explicit logic to justify the existence of OR-form states. This is evident in the *apparatus problem*:** an important part of the measurement problem.

Just as we might assert that all things ought to be purely OR-form (i.e., naive realism), from a quantum mechanical perspective, the universe ought to be purely and only AND-form: bereft of any OR-form contaminants or an OR-form apparatus; bereft of the possibility that we might configure the apparatus to an OR-form single-slit or double-slit configuration. By rights, upon isolation, the barrier ought to dissolve into an AND-form superposition of its alternative barrier-settings, and subsequently resolve into any one of the alternative settings on a purely quantum probabilistic basis, without concern to consistency or experimental initial conditions: i.e., the apparatus. Thus, **generic quantum mechanics must treat the apparatus as an AND-form state and cannot furnish any necessary compelling reason why we must treat the apparatus as an OR-form state.**

However, practically, we *do and must* treat the apparatus as an OR-form system. Yet, this is something we must bring to the fold from *outside* of the logic and structure of generic quantum mechanics. The science *does not* and *cannot* furnish us with any compelling reason why we must have OR-form variables or equipment.

This is a serious problem: **Systems *must* assume OR-form characteristics in order that physically meaningful experiments become possible,** such as *en masse* OR-form outcomes that bring into evidence self-interference, the interference pattern, and permit the inference of the superposition principle, *and* render possible meaningful statements about the 'measurement problem' itself.

This, and the fact that an apparatus can exist and function in support of quantum mechanical conclusions, indicates that the universe is not a pure AND-form state of affairs: that it is not purely or only quantum mechanical. It not only consists of OR-form states, but such states, or their *ultimate* basis, somehow survive and are retained by the physical order against the vagaries of the quantum mechanical AND, and the AND-to-OR collapse process, and its attendant quantum indeterminate sortology. **In other words, OR-form states, or their ultimate bases, are somehow not 'washed out' from the physical order by either AND-form realities or by quantum indeterminate processes: Somehow these are retained in the order of nature.**

But how is this possible?

*

41

We have already intimated the reality of abstract memory in nature. But the retention of information, of the past, and of the experimental apparatus as a set of initial conditions… as abstract memory… is *not* explicated by quantum mechanics: abstract memory in nature does not seem to follow ineluctably from the intrinsic logic and structure of generic quantum mechanics. Yet, without such memory, initial conditions (the apparatus) could not abide, and quantum mechanics as we know it would not be tenable, even though it cannot account for the very abstract memory upon which quantum mechanics critically depends. **What is missing in generic quantum mechanics is a conclusive and decisive argument for abstract memory in nature. To put the key part of the solution succinctly,** *the physical universe has memory…* **and memory is critical to any solution to the apparatus problem.** The physical order does not conform to an *Amnesic Universe thesis*, and even less to an *Alzheimer's Universe*.

*

How do we obtain preliminary demonstration of the existence of abstract memory in nature?

In the two-slit experiment, a total of four combinations for the barrier-setting are possible: These are… "both slits open"; "left slit open"; "right slit open"; and 'both slits closed'. We envision that, upon isolation, the barrier *must* transform into an AND-form superposition state for all four future-possible barrier-settings. This is per isolation, the consequent deference of the state of the barrier to the future per enforced isolation, and the AND-form character of that future and of the isolated barrier, as demanded by the firewall principle that wards against signals from the future and the usurping of causality and the principle of conservation. Hence, the barrier *must* dissolve into an AND-form superpose of possibilities, and, in that interim of isolation, it cannot abide as an OR-form state-of-affairs. **If this is so, then the experimental setup decided by the experimenter ought not to survive the interim of physical isolation and would reconstitute at the end of isolation purely per quantum happenstance.** Consequently, at the end of isolation, the kind of distribution pattern we might obtain at the detector ought to be fated to which barrier setting was quantum indeterminately resolved into actuality, purely on a one-in-four basis, regardless of how we configured the barrier initially… with a 75% likelihood of inconsistency with our initial setup.

Yet, in the real world, save by the intercession of contingency, such as analogised by the action of a mischievous laboratory cat that disrupts the experiment, the setting on the barrier is *not* subsequently erased or scrambled by quantum mechanical processes, even if quantum depositions on the detector *are* undeniably quantum indeterminate. **The fact that the barrier is not subject to or fated to quantum indeterminate scrambling, or even to erasure in the interim of isolation, despite its necessary quantum mechanisation in that interim of isolation, clearly implies the retention of the barrier-setting as a critical piece of information: a retention that survives isolation, with implied solutions to the apparatus problem based on informational retention… or, simply,** *abstract memory.*

Again, the paradox here is that generic quantum mechanics implicitly needs physical abstract memory in order to deliver the kinds of results it has achieved. Yet, generic quantum mechanics cannot furnish the reality of such memory from a pure AND-form perspective: Hence, the need for quantum mechanics-ii…for quantum mechanics *with* abstract memory.

One possible counter to abstract memory is the Manyworlds interpretation and the attendant hypothesis of information-erasure and informational nihilism. It will be the task of Book-III to show conclusively why such a move is forlorn, and, through it, augment the case for abstract memory.

*

By Book-III, as precursor to mnemonic implications of the growing block universe and from de-spatialisation and temporisation, Alt-R will establish the physical reality of abstract memory: entailing the retention of critical informational states outside of putative radiation, matter, 'space' and time. The undeniable reality of memory is clearly intimated in the apparatus paradox just adumbrated, as well as by implied information-retention per facile information-matter dualism from **0.04**. Indeed, abstract memory is almost interchangeable with facile information-matter dualism: The fact of the persistence and retention of information pertinent to initial conditions in basic physical experiments, and even in the process of photography, necessarily implies the retention of information *as* memory *outside* of present-continuous 'materialised' radiation, matter and 'space', despite quantum mechanical AND-form interims.

Our case for abstract memory will constitute one of the major ingredients to the solution of the apparatus problem; part of the larger solution to the AND-to-OR collapse problem as whole, with implications to the retention of memory vis-à-vis the brain *and* to anticipated Cartesian mind-brain relations.

By Book-IV, the reality of abstract memory, its clear pertinence to the apparatus problem, hence the case for the abstract physical basis for retention of experimental initial conditions, will be furnished in succinct form by de-spatialisation and as the direct seamless consequences of de-spatialisation. There is no space. It follows that the apparatus has no 'location' as a material corpus in space: The apparatus is *not* a 'distribution in space': It is *a pure distribution in time…* constituted as the abstract past. As such, **the apparatus resonates from the past and delimits or decoheres the evolution and AND-to-OR collapse of the experimental future into possibilities and outcomes that are consistent with the pre-set initial conditions, or with the state of the apparatus ultimately retained as abstract memory.** This resonation of the apparatus is *not* through or across 'space', and it is not mediated by contiguous impact causality between particles. Instead, it is a resonation purely across time, without contact or impact causality. The apparatus manifests itself, *not* as 'arranged matter in space', but as the recollection of the past across time, as the consistent decoherence of the future, re-materialised via subsequent AND-to-OR collapse or time into the endurance, or at least perdurance.

0.26: EPR Paradox, Information & Alt-R

The Einstein Podolsky Rosen experiment, or EPR, was first experimentally investigated by Alan Aspect in 1981[11] **EPR demonstrates the reality of apparent instantaneous 'spooky action at a distance'**. As such, it apparently violates a principle from Einstein's relativity that physical information, or, simply, the consequences of causality, cannot be mediated at faster than light, much less instantaneously. Stating the issue in terms of the speed of light limit and its 'violation' describes the mere surface: At a deeper level, the problem is one of implied breakdown of causality.

We have already explored the problem of causality in the context of the cause of AND-to-OR collapse and measurement problem, and how both implicitly constitute the crisis of the materialist closed input-operation-output schema for causality. In the context of EPR paradox, causality can be further understood as the principle that, in a flight from London to New York, one cannot arrive at New York before having departed from London. This is consonant with the **firewall principle** behind AND-to-OR collapse and time itself and **the growing block universe** approach to 'spacetime': The future cannot be described as a state of resolved affairs before it gets to happen, otherwise we would obtain 'signals from the future' *and* the violation of the principle of conservation of energy-matter. Hence, the firewall principle: wherein a system in isolation, hence a system deferred to the future, cannot be constituted as an OR-form state, but *must* be configured as an AND-form state of potentiality... until, at a potential future-moment, it undergoes AND-to-OR collapse into an OR-form realised event.

Does action at a distance from EPR violate the principle of causality and the firewall principle?

Let us take a closer look at the nature of physical signals conceived in Einstein's Special Relativity. Therein, any impossible signal said to travel at 'faster than the light' (sometimes speculated to happen in EPR) would constitute a signal that purportedly travels backwards in time: a signal from the future, with all the violation of the principles of order and causality implied. Such violations of causality, framed in terms of the constancy of the speed of light, would not simply imply putative reversal of time, but undermine the constitution of events about which all observers must agree, whatever their frame of reference or physical state: (See **Fig. P1.10**, for an explanation that involves speculative 'faster than light' signalling).

Insofar as EPR entails instantaneous 'spooky action at a distance', and insofar as this ought to entail 'faster than light' signalling, EPR ought to imply the total breakdown of causality as illustrated in **Fig. P1.10**, supposedly *with* 'signals from the future', *and* attendant violations of the principle of conservation.

Conventional circumvention of implied breakdown of causality in EPR posits that no meaningful physical information can be transacted in EPR. This is per the information-scrambling role of quantum indeterminacy entailed in, say, spin-resolution of entangled particles. But is this explanation sound?

Envision a pair of photons originating from a common particle decay. When the source particle decays, each photon displaces in opposing directions: to the 'left' and 'right', respectively... at the speed of light. So long as experimental isolation remains in effect, the photon travelling toward the left-side will constitute a unitary AND-form superposition of "spin-up" AND "spin-down". The converse will hold true for the right-side photon: a converse AND-form superposition of "spin-down" AND "spin-up".

When we finally make a measurement on the left particle and end isolation, the left-side might AND-to-OR collapse into, say, the OR-form "spin-up". It has a fifty-fifty chance of doing so. Before any physical signal at the speed of light has had time to reach it from the left-side particle, the right-side particle will instantaneously AND-to-OR collapse into an opposite corollary... the OR-form of quantum "spin-down".

To clarify, in EPR, every time we obtain "spin-up" on the left-side, we get a *simultaneous* and instantaneous correlate "spin-down" on the right-side. Conversely, when we get "spin-down" on the left-side we get "spin-up" on the right-side. **In short, spin is conserved. Somehow, mutual consistency and conformity in the form of the principle of the conservation of spin will be enforced on the outcomes, despite there being no physically permissible signal between the far-flung particles to mediate or enforce such a consistency.**

How does the particle on the right-side 'know' that it should resolve into a consistent 'spin-down' in order to conserve spin, given that no signal, causal effect, or physical information could have reached the right-side particle from the left-side particle?

As stated, **the conventional solution from quantum indeterminacy holds that no meaningful information can pass from the left-side to the right-side particle in EPR, given the ineliminable effect of information-scrambling quantum indeterminacy.** This notion can be clarified in the forlorn generic attempt to exploit EPR for putative 'faster than light' communication. To this end, let us imagine a device that employs the simple Morse code system of DOTs and DASHes: We correlate these to specific spin states of entangled couples in the forlorn hope of using these as means for instantaneous transmission and communication of whatever strings of DOTs and DASHes we might constitute as a message. We place the transmitter on Earth (the left-side) and the receiver on Mars (the right-side): We then make a measurement on the transmitter Earth-side particle, resolving its spin. A correlate resolution of spin will instantaneously occur on the Martian 'right-side' receiver.

For meaningful communication to be transacted from Earth to Mars, we must be able to control with certainty the specific spin outcome generated on Earth, and thereby determinately control the correlate spin-outcome on Mars, consistent with whatever spin couples we formally agree to code as DOT or DASH: Thus, if we could control the outcome into 'spin-up' on Earth, its correlate would

---

[11]Experimental Realization of Einstein-Podolsky-Rosen-Bohm Gadenkenexperiment: A New Violation of Bell's Inequalities. A. Aspect, P. Grangier, and G. Roger. Physical review Letters, Vol 49, Iss.2, pp.91-94 (1982)

imply the transmission and reception of "DOT" on Mars. If we could control the spin-outcome on Earth into 'spin-down', its correlate on Mars would be read as "DASH". If we could construct a large array of such EPR quantum-coupled devices and distribute these to Earth and Mars, and use these in accordance with the agreed Morse code spin-correlates, we might hope to communicate DOTs or DASHes from Earth transmitters to Martian receivers at faster-than-light... indeed, *instantaneously*.

What is it that prevents such a scheme from working? Quantum indeterminacy is ineliminable. Hence, we cannot determinately control spin-outcomes on the Earth left-side. Thus, on Earth, we could never be sure of a DOT when we want it, or a DASH when we want it. Quantum indeterminacy will defeat *any* attempt to orchestrate the outcome on the Earth-side. An equal scramble will obtain on the Mars-side. Thus, **in EPR, intrinsic and ineliminable quantum indeterminacy appears to scramble any hope for causally controllable outcomes:** *This* **is what is meant when it is asserted that EPR 'action at a distance' cannot allow the transmission of meaningful information.**

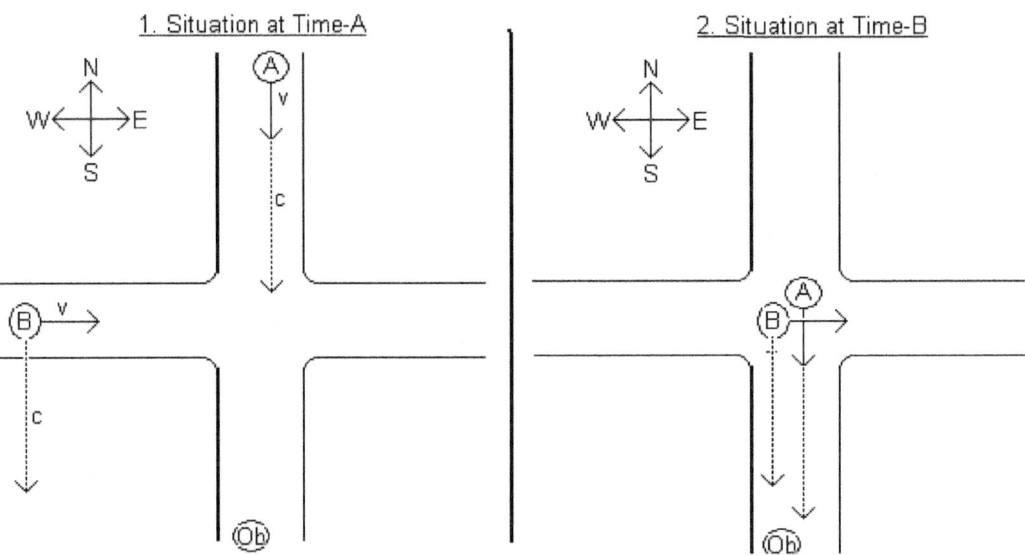

Fig. P1.10 Faster-than-light signals and the constitution of events
Notes: The constitution of events must be the same for all observers, even with the relativity of simultaneity and attendant disagreement about the order of events. **Diagram-1** depicts two moving observers, **A** and **B**, destined to collide at the crossroad: Another observer, **Ob**, located at the end of the southward road of the crossroad, must rely on the light signals (the dotted arrows) emitted from **A** and **B** in order to observe the expected collision of **A** with **B** at the crossroads. Let us assume that the speed of light is modified and increased by the motion of **A**, such that the signal from **A** becomes C + V: resulting in a faster-than-light signal from **A**. Since **B** does not have a southward component to its motion, the southward light-signal from **B** is not expected to speed up to greater than C. The signal from **B** will reach **Ob** at the speed of light. But the signal from **A** will reach **Ob** at the speed of light plus V. In **Diagram-2**, **A** and **B** finally meet and collide at the crossroads: As far as **A** and **B** is concerned, the collision has undeniably occurred. However, **Ob** disagrees with the composition of facts accounted for by **A** and **B**. Indeed, according to **Ob**, **A** passed safely through the junction, followed moments later by **B**: "**A** and **B** just missed each other. There was no collision". This is because the speeded-up light signal from **A** got to **Ob** before the non-modified light signal from **B**, implying that **A** got to the junction just before **B** and passed safely through. This illustrates that faster than light signals would lead to the breakdown of consistency and produce tandem breakdown in causality, insofar as causality requires consistency and agreement about the composition of facts between all observers: Did **A** and **B** collide, or did they just miss each other? If the notion of faster-than-light immediately leads to the breakdown of consistency and causality in the way just described, faster than light in EPR ought to accomplish the same breakdown.

*

However, the case is not quite closed. Notice that the utilisation of the information-scrambling role of quantum indeterminacy in EPR contradicts outcomes produced by the same sort of quantum indeterminacy in non-EPR situations. **In non-EPR situations, we discovered the persistence of physically meaningful information in the endurance and perdurance of the identity and composition of things over long stretches of time...** *despite* **the ineliminable operation of 'scrambling' quantum indeterminacy,** at all scales and in all AND-to-OR resolution processes in the developmental unfolding of structures and identities across time.

QUANTUM MECHANICS AND MIND

If quantum indeterminacy really was the basis for prohibiting the transmission of meaningful information in EPR, **the formation of any image in photography ought to be fated to the same information-scrambling from quantum indeterminacy as asserted to hold in EPR**. The photo-image ought not to form, and it ought to be as impossible as Morse-code communication using spin-entangled couples in EPR. Yet, in experience, quantum indeterminacy does not at all prohibit the emergence of coherent photo-images, nor prohibit the manifestation of meaningful information in any other context.

Although processes involved in non-EPR situations and in photography do not occur at faster than light, they are none the less subject to ineliminable quantum indeterminacy at the critical resolution-point. In other words, equally ineliminable quantum indeterminacy in non-EPR situations does not scramble the global formation and manifestation of information, despite its undeniable operation: Yet, as we have seen, EPR situations are asserted as exceptions to this.

*

Clearly, something is amiss here: Does quantum indeterminacy scramble information or does it not? Previously in **0.23**, we intimated that **quantum indeterminate processes are funnelled into ordained, fated and fully determined global pattern-outcomes, despite quantum indeterminacy. Thus, quantum indeterminacy is *not* inimical to information**. If quantum indeterminacy cannot ultimately prohibit the formation and re-formation of meaningful information complexes in non-EPR situations, why should we believe things to be different in full-fledged EPR situations? If quantum indeterminacy in non-EPR situations somehow allows for the manifestation and persistence of pattern and information, what reason is there to suppose that it cannot do so in EPR-like situations? Either quantum indeterminacy is to be treated as an information-scrambling factor in *all* situations, including EPR... or in none, including EPR.

Yet, it is certain that, at least in conventional terms previously outlined, EPR cannot be exploited for faster than light communication, simply because we cannot control code-pertinent spin-outcomes on either end; no more than we can determinately control the outcome of *any* quantum indeterminate process in non-EPR frameworks: for example, in two slit experiments and in photography.

What about the 'conservation of spin'? **The enforcement of the conservation of spin constitutes the minimal concession to information-communication in EPR. The requirement that this principle *must* somehow be imposed on quantum-entangled couples despite quantum indeterminacy shows that quantum indeterminacy in EPR cannot be used as a sufficient argument against the transaction of meaningful information**. That is, we find that meaningful information *is* being transacted in EPR, despite quantum indeterminacy: The conservation of spin *is* the meaningful information being transacted and enacted in EPR.

*

The above contradictory-seeming discussion shows that there is far more going on in both EPR and in non-EPR situations than hitherto appreciated: It all revolves around the correct appraisal and definition of information and how it relates to the physical order. Indeed, facile information-matter dualism, and many other things so far intimated, have a role to play in resolving such problems.

First, as was clarified previously, in EPR, what we have is the enforcement of critical information pertaining to physical law: namely, the conservation of spin. This enforcement is not enacted or mediated by exchanges between the particles involved, and it is not mediated or affected via radiation, matter, or across space. This has implications to the ontology of physical law. Indeed, EPR shows that abstract-immaterial physical law (e.g. the conservation of spin) *is* being enforced on radiation, matter and putative 'space'... from *outside* of radiation, matter and 'space'. **What EPR demonstrates is the dualism of extant abstract physical laws vis-à-vis radiation, matter and 'space', and, with it, the enforcement of otherwise extant physical law on radiation, matter, 'space', and time... from outside of radiation, matter, 'space', and time**.

Yet, this does not directly clarify how quantum indeterminacy can and does act as an information-scrambler in EPR, but not in general non-EPR situations. Part of the answer to the latter paradox was presented in summary-form in the previous essay on the stochastic-coherence paradox (**0.22**): Therein, the solution to the stochastic-coherence paradox was unravelled by reframing quantum indeterminacy to abstract memory and future potentiality, and by integrating these to facile information-matter dualism amongst other anticipated discoveries, such as implications by de-spatialisation.

Recall that, in non-EPR situations, even per facile information-matter dualism, information can dissociate from matter, radiation and 'space' into an abstract form. At the point of restitution to radiation, matter and 'space', information circumvents the information scrambling effects arising from quantum indeterminacy by function of the fact that these are funnelled into the global outcomes through a three-phase developmental process, without requiring any deterministic orchestration or 'hidden gears'. We can provisionally model abstract information (both memory and futurity) as either a carry-over from the past (per the postulate for abstract memory) or as a never-before realised future potentiality, consistent with our case for the objective reality of the future and the *raison d'etre* of quantum AND-form logic. These abstract states of memory or potentiality will funnel quantum indeterminate terms into ordained fated global outcomes, without any need of orchestration or obviation of quantum indeterminacy.

In EPR, ineliminable objective quantum indeterminacy abides, but it is funnelled by an abstraction: namely, the abstract law pertaining to the conservation of spin. The law is inherent to the structure of the wavefunction for quantum entangled states, with the principle imposing itself across very large distances, instantaneously, without recourse to anything more than a single AND-to-OR collapse-process, despite the 'speed of light' limit. Abstract dualistic law accomplishes this in such a way that a left-side measurement on a quantum entangled couple is guaranteed to yield a right-side spin-outcome that conforms to the conservation of quantum spin, even if the putative spatial separation between the terms might equal a million parsecs, and even when there is no orchestration and determination of the spin outcomes.

*

However, the sorts of abstraction that can burn through AND-to-OR wavefunction collapse *and* attendant quantum indeterminism, and do so across vast distances, need not be restricted to the enforcement of physical law. More complex types of information, acting as **abstract states operating across vast domains of 'space' and time, can also funnel *en masse* quantum indeterminate terms into globally coherent wholes and outcomes, and even into cosmic-spanning wholes and outcomes, even though the quantum indeterminate constituent terms of such wholes might be separated from each other by thousands, if not millions, of parsecs**, superseding generic 'common sense' assumptions about the transmission of information that must supposedly happen 'across space' at or below the 'speed of light ', **allowing for coherence and continuity of large-scale patterns and structures by means effectively at faster than light, if not instantaneously.**

Such claims appear outlandish… until we introduce de-spatialisation. **There is no space. Therefore, information and its transmission does not transpire across space, much less assume a speed-to-distance form, or transpire at or below a putative 'speed of light'.** Instead, we must relate information to the physical order purely in terms of time.

In simplest terms, if we take the quantum indeterminate incidences that make up manifest information and pattern, and treat these as 'pixels', theses must inter-relate to each other as distributions in time, not as distributions in space, and as events structured according to delayed choice time-interval relations.

By way of a simple example, consider a bell curve distribution formed by any number of consecutive or parallel quantum mechanised die-rolls. *En masse,* these will inexorably culminate into the expected bell-curve distribution, rendering it manifest or 'materialised'. But there is no 'communication' or any secret co-ordination between the quantum mechanised die-rolls and their outcomes. These are seemingly a-causally funnelled and *inexorably* culminate into the fated and ordained bell-curve distribution.

Now, replace the bell curve with the delimits that attend quantum entangled couples in EPR and attendant enforcement of the law of the conservation of spin. The outcomes cannot help but materialise the conservation of spin, inexorably.

Of course, in non-EPR situations, we replace the bell curve distribution with the complex wavefunction that attends, say, an intricate photographic image: The quantum indeterminate terms must inexorably culminate into the expected image, and will do so in a totally a-causal way, without secret orchestration or co-ordination. And because these are objectively indeterminate, their interrelations remain a-causal, just as is true in the indeterminate resolution of quantum mechanised dice in the bell-curve scenario and spin-outcomes in EPR.

The complex wavefunctions that pertains to photography constitutes an instantaneous whole, as does the complex wavefunction that pertains to the basis of the bell curve distribution of quantum mechanised die-outcomes. The quantum indeterminate terms therein do not communicate or co-ordinate with each other at all, much less mediate such co-ordination at the 'speed of light', simply because they are a-causally related.

At the cosmic scale, we replace the complex wavefunction pertaining to our photographic image with a complex wavefunction that attends, say, whole galaxies, or even the universe itself. The quantum indeterminate terms, as the 'pixels' of such vast patterns, will inexorably culminate into the materialisation of these vast patterns, in a totally a-causally way, without speed of light communication or secret orchestration, per function of objective quantum indeterminacy, and effectively 'at faster than light'.

What happens when we recapitulate cosmic structures into states distributed and organised in pure time, without space? **With the fact of de-spatialisation, complex structures, regardless of their 'size', are no longer 'distributions in space': nor do these require that their terms communicate and co-ordinate with each other at or below the speed of light. Instead, with de-spatialisation and temporisation, these become combinations of the abstract past (memory) and abstract potentiality (the future), brought into manifestation through universe-scale processes of AND-to-OR wavefunction collapse or time that operate on wavefunctions that are constituted as instantaneous wholes**, with a-causal quantum indeterminacy.

Note that, with de-spatialisation, temporisation, and the recapitulation of the wavefunction as an instantaneous whole, **the relation of information to the physical order is totally transformed**: Our generic notions of 'information distributed in space and matter', supposedly subject to the 'speed of light', collapse in the following ways.

- **First, the relation of information to the physical order collapsed into information-matter dualism** and its bizarre dissociation of information from radiation, matter, and space.

- **Second, information-matter dualism has now in turn collapsed into a framework comprised of de-spatialised information distributed in time; made manifest in totally a-causal ways from the collapse of instantaneous wholes, circumventing the spatial notion of the mediation of causality and information at or below the 'speed of light'.**

Thus, the form of information has superseded spatial notions, and we have broken from the spatial mediation of information at some 'speed'… while our notion of how information relates to the physical order and to quantum indeterminate processes has transformed into a form beyond ready recognition, courtesy of Alt-R.

*

The above is but a presage to our full exposition of the growing block universe and the overhaul of 'spacetime' accordingly, culminating into the intermediate model spacetime in Book-II, which will in turn be superseded by a pure temporised spacetime, *without* space, in Book-IV. Critical ingredients to these revolutionary insights will be the *nested futures* perspective, *counterfactual causality* and *grad decoherence*. These insights are introduced below.

QUANTUM MECHANICS AND MIND

0.27: Delayed Choice, Nested Futures, the Unitary Universe, & Alt-R

John Wheeler's Delayed Choice[12] is an unexpected seam of gold for the Alt-R perspective. Based on an alternative scrutiny of Wheeler's experiment, Alt-R will assert that AND-form states are not only composed of immanent futures but the immanent futures in turn nest subsequent futures. Hence, the revolutionary *nested futures* view of the quantum mechanical wave.

In Book-II, the intermediate model of spacetime will emerge from the fact that generic spacetime models possess no provision for the future, and that generic models constitute block universe systems, as opposed to growing block models. Insofar as future-possible events must be configured in AND-form, a spacetime model that explicitly incorporates the future, hence constitutes a growing block universe, must model a pure quantum mechanical spacetime domain. The absence of the future in generic spacetime models frustrates the want of a unification of the classical-relativistic and quantum mechanical descriptions of physics.

How do we justify the existence of this nested futures wavefunction and the attendant growing block future-form spacetime domain? **John Wheeler's delayed choice constitutes the justification-basis of the nested futures system**.

The moment we grasp the AND-form state as the ontological manifestation of the future, we must incorporate that future into spacetime models as the **domain of spacetime in potentiality**: an AND-form system of future possibilities that elaborate to the indefinite future. This domain must comprise not only the immanent futures, but must configure as a complex of nested futures attached to the immanent futures. One implication is that the universal wavefunction that incorporates the total future must re-structure as an *instantaneous whole* throughout the total future, to the infinitely removed future: That is, per events that come into being out of future potentiality, and the consistency that the total future must exhibit with respect to old and new events, the total future must undergo **grand decoherence**, with remarkable consequences that presage quantum mechanical solutions to inertia and other physics mysteries.

With the advent of de-spatialisation, the intermediate model of spacetime to which the nested futures perspective will relate, will in turn give way to the superlative Bergson-Whitehead amalgam (Book-IV): spacetime *without* space. Nested futures and grand decoherence will survive into the Bergson-Whitehead amalgam.

*

What is the relevance of the anticipated nested futures view of both the quantum mechanical wave and the future-form spacetime? **The nested futures perspective leads to non-local and transtemporal counterfactual causality or grand decoherence**. The case for non-local counterfactual causality is distinct from non-locality evinced in EPR. In both, measurement in one locale or moment has an instantaneous effect on another, across a 'distance' that no signal could have had time enough or be swift enough to traverse. In both, the pertinent frames are entangled or in unity with every other frame, despite the putative speed of light limit. In other words, **there is a unitary binding of the physical universe, of a form in addition to but distinct from the transtemporal enforcement of physical law exemplified in EPR**.

Consider the following thought experiment in which an extraterrestrial in orbit about Alpha Centauri attempts to transmit the Encyclopaedia Galactica to Earth. Alpha Centauri has just exploded: The supernova shock wave is about to smash our alien friend to oblivion. Our friend has a fifty-fifty chance of hitting the transmission button before being destroyed. At that juncture, the AND-form wavefunction of futures attached to our alien friend will configure into two principal *imminent* possibility sets: "transmission successful" AND "transmission failed".

Observers on Earth will not see the explosion of Alpha Centauri for another four or so years: Assuming the ontology of 'space', it will take light from Alpha Centauri some four years to reach Earth. Observers on Earth also have a future-form wavefunction attached to their frame. This will include both immanent possibilities *and* their attached nested futures. The possibility tree so-formed will project and branch out up to the pertinent four-year future limit: the time required for putative signal-events to reach the Earth from Alpha Centauri.

The Earth-frame of nested futures, or that part of it covering the future up to the said four years, must instantaneously decohere into a consistent form upon the resolution of events at Alpha Centauri, despite the putative speed of light limit. Remember, our extraterrestrial friend has a 50-50 chance of hitting the transmit button. Before this happens, the Earth-frame wavefunction has two main alternate nested futures: One branch leads to the future observation of the supernova *and* the reception of the Encyclopaedia Galactica in four years' time. The other nested future branch potentialises for the future observation of the supernova, but *without* reception of the Encyclopaedia Galactica. Both sets of futures will be consistent with whatever gets to happen to our extraterrestrial friend at Alpha Centauri. But the sets will be mutually incompatible. Thus, **if our alien friend successfully transmits the Encyclopaedia, all the nested futures attached to the non-transmission alternative will be instantaneously erased from *all* frames, including from the Earth frame: The Earth-frame wavefunction will instantaneously restructure and decohere *only* to those futures that are consistent with that far-away successful transmission. On the other hand, if our alien friend fails to transmit, the branch entailing 'transmission' will be rendered into non-consistent nested futures: it will be erased**. Hence, the future permitted to our Earth-observers vis-à-vis events related to Alpha Centauri, but otherwise destined to transpire in the Earth frame in four-years' time, will be subject to instantaneous consistency-based restructuring per quantum counterfactuals, leading to **grand decoherence of the joint nested future wavefunction that unifies both the Earth-frame and the Alpha Centauri frame**.

---

[12]Mathematical foundations of Quantum Theory, edited by A.R. Maslow, academic press, 1978, pp 39

However, the information 'exchanged' in this non-local transtemporal counterfactual way is *not* substantive information: It is about *what did not happen,* and the elimination of what cannot transpire in the future. As such, somewhat like EPR, it cannot be used for the communication of substantive information. Even so, **the nested futures view implies the counterfactual unitary binding and *grand decoherence* of the whole universe onto itself: the unitary binding of every location and moment with every other location and moment through their common convergent futures, no matter what their relative putative 'distance' or the delayed-choice time-intervals that separate them**. In other words, the notion of the instantaneous whole, and the characterisation of the growing block future and *all* subsidiary wavefunctions as instantaneous wholes, as espoused in **0.26**, is a truism of nature, *not* an outlandish idea.

This insight comprises one of the decisive innovations of Alt-R, on a par with de-spatialisation itself. It is also a **basis for reviving an alternative revamped version of Mach's principle: an important step in the long-elusive quantum mechanical growing block solution to inertia**: Through non-local transtemporal counterfactual causality and grand decoherence just outlined, matter here 'knows' what matter elsewhere is *not doing*. The future-possible motion and momentum of matter 'here' will be restructured and adjusted per attendant counterfactuals-based consistency rules vis-à-vis matter elsewhere and per the futures no longer permitted to happen in their common convergent futures. **The quantum mechanical basis for inertia also constitutes a partial contribution to prospective quantum gravitation, as well as a primer to a quantum mechanical theory of the permanent magnet**.

With prospective de-spatialisation obtained by Book-IV, 'non-locality' will give way to pure transtemporality: Thus, the nested futures perspective, counterfactual causality and grand decoherence, will survive into the Bergson-Whitehead amalgam: into spacetime *without* space.

0.28: The Uncertainty Principle & Alt-R

We now come to the uncertainty principle: one of the most popular quantum mysteries, though less significant than one might suppose. The clarification of the uncertainty principle from Alt-R follows seamlessly from the *raison d'etre* of AND-form logic and the quantum wave, grasped as the ontic manifestation of the future.

The uncertainty principle states that one cannot resolve accurately both the position and the momentum (i.e. the velocity times the mass) of a particle at the same time: **If we resolve the position of the particle we end with calculable intrinsic non-resolution of its momentum. But if we resolve the momentum of that particle, we end with related calculable non-resolution of its position.**

Despite the hyperbole that surrounds it, the uncertainty principle remains the most obscure of the quantum mysteries because its proper account is rendered in inaccessible mathematical form. This has meant that the *raison d'etre* for the uncertainty principle has remained inaccessible to a non-mathematical public.

The reason why we cannot resolve both the position and the momentum of a particle at the same time is for essentially similar reasons that one cannot walk forwards and backwards at the same time: Uncertainty is per the structural limits of nature: limits enforced despite our ideas and wishes to have it otherwise.

What we need is **an explanation of the uncertainty principle derived from a proper understanding of the *raison d'etre* of the quantum mechanical-AND.** This will be achieved in sufficient detail in Book-I and in successive books. To glimpse the answer, we first ignore implications from de-spatialisation, which undermines the whole notion of 'position' and its uncertainty (more on this later). When we resolve putative position accurately (i.e., when we use very short wavelength inputs of observation such as to confine our particle to a very small area) we defer and delay the resolution of our particle's momentum to the future. Hence, its momentum can only be described as a future-form state of as-yet not-happened AND-form momentum-possibilities. It is *not* the case that the momentum so deferred remains inherently concrete or resolved. Not at all: The momentum is rendered into a pure as-yet not-happened set of alternative future possibilities, all rendered in AND-form. However, the AND-form futurity, as a spread of momentum-possibilities, *is* objectively real, because the future is objectively real.

The converse is also true: When we resolve momentum accurately (i.e. when we use long wavelength inputs of low energy so as not to disturb the system too much), we must defer position-resolution to the future. Deferred to the future, the position of the particle must assume an as-yet not-resolved AND-form spread of alternative position-possibilities. The uncertainty is *not* per subjective uncertainty: The particle has no 'objective' or resolved position. On the contrary, position is deferred to the future and, as such, remains totally non-resolved: configured as a collection of future-potential as-yet not-happened positions, all rendered in AND-form. Again, the AND-form futurity as a spread of position-possibilities *is* objectively real because the future is objectively real.

The reason why we must always defer one of the non-commutative attributes to the future arises from the structural requirements that regulate the type of observational input required to resolve one or the other of the attributes: Hence, **position-resolution requires short-wavelength high-frequency inputs. Momentum-resolution requires long-wavelength low-frequency inputs. If we try to use both at the same time, we end up with no accurate resolution of neither position nor momentum. When we use one input for one attribute, we must necessarily defer the resolution of the other to the future... into AND-form rendition. The deferred attribute has no OR-form resolved characteristics: it is purely in an AND-form state. That is, it is not the case that the deferred attribute has a defined singular value, merely hidden from view. Instead, it is reduced to a set of possibilities, while the other attribute gains actuality.**

Of course, when we use high-frequency inputs to resolve position, the high frequency implies high energy, and a greater 'kick to the system'. Consequently, the range of future-potential not-yet happened momentum values in AND-form will be large. i.e. the uncertainty about the as-yet not happened future state of the momentum will be large.

Conversely, to resolve momentum accurately we use low-frequency long-wavelength observational inputs: i.e. low energy 'kick' with consequent reduced range of future momentum possibilities, or higher resolution of momentum. But this gives a future-potential position-range proportionate to the area of the inputted large wavelength. The attendant larger area of confinement for the particle constitutes a larger range of position-possibilities: hence the basis for uncertainty about not-yet realised AND-form future position.

Our 'deference to the future' explanation of the uncertainty principle, founded on the *raison d'etre* of the quantum mechanical AND as the ontic objective future, constitutes the heart of the uncertainty principle as grasped from Alt-R. It is now clear that 'uncertainty' has nothing to do with the 'disappearance of reality', or 'with ultra-subjectivity', or with the notion that 'everything is in the mind'; all born out of obtuse interpretations of the word 'uncertainty' attached to the principle.

With de-spatialisation and consequent dissolution of the notion of 'position', we must reform the uncertainty principle. Given the demise of space, we can no longer speak of 'position' or *its* deference to the future. However, there is another form of the uncertainty principle: one that relates energy to time. **Anticipated de-spatialisation will simply require that we supplant the position and momentum-based variant of uncertainty with the energy-time variant of the same.**

SPACETIME, MEMORY & MIND

0.29: Primitive Unification:
The Quantum-Relativistic Incorporative Intermediate Model of Spacetime

The Alternative Realist interpretation accommodates the established facts of quantum mechanics: It accepts AND-form logic synonymous with the superposition principle; it accepts quantum indeterminacy; it accepts EPR and 'action at a distance'; and it accepts the uncertainty principle and the measurement and apparatus problems.

It is generally held that all the above-stated issues render quantum mechanics permanently at odds with classical-relativistic physics, and even a-causal. Yet, **Alt-R asserts deep harmony between the quantum description and the classical-relativistic descriptions of nature**. **In espousing for the growing block universe approach, Alt-R asserts the unity of AND-form and OR-form realities into a quantum-relativistic incorporative spacetime model:** an intermediate model that will itself be superseded per de-spatialisation by Book-IV.

Generic physics and quantum mechanics implicitly posit in favour of an amnesic universe: hence, the implicit *Amnesic Universe thesis*. Therein, the universe is dispossessed of abstract memory, or of any possibility for the retention of the past in dissociated form from putative 'materialised' radiation, matter, and space. At best, generic science and quantum mechanics seek to reduce information and memory to hands-on material trace-states and cannot conceive information as abstract and extant, much less admit to any causal role for abstract memory in the generation of outcomes from AND-to-OR collapse or time. In short, generic physics has a memory problem, compounded by a future-problem. Both problems are inherent to generic block spacetime models. That is, just as generic spacetime models have no place for the ontic future and do not incorporate the AND-form quantum mechanical realities explicitly, these also fail to furnish a domain of memory in which realisations from AND-to-OR collapse and time are subsequently retained.

Indeed, the necessity for a mnemonic universe emerges most clearly in the apparatus problem, and in how the memory problem of generic quantum mechanics is rendered evident in the fact that the science cannot furnish any necessary reason or basis for the retention of the apparatus as a set of initial conditions in memory *and* physically justify memory retention in abstract form (see **0.25**). Yet, the critical viability of quantum mechanics was shown to rests upon the necessity of retention of, and dissociated carry-over of, the experimental conditions, from one moment to a succeeding moment, across experimental isolation… despite the seeming scramble from quantum indeterminism. Thus, despite itself, generic quantum mechanics requires memory: a requirement it cannot furnish to itself in its generic form.

It is hardly a coincidence that generic spacetime models follow suit the general obviation of abstract memory across physics and, with full self-contradiction, in generic quantum mechanics. The neutering of a possible active past in generic spacetime models leads to the distorted apprehension of spacetime. While generic models assert for the amalgamation of space and time, these do not incorporate the past-form of that spacetime in any causally active way. Similarly, the quantum mechanical aspect of spacetime, the future as explicit ontic potentiality, is *not* a discernible domain in generic models of spacetime. Therefore, **generic models of spacetime amount to an understanding of spacetime that is strangely bereft of both the causally active past constituted as memory *and* equally bereft of an AND-form quantum mechanical futurity**.

Nor is the process of AND-to-OR wavefunction collapse explicitly grasped as time itself in generic spacetime models, much less incorporated as such. Instead, the models entertain a geometrical 'fourth dimension' stand-in for 'time'... *not* at all synonymous with the process of AND-to-OR collapse. Yet, AND-to-OR wavefunction collapse *is* time, grasped as the process of 'coming into being': grasped as the conversion of future potentials into OR-form actua, the basis of which then get relegated to a spacetime domain of memory. In short, **time is *not* a fourth dimension at all: time is synonymous with perpetual wavefunction collapse.**

Generic spacetime models, together with the implicit postulate for the amnesic universe, lies at the heart of the long-standing divide between quantum descriptions versus classical-relativistic descriptions of nature: between AND-form and OR-form logic. Once corrected, we not only obtain a more complete growing block model of spacetime, but we obtain integration and *unification* of AND-form quantum descriptions with OR-form relativistic-classical descriptions of nature. In short, we obtain the primitive unification of classical relativistic physics with quantum mechanics.

We reiterate that the AND-form quantum mechanical state is nothing less than the physical phenomenalisation of the future, and the future as ontically real. The wavefunction is the implicit culminating mathematical attempt to grapple with the future as ontic and reality. This understanding will be augmented in Book-II based on John Wheeler's delayed choice experiment, culminating into **the nested futures model** of the AND-form quantum mechanical state: the basis for non-local counterfactual causality or grand decoherence (see **0.27**): all indispensable to the presage to prospective growing block quantum mechanical theories of magnetism, inertia, and gravitation. So understood, the nested futures view will force the incorporation, not just of immanent futures, but nested futures into spacetime models. This nested futures domain will constitute as the domain of **spacetime in potentiality**.

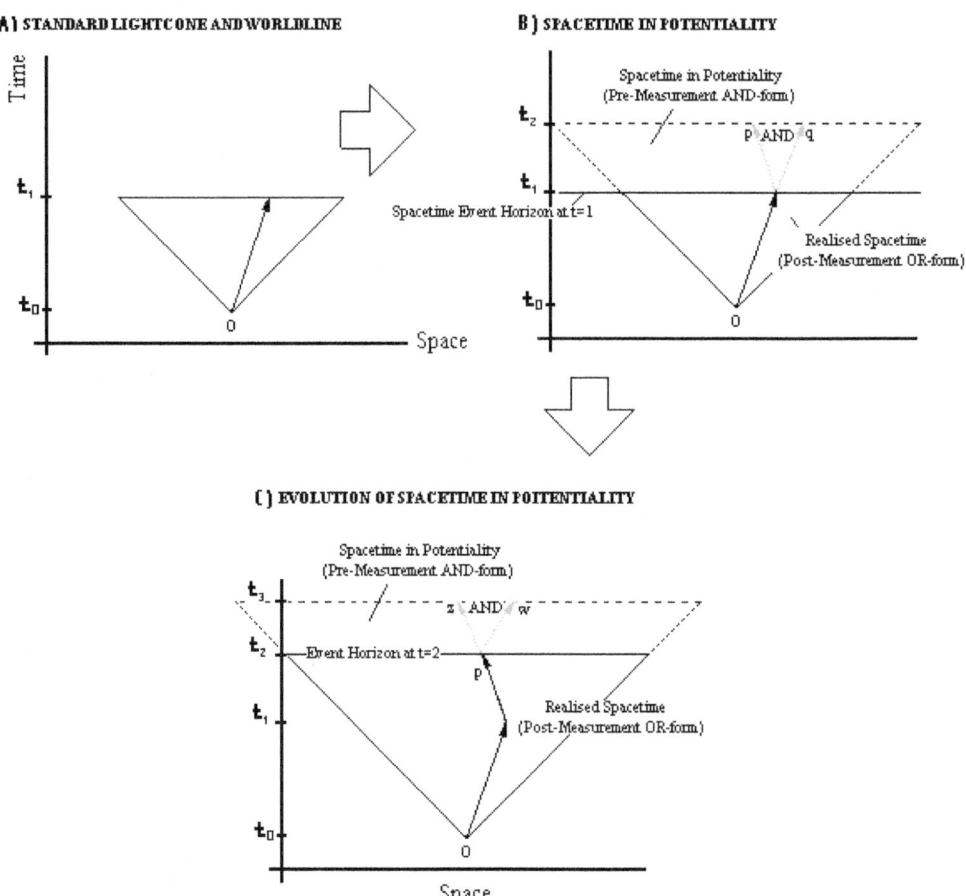

Fig. P1.11: The quantum-relativistic incorporative intermediate model spacetime
Notes: Diagram-A: a generic spacetime diagram for a universe with one lightcone and worldline; origin at O; progressing from t_0 to t_1. Such models are block models and purely OR-form, and do not include quantum mechanical realities vis-à-vis spacetime. Diagram-B corrects this omission: It extends the lightcone with dashed lines into a potential lightcone region stretching from t_1 to potential t_2, cut by an 'event horizon' vis-à-vis 'realised spacetime' 'below' it, and the AND-form quantum mechanical 'spacetime in potentiality' above it: This forms the growing block system. Realised spacetime contains the realised part of the lightcone and a single realised worldline terminating at the event horizon at t_1. At that point, we find two quantum mechanical future-form not-yet realised potential worldlines comprised of {p AND q}: Either potentially terminates at future-potential t_2. In diagram-C the event horizon advances to t_2, with 'wavefunction collapse' of {p AND q} into, say, the OR-form worldline-p. This latter is in turn potentialised with its own nested not-yet realised possibilities, i.e. {z AND w}. This is within 'spacetime in potentiality' above the event horizon at t_2. The event horizon is expected to progress from t_2 to t_3: either potential {z} or {w} will become OR-form: expanding the domain of realised spacetime. In this way, the intermediate model spacetime consolidates and *unifies* quantum descriptions of nature with classical-relativistic descriptions of the same, albeit in a primitive way.

Why not also incorporate the past as an abstract memory-repository: a domain constituted as the history and unitary sum of all past states? Subsequent AND-to-OR collapse of the future-form spacetime must render the past both causally active and complicit in the perpetual AND-to-OR decay of AND-form spacetime in potentiality into new OR-form outcomes and events, whose abstract basis is subsequently relegated to and retained within the domain of realised spacetime as new memory. Hence, **the incorporation of both quantum and classical-relativistic descriptions of nature into a unifying quantum-relativistic incorporative growing block spacetime model, and, with it, the primitive unification of quantum and classical-relativistic physics.**

Fig. P1.11 presents didactic illustrations leading to just such a spacetime model; one that consist of three domains: **The domain of spacetime in potentiality** constitutes the AND-form future of spacetime: the ontically real nested futures system of as-yet not-realised possibilities. **The domain of mnemonic realised spacetime** constitutes the repository for the basis of past OR-form events: i.e., abstract memory. Finally, **the event horizon** constitutes the relative present continuous 'cut' between the two domains: a 'surface' or 'skin' that we tend to confuse for the whole of the physical order. The event horizon is comprised of the most recent OR-form outcomes *and* the most immanent AND-form possibilities. As such, the event horizon constitutes the perpetually 'moving' boundary, or the *event horizon*, that resides between and separates the two spacetime domains. All of this is culminated in **Fig P1.11-C**.

Spacetime in potentiality is subject to partial-perpetual AND-to-OR collapse, or simply, quantised leaps in *time*. i.e. the temporal version of the essential 'quantum transition'. In other words, time is portrayed as quantised. This leads to the 'upward' progression of the *event horizon,* obtained through perpetual broken-discontinuous 'quantum leaps in time'. This contrasts with the generic conception of wavefunction collapse, conceived as involving one-time comprehensive AND-to-OR collapse processes.

Spacetime in potentiality is implicitly a system of nested futures, although only the immanent futures, comprised of potential worldlines, {p AND q}, and, by **Fig P1.11-C**, {x AND y} are rendered explicit in the given diagrams.

As the tableau for the most recent leading OR-form outcomes, the event horizon need not be flat: It could be 'curvy'. Nor does the event horizon contradict the relativity of simultaneity from Special Relativity: i.e. it is *not* a claim for absolute time. The event horizon, as a critical feature of the intermediate model spacetime, constitutes the frontier at which continuous partial-perpetual universe-scale AND-to-OR wavefunction collapse (time) takes place. This forgoes the need to model time as a 'fourth dimension', although the time axis is retained in the diagrams, and the progression of the event horizon and time is modelled along it. Thus, **time is the partial and perpetual universe-scale AND-to-OR collapse of 'spacetime in potentiality' into 'realised spacetime', on and through the 'upward' progress of the event horizon.**

Note how the quantum mechanical world embodied as spacetime in potentiality perfectly and seamlessly fits together with the domain of mnemonic realised spacetime, and does so through the process of AND-to-OR collapse, synonymous with the progression of the event horizon, in turn synonymous with *time* itself. In other words, we obtain the primitive unification of classical and quantum descriptions of nature: i.e. The quantum-relativistic incorporative intermediate model spacetime.

With de-spatialisation by Book-IV, our incorporation of the quantum-relativistic domains, together with the primitive unification obtained, will survive into the Bergson-Whitehead amalgam: into spacetime *without* space.

*

With **the intermediate spacetime model, facile information-matter dualism adumbrated in 0.01 to 0.04 turns out to be an aspect of the broken-discontinuous progression of the event horizon**: Manifest OR-form information, as resolved radiation, matter and 'space', constituted as the first event horizon 'frame', dissolves and gives way into a pure AND-form quantum mechanical interval. The information is restored to the radiation, matter, and 'space' through subsequent AND-to-OR collapse (time) at the succeeding event horizon frame, so generating the facts of facile information-matter dualism. In this way, information-matter dualism is incorporated into the anticipated intermediate spacetime model (and to its superlative form in the superseding Bergson-Whitehead amalgam) as an aspect of the dynamics of spacetime, or as an aspect of the incorporation of abstract memory, future possibility, and time.

0.30: Primitive Unification, Spacetime & Mind

It turns out that the mind-brain relation can also be incorporated into the intermediate model of spacetime, at the event horizon. Hence, mind and memory, hitherto assumed or when assumed to be embodied in the material corpus of the brain, appear to dissociate from that material corpus in the transition of a first event horizon to a succeeding one. Thereafter, following a quantum mechanical interval between the event horizons and subsequent AND-to-OR collapse (time), mind and memory, or at least their consequences, are restored to the brain, albeit with state-change affected at the succeeding event horizon. In short, **a preliminary growing block intermediate spacetime model of memory-brain and mind-brain relations characterises the dissociation of mind and memory from brains, with the latter confined to the event horizon, in the succession of event horizons**. This is consistent with the claims from facile information-matter dualism about the same (see **0.05**).

From our intermediate spacetime, memory is retained, *not* in brains but within the domain of realised spacetime 'below' and extant the event horizon and its momentarily OR-form materialised brain. Consequently, **the recollection of memory by mind must involve a retrieval of that memory from the domain of realised spacetime, *not* from the brain confined to the event horizon**. For this to be possible, the mind as the decider and agency of recollection, insofar as it must reach into memory, cannot itself be confined to the event horizon or its brain. If it were so confined, it could not reach to and retrieve memory from the domain of realised spacetime 'below' and extant the event horizon and its brain.

Moreover, for mind to affect its intents, it must do so by selecting for that intension an appropriate future possibility-set embedded within the domain of spacetime in potentiality. That is, **the mind must be able to 'reach into the future' and select from that future potentiality what it wants to realise.** This also implies that mind cannot be restricted or confined to the event horizon and its brain. Otherwise, mind could no more reach into the future than it could reach into the past, if confined to the event horizon and the brain materialised upon it.

Paradoxically, memory expression is always a possibility that resides in the future: That is, memory-recollection and expression is potentialised as a future-possible expression or event: To express recollection of memory, mind must select an appropriate memory recollection-pertinent possibility function from its future potentialities; the AND-to-OR collapse of which then generates the expression of that recollection. That is, the recollection of the past is realised through the AND-to-OR collapse of the future and, in this sense, strangely, memory of the past is potentialised to the future, even while the memory itself is retained in abstract form within realised spacetime, despite memory-expression potentialised within spacetime in potentiality. These contradictory-seeming realities will be fully clarified in Book-V.

Finally, the mind must be able to accomplish all the above from between the event horizons, from the 'gaps in time' in the succession of event horizons: Detail of the attendant process must also await Book-V. Thus, **the primitive unification of physics in the growing block universe does not only entail the incorporation of quantum and relativistic descriptions into unity, but also incorporates the processes of mind, memory, and brains.**

However, with the advent of the Bergson-Whitehead amalgam by Book-IV, all the above will abide… but *without* space. **Thus, the ontics of de-spatialisation compels automatic Cartesian dualism: first by usurping the notion that memory is a spatial distribution retained in spatial brains, and second, by unravelling the notion that mind and consciousness itself is produced by brains.** Given that there is no space, there can be no such thing as a 'location' for mind and consciousness, or of any *where*-state for memory-retrieval processes, or even for memory itself. Hence, de-spatialisation will get rid of Rene Descartes' *minor error*: i.e., the notion of *res extensa*.

There is no space. It follows that neither the brain itself, nor the manifest material world, can be modelled as *extensa*. Equally, Cartesian non-extended mind, or *res cogita*, will have no problem in interfacing with a non-extended brain and world, given de-spatialisation, and given that the interface must now assume a pure non-contiguous temporal form. The causal relation between mind and brain is a relation unfolding in and through time. As such, it cannot be mediated by 'material' contact causality, given that there is no 'location' or 'point' for any 'spatial contact' in nature. That is, contiguous or contact causality unravels per de-spatialisation and pure temporisation of the physical order and of mind-brain relations. Automatic Cartesianism follows suit.

0.31: Comprehensive Summary of Preliminary Part-II

What is wave-particle dualism? What is the *raison d'etre* of the quantum mechanical wave and the superposition principle?

- The simple distinction between **AND-form logic versus OR-form logic is synonymous with wave-particle duality.** (See **0.09**).
  - According to OR-form logic, a particle or object is expected to be located at either one position OR at another: never at both simultaneously: It must either be 'here' OR 'there'. *Never* here AND there.
  - Yet, quantum mechanics demands the equal reality of AND-form logic: wherein a particle *can* simultaneously occupy one position AND another, forming a superposition of such. The same holds for momentum, spin, etc.
  - AND-form logic stands fully experimentally confirmed by the double-slit experiment. Consequently, the reality of AND-form logic in nature, complemented by an equally undeniable OR-form logic, constitutes AND-OR duality, synonymous with *wave-particle duality.*
  - Facile information-matter dualism arises from wave-particle AND-OR dualism in a seamless way: They are almost synonymous.

What does the AND-form quantum mechanical wave constitute?
  - The future is objectively real and 'out there': the future is *not bloss* subjective. **The AND-form quantum mechanical wave *is* the physical manifestation of the ontically real future.** (See **0.10**).
  - The distinction and cut between the OR-form past and AND-form potentiality constitutes **the firewall principle of quantum mechanics (0.15)**, per the intertwine of causality and the principle of conservation of energy-matter. Hence...
    - Any attempt to treat the attributes of a quantum object, or any object at whatever scale, as if it were comprised purely and *only* of OR-form resolved events (naive realism) before their AND-to-OR collapse out of AND-form futurity (hence, any attempt to treat the future as if it had already transpired into resolved events, such as to constitute a 'block universe') would entail the reception of 'signals from the future', with violations of the principle of causality *and* the violation of the principle of conservation of energy-matter.
    - **A block universe without a future, or with the future fully OR-form resolved, could never produce evidence of self-interference or interference patterns in the double-slit experiment: pre-resolved paths cannot accomplish quantum interference patterns. Therefore, the universe is a *growing block universe* with an ontic and objective future 'out there'.**

○ Insofar as AND-form and OR-form descriptions of nature are in fact mutually compatible and harmonious, as clearly intimated and necessitated by the firewall principle, we discover an unexpected harmony and **primitive unification of quantum mechanical descriptions of nature with those of classical-relativistic descriptions** of the same.

Preliminary assertions can be made about the relation of quantum phenomena to mind and to scale-issues (see **0.12**): Both are often connected, given that those who espouse against quantum mind theories do so by restricting quantum mechanics to the very small, excluding it from macro-scale brain. By Book-V, Alt-R will furnish proof that...

- **Mind or consciousness does not bring about AND-to-OR wavefunction collapse**: This comes about spontaneously per general causality that subsumes and transcends both mind and 'matter'.

- **Mind or consciousness does not constitute the secret orchestrator of quantum indeterminism**: Nor does it alter the probability of quantum indeterminate outcomes in the brain.

- Both of the above espouse against certain types of quantum mind theories, especially idealistic ones.

As for the issue of scale, **quantum phenomena *do* manifest at the macro-scale**. The very structure of the evidence for the superposition principle in double-slit experiments demonstrate this in an obvious way. Quantum AND-form logic and quantum indeterminacy are *not* scale-restricted, and they are not excluded from the brain. Thus...

- **It is only per *en masse* OR-form particles depositions obtained *at the macro-scale* that we obtain macro-scale evidence of the superposition principle and of AND-form self-interference in the macro-scale interference pattern** in the two-slit experiment.

- One other critical reason why quantum mechanics is not scale-restricted is per the ***firewall principle***. Therein, **causality and the principle of conservation demand that AND-form quantum phenomena (i.e., the future-form descriptions of systems) abide *at all scales* and at the macro-scale, as much as they do at the micro-scale, as is required by the growing block universe approach**. Otherwise, it would be possible to obtain 'signals from the future' at the macro-scale. To firewall against this, **the futurity of the macro-scale *must* configure as an AND-form state...** as a quantum mechanical state... just as it does at the micro-scale. **The macro-scale brain is not going to be exempt from this truism.** Hence, AND-form logic must also apply to the macro-scale brain, and implies the necessity of a growing block quantum mechanical model of brains.

The double-slit experiment constitutes the core experiment of quantum mechanics and demonstrates wave-particle dualism (see 0.13): An emitter injects either a single quantum of radiation or a 'particle' of de Broglie matter: This passes across a barrier with either one slit or two slits open: The outcome manifests at the detector.

Two standard claims are often asserted from the two-slit experiment:

- It is often claimed that **the single-slit experiment emphasises the particle-like aspect of the wave-particle duality: That it highlights the OR-form of radiation or matter**, from which OR-form 'which way' information could supposedly be inferred about the particle's prior state in isolation.

- It is also claimed that **the experiment with two open slits emphasises the wave-aspect of the wave-particle duality: the AND-form aspect of radiation or matter**. When both slits are open, the one particle apparently passes through both slits. Thereafter, it constructively and destructively self-interferes, eventually forming an interference pattern at the detector.

- However,...

The non-standard appraisal of the double-slit experiment (0.14) questions the notion that the single-slit exclusively resolves for the particle-like OR-form aspect of radiation and matter, or that it can furnish 'which way' information. It espouses that both the single *and* double-slit experiments involve AND-form *and* OR-form logic in observationally simultaneous ways, culminating into a **new complementarity thesis** of quantum mechanics. Why?

- **OR-form 'particle' realities are required to bring into evidence AND-form wave-like realities at the detector in both single and double-slit experiments. This happens observationally simultaneously.**

  - In the double-slit, the interference pattern (AND-form) is evinced by macro-scale *en masse* quantum depositions (OR-form). Without the particles, we cannot obtain evidence of wave-interference.

  - **The same holds in the single-slit experiment, except that no interference is produced**. Yet, the specific distribution of particles therein is generated by an AND-form wave that designs for its specific macro-scale distribution of OR-form particles.

  - Hence, **OR-form *and* AND-form realities are *always* observationally simultaneous at the detector, at least at the end of isolation**, or at the post-experimental phase, in both single and double-slit experiments.

  - **Generic complementarity is saved by recognising that**, while OR-form and AND-form logics are always in observational simultaneity at the post-experiment phase, **in the interim of experimental isolation, only AND-form logic abides in both single and double-slit experiments: It is not possible to assign any sort of OR-form 'which way' state to the interim of isolation, even in single-slit experiments.** The presumed 'which way' states in the interim of isolation do *not* constitute OR-form resolved 'paths', even in single-slit experiments.

The resulting **new complementarity thesis** asserts that…

- **At end of isolation, OR-form and AND-form realities are observationally simultaneous and occur together, but this cannot permit any inference of any prior OR-form 'which way' path in the interim of isolation,** even in single-slit conditions…and…

- **In the interim of isolation, the emitted 'particle' constitutes a pure AND-form superposition state of possible, *not* realised, 'which way' paths.**

- All of this assumes the ontology of 'space' and 'path'. With de-spatialisation, these are at best transformed into AND-form potentials for 'which moments' organised into potential delayed choice time-intervals.

Also, **the firewall principle (0.15) absolutely necessitates that we treat the interim of isolation as a pure AND-form state, lest we entertain 'signals from the future'.** Hence, we cannot infer any OR-form 'which way' or 'which moment' information about the interim of isolation, even in single-slit experiments. Thus…

- **When we place any system into isolation, we defer both it and its resolution to the ontically real future.** Hence, the system must assume a pure as-yet not-realised futures, rendered into an AND-form superposition state.

- **To treat the future *as if* it was an OR-form realisation (to treat the interim of experimental isolation as if it entailed any OR-form 'which way' information)** would be to treat it as if it was resolved into concrete events *before* it transpired into such. To prove this **would require 'signals from the future'.**

- Both **causality and the principle of conservation intertwine and enforce a firewall principle that asserts that you cannot gain concrete OR-form signals or physical effects from the future, or from a future-deferred experimentally isolated system, including from single-slit experiments.**

Now we can resolve **the Afshar controversy, premised on the claim that we can obtain OR-form 'which way' information about an AND-form system by generating a single-slit outcome from a two-slit arrangement (0.16).** Our non-standard appraisal of the two-slit experiment, the new complementarity thesis, combined to the firewall principle, assert that the AND-form state *must* abide in the interim of isolation, even in the Afshar experiment **(0.17)**. Thus...

- **The non-standard appraisal asserts that the single-slit experiment also entails AND-form logic. Thus, obtaining a single-slit outcome from a two-slit arrangement cannot avail against the assertion that AND-form logic precedes the OR-form resolutions of outcomes,** even in the specific arrangement of the Afshar experiment. Hence, the complementarity principle remains intact.

- **The new complementarity thesis asserts that OR-form and AND-form realities are observationally simultaneous at the end of isolation, even in single-slit experiments, and even in single-slit outcomes obtained from a double-slit arrangement** in the Afshar experiment.

- **The firewall principle of quantum mechanics** asserts that, even when we arrange to obtain a single-slit outcome from a double-slit arrangement, in the interim of its isolation, the system *must* be deferred to the future and must remain in quantum AND-form. **In the Afshar's experiment we cannot attribute any OR-form 'which way' information about the interim of isolation, unless one is willing to entertain 'signals from the future' in violation of the principles of causality *and* the conservation of energy.**

- Finally, **anticipatory de-spatialisation itself renders null and void notions of 'paths' and 'which way' states,** (see **0.18**) whether in OR-form or AND-form terms. The best one can assume is a pure temporal 'which moment', but never as an OR-form state in the interim of isolation, as is demanded by both the non-standard appraisal and the firewall principle.

*

What causes AND-to-OR wavefunction collapse? Are there hidden gears behind this? This is **the missing gears problem of wavefunction collapse** (See **0.20**). How will Alt-R solve the problem in Book-III and beyond?

- Recall that **AND-to-OR wavefunction collapse is synonymous with time.**

- Also consider the postulate that the future is an AND-form state comprised of both immanent futures *and* nested futures that elaborate to and totally converge at the indefinite future. This implies that wavefunction collapse 'here' must instantaneously decohere and restructure the infinitely many nested future possibilities 'over there' to consistent-only possibilities: i.e., counterfactual causality and **grand decoherence.**

- **Hence, any 'hidden gears' purported to counterfactual eliminate the infinitely many non-consistent futures and bring about the attendant grand decoherence of the universal wavefunction, could never overcome the attendant infinities without crashing into an insurmountable halting problem… and time would 'freeze up'.**

- It follows that, **whatever brings about AND-to-OR collapse and time is not going to be a definable 'hidden gears'. Only a structure-less ineffable causality can sort out and eliminate infinite non-consistent nested futures without crashing into insurmountable infinities and halting problems.**

- Gear-less wavefunction collapse is independently corroborated by renormalisation issues in particle interaction: **If particle interaction presupposes an interval between particle inputs and outputs comprised of infinitely many particle interaction-steps, any 'gears' to sort through these would crash into an insurmountable halting problem. The process could never obtain a final particle output. Particle interaction would 'freeze up'.**

- The solution is to grasp that **the process of AND-to-OR collapse or time that issues particle inputs into outputs simply 'leap over' the intermediate infinities**... if only because AND-to-OR collapse and time itself is driven by a non-gears based ineffable causality. This is the 'quantum leap in time' solution to particle interaction.

- De-spatialisation also corroborates gear-less wavefunction collapse and the 'quantum leap in time' solution: **There is no 'space'. Therefore, there can be no 'locations' or 'spatial points' at which purported particle-contact or impact take place.**

- Since missing gears is synonymous with the missing middle-term 'operation' in IOO-schema causality central to materialism, the crisis of causality in AND-to-OR collapse and time is simultaneously the crisis of materialist causality.

- **Mind brain causality must also be a gear-less ineffable process, fashioning for revived Cartesian dualism.** If the very process of AND-to-OR collapse and time that resolves successive moments of the brain out of AND-form future-potentialities is a gear-less ineffable process, so is the mind process.

*

The stochastic-coherence paradox (see 0.22) constitutes the problem of 'impossible' order from apparent quantum mechanical chaos. **If quantum indeterminacy is objectively real, any sort of coherence, consistency and pattern-persistence, or even 'illusory' functional self-determining mind and consciousness vis-à-vis brain and world, ought to be totally untenable.** How will Alt-R solve this great problem?

- **Book-III will make the case for the objectivity of quantum indeterminism:**
  - Therein, we will take some key process in which randomness or quantum indeterminism is presumed to operate and systematically generate for it wholly deterministic orchestrating scripts.
  - **Script-operations, when hidden from view, will appear to engender randomness or quantum indeterminacy,** even though the scripts proceed in a wholly deterministic way, constituting putative 'hidden gears' behind facile quantum indeterminacy.
  - It will be shown that **each script, from the very large number of such, are equally adept at emulating facile randomness. There is no one exclusive script,** and there exist no structural or qualitatively unique standard for discriminating for such an exclusive script from among the very large number of such.
  - Hence, **the only recourse is random selection from among the very large number of equally apt scripts. Thus, randomness cannot be eliminated: There are no hidden gears behind quantum indeterminism: Hence, quantum indeterminism is objectively real.**

- The objective reality of quantum indeterminism can only retrench the seeming impossibility of pattern and persistence. Yet, **Alt-R promises to save 'order' from chaos in the following way**:
  - Patterns, such as the materialised bell-curve distribution, the interference pattern in the two-slit experiment, the developed photographic image, or the materialised universal wavefunction that constitute the physical universe, inexorably emerge out of *en masse* quantum indeterminate outcomes. Thus, relative **order will emerge *despite* objective quantum 'chaos' *so long as we identify order with the abstract complex wavefunction itself.*** This latter will force a reframing of facile information-matter dualism, which does not identify information with the complex wavefunction, while it deems information as something that must dissociate from that complex wavefunction.
  - **Quantum indeterminate processes unfold through a three-phase developmental process, itself bounded by the abstract complex wavefunction. The latter frames the natural odds in its own favour, for the sake of its own inexorable materialisation, despite objective quantum indeterminacy.** This is the central insight entailed in the solution to the stochastic coherence paradox by Alt-R.
  - **A complementary solution to the stochastic coherence paradox also entails the reality of abstract memory** (Books III and IV) **entailing the *retention* of past complex wavefunctions.** Memories also funnel *en masse* quantum indeterminate terms into enduring and perduring patterns, manifesting key features of facile information-matter dualism.
  - Indeed, **anticipated de-spatialisation from Book-IV will force a pure temporal-based organisation of information.** Therein, what appear to be 'objects in space' turn out to be resonations, *not* across 'space', but *across time*. That is, **de-spatialisation itself will seamlessly and perfectly culminate into abstract memory, and automatically prove the retention of past wavefunctions *as* memory.**

*

The most obvious problem posed to mind-brain causality derives from quantum indeterminacy and the stochastic-coherence paradox: (see 0.23). Alt-R will first confront the much-touted claim that quantum mechanics has no relevance to the macro-scale or to

55

the brain, by showing that AND-form logic and quantum indeterminism are not scale-restricted and apply to the macro-scale, at least as much as they do to the micro-scale. How?

- A double-slit experiment involving very short wavelength waves could confine the experiment to the near micro-scale. **But the utilisation of very long wavelength (low frequency) radiation and matter will manifest pertinent quantum effects, including quantum interference and quantum indeterminacy, at the macro-scale**. It follows that, in terms of wavelength considerations, quantum phenomena are not restricted only to the very small or very-short wavelength domain. Ultimately, there can be no scale-restriction of quantum phenomena.

- **The three-phase development of a photographic image (a macro-scale frame) demonstrates that quantum indeterminacy cannot be abstracted out of the macro-scale framework or from the brain**:
  - Herein resides the solution to how mind overcomes quantum indeterminacy in brains: **Like the photographic film, the brain, as a hot-body macro-scale framework for the three-phase development process, is not free of quantum indeterminacy any more than is the process of hot-body macro-scale photographic development. Hence, coherent brain activity will emerge out of *en masse* quantum indeterminacy, as do coherent photographic images obtain from the same.** Indeed, both quantum interference and indeterminacy must be present in the macro-scale brain, as much as these obtain in macro-scale photography.

*

Irreversibility or time-asymmetry through AND-to-OR wavefunction collapse and time contrast with time-symmetric timeless physical laws (see **0.24**). Yet, physical laws are evinced by time-irreversible time-asymmetric outcomes obtained from AND-to-OR collapse or time. Alt-R will solve this conundrum thus:

- **The problem of irreversibility and the question of time-symmetric physical law in the context of AND-to-OR collapse will be partly solved by the *quantum theory of entropy*** (i.e., the theory of potential entropy) in Book-I.

- Second, **both the irreversibility question and the time-symmetry question of physical law will be fated to the explication of what causes AND-to-OR collapse**, pertinent to the quantum theory of entropy. In brief, Alt-R will address these problems per the following in summary...
  - **Insofar as AND-to-OR collapse *is* the conversion of future-form potentiality into OR-form events, AND-to-OR collapse must be synonymous with time itself**. This is readily apparent from the ontic reality of the future per the *raison d'etre* of AND-form logic and attendant firewall principle.
  - Hence, **the question of the causality responsible for AND-to-OR collapse is essentially identical to the question of causality responsible for *time* itself.** (Book-IV).
  - It will turn out that **neither non-mental hidden gears nor consciousness is causally responsible for time. The notion that mind is the drive behind AND-to-OR collapse and time will be shown to be as dubious as resort to dumb 'hidden gears' in the same:** (Book-V).

*

Generic quantum mechanics does not furnish necessary or compelling reasons to treat the apparatus as an OR-form state, in the sense that the choice to do so is imposed by the experimenter on a pragmatic basis, given that generic quantum mechanics is purely about AND-form states. Yet, experimental systems *must* assume OR-form characteristics, or the basis for this must be retained somehow, despite the exclusive AND-form bias of quantum mechanics, in order that physically meaningful experiments become tenable and quantum mechanics as we know it become viable. Hence, the *apparatus problem*: (see **0.25**) How will Alt-R solve this problem? The solution is abstract memory.

- **What is missing in generic quantum mechanics is a theory of abstract memory in nature**. Alt-R must demonstrate that **the physical universe has memory**.

- **The paradox here is that generic quantum mechanics implicitly needs physical memory, or the retention of initial experimental conditions despite isolation, in order to deliver the kinds of experimental results it has achieved. The physical retention, as memory, of the experimental apparatus will be furnished by de-spatialisation.**

- The 'both slits open' state, retained in memory as a complex wavefunction in memory, delimits the subsequent evolution of the future and decoheres that future only to possibilities that are consistent with the "two slits open" condition, or any other initial condition, eliminating the alternatives. **The resonation of the initial condition, or of the apparatus, to the future is something accomplished purely across time, given de-spatialisation and the pure temporisation of physics**. This is also consistent with implications from facile information-matter dualism, consonant with abstract memory theory.

*

The Einstein Podolski Rosen experiment (EPR) demonstrates 'spooky action at a distance'. In terms of quantum spin, when we get 'spin-up' on a left-side particle, we get an instantaneous 'spin-down' on the right-side. i.e., spin is conserved. **The conservation of spin is enforced, even while there are no physical signals between the particles to mediate and enforce it**. Yet, the generic solution to EPR holds that no meaningful information can pass from the left-side to the right-side particle, due to ineliminable scrambling from quantum indeterminacy, and that EPR 'action at a distance' cannot entail the transmission of meaningful information (see **0.26**). However, ...

- **First, the enforcement of the conservation of spin in EPR itself constitutes minimal concession to meaningful information in EPR**, given the requirement that this principle *must* be imposed on quantum entangled couples, despite indeterminacy. **This shows that ineliminable quantum indeterminacy in EPR cannot be used as a basis for any final argument against the transaction of meaningful information in EPR.**

- **What EPR actually demonstrates is the dualism of abstract physical laws vis-à-vis radiation, matter and 'space'**, and, with it, the enforcement of extant physical law from 'outside' of radiation, matter, and time. In EPR, this is evinced as the enforcement of the conservation of quantum spin.

- In non-EPR situations, such as photography, we find the persistence of physically meaningful information and the endurance and perdurance of identity, *despite* ineliminable quantum indeterminacy. **The formation of an image in photography ought to be fated to the same information-scrambling from quantum indeterminacy as asserted to hold in EPR.** Yet, equally ineliminable quantum indeterminacy in non-EPR situations do not scramble the global formation of a photographic image, for example.

- **The second part of the solution is to grasp that, in non-EPR, quantum indeterminate processes are funnelled into pre-ordained fully determined global pattern-outcomes, despite quantum indeterminacy, and per Alt-R's solution to the stochastic coherence problem... to the effect that quantum indeterminacy is not inimical to information**.

- **The third part of the solution is to realise that information in the physical order is being imposed and manifested in ways that defy convention**: Complex wavefunctions, laid out across vast domains of 'space' and time, as instantaneous wholes of future potentiality, can funnel *en masse* quantum indeterminate terms into globally coherent wholes, and even into universe-spanning wholes, even though the quantum indeterminate terms are separated by thousands, if not millions of parsecs. Yet, these culminate into coherent large-scale patterns; forming wholes that persist over long periods of time. The role of instantaneous wholes is exemplified by complex abstract wavefunctions in photography; for whole galaxies, and by the universe-scale wavefunction itself. **The quantum indeterminate terms, as 'pixels', inexorably culminate into the materialisation of such instantaneous wholes, *without* speed of light communication and *without* any secret orchestration of quantum indeterminate terms.**

- **The fourth solution emerges out of de-spatialisation, which implies that the quanta that materialise instantaneous wholes cannot be characterised as 'distributions in space', whose terms must supposedly communicate with each other at or below the speed of light.**

*

Part of **Alt-R's overhaul of putative spacetime will emerge out of the nested futures perspective, in turn derived from John Wheeler's delayed choice**. The nested futures perspective leads to non-local transtemporal **counterfactual causality** and **grand decoherence**. (See **0.27**) Thus...

- Wavefunctions and frames of reference do not exhibit only immanent AND-form futures, but also embed subsequent futures attached to the immanent futures, forming **a system of nested futures that elaborate to the indefinite potential future.**

- This obtains **unitary non-local and transtemporal binding of the physical universe, or of one frame on the event horizon to every other frame along the same event horizon, per function of counterfactual processing and grand decoherence of nested futures**: This constitutes a unitarity related to but distinct from unitarity evinced in non-locality in EPR.

- In the case of a hypothetical relation between an Earth-frame and an Alpha Centauri frame over four lightyears distant from Earth, the futures attached to each frame, though separate along the event horizon, must yet converge in the future. i.e., observers on Earth are potentialised to observe events at Alpha Centauri in four years' time. This is per the nested futures structure that co-join both frames in the future, in four years' time. The respective futures up to the said four years must instantaneously decohere to consistent-only possibilities the moment key OR-form events transpire at Alpha Centauri: **If an extraterrestrial at Alpha Centauri successfully transmits the Encyclopaedia Galactica before its world is destroyed by a supernova, all nested futures related to the alternative 'failed transmission' will be erased from all nested future frames, including those inherent to the Earth-frame of future potential signals that might be observed or detected on Earth four years in the future.**

- Hence, the nested futures view implies **the instantaneous 'non-local' counterfactual unitary binding and <u>*grand decoherence*</u> of the whole universe onto itself**... no matter what the putative 'distances' between all of its frames or, more appropriately, no matter the delayed choice time-interval relations that separate the frames and events: All frames converge to a common and same future, especially at the indefinite or infinitely removed future, and the Earth frame is certainly convergent with the Alpha Centauri frame in terms of their common future in four years' time. What happens at Centauri will decohere permissible possibilities into consistent-only possibilities over here, in an instantaneous 'non-local' or transtemporal way.

57

- Counterfactuals and grand decoherence constitute the basis for a **revived Mach's principle; the quantum mechanical solution to inertia** and a primer to the **quantum mechanical solution of the permanent magnet**: (see Book-II).

*

The uncertainty principle in a nutshell: If we accurately resolve the position of a particle, we end in the calculable intrinsic non-resolution of its momentum. If we resolve the momentum of that particle, we end in related calculable non-resolution of its position. **In Alt-R, we need an explanation of the uncertainty principle derived from the *raison d'etre* of the quantum mechanical-AND**: (see **0.28**) Thus...

- **The quantum mechanical wave is the objective manifestation of the future**, and any attribute deferred to the future must remain non-OR-form or non-resolved. i.e., in an AND-form configuration.

- Momentum requires low-frequency observational inputs. Position requires high-frequency observational inputs. We cannot bombard a system with both low and high frequency inputs: This would furnish an ambiguous resolution of both momentum and position.

- **Per Alt-R, when we resolve in favour of position, we defer momentum to the future**: The momentum so-deferred is rendered into a wide range of as-yet not-happened AND-form future momentum-possibilities. **Note that momentum so deferred cannot be 'objective'… it cannot be a resolved OR-form state.**

- **Per Alt-R, when we resolve momentum, we defer the resolution of position to the future.** The position so deferred to the future configures as a wide range of future-potential as-yet not-happened possible positions configured in AND-form. **Note that position so deferred cannot be 'objective'… it cannot be a resolved OR-form state.**

- **With de-spatialisation, and the dissolution of the notion of 'position', we must reform the uncertainty principle**: This can be done per the adoption of the variant of uncertainty configured as energy-time uncertainty.

*

In Book-II, **Alt-R will obtain the intermediate model spacetime: a growing block model that incorporates the primitive unification of classical-relativistic and quantum mechanical descriptions** of nature: (see **0.29**). Thus…

- **Conventional models of spacetime are bereft of both a causally active past (retained as memory) *and* an AND-form quantum mechanical future**, and do not even explicitly incorporate the process of AND-to-OR wavefunction collapse, or time itself.

- The incorporation of the future into spacetime entails an AND-form domain called **spacetime in potentiality**. The future therein is comprised of immanent futures and nested futures that elaborate to the infinitely removed future.

- In turn, the active past is incorporated into spacetime as the mnemonic **domain of realised spacetime**. Once events come into being via AND-to-OR collapse and time, their basis is relegated to and retained within this domain as abstract memory.

- **The event horizon** constitutes the 'cut' between the two domains: a perpetually 'moving' boundary between the two; comprised of the most recent OR-form outcomes *and* immanent AND-form futures.

- **Time is synonymous with the AND-to-OR wavefunction collapse of 'spacetime in potentiality' into 'realised spacetime'** and attends the progression of the event horizon. The identification of time with wavefunction collapse itself, and its incorporation into spacetime models, forgoes the need to model time as a 'fourth dimension'.

- **Facile information-matter dualism turns out to be an aspect of the broken-discontinuous progression of the event horizon.** The processes of the event horizon are readily identifiable with the processes of facile information-matter dualism, described in **0.29**.

- **As a growing block universe model, the intermediate model of spacetime and its tripartite structure constitutes the explicit primitive unification of OR-form classical-relativistic with AND-form quantum mechanical descriptions of nature**: the primitive unification of OR-form and AND-form logics.

- By Book-IV, **the intermediate model will itself be succeeded and superseded by the Bergson-Whitehead amalgam**: i.e., **spacetime without space**. The tripartite structure described above, the synonymity of time with AND-to-OR collapse, and the attendant broken-discontinuous processes of the event horizon, together with information-matter dualism itself, will survive into the Bergson-Whitehead amalgam..

*

Unification of classical and quantum mechanical descriptions of nature also incorporate mind and memory (see **0.30**):

- Per anticipated intermediate spacetime model, **the recollection of memory by mind must involve memory retrieval from the domain of realised spacetime below the event horizon, *not* from the brain restricted to the event horizon.**

- **To realise a future outcome, the mind must be able to 'reach into the future' and select from its potentiality the goal it seeks to realise and express through brains.**

- In both cases, **the mind cannot be reduced to or confined to the event horizon and the brain therein.** The mind must be extant of both memory recollection and future affectation. Hence, revived Cartesian dualism.

- **Anticipated de-spatialisation (Book-IV) will usurp the notion of memory as any sort of 'spatial distribution' in brains,** simply because there is no space. **De-spatialisation will also unravel the notion that 'illusory' mind, or consciousness is a thing that arises from or produced at a 'spatial location', or from and at brains**.

PRELIMINARY PART III :
MATERIALISM AND CLASSICAL PHYSICS

0.32: Materialism and the Claims of Natural Science: Work Function Zero Processes

According to culturally hegemonic materialism, there is only 'matter in space'. All things arise from the interactions of matter in space. It follows that any claim for the existence of abstract-immaterial mind must be dismissed as 'religion-inspired'. Thus, according to Daniel Dennett…

> The prevailing wisdom, variously expressed and argued for, is *materialism*: there is only one sort of stuff, namely matter –the physical stuff of physics, chemistry, and physiology- and the mind is somehow nothing but a physical phenomenon. In short, the mind is the brain [13]

Also, according to Edelman…

> Any adequate theory of brain function must include a scientific model of consciousness, but to be scientifically acceptable it also must avoid the Cartesian dilemma. In other words, it must be uncompromisingly physical. [14]

Notice the equivocation of "scientific" with materialism. Notice how physicalism is treated as if interchangeable with materialism. Therein, there can be no possibility of a dualistic approach to the mind-brain. Edelman continued:

> Scientific epistemology must confront the issue of consciousness in terms of evolution, development, brain structure, and the physical order as we know it. If the confrontation is to remain in the scientific domain, a dualistic solution or any other Cartesian empiricism cannot be countenanced, [being] often accompanied by what might be called Cartesian shame. [15]

One wonders what Dennett, Edelman, *et al* would make of the Alternative Realist thesis. How would they consolidate their materialism with the dissociation of information from radiation, matter and 'space'; with the abstract extant and dualistic character of physical law per the conservation of spin in EPR; or with the discovery that we do not inhabit a physics of 'objects in space', but a de-spatialised purely temporised order of information organised into abstract memory and future-potentiality? Indeed, these necessitate the pure temporal integration of mind and consciousness to a de-spatialised growing block system of memory and possibility: a daz that transcends 'objects in space' as much as it undercuts any 'spaceballs' notion of contiguous impact causality. Alt-R usurps materialism and compels for shameless Cartesianism.

Indeed, **even in classical physics, materialism was at best a *prima facie* parasite ideology: It enjoyed a precarious relation to naturalism and science: an EPR–based analysis of physical law is not necessary to arrive at similar dualistic conclusion about physical law vis-à-vis 'matter in space':** The fact is obvious from the supposed time-symmetric character of physical law: a feat only plausible for an abstraction extant time itself ; harking to an ontology that supersedes apparent 'matter in space': a view that will gain full saliency by Book-I, but it is obvious even from the content of our Preliminary.

Materialist ontology asserts for closed causality: i.e., closed to 'outside' influence, especially from abstract-immaterial extra ontological factors, despite implications from the character of physical law; even when we ignore implications about physical law from EPR. **Hence materialism must account for causality in terms of specifiable face-front spatially contiguous states in contact.** Thus, when materialism speaks of 'natural physical causes', and about causality in general, it expects materially specifiable contact and impact relations between definable variables, events, or 'matter', **structured into input-operation-output schemas (IOO). Insofar as this must involve causality through seamless-continuous contiguity across 'space', it must constitute a totally closed system, with no possibility of extant extra variables, or of any extra-ontology, despite the character of physical laws. If true, materialist closed causality and closed ontology must also apply to the mind-brain relation, as much as it supposedly applies to all else.** As Dennett stated it…

> Let us concentrate on the returned signals, the directives from mind to brain. These, *ex hypothesi,* are not physical; they are not light waves or sound waves or cosmic rays or streams of subatomic particles. No physical energy or mass is associated with them. How, then, do they get to make a difference to what happens in the brain cells they must affect, if the mind is to have any influence over the body? A fundamental principle of physics is that any change in the trajectory of any physical entity is an

---

[13] D. C. Dennett, Consciousness Explained, 1991, p.33

[14] G.M. Edelman, The Remembered Present, 1989; p. 10

[15] G.M. Edelman, The Remembered Present, 1989; p. 278

acceleration requiring the expenditure of energy, and where is this energy to come from? It is this principle of conservation of energy that accounts for the physical impossibility of 'perpetual motion machines', and the same principle is apparently violated by dualism. This confrontation between quite standard physics and dualism has been endlessly discussed since Descartes's own day and is widely regarded as the inescapable and fatal flaw of dualism.[16]

Thus, we stumble on **the materialist standard abuse of the principle of conservation of energy-matter, arrayed against dualism**; founded on a lack of awareness of work function zero processes across nature., which we will adumbrate shortly.

However, the materialist presuppositions about the nature of causality cannot survive actual physicalism, as was made obvious in Part-I and Part-II, even when we ignore implications to the materialist presumptions about closed ontology and closed IOO-schemas: implications that emerge from the measurement problem of quantum mechanics, and from the preliminary case for the character of information and memory; or from the abstract dualistic character of physical law amplified in EPR; or from all other discoveries obtained in the Alternative Realist thesis; and most especially from de-spatialisation, temporisation, and the collapse of contiguous causality.

*

Without undermining principles of conservation, certain processes in nature fall under the rubric of *work function zero*. Work function zero processes raise doubts about the materialist conception of causality.

The consummate classical physics exemplification of work function zero, which at first appears harmless to materialism, is salient in **the principle of inertia**: Therein, a material body will coast in space in a state of uniform or unchanging motion, and in perpetuity, without requiring any continuous application of force, power or work: a state in perpetual motion, until such time as it encounters an impinging force that accelerates or decelerates it; perhaps also changing its direction.

In the case of inertia, the left term in the equation, $\mathbf{F} = \mathbf{ma}$ (or *force = mass* times *acceleration*) equals zero…simply because acceleration remains zero. The body is as good as being at rest, even though it is undergoing perpetual motion and change. Thus, a mass in uniform motion is apparently indistinguishable from a mass at rest under $\mathbf{F} = 0$, even though it is consummately distinct from a body at rest. In other words, *work*, energy, or power requirements to bring about the continuous *uniform* perpetual motion of an inertial body always equal zero: hence *work function zero*. This will remain so long as there is no inexplicable change in direction or velocity of the mass.

However, **gravitational phenomena also constitute implicit work function zero processes**. The orbital motion of the Moon about the Earth, or the acceleration of a body under Earth's gravity, or the motion of cascading water through a hydroelectric dam by **gravitation… involves changes in the state and trajectory of the affected material, with acceleration of that material in a form almost interchangeable with an inertial body in perpetual uniform motion, but *with* non-uniform motion or acceleration… *without* the violation of principles of conservation**.

In gravitation, we do not only obtain the acceleration of objects out of seeming nothing but we can also release the putative potential energy-store from the matter under its influence, such as in water flows in dams and in gravity-driven stellar fusion… again *without* the violation of principles of conservation. **At no point in any of these phenomena is the sum of energy-matter altered or increased, even though the induction of state-change and energy-release *is* obtained through the application of essentially *nothing***. In other words, real physical causality has a very different form from that imagined in materialism, or by Dennett and Edelman et al. The issue is not only salient in gravitational phenomena.

At this juncture, we cannot help but resort to Alt-R: None of the above should come as a surprise: **De-spatialisation implies that there can be no contact causality: There is no space. Therefore, there are no 'locations' at which gravitational influence could be realised through any contiguity or contact-based causality. Consequently, the causal mediation of gravitational influence upon putative matter cannot entail impact-mediated 'force', energy, or power, much less involve force 'carrying' particles**.

Moreover, **all change, including acceleration and energy-release due to gravity, must involve gear-less AND-to-OR collapse processes (time)** through which we obtain the set of successive events for which we infer gravitational acceleration and stellar fusion: Recall from **0.20** that **AND-to-OR wavefunction collapse and time is *not* realised through a gears-based IOO-schema**. Thus, the AND-to-OR collapse that evinces gravitational acceleration or orbital motion is not going to be exempt from this, and **gravitational causality, subsumed to ontologically primary AND-to-OR collapse and time, must also constitute a gear-less process**.

What happens when we attach putative energy relations to AND-to-OR collapse process that attends gravitation? One can state that **the total energy after AND-to-OR collapse will be the same as the total energy before AND-to-OR collapse attendant gravitational acceleration or induced nuclear fusion** in stars, and conservation principles will remain intact. **Yet, the AND-to-OR time process that preserves the principle of conservation and evinces gravitation is not itself an energy or power-driven process**. That is, you do not obtain AND-to-OR collapse and time by 'shining particles on time' or applying force, energy, or power upon time. Succinctly, **the AND-to-OR time process, or simply wavefunction collapse, constitutes a work function zero process**. It also follows that the gravitational phenomena subsumed to AND-to-OR collapse and time, and the change affected thus, is also necessarily a work function zero process, because it was brought about by AND-to-OR collapse, which is *itself* a work-function zero process. Yet, notice again that conservation principles abide despite the ubiquity of work function zero process and attendant 'weird' causality.

---

[16]D.C. Dennett, *Consciousness Explained*. 1991: p35

In short, **the notion that you cannot obtain change without an energy-input, or that such a feat would entail the violation of principles of conservation... is false**.

These insights are critically relevant and inimical to the materialist expectations about noetic causality vis-à-vis the brain: (See previous quote from Dennett). **The Alt-R contention is that mind-brain causality is also a work function zero process: one that exploits ontologically primary work function zero gear-less AND-to-OR collapse and time vis-à-vis the brain.** This segues seamlessly with the contention in **0.06** ...to the effect that mind employs non-clockable processes not possible to embody as brain structure. Revived Cartesian dualism inexorably follows.

0.33: Gravitation & Work Function Zero

Let us return to the pure classical physics format to further augment our understanding of work function zero processes without resort to Alt-R. Here we will look at how gravity is generated by matter in classical physics, followed by the augmentation of gravitationally induced acceleration. Both expose the reality of the work function zero format in nature.

Putative matter apparently generates gravity: but from what? We might first conceive gravity as a classical 'force field' or energy field. But where does this force or energy come from? Insofar as matter seemingly generates it, it is tempting to think that the energy of the gravity field must be 'bled out' from its engendering matter.

A more sophisticated version of the principle of conservation holds that matter and energy are interchangeable, or that matter is 'congealed energy': a 'battery' for energy. Therein, the sum of energy and matter must remain constant. Thus, if a certain quantity of matter is converted into radiation-energy, or vice-versa, the sum total of energy at the end of the process must remain unchanged from its quantity at the beginning of that process. This is encapsulated in the famous equation, $E = mc^2$.

If a gravitational field were treated as some sort of force field or energy field, one might imagine that its energy E content must issue out from its engendering matter m. Since the sum of energy and matter must remain unchanged, the gravity-engendering matter must eventually run out, as would a battery run of its power; unless we are willing to countermand principles of conservation or abandon the notion that the sum of energy and matter must remain unchanged.

As far as is known, matter can generate a gravitational field in perpetuity, without ever running out of or depleting its congealed energy-matter content, *ceteris paribus* with respect to entropy, etc. Indeed, when the gravity field is treated as an energy field, there is not enough energy content in the gravity-generating matter to account for the total energy constituting the gravity field hitherto generated. Note that such fields extend and expand outward indefinitely from their material 'source': In Earth's case, its gravity field has been expanding in all directions for some five-billion years. Even with diminution over 'distance' and time, imagine how much total energy this field must now possess, *if* it is to be treated as an 'energy field': There is not enough congealed energy-content in the material of the Earth to account for this colossal quantity.

When we notice the obvious fact that the gravity-generating matter doesn't appear to run out, its gravity field appears to arise *out of nothing*: equivalent to conceiving it as if it were 'energy out of nothing', despite rules of proportionality of the gravity to the associated mass. But this ought to lead to the violation of principles of conservation, *if and only if* we insist that a gravity field *must* be treated as some sort of 'energy field', and that its causal power to accelerate or decelerate matter and radiation *must* involve the expenditure of energy or power, consistent with materialist expectations.

Again, perhaps the solution requires that we stop treating gravity fields as if these constituted energy fields. Instead, we could treat gravitational fields as the curvature of the geometry of spacetime, generated by spacetime-curving matter. Einstein's spacetime curvatures are abstractions that might easily smack of immaterialism: They *are* physically immaterial in character; even though they do not violate the principle of conservation or, as far as is known, add or subtract from the total putative energy-matter content embedded within the spacetime curve.

Let us be generous and assume that some sort of material causality is involved in the generation of the spacetime curve by the gravitating matter. We run into the same problem as before: Since matter 'tells' spacetime how to curve, does the generating matter accomplish this by doing work? If so, from where is this energy or work to come from? Will the generating matter obtain it from its own content? If the generation of the spacetime curvature requires the expenditure of energy from the gravitating matter itself, or its conversion to gravitation by some variant involving $E = mc^2$, there will not be enough congealed energy in the matter to account for the total gravity field hitherto generated. Are we to contrive a 'hidden energy' to account for this? Are we again to brook potential violations of the principles of conservation?

Why not simply admit that, **however we choose to describe it, gravitational fields, associated with putative matter, do not arise from a matter-energy based physical causality, and that such fields do not actually consist of matter, or of radiation, or of *any* sort of 'energy': The generation of gravitation, curve or other, is a work function zero process. The nature of a gravity field, whether conceived as an 'energy field' or as 'spacetime curvature', is phenomenally abstract. In other words, gravitation clearly evinces the reality of a wholly physical and entirely *immaterial* state and causality: a work function zero-based causality.** Such an admission is the only way to preserve principles of conservation in the framework of gravitational phenomena.

*

We now return to the induction of change and acceleration by gravity, distinct from the question of how the gravity is generated. To this end, consider a cup at relative rest on the surface of a table. There is no force acting on it: It is equivalent to an inertial object. That is, the term **a** in $F = ma$ appears to equal zero. Thus, **F** is also zero: i.e. The work function equals zero. Yet, the cup is an object

61

on the surface of the Earth, which is rotating. In the latter framework, on account of the perpetual change in the direction of its motion due to the spin of the Earth, the cup is undergoing continuous change in its direction: i.e., *acceleration*. Yet, we find that there is no *accountable* force **F** acting upon our cup to bring this about, whether we use the centrifugal approach or the countervailing factor from gravity, or both acting on our cup. (We have already noted the problem of treating the gravity as an energy field' or its generation as an energy driven process).

We also find that **a** is not zero: A positive value for **a** *is* obtained, even though **F** is apparently zero.

We can generalise the same to the orbit of the Moon about the Earth. The Moon operates in essentially the same way as our cup, albeit from a considerable altitude vis-à-vis the surface of the Earth. The Moon undergoes a slightly elliptic orbit about the Earth; perpetually changing its direction of motion throughout its orbit; thus, undergoing acceleration: All based on a work function zero gravitational process, *without the violation of principles of conservation.*

Only if we change the orbit of the Moon to a higher or a lower orbit about the Earth would we need to input any energy, power, or force. Only then would the work function become non-zero: (The same holds for the cup). Otherwise, without requiring any input of energy or power, the Moon will continue in its orbit in *perpetuity*, with perpetual acceleration and motion, as will the cup. As such, **the perpetual orbit of the Moon is almost equivalent to the motion of an inertial body: Just like the inertial body in perpetual motion, the Moon does not require a perpetual input of force, energy, or power to affect its motion, even though it *is* undergoing perpetual *acceleration*.**

If in the want to account for such processes we resort to some sort of implicit energy or power-input, then we risk upsetting the accounting books, as was the case in the want to account for the generation of the gravity field by the same: i.e., we end up violating principles of conservation. Where would such an implicit energy, power or force come from? **Again, in order to retain the principles of conservation, especially in gravitational phenomena, we need work function zero processes** and, with these, the induction of motion, acceleration, energy-release, and state-change... *without* material or radiational inputs...*without* work... but *without the violation of conservation principles.*

Let us further augment our insight: Imagine a tunnel cut right through the Earth. We drop a material object into it, which accelerates through the tunnel by the effect of the Earth's gravitation. The object undergoes free-fall: It does not experience force or acceleration, even though it *is* undergoing acceleration: An equivalent of this takes place in astronaut training, which involves a plane taking a nosedive from high altitude in such a way as to nullify the effect of gravity: Gravity is still in force, yet the astronauts feel no force or pull: they are indeed in free fall.

Returning to the tunnel, reaching the very centre of the Earth, our object begins to decelerate under the same gravitational power. It emerges at the other end of the tunnel cut through the Earth. There, the object briefly halts, only to reverse its course and fall back through the Earth... eventually re-emerging at the original tunnel-end.

All things being equal, our object is destined to oscillate from one tunnel-end to the other, *in perpetuity*. In effect, with gravity, we obtain virtual perpetual motion in yo-yo form... so long as the object does not touch or skirt the walls of our tunnel, and nothing impedes it on its unobstructed course.

The behaviour of Earth's Moon cannot obviously be described in the same way, save that it is the circularised or elliptical version of the same yo-yo. The operating principle is the same in both instances. In both, we obtain perpetual motion by gravitation, without requiring *any* energy or work, *ceteris paribus*. Again, it is only when we try to shift the orbit of the Moon about the Earth to, say, a closer orbit, or attempt to speed up or slow down our falling object through the Earth's tunnel, that we must expend energy or power, which must be obtained from some specifiable source; and wherein the work function can no longer equal zero. But, left to itself, the Moon will go about its expected orbit about the Earth in perpetuity, in free-fall, just as our object will oscillate through the Earth's tunnel in perpetuity, in free-fall, even though in both cases the work function for describing either will equal zero, and no force or energy will be expended, despite the reality of perpetual acceleration in both instances.

Again, the inevitable perpetual change in the direction of the Moon's motion, from a simple straight-line inertial state into an almost circular-elliptical almost-inertial orbit under gravity, necessarily implies *acceleration*: all of it achieved through work function zero, *without* the expenditure of work: Per **F = ma**, it is as if the acceleration **a** is absent, even though obviously present; and only mass **m** appears to apply. Thus, despite observable reality, the Moon remains effectively at rest (or in free-fall) even while it is perpetually accelerating (but at rest?)... so long as its orbit is not disturbed to a lower or higher orbit.

*

What is the significance of gravity-driven perpetual motion to claims from materialism? Recall from Dennett:

> ...A fundamental principle of physics is that any change in the trajectory of any physical entity is an acceleration requiring the expenditure of energy, and where is this energy to come from? It is this principle of conservation of energy that accounts for the physical impossibility of 'perpetual motion machines'...

Yet, as the above cases show, we obtain precisely just such 'impossible' perpetual motion and acceleration in gravitation, *without* any expenditure of energy or power: a *work function zero* process. We must either assume that the gravity field induces perpetual motion and acceleration by expending some sort of 'hidden energy' (a contrivance), or resort to outright violations of the principle of conservation. Alternatively, we must resort to the notion that perpetual motion by gravitation is something induced by immaterial causality, *without* altering the quantity of energy-matter in the systems that it changes. In short, **if we want to preserve principles of conservation, we must reject materialism**.

Perhaps the starkest example of how principles of conservation are retained under gravitation is evinced in stellar nuclear fusion. The material of a star, such as that of the Sun, conglomerates by gravitation: also, a work function zero process. At some point, under the same gravitation, the material at the core of the sun begins to fuse to form more complex elements; the by-products of which are heat, light and the various forms of radiation.

Per $E = mc^2$, if we sum the matter-content of the pre-fusion reactants and compare this to the post-fusion products, we discover that no change in the energy content has taken place: i.e. the principle of conservation abides.

How is it possible for nuclear fusion, ubiquitous throughout the universe, to be induced by gravity, all without the expenditure of energy, and without violating principles of conservation? Our answer is obvious: The induction of nuclear fusion by gravitation requires no energy expenditure and no violation of conservation principles, simply because it is wholly a work function zero process. Insist otherwise and you *will* get the violation of the principle of conservation. **In short, materialist causality is false because it violates principles of conservation. An ironic conclusion.**

What about entropy? In gravity-driven nuclear fusion in stars, the total quantity of energy-matter does not change. Even so, the significant dissipation of that energy as heat, light, and the various forms of radiation into non-useful forms, culminates into the entropy of energy.

Now, in the absence of gravity, one would obtain a lower level of entropy from the same material: The material would mostly sit there, do nothing, and not dissipate at the usual rate. But, in the presence of work function zero gravitation, one obtains accelerated entropy. In other words, **the price paid by nature for utilising work function zero gravitational induction of change is *always* a much greater level of entropy than would be produced otherwise.** This fact also holds true in the hydro-electrical generation of power in dams. It is also implicitly true in the case of an object 'falling' under the influence of gravity, or in the orbit of the Moon about the Earth, or in orbital slingshot techniques used by various space agencies for deploying space probes.

In the latter case, every time we slingshot a probe using the gravitational acceleration of, say, Jupiter, Jupiter's orbit about the Sun is reduced to a calculable lower altitude about the Sun. This is often interpreted in energy-terms. Yet, both the generation of gravitation by the Sun and by Jupiter, and the gravitational acceleration afforded in the slingshot, proceed on a work function zero basis, and for reasons already clarified. The decay of the orbit of Jupiter per the slingshot effect is a case of entropy, not an argument for any energy-driven gravity process, much less for a process that involves the violation of principles of conservation.

*

We have already stated Dennett's generic view on the alleged 'impossibility' of perpetual motion. Dennett also purports the impossibility of the dualism of mind vis-à-vis the brain. We reiterate thus:

> Let us concentrate on the returned signals, the directives from mind to brain. These, *ex hypothesi,* are not physical; they are not light waves or sound waves or cosmic rays or streams of subatomic particles. No physical energy or mass is associated with them. How, then, do they get to make a difference to what happens in the brain cells they must affect, if the mind is to have any influence over the body?

One may as well pose the same question about gravitation: How does gravitation influence the orbit and acceleration of matter, initiate nuclear fusion, and realise effective and 'impossible' perpetual motion and *change*, even though no physical energy or mass is expended or involved in gravitational induction of state-change, or even in the formation of the gravity field itself?

Mind induces the activity of the brain, and changes brain-states from moment to a succeeding moment, by means of work function zero process, without violating conservation principles, as also found in nature. As such, abstract mind is wholly irreducible to the brain, in the same way that other abstractions, such as physical laws, remain wholly irreducible to matter, radiation and space.

Historically, **the key materialist argument against mind-brain dualism was based on the much-abused principles of conservation, married to the false notion that causality in nature can only come in the form of materially specifiable inputs manifested through spatial contiguity: all according to IOO closed causality and closed ontology.** Yet, even from a classical physics frame, **we find abstract principle-driven causality (physical laws) *and* gravitational processes that unfold on a work function zero basis, all *without* contiguity or contact... *without* the violation of principles of conservation. The same can be postulated about mind-to-brain causality, without contradiction of nature's 'quite standard physics'.**

0.34: The Future of Gravitation

Any hope that a future gravitation theory to replace General Relativity might be reworked in line with materialism, or in such a way as to save IOO schema closed causality and spatial impact causality, is totally forlorn. Ultimately, gravity and quantum gravity must involve the incorporation and primitive unification of classical and quantum relativistic descriptions of nature, obtained by Alt-R via the growing block intermediate model of spacetime (Book-II)... itself superseded by the de-spatialised temporised Bergson-Whitehead amalgam in Book-IV.

The key to future gravitation is de-spatialisation: So long as we assume the false notion of 'space', gravitation will be erroneously modelled, either as a contiguously operating 'force field' in space, or else as a contiguously generated 'curvature of spacetime'. However, with de-spatialisation, the notion of 'space' cannot be saved. Only *time* abides. Therefore, **the notion of 'matter**

63

in space' must give way to information organised in time, processed by means of a contact-less non-contiguous causality. Both obviate spatial 'force fields' as well as 'spacetime curves'. The spatial aspect of the 'curve' in spacetime will be superseded by spacetime *without* 'space' in the Bergson Whitehead amalgam.

By Book-IV, we will discover that gravity is a memory effect incorporated to a de-spatialised purely temporised physics: Therein, gravity will be posited as generated by the superpose of the abstract past (memory) retained within the Bergson-facet of the temporal Bergson-Whitehead amalgam.

That gravitation is a function of the past, a memory effect, is obvious from the influence of the Moon upon the Earth and, among other things, the consequent tidal bulge: Even assuming space, the Moon's pull upon the waters of the Earth generates the bulge and the associated tidal effects. Yet, again assuming space, the Moon's influence obtains from the Moon as it was 1.5 seconds ago in its past, given the spatial putative 'speed of light' transmission of its effect. That is, the Moon's effect upon the Earth is an effect of the Moon's past. Per de-spatialisation, we now remove space itself, but retain the effect of that past, consonant with the relativity of simultaneity: but now modelled as a de-spatialised pure temporised abstract memory effect. Thus, we arrive at gravitational influence as a non-contiguous resonation, purely across time, from the abstract past-state of the physical universe upon the relative present frame: Therein, the past as memory from the Moon resonates, structures and delimits the probability structure of the Earth's complex wavefunction, which then undergoes AND-to-OR collapse, biasing its probability structure to a form consistent with the facts of the Moon's past. The outcomes generated from this AND-to-OR collapse evinces as 'acceleration', as 'orbital motion', and as 'tidal effects' under gravity. But this influence is not a function of a single past state. It is a function of the superpose of all past states upon any relative present frame.

However, **the superpose of the sum of abstract memory, all without space, is obtained through a work function zero process**: That is, the superpose of the past *en sum* does not require an energy-power input to realise. **Hence, gravitation, as the superpose of the *en sum* abstract past *really is* a work function zero process, as is its ontologically primary AND-to-OR wavefunction collapse (time) through which said gravitational outcomes are realised. Therein, the principle of conservation is *not* violated and remains intact.** In other words, the overlooked reality of work function zero processes in generic gravitation, obvious even from a classical physics framework, will simply carry over into the presage to the new theory of gravity and quantum gravity, to be developed in Book-IV. Therefore, the future development of gravity theory is expected to be wholly inimical to materialist presuppositions.

0.35: Unauthorised History of Materialism from Within Natural Science

Let us recapitulate how causality and 'particle interaction' *actually* unfold. According to generic quantum mechanics, **given the wave-like nature of 'particles', when two particles 'interact', a non-directly observable superpose of their respective possibility wavefunctions takes place**. The waves involved constitute probability amplitude-sets for futures that the respective particles might get to assume. i.e., the 'interaction' is a combining of their abstract non-directly observable not-yet happened futures-sets. **Thus, even from generic quantum mechanics, 'particle interaction' does not involve a literal 'spaceballs' impact or contact. Instead, it involves a combine of abstract not-yet transpired futures and probabilities, forming a new superpose of futures and probabilities.**

As stated, this superpose is not directly observable, and for reasons given in Part-I (see **0.03** to **0.04**). **At some point, the joint superposition of abstractions undergoes AND-to-OR collapse. This process of time does not derive from hidden gears**, as we shall conclude in Book-III and have summarised in **0.20**: Thus, AND-to-OR collapse is *not* subject to any sort of input-operation-output (IOO) schema, given its gearless character. It is not brought about by an input of energy, power, or force, much less one mediated through contact causality across 'space', much less carried by 'particles'.

On the false presumption of 'space', we could give the involved superposes a spatial and wavelength-based description, as is done in generic quantum mechanics. The result would be no different. However, given anticipated de-spatialisation, we must furnish only a pure time-based or frequency-only model to particle superposes, both before their combination and after. Therein, with de-spatialisation and pure temporisation, there can be no 'spatial contact' between the particles. With de-spatialisation, contact-less non-contiguous interaction becomes obvious and tautological.

Of course, a pure frequency-based account of particle-interaction constitutes a purely temporal approach; an oracle of the rate of future potential probable events. Again, this pure temporal account of the process of 'particle interaction' obviates any recourse to impact or contact-causality in 'space' and relates the whole process purely to time. Such a requirement is obvious from de-spatialisation.

We must now reiterate that **the AND-to-OR wavefunction collapse (time) of the combined possibility-superpose of our 'particles' constitutes a work function zero process**. Now, pragmatic 'energy relations' and particle interactions are brought into apparent realisation in and through ontologically primary work function zero wavefunction collapse and time, and conservation principles inferred from the outcomes certainly abide. But **wavefunction collapse itself, and time itself, is *not* brought about by any energy or work-driven process or impact**. We shall discover that **wavefunction collapse and time are brought about by a gearless process: necessarily a work function zero process**.

This, then, is how *real causality* in the physical order proceeds. One should have no issue with the idea that a similar form of causality must attend gravity, if not mind-brain causality: Indeed, why should mind-brain causality be the exception to the de-spatialised, contact-less and gearless nature of nature, of time, of wavefunction collapse, and of general causality?

Notice that such conclusions are obvious even from generic quantum mechanics. Yet, pleonastic and extraneous materialist notions are unambiguously foisted on physics, as evinced in the confident claims of non-physicist Dennett and Edelman *et al*, if not by the physics community itself, strangely enough.

QUANTUM MECHANICS AND MIND

*

The decay of materialist notions about the nature of causality, and of materialism itself, was almost evident from the beginning of modern physics. Galilean physics from early to the mid seventeenth century may have accorded well with the notion of 'spaceballs' and contact causality. Yet, by the end of that century, **in Newton, we obtained the first instance of contact-less causality in gravitational 'action at a distance'. Less well known is the abstraction of 'force' from the Newtonian system of equations, achieved by Ernst Mach,** ironically an ardent materialist, who posited that 'force' in Newton was pleonastic, facile, or tautological, or simply pragmatic and not ontologically real[17]. But this insight fell to the wayside, given the conceptual and pragmatic convenience of 'force'. (A similar judgement could be made about energy or work in lieu of 'force': More on this will be said in Book-IV and V).

By the Nineteenth century, the advent of a field-based approach in physics, such as in early models of electromagnetism by Maxwell *et al*, did not necessarily imply the collapse of materialism. Yet, 'fields' (what are they?) surely hinted at ephemeralisation in physics, especially in matters of physical contiguity and contact. Indeed, **what exactly is engaging in 'contact' in a field, *especially* if the terms influenced by it themselves turn out to be nothing but fields**?

Such problems were rendered more salient by the discovery of the early wave theory of matter. Despite Newton's preference for corpuscular or particle-like matter, by the Nineteenth century, the famous Thomas Young's experiment of 1801, the prototype of the modern two-slit experiment, forced the reality of a continuous and wave-like version of matter and matter-interaction. Therein, the wave model envisioned matter as a mechanical wave, not as a quantum mechanical one. Yet, it created serious problems. How do you get mechanical waves to interact? **When two mechanical waves in, say, a pool of water approach and coincide, they converge and pass through each other with impunity. It is as if no interaction had taken place.** If matter is a wave, the larger waves seen on the surface of a pool of water could only be made up of smaller wavelets. But these will also pass through each other with impunity, without apparent interaction. So how do you obtain interaction or *contiguity* via waves?

It ought now to be obvious that causality was always a problem in physics and it *never* conformed to a simple materialist contiguous impact-causality or IOO-schema. This latter was always an *a priori* assumption foisted on science, *not* a fact of science. The ideas that culminate into the philosophy of Alt-R are rooted in the long-standing problems of causality that, if we should admit to their history, as we must, these can only expose the historic and ontological precarity of *a priori materialism* vis-à-vis science. This insight renders totally suspect the common adage that 'materialism is the best possible hypothesis of the physical world', or that physics and naturalism are the same as materialism: or that 'materialism is science… and science, materialism'.

*

One might be forgiven for thinking that modern physics in the Twentieth century had solved these problems, especially with the advent of quantum mechanics. **The 20th Century accomplished the replacement of putative mechanical waves with quantum mechanical probability waves**: Converging and superposed quantum waves must now AND-to-OR collapse into discrete events, which in turn collectively evince wave-particle duality. The discrete outcomes obtained give the false impression of a world made up of spatial 'particles': a belief made worse by particle physics: a specialism in need of comprehensive de-spatialisation and temporisation.

Particle physics must surely constitute proof that the world *is* made up of point-like states and spatial particles or blocks, and that 'spaceballs' collide and produce all the phenomena we know of, including the *illusion* of consciousness and mind. Depictions of particle interactions in Feynman diagrams and in old particle gas chamber trails certainly reinforces this belief… until we look closely and crash into infinities in particle-interactions... and *renormalisation*.

Infinities and renormalisation pre-existed quantum mechanics and quantum field theory: In the 19th century, **it was found that, for a spherical particle with mass, as its radius approached point-like zero, its inertia shot to infinity. But how do you get one particle to move another if the latter's inertia is set to infinity?** Such problems are artefacts of the presumption of matter as spatial; as 'spaceballs', or else as point-like... combined to the dubious notion that causality must be mediated by particles, by way of spatial contact and impact.

When such infinities emerged, these were subtracted from the equations through renormalisation. Renormalisation permitted calculations to work, but did not resolve, much less recognise that such artefacts derived from the false assumption of 'space'.

Similar problems re-emerged in quantum field theory in the 20th Century, involving infinities in particle-interaction. Again, by the 1950s, renormalisation solved the calculation problem that arose from infinities, but belied the very dubious idea of impact causality and attendant IOO-schemas. It did not foster the elimination of 'space' itself, or its recognition as a pleonastic conceptual artefact. **If physical causality truly proceeded via contact causality in 'space', in the face of the infinities inherent to such a vision, and even when the infinities are abstracted out of the equations and the calculations are made to work, if the objective process of particle interaction really entailed contiguity, such causality would crash into the said infinities; into insurmountable halting problems. Consequently, the outcome or output could never be realised, and AND-to-OR collapse, or time itself, would 'freeze up'.**

In any case, our belated de-spatialisation of physics, which should have emerged from the usurping of absolute time in Einstein's relativity and from the structural delimits of quantum mechanics against the direct observability of the 'motion' of quantum mechanical waves, combined to the impossibility of seamless-continuous physical inputs and observational outputs, with attendant broken discontinuous motion, information and time... should have rendered the whole notion of 'spatial particles' and 'contact-causality' bunk.

---

[17]Ernst Mach, Mechanics in their Development, Presented Historically and Critically, Akademie-Verlag, Berlin, 1988, section 2.7.

All of this is obvious even from facile information-matter dualism (see Part-I **0.01** to **0.04**) and from other findings previously adumbrated in Part-II and now in Part-III.

Hence, *there is no space*. **Therefore, there are no 'locations' at which purported particle-contact can take place, much less exhibit spatial point-like entities, or even spatial block 'spaceballs'**. Instead, particles, though real enough in their de-spatialised pure temporal and frequency-based forms (i.e. as temporal *durata*, and no longer *atomos*), interact in the way first described in this essay (above), with the added proviso that the AND-to-OR collapse-process that generates the outputs proceeds in the fashion of a 'quantum leap in time': Consequently, **nature 'leaps over' the infinities inherent to the interval that resides between the input and the output, obtaining the particle-output *without* impact, without contact, without 'interaction'... without the mediation of 'force', power or energy**. That is, **nature renormalises herself by forgoing 'space'... and by temporally 'leaping over' her infinities...** by unfolding possibilities into actualities **through a broken-discontinuous gearless de-spatialised time-process**.

The same sort of causality must be posited in mind-brain causality, leading to the revival of Cartesian dualism.

*

While the association and identification of materialism with naturalism, physicalism and science is dubious at best, there *is* an aspect to the physical sciences that renders it most susceptible to a materialist foisting. There is a tendency in human beings to the effect that, if one invents the hammer, then all things are declared nails. Things that do not fit the nail-schema are either declared nails in disguise or, if this will not work, dismissed as 'illusions'... 'not parts of real existence': hence, consciousness is declared an 'illusion'. The assumption that 'all is nails', or that there might be more to the universe than nails, will go unquestioned.

In a similar vein, **the moment humankind developed a mathematical physics and mechanics of cannonballs, all of ontology was declared 'spaceballs'** or 'matter', despite the undeniable reality of abstract immaterial physical laws. This co-joined with another facet of the same problem: **The triumph of mathematics in physics encouraged the claim that only quantifiable things are real**. This latter was institutionalised in John Locke's epistemology and ontology, only to be augmented and retrenched in David Hume's extension of the same to morals.

Galvanised by the mathematical physics of Isaac Newton, Locke asserted that 'primary qualities', composed of extension, quantity, and number, must be differentiated from secondary qualities, such as colour, texture, flavour etc. Locke declared the secondary qualities as 'not parts of real existence': supposedly imposed by the mind on a world ultimately devoid of these and ultimately comprised only of primary qualities. However, the mind itself was also asserted by Locke a product of brains, ultimately composed of primary qualities. In that case, it ought to be impossible for primary-state brains to produce non-real secondary qualities, much less impose these on sensory experience: a silly conundrum, and the original 'hard problem' that materialist philosophy of mind, in turn galvanised by its anti-Cartesian credentials, has failed to resolve.

However, empirically and historically, the differentiation of primary-quantifiable and secondary-qualitative categories, and the assertion that the latter are not parts of real existence, is an error. It is purely an *assumption* that is empirically unjustifiable, and historio-empirically falsifiable.

Historically and empirically, the quantities of physics in their inception were grasped either as countable qualities, such as the spectrum of countable colours evinced by the refraction of white light by a prism... or as quantitative relations inferred from qualitative differences, such as the relations between lines in absorption spectra from which one can infer the mathematical model of the electronic configuration of the hydrogen atom. Indeed, **empirically, the universe is comprised of either countable qualities or of qualitative differences that interrelate in quantifiable ways. That is, we *always* start from necessarily ontically real qualities and their differences and, from these, infer the embedded quantities upon which we then derive, deduct and induct a mathematical physics. It is *never* the case that we infer the qualities by starting from the quantities: This latter claim is simply historically, experientially and experimentally untenable and impossible.**

This insight also falsifies the notion often touted that... 'only frequencies are real, but the colours are subjectively imposed': In such claims, we clearly find the ghost of Locke. Yet, historically and empirically, we had to start with the colours before we could count and infer the quantities that then make up the mathematical physics of light and radiation, to which we rightly attach 'frequency' measures. These are real enough, but they are *embedded* and inferred states: Succinctly, these are embedded in and inferred from a qualitative milieu... implying that the universe is primarily qualitative, and its quantifiables are real but *embedded* features of qualities, and are fashioned from qualities. Otherwise, we could not have arrived at the quantities, or attain the physics.

Historically, it was never the case that we arrived at quantities first before we inferred the qualities. **Physics began with the experience of qualities, and only later extrapolated the countable and quantifiable relations between the qualities. Even if the pragmatics of physics later required that, once we inferred the quantities, we forget about the qualities and their empirical, historical and *ontological* primacy, their relegation thus is a convenient pragmatics, *not* a basis for ontologically false assertions for a purely quantitative universe**. That is, the pragmatic move cannot obviate the historical and empirical roots of mathematical physics as *embedded ontics* within a grander primarily qualitative universe: The move cannot obviate **the reality of a primary qualitative and superlative universe that embeds and supports secondary mathematical physics, wherein the former exceeds and supersedes the latter, and cannot therefore be reduced to a physics of quantities**.

To reiterate, historio-empirically, the universe and existence remain primarily qualitative, *not* quantitative. Quantifiables certainly exist: They *are* parts of real existence, albeit as *embedded* states. Indeed, quantifiables at the core of mathematical physics are real, but *only* as embedded and subsumed states of an ontologically primary *qualitative* universe.

66

QUANTUM MECHANICS AND MIND

In short, the real universe has no provision for the notion that 'quantities turn into qualities'. Yet, the pragmatics of forgetting the qualities, or relegating these to the background, easily led to **the erroneous assumption of the absence of qualities from reality, and then degenerated into an acid-stripped ontology permissive of dubious materialism and *its* unwarranted foisting on naturalism.**

*

It is also worth mentioning **the problem of spectacle** in how materialism came to be foisted on physics, despite its stated precarious relation: It cannot be sufficiently avowed that mathematics, physics and engineering have achieved great strides, while other human concerns, such as ethics, culture, and politics (the qualitative domain) remain in a permanent dark age. Of course, the said advancement was inevitable given that mathematics, physic and engineering contend with the most primitive aspects of existence, necessarily embedded in a qualitative and ultimately superlative universe that exceeds and supersedes mathematics, physics and engineering. But **the spectacular advancements secured by mathematics, physics and engineering have galvanised the rather silly view that all aspects of human existence ought to be reduced to mathematics, physics and engineering (i.e. the hammer and nails problem, again) making for a necronic mechano-nihilistic world-spanning culture**... made only worse by the assumption, explicit in Locke *et al*, and implicit in the local physics department, that only primary-quantitative states possess real existence. This, combined with the said spectacle, has render it exceedingly difficult to make advancements in qualitative domains: It has crippled humankind's ability to make any headway in ethics, politics, aesthetics, and especially in our understanding of mind and consciousness.

If we now add to this the presumption that all is 'spaceballs'; that only quantifiables exist; and that causality itself is a closed contiguous input-operation-output schema subsumed to a closed necronic mechano-nihilistic ontology, wherein agency is at best an 'illusion', combined to the mad and dismal promissory apotheosis of ever-elusive AI-gizmo... it is plain to see that humankind is in serious trouble.

The tendency in physics to confuse its pragmatics for claims on ontology and existence or, at the very least, the agnosticism and indifference of the physics department to qualitative and superlative realities, which it is in any case ultimately *not* qualified to make ontological judgments about (given that the qualitative and superlative universe exceeds and supersedes the mathematical physics it obviously embeds), makes it susceptible to ready exploitation; making for a necronic mechano-nihilistic daz, upon which materialism readily thrives as now-hegemonic parasitical anti-philosophy.

*

With the demise of materialist conception of causality in physics generally, and with due consideration to the ontological and cultural fallout from dismal materialism, mind-brain causality of the Cartesian form can no longer be dismissed out of hand. Indeed, Cartesian mind-brain dualism is the inexorable conclusion, given the general de-spatialisation and temporisation of nature and causality, the demise of impact causality, and the saliency of work function zero processes; especially in gear-less work function zero AND-to-OR collapse or time, which subsumes gravitational causality as it surely subsumes mind-brain causality itself.

0.36: Summary of Part-III

We can summarise Part-III thus:

- The notion that causality *must* require specifiable inputs of energy-matter was never wholly true even in classical physics. (See **0.32**)

- Materialists have only ever abused principles of conservation.

- **Work function zero processes are ubiquitous throughout nature and consistent with principles of conservation.** Thus, the conjecture for a work function zero-basis for mind-brain causality remains more than plausible: it is almost a certainty, given that we cannot resort to misconceived materialist notions of nature to argue against it. (See **0.33** and **0.34**)

- **An unauthorised history of materialism in natural science demonstrates that materialism was in decay almost from the very inception of natural science**, and always held a precarious position therein. **Materialism, and the supposition of 'space' with it, has caused problems and artefacts in physics that can only be overcome through de-spatialisation, amongst other things. Hence, the prospective demise of materialism** from within physics and naturalism, despite its unwarranted salience in all manner of fields, including brain research. (See **0.35**).

Materialism claims that 'everything is matter in space' or that all things arise from the interaction of matter in space... on the supposition of contiguity or contact causality between matter in space. In its succinct form, materialism posits an input-operation-output schema (IOO) for all of causality and ontology. The inputs, operations and outputs are, in principle, fully accountable and definable. Thus, in principle, all is fully explainable without resort to any extra variables (mind?) or extra ontology (God?). Hence causality is declared closed, and with it, ontology also closed... despite immaterial and time-independent physical laws... and despite all the problems that historically attended the assumption of materialism: (see the unauthorised history in.**0.35**).

It turns out that, per Alt-R, with de-spatialisation, the notion of contact or impact causality must totally break down. De-spatialisation takes us from 'matter in space' into distributions of memory and possibility in pure time, *without* space. There is no space. Therefore, there are no points of contact or impact for causality to proceed in the fashion espoused by materialist contiguity: the very assumption that led to infinities and renormalisation.

Again, per Alt-R, it also turns out that time is not a 'fourth dimension'. Instead, time is synonymous with partial-perpetual broken-discontinuous AND-to-OR wavefunction collapse, implicit to facile information-matter dualism uncovered in Part-I. Time as wavefunction collapse can be incorporated into a growing block intermediate model spacetime, later superseded by 'spacetime' *without* space. Moreover, the process of AND-to-OR collapse proves to be gear-less. Thus, gear-less time transits one state into a succeeding state, in and through which putative physical inputs are converted into outputs... without any gears.... without any requirement of a definable 'operation' term, or even 'contact'. Hence, the middle term in IOO schema causality unravels. Co-join this to a similar conclusion implied by contactless causality per de-spatialisation, and the case against materialism is complete.

Consequently, the notion of closed causality *and* ontology is also broken. Gear-less time and causality design for both open causality *and* supports the real existence of a grander ontology that supersedes and subsumes even de-spatialised memory, possibility, and time. Therein, abstract mind enters the fray, and can decide the fate of brain and world with near impunity, presumably on a work function zero basis, without the violation of principles of conservation.

In short, Alt-R culminates into revived Cartesian dualism, albeit without *res extensa*, or 'space'.

PRELIMINARY
PART-IV:
MATERIALISM AND THE EXPERIMENTAL INVESTIGATION OF THE BRAIN

0.37: Aims

Materialism constitutes the background to contemporary experimental investigations of the mind-brain relation. **Materialism asserts that the mind is nothing but the brain**: **There is no extant immaterial mind**. Confidence in this assertion does not arise from the scientific veracity of materialism, given the precarious relation of materialism vis-à-vis physics, adumbrated in Part-III and by the Alternative Realist thesis (Alt-R). **The dominance of materialism is a consequence of a runaway ideology**, unaware of the serious weaknesses of its assumptions about the nature of nature in the face of long-running problems that beset physics; precarity obvious, even from within classical physics (see Part-III **0.32** to **0.35**). Together with subsequent de-spatialisation and the breakdown of both impact causality and the demise of 'hidden gears' in wavefunction collapse or time, work function zero processes necessarily defy generic materialist notions of causality, even while principles of conservation much abused by materialists survive the de-materialisation of physics. Yet, an almost unconscious **confidence in materialism vis-à-vis physics has led to the unconditional confidence of neuroscience and brain-science in forlorn *a priori* materialism, and the foisting of materialism upon generic interpretations of experimental facts garnered about the brain**. Among other things, this involves a form of confirmation bias, no different from the fallacy of imputing God into every natural fact and then interpreting or presenting such facts as *if in* proof of God.

The dominance of forlorn materialism in the philosophy of mind, in neuroscience and brain science generally, entails the following fallacy-chain, culminating into what is essentially a post hoc fallacy: one that unconsciously twists facts garnered in experiment in support and supposition of materialism:

- **The experimentalists assume the scientific veracity of materialism as 'obvious', or as a 'given from physics'**: One could call this 'the physics fallacy': We reiterate the many converging challenges from old, new, existing, and anticipated physical discoveries (see Part-III and **0.35** for the summary). These render materialism wholly untenable. Yet, galvanised by 'physics', the presumed veracity of materialism purports that mind and memory is nothing but the brain. *Eliminative materialism*, e*piphenomenalism*, and the newer *emergentism*, which precede and then condition the views of brain and neuroscience, allege the primacy of brain structure and activity over 'produced' and otherwise causally neutered 'illusionary' consciousness. Epiphenomenalism and emergentism constitute default forms of eliminative materialism: i.e., these deny the causal mind vis-à-vis the brain... supposedly in order to save the conservation principle. (See Part-III **0.33** for why this is a dubious contention).

- **Alternative non-materialist views are automatically dismissed, or not even thought of on the basis of point-1**. When challenged, predictably, the principle of conservation is invoked and abused in the manner described in Part-III: For example, the reality of work function zero processes is completely ignored or not know of. Alternatives such as dualism are dismissed as 'religion inspired'... 'superstitious'... 'metaphysical', etc.

- **The assumption of materialism is loaded into the facts experimentally garnered about the brain**. This loading is on the basis of step-1 and step-2. In short, materialism is 'true': The mind is the brain: Any facts uncovered *must* be framed in these terms... or else, 'religion!'.

- **The facts uncovered by experiment are then presented *as if* in 'proof' of the loaded materialist assumptions**, based on the *a priori* materialist interpretation of the facts. (This is the post hoc fallacy in its saliency: A occurred, then B occurred. Therefore, A caused B: e.g., Brain activity occurred. Then consciousness occurred. Therefore, brain activity caused consciousness). For example, Amnesia is oft posited as 'proof' that memory is stored in brain, 'obviously'. That the brain might be a means of expression of memories as opposed to a store of memories, in the same way that a computer-keyboard is a means for expressing outputs and not a storage-milieu for outputs, is not considered. Why should it be? Materialism is 'true'. Therefore, memories are stored in the brain. Hence, when damage to the brain leads to loss in the expression of memories, this is then presented *as if* 'proof' that memories are stored in the brain. Another example: The outcomes of the experiments of Benjamin Libet are oft presented as 'proof' that volitional consciousness is an illusion. This is on the core assumption that the brain activity generates the consciousness. The alternative possibility, that brain-activity might follow *after* causal intercession by independent extant minding, in the same sense that computer keyboard activity follows on *after* the decisions made by an extant end-user, and that mind might not at all coincide in either time or 'space' with brain activity, is not proposed, even as hypothesis: Why should it be? Materialism is 'true'. Therefore, the consciousness *must* be produced by brain activity.

In this concluding part to our Preliminary, our aim is to break the above fallacy-chain. By attacking materialism based on old, contemporary and anticipated science, **we have already usurped step-1 in the chain:** The presumption of *a priori* materialism, or the dubious notion that materialism is justified by physics, is rendered totally suspect, to say the least. We must also usurp fallacy step-2 and step-3 of the fallacy-chain. **Against step-2, we will pose an alternative dualist hypothesis for amnesia and other claims, based on implicit default crucial experiments**: We will show that dualism is equally consistent with the facts experimentally garnered, if not superior to materialism in saving those facts.

*

It is not a matter of passing interest that the default crucial experiments scrutinised throughout Part-IV were never carried out with dualism in mind: They were designed with the implicit or explicit assumption of materialism: all the more interesting, given that the experiments invariably point to dualistic conclusions, counter-materialism.

The said experiments are divided into two sorts: **Memory-side experiments** entail the investigation of memory vis-à-vis the brain: These typically investigate procedural memories. The classical experiments of Karl Lashley and the later experiments of Rose and Harding *et al*, are examples. But there is an overlap between memory and minding. Thus, memory-side experiments invariably shed light on the mind-side aspect of the conundrum.

Mind-side experiments grapple directly with consciousness vis-à-vis brain activity. The experiments of Benjamin Libet comprise a beginning-point and prime example of mind-side experimentation. Therein, materialism, and the notion that the brain produces consciousness, was assumed as a given. Libet hypothesised a 'mind field' approach to address the problems wrought by his *a priori* materialism.

When we apply the counter-materialist assumption to mind-side experiments, we will find consistency with the facts superior to those obtained from the assumption of materialism. And when we reframe such experiments in terms of the logic of the crucial experiment, in which contesting assumptions are married in such a way that the experiment tends to eliminate one assumption and substantiate the other, we obtain consequences that clearly favour dualism.

Of course, dualism is the foregone conclusion forced upon us by Alt-R itself, most explicitly by de-spatialisation: There is no space. Therefore, there is no spatial location for memory or for mind in the brain, much less any spatial locus or corpus that retains or retrieves memory through contiguous causal means, or one that could produce mind and consciousness itself. De-spatialisation implies the pure temporisation of the whole physical order, including the brain itself... including memory-brain and mind-brain relations. Hence, memory and mind must be reframed in terms of pure delayed choice time relations between mind, memory, brains, and the world: relations that must unfold through the process of gear-less work function zero AND-to-OR collapse and time, whose work function zero and gear-less character independently usurp the materialist false contention on causality and on memory-brain and mind-brain relations, inexorably culminating into Cartesian revival.

MEMORY-SIDE INVESTIGATIONS OF THE MIND-BRAIN

0.38: The Experiments of Karl Lashley

In a series of experiments involving rats, chimps, and monkeys, carried out over a period of some thirty years ending in the 1960s, Karl Lashley sought to locate memory 'engrams' (succinctly, procedural memories) in the brain[18]. Thus...

> The original programme of research looked toward the tracing of conditioned-reflex arcs through the cortex, as the spinal paths of simple reflexes seems to have been traced through the cord.

The experiments ought to have established the veracity of the materialist postulate on memory. However, according to Lashley himself...

> ...The experimental findings have never fitted into such scheme. Rather.... [*the experiments*]...have emphasised the unitary character of every habit, the impossibility of stating any learning as concatenations of reflexes, and the participation of large masses of nervous tissue in the functions rather than the development of restricted conduction paths.

---

[18]Lashley, K.S. 1929 Brain Mechanism and Intelligence. Chicago: Chicago University press...1950. In search of the engrams. Symposium of the Society for Experimental Biology 4:454-483

Thus, Lashley experiments never identified memories with 'restricted conduction paths' or reflexes. Thus, **procedural memories could not be identified with any specific or localisable material structure in brains**: precisely the thing one would expect of spatially localised hands-on in-brain material memory.

We shall soon discover that, once procedural memories are acquired, the ability to re-enact these will return, even when the putative structures are surgically destroyed: **How is it possible for surgically destroyed memory to return when, as the dogma would have it, the materially embodied memory has been irretrievably lost per the said destruction?**

In response to the perplexing results of his experiments, Lashley resorted to an approach entirely unsurprising. **To save the assumption of material memories, Lashley subscribed to a view later developed into the holonomic postulate on memory**: i.e., materially delocalised or 'clouded' memory in brains, as opposed to materially localised engrams.

In Lashley's experiments, prior to surgery, test animals were subject to some sort of specific rote training: e.g., puzzle solving. With procedural memory 'engrams' presumably formed up as nerve tracts in brains, the surgical destruction of these would follow. Post surgery, the animals were presented with the puzzle for which they had been trained. The initial expectation was that, per the said destruction, the ability of the animals to enact procedural memories *and* carry out the tasks ought to have been permanently lost.

In one set of experiments involving rats, the animals were trained to respond to lights in certain ways. Surgery was then carried out to remove the relevant nervous tracts. In some versions of the same experiment, the entirety of the motor cortex was removed. However: in both versions, upon recovery from surgery, the rats showed no fall or reduction in performance.

The fact of the recovery of 'destroyed' procedural memories must imply that the memories were not ultimately fated to the destroyed structures, let alone to the entirety of the motor cortex. The pertinent memories must have been retained in some fashion unrelated to the surgically targeted material of the brain or of the motor cortex.

In another set of experiments, monkeys were trained to open or close latch boxes relative to certain stimuli. Much of their motor cortex was removed. Paralysis followed immediately, with an eight-to-twelve-week recovery. With the reintroduction of the puzzle boxes, the monkeys were able to perform as well as they had done before surgery: That is, despite the irreversible destruction of putative procedural memory structures, procedural memory and ability inexplicably returned.

Certain versions of experiments even entailed deep incisions into the cerebral cortex and its cross connections. Other versions entailed the removal of the cerebellum. Yet, in all cases, procedural ability, hence implied memory, returned to the test animals.

Lashley sought to explain these puzzling findings by conjecturing for redundant reduplication of memories: the precursor to the later materialistic holonomic conjectures of Karl Pribram. Thus…

…The characteristics of the nervous network are such that when it is subject to any pattern of excitation, it may develop a pattern of activity, reduplicated through the entire functional area by spread of excitation, such as the surface of a liquid develops an interference pattern of spreading waves when it is disturbed at several points.

Implicit to Lashley's account is the belief that memory *must* somehow be located and reduce to the material corpus of the brain. This obviously forgoes the need for any 'superstitious' solutions, such as required by dualism. Faced with the failure to account for memory as specifiable localisable material structures and pathways in the brain, circumvention of unpalatable alternatives must come into operation, as demanded by ideology. Hence, holonomic memory theory.

0.39: The Holonomic View of Memory & Experimental Investigation

Subsequent experimental investigations into the relationship between memory and brains has undermined the 'filing cabinet' approach to memory; wherein memories are characterised as stored under specific localisable addresses in the brain; equivalent to the same as found in computers; with assumed equivalence and interchangeability between computers and brains.

Experiments beginning with those of Karl Lashley have shown that memories vis-à-vis the brain are not deposited in localised form at spatial 'addresses', much less in stable physical locations, as the 'filing cabinet' approach demands. The response led to the holonomic paradigm; a derivative of Karl Lashley's own approach. Thus, according to Boycott,…'Memory is both everywhere and nowhere in particular'.[19]

By extending the original conception of delocalised memory originally hypothesised by Karl Lashley, others, such as **Karl Pribram and Wilber, have developed the sophisticated, albeit materialistic, holonomic theories of memory[20].** This has the advantage conveyed by all holonomes and 'holograms': If one destroys, say, ninety-nine percent of a holographic picture, it is possible to reconstruct the whole image from the residual one-percent part. (See **Fig. P1.12** for a basic illustration of the hologram). On the same logic, materialistic memory conceived in the holographic-holonomic form, as memory 'clouded' across the whole of the brain or across pertinent functional areas, would have the advantage that, **if we could surgically destroy ninety-nine percent of that functional area, we ought to be able to recover the 'lost' memories from the remaining one-percent residue.** In other words, the ability of holograms to preserve information in the way just described appeared consistent with Karl Lashley's own experimental discovery: namely, the

---

[19]1.Boycott, 1965, p48 (Boycott, B.B. 1965. Learning in the Octopus. Scientific American 212(3):42-50

[20]Pribram, K.H. 1971. Languages of the Brain. Englewood Cliffs: Prentice Hall…Wilber, K., ed. 1982. The Holographic Paradigm and Other Paradoxes. Boulder: Shambala

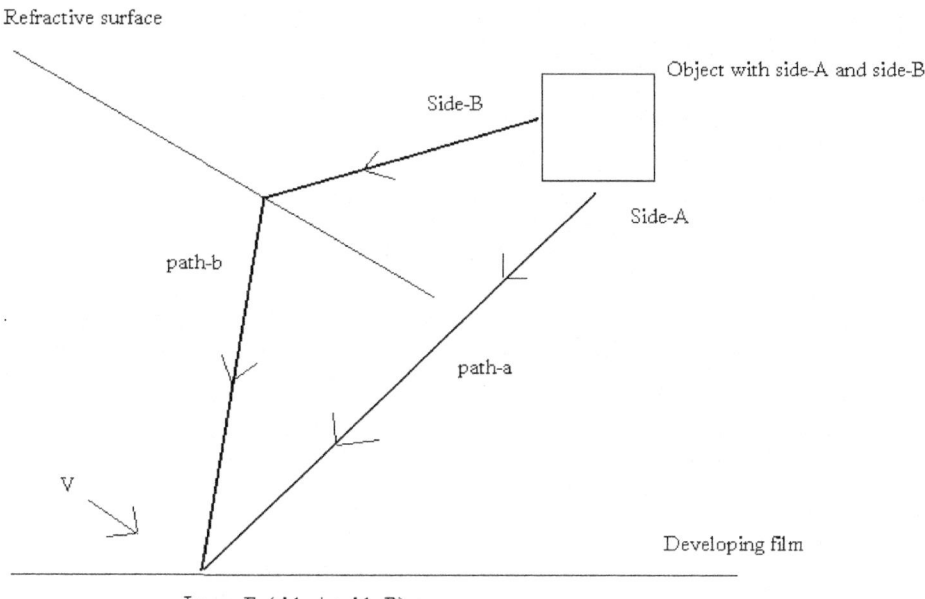

Image-E:{side-A+side-B}

Fig. P1.12: Elementary illustration of the hologram
Notes: The hologram depicted is of an object with sides **A** and **B**. An image from **side-A** follows **path-a** to **site-E** on the developing film. The image from **side-B** follows **path-b**, which is refracted by a transparent surface to arrive at site **E**. The image of both **side-A** and **side-B** are therefore developed at **E**. To an observer at **V** the image manifested on the film at **E** will be a three-dimensional image, exhibiting both side-A and B. The same three-dimensional image is being formed across the whole developing film: Hence, at successive points to the right of **E**, we will obtain combinations of **side-A** and **side-B**, formed by different paths and at different times. Now, if we destroy almost the whole photographic image, except **site-E** it will be possible to recover the three-dimensional image of our object from just **E**, or from any other fragment-site along the film. It has been hypothesised that memory in brains might be stored in just the same way, albeit in material form. As with any hologram, if we destroy most of the brain, but leave a fragment residue, it ought to be possible to recover the whole of memory from the surviving fragment, so explaining the anomalies uncovered by Lashley *et al*. However, with anticipatory de-spatialisation and temporisation, the refracted path 'distances' that converge to constitute the hologram at **E**, and at all successive points thereafter, are replaced by de-spatialised purely temporal delayed choice time-relations that incorporate the complex past as a resonation and complex superposition of all pertinent past states belonging to the original 'object', all resonating across time, *without* spatial contiguity or impact, focused upon the relative present moment that constitute the developing film at **E**. This obviates the need for material memories distributed in 'space', or in a spatial brain. Thus, we espouse a de-spatialised, immaterial, purely temporal holonome of memory that includes, but ultimately exceeds, the brain.

gross resilience to surgical destruction of procedural memories, and the tenacious ability of memory to recover, despite targeted gross surgical destruction.

With the holonomic paradigm of Karl Pribram, the materialist view on memory appeared saved. Yet, this move constitutes a final gasp of the materialistic memory theory.

We have no objection to a dualistic version of holonomic memory. The very possibility for this is tacitly supported by Alt-R; especially by de-spatialisation and temporisation: wherein all corpuses, including the brain, must be re-framed, *not* as 'matter in space', but as abstract de-spatialised information distributed in pure time. Therein, the recollection of memory cannot help but constitute a process realised through non-contiguous or contact-less resonation in and across pure time: Therein, obviously, memory could *not* be 'stored in spatial addresses' in the brain. **A dualistic temporal holonomic theory makes for memory 'clouded',** *not* **across space, but across time: a temporal holonomic memory theory: an immaterialistic Cartesian dualistic understanding of memory**.

It so happens that there are experiments that control for *and* against the materialistic version of the holonomic explanation of memories, even if the outcome from such experiments is obviously unintended and arrived at by default: Such experiments inadvertently falsify the materialistic version of the holonomic view *and* verify the dualistic view on memory.

0.40: Default Crucial Experiments on Memory & the Materialist Paradigm

The experiments of Karl Lashley lacked precision. Not so in the later experiments of Rose (1981; 1984); Rose and Harding (1984); Rose and Csillag (1985) and Horn (1986). These entailed the injection of short-lived radioactive substances into the bloodstream of the test animals. Localised radioactive sites formed up in brains: i.e., the location of putative memories, revealed by *positron emission tomography*: otherwise known as the **glucose pump method**.

Peter B Lloyd describes the method thus:

> …radioactive glucose is injected into the bloodstream. Since glucose serves as fuel for the cells in the human body, it is absorbed more by brain cells that are currently active than by idle ones. Minute particles are emitted by the radioactive glucose and pass straight through the body. They are detected by sensitive electronic devices, whose readings are automatically fed into a digital computer…. the computer works out where in three-dimensional space the radioactive sugar is being used most rapidly. Those are the sites where brain cells are busy…[21]

The acquisition of procedural ability must surely entail the formation of structures *in* the brain: with putative material memory states evinced as new cellular connections, growths, and structural developments. These must incorporate the very radioactive glucose injected per the glucose pump method. **The newly formed radioactive sites in brains, the presumed sites for new procedural memories associated with the new skills acquired, will then be visible to appropriate detection procedures; permitting their precise location and surgical removal** at much greater precision than was possible in Lashley's experiments.

*

It so happens that **experiments that entail precise surgical removal of presumed memories via the radioactive glucose pump method control for and simultaneously test for the materialistic holonomic memory theory, as much as test for localised engrams:** Memories deposited in holonomic form in brains ought to be deposited, if not across the whole of the brain, then at least across a pertinent functional area. The corollary radioactive glucose ought then to cloud *throughout* the brain or throughout the pertinent functional area. But experimental results proved the contrary: **There is *no* evidence of generalised spread and incorporation of radioactive glucose; not even across functional areas. Instead, the radioactivity is of the highly localised sort: consistent with the classical memory engrams of Lashley: an invalidation of materialistic holonomic approach**.

The experimental designs of Rose, Harding, *et al* integrate the two opposing materialist assumptions about memories, even if inadvertently or by default. Their experiments unintentionally test *both* contentions against each other: falsifying the holonomic assumption by obtaining the absence of the said radioactive clouding, albeit with *apparent* substantiation of the localised engram memory.

But what happens when the localised structures are surgically destroyed, courtesy of the precision from the glucose pump method?

In the experiments of Rose and Harding, just one day after hatching, the test chicks were subject to simple training procedures, accompanied by the injection of radioactive glucose into their blood streams per the glucose pump method. Some of this radioactivity was incorporated into the newly formed structures in the forebrain of the chicks; in areas of their brains associated with learning of the pertinent procedures. i.e., the left hemisphere. These radioactive locales, presumably the material deposits of memory associated with the training, were then surgically destroyed.

Following surgery, the same chicks were exposed to the stimuli associated with their previous training. The chicks performed normally, indicating that **the putative localised memories that were surgically destroyed had somehow returned**. Thus, even with precision-targeted surgical destruction, the memories were not lost. **Memory proved to be *not* ultimately dependent on the destroyed localised material structures identified by the glucose pump method**.

What do such results imply? On the one hand the targeting of ostensive memories via radioactive materials appears to favour localised memory, given that we do not obtain any evidence for clouded radioactivity or clouded memory across pertinent functional areas: expected per the materialistic holonomic view. Indeed, the experiments of Rose and Harding do not entail the surgical removal of whole functional areas, which is what one would expect if radioactivity and correlate memory were clouded throughout pertinent functional areas.

On the other hand, with de-localised clouded memory now falsified, the surgical destruction of the apparent localised memory structures ought to have led to the permanent and irretrievable loss of memory, but did not do so.

In short, we have two results from the inadvertent default crucial experiments of Rose and Harding, which implicitly incorporate the two contrasting materialistic assumptions about memory into the very structure and process of the experimentation:

- **The first outcome falsifies the holographic paradigm on memor**y, given that the pertinent radioactivity is *neither* found clouded across either a pertinent functional area, much less throughout the brain.

- **The second tandem outcome falsifies localised putative memory engrams**: The surgical destruction of ostensive localised memory-sites do not ultimately prevent the return of memory, clearly implying that memory is not fated to, much less materially identifiable with, the destroyed localised structures.

---

[21]Is the Mind Physical? Dissecting Conscious Brain Tissue; an article published in Philosophy Now, No.6, 1993

The conclusion is obvious: **The retention of memory in *any* fashion, localised or clouded, in and as the material of brains, is not tenable**. Upon the factual-experimental rejection of *both* the materialist holonomic paradigm *and* classical materialist localised engram memory, **the results favour a form of memory *not* retained in brains**: memory ultimately retained at a remove from brains. In other words, the default crucial experiment favour **the irreducibility and dualism of memories vis-à-vis the brain**: or **memory-brain dualism**.

*

How is this possible? Independent of the experimental outcomes described, the background assumption of materialism, loaded to the facts uncovered in such experiments, is an error, as can be grasped from fundamental Alt-R considerations intimated from Part-I to III: We need a post-materialist approach to memory-brain conundrums, of a form encapsulated per the key assertions of Alt-R. The most prominent Alt-R assertion directly pertinent to the memory-side investigation of the brain is...

- **De-spatialisation: There is no space. Therefore, there are no 'locations' at which anything, including memories, could be stored**. Memories are retained extant the brain because these are distributed purely across time, and relate to the brain and mind per pure non-contiguous time-relations: De-spatialisation and attendant temporisation forces this conclusion inexorably. **The putative surgical destruction of sites in the brain in the relative 'now' cannot imply the destruction of memories extant that brain, which are at a temporal remove from that brain**, retained within abstract 'realised spacetime'… retained within the growing block Bergson-facet of the Bergson-Whitehead amalgam, as we shall discover by Book-IV.

0.41: Candidates for the Materialistic Conception of Memory

With our dual falsification of both localised *and* clouded contentions on memory, combined to implications from de-spatialisation, materialist views on memory are rendered precarious, to say the least. Even so, by compiling the list of long-abandoned or otherwise forlorn candidates for memory, and by rejecting these explicitly, we can only augment our own case. To this end, we must first look at the long-abandoned materialist candidate for memory proposed by James C. McConnell: **memory ribonucleic acid, or m-RNA**[22].

Therein mRNA would need to be synthesised in association with pertinent memory-forming stimuli. However, the synthesis of memory-pertinent m-RNA could not avoid 'spatial' correlation with radioactive glucose from the glucose pump: When the glucose pump method is used, radioactive clusters would be expected to spatially associate with on-site metabolic processes of m-RNA formation. If the holonomic theory applied, we would expect radioactive glucose associated with formed mRNA to cloud throughout the relevant functional area. Such a thing is *not* observed in brains or across relevant functional areas. Only localised and spatially restricted areas of radioactivity form up. Yet, their surgical destruction do not prevent the recollection of the 'destroyed' memories.

What about **patterns of synaptic connections** and the formation of such as the material-structural candidates for in-brain memories? The holonomic approach to synaptic connection-patterns would require the reduplication of same-pattern synaptic connections throughout a functional area, or even throughout the brain as a whole: (See **Fig. P1.13** for this and various other in-brain memory candidate structures). This variant of memory suffers the same fate as defunct mRNA: The injected radioactive glucose from the glucose pump would be expected to spatially cloud throughout the relevant functional area of the brain. But this is not observed in the experiments of Rose, Harding *et al*, although localised structures of radioactive glucose *are* observed. These could be supposed to entail the new synaptic connections and new memories, and their surgical destruction ought to lead to the permanent loss of memory. Yet, this does not happen. Hence, synaptic connections and patterns, real enough and critical to mind-brain operations, and certainly correlated with memories, do not constitute the final repositories of real memories.

Let us assume that **microtubular structures within cytoskeletal strands** constitute memory candidates. Microtubular structures are found within dendritic 'wiring' that physically link neurons together. They also terminate within bouton structures which contain synaptic vesicles: (see **Fig. P1.13**). Roger Penrose *et al* have proposed the possibility that quantum coherent phenomena, possibly related to minding, might occur within such structures[23]. Consider that single-cell organisms (i.e., the protozoa) appear to engage in complex self-guiding behaviours, even though these do not possess any sort of neural nets to realise such behaviours. However, they *do* possess the primitive equivalent of a sub-cellular neural network: one comprised of cytoskeletal and microtubular organelles. Hence, there is a complicity of such structures to the said complex behaviours, and with putative mind at the most rudimentary pre-conscious level[24].

Perhaps just such cytoskeletal microtubular complexes constitute the candidates for material memories. If so, any newly acquired memories would be expected to correlate with growth and development of just such structures, and with accretions of radioactive glucose from the glucose pump that attend their formations. The surgical destruction of these, and the subsequent recollection of memories despite their destruction, invalidates the implicit postulate that cytoskeletal microtubular structures constitute the material form of memories. Again, the radioactivity does not cloud throughout a functional area that might subsume such putative cytoskeletal subcellular memories.

---

[22]J.V. McConnell, (1962) Memory transfer through cannibalism in planarium, J. Neuropsychiat. 3 suppl 1 542-548

[23]Shadows of the Mind. Roger Penrose, Vintage 1995 p367

[24]Ultimate Computing: Biomolecular Consciousness and Nano Technology, Stuart Hameroff, North-Holland, Amsterdam 1987

On the other hand, in the framework of the delocalised holonomic version of cytoskeletal microtubular memories, the memory-specific cytoskeletal growth would need to be reduplicated and clouded to many neurons and cellular bodies throughout pertinent functional areas, with radioactive glucose clouded across the whole. Again, no such effect is observed. Only localised structures are formed. These latter may well entail a cytoskeletal microtubular basis and may well play an associative role with real memory. Yet, their surgical destruction does not lead to the erasure of memories. Therefore, cytoskeletal microtubular structures, though possibly involved in memory, cannot be the ultimate repositories of memory.

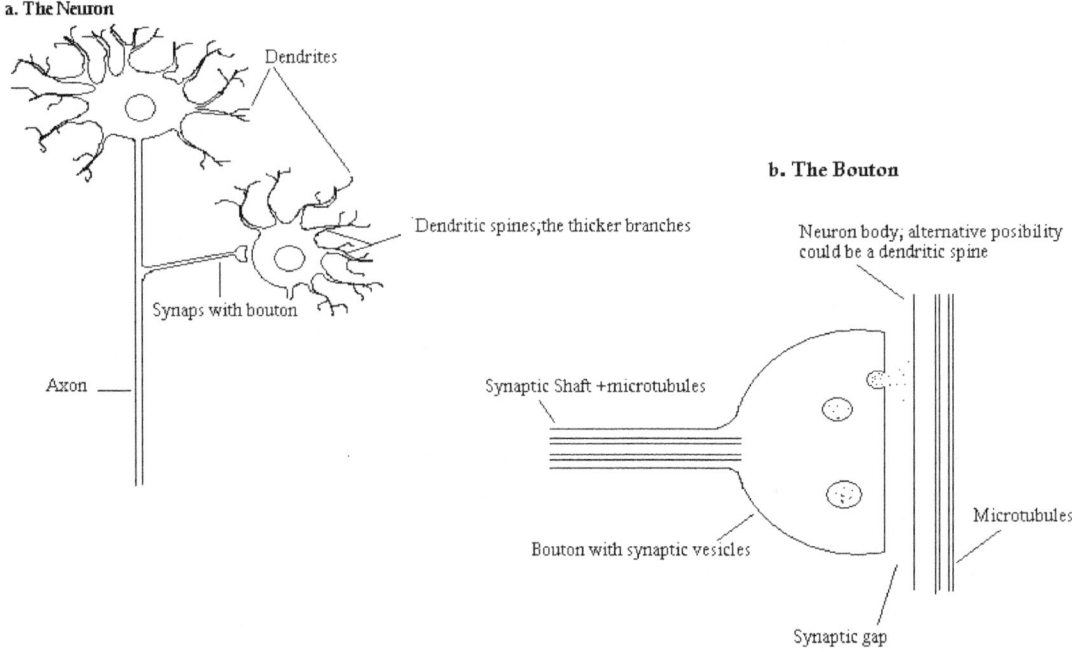

Fig. P1.13 Candidate memory structures
Notes: The illustration is divided into two: illustration **(a)** shows a larger conglomerate of neurons connected through axons and synapses. Illustration **(b)** zooms in to the synaptic shaft with boutons and synaptic vesicles. Those who insist on some sort of reduction and identity of memories to material structures in brains must choose some or all the above as memory-candidates.

*

According to the late John Eccles, memory is not reducible to brains at all: It manifests in the form of what he called *psychons*, thought to be ultimately associated with the fine structures and processes that *are* located within the brain: i.e., the *dendrons*. The most critical structure pertaining to dendrons are the **complexes of vesicular formations located within synaptic boutons** (see **Fig. P1.13** for clarification). According to Eccles, through the psychons, mind associates with and induces, but does not itself reduce to, the complex of vesicular structures; triggering and controlling neurotransmitter discharges into the pre-synaptic vesicular grid (again, see **Fig P1.13** for clarification). In his book, *How the Self Controls Its Brain*, minding and psychon-related quantum mechanical processes are postulated to interact at the pre-synaptic vesicular grid: The sensitivity and possibility of vesicular discharge is hypothesised as subject to control by the psychons and the dualistic interaction mind-brain process behind these: The dualistic interaction supposedly adjusts the quantum probability of vesicular discharge per the intents of mind.

The claims of Eccles as they relate to pre-synaptic vesicles correlate with findings from research: According to Rose[25], as a supplementary to experiments associated with memory acquisition in chicks, and according to fine-structure investigation, the acquisition of memory turns out to be quantitatively correlated with detectable changes in the number of vesicle formations within boutons.

Assuming a pure a materialist turn on the claims of Eccles, let us correlate memories with the pre-synaptic vesicles themselves: their formation and their patterns of distribution and discharge. Therein, radioactive glucose from the glucose pump ought to spatially correlate with the formation of newly formed pre-synaptic vesicles and grids, which are expected to correlate with newly formed

---

[25]Rose, 1986

memories. If these are being surgically destroyed in the sorts of experiments carried out by Rose and Harding *et al*, the memories inexplicably return. Therefore, these cannot be the ultimate repositories of memory. As for the holonomic alternative that entails the clouding of synaptic same-vesicular grid-patterns throughout pertinent functional areas, this is *not* observed. Thus, memories cannot ultimately be identified with synaptic vesicles and patterns.

*

In his intriguing development of the holonomic paradigm of memory, Karl Pribram focused on visual information and memory: Therein, visual memory is thought to be vested in distributed or clouded form over large numbers of dendritic bodies associated with neurons and with dendritic processes associated with visual inputs. This idea was explored in Pribram's book, *Brain and Perception: Holonomy and Structure in Figural Processing.* (For clarification of dendrites, see **Fig. P1.13**).

Karl Pribram's dendritic-based postulate appears to tie closely with the previously discussed cytoskeletal microtubular candidate for memory. The later structures are found formed and embedded within the dendrites: Therein, visual memory would be expected to entail molecular-cellular activity and growth of dendritic structures and embedded potential memory candidates.

If dendritic structures and processes *are* the candidates for the materialistic version of holonomic visual memory, and if such structures are being reduplicated across relevant functional areas, then, in the pertinent experiments, these ought also to incorporate the radioactive glucose from the glucose pump, thus manifest radioactivity clouded across the whole of the pertinent functional area.

Alternatively, if localised structures were rendered evident using the glucose pump, these ought to be the real sites of memories in the brain, falsifying the *materialistic* variant of holonomic theory of memory (...but *not* the de-spatialised pure temporal version of the same).

However, no glucose pump method and test of the Pribram's holonomic hypothesis for visual memory has been carried out; at least, not directly. Yet, as we have already discovered, pertinent experiments *have* been carried out inadvertently by Rose, Harding, and others: The training of chicks must surely entail visual information as critical components of acquired abilities and memories. Thus, the retention of the pertinent visual information must also have occurred in the brains of the chicks. If the materialist holonomic paradigm were true, delocalised or generalised radioactivity in pertinent functional areas in chick brains, associated with visual processing and visual memory, should have been observed. No such delocalised spread of radioactivity was observed. Therefore, it is unlikely that any future explicit experimental investigation of the visual cortex using the glucose pump method will yield different result.

*

The reader should note that **there is no reason why we cannot develop an alternative dualistic holonomic theory of memory**: Applying de-spatialisation and using the anticipated Bergson-Whitehead amalgam successor to 'spacetime' in Book-IV, we will discover that memories ultimately prove to be abstract states distributed in time; retained at a physical remove from the present continuous brain within the mnemonic Bergson facet of the growing block Bergson-Whitehead amalgam. This is true of information and memory across the whole of nature: Memory vis-à-vis the brain is not going to be exempt from this general truth. Once we appreciate a de-spatialised temporised form of memory as information distributed in and through time, we automatically arrive at a temporal holonome of information and memory... *without* 'space', and *without* 'locations': We arrive at fully de-spatialised, temporised, and dematerialised memory theory.

Indeed, we need only reframe the hologram depicted in **Fig. P1.12** in terms of delayed choice time-interval relations demanded per de-spatialisation; wherein the 'distances' covered by the paths of refracted light that converge to constitute the hologram of the original 'object' are replaced by de-spatialised pure delayed choice time-relations, which get to incorporate a complex distribution of the past into a relative present moment. Therein, forlorn 'material' correlates of memory need no longer be posited as the memories, but only as states of the relative present that resonate with the complex abstract past (i.e., the real memory), itself structured and organised into a purely de-spatialised temporised form.

Note that our de-spatialised purely temporal holonomic contention on memory is inexorably dualistic memory: It designs for revived Cartesian dualism.

AMNESIA & MATERIALISM

0.42: Amnesia: Implications to the Nature of Memory

If memories are not in the brain, how is one to account for amnesia, especially permanent amnesia? Surely, permanent amnesia indicates a strong relation between memory and the structures of the brain: It ought to validate the claim for an outright identity of memory with brain structures... *for* materialism and *against* dualism. It turns out that amnesia is as useless to the materialist contention on memory as it is to the dualist approach: **Amnesia is consistent with *both* the materialist *and* dualist point of view on memory and brain relations**. However, it also follows that the facts of amnesia cannot be contrived to support an exclusive materialist position on memory.

Insofar as materialism *assumes* identification and reduction of memories to specific brain structures, the disruption and disorganisation of these at a given threshold must surely lead to the loss of the ability to recollect memories. Hence, permanent amnesia. Indeed, the assumption of materialism appears to save the evidence from amnesia.

However, the facts of amnesia are equally consistent with memory-brain dualism. This consistency does not constitute proof of dualism. But a dualist account of amnesia, if more than plausible, can break the exclusive hold of materialism in amnesia, exposing it

75

as an assumption loaded into the facts of amnesia, but *not* a fact of amnesia. To this end let us, assume the reality of space, and use John Eccles's scheme from our previous analysis, wherein non-material abstract *psychons*, comprising real memories, are grasped as dualistic and irreducible with respect to in-brain material *dendrons*. Therein, the abstract psychons will need the material dendrons because it is through the dendrons that the psychons get expressed.

Using **the keyboard metaphor**, the material dendrons can be likened to the keys on a computer keyboard device. Per Eccles' dualism, the end-user mind, together with the real memories, resides at a remove from the 'keyboard' brain comprised of dendrons: The abstract psychons reside at the level of the dualistic-irreducible end-user and constitute the real 'location' of the memories. The in-brain dendrons associated with these, like keys on a keyboard, constitute the mere means of memory expression. Framed in these terms, **materialism confuses and identifies the means of expression for memories (the in-brain 'keyboard keys') with the memories themselves, if not with the end-user itself.**

In Eccles' scheme, a continuous association of dualistic memories with the dendrons is formed. The continuous exploitation of this association by the abstract mind through the regular activation of the said dendrons must necessarily entail the formation of strong historical dependency between dualistic-irreducible real-memory psychons versus their expression-facilitating 'keyboard' dendrons. The association is expected to be consistent in 'spatial' and structural terms, even if the psychons of real memories are ultimately dualistic and irreducible vis-à-vis the said dendrons.

However, with physical destruction or serious disruption of the said dendrons by, say, impact affecting the brain, or through Ewen Cameron's annihilation ECT methods, the capacity to express the psychons of memory will appear lost, even though the real memories (the psychons *not* identifiable with the dendrons) are not lost at all and remain independently retained vis-à-vis the brain and its now-destroyed dendrons.

It is tautologically true that we need to become conscious of the expression of our memories in order to claim that we have them, and attest to the fact that we have recollected them. To recollect them, we need the in-brain dendronic means with which to express them, in the same sense that the end-user needs keyboard keys in order to express sought outputs. **The destruction or serious disruption of the means of expression (the dendrons; the keyboard keys) must imply the respective destruction or serious disruption of the process of recollection itself, dependent as this is on the possibility of expression**. When the destruction or disruption is at a level above a critical threshold, the means of expression is lost. So also is the possibility for recollection, dependent as this is on the possibility of expression. **Hence, the dualistic account of amnesia.**

Our point is simple: Amnesia lends itself as easily to a dualist description as it does to a materialist one: The same effect can be consistently reasoned from within a dualistic framework. Unfortunately, materialism is so entrenched that only the materialist perspective on amnesia predominates. This leads to a dogmatic unreflective supposition, contrived as fact, that the facts of amnesia supposedly validate the materialist interpretation exclusively. Hence, the loading of the assumption of materialism to the facts of amnesia, and the presentation of those facts *as if* in 'proof' of that assumption.

The notion that amnesia somehow vindicates the materialistic view of memory was doubted by Karl Lashley: a fanatical materialist. Thus...

> I believe the evidence strongly favours the view that amnesia from brain injury rarely, if ever, is due to the destruction of specific memory traces. Rather, the amnesia represents a lowered level of vigilance, a greater difficulty in activating the organised pattern of traces, or a disturbance of some broader system of organised functions[26].

Lashley did not abandon materialism, but only questioned the presumed final loss of memory. In our case, from Alt-R and dualism, abstract memories are not in the first place in brains, given that there is no 'space'. There are no 'locations' at which memory can be stored in anything, including brains, much less be susceptible to surgical destruction, or to physical impact causative of amnesia.

0.43: Mystery of the Hippocampus

Before we explore the mystery of the hippocampus, let us enumerate the three categories of memory. These are...

- **Procedural memories**: the kind one utilises to drive a car or write one's name.
- **Declarative-factual memories**: matters about which we can make declarations. i.e., "Roses are red"; "The sky is blue": "The road bends south." ... etc.
- **Episodic-autobiographical memories:** pertain to personal history and life experiences. e.g., "I loved him, but he didn't even notice me"; childhood memories; emotive memories, etc.

In much of the preceding material, we focused on procedural memories: the experiments of Lashley, Rose and Harding *et al*. The mystery of the hippocampus opens the matter to the other types of memory vis-à-vis the brain.

David Chalmers 'hard problem' approach[27] is highly pertinent: Memory-brain claims are either easy problems or hard problems. **Procedural memories constitute easy problems**: These entail *pull-push, on-off* or *binary* (i.e. thermostat-like) forms of structure and activity: In a car, pressing down on the accelerator while lifting one's foot off the break requires that certain muscles contract and others

---

[26]1950. In search of the engrams. *Symposium of the Society for Experimental Biology* 4:454-483

[27]Chalmers, David (1995)."Facing up to the problem of consciousness".Journal of Consciousness Studies.2(3): 200–219.

expand, to which we could find corresponding neural structures with neural discharges on a "fire" versus "no-fire", on a push-pull basis, or on-off basis. These are easier to infer and map as putative physical states, at least in principle.

In the context of integrated synapses, the neurons either fire or they do not fire: They either 'pull' or switch 'on', or otherwise. The same sorts of neural analogues might be utilised by chicks in the experiments of Rose and Harding *et al*. As such, we could know and see what procedural memories in brains might look like by means of literal mapping, and we might hope to find such structures in the *Striatum* and in the *Basal Ganglia*: (See **Fig. P1.14**).

However, the default crucial experiments of Rose, Harding *et al* demonstrate that the real memories behind these are not the mappable putative neural and synaptic arrangements (see **0.40**). Thus, even apparent easy-problem memory-brain issues turn into seriously conundrums, but only if we adhere to the forlorn materialist framework.

What about ostensive hard problems memories? **Declarative memories have easy-problem and hard problem aspects. The hard problem aspect involves the problem of qualities, which cannot be constituted or designed as material pull-push, on-off schemas,** even if their expressions comprise of easy-problem procedural structures: For example, one could in principle map out the neurological pull-push and on-off complexes to express the declaration, "Roses are red"… "The tea was sweet"… etc. Yet, it is impossible to grasp what the cognition of 'red' or 'sweet' would look like in terms of pull-push and on-off in-brain schema. **The structural account of their means of expression is an easy problem. However, 'red' and 'sweet' do not have the same form and identity as their procedural schemas of expression.** i.e., 'red' and 'sweet' does not have an on/off structural analogue writ in terms of excited or inhibited neurons, even though such schemes must apply to their means of expression. Thus, 'red' and 'sweet' at the qualitative apprehension and cognition-level must constitute hard problems.

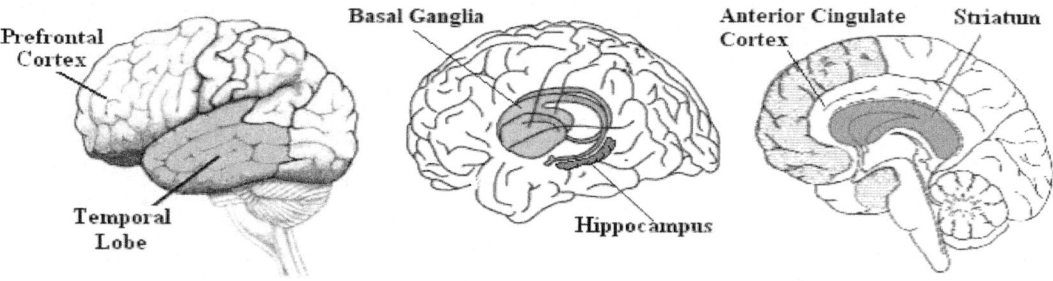

Fig. P1.14 Some functional areas and lobes of the brain
Notes: The illustrations depict some of the brain structures pertinent to issues raised in preceding and forthcoming essays. The prefrontal cortex, which includes the middle prefrontal cortex, is central to the experiments of David Nutt. The middle prefrontal cortex integrates the other lobes of the prefrontal cortex, and it is correlated with planning, decision making, social behaviour and personality. The temporal lobe was an area of concern in the experiments of Wilfred Penfield in his Montreal Protocol: The structure correlates with visual memories, emotional expression, and language understanding. Note that the Hippocampus is depicted as an embedded structure: It correlates with the formation of episodic and autobiographical memories and navigational place-memory. The basal ganglia is embedded within the hippocampus and, together with the Striatum, correlates with procedural types of memory we previously explored in the memory-side investigations of the brain.

With autobiographical-historic memories, the hard problem is rendered undeniably salient: It is impossible to grasp in any neural and structural on-off or push-pull schema what historical and episodic 'feelings' look like. The pure procedural in-brain neural schema for 'I once fell in love with Lucy…' will not have the same form as the qualitative, *I once fell in love with Lucy*. For example, what 'love' looks like in the qualitative sense is wholly unrelated to what the expression-schema for the utterance "love" looks like in terms of pull-push and on-off neural schema. Indeed, the map of the neural schema for the expression 'I once fell in love with Lucy…' is going to have a completely different structural design compared to "*Odnazhdy ya vlyubilsya v Lyusi…*" (in Russian)… even though both constitute the exact same quality and meaning. The same can be paralleled by the obvious difference between the meanings in our minds (at the end-user level) versus pertinent comparative keyboard keys and keystrokes on the computer keyboards.

To augment, consider the apprehension of *infinity* versus the bloss neural schema for the procedural expression of 'infinity'. The pertinent neural schema for the expression does not constitute the abstract meaning *infinity* and it is not interchangeable with it. Indeed, we boldly state that the said procedural neural schema does not actually constitute the memory of *infinity*. What would the real memory of infinity look like? The question is further complicated by the fact that we do not arrive at the idea of infinity from sense-data or 'material experience', or from environmental 'inputs'. The empirical-sensate world does not furnish an analogue for *infinity*. Nor can the finite brain.

Succinctly, **if we sought to model the apprehension of infinity and its subsequent in-brain memory as a material neural schema, it will not be enough for it to constitute a mere means of expression for the word and utterance 'infinity'. It would need to constitute a counting procedure for infinity, manifested as a procedural schema comprised of pull-push and on-off switches. No such schema can exist: It is not possible for any finite in-brain procedure to count and 'compute' the meaning and**

apprehension of infinity. Such a structure would crash into a non-terminating problem. The problem of the qualitative apprehension of infinity constitutes an insurmountable hard problem.

Recall from **0.06** how the problem posed by infinities. Later, we will discover that infinity extends beyond mathematical context into basic dictionary definitions; such as "to run". These have an unbound and indefinite character, putting the infinity to the infinitive without committing any equivocation fallacy. Also consider the infinitive "to be" as it pertains to *being*: of most directly pertinence to autobiographical memory; if only because "to be" and *being* constitutes the essential milieu to perduring identity and auto-biography and episodic memory pertinent to memory often associated with the hippocampus. Thus, the hard problem of infinity cannot be excluded from considerations of auto-biographical episodic memory attendant the hippocampus.

*

Relating memory types to the brain, the *medial temporal lobe* (See **Fig. P1.14**) appears to deal with declarative or factual memories. Episodic or autobiographical memories correlated to the peculiar structure called the *hippocampus*. And it is the hippocampus and episodic-autobiographical memories that directly concerns us. We must keep in mind the clear differentiation between easy problem procedural memory components of memory and hard problem qualitative ones. And we must assert the inequivalence between their material neural expression-schemas versus their noetic qualitative apprehension presupposed to and non-reducible to the former schema. This distinction must carry into our examination of the mystery of the hippocampus and amnesia as it relates to episodic-autobiographical memories.

The conventional accounts claim that **episodic-autobiographical memories are first formed within the hippocampus** (See **Fig. P1.14**). **These supposedly migrate out from the hippocampus to elsewhere in the brain**[28]. While the hippocampus appears essential to the initial formation of episodic autobiographical memories, the hippocampus is not *ultimately* required for their recollection and expression. This is augmented by the fact that, in specific cases, damage to the hippocampus in both hemispheres will not prohibit recall, even though it *will* prohibit the formation of new episodic memories: The condition is called Anterograde amnesia.

The problem here is not that the hippocampus is required for the initial formation of episodic memories, although the fact cannot automatically imply the veracity of materialist view of memory, given that the dualistic approach to the brain as the means of expression of otherwise extant autobiographical memories can also be consistent with known facts. That is, we can load the assumption of dualism to our interpretation of the facts pertaining to the hippocampus and get away with it, even though it is taboo to do so per the dominant materialist hegemony. **Materialism has no exclusive privilege vis-à-vis episodic-biographical memories, especially given the hard problems previously detailed**.

*

Mind, even a dualistic and ultimately irreducible one, has as much a need for functionally specialised brain as does the end user for the keyboard: primarily as means of expression. That the brain should be divided into different functional areas or modules through which it can be utilised to expresses different capabilities and memories does not validate materialism exclusively, save by culturally sanctioned assumption-loading. A dualistic-irreducible mind cannot obviously use same-areas of the brain for different purposes of expression: The dualistic non-reducible and extant mind must also require different areas of the brain for the expression of different things. The problem with generic materialist accounts is the *assumption* that memories are stored in the brain, and that the damage to the hippocampus, and with memory-recall despite such damage, supposedly implying that the memories must have migrated to other locations in brains outside the hippocampus.

The claim for such migratory memory *assumes* that the schema for the expression of episodic, autobiographic, and emotive memories have reformed elsewhere in the brain. But this cannot preclude dualism and non-reducibility of memory: From dualism, the brain is principally a means of expression of memory, not its ultimate store. The memories are extant the brain: a possibility also forced by hard problems as pertains to real memory versus its mere schema of expression, highlighted by the problem of *infinity* versus its mere schema of expression. That the means of expression of memory might be reformed elsewhere in the brain and not in the hippocampus, is not a problem that excludes dualism or favours materialism exclusively.

Thus, as with procedural memories inferred from the default crucial experiments of Rose and Harding, **from dualism, the reason why old autobiographical memories can be recalled despite damage to the hippocampus is because these memories are ultimately retained outside of the brain, *not* in the brain. Their 'migration' to other parts of the brain, as the mere migration of the means of expression, can be easily accommodated by dualism, given that the real memories, as hard problem qualitative states and infinities, cannot in the first place be reduced to brain-states or to mere schemas of expression.**

*

There is a good amount of evidence to suggest that the hippocampus is also involved in 'place' memory and navigation. Evidence for this has been found in rats[29], monkeys and in humans. Pertinent neural activity in the form of bursts of action potentials, characterised as 'place fields' within the hippocampus, occur when test animals pass through an environment: The bursts are observed to happen in relation to the place a specimen may be looking at, by the direction of its displacement, or by the destination toward which the specimen is moving.

---

[28]Squire, LR; Schacter DL (2002). The Neuropsychology of Memory. Guilford Press: Chapter 1

[29]Morris, RGM; Garrud P; Rawlins JNP' O'Keefe J (1982)." Place navigation impaired in rats with hippocampal lesions". Nature 297 (5808): 681-83

Again, it is tempting to assume that this must be evidence for memory-storage within the brain, consistent with materialism. However, as with implications from the surgical destruction of procedural memories in the experiments of Rose, Harding *et al*, one should not expect the surgical destruction of 'place fields' from within the hippocampus to obtain any different an outcome.

The utilisation of the hippocampus for 'place memory' and navigation is certainly consistent with a materialist hypothesis. Yet, it is equally consistent with the dualist one: Again, the dualistic mind needs specialised structures of the brain through which to express its control over the body in relation to 'location and place'. Hence, the existence of 'place memory' schemas do not imply materialism, nor exclude dualism. For example, the dualistic mind might require material procedural schema in the hippocampus for, "Place-A: turn right". Equivalent to arrow keys on a keyboard (No one would suggest that the end-user's decision to use arrow keys is reducible or identifiable with the activity of the arrow keys). The dualistic mind will seek to keep this potential within the brain in readiness and not have it reformed again and again, or every moment Place-A is to be visited. Thus, place-memory functions and navigational functions of the hippocampus cannot be misconstrued as automatic validations of the materialist contention. The materialist contention is entirely assumptional, and we achieve no different a result when we load dualism to the same facts and present *it* as the certain conclusion. Yet, given the hard problem issues that beset the want to relate qualitative states to their mere expression schema, in this instance, as in all others, dualism has an easier time accommodating the facts than materialism.

*

Finally, per de-spatialisation and other anticipated findings from Alt-R, and from the unauthorised history of physics (see Part-III), clearly, materialism is dubious at best. So is its assumption-loading to the facts of amnesia and to theories of the hippocampus. Even if schemas for expression of autobiographical and emotional memories could be contrived as the real memories per the usual materialist expectation, de-spatialisation and temporisation alone forces these, and their putative retrieval processes, to transform into resonations organised and distributed in and across time, *without* 'space'...*without* contiguity or impact.... *without* resort to input-operation-output schemas... all based on work function zero processes.

MIND-SIDE INVESTIGATIONS OF THE BRAIN

0.44: The Experiments of Benjamin Libet

Insofar as minding is necessarily linked to the apprehension and articulation of memories, inferred memory-brain dualism must imply mind-brain dualism. At the minimum, and in the framework of de-spatialisation and temporisation of the physical order, temporally dissociated dualistic memories must entail a recalling mind that must 'reach out' to the memories temporally extant the brain, itself becoming extant that brain, even if it began in the brain. Hence, we can use memory-side conclusions to infer the reality of mind-brain dualism. **Yet, we need not confine ourselves to implicit memory-side arguments to assert mind-brain dualism.** For example, it is possible to use electroencephalogram or EEG procedures to measure the time-relation between brain activity and consciousness, or even putative volitional consciousness. This was carried out by Benjamin Libet in the 1970s[30] on the materialist assumption that conscious volition, the consciousness of a thing, or of *anything*, is a thing produced by brain activity. The contrary dualist view to the effect that the mind is not the brain and cannot arise at the same time as observable brain activity (and that the latter no more plausible than the claim that the end-user arises out of his keyboard activity) was not entertained, even as a side-hypothesis. In short, the Libet experiments loaded the assumption of materialism to the facts garnered, and then posed the garnered facts *as if* in 'proof' of the loaded assumption, without due consideration of alternatives.

Libet discovered that the report of conscious volition (and, presumably, all forms of consciousness) occurs at a measurable time-delay *after* brain activity. Materialism is assumed. Per the latter, it appears that conscious volition occurs *only* after the preceding and producing non-conscious brain activity, readiness potential, or *Bereitschaftspotential*... some three hundred milliseconds *before* the test-subjects report first awareness of the volition. At first, this fact appears to portend serious problems to the notion of conscious volition and free will, if not to consciousness itself: **It appears that the brain has already made the non-conscious decision to act before we have consciously decided to act, implying that both volitional consciousness and free will are apparently illusory.** The greater problem is that, **together with free will, consciousness and mind appear eliminated in causal terms.** The apparent 'fact' that consciousness has now become an output of non-conscious brain activity, and *not* the causal generator of that brain activity, must surely imply that consciousness generally, and not just volitional consciousness, must be an outright illusion, insofar as it has no causal efficacy, and cannot be part of any real ontology: i.e. A thing that has no causal power *effectively* does not exist.

However, all of the above assumes that the conscious volition and consciousness generally is something 'produced' by brain activity, equivalent to the claim that the end-user is produced by his keyboard activity (an absurd notion). Libet-type interpretations and conclusions are artefacts of this uncritical assumption: i.e., fallacy chain, step-1: (see **0.37**).

*

Let us now postulate the dualistic interpretation of the Libet experiments and its facts. **Dualism holds that** consciousness and minding are, as is true of memories, ultimately not at all in the brain: and that **consciousness and minding do not arise from or**

---

[30]Libet, Benjamin; Gleason, Curtis A.; Wright, Elwood W.; Pearl, Dennis K. (1983. "Time of Conscious Intention to Act in Relation to Onset of Cerebral Activity (readiness-Potential) –The Unconscious Initiation of Freely Voluntary Act". Brain 106: 623-642

79

through the activity of the brain, even though consciousness and minding *are* expressed by means of transitive brain structure and activity, in the same way that an extant end-user needs and initiates keyboard activity to generate outputs. This view is automatically furnished when we model both memory recollection and the expressions of consciousness in terms of a de-spatialised and temporised physics, brain, and mind. However, we must put aside de-spatialisation for now and entertain dualism as just an assumption, a hypothesis, in order to find out if, as a contending assumption, it can save the same evidence as the materialist assumption.

In terms of the dualistic account, brain activity would constitute an intermediary or transitive state to two modes: **In the first mode**, mind would trigger brain activity in order to bring about or express its intentions. **The decision and initiation of brain activity by mind would take place *before* the subsequent brain activity: Indeed, the preceding minding process could not be evinced as any brain activity, or even as *any* even**t: Analogically, the end-user's decision to initiate keyboard activity precedes the keyboard activity and is not evinced as keyboard activity. From dualism, the mind is not the brain: it is extant the brain in the same sense that an end-user is extant the keyboard.

Per the assumption of dualism, the extant mind's antecedent decision initiates the brain activity. The readiness potential in the brain builds up to a threshold. It is then released. The intents of that volition are expressed through the consequent brain activity and the orchestration of the whole nervous system. Note that, **at no point in this scheme is mind assumed to be in synchronicity with subsequent brain activity or with any other subsequent brain-event**. Therefore, **the putative moment of volition, insofar as it precedes and does not involve brain activity, and is not in the first place evinced as such, cannot be synchronously timed or clocked as brain activity** or as any process evinced in brains.

In short, the want to time the moment at which mind initiates brain activity (the want to time the moment of volition) simply because the materialist supposes that it is produced by and ought to be synchronous with brain activity, is in error.

The second mode is the feedback mode: Therein, **incoming externally sourced stimuli trigger intermediary brain activity. The attendant brain activity then conveys the informational content of the external stimulus to the dualistic mind**. But the mind is extant the brain, and hence extant the pertinent brain activity. Hence, **non-conscious stimuli generate non-conscious brain activity. But, at some succeeding moment thereafter, the activity is received or witnessed by consciousness… witnessed temporally *after* brain activity**.

We reiterate that, in the Libet experiment framed per dualism, mind and conscious volition are processes that take place extant of the brain: In the first mode, directly relevant to the issue of conscious volition, active mind temporally precedes subsequent brain activity: The brain activity does not produce the deciding mind. Therefore, consciousness need not coincide with brain activity, either in 'space' or especially in time.

For one to become conscious of the volitional 'wiggle of the little finger', the volitional causality must decide this temporally before the non-conscious *Bereitschaftspotential* builds up in the brain. The build-up is formed and then released, culminating into one's non-conscious 'wiggling finger'. This fact must then feedback through the senses into the brain, producing the pertinent non-conscious feedback second-mode brain activity. A moment *after* this non-conscious feedback brain activity, one becomes conscious of the volitional act, but always *in retrospect*…of the volitional minding that temporally preceded the brain activity itself.

Thus, according to dualism, neither the causal triggering that instigates the brain activity, nor the subsequent conscious moment witnessing the results or the facts of its past volitional act, occur as brain activity: **Mentality happens temporally *before* brain activity and temporally *after* brain activity, respectively. Mentality and, with it, consciousness, or volitional consciousness, cannot be constituted as a set of brain-events**, at least from the dualist framework.

To clarify and reiterate by analogy, we expect to find the same sort of time-relation between an end-user and its computer keyboard device: The initiation inherent to the end-user happens *before* keyboard activity; *not* at the same time as. The subsequent awareness of that keyboard activity, as a set of past events, comes to the attention of consciousness of the end-user *after* the pertinent keyboard activity, *not* at the same time as. In a similar vein, according to dualism, mind and consciousness do not arise because of keyboard-like brain activity, given that the end-user mind and *post facto* consciousness is non-identifiable with and non-reducible to the keyboard brain activity. Therefore, it should be of no surprise that consciousness, especially volitional consciousness, is experimentally found *not* to occur at the same time as its keyboard brain activity, and that consciousness and brain activity are necessarily a-synchronous: the Libet finding, in a nutshell.

The real problem exposed by Libet's experiment is *not* the elimination of volitional consciousness, but the reality of contrived artefacts and conclusions about the relation of mind and consciousness vis-à-vis brain activity, concocted on the unchecked assumption of materialism.

*

The results of Libet's experiment are not problematic to free will, nor to consciousness, unless one assumes forlorn materialism. Indeed, **if minding and consciousness could not be anything but brain activity (i.e., if materialism was true) then consciousness ought to be perfectly synchronous with the brain activity that produced it**. Yet, the experiments of Libet show that consciousness is *not* synchronous with brain activity, but a-synchronous. **If consciousness does not occur at the same time as brain activity, then it cannot occur at the same 'place' as brain activity**, on the assumption of 'space'. **Therefore, it cannot be produced by that brain activity. Therefore, the mind is not the brain**. This ought to favour dualism, given that dualism saves the evidence garnered from the Libet experiment in a far more effective way than materialism: That is, **the observed a-synchronicity arises seamlessly from dualism, but it is problematic for materialism**.

80

Of course, de-spatialisation, temporisation and other findings from Alt-R render dualism the inexorable fact of mind-brain relations, beyond mere assumption or hypothesis... to the demise of materialism.

0.45: Hidden Crucial Experiment: The Libet-Nutt Bridge

We need another crucial experiment: one that directly attacks the assumption that consciousness is produced by brain activity, independent of the form implicitly furnished by Libet, but incorporating the findings secured by Libet. Just such a crucial experiment, another default experiment, was furnished by Dr. David Nutt *et al*[31].

In his experiments, subjects were linked to an EEG (electroencephalogram) and an MEG (magnetoencephalogram), and, more critically, to an MRI (magnetic resonance imaging) scanner. Around thirty subjects were intravenously injected with the active ingredient of the psychedelic psilocybin, **with the aim of assessing the impact of the psychoactive material on the medial prefrontal cortex and the anterior cingulated cortex of the brain**: (See **Fig. P1.15** above). Both integrate other parts of the brain and are posited as responsible for the processing of hard problem states of emotions, introspection, and cognition. (See **0.43** for the problem posed by hard problem approaches to memory in the context of the hippocampus). Per the materialist assumption, the said structures constitute leading candidates for a 'hub of consciousness': one that integrates all other functions and processes of the brain into a coherent unitary whole.

From past experiments involving similar psychedelic drugs, it was expected that both stated brain structures must become overactive. **Given vivid and powerful psychedelic experiences, these must surely involve far more brain activity than mundane experiences: the brain ought to 'light up like a Christmas tree'.**

David Nutt *et al* obtained outcomes that not only contradicted past experiment but uncovered the opposite: EEG, MEG, and MRI scans of brain activity in the stated regions showed no activity at all. **The brains of the subjects were running cold, dark, and silent, equivalent to the unconscious state**. Yet, the subjects were anything but unconscious: They were deeply immersed in a powerful psychedelic trip.

In short, what the **Nutt's experiment does is attack, even if inadvertently, the ready materialist notion that 'brain activity produces consciousnesses'. It does so by showing that vivid psychedelic consciousness can occur** *prima face* **even in the absence of brain activity**.

*

What is the relevance of the experiments of Nutt to those of Benjamin Libet? Recall that Libet ran his experiments on the assumption that brain activity 'produces' consciousness; specifically, volitional consciousness. What if we were to run a Libet-type experiment on Nutt's experimental subjects? What if we sought to time the moment of their consciousness of their psychedelic experiences vis-à-vis their brain activity? **If, per materialism, we assumed a strict causal relation between brain activity and 'produced' psychedelic consciousness, we would have to assert that, in Nutt's subjects, on account of the lack of brain activity in the stated structures, there ought to be** *no* **consciousness at all. Yet, there** *is* **consciousness, although it does not seem to require brain activity**.

However, past experiments utilising psychedelic substances *have* evinced over-excited brain activity, such as in the experiments of Dr Franz Vollenweider. **Yet the experiments of Nutt have ontological primacy and demonstrate the absence of brain activity, even while vivid consciousness is still in effect**: This is such that, even if other experiments involving psychedelics evince brain activity, the results from the experiments of Nutt ontologically override findings from the other experiments: Just because the assumption of the primacy of brain activity vis-à-vis consciousness (as the supposed 'producer' of consciousness) is not disturbed by the evidence of brain activity in other experiments, this does not obviate the experiment that *does* disturb that assumption, as do the experiments of Nutt *et al*: **If you can demonstrate consciousness** *without* **brain activity in one domain or experiment, the evidence of brain activity** *and* **consciousness in another domain or experiment can no longer be construed as absolutely requisite to the possibility of consciousness, even if such brain activity is undeniably in evidence in the latter case**. Indeed, if one were to insist that one cannot obtain consciousness without the 'producing' brain activity, then Nutt's experiments ought not to evince *any* consciousness, given the absence of brain activity. **Hence, brain activity may well typically correlate with consciousness, but it is** *not* **necessary to consciousness**, per the findings of Nutt *et al*.

Of course, in its own way, the experiments of Benjamin Libet also recommend the same conclusion: albeit obscured by the emphasis on volitional freedom, the precarity of which turns out to be an artefact born from the imposed assumption that brain activity produces consciousness.

*

Consider the approach entertained by Roger Penrose and Stuart Hammeroff, independent of Nutt's experimental findings: that the real brain activity, one supposedly producing consciousness, might be occurring at a lower level in the brain: through fundamental spacetime geometric collapse and associated quantum computations at the cytoskeletal and microtubular sub-cellular level (see **0.41**). Thus, the purported consciousness-producing activity would be at a level lower than the baselines of MRI, EEG and MEG scanners used in the experiments of Nutt.

Could Hameroff's approach obviate the findings from the experiments of Nutt?

---

[31]David J. Nutt, Carhart-Harris RL, Erritzoe D, Williams T, et al. (February 2012). "Neural correlates of the psychedelic state as determined by fMRI studies with psilocybin" Proc. Natl. Acad. Sci. U.S.A. 109 (6):2138-43

If 'real' brain activity producing the psychedelic consciousness, or any consciousness, is taking place at a lower unobservable level vis-à-vis pertinent instruments, serious difficulties must follow: **If real brain activity is not the manifest brain activity, then what is the purpose of manifest brain activity? Why should the brain bother to produce it?** Second, **if the ostensive brain activity, including readiness potentials in Libet's experiments, are not the real producers of volitional or of *any* consciousness, then hitherto interpretations of the Libet experiment, and generic materialist assertions born of these, if not from the whole of brain science and neuroscience based on generic brain activity, must now be questioned.** We have no qualms with such difficulties: since these render materialism even more implausible.

*

The puzzling facet of the Libet-Nutt bridge is resolved the moment one grasps the fundamental physical framework in terms of anticipated de-spatialisation and purely temporised growing block Bergson-Whitehead amalgam: i.e. spacetime *without* 'space': at which point we must completely re-frame and overhaul our ideas about mind-brain relations into alternative inexorably dualistic forms.

There is no space. Therefore, the want or expectation that we must model brain activity as a spatial distribution, *and* associate mind, consciousness and memory-recollection to spatial processes correlate with brain activity (now itself de-spatialised) is rendered wholly forlorn: We can no longer frame the brain as a material corpus in 'space', but only as a complex information structured in pure time, subsumed to a grander physical order: itself structured as a combine of past-realised and future-potential information distributed in pure time. In such a framework, one is forced to give up forlorn attempts to 'locate' mind and consciousness in the brain (i.e., as 'brain activity' in space) simply because there *is* no 'space' or 'location' as such... for *anything*, including the brain or the mind.

We must instead relate mind and consciousness vis-à-vis the brain, itself a de-spatialised temporised complex of events, in terms of pure time-relations that force one to adopt consciousness and mind as necessarily temporally antecedent, precedent, and *always extant* brain-events and world. **That is, in a pure de-spatialised order of nature, nous must either temporally precede the manifestation and resolution of any AND-form future-potential world and brain-events, and therefore cannot be identified with or reduced to such future-potential brain-events... or it must be temporally antecedent any realised past brain-events: again, without reducing to or identifying with these.** This expectation is perfectly consistent with the findings from Libet and Nutt... and perfectly consistent with mind-brain dualism imposed by a de-spatialised radicalised ontology of physics.

0.46: The Montréal Procedure & the Unity of Brain Activity

If materialism held true, then consciousness ought to be investigable as something generated by an in-brain corpus or functional area: a hub of consciousness: perhaps the medial prefrontal cortex and the anterior cingulated cortex. Yet, as was discovered in the Libet-Nutt Bridge (see **0.45**) psychedelic consciousness abides even when these structures are running cold: This calls into question the notion that consciousness is a something produced by *any* brain activity, much less by a hub for it. Yet, there are older investigations that also question this presumption: **In Wilder Penfield's Montréal procedure from the 1930s, the electrical stimulation of parts of the brain led to the elicitation of memories, even if in a non-volitional manner: Yet Penfield found no basis or corpus for decision-making or for unitary binding in the brain**[32].

Penfield's experiments tend to be portrayed as *prima facie* 'proof' that memory must be in the brain. Yet, per dualism, electrically stimulated tissues in the Montréal procedure are nothing more than 'keyboard' structures with which dualistic-irreducible memories, perhaps Eccles' hypothetical psychons, relate or correlate: **Per the findings of Libet, and consistent with the a-synchronicity established therein, even non-volitional memory elicited in the Montréal procedure is simply not going to occur at the same time as the electrical stimulation of that activity, no more than it could occur at the same time as mind-triggered brain activity**: If these do not occur at the same time they cannot be said to occur at the same 'place', even ignoring de-spatialisation and assuming the ontics of 'space'. **Hence, consciousness of memories in the Montréal procedure occurs 'outside' of, and inevitably *after*, the electrical stimulation, vis-à-vis the witnessing consciousness, which comes *after* (and not from) the stimulated brain activity, as purported by our dualistic analysis of the results of the Libet experiment.**

Furthermore, **memory-side investigations from Lashley, Rose and Harding *et al* show that memories are not *ultimately* identifiable with *any* structure or process in or of the brain. When we add the fact that consciousness need not even require brain activity, as found in the experiments of Nutt, the case against the materialist contention is seemingly concluded**, even if memories are found to correlate with attendant brain structures and utilise these as means of expression, as 'keyboard' terms. **Hence, electrical stimulation of brain tissue only elicits dualistically associated memories:** memories that are not ultimately identifiable with, or reducible to the stimulated structures, or with attendant brain activity.

It follows that **the results from non-volitional electrical stimulation in the Montréal procedures are equally consistent with the dualistic view**: Dualism implies that the electrical stimulation of brain tissue must lead to an involuntary trigger of dualistically associated memories (Eccles' psychons, perhaps) associated with the stimulated brain structures (Eccles' dendrons).

Thus, the balance of evidence so far presented favours dualism, *not* materialism. At the very least, materialism has no special privileged or exclusive claim over the outcomes from the Montréal Procedure, when dualism can save the same evidence just as well, if not more efficiently.

*

---

[32]The Mystery of the Mind: A critical Study of Consciousness and the Human Brain. Penfield, Wilder. Princeton university Press, 1975

82

QUANTUM MECHANICS AND MIND

If the consciousness of the non-volitional electrically stimulated dualistic memory was being produced by a specialised part of the brain (ideally, by the medial prefrontal cortex or the anterior cingulated cortex, or both) then the stimulated part ought first to send a traceable neural signal to that consciousness-producing hub in order to communicate the stimulation to that part, before one became conscious of the stimulated memory. Using positron emission tomography, it should be possible to observe such signals sent from the electrically stimulated putative memory-site to the consciousness-producing hub, and thus trace down the physical location of the mechanisms of minding and consciousness in the brain, assuming materialism held true. But if dualism holds true, no such corpus or 'hub' need exist: The minding of memory would take place 'outside' the brain without requiring any specialised mind-producing corpus… in the same way that the end-users mathematical mind utilises but does not reduce to the keyboard numeric keypad.

Nutt's experiment clearly shows that consciousness need not require brain activity at all, with the consequence that there is no 'hub of consciousness', especially at those functional areas where it is expected to be found: The functional areas were found to run cold.

What about the normal volitional trigger of memories without artificial stimulation? This would demand greater complications to the scheme: **Per materialism, a consciousness producing corpus would need to transmit a traceable signal to the purported location of the memory. The memory thus triggered would then generate a feedback signal. This latter would propagate back to the consciousness-producing corpus to register in consciousness**, per an anticipated and expected hub-and-strand pattern: an effect that ought to be easily detectable through the glucose pump method, positron emission tomography, or by other advanced realtime brain scanning methods. **Yet, despite the use of said techniques for decades, and especially in view of the Nutt experiments, no such 'hub-and-strand' effect has been observed**.

Moreover, if the memory sought by a putative consciousness-producing corpus is in the brain, but at a spatial remove from that consciousness-producing corpus, and if the latter needs to send a neural signal to and from that memory, how will it know which part of the brain contains the sought memory? How could the consciousness hub know to which part of the brain to send its retrieval-signal? **How could the consciousness hub know which memory it needs to trigger in order to acquire a consciousness of it, if the pertinent memory is *not* coincident with the consciousness-producing corpus to begin with, but spatially located elsewhere in the brain? But if the consciousness producing domain already possesses the memory within its own confines, why would it need to trigger and retrieve the same from some other location in the brain**?

Such problems cease the moment we admit that consciousness and the coherent control of the brain is something imposed from a minding process dualistically irreducible to and extant that brain: captured in the analogy of how and end-user imposes unity to its keyboard activity, the unity and orchestration of which does not derive from the keyboard activity itself, or from any structural procedural feature of the keyboard. If so, there is no need for a consciousness-producing specialised corpus or hub in brains. Consequently, such a hub has never been observed.

Curiously, and as split-brain patients demonstrate (see **0.48**), the physical integration and unitary binding of the brain is necessary if the minding process is to impose unitary control over its brain. Yet, as argued, this is never accomplished though any specialised consciousness and mind-producing corpus: No such thing has been observed in any of the myriad investigations of the brain by the application of *any* method: Let us not forget that the mind-side Libet-Nutt bridge (**0.45**), combined with decisive memory-side facts that call into doubt both localised engram memories as well as clouded holonomic material memories: These seriously undermine the want or possibility of such a corpus.

In the Montréal procedure, what seems to happen is that the non-volitional consciousness related to the location electrically stimulated is not generated by any adjacent specialised consciousness-producing corpus, but arises *dualistically* vis-à-vis the very part of the brain electrically stimulated. **In the Montréal procedure, the non-volitional consciousness of the elicited memories anywhere in the brain will not be produced by *any* specialised part of that brain, much less be mediated by some non-existent neural signal that needs to displace to a putative specialised consciousness-producing corpus. The consciousness will simply arise relationally (albeit dualistically and, per Libet, a-synchronously) vis-à-vis the region non-volitionally stimulated. The point? What holds true for non-volitionally stimulated consciousness of memory per the Montreal procedure must also hold true vis-à-vis volitionally triggered memory.**

*

It is of no passing interest that the brain possesses specialised parts or 'functional areas'. These pertain to, say, speech, mathematical abilities, motor-neural skills, musical abilities. We have already explored the fact that procedural memories, declarative memories and autobiographical-historical and episodic memories, are expressed through different structures. Such functional areas can be likened to the specialised keypads of a computer keyboard device: The dualistic-irreducible mind certainly needs in-brain specialised functional areas to output, say, mathematical expressions, in the same sense that an end-user needs a numeric keypad and cannot express these through the alphabetic keypad. Yet, real mathematical insight will not ultimately reside in the pertinent functional areas of the brain, no more than these reside in the keyboard numeric keypad vis-à-vis the end-user. Memories and abilities do not arise from the keyboard brain; no more than consciousness itself arises from any hub or in-brain corpus, given the implications of the Montréal procedure in this essay, combined to obvious inferences from preceding material.

That the brain has no special functional area to create the unitary binding over brain activity is of no surprise: The end-user has unitary control over the keyboard without reducing to any part of his keyboard. There is no consciousness or unitary control-producing keypad. **The unitary control of the keyboard-brain is similarly imposed on that brain from 'outside' of it, by an end-user dualistic mind. The latter exploits information-matter dualism (0.01 to 0.04 and 0.05), the structure of intermediate spacetime (0.29 to**

0.30), and, ultimately, the de-spatialised temporised structure and processes of the Bergson-Whitehead amalgam (the successor to spacetime): all on a work function zero basis (see 0.32).

The puzzling facts of the Montréal procedure, and the inability to find a locus for consciousness in the brain, is resolved the moment we grasp the brain in pure temporal de-spatialised terms, and hence give up the notion of the brain as an 'object in space'. By Book-IV we will discover that de-spatialisation emerges thus:

- **The notion of 'space' implicitly broke down with the abandonment of the notion of absolute time by Einstein in 1905**: i.e. It turns out that absolute time is implicitly synonymous with 'space'.

- **The notion of 'space' cannot be empirically justified even per generic quantum mechanics**: per the impossibility of seamless-continuous inputs and feedback-signals required to prove seamless-continuous motion and, with it, space itself.

- Per Alt-R and nascent quantum mechanics-ii, **if we insist on keeping the notion of 'space', the best we will obtain is the spatially static quantum mechanical wave**, embedded within an as-yet not-realised quantum mechanical future-form 'space': i.e. basically, an as-yet not-existent 'space'; and hence a nonsensical notion of space... in and across which nothing can displace... simply because it does not yet exist to allow such displacement.

- Per Alt-R, **a specialised application of John Wheeler's delayed choice (one devised to analyse space) totally unravels, first the notion of the motion of quantum mechanical waves, and finally the very notion of space itself**.

- With the notion of 'space' rendered pleonastic, **we need only map events purely in terms of *time, without* 'space'**: i.e., in terms of pure delayed choice time-interval relations.

Consequently, **we need no longer frame the brain as a 'material corpus in space', but only as a complex de-spatialised temporal distribution of the past, in turn structured purely as a distribution in time, combined to an as-yet not-realised future potential brain... which is also de-spatialised**. This is the growing block temporised approach, universally true for the physical order, with respect to which the brain is not going to be the special exemption. Hence, we must abandon the want to 'locate' mind and consciousness in the brain, simply because there *is* no space or 'location' for such things, or for *anything*, including the brain itself. Instead, we must 'locate' both mind and its unitary binding power over brain-events and world-events at a dualistic temporised remove from the brain. It follows that we need not and cannot reduce the unitary control of brain activity to an in-brain corpus. **Hence, mind-brain dualism follows suit.**

0.47: Annihilation Electro-Convulsive Shock & De-Patterning:
The Locus of Consciousness & the Penfield-Cameron Bridge

We now come to a*nnihilation electro-convulsive shock*: a procedure inspired by the prior humanistic work of Wilder Penfield in the 1950s. Donald Ewen Cameron subjected the brains of his patient's (or victim's) to high voltage current of around 800 mils of amperage, for durations lasting between half a second to 1.5 seconds; carried out multiple times until total 'de-patterning' was achieved: Essentially, this amounted to effective memory-wipe and induced irreversible amnesia.

Cameron hoped he could re-program his memory-wiped 'patients' by means of recorded instructions: i.e., so-called 'psychic driving'[33]. This failed.

Annihilation ECS proved to be capable of total memory-wipe. Note that both materialistic and dualistic approaches can accommodate this fact, as was clarified in preceding material. Indeed, dualism can handle the evidence more efficiently: Thus, one must not misrepresent the results from annihilation ECS as somehow affirming the materialist view on memory-brain relations. In dualism, per our own scheme, and per the scheme described by John Eccles, it is expected that memories (the hypothetical psychons) dualistically relate to specific 'locales' of the brain (the dendrons) *without* reducing to the latter. Since different dualistic-irreducible memories cannot be expressed through the same dendrons, different memories will require different dendrons and complexes of such for their expression.

It follows that the sustained disruption of dendrons by means of Cameron's procedures will necessarily destroy the capacity of the mind to express its otherwise dualistic-irreducible memories (it's psychons).

The logic is simply this: *No expression, no recollection*. Hence, the total effective loss of memories will follow despite the ultimate reality that, per dualism, memories are retained intact 'elsewhere', but are no longer expressible.

However, **the critical point, easily obscured from the apparent destruction of memory, was that Cameron's *annihilation ECS could not result in the annihilation of nous itself.***

*

Insofar as consciousness must relate to its memories in order to successfully transact with the world, the total annihilation of the expressibility of memories, necessarily implying the impossibility of conscious recollection tandem to such expression, ought also to cripple the conscious mind's power to transact knowledgeably and sensibly with the world. This much turned out to be true in what Cameron finally 'achieved'. Yet, the unitary nous and the binding of the brain that attends it (but not the biographic identity), albeit conditionally crippled per the 'memory wipe', survived annihilation ECS. This reality sits well with the findings of Libet and with

---

[33] Marks, John (1979). The Search for the Manchurian Candidate. New York: Times Books. Pp. 140-150

implications derived from the Libet-Nutt Bridge previously detailed. In an entirely notorious and unethical way, **Cameron's work completed the work of Penfield: augmenting and validating his view that the basis of nous, and the basis of the unitary binding of the brain, could not be located in the brain. Hence, the Penfield-Cameron Bridge.**

What if we applied Pribram's forlorn materialist holonomic approach to the putative production of consciousness itself? What if we claimed that consciousness is produced by a same-mechanism reduplicated and clouded across the brain, or at least across, say, the temporal lobe? Cameron's destruction of these at many sites would presumably miss some of the re-duplicated consciousness-producing material: The consciousness could then be reconstructed from such missed sites, in the same way that, in holograms, we can reconstruct a whole picture from any small surviving sample. But if this could hold for consciousness it ought also to hold for memory: both supposedly clouded throughout the temporal lobe and reconstituted from samples missed by Cameron's destructive procedures. Yet, the memories, succinctly, their means of expression, are effectively destroyed by Cameron's methods.

How could it be that unitary consciousness can be 'reconstructed' despite Cameron's destruction, but memory cannot be so reconstructed? (The latter is *not* clouded in brains, as was shown in 0.40). If the holonomic approach appears to fail at the memory-side, there is no reason to suppose it likely to possess any validity on the mind-side, or in matters of the unitary binding of brains, or in survival of unitary consciousness, despite annihilation-ECS.

Framed in terms of dualism, the same facts from Cameron's 'experiments' can be interpreted to mean that **the minding process responsible for consciousness is not at all located or reduced to any specific part of the brain and cannot be destroyed in brain sites irreversibly disrupted by annihilation-ECS**: Hence, in annihilation-ECS, the content of consciousness is apparently wiped out, but *consciousness in itself* is unaffected. Moreover, the full unitary binding of the brain remains intact.

In terms of mind-brain dualism, while memory itself requires specific structures through which it must be expressed, mind and consciousness itself remains wholly explicitly irreducible and extant: That is, annihilation-ECS cannot erase nous and unitary binding itself, given that nous is not in the first place produced by any specialised corpus in the brain, much less clouded to the brain per Pribram's materialist holonomic scheme.

When we utilise memory-side analysis from the re-appraisal of the experiments of Rose and Harding (**0.40**) to raise doubts about the notion that memories are ultimately reducible to brain-structures; and when we combine these to the undermining of the materialist notion that consciousness is produced by brain structures via the case from the Libet-Nutt Bridge, it is fair to conclude that **dualism, *not* materialism, must have the final say on how we make sense of the outcomes from Cameron's annihilation-ECS.**

0.48: The Split-Brain Conundrum: How De-Spatialisation can Explain

We must now relate the unitary binding problem posed by Penfield *et al* to observations obtained from split-brain patients. The 1940s saw brief indulgence in the use of *corpus callosotomy*: the surgical separation of the two hemispheres of the walnut-like brain, aimed at the reduction of epileptic seizures[34]. It produced odd outcomes and conclusions seemingly in favour of split-consciousness. **With the surgical splitting of the brain, there is the expectation that two distinct individuals and identities must now exist; with each confined to its own hemisphere: i.e., split consciousness.** From it, it might be argued that consciousness *must* be produced by the brain, and that, if this were not so, the observed split of the original consciousness into two could not transpire.

From a materialist viewpoint, whatever is responsible for the production of consciousness in the brain must also split into two, per each hemisphere. Of course, if mind-brain dualism held true, the split-consciousness could not occur.

In the older data, split consciousness appeared to be proven, accompanied by certain anomalies: for example, **under normal conditions, split-brain patients did not report any split in identity and would not end in loggerheads with themselves: The identity remained unitary, and by implication, the consciousness also.** That the split-brain brain must also act in unitary fashion under normal conditions was inferred, despite the surgical split. **Yet, split identity appeared to be obtained if experimental blinders and restrictions were applied, and one could consult one hemisphere without the inclusion or knowledge of the other**[35].

However, **recent observations by Yair Pinto *et al* have contradicted the older findings that appeared to favour the splitting of consciousness**[36]. **Pinto's experiment demonstrated the reality of unitary consciousness and binding of both brain-hemispheres, despite surgical splitting, and despite the enhancement of the split by experimental arrangement.** Thus, both halves of the split brain appear to concert into unity.

*

How could a split-brain cohere into a single co-ordinated brain and retain unitary identity, even in the absence of experimental contrivances to exploit the split? **How could the separated halves become subject to unitary control that ought no longer to be possible** even when the experimental separations and blinders are removed? This is the central conundrum found in the old data. The

---

[34]Mathews, Marlon S.; Linskey, Mark E.; Binder, Devin K. (2008-02-29). "William P. van Wagenen and the first corpus callosotomies for epilepsy". *Journal of Neurosurgery*. 108 (3): 608–613.

[35]Wilson et al. 1977; Gazzaniga, LeDoux, and Wilson 1977

[36]Yair Pinto, David A. Neville, Marte Otten, Paul M. Corballis, Victor A.F. Lamme, Edward H.F. de Haan, Nicoletta Foschi, Mara Fabri (2016): 'Split brain: divided perception but undivided consciousness' in *Brain*.

unitary control of the brain by unitary identity abides: As stated, Yarn Pinto's findings concur with this, demonstrating unitary coherence of the brain and of identity despite the split, even when the split is enhanced by the experimental arrangement.

Again, how is any of this possible? **The key is de-spatialisation**.

*

In a physics that assumes space and 'the distribution of matter in space', cutting a block of cheese into two and separating the pieces across space does not appear to present any obvious problems: The same ought to hold in the surgical splitting of the brain. However, even in a physics that assumes the ontic reality of space, the two halves of the former cheese-block, now apparently separate, behave in inexplicable ways: Without obvious material mediation, the two halves gravitate toward each other into unity.

There is no denial of gravity as a real phenomenon. Yet, as we saw in **0.34**, hitherto, gravitation defies explanation. Also, while Newton's gravity, based on 'force', is now understood as having pragmatic value only, and was subsequently succeeded by General Relativity (GR), the search for ever-elusive quantum gravity admits that GR must itself be wrong: Quantum gravity seeks to replace GR, simply because GR cannot accommodate any quantum mechanical description.

With anticipated de-spatialisation, the notion of a 'curve in space' disappears per the ontic elimination of space itself, and we can then presage the overall character of quantum gravity in a de-spatialised purely temporised growing block universe: (see Book-IV).

However, unitary coherence of identity and consciousness in split-brain patients is not due to any gravitational relation between the two hemispheres. Gravitation is but a small and distinct aspect of implications that follow on from de-spatialisation, temporisation and the growing block approach.

How does de-spatialisation solve the problem posed by split-brain phenomena *and* explain the continuation of unitary concert of brains and of identity and consciousness, despite the surgical split? There is no space. Only time is real. Therefore, information, as both memory and possibility, remains a thing purely distributed in time. This new truism holds true for the cheese-block as much as it does to brains: The cheese block, and the brain itself, together with the larger order of nature in which these are embedded, constitutes a wholly de-spatialised complex distribution of information in time. Thus, we drop all the distance relations between the internal components of the cheese-block or the brain to itself, and we drop all the external distance-relations through which these relate to other 'objects' in the physical order. We replace all of these with pure time-relations: namely, delayed choice time relations... *without* space.

Again, it is easy to conceive the cutting of a spatial cheese-block into two. But now treat the cheese-block as a complex distribution in time and see if we can cut such a temporal distribution into two. Again, we can easily conceive cutting a spatial extension into two. But cutting a temporal distribution into two runs into immediate implausibility. In the case of the cheese-block, we ostensibly cut it into two, but we cannot ultimately separate either block from their common past: a relation not mediated in or across space (which does not exist), but one established in and across pure time; a relation of the abstract past to the relative present. Nor can we separate the now-cut cheese blocks from their common future potentiality or from their ontically real common futures. This is because we inhabit a growing block model universe.

Now treat the brain as a complex pure distribution of information in time within a de-spatialised purely temporised daz. Try to sever the brain from its past-history, from the larger order of information related to it in and across time, resonating from its past... a relation *not* rendered as a contiguous 'material exchange', or as 'information mediated by radiation or matter', but as a pure time-relation and abstract mnemonic superposition.

Even as we ostensibly cut the brain into two, from this new a superlative understanding wrought on by de-spatialisation and temporisation, we cannot sever the brain from its prior memory in time, nor from the larger order of memory in which it is embedded in a de-spatialised way... nor indeed from the common future-potentiality of the two hemispheres, in which both hemispheres will undergo grand decoherence to consistency in light of their common past in memory, *and* resolve into unitary activity thus.

Consider the mundane fact that that we cannot shield against gravitation. We can stop all manner of putative radiation and matter by means of apparent material barriers or other appropriate shielding and segregation. But we cannot block gravitational influences by such means. Why? This is because, in a de-spatialised purely temporised universe, gravity is an abstract memory effect: one realised per function of a pure temporal resonation of the past in and across time, *without* space. To block a gravitational effect would require that we sever an object from both its own past and from the larger past generally. But, as is already obvious, we cannot cut a time-relation into two: we cannot sever a system composed of pure distributions in time from its past, or from time, or even from their common futures. We cannot therefore block that past. We cannot therefore block memory effects. Hence, we cannot block gravitation.

The implications are obvious. The reason we obtain unitary brains, identity and consciousness in split-brain cases is per function of the fact that cutting the brain into two cannot cut off either hemisphere from their common past, which is retained as abstract memory in and across time, and which superposes upon the present brain, split or not, by means of a non-contiguous immaterial resonation across time.

What if the two halves of the brain were grown and developed separately from the very beginning: effectively as two distinct brains, albeit with reduced and absent functions? Never having formed together, we would indeed expect two distinct identities and consciousness per each hemisphere, in the same fashion that we obtain distinct identities and nous per two distinct whole brains.

What if the brain, split or not, were subject to gross continuous or systematic disruption per Cameron's annihilation-ECS? (See **0.47**). Therein, the ready ability of an otherwise unitary mind to utilise its means of expression in and through the brain would be broken. The identity would lose the ability to express itself through that brain. Therefore, the identity would be effectively lost. Yet, the pure unitary consciousness, or the possibility of unitary experience independent of identity, could no more disappear that it does in split brain patients. This pure unitary consciousness would quickly develop a new history of experiences… a new identity.

However, notice that **the ultimate point of all these remarkable implication from de-spatialisation vis-à-vis unitary brains, identity and consciousness in split-brain cases, is the culmination into memory-brain and mind-brain dualism.**

0.49: Comprehensive Summary of Part-IV

Part-IV deserves a comprehensive summary, given its controversial claims. But it will be brief about the memory-side aspect.

Memory-side problem in summary: The general and popular impression is that experiments in brain science and neuroscience 'prove' that memories are in brains, and that the mind is nothing but the brain. This is not so, save for the fact that **the assumption of materialism is loaded to the facts gathered from experiment, and then the facts are presented** *as if* **in 'proof' of the materialist contention,** (see **0.37**) even though dualism could save the same facts far more effectively, and even while the facts favour dualism.

- The consummate materialist **Karl Lashley searched for in-brain memory engrams over many decades. He failed** and was forced to conclude that memories must be distributed across the brain or functional areas: Thus, the first variant of holonomic or 'clouded' memory theory, albeit in the materialist mode (see **0.38**).

- **Karl Pribram later developed the intriguing holonomic theory of memory, elaborating on Karl Lashley's distributed memory**. It also assumed memories are materially stored in brains but clouded across it in the fashion of a hologram or holonome (see **0.39**).

- In the 1980s Rose and Harding performed experiments in which apparent localised memory-formations (presumably engrams) revealed by the glucose pump method would be surgically destroyed… only for attendant ability (and memory) to return regardless. **The experiment inadvertently falsified the assumption that the localised structures constituted memory,** otherwise the memories would have been permanently destroyed upon surgical destruction (see **0.40**).

- On the other hand, **the same experiments by Rose and Harding inadvertently falsified the materialist variant of the holonomic memory theory**: The glucose pump method did not reveal any generalised radioactivity, as would be expected if memory was reduplicated and clouded throughout pertinent functional areas of the brain (see **0.40**).

- **Thus, the experiments of Rose and Harding constituted a default crucial experiment, falsifying both materialist contentions about the nature of memory.**

- **By implication of the default findings of Rose and Harding, cases of amnesia do not constitute ready or automatic proof that memories are in the brain,** as Karl Lashley himself admitted, and in relation to which he posited his later distributed form of memory: the precursor to holonomic memory inadvertently falsified by Rose and Harding. Indeed, **amnesia is consistent with dualism,** insofar as material structures in brain can be treated as mere intermediaries and means of expressions of memories that are 'elsewhere', analogised by the end-user and his memories vis-à-vis his keyboard keys. Destroy the keyboard keys, you do not destroy the end-user's memories (see **0.42**).

- Alternative materialist contentions about how and in what form memories are stored in brains, even though some supersede the neuron, cannot escape the implications of the glucose pump method and of outcomes from default crucial experiments inadvertently carried out by Rose and Harding et al (See **0.41**).

- Autobiographical-episodic and emotive memories, presumably formed in the hippocampus and later presumed to migrate elsewhere, present a grander conundrum than the 'easy problem' procedural memories (see **0.43**) inadvertently and obliquely attacked by the experiments of Rose and Harding. **While procedural memories could be supposed as in-brain on-off, push-pull complexes of neuronal connections and activities, when it comes to hard problems, it is impossible to conceive how 'sweet', red', 'I love Lucy', and especially 'infinity' might be constituted as neural sets,** or as any other brain-structure: What does 'love' look like in neural terms? More especially, what does infinity look like as neural memory or 'code'? **In the case of infinity, the finite brain cannot compute it without crashing into halting problems. Indeed, where and in what form exactly would the brain store the outcome of such a computation**?

- In any case, **even if correlate structures to such hard problem states** *do* **form up in the brain or in the hippocampus, it is reasonable to assert that, if subject to the sort of experimentation carried out by Rose and Harding, the default crucial experimental conclusion would be the same: i.e., memories are not in brains.**

*

The mind-side investigations of the mind-brain problem, with obvious tandem conclusions about the nature of memory, remain critical and deserve comprehensive summarisation: **If mind and consciousness is not in the brain, then the memory attended by mind and consciousness is no more likely to be in the brain than is the attending nous itself.** Pure mind-side investigations can corroborate this assertion.

The Libet experiment and implications from a-synchronicity of consciousness with brain activity: (**0.44**). In the Libet experiments, **conscious volition (by implication, all types of consciousness) occur at a measurable time-delay** *after* **brain activity: thus, consciousness-brain a-synchronicity.** Per materialism, the conscious volition of a thing occurs as the output of a preceding non-conscious brain activity. Hence, the materialist conclusion asserts volition, if not consciousness itself, to be 'illusory'. However, from a dualistic perspective...

- The **a-synchronicity observed in the Libet experiment is a thing** *expected* **from dualism, but totally problematic to materialism**: If minding and consciousness could not be anything but brain activity, then consciousness ought to be perfectly synchronous with brain activity. But it is not. If consciousness does not occur at the same time as brain activity, then it cannot occur at the same 'place' as brain activity: Thus, the a-synchronicity is seamlessly natural to the dualist approach:
  - Per dualism, the **decision and initiation of brain activity by extant mind would take place temporally** *before* **the subsequent brain activity** and could not be evinced or clocked as the brain activity, or as *any* activity or event.
  - Therefore, **the putative moment of volition, insofar as it temporally precedes and does not involve brain activity, and cannot be evinced as such, cannot be synchronous to brain activity. Hence, the observed a-synchronicity in Libet experiments.**
  - Per dualism, externally sourced stimuli trigger intermediary brain activity. The activity is received by the extant nous *after* that brain activity. **The attendant extant consciousness of said events must also occur after environmentally induced brain activity** *and* **after brain activity.**
  - Thus, per dualism, **mentality happens** *before* **brain activity and** *after* **brain activity, respectively**. Mentality and, with it, **consciousness or volitional consciousness, cannot be produced by the middle-term intermediary brain activity.**
  - In short, **a-synchronicity in evidence seamlessly arises from dualism.**

*

Implications from the experiments of David Nutt combined to findings from Libet: (**0.45**) In the experiments of Nutt *et al*, subjects were injected with psychedelic psilocybin, with the aim of assessing the impact of the psychoactive material on the medial prefrontal cortex and the anterior cingulated cortex of the brain: candidates for a 'hub of consciousness' and for coherent unitary brain control. The brain ought to have gone into over-activity. However, ...

- The stated regions showed no activity at all. **The brains of the subjects were running cold, dark, and silent, equivalent to the unconscious state.**
- **The finding inadvertently attacks the notion that 'brain activity produces consciousnesses'**, given that vivid psychedelic consciousness seems to occur even in the absence of brain activity. **Consciousness does not seem to require brain activity.**
- If one can demonstrate consciousness *without* brain activity in one domain or experiment, the evidence of brain activity *and* consciousness in another can no longer be construed as requisite to the possibility of consciousness, even if brain activity is evident in the latter case: That is, **brain activity is correlate with consciousness, but** *not* **necessary to consciousness.**

In relation to the Libet experiment...

- If ostensive brain activity is not the real producer of volitional or *any* consciousness, as inadvertently revealed by Nutt *et al*, then **hitherto interpretations of the Libet experiment, and generic materialist assertions from these, if not from the whole of brain science and neuroscience, are immediately rendered suspect.**
- A dualistic solution will arise from the finding that, **in a pure de-spatialised and temporised physics, nous must either temporally precede the manifestation and resolution of world and brain-events, or it must be temporally antecedent any past-constitution of world and brain-events, without reducing to brain or world events.** In either case, consciousness is not going to be synchronous with brain activity, or arise from brain activity or, as Nutt has shown, even require brain activity. **The dualism of consciousness vis-à-vis the brain must follow.**

*

Implications from the Montreal procedure and the Libet experiments combined: (**0.46**) In the Montréal procedure, Penfield elicited non-volitional memory recollection. **Penfield found no basis or corpus for decision-making or for the unitary binding in the brain.** We summarise both facets of the implications from the Montreal procedure in the following. In the case of elicited memories...

- Per findings from Libet's experiments, the consciousness of non-volitional memory elicited in the Montréal procedure is simply not going to occur at the same time as the electrical stimulation of the attendant brain activity. **The consciousness of memories in the Montréal procedure must occur 'outside' of, and inevitably temporally** *after* **the electrical stimulation vis-à-vis the witnessing consciousness that comes after (and not from) the brain activity.** Also...
- Memory-side investigations from Lashley, Rose and Harding *et al* showed that memories are not *ultimately* identifiable with *any* structure or process in or of the brain. Hence, **electrical stimulation of brain tissue in the Montreal procedure only elicits the mere means of expression of otherwise dualistically associated extant memories.**

In the case of the unitary binding of brains and how this is produced...

- Per materialism, a consciousness-producing corpus would need to transmit a traceable signal to the location of the memory. The memory thus triggered would then generate a feedback signal. This latter would need to propagate back to the consciousness-producing corpus in order to register in consciousness. **No such effect is observed.**

- On the assumption of a consciousness and unitary co-ordinating corpus, how would such a corpus know which part of the brain contains the sought memory, if the pertinent memory is *not* coincident with the consciousness-producing corpus itself, but located elsewhere in the brain?

- In the Montréal procedure, **the non-volitional consciousness of the elicited memories anywhere in the brain is not produced by *any* specialised part of that brain. The consciousness simply arises relationally (albeit dualistically and, per Libet, a-synchronously) vis-à-vis the stimulated region.**

- **This must also hold true for volitionally triggered memory: it is not produced at a special corpus, but from 'outside' the brain and the region volitionally stimulated.**

- **The unitary control of the keyboard-brain is similarly imposed on that brain from 'outside' of it,** by an end-user extant dualistic minding. This exploits information-matter dualism, and, ultimately, the de-spatialised pure temporised structure and processes of nature and the growing block universe; all on a work function zero basis.

- Hence, mind-brain dualism.

*

Implications from Donald Ewen Cameron's annihilation ECS (0.47) proved to be capable of achieving total memory-wipe. But the critical point, easily obscured from the apparent destruction of memory, was that Cameron's **annihilation ECS could not result in the annihilation of consciousness itself.** Cameron's work also inadvertently completed the work of Penfield: **augmenting and validating Penfield's view that consciousness, and the basis of the unitary binding of the brain, could not be in the brain.** But how could it be that unitary consciousness can abide despite Cameron's annihilation ECS, while memory cannot?

- **The minding process responsible for consciousness is not at all located or reduced to any specific part of the brain and cannot be destroyed in brain-sites irreversibly disrupted by annihilation-ECS.**

- Dualism, *not* materialism, must have the final say on how we make sense of outcomes from Cameron's annihilation-ECS.

- The loss of memory is due to permanent irreversible disruption of the means of expression; the brain-equivalent of keyboard terms that the end-user mind needs to express its memories. The ECS disruption of such readiness states implies the disruption of the means of expression. This in turn means the impossibility of expression. Effectively… no expression, no recollection. Yet, the consciousness, the "I" … abides, since it is not produced by the readiness states, no more than the analogous end-user is produced by the keyboard keys.

*

Implications from split-brain experiments of Pinto *et al*: (0.48) With the surgical splitting of the brain, there is the expectation that two distinct individuals and identities must now exist; with each confined to its own hemisphere: i.e., split consciousness. What happens is a contradiction:

- **Split-brain patients do not report any split in identity** and never get into loggerheads with themselves. **The identity remains unitary despite the split,** and, by implication, the consciousness must also remain unitary.

- **Yet, seeming split identity could be obtained if, under experimental conditions**, one could consult one brain hemisphere without the inclusion or knowledge of the other.

- **But recent experiments by Yair Pinto *et al* demonstrate unitary consciousness and binding of the brain *despite* surgical splitting** and despite the enhancement of that split by the experimental arrangement.

- **The key to unravelling this conundrum is de-spatialisation:**
  - There is no space: **The brain is not a distribution of matter in space but a distribution of information and memory in pure time, itself embedded in a grander system of information and memory organised in time.**
  - It is easy to conceive cutting an object in space into two separate halves. But we cannot in truth cut a distribution in time into two halves. How does one 'cut' time? Ostensibly, **surgically splitting the brain into two cannot sever it from its whole history of abstract memory in which it is embedded in de-spatialised purely temporised form.**
  - The reason we obtain unitary identity and consciousness in split-brain cases is per function of the fact that **cutting the brain into two cannot imply the separation of either half from a common past through which the original identity (hence the unitary singular consciousness) of the patient was formed, and in which it is perhaps retained as a de-spatialised distribution in time**: a unitary identity not reducible to or identifiable with the manifest relative present brain, whether or not the latter is split.
  - **The unitary identity is extant the split brain, and can co-ordinate that brain in a unitary way, despite the split. Cartesian dualism follows**, albeit, without *res extensa*.

PRELIMINARY FINAL DISCUSSION: MIND & LEGITIMATE PHYSICALISM

The outcomes obtained from the experiments of Rose and Harding constituted inadvertent crucial experiments that pitted the materialist localised engram notion of memory against the materialist holonomic version. The implicit and inadvertent crucial experiment falsified both, usurping the notion that memories are deposited in brains. For its part, the Libet-Nutt Bridge constituted another inadvertent crucial experiment: showing that vivid consciousness *can* transpire even in the absence of brain activity: breaking the presumed set relation between consciousness and its 'causative' brain activity; augmented by the a-synchronicity between consciousness and brain activity revealed in the experiments of Libet that were implicit to the experiments performed by Nutt *et al.*

Finally, with the hitherto failure to find a hub to produce consciousness, or for any basis for the unitary binding of brains from within the brain, the Penfield-Cameron Bridge constituted another inadvertent crucial experiment: one that tested the association of consciousness with memory and with the general unitary binding of brains. Therein, the failure to find a hub to produce consciousness in the brain (from Penfield), combined to the fact that, in the Libet-Nutt Bridge, consciousness can transpire in the absence of any brain activity, almost completes the mind-side case against the materialist dogma. But completion is obtained with the outright crucial experiment of Pinto *et al* and the remarkable solution to the split-brain conundrum furnished by de-spatialisation... from the fact that, in a de-spatialised pure temporal physics, we cannot ultimately 'cut' or 'split' any temporal distribution, much less the brain recapitulated as a temporal distribution, or from the larger whole of de-spatialised memory-in-time in which the temporal brain is embedded,

Hence, the ready 'scientific' presumption in favour of materialism in matters of memory and consciousness vis-à-vis the brain, is no longer tenable. At the least, one must concede that we cannot automatically assume the veracity of materialist contention, if at all. Indeed, the evidence favours dualism. Yet, groundless proclamations portray experimental facts from neurology and brain science *as if* these support the materialist contention. This is not merely because of the stated fallacy-chain (see **0.37**): Materialism is not merely an idea, it is culture-defining ideology: the implicit civic religion and core of an establishment civilisation that almost bestrides the world. This has two consequences: First, through portraying a false consensus built mainly on ideology and blind sight, it disables any honest evaluation of uncovered facts (e.g., assumption loading: again, see **0.37**). Hence, evidence contrary to the materialist ideology and possibly in favour of dualism is liable to instant dismissal, if noticed at all.

Second, the presumption of unquestioned materialism leads to a thing very dangerous to human progress: Materialism constitutes an ideological, irrational, and wholly erroneous postulate about the ultimate ontology of nature, the universe, and mind; positing itself as 'scientific', or even as the defender or basis of 'rationality' itself. Materialism is ultimately as anti-scientific and as irrational as any religion, as all *a priori* beliefs tend to be. As such, it cripples serious study and advancement *from within* the scientific establishment. Irrational forces outside of science pose enough of a threat. Recalcitrant irrationality from within the scientific establishment virtually cripples the advancement of humankind, entirely dependent as such advancement is on our discovery about what the universe is really like... what life is... and what *we* are.

*

Despite the nature of the evidence inadvertently uncovered, Rose and Harding may well insist, even today, that memory must somehow be stored in the material of the brain. As far as one knows, the fact that their experimental approach implicitly pitted the engram versus the holonomic paradigm of memory, and that it falsified both, was never posed by Rose and Harding, or by anyone else since. That the total implication of all of this must be the comprehensive undermining of any materialist conception of memory remains obscured, despite the obviousness of the evidence. Any dualistic counter-hypothesis, or any other non-materialist hypothesis, is *not* posed, not even as a false profile against which to bounce facts and ideas. Dualism is *prima facie* considered false, 'anti-scientific'... 'religion-inspired', and irrational. Ultimately, materialism is the only avenue permitted.

Another example: Benjamin Libet had misgivings about the supposed detrimental implications of his findings to volitional consciously directed will: an artefact that emerged from the presupposition that consciousness is produced by the brain: an assumption never questioned. Hence, to save the possibility of conscious volition, Libet developed a 'mind field theory', whose only real merit is its consistency with established materialist and anti-Cartesian presuppositions[37]. Yet, the fact that the solution to the problem was always clear in the very grain of the facts uncovered (namely, that consciousness does not occur at the same time as brain activity, and therefore cannot occur at the same 'place' as brain activity) did not emerge even as a hypothetical stance... because such a hypothesis is obviously dualistic. Yet, minding must either occur temporally *before* brain activity or, in the case of requisite feedback, it must occur temporally *after* brain activity, but never at the same time as brain activity (the central finding of Libet), and *never* identifiable with or reducible to brain activity in time, as compelled by a-synchronicity. The dualism of consciousness vis-à-vis the brain must surely follow.

When we now bring in memory-side conclusions in favour of dualism and combine these to the facts uncovered by Libet and others, dualism becomes the only conclusion permissible per the total evidence, and that memory is no more 'in the brain' than is mind and consciousness.

*

In order to maintain materialism against the grain of reality, or else to avoid the obvious dualistic conclusions implied by the evidence, one approach to the findings from Libet consist in retro-causality: wherein the brain operates on itself *from the future*: presumably through some physical signal that travels backwards in time. Retro-causality was proposed as part of the Penrose-Hameroff hypothesis of *orchestrated objective reduction*. A similar notion might also be imagined in terms of the transactional interpretation of quantum mechanics[38]. However, the brain acting upon itself from the future in the way imagined would imply the violation of principles

---

[37]Benjamin Libet, Mind time: The temporal factor in consciousness, Perspectives in Cognitive Neuroscience. Harvard University Press, 2004. pp 157-184

[38]The Transactional Interpretation of Quantum Mechanics by John Cramer. Reviews of Modern Physics 58.647-688, July (1986)

of causality *and* the principle of conservation, as all 'signals from the future' would imply. In any case, even if one could circumvent such violations, retro-causality cannot be tested. By its very nature, it is conveniently beyond experimental investigation. Like materialism, retro-causality can be *assumed* and loaded onto the facts. The facts will then appear consistent with the contention.

Such dubious approaches illustrate the operation of a dominant implicit motto: *Anything goes, no matter how crazy it is, so long as it is not dualism.* Radical though such speculations appear, as is the case with Pribram's materialist holonomic theory on memory, they are inadvertently conservative and reactionary: arising from an automatic almost-unconscious aversion against obvious dualism.

Then there is the more mundane non-conscious ideological abuses of evidence that we have exposed: i.e., the fallacy-chain. These perpetuate the materialist ideology by loading materialist assumptions to uncovered facts, and then presenting those facts *as if* in 'proof' of the materialist assumption: such as per the facts of amnesia; or the claim that the materialist view is 'proven' from the electrical stimulation of brain tissue in Penfield's Montréal protocol; or the notorious assumption-loading to the experimental findings of Libet. All such facts are unquestioningly presented as 'obvious proofs' of the materialist contention on the nature of memory and consciousness, even though the same facts are equally consistent with dualism, if not better served by dualism, as we have shown. The claims made in favour of materialism are made as if conclusive and, like all other ideology-driven proclamations of 'evidence', consciously or unconsciously end all further discussion or examination of the same facts. The dismal situation gains succour from the false association of both physicalism and naturalism with materialism. This is highly dubious per the following intimations and discoveries that will only gain augmentation and amplification in the books that follow the Preliminary. Thus...

- **Facile information-matter dualism** leads to the breakdown of the association of information vis-à-vis radiation, matter and 'space' (see **0.01 to 0.04**: to be further augmented in Book-I).

- **The breakdown of the materialist conception of causality in the face of the measurement problem** (see **0.20 to 0.21**, and further developed throughout this work).

- **The anticipated structure-less form of causality responsible for both AND-to-OR wavefunction collapse (i.e. time itself) and in the quantum indeterminate resolution of outcomes**; implying a form of causality that is extant both 'matter in space' and the brain (see **0.20** and **0.22 to 0.23** fully developed in Book-III).

- **The intermediate quantum-relativistic incorporative spacetime model** whose larger ontology exceeds, supersedes and is *extant* the facile notion of 'matter in space' or 'brains in space'. (See **0.29 to 0.30**, and fully developed in Book-II).

- **De-spatialisation, temporisation, and the unravelling of the notion of 'space'** in the anticipated successor to the intermediate model spacetime: namely, the Bergson-Whitehead amalgam... and the incontrovertible **necessity of grasping all things, including the brain, as complexes of information distributed in pure time, without 'space'**: forcing conclusions in favour of mind-brain dualism (see split-brain conundrum per **0.48**): Indeed, de-spatialisation makes for a mind-brain theory without the problematic notion of *res extensa*, or 'space' in classical Cartesian dualism (see **0.29 to 0.30**, but fully developed in Book-IV).

Book-I through to Book-III will develop the solid basis for these findings. By Book-III and IV we will have expunged physics of the notion of 'space' based on the following...

- **The notion of 'space' cannot be empirically justified from the structure of quantum mechanics**.

- **The notion of 'space' can be fully dispensed with**: We need only map events purely in terms of time, *without* 'space': i.e., delayed choice time-interval relations.

- **The best we can obtain are spatially static quantum mechanical waves** embedded within an as-yet not-realised quantum mechanical future-form Hilbert domain: effectively a non-space, or nonsense. Finally...

- **A specific application of John Wheeler's delayed choice experiment to analyse and attack both the notion of the motion of quantum mechanical waves *and* the notion of their 'space'** in and through which such purported motion supposedly occurs, **usurps the viability of both**.

The usurping of the notion of 'space' alone derails any materialist contention about causality and in the mediation of causality through contact-points in space, supposedly by spatial matter and radiation in motion and in 'collision'. De-spatialisation also usurps the expectation that any mind-brain causality must also proceed through material transactions and contact in a 'brain-in-space'. **By Book-IV we will fully articulate just what it is about the brain that allows for dualistic mind-brain interface, as part of a Cartesian revival, without *res extensa*.** Therein, we will discover that we need not reduce consciousness and mind to brain activity at all, much less assume it to be 'created' by any part or corpus of that brain.

In short, the mind is not the brain. Cartesian dualism follows suit.

BOOK-I:
INFORMATION-MATTER RELATIONS IN DEFINITION

CONTENTS

BOOK-I:
INFORMATION-MATTER RELATIONS IN DEFINITION

AIMS OF BOOK-I

Recall from Preliminary **0.01** to **0.04** that information-matter dualism involves the apparent physical dissociation of information from radiation, matter and space, *and* the retention of information in abstract remove from the same: the preliminary case for the existence of abstract memory in our explorations. The pivotal aim of **Book-I is the deeper case for information-matter dualism**. To this end, it will scaffold for a more refined evolving and working definition of physical information, one conducive of a deeper account of information-matter dualism. However, we forewarn that information-matter dualism itself is only the surface of a more radical, purely temporised and de-spatialised ontology, framed to the growing block universe, furnished by the Alternative Realist Thesis (Alt-R). Consequently, information-matter dualism will itself be superseded by the structure and processes of the growing block universe, even if its essential features, such as effective information-dissociation and abstract memory, will abide. The supersession of information matter dualism will be fully accomplished by Book-V, and much of Book-I will only seek a deeper capitulation of surface information matter dualism on the assumption of the ontic reality of space, and by directly utilisation of generic quantum mechanics.

*

Book-I Part-I will grapple with the definition of physical information requisite to the deeper view of information-matter dualism. To this end, Part-I will innovate information-entropy relations. It will do so by developing the quantum mechanical theory of **entropy of future-potential information** and, with it, furnish tandem **entropy of memory**: both of which are implicit to information-matter dualism and Alt-R; precipitating solutions to various problems, such as the irreversibility and time; furnishing a provisional definition of time itself beyond its supposition as the 'fourth dimension'; and securing the harmonisation of apparent time-symmetric physical laws with time-asymmetric change: with partial solutions to the perennial problem of permanence and change. Indeed, we will arrive at the remarkable conclusion that **physical laws are time-asymmetric, *not* time-symmetric, against the staple claims of generic physics.**

Part-II will be pivotal to Book-I: It will offer four of the **five principal arguments for information-matter dualism** beyond the facile case furnished in the Preliminary, and explicitly ground it in quantum mechanics: in AND-OR dualism; in quantum indeterminism; and other staple quantum mechanics features.

Part-III will justify the reality of quantum mechanical **AND-form logic as part of the objective ontic reality of the future: the future phenomenalised per AND-form logic, phenomenalised as the quantum mechanical wave**. We shall discover that the ontic reality of the future is **demanded by the firewall principle** that, within the framework of the growing block universe, demarcates the realised past from the potential future; a demarcation which emerges from the intertwine between causality and the principle of the conservation of energy-matter.

Insofar as putative information-matter dualism arises from the ineliminable reality of AND-form logic, and insofar as the ontic reality of AND-form logic is forced by the intertwine of causality and the conservation principle, from the firewall principle that these latter constitute, the **fifth argument for information-matter dualism will emerge: information-matter dualism per the firewall principle**: information-matter dualism from the intertwine of causality and the principle of conservation.

Part-IV will list key spin-offs from the five arguments for information-matter dualism: **Decoupling theory** will relate information to the putative speed of light itself, centred on the anomalous-seeming experiments of Gunter Nimtz. Decoupling theory will augment our case for abstract memory and for dualistic mind-brain relations. It will also advance our understanding of information and augment the case for information-matter dualism.

Finally, Part-V will posit the clear distinction between **noetic information** specific to mind, consciousness and cognition, versus information specific to the physical order: Succinctly, via **rigging theory**, combined with key hard problems of consciousness, Part-V will develop an EPR-based thought experiment to shed light on **the code-information-distinction**; an augmentation of John Searle's Chinese Room experiment, culminating into **the case against strong artificial intelligence**, with unexpected outcomes that presage mind-brain dualism anticipated in Book-V.

BOOK-I PART-I:
SEARCH FOR THE DEFINITION OF PHYSICAL INFORMATION: INFORMATION, ENTROPY & THE "ARROW OF TIME"

1.01: Aims of Part-I

Before developing the case for information-matter dualism in Part-II, we must furnish a **working definition of physical information**. This will involve a brief history of both pre-scientific and modern ideas on information, culminating into a **critique of**

mereological nihilism, which rejects the ontology of both wholes and of information. The search for the definition of information will reach completion by Book-III and IV. In Part-I, our evolving definition will incorporate such things as **read-only versus read-write categories of information**, combined to the dualism of **AND-form versus OR-form information**: both relevant to quantum mechanics and to information-matter dualism.

In the course of Part-I, we will also integrate to our categories of information **information-entropy relations**, with the specific aim of combining and harmonising generic definitions of entropy, information-matter dualism, and generic quantum mechanics. The centre piece will be **entropy of future-potential information**. The reasoning will be thus: If the quantum wave *is* the phenomenalisation of an ontically real future, and if wavefunction collapse *is* synonymous with time itself, the perpetual decay of the future must constitute the perpetual entropy of the ontically real future: constituting the perpetual entropy of future-potential information via AND-to-OR collapse. In short, ultimate entropy entails the irreversible decay of the ontic future.

By **harmonising entropy with information-matter dualism**, we will obtain unexpected solutions to the perennial problem of permanence and change, resolving the relation between apparent time-symmetric physical laws versus time-asymmetric phenomena; establishing **a firmer basis for the perennial 'arrow of time'** than that obtained from statistical generic entropy. Through these developments, we will obtain an unexpected conclusion for **time-asymmetric physical law, against the generic claim for time symmetric laws**.

1.02: The Definition of Physical Information:
The Historical Background to Information-Matter Dualism

We now reiterate in brief the case for facile information-matter dualism first adumbrated in the Preliminary. Therein, we discovered that, in nature, there is no such thing as seamless-continuous motion or trajectory: Such a feat would require observational intrinsic zero wavelength infinite frequency inputs of radiation and energy. Also, quantum mechanics demands that both the observation of motion *and* motion in itself cannot be seamless-continuous and must be broken into 'slices', 'frames' or 'snapshots' across putative space and time, interspersed with gaps about which no observation can be made and to which no resolved attribution can be posited.

What holds true for motion must also hold true for the transportation of physical information across putative space by radiation and matter, under motion or under time-evolution. The information thus 'carried' is not rendered in seamless-continuous form across 'space': Consequently, the information 'carried' must dissociate from radiation, matter, and space, across pertinent gaps, only to re-manifest amidst the same in the form of a finite number of broken-discontinuous 'snapshots' or OR-form resolutions. Across the intervals or gaps between these 'snapshots', the information apparently dissociates out of quantum mechanical radiation and de Broglie matter into an extant form. Hence, the physical 'transportation' of information across space is dualistic and dissociative vis-à-vis putative 'carrying' radiation and matter.

Note that, in facile information-matter dualism furnished in the Preliminary (see **0.01** to **0.04**) information was defined by reference to specific examples: such as a photographic image manifested by light radiation; the electron micrograph image evinced by *en masse* electrons; the snowflake pattern exhibited by a drifting corpus of matter; and as the organisational content manifested in and as the material of the brain. Also, in the Preliminary, information was exemplified by say, resolved quantum spin-states assumed by quantum entangled couples in EPR experiments. Indeed, the principle of the conservation of spin itself was asserted as an example of fundamental information, exemplifying dualistic Platonic-Cartesian physical law.

However, what the Preliminary lacked was a more comprehensive working general definition of information: a lack we can get away with in most cases, insofar as we can state a specific configuration or concern as constitutive of information. We must seek greater clarity, and the following historical background and evolving definition of information should help amplify further the nature and definition of information.

*

In the most general terms, **physical information can be defined as that which informs structure and pattern amidst putative radiation, matter, and space**. Information becomes tangible through apparent embodiment in radiation and matter. This makes information distinct from and unlike the 'substance' or matter through which it becomes manifest and which it informs. We need only grasp the critical distinction that holds between, say, a pile of bricks that form a house versus the plan, design or blueprint that organises, informs, and forms that pile into a house. In that context, information is discernible as the abstract-immaterial design and organisation that operates over the bricks in order to bring into effect the house. Thus, **information is the architectonic immaterial principle, or over-arching abstraction, distinct from the constituent elements:** the bricks, the mortar, cement, girders, etc… that form the house, distinct from the 'matter' subsumed to its informative and formative powers.

The mere distinction of information versus 'stuff' does not automatically lead to information-matter dualism of a form in affinity with our facile information-matter dualism introduced in the Preliminary. It merely makes for a limited classical Aristotelian understanding of information-matter relations, implicit to contemporary natural science, especially in those sciences in which the subject of information and complexity is earnestly explored: such as in genetics; in chaology; and in complexity theory. **In classical Aristotelian dualism, form, and substance (i.e., information and matter) are asserted to be inseparable.** According to Aristotle, information and matter must be grasped as **incapable of existing or subsisting independently, albeit discernible as a distinct aspect of a non-dissociable dyad.** As such, Aristotelian dualism cannot allow for a radical dualism *with* the dissociation of information from matter, or of the sort argued for in facile information-matter dualism. (See Preliminary: **0.01** to **0.04**).

QUANTUM MECHANICS AND MIND

The only dualism that permits the dissociation of information from matter was historically prototyped in Plato's theory of forms. For Plato, matter, or 'appearance', was held to 'partake' in superlative abstract principles: the Platonic forms. Therein, **the forms supersede and cannot be identified with 'appearance' or matter so informed.** Accordingly, matter attains intelligibility, and is entirely dependent on abstract-immaterial Platonic forms. Hence, **the form (hence the information) can subsist and exist in the abstract, independent of 'appearance' or matter... and can separate out from matter into pure abstract form.** Consequently, Plato held information, the forms, to be immune to the vagaries of change, decay, or *time*: To Plato, information, and the intellect that apprehends the forms, and is akin to the forms, must transcend and survive material change, decay, and dissolution.

Plato's view of information constituted the first truly comprehensive attempt, at least in the White Sea world, at a solution to the perennial problem of permanence and change: a concern that survives even into contemporary natural science. For example, the problem resurfaces in the question of supposed time-symmetric reversibility of physical laws and in the context of physical irreversibility; in physical entropy, change and time. All of this falls under the rubric of the question of permanence and change, reframed to modern physics terms.

*

What possible value could archaic Aristotelian and Platonic views on information-matter dualism have vis-à-vis contemporary natural science, or even Alt-R? Aristotelian and Platonic views are of no less value to science than the much-vaunted ancient atomic theories of Epicurus and Democritus. Of course, the latter, in their original and naïve forms, do not meet contemporary standards. Yet, their historic value is appreciated, often in over-stated fashion, as precursors to present-day chemistry, atomic and particle physics.

The archaic informational theories of Aristotle and Plato must be rightly appreciated as historically prototypical to the concerns of contemporary natural science. As already intimated, Aristotle's view of information-matter relations is implicit to most perspectives on physical information in biology and in cybernetics. Although contemporaries would be loath to admit to Aristotelianism within natural science, this denial is more a consequence of a historic hangover from the old demoded conflict between scholasticism (mostly founded on abuse of Aristotle) versus the then-nascent Galilean physics: The latter undermined Aristotelianism by rightly attacking Aristotle's erroneous conceptions motion. Finally, insofar as natural science must entertain a physics of information, it does so in implicit conformity to the Aristotelian conception of information-matter relations: conceiving physical information as bounded to and inseparable from 'carrying' or exhibiting radiation, matter and space. While no one would rightly endorse Aristotelian ideas of motion, the implicit position on information-matter relations in modern science is, by and large, a modernised version of the Aristotelian form and substance relation, as we shall discover.

If contemporary science is loath to openly admit any concession to Aristotelian information-matter dualism despite relying on it implicitly, it is liable to turn apoplectic vis-à-vis any concession to Platonic dualism. Yet, contemporary science implicitly concedes to Platonic dualism, succinctly in the context of abstract-immaterial physical laws vis-à-vis radiation matter and space… or, as Plato might have characterised it, "…timeless abstract laws versus the domain of appearance, substance and change"… even to the point that physical laws are characterised as time-symmetric… able to operate from past-to-future as well as from future-to-past: This is an error, as we shall discover in **1.12**. Yet the Platonism implicit to that error is obvious and abides: Physical laws *are* Platonic (see **1.13**).

However, it is through our case for information-matter dualism, even the facile variant from the Preliminary, that a more profound relation of information to radiation, matter and space is uncovered: one that favours the form of dualism espoused by Plato.

*

Epicurus and Democritus are historically recognised and over-vaunted as having anticipated the physics of the atom, of nuclear physics and particle physics. Given the reality even of facile information-matter dualism, and the implicit Platonism of physical laws, we must now forge a more complex picture of physics: one that includes, but radically supersedes, Epicurus and Democritus, and makes appropriate concessions to Aristotle and Plato. Indeed, the Platonic and Aristotelian anticipation of a physics of information, of which the Platonic variant turns out to be the more important one, promises to be far greater in importance to the future of natural science than the contributions of Epicurus and Democritus. It was Plato who came closest to the truth about the nature of information and its relation to radiation, matter, and space, even if his methodological *a priorism*, and, with it, his apparent dismissal of the empirical world, must remain suspect. Our forthcoming five physical arguments for information-matter dualism will incorporate but transcend the Aristotelian dualism in support of Platonic dualism, which entails the outright physical dissociation and independent retention of information from putative radiation, matter, and space.

With de-spatialisation in Book-IV, and the growing block universe framework that attends it, our case for information-matter dualism and Platonic dualism will be complete.

1.03: Definition of Information & the Problem of Mereological Nihilism

With the general definition of information as 'that which informs', and with the roots of the idea of information prototyped in Aristotle and Plato, and with their historic relevance to contemporary natural sciences appropriately appreciated, we must now address the ontology of information: **Is information something that really exists, or is it a *bloss* subjective illusion, not part of real existence**? Mereological nihilism, asserts wholes to be illusions. Consequently, that which informs (i.e., the abstract wholes that inform and form up parts into wholes) must also be illusory.

Mereology is the study of the relation of parts to wholes. As such, it is critical to information issues insofar as information is broadly understood as 'that which informs' and integrates the parts to other parts so as to form wholes.

97

In its Aristotelian variant, mereology assumes that the whole that relates and informs the parts cannot separate out from the parts (the substance) interrelated into and by the whole. The Platonic variant on mereology makes the further claim that the whole that relates and informs the parts to each other and to the whole *can* separate out from and subsist independently of the parts that collectively manifest that whole. Thus, the blueprint for, say, a house, is not merely intellectually discernible from the parts that make up that house, but, in the Platonic understanding, it is possessed of an ontology that can subsist at a remove from the parts and from the manifest house.

With anticipated de-spatialisation, both the Aristotelian and especially the Platonic claims on mereology become obvious, but only on the assumption that information has real ontology. De-spatialisation will reframe the understanding of wholes, from distributions and relations actuated in 'space' to relations and entities distributed only and purely in and across time… wherein wholes and identities transform into projections in and across time, can endure and perdure as such, and can dissociate from any relative present state of most recent OR-form manifestation into an extant form; succinctly, into abstract memory… all on the assumption of the ontic reality of information.

However, mereological nihilism, born out of materialist philosophy, and subject to a heavy spatial bias, claims that only the basic building blocks of the physical universe are real (or partology), whatever these might turn out to be; and that the apparent wholes built up out of these are not real and are not parts of real existence. In short, whether conceived in its Aristotelian variant or in its Platonic variant, **according to mereological nihilism, information is not part of real ontology or existence: 'that which informs' does not really exist. Consequently, both information-matter dualism and prospective temporal abstract memory is seemingly usurped by mereological nihilism.** Indeed, in Book-III, mereological nihilism will combine with the Manyworlds Interpretation of quantum mechanics to form an anticipated ultimate nihilism in rejection of information-matter dualism, the very ontology of information, and abstract memory. But it will fail.

Mereological nihilism can be traced back to the atomic philosophies of Epicurus and Democritus: These atomic philosophies claimed that the only real existents are *atomos* and void, and that the *apparent* patterns or mereological states that arise purely through their contingent relations, are mere appearances: not fundamental to existence: part of a transient perishable phantomatic shadow-order, manifested through the mindless purposeless motion of *atomos*.

There is close affinity between the *atomos* of ancient materialism versus contemporary mereological nihilism at the background of eliminative and other materialist views on mind-brain relations: Since only the atoms or fundamental particle building blocks are real, and wholes are not parts of real existence, mind as mere mereology emerging from *atomos* is also not real… or so it could be argued. Clearly, non-nihilistic Platonic-Cartesian mereology could not survive either classical or modern materialist mereological nihilism, if these held true.

*

Mereological nihilism is easiest to grasp per artificial objects. One can assert that a chair does not really exist: The chair is neither necessary nor compelled by its constituent building blocks, atomos, or fundamental particles: the collection of electrons, protons, neutrons, and other fundamentals that make it up. Nor is the chair necessary or compelled by any universal physical laws that govern the said fundamentals. The same fundamental components and laws could easily constitute any other *apparent* form, pattern or whole. The form or whole would be contingent, and ontically pleonastic thus.

For the same reasons, a house, as a compound of its building blocks is non-existent for essentially the same reasons as holds for the chair. Building blocks could as easily admit of any other apparent purely contingent and pleonastic form other than the house, itself ultimately indistinguishable from a rubble pile of bricks: The 'house'… is an illusion.

Per the same reasoning, a message coded in and transported by electromagnetic radiation or radio carrier-wave is not there. The only real existents are the electromagnetic photons and pertinent physical laws. The apparent 'message' is not a necessary or compelled outcome of the real existent electromagnetic constituents: The 'message' does not actually exist.

*

Is mereological nihilism in error? The form of the chair *is* really-existent: Its form predicates a certain range of configurations for physical laws, and, in doing so, configures unique consequences (future possibilities) specific to chairs. That is, the 'illusory' form of the chair structures basic physical laws in such a way that some possibilities are plausible upon it, while others are not. Indeed, using the same mass and materials that make up the chair, and obeying the same laws, we might construct a spear instead of a chair? While both can be used as weapons, per function of the way their respective forms structure putative physical principles and constituents, they cannot be used as a weapon in the same way, or with exactly the same efficiency and effect. With a spear, one could skewer a target, because the spear structures pertinent physical laws and constituents into a form that allows for such a possibility. But the chair cannot because it structures physical principles and constituent fundamentals entirely differently vis-à-vis how a spear accomplishes it.

How is it possible for an illusion, a non-existent whole and form, to structure physical principles and constituents, and delimit the range of possibilities, one way and not another, as is obvious from the contrast between the spear versus the chair? The answer is simple: the 'illusion' is not an illusion at all. The chair is real by virtue of the fact that, even though its form is not compelled by the basic constituents and physical laws, but permitted by the same, and certainly not prohibited by the same, it has the power to structure, form and *inform* the laws and constituents in a unique and specific way vis-à-vis other forms.

If houses do not really exist then why did the residents of London, Berlin, Stalingrad, and Hiroshima go to the bother of rebuilding these after they were turned into rubble in WWII? If the illusory 'house' is 'not really there', and indistinguishable from rubble, then rubble ought to have been as good as the illusion. The answer is simple: The form of the 'house' structures constituents and pertinent physical laws in such a way as to permit possibilities not obtainable from mere rubble: these being the broad range of real advantages that we obtain from houses, but not from rubble. The form of the house, though not compelled by its constituents and physical law, but not prohibited by these, is therefore real.

In a similar vein, according to mereological nihilism, a message supposedly associated in a radio carrier-wave does not really exist, simply because it is not compelled by the fundamental electromagnetic constituent photons or laws. However, assuming mereological nihilism, what happens when we confront the claim with, say, the American victory at the Battle of Midway, in WWII, 1942, in which the breaking of Japanese communications codes proved decisive? By breaking the code, US military intelligence read the secret radio messages of the Imperial Japanese Navy and conjured a trap for their carriers[39]. Self-consistent mereological nihilism ought to assert that the secret Japanese radio code-states could not have been associated with the electromagnetic radio waves, because these were never part of the real existence of electromagnetism or compelled by electromagnetism.

Notwithstanding information-matter dualism, the critical Japanese code-states were either associated with the pertinent radio signals and constituents in an appropriate form, or not: If not, then the encryptions could not really have been part of real existence. Thus, the American claim to have 'decoded' these was as much an illusion as the prior encryption by the Japanese, with any subsequent associations reduced to extreme improbability: assuming mereological nihilism is correct.

At this juncture, we must introduce the code-information distinction: a distinction we will directly grapple with in Part-V. Therein, it is recognised that the real information, the realia, are not of the same form as the code-states upon which they are foisted. In short, code is not information. We foist meanings on agreed-upon codes and then communicate the information by transacting codes… even though the code is not in some literal sense the information or realia. For example, the word "apple" is not the real apple: the word, the code, is of an improper form to the realia. Even so, the meanings we want to communicate, combined to the code-information association we agree upon and form a shared consensus upon necessarily structures the codes conveyed into a suitable organisation. In a similar vein, the information, "AF was short on water", immanent to the Battle of Midway in 1942, was not literally in the intercepted and decoded Japanese message. Yet, the meanings and code-associations the Japanese agreed upon structured the transmitted and intercepted electromagnetic radio transmissions into a conformal form. The transmission could not have assumed any other form at that point: Thus, the information structured the code and constituted the effected form of the Japanese radio signal. That is, the whole, the mereology, conditioned and organised the parts, which then proved critical to the decoding operation and the battle of Midway.

*

From the given examples, the claims of mereological nihilism simply do not measure up against the complexities and actualities of the real world. For mereological nihilism to be true, for it to be consistent with the way we really experience things, it would require the operation of extreme improbability, if not the miraculous. A commensurable world constructed on miraculous improbabilities would soon breakdown: the odds would very rapidly build up against it. The extreme consistency and commensurability of the *real* world must raise doubts about mereological nihilism, if not about *all* nihilism.

If mereological nihilism cannot square with the complicated reality of even artificial mereologies, if it cannot eliminate mereology even in artificial forms not compelled by nature but permitted by nature, it follows that information *really does* exist, especially in the context of natural wholes: At a minimum, and putting aside radical Platonism, at least an Aristotelian relationship between information and substance must be conceded in all mereology, including in both natural and artificial wholes.

To rephrase Aristotle… *Without substance, form cannot exist. Without form, substance is nothing*.

*

The position against mereological nihilism can be crystallised into three arguments: The first is **the argument from improbable commensurability**: This was exemplified by the example of code breaking operation in the Battle of Midway, 1942 (see above). We could also include similar codebreaking by Bletchley Park during WWII.

The second is the argument from intrinsic consequential inequivalence: It was presented in the previous contrast between the chair versus the spear and is also subsumed to the argument from improbable commensurability. The third is the argument from generic quantum mechanics: more on this later.

The argument from intrinsic consequential inequivalence can be further augmented by the following: Suppose we had a small quantity of steel. From it, we could construct one of two objects: A hollow sphere made of a very thin shell of steel, or a steel dart. To mereological nihilism, these forms are illusions and do not exist: Only the fundamental constituents and laws that make up and govern the quantity of steel comprises the real existents: The form of the sphere, or the form of the dart, is not compelled by the fundamental constituents and laws. No matter how closely we examine these, we will never find any necessary logical, rational, or physical grounds for claiming the reality of either the form of the sphere or the form of the dart. Thus, neither the ball nor the dart objectively exists: They are mereological illusions.

However, if we throw a hollow steel sphere in such a way as to strike another's skull, it is doubtful it would cause any grievous or mortal harm, per function of how the form of the sphere structures and delimits the constituent and physical laws, and attendant horizon of possibilities and effects. If we construct a dart from the same quantity of steel, the anticipated outcomes obtained from it will be very different from those obtained from the sphere: real harm, or even death, is likely. The forms are *not* mereological illusions.

---

[39] Joe Rochefort's War: The Odyssey of the Codebreaker Who Outwitted Yamamoto at Midway, Elliot Carlson, Naval Institute Press 2013

The same must hold true in the case of a complex object, like a wristwatch: The form of the watch, hence the watch itself, is a real existent. It is not a mereological illusion. The same constituents and laws, transformed into a mere pile of minerals, or some other form, will not yield the same horizon or possibility frontier. One cannot 'tell the time' from a pile of minerals, even if these are indistinguishable in type and quantity, and obey and entail the same laws, as does the 'illusory' wristwatch.

Thus, we have **the argument from intrinsic consequentiality inequivalence**, which asserts that anticipated outcome **possibilities from the one form are not interchangeable with that obtained from another form, even when both forms assume the same type and quantity of fundamental real existing constituents,** *and* **employ the same physical laws; and even when these are not ontically compelled by those laws and fundamentals. Therefore, the whole and its form, and hence the mereology of that form, is real.**

*

If we ignore Alt-R and put aside implications from de-spatialisation, generic quantum mechanics itself can raise doubts about mereological nihilism. Thus, we have a third argument against mereological nihilism: one that implicitly incorporates the previous two arguments: Namely, the argument from generic quantum mechanics, or, succinctly, **the argument from instantaneous wholes**.

Recall that mereological nihilism denies the ontic reality of the whole, and even more so the notion that this whole can form and inform the parts. But if we explore the relation between, on the one hand, the abstract quantum mechanical wave and wavefunction versus, on the other hand, the *en masse* quantum indeterminate terms that the wave funnels into the simplest wholes, such as the interference-pattern evinced in the two-slit experiment, we discover that an abstract whole, a pattern-potentiality, *does* organise its fundamental parts: a fact often obscured by the quantum indeterminacy entailed in the process.

Clearly, the interference-pattern is strictly conditional to the two open slits: we cannot obtain it from a single open slit. If the whole interference-pattern manifested in bottom-up fashion from the discrete *en masse* quantum depositions, and only the constituent quantum indeterminate depositions were real, then the two-slit outcome would be indistinguishable from the one-slit outcome, or from any mere amorphous set of alternative outcomes... given that the outcomes could not be part of real existence; given that the mereological arrangement of the barrier with one or two slits could not be part of real existence, much less distinguishable or capable of generating alternative distinct global outcomes at the detector.

This observation comes into its own in photography, wherein a complex image develops out of an abstract complex quantum mechanical wave and wavefunction: an *instantaneous whole* in quantum mechanical abstract form. Therein, **the photographic image does not arise from the quantum indeterminate depositions, bottom-up. Instead, as is also true in the two-slit formation of interference patterns, the quantum indeterminate terms are inexorably funnelled and fated to render the pre-existing abstract instantaneous whole into saliency, top-down.** But the instantaneous whole is not a state in absolute time: Given its AND-form state and its status as future-potentiality, it does not constitute realised events in a common present moment, and the quanta of events that decay out of it are interrelated per the relativity of simultaneity.

While the quantum depositions that collectively make up the interference pattern or the photographic image *are* objectively quantum indeterminate, as argued in the Preliminary (see **0.22**) and as will be assert in in Book-III, these are inexorably funnelled to a pre-fated global outcome by the abstract quantum mechanical wave-state. Indeed, per each quantum deposition destined to collectively evince the interference pattern or the photographic image, **we envision a Manyworlds scenario of the two-slit experiment, wherein the universe splits into as many possible alternative histories and endpoints as permitted by the complex wavefunction involved. Yet, each seemingly unique history could not help but culminate into the same interference-pattern or photographic image**. The key insight here is that **the abstract instantaneous whole that organises the quantum discrete parts exists and** *pre-exists* **its constituents: the mereology identified** *as* **the abstract complex wave and wavefunction is ontically real and** *precedes* **its parts, either in the Aristotelian form-substance fashion, or in the more hidebound Platonic form fashion.**

Indeed, what exactly would one obtain from the collective of quantum depositions in the absence of any funnelling abstract wavefunction? One would obtain perfect mereological nihilism: an amorphous pile of discrete depositions *without* global pattern or structure. But, in the presence of abstract wavefunctions, we get the exact opposite of mereological nihilism: **The quantum mechanical wave-pattern and wavefunction constitutes an instantaneous abstract whole; one distinct from the materialised global outcome of discrete partology that it funnels into evidence** (see Preliminary **0.22** and **0.23,** but *especially* **0.26** and **0.27** for the role of instantaneous wholes in the growing block universe).

The same can be said about what might well be the ultimate whole: namely, the universe-scale wavefunction that fates the pattern of the whole universe out of its constituent ultimately quantum indeterminate partology. This unique pattern-outcome would obtain in every culmination of every Manyworlds history rooted in the same 'instantaneous whole' or universal wavefunction, *ceteris paribus*.

*

The discerning reader will notice a seeming contradiction between the above declaration about complex wavefunction instantaneous wholes versus the status of information in facile information-matter dualism furnished in the Preliminary, which assumed real information to be purely OR-form in character and sought to explain its dissociation and retention vis-à-vis incompatible AND-form radiation and matter. This seeming contradiction will be resolved by Book-III and IV, wherein the same essential dualism from facile information matter dualism will re-emerge, but in a very different form: succinctly, as part of the growing block model universe... and as the dissociation of the past (abstract memory) from the AND-form future. The reformation of spacetime in Book-II, and its supersession in turn by anticipated de-spatialisation in Book-IV, will be key to this transformation and supersession of information-matter dualism: resolving the stated contradiction.

100

Even so, consider that **the critical element throughout is the abstract instantaneous whole, now intimated as the complex wavefunction itself. Hence, wholes matter. In short, information matters: That which informs and forms is not a mereological illusion. The whole is *real* … and mereological nihilism is false.**

We shall return to this theme in Book-III, wherein mereological nihilism will combine with Manyworlds and other claims to posit the ultimate nihilistic case against both information and abstract memory, culminating into unexpected conclusions that are totally inimical to *any* and all forms of nihilism… *and* materialism.

1.04: Definitions of Information: Mereological Categories

Mereology is real: Some forms are compelled by their fundamental laws and constituents: e.g., salt crystals, the structure of diamond, and other natural forms. Others, like chairs, houses, and radio-messages, are *not* compelled by their contingent fundamental constituents and laws, though they are no less real per their unique and inequivalent consequences and possibilities. Thus, we have two sorts of wholes or informational states: **compelled mereologies** versus **flexible mereologies**.

An example of compelled mereology is given by the simplest of crystals: The salt crystal really is a product of necessity, compelled by its fundamental micro-constituents and laws on a ground-up basis. As such, the large-scale mereology of the salt crystal mirrors the micro-mereology from its constituents and laws.

In contrast, flexible mereologies are *not* compelled in any necessary way by their fundamental constituent and laws: Chairs, houses and radio messages are the given examples. We have already established their objective realities per improbable commensurability and consequential inequivalence.

However, a third category of mereology, **intermediate mereologies**, lie between compelled and flexible categories. For example, a snowflake pattern certainly is an outcome born of necessity: it is compelled per fundamental constituents and laws, ground-up. Yet, for the same amount of water molecules, manifold numbers of alternate possible snowflake patterns are possible: a broad range of multiple-minima constitute the possibility-horizon for snowflake patterns. None of these can be dismissed as perceptual illusions just because they are not bourn of or totally reducible to the necessity of pertinent fundamental constituents and laws on a ground-up basis.

Consummate intermediate mereology is perhaps best exemplified by living forms and wholes. For example, the naïve attempt to reduce and explain all of morphology, physiology and even psychology in terms of the human genome confronts the fact that the human genome is largely a molecular parts-list with in-built regulatory on-off switches. Though critically indispensable to the functioning of the human organism, the human genome and its contents are not capable of accounting for the whole of the morphology of the human organism. Morphology is not reducible to the genome, although the genome is certainly indispensable to that morphology.

Ultimately, with radical information-matter dualism, itself to be superseded by implications from anticipated de-spatialisation and temporisation in the framework of the growing block universe in Book-IV, it turns out that mereology, or information… or abstract instantaneous wholes… can dissociate and persist as extant abstract memory vis-à-vis radiation, matter, and 'space', and can certainly dissociate from the very materialised patterns they evince.

1.05: Read-Only and Read-Write Physical Information

To further complete our attempt to define an evolving working definition of information, we must borrow terms from computer and information technologies. Thus, we will distinguish between **read-only** versus **read-write** informational states.

The kinds of physical information constitutive of read-only states are of the sort that, as far as is known, appear to remain immutable: The consummate examples are abstract physical principles and laws: inferred through the investigation and behaviour of 'matter', but never *as* matter… inferred as abstractions *without* extension or space; and even as timeless. Hence, time-symmetric physical laws. Indeed, physical laws and principles are dualistic Platonic-Cartesian realities, as was stated in the Preliminary, insofar as these cannot be weighed, physically divided, or precipitated into a test-tube, and cannot be described as spatially extended, and obviously operate across nature... while they operate *from nowhere and everywhere.*

Fundamental Platonic-Cartesian realities, or read-only realities, are unalterable and immutable by any *know* means. Of course, one cannot be absolutely certain whether physical laws are subject to mutability, evolution or even succession in some ultimate sense. Even if proven to be mutable, their immaterial Platonic-Cartesian characteristics, their Platonic dualism vis-à-vis matter, radiation, and space, cannot be denied: (See Preliminary: **0.26**)

The evidence for immaterial physical laws vis-à-vis matter, radiation and space constitutes the case for fundamental information-matter dualism operative across the whole of nature: The very Platonic-Cartesian character of natural principles implies, *not* an Aristotelian distinction of principle from *extensa*, but their Platonic and Cartesian distinction; entailing the literal dissociation of natural principles and laws from *extensa*, implying their immateriality and independent subsistence from *extensa*.

*

In addition to Platonic-Cartesian read-only informational states, we find clear evidence for plethora of inexhaustible *mutable* patterns assumed by putative matter, radiation and space that incorporate read-only Platonic-Cartesian laws. Such transient **read-write** states are composed of flexible and intermediate mereological categories: (see **1.04**).

Our case for information-matter dualism will focus on the dualism that resides between transient read-write information versus radiation, matter, and space. **It is read-write states that explicitly dissociate from radiation, matter, and space, and are subsequently retained in abstract form as abstract memory at physical remove from radiation, matter, and space.**

One snowflake pattern could easily replace and succeed almost any other snowflake pattern, of which there exist a huge number of unique examples and possibilities, even with the same number of constituent water molecules and unchanged pertinent physical laws. The fact that one snowflake pattern is as good as any other shows that such wholes, as organisational contents, or pattern-contents, comprise domains that are alterable and mutable: comprising intermediate mereological read-write informational states.

Another example: Consider read-write informational states in the context of a developing photographic film. Obviously, the light sensitive silver-based compounds that coat the surfaces of old-school photographic films are not collectively biased toward any exclusive photographic image, and a huge number of alternate unique images are possible: That is, the system is open to alternative wholes. The image that finally forms on the film constitutes the sort of information we have in mind when we speak of read-write physical information and intermediate mereologies. Such read-write states are physically distinct from the constituent silver-based building blocks that get to manifest them and could be substituted with any other equally possible read-write image. Therefore, unlike read-only natural laws and principles, read-write states are indeed alterable, mutable, and replaceable by other such states.

Nature can be likened to a vast photographic film-surface: albeit three-dimensional or stereotemporal, but one that can perpetually revert to a stand-by state per alternative read-write patterns that might succeed previous formations, within the limits of read-only physical laws. Indeed, in the context of radical information-matter dualism, our aim will be to show that captured read-write patterns of the universe can wholly or partially dissociate and can be retained in abstract form from natures 'canvas', without erasure, in the form of abstract memory. These memories fade out in terms of their power or influence over subsequent formations or events. Consistent with the growing block model, this implies the reality of universe-scale abstract memory, comprised of an accumulating 'presence of the past'[40]: a domain for the sum of the past, retained in abstract form at a remove from the manifest materialised order.

*

We now have a general useful definition of information as 'that which informs' radiation, matter, and putative space. These are subdivided into *read-only* and *read-write* types, consistent with *compelled, intermediate* and *flexible* mereological categories. We have also sought to distinguish between Aristotelian forms of informational in contrast to Platonic forms of the same: In our quest for the definition of information, our openly stated aim has been to extend the Platonic view about information-matter relations to the larger read-write domain, but within the context of information-matter dualism, prior to the similar and more radical implications from anticipated de-spatialisation, temporisation and the growing block reform of generic spacetime.

1.06: Definition of Physical Information from Wave-Particle Dualism: AND-Form & OR-Form Categories of Physical Information

The incorporation of read-only and read-write categories of information, together with compelled case for intermediate and flexible mereological categories, must be complemented by two further categories, both inferred directly from wave-particle dualism: namely, **AND-form versus OR-form categories**. Taking quantum spin as an example, the *pre-measurement* quantum mechanical description of a particle's spin will constitute a unitary superposition of both "spin-up" AND "spin-down" before it is resolved into a realised event. With the application of physical inputs that supposedly disturb a hitherto isolation-state, a mysterious conversion takes place: The unitary superposition "spin-up" AND "spin-down" collapses into a realised OR-form outcome: either spin-up OR spin-down... but not both. This underpins AND-OR dualism synonymous with generic wave-particle dualism: a dualism between AND-form potential information and OR-form resolved information, requisite to our own facile-level information-matter dualism and to the subsequent growing block model of the universe. Thus, **AND-form potential information constitutes the quantum mechanical waveform of physical information: the sum of *potential* attribute-states and possibilities that a physical entity, in its pre-measurement and *pre-event* quantum mechanical future-form, might come to acquire in concrete form at some future moment or resolution.**

On the other hand, **OR-form physical information consists only of post-measurement or post-event states that a system has come to assume through a process that collapses AND-form states into OR-form outcomes.**

*

In the context of radical information-matter dualism, we must specifically concern ourselves with the relation to radiation and matter of read-write types of physical information, and are less concerned with read-only Platonic-Cartesian physical laws. To this end, we must now reframe read-write physical information to the quantum mechanical AND-form and OR-form forms, i.e., to AND-OR dualism.

Recall from our EPR-based analysis first introduced in the Preliminary, wherein the principle of conservation of spin constitutes a read-only abstract Platonic-Cartesian physical law (See Preliminary **0.26**). Yet, the materialised OR-form spin-state that a particle might acquire therein will constitute a read-write informational state, insofar as any subsequent measurement on the same particle might subsequently resolve into OR-form of "spin-down". Thus, the spin-attribute of a particle will vary: Hence, quantum spin is not a read-only state, but a read-write informational state.

---

[40] *The Presence of the Past*: A book of the same name by Rupert Sheldrake advocates essentially abstract memory, but per the 'hypothesis of formative causation'. Sheldrake is often dismissed. Yet, if we take his 'morphic resonance' across time, but simply remove space, as we shall do by Book-IV, the two approaches converge. The only factor lacking in Sheldrake's account is explicit de-spatialisation and temporisation of physics.

102

The same holds for putative 'position': Following one measurement, a particle may be resolved into an OR-form state at a putative x-y-z co-ordinate position. A subsequent measurement might resolve it into a different x^1-y^1-z^1 co-ordinate. Subject to change and mutability, the positional attribute of a particle is also read-write information.

Stated in terms of AND-OR dualism, read-write physical information comes in two forms: either as **pre-measurement AND form read-write potential information** or as a **post-measurement realised OR-form read-write information**. Per the lay-formalism first introduced in the Preliminary, quantum spin in its pre-measurement AND-form read-write state is describable as…

Particle spin (Pre-Measurement): {"spin-up" **AND** "spin-down" | all are future-potentials}

However, the post-measurement expression of this will assume the mutually exclusive form…

Particle spins (Post-Measurement): {"spin-up" **OR** "spin-down" | only one of these is realised}

Thus, read-write physical information may assume either a pre-measurement pre-event AND-form quantum mechanical state or a post-measurement OR-form variant. These will not occur simultaneously, as both old and new complementarity principles testify (see Preliminary **0.13** to **0.17**) even though, post measurement, at the two-slit experiment detector, both AND-form and OR-form realities will be found to be observationally simultaneous, in the sense that the evidence for AND-form wave-interference resolved as the interference pattern at the detector requires that that this is formed out of and evinced by *en masse* OR-form particle-depositions.

In the context of our present concern with information-matter dualism, and in the context of the same per our anticipated intermediate spacetime model in Book-II, and its de-spatialised temporised successor in Book-IV, we are principally concerned with the fate, dissociation and restitution of read-write OR-form informational states, and the question of where and in what form these are retained subsequent to their dissociation. We are also concerned with whether these are subject to complete erasure from the physical order upon their dissociation: a concern pertinent to the question of abstract memory in nature.

Generally, **AND-form states comprise pure *potential information*** devoid of any 'real', realised or resolved OR-form states. Thus, with the physical conversion of AND-form possibilities into OR-form realities, **with wavefunction collapse, we enter the domain of *realised information***: the emergence of realised events.

It is per the subsequent fate, retention, and restitution of realised information that we infer both abstract memory in nature and information-matter dualism generally: Both segue to the growing block universe and the reform of spacetime.

*

In the next step, we must examine complex forms of AND-form and OR-form read-write states that involve *en masse* 'particles', such as in the formation of a snowflake; in the formation of a photographic image; etc. (i.e., in intermediate and flexible mereologies, respectively). In the case of the formation of the snowflake, and over-simplifying it to its essentials, the pre-measurement description of the snowflake is a potentiality expressed as an AND-form unitary superposition of **x**-number of unique *future-potential* snowflake patterns eligible to form up. Using our lay formalism…

Snowflake (Pre-Measurement): {pattern-1 **AND** pattern-2 **AND**... **AND** pattern-x | assumes all as future-potentials per combinations of **n**-water molecules}

This pre-measurement AND-form potentiality for snowflake patterns does not and cannot constitute a repository for any *real* or realised snowflake: The AND-form complex does not constitute or 'carry' any *real* or realised OR-form snowflake pattern. If it did, it would no longer constitute an AND-form future potentiality for such. Thus, the quantum mechanical de Broglie matter-waves associated with snowflake patterns are incapable of acting as repositories or carriers of realised OR-form snowflake patterns: As repeatedly stated, this sort of insight implies information-matter dualism.

A similar logic also applies in the formation of a photographic image in its pre-measurement state. Before light radiation is allowed to resolve into OR-form depositions on the photographic film, the film might be described as an AND-form future-potentiality for alternative image-possibilities (**x**-number) permitted by **n**-number of silver-based light sensitive compounds arrayed on the surface of the pre-developed film. Thus…

Photo-film (Pre-Measurement): {image-1 **AND** image-2 **AND**…**AND** image-x | assumes all as future-potentials per combinations of **n** photo-sensitive molecules arrayed on a surface}

Obviously, the unitary superposition of images in potentiality is not a repository or a carrier of any realised OR-form photographic image. Post-measurement, the unitary superposition of potential images in the quantum mechanical AND-form state will resolve into a specific actualised image out of the welter of future-potential images. It is the subsequent dissociation, retention and restitution of the actualised OR-form read-write image that concerns us in our pursuit of information-matter dualism and abstract memory theory.

*

We can generalise the above to the whole of the physical universe. In its pre-measurement or pre-resolution AND-form state, the whole physical universe, grasped as comprised of **n** number of basic entities or constituents, will constitute a quantum mechanical AND-form unitary superposition… or Hugh Everett's universe-scale wavefunction... for all as-yet not-realised **x**-number of alternative *potential* universe-scale configurations. Thus…

Whole universe (Pre-Measurement): {config-1 **AND** config-2 **AND**…**AND** config-x | assumes all as potentials from the combinations of **n**-basic constituents}

This pre-measurement perennial AND-form wavefunction for the whole universe cannot constitute a repository of any actualised OR-form configuration of the universe: Upon AND-to-OR wavefunction collapse, the quantum mechanical AND-form of future potential configurations for the whole universe will resolve into just one *real* read-write universe-scale configuration, notwithstanding Manyworlds. Again, it is the subsequent fate of this latter read-write information that concerns us in our pursuit of information-matter dualism and attendant memory theory.

1.07: Information & Entropy: The Entropy of Future-Potential Information

Entropy, from *the second law of thermodynamics*, is the measure of disorder in a physical system, understood in two interrelated ways as entropy of energy and entropy of information. **In entropy of energy, disorder is understood in terms of the dispersal of the energy content of a physical system over time, and its transformation into relatively useless form.** For example, a hot cup of coffee will cool down, losing its heat-energy to the larger environment. This loss does not entail the destruction of energy, but its conversion into a form that cannot be readily recovered and re-used, save at the cost of even more energy expense, and even greater energy dispersal.

In entropy of information, disorder is understood in terms of the degeneration of a higher complexity system into a lower complexity system: a reduction in and loss of information. A typical example consists in a relatively complex chemical compound decomposing into its more basic constituents. The tendency toward this decomposition is, in putative energy terms, often easier to obtain and less costly than the energy cost required for the formation of the complex compound. The corollary is that increase in the entropy of energy often entails a reduction in the informational content of that system: a reduction from a more complex configuration to a less complex one. In terms of information, this constitutes an increase in the entropy of information: a reduction and degeneration in the information-content of the system, wherein information is taken to be synonymous with organisational complexity.

Under unique conditions, it is possible to obtain an increase in the organisational complexity or informational content of a system. This is always exacted at the cost of tandem increase in the total entropy of energy. For example, the development and brief continuation of a living organism entails a superlative increase in the organisational complexity and attendant informational content that defines that organism, amounting to a superlative reduction in the entropy of information. However, this reduction in informational entropy, attained principally through anabolic processes that synthesise and maintain complex structures from simpler ones, will always entail an increased in the use and dissipation of energy, and, with it, an increase in the entropy of energy. Therefore, a brief local reduction in informational entropy (or increase in organisational complexity and order accompanied by the emergence, development and perdurance of a living organism) will *always* entail a marked increase in the entropy of energy and its dispersal into less useful forms. This insight puts a lie on the erroneous notion, oft abused by so-called creationists, that living systems are somehow anomalous to a universe where entropy tends towards inexorable increase.

Having defined and clarified entropy in both energy and in informational terms, we must now recuperate entropy within the context of our own definitions of information, and clarify how entropy is pertinent to information-matter dualism and attendant abstract memory theory.to

*

We first start with **entropy of future-potential information**: In the case of a complex system, the quantum mechanical AND-form state of that system is the unitary superposition of all *potential* organisational configurations that that system might get to assume *in the future*. In short, **the AND-form description of a system represents its total *future-potential information*.** This quantity of potential information is ultimately subject to entropy: i.e., all things being equal, the entropy of future-potential information must *always* increase. The act of measurement, or attendant **AND-to-OR wavefunction collapse, hence the collapse from a domain of pre-event future possibilities into OR-form actua, must** *always* **lead to a reduction in the future-potential informational content of the system**, whether it is a local AND-form system or the universe-scale AND-form wavefunction itself.

In its pre-measurement or pre-collapse state, an AND-form quantum mechanical wave of potential information remains unchanged, ignoring quantum decoherence (see later), or all things being equal. Hence its potential informational entropy is conserved: again, all things being equal, or so long as isolation holds. In such a suspended state, the system (the future) is not subject to entropy. However, **with 'measurement' and AND-to-OR collapse, the prior AND-form future-potential informational state will decay into a unique actualised OR-form state. It will degenerate from a higher future-potential informational state to a lower-level future-potential information,** or even a highly idealised fully exhausted maximum future-potential entropy state, insofar as all of a system's future-potentials might well become eliminated. A fully exhausted futurity would be subject to the total *irreversible* loss of all future potential informational states; degenerating into a state of *absolute* entropy of future-potential information, assuming there are no subsequent AND-form futures available to the system for subsequent AND-to-OR collapse: a speculation we will return to in Book-II when we develop the nested futures perspective of the quantum mechanical wave.

In more complex systems, the wavefunction of future-potential informational states may decohere without total conversion. In decohering systems, wherein some factor or variable reduces the scope of future possibilities available to the system, there will be no AND-to-OR conversion as such, while the entropy of potential information from decoherence will be proportionate only to the future potential states erased by the decoherence, minus the intact AND-form future-potentialities that remain.

Generally, we may state that the entropy of future-potential information of the whole universe must always increase with every instance and sequence of universe-scale AND-to-OR *partial* wavefunction collapse or decoherence of the universe-scale wavefunction.

The increase in the entropy of future-potential information must also be accompanied by the increase in the putative entropy of energy. This might also be accompanied by a brief reduction in generic entropy of information, with tandem increase in OR-form organisational complexity, such as evinced in living forms. Yet any reduction in informational entropy (or increase in organisational content) will always be accompanied by both increase in the entropy of energy *and* increase in the entropy of future-potential information. Thus, increase in OR-form organisational complexity in living forms is an outcome *always* obtained at the cost of the increase in the entropy of future-potential information, especially given the irreversible erasure of alternate future-potentials entailed in both AND-to-OR collapse of that future *and* in its likely quantum decoherence in the absence of outright collapse.

1.08: Entropy of Abstract Memory & Entropy of Future-Potential Information

We must now integrate memory and entropy into a complementary theory of **entropy of abstract memory**. To this end, we assume information-matter dualism as a given. What are the implications of putative information-matter dualism to entropy? **Per information matter dualism, and its tacit admission to abstract memory, the dissociation of information does not imply erasure, and the final failure to re-associate does not obviate the independent retention of that information in extant abstract form**. In short, abstract memory attendant information-matter dualism changes the rules of entropy.

The implication from abstract memory is that information ultimately survives its eventual failure to re-associate with matter: It is retained and kept in extant form as abstract memory from all subsequent AND-to-OR collapsed OR-form resolutions of radiation, matter, and 'space'. **Only the ability of abstract memory to re-associate with radiation, matter and 'space' is physically hampered by inexorable entropy. Otherwise, it survives and abides as abstract memory. Indeed, in nature, there must be a build-up and accumulation of memory over time.** This view is consistent with the growing block universe destined to emerge by Book-II and especially in Book-IV, wherein events decay out of an ontically real future domain and, as such, lead to the accumulation of the past. Hence, the accumulation of memory, regardless of whether it is re-materialised in the future or fails to do so.

The implied build-up of memory, in whatever form it is retained, ought to imply a perpetual increase in the abstract informational content of the whole universe: grasped as an increase in the stock of abstract informational states pertaining to the past, with attendant decrease in the entropy of information from memory.

While the retained informational content of the universe must increase, and must surely imply decrease in the entropy of information per increase in information per memory, this cannot obviate the overall tendency toward the increase in generic entropy, much less obviate the increase in the entropy of future-potential information in both local and universe-scale terms.

*

There are several solutions that harmonise the accumulation in memory with the increase in future-potential entropy. The first solution comes from **the primacy-recency rule** and the diminution of the power of memory to re-associate and organise matter and radiation: so contributing to the increase in the entropy of abstract memory.

The oldest memories derived from past AND-to-OR collapse-processes will constitute the primary memories. **Memories in primacy are comparatively weaker in their power to inform and form subsequent leading OR-form resolutions of organisation and pattern. This is because these are temporally the most removed from leading resolution processes.** On the other hand, the newest memories, obtained from the most recent AND-to-OR collapse-processes, are much less removed in time from the latest OR-form resolutions. Hence, **memories in recency have greater informative and formative power over subsequent resolutions of organisation and pattern compared to memories in primacy... because these are less removed in time and more recent**.

The power of memory to influence and organise the outcomes obtained from subsequent AND-to-OR collapse is per function of **the inverse-square law of time**: Succinctly, this is the square of the time-interval that separates the moment of AND-to-OR collapse vis-à-vis the past memory, all things being equal with respect to such things as time dilation. Hence, the longer the time-interval, the weaker the memory. And the shorter the time-interval the stronger the memory in its power to structure the formation of subsequent events out of generative AND-to-OR collapse.

The primacy-recency rule, combined to the inverse square law of time, makes for the **entropy of abstract memory: The increase in information per memory over time is compensated by the perpetual weakening of the total power of memory to influence the formation of new events and patterns in subsequent AND-to-OR collapse-processes of the universe-scale wavefunction.** Put another way, even as the memory-content of the universe increases, the memory-content of the universe recedes in its power to inform and form leading OR-form outcomes: This makes for entropy of abstract memory.

Let us take a closer look at implications directly from primacy-recency rules: **As memories in primacy recede in time and are weakened, these are saturated by newly formed memories in recency, obscuring and subsuming the older memories**. Over time, memories in primacy must fade out and, even though ineradicable, must effectively 'drop out' from any discernible influence over new events and patterns. **But the weakening of memories in primacy is also due to their recession in time** vis-à-vis leading events and patterns. **That is, primacy-recency rules work together with an inverse-square law of time vis-à-vis memories in primacy in order to affect increase in the entropy of memory.**

*

105

What is the relationship between entropy of abstract memory of the past... a past that accumulates even in the face of the previously described primacy-recency rules... to the entropy of future potential information, which increases per AND-to-OR wavefunction collapse-process? **First, the accumulation of information as memory is always obtained in and through AND-to-OR wavefunction collapse**: The accumulation of memory is *always* engendered by the tandem increase in the entropy of future-potential information.

Second, and interrelated to the first, **the decrease in the entropy of information per the accumulation of abstract memory is *always* achieved at the expense of critical and irreversible increase in the entropy of future-potential information** in and from each instance and sequence of AND-to-OR collapse.

Whichever way we look at the matter, entropy always tends toward an increase: Even the initial decrease in the entropy of information from the accumulation of memory is offset by the tandem increase obtained in the entropy of future-potential information.

When we incorporate the time-recession of memory through the inverse square law of time, and add consequent entropy of memory to the fray, our assertion for the overall increase in entropy in the universe is secured.

1.09: Entropy of Future-Potential Information, AND-to-OR Wavefunction Collapse and the Arrow of Time

There is an established relation between the metaphorical 'arrow of time' and entropy. The relation asserts that **the 'arrow of time' is a function of entropy and points toward increasing entropy**. Presumably, time-reversal would entail the perpetual decrease in entropy and increase in order.

In conventional terms, whether approached in terms of energy or information, **generic entropy tends towards an increase and is generally irreversible**, with local reduction in the generic entropy of information (as is evinced in living systems) accomplished only at the expense of and offset by even greater increase in the generic entropy of energy. This irreversible increase in generic entropy is consistent with information-entropy relations from facile information-matter dualism and the attendant case for abstract memory, regardless of the dissociability of information; and regardless of the accumulation of abstract memory (see **1.08**).

What better indicator of irreversible change than irreversible and inexorable entropy? What better intrinsic indicator of time than entropy itself; with its figurative 'arrow' always pointing toward perpetual increase in entropy? What alternative could one recommend? Indeed, we could not base the 'arrow of time' on purported time-symmetric physical laws, given that laws are *presumed* to work equally well in both time-directions under increasing entropy as well as under improbable decreasing entropy.

However, there *is* an alternative basis for discerning the objective reality of irreversible change, time, and time-asymmetry, or the 'arrow of time': one that incorporates *and* supersedes generic entropy-based approaches to the same. We assert that the arrow of time is a function of two interrelated things: **First, AND-to-OR polarity in ontically primary AND-to-OR wavefunction collapse *is* the basis of the 'arrow of time'** and supersedes conventional statistics-based entropy conception of time-asymmetry and irreversibility. Indeed, **AND-to-OR collapse *is* time, and always proceeds from AND-to-OR, *never* from OR-to-AND**.

Second, AND-to-OR collapse *always* culminates into an irreversible increase in the entropy of future-potential information. Since time must always proceed from AND-to-OR, but never OR-to-AND, it inescapably follows that time always proceeds from a state of lower entropy of future-potential information (a level of greater future-potentiality) to a state of higher level of entropy of future potential information (to decrease and decay in future-potentiality).

Notice that we have superseded the generic postulate that both time and entropy are discerned from the mere statistical disposition to an increase in generic entropy of energy and entropy of information. We have moved to a clearer and more certain basis for 'time's arrow', now linked directly to AND-to-OR directionality and the tandem non-statistical increase in the entropy of future-potential information.

Our re-basing of the arrow of time on ontologically primary and superior AND-to-OR collapse is not at all inconsistent with conventional entropy, given that **irreversible AND-to-OR collapse *is* the process through which the increase in generic energy and informational entropy manifests in OR-form outcomes**. Hence, the generic increase in entropy is always tandem with, but ontologically *subsumed* to, the increase in ontologically primary entropy of future potential information.

But generic entropy is always secondary and not fundamental: **generic entropy is not ontologically primary**.

Given its ontological primacy over generic entropy, the ontology of AND-to-OR wavefunction collapse must subsume generic entropy and succeed the latter as the *true* basis for change, irreversibility, time-asymmetry, and the figurative 'arrow of time'.

Indeed, **the process of AND-to-OR wavefunction collapse is synonymous with time itself. Wavefunction collapse *is* time**. Time is not a fourth dimension: a conclusion we will clarify in Book-II in the intermediate model of spacetime. Instead, time *is* the very process of perpetual wavefunction collapse... through which future-potentiality and possibility irreversibly decay into actuality in the genesis of new events. Hence, the 'arrow of time', time-asymmetry and irreversibility, *must* be sought strictly in the process of ontologically primary AND-to-OR collapse or time, not in generic statistical entropy, which is a secondary non-fundamental residue. To augment this claim, note that **AND-to-OR collapse itself (time itself) does not depend on or arise from generic entropy as a mere 'statistical disposition'. But the probability of generic entropy of energy and information *does* arise, as a secondary and ontologically subsumed feature of ontologically primary AND-to-OR collapse and time**. If there is no primary AND-to-OR collapse, then there can be no succession of events from which generic entropy could be garnered. Even if generic entropy decreased, as a mere secondary effect, the entropy of potential information will always increase, regardless of what happens to generic entropy, and the whole process and outcome will manifest through ontologically primary AND-to-OR collapse, which can never proceed from OR-to-AND.

Implicit to all of this is the entropy of memory (see **1.08**), also subordinated to ontologically primary AND-to-OR process synonymous with time itself.

1.10: AND-to-OR Wavefunction Collapse as Time,
The Arrow of Time & the Structure of Spacetime

There are two basic states that any physical system might occupy: A pre-measurement or pre-AND-to-OR collapse-state versus a post-measurement post-AND-to-OR collapse state. In its pre-measurement pre-AND-to-OR collapse state, a physical system is in unitary AND-form superposition of future-potentials. It is then disturbed by hitherto mysterious measurement processes, only to collapse into an OR-form state, wherein only one of the former future-potentials is realised as an OR-form event or outcome.

Another more lucid way of describing the above is thus: A physical system may occupy one of two states: First, a **pre-event state** of pure future possibilities in unitary AND-form, the possibilities of which are not-yet realised and cannot be attributed the status of realised events, save by some impossible 'signal from the future' in violation of the principle of causality and the principle of conservation. But the other state that a system might occupy is, of course, the **post-event state** composed of just the one OR-form possibility, now realised as a concrete event: no longer a possibility but an actuality.

As stated repeatedly, the AND-OR distinction, and the process of AND-to-OR collapse itself, are not merely linked with time: Indeed, AND-to-OR collapse *is* the process of time, and most certainly constitutes the proper basis for its 'arrow'. But both AND-OR dualism *and* AND-to-OR collapse also constitute the framework for the structure and processes of the growing block universe or the intermediate model of spacetime anticipated in Book-II, which will itself be superseded by Book-IV per de-spatialisation and temporisation.

In the anticipated growing block intermediate model spacetime, the whole of **spacetime is divided into three basic domains**, which includes the 'event horizon'. (The model was briefly introduced in the Preliminary: **0.29 and 0.30**). The first domain is the pre-measurement and pre-event quantum mechanical **spacetime in potentiality: comprised of all future-potential possible spacetime configurations for events; all rendered in AND-form unitary superposition**. Attendant AND-to-OR collapse-processes (time) will decay the future potentialities belonging to spacetime in potentiality into realised **OR-form events on or along the event horizon**, which constitutes the first-line post event domain and 'cut' between two major domains of spacetime. These events, or their basis, whatever it may turn out to be, is relegated to the mnemonic domain of **realised spacetime: the post-measurement or post-event domain for abstract memory**. The decay of AND-form spacetime in potentiality into realised OR-form outcomes and events on the event horizon is obtained through universe-scale AND-to-OR wavefunction collapse: *time*.

The AND-form domain of spacetime in potentiality is ontologically primary to the OR-form event horizon and to the accumulation of memory within realised spacetime 'below' the event horizon: Realised spacetime (the past *en sum*) decays out of pre-existent ontologically prior spacetime in potentiality (the future). In other words, the future is ontically primary and requisite to the present and past: Spacetime in potentiality could never arise out of OR-form events, or out of realised spacetime. But the later *need* spacetime in potentiality as their ontic precursor. Put another way, per the ontic primacy of AND-form spacetime in potentiality, the AND-to-OR directionality or polarity entailed in the collapse-process (time) which generates new OR-form events and secures their retention as memory within realised spacetime, *is* the direction-determinant of time or the 'arrow of time': the basis of time-asymmetry and irreversibility.

Moreover, it is only with AND-to-OR collapse that irreversible processes and outcomes can be said to take place. **Through the process of AND-to-OR collapse and time, the alternative possibilities belonging to the ontologically prior AND-form spacetime in potentiality become irreversibly and uniquely erased, with attendant irreversible increase in the entropy of future-potential information** and the generation of new OR-form events... involving tandem irreversible and unique qualitative and quantitative transformation of the universe.

The 'arrow of time' therefore arises in a necessary way from the asymmetric structure and composition of growing block spacetime and its event-generating processes: a spacetime asymmetry made possible by the co-joining of diametrically opposed ontically primary AND-form and subsidiary OR-form domains, with the 'arrow of time' inextricable to the decay of OR-form events out of ontologically primary AND-form spacetime in potentiality.

*

The universe-scale AND-to-OR collapse of 'spacetime in potentiality' *is* time. It is precisely because time arises from the process of AND-to-OR collapse-processes, and always proceeds from AND-to-OR, *never* from OR-to-AND, that time has its characteristic 'arrow' or direction. This insight renders void the usual way we describe time... as something that proceeds from past to future. On the contrary: time *always* proceeds from AND-to-OR: from an ontologically primary spacetime in potentiality (the physical manifestation of the future as ontically real) to a post-collapse event horizon of realised OR-form events, the basis for which is finally relegated to realised spacetime as abstract memory.

Put another way, **time proceeds from possibility to actuality: In effect, time proceeds from *future*-possibility to past-outcome... *from future to past... not* from past to future, as is commonly and erroneously assumed.**

The finding is not trivial: It constitutes the basis for the resolution of the *raison d'etre* of the quantum mechanical wave and its defining AND-form logic. It is also the basis for resolving problems of fundamental cosmology pertaining to the origins of the universe: It turns out that, with or without the Big Bang, the error of placing the origin of the universe in the remote past, which conforms to the erroneous characterisation of time and causality as 'flowing from past to future', must now be replaced: **With or without the Big Bang, the manifest universe perpetually emerges out of an ontically prior quantum mechanical domain of 'spacetime in potentiality': i.e. the universe decays out of and originates... from the *future*.** Therefore, the *penultimate* origin cannot issue from a creation-event placed in some remote past, but from a vast ocean of inexhaustible potential information that elaborates to the infinitely removed future; a 'cosmic ocean' of futurity essentially inexhaustible and requisite to the very possibility of AND-to-OR collapse and time itself: a future potentiality that is itself beyond the need of an origin-point or a creation-event.

1.11: Physical Laws in the Context of
AND-to-OR Collapse & Entropy of Future-Potential Information

We will now integrate physical law to time, recuperated as inherently time-asymmetric AND-to-OR collapse and with the entropy of future-potential information; all within the framework of the growing block structure of spacetime.

Clearly, **one cannot base the objective reality of time on putative time-symmetric physical laws that are alleged to operate equally well, in either 'direction' of time. Nor can one argue for the objective reality of causality based on putative time symmetric laws that supposedly cannot distinguish change and irreversibility**. If causality asserts that the future cannot transpire as a concrete event before the past, and that, in a flight from London to New York, one cannot arrive at New York before first departing from London, the objective reality of causality itself cannot be derived from putative time-symmetric physical laws, since these would apparently remain identical in form even if we could arrive at New York before departing from London; even if time and causality so understood were to completely breakdown.

As is clearly compelled by their apparent time-symmetric characteristics, abstract physical laws are incontrovertibly *timeless* Platonic-Cartesian principles. Timeless, immaterial architectonic physical laws present an immediate philosophic problem to science, insofar as their apparent time-symmetry, together with their consummate timeless natures, collide with a natural order characterised by undeniable irreversibility and time-asymmetry.

The conventional response to this problem has been to subsume timeless physical laws to the context of change and irreversibility grasped from the framework of merely statistical generic entropy: However, this is inadequate in the face of the ontological primacy of AND-to-OR collapse vis-à-vis generic entropy itself, as we discovered in **1.10**.

What if we place putative timeless physical laws into the context of change and irreversibility grasped in terms of local and universe-scale AND-to-OR collapse processes? This is a surer path to the conciliation of supposedly time-symmetric timeless physical laws with undeniable time-asymmetry and irreversible change: leading to certain solutions to the problem of how one can harmonise the real ontology of timeless Platonic-Cartesian physical laws with the objective reality of temporal and irreversible change: this being the modern physics-version of the old problem of permanence and change from vintage ontology and classical philosophy.

*

No matter how time-symmetric physical laws appear to be, their operation must always be incorporated to the AND-to-OR collapse process synonymous with time: to the process that brings into evidence the operation of the very physical laws that are otherwise timeless, accompanied by the tandem increase in the entropy of future-potential information.

It follows that, *always, physical laws must be framed to a milieu that proceeds from AND-to-OR, never from OR-to-AND: always from within the framework of an irreversible 'arrow of time' set by irrevocable AND-to-OR directionality; accompanied by the increase in the entropy of future-potential information.*

As we shall find in Book-II, the whole of spacetime incorporates this principle by means of its three domains: An apparently inexhaustible system of future possibilities embodied as 'spacetime in potentiality' perpetually decays into realised events, which relegate to memory in the domain of 'realised spacetime', thus making evident the operation and enforcement of physical law in the two ways highlighted above.

What would happen if we framed physical laws to a hypothetical collapse-inversion or time-reversal? Therein, AND-to-OR directionality would be reversed into OR-to-AND de-collapse, and the structure of spacetime anticipated in Book-II would be turned upside down. The **OR-to-AND de-collapse process would entail the *de-evidencing* of pertinent events that, in the first place, bring into evidence the operation of otherwise timeless physical laws. Thus, physical law would literally *fall out of evidence* through OR-to-AND de-collapse or 'time-reversal'.**

Timeless and eternal they may well be, physical laws can *only* ever come into evidence in the form we understand them under time-process that *always* proceed from AND-to-OR: The evidence of physical law is possible *only* under AND-to-OR directionality. Moreover, the tandem evincing of physical laws must also *always* correlate with the tandem increase in the entropy of future-potential information; if only because this increase is inevitable from the fixed non-reversible AND-to-OR directionality or 'arrow of time'.

1.12: The Time-Asymmetry of Physical Laws: Physical Laws do not Operate in all Time-Directions

One surprising implication is that supposedly **time-symmetric timeless physical laws are effectively time-asymmetric, even though they *are* timeless and eternal *in themselves***. In other words, physical laws are in a sense Aristotelian... in the sense that law *without* the substance (without the events) to render law evident is effectively a non-law; a nothing. But the same law is in another sense Platonic: i.e., independent of substance (the events) that bring its operation into saliency, because law does not itself change per function of the events, or from law's time-asymmetric relation to those events, or in relation to the process of AND-to-OR collapse and time that realises law in and through events but does not define physical law's ontology. Succinctly, **physical laws are effectively time-asymmetric because they *always* manifest in and through the irreversible increase in the entropy of future-potential information, per time-processes that *always* proceed according to an asymmetric AND-to-OR directionality, and in the strict sense that you cannot obtain physical laws via OR-to-AND time-reversal.**

*

Physical laws are time-asymmetric, at least in the empirical observational sense of their observed and inferred operation, but *not* so in the final ultimate ontological sense. We reiterate that physical laws can only operate in and through AND-to-OR collapse (time), and always amid the increase in entropy of future-potential information. **The alternative would entail the absurdity of physical laws that 'fall out of evidence' through time-reversal via OR-to-AND: a feat absurd and contradicted by experience. Hence, the evidence is for time-asymmetric physical laws.**

To demonstrate, consider the conservation of spin in the context of EPR (Einstein Podolski Rosen) experiment. Recall that EPR was first introduced in the Preliminary (see **0.26**) to demonstrates the timeless transtemporal ontology and dualism of Platonic-Cartesian physical law. Even so, let us embed the resolution of spin in quantum entangled couples to the framework of AND-to-OR time directionality and the tandem irreversible increase in the entropy of future-potential information.

Of course, the resolution of quantum spin entailed in quantum entangled particles in EPR is a process whereby the principle of the conservation of spin, as law, comes into evidence as an inferred abstract law that structures and delimits pertinent empirical outcomes, even while the law itself is not any sort of empirical 'stuff' and enjoys an ultimate extant ontology versus the empirical outcomes that it structures and delimits.

What begins as an intact future-potential informational state comprised of all the potential spin-correlates in unitary AND-form superposition must AND-to-OR collapse into a single OR-form spin-correlate relation. The alternative potential spin-correlate is erased. Hence, we obtain an increase in the entropy of future potential information in terms of the erasure of the foregone potential of spin correlates, attended by the conservation of spin. Thus, the evidencing of an otherwise timeless principle of the conservation of spin in EPR is obtained through a time-asymmetric AND-to-OR event-outcome, accompanied by the pertinent increase in the entropy of future-potential information. That is, **we need the asymmetric irreversible event from which to infer the transtemporal and timeless conservation principle of quantum spin. That is, we gain evidence of timeless physical law amid time-asymmetric time and events. Hence, we obtain the time-asymmetric operation of physical law.**

In other words, operationally, physical laws are time-asymmetric and always operate from AND-to-OR, and never from OR-to-AND.

*

If physical laws were truly time-symmetric, they would be able to operate in any time direction. But what would happen to physical law if we reversed time from AND-to-OR to OR-to-AND de-collapse? Through OR-to-AND de-collapse, the entropy of potential information would decrease as the OR-form state of resolved spin-correlates would fall out of evidence and revert back to pre-event AND-form potentiality. That is, the OR-form spin correlate would 'fall out of evidence', become 'un-evidenced', as these fell out of the domain of realised events altogether. **Insofar as the principle of the conservation of spin can only be brought into evidence *through* the formation of an OR-form spin-correlate via AND-to-OR collapse, the falling out of evidence of the OR-form spin-correlates via time-reversal OR-to-AND de-collapse would constitute the falling out of evidence of the principle of the conservation of spin itself**: That is, OR-to-AND de-collapse and time reversal would bring about the 'falling out of evidence' of physical law. But no one could bear witness to any of this, since the evidence, experience and consciousness of the scientist (or the data recorded in any a-noetic device) would also fall out of evidence in tandem per OR-to-AND de-collapse; a process of 'un-science'.

Physical laws are only operationally relevant and evinced in the AND-to-OR direction. Therefore, physical laws are operationally time-asymmetric and operate *only* in the AND-to-OR direction, and cannot be meaningfully spoken about under OR-to-AND de-collapse or time-reversal.

Simply put, physical laws do not operate in both directions of time. Physical laws operate in only one time-direction: that is, AND-to-OR collapse. Therefore, physical laws are empirically, evidentially and operationally time-asymmetric, *not* time-symmetric, as convention presumes.

1.13: Reiterating the Transtemporal Platonic Ontology of Physical Laws
Without Contradiction to the Operational Time-Asymmetry of Physical Laws

How do we know that time-asymmetric physical laws are timeless and eternal as much as they are time-asymmetric? The first evidence comes from the fact that physical laws are different from the states they configure and delimit, even if the effect of law is always through a process of time-asymmetric AND-to-OR collapse. The principal case for the timeless nature of physical law was first presented in the Preliminary (**0.26**): Therein, **our EPR-based approach demonstrated that the enforcement of the conservation of spin (and the time-asymmetric milieu in and through which the law is evinced) is not causally mediated across putative 'space' by means of a material 'signal'. Instead, it is enforced from outside of pertinent quantum entangled particles... from outside of 'space' and time.** The implication is that physical law and principle is dualistic, 'non-local', trans-spatial, and ultimately *transtemporal*, despite the time-asymmetric milieu through which law is rendered evident and affective.

109

In the Preliminary, we anticipated the dissolution of the notion of 'space' from ontology of the physics. The case for de-spatialisation will be fully made by Book-IV through a unique application of John Wheeler's delayed choice approach applied to the analysis of space itself. **The implication from de-spatialisation is that there are no 'objects in space', and only time is real: that corpuses of events are distributed in and structured by pure time-relations** or delayed choice time-interval relations.

By Book-IV, in the face of de-spatialisation and the unravelling of the notion of 'motion', we will re-work the concept of the 'speed of light' and re-define it as an invariant *delayed choice time-interval standard* that resides between a minimum of two events. At this juncture, we simply assert that, in the context of quantum entangled couples in EPR, the principle of the conservation of spin is enforced upon quantum entangled couples in a transtemporal way: i.e., in a fashion that does not require the mediation of the principle across time by means of some putative 'signal-bearing particle'. That is, even with anticipated de-spatialisation, and even with the fact that only time is real, **the being of the principle of conservation of spin instantiated between two entangled couples does not require time**, no more than it requires mediation or displacement of causality-bearing particles through non-existent 'space'.

Hence, in EPR, the principle of conservation of spin is *transtemporal* and *extant* the quantum entangled couples whose resolved spin-attributes it delimits in line with its specific principle. Its final ontology is superlative and extant vis-à-vis the states that it brings into ordered manifestation.

Insofar as this is true, and it is *incontrovertibly* true, the principle of the conservation of spin is not only an abstract-immaterial Platonic-Cartesian state, but, as such, it is not at all subject to modification and mutability by any AND-to-OR collapse process, or by the increase in the entropy of future potential information through which it is brought into evidence. That is, as undeniably time-asymmetric as it and other laws are in the empirical and operational sense, the principle of conservation of spin does not degrade, decay or mutate through AND-to-OR collapse and time… because **the principle ultimately belongs to an ontology above, beyond and extant the domain of "matter in space"; beyond and extant events in time: unchangeable, immutable… perhaps *eternal*.**

This does not mean that laws could not drop out of the domain of 'matter in space' and time, for whatever reason this might occur. But laws would not cease to exist because of time, 'space' and events. Indeed, even **if the process of AND-to-OR collapse could reverse into OR-to-AND 'de-collapse', physical laws would certainly fall out of evidence from memory and time and cease to be causally pertinent. Yet, the same laws could not fall out from the ultimate ontology that resides beyond future-potentiality, memory and time**. Thus, physical laws are indeed relationally eternal and timeless vis-à-vis the domain of memory, possibility, and time, even under absurd OR-to-AND de-collapse and time-reversal.

*

What holds true for the principle of the conservation of spin holds true to all physical principle or law: All physical laws are time asymmetric: All physical laws are abstract-immaterial: All natural laws are Platonic-Cartesian: All are dualistic and extant vis-à-vis 'spacetime'… extant vis-à-vis potentiality, memory, and time. Hence, physical laws *are* timeless and eternal at the same instant they are operationally and empirically time-asymmetric.

1.14: Part-I Summary & Conclusion

The **facile proof of information-matter dualism,** first adumbrated in the Preliminary, is summarised thus: There is no such thing as seamless-continuous motion or trajectory: Such a feat would require observational and intrinsic zero-wavelength inputs of infinite frequency and energy. Also, quantum mechanics demands that both the observation of motion, and the motion in itself, cannot be seamless-continuous, and must be quantised into discontinuous 'slices', 'frames' or 'snapshots' of observations across putative space and time. What holds true for motion must also hold true in the transportation of physical information across putative space by 'carrying' radiation and matter. The information is not 'carried' in a seamless-continuous form across space. The information apparently dissociates from radiation, matter and space, only to re-manifest in the midst of finite sequence of 'snapshots' composed of OR-form resolutions. Between these 'snapshots', the information apparently dissociates out into extant form, implying abstract memory.

Now, the summary of our attempt to define information:

- **The broadest definition of information is, "… that which informs and forms"** putative radiation and matter.

- **Aristotelian and Platonic background to information-matter dualism (1.02):** Information-matter dualism was anticipated in the philosophies of both Aristotle and Plato. For Aristotle there can be no dissociation of information from matter. Even if openly denied, Aristotelian dualism is implicit to information-matter considerations in contemporary science. Platonic dualism enjoys a latent standing within science: occasionally conceded in claims about abstract physical laws. With the advent of information-matter dualism, it is the Platonic understanding of dualism that gains ascendancy: it anticipates the dissociation of information from radiation, matter, and putative space evinced in facile information-matter dualism.

- **Overcoming mereological nihilism (1.03):** Mereological nihilism denies the real existence of wholes that form and inform the partology of matter. Mereological nihilism fails on account of **the argument from improbable commensurability** and **the argument from intrinsic consequential inequivalence**. In the first instance, communication, consensus, and coherent experience about anything would be rendered impossible. The case from consequential inequivalence is especially salient: The same amount of matter formed into a dart, but re-formed into a hollow sphere thereafter, will produce entirely different consequential outcomes versus a target: a consequence of the form or whole that informs the same quantity of matter. But the clincher is **the argument from generic quantum mechanics *instantaneous wholes*:** Therein, complex wavefunctions involved in, say, photography, are decisive in the pattern and consequence they ultimately manifest on the film, while the

110

quantum indeterminate 'parts' that manifest the image are funnelled into that image by the complex wavefunction: by the whole: *an instantaneous whole*. Hence *the whole precedes the partology*. The apparent contradiction between this and the facile claims of information-matter dualism, which initially presumes real information to be purely OR-form, will be resolved by Book-II and IV as part of the dissociation of the past into abstract memory out of the future. Hence, **the abstract wavefunction, the *instantaneous whole*, is real. Consequently, information synonymous with wholes and their identities cannot be dismissed as mereological illusions.** The claim for the physical dissociation of information, of wholes, from radiation, matter and 'space' is also ontologically real. Hence, we may speak of mereology, and of information, as being either **compelled** by its fundamental constituents and law, or as **flexible** with respect to those fundamental constituents and laws **(1.04)**. Much of nature is replete with **intermediate** types of mereologies or information which have both compelled as well as flexible relations vis-à-vis fundamental constituents and laws.

- **The definition of information in terms of <u>read-only</u> and <u>read-write</u> categories (1.05)**. The mereological account of form and information can be further elaborated into read-only and read-write informational states, wherein read-only types of information are the immutable forms exemplified by physical laws, while read-write types of information constitute the mutable and changeable types of form and information. Much of information-matter dualism concerns itself with the dissociation and independent retention of read-write information vis-à-vis radiation, matter, and 'space'.

- **<u>AND-form</u> versus <u>OR-form</u> information (1.06)**. We may further distinguish read-write types of information in terms of AND-form versus OR-form informational states. Thus AND-form information is quantum mechanical future-potential information: It is pre-measurement or pre-event information. As such, it does not constitute realised information. The process of 'measurement', of AND-to-OR collapse, converts AND-form potential information into an OR-form *event*: into realised information. This duality translates into a reformed intermediate spacetime model with two key domains: an AND-form quantum mechanical **spacetime in potentiality** versus a domain of **realised spacetime**. In information-matter dualism, the basis of OR-form information, whatever this turns out to be, survives its dissociation and is retained in the mnemonic domain of realised spacetime as abstract memory.

- **Harmonisation of information-matter dualism with entropy considerations**: Generic entropy is superseded by implications from information-matter dualism and abstract memory, in that both compel for the accumulation of the past in extant form vis-à-vis radiation, matter and 'space'. A feasible reduction in entropy on the basis of accumulating memory or decrease in the **entropy of memory** is offset by irreversible increase in the **entropy of future-potential information** (or the entropy of the future) whatever else might happen to generic statistical-based entropy. When we also consider **primacy-recency rules vis-à-vis the structure of memory** and combine this to AND-to-OR collapse as definitive of the 'arrow of time', we obtain harmony between the accumulation of memory on the one hand and the increase in entropy of the future-potential information on the other. Insofar as generic entropy is physically manifested through fundamental ontologically primary AND-to-OR collapse or time, and this entails the irreversible decay and erasure of all or most of the AND-form future potential informational state, **we obtain irreversible increase in the entropy of future-potential information despite the improbable reduction in generic entropy of energy**.

- **The true basis for the 'arrow of time' and the time-asymmetry of physical laws (1.07 to 1.12)**. Insofar as increase in generic entropy of energy and information is nothing more than a statistical disposition, while ontologically primary AND-to-OR wavefunction collapse and attendant increase in the entropy of future-potential information is a matter of certainty and ontic primacy, **generic entropy must be subsumed to ontologically primary entropy per fundamental and subsuming AND-to-OR collapse. (1.07)**. This also opens the way to **the entropy of memory per the inverse square law of time (1.08), hence the weakening and effective entropy of memory per time**. As a result, the conundrum between permanence versus change, irreversibility, the time-symmetry of physical laws versus the time-asymmetry per the 'arrow of time', can finally be solved. Indeed, **the 'arrow of time' is to be grasped in terms of the ontologically primary event-generating AND-to-OR wavefunction collapse process, which *is* time (1.09)** ... which *always* proceeds from AND-to-OR, *never* from OR-to-AND. Thus, time proceeds from a state of low entropy of future-potential information to higher entropy of potential information. Insofar as physical laws only ever become manifest amidst complexes of events resolved through AND-to-OR collapse, it follows that **physical laws can only operate per AND-to-OR directionality, not OR-to-AND. In the latter case, physical law would 'fall out of evidence'. Therefore, physical laws are not time-symmetric: (1.11 to 1.12)**. Therefore, *physical laws are operationally and empirically time-asymmetric.* All of this is expected to be incorporated into the intermediate model of spacetime (Book-II) and its de-spatialised temporised successor in Book IV: **The tripartite structure of these growing block models of spacetime incorporate time as AND-to-OR collapse, its AND-to-OR directionality, and the time-asymmetry of otherwise timeless physical laws (1.10)**.

In short... information is both definable and real: information-matter dualism is undeniable, even if it is to be superseded by implications from anticipated de-spatialisation in the framework of the growing block model. Conclusions from information-matter dualism are in harmony with generic ideas of entropy, although these are superseded by and subsumed to superlative entropy considerations per the ontic reality of the future as potential information, and from the identification of time with AND-to-OR collapse.

Finally, the problem of the 'arrow of time', and of the relation between supposed time-symmetric laws and time-asymmetric universe can finally be solved: Time and its 'arrow' are operationally and evidentially fated to AND-to-OR collapse and to the tandem increase in the entropy of future-potential information: Physical laws evinced through both are necessarily time-asymmetric, *not* time-symmetric.

BOOK-I PART-II:
THE FOUR-FOLD PROOFS OF
INFORMATION-MATTER DUALISM & THE APAGOGIC CASE FOR THE ONTOLOGY
OF INFORMATION

1.15: Aims of Part-II

Part-II constitutes the deeper scrutiny of information-matter relations requisite to the further development of the Alternative Realist thesis (Alt-R). It will rely far more on staple quantum mechanics than the facile case for the same espoused in the Preliminary (see **0.01** to **0.04**), which was based primarily on the critique of seamless-continuous motion. While Part-II seeks a deeper basis for information-matter dualism, it will become obvious that **the overriding concern is the ontic reality of information itself**, expressed as a series of apagogic arguments vis-à-vis information and against mereological nihilism; an apagogic approach that will not reach full completion until Book-III.

Part-II will begin with a brief **reprise of the double-slit experiment recapitulated in its non-standard evaluation**. This was first detailed in the Preliminary (see **0.13, 0.14** and **0.15**). The non-standard evaluation will be **followed by our four proofs of information matter dualism**. A **fifth proof** will emerge from the ontology of AND-form logic and the firewall principle in Part-III: The proofs will **focus on information-radiation dualism vis-à-vis radiation** in photography under ideal conditions and, finally, under realistic conditions. What we discover therein will be **generalised to information vis-à-vis 'carrying' de Broglie matter-waves**. Indeed, the fourth apagogic proof of information-matter dualism will directly relate information to de Broglie matter, insofar as its primary focus is the information pertaining to the material barrier featured in the two-slit experiment.

The discourse on the various definitions and categories of information formed in Part-I, including the reprise of entropy of future potential information *and* the entropy of memory, will be implicit to our four proofs of information-matter dualism. The proofs will assume quantum mechanical radiation and de Broglie matter-waves as spatially dynamic states that supposedly undulate and displace across putative 'space': This is also assumed in conventional generic quantum mechanics. But the assumption must eventually succumb to de-spatialisation as we discover reasons to abandon space. Indeed, Book-II will argue that, at best, quantum mechanical radiation and matter can only be modelled as spatially static. If the quantum mechanical wave constitutes the ontic reality of the future, then it is obvious that the future cannot be treated as 'moving', but only as a static state of nested future possibilities. By Book-IV, we will have completely usurped the notion of 'space' itself, using a variant of John Wheeler's delayed choice that attacks both the notion of the motion of the quantum wave *and* the 'space' across which it is purported to move.

The anticipated case for spatially static quantum mechanical waves in Book-II, and the even more revolutionary de-spatialisation furnished in Book-IV, must surely overhaul quantum mechanics and, with it, both the facile and the deeper argument for information matter dualism, leading to a more exotic and deeper understanding of both quantum mechanics and information-matter relations, as part of a superseding growing block system. Even so, the four proofs of information-matter dualism espoused in Part-II (and the fifth proof in Part-III) are sound and valid on the assumption of space and motion, which are treated as ontically real or axiomatic. The choices are…we either impale materialism per information-matter dualism that presupposes the ontic reality of space and motion… or we impale materialism per later de-spatialisation and the consequent temporisation that supersedes information-matter dualism, but which leads to the very same dissociation of information into abstract memory found in information-matter dualism.

1.16: Reprise of the Double-Slit Experiment:
Significance of the Non-Standard Evaluation to Information-Matter Dualism

What follows is a brief reprise of the double-slit experiment, especially per the non-standard appraisal (Preliminary **0.14**) which resides at the heart of the new complementarity thesis.

Let us recall the three main components of the experiment: i.e., the emitter; the detector; and the intervening barrier. Typically, the barrier can be set to one of two 'one slit open' states, or to a 'two slits open' form: The null setting, in which both slits are closed is typically ignored. The emitter may be designed to issue light radiation (photons) single quanta at a time, or de Broglie matter-waves, such as electrons, one 'particle' at a time. In the case of electromagnetic light radiation, a simple developing photographic film will suffice as the detector, or an electron micrograph film in experiments with de Broglie electrons.

Released one 'particle' or one quantum at a time, per the assumption of space and motion, the emitted particle will supposedly displace toward the barrier. When both slits are open, the particle will allegedly pass through both slits and proceed to the detector, becoming incident at the detector as a localised particle-like deposition. It is on the basis of the global pattern evinced by *en masse* depositions that we garner conclusions about radiation and matter as both particle-like and wave-like.

QUANTUM MECHANICS AND MIND

When two slits are open, the pattern formed by the *en masse* quantum depositions will evidence an interference pattern. Presumably, each quantum wave passes through both slits and splits into two independent wavelets. Subsequently, on approach to the detector, the wavelets meet and interfere constructively in some places and destructively in others to form the characteristic interference pattern at the detector. Given that the wavelets per each open slit derive from one original wave per just one quantum particle released at the emitter, the pattern thus formed from the *en masse* particle-depositions will be rightly inferred as the result of the self-interference of each particle with itself, possible only if the particle is wave-like.

Note again that **it requires *en masse* particle depositions to manifest and bring into evidence the reality of self-interference through the so-formed interference pattern at the detector. Hence, particle-wave duality is always observationally simultaneous at the detector, in that we need particles to manifest the former reality of the waveform at the point of final observation, and both particle and wave-like outcomes are simultaneous at the point of observation.**

*

Per generic quantum mechanics, the emitted quanta of radiation or matter begins as a particle, allegedly propagates as a wave in the usual *imagined* pseudo-realtime way; passes through the barrier; constructively and destructively self-interferes on its way to the detector; and, finally, AND-to-OR collapses into a particle at the detector-surface. To state that the particle self-interfered is to state that it occupied multiple putative *potential* positions, rendered in AND-form superposition, not just at the two open slits but throughout its purported approach to the barrier and thereafter.

On its way to the detector, now as two wavelets, the particle will assume, not just one *potential* 'which way' path but a superposition AND-form state of many such potential paths. Self-interference implies that, in some places, coinciding *potential* 'which way' paths embedded in the now-overlapping wavelets will cancel each other, or otherwise reinforce the probability of a 'which way' path. Of course, all of this assumes the real ontology of 'space', 'position', and, with it, the reality of the notion of the motion of the quantum mechanical wave.

The quantum mechanical wave-state for *potential* particle-position will constitute an AND-form unitary superposition. Thus…

… {position-1 AND position-2 AND position-3 AND…AND position-n}

This will AND-to-OR collapse into a localised OR-form 'particle' at the detector, and, collectively with others, manifest the relic of self-interference by forming the expected interference pattern at the detector: so constituting the physical-experimental evidence for self-interference, the reality of AND-form logic, and the superposition principle.

*

The generic and popular accounts of the two-slit experiment, or the standard evaluation, states that, when only one slit is open, the distribution pattern formed at the detector evinces only the particle-like nature of the wave-particle duality, with *presumed* definite OR-form 'which way' positions and paths convergent and divergent from that single open slit. Therein, it is typically purported that particles begin as particles, presumably travel as particles (with definite OR-form 'which way' paths) and arrive as such at the open slits or holes in the barrier and beyond, and finally deposit as particles on the detector, with no AND-form wave-like intermediate states: all supposedly on account of the fact that an interference pattern *does not* form at the detector.

On the other hand, under two-slit conditions, the same radiation or matter exhibits clear AND-form characteristics per expected self-interfering wave-like behaviour, evinced in the formation of the interference pattern at the detector; not possible to obtain under one-slit conditions. **Supposedly, the two-slit condition exhibits and permits only AND-form realities, both per the fact of the interference pattern, and from the fact that it is not possible to infer any definite OR-form 'which way' path for the particle in any interim state before its deposition on the detector.**

However, **the same can be asserted in the case of the one-slit experiment, given that there is no justifiable basis for assuming that, under isolation, when all observational inputs are abstracted from the system, the potential paths that converge upon and diverge from the single-open slit are in any sense resolved OR-form states and paths: It could equally be argued that these convergent and divergent 'paths' are pure potential AND-form, *not* realised paths… *not* paths at all),** as we shall reiterate below.

*

The standard evaluation was challenged by our non-standard appraisal of the two-slit experiment in the Preliminary (see **0.14**). **The non-standard evaluation makes the assertion that both single-slit *and* double-slit experiments exhibit both AND-form *and* OR-form realities, and that AND-form logic is not restricted to the double-slit experiment. Critically, AND-form logic applies in both experiments to an equal degree, even in the single-slit experiment. The only concession is that the two-slit variant exhibits the interference pattern at the detector. But this is made manifest through OR-form particle-depositions at the detector: You need OR-form 'particle' depositions to render evident AND-form 'wave' realities and the interference patterns. Hence, both OR-form and AND-form realities are observationally simultaneous.**

In the single-slit variant, there is no interference pattern formed at the detector. Yet, the distribution pattern formed therein is per function of quantum probabilities that attend the AND-form wave structurally delimited by the single open slit. The quantum probability that attends the AND-form wave is reflected in the *en masse* OR-form 'particle' depositions that it imposes and manifests at the detector. In short, **the single-slit experiment does not exhibit exclusive OR-form particle formations but, through such formations, exhibits the probability structure per function of an AND-form wave imposed on the detector. Indeed, there is no definite OR-form**

113

'which way' information to obtain in the single-slit experiment, given that the convergent and divergent potential 'paths' at the single slit (or single hole) are purely AND-form potentials... *not* OR-form realised paths.

Moreover, direct observational evidence to the contrary to prove that there ever was *any* realised OR-form path under conditions of isolation in the single-slit experiment is not obtainable. Even if we place a detector at the single open slit and resolved the particle there into an OR-form state, and even if we subsequently obtained the single-slit distribution at the detector, this could not constitute direct evidential proof that the particle assumed any OR-form attribute subsequent to the single hole, or assumed any definite OR-form 'which way' path thereafter. Indeed, *in the absence of direct observational evidence under conditions of isolation*, one could not obviate the assertion that the distribution pattern obtained when the single-slit condition applies could also be obtained by a pure AND-form superpose of not-realised (but non-interfering) 'path-potentials' bereft of any definite OR-form characteristics or paths. Thus, you do not need to assume definite OR-from paths to obtain the single-slit particle distribution, and can obtain the same by assuming the isolation-interim to be AND-form throughout.

This insight constitutes the basis for the new complementarity thesis (see Preliminary **0.16** and **0.17**) wherein, per generic complementarity, **in both single and double-slit experiments, the system develops as a pure AND-form wave; collapses into an OR-form outcome at the detector; while the OR-form depositions that render the preceding AND-form wave, by reflecting the probability distribution of that AND-form wave, become observationally simultaneously at the detector. Again, this holds in both experiments, regardless of whether a single-slit or both slits are open; or whether we obtain an interference pattern**.

Therefore, it is false to assert that the single-slit experiment evinces only the particle reality, or even exhibit definite 'which way' paths, or that the two-slit experiment exhibits only AND-form realities with only pure potentials of not-realised 'which way' paths. Both experiments exhibit AND-form *and* OR-form realities, while *both* exhibit only *potential* paths rendered in AND-form, *never* as OR-form paths. Succinctly, AND-form logic, and AND-OR dualism, is evident in both single and double-slit conditions, save that the interference pattern is exclusive only to the two-slit condition.

*

What is the relevance of any of this to information-matter dualism? In its facile form described in the Preliminary, information matter dualism held that information dissociates from its 'carrying' radiation, matter and space. This assumes the ontic reality of space. It also assumes that information is purely OR-form in character: This was questioned in **1.13** (i.e., generic quantum mechanics against mereological nihilism). It was also questioned in the Preliminaries in relation to the first introduction of instantaneous wholes (**0.26** and **0.27**). It is to be fully questioned in successive Books. Yet, the essential claim from information-matter dualism abide: realised information cannot be carried or retained by AND-form radiation or by de Broglie matter.

Now, if we also grasp the AND-form radiation and matter as the objective future-potential state of a system then, obviously, the past and present, as resolved events, cannot be carried in or by the future, given that the latter constitutes a pure potentiality for future events that have not yet happened, and, as such, cannot constitute a repository or a 'carrier' or a 'transport' for the past. The information (of the past) is therefore carried 'elsewhere', not in or by the AND-form futurity. **The point of the non-standard evaluation of the double slit experiment is that it removes any hope that there could be** *any* **preceding resolved OR-form paths in** *any* **process placed in isolation (deferred to the future), by asserting that, even the vaunted single-slit condition cannot furnish OR-form paths**. This was made evident in our critique of Afshar's controversy (see Preliminary **0.17**). Therein, Afshar essentially claimed to have obtained single-slit OR-form 'which way' path-outcomes from double-slit conditions: He obtained this from two-slit conditions that ought to be bereft of such information. Yet, as we had argued in the Preliminary and have summarised above, while it true that the two-slit condition ought not to furnish any 'which way' information, even the single slit experiment cannot do so. And obtaining a single-slit outcome from a two-slit condition does not imply any OR-form 'which way' path: At best, it can only render a pure not-yet happened superpose state of convergent and divergent AND-form *potential* paths: These are *not* resolved 'which way' paths in any sense... in the sense that the future *en toto* cannot be treated as a resolved set of OR-form events or paths, and isolation in both two-slit *and* one-slit experiments (and in the Afshar experiment) secures just such a non-resolved future, and no realised 'which way' paths.

The Afshar experiment is no *more* capable of obtaining definite 'which way' information about the interim of isolation than is the mundane single-slit experiment, even if Afshar did obtained single-slit outcomes from two-slit conditions.

Through our approach to the Afshar controversy, we saved the principle of AND-OR complementarity, albeit as a new thesis that must admit to the observational simultaneity of both AND-form *and* OR-form realities at the detector, at least in post-isolation: wherein AND-form realities are evinced by the *en masse* OR-form states in an observationally simultaneous way, ultimately in both single *and* double-slit conditions.

The point is simply this: In the interim of isolation, concrete OR-form information will dissociate from its AND-form radiation, matter, and 'space', because our non-standard evaluation shows that the radiation remains purely AND-form (or future form) under *both* single and double-slit conditions, so long as isolation abides. Thus, AND-form logic cannot be eliminated or obviated even under single-slit conditions. It necessarily follows that any OR-form or realised information cannot be carried by or transported in the interim of isolation in single slit, double slit, or any other process subject to isolation, including the remarkable experiment carried out by Afshar. Therefore, any information at the initial point must dissociated from and across that AND-form interval of isolation, even in the single-slit experiment, and even in Afshar's experiment.

In short, the conclusion of the non-standard evaluation supports information-matter dualism.

QUANTUM MECHANICS AND MIND

1.17: The Causality Argument for AND-form Logic:
Implications to the Veracity of Information-Matter Dualism

Our non-standard evaluation of the double-slit experiment, and the conclusion in favour of information-matter dualism therein, is further augmented by another fundamental finding: namely, the causality argument for the ontology of AND-form logic, which derives from the ontological meaning of AND-form logic and of quantum mechanical waves. This envisions the quantum mechanical wave as the ontic expression of the objective future. The causality argument also justifies information-matter dualism and prefigures the fifth proof for the same in Part-III.

The causality argument holds that, in the pre-measurement pre-event state, or in the interim of isolation of *any* system or process (including in the single-slit experiment), we can only speak of pure potentialities or possibilities that a system might assume at some future moment. These are necessarily in AND-form superposition. Consequently, **to speak of the AND-form future, but treat it *as if* it constituted an OR-form state, exemplified by the typical assumption of OR-form 'which way' information, under conditions of isolation in single-slit experiments and in Afshar's experiment, is dubious**. If it could be practically achieved, if we could treat the future as if it constituted a resolved event-set before it AND-to-OR collapsed into such, **the proof would constitute the reception of 'signals from the future' and, with it, the violation of the principle of causality. It would also entail the tandem violation of the principle of conservation of energy-matter entwined with the principle of causality**.

In its simplest form, if one could receive signals from one's own future, now constituted as a resolved set of events, also constituted by energy-matter, where is the extra energy-matter going to come from? The supposition that the future (or any isolated system) might exist as a realised state-of-affairs (as OR-form) implicitly supposes the creation of above-unity energy and matter to constitute that future. Again, where is this to come from? Hence, the violation of the principle of conservation, inexorable upon the erroneous treatment of an isolated system *as if* it constituted OR-form states.

It is precisely because of the demands of causality entwined with the principle of conservation that nature must exhibit AND-form quantum mechanical characteristics for systems and processes placed in isolation. **It is because of isolation that the system and process (including in the single-slit experiment) is deferred to the future, and, in its interim of isolation, could *only* constitute *the future*: a set of as-yet not-happened possibilities that are necessarily rendered in AND-form, even when the AND-form future possibilities are restricted and forced to a single-open slit or hole**. This constitutes *the firewall principle* which wards against the possibility and absurdity of 'signals from the future'. It was first adumbrated in the Preliminary in **0.15**.

Hence, in the interim of isolation, the AND-form 'path' possibilities constituting the total AND-form state will converge at the single open slit, but *only as pure future possibilities... not* as OR-form realised converging paths. Beyond the open slit, the AND-form wave will then constitute diverging potential 'paths', again *as future possibilities... not* as OR-form realised paths. To assume otherwise would entail the treatment of a pure future-form pre-event possibility-state *as if* it constituted resolved OR-form realised events or outcomes... with implied 'signals from the future', and, with it, the violation of th firewall principle..

Therefore, the principle of causality demands that radiation and matter under isolation, even under single-slit conditions, and even in the Afshar experiment, must remain wave-like and AND-form, as is also true when under two-slit conditions. Therefore, the single-slit condition is not capable of emphasising exclusively the particle-like facet of the wave-particle duality. In the isolation-interim, the single-slit condition *must* exclusively exhibit the reality of the AND-form wave facet of the wave-particle duality, as must the Afshar experiment.

*

What is the relevance of the firewall principle and the causality argument to prospective information-matter dualism? The case regarding information-matter dualism critically rests on whether it is possible to completely eliminate AND-form wave-like characteristics from *any* physical system or process: tantamount to the elimination of the future. If this could be done, it would eliminate any possibility of information-matter dualism: Indeed, the feat would constitute proof of seamless-continuous motion, which was undermined in the Preliminary (see **0.2** and **0.3**).

However, **the causality argument and the firewall principle independently and permanently undermine any prospect of the elimination of AND-form intervals (the future) from *any* system or process under isolation, unless one is willing to entertain 'signals from the future'** and the usurping of the firewall principle..

Insofar as the viability of information-matter dualism is critically fated to our inability to rid any physical system of its inherent ultimate AND-form characteristics, and insofar as this is not permissible per the demands of causality intertwined with the conservation principle (hence, the firewall principle), the apparent dissociation of OR-form information from AND-form 'carrying' radiation and matter must follow inexorably.

In short, information-matter dualism *is* indubitably true, and is demanded by causality, the principle of conservation of energy-matter, and the firewall principle that incorporates these... against 'signals from the future', and *for* the ontic reality of that future as the decisive component of our impending growing block universe model.

1.18: The Four-Fold Proofs of Information-Radiation Dualism in a Nutshell

We will now present four proofs of information-matter dualism, first in the simpler context of radiation and photography, later generalised to de Broglie matter. There are four distinct factors behind the four proofs for the dissociation of information from radiation and matter: The first can be summarised as follows: In all accounts of the double-slit experiment with two-slits open, radiation assertedly propagates from emitter to the detector in pure quantum mechanical AND-form state: a form incapable of carrying or retaining *any* resolved OR-form information. It is this structural incompatibility and difference between the AND-form character of the 'carrying' quantum mechanical waves of radiation, versus the *presumed* OR-form character of the 'carried' realised information, that forces the conclusion that OR-form information *must* dissociate from the 'transporting' AND-form radiation throughout the interim of isolation, in which it is alleged to be 'carried' by that radiation. Generalised to the context of photography, the AND-form light radiation alleged to 'carry' the photographic image from its point-origin to the photographic film-surface *ought not* to be any sort of carrier of the OR-form photographic image. Hence, **given AND-OR dualism, in the interim of its 'transportation' from origin to film, the OR-form photographic image must dissociate from the AND-form radiation. Hence, information-radiation dualism**.

*

The second of the four-fold proofs can be summarised as follows: In the non-standard account of the double-slit and single-slit experiment reprised in **1.16**, combined with the firewall principle reprised in **1.17** (which forces the AND-form state to remain in its AND-form state, as demanded by causality intertwined with the principle of conservation) it was found that, even in the single-slit experiment, the AND-form character of the radiation cannot be eliminated, and no resolved OR-form 'which way' path or information could be inferred or can exist in that interim of isolation, even if possibilities are reduced to a single open slit or hole. The radiation will purportedly propagate from emitter, through the barrier and to the detector, as a pure AND-form quantum mechanical wave: as the pure future.

Upon the enforcement of isolation, a system or process is necessarily relegated to the future, into a not-yet realised pure possibility state. As such, it is firewalled from us per the demands of causality intertwined with principles of conservation, which wards against 'signals from the future'. Thus, we cannot rid the system or process of its AND-form quantum mechanical characteristics, even when we set up the system and process in such a way as to emphasise *only* its OR-form 'particle' characteristics: i.e., the single-slit condition.

Applying these to our photography context, the fact that we cannot ultimately or completely rid a system of its quantum mechanical AND-form characteristics (the future) necessarily implies the dualism of the photographic image vis-à-vis its 'carrying' AND-form light radiation even when photography unfolds under non-ideal conditions that reduce, but cannot eliminate, AND-form intervals. It follows that the image must physically dissociate from its 'carrying' and 'transporting' AND-form light radiation. This dissociation must begin at the source. The information must remain at a dissociated remove from the AND-form light radiation throughout the interim of the latter's putative journey from source to film-surface.

In short, **information-radiation dualism applies in photography, to whatever extent the radiation remains in its AND-form state, and to the extent that the AND-form character of it cannot be completely eliminated from the system, as demanded by the non-standard evaluation, causality, principles of conservation, or the firewall principle**.

*

The third proof derives from the role and implications of quantum indeterminacy, which operates on the 'carrying' light radiation incident on the developing photographic film. Under non-ideal conditions, quantum indeterminacy must surely operate throughout the putative displacement of the radiation from source to film-surface. Recall that, when AND-form radiation undergoes AND-to-OR collapse into a particle-like OR-form incident at the detector or photographic film-surface, the process *always* entails quantum indeterminacy. This ought to act as an information-scrambling process vis-à-vis any resolved OR-form information (i.e., the photographic image) that one erroneously supposes must have been carried by the preceding AND-form light radiation.

The operation of quantum indeterminacy alone ought to destroy any hope we might otherwise entertain that the AND-form light radiation must be the physical 'carrier' and repository of the OR-form photographic image. **The *assumed* information-scrambling effect of quantum indeterminacy serves to augment the first and second proofs of information-matter dualism, while it also constitutes an independent basis for the dissociation of information from its 'carrying' quantum mechanical radiation**. Whatever the information does, it must somehow circumvent the *presumed* information-scrambling effects of quantum indeterminacy. This implies effective dissociation of information from its radiation.

*

The fourth proof consist in the *apagogic argument from persistence*. **Despite the dissociation of information per the preceding three proofs of dualism, the photographic image will yet manifest on the film-surface**. The OR-form photographic image, the realised information, or the basis thereof, somehow persists vis-à-vis the system and process, even when it ought to be permanently erased upon its physically dissociation in the AND-form isolation-interim, if not by the information-scrambling effects of the quantum indeterminacy that attends AND-to-OR collapse of the image upon the film-surface. This 'miraculous' manifestation of the image at the film-surface indicates that the real information constituting the image, or its ultimate basis (whatever that may be) cannot be physically erased, despite the said OR-AND dissociation, and despite subsequent apparent scrambling from quantum indeterminacy.

The information, in its OR-form configuration, or whatever its basis turns out to be, is placed at a physical remove from the AND-form light radiation. As such, it is retained by the physical order in some peculiar way: We will need de-spatialisation and particle-wave dualism reframed into past-future dualism, combined with other radical findings of the growing block universe, to infer what this peculiar dissociation truly looks like.

From its state of remove from the AND-form light radiation, **the information subsequently re-associates or re-merges (somehow) with that radiation at its point-of-incidence on the surface of the photographic film, and abides despite the operation of *apparent* information-scrambling quantum indeterminacy**. Hence, information-radiation dualism per information-persistence must follow inexorably.

116

Let us augment thus: **Unless information really *is* retained intact at an abstract-immaterial remove from 'carrying' radiation, and can thus persist, and even survive quantum indeterminacy, quantum science as we know it would become untenable.** We need only imagine what would happen in a double-slit experiment in which total informational erasure was hypothetically entertained, and the setting on the barrier itself became subject to erasure and scrambling by quantum indeterminate resolution. Therein, and assuming any semblance of the barrier could survive total erasure, the barrier setting would be resolved in a purely probabilistic way... in a way uncorrelated with the original barrier-setting, save by a 1-in-4 probability... totally inconsistent with how the real experiment works and for quantum mechanics as we know it to be tenable. Thus, if erasure of information is assumed in a strict and total way, the double-slit experiment as we know it would become practically untenable. Consequently, staple conclusions foundational to quantum mechanics could not be secured, and quantum science as we know it would become impossible.

In short, **the very viability and possibility of quantum mechanical science *absolutely requires* the non-erasure of critical information, despite its dissociation and apparent scrambling. This insight augments and supports the persistence argument in a critical way: one that will become fully salient by Book-II in the preliminary case for abstract memory.**

1.19: Proof-I: Argument from AND-Form Logic: Information-Radiation Dualism in Photography

Let us recall the distinction between *potential information* and *realised information* from **1.06**. Therein we discovered that quantum mechanical waves comprise the AND-form potential informational content of a physical system: i.e., the future possibilities of that system. For example, until the end of isolation, quantum spin possibilities (the future) must be comprised of as-yet not-realised superpose of "spin-up" AND "spin-down". With the end of isolation at some future moment, this must AND-to-OR collapse into one or the other OR-form of spin, even if in a quantum indeterminate way.

What holds true for spin also holds true for putative 'position': Assuming an over-simplified case limited to just two alternate future possible positions that a single photon of light might assume at some future time, its future per isolation will be comprised of "position-1" AND "position-2". With the end isolation and AND-to-OR collapse, this might decay into the OR-form of one or the other position, albeit in a quantum indeterminate way.

Note that, throughout the above, we are assuming the ontology of 'position', of 'which way' states, and of 'space' itself: All of these are finally debatable. For brevity, we apply these assumptions with the full understanding that physical ontology does not ultimately support these. Thus, assuming the ontology of 'position', how do we really know that positional information, hence the photographic image fated to the *en masse* OR-form positional factors, cannot be carried by the group quantum mechanical light radiation pertinent to the image? We know this from the fact that **the effective dissociation of realised information from the future-potential AND-form radiation in the interim of isolation, is certain: It remains certain regardless of any group form characteristic of the AND-form quantum waves involved.** We possess this certainty from the double-slit experiment and from the non-standard evaluation of that experiment, summarised in **1.16**.

*

What is the relevance of the above to information-radiation dualism in photography? To address this, we now carry the findings summarised in **1.16** to our photography context. **Let us assume photography under ideal conditions, wherein the intervening 'space' between the photographed object and the film-surface is comprised of a pure vacuum and constitutes an AND-form interval in near-perfect isolation, as is typically obtained in single and double-slit experiments**, from which *all* conceivable extraneous physical-observational inputs are almost perfectly and completely removed. Indeed, it is possible to develop a photographic film in as pure a vacuum as one could physically obtain. Real photography rarely unfolds in this ideal way. Even so, we must first explore ideal photography before we examine realistic photography.

The same principles that are found in the ideal double-slit experiment must also apply to individual and group light-radiation involved in photography and to the process of 'transmission' of a photographic image under ideal conditions. **The photo-image from origin to film-surface must abstract out of the group of AND-form 'transporting' light radiation, in the same way that definite OR-form 'which way' information abstracts out of AND-form radiation in both the single and the double-slit experiment. This abstraction cannot entail the erasure of the image**, given the fact that, at the film-surface, the photographic image (the information) is realised, and the information is thus recovered. In photography, **the information dissociates from the radiation, with the latter configured as a future-potential AND-form state in the interim of isolation. It is then abstracted from that radiation. It is then subsequently restored to it at the film-surface. Hence, proof-i of information-radiation dualism.**

The fact that *any* image, or the correct image, develops at the film-surface clearly implies that its pertinent information must somehow be retained by the physical order, despite the AND-form character of the 'carrying' light radiation in the interim of isolation. It remains in this physically dissociated form vis-à-vis that radiation until its re-merger with it at the film-surface at the end of isolation.

1.20: Proof-II: AND-Form Ineliminability, Information-Radiation Dualism
And the Non-Standard Evaluation of the Double-Slit Experiment

The second of our four proofs of information-radiation dualism is based directly on the non-standard appraisal of the double-slit experiment. In the most general sense, the issue boils down to the question of whether we can ever completely attenuate the AND-form quantum mechanical character of radiation and matter and accomplish this in such a way as to assert a permanent merger between information, radiation and matter. Such a feat would permanently rid any system or process of *any* possibility for informational

117

dissociation and dualism. **The non-standard account of the double-slit experiment shows us that, from source to detector, or to film-surface, we cannot ultimately get rid of the future-potential AND-form character of radiation in any absolute way**. Recall from **0.16** how, in both single and double-slit conditions, we need *en masse* particle-like depositions to materialise into evidence the wave-like reality, but in an observationally simultaneous way with the OR-form reality. Under two-slit conditions, the *en masse* particles render the specific probability distribution that attends wave-interference and self-interference. Under the one-slit condition, we cannot obtain self-interference, although, the wave facet of the wave-particle duality *is* clearly manifest at the detector in and through the *en masse* particle OR-form depositions, the distribution of which abides to the probability structure specific to the non-self-interfering quantum mechanical wave. The point is that **the AND-form wave-facet of the duality is evident in both experiments, and it is not a peculiarity of the experiment with two-slits open**.

This observational simultaneity applies only at post-measurement and at post-isolation: **In both single and two-slit experiments, the preceding interim of isolation is purely AND-form, and the single-slit experiment does not furnish any OR-form realised 'which way' information pertaining to the interim of isolation**, despite the restriction of the AND-form state to convergence and divergence at a single open slit (or hole): But this is a convergence and divergence of pure possibility in AND-form, and cannot be attributed the status of a concrete resolved 'path'. Thus, we get a definite and non-contrived tilt in favour of radiation and matter as something that must behave as a pure future-potential AND-form quantum mechanical wave under *all* isolation-conditions: Only at the end of isolation, and post-measurement, will AND-form radiation and matter finally AND-to-OR collapse and exhibit the OR-form particle-like facet of the wave-particle duality, and, through it, furnish observationally simultaneous evidence for the interim AND-form wave through subsequent *en masse* OR-form particles.

*

In its specific pertinence to information-radiation dualism in photography, the non-standard evaluation shows that it is not possible to eliminate the interim AND-form quantum mechanical nature of radiation under any circumstances, even under single-slit or equivalent conditions. The same must abide in photography and across nature: Thus, in photography, the AND-form quantum mechanical character of light radiation displacing from source to film-surface cannot be eliminated, **Therefore, the interim AND-form light radiation cannot constitute or transport the previously resolved OR-form information or photographic image**. It follows that the real information, the image, must *effectively* dissociate from the displacing quantum mechanical light radiation: It must dissociate, just after its issuance from its source, and then remain dissociate throughout the interim from source to film-surface. The image must remain dissociated to whatever extent the future-potential AND-form characteristics of the 'carrying' radiation abides throughout the interim of isolation; perfectly under idealised conditions and imperfectly or decohered under imperfect isolation… **until the photo-image re-merges with the radiation per AND-to-OR collapse into OR-form realised information on the film-surface.**

*

Insofar as the quantum mechanical character of radiation is essentially the AND-form future-potential information state of that radiation, and insofar as this cannot be a carrier of any kind of OR-form realised (past) information, **our inability to eliminate the quantum mechanical character of radiation in photography necessarily compels for effective information-radiation dualism.** Of course, this can be generalised to any system or process involving radiation beyond photography. Hence, proof-ii of information radiation dualism.

1.21: Proof-III of Information-Radiation Dualism: Argument from Quantum Indeterminism

Information-scrambling from quantum indeterminism vis-à-vis AND-to-OR collapse and the formation of the photographic image must also apply. If so, how can one obtain any pattern or image-formation out of quantum chaos? This is a core mystery of quantum mechanics that we will fully resolve by Book-III: A summary of the expected solution was given in the Preliminary (see **0.22**): It entails very different implications to how information relates to quantum indeterminacy versus the sort of facile assumptions here entertained. Yet, for all its radical differences, what we will garner in Book-III and IV will entail *essentially* the same implications from proof-iii: namely, information-radiation dualism, albeit rendered in alternative radical form by Book-III and IV.

Let us first clarify what is meant by 'information-scrambling effect' of quantum indeterminism: It boils down to the scrambling of information-containing signal by a randomising process that operates on that signal at some critical point of its progress across 'space' or at the point of its reception. For example, treating the radio wave as if it were the literal 'carrier' of information, we insert a randomiser at the receiver: one that randomly re-arranges its constituents; scrambling the original signal into something different from the original.

An analogous scrambling process is expected to occur throughout the development of a photographic film, and at the culminating AND-to-OR collapse and formation of the photographic image on the film. **The future-potential AND-form group of quantum mechanical light radiation becomes incident on the film-surface. At which point AND-to-OR collapse processes unfold, transforming the radiation into photon-particle depositions. The photons are expected to deposit on the film in a completely random and haphazard way,** *without* any relation to the image that must finally emerge out of such quantum chaos. **Yet, paradoxically, we** *will* **obtain a coherent image out of the quantum-indeterminate scramble: the** *right* **image at that**, when we ought *not* to obtain *any* sort of image at all, save mere white noise.

The OR-form incidences that deposit on the film-surface, and the issue of where and in what order these deposit on that surface, is not susceptible to any kind of deterministic orchestration or technological control (as illustrated in EPR vis-à-vis quantum spin). The process of photon-deposition is entirely randomised. But processes subject to quantum indeterminacy ought not act

118

as viable carriers of, or depositories for, any sort of information, much less permit the survival of any coherent OR-form informational state, such as the photographic image. Yet, the restitution of information (the formation or re-formation of the photographic image on the film-surface) necessarily implies that the information was *somehow* retained by the system, even if it was not apparently transported in and by the future-potential AND-form quantum mechanical radiation or expressed through its subsequent information-scrambled quantum indeterminate processing at the film-surface.

The information, the image, seems be retained by the physical order in a dissociated abstract form, at an *apparent* physical remove from both the AND-form quantum mechanical radiation as well as from, or despite the operation of, the quantum indeterminate processes at its AND-to-OR collapse. This remove seems to lead to inescapable information-radiation dualism, even if from the specific vantage-point given to us by quantum indeterminism. Hence, proof-iii of information-radiation dualism.

1.22: The Apagogic Proof-IV-a:
Argument from Informational Persistence, Fundamental Non-Erasure
And the Viability of Quantum Mechanics

So far, we have presented three interrelated but distinct physical arguments in proof of information-radiation dualism. All of these presuppose the survival and persistence of OR-form information vis-à-vis the AND-form quantum mechanical wave and attendant quantum indeterminate process. This insight leads to proof-iv of information-matter dualism per the non-eradicability and retention of information: in short, *abstract memory*.

The persistence of information across nature, hence information in abstract memory, is evinced by the repeat re-manifestation, endurance, perdurance and continuity of identities and mereologies, despite seeming inimical quantum mechanical processes, and despite even generic entropy.

However, **instead of the retention of information and abstract memory, one could seek an alternative interpretation that entails the total erasure of information upon its dissociation from radiation and matter, *and* add the vagaries from quantum indeterminacy,** if not argue against the very ontology of information itself per the anticipated co-joining of the assumption of erasure with the Manyworlds interpretation of quantum mechanics. This latter is to be hammered out in Book-III.

If such absolute erasure held true, the manifestation of the photographic image on the film-surface, in a form residually true to the originary-source, and the repeat re-manifestation of patterns in nature in general, would be a product of the miraculous. We will present and solve against anticipated erasure hypothesis co-joined to Manyworlds in Book-III. However, **the consequent erasure and nihilism of information would render impossible the foundational inferences and conclusions of quantum mechanics itself; rendering the science in the form we know it impossible.** This insight constitutes a key apagogic argument for abstract memory, in that the very viability of quantum mechanics itself requires and favours abstract memory and attendant information-matter dualism.

In order **for quantum mechanics to be possible as a science and as an empirically justified set of conclusions, the basis for information and its critical retention and survival must abide, whatever the ultimate form or basis by which this is achieved.**

1.23: Proof-IV-b of Information-Matter Dualism:
Causality Argument, Limited Erasure & Information-Retention

For quantum mechanics to be true, in the double-slit experiment, the key variable that must survive and persist the interim of isolation and subsequent quantum indeterminism is the barrier, whatever its setting: the information pertaining to the barrier-setup must dissociate from its 'carrying' milieu in the same way and for the same reason as the photographic image dissociates from its 'carrying' radiation upon the enforcement of isolation, and upon the consequent deference of both to the abstract future. Per its dissociation, the barrier setting cannot be treated as something erased, despite isolation. Thus, upon the enforcement of isolation, we allege that the barrier might transform into a superpose of its alternative settings: Thus....

Status of Barrier: {"only slit-one open" **AND** "only slit-two open" **AND** "both slits open" **AND** "both slits closed"}

The initial barrier-setting, whatever this was, is expected to dissociate and be retained independent of the AND-form superposition state just described.

It cannot be sufficiently avowed that isolation *must* imply the quantum mechanisation of the whole system, including the barrier itself, into an AND-form future potentiality and futurisation, as described above. Also recall that, to assert otherwise, or to assert that the system and the barrier must remain OR-form despite enforcement of isolation, requires empirical proof, *not* possible to obtain under conditions of isolation, save through 'signals from the future', and, with it, the violation of causality and the contravention of the principle of conservation of energy-matter: (see **1.17** and in the Preliminary: **0.15**). As such, upon isolation and deference to the future, that future must now be open to all alternatives possibilities for the barrier, and its potentiality must assume an AND-form superpose of these. The barrier must remain in this AND-form state until the end of isolation and attendant AND-to-OR collapse.

However, **the outcome from the pertinent AND-to-OR collapse cannot be a matter of 0.25 probability for any of the four alternative barrier-settings, despite the AND-form futurisation of the barrier... in the same way that it does not matter how quantum mechanical the 'image-carrying' light radiation becomes upon similar physical isolation. As a piece of information, the original barrier-setting, just as was true of the photo-image, *must* be restored at the end of isolation** into the OR-form of "both

119

slits open"... baring contingency that might have it otherwise, *ceteris paribus*. In which case, throughout the interim of isolation, the initial setting was retained independently of the AND-form state of the barrier in isolation, and even from the subsequent quantum indeterminate processes of resolution of the whole barrier at the end of isolation. Hence, we again arrive at information-matter dualism and attendant abstract memory.

*

The initial OR-form state of the barrier, as critical information, is either somehow retained at a remove from the AND-form isolation-interim, and hence *persists* in the physical order in just such a state, or, per the demands of the anticipated erasure hypothesis, it must be totally erased from the physical order. But, **if the original setting on the barrier is erased upon physical-experimental isolation, and if this erasure is so total as to completely wipe out the initial setting of the barrier, then, at the end of isolation, the barrier could AND-to-OR resolve into any of the four potential settings available to it on a 1-in-4 or 0.25 probability basis: This is wholly inconsistent with the real world. Hence, the apagogic principle of the retention of information as abstract memory**.

In the real world, the original or initial barrier-setting, despite its dissociation, will be restored at the end of isolation, and structure subsequent AND-to-OR outcomes to produce, say, the interference pattern consistent with the "both slits open" setting: The real world implicitly exhibits *memory*, despite quantum mechanisation and quantum indeterminacy in the interim and in the process of AND-to-OR resolution. Hence, **apagogic reference to the real-world renders doubt upon the anticipated erasure hypothesis, and preserves the ontology of information in both information-radiation and information-matter dualism**.

*

Let us view in detail the apagogic argument at the centre of our case for abstract memory. If erasure were to hold true then information nihilism would also hold true: Having initially fixed the barrier to the OR-form "both slits open" setting, the complete erasure of that setting upon isolation would require that the barrier subsequently resolve into "only slit-one open", or into its opposite... or into "both slit open"... or even into "both slits closed". This would happen purely on a quantum indeterminate 0.25 probability.

Even if the outcome obtained at the detector was found to be consistent with the initial setting of the barrier to "both slits open", its restoration at the end of isolation would be a matter of pure 1-in-4 coincidence, brought about by quantum indeterminate chance: *not* by experimental design and control: *not* per memory. **If erasure held true, quantum indeterminacy could never allow the experimental practical determination of the barrier-setting in any short-run or long-running sequence of experiments**.

Moreover, if the barrier-setting was not retained and was subsequently resolved in a quantum indeterminate way, the same would apply to the global outcome obtained at the detector, and major inconsistencies between initial conditions and outcomes would emerge: For example, we might end up with the expected interference pattern, even if the barrier subsequently resolved into a single-slit configuration: a contradiction. That is, **if erasure held true, there could be no meaningful correlation and consistency between the initial setting versus the AND-to-OR restored state of the barrier and pattern-outcome on the detector**.

We immediately grasp that the assumption of erasure and information nihilism is wholly inconsistent with experience. The barrier state is *not* resolved per a 1-in-4 quantum indeterminate sorting process. The outcome on the detector is *not* purely contingent and irrelevant to the barrier-setting.

Once the barrier is set to a certain configuration, it dissociates from and persists, extant the interim of isolation, despite the automatic deference of the barrier to the future into its AND-form state of alternative barrier settings. At the end of isolation, the original setting is recovered while the detector-outcome that emerges is fully consistent with the initial barrier-setting. Otherwise, quantum mechanics as we know it would be impossible.

In short, the viability of quantum mechanics presupposes abstract memory and the retention of critical information in dissociated form from quantum mechanised and futurised radiation and matter: i.e., information-radiation and information-matter dualism. That is, **erasure hypothesis is not tenable. Therefore, abstract memory, and attendant information-radiation and information-matter dualism... must abide. Hence, proof-iv-b**.

1.24: The Apagogic Proof-IV-c: Self-Voiding Erasure: Proof of Information-Radiation Dualism

One could argue that, **even with the assumption of erasure and the rejection of the ontology of information, and despite the claims of essay 1.23, outcomes obtained at the detector could yet furnish quantum mechanics**, albeit not in the form normally obtained. To explain by example, if we set up the barrier to "both slits open", but at the end of the experiment we obtained an outcome specific to the "one-slit open" outcome, this could yet permit conclusions in favour of AND-form wave-particle duality per function of the non-standard appraisal of the double-slit experiment, wherein single-slit outcomes can and *do* manifest evidence for AND-form waves: (see **1.20**).

Another example: If instead we set the barrier to the "one-slit open" state, but subsequently obtained outcomes characteristic of "two slits open", we could yet infer the reality of self-interference, AND-form logic, and reach conclusions for particle-wave duality.

In either case, **it might seem reasonable to suppose that the original barrier setting need not at all persist in, or be retained by, the physical order... much less require requisite dissociation, or even less require the real ontology of information itself**. The initial barrier-setting could be fully erased, with no persistence or abstract memory, and with subsequent barrier-setting resolved on a pure quantum indeterminate 1-in-4 probability basis at the conclusion of the experiment. From this, quantum mechanical-like conclusions could still be obtained, even if in an obviously weird way, and inconsonant to real-world experience. Therein lies the problem with the approach: it fails to abide with really-existing experience and science.

*

120

We cannot eat our cake and have it too i.e. **If erasure applies to the barrier, it cannot be excluded from the rest of nature or experience: it must operate at every point and facet of the experiment across the physical order, and across all experience... on an all-or-nothing basis**: *Everything* subsequent to *all* initial conditions and settings, *in all systems and processes*, would be sorted and resolved on a pure quantum indeterminate basis, *without* the possibility of retention of information as memory...indeed, *without* the ontology of information. Full information nihilism would reign. Hence, *everything* would be erased. Not even reportable experience about double-slit experiments, the memories of the experimenters, or anything else, could survive the presumed nihilism of information: a point we shall return to in our attack against anticipatory informational and memory nihilism in Book-III.

The apagogic implication is simple: **The notion that we could obtain some sort of quantum mechanical science despite erasure *and* full ontological nihilism of information, is totally untenable. Indeed, it would not be possible to obtain *any* outcome in proof of *any* quantum mechanical realities, or of anything else, under conditions of all-or-nothing erasure and information nihilism. Indeed, the outcomes from which we might infer such realities, the critical informational states, if not experience itself, would be totally erased.** To illustrate, suppose a physicist performed a double-slit experiment in the nonsense universe in which erasure and nihilism applied on an all-or-nothing basis, as it must, with no possibility of retention of dissociated information in abstract form, and no possibility of its restitution at the end of isolation. At the end of the experiment, suppose the physicist obtained an interference pattern in consummate proof of self-interference. The physicist could not control the barrier setting, given total erasure. Therefore, the outcome from which the physicist obtained the said result would be resolved purely on a quantum indeterminate probabilistic a-causal basis, without any meaningful relation to a now-vaporous set of initial conditions. Yet, the 'proof' obtained about self-interference and the reality of quantum waves would be of no use to science unless the experimenter could present the material to peers: The experimenter must accumulate evidential exhibits about the self-interference; the implied reality of AND-form waves... with notes and data about the original barrier setting, the method, the lack of control over the barrier (analogous to the inability to control spin-resolution in EPR), and the observation that what happens to the barrier was a function of subsequent pure quantum indeterminate sorting. All of this would need to be placed in storage for subsequent peer review.

However, erasure must apply on an all-or-nothing basis: We cannot assert erasure at one point, only to suspend it per convenience at some other point. Hence, come the morning, our physicists will open the storage box… only to find that its contents have been erased, or else AND-to-OR transformed into, say, a fruit cake.. or into *nothing*. This is expected given nihilism and the operation of erasure and subsequent pure quantum indeterminate sorting... both of which must apply throughout, and for the same 'reason' that it supposedly applies to the barrier-setting and to the whole ontology of information.

Note the obvious: *This* is not the universe we inhabit. Total erasure does not abide in the real-verse. In our universe, critical information survives: Information really *is* part of real ontology.

*

Why stop there? Why not extend erasure to the physicist; to the peers; to the laboratory; to the academy as a whole? Therein, there will be no certainty that the laboratory scientist from the preceding moment will remain laboratory scientists at a succeeding moment. There is no insurance that they might not turn into a collection of pumpkins the next moment... or, perhaps into *nothing* thereafter. Erasure must imply the total impossibility of the continuation of *any* kind of communicable and consistent experience or basis. Erasure would render impossible the communication of *any* facts about science, or about *anything*.

It would be disingenuous to suppose erasure to operate selectively: it must operate on an all-or-nothing basis. If we are going to deny the ontology of information itself and, with it, dismiss the possibility of its dissociation, retention and survival, despite or because of AND-form interims and quantum indeterminate processing, the possibility for proving, communicating and investigating quantum mechanical realities, or even the possibility of founding quantum mechanics itself, would be rendered impossible.

Clearly, **the very possibility of quantum science is *absolutely fated* to the ontological reality and survival of critical information across time, albeit as dissociated abstract memory… viable *only* in a universe where erasure does not apply on an all-or-nothing basis, despite AND-form interims and futurisation, and despite putative information-scrambling quantum indeterminacy**.

Moreover, **since we cannot obviate AND-form logic, futurisation, the firewall principle and subsequent quantum indeterminacy, it follows that we cannot circumvent informational dissociation, its retention as abstract memory, and consequent information-radiation and matter dualism… no more than we could obviate the facts of quantum mechanics itself and *its* total dependence on abstract memory and effective information-radiation and information-matter dualism.** Hence, the completion of proof-iv.

<u>1.25: Photography Under Non-Ideal Conditions & Implications from the Quantum Zeno Effect</u>

We explored information-radiation dualism in the context of photography (proofs i through to iii) principally under ideal conditions involving the perfect withdrawal of system-disturbing and decohering inputs, under perfect isolation in pure vacuum. Could information-radiation dualism also arise in photography carried out in non-ideal conditions? **Is it possible to saturate a system under initial isolation to such an extent that its isolation can be completely eliminated, and so obtain an absolute fusion between information on the one hand and the 'carrying' radiation and matter on the other? There exists a physical procedure that could bring us close to this goal: This is the Quantum Zeno effect**, and it was first grasped by Alan Turing and later fully developed by A. Degasperis and G. Sudarshan. To quote Alan Turing…

121

It is easy to show using standard theory that if a system starts in an eigenstate of some observable, and measurements are made of that observable N times a second, then, even if the state is not a stationary one, the probability that the system will be in the same state after, say, one second, tends to one as N tends to infinity; that is, that continued observations will prevent motion.[41]

Although N (the number of observations) might tend to infinity, it cannot obtain infinity, as was previously discussed in the facile case for information-matter dualism, and on the grounds of the impossibility of seamless-continuous motion: (see Preliminary: **0.2** to **0.4**). Indeed, if N obtained infinity, frequency would also obtain infinity… and so would putative energy… and we would be back to the absurd world of infinite power sources and ultraviolet catastrophes. Thus, the want to rid a system so completely of its quantum mechanical AND-form characteristics through the Quantum Zeno effect, hence, to rid it of all facets of its isolation in some absolute way, must remain as forlorn as the want of a complete erasure of critical information pertaining to the initial setting on the barrier envisaged in **1.24**.

Consequently, **it is simply physically impossible to render a system into a pure OR-form seamless-continuous state, given the limits of the Quantum Zeno effect, and given the ineliminability of AND-form logic. Therefore, it is impossible to obviate information-radiation dualism consequent upon ineliminable AND-form logic.**

On the other hand, by accentuating the AND-form quantum mechanical character of a system under ideal conditions of perfect isolation, hence by reducing to the absolute minimum the Quantum Zeno effect, and by highlighting the relation between physical information and quantum mechanical light radiation under just such conditions, we obtain the effective physical dissociation of information from the now-futurised AND-form radiation in the isolation-interim, and, with it, established the reality of information radiation dualism.

We can also assert that perfect isolation is impossible, and that nature can never permit the perfect amplification of the AND-form quantum mechanical to the point of achieving total dissociation or even total erasure, assuming the apagogic impossibility of erasure. That is, perfect isolation cannot be achieved, no more than perfect saturation through the Quantum Zeno effect could be achieved. Consequently, the system under optimum isolation must be subject to consequent quantum decoherence. i.e., an imperfect, reduced AND-form state per OR-form contaminants. Indeed, **this insight is admitted in generic quantum mechanics under the rubric of** *decoherence* first developed by H. Dieter Zeh[42].

In its own way, decoherence constitutes admission to a minimal implicit Quantum Zeno effect: to the effect that a system or process cannot be purely AND-form and must also possess OR-form contaminants: i.e., unintentional, and ultimately ineradicable system disturbing inputs in the form of, say, incident cosmic rays, lone photons, or even vacuum fluctuations. These will *decohere* an otherwise ideal AND-form unitary state, but will fall short of bringing about total AND-to-OR collapse.

The world we experience, one in which AND-form quantum phenomena do not appear to be obvious, but from which they are provably ineliminable even at the macro-scale, can be understood as the result of a generalised, but *not* absolute, implicit Quantum Zeno effect and attendant all-scale rapid decoherence. Therein, photography under normal non-ideal conditions will entail some minimal realisation of the Quantum Zeno effect, delimited to some degree by the isolation enforced by the camera chamber, amongst other things. Hence, all things being equal, in photography under non-ideal conditions, the fullest possible decoherence will apply to the light radiation involved in photography.

Yet, just as we cannot completely erase OR-form contaminants and eliminate the said decoherence, neither can we eliminate from the process of photography the *inevitable* AND-form quantum mechanical interims and states, even if some concession to minimal generalised Quantum Zeno-like effect, or even a full Quantum Zeno effect, might apply. Hence, even under non-ideal photography, the information or image *must* dissociate from the ineliminable AND-form interims inherent to the whole process.

*

To reiterate, photography carried out under non-ideal conditions entails an imperfect interfering and decohering medium. For reasons given, this imperfect medium cannot entail a perfect Quantum Zeno effect or absolute decoherence. It follows that the quantum mechanical character of the pertinent light radiation, and a degree of AND-form isolation, must remain ineliminable from the system and process of photography, even under non-ideal conditions. Therefore, the effective dissociation and dualism of the image vis-à-vis its 'carrying' radiation, will abide. Thus, information-radiation dualism in photography cannot be obviated and must abide, with implied abstract memory and information-radiation dualism.

Insofar as the AND-form quantum mechanical character of radiation cannot be eliminated in any absolute way, it follows that we cannot rid physical systems of instances of apparent dissociation of information. Therefore, we must again concede in favour of information-radiation dualism… *and* **implied abstract memory.**

1.26: Information-Matter Dualism & Heisenberg's Uncertainty Principle

A final consideration must be brought to bear to complete our assertion that, in photography, and across nature generally, it is not possible to eliminate the AND-form quantum mechanical character of radiation and matter, and, consequently, it is impossible to

---

[41]Sudarshan, E.C.G; Misra, B. (1977) "The Zeno's paradox in quantum theory". Journal of Mathematical Physics 18(4): 756-763

[42]Dieter Zeh. H. (1970) "On the interpretation of measurement in quantum theory". Foundations of Physics, Volume 1, Issue 1, p 69-76

eliminate attendant information-radiation and information-matter dualism from any system or across nature. This derives from Heisenberg's uncertainty principle.

In its usual definition, **the uncertainty principle states that, the more accurately we measure the putative position of a particle through, say, short-wavelength high-frequency inputs, the greater the intrinsic uncertainty obtained about its momentum (i.e., its mass times velocity). Conversely, the more accurately we measure the momentum of a particle by means of long-wavelength low-frequency inputs, the greater the intrinsic uncertainty garnered about its position.** In other words, we cannot measure accurately both the position and the momentum of a particle at the same time.

Stated succinctly in our own terms, this implies that, **to resolve the putative position of a photon of light accurately would entail that we reduce as much as possible the position-relevant AND-form state of that particle**: That is, we must 'measure' or convert the AND-form state as it pertains to the superposition of potential positions that the particle *might* assume in its future to an AND-form state that approaches absolute resolution of position into one position. The accuracy obtained will only be as good as the resolution power of the physical observational input used: The smaller the wavelength and higher the frequency of our physical input, the more spatially and temporally confined the set of potentialities belonging to the particle, and the more accurate will be its positional resolution upon subsequent AND-to-OR collapse. If we could devise a zero-wavelength and infinite frequency observational input, the AND-form quantum mechanical character of the radiation as it pertains to photon position would be eliminated: We would obtain absolute resolution of position, and, with it, the absolute elimination of information-matter dualism pertinent to position.

Now, according to the uncertainty principle, the less AND-form quantum mechanical position becomes, the more quantum mechanical momentum becomes, and vice versa. In other words, even if we could eliminate the quantum mechanical character of light radiation as it pertains to putative positional information, hence eliminate information-radiation dualism pertinent to position, this would only amplify the AND-form character of its momentum. It follows that, even if we could resolve putative positional information absolutely by the application of an impossible zero-wavelength infinite-frequency input, hence eliminate information-radiation dualism, the uncertainty about the momentum of the radiation would then become magnified to infinity.

In conclusion, **the reduction of the quantum mechanical description of one attribute of a system (i.e., putative position) would only serve to create the conditions for the magnification of the quantum mechanical character of the other physical attribute (i.e., momentum)**.

It follows that the AND-form quantum mechanical character of a physical system cannot be eliminated vis-à-vis the other variable, given the non-commutative character of position versus momentum... and from the fact that the attempt to eliminate it for one attribute will only magnify it per the other attribute.

In the photography context, we imagine a way to eliminate the quantum AND-form character of the positional state of the propagating light radiation by means of physical-observational inputs of zero-wavelength. Even if this could be accomplished for position, it could not be simultaneously achieved for momentum: (Yes, even mass-less photons have momentum. Particles with mass, such as electrons, hence 'matter', explicitly possess momentum attributes that are non-commutative with their putative positional attributes). **It follows that, some significant attribute of radiation or matter must always remain in AND-form, no matter what we do about the other attribute.** Indeed, something of *any* system must remain quantum mechanical per the implications of Heisenberg's uncertainty. This applies as much to the intrinsic wavelength of radiation and matter as it does to the extrinsic observational inputs used to garner information about radiation and matter.

Since we cannot eliminate the AND-form state of a system per Heisenberg's uncertainty principle, we cannot eliminate attendant putative dissociation of information from AND-form radiation or matter. Information-radiation and information matter dualism necessarily follow.

*

Of course, the above thought experiment supposed that we could apply zero-wavelength observational inputs to resolve the position attribute absolutely. In the first place, zero-wavelength observational inputs are impossible: Such impossibilities would also possess infinite frequency and constitute infinite energy inputs. In the second case, if we could work the impossible, the absolute resolution of position would shoot the potential momentum attribute to infinity. This is also impossible. Even so, on the notion of the impossible, per Heisenberg's uncertainty, one attribute of the resolved system would yet remain AND-form quantum mechanical. It follows that, per uncertainty, we could not ultimately eliminate the AND-form character of a system.

Information-radiation and matter dualism, premised on the ineliminability of AND-form logic, must yet again abide.

1.27: Generalisation of the Four-Fold Proofs to Matter: Information-Matter Dualism

In electron micrography, it is the electron de Broglie matter-waves that supposedly propagate from source-object to an electron micrograph film-surface and manifest an image on that surface. Given the essential similarity of this to photography with light radiation, **the four-fold proofs and conclusion garnered for radiation in photography apply in the same way to de Broglie matter** such as evinced in the development of an electron micrograph. Thus, by generalisation, *any* de Broglie matter, whether under ideal or non-ideal conditions, will exhibit information-matter dualism: **We need not repeat in detail the four-fold proofs pertaining to information radiation dualism: we simply transplant these vis-à-vis de Broglie matter.** In any case, the fourth proof directly relates to the case for information-matter dualism.

A key difference between radiation and matter resides in the fact that, within given frames of reference, material corpuses can usually be treated as if at relative rest, while light radiation especially cannot usually be treated in such a way. Matter-waves can also be treated as capable of undergoing displacement, albeit at less than the speed of light.

How do we generalise information-matter dualism to matter and mereology at relative rest? This task was accomplished in the previous examination of information as it relates to the barrier-setting in double-slit experiment (proof-iv from **1.22** to **1.24**). The barrier constitutes a material corpus at relative rest: one that apparently undergoes only time-evolution. The dissociation of information from the material of the barrier upon experimental isolation and futurisation can be modelled as a dissociation in time; part of the time evolution of matter or barrier at relative rest.

Also, as part of information-matter dualism, the fourth proof established the non-erasure of the dissociated critical information pertaining to the barrier-setting ; an important segue to the case against anticipated erasure hypothesis and information nihilism, to be featured in Book-III. Therein, it was shown that, such critical information must survive both dissociation *and* scrambling by quantum indeterminacy. Hence, abstract memory.

What holds true for the barrier must equally apply to the emitter, to the detector, to the whole apparatus: These are critical informational states in their own right and cannot be subject to erasure: All are subject to information-matter dualism, wherein information will effectively dissociate from ineliminable AND-form interims to whatever extent isolation abides, given that isolation cannot be eliminated perfectly per the delimits of the Quantum Zeno effect (**1.25**).

This insight also contributes to the solution to the apparatus problem of quantum mechanics, introduced in the Preliminary (see **0.25**): The apparatus cannot be erased and, in the way described for information entailed in photography and the barrier, circumvents undeniable quantum mechanisation and quantum indeterminacy.

Of course, with or without motion, **information-matter dualism per the four proofs abides generally and equally vis-à-vis *all* matter... to the clover leaf; to the drifting snowflake; to worlds, stars, and galaxies... and to the brain itself**.

1.28: Final Notes on Information-Matter Dualism

Our deeper four-fold proof of information-matter dualism, as with the facile case adumbrated in the Preliminary (**0.1** to **0.4**) presumes the reality of 'space' and admits to the notion of 'motion'. It also assumes that information is typically OR-form, while the AND-form state constitutes pure future-potentialities that are rightly judged as structurally and logically distinct from and incapable of acting as the carriers or repositories for OR-form information. Within the stated assumptions, attendant effective information-dissociation, the independent subsistence of information from putative radiation, matter and 'space', or abstract memory, holds true.

However, the anticipated evolution of spacetime in Book-II, and its later supersession by the superlative Bergson-Whitehead amalgam per de-spatialisation and temporisation in Book-IV, will call into question both the notion of 'motion' *and* the ontology of the 'space' across which AND-form radiation purportedly displaces and from which information effectively dissociates.

Moreover, as we saw from the quantum mechanical argument against mereological nihilism in Part-I (**1.03**), and against information nihilism from the previous set of apagogic arguments (**1.22** to **1.24**) we must transcend the generic limited understanding of information. **As we develop de-spatialisation, and temporisation the fact that both usurp notions of the 'distribution of matter in space' and succeed this with the distribution of information in time (without 'space') must seamlessly lead to the development of abstract memory and the de-spatialised Bergson-Whitehead amalgam. De-spatialisation, temporisation and the growing block Bergson-Whitehead amalgam replacement to 'spacetime' will incorporate and transcend information-matter dualism in its present form, even as the latter's essential characteristics, such as dissociation and extant retention of information as abstract memory, will yet abide, albeit recuperated into radically different forms.**

1.29: Summary & Conclusion: Part-II

Below is a summarisation of the four-fold proofs of information-matter dualism: proofs that go beyond Aristotelian form-substance dualism and entail the full physical dissociation of information from radiation, matter and 'space', consonant with Platonic conceptions of information-matter relations... all of it secured on the basis and stricture of quantum mechanics recuperated in Alt-R.

- **Proof-I: Argument from AND-form logic (1.19):** This is the most basic of the proofs: Therein, OR-form information is distinct from AND-form potential information. The latter is structurally inimical to OR-form information. Thus, OR-form information cannot be retained in or transported by AND-form radiation and matter. The information 'carried' by AND-form radiation or matter (the photo-image, the electron micrograph image, or the 'material' state of the barrier) must, in the interim of isolation, dissociate from and survive the AND-form state. Its survival is empirically proven by the fact of its re-manifestation at the end of isolation via AND-to-OR collapse. Information can only survive the interim of isolation if it is retained at a physical remove from its AND-form quantum mechanical radiation and matter. Information-matter dualism follows, as does abstract memory, even if by inference.

- **Proof-II: Ineliminability of AND-form logic from the non-standard evaluation implies dualism (1.20):** In the non-standard evaluation of the double-slit experiment, even under single-slit conditions, the future-potential AND-form state abides throughout isolation... because isolation implies the deferment of the system to the future, and **futurisation** into AND-form superposition. While only the two-slit experiment variant exhibits wave-interference, it is erroneous to suppose that the single-slit exclusively emphasises the OR-form 'particle' facet of the wave-particle or AND-OR duality. **Indeed, the single-slit** experiment exhibits *both* aspects of the wave-particle duality in an observationally simultaneous way, just as do two-slit experiments, given that the distribution-structure of particle-depositions exhibited in the single-slit experiments can equally be formed by AND-form quantum mechanical waves that control for the probability of said depositions, albeit *without* interference. In any case, the principle of causality necessitates that, whatever the barrier-setup, the interim radiation or matter before the end of isolation *must* be treated as an AND-form future potentiality, or else we entertain 'signals from the future' in contravention of causality and the conservation principle. Per implied **firewall principle**, the non-standard evaluation of the double-slit experiment establishes the ineliminability of AND-form characteristics of systems in the interim of isolation, even when these operate under single-slit conditions that appear to attenuate AND-form logic. Given the ineliminability of the AND-form interims per the non-standard evaluation *and* the firewall principle, we cannot entertain the notion that OR-form information could reside or be carried in AND-form interims of radiation or matter: The dissociation of information, hence information-matter dualism, necessarily follows.

- **Proof-III: Quantum indeterminism implies information-matter dualism (1.21)**: The end of isolation entails that AND-form quantum mechanical radiation or matter collapse into OR-form realised outcomes that restore the original (or a consistent residue of) OR-form setup or initial condition. The process of AND-to-OR collapse coincident with this entails quantum indeterminate processes that ought to scramble, if not entirely erase, the OR-form information. Yet, quantum indeterminate processes paradoxically manifest and re-manifest the information, *without* erasure. If we *assume* that the information cannot be mediated into saliency through scrambling quantum indeterminate process, yet information abides despite such quantum indeterminism, then information *must* dissociate *and* be retained at a physical remove and survive, not only its 'carrying' quantum mechanical AND-form state, but the very quantum indeterminate process entailed in its subsequent AND-to-OR collapse and restoration. Information-matter dualism necessarily follows, as does abstract memory.

- **Proof-IV: Three apagogic arguments from persistence, fundamental non-erasure, and the viability of quantum mechanics, validate information-matter dualism (1.22 to 1.24)**: Using all three arguments as background, one could assert for a total erasure of information from the physical order per function of inimical AND-form logic *and* information-scrambling quantum indeterminacy. Hence, the anticipated erasure hypothesis and information nihilism; both of which are to be fully developed in Book-III. However, erasure is untenable: information is clearly retained at a remove from radiation, matter, and 'space'. If it were otherwise, apagogically, critical informational states, such as the barrier-setting in the two-slit experiment, and the very possibility of quantum mechanical facts in the form known, would become untenable. Indeed, any attempt to generalise information and memory erasure and the nihilism of information vis-à-vis nature, experience, and universe... a feat that must be accomplished on an all-or-nothing basis if it is to claim any merit... would culminate into a universe in which *no* experiment, science or experience is possible. Thus, anticipated erasure and information nihilism is dubious. Hence, information *must* be retained at a remove from radiation, matter, and 'space', *and* survive its dissociation… *without* erasure. Hence, abstract memory and information-matter dualism, and the ontic reality of information itself, most certainly abide.

- **The Quantum Zeno effect and the ineliminability of AND-form logic implies information-matter dualism (1.25)**: It is possible to so saturate a system with observational inputs so as to almost eliminate its AND-form isolation completely, and almost obtain the absolute fusion between information on the one hand and the 'carrying' radiation and matter on the other. This is the Quantum Zeno effect, first espoused by Alan Turing. Unfortunately, or fortunately, the Zeno effect cannot employ or reduce observables to zero-wavelength infinite frequency outcomes: a feat necessary for the total fusion and the total elimination of AND-form interims from a subject system. Insofar as the Quantum Zeno effect cannot eliminate AND-form interims from any system in some total or ultimate way, information matter dualism consequent upon AND-form logic must abide: Both the dissociation of information from the system *and* the retention of this as abstract memory must inexorably follow.

- **Heisenberg's uncertainty as an independent basis for information-matter dualism (1.26)**: The ineliminability of AND-form logic can also be independently supported from Heisenberg's uncertainty. To say that we cannot resolve the position and momentum of a particle at the same time implies that we cannot reduce the AND-form state pertinent to position *and* reduce the AND-form state pertinent to momentum at the same time: In short, we can never eliminate the AND-form character of a system in one respect (the momentum), even if we could totally eliminate it for the other (position). Insofar as putative information-matter dualism is apparently based on the ineliminability of AND-form logic, the ineliminability of the same per Heisenberg's uncertainty implies the independent tacit validation of information-matter dualism from Heisenberg's uncertainty.

BOOK-I PART-III:
ONTOLOGY OF THE QUANTUM MECHANICAL "AND" & THE FIFTH PROOF OF INFORMATION-MATTER DUALISM

1.30: Aims of Part-III

In furtherance of the Alternative Realist thesis (Alt-R) and our advancement of our understanding of information, the core theme throughout Part-III will comprise the recapitulation and deeper insight into the assertion that **AND-form reality constitutes the physical phenomenalisation and ontic reality of the future; part of the growing block universe approach, forced upon us by the intertwine of causality and the principle of conservation (i.e., the firewall principle)**.

Implicit to this will be the tandem insight that AND-form logic operates at all scales: and that quantum mechanics is *not* confined only to the micro-scale or to short time-interval physics; and that, whatever the scale, the ontic reality of the AND-form future cannot be treated as a resolved set of events, unless one entertains 'signals from the future', with consequent contravention of both causality and principles of conservation.

All of this implies the ineliminability of AND-form logic from nature... at all scales. Hence, the compelled dissociation of information from AND-form radiation, matter, and 'space'... *again*, at all scales. **Hence, the fifth proof of information-matter dualism, founded on the ontic reality of AND-form futurity as demanded by the all-scale *firewall principle*.**

The ontological account of AND-form logic in terms of the principles of causality and conservation will also furnish a **natural and seamless clarification of Heisenberg's uncertainty principle**: From the structural restriction imposed by the observational input, we cannot measure the position and momentum of a particle accurately at the same time: **When we use one sort of input for one attribute, we structurally defer the resolution of the other attribute to the AND-form future**, which is quantum mechanised or futurised into an AND-form 'smear' of future possibilities... bereft of any 'objective' OR-form attributes or resolved state. In short, **Heisenberg's uncertainty is per function of the ontic reality of the future**, combined to the structural delimits of the measurement process that must defer one non-commutative attribute to that future and into a 'smear state' of as-yet not-happened possibilities.

These insights will segue into the reiteration and explicit articulation of the Alt-R approach to time and causality: wherein AND-to-OR wavefunction collapse constitutes time itself; wherein time is *not* a 'fourth dimension'. As oft repeated, the process of **AND-to-OR collapse or time also leads to the accumulation of the past as abstract memory**, necessarily in dissociated form from the event horizon, in turn comprised of the most recent OR-form resolutions. The past as abstract memory delimits the possibilities that the AND-form future is permitted to assume (i.e., abstract memory decoheres the future): This memory effect must be incorporated into the anticipated intermediate spacetime model in Book-II and to its de-spatialised and purely temporised successor in Book-IV.

The process of **AND-to-OR collapse, synonymous with time itself, is a structure-less and gear-less process** that must defy rendition into closed-causal input-operation-output schemas, as well as obviate contiguous or contact causality, in tandem with anticipated de-spatialisation and temporisation. All of this will herald the demise of materialism and presage the intermediate model of spacetime in Book-II: with its nested futures model of the future; its counterfactual causality and attendant process of grand decoherence. When combined with the later de-spatialisation in Book-IV, the transcendence of the intermediate model by the Bergson-Whitehead amalgam, or spacetime *without* space, will be at hand, as will the final demise of materialism and the revival of Cartesian dualism.

1.31: Recapitulating the Causality-Aspect of the Firewall Principle
And the *raison d'etre* of the Quantum Mechanical "AND"

If the sort of realist view espoused by Einstein *et al* held true... if AND-form quantum mechanical states and the superposition principle turned out to be false, and if the natural order in its pre-measurement or pre-event state constituted as a pure OR-form state... causality would be rendered untenable.

From the Alterative Realist approach, one of the two critical reasons why AND-form quantum mechanical realities pervade across nature is per function of the principle of causality. Without AND-form logic, causality would crash. Conversely, without the principle of causality, there would be no AND-form logic, and no quantum mechanics.

Contrary to the philosophy of *presentism*, which denies the ontic reality of both the past and the future in favour *only* of the present, and contrary to *block theory* which asserts that the future is pre-formed into a fully resolved concrete OR-form state of events, causality demands that, until events get to happen, they *must* be suspended into pure as-yet not-realised AND-form future-potentialities. As such, the future cannot be treated as a resolved OR-form set of events, and until AND-to-OR wavefunction collapse happens (until time happens) the future must exist suspended in its AND-form superposition quantum mechanical state and form.

As part of the firewall principle, the causality argument from Alt-R is consistent with the *growing block theory of time*, promulgated by Michael Tooley[43], for example. But the same was also presaged in the views of North Whitehead, with the difference that Tooley denies the ontic reality of the future, whereas Whitehead assumed its onticity, as we do. To the growing block universe, Alt-R furnishes a quantum mechanical justification for the ontic reality of the not-yet happened future: It accomplishes this in part through the intertwine of causality and conservation principle amalgamated into the *firewall principle*. Our aim in this essay is to hammer out the ontic reality

---

[43]Tooley, Michael (1997). *Time, Tense, and Causation*. Oxford University Press

of AND-form logic and of the ontic reality of the future, and, with it, validate the growing block theory in the form that Whitehead envisioned it.

Indeed, as a first salvo and obvious conclusion, in the block universe, the future is pre-resolved into definite 'paths' and 'which way' states. Hence, **if the block universe held true, the two-slit experiment could never yield evidence for self-interference, nor the formation of interference patterns, given that these cannot be furnished by pre-resolved definite OR-form 'paths' but require alternative AND-form possibilities for self-interference. In real two-slit experiments, we** *do* **obtain self-interference and macro-scale interference patterns. This can only happen if the future ontically exists as an objective state of the universe... as as-yet to happen possibilities and futurities that permit self-interference. In short, AND-form logic, the quantum wave, and wavefunctions really do pertain to an ontic future, and the growing block model** *is* **the true depiction of our universe.**

*

If we could treat the future as if it constituted a fully resolved OR-form event-state before it transpired, as block theory would demand, and concede that this would require empirical evidence, and not be simply assumed on faith, the claim would require 'signals from the future' from that pre-resolved future.

In a similar vein, if we insist, as the naïve realists approach does, that a particle's given attribute must always constitute as an 'objective' OR-form state , and must do so even under isolation and the withdrawal of all observational inputs, and that it must always remain OR-form even before requisite AND-to-OR collapse gets to happen, in proof, this will also require 'signals from the future' in travesty of causality. If one must assert that the future has already transpired per the naive realist assumption that all things must be 'objective' OR-form, even in the absence of observation and in isolation, and even before these have transpired into realised events... this would also imply and require empirical evidence about pre-resolved futures. Hence, naive realism inadvertently requires 'signals from the future' and a block universe inimical to the viability of causality, quantum interference, *and* quantum mechanics.

Clearly, for causality to be tenable, the process of AND-to-OR collapse, and its AND-to-OR directionality and polarity as the true basis of the 'arrow of time' (see **1.09**) cannot be reversed, circumvented, or voided. Hence, future possibility must remain configured in its AND-form potentiality state until AND-to-OR collapse catches up to it. At which point it is permitted to collapse into an OR-form realised outcome. So long as isolation holds, so long as the AND-to-OR collapse process has not caught up with that futurity, and so long as a system in isolation is synonymous to a system deferred to the future, the said future possibility cannot be treated as an 'objective' OR-form state. It must indeed remain in its limbo AND-form state as a pure not-yet happened future possibility and ontic futurity.

This truth must hold at *all* scales, not just at the micro-scale or for short time intervals. This **all-scale theory of AND-form logic, hence all-scale quantum mechanics**, is crucial to the further development of Alt-R, given that it is often erroneously asserted that quantum mechanics applies only to the very small, and, by implication, only to very short time-interval relations... and that quantum mechanics and AND-form logic is supposedly excluded from the macro-scale domain.

Even if we disregard the strict demands from causality, we can yet directly infer from experiment and experience that both quantum indeterminate processes *and* self-interference... hence **the manifestation of the quantum mechanical "AND" at the macro-scale... can be vouched for in the double-slit experiment through macro-scale outcomes evinced on the macro-scale detector. The macro-scale culmination reveals both OR-form** *en masse* **particles** *and* **AND-form waves in an observationally simultaneous macro-scale way, constituted** *as* **the macro-scale interference pattern**. It follows that, AND-form logic, at least in its consequences and post AND-to-OR saliency, is not restricted to the micro-scale.

This assertion is consistent with expectations from growing block theory: Indeed, insofar as the process of AND-to-OR collapse is synonymous with 'coming into being' and time itself, the forlorn exclusion of the AND-form future from the macro scale, or the rejection of the ontic reality of the future as the AND-form potentiality state, would render time itself impossible, in that there would be no need for AND-to-OR collapse to transpire, and synonymous time could not possess onticity.

*

In a block universe in which the future is pre-resolved and has fully transpired, in a flight from London to New York, it would be possible for one to arrive at Kennedy Airport before having departed from Heathrow. If, per the block universe thesis, AND-form futurity could not exist, and the macro-scale future was treated as a fully resolved OR-form corpus before having transpired into such, one could then confirm the above travesty by receiving a telephone call in London Heathrow from a future version of oneself at the future Kennedy Airport, New York.

In a block universe *without* AND-form realities, is such an already-constituted OR-form future compartmentalised and isolated from us, or is it disclosed via 'signals from the future'? Either we can obtain signals from the fully resolved future, or we cannot. **If we cannot obtain 'signals from the future', we would need to reject the viability of block theory: not simply because it cannot be falsified but, in the absence of any 'signals from the future', the block universe could never be substantiated**.

On the other hand, to reiterate, in a block universe, **wave-interference, let alone self-interference attendant the double-slit experiment could not arise, since all things would be OR-form resolved, and given that only AND-form states produce self-interference. The fact that we can obtain interference patterns in the double-slit experiment speaks volumes against the block universe approach; against the notion of a pre-resolved future**. It inexorably leads to the ontic reality of AND-form futures and secures the reality of the growing block universe. Thus, the scenario above could never transpire, and it is not an accident that we do not receive telephone calls from our future selves from any future Kennedy Airport, New York.

127

The future entailed in any flight from London to New York would be subject to the same principle of isolation as found to operate in the double-slit experiment, and as espoused in the four proofs of information-matter dualism: **By definition, the future is the domain from which all observational inputs and feedback signals are withdrawn: That is, the future is generally equivalent to an experimental isolation-interim. This certainly applies to the macro-scale barrier in the double-slit experiment upon the enforcement of isolation.**

Returning to the scenario involving the flight from London to New York, **all possibilities... from the immanent boarding of the plane at the immanent-future London-Heathrow, to the later future-potential arrival at New York... are objectively isolated from observation and from any observational inputs at present-moment Heathrow: As such, the future *is* equivalent to the interim state of isolation as found to apply in the double-slit experiment.** This is true **if for no other reason than the fact that, even the requisite observational inputs to bring into saliency the not-yet transpired events have not yet transpired in their own right, and could not facilitate the observation of the future or any 'signals from the future'.**

Hence, you cannot arrive at New York before departing from London... and you cannot receive telephone calls (the equivalent of observational inputs and feedback that have not yet formed up) from your future-self at New York. Hence, unless we contrive 'signals from the future', the growing block universe must abide.

Implicit to the above, **it also follows that AND-to-OR collapse, as the critical process of 'coming into being' of events out of the AND-form future, and as the process through which the growing block universe gets to 'grow'...** *really does* constitute time. That is, AND-to-OR wavefunction collapse *is* time and, grasped as such, supersedes the generic conception of it as a 'fourth dimension'.

1.32: Recapitulating the Principle of Conservation Aspect of the Firewall Principle
And the *raison d'etre* of the Quantum Mechanical "AND"

If causality could be violated at both the macro-scale and at the micro-scale, and if we could receive 'signals from the future', and if the block model of the universe held true, what would happen to the principle of conservation? What would happen to the sum of energy-matter epitomised in $E = mc^2$? Would the sum of energy-matter remain constant? It turns out that the putative principle of conservation of energy-matter is inseparable and mutually intertwined with the principle of causality: Both are aspects of the same one thing. Together, they constitute *the firewall principle*. This keeps 'signals from the future' at bay.

Recall from **1.31** how in a flight from London to New York, wherein one arrived at New York before having departed from London, led to the violation of causality, compounded by a telephone 'signal from the future': from a future-self at New York to a present-self yet to embark on the plane at London-Heathrow. But where will all the extra putative energy-matter to constitute the signals from the future, the future-passengers, *and* the future plane at New York come from? Indeed, there are now two of each of Heathrow and Kennedy Airports: These also need putative energy-matter to embody them. Indeed, there are now two Earths, two Moons, and two sets of solar systems, if not two 'whole universes'; all of which require energy-matter to constitute them.

If the future could transpire before AND-to-OR collapse processes caught up with it, or if the future had already transpired in fully realised form as is demanded by the block model approach, and if we could gain empirical evidence and physical effects from that future via 'signals from the future', this would inexorably lead to the violation of the principle of conservation: There would be uncontrollable multiplications of energy-matter, presumably created out of nothing, to embody every future state of any and all pertinent systems, insofar as we would need to treat these as pre-resolved OR-form fully 'materialised' events, and as embodied in energy-matter, *whatever the scale.*

It follows that the future cannot be treated as a resolved set of events: not simply per the demands of causality, but per the demands of an intertwined principle of conservation of energy-matter.

Note that any system in the future, even a micro-second from now, even five minutes from now, is effectively suspended in a state of interim AND-form isolation: an isolation-interval equal to a micro-second or to five minutes, respectively. Otherwise, the Moon five minutes from now would be treated as a concretised fully materialised corpus of events... so also all the Moon's tidal effects upon the Earth pertinent to that five-minute interval. In such a case, we would be forced to dismiss the principle of conservation, at least as much as causality itself... although the Earth could not possibly survive the tidal effects from the Moon five minutes from now combined to the same effects from the same Moon at each instant of its resolved future-state block. The fact that the Earth abides speaks against the block universe approach and in favour of the ontic reality of AND-form futurity, *at all scales.*

*

We now return to the issue of scale as it applies to AND-form superpositions and related conservation principles: **To assert that AND-form logic and the superposition principle are restricted only to the micro-scale is to restrict both causality *and* the principle of conservation to a micro-scale domain of short time-interval phenomena.** On that account, cause and effect would need to be excluded from the macro-scale, with tandem contravention of the conservation of energy-matter at the macro-scale. **If AND-form logic did not apply at the macro-scale world, we would inhabit a very different physical order: one with 'signals from the future': one with violations of the principle of conservation at every moment and turn. Hence, the belief that quantum phenomena *only* apply to the very small is rendered false, given the normalcy of the macro-scale, as testified by experience.**

Clearly, the real universe and experience at all scales does not typically exhibit either the violation of causality or of the usurping of the principle of conservation of energy-matter, as would be expected from the dubious restriction of AND-form logic to the micro-scale, or by its abstraction from reality as whole by the block model approach that treats the future at *any* scale as a fully pre-realised

OR-form daz. Hence, AND-form logic *indeed* applies to the macro-scale and to long time-interval processes, as much as it does to the micro-scale short time-interval domains.

As to why macro-scale billiard balls do not behave like quantum objects, there is an explanation to this that is wholly consistent with the assertion that AND-form logic abides at the macro-scale. However, the proper answer will need to await further developments, starting with Book-II and the case for grand decoherence.

If AND-form logic applies, not just to the very small but to the macro-world (as it must) it must also apply to the barrier in the macro-scale double-slit experiment in its state of isolation; to the macro-scale development process in photography; to the macro-scale drift of the snowflake; to the macro-scale flight from London to New York; to the universe as a whole… and to the brain itself.

From the firewall principle per the intertwine of causality and the conservation principle, and its incorporation to the scale-issue in support of the ontic reality of the not-yet resolved all-scale future, we arrive at the *raison d'etre* of the quantum mechanical "AND" and the superposition principle. **Why does AND-form logic exist in nature? It exists because the principle of causality, entwined with the principle of conservation, must *demand* the reality of AND-form logic. Natural order at all scales without AND-form logic would imply, not simply the contravention of causality, but the tandem violation of principle of conservation. Hence, we are forced to permanently give up on the naïve realist approach to quantum mystery and, with it, reject the forlorn block universe approach, which seek to treat even isolated micro-scale systems as 'objective' or OR-form.**

1.33: Proof-V of Information-Matter Dualism
From the Ontic Reality of the Quantum Mechanical "AND" & the Firewall Principle

We have clarified that the ontic reality of the future is the *raison d'etre* for the quantum mechanical "AND": The future is ontically real, *at all scales*… consistent with the growing block approach… as demanded by the firewall principle that preserves both causality and the principle of conservation.

The principle of causality played a key role in the apagogic arguments of the fourth proof of information-matter dualism presented in Part-II (**1.22** to **1.24**). It held that isolation must engender the AND-form quantum mechanisation and futurisation of the barrier-state from its initial OR-form configuration into its AND-form superposition of future possibilities, per the strict demands of the principle of causality operating as part of the firewall principle. Therein, the initial OR-form information pertaining to the barrier-setting, which constitutes a critical piece of information without which quantum mechanics as we know it would be rendered untenable, must *effectively* dissociate from the quantum mechanised barrier upon experimental isolation, with the information so dissociated retained as abstract memory extant the quantum mechanised radiation, matter and 'space' pertaining to the interim of isolation.

Just as causality implies the inescapable necessity of AND-form logic, of the quantum mechanisation of the barrier, and the apparent dissociation of information from matter in the interim of isolation, the intertwined principle of conservation, hence the firewall principle *en toto*, must also imply the same inescapable necessity of AND-form logic and the dissociation of information from matter. Hence, information-matter dualism follows yet again from the all-scale firewall principle, now co-joined with the growing block universe approach; both constituting the fifth proof of information-matter dualism.

1.34: Elucidating Heisenberg's Uncertainty Principle:
Application of the Ontology of the Quantum Mechanical "AND"

The firewall principle can help elucidate the Heisenberg's uncertainty principle in a unique and accessible way. **Heisenberg's uncertainty states that two attributes of any particle (specifically, the position verses the momentum of that particle) cannot be resolved accurately at the same time.** Succinctly, we cannot realise the OR-form particle-position at the same time as the OR-form resolution of particle-momentum (i.e. mass times velocity). If we resolved the position of the particle accurately, hence reduced its positional AND-form state to a minimum, this would only serve to increase the AND-form state of its momentum. Conversely, if we reduced the AND-form of momentum to the minimum, this would only serve to increase the AND-form state of its position. Thus, the resolution of the particle's position and momentum are mutually exclusive.

The 'uncertainty' refers to the fact that, as we reduce the range of AND-form possibilities for one attribute, we enlarge the AND-form range of possibilities for the other attribute. The enlarged range of the latter constitutes the 'uncertainty'.

The generic reason for Heisenberg's uncertainty springs from the mathematical non-commutativity of position versus momentum. Indeed, we ought to call the principle *the Heisenberg's non-commutativity principle*. Hence, position and momentum are non-commutative. To illustrate, consider commutativity and non-commutativity in elementary multiplication and division: Multiplication is commutative because 3×2 and 2×3 both produce 6. The order in which we multiply the inputs is not important or decisive to the output. But division is non-commutative: i.e., 3 ÷ 2 does not give the same quotient as 2 ÷ 3. The relation between the measure of position verses the measure of momentum can be grasped in terms of the same sort of logic: the relation is non-commutative.

However, mathematical non-commutativity, while easy to grasp in the given simple analogy, is mathematically hypertrophic in Heisenberg's uncertainty. Consequently, outside the physics community, the matter is subject to obscurity and confusion. This lends to dubious idealist notions of 'reality as illusion', or even the 'consciousness-generated universe', if not postmodern obscurantism of the sort exposed in the Sokal Affair[44].

---

[44]Sokal A *1996) "Transgressing the Boundaries: Toward a Transformative Hermeneutics of Quantum Gravity". Social text 46/47 (46/47): 217-252

Thankfully, there *is* an entirely non-mathematical and wholly accessible way we can grasp the principle. This is per the firewall principle in justification of AND-form logic grasped in the context of wavelength and frequency consideration vis-à-vis observational inputs.

*

A popular and somewhat erroneous account of Heisenberg's uncertainty principle[45] can help scaffold for the sort of elucidation we seek. Therein, to resolve the position of a particle, we must input to it short-wavelength or high-frequency observational inputs: These are also high-energy inputs. Hence, considerations of resolution power proportionate to wavelength obvious in light microscopy apply to position-measurements of a particle. Therein, the shorter the wavelength or higher the frequency of the input of light used, the higher the definition and resolution obtained. Thus, in light microscopy, we must at least use a wavelength of light equal to the length of the object we want to observe. If the object is smaller, we must use a shorter wavelength of light. The same basic rule applies to the observation of a particle's position: the spatial wavelength confines the position to a smaller area and furnishes greater accuracy in positional resolution per the confinement. But note that the shorter the wavelength, the higher the frequency. Hence, the higher the energy of the observational input.

An absolute position-resolution would require a zero-wavelength input to furnish a zero-area confinement of the particle. But the frequency and energy of such an input would shoot to infinity.

Now, the application of normal non-zero short-wavelength high-frequency (hence high energy) observational inputs will give a 'kick' to the particle, forcing a broader range of AND-form future-possible momentum values. Hence, uncertainty in the momentum will emerge, expressed as the enlarged range of AND-form momentum possibilities consequent upon the high energy 'kick'. This will contrast with the more precise resolution or smaller restriction and confinement of AND-form position-possibilities of the same particle, given the shortened wavelength and reduced area of confinement of the particle.

However, while the above explanation is correct up to a point, it fails by leaving room to the dubious notion that the uncertainty about momentum is merely apparent; *not* objective, and that the particle might somehow continue to possess a fully resolved OR-form momentum despite our uncertainty, and despite our want to express it as an AND-form superposition of broad-range momentum values.

Conversely, if we resolved the momentum of our particle more accurately, we must input long-wavelength low-frequency (low energy) observational inputs. This will give the particle a much smaller 'kick', reducing and restricting its possible momentum-possibilities to a much smaller AND-form spread. Hence, more accuracy vis-à-vis momentum. However, the long-wavelength input will now impose a much larger spatial area of confinement on our particle: The particle's possible positions will now assume a larger range of AND-form positional possibilities. Hence, uncertainty in position.

Again, this explanation fails because it appears to permit the notion that the uncertainty about position is merely apparent, and that, ultimately, the particle has only one definite implicit OR-form position, not an AND-form smear of such.

Why is it not possible to measure both momentum and position at the same time? **If we sought to measure momentum and position at the same time, we would need to use both short-wavelength high-frequency *and* long-wavelength low-frequency inputs at the same time. But these would intermingle, giving a sum-input that is neither clearly momentum-biased nor position-biased**. In short, the reason why we cannot measure position and momentum at the same time is for objective physical-structural reasons that derive from wavelength and frequency considerations, and the incompatibility of, and different outcomes from, the use of one versus the other. Thus, if we want to emphasise position, we must use the shortest possible wavelength or high-frequency inputs and exclude long wavelength low-frequency ones. But this will produce uncertainty in momentum. Conversely, if we want to measure momentum, we need to use long-wavelength low-frequency inputs to the exclusion of short-wavelength high-frequency ones. But this produces uncertainty in position.

*

When we resolve for position via short-wavelength high-frequency inputs and reduce the AND-form state for position to a much smaller range of possibilities, we bias the subsequent AND-to-OR collapse process in favour of position, which will now resolve into an OR-form position (typically the highest probability position) within the reduced area of confinement. We will be able to predict this with greater certainty. And because we use short-wavelength inputs only, we must structurally defer the resolution of momentum... to the *future*. That is, **by resolving position accurately, we do not simply increase the range of possible momenta in future-potential AND-form, but objectively relegate momentum into the not-yet happened futurity, wherein momentum is *not* an 'objective OR-form resolved state'**, and *cannot* be so, unless we entertain 'signals from the future' and contravene the firewall principle. Hence, the AND-to-OR collapse will yield position with great certainty, while momentum will be deferred to the status of a future non-event, without any 'objective' OR-form momentum value. **Indeed, per the firewall principle, to assume that momentum so deferred must have a definite OR-form value would be assume that the future has transpired before AND-to-OR collapse has caught up to it: with 'signals from the future' and all the implications that attend such an absurdity.**

When we do the same but use long-wavelength low-frequency and low energy observational inputs **to resolve momentum accurately, we defer position to the future instead: to a wide range of alternative not-yet happened positional futures.** Subsequent, AND-to-OR collapse will now resolve momentum into an OR-form event, but from a small range of possible momentum values... while position will be deferred to the future and remain in a not-yet event-state that has no 'objective' concrete characteristics. **Again,**

---

[45]*The Quantum Universe*, Tony Hey and Patrick Walters, Cambridge University Press, 1987, pg 16-18

to assume that position is resolved regardless of its deference to the future would countenance 'signals from the future' and the violation of the firewall principle.

*

In summary, the non-commutativity in Heisenberg's uncertainty arises from the necessary structural deference of one or the other attribute to the future. This is per implications from strictly short-wavelength or strictly long-wavelength inputs, on the understanding that the future is ontically real, and per the operation of the firewall principle that protects causality and principles of conservation. This wards against 'signals from the future'… wards against the notion of a resolved momentum otherwise deferred to the future, or of position otherwise deferred to the future.

1.35: Anticipating the Alternative Realist Theory of Causality & Time

All the above constitutes the foundation for a form of causality that supersedes and renders doubt on the materialist contention on the same. Hence, our Alternative Realist take on causality is premised on the ontic reality of the future, exhibited as AND-form logic and phenomenalised as the quantum mechanical wave: consistent with the growing block theory of the universe.

How does the ontic future decay into realised events? By Book-II, in the context of the intermediate model of spacetime, we shall refine our understanding that time is not a 'fourth dimension'. Instead, **time is AND-to-OR wavefunction collapse**; the process through which future possibilities or potentialities irreversibly decay into realised OR-form events. This was obvious in **1.07** and it led to the quantum mechanical theory of entropy: namely, the *entropy of future-potential information*. Therein, **the 'arrow of time' was found always to proceed from AND-to-OR, never from OR-to-AND**. In turn, this called into question generic time-symmetry of physical laws (**1.12**) which must evince through AND-to-OR collapse and cannot operate from OR-to-AND, as forlorn conventional belief in the time-symmetry of physical laws assumes. It also broke the supposition of the reversibility of time, in that time cannot be reversed into OR-to-AND. This AND-to-OR polarity of time furnished irreversibility to time *and* to tandem causality, in that **causality itself must always proceed from AND-to-OR… from future-potentiality for events to new events: never from OR-to-AND**. These insights will survive into the supersession of the intermediate model spacetime and into the de-spatialised and temporised Bergson-Whitehead amalgam in Book-IV: i.e., spacetime *without* space.

*

Previously, from both facile and the later deeper account of information-matter dualism in Part-II, we concluded in the reality of abstract memory extant matter, radiation, and 'space'. Thus, perpetual **AND-to-OR collapse must also lead to the accumulation of the past in the form of accumulating abstract memory**, which must physically dissociate from the event horizon of the most recent OR-form events of 'matter in space'... into a form distinct from both OR-form resolutions at the said event horizon *and* from the pure as-yet not-realised AND-form futurity (see **1.08**). The answer to the specific form in which abstract memory is constituted must await developments leading into Book-IV. It is sufficient to state that **abstract memory is not 'carried' or 'transported', or 'stored' in matter, radiation, and space**. This is consistent with the claims of both facile and the deeper account of information-matter dualism, and from the entropy of abstract memory in **1.08**.

Of course, **dissociated information, constituted as abstract memory, undermines materialist contentions on information in general and memory specifically**, especially in relation to the brain, with obvious consequences to be fully adumbrated in Book-V.

The past retained as abstract memory delimits the possibilities that the as-yet not-realised AND-form future is permitted to assume. In this way, abstract memory contributes to the structure of causality. By Book-II we shall discover that **resolved events, including the retained past itself, force the effective non-local counterfactual processing and restructuring of the AND-form future, which must in turn elaborate to the indefinite potential future. Counterfactuals then generate the instantaneous grand decoherence of the total future into consistent-only alternative futures.** This important insight will revive Mach's principle and will lead to primers and the presage to the quantum mechanical growing block solution to inertia, to magnetism, and to gravity itself.

Of course, counterfactual processing and grand decoherence attendant AND-to-OR collapse constitute a critical aspect of causality: **Insofar as counterfactual causality can circumvent the inevitable infinities to affect said grand decoherence of the total future, this implies a form of core causality that defies account in terms of a 'gears' behind both it and its attendant AND-to-OR collapse time-process.** Hence, both counterfactual causality *and* AND-to-OR collapse (time) is necessarily gear-less and structureless, and must break with materialist contentions in expectation of gears, mechanism, or clockable iterational input-operation-output schemas behind AND-to-OR collapse and time, and behind grand decoherence itself.

Insofar as AND-to-OR collapse and time constitute the transition and evolution of any system from one state to a successor state, and does so by means of a gear-less and structure-less process, **this necessarily circumvents the materialist expectations for contiguity, contact and impact causality between a first state and its successor-state.** The gear-less nature of AND-to-OR collapse and time implies that there is no definable operation to tie putative inputs to outputs in any input-operation-output or IOO-schema. Therefore, there is no contact, contiguity or impact involved in the transition from inputs to outputs: **There is no 'mechanism'… and causality is not a mechanical or clockable process. Ultimately, we inhabit a non-mechanical universe**, at least at the ontologically primary level and in the process of AND-to-OR collapse and time, and in the attendant grand decoherence of the total future per function of abstract memory.

By Book-IV, de-spatialisation itself will constitute a completely independent basis and corroboration of the above-claim: **There is no space. Only time is real. Therefore, there is no possibility of impact or contact between putative matter in the character of**

131

causality: The 'spaceballs' theory of causality is bunk. Therefore, in a de-spatialised universe, causality cannot be mediated through impact or contact in or across 'space', since there are no spatial points or positions at which such contact or impact could transpire. This confirms the claim from the approach garnered from gear-less counterfactual-driven grand decoherence, and equally so from gear-less tandem AND-to-OR collapse and time.

1.36: Summary of Part-III

- **AND-form logic is justified by the principle of causality (1.31):** The imposition of isolation on a system defers its resolution to the *future:* a future that cannot be defined as a resolved OR-form event-set and which must assume a pure AND-form character. To treat an isolated system as if it constituted a pre-resolved OR-form state is tantamount to 'signals from the future', usurping causality. The integrity of causality demands and necessitates AND-form phenomena, *at all scales.* i.e., causality demands quantum mechanics, insofar as AND-form realities are central to quantum mechanics... *at all scales.*

- **AND-form logic is justified by the putative principle of conservation of energy-matter (1.32):** The principle of conservation of energy-matter is entwined to causality as part of the *firewall principle*: The future five minutes from now can only be described as an AND-form state of unrealised futures. Otherwise, we contravene the conservation of energy-matter via 'signals from the future': Where is the extra energy-matter to constitute the Moon five minutes from now going to come form, if it really is pre-realised into OR-form terms five minutes from now, as is demanded by block theory? Hence, the principle of conservation necessitates AND-form phenomena *at all scales.* Insofar as the principle of conservation demands AND-form logic, the principle also justifies quantum mechanics and the growing block model.

- **The principle of causality, combined with the principle of conservation (the firewall principle), constitute the *raison d'etre* for AND-form logic: the ontic reality of the future (1.32):** Both causality and the principle of conservation constitute the firewall and 'cut' (the event horizon) between all that has transpired (memory) per hitherto AND-to-OR collapse... versus all potentiality yet to happen (the future). Thus, the firewall principle constitutes the natural basis for universe-scale isolation of the total future from the past *and* from the relative present. This constitutes the growing block model of the universe and necessitates the ontic reality of the future as an AND-form complex or domain.

- **The growing block universe (*with* an ontic future) constitutes the true depiction of the universe, while the block universe is false (1.31):** If the block universe held true, the two-slit experiment could never yield evidence for self-interference or the interference pattern. These cannot be furnished from or emerge out of pre-resolved definite OR-form 'paths' or from a pre-resolved OR-form future. In two-slit experiments, we *do* obtain macro-evidence for self-interference and interference patterns. This can only happen if the future ontically exists as an objective. In short, AND-form logic, the quantum wave and wavefunctions really do pertain to an ontic existent future, and the growing block approach constitutes the true depiction of the universe.

- **The principle of causality, combined to the principle of conservation (together constituting the firewall principle), constitute the basis for <u>proof-v of information-matter dualism</u> (1.33)** … insofar as both force the ontology of AND-form logic and the ontic reality of the future. This implies the effective dissociability of information from AND-form radiation, matter, and 'space', simply because the past, or memory, cannot obviously be transported in the as-yet not-happened future. Thus, information-matter dualism is the conclusion forced by causality and the principle of conservation per the growing block model.

- **The *raison d'etre* for AND-form logic as the ontic reality of the future, elucidates Heisenberg's uncertainty (1.34):** When we resolve position, we structurally defer momentum-resolution to the AND-form future, which must remain suspended in an AND-form as a not-yet realised non-event future. When we resolve momentum, we defer the position-resolution to the future into a not-yet realised non-event. We cannot treat the deferred attribute as an OR-form state, unless we presume 'signals from the future' in violation of the firewall principle.

- **The true character of causality undermines materialist causality.** Insofar as time itself is AND-to-OR wavefunction collapse, and always proceeds in a time-asymmetric way from AND-to-OR, never from OR-to-AND, causality entwined with time must be structured in the same way. Also, AND-to-OR collapse and time is anticipated to be a gear-less non-clockable process, undermining the want of a definable 'operation' to mediate inputs to outputs. This assertion is augmented by anticipated de-spatialisation: There is no space. Hence, there is no 'point-of-contact' for the contiguous mediation of causality, or 'spaceballs' theory. Therefore, causality cannot be mediated in contiguous clockable fashion espoused in materialism.

BOOK-I PART-IV:
IMPLICATIONS FROM INFORMATION-MATTER DUALISM:
DECOUPLING THEORY

1.37: Aims of Part-IV

Part-IV has three main goals requisite to the further advancement of our Alternative Realist thesis (Alt-R), the nature of information, and the character of information-matter dualism itself. The first is **decoupling theory**: Insofar as information-radiation and information matter dualism hold effectively true, and information really does effectively dissociate from radiation, matter and 'space', it follows that the very integrity of information cannot be fated to the 'transporting' radiation or matter; even if the later should appear to behave in a highly anomalous way, or if the 'transporting' radiation appeared to propagate at 4.7 times in excess of the speed of light, as was found in the experiments of Gunter Nimtz[46]. The integrity of the dissociated and decoupled information will remain intact, despite the 'faster than light' radiation, and no violation of causality will have occurred, despite presumptions to the contrary. This is **the meaning of decoupling theory: i.e., the decoupling of information, order, and causality, not only from AND-form matter and radiation, but from the very putative propagation constant; the speed of light**.

Our aim will be to highlight this bizarre finding in the context of the controversial Gunter Nimtz's quantum tunnelling experiments, which appeared to entail the transportation of the musical composition, Mozart-40, at speeds up to 4.7 times the speed of light. We shall argue that effective information-matter dualism allows for such effects, but does so without violation of causality or the facile 'speed of light' limit, despite effective faster than light radiation.

A key aspect to all of this, but one that will not be explored fully in Part-IV, and one that must await Book-IV, is de-spatialisation: There is no space, much less putative 'motion' therein. What we call 'motion' is a thing we foist on a particular type of temporal succession of events. The insight from de-spatialisation, and against literal 'motion', has direct consequences to the 'speed of light'… which obviously presupposes the ontic reality of space. With de-spatialisation, the speed of light must be succeeded and replaced: namely, by a pure temporal account of the same, recuperated as the *delayed choice time-interval standard*. Therein, both the form of information-matter dualism *and* the relation of information to the pure temporal succession of events organised by the said-standard, assumes a form that supersedes the notion of an 'information-carrying' light-radiation that supposedly 'transports' the said information at the putative speed of light. Therein also lies a clue to why the basis for the integrity of information and causality must obviously decouple from the putative 'speed of light' notion.

Of course, the point of Part-IV is to accomplish all of this without relying on de-spatialisation, and by relying on and grappling with the anomalous results of the Nimtz experiments almost solely on the basis of information-matter dualism hitherto developed.

Our second aim will be **the recapitulation of abstract memory theory**, as was inferred in the apagogic non-erasure argument entailed in proof-iv of information-matter dualism (see **1.22** to **1.24**) … **now to be augmented by decoupling theory**. The retention of information at a dissociated remove from putative radiation, matter, and 'space', combined to the non-erasure of critical information, necessarily implies abstract memory: wherein memories are retained, *not* within complexes of manifest radiation and matter, but at abstract physical remove from radiation and matter in 'space'. Of course, with anticipated de-spatialisation in Book-IV, abstract memory will emerge seamlessly from the fact that information inter-relates purely across time and, consequently, it is obviously dissociated from leading events: a dissociation expressed in terms of time, augmented by the fact that, in a de-spatialised universe, there is no place or 'location' for information, much less for radiation or matter. Yet, prior to de-spatialisation, decoupling theory itself augments abstract memory in a seamless way. The decoupling of Mozart-40 from quantum tunnelled microwaves in the Nimtz experiments necessarily implies the abstraction of M-40 into abstract memory vis-à-vis the 'transporting' faster-than-light microwave radiation and 'space'.

The tacit implication from all of this is memory-brain and mind-brain dualism. The dualism of mind and memory vis-à-vis the brain, and the radical-remove of both memory and mind from the brain, must inexorably follow from both information-matter dualism *and* abstract memory, and it must gain augmentation from decoupling theory before we arrive at obvious implications from anticipated de-spatialisation and temporisation.

1.38: Decoupling Theory: Gunter Nimtz's Tunnelling Experiments & the Problem of Causality

Decoupling theory emerges from the application of information-matter dualism to the experimental findings secured by Gunter Nimtz: It posits the now-justifiable dissociation of information *from* 'transporting' quantum mechanical radiation, which normally propagates at the putative 'speed of light'.

First, it must be point out that, as Einstein readily admitted, while the speed of light must be typically treated as a constant, this is strictly true within the idealised domain of Special Relativity (SR). It proves not to be so constant in the context of General Relativity (GR). Einstein himself emphasised this fact in terms of the curving of 'light rays' manifested by gravitational fields. Thus…

…our results show that, according to the general theory of relativity, the law of the constancy of the speed of light *in vacuo*, which constitutes one of the two fundamental assumptions in the special theory of relativity… cannot claim any unlimited validity. A curvature of rays of light […*about a gravitating body*…] can only take place when the velocity of the propagation of light varies with position. Now we might think that as a consequence of this, the special theory of relativity and with it the whole theory of relativity would be laid in the dust. But, in reality, this is not the case. We can only conclude that the special theory of relativity cannot claim an unlimited domain of validity; its results hold only so long as we are able to disregard the influences of gravitational fields on the phenomena (*e.g.,* of light).[47]

[46]Enders, A; Nimtz, G. (1992). "On superluminal barrier traversal". *J Phys*. I France 2 (9): 1693-16

[47]*Relativity*: The Special and the General Theory, Albert Einstein, Routledge 1993, pg 76

In other words, within a gravitational context, the speed of light is no longer a constant. It also follows that the universe described in Einstein's SR is indeed very special: an idealised universe of approximate validity and convenience, wherein bodies and reference frames enjoy uniform non-accelerated motion. In the real universe, motion is typically non-uniform per gravitational effects. Insofar as this is true, and insofar as gravitation is ubiquitous, the putative propagation constant for light is no longer strictly a constant; although it may be treated as such approximately and, all things being equal, in appropriate frames of reference.

The speed of light appears to be non-constant in experimentations involving quantum tunnelling procedures using microwave radiation: specifically, in the experiments of Gunter Nimtz. Therein quantum tunnelled microwaves have been used to 'transport' the musical composition Mozart-40… at 4.7 times the speed of light.

However, **the acceleration of light by any means ought to lead to serious violations of causality**: a breakdown in the constitution of events about which all observers must agree, whatever their reference frame (See **Fig.1.01** for illustration: a recapitulation of the same illustration from the Preliminary: **0.20**).

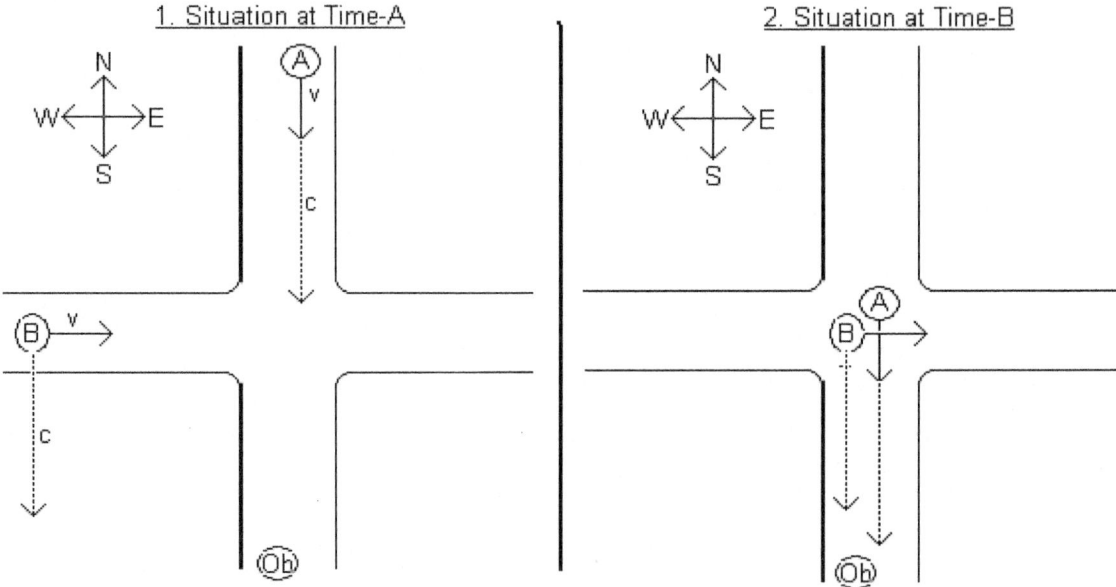

Fig.1.01 Faster-than-light signals and the constitution of events
Notes: **Diagram-1** depicts two moving observers, **A** and **B**, destined to collide at a crossroad. Another observer, **Ob**, located at the end of the southward road of the crossroad, must rely on the light signals (the dotted arrows) emitted from **A** and **B** to observe the expected collision of **A** and **B** at the crossroads. Let us assume that the speed of light is modified by the motion of **A**, such that the effective speed of light of the signal from **A** becomes C + V: (the sum of velocity of **A** and the speed of light, C). Thus, faster-than-light signalling. Since **B** does not have a southward component to its motion, the southward light signal from it will not be modified by its eastward motion. The signal from **B** will reach **Ob** at the speed of light. But the signal from **A** will reach **Ob** at the speed of light plus V. In **Diagram-2**, **A** and **B** finally collide at the crossroads: As far as **A** and **B** is concerned, the collision certainly occurred. However, **Ob** disagrees with the composition of facts accounted by **A** and **B**. According to **Ob**, **A** has passed through the junction without collision, followed moments later by **B**, which misses **A**. This is because the speeded-up light signal from **A** got to **Ob** before the non-modified light-signal from **B**, implying that **A** got to the junction just before **B** and passed safely through without colliding with **B**. This illustrates the breakdown of consistency in the composition of facts engendered by faster-than-light signalling, with tandem breakdown in causality. The integrity of the latter requires consistency and agreement about the composition of facts by all observers. The scenario suggests that faster-than-light signalling would lead to the breakdown in the consistency and into disagreement about the composition of facts and events, with breakdown in causality. Faster-than-light in EPR or in the quantum tunnelling experiments of Gunter Nimtz ought to lead to the same sort of breakdown.

Moreover, per the mathematics of SR, **faster than light signals ought to lead to time-reversal and, with it, the absurdity of 'signals from the future'**, the possibility of which we have challenged on the basis of the firewall principle in Part-III (**1.32** to **1.33**).

QUANTUM MECHANICS AND MIND

The first critical solution to these contradictions emerges from the fact that 'the arrow of time' is based, not on the putative speed of the radiation or even on the speed of light. Instead, it is purely per function of ontologically primary AND-to-OR directionality of wavefunction collapse, synonymous with time. The ontologically primary process that generates realised events out of future potentiality always proceeds from AND-to-OR, never from OR-to-AND. **So long as wavefunction collapse and time is always from AND-to-OR, the 'speed' of the light radiation is irrelevant: it cannot alter or reverse the AND-to-OR directionality of time into OR-to-AND time-reversal**. Indeed, the observation of a 'faster than light' effect requires that the potentiality for this collapses out of the futurity of AND-form superposition into a succession of OR-form events (a succession of OR-form photon resolutions) to which we foist the violation of the cosmic speed limit: As such, a faster than light effect, insofar as it presupposes ontically primary AND-to-OR directionality to time, cannot evince or indicate OR-to-AND time-reversal.

The second solution resides in information-matter dualism itself, and the effective dissociation of information that it furnishes: **The content and constitution of events is preserved through the effective dissociation of information from light radiation, and the information is _not_ being carried in the radiation, whatever its 'speed'**. Hence, in the Gunter Nimtz experiments, Mozart-40 dissociates from the radiation into an abstract form (into memory retained independently of the quantum tunnelled microwave 'carrier-radiation' of M40) regardless of whether the radiation is tunnelled to 4.7 times the speed of light, or else proceeds at the normal cosmic speed limit.

*

Recall from Part-I (see **1.09 to 1.10**) that the perennial 'arrow of time' is partially per the increase in future-potential entropy of information attendant ontologically primary AND-to-OR collapse-directionality. Even if AND-to-OR evinces a succession of events from which 'faster than light radiation' is inferred, the increase in the entropy of future-potential information will yet transpire. Thus, the anomalies of Nimtz will make no difference to this or to time's arrow so conceived, or to the integrity of information or causality that attends the process. Indeed, **causality can only be violated if and only if the AND-to-OR directionality were reversed into OR-to-AND**. The mere acceleration of electromagnetic radiation to 4.7 times the speed of light, as is apparently achieved in the experiments of Nimtz, cannot accomplish the reversal of time into OR-to-AND, and, as stated, the anomaly itself requires AND-to-OR collapse processes to transpire it into evidence.

Let us augment with the following apagogic argument: In order to make an empirical claim that light radiation is being made to travel 'faster than light', we must confirm this as a fact in the midst of events constituted as OR-form outcomes. For such OR-form complexes to render evident 'faster-than-light radiation', the effect must first decay out of an _a priori_ AND-form future potentiality for it. The corpus of events that make 'faster than light' radiation retrospectively evident emerges out of ontologically primary AND-to-OR collapse which, in the first place, gets to define both the 'arrow of time' as well as delimit causality itself, _and_ render evident 'faster than light' as a purported effect.

Indeed, AND-to-OR collapse does not itself arise from the speed of light, much less from putative faster-than-light effects: Such effects are ontologically secondary and subsidiary to the ontological primary AND-to-OR collapse-process (time) that evinces these. That is, we infer the generic speed-of-light from the temporal succession of at least two events and the time-interval demarcated by these. From these we conclude in the reality of a 'speed of light'. The said two successive events are brought into realisation out of future-potentiality through ontologically primary AND-to-OR collapse. But AND-to OR collapse is not brought about by the subsidiary events from which we infer the 'speed of light'. Again, AND-to-OR collapse is not brought about by a speed of light effect, while the speed of light effect _is_ brought about by AND-to-OR collapse. In the case of anomalous faster-than-light effects evinced in the Nimtz experiment, such anomalies also require the minimum of two successive OR-form events, used as time-markers from which to infer the anomaly… both brought about by ontologically primary AND-to-OR collapse and time.

If the evidence for faster than light could bring about reversal into OR-to-AND, we would obtain paradoxical reversal of outcomes: one in which the two minimum successive events required to infer the fact of the faster-than-light anomaly would fall out of evidence, or would 'un-happen' through OR-to-AND de-collapse: The two essential events would de-convert from OR-form back to not-yet happened AND-form... from evidence to 'out of evidence', ending in a process of 'un-science'... and _nonsense_.

*

Could putative faster-than-light radiation constitute a 'signal from the future'? One speculation holds that faster than light constitute signals that ravel backwards in time: the fabled _tachyons_ from the future[48]. When we enforce isolation on a system, the system becomes an AND-form quantum mechanical state, as was inferred for the barrier in the double-slit experiment. This futurisation is enforced by the demands from causality entwined with the principle of conservation: the firewall principle. Thus, AND-form logic rules throughout the interim of isolation... until AND-to-OR collapse furnishes its end. Hence, we cannot treat the isolated system as comprised of any realised OR-form events or resolved outcomes, until isolation ends per AND-to-OR collapse. To put it another way, we cannot treat the AND-form future as if it happened before it transpired, or obtain 'signals from the future'.

Again, could putative faster-than-light radiation constitute a 'signal from the future'? In the light of the preceding reasoning, let us put the question another way: Could a putative faster than light radiation, such as from the tunnelling experiments of Gunter Nimtz,

---

[48]Feinberg, G. (1967). "Possibility of faster-than-light particles". _Physical Review_. 159 (5): 1089–1105

135

provide information about an isolated system, or from the firewalled as-yet not-happened future, by means involving a process other than the process of AND-to-OR collapse? The answer is an emphatic, *no*. As previously stated, any effect, including putative faster-than-light, could only come into evidence via the temporal succession of OR-form events, *always* brought about by AND-to-OR collapse. Again, faster-than-light effects must pre-exist as pure AND-form potentialities within the as-yet not-realised future. Again, it is through the collapse of this future rendered in AND-form potentiality that we obtain the OR-form outcomes from which we infer 'faster-than-light'. So long as the very possibility of putative faster-than-light depends on the ontologically primary AND-to-OR collapse-process or time, its anomalous realisation can only confirm ontologically primary AND-to-OR directionality of time and causality. **So long as apparent faster-than-light needs and conforms to AND-to-OR directionality, the superluminal effect in itself cannot bring about evidence for any event before it has transpired. Thus, faster than light effects cannot evince 'signals from the future'. That is, evidence for faster-than-light could not constitute 'signals from the future' in violation of causality, much less evince OR-to-AND time reversal, notwithstanding the un-science and non-sense that both would entail.**

*

However, there is one factor that we must not overlook and must reiterate in support of the above conclusions: **Per effective information-matter dualism, information is not and cannot be carried or transported by light radiation, no more than by de Broglie matter**, as we have already established in Part-II and III, even if, per prospective de-spatialisation and other developments, information-matter dualism must itself be superseded by a radically different form of the same. Yet, **the firewall principle demands the reality of effective information-radiation dualism,** notwithstanding later de-spatialisation and temporisation. With it, the decoupling of information from radiation, regardless of whether such radiation is subject to a propagation constant, or is pushed to 4.7 times the speed of light in the Gunter Nimtz experiment, it cannot lead to any violation of causality, or conduce OR-to-AND time-reversal, much less furnish 'signals from the future'. In any case, Mozart-40 is *not* carried in the quantum tunnelled radiation to begin with, no more than it is carried by normal light.

Of course, the faster-than-light radiation will eventually get to manifest the information, or M40, as an OR-form corpus, always through ontologically primary AND-to-OR collapse... through the very process that defines, not only the 'arrow' or directionality of time, but constitutes time itself: the process of 'coming into being': the process that cannot be reversed into OR-to-AND de-collapse, much less permit 'signals from the future' in violation of causality or usurp the integrity of information.

*

With these insights, we obtain decoupling theory: Thus...

- Information is decoupled from radiation and matter because of information-matter and information-radiation dualism.

- Insofar as it is decoupled from the radiation, information must also be decoupled from the putative 'speed' of the radiation: specifically, from the generic propagation constant; the 'speed of light' limit.

- The integrity of information and causality cannot be overturned by *apparent* violations of the speed of light limit, given that this critically depends on...
  - ontologically primary AND-to-OR collapse and time, which defines the 'arrow of time' and is the only means through which apparent faster-than-light effects could come into apparent evidence.

- The evidence for 'faster than light' cannot constitute evidence for OR-to-AND de-collapse, or time-reversal, much less entail 'signals from the future'.

1.39: Gunter Nimtz Experiment: The Experimental Setup & Process

Gunter Nimtz's quantum tunnelling experiment is described in **Fig 1.02** and **Fig 1.03. The experiment entails the effective transmission of information (the musical composition, Mozart-40) by way of microwave radiation tunnelled to 4.7 times faster than light. This feat is accomplished by utilising quantum tunnelling procedures and 'barrier penetration'.** Barrier penetration was historically inferred by Friedrich Herman Hund and later independently discovered by Leonid Mandelstam and Mikhail Leontovich[49]. Such 'faster than light' effects ought not to be physically possible. Yet, these and the apparent transmission of physical information with these, *are* possible, as subsequently demonstrated in the experiments carried out by Gunter Nimtz.

However, the anomalous results therein do not imply the violation of causality, for reasons already outlined in our preliminary to decoupling theory in **1.38: Information in the form of M40 is not in the first place physically carried by the quantum tunnelled microwave radiation. Instead, per effective information-radiation dualism, M40 dissociates and *decouples* from the quantum tunnelled microwave radiation and from its 'speed'**, and it is retained at a physical remove, in abstract form, from that radiation and its speed.

If Mozart-40 is decoupled from its generic 'carrying' radiation, which normally propagates at the normal speed of light, it can equally become decoupled from a 'carrying' quantum tunnelled microwave radiation displacing at 4.7 times faster than light.

If it should ever become experimentally possible to quantum tunnel de Broglie matter at faster than light, and 'transmit' M40 using matter instead of radiation, decoupling might yet abide: As is true with respect to M40 vis-à-vis microwave radiation, *any* information

---

[49]Mandelstam, L.; Leontowitsch, M. (1928). "Zur Theorie der Schrödingerschen Gleichung".*Zeitschrift für Physik*. 47 (1–2): 131–136.

136

associated with, but not carried by putative quantum tunnelled de Broglie matter, would be entirely removed from the tunnelled faster than light matter, at least in the interim of isolation.

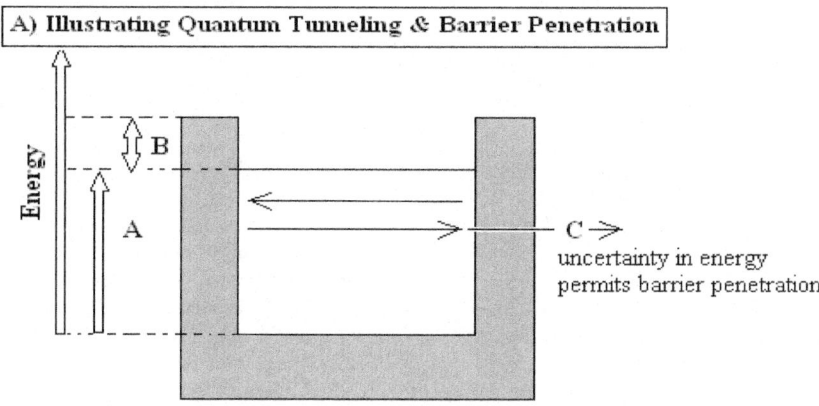

Fig: 1.02: Barrier penetration and the Gunter Nimtz experiment
Notes: The didactic depiction clarifies the meaning of 'barrier penetration' at the heart of the experiments of Nimtz. A particle of radiation or matter is confined to and trapped within a cavity (shaded grey) at an energy level, energy-A. For it to escape the cavity, it would require an additional amount of energy; energy-B. Classically, the radiation or matter so-confined could never escape the cavity. However, a version of Heisenberg's uncertainty relates potential energy to time: It asserts that, as we reduce time to very short intervals, the AND-form potential energy possibilities for the 'trapped' radiation or matter will broaden to include the additional energy-B. This will permit barrier penetration of the radiation, even if this would be a low-probability outcome. With subsequent AND-to-OR wavefunction collapse, such barrier penetration may well become evident, and the trapped radiation or matter might penetrate the barrier and escape the cavity. The thicker the barrier the lower the probability of penetration.

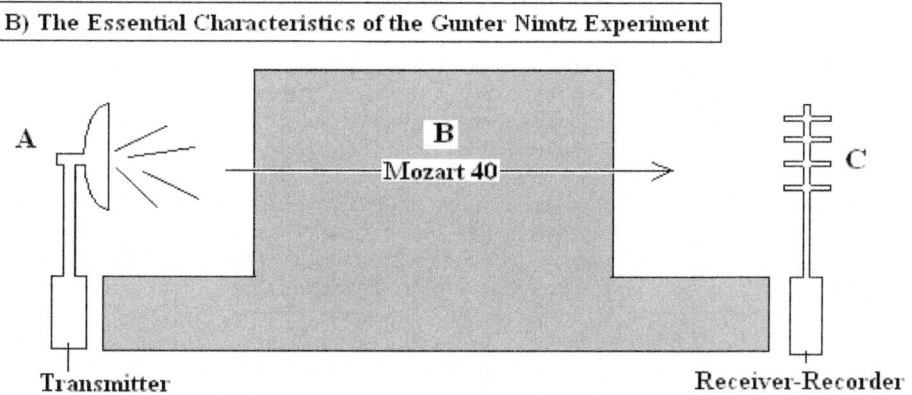

Fig: 1.03: Quantum tunnelling and the Gunter Nimtz experiment
Notes: The didactic illustration depicts the *essentials* of the quantum tunnelling experiment of Gunter Nimtz: Mozart-40 is transmitted at **A** using microwaves. These penetrate the barrier (**B**) by exploiting the process of barrier penetration depicted in **Fig.1.02**, deriving from Heisenberg's energy-time uncertainty relation. Mozart-40 is received and recorded at **C**. The microwave supposedly 'carrying' Mozart-40 has to penetrate **B** and displace a distance equivalent to light radiation travelling at speeds 4.7 times the speed of light. Hence the faster-than-light anomaly evinced in the Nimtz experiment.

We anticipate the further evolution of decoupling theory in the face of subsequent discoveries in Book-II to Book-IV: **The ontic reality of space and motion are assumed in the experiments of Gunter Nimtz and in our own first iteration of decoupling theory herein. Yet, when we get to the deeper understanding of information-matter dualism by Book-II, we will discover that, ultimately, it is *not* correct to treat quantum mechanical matter and radiation, tunnelled or otherwise, as spatially dynamic displacing states: Even on the assumption of the ontic reality of 'space', at best, by Book-II, we will discover that radiation and matter prove to be spatially static states** comprised of equally static nested futures that elaborate to the indefinite future. As such, de Broglie matter and radiation do not in truth displace at the speed of light, much less at 'faster-than-light', or at *any* 'speed'. Thus, core discoveries promulgated in Book-II will put an entirely different perspective on decoupling theory and precipitate the further evolution of information-matter dualism itself, reframed in terms of spatially static quantum mechanical AND-form waves; part of the growing block model that supersedes 'space and motion'.

This anticipated evolved appreciation of decoupling theory will not end in Book-II: An even more exotic development of decoupling theory will emerge by Book-III and IV, comprised of a series of iterations of decoupling theory framed to the solution to the stochastic coherence paradox and quantum indeterminacy, and per subsequent de-spatialisation and temporisation. **The very abolishment of space itself will precipitate a deeper understanding of nature, including information-matter dualism itself. De-spatialisation by Book-IV will radically transform even the notion of the 'spatially static quantum wave' furnished in Book-II: It will culminate into decoupling theory that relates information to the physical order purely in de-spatialised temporal form, involving the replacement of the 'speed of light'** with the *delayed choice time-interval standard.*

1.40: Minkowski Spacetime Diagrams and the Gunter Nimtz Experiments

We must now harmonise the findings of Nimtz with Special Relativity (SR). The methodology assumes two forms. The first arises from the stipulations of decoupling theory previously given: The decoupling of the integrity of information and causality from the purported 'carrying' radiation, and even from the propagation constant of electromagnetic and light radiation, is the first solution (see **1.37** for details). The second requires that we consider the critical relations and observations obtained in the Nimtz experiment within the framework of Minkowski spacetime diagrams. This essay is concerned with the second method, to which the assertions from **1.37** are inherent.

In a series of Minkowski diagrams, we will imagine Nimtz performing his experiment using the procedures depicted in **Fig.1.03**. Typically, Nimtz listens to M40 after its reception in order to confirm both its transmission and reception. He can only accomplish this in various unique ways, each depicted as a unique scenario that incorporates both information-matter dualism *and* decoupling. The scenarios will help us find out whether there can be any serious violations of the postulates of Einstein's SR: Succinctly, we must find out whether the M40 signal was received by Nimtz from either within pertinent lightcones attached to the critical events of the experiment, or from regions that reside outside of the critical lightcones: i.e., the 'elsewhere' regions. **Per any given scenario, if we find that Nimtz received M40 within a pertinent critical lightcone that forms up at the putative speed of light, the event will be in harmony with SR and the speed-of-light limit, despite apparent information exchange at 4.7 times the speed of light through quantum tunnelling.**

On the other hand, in any given scenario, **if we discover that Nimtz received M40 in an area 'outside' of some pertinent lightcone, or from 'elsewhere', we could then obtain the subversion of SR and the speed-of-light limit.**

In the following essays, **we endeavour to show that M40 and attendant critical events are received and processed by Nimtz within pertinent lightcones, never from 'elsewhere',** despite 4.7 times the speed of light of the 'transmission' of pertinent events. **This insight will become fully salient when we develop the *maincone approach*** to the anomalies of the Nimtz experiments.

We must first recapitulate the Minkowski spacetime approach involving lightcones, worldlines, and the 'elsewhere': This is accomplished in **Fig. 1.04** and **Fig. 1.05**: These are recapitulations of illustrations first presented in the Preliminary. The Minkowski spacetime diagrams in **Fig. 1.04** develop the idea of the 'lightcone'. To this end, we imagine light emitted in all directions from a point-source or *origin*. Over time, this will propagate in all directions, describing a perpetually expanding 'sphere of light' of diameter measurable in appropriate units. Since the light will propagate at the speed of light c, the diameter of the sphere it will describe at time-t will be the speed of light c times t, or ct. By t_2, the sphere of light will have expanded to a larger diameter, ct_2. To describe a lightcone, we use the depicted 'light-spheres' as reference. In relation to the first 'light-sphere' in **Fig. 1.04, diagram-A**, below it, we define a graph with a vertical axis for time and a horizontal axis for space: the Minkowski spacetime diagram. At time $t = 0$ we designate the point-of-origin for the pertinent lightcone. At $t = 1$, we draw out the diameter of the light-sphere so far formed and then join its ends to form the triangular cone: i.e., the 'lightcone' originating at $t = 0$. We do the same in relation to the second larger 'light-sphere' in **diagram B**, describing a lightcone that expands as the 'light sphere' expands. The 45-degree angle to the vertical of the lightcone represents the graphical analogue for the speed-of-light limit. A larger angle would constitute an analogue for radiation that propagates at much faster than light, so describing a broader lightcone. By logic, an instantaneous signal from one point in space to another would require a purely horizontal line: a cone flattened to the horizontal.

Things that happen at or to the limit of the speed of light transpire within the boundary of, or inside of, the lightcone: within the interiority of that cone. Transmissions that transpire faster than light would reside 'outside' the region specified by the lightcones: i.e., the *elsewhere*. Only when observed events confirmed to happen *elsewhere* do we end in the violations of SR and causality.

Do quantum tunnelled transmissions of M40 in the experiments of Gunter Nimtz imply such 'elsewhere' states?

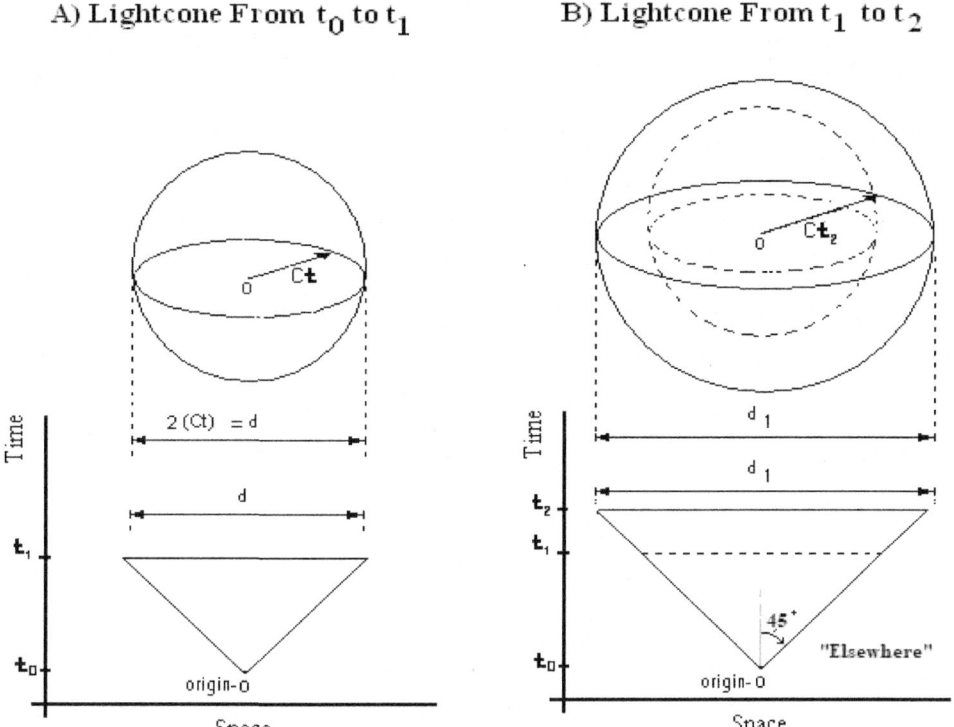

Fig: 1.04 Reading spacetime diagrams: illustrating lightcones

Notes: In the top part of **diagram-A**, in the 'light-sphere', light radiation originating from **origin-O** propagates in all directions at the speed of light **c** in time t_0 to t_1: The light radiation describes a sphere whose radius is **ct** and whose diameter **d** is given by **2(ct)**. In the Minkowski spacetime diagram correlated with the 'lightsphere' (bottom depiction)) in **diagram-A**, diameter **d** describes the top horizontal line of an upside-down triangle called the lightcone: It is simply the diameter of the sphere seen edge-on, formed up by the light radiation originating in **origin-O**, thus forming the lightcone for an expanding 'sphere of light' from time t_0 to t_1. The 45-degree of the lines from **origin-O** analogues the speed of light. Anything greater than 45-degrees would describe flatter cones and analogue radiation that travels at faster-than-light. Anything less than 45-degrees analogue radiation or matter travelling at less than light (These apply to 'worldlines' as we shall later see). In **diagram-B**, t_1 to t_2 describes a larger expanded 'lightsphere' per ct_2, and a new enlarged lightcone of diameter d_1 given by $2(ct_2)$.

What about worldlines? Worldlines are diagrammatically illustrated in **Fig. 1.05**. A body relatively stationary describes a pure vertical worldline (**diagram-A**). In **diagram-B**, as the light radiation propagates in all directions from its point of origin, from the same origin, an object travelling at much less than the speed of light heads out in a given direction at a uniform velocity **v**. By the time the lightsphere and lightcone obtain the diameter given at **t = 1**, our object will have travelled a distance given by **v** times *time*: In **Fig. 1.05**, this is given as **vt**. In the pertinent Minkowski diagram below this, we draw **vt** as a worldline from point-origin, which intersects an implicit horizontal line at **t = 1**. This worldline represents the object undergoing its stated uniform motion at vt: (See **diagram-B, Fig. 1.05** for clarification. Also see **diagram-C**, which accounts for acceleration in terms of worldlines). The worldline is the spacetime 'path' or *history* of the stated object travelling at uniform velocity **v** in time **t = 0** to **t = 1**. The angle of the worldline is confined to the region within the lightcone: It is less than the speed of light.

*

Having clarified the meaning of lightcones, worldlines and the logic of Minkowski spacetime diagrams, we can model the four possible scenarios that entail the apparent transmission of quantum tunnelled Mozart-40 at faster than light, as made evident in the Nimtz experiments, in supposed violation of the basic postulates of Einstein's SR. Let us briefly summarise each of the four scenarios: Thus…

- **First scenario:** After sending M-40 to the recording device by means of quantum-tunnelled microwaves at 4.7 times the speed of light, Nimtz must physically move to that recording device in order to play and confirm the reception of M-40.

139

- **Second scenario:** Nimtz quantum-tunnels M-40 to the recording device at 4.7 times the speed of light. Instead of physically moving to it, he uses a normal 'speed-of light' signal issued from the recording device *after* it receives and plays M40, so confirming the transmission and reception of M40 at 4.7 times the speed of light.

- **Third scenario:** This incorporates scenario-2, but Nimtz sends the play-trigger signal to the play-button on the recorder, *not* using normal radiation, but as a quantum-tunnelled signal, timed to arrive at the recorder-player immediately *after* the arrival of quantum tunnelled M40.

- **Fourth scenario:** Nimtz sends both M40 and the play-trigger signal to a first recorder by means of quantum tunnelled microwaves. This is linked to another quantum tunnelling device. Once the first recorder receives the Mozart-40 and, with it, the pertinent play-trigger signal, it re-transmits both signals back to Nimtz by means of a second quantum tunnelled microwave 'carrier' signal: Both signals get to the second recorder-player device attached to Nimtz.

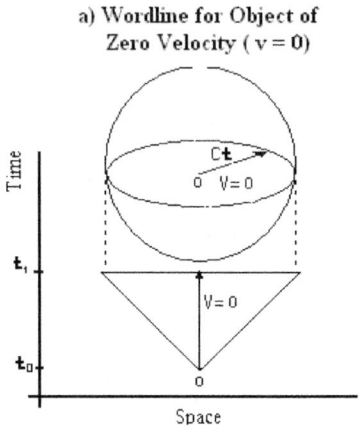

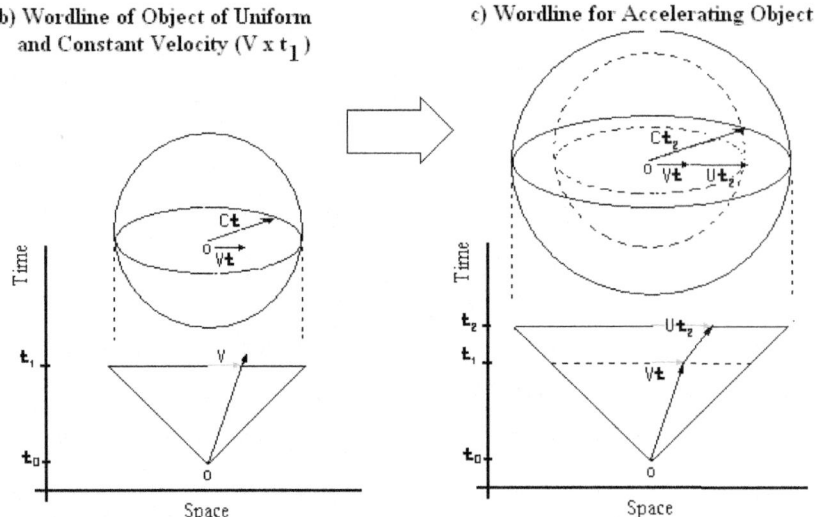

Fig.1.05: Spacetime diagrams illustrating worldlines.

Notes: Using our previous schema for 'lightspheres' and lightcones as the framework, embedded within the lightsphere and the lightcone diagram in **(a)**, an object of zero velocity will describe a simple vertical worldline (i.e., the line and arrow from t_0 to t_1) from **origin-O** and within the bounds of the lightcone attached to **origin-O**: In diagram **(b)** we describe the worldline from t_0 to t_1 of an object at uniform unchanging velocity vt. This is shown as a vector within the lightsphere and is also included as a clarifying grey vector line **v** within the lightcone at t_1: a worldline from **origin-0** offset to the right of the vertical by **v**. Diagram **(c)** depicts a change in velocity (i.e., acceleration) of the object at t_1, from vt to ut and a new adjoining worldline from t_1 to t_2.

Through these scenarios, we will discover that only the fourth scenario might subvert the claims of Special Relativity; corrected the moment we embed all its pertinent events and outcomes within what we refer to as the *maincone*: **The maincone is a lightcone in which all the various apparent faster than light anomalous-seeming outcomes and events are embedded. It is ultimately by reference to the maincone that we decide whether the anomalies truly subvert the basic postulates of Special Relativity and causality.** Thus, we will discover that the transmission of information through quantum tunnelled faster-than light microwaves does not subvert SR, insofar as the final reception of any meaningful signal or coherent information always manifests within pertinent lightcones or, per the final and fourth scenario, within an all-embedding maincone: never 'elsewhere'.

Throughout each scenario, **especially vis-à-vis the maincone, we will reiterate implications deriving from information-radiation dualism and from tandem decoupling, justified by the principle of causality and the firewall principle implicit to the process of AND-to-OR collapse and time that, in the first place, brings into evidence the outcomes of the four scenarios within their respective cones and maincones.**

1.41: Harmonizing Gunter Nimtz's Quantum Tunnelling Experiment with Special Relativity

The first scenario entails the quantum tunnelled transmission of Mozart-40 at faster than light to a distant recorder-player device. For Nimtz to prove that Mozart-40 has been transmitted thus, he must physically displace from the transmission-point to the location of the recorder-player. Once there, he must press the play-button. Nimtz will displace to the recorder-player at speeds much less than that of light. Consequently, the whole process, including the said displacement, the press of the play button, and the play of M40, will be embedded within a lightcone originating at Nimtz's original position and encompassing all subsequent positions of the transmitter, recorder, player and the M40 play: See **Fig 1.06** for the clarifying illustrations.

Thus, even when we treat M40 as if it was carried, displaced and deposited upon the recorder-player device by means of the faster than light quantum tunnelled microwave radiation, there will be no contravention of the postulates of Special Relativity, given the fact that, by the time Nimtz has displaced to the recorder and has played M40, the 'elsewhere' events presumably formed up by the initial faster-than-light transmission vis-à-vis the recorder-player will have 'moved' into a common embedding lightcone: one centred on and encompassing Nimtz himself: The subsequent play of Mozart-40 will constitute an event experienced and observed *within* that encompassing lightcone, not outside of it; *not* 'elsewhere'. Thus, the postulates of SR remain intact.

We must also keep in mind that the supposed transmission of M40 at faster than light is inferred *as if* it constituted an 'elsewhere' corpus of events, even though it had not been directly observed as such. Thus, we never get to observe the supposed faster than light transmission within any sort of 'elsewhere' as a corpus of directly observable events. Indeed, all subsequent reportable events will occur within a lightcone attached to and embedding Nimtz's and subsuming all, even if parts of these events are inferred (in retrospect) to have happened at faster than light, *without* any direct observation. Hence no subversion of SR will have occurred.

To round up, let us describe this **first scenario in detail** by reference to **Fig. 1.06**. First, Gunter Nimtz transmits M40 via quantum tunnelled microwaves at 4.7 times faster than light, from t_0. At the same moment that the transmission is triggered, from the same origin at t_0, normal light begins to radiate in all directions at the speed of light, forming its perpetually expanding lightcone. By the time the quantum tunnelled faster-than-light M40 reaches the recorder-player, the lightcone pertinent to the normal light will have expanded to t_1, while the quantum tunnelled M40 will have arrived at the recorder-player at point **Ee**: a putative 'elsewhere-event' that apparently resides outside the said lightcone, but is *not* directly observed or observable by Nimtz . Indeed, **Ee** cannot be observed at all.

As all of the above is happening, Nimtz begins his displacement toward the recorder-player. His displacement is described by a worldline with velocity **v**, which is much less than the speed of light. Notice that Nimtz, synonymous with that worldline, is embedded *within* the pertinent normal light lightcone.

Nimtz finally arrives at the recorder-player, say, at t_4. As this happens, the recorder-player, originally at **Ee**, will itself describe a vertical worldline: one that intersects Gunter's worldline at t_4. At which point, **Ee** can no longer be described as 'elsewhere': it has 'moved' into, and is now *subsumed* to, the domain of the normal lightcone by t_4. At t_4, Nimtz presses the play-button, confirming the reception of M-40 and the successful transmission of M40 at 4.7 times 'faster than light'. Yet, this confirmation occurs within the pertinent lightcone that, by t_4, embeds within it Nimtz, the recorder-player (no longer 'elsewhere' by that point) *and* the M40 play.

Recall that information-matter dualism is inextricably foundational to decoupling theory. Recall that the very dissociation of information from radiation generally, and from the manipulated microwaves in Nimtz' experiment specifically (hence the non-dependency of the putative transfer of Mozart-40 on any 'carrying' radiation, even one quantum tunnelled to 4.7 times 'faster than light') can only serve to augment the assertion that no subversion of SR could have transpired. Information-matter dualism and decoupling theory leads to, and justifies, the physical decoupling of information from quantum mechanical radiation, no matter how fast the latter might be pushed above the speed of light, notwithstanding consequences from static quantum mechanical waves and from anticipated de-spatialisation.

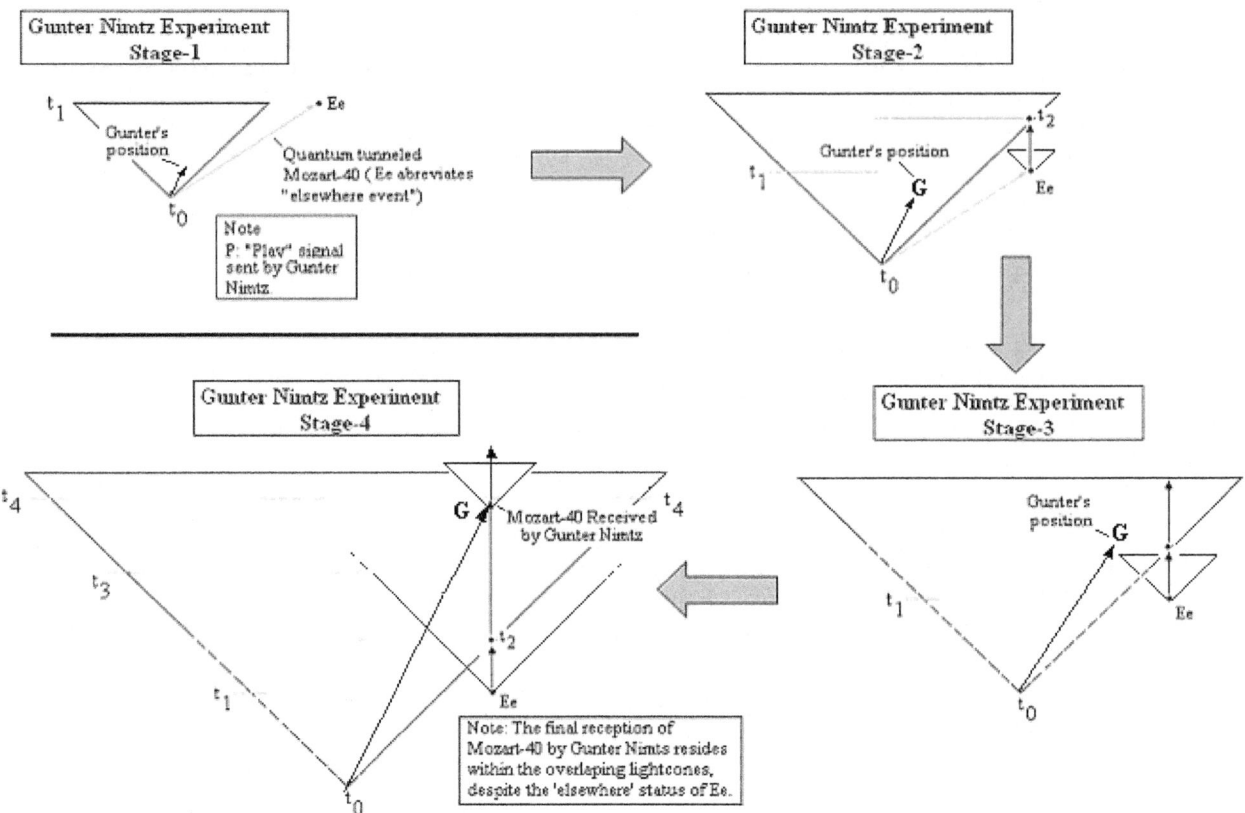

Fig.1.06: Illustrating the first scenario: Gunter Nimtz experiment

Notes: A didactic illustration of why the Nimtz experiment does not undermine Special Relativity: The first scenario. At **Stage-1** Gunter Nimtz transmits M-40 via quantum tunnelled microwaves at 4.7 times faster-than-light from transmitter at t_0. As the transmission is triggered at origin t_0, normal light radiation begins to radiate, forming its own expanding lightcone. By **Stage-2** the quantum tunnelled M-40 reaches the recorder-player, while the normal lightcone expands to t_1: The quantum tunnelled M-40 arrives at the recorder-player at point **Ee**: the 'elsewhere' point that resides outside the generated lightcone. But it is not directly observed by Gunter Nimtz or by anyone else: Therefore, no actual contravention of SR takes place. From Stage-1 to **Stage-3**, Nimtz displaces toward the recorder-player. This is described as **worldline-G**, at less than the speed of light. Note that Nimtz, synonymous with the **worldline-G**, is embedded within the pertinent normal lightcone. Nimtz finally arrives at the recorder-player say, at t_4. The recorder-player, originally at **Ee** itself describes a vertical worldline: one that meets with Nimtz' **worldline-G** at t_4, and it is now within the lightcone originating at t_0 … and no longer 'elsewhere'. Nimtz now presses the play button, confirming the reception of M-40 and, indeed, its prior transmission at 4.7 times faster than light. Yet, this confirmation occurs within the pertinent lightcone that, by t_4, embeds within it both Nimtz, the recorder-player (no longer 'elsewhere') and M40 play.

*

    The second scenario entails almost the same process of subsummation as the first, except that, by stage-2, Nimtz need not displace toward the recorder-player. Instead, Nimtz sends a normal speed-of-light trigger-signal to the device in order to initiate the play of the previous faster-than light transmitted Mozart-40. (See **Fig. 1.07**).

    Therein, M40 is quantum tunnelled at faster than light and reaches the recorder-player, well before any normal signal could do the same. This is constituted as an 'elsewhere event' Ee: outside the lightcone attached to Nimtz, in apparent contravention of SR, even though it is never directly observed as an elsewhere-event, and cannot be asserted to have occurred in concrete empirical event-terms as such by *any* observer in *any* frame.

    While M40 is being quantum tunnelled toward the recorder-player, Nimtz sends a second play-trigger signal toward the same recorder-player. This is sent at the normal speed of light. It travels along the edge of its pertinent lightcone and is denoted **signal-P**. It reaches the recorder-player a moment just *after* the faster-than-light M40 reaches it. Consequently, the play-button is triggered by **signal-P**: M40 is played.

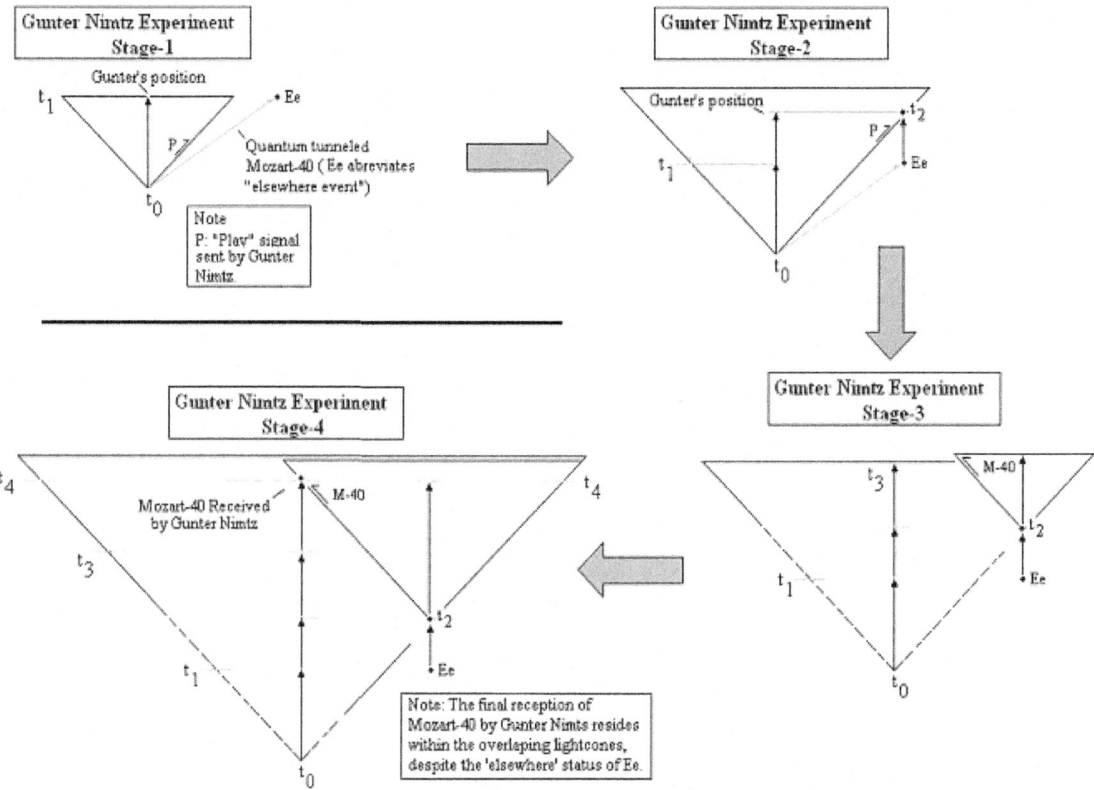

Fig. 1.07 Nimtz's quantum tunnelling experiment involving a normal "play" signal

Notes: In **Stage-1**, Gunter Nimtz transmits Mozart-40 by means of the quantum tunnelling procedure first shown in **Fig. 1.03**. The worldline for the transmission begins at origin t_0 and terminates as an apparent elsewhere event, **Ee**, at t_1. Simultaneously, Nimtz sends an electromagnetic play-trigger signal toward the recorder-player coincident with **Ee**. This signal must travel at the speed of light and, in **Stage-1**, is denoted as **P**, along the lightcone. At **Stage-2**, by t_2, the play-trigger signal **P** reaches the recorder-player and triggers the play-button, tying **Ee** to signal-**P** at t_2. Thus, **Ee** is no longer 'elsewhere' and it is incident to the lightcone and with signal **P**. At **Stage-3**, Mozart-40 begins its journey towards Nimtz as a radio-signal at the speed of light: Gunter is located along the vertical worldline attached to origin-point t_0. The Mozart-40 transmitted from the recorder-player begins to form a normal lightcone, along the edges of which Mozart-40 propagates toward Nimtz at the speed of light. By **Stage-4** and t_4, Nimtz receives the M-40 signal: However, notice that the reception at t_4 is embedded within the larger lightcone originating at t_0. Insofar as this holds true, all things remain consistent with the demands of Special Relativity, even though Mozart-40 was transmitted to the recorder-receiver at **Ee** apparently at much faster than the speed of light.

Let us assume that, from stage-3 to stage-4, sound signals propagate toward Gunter Nimtz at the speed of light (this is absurd, but the notion of it will help us prove our point). Let us denote this incoming music as M40. It eventually reaches Nimtz, who now enjoins at t_4. Note that the event will happen within the principal lightcone issued from t_0. M40 is not experienced as an 'elsewhere event' at that juncture, and SR is saved.

We have discovered that, per the dominating and subsuming lightcone, the first scenario does *not* imply the contravention of SR. The embedding lightcone from t_0, evinced in both the first and second scenarios, saves SR. Yet, the question remains: In both the first and second scenarios, was M40 transmitted in the quantum tunnelled light radiation in the first place? To address this, we must again apply information-matter dualism and the understanding of causality based on AND-to-OR collapse and directionality.

From information-matter dualism, information is not in the first placed carried by quantum tunnelled 'faster-than-light' radiation, and both it and causality are not fated to the propagation constant for electromagnetic radiation: i.e. the speed of light. (see **1.38**)

Also, the AND-to-OR collapse process and time itself, the very process that saves causality, and which brings into evidence the very outcomes of Nimtz's experiment and our derivative scenarios, is ontologically primary to both normal speed-of-light *and* to the faster-than-light events, *and* constitutes the proper basis for causality. Neither AND-to-OR collapse and time, nor causality, are fated to the propagation constant for electromagnetic radiation, which itself depends on AND-to-OR collapse to bring it into evidence or inference. The process of AND-to-OR collapse would happen even if the events it brings into the fray permit inference in favour of faster than light radiation. But this could not lead to OR-to-AND de-collapse or time-reversal.

Indeed, **each stage involved in scenario-1 and scenario-2 entail sequences of AND-to-OR collapse processes that bring into being the evolution of all lightcones, worldlines, displacements and events**… of light, of the quantum tunnelled microwaves… of Gunter Nimtz himself… of M40. And, as was shown in the illustrations, none entail direct observation of *any* elsewhere-events, and **there is no OR-to-AND time reversal… no 'signals from the future': i.e., events do not arise before they transpire per attendant AND-to-OR collapse.**

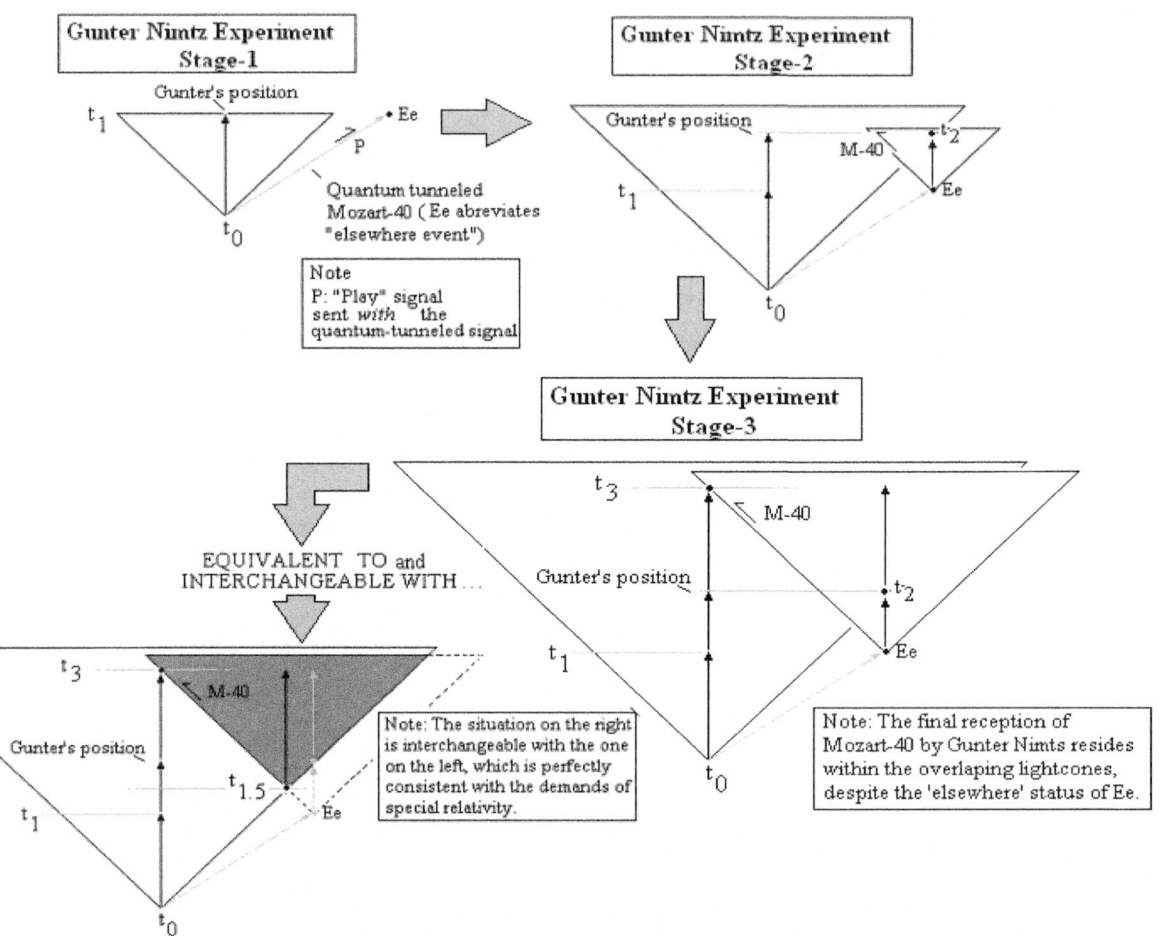

Fig. 1.08 Nimtz's experiment involving a quantum-tunnelled "play" signal
Notes: In **Stage-1**, the play-trigger signal-P is quantum tunnelled to the first recorder-player at **Ee** together with Mozart-40. Both reach the recorder-player at t_1. At **Stage-2**, M40 begins its back-transmission toward the second recorder-player, beginning its journey at t_1, and generating its own lightcone originating at **Ee**. It gets to the second recorder-player and Nimtz by t_3, by **Stage-3**. The reception of M40 by Nimtz has occurred within the larger lightcone in which he is embedded: From his frame, within his lightcone, events have not reached him from any 'elsewhere', and he cannot know of any such event, even though faster-than-light components were involved. Also note that the situation portrayed in the first diagram in **Stage-3** is interchangeable with and equivalent to the scenario below, in which the M40 originates at $t_{1.5}$, consistent with the speed of light limit: That is, the dark-shaded lightcone is perfectly embedded and contained within the larger lightcone attached to Nimtz. Insofar as any scenario entailing apparent violations of the speed-of-light limit can be interchanged with one that does not violate that limit, *and* is capable of producing exactly the same outcomes, then no subversion of Special Relativity can be said to have transpired: This is the **interchangeability rule**, useful in types of experiments and anomalies carried out by Nimtz, or in any situation that appears to entail the violation of the speed-of-light limit.

Notice that, from the preceding scenarios, we obtain a basic assertion: **Even if a scenario entails apparent faster-than-light components that culminate into apparent elsewhere-events, if the elsewhere-events eventually transform into directly observable events and outcomes *within* a common lightcone in which both the corpus of events *and* the observer are embedded. Hence, the pertinent events and outcomes, and their subsequent observations, cannot be of, or pertain to any 'elsewhere'. Therefore, there can be no violation of SR, despite putative quantum tunnelled faster-than-light 'transmission'.**

Now we come to **the third scenario**, illustrated in **Fig. 1.08**. The scenario develops the *interchangeability rule*: Insofar as the outcomes of any scenario entailing processes in violation of the speed-of-light limit could be interchanged with scenarios that could produce the same outcome *without* apparent violations of that limit, and insofar as no putative 'elsewhere event' is directly observed or observable, then no subversion of the speed-of-light limit, much less any undermining of Special Relativity, can be asserted to have transpired.

This interchangeability rule specifies that, only those scenarios that entail faster-than-light processes that cannot be interchanged with equivalent Special Relativity-consistent scenarios (hence, only those scenarios that involve directly observable elsewhere-events that have not been brought into the domain of any common embedding lightcone, but can yet be observed) could constitute genuine cases of the violation of the speed-of-light limit and the contravention of SR.

However, we must point out again that, insofar as information-radiation and information-matter dualism both entail the dissociation of information from radiation and matter, putative faster-than-light radiation or matter could not in the first place act as a 'carrier' of information: That is, in all scenarios, Mozart-40 is decoupled from the quantum tunnelled faster-than-light radiation, notwithstanding other assertions we have made about the integrity of causality and time vis-à-vis radiation and the cosmic speed limit.

The third scenario in detail entails, *not* merely the quantum tunnelled faster-than-light transmission of Mozart-40 to a recorder player at **Ee,** but the utilisation of point **Ee** as a conduit for a similar quantum tunnelled faster-than-light return-signal: one that shunts M40, together with its own play-trigger and signal back to a second recorder-player directly attached to Nimtz. The scenario is depicted in **Fig 1.08**.

As can be appreciated by inspection of the final diagrams for stage-3 and stage-4 in **Fig.1.08**, triggered by the incoming putative faster-than-light signal, the play-button located on the recorder-player attached to Nimtz gets to play M40, but does so within the lightcone in which Nimtz is embedded. In other words, despite faster than light shunting of M40 to **Ee** and back to Nimtz (i.e., events that are *not* directly observed or observable by Nimtz) all observed outcomes or events pertinent to the *final* reception and play of Mozart-40 will have occurred *within* the lightcone to which both Nimtz, together with the second recorder-player, remain confined. It is as if the signal was sent directly to the second recorder-player, but for a time-delay lasting from t_1 to t_3: an *interchangeability* that hardly constitutes any violation of causality or time. In short, the pertinent elsewhere-events are not directly observable, even when faster-than-light quantum tunnelled signals are involved in the transmission *and* reception-steps. Therefore, no subversion of SR can take place.

Indeed, suppose that the play-trigger on a music device was linked to a complex relay system, involving several shunting-steps through quantum tunnelled faster-than-light means. Let us imagine that this spanned the entire solar system, if not the whole Milkyway galaxy. The culminating play of M-40 would merely constitute the final concluding-point of this quantum tunnelled faster than light complex relay system. Let us assume that the whole process of relays took one minute to conclude. All that one could empirically assert in terms of directly observable and concrete events would be a one-minute delay between the depression of the play button and the play of Mozart-40: One could not use this observation, or use the delay itself, as any basis for concluding in *any* sort of violation or subversion of generic SR. Effectively, such a violation or subversion did not transpire in scenario-3 per any directly observable and accounted-for elsewhere-event. Indeed, **the situation is perfectly interchangeable with a scenario that might entail a mere normal non-faster than light circuitry inbuilt to the device, but with an inbuilt one-minute delay between its trigger and the play of M40.**

In other words, **we can apply the interchangeability rule if the outcomes of any scenario supposedly entailing violations of the speed of light could be interchanged with a scenarios that can produce the same outcomes *without* that apparent violation, and wherein no 'elsewhere event' is directly observed or observable, and no subversion of the speed of light limit has occurred,** and SR abides.

Of course, when we add our own decoupling theory per information-matter dualism, *and* add the fact that causality is *not* in the first place fated to a propagation constant for light, but to the AND-to-OR directionality ontically *a priori* the manifestation of the propagation constant, or even *a priori* its 'violations', we then obtain an additional and independent argument for the non-violation of SR and its pleonastic speed-of-light limit.

*

Only with **the fourth scenario** depicted in **Fig. 1.09** do we come close to the apparent violation of the speed-of-light limit. The scenario has the same form as the third: except that an assistant is now placed at **Ee**: The events pertaining to **Ee** must now become directly observable by the Nimtz's assistant.

Nimtz remains on planet Earth while his assistant travels to Mars, where **Ee** will form. With his assistant now at point-**Ee**, Nimtz uses a quantum tunnelled faster-than-light signal to mediate M40 to his assistant on Mars. The reception manifests as an observable event at putative **Ee**. Nimtz's assistant uses a similar quantum tunnelling procedure to send a faster-than-light response-signal back to Nimtz on Earth: The whole chain involving the transmission, reception, and response between Nimtz and his assistant will unfold at speeds much faster than light, and **Ee** will be apparently directly observable, at least to Nimtz's assistant.

145

Surely, now we have a real violation of the faster-than-light limit to which the interchangeability rule cannot apply, and **Ee** appears no longer to enter and subsume within any pertinent lightcone… or so it would appear. But we must take a larger-scale look.

Consider that Nimtz's assistant would first need to travel to Mars. He will do so at speeds much less than that of light. The worldlines of both Nimtz and his assistant will therefore reside within the confines of a common embedding **principal maincone** originating at **t = 0**. This marks the beginning-point of the assistant's journey to Mars, and embeds both the beginning-point and the assistant's arrival-moment on Mars. It also embeds the time spent on Mars. The principal maincone will also embed every subsequent quantum tunnelled transaction between Nimtz and his assistant. Insofar as these are embedded within the principal maincone, any putative or potential **Ee** must also eventually 'move' into and emerge as an event within the principal maincone.

It is the principal maincone that demarcates the possibilities and limits for the manifestation of information: Any information received and manifested within its bounds constitutes a signal or event that does not undermine the strictures of SR, even if such events should coincide with putative 'elsewhere' points in relation to the lesser lightcones subsumed to and embedded within the larger all-encompassing principal maincone. By such means, these are no longer 'elsewhere' but within the principal maincone. It is this role of the principal maincone, *and* its combination to decoupling theory, which saves SR from apparent faster-than-light chaos.

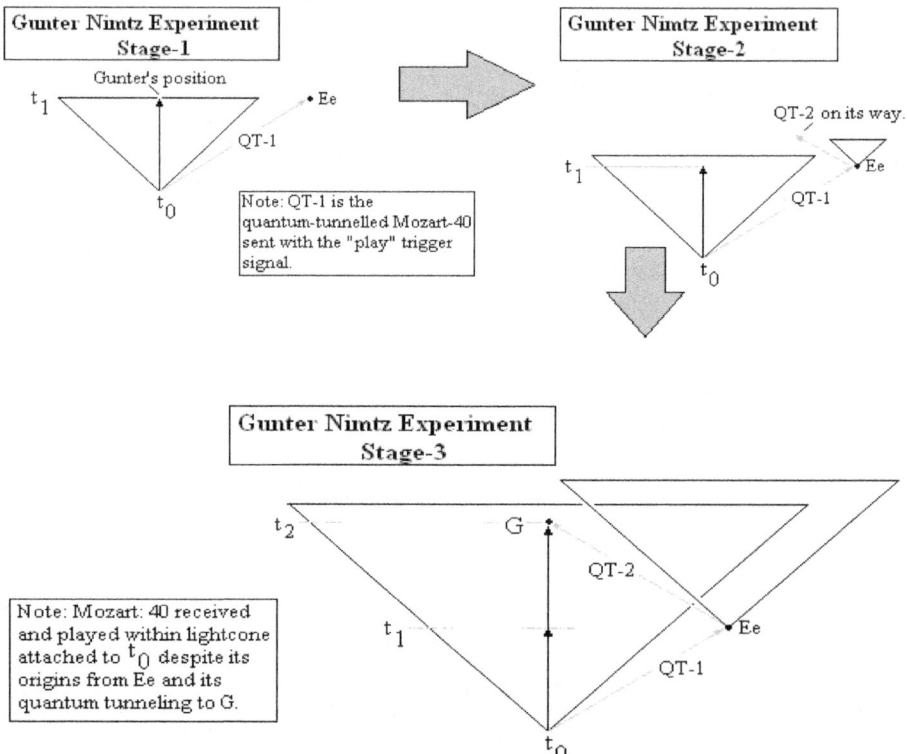

Fig.1.09 Gunter Nimtz's experiment with quantum tunnelling circuitry
Notes: An imaginary experiment entailing the quantum tunnelled faster than light transmission of M-40 to **Ee** at **Stage-1** by means of **QT-1**. At **Stage-2**, M-40 is then quantum tunnels back to Nimtz via **QT-2**: also faster-than-light. However, by **Stage-3**, for Nimtz located at the final point of reception of M-40, all that he observes is the trigger, followed by a momentary delay from t_0 to t_2, followed by the subsequent play of M40. Nimtz cannot report **Ee** directly. Yet, his assistant at **Ee** could verify that **Ee** was involved as a shunt by clocking the fact that M40 *was* received at **Ee** at faster-than-light from the Earth. A seeming violation of the speed-of-light limit will occur. Even so, recall that what preserves causality *and* the integrity of information is the ontological primacy of AND-to-OR collapse, combined to the fact that per information-matter dualism, M40, dissociates from the pertinent radiation, whatever the latter's speed. In any case, there is a solution to the conundrum based on the maincone principle, described in a subsequent illustration.

Recall yet again that information is not ultimately retained in, or even carried and communicated by, putative radiation and matter: It *effectively* dissociates from both. Causality and order are not fated to an electromagnetic propagation constant, or even undermined by its supposed violation, given the said dissociation. And it is the ontological primacy of AND-to-OR time that matters in the preservation and integrity of causality and information: unaltered by 'faster-than-light' effects (see **1.38** for the comprehensive account of these insights).

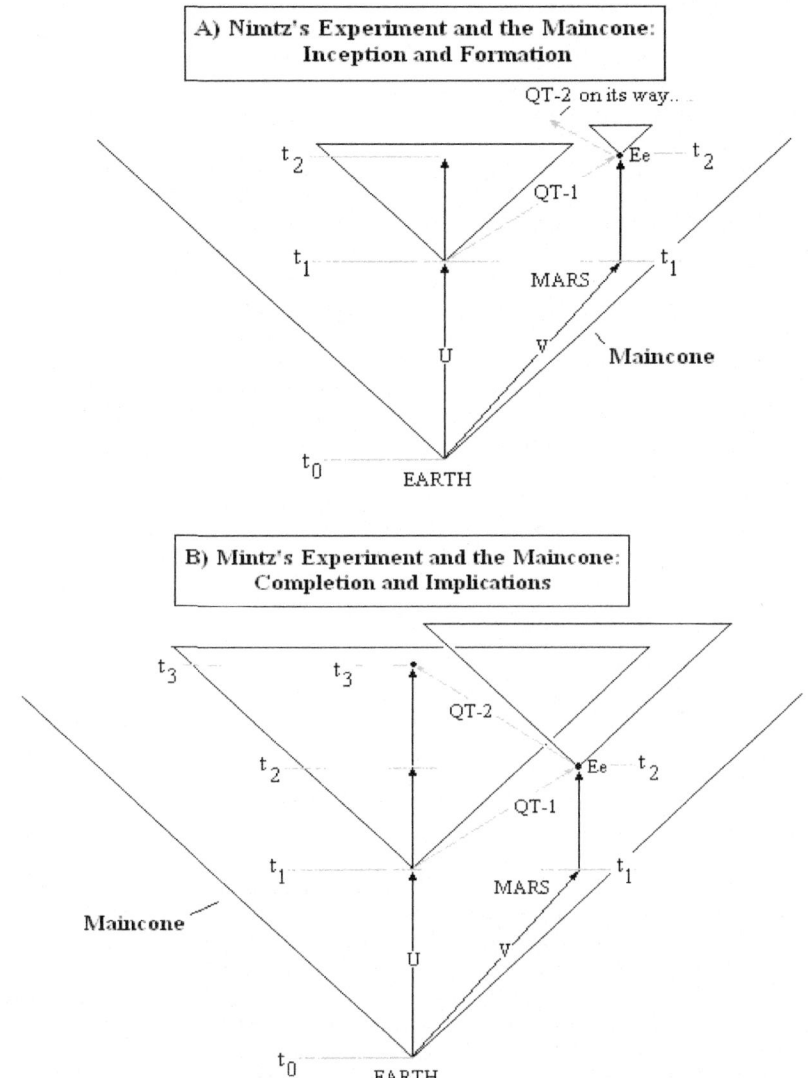

Fig.1.10: Maincone approach and the harmonisation of Nimtz experiment with Special Relativity
Notes: Nimtz remains on Earth. His assistant is sent to Mars at velocity-**v**: His departure at **t₀** is embedded in an encapsulating lightcone; the **principal maincone**. This encapsulates subsequent lightcones and 'anomalous' faster-than-light relations within its bounds. Upon arrival at Mars, the assistant establishes quantum tunnelling circuitry between Earth and Mars by **t₁**. The assistant awaits a faster-than-light quantum tunnelled signal, **QT-1** from Nimtz, from Earth… and receives it by **t₂** as event **Ee**. per which the assistant immediately shunts a faster-than-light quantum tunnelled, **QT-2**. This reaches Nimtz by **t₃**. If the scenario were to unfold without the encapsulating maincone, Special Relativity and causality would be undermined. Fortunately, the existence of the encapsulating maincone, combined to information-dissociation per information-matter dualism, prevents the travesty. Per the maincone, M40 is present and dissociates at **t₀**; M40 is present in dissociated or clouded form throughout the maincone, and re-associates via AND-to-OR collapse at **t₂** and **t₃**. M40 is not being carried by the quantum tunnelled or other radiation, nor is M40 fated to radiation, whatever its speed.

These insights force radical reconsiderations about the manner and form of the presence of information vis-à-vis systems and relations. Per information-matter dualism and decoupling, the dissociation and retention of information is extant the 'carriers'. Per information-dissociation, **the information communicated between Nimtz and his assistant at 'faster than light' is potentially present to the *whole* system demarcated by the whole principal maincone, as an effectively de-localised or clouded form of information... and it will be re-manifested in concrete 'localised' OR-form per the grander all-encompassing and ontically primary AND-to-OR collapse process, which will reinstate that information back into its localised form *within* the maincone.**

Per the illustrations presented in **Fig.1.09**, Nimtz's assistant will arrive on Mars at t_1 and established part of the quantum tunnelling circuitry that resembles the setup depicted in **Fig. 1.03**. Nimtz on Earth will then send his faster-than-light quantum tunnelled signal, **QT1** to Mars. The point of reception at t_2 will be **Ee**, at least as far as the lightcone attached to Nimtz is concerned: To Nimtz, **Ee** will be 'elsewhere' on Mars.

Upon the reception of Nimitz's quantum tunnelled faster-than-light signal at **Ee**, his assistant on Mars will immediately send a similar faster-than-light quantum tunnelled response, **QT2**, to Earth, which will be received by Nimtz by t_3. The response-signal **QT2** will reach Nimtz on Earth well before any normal speed-of-light radiation signal can reach him from the same, originating at t_2 on Mars: This is illustrated as the evolving lightcone attached to the assistant at **Ee** in stage-2 and stage-3.

Thus, in **Fig 1.09** we have a genuine-seeming case for the subversion of SR. Yet, the solution is implicit to and fully demonstrated in **Fig. 1.10**. Therein, the whole system of 'violations' are transpiring within a principal maincone originating at t_0. Indeed, all previous scenarios also transpire within an implicit principal maincone that subsumes and integrates *all* subsequent anomalous-seeming 'faster-than-light' outcomes.

Why is the maincone important? While preceding conclusions and solutions from non-observable 'elsewhere events', subsummation and interchangeability rules do not rely on or combine directly with information-matter dualism and the dissociation of M40, the maincone approach does exactly that. Therein, M40 is present at t_0 as part of an initial set of conditions. It dissociates from the radiation pertinent to the principal maincone at t_0. It remains present in dissociated form to pertinent systems embedded *within* the evolving maincone that encompasses Nimtz, the travels of his assistant, Ee, and all subsequent actions. It re-associates and re-merges to materialisation vis-à-vis the system at t_3. Thus, M40 is dissociated, decoupled, and present in abstract 'clouded' form vis-à-vis the total region defined by the whole principal maincone, and it is re-materialised at the subsequent pertinent moment with radiation and matter.

Again, it bears reiteration *ad nauseam*, that principles that harmonise outcomes with causality rest, *not* in the dubious purported mechanisms of 'spatial transmission of information' by radiation or matter, much less with the putative propagation constant for light. The principles that harmonise event-generation with causality belong to the ontologically primary process of AND-to-OR collapse behind and underpinning the 'transmissions of signals', at the speed of light, or even at 'faster-than-light'; and even in the manifestation of the speed of light in the formation of the principal maincone, within which M40 is retained in dissociated and clouded decoupled form: again, see **1.38**.

1.42: Decoupling Theory & Abstract Memory

There are two distinct conceptions of memory: The conventional conception is implicitly Aristotelian, despite its materialist pretensions. But, like materialism, the Aristotelian view cannot conceive of information as capable of physical separation from either radiation, matter, or from 'space'. In contrast, and per our Alternative Realist perspective, we have the Platonic conception of memory. **Abstract memory asserts that the informative power of nature can fully dissociate out from radiation, matter, and 'space', and even from the 'speed of light' limit, *and* survive such a dissociation and decoupling in extant abstract-immaterial form.** This is the form of memory vindicated by information-matter dualism, notwithstanding its own expected eventual succession by an even more exotic form per de-spatialisation and temporisation in Book-IV. In any case, **the Alt-R case for abstract memory is further augmented by decoupling theory and the maincone approach, combined to informational dissociation** discussed in **1.41**.

As we further develop abstract memory theory in Book-II through Book-IV, we will discover that much of the informational content of the universe exists as an abstract superpose of past events in holonomic form: retained at physical remove from manifest OR-form radiation, matter, and 'space'. Indeed, the very possibility of quantum mechanics and experimentation depends on abstract memory, as was espoused in our apagogic arguments presented on **1.22** to **1.24**. Clearly, abstract memory, together with tandem information-matter dualism *and* decoupling theory, applies to the four scenarios that harmonise the anomalous findings of Nimtz with the strictures of Special Relativity, and co-joins with decoupling theory, *especially* in scenario-4: wherein **the dissociation of M40 within the whole of the principal maincone constitutes a case for abstract memory**, as much as it does for information-matter dualism.

Abstract memory, though apparently hidden from immediate view, through successive AND-to-OR collapse-processes and time, gets to play a critical part in the endurance, perdurance, persistence, formation and re-formation of pattern-structures and identities. **In Book-II, we will incorporate this claim in explicit form to the intermediate spacetime model, wherein abstract memory resides in and constitutes the mnemonic domain of realised spacetime.** By Book-IV, in the face of de-spatialisation and other things, the intermediate spacetime model will itself be succeeded by the Bergson-Whitehead amalgam, or spacetime *without* space... in which information is finally understood as organised and distributed purely in and by time, *without* 'space'... therefore, necessarily abstract and dissociated in essentially the same form promulgated in information-matter dualism and in decoupling theory, scenario-4. **By Book-IV, we will have arrived at the purest definition of what is meant by abstract memory: namely, information organised in and by delayed choice time-intervals... by time... *without* space, with direct pertinence to the principal maincone from the fourth** scenario of the Nimtz experiment, and to the form in which information pertaining to M40 and much else besides dissociates, delocalises and clouds to that maincone, and across nature.

Also of direct pertinence, **by Book-II, we will have revamped the clouding of memory vis-à-vis the future as part of counterfactual causality and grand decoherence… wherein the past eliminates from all possible futures within future-potential lightcones** *all* **non-consistent futures vis-à-vis the abstract past. Hence, Gunter Nimtz could 'receive' only Mozart-40 because all possible futures had decohered to consistent-only outcomes: to outcomes consistent with the mnemonic fact of M40 retained within the maincone… within mnemonic realised spacetime… within the Bergson facet of abstract memory.**

But, most remarkable of all, the dualism of information and matter, and the decoupling and clouding of information that attends it, which holds true throughout nature, must also hold true vis-à-vis the brain, given that the brain itself is not going to be exempt from this general truth. On the assumption that both memory and nous are typically *presumed* to be constituted as in-brain physical categories, our case throughout Book-I in favour of information-matter dualism, decoupling theory, the clouding of memory and abstract memory, must necessarily culminate into memory-brain and mind-brain dualism… and into Cartesian revival.

1.43: Summary of Part-IV

The summary and conclusion to Part-IV is thus:

- **Decoupling theory**: Information-matter dualism abides. Hence, information and causality is decoupled from matter, radiation, 'space', *and* from the speed of light. This implies that apparent faster-than-light radiation is of no consequence to the integrity of information, which dissociates out of the radiation and is not 'carried' by it. The viability of causality is not dependent on the 'carrying' radiation or matter, no more than it is on the speed-of-light , given that causality and the 'arrow of time' is solely dependent on ontologically primary AND-to-OR collapse that, in the first place, brings into evidence both radiation at speed-of-light *and* apparent faster-than-light, clarifying the anomalies of the Nimtz experiments.

- **Solutions to the anomalies of the Nimtz Experiment**: These are distributed into four scenarios in **1.41**.

- **The dissociation of information from apparent faster-than-light radiation implies that information is *not* fated to the radiation or to its motion, whatever its speed**. This applies in all four scenarios in **1.41**.

- Causality, time, and its 'arrow' is fated to and synonymous with AND-to-OR collapse. **Even when AND-to-OR collapse evinces events in apparent violation of the speed-of-light-limit, this could not alter AND-to-OR directionality or usurp causality, nor bring about OR-to-AND time-reversal: (1.38).** This insight is implicit to all four scenarios.

- **The worldline of an elsewhere-event Ee in supposed violation of the speed-of-light limit, will, in time, 'move' inside some critical lightcone, wherein it will be observed as a normal event. (Scenario-1). (1.41).**

- **Elsewhere-events are never directly observed as such: As such, elsewhere-events cannot constitute the violation of SR postulates.** This applies throughout most scenarios, except scenario-3 and 4, or seemingly so. (**1.41**).

- Per **the interchangeability rule**, in a scenario involving the ultimately non-observable violation of the speed-of-light limit in the formation of a supposed elsewhere-event, **if the 'violating' event, and all pertinent relations, are effectively interchangeable with an equivalent arrangement that does not violate the postulates of Special Relativity**, then no violation of the speed of light limit can be claimed. See scenario-3 (**1.41**).

- **Maincone theory**: This is presented in scenario-4. All subsequent events, including those that apparently violate the speed of light limit, unfold from a common origin t_0 and are subsumed within an all-encompassing lightcone originating at t_0, (namely, the principal maincone) that exhibits the said events, ultimately as ordinary non-violating events. **Per information-matter dualism and decoupling, information and M40 itself dissociates at t_0 and remains dissociated throughout the region defined by that principal maincone: It is not fated to apparent 'faster-than-light' events within that maincone.** The claim for the dissociation of information and its de-localisation and clouding throughout the maincone will follow seamlessly from de-spatialisation in Book-IV: De-spatialisation attacks the whole notion of information transportation in and by radiation and matter 'in space', whatever its 'speed'. De-spatialisation renders information into a pure abstraction distribution in time, *without* mediation or resonation across space: necessarily dissociated and abstract in its form.

Mind-brain dualism: What holds true for putative 'matter in space' holds true for the brain: the dissociation of information, implications from decoupling, and all other anticipated radical findings, imply the dissociation of memory and nous from brains, even or especially when, per materialism, we erroneously assume that both must identify and reduce to hands-on structures and processes of the brain. In the face of information-matter dualism, decoupling theory and the clouding of physical information vis-à-vis maincones, the brain is not going to be an exemption, with obvious Cartesian implications.

EMRE ASENA

BOOK-I PART-V:
INFORMATION, MACHINES AND MINDS:
RIGGING THEORY AND NOETIC INFORMATION

1.44: Aims of Part-V

The dualism of information and matter, which holds *effectively* true throughout physical nature, even if destined to be superseded by a more exotic form per anticipated de-spatialisation and other findings, must also hold true vis-à-vis the brain: The brain is not going to be exempt from such truth. Hence, on the dubious materialist assumption that both memory and mind is constituted as in-brain material trace-states, information-matter dualism must necessarily lead to default mind-brain dualism, given the expected dissociation into abstract memory of the in-brain material embodiments of memory and mind. Yet, this contention assumes that mind, cognition, and memory, are like physical informational states; perhaps initially constituted as neural formations, patterns, and processes, whatever their fate in the face of dissociation per information-matter dualism, decoupling theory, and other radical considerations. Although the assumption could not obviate consequent default mind-brain dualism, it is by no means a given that consciousness and cognition, necessarily look like the ostensive in-brain material structures and processes. Indeed, Preliminary Part-IV raised memory-side and mind-side experimental and empirical reasons why we must suspect (to say the least) any presumed initial materialist reduction of memory, mind, and cognition to in-brain structures. Essay **0.6** and especially key passages in **0.43**, if reprised, are sufficient to lay bare the suspect nature of the notion that structures of the brain could constitute noetic memory or even cognition, and why mind, consciousness and cognition must be of a radically different order vis-à-vis in-brain structures and processes: suspicions and reasons destined to be fully vindicated by Book-V.

To appreciate what we mean by noetic information, and its distinction from physical information, we refer directly to material from **0.6** (Preliminary) and focus on the **operational definition of mind**; wherein the ***mind is defined in terms of what it does.*** Therein, we need only choose a specific hard problem and we find that it engenders serious challenges to the notion that mind, cognition and memory are in-brain structures or process, notwithstanding implications from information-matter dualism, decoupling, and other findings.

Renormalisation in quantum field theory is one such hard problem. Recall that, therein, when two particles approach, these purportedly interact by exchanging mediating particles. But the interaction is wrought by infinities, and calculation becomes impossible. To overcome this, physics turns to renormalisation: the subtraction of the infinities. Whether this is a contrivance or a perfectly justifiable mode, renormalisation raises key hard-problem questions:

- **What sort of causality could operate within the brain (assuming materialism) such as to allow renormalisation?** To renormalise, the physicist obviously needs to apprehend infinity.

- **From where does the physicist get the idea of 'infinity'? It cannot come from sense-experience:** There is nothing in the sensate-empirical order that constitutes an analogue for infinity. Sensate experience is always experience of the finite.

- **Does the physicist's brain generate the idea of infinity through an in-brain clockable process**, so as to permit the possibility of renormalisation?

The attempt to grapple with these questions leads to two problems:

- **The problem of memory: How do we store the idea of 'infinity' in the finite structure and capacity of the brain?** What would the 'memory of infinity' look like in biochemical and neural terms... or as in-brain 'code'?

- **The problem of process: How could the finite iterational and presumably clockable processes of the brain generate the apprehension of 'infinity' without crashing into insurmountable non-terminating or halting problems?** This forces the intuition that *the apprehension of infinity must entail a non-clockable process*: one that must defy rendition in brains or execution by any in-brain clockable process.

Such problems cannot be waved away by resort to 'complexity': How complex must a clockable in-brain process and structure be to 'compute' and store infinity as memory? Nor will the problem go away by dismissing 'infinity' as a *bloss* subjective idea, much less an 'illusion': The latter would render infinity *not* an objective state of the world or of brains, which is precisely our point. But the apprehended 'illusion' and 'fiction' of infinity must yet demand explanation in terms of real existence and the structures of the brain, at least insofar as materialism requires it. Nor could infinity be dismissed as a 'false' made-up idea: Even if 'false', it must yet defy clockability and rendition as in-brain structure: The problem of memory and the problem of process will yet abide.

In all cases, for generic quantum field theory to work, the intuition about infinities central to renormalisation is indispensable, subjective or otherwise, 'illusion' or otherwise, 'false' or otherwise.

Moreover, quantum filed theory underpins quantum electronics and myriad practical applications, and never mind infinities entailed in both infinitesimals and limits in old and new calculus and across modern engineering. All of these, and ostensive material civilisation built upon these, critically depend on our ability to grasp the 'false' idea of infinity.

The hard problem is not confined to mathematical physics and engineering: It is also exhibited in basic dictionary definitions and day-to-day meaning-extraction. The infinitive 'to run' most certainly entails infinities: This is not an equivocation fallacy. Indeed, 'to run' is an indefinite statement: Run? For how long? Given its unboundedness, 'to run' entails infinity.

QUANTUM MECHANICS AND MIND

How does the brain use a clockable process to arrive at, and then store… "to run"… within the finite brain? What does the memory of, "to run", look like within brains?

The infinitive "to be" is liable to cause even greater palpitations, and we will come back to it at the end of Part-V.

In short, the basic taken-for-granted daily meaning-extraction and communication itself defies explication in terms of finite structures and clockable processes of the brain, even though, per the keyboard-brain approach espoused throughout Part-IV of the Preliminary, the brain and its functional areas are certainly needed as a means of expression of otherwise non-clockable processes of intuitions, ideas, intentions… or, simply, of noetic information.

*

Renormalisation and infinities, as consummate hard problems, smack of mind-brain dualism and intimate inescapable radical conclusions about the nature of causality pertinent to mind and memory by **forcing the distinction between *noetic information* versus the usual sorts of physical finite-state information** that pertain to physical events, initial conditions, to AND-to-OR outcomes and to our nascent growing block universe. Indeed, noetic information in its most explicit form entails the apprehension of infinities, in a fashion that must necessarily break materialist conceptions of memory supposedly stored or traced within the finite brain, it must also challenge the capacity of the finite brain to carry out clockable processes to 'compute' infinities and renormalisation procedures: forcing conclusions in favour of mind-brain dualism involving non-clockable processes that cannot be evinced in brains, even if brains are indispensable as means of expression for non-clockable thought and mind.

In short, **it is obvious that the apprehension of infinity is a non-clockable process and, as such, does not entail *any* clockable process writ brains… much less 'storage' as a finite-structure comprised of in-brain 'code' or material corpus**. The non-clockable process that permits the apprehension of infinities and can permit the possibility of the renormalising quantum mechanist, is necessarily *extant* the brain. **Mind-brain dualism necessarily follows**, of a form different even from the default version implied by information matter dualism, decoupling and other findings that span this work.

*

What holds true for brains must also hold true for general purpose machines, computers and putative strong artificial intelligence (strong-AI). The expectation in strong-AI, to the effect that computers might one day accomplish what the brain supposedly does, is not going to progress well, given that computers suffer the same problem in terms of finite capacities and clockable processes as do brains, when both are confronted with hard problems that exhibit infinities, and wherein both require concession to non-clockable processes that defy rendition in or as any putative structure and process, whether in brains or in computers.

The historic association between brains and computers is, of course, founded on *a priori* materialism: The materialist notion that the brain produces mind and consciousness (assuming these are not dismissed as illusions in certain materialist approaches) merely constitutes AI-organic. Thus, **assume that some sort of hands-on in-brain structural iteration and mechanics can constitute memory and even 'excrete' mind, consciousness, and cognition, it will be but a single small step to suppose that alternative hands-on material structures and mechanics, embedded within inorganic Babbage engines or in modern silicon-based electronic computers, could accomplish the same**. Thus, original AI-organic gave way and gave succour to AI-inorganic, whether silicon based or other: Hence, the contemporary basic postulate behind strong-AI.

However, when both brains and computers are confronted with appropriate hard problems, both 'crash', so to speak. To augment, the question… "What does the apprehension of infinity look like in and as in-brain structure?" … is essentially interchangeable with… "What does infinity look like in terms of machine-state or 'code' in computers?"… and… "How does the brain or the computer circumvent halting problems when it is confronted with the need to accomplish the computation of infinity by means of an inevitable clockable process, *and … somehow*... store the result as memory?". The insurmountable problems posed by the question will not go away by resort to 'complexity, or even prospective quantum computation in brains or in future computers.

For our purposes, we shall reduce the matter in hand to the simple 'code-information distinction'. We will assert that code (in machines or in brains) is *not* information... and information… is *not* code. Again, the actual apprehension of infinity by mind… the *realia* of that apprehension… involves a non-clockable process: one that does not require the machinery and clockable processes of the brain. The same holds with respect to memory of non-clockable intuitions. Hence, the reason why computers will never be able to accomplish even what the brain cannot accomplish is because computers, like brains, *cannot* carry out pertinent non-clockable processes, much less get to store non-clockable cognitions (or noetic information) as 'codes' or machine-states. **The code-information distinction clarifies and augments what is meant by noetic information: Noetic information is non-clockable and non-codable intuition, involving infinities or infinity-laden ideas**, whose reality is necessarily conducive to Cartesian mind-brain dualism.

*

The aim of Part-V will be to establish the code-information distinction as fact, conducive of mind-brain dualism. Following a brief history of the **code-information distinction**, itself followed by **rigging theory** to explain how putative artificial intelligence systems actually work, we will apply the Einstein Podolsky Rosen (EPR) relations in a radical scheme that, through **superlative rigging**, will establish both the code-information distinction *and* rigging theory as indisputable facts, permanently calling into question strong-AI, while firmly establishing noetic information as a truly distinct form of information, with attendant mind-brain dualism, against the grain of *a priori* materialism.

The question of *a priori* materialism in relation to physics, and the false confusion of materialism with physicalism, was dealt with in Preliminary Part-III; especially in essay **0.35**. The latter adumbrated the unauthorised history of materialism in the context of physics.

151

It exposed the precarious relation of materialism to physics: an important consideration, given that **the belief that the brain can compute cognition, and the generalisation of this to the notion that 'computers can think', gains succour from background *a priori* materialism: This presupposes that the brain employs a materialistic closed-causal input-operation-output schema to generate mind and thought, and that this could be replicated in other material mediums: namely Babbage engines and, later, in contemporary electronic computers.**

Of course, given implications from information-matter dualism itself, and its succession in turn by de-spatialisation, temporisation, and other findings central to our Alternative Realist thesis (Alt-R), these considerations also undermine materialism. By Book-IV, anticipated de-spatialisation will force physical relations to be expressed purely in terms of time-relations between abstract states that can no longer be defined as 'objects or points in space': This must also apply to memory, cognition, consciousness, and mind vis-à-vis the brain. **Simply put, there is no space. Therefore, memories and minds are not 'spatial objects in space' and cannot be rendered 'in' brains, especially given that the brain itself can no longer be modelled as a spatial corpus.** And, in terms of time-relations alone, the mind turns out to be extant and dissociated vis-à-vis its brain, in the sense that a generic end-user is extant its keyboard.

Moreover, by Book-IV, **the process of AND-to-OR wavefunction collapse, or time itself, will be shown to be ontologically primary and presupposed to the succession of events pertaining to the physical order as well as to brains, whose succession of events constituted as brain activity might be naively supposed to constitute 'thought' and 'memory'. The process of AND to or collapse and time proceeds on a gear-less work function zero basis: a basis that breaks the materialist closed-causality IOO input-operation-output schema generically employed to explicate the physical order, and especially the workings of the brain**.

The perspective from anticipated de-spatialisation and from gear-less AND-to-OR collapse and time constitute distinct considerations that doom the materialist contention on minds, brains and computers: an assertion to be developed in the succession of books that constitute this work.

1.45: Lovelace, Searle & the Code-Information Distinction: Code is *not* Information

As far as is certain, the first historic instance of the **code-information distinction** was made by Ada Lovelace in the 19th Century, in the context of the Babbage analytical engine: the forerunner to the modern computer. Lovelace explicitly understood that symbols or codes do not constitute the meanings, (i.e. the proper informational states) or the *realia* that we otherwise correlate with and foist on symbols. Thus…

Considered under the most general point of view, the essential object of the machine being to calculate, according to the laws dictated to it, the values of numerical coefficients which it is then to distribute appropriately on the columns which represent the variables, it follows that *the interpretation of the formulae and of results is beyond its province, unless indeed this interpretation be itself susceptible to expression by means of the symbols which the machine employs*. Thus, although it is not itself the being that reflects, it must yet be considered as the being which executes the conceptions of intelligence.[50]
(my italics)

If the symbols could constitute the 'interpretations', the meanings, the realia, then the analytical engine, and computers generally, would possess the power to reflect on, as opposed to merely execute, the 'conceptions of intelligence'.

It could be argued that the **code-information confusion** springs from ordinary language-use, which then contaminates our views about code-processing devices, such as Babbage engines and modern computers: wherein the proper meanings (the *realia*) are so enmeshed with the symbols, codes, or words of the language utilised, that it is almost as if the codes themselves are the meanings. It is almost assumed that the meanings are intrinsic to the words, symbols, and codes, and *not* extrinsic, imposed or foisted on these: namely, by us, and upon consensus.

All that has happened is **that we have mutually agreed to *rig* words or symbols, different in form from the meanings-proper or from the realia themselves.** There may well be as many words for 'apple' as there are human languages. Yet the real apple, or the *realia* apple, has no name or word or symbol: The realia is nameless: it is the apple *in-itself*, which cannot be substituted with any improper form (i.e., the code, symbol, or word). You cannot eat the word 'apple' and obtain the same result.

The other culprit behind the information-code confusion is **likely blindness about the infinities involved in meanings (e.g., "to run" … "to be"). These imply memory that cannot be constituted as a finite structure and cannot be substituted with or replaced by a finite symbol. It also implies an inescapable non-clockable process of cognition and apprehension that defies in-brain time-bound processes and structures, or equivalent structures and processes in computers.**

A clearer expression of the code-information distinction can be found in the assertions of John R. Searle. Thus...

Let us program our favourite PDP-10 computer with a formal program that simulates thirst. We can even program it to print out at the end "Boy, am I thirsty!" or "Won't someone please give me a drink?" etc. Now would anyone suppose that we thereby have even the slightest reason to suppose that the computer is literally thirsty?.. The answer, alas, is that a large number of people are committed to an ideology that requires them to believe just that.

---

[50]Menabrea, L.F., *Taylor's Scientific Memoirs, Vol. II*, pp, 666-731, trans and annotated by the Countess of Lovelace; Bibliotheque Universelle 82:10, 1842; reprinted in Bowden, B.V., *Faster than Thought* (London: Sir Isaac Pitman and Sons, 1953)

And…

So let us imagine our thirst-stimulating program running on a computer made entirely of old beer cans, millions (or billions) of old beer cans that are rigged up to levers and powered by windmills. We can imagine that the program simulates the neuron firings at the synapses by having beer cans bang into each other, thus achieving a strict correspondence between neuron-firings and beer-can bangings. At the end of the sequence a beer can pops upon which is written "I am thirsty" Now, to repeat the question, does anyone suppose that this Rube Goldberg apparatus is literally thirsty in the sense in which you and I are?[51]

Clearly, beer-can 'bangings' are not the proper forms for neuron discharges, no more than are electronic logic and switch-gates, machine states and codes in computers, and even less the symbol-outputs generated by these. Of course, Searle's assumption here, part of his biological naturalism approach to the mind-brain, is that neuron discharges might ultimately constitute the proper forms for meanings, feelings and cognitions: a matter we questioned in **1.44** and throughout Part-IV of the Preliminary.

The most lucid expression of the code-information distinction made by Searle can be found in the following: To the question of whether a computer program can think, Searle asserts….

…the answer...is no.
Why not?
Because the formal symbol manipulations by themselves don't have any intentionality; they are quite meaningless; *they aren't even symbol manipulations since the symbols don't symbolise anything.* In the linguistic jargon, they have only a syntax but no semantics. *Such intentionality as computers appears to have is solely in the minds of those who program them and those who use them, those who send in the input and those who interpret the output.*[52]
(My italics)

To put the matter in our terms, **the information, the noetic information, is extrinsic and foisted, by us, on the internal and outputted machine states, symbols and processes: The information so foisted is *not* intrinsic to the device, to its processes or to its output… because the codes, symbols or machine-states are *not* of the same or proper form as the meanings, the intentions, or the *realia* presupposed to them or foisted upon them.**

Searle's Chinese Room thought experiment set out to prove this. In the experiment described in the same paper from which the above-quote was taken, Searle imagined himself as a symbol-processing program, sitting in a room, the Chinese Room, and following mechanical instructions written in English for manipulating Chinese symbols… in the same sense that a computer executes a program to do the same, but with instructions written in machine language. Thus, Searle appeared to 'understand' Chinese by following the instructions, but in truth could not achieve any understanding… because Searle does not speak or understand in Chinese. In the same vein, neither does a computer that merely executes the same program to manipulate Chinese symbols.

The conclusion was that a mere program cannot attain understanding... of Chinese, or of any other informational process. In fact, we assert that the computer is not an 'information processor' at all; and that this is a misnomer, as indicated by Searle: '...the symbols don't symbolise anything'.

Indeed, per the code-information distinction, the symbols are informationally empty, and their 'processing' equally void.

Unfortunately, elsewhere in the same paper, but consistent with Searle's biological naturalism approach, Searle left his position vulnerable. Thus…

Could a machine think?
The answer is, obviously, yes. We are precisely just such machines.

Searle's concession is typically seen as effective capitulation to the notion that machines could think, *somehow:* It appears to undermine his otherwise decisive and conclusive Chinese Room approach. This is because of a central fallacy: We are supposedly organic machines, or so materialism will assert, as does Searle. If so, non-organic machines could ultimately accomplish the same things as organic machines: Indeed, why not? That is, AI-inorganic ought to be possible for the same reason that AI-organic is possible. And Searle's Chinese Room experiment does not appear to undermine *this* assumption. In his capitulation, Searle left room for both materialism and the AI-believer to, as it were, 'escape the Chinese Room'.

However, *a priori* materialism that succours confidence in materialism, in AI-organic, and aspires to AI-inorganic, is itself totally suspect, as was shown in Preliminary Part-III, as asserted in **1.44**, and as will be established conclusively by Book-V.

Nor is passing the Turing test sufficient to put such doubts to rest. The Turing test, wherein a machine successfully imitates what a human being could accomplish in a limited purview or context, even one that can be passed, does not address the code-information distinction, and it does not overcome the hard problems that emerge from infinities and from attendant non-clockability.

---

[51] John R. Searle. "The Myth of the Computer." in the *New York Review of Books.* 1982

[52] John R. Searle, 'Minds, Brains, and Programs', Behavioural and Brain Sciences 3 (1980)

1.46: Facile Rigging Theory: Passing the Turing Test

The following constitutes a simple system designed to easily pass the Turing test or imitation test. The overall design is depicted in **Fig.1.11**. Rigging theory asserts that, from input to operation, through to output, **the rigged system is informationally void, and it only consists of improper forms: That is, the terms of the system have no bearing on the proper forms of the information foisted upon it through deliberate rigging on the part of programmers and end-users: a fact obscured by the code-information confusion, or naivete in the supposition that code-states constitute informational states or the realia.** While the core fallacy is obvious from the code-information distinction espoused in **1.45**, and equally obvious from the hard problem imposed by infinities and non-clockability reprised in **1.44**, it will also become obvious from both facile rigging here-espoused, and especially from superlative rigging based on the utilisation of EPR relations and transactions, to be adumbrated in the forthcoming essays: **1.47** through to **1.50**.

In the simple rigged system depicted in **Fig.1.11**, the left input-column constitutes every conceivable interrogative that one could pose to the device about the infinity or otherwise of whole number integers. The number of such interrogatives are finite, given that there are only so many ways one may pose the same question in the English language or any language. The end-user poses the question. The device matches the question to a memory bank of a finite number of English-language interrogatives, 'recognising' the question thus posed via matching.

On the right-output column, we have every conceivable correct same-answer output vis-à-vis the inputted interrogatives. Again, there are only a finite number of possible same-answer outputs in the English language.

In the intermediating centre that resides between the input-column and the output-column, an embedded algorithm selects the same-answer output from the list of possibilities pre-rigged in the output-column on the right.

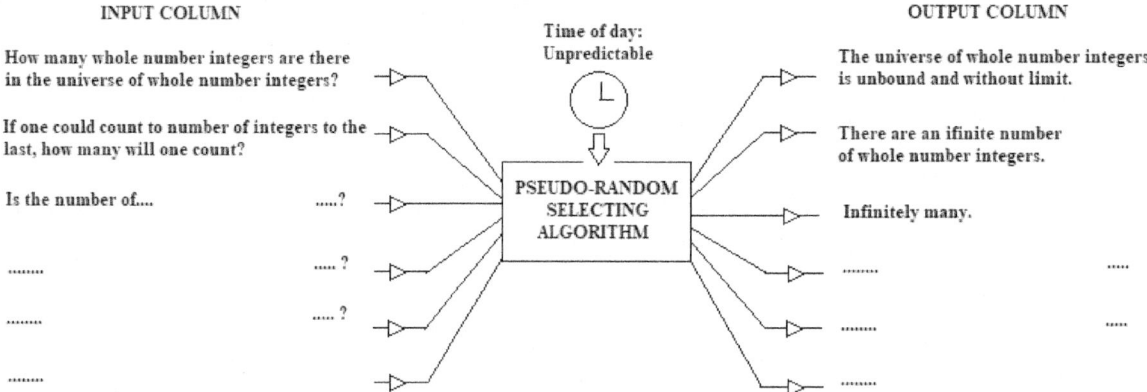

Fig. 1.11 A simple rigged system for "the infinity of integers" passes the Turing test

Notes: In the simple rigged system depicted above, an end-user can pose any same-question alternatives about the limit of whole integer numbers: Only a finite number of such questions is possible in the English language or in any language, all of which are listed in the input column (for brevity, not all are shown). The answer expressed by the machine at the output column could be any of the listed therein. One same-answer alternative listed in the output column is selected by a pseudo-random algorithm. The algorithm is of the form that operates in coin-driven gambling machines: It operates by incorporating an unpredictable variable to the calculation performed, such as the time of day at which the end-user poses the question. The listed answers are assigned ranges of values calculated by the algorithm (not shown). If the value produced by the algorithm lies within an appropriate range, the pertinent output linked to that range is selected and expressed. In every instance, our machine will pass the Turing test: it will imitate the correct answer (whatever this turns out to be) to any one of the same-question alternatives posed. The answer will be indistinguishable from one that could be furnished by a human responder. The machine is even 'creative' (per the pseudo-random algorithm) in that it does not always produce the same form of the answer. It is also 'intelligent', albeit in the limited purview sense, and more remarkable for the fact that it appears to circumvent a key hard problem, but can address questions about whole integer infinities, *without* crashing into any insurmountable halting problem.

The said algorithm works on a pseudo-random basis: It is entirely deterministic. Yet, the time-of-day at which the end-user might input the interrogative cannot be determine beforehand. Thus, the time-of-day constitutes the random variable. When the question is posed, an in-built clock is consulted, and the time of day is feed into the algorithm. Per the value-ranges pre-assigned to the answers, modified only by the random time-of-day, one of the listed same-answer will be outputted on a randomised basis. (This method is utilised in gambling machines, wherein the time-of-day, the unpredictable variable, is the basis for the said generating random sequences of fruit symbol sequence on the machine-display).

The point of including the pseudo-random output-selector is to convey the impression of machine 'creativity' to the naive end-user: there is no 'creativity' involved, no more than can be found in any fruit machine. To further augment the impression of 'creativity', the previous value calculated by the algorithm could be kept in memory and cycled back into the next round of calculation and selection; circumventing the risk that a clever end-user might choose exactly the same time-of-day and get the same-answer: exposing our trick.

The end-user's question could be… "How many whole integers are there in the universe of whole number integers?" … or… "If one could count the number of integers to the very last, how many will one count?". The answer outputted by the machine could be… "There are an infinite number of integers" …or… "The universe of whole number integers is unbound and without limit" … or "Infinitely many" etc... depending on the time-of-day fed into the pseudo-random selector-algorithm.

Notice that, in every instance it is consulted, **the device passes the Turing test**: That is, it *imitates* the correct answer, apparently indistinguishable from the same that might be furnished by a human. It is even 'creative' per the pseudo-random trickery described. The device is effectively 'intelligent', albeit in the limited purview sense: in the sense that the device can only be consulted about the infinity of whole number integers, and nothing else. **Our device is even more remarkable for the fact that it appears to circumvent a key hard problem: It can address questions about infinities without crashing into non-terminating halting problems: Of course, the device cannot count the whole integers to infinity, much less store the result as memory. It does not need to do so. So long as it furnishes the correct answers, it** *appears* **indistinguishable from a human that claims to possess non-clockable intuitions about infinities**. Indeed, if we placed a screen between the end-user and the device, and added a human voice generator to furnish the outputs, the end-user might not be able to tell the difference.

As admitted by John Searle, some argue that the machine depicted in **Fig.1.11** is literally thinking, and that it has full intelligence and cognition… because human intelligence and cognition consist of the same symbol-crunching mechanical processes, albeit writ brains. We are merely AI-organic… and what is possible for AI-organic (us) ought to be possible for AI-inorganic (gizmo) in essentially the same way and per the same process: a processes born of the materialist depiction of the physical order.

We beg to differ: Materialism is wholly questionable, to say the least (see Preliminary Part-III for the historical decay of materialism, and **1.45** above). Symbols, codes, machine-states, as well as hands-on brain-structures and processes, are *not* the proper forms of noetic information: The realia for infinity is not a symbol or word, no more than it is a neural strand or discharge-pattern. And the intuition about infinities inherent to definitions and meanings involve processes that are entirely non-clockable and cannot be captured in and as any code or clockable structure, whether in computers or in brains, or even as neural strands. As such, cognition cannot be rendered as brain-writ structure, or machine-state, or as any sort of clockable process, either in brains or in gizmo.

The test described in **Fig.1.11** exemplifies facile rigging. The sum of possible finite questions is *rigged* to the sum of possible finite answers by a mechanical intermediary. Yet, the proper form of the inputted questions is absent from the left column, per the code information distinction. The same follows for the outputs: also comprised of improper forms. Recall that the code, the symbol, and the machine-level state do not constitute the realia, much less the noetic information. Relative to noetic information, from input to output, the system is informationally empty. The rigged system certainly consists of physical states, and these may as well be constituted by Searle's 'banging beer cans' (see **1.45**), or the meaningless informationally void Chinese ideograms featured in Searle's Chinese Room experiment. But it takes real naivete to capitulate to the code-information confusion, and treat the inputs, machine-churnings, and the outputs of the system *as if* these constituted the realia, or even 'thinking' itself.

*

Within the limited purview of its task, the rigging in the system illustrated in **Fig.1.11** pertains to a perfectly definable domain. There are other perfectly definable domains: such as exemplified by the tic-tac-toe domain (or 'noughts and crosses') and the Chess domain. It is precisely because of their perfect definability that these were the first to be mechanised, even before the advent of modern electronics and later computers.

These domains are perfectly definable in the sense that the number of position-possibilities or 'squares' in Chess and in tic-tact-toe are fully countable, and the noughts, crosses or units related to these can be recapitulated at the machine-level improper forms as so many 'orbits of a seed', and, per the limits on such orbits, as 'rules of motion'. Thus, how any seed is permitted to orbit (such as the seed that pertains to the Pawn or the Bishop in Chess) can be specified and delimited at the machine-level. Hence, in principle, if not in practice, the total possible orbits of every seed can be calculated and embedded into the machine. The 'winning' configurations and arrangements, as 'definitions of goodness', can also be embedded into the machine as relative combinations of seed-orbits that constitute "pawn takes bishop"; "mate"; and "checkmate", etc. Indeed, these, and the total sum of combinatorial possibilities for Chess, are finite: part of what makes Chess a perfectly definable domain, at least in principle. The combinatorial sums for Chess could be calculated: an extremely large but finite number. All these combinatorial states could in principle be embedded in a machine to surpass even Deep Blue[53] and its successors. Hence, the prospective *God-Machine of Chess*. Yet, destined to be a complete moron; nothing more than a perfectly rigged perfectly definable domain, devoid of noetic information and devoid of fundamental non-clockable noetic processes that make both animals and humans distinct from inanimate objects and machines.

That such machines could surpass human players in Chess and in other domain-definable games of opportunistic exploitation, exclusion, elimination, and amassment (i.e., 'competition'… or *sociopathy*) is not in question. Indeed, notwithstanding the God-Machine of Chess, this has already transpired in the case of Kasparov versus Deep Blue. That such devices might be implemented in industry, commerce and in the military, and pass the Turing test in the same sense that our rigged system in **Fig.1.11** does so, is also

---

[53]Warwick, Kevin (29 July 2017). "A Brief History of Deep Blue, IBM's Chess Computer". *Mental Floss*.

not in question, albeit on the strict condition that the pertinent gizmo are limited purview, or perfectly definable, or substantially definable domains, such as found in fly-by-wire flight computers in modern aircraft.

Of course, not all gaming, industrial and command-control domains are perfectly definable. Many domains are simply admixtures of the partially definable and mostly ambiguous. This is obvious in commercial computer games and in automated self-driving vehicles, for the perfect operation of which the sum of possibilities cannot be fully anticipated or embedded as machine states. Nor can domain definability or ambiguity be solved by greater memory capacity or increased processing speeds.

To the extent that domain definability remains ambiguous save in rear cases such as Chess, computers and industrial automation, computers cannot totally replace human agency, except in the most limited cases where perfect or good-enough domain-definability is obtainable, or in the most simplest of tasks in industry, manufacturing and services, and secured on the basis of two centuries of human de-skilling, courtesy of the antecedent divisions of labour exemplified by Adam Smith, combined to subsequent Taylorism and Fordism, and into the computer revolution since at least the 1970s. Hence, much of industrial and manufacturing work, the service sector, and significant parts of warehousing, could be automated and computer driven. It might be possible to progress into a post-work world by accomplishing for industry, manufacturing, logistics and commerce what was achieved in agriculture: Once the province of 95% of any population, now involving 3% at most.

However, in instances where non-clockable processing is required and remains critical to various endeavours, the machine is *almost* destined to remain subordinate to the human. Of course, fantasies about von Neumann's Technological Singularity[54], or human-to-machine download, or the ever elusive 'Skynet coup d'état', will remain *fantasies*: potentially dangerous ones that obviate the actual risks from such devices, while they galvanise superficial moral and existential hysteria.

1.47: Reprise: On The Impossibility of Instantaneous Transaction
Of Meaningful Information Through EPR

As prelude to our superlative EPR-based version of rigging, we must first reprise the Einstein Podolsky Rosen experiment and the impossibility of using 'spooky action at a distance' for the transaction of meaningful information. Insofar as EPR entails instantaneous effects, EPR ought to imply faster-than-light causality and communication, in contravention of Special Relativity and the speed-of-light limit. Conventionally, this problem is circumvented through the assertion that no meaningful information can be transacted in EPR, courtesy of the information-scrambling role of quantum indeterminacy entailed in the spin-resolution of entangled particles. The only exception consists in the implicit enforcement of the conservation of quantum spin: part of transtemporal enforcement of physical law. But this is of no relevance to substantive communication of information: (see Preliminary: **0.26**).

To grasp why faster-than-light or instantaneous communication of meaningful information is impossible via EPR, we must recapitulate the reasons previously adumbrated in **0.26**. To that end, envision a pair of photons originating from a single particle-decay. From the decay, each photon displaces in opposing directions at the speed of light. So long as experimental isolation remains in effect, the photon travelling to the left will constitute a unitary AND-form superposition of "spin-up" AND "spin-down". The converse will hold for the right-side photon: a corollary AND-form superposition of "spin-down" AND "spin-up".

When we finally make a measurement on the left particle and end its isolation, the left-side AND-form state might AND-to-OR collapse into, say, the OR-form "spin-up". It has a fifty-fifty chance of doing so. Before any physical signal at the speed of light has had time to reach it from the now-resolved left-side particle, the right-side particle will *instantaneously* AND-to-OR collapse into the opposite OR-form of "spin-down".

To clarify, **in EPR, every time we obtain "spin-up" on the left-side, we get a simultaneous and instantaneous correlate "spin-down" on the right-side. Conversely, when we get "spin-down" on the left-side, we get "spin-up" on the right-side.**

Could this process be used for directed instantaneous communication of meaningful information? As stated, **the generic assertion holds that no meaningful information can pass from the left-side to the right-side particle in EPR, given ineliminable quantum indeterminacy.** To this end, let us imagine a device that employs the simple Morse code system of DOTs and DASHes: Recall that DOT and DASH are contrived improper forms vis-à-vis the proper form realia of ideas we want to communicate using these. Hence, the message, "I am thirsty", can certainty be coded as DOTs and DASHes via Morse code, but the latter do not constitute the realia pertaining to "I am thirsty". Similarly, the DOTs and DASHes could in turn be coded as quantum-entangled spin-correlates. But the spin-correlates are no more DOTs and DASHes than they are the proper forms or the realia pertaining to "I am thirsty". For brevity, we will ignore the fact that the English language itself is a contrivance whose terms are of improper form to realia pertaining to the realia "I am thirsty".

Somewhere between Earth and Mars we produce quantum entangled spin-sates. We trap each entangled photon produced by the single-particle decay within quantum-suspending idler boxes, similar to those employed in nascent quantum radar systems: with one idler box for each of the particles constituting the entangled couple. The idler boxes are specifically designed to preserve the isolation and AND-form characteristics of the trapped entangled photons.

We then construct arrays of entangled transmitter-receiver idler boxes: one set for DOTs and another set for DASHes. We then move the transmitter-end arrays to Earth (the left-side) and the receiver-ends to Mars (the right-side). We then make a measurement on

---

[54] *The Technological Singularity* by Murray Shanahan, (MIT Press, 2015), page 233

the transmitter Earth-side particle by breaking into its transmitter-end idler box, resolving its spin into an OR-form outcome. A correlate opposite OR-form counter-spin will be instantaneously realised on the Martian 'right-side' receiver-end idler box.

In the system just described, for meaningful communication to be transacted from Earth to Mars, we must control and orchestrate with certainty the specific spin-outcome generated on Earth, and thereby determinately control the correlate spin-outcome on Mars, consistent with whatever spin correlates we formally agree to foist with DOT or DASH at the first-level rigging. Thus, if we could control and direct the outcome into the OR-form of 'spin-up' on Earth, its correlate could imply the transmission and reception of a rigged and foisted DOT on Mars. Conversely, if we could control the spin-outcome on Earth into 'spin-down', its correlate on Mars would then be read as DASH, again upon prior agreement about what we foist and rig upon quantum spin-correlates.

Could we realise instantaneous faster-than-light meaningful communication by such means, or the *ansible* of science fiction?

Quantum indeterminacy is ineliminable. Consequently, we cannot determinately control and orchestrate spin-outcomes on the Earth-side. Thus, we could never be sure of the transmission of a foisted DOT when we seek it, or a rigged DASH when we want it. Quantum indeterminacy will defeat any attempt to orchestrate the outcome on the Earth-side, and it will equally scramble what we want to transmit to the Mars-side.

Note that, in place of DOT and DASH, we could easily substitute foisted **0** and **1**, or equivalent machine-states (logic gates and transistors) for digital processing. In principle, we could even send complex instructions and imagery by such means, if only we could make EPR 'spooky action' work in an orchestrated way. But, given quantum indeterminacy, we cannot exploit EPR in the way described. Therefore, we cannot send information by means of an EPR-based 'ansible'... or *not in the way so far portrayed*.

Thus, **in EPR, intrinsic and ineliminable quantum indeterminacy appears to scramble any hope for causally controllable outcomes requisite to communication.** *This* **is what is meant when it is asserted that EPR 'action at a distance' cannot allow faster-than-light or instantaneous transmission of meaningful information.**

1.48: Superlative Rigging: EPR-Circuitry & the Transmission of Quantum Noise

Is it necessary that we determinately orchestrate the AND-to-OR resolution of spin correlates? If we wish to directly impose, foist, and rig on quantum spin outcomes some sort of Morse, binary, or other code, the control and obviation of quantum indeterminacy ought to be necessary. But quantum indeterminacy is impossible to obviate. Therefore, we cannot control and direct quantum spin-outcomes we want to obtain on the receiver-end idler box on Mars. Therefore, we will not be able to transmit our imposed code; whether this is comprised of a foisted DOT, a DASH, or a **0**, or a **1**; or even the complete sentence... "There are infinitely many whole number integers in the universe of numbers".

But is it at all necessary that we foist code *directly* onto specific spin-correlates? **What if we constructed and placed on Mars a non-discriminating receiver-end idler box (see Fig. 1.12), constructed in such a way that, whatever the OR-form spin outcome resolved within it, it will trip a circuit and output whatever code or message we pre-rigged to the idler box, but *did not* rig upon the spin correlates within it?** Therein, if the spin-outcome inside the non-discriminating receiver-end idler box is resolved into OR-form "spin-up", this will trip a circuit that will flash the foisted message... "There are infinitely many whole integers in the universe of numbers" ... on a screen attached to the outside of the Mars idler box. On the other hand, if the quantum spin inside the box on Mars is resolved into the opposite OR-form "spin-down", this will *again* trip the same message... That is, the same message will be 'received' on Mars, whatever the spin-outcome obtained, *without* the need to determinately orchestrate or control quantum spin-outcomes in entangled states: (see Fig.1.13).**

The above assumes that we can construct receiver-end idler boxes that, upon the spin-resolution of the particle within it, whatever the spin thus resolved, the AND-to-OR resolution of spin will break the imposed isolation of the box from inside out, and, in the process, trip the appropriate circuit and flash the pre-rigged message.

Purely for the sake of argument, let us assume that such an idler box could be constructed: We are not certain of the technical possibility, but let us indulge. In that case, so long as we have entangled and paired the receiver-end idler box on Mars to a transmitter end idler box on Earth, and we have agreed to pre-rig and foist upon the system the message that will be flashed on the Mars-end, we need only tamper with the Earth-end box, break its isolation, and quantum indeterminately AND-to-OR collapse the particle inside it into any of its possible spin-outcomes. Instantaneously, without the requirement of any physical signal, and without any need of orchestration or obviation of quantum indeterminacy, the particle inside the receiver-end idler-box on Mars will resolve into the opposing OR-form spin (any), break the isolation enforced upon it from inside-out, and trip the normal circuit attached to the box. And, on its outside, flash the message... "There are infinitely many whole integers in the universe of numbers": (see **Fig.1.13**).

Note that, **in this arrangement, there is no requirement to deterministically direct or obviate quantum indeterminacy. Indeed, we need not impose or foist** *any* **code directly to specific spin-possibilities: The spin-outcomes are** *not* **the constituents of any 'message'**, and quantum indeterminacy can operate in its usual inexorable fashion. Yet, we have apparently 'transmitted' the message, "There are infinitely many whole integers in the universe of numbers" ... to recipients on Mars.

We could instead rig our arrangement to flash DOT or DASH on the outside of the receiver idler box on Mars. Or we could instead rig it to flash 0 or 1 (See **Fig.1.14**). Either possibility would entail the same form of EPR circuitry and operation. There will be no need to orchestrate or obviate the quantum indeterminacy, given that the functioning of our putative *ansible* is not dependent on the attainment of specific spin-outcomes.

157

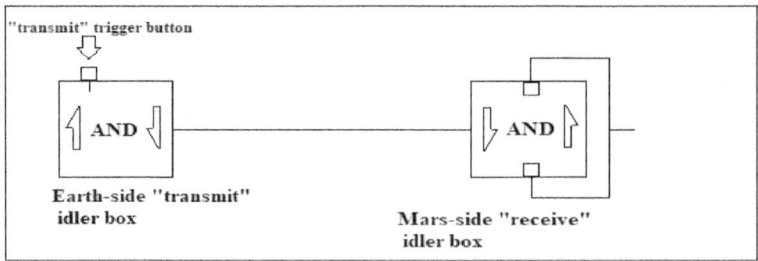

Fig. 1.12: Non-discriminating EPR-circuitry

Notes: The depiction above assumes that we can construct a receiver-end idler box (Mars-side) that, upon the OR-form spin-resolution of the particle within it, and whatever the spin resolved, it will break the imposed isolation from the idler box from inside-out... and trip the appropriate circuit that will flash a pre-attached pre-rigged message, agreed upon by the designer and end-user. We need only tamper with the Earth-end transmission idler box, break its isolation, AND-to-OR collapse the particle inside it into any of its possible spin-outcomes... and instantaneously OR-form resolve the particle inside the receiver-end idler-box on Mars. Whichever spin outcome the receiver end-particle resolves into, it will break the isolation enforced upon it *inside-out*, trip the normal circuitry attached to the Mars box, and, on the box-outside, flash the message previously rigged to it... *without objectively transmitting it.*

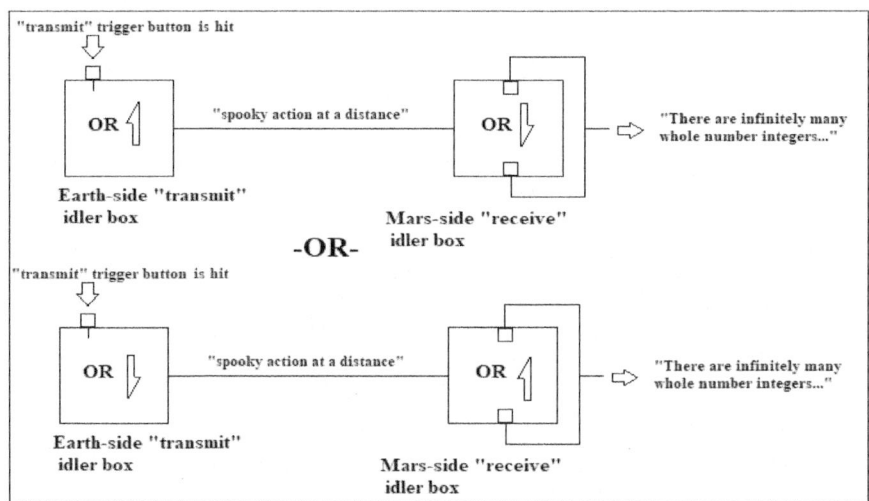

Fig.1.13: Instantaneous transmission of a "message" using non-discriminatory EPR-circuitry

Notes: The Mars-side idler box is non-discriminatory. i.e. whatever the spin outcome resolved within it, this will break the isolation enforced by the idler-box from inside-out. Thus, we need only tamper with the Earth-end box, break its isolation, and quantum indeterminately AND-to-OR collapse the particle inside into any of its possible spin-outcomes. Instantaneously, the particle inside the receiver-end idler-box on Mars will resolve into the opposing OR-form spin, break the isolation enforced upon it from inside-out, trip the normal circuit attached to that box, and, on its outside, flash the message previously agreed to and rigged upon it... *without* objectively transmitting it: The message will be pre-rigged to... "There are infinitely many whole number integers in the universe of numbers". Note that there will be no requirement to deterministically direct or obviate the quantum indeterminacy involved. Indeed, we need not impose or foist *any* code directly to any specific spin-possibilities, since these are *not* the constituents of the 'message'. Yet, we have apparently *instantaneously* 'transmitted the message' to recipients on Mars, *without* accomplishing any such thing, The form of EPR circuitry envisioned here could only mediate pure quantum noise: there can be no intrinsic 'message' or information transmitted owing to such quantum noise. Yet, the superlative rigging depicted could be used to fashion the same form of device envisioned in **Fig 1.11**. Such a device, operating using EPR-circuitry and nothing but quantum noise, could 'pass the Turing Test'. Yet, informationally, it would be as dumb as a doornail, given that all that would be transacted by the device at the base-objective level must consist of nothing but quantum noise.

For effective utilisation of this system, we would need large numbers of DOT-rigged and DASH-rigged boxes (or **(0)**-rigged and **(1)**-rigged idler boxes) arranged into arrays, with entangled idler box arrays sent and distributed to Earth and Mars, respectively. Second, we would need to code the English language either in Morse code or in terms of a binary code **0**s and **1**s: all on a pre-agreed codebook shared between the Martian astronauts and personnel on Earth. Using the appropriate agreed upon code, we could then 'transmit' the message… "There are infinitely many whole number integers in the universe of numbers" … through a series of DOTs and DASHes, or 0s and 1s… at faster-than-light. Indeed, instantaneously.

Or so it would appear… except that, **upon closer inspection, there will be no *intrinsic* transmission of *any* meaningful information. First, the message, the information, *the realia*, could not be intrinsic to the spin-correlates resolved: Recall from 1.47 that meaningful information cannot be transmitted in EPR because of the scramble from ineliminable quantum indeterminacy. Second, even if we could orchestrate and obviate quantum indeterminacy, the mere spin-correlates remain improper forms to the DOTs, DASHes, or the 0s and 1s, at the pertinent levels of rigging. Even the DOTs and DASHes, and 0s and 1s, are ultimately improper forms to the proper form of their own realia, and the message that these apparently code at a next level of rigging.**

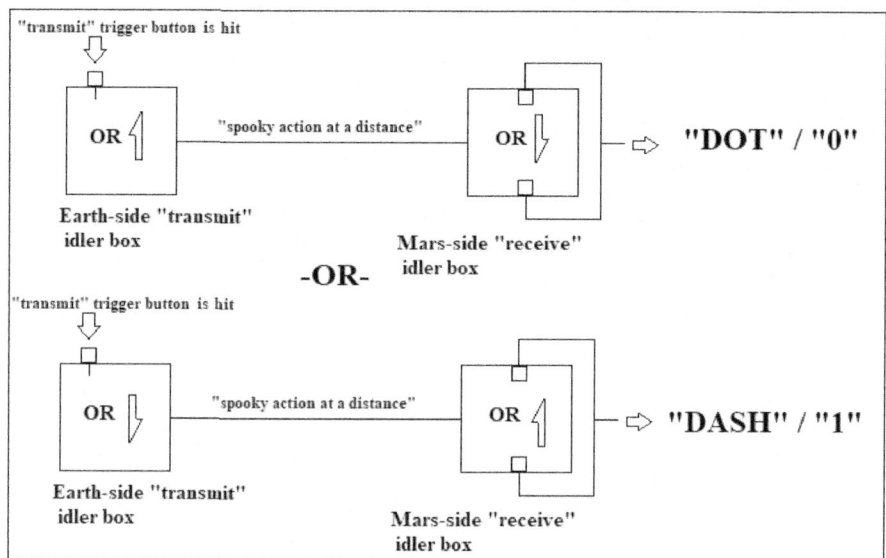

Fig. 1.14: Setup for the transmission of morse code or binary code via non-discriminating EPR-circuitry
Notes: Again, the Mars-side idler box is non-discriminatory. Whatever the spin outcome resolved within it, this will break the isolation enforced from inside-out: We need only tamper with the Earth-end box, break its isolation, and quantum indeterminately AND-to-OR collapse the particle inside it into any of its possible spin-outcomes. Instantaneously, the particle inside the receiver-end idler-box on Mars will resolve into the opposing OR-form spin, break the isolation enforced upon it from inside-out, trip the normal circuit attached to that box, and, on its outside, flash the message rigged upon it, *without* transmitting it. This time the 'message' could comprise of Morse code or binary code terms: i.e., DOT or DASH… 0 or 1. The latter could be used to operate computational devices on Mars, feeding instructions to these instantaneously from Earth. Notice that there will be no requirement to deterministically direct or obviate the quantum indeterminacy involved. Indeed, we need not impose or foist *any* code directly to specific spin-possibilities, since these are *not* the constituents of any 'message'. Yet, we have apparently 'transmitted the message': in this case, a DOT or a DASH… or a 1, or a 0… to recipients on Mars. In truth, we have accomplished no such thing: At the base-objective level, the device mediated nothing but quantum noise. Obviously quantum noise cannot not constitute code, and it is totally of improper form to any code, idea or realia. In truth, the designers and end-users have pre-agreed to foist ideas upon key machine-outputs and states: The impositions are not intrinsic to the system and depend on the code-information confusion fallacy committed on the part of both designers and end-users. An AI device (see **Fig. 1.11**) that runs on EPR circuitry would involve the mediation of mere quantum noise, not even codes, and *certainly not* ideas: It would be as dumb as a doornail. Yet, such a device could pass the Turing Test, despite its basis in mere quantum noise. This shows that the Turing Test is *not* a valid basis for attributing intelligence to any device. While generic AI-devices do not necessarily involve the transaction of quantum noise, what they *do* transact are improper forms to the ideas and the realia we foist on the outputted states, or when we committing the code-information confusion fallacy.

All of this assumes that the symbolic codes (the words) of the English language themselves constitute the proper forms for the realia and meanings: They do not. But, even when we ignore the rigging contrived at all levels, and seeing through the insidious code-information confusion, **our ultimate point is that, at the base-objective level, what we *actually* 'transmit' between Earth and**

Mars will be comprised of *nothing but quantum scramble or quantum noise*: **no meaningful information. Yet, the said message will be 'transmitted'.**

The whole system is pre-rigged from the beginning. The message or information is not at all intrinsic to the system, or even to the codes utilised and flashed by it. **All that has happened is that we had agreed on a contrived association of proper meanings (the realia) with non-proper forms: the codes, the symbols, etc that do not and cannot constitute the realia, much less elicit these. We then married these to a peculiar spooky cause-effect system involving EPR 'action at a distance' that, whatever its indeterminate and scrambled outcome at the base-objective level, and despite the fact that it mediates nothing but quantum noise without meaningful information, it is yet rigged and contrived to prompt a 'code',** *itself* **informationally empty.**

So long as we do not fall for the notorious information-code confusion, and can distinguish information from mere code, we will see through the trickery, even as we might utilise it for effective faster-than-light and instantaneous 'communications'… *if* and *only if* the appropriate idler box could be constructed on the receiver-end: a great technical hurdle: perhaps impossible to achieve.

1.49: EPR Circuits Applied to Logic Gates

The point of the above exercise was to devise a means through which we can demonstrate the reality of the code-information distinction without ambiguity, and, with it, the reality of contrivance of rigging in putative strong-AI. In this way, we expose the notorious code-information confusion, fallacy (perhaps even fraud) that resides at the heart of strong-AI. To further this goal, **we will apply the sort of EPR-circuitry developed in 1.48 and construct 'spooky action' driven logic-gates that could operate within real computers; in our integer number test from Fig.1.11; and in Searle's Chinese Room experiment; or in any putative system capable of passing the Turing Test.**

First, we must clarify the use of terms "AND" and "OR" in the context of logic gates. We have used the terms in AND-OR dualism or wave-particle dualism and in AND-to-OR wavefunction collapse. But these should not be confused with the same used in logic gates, such as in AND-gates and OR-gates: See **Fig.1.15 to Fig. 1.17.**

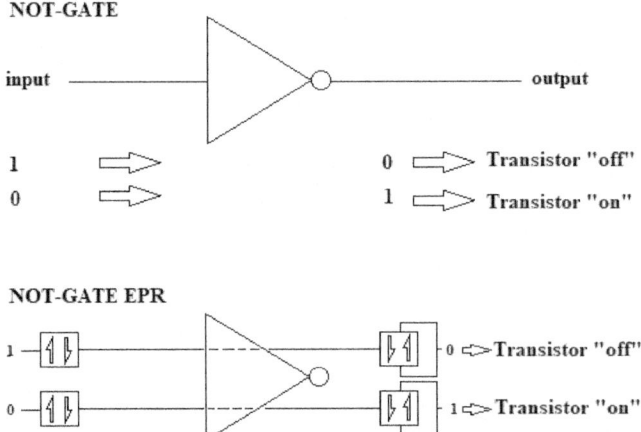

Fig. 1.15: The NOT-gate and NOT-gate with non-discriminating EPR-circuitry
Notes: We depict the simplest logic gate: the NOT-gate, which invert inputs to their opposite: i.e., input **1** into output **0**, such that the transistor switches "off". But when input **0** is inverted to output **1**, the transistor is switched "on": (see the 'truth table' below the top depiction). In the EPR-mediated version of the NOT-gate, we construct two pairs of idler boxes with quantum entangled pairs. Per each pair of boxes, the left idler-box constitutes the transmitter-end and the right idler the non-discriminating receiver-end. One of the pairs will be rigged to 'transmit' **1**, although this is merely foisted on the idler box pair and is not an intrinsic attribute of the quantum entangled pair, given the code-information distinction. The other pair will be rigged to 'transmit' **0**. To input and transmit **1**, we tamper with the left idler box and we AND-to-OR collapse the particle therein, which instantaneously AND-to-OR collapses the particle inside the right receiver-end idler box to an opposite spin. Whatever the OR-form of spin resolved therein, it will trip a normal circuit whose external output will be to 'switch off' the transistor attached to it. i.e., input-**1** will be inverted to output-**0**, and the transistor will switch 'off'. If we input **0** instead, the opposite outcome will be obtained: the transistor will switch to 'on'. In this way, we obtain a NOT-gate driven purely by EPR 'spooky action' and quantum noise. At the base-objective level, the realia of **0** or **1** (or even the codes terms improper to the realia) are *not* transmitted: Only quantum noise is transmitted. Yet, we get a device that can 'transmit' "**1**" and "**0**", instantaneously to recipients on Mars, and garner *instantaneous* control over transistor circuitry on Mars from Earth.

Logic gates are input-output relations implemented in modern electronic computers, although they can also be applied in water-driven computers and other mechanical devices. There are a total of seven different logic gates: we will illustrate three basic examples only: namely the NOT-gate, the AND-gate and the OR-gate. In modern computers, logic gates are constructed out of transistors (essentially "**on**" or "**off**" switches, in lieu of "**1**" and "**0**"). Logic gates can be combined into ensembles to execute complex protocols.

Notice how the NOT-gate in **Fig.1.15** (bottom) is also duplicated and portrayed as an EPR-circuit system involving quantum-spin idler boxes first described in **1.48**. Any computational device could in principle be constructed out of such putative EPR-circuitry to mediated outcomes via 'spooky action'. Indeed, given EPR 'spooky action', we need not localise the logic gates in any specific contiguous spatial way. Instead, we could scatter these randomly across lightyears (in principle). Yet, their rigged instantaneous relations will be preserved and successfully execute any viable program, despite radical spatial scattering.

Let us examine in detail the simplest logic gate: the NOT-gate portrayed in **Fig.1.15**: (top depiction). The NOT-gate inverts the input to its opposite: i.e. If the input is **1**, the output will be **0**, and the transistor affected will switch off or remain off. If the input is **0**, the output will be **1**, and the transistor will switch to "on".

How do we render NOT-gates using EPR-circuitry? **We construct two pairs of idler boxes** in which quantum entangled pairs of particles are contained: see **Fig.1.15** (bottom depiction). **Per each pair, the left idler-box will constitute the transmitter-end, while the right idler-box will comprise the receiver-end: These involve pertinent non-discriminating idler boxes,** as depicted in **1.48, Fig. 1.12**.

One of the pairs on the right-hand idlers will be rigged to 'transmit' **1**, although the "**1**" is an improper form code and, as such, it is merely foisted on the pair: It is not an intrinsic attribute of that pair. The other idler box pair will be rigged to 'transmit' **0**.

To input "1", we tamper with one of the left-hand pair of pertinent idler boxes. This leads to the AND-to-OR collapse of its content into one of two possible OR-form spin-states. Instantaneously, the particle within the right receiver-end idler box will AND-to-OR collapse into the opposite spin-outcome. Whatever the OR-form of spin resolved at the receiver-end, it will trip a normal circuit whose external output will 'switch off' the normal transistor wired to the box. i.e. the input-**1** will be inverted to output-**0**, and the transistor will switch 'off'.

If we input **0** instead, the opposite outcome will be obtained, in the same way via the other pair of entangled idler boxes; The transistor will be commanded to switch 'on'.

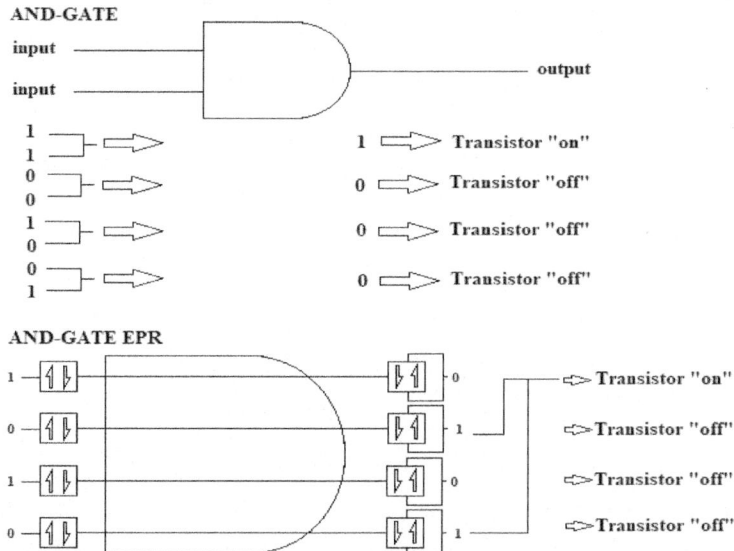

Fig.1.16: AND-gate and AND-gate with non-discriminating EPR-circuitry
Notes: The depicted AND-gate (top) accepts two inputs. If the inputs are **1** AND **1**, the output will be **1**: the transistor will switch 'on'. All other input-combinations will output **0** and switch the transistor 'off'. Again, we could construct an AND-gate using EPR-circuitry (bottom), with receiver idler boxes set on Mars. The system will have two sets of pairs of EPR circuitry to 'transmit' **0** *and* **1** input combinations. Again **1** AND **1** will switch the transistor "on"... all other combinations will output "off". At the base-objective level, only quantum noise is being transmitted: Neither the code-states not the realia of **0** or **1** could be transmitted: Again, no meaningful information could be mediated, given the operation of quantum noise. Yet, again, we have contrived an instantaneous a-causal processes, based purely on quantum noise, which can instantaneously affect transistors on Mars to either switch 'off 'or to switch 'on'.

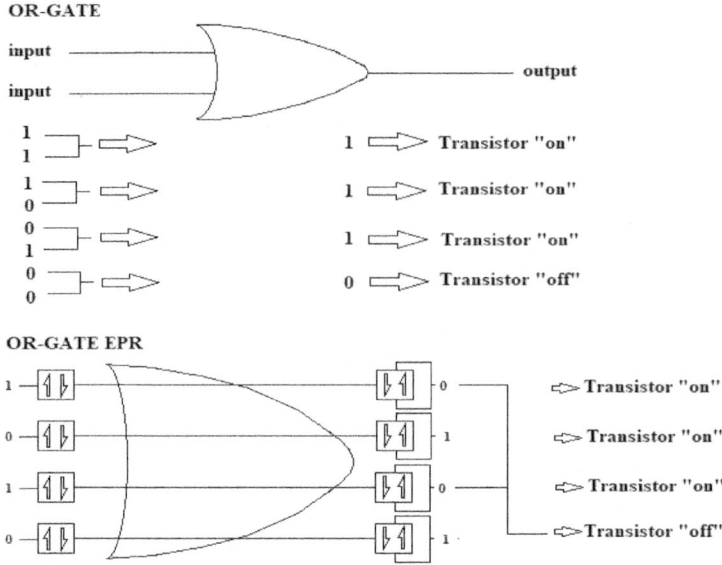

Fig. 1.17: OR-gate and OR-gate with non-discriminating EPR-circuitry
Notes: The depicted OR-gate (top) accepts two inputs. If the inputs are **0** AND **0**, the output will be **0**: the transistor will switch "off". All other input-combinations will output **0**.and will switch 'on' the transistor. Again, we could construct an OR-gate using EPR-circuitry, involving the mediation of quantum noise (bottom diagram). We will have two sets of pairs of EPR circuitry to 'transmit' **0** *and* **1** input combinations. Again, **0** AND **0** will switch the transistor on Mars "off" … all other combinations will switch it "on". As before, at the base-objective level, the realia of **0** or **1** *cannot* be transmitted: Again, all we have achieved is the 'transmission' of quantum noise: No meaningful information is mediated. Yet, once again, we have contrived a device for the instantaneous 'transmission' of control signals to direct transistor circuitry on Mars.

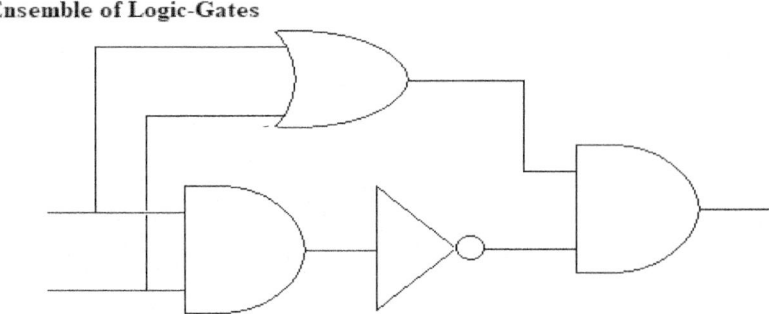

Fig.1.18 An ensemble of logic gates
Notes: Typically, different logic gates are put together into ensembles that can perform more complex conditional protocols or programs. The above depicts a non-EPR ensemble. An EPR-based quantum noise-mediating ensemble would be comprised of non-discriminatory EPR-circuitry, with non-discriminatory receiver-end idler boxes, which switch complex sets of transistors on Mars on or off; all via instantaneous 'spooky action at a distance'. Indeed, the components need not coincide in putative space and could be scattered lightyears apart, yet still direct transistor outcomes on Mars, at least in principle. No doubt, the sort of simple AI-device depicted in **Fig. 1.11**, or even an EPR quantum noise mediated chess machine, could operate between Earth and Mars instantaneously via the mediation of pure quantum noise, void of any intrinsic information, *and* pass the much-vaunted Turing Test, and even beat Kasparov at chess. Yet, the device would be as dumb as a doornail, given that it is comprised of pure quantum noise, albeit with a touch of superlative rigging, co-joined to the fallacy of code-information confusion.

In this way, we obtain a NOT-gate driven purely by EPR 'spooky action at a distance': mediated by pure quantum noise, *without* **any orchestration of spin-outcomes**. Again, at the base-objective level, the realia of **0** or **1** will *not* be transmitted: All we can ever obtain is the 'transmission' of quantum noise: no meaningful information. Yet, we have contrived and foisted upon an instantaneous a-causal processes the associated code-states of "**1**" and "**0**", with real consequences that must either switch off the transistor or switch on the same, by means of 'action at a distance' and in an orchestrated way, without obviating quantum noise, but involving the transmission of *nothing but* quantum noise.

However, so long as we are clear about the code-information distinction *and* can reign in our disposition to the usual naivete toward the insidious and notorious code-information confusion, we will not be fooled into thinking or assuming that *any* real transmission of information has transpired.

In **Fig.1.16** above we apply the same arrangement to the AND-gate. In **Fig.1.17** we apply the same treatment to the OR-gate. Both have two pairs of EPR-circuitry instead of just the one pair exhibited in the NOT-gate in **Fig. 1.15**. The AND-gate accepts two inputs. If the inputs are **1** AND **1**, the output will be **1** (switch transistor to 'on'). If the inputs are **0** AND **0**, the output will be **0**: (switch transistor to 'off'). All other input-combinations will output **0**. (For clarification, see attached truth table in **Fig. 1.16)**.

In the case of the OR-gate depicted in **Fig.1.17**, if the two inputs are **0** AND **0**, the output will be **0**: (transistor 'off'). All other combinations of inputs will generate output-1 ('on'), as summarised in the truth table in **Fig.1.17**,

The EPR-circuit versions for both the AND-gate and OR-gate are also portrayed in **Fig.1.16** and **1.17**, albeit using two pairs of EPR-circuitry. These also produce the expected outcomes, but involve instantaneous spooky action at a distance, through the mediation of pure quantum noise. **Again, at the base-objective level, in the operation of AND-gates and OR-gates mediated by EPR-circuitry, all that is transacted consists in quantum indeterminate noise: no meaningful information is being transmitted**.

Now, we could arrange two NOT-gates to feed **1** or **0** inputs respectively into either an AND-gate or some other gate designed to accept two inputs. Indeed, we could create elaborate combinations and ensembles of logic gates to execute extremely complex protocols via EPR-circuitry: all mediated by spooky instantaneous transactions that consist of nothing but quantum indeterminate noise (see **Fig. 1.18** for a simple depiction of an ensemble of logic gates, but with EPR-circuitry) Yet, assuming that such non-discriminating EPR-circuitry could work, the trick of transmitting ideas through quantum noise would also 'work' because information is not intrinsic to the process, but foisted and rigged upon the improper terms of the system, by the end-users at both the 'transmission' and 'reception' frames: **We have imposed and foisted improper forms on a noise-mediated process. The realia are** *not* **intrinsic. This fact is rendered salient by the truth that, what** *is* **intrinsic to the system is nothing but quantum noise: no meaningful information**.

Again, we could scatter and cloud the EPR quantum noise-mediated logic gates that make up the total ensemble across many lightyears. The scheme will yet 'work': Distances do not matter where 'spooky action at a distance' and quantum noise is concerned. And, **again, there will be no requirement to orchestrate and obviate quantum indeterminism. Indeed, we rely on the impossibility of such a feat in order to assert that all that can be transacted at the base-objective level is nothing but quantum noise: no intrinsic meaningful information**.

1.50: Superlative Rigging with Logic Gates: Passing the Turing Test

Let us return to the limited purview AI-system devoted to whole number integers and infinity. This was described in **1.46** and illustrated in **Fig.1.11**. Recall that such a system is expected to pass the Turing test. Recall that, at the input-side, we present all possible finite number English language same-questions that might be posed about whole integers and their range. Recall that, on the output side, we placed all possible finite addresses to the questions posed on the input side: all variations of same-answer English language possibilities: all correctly stating that "there are infinitely many whole number integers". Recall that, at the central mediating area, we placed the requisite protocols for generating the outputs, combined to a pseudo-random processor that, using the time of the day as an unpredictable variable, selected the output in a seemingly 'creative' way.

We will now reconstitute this system using non-discriminatory EPR-circuitry described in 1.48; again, cast across the Earth and Mars framework. We place the input-side question-set and the end-user on Earth., The Earth side is integrated to the Mars-side via quantum noise mediating EPR-circuitry. We place the core processor and the data store of 'answers', and the pseudo-random selector, on Mars. We add a second set of EPR-circuitry to mediate the 'answer' so-processed back to Earth.

The Earth-side poses one of the appropriate questions from the list of such: e.g. "How many whole number integers are there in the universe of numbers?" This AND-to-OR resolves quantum spin inside a pertinent Earth-side idler box. This instantaneously triggers the appropriate EPR-circuitry on the Mars-side, which in turn triggers the pseudo-random selection for one of the appropriate 'answers'. By means of similar EPR-circuitry, but entangled from Mars to Earth, and via pure quantum noise, the selected answer is 'transmit' to Earth and received by the end-user on Earth. It reads… "There are an infinite number of whole integers". **Clearly, this superlatively rigged system will predictably pass the Turing test, but with no possibility of any** *real* **transaction of meaningful information: only quantum noise**.

What would happen if we replaced this system with a more elaborate one that directly utilises binary "1s" and "0s" 'transmitted' via EPR-mediated logic-gates? These would select and 'transmit' either "1s" or "0s", in the fashion explored in **1.49**. While this would be closer to what actual computers do, it would not make the slightest difference: The system would again pass the Turing test, but based on a different layering of rigging. Again, at the base-objective level, no information can be mediated: *only* quantum indeterminate noise… further exacerbated by the various levels and layers of rigging of improper forms foisted upon nothing but quantum noise; rendered superlative by the incorporation of remarkable quantum noise EPR-circuitry and spooky action at a distance.

*

What if we applied the same EPR-circuitry and noise-mediated spooky action process to Searle's Chinese Room experiment? Perhaps we might place Searle and the Chinese room on Mars, and get him to process the syntactic instructions 'sent' to him from Earth by means of EPR-circuitry and the instantaneous transaction of objectively meaningless quantum noise. We could even replace Searle with EPR-circuit and quantum noise-mediated ensembles of logic gates and transistors embodying the syntactic rules of Chinese: That is, we finally replace Searle with a computational device, but one constructed out of EPR-circuitry and the transaction of mere quantum noise. Moreover, we could scatter the pertinent components lightyears apart.

Again, at the base-objective level, only quantum indeterminate noise will transpire, exacerbated by the various levels of rigging, but rendered superlative by the incorporation of 'spooky action at a distance' and the mediation of meaningless quantum noise.

*

What if we replayed Kasparov versus Blue, but this time, we placed Kasparov on Mars; recuperated Blue into EPR-circuitry, and then scattered its pertinent components across lightyears; all mediated by non-discriminatory EPR-circuit logic-gates integrated through spooky action, and the transmission of mere quantum noise, with no objective meaningful information? Again, Blue would be as informationally empty as the systems previously described, even with rigging and the naivete of the code-information confusion. Yet, such a device might yet beat Kasparov again, @prive' itself a superior sociopath thus… and pass the Turing test.

What if we finally succeeded Blue with the God-of Chess machine, but mediated by non-discriminatory EPR-circuitry and quantum noise? Again, it would constitute an informationally void system, given base-objective-level quantum noise, notwithstanding the code-information confusion on the part of Kasparov and the cheering AI-enthusiast audience.

1.51: Implications from Superlative Rigging to Strong-AI

The implications from rigging, especially from superlative rigging, as espoused in **1.46** through to **1.50**, are obvious: **Code does *not* constitute information. Therefore, machine states, grasped in terms of logic gates and transistor arrays, even if these could be refashioned into quantum computational states, are improper to the proper forms peculiar to the realia foisted upon them. Such foisting, in its fallacious code-information confusion form, is especially exposed by our pure quantum noise-based mediation and machination of rigging, wherein the base-objective condition consists in nothing but pure quantum noise… *not* information**. This fact is often obscured by several overlays of rigging that foist ideas and codes upon improper forms per the insidious code-information confusion that then prompt the naive to assign real information to machine-states and their symbol-outputs… akin to an ignorant ancient Egyptian desert-dweller's astonishment that "…the stone speaketh!", because the local priest read out the hieroglyphs etched onto the stone work. If the same hieroglyphs could be animated and rigged using techniques employed in present-day computers, our desert-dweller friend would be even more impressed. The AI-believer is but a new variant of our old desert dweller.

But information is *not* code. This fact is especially salient in the case of realia pertaining to infinities.

The same must apply to any notion that the universe could be embodied in or as a simulation, as espoused by Niklas Boström[55]… or the related dubious notion that the universe *is* a simulation. The realia that constitute the world and universe, when rendered into machine-state form, are transformed into improper forms. The realia, the proper forms, cannot be constituted into improper forms without losing their identity. Simulations are based on trickery, or rigging of realia with improper forms, combined to the end-user's insidious disposition toward the code-information confusion fallacy. In principle, such tricks could be achieved via superlative rigging involving EPR-circuitry, spooky instantaneous action at a distance-driven logic-gate processing (**1.48** and **1.49**) … all achieved through the mediation of pure quantum noise: no information.

This brings us to the even more dubious promise that, through so-called whole brain emulation, we might one day upload ourselves into simulated worlds, perhaps to circumvent the heat death of the universe[56], or else to obtain immortality[57]. Such necronic mechano-technocratic and nihilistic fantasies immediately fail when confronted with the code-information distinction, especially in the form rendered salient by our superlative rigging, based on the utilisation of EPR-circuitry and pure quantum noise. Indeed, how could a downloaded human being, now a mere machine-state composed of improper forms, retain its proper form, *and* grasp non-clockable infinities, or even dictionary definitions, such as the infinitive "to run", or "to be"? How will an uploaded renormalising quantum mechanist, now a simulation, handle infinities and normalisation? (See **1.44** and Preliminary: **0.6**). How could non-clockable processes ever be clocked by any machine, with or without quantum noise-based superlative rigging, much less via logic-gate ensembles?

We come full circle: **If non-clockable processes peculiar to mind, consciousness and thinking cannot be rendered in or as improper form machine-states, or logic gates and codes, nor could these be rendered in the material corpus of the brain, much less be generated by brains and *its* peculiar improper forms.**

---

[55]Bostrom, Nick (2003). "Are You Living in a Computer Simulation?". *Philosophical Quarterly*. 53 (211): 243–255

[56]Freeman J. Dyson, "Time without end: Physics and biology in an open universe," *Reviews of Modern Physics*, Vol. 51, Issue 3 (July 1979), pp. 447-460

[57]Martin GM (1971). "Brief proposal on immortality: an interim solution". *Perspectives in Biology and Medicine*. 14 (2): 339–340

QUANTUM MECHANICS AND MIND

Recall that strong-AI is merely organic-AI writ inorganic. Strong-AI begins with the notion that human beings are nothing but machines (and, by the time of Babbage, nothing but computers), and the reification of this dubious notion, through the intermediary of the code-information confusion and the trickery of rigging, into the notion that a machine could in its turn effectively become a conscious being: a thinking intelligence. Yet, the forms that constitute the states and processes of the brain, whether neural or subcellular, are equivalent to machine-states, logic gates and on-off transistors: Hence, a neuron either fires or it does not fire. At the sub-cellular level, Tubulin either enjoins to one molecular conformity ("on"), or the other: ("off"). A transistor or light bulb either lights up or it does not. **These cannot generate, 'compute', much less 'store' non-clockable realia and meanings, much less transform code into information, given the improper form of the former vis-à-vis the realia, *especially* vis-à-vis the realia of infinity. Indeed, what holds true for machines also holds true for brains...** in the same sense that Ada Lovelace suggested about the Babbage engines, highlighted in the quote from **1.45**: Hence, applying Lovelace's assertion about machines, but to the brain, the brain...

> *... although it is not itself the being that reflects, it must yet be considered as the being which executes the conceptions of intelligence.*

*

Again, if we employed logic gates and attendant transistor ensembles based on our EPR-circuitry involving the instantaneous mediation of pure quantum noise to affect the simple device originally depicted in **Fig. 1.11** (**1.46**) and fully described in **1.50**, the device would pass the much vaunted and over-conflated Turing test. Yet, the device would intermediate nothing but quantum noise, totally inimical to any sort of information, and even inimical to improper-form codes... so demonstrating that an informationally void system could pass the Turing test... so rendering worthless the value of the Turing test, while rendering the trickery of rigging fully salient.

The same conclusion must be reached about the same device based on non-EPR normal circuitry. The device fulfils its trickery purely on the basis of extrinsic impositions of meaning and information by the end-users on pre-rigged improper form machine-states and outputs. The information so imposed per the code-information confusion is no more intrinsic to the device and *its* states than is the supposed 'intelligence' foisted upon it but absent from it. Indeed, the issue is not that the system can pass the Turing test, but that the end-user and the AI-believer *failed* the simpler code-information fallacy test: In effect, like our ignorant friend desert dweller from the previous essay, both assert, 'the stone speaketh!'.

Again, what would happen if we reconstituted Deep Blue, or even our hypothetical God of Chess machine, on the basis of remarkable EPR-circuitry, whose inputs, operations, and outputs involved nothing but the instantaneous transaction of quantum noise? The answer is obvious: the machine might beat Kasparov again... and pass the Turing test thus. Yet, the base-objective level quantum noise, with or without the code-information distinction, is totally inimical to any sort of information, even improper-form codes. Therefore, again, the value of the Turing test, and Blue passing it, is rendered moot and void.

Again, the same conclusion must be reached about the actual Deep Blue device, which is not based on EPR-circuitry or quantum noise: The code-information distinction yet abides vis-à-vis really-existing Deep Blue. Yet, the device can pass the Turing test purely on the basis of extrinsic impositions of meaning and information on a pre-rigged set of states and outputs devoid of those meanings... because code is *not* information. Again, the information so imposed per the code-information confusion is no more intrinsic to the really existing Deep Blue device and *its* states than is the supposed 'intelligence' foisted upon it.

Again, the issue is not that Deep Blue can pass the Turing test, but that both Kasparov and the audience of AI-believers failed theirs: They confused code for information and failed the code-information fallacy test.

Moreover, what conclusions can we draw from Searle's Chinese Room experiment when it is recuperated in terms of EPR-circuitry and the transaction of pure quantum noise. Clearly, we could only confirm Searle's original conclusion, now augmented on the basis of superlative rigging and quantum noise, inimical to any sort of information or realia, even with respect to improper-form code. Again, it makes little difference if the Chinese room experiment is constructed out of non-EPR circuitry absent of quantum noise: The code-information distinction will abide, and, in their failure to pass the more elementary test (namely, the recognition of the code-information distinction) both the end-user and the AI-believer will exemplify the general failure of humankind to come to terms, with complex machines *and* its own identity as more-than-machine.

*

Our superlative rigging renders incontrovertibly salient the code-information distinction and the fallacy of the code-information confusion at the centre of strong-AI. It does so by introducing a system and process that operates purely on quantum noise... to which not even improper-form code could be attributed, never mind information... never mind realia... and even less 'intelligence'. This is aside the preceding non-clockability thesis and the claims from hard problems based on the intuition of infinities and infinitives, both beyond the purview of any finite-state machine or clockable operation.

Superlative rigging on the basis of quantum noise is conclusive about the impossibility of strong-AI, even when we ignore problems of clockability and infinities... notwithstanding the prospective revival of Cartesian dualism per anticipated de-spatialisation, temporisation and the growing block approach, all of which render the notion of AI-organic historically presupposed to AI-inorganic totally void, even when we ignore the demise of materialist ontology historically presupposed to both; a fact and reality we can no longer ignore.

165

1.52: Harmonising Superlative Rigging with Special Relativity:
Return to the Maincone Approach & the 'Pre-Tee-Zero' Contention

Putative devices based on superlative rigging and involving promissory EPR-circuitry entailing noise-mediated instantaneous a-causality, appear to violate Special Relativity (SR) and the speed of light limit. Such putative devices might even appear to violate the prohibition against meaningful information in EPR, but only for those who cannot overcome the code-information confusion, and invariably believe that code (which is *not* information) is being instantaneously transmitted in such systems.

There are two main reasons why the speed of light limit is not violated in our superlative rigging scenarios and thought experiments: First, at the base-objective level, there is only quantum noise. Therefore, there is no transaction of meaningful information, even if the resolution of pertinent spin-correlates and quantum noise occurs instantaneously or at faster than light.

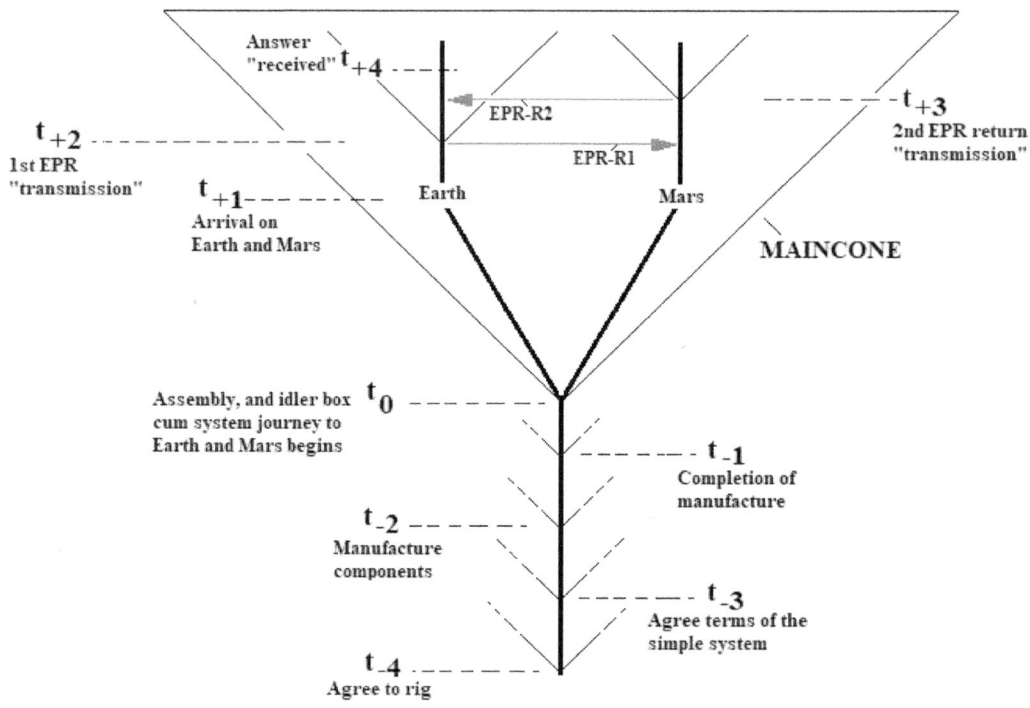

Fig. 1.19: "Pre-Tee-Zero": Maincone solution and SR-harmonisation of EPR-mediated Turing test
Notes: In the didactic illustration above, we follow the path of a principal worldline and the time-marked events along it. At **t-4**, we agree to develop a rigged system and set out our plans. At time **t-3** we agree on the English language terms for the input and output-side of our system (the first layer of rigging) and the meanings (the realia) foisted or imposed on these. We could use Mores code or binary coding, but instead use simple English prompts and outputs as featured in the preceding **Fig. 1.11**. Our system will involve superlative rigging with EPR-circuitry and the mediation of pure quantum noise. At **t-2**, we manufacture the components, and by **t-1** we manufacture attendant EPR circuitry with idler boxes, forming pertinent EPR input and output arrays. At **t-0** the journey begins: the idler box arrays are sent to Earth and Mars, respectively… following diverging worldlines originating at t_0: (This is also the moment the maincone begins to form up). By **t+1** the idler box arrays arrive at Earth and Mars, respectively. Note that the whole setup resides within the maincone. By **t+2** Earth chooses the question-form it wants to 'transmit' and triggers the appropriate transmitter-end idler box. Note that an embedded lightcone begins to emanate from **t+2**, the first receiver-end idler box on Mars is also instantaneously triggered via EPR-R1. No meaningful information is transacted via EPR-R1: only quantum noise. EPR-R1 triggers the receiver-end idler box on Mars, which pre-selects a same-answer alternative response to the 'incoming' question. By time **t+3**, the selected answer triggers another EPR-circuit which trips the idler box on Earth via EPR-R2 from Mars to Earth. The 'transmission' is instantaneous, but no meaningful information is sent; only quantum noise. This answer selected on Mars is outputted and 'received' on Earth at **t+4**. Note again that the Mars-to-Earth EPR-relations convey no meaningful information, save quantum noise. Even if the process could entail the transaction of meaningful information despite the base-objective pure quantum noise, and despite all the layers of rigging and trickery per the code-information distinction, the information would be subject to decoupling (see Part-IV) and, with it, dissociation from putative radiation, matter and space before or pre-t_0: Hence, "Pre-Tee-Zero".

QUANTUM MECHANICS AND MIND

The second reason why no violation of SR could occur resides in the fact that the foisting and rigging of informationally empty codes upon base-objective quantum noise cannot constitute any basis for meaningful information in its own right: All we can assert are layers of rigging… of improper forms foisted on realia, that are not intrinsic to or present in the system, and even **less transacted by it**. Only if end-users inadvertently commit the fallacy of the code-information confusion will they presume a violation of the speed of light limit, or arrive at the erroneous conclusion that meaningful information has been transacted via EPR instantaneous action, despite the saliency of quantum noise at the base-objective level.

A third reason why the violation of the speed of light limit is not committed by any of our thought experiments can be clarified via the maincone approach, in a form similar to the maincone approach in decoupling theory espoused in 1.41: See the fourth scenario therein. Indeed, the maincone approach independently clarifies and augments the nature of superlative rigging, and thus comprises the last item in superlative rigging theory: i.e., the 'pre-tee-zero' contention

The pertinent form of the maincone approach is illustrated in **Fig.1.19**. Therein, we follow the path of the principal worldline as it pertains to the quantum noise-driven rigged system from **Fig1.11**. We number the events along that worldline all the way to time **t 0**. At **t -4**, the experimenters agree to develop the rigged system and set out their plans. At time **t -3** the parties agree on the use of English language terms, both in the input-question and output-answer side of their system (the first layer of rigging) and the meanings (the realia) that will be foisted on these, but fully appreciating the code-information distinction and openly declaring that the realia, the meanings, are *not* intrinsic to the utilised codes. Of course, the experimenters could apply another layer of rigging, wherein the English language is itself coded via Morse code or via binary coding using **1** and **0**. But it would not make the slightest difference: the code-information distinction abides, only complicated by another layer of rigging and foisting per Morse-code or binary coding.

To implement superlative rigging involving instantaneous action via quantum noise, the designers must implement appropriate logic gates and transistor arrays to facilitate the experiment: In superlative rigging, these will involve EPR-circuitry with non-discriminatory receiver-end idler box termini. Thus, at **t -2**, the experimenters manufacture the components of their *ansible* and, by time **t-1** complete and assemble the components into the required input and output arrays.

Finally, at **t 0**, the journey begins: The appropriate space vehicles haul the transmission-side and reception-side idler boxes and arrays to Earth and Mars, respectively: This is indicated by the diverging pair of worldlines in **Fig.1.19**, at **t 0**, which indicate travel at less than the speed of light.

Notice that the maincone also begins at **t-0**, although its informational content was incorporated to it by all the preceding lightcones pertinent to **t-4** to **t-1**: These lie along the principal worldline and are shown only partially using hatched inscriptions.

By **t +1** the equipment and attached idler boxes will have arrived at their respective stations on Earth and Mars. Note that the whole setup resides within the maincone originating in **t-0**.

By **t+2** personnel on Earth will have decided and selected the question-form they want to input and 'transmit'. They will have triggered the appropriate Earth-end idler 'transmission' box. Note that an embedded lightcone begins to emanate at **t +2** from the Earth frame. The trigger of the Earth-side idler box instantaneously triggers the first non-discriminatory receiver-end idler box on Mars, which also happens at **t +2**: (see grey horizontal line with arrow denoted EPR-Relation1, or **EPR-R1**). But no meaningful information is transacted across this **EPR-R1** as explained from 1.48 to 1.50: only quantum noise.

The trigger of the receiver-end idler box by **EPR-R1** on Mars selects one of the same-answer responses to the 'incoming' question.

By time **t +3**, the selected answer on Mars triggers another EPR-circuit, quantum entangled to a non-discriminatory receiver idler box located on Earth. Thus, another instantaneous trigger transpires, this time from Mars to Earth at **t+3**: (See second grey horizontal line, **EPR-R2**). This outputs the 'answer' on Earth: It is 'received' by personnel on Earth at **t +4**. Note again that the Mars-to-Earth horizontal transaction rendered along **EPR-R2** involves no meaningful information but, at the base-objective level, involves *only* quantum noise.

Even if the process somehow involved the transaction of meaningful information despite the total dominance of base-objective pure quantum noise along **EPR-R1** and **R2**, and despite the several layers of rigging, and the superlative rigging and trickery exploited the insidious fallacy of the code-information confusion, the information would be subject to dissociation per information-radiation dualism (Part-II and III) and attendant decoupling (see Part-IV) and, with it, the dissociation of information from putative radiation, matter and space. This would happen before, or pre-**t -0** (hence "Pre-Tee-Zero": or "pre-time-zero"). Such decoupling would occur per information-matter dualism and the delocalisation or clouding of the dissociated information *throughout* the maincone, as was posited in the case of Mozart-40 'transmitted' at 4.7 times 'faster than light' in the Nimtz experiment and in scenario-4 (see **1.41**).

By Book-II we will revamp this informational decoupling, dissociating, and clouding as a part-consequence of counterfactual causality and grand decoherence; wherein the "pre-**t 0**" past eliminates all non-consistent nested futures within the future-potential maincone: Only those futures consistent with the past can subsequently AND-to-OR resolve into OR-form events. These must necessarily be consistent or conformal with the past. Hence, Gunter Nimtz could only 'receive' Mozart-40 in the future, regardless of the putative 'transmission' of M40 at 4.7 times faster than light, because the future was forced to decohere in favour only of M-40. In a similar vein, in our superlatively rigged system, even assuming the impossible notion of any transaction involving meaningful information, we could only obtain the rigged outcomes despite spooky action at a distance, simply because all possible futures must decohere to consistent-only futures conformal to the rigged terms and outcomes, and the shared code-information fallacy therein.

1.53: Technological Perils… and Possibilities

Is it technically possible to build a real-world working *ansible* that utilises non-discriminatory EPR-circuitry and quantum noise, as depicted in the systems portrayed in **1.48** through to **1.51**? Even with such a technology, we could never realise impossible transaction of real information instantaneously or at faster-than-light: We could only ever 'transmit' meaningless quantum noise. Indeed, we could only ever 'transmit' rigged and prompted informationally empty code-states, *never* realia. Yet, such a feat would contrive a thing as good as 'faster-than-light communication', even though the transaction of meaningful information could not transpire at the base-objective level composed purely of quantum noise.

Recall from **1.48** to **1.51** that the envisioned devices would not require the suppression of quantum indeterminacy or involve any deterministic orchestration of spin-outcomes into 'meaningful information' outputs. The devices would only require that our superlatively rigged and totally informationally empty code-states be instantaneously activated at the receiver-end on Mars, triggered by pure quantum noise, itself devoid of information or even of mere code. To realise such a feat technologically, we would need quantum circuitry that can exhibit and preserve the isolation and AND-form suspension of quantum spin potentials for quantum entangled couples, captured and suspended within transmitter-end and receiver-end idler boxes.

Are such idler boxes technically possible? Nascent quantum radar apparently features the use of such idlers, although the technique is classified[58]. **The device that holds the microwave idler beam featured in nascent experimental quantum radar might be modified into an appropriate form and applied to the construction of requisite idler boxes pertinent to our ansible. This would require that the idler box on the receiver-end both preserve the AND-form spin state of the particle within it *and*, upon the latter's resolution into an OR-form outcome, break the isolation enforced by that idler from inside-out.** That is, the receiver-end idler box must be constructed in such a way that the resolution of the OR-form outcome within it must itself break the isolation enforced, from inside-out, *and* trip the normal circuitry attached to that box to affect whatever informationally empty improper-form code-output rigged upon it by designers, end-users, and otherwise foolish people who confuse code for information.

Can such an idler box be constructed? The answer to this question is beyond the purview of this work. If it is possible to construct such a device, then instantaneous faster-than-light 'communication' of informationally empty states, utilising superlative rigging and quantum noise… and attendant *fools*…would become possible, notwithstanding practical proof of the code-information distinction per base-objective level quantum noise, both of which prohibit *any* possibility of the transaction of intrinsic meaningful or real information, while both permanently invalidate the claims of strong-AI exploitative of the code-information confusion and fallacy.

If receiver-end idler boxes prove to be practically impossible, our though experiments yet furnish the requisite purely didactic apagogic case against the claims of strong artificial intelligence, no less true than what might be obtained if requisite receiver-end idler boxes prove to be practically possible. In short, **it does not matter if idler-boxes cannot be constructed to practically falsify the claims of strong-AI: Our exercise has didactic and apagogic value: good enough as such in the falsification of strong-AI**.

*

However, let us assume for the sake of argument that the construction of the appropriate idler boxes *is* technically possible, and practicable enough for applications beyond the falsification of strong-AI. What would be the possibilities, limits, and perils of such a technology?

First, the limits: Unlike normal means of communication, our ansible would be restricted to finite capacity: As we use up the finite number of quantum entangled idler boxed pairs, the system would run out of capacity. This could be partially compensated for by miniaturisation and small volume idler boxes (micro, or even nano-size) combined to large numbers of such arranged into appropriate arrays, and a permanent industrial base to resupply new idler boxes as we exhaust stocks.

Let us now examine the possibilities. It would be pointless to use our ansible for normal communication across Earth and even to and from high orbit, given that electromagnetic-based cable and satellite-mediated speed of light communication is more than sufficient for most purposes, even for very complex communication. Even so, the ansible could be utilised in critical military and state communications, with assured near-perfect security, given that no physical mediation or transaction is entailed by quantum noise, and the attendant 'signals' could neither be intercepted nor tampered with.

One legitimate area of use might be in the command and control of nuclear intercontinental ballistic missile submarines: overcoming the prospect of unauthorised launch. In principle, via remote control by ansible, even the launch process could be taken away from the submarine officers and placed directly in command centres on land. In principle, one could also envision fully remote-controlled crewless missile submarines. Therein lies the peril of such a technology: namely, a new generation of non-jammable warfare drones upon and under the sea, on land, in the air, if not in space.

The obvious peril is that, with effective ansibles, states and militaries might be tempted to remove direct human presence from the battlespace and engage in wars of indulgence: A similar problem occurs with respect to present-day military drones, but the tendency is much less than it would be with non-jammable ansibles. Lulled into a false sense of immunity and invulnerability, the technology might exacerbate warfare; forcing equally capable adversaries to develop similar means, with similar illusions in an accelerated arms race. The technology would likely force weaker adversaries into even more desperate forms of asymmetric warfare, as those with the technology might become even more aggressive and reckless vis-à-vis captive countries and peoples… who might possibly resort to

---

[58] Pirandola, S; Bardhan, B. R.; Gehring, T.; Weedbrook, C.; Lloyd, S. (2018). "Advances in photonic quantum sensing". *Nature Photonics*. 12 (12): 724–733

the use of 'counter-coward' weapons of mass destruction to 'balance' the score. That is, the presumed 'invulnerability' and 'immunity' that might arise from the technology would likely produce a diametrically opposite result.

Unfortunately, existing civil and military authorities and mentalities cannot be trusted. Associated host-populations have proven to be equally insane, if not totally oblivious to even present-day relatively mundane forms of similar dangers.

However, in a saner world, the most legitimate application of the ansible could be in deep-sea mining and especially in space mining, industrialisation, and exploration. Both deep-sea and space mining present extreme difficulties and hazards, if direct human presence is required in either. In deep-sea mining, one could forgo the complexities and costs associated with heavy-duty control cables required in remote mining by usual speed-of-light means of communication. The ansible would obviate the need for direct human presence in deep-sea mining, reducing both complexity and expense required in life-support, hazard pay, inevitable medical complications and expenses, and casualties.

All of the stated benefits would especially apply to space mining. For space mining, it would be ideal if both the equipment and the ansibles were manufactured *in situ* and in proximity to space resources, reducing the need or frequency of costly Earth-to-space launches and attendant pollution, save in the initial investment required to deploy the starting infrastructure, mining, and manufacturing.

With the long-term transfer of most mining and manufacturing to space itself, this would open the way, not only to space exploration beyond the present-day limits and costs, but might segue into the construction of O'Neill structures by remote means via ansible, culminating into the colonisation of the solar system[59], *assuming* such structures prove to be habitable or worth living in. Kardashev-1 civilisation would be at hand.

1.54: Noetic Information & Dualism from the Critique of Strong-AI

Returning to the principal theme, noetic information is distinct from physical information. Rigging, especially superlative rigging via quantum noise in critique of strong-AI in **1.48** to **1.51**, constitutes the necessary milieu for establishing the distinction without ambiguity.

Per generic object-subject relations, consciousness attends a-noetic events that are 'out there'. Thus. **when we experience the events of the world, such as... a man running down the road... and seek to encapsulate the events of that world in the definitions of language, such as via the infinitive, 'to run', this entails two components: a component that refers to the world 'out there' or the event or the object *in itself* ... versus the hard-problem component that supersedes that world, even while it is enmeshed with it, rendered salient by the infinity implied by 'to run'. Thus, on the one hand, we have the finite events and 'objects' of the world (the running man, 'out there'), and, post facto, the retrospective cognition of the 'running man': a cognition that involves infinities that supersede the original finite world and 'object'**, if only because the world 'out there' always manifests upon our awareness and consciousness in retrospect, and always as a finite state. We never intuit infinity from the finite sensate-empirical world 'out there', or from the 'running man'. Our intuition of infinity, inherent to the infinitive, 'to run'... and inherent to our cognition about the world 'out there'... requires an extra-ontology enmeshed with but exceeding and superseding the finite world. It follows that the subject is not reducible, identifiable or accountable with or as the object, and subject-object relations are ontologically real, *not* illusions of folk psychology, and not collapsible to one or the other term in the object-subject relation in forlorn materialism and idealism.

The problem with materialism is that it seeks to trace the causality and ontology of our cognition of the infinitive 'to run', and of infinities in general, if not the whole of cognition, to the finite world: Succinctly or transitively, to the finite brain... as generated by the hands-on finite structures writ in brains. That is, materialism seeks to reduce the subject to the object-brain. But the finite brain, in its mode as a keyboard complex, though needed as the facility to express mind's declarations and its agency in the world, cannot produce the intuition of infinity, much less the cognition 'to run', much less minded agency or the subject... given the inherent finite nature of the brain. And even less is the finite brain capable of storing its intuitions and cognitions as finite 'code' in the improper forms of in-brain material trace-states, given the infinities inherent to the intuitions and cognitions peculiar to the subject, and their differentiation from the finite structure of putative improper-form in-brain code-states.

Succinctly, the problem of materialism is that it assumes it can constitute the realia of infinity and our intuition of it in the improper forms of the finite brain. It then generalises the same fantastical supposition to the improper forms exhibited in finite machines and computers: That is, it starts with dubious AI-organic, only to degenerate into AI-inorganic. Thus, in a similar context, and without any embarrassment...

A robot is an intelligent system equipped with sensory capabilities, operating in an environment similar to the everyday world inhabited by *human robots*.[60]
(my italics)

This sort of doubtful claim is compounded by the necessary illusion entailed in efficient language formation and use: one that requires an almost inbuilt predilection to confuse code with information, and, with it, espouse the fallacious idea that improper forms (the code

---

[59] O'Neill, Gerard K. (1977). *The High Frontier: Human Colonies in Space*. New York: William Morrow & Company
[60] Hayes, P.J., 'The Frame Problem and Related Problems in Artificial Intelligence', in Artificial and Human Thinking: Proceedings of a Nato Symposium entitled 'Human Thinking: Computer Techniques for its Evaluation', A. Elithor & D. Jones (eds), Amsterdam, Elsevier, 1973

or the word 'apple') could constitute the realia (the real apple), culminating into the absurd apotheosis of strong-AI… notwithstanding inherent hard problems posed by infinities and attendant non-clockability.

Shorter still, materialism assumes that physical information constituting the world and the brain itself, can constitute mental information, or noetic information, without resort to any 'religious' extra-ontology of the sort espoused by Plato and Descartes.

Yet, information that pertains to nous, though certainly enmeshed with the domain of physical information, cannot reduce to physical information, notwithstanding the independent implications from information-matter dualism. Information that pertains to mind requires an abstract extra-ontology… a noosphere[61]**… as demanded by the onticity of world-exceeding and brain-exceeding infinities and attendant Cartesian minds**.

Infinities constitute the clearest hard problem for why we need a noetic extra-ontology. But so does rigging, especially superlative rigging: The improper forms of the brain cannot constitute the realia of the larger world: The realia of apple 'out there' cannot be elicited by any brain-writ neurological improper form, for essentially the same reason that the realia of apple cannot be constituted as the word 'apple', much less constituted as a specific ensemble of logic-gates and transistors whose output is rigged to prompt 'apple'. Thus, the same realia that comprise the larger world cannot be reconstituted and simulated as ensembles of logic-gates, transistors, codes, or simply machine-states in computers, much less as interchangeable analogues in brains. The limitations that hold for brains hold for computers, and vice versa. Hence, AI-inorganic is no more plausible than the original root AI-organic… because the differentiation of noetic information vis-à-vis the physical information that constitutes world, the brain, and even computers… is *real*. **The distinction gains full saliency with superlative rigging via the application of EPR-circuitry and rigged 'spooky action at a distance', and the transaction of pure quantum noise: the ultimate improper form inimical even to the improper forms that comprise codes, logic gates and transistors foisted upon it, and never mind noetic abstract ideas and infinities**.

However, when we seek to intuit mind, consciousness and awareness itself… when we intuit the very agency and power that can apprehend infinities, and can cognitise these into basic definitions, we arrive at 'to be': We arrive at awareness, consciousness and mind *without* infinitely recursive self-reference: We arrive at noetic being that, though it is wholly dependent on time, it yet transcends time… and points to an extra-ontology that is as timeless and as abstract as Plato's Good, or the abstract world of time-asymmetric physical laws (see **1.13**).

1.55: Computers, Brains & the Fallacy of Emergence Theory: Final Thoughts on Noetic Information

Emergence is a recent development within science and in the philosophy of mind. **Emergence theory is the next attraction, given the failures of identity theory, eliminative materialism, and epiphenomenalism**. Moreover, emergence could be used to rationalise away our conclusive findings secured throughout Part-V. Therefore, we must grapple with it.

Originally espoused by Edwin G. Boring in 1933, **identity theory, or type physicalism, asserted that mental events can be reduced and identified with physical events or 'correlates' in brains**. This view collides with our own code-information distinction, wherein the realia pertaining to noetic information are confused with improper forms evinced as codes, symbols and correlated brain structures. The code-information confusion was succinctly exposed by the hard problems that attend the intuition of infinity and non-clockable cognition and consciousness (**1.45**). It was also especially undercut by our superlative rigging theory (**1.47** to **1.50**). Both the code-information confusion *and* superlative rigging are pertinent insofar as what is foisted in AI-inorganic is foisted vis-à-vis the brain, in AI-organic, and in identity theory vis-à-vis the improper form of brain structures and processes. But identity theory was also undermined by both memory and mind-side investigations furnished in the Preliminary (Part-IV) and by the precarity of materialism vis-à-vis generic physics, and the demise of the false equivalence of naturalism with materialism (Preliminary Part-III), notwithstanding distinct conclusions from Alt-R born out of anticipated de-spatialisation, temporisation and the growing block universe from Book-II onwards.

Espoused by Paul and Patricia Churchland, **eliminative materialism asserted that presumed mental states do not exist at all. Their approach sought to eliminate these through the development of a language of description of the brain that would completely circumvent resort to the terms of 'folk psychology'**: terms that espouse presumably non-existent states as consciousness, intentionality, will, etc. Therein, the brain was asserted to constitute a pure physics-milieu…a fully deterministic state (closed causality and closed ontology) and, as such, the *elimination* of the possibility of intentionality and nous.

Of course, both identity theory and eliminative materialism presupposed materialism. Yet, root materialism is dubious at best (see Preliminary Part-III: **0.32** to **0.36**… and the whole of Preliminary Part-IV; notwithstanding the case for gearless AND-to-OR collapse and time; or implications from de-spatialisation and the attendant demise of both IOO-schema causality and contiguous causality; or the reality of work function zero processes in nature and the anticipated abstraction and pleonastics of energy concepts from the ontology of physics… all to be realised in the intermediate model of spacetime and its de-spatialised successor. Of course, eliminative materialism cannot accommodate hard problems that arise from the intuition of infinities and attendant non-clockability, the basis of which must supersede the finite brain and its clockable processes, beyond the purview of any code-processing computational device.

A priori materialism also undergirds epiphenomenalist theories, such as those advanced by Keith Campbell in the 1970s, although originally conceived by Thomas Huxley (1874). Essentially, **epiphenomenalism supposes that the mind is to the brain what**

---

[61] We take the restricted quasi-materialistic view of Vladimir Vernadsky, but even go beyond Teilhard de Chardin: In our use, the *noosphere* constitutes the collective sphere of nous extant and parallel the growing block system of a-noetic memory, possibility and time.

magnetism is to electricity: two aspects of the same thing. Hence, the brain supposedly generates the mind in the same way that an electric current generates a magnetic field. Yet, upon closer inspection, the mind thus generated is causally neutered, unlike magnetic fields vis-à-vis electricity, or electricity versus magnetic fields, which tend to mutual causal efficacy: Nous is at best causally passive with respect to the brain which supposedly generates it , while it cannot in turn influence or control its brain, but its brain can influence and control it.

Why is mind neutered in epiphenomenalism? This is for the same reason that it is neutered in all materialist philosophy of mind: Therein, the possibility of mind-brain causality would supposedly violate principles of conservation. Again, materialisms, and the confusion of much-abused conservation principles with both closed causality and closed ontology and IOO-schema impact causality, is presupposed in epiphenomenalism as much as it is in identity and eliminative theories, all without regard or awareness of work function zero processes (see 0.32 and 0.33). For example, time, grasped as the process of AND-to-OR wavefunction collapse turns out to be a gear-less process that proceeds on a work function zero basis: i.e., time does not require an input of energy or power to make it happen, much less mediation by a 'stream of particles' that must 'collide' with time to 'push it over' to a successor-state. Moreover, insofar as putative energy concepts and transactions presuppose ontologically primary time, without which these could not become apparently evident, and insofar as AND-to-OR collapse and time are gearless work function zero processes ontologically primary to putative energy relations, state-change can transpire with work function zero time and *without* the requirement of energy concepts. Hence, energy concepts can be abstracted out of physical ontology and declared pleonastic... in the same way that Ernst Mach[62] sought to abstract 'force' out of Newtonian equations and asserted force as *pleonastic*, but far more profound in implication. Thus, in the context of gear-less work function zero time, both energy concepts and notions of causality that presuppose the mediation and transfer of energy or work turn out to be *pleonastic*, and ultimately not part of physical ontology (see Book-II 2.39).

If energy concepts and associated work functions in contemporary forms of conservation principles are not real aspects of nature, and if time itself, which mediates possibility into actuality and state-change via AND-to-OR collapse is a work function zero process, then so is the causal process pertinent to mind's causal control of its brain. These insights make nonsense out of the much vaunted abuse of the conservation principle, and render mute the claims of epiphenomenalism and other materialist theories, given that *all* presuppose and abuse the said conservation principle. In any case, insofar as epiphenomenalism seeks to causally neuter mind in the want to save conservation principles, it constitutes effective eliminative theory disguised as something other than.

∗

Now we come to emergence theory, or supervenience theory, espoused by Jeffrey Goldstein in at least its physics form: also undergirded by the presumption of materialism and attendant dubious notions of closed ontology, closed causality IOO-schemas, and contiguous causality. No extra-ontology above 'matter in space' is permitted.

In emergence theory, per complexity, pattern is said to arise from the self-organising tendency of matter. In some sense we come full circle, back to mereological nihilism (see 1.03): A pattern etched on sand does not and need to arise from the sand grains on a ground-up basis: The pattern is not a given of its constituent parts. Yet, the pattern exists as a supervenience. Similarly, mind and consciousness can associate vis-à-vis the self-organising brain, just as the sand-pattern could relate to sand...as part of a putative evolutionary process that supposedly leads to noogenesis: the evolutionary emergence of mind.

Once could presumably espouse the same in the context of AI-inorganic under superlative rigging: the notion that intelligence could emerge out of even quantum noise: See (1.47 to 1.50). Indeed, the closest that emergence theory has come to such a dubious proposal is found in so-called *Boids*: swarms of artificial lifeforms constituting a 'swarm intelligence'. The idea was developed by Craig Reynolds in 1986. One could argue that mind is an emergent supervenient 'swarm intelligence'... an emergent state of the swarm brain, perhaps circumventing the code-information distinction, or the fact that the structures of the brain are improper to the forms of realia, such as non-clockable infinities inherent to and definitive of noetic information. Once could also transplant such ideas to AI-inorganic in general, in the context of computers that supposedly evince 'machine intelligence' as an emergent property. All of this can easily be subsumed to the dubious notion that the universe (and the mind itself) could be simulated or uploaded into a computational device, or that such things could *emerge* from computers, per the latter's spontaneous self-organisation... as *supervenience*.

Again, the core fallacy throughout emergence theory is the presumption of the veracity of materialism and the supposition that materialism is the truth about ontology, nature and existence.

∗

Why is emergence theory erroneous? Recall the quantum mechanical case against mereological nihilism from 1.03 and in the Preliminary essays 0.26 to 0.27, which espoused the case for 'instantaneous wholes'. Recall from 1.03 the relation between the abstract complex quantum mechanical wave or wavefunction, which constitutes an abstract instantaneous whole, versus the quantum indeterminate terms (the parts, the 'material' constituents) that the complex wavefunction funnels into a global pattern. We saw the operation of instantaneous wholes in the formation of interference patterns in two-slit experiments, and especially in the image formed out of quantum indeterminate photon-depositions in photography. Therein, we discovered that abstract wholes, as pattern-potentialities, *precede* the parts, the quantum depositions and indeterminate processes and their organisation: The whole precedes the parts as a set of coherent future-potentialities, or, with the advent of anticipated abstract memory, as abstract memories that work as instantaneous wholes vis-à-vis the quantum indeterminate partology.

---

[62]Ernst Mach, Mechanics in their Development, Presented Historically and Critically, Akademie-Verlag, Berlin, 1988, section 2.7

171

Now, this does not imply that the wavefunction, as an instantaneous whole, constitutes a state in absolute time: Given its AND-form state and its status as future-potentiality, it does not constitute a set of realised events rendered into a common present moment. Indeed, the events that subsequently decay out of it are interrelated by the relativity of simultaneity, without absolute time.

The point is that the whole is *not* a supervenience that arises from the parts: The photographic image, or the interference pattern evinced on the two-slit detector screen, do not constitute a supervenience 'emerging' out of random quantum depositions or 'parts'. On the contrary: the parts are funnelled and organised by the instantaneous whole constituted as a complex wavefunction. Emergence theory is fallacious because it seeks to invert how things actually happen in nature. Simply put, emergence theory is science fiction, as is the whole gamut of mereological nihilism.

Faced with the primacy of instantaneous wholes vis-à-vis parts, and the demise of mereological nihilism that accompanies it, at the very least, one must concede to an Aristotelian vision of abstract form, wherein wholes precede and organise 'substance', but without separation from substance. Although it is ultimately Plato that turns out to be vindicated, given the full dissociation and decoupling of the abstract whole from its totality of parts: (See Book-II Part-II to III).

As was argued in the Preliminary (see **0.22**) and as will be argued in detail in Book-III, quantum indeterminate terms are inexorably funnelled to their fated global outcome by the very probability structure of the abstract complex quantum mechanical wave and wavefunction, which functions as an instantaneous whole. Therein, it is obvious that the abstract whole, the complex wavefunction, *does not* emerge from its parts through 'self-organising'. **In its future-form, the whole *absolutely* precedes its AND-to-OR collapse into materialised pattern... simply because the growing block model demands that the future always precedes the partology of events that it funnels into global coherent *en masse* outcomes. Thus, the parts are finally globally fated into wholes by the ontically primary abstract instantaneous wholes, not the other way round. In short, again, emergence theory is back-to-front, and erroneous**.

Indeed, if we could envision a Manyworlds scenario wherein the universe split into as many possible alternative histories comprised of the putative 'paths' and deposition-points for all the photons involved in photography, each apparently distinct history and distinct universe must globally culminate into the same final photographic image, obviating the usefulness of the much-vaunted Manyworlds interpretation, but also putting an end to dubious emergence theory as it pertains to pattern and structure in nature. This must equally apply to the dubious notion that intelligence emerges out of Boids 'swarms'. It especially casts permanent doubt on the notion that mind, consciousness and cognition emerge as supervenient from the constituent activity of brains. Indeed, what controls and coheres brain activity turns out to be the abstract instantaneous whole imposed on the brain, either out of future potentiality or from memory, by extant Cartesian mind: That is, mind selects which abstract complex wavefunction or 'instantaneous wholes' get to develop the activity of the whole brain. The mind accomplishes its selection of the complex wavefunction by means of a work function zero process: a process to be fully explained by Book-V.

*

A qualitative variant of emergence theory, part of *non-reductive physicalism*, was also espoused by John Searle, in his attempt to grapple with how intentionality and mind (noetic information, as we have termed these) could *emerge* from the obvious improper-form 'mush' that constitutes brains. Our clear contention is that it cannot, and that wholes do not emerge from parts in any case, as was just amplified, and as will be justified fully in successive books that comprised this work.

Searle termed his view **biological naturalism***:* Therein, in his book, *The Rediscovery of Mind*, Searle claimed that the mind *is* produced by the brain by some natural complex process that we have yet to fathom, and that... **"consciousness is just an ordinary biological feature of the world"** ...even if what it thus produces is non-reducible to brains. A useful analogy here is between what water qualitatively is at our level versus what water is at the molecular and atomic level. **The familiar qualities of water at the higher level do not appear to be available as properties at the lower molecular-atomic level. In a similar vein, according to Searle, the mind is to the brain what the qualitative state of water is at its higher level vis-à-vis its lower molecular-atomic level, *and* non-reducible to that level, but presumably produced by that lower level.** This supposedly forgoes Cartesian substance dualism, or the idea that there is a matter-substance versus a distinct mind-substance.

Searle's approach is often charged with property dualism: the notion that there is only one matter, but that matter can have different non-reductive properties at different levels.

*

The core premise in Searle, and in the debates that surrounds it, is historically rooted in the differentiation of primary qualities from secondary qualities that came into its institutionalised form in the ontology and epistemology of John Locke: articulated in Book-II of *An Essay Concerning the Human Understanding* (1689-90). Therein, galvanised by the mathematical physics of Isaac Newton, Locke asserted that primary qualities composed of extension, quantity and number, must be differentiated from secondary qualities, such as colour, texture, flavour etc. **According to Locke, secondary qualities are not parts of real existence: these are imposed by the mind on a world ultimately comprised only of primary qualities: the quantifiables. However, if the mind must itself be a product ultimately of primary qualities, it ought to be impossible for primary quality brains to produce non-real secondary qualities, much less get the brain to impose these on sensory experience.**

One can see within this tiresome conundrum all the different paths that led to later-day identity theory, eliminative materialism, epiphenomenalism and, of course emergence theory, as well as to Searle's biological naturalism. These are not merely reactions against Cartesian dualism (which turns out to be true, in any case) but are attempts or reactions to clear up the mess created (or institutionalised)

172

by John Locke *et al*. The mess Locke helped created was later entrenched by David Hume and retrenched by Immanuel Kant, per his 'categories of the mind' imposed on experience… upon a world devoid of said categories.

However, as we discovered in Preliminary essay **0.35, empirically, the differentiation of primary and secondary qualities, and the assertion that the latter do not constitute real ontology or existence, is in error**. It is based purely on assumption; one empirically unjustifiable and historically falsifiable.

Historically and empirically, the quantities of physics were inferred either as countable qualities, such as the spectrum of countable colours evinced by the refraction of white light by a prism, or as quantitative relations between qualitative states or qualitative differences, such as exemplified in the relations between lines in absorption spectra from which one can infer the mathematical model of the electronic configuration of the atom. In summary, empirically, the universe is comprised of either countable qualities or their qualitative differences in interrelation. We always start from the qualities and their differences and, from these, infer the quantities… upon which we then constitute a mathematical physics. It is *never* the case that we infer the **qualities by starting from the quantities: This is simply historically and experientially impossible**: and the admission to the truism falsifies the notion often touted that 'only frequencies are real, but the colours are subjectively imposed' (the ghost of Locke, again).

No. The colours are real. From the countability and quantifiable relations embedded in and between qualities, we then infer 'frequency' or 'wavelength' measures: The quantifiables (the frequency) so inferred are *also* real, but *only* as *embedded* ontics subsumed to a larger ontologically primarily qualitative milieu.

Historically, physics began with the qualities and extrapolated the countable and quantifiable relations of and between these, even if **the pragmatics of physics required that, once it had grasped the inferred embedded quantities, it relegated the qualities to the background. Yet, such pragmatics is** *not* **a sound basis for assertions about ontology, and cannot be abused to obviate the historical and empirical roots of mathematical physics embedded in and subsumed to what is primarily a qualitative universe: one that ontically exceeds and supersedes the quantifiable universe of mathematical physics in its scope and being.**

*

The universe and existence remain primarily qualitative, *not* **quantitative. Quantifiables certainly exist: They** *are* **parts of real existence. Yet, quantifiables at the core of mathematical physics are real only as embedded and subsumed states of a grander qualitative universe**. And the existential preoccupations of mind, consciousness, cognition, and experience (its *pilgrimage*, so to speak) hence, almost the whole spectrum of noetic experience and existence, involves engagement with ontologically primary qualities over and above embedded, sustained, but superseded quantities.

The qualitative and superlative universe subsumes quantifiables, supports their existentiality, but cannot be reduce to quantities, or be accountable as these… much less arise or 'emerge' from the quantifiables as supervenient to quantities. And herein lies the problem or presumption of John Searle *et al*, implicitly or explicitly steeped as their views are in the historical errors of John Locke *et al*.

In criticism of the biological naturalism or qualitative emergence theory of Searle, the vaunted quantifiable lower level states of the brain are, like the rest of the quantifiable universe, subsumed to an overarching and embedding qualitative ontology that itself transcends, not just the lower level quantifiables of brain and world, but even the basic qualities at intermediate levels via superlatives, and, beyond superlatives, by metalatives that make for experiential-existential breakthroughs, such as the intuition into infinities and "to be"; the exemplars of the simplest of metalatives. **The lower levels of the brain cannot generate qualities, superlatives, much less metalatives. Indeed, the lower-level processes are not required to generate these, given that the lower-levels are subsumed to, organised by and realised, not merely by abstract instantaneous wholes (see photography example above in criticism of forlorn emergence) but by the non-mathematical qualitative, superlative and metalative wholes, and a universe that subsumes the quantifiables** *and* **existentially sustains them as its ontic supplicants.**

1.56: Summary of Part-V

- **The code-information distinction:** The distinction can be traced to the views of Ada Lovelace in the 19th century and is salient in the views of John Searle in the 20th century. Code or symbols are not the proper forms of information that they are purported to 'represent' or elicit per the code-information confusion. (**1.45**).

- **Putative simple AI-systems work purely based on rigging and the fallacy of the code-information confusion (1.46):** Simple artificial intelligence systems exploit the information-code confusion and the naivete of the end-user and AI-believer. In perfectly definable domains, wherein the finite combinatorial possibilities can be calculated without ambiguity (tic-tac-toe and chess exemplify such domains) putative attendant AI can pass the Turing test. Yet, such systems, and computers cannot exhibit or process realia or real information. They can only process codes that are improper in form to the realia. Hence, the code-information distinction. The distinction and the code-information fallacy is rendered undeniable in systems employing rigging (**1.46**) and especially superlative rigging, involving EPR- circuitry and the mediation of pure quantum noise, inimical even to improper-form codes.

- **In superlative rigging**, it is hypothetically possible to construct logic gates and transistor-ensembles, and input and output devices based on these, whose basic causality is mediated by EPR-circuitry involving instantaneous 'action at a distance' and entailing the base-objective mediation of pure quantum indeterminate noise: no information. **What is ultimately mediated in such systems is** *not* **information, and not even improper-form codes, but pure quantum noise devoid of information**

173

(1.47 to 1.49). Yet, putative AI-systems employing superlative rigging and pure quantum noise will pass the Turing test, even though no meaningful information can be transacted via such systems, and wherein the objective transaction involves nothing but pure quantum noise inimical to both realia and code. (1.49 to 1.50).

- **The 'Turing test' should be appropriately renamed the 'rigging test', and its veracity permanently doubted.**

- **Noetic information is distinct from physical information (1.51, 1.54, 1.55):** Noetic information is evinced in the cognition of infinitives and the intuition of infinities (consummate hard problem of consciousness) versus the clockable finite brain and world incapable of generating such cognitions and intuitions without crashing into insurmountable halting problems. The insight into hard problems from the intuition of infinities is augmented by the code-information distinction and divide: The intuition into infinities does not merely defy clockable finite processes in computers and brains, but per the fact that the realia of infinities cannot be captured in or as code, or embedded in a machine, or elicited by the output of mere code or symbol. All of this is rendered salient by the highlighting of the code-information distinction in both rigging *and* superlative rigging via quantum noise.

- **What applies to putative AI applies to the brain: AI-organic is as untenable as AI-inorganic**: The brain is no more capable of eliciting realia or infinities, much less being-in-awareness and consciousness presupposed in the intuition of infinities, given the improper forms of the brain vis-à-vis the realia, especially with respect to infinities. It follows that, like computers, the brain…

  > … *although it is not itself the being that reflects, it must yet be considered as the being which executes the conceptions of intelligence.* (Ada Lovelace)

- **Technological spin-offs and perils from the employment of hypothetical EPR-circuitry might be achievable (1.53)** if receiver-end idler boxes can preserve the AND-form suspension of an entangled particle, and, upon the latter's AND-to-OR spin-resolution, break the enforced isolation from inside-out of the idler box. The attendant 'ansible' technology could revolutionise space industrialisation, exploration, and colonisation, but it could also lead to major instabilities and insecurities by fostering a dangerous revolution in military means.

- **Critique of emergence theory**: This view holds that consciousness emerges as a supervenient state out of lower level brain or, in AI, perhaps from a Boid-swarm. Supposedly, it is all a matter of complexity. In short, emergence theory presumes that wholes emerge out of parts. This is contrary to the evidence of nature: In nature, abstract complex quantum mechanical wavefunctions, as abstract instantaneous wholes, AND-to-OR collapse and funnel otherwise quantum indeterminate 'parts' into materialised wholes. That is, **the whole precedes the parts**, whether these are wholes configured as future-potentials or as abstract memory. The whole organises the parts: The whole is *not* supervenient to the parts. This is obvious in the development of a photographic image, and even in the development of the interference pattern in the two-slit experiment. It is most certainly true across nature. The same must also hold with respect to whole-brain activity and unity: also regulated by abstract instantaneous wholes, imposed on the brain by the Cartesian mind. The abstract instantaneous whole so imposed then funnels the quantum indeterminate constituent partology into coherent global brain activity, *not* the other way round. In short, the 'parts' emerge and are organised by the abstract whole, against the grain of emergence and supervenience.

- **Problems of noetic information from the division of ontology into primary-quantifiable versus secondary-qualitative categories**. Critical noetic information must be comprised of qualitative states… of superlatives (aesthetics and aretaic), and even metalatives (the simplest are infinities and the infinitive "to be", or Being). But qualitative ontology cannot be furnished by a purely quantitative mathematical universe. Insofar as the brain is misconstrued as a pure primary-quantitative Lockean corpus, it cannot generate qualitative sense experience, or even impose these on the world in the fashion entertained by Kant, and even less generate superlatives and metalatives. The problem was first institutionalised in the views of Locke. Yet, the quantifiables that render mathematical physics possible were historically and empirically inferred either as countable qualities or as quantifiable relations that reside between qualities. That is, the quantities of physics are *embedded* in architectonic superseding qualitative states and wholes. Thus, the universe is primarily a qualitative whole, with embedded supplicant quantifiables that render mathematical physics possible. Yet, the larger qualitative universe and whole irreducibly and ontically exceeds and supersedes really-existing embedded quantities and the mathematical physics these sustain. Qualitative experience is thus possible because the universe is primarily qualitative, if not superlative, if not more.

KEY INNOVATIONS EXHIBITED IN BOOK-I

Several innovative ideas and breakthroughs are exhibited in Book-I: These are…

- **The deeper account of information-matter dualism based on generic quantum mechanics**: The central aim of Book-I constituted the case for the dissociation of information from radiation, matter and 'space' per the five principal proofs based principally on generic quantum mechanics.

- **The wavefunction describes the future: The future is phenomenalised as the quantum mechanical wave: The future is ontically real and 'out there'**: The AND-form state of a system constitutes the future-form of that system; hence the *raison d'etre* for why nature exhibits AND-form logic and the superposition principle. Per **the firewall principle**, causality intertwined with the principle of conservation demands the ontic reality of the future and the attendant growing block model of the universe. Otherwise, a block universe in which the future is fully resolved would end in 'signals from the future' and, with it, the violation of causality and principles of conservation of energy-matter.

- **Time is synonymous with AND-to-OR wavefunction collapse.** Time is not a 'fourth dimension': AND-to-OR wavefunction collapse *is* time. This necessitates a division of spacetime into an AND-form domain of future potentiality versus a memory-domain of retained past events: The attendant growing block model will be introduced in Book-II.

- **Entropy of future-potential information and AND-to-OR collapse and time:** The increase in the entropy of future potential information per the AND-to-OR perpetual decay of AND-form future-potentiality, constitutes the primary basis for entropy. Thus, AND-to-OR collapse is ontologically superlative and subsuming of merely statistical generic entropy. The latter permits the dubious notion of the reversibility of the 'arrow of time' that supposedly attends the improbable decrease in generic entropy. Hence, AND-to-OR collapse and time constitutes the proper ontological basis for the irreversible 'arrow of time: The 'arrow of time' *always* proceeds from AND-to-OR, never from OR-to-AND: Hence, irreversibility and the 'arrow of time'... from low to high entropy of future-potential information... and even when occasionally accompanied by low probability decrease in generic statistical entropy... cannot imply 'time reversal', given that such instances cannot reverse AND-to-OR into retrotemporal OR-to-AND, or emerge into evidence through AND-to-OR collapse. Hence, **increase in entropy of future-potential information is the only basis for time-asymmetry and the 'arrow of time'.**

- **The manifested universe decays and originates out of the future: The universe emerges out of the future, *not* from a cosmic event in remote past.**: With or without the Big Bang, consistent with growing block theory, in its penultimate origin, the manifest universe perpetually decays out of AND-form future potentiality. The universe emerged out of future-potential spacetime, which in turn elaborates inexhaustibly to the infinitely removed future. Indeed, origin cannot ultimately reside in a Big Bang creation-event in the remote past, even if the Big Bang transpired. Origin is always out of future potentiality. Put another way, the Big Bang monoblock must also incorporate future potentiality. So incorporated, the new monoblock must be placed, not in the remote past, but constituted as the block of future potentiality out of which the universe...and perhaps the Big bang itself...perpetually decays into issuance.

- **The time-asymmetry of physical laws**: The empirical coming into evidence of physical law is always via AND-to-OR collapse and time, and always accompanied by the tandem irreversible increase in the entropy of future-potential information. It follows that physical laws are time-asymmetric, *not* time-symmetric. The latter is not viable. Why? Physical laws under OR-to-AND time-reversal would 'fall out of evidence', while their effects and our experience of these would 'un-happened', and we would end up in 'un-science'... or *nonsense*. Thus, physical laws do not look the same in any time-direction, but look the way we know them *only* in the AND-to-OR time-direction. Yet, although time-asymmetric in the way just described, physical laws are also ultimately eternal and timeless. i.e., Physical laws participate in time and entropy in an exclusively time-asymmetric way. Yet, physical laws are timeless and extant both AND-to-OR collapse *and* time, its 'arrow', *and* vis-à-vis irreversible entropy of future-potentials. This is inferred from the fact that AND-to-OR collapse and entropy does not change or mute the character of time-asymmetric physical law.

- **Abstract physical memory**: Critical informational states, such as the setting on the barrier in the two-slit experiment, must survive isolation, the implied quantum mechanisation of the barrier into AND-form possibilities, and the subsequent scrambling by quantum indeterminism at the end of isolation. The survival is consistent with information-matter dualism, wherein the critical information pertaining to the barrier effectively dissociates and is retained as abstract memory. As critical information, the barrier setting is not erased, otherwise, per erasure on an all-or-nothing basis, experience, or physics itself, including quantum mechanics, would become untenable. Succinctly, memory is directly relevant to the two-slit experiment and pertains to the survival (in memory) of the barrier setting, which then permits quantum mechanical facts and conclusions to be drawn. In short, **the viability of know quantum mechanics critically requires abstract memory**.

- **Decoupling Theory**: From information-matter dualism emerges the *decoupling* of information and causality from radiation, matter, *and* from the speed of light. Decoupling accounts for the anomalous results of the experiments of Gunter Nimtz, wherein Mozart-40 is apparently quantum tunnelled and 'transmitted' at 4.7 times faster than light. Yet, both decoupling itself and the experimental outcomes from Nimtz can be harmonised with Special Relativity, ultimately through the co-joining of decoupling and information-matter dualism, with anticipated abstract memory co-joined to *a maincone approach*.

- **Noetic information pertaining to mind is distinct from generic physical information constitutive of world, brains and machines: The insight fashions for both mind-brain dualism and the critique of strong-AI.** The character of noetic information is readily discerned from hard problems evinced in the intuition of infinities as well as from the saliency of the code-information distinction (and confusion) in putative 'artificial intelligence'.. in turn highlighted via superlative rigging involving putative AI-systems driven by the mediation of pure quantum noise (via EPR-circuitry)... *without* information. A putative AI-system based on such circuitry cannot be said to process information because it consists in the mediation of pure quantum noise inimical to *any* information... and wherein the claimed 'intelligence' of the AI system (absent from the quantum noise-driven system in as much as information is absent from the same) is a fallacious attribution to that system from the imposition upon it by end-users, per the fallacy of the code-information confusion. Aside from the critique of strong-AI that this furnishes, superlative rigging based on quantum noise might well garner technological applications, perils and advantages, if key technical problems can be overcome.

- **The crisis of materialism and inexorable mind-brain dualism**: Information-matter dualism, decoupling theory, and implied abstract physical memory, are central to the demise of materialism vis-à-vis memory-brain and mind-brain relations. But the demise of materialism also follows from the critique of strong-AI on the basis of superlative rigging, the code-information distinction, and from the distinction and definition of noetic information thus enhanced.

BOOK-II:
ONTOLOGY OF THE QUANTUM MECHANICAL "AND"
AND THE STRUCTURE OF SPACETIME

CONTENTS

Page

BOOK-II
ONTOLOGY OF THE QUANTUM MECHANICAL "AND"
AND THE STRUCTURE OF SPACETIME

AIMS OF BOOK-II

What is the physical meaning of the quantum mechanical wave? Why does nature exhibit AND-form logic in an apparent incompatible way with equally evident classical particle-like OR-form logic? The Preliminary and Book-I introduced the *raison d'etre* of the quantum mechanical wave. Book-II will fully justify its ontology, culminating into a growing block reform of spacetime and the attendant primitive unification of quantum and classical-relativistic physics. Thus, Book-II will constitute the watershed in the Alternative Realist interpretation or thesis: (Alt-R).

Part-I will review the erroneous view of the quantum wave originally espoused by the Astronomer James Jeans. In the Preliminary essays, **0.9** to **0.10**. Jeans applied a navigational dead-reckoning analogy to explain away AND-form logic and to declare the quantum wave illusory. Part-I will reprise our own forecasting analogy involving the forecast of all future possibilities for a ship that has not yet left port. By such means, we will reiterate that **the AND-form quantum mechanical wave is the physical manifestation and phenomenalisation of the ontically real future**. We will also reiterate causality-based and other supplementary arguments, culminating into the recapitulation of the **firewall principle**: The *raison d'etre* of the quantum mechanical wave based on the intertwine of causality and the principle of conservation: the firewall wards against 'signals from the future' and against the breakdown of causality and order.

The quantum wave, grasped as the ontic future, culminates into the reform of spacetime: one that will incorporate the quantum mechanical ontic reality of the future into a growing block spacetime model. From within the growing block **intermediate spacetime model**, the AND-form quantum wave, and its attendant universal wavefunction, will constitute the domain of **spacetime in potentiality: the future-form of spacetime**, configured as an AND-form complex delineated from **the domain of realised spacetime; the mnemonic past**. The domains are separated by **the event horizon: the cut or frontier between the two domains**. Through the progression of the event horizon, we obtain the process of *coming into being...* of OR-form events out of the decay of AND-form spacetime in potentiality, or AND-to-OR wavefunction collapse synonymous with time itself. We will discover that, once formed, the abstract basis of OR-form realised events, whatever this turns out to be, relegate to and are retained within the mnemonic domain of realised spacetime, constituting the growing past retained in an accumulating abstract memory.

Critically, the intermediate model of spacetime will incorporate **time as AND-to-OR collapse;** *not* as a fourth dimension: Thus, the notion of time as a fourth dimension will be obviated. With time grasped as synonymous with AND-to-OR collapse, the **justification of the quantum wave as the ontic future will emerge seamlessly from the structures and processes of the intermediate model of spacetime:** constituting the third and final element of the case for the ontic reality of the future and of the future as the quantum mechanical wave.

The intermediate model of spacetime will realise the primitive unification of classical and quantum mechanical descriptions of nature: one that falls short of, but scaffolds for, prospective quantum gravity. Yet, with anticipated de-spatialisation in Book-IV, the intermediate model will itself give way to a successor spacetime model: spacetime *without* space. In Part-I, the firewall principle, co-joined to the intermediate model spacetime, will integrate **the explanation of Heisenberg's uncertainty from Book-I; directly reframed to and restated in terms of the structures and processes of the intermediate spacetime model**.

Our growing block reform of spacetime will prove mutual with information-matter dualism and decoupling theory from the preceding Book-I. Hence, **the intermediate model spacetime will exhibit information-matter dualism *and* decoupling as inherent to its key structures and processes. Information-matter dualism will prove to be the mirror-image of the tripartite structure of spacetime**.

Part-I will also introduce the key finding that scaffolds toward de-spatialisation: namely, **the spatially static quantum mechanical wave**, which is obvious even from the structure of the intermediate model of spacetime (i.e., Obviously the future, grasped as the domain of spacetime in potentiality, does not itself 'move'. Therefore, quantum waves synonymous with that ontic future cannot 'move' and are consequently spatially static).

In Part-II, **the spatially static quantum wave will emerge from the delayed choice analysis of motion... in the invalidation of motion**. Simply put, quantum waves do not move. Instead, they AND-to-OR decay into successions of OR-form events that we then 'interpret' as evidence for motion, or upon which we foist the *assumption* of motion. It follows that there is no point to the 'space' requisite to such 'motion'. Hence, the intimation of de-spatialisation from spatially static quantum mechanical waves.

Part-II will further develop the **domain of spacetime in potentiality into a complex system of nested futures**, comprised, not just of immanent AND-form futures, but of a more complex wavefunction that nests to each immanent future the unique subsequent futures and *their* wavefunctions. That is, we will arrive at a wavefunction that nests within its future milieu the wavefunctions that belong to subsequent futures. In this way, we will obtain **a superlative futures-within-futures model (or 'wavefunctions within wavefunctions' approach) of the quantum mechanical wave, wavefunction, and spacetime in potentiality**. This nested futures view of the quantum mechanical wave was always implicitly to John Wheeler's delayed choice approach. **We will exploit Wheeler's delayed choice to validate the nested futures view** and generalise it to spacetime in potentiality.

Conclusions obtained from delayed choice and the nested futures view will engender important distinctions between **closed-choice and open-choice quantum mechanical systems**. Closed-choice systems are essentially generic quantum waves that possess only immanent futures. Open-choice states are quantum waves with nested futures. We will also obtain the basis for **wavefunction collapse, *not* as a one-step one-time all-comprehensive collapse, but as a partial and perpetually decaying future spacetime domain**. This partial-perpetual AND-to-OR collapse of spacetime in potentiality was mirrored in the case for 'broken-discontinuous motion' in the Preliminary (**0.3** to **0.4**) requisite to information-matter dualism; in turn synonymous with the marginal broken and discontinuous progressions of the event horizon featured in the intermediate spacetime model, attendant partial-perpetual collapse.

We will also establish the **presage to the reform of the Schrödinger wave-equation in the light of the nested futures view; in view of the spatially static quantum wave; and per the partial-perpetual nature of gear-less AND-to-OR wavefunction collapse**.

Finally, in Part-III, we will use the advances secured in the preceding parts of Book-II to develop **the categories of nested futures, constitutive of spacetime in potentiality**. This will lead to **quantum erasure** in the context of nested futures and non-local counterfactual causality, and consequent **unitary grand decoherence of spacetime in potentiality**: a form of non-local causality superlative to the form exhibited in EPR, generalised to the whole intermediate spacetime in potentiality, in turn treated as an 'instantaneous whole'. These advances will constitute **the revival of Mach's principle re-framed as grand decoherence, foundational to the presage to the growing block quantum mechanical solution to inertia, and to a similar quantum mechanical solution to the permanent magnet**. Beyond these, we will secure foundational solutions to problems attendant zero-point vacuum fluctuations and electron particle-spin.

Basic energy concepts will also be reformed through their reframing to the growing block intermediate model spacetime. Therein, potential energy will become inherent to spacetime in potentiality (the future) and kinetic energy will be shown to be inherent to the transition and succession of the event horizon. Yet, upon the fact that AND-to-OR collapse constitutes a gear-less work function zero process presupposed to and responsible for bringing about putative energy relations, but itself *not* driven by energy or work, must imply that **energy concepts and relations can be abstracted out of the ontology of physics entirely, and declared pleonastic**; augmenting the conclusion that inertia, the behaviour of the permanent magnet, and much else, are indeed ultimately driven by work function zero processes, albeit *without* the violation of the principle of conservation; with similar implications to the affectation of gravity.

Part-III will **revamp the model of spacetime curvatures and gravitational phenomena per the structures and processes of the growing block intermediate model spacetime:** Therein, we will presage how **curvy event horizons can better model gravitational phenomena**. But we will also integrate gravity to the process of grand decoherence: our first presage to prospective quantum gravity in Alt-R. Other significant highlights from Part-III will include **the objective reality of time** and the **critique of the idea that the universe is a computer**, challenged on the basis of the inherent infinities involved in grand decoherence and the insurmountable non-terminating problem that this poses to any iterative mechanical or computational processing of spacetime and AND-to-OR collapse, or time, requisite to the hypothesis of the computational universe.

These discoveries will then support anticipated mind-brain dualism, with attendant Cartesian revival, undermining the materialist contention on the nature of causality, of the physical order, and specifically of the brain. Indeed, gear-less other-than mechanical time, with attendant generalised gear-less other-than mechanical causality, necessarily implies gear-less other-than mechanical Cartesian causality vis-à-vis the brain, transgressing and transcending the forlorn notion of the 'mind machine', or the supposition that thought must be computation.

BOOK-II PART-I:
ON THE ONTOLOGY OF THE QUANTUM MECHANICAL WAVE
AND THE INTERMEDIATE MODEL
QUANTUM-RELATIVISTIC INCORPORATIVE SPACETIME

2.01: Aims of Part-I

The main aim of Part-I is the completion of the answer to why nature displays AND-form logic and quantum mechanical wave-like phenomena. The answer, prefigured in the Preliminary and in Book-I, will advance our understanding of spacetime, transforming it from a generic block theory model devoid of an ontic future to a growing block system that incorporates the future as the quantum mechanical 'AND' domain. This will also bring into unification the quantum mechanical AND-form with classical-relativistic OR-form descriptions of nature: resolving the contradiction said to reside between these. The solution to the physical mystery of AND-form quantum mechanical logic will prove to be perfectly consistent with information-matter dualism: ultimately justifying it as a mirrored effect of the structure of spacetime itself, exceeding the proofs furnished in Book-I.

Our first task will be the full presentation of the view of the quantum wave articulated by James Jeans, which applies a navigational dead-reckoning analogy to debunk the physical reality of the quantum wave. Our counter to Jeans' view, based on navigational future forecasting, will reiterate the *raison d'etre* of the quantum mechanical wave as the physical phenomenalisation of the ontically real future; addressing why nature exhibits AND-form logic. A detailed reiteration of previous material, this will constitute the prelude to deeper insights: It turns out that the ontically real futurity can be seamlessly incorporated into models of spacetime, hitherto depicted *only* in terms of the classical OR-form logic-based block model. The resulting growing block intermediate model spacetime will revolutionises our understanding of spacetime, even though it will constitute but a transitive development that presages the greater revolution furnished in Book-IV, but leading to the primitive unification of AND-form quantum mechanical descriptions of nature with OR-form classical and relativistic descriptions of the same.

By incorporating only OR-form descriptions of spacetime, **generic models of spacetime present models strangely bereft of the future: an erroneous block-theory approach, absent of any open future possibility or potentiality for the genesis of new events. The quantum mechanical future, once incorporated into the structure of spacetime, culminates into the primitive unification of AND-form quantum mechanical descriptions of nature with classical OR-form description of the same, capitulated in an intermediate model spacetime: a growing block model.**

Part-I will argue that the exclusion of quantum mechanical realities from spacetime models constituted an historic error: one that our intermediate model of spacetime will partially correct. Through it, we will instantly grasp that the quantum mechanical and the classical are not mutually contradictory or disharmonious. Other advances will follow on from these: **The causality argument intertwined with the principle of conservation, such as to constitute the firewall principle, will be seamlessly incorporated into and emerge out of the intermediate spacetime model, as will static quantum mechanical waves**... as will the further clarification of Heisenberg's uncertainty... as will revamped information-matter dualism and decoupling theory recapitulated in terms of intermediate spacetime structure.

2.02: The Quantum Mechanical Wave According to James Jeans

The following is the full version of the same analogy first presented in summary-from in Preliminary **0.10**. It was espoused by the astronomer James Jeans: part of his attempt to debunk the physical reality of the quantum mechanical wave. While the attempt was forlorn, it provides a useful springboard toward a valid resolution of the *raison d'etre* of quantum mechanical "AND" and of quantum waves. Thus, according to Jeans...

Imagine a ship crossing from New York to Southampton. The first day out the ship's position would normally be determined by taking the altitude of the sun at noon; the navigation officer would then mark the position on the ship chart.

If the sky was too cloudy for the sun to be seen, it might be necessary to calculate an approximate position by dead reckoning; the officer would know the approximate speed of the ship, or the distance it has travelled through the water as recorded by the log and could make a rough allowance for the motion of superimposed currents in the sea. He might in this way be able to fix his position to within, say 5 miles. He could not mark a cross on his chart to fix his position, but might draw a circle 5 miles in diameter; this like the waves of the undulatory theory would represent his knowledge of his position[63]

By the 'undulatory theory', Jeans was referring to the quantum mechanical wave-model for radiation and matter, which he took to be a consequence of our ignorance and uncertainty pertaining to the position of a particle. Jeans sought to treat the positional information of the particle in a similar way to how one treats an accumulating ship's positional uncertainty under conditions of navigational dead reckoning. Thus...

As the ship progressed on its journey, we can picture this circle travelling over the chart like a wave travelling through space, at a speed representing the speed of the ship. As the new uncertainties accumulated, the circle would continually increase in size. If the sun were still invisible on the next day, it might be necessary to indicate the ship's position by a circle 10 miles in diameter. If the sun could not be seen throughout the voyage, the uncertainty as to the ship's position would continue to increase, until, by the time the ship was close to land, it might have been represented by a circle 50 miles in diameter.

Suppose that when such a circle had been marked on the chart, half of it was found to lie over the Cornish coast. As the ship could not be on land, this half of the circle would at once be ruled out; this bit of knowledge would at once reduce the extent of the uncertainty by half... If the Lizard was sighted a few moments later, the further knowledge thus provided would reduce uncertainty principally to zero, and the ship's position could now be marked by a point.

This analogy explains a more general orientation in other respects. We know how in practical life one uncertainty leads to another; for instance, the uncertainty which prevailed as to the ship's position when it was one day out continually increased; this uncertainty made it impossible to allow exactly for the currents encountered on the second day's run, and as the voyage proceeded uncertainty was piled on uncertainty. The wave-picture of radiation faithfully reproduces this cumulative uncertainty in knowledge, because it is an inherent property of a group of waves always to spread out, and so to occupy more space.

In this analogy the ship represents a photon [i.e. a particle of light radiation], the sea represents the space in which the photon, and the land represents the barriers which prevents the photon moving through the whole space. The sea, land, ship and photons all exist and move in the ordinary space of everyday life: indeed this is what we mean by ordinary space –space in which we

---

[63]James Jeans, Physics and Philosophy, Cambridge: Cambridge University Press, 1942

see things through the impact of photons on our retina, and travel by ship. But the waves which represent the navigator's knowledge of his ship's position do not travel through ordinary space, but over a nautical chart which is a sort of diagrammatical representation of ordinary space.

In precisely the same way, the space traversed by those waves which represent our knowledge of photons is not ordinary space but a mathematical representation of ordinary space…In brief, the space of photons is ordinary physical space, while the space traversed by the waves of the undulatory theory is a conceptual space. Indeed, it must be, since the waves, as we have seen, are mere mental constructs and possess no physical existence.

James Jean's view can be summarised thus: The expanding dead-reckoning circle represents the accumulated uncertainty pertaining to ship's position: i.e., the navigator's informational ignorance of ship's position over time. As such, it does not represent any intrinsic objective uncertainty of the real ship's position. The ship's position is real and certain, and only the navigator's ignorance of an exact determination of position engenders uncertainty. This uncertainty is purely in the mind of the navigator. From the navigator's point of view, we might assert that the ship occupies *all* the possible positions within the navigation circle; and that it represents an AND-form unitary superposition of all these positions within the bounds of that circle.

Assuming the oversimplification of just ten possible positions within the navigation circle, by time t_n, the unitary superposition for the ship may be described as…

Ship (position at t_n): {position-1 **AND** position-2 **AND**…., **AND** position-10:
such that the ship assumes all positions}

Or the same shortened to…

Ship (t_n): {**[AND]** position-1, position-2…., position-10 | **ship assumes all positions**}

However, this does not appear to be true of the ship: All that the navigator needs to do is look outside any porthole and grasp the fact that the ship has only one real position, *not* a unitary "smear" of ten distinct ones. Even so, the navigation circle will not go away; this is because it is not based on the actual position of the ship, but on the inevitable informational ignorance of the navigator about exact ship's position.

The navigator is therefore correct to assume that, despite what his navigation circle states, the position of his ship is not in reality an AND-form superposition of ten positional possibilities by time t_n. Realistically, the position of his ship can only be a collection of mutually exclusive positions, with only one *real* position, in concordance with OR-form logic. Thus, the navigation circle represents…

Ship (position t_n): {position-1 **OR** position-2 **OR**…, **OR** position-10: **such that only one position is real**}

…shortened to…

Ship (t_n): { **[OR]** position-1, position-2,…., position-10 | **only one position is real**}

On this account, the AND-form description of the ship is an illusion: an artefact of method and ignorance: The ship is constituted purely into an OR-form state.

*

Is the quantum mechanical AND-form logic for positional information a product of our ignorance in essentially the same way as found in navigational dead reckoning? This is what Jeans had in mind. Yet, Jeans' view will not work: **Recall from the Preliminary (0.13 to 0.14) that, when a particle of radiation or de Broglie matter is made to pass across a barrier with two slits open, it exhibits self-interference, inferred from the interference pattern formed at the detector. If putative particle position and path was OR-form throughout, self-interference could not happen, and we could not obtain the interference pattern at the detector. This clearly implies the objective reality and operation of AND-form logic. Hence, AND-form logic is *not* an illusion**, even with the eventual usurping in Book-IV of the notion of space and of 'which way' paths, and the reformulation of these in pure temporal terms, or into AND-form superposes of potential 'which moments'. Therein, we shall find that AND-form logic and the superposition principle must yet abide as phenomenal and objectively real.

Thus, contrary to Jeans' view, **the quantum mechanical wave, hence AND-form logic, is physically real: it is not a *bloss* subjective construct born of human ignorance, much less an artefact of method**.

But why does nature exhibit AND-form logic and the quantum mechanical wave?

2.03: Ontology of the Quantum Wave: Reprise of the Future-Form Argument

We need only imagine a non-standard navigational description for a ship that, as yet, has not left port. This was first presented in the Preliminary **0.10** and is reiterated in summary-form below.

It is WW-II. A U-boat is operating in the Atlantic. Part of its mission involves a *forecast* of the possible future positions that a target ship, at present anchored at port, *might* get to assume in the *future*. Intelligence suggests that the ship will most likely take the direct course from New York to Southampton. The navigator draws this path on the map and designates it as the most probable *future* position set for the enemy ship: i.e., **possibility-1**. A more comprehensive forecast must also assume that the ship might not depart from port at all (**possibility-2**): the second most probable future. According to the same intelligence, the navigator on board the target ship is a drunkard: Thus, a third future possibility: i.e., the meandering drunken sea-voyage: (i.e., **possibility-3**). We may incorporate other possible future ship plots. For example: the ship might never get to arrive at its destination due to iceberg hazards: (**possibility-4**). Other legitimate future paths may also be included, whatever these might be: Hence, **possibility-5**, AND **possibility-6**... AND **possibility-n**.

In descending order of probability, the set of *future* potential paths for the target ship might be expressed as....

Ship at New York (future):
{Possibility-1 **AND** Possibility-2 **AND** 3...**AND**..., **AND** Possibility-n: | **ship assumes all; most to least probable**}

...shortened to...

Ship at New York (future):
{**[AND]** Possibility-1, Possibility-2,3,4,5,6, 7..., Possibility-n: | **ship assumes all; most to least probable**}

This is an AND-form unitary superposition for navigational possibilities that the target ship *might* assume...*in the future*: essentially identical to the AND-form superposition found in quantum mechanical particles. In short, **forecasting as it applies to futures entails precisely the same form of AND-form logic we find in quantum mechanical systems**. The futurity for radiation or matter, hence their AND-form description, really is physically true and objective: a real physical character that, per the self-interference and subsequent interference patterns, *does* manifest as an objective reality 'out there'. Thus, **the futurity of a particle constitutes a physical reality, and the AND-form quantum mechanical wave is the physical phenomenalisation of that ontic futurity, evinced by self-interference and in interference patterns featured in two-slit experiments**.

*

The naïve realist's tacit insistence on the exclusive reality of OR-form logic, and the alleged impossibility and incompatibility of AND-form realities with OR-form realities, has historically produced a wholly contrived set of contradictions between quantum mechanical AND-form based descriptions of nature versus classical-relativistic OR-form descriptions of the same. Yet, the *raison d'etre* of the quantum mechanical wave as the physical phenomenalisation of the future reveals an ultimate mutual harmony and mutual necessity between the descriptions: one that leads to the primitive unification of the quantum and classical descriptions of nature.

The OR-form classical-relativistic description of nature remains valid for states that have come into definite physical resolution and have decayed out of the potential AND-form futurity. On the other hand, the AND-form description of nature remains valid insofar as it applies and encapsulates those parts of the physical order that have not yet transpired into OR-form events. These futures can only be exhibited as spacetime potentialities or spacetime futurities, rendered in AND-form. Restating all of this in explicit spacetime terms, states of spacetime that have not-yet come into OR-form realisation must be grasped as harmoniously integrated with those parts of spacetime that *have* come into OR-form realisation, in the same way that the idea of the future in no way contradicts either the present or the past, but remains both distinct from and complementary to the present and the past. In this way, **we obtain the harmony and integration of AND-form (future) with OR-form (past) descriptions of nature: A fundamental primitive unification between the quantum mechanical and the classical must seamlessly follow and culminates into a growing block intermediate model spacetime, on the strict provision that we recognise the ontic reality of the future and explicitly incorporate both past and future into our spacetime models as distinct domains.**

2.04: Deficits of Generic Models of Spacetime: Precursors to the Intermediate Model Spacetime

Quantum mechanical AND-form states must constitute the futurity of spacetime, or that part of spacetime comprised of the set of possible as-yet not-realised future events, constituted in AND-form superposition; helping resolve the *raison d'etre* of the quantum mechanical wave while incorporating the insight into spacetime as a domain called *spacetime in potentiality*. Unfortunately, the insight was not historically recognised, despite key contributions from North Whitehead at the dawn of quantum science (to be articulated in Book-IV). Thus, **in generic spacetime models, the future is not incorporated into spacetime**.

The historic oversight may well partly arise from the fact that the quantum mechanical wave is often described as a spatially dynamic phenomenon; the physical basis for force-carrying and information-carrying radiation and matter; or as a something that supposedly displaces across putative space. So characterised, **one will typically attempt to model the quantum mechanical wave as a phenomenon displacing and transpiring *within* spacetime, *not* as a precursor or future potentiality for spacetime: not as a spacetime domain for future events that must AND-to-OR decay into realisation**. Amongst other things, this probably helped obscure and confuse the physical meaning of the quantum mechanical wave. Yet, the notion that the quantum wave undergoes spatial displacement is dubious, as will be establish when we present reasons for why it is spatially static (see Part-II **2.23** to **2.26**), even assuming 'space' itself is part of physical ontology and not pleonastic. But if the ' notion of the motion' of quantum wave radiation and matter is to be doubted, what need of the 'space' requisite to such motion?

The said oversight to grasp the *raison d'etre* of the quantum mechanical wave led into models of spacetime strangely bereft, not only of a quantum mechanical domain of open-ended not-yet formed futurities in a growing block universe but, more subtly, devoid of time itself: Describing time as a 'fourth dimension' obscures the very absence of futurity and time in generic spacetime models. Indeed, **the idea of the 'fourth dimension' does not equate to an explicit rendition of an AND-form domain of spacetime (the ontic future), much less concede to time as synonymous with the AND-to-OR wavefunction itself, entailed in decay of futurity into actuality; with the very process absent from generic Minkowski and the occasional Liodel spacetime models.**

185

The generic model spacetime is exemplified in the diagrams in **Fig. 2.00-A** and **B**. In **A** (top), **O** is the origin of the light radiation which displaces in all directions at the speed of light **c**. In each time **t**, it will displace a distance **r** from **O**, given by **r = c x t**. By time **t**, it will describe a 'light-sphere' whose radius is the distance travelled by light per **c x t = r**; and whose diameter is **2 x r = d**: i.e. (**c x t**)^2 gives diameter **d**.

All of this is reconstituted as a Minkowski spacetime diagram in **Fig. 2.00-A- Bottom**: Therein, the vertical y-axis stands for time, and is divided into equal intervals from t_0, to t_1, t_2,... and so on, to the infinitely removed future. The horizontal x-axis stands for space. Thus, origin-**O** must rest along space at time t_0. At t_0 the light radiation has not yet set out on its journey. But by time t_1 the light radiation will have propagated in all directions to describe the previous 'light-sphere' whose diameter-**d**, lies horizontally along t_1. By joining the two ends of **d** to **O**, we describe an upside-down triangle: i.e., the lightcone.

The lightcone is merely an alternative description of the expanding light-sphere centred at **O**: except that the diameter **d** at t_1 of the sphere now constitutes the top of an upside-down equilateral triangle, while the origin **O** of the sphere becomes the bottom vertex of that upside-down triangle.

The illustration in **Fig. 2.00-B** (**top** and **bottom**) follows the same logic, but with an additional passage of time from t_1 to t_2: The light radiation will have propagated further than **d**. By t_2 it will describe a larger light-sphere with a diameter d_1. We now place this larger diameter d_1 horizontally along t_2 and connect the ends of d_1 to origin-**O**. We obtain a larger lightcone. Note that the dashed horizontal line embedded within this larger lightcone is the old diameter given by the smaller light-sphere and lightcone per **d**. The lightcone, as with the light-sphere originating at **O**, may expand in perpetuity. As the light-sphere expands in time, its associated lightcone also enlarges.

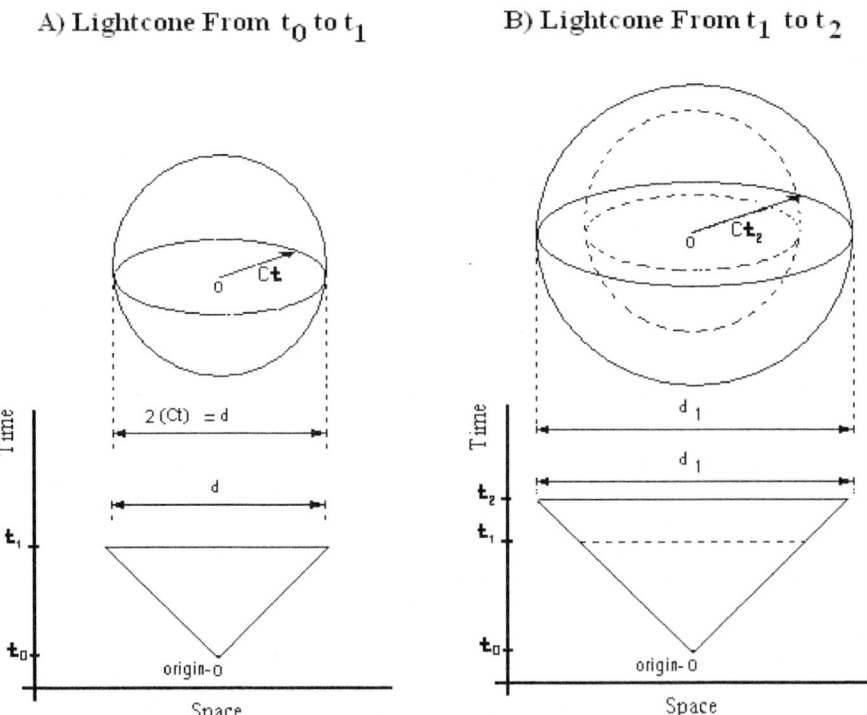

A) Lightcone From t_0 to t_1 **B) Lightcone From t_1 to t_2**

Fig: 2.00: Spacetime diagrams illustrating lightcones
Notes: In the top depiction **diagram-A**, light radiation originating from **origin-O** propagates in all directions at the speed of light **c**, in time t_0 to t_1. It describes a light-sphere whose radius is **ct** and whose diameter **d** is given by **2(ct)**. In the Minkowski spacetime depiction below the light-sphere in **diagram-A**, diameter-**d** describes the top horizontal line of an upside-down triangle: i.e., the "lightcone": This is simply the diameter of the sphere seen edge-on, formed by the light radiation originating from **origin-O**, from t_0 to t_1. The 45-degree lines imply that the light radiation is travelling at the speed of light. Any angle greater that 45-degrees would imply radiation travelling at faster-than-light: Less than 45-degrees would imply radiation or matter travelling at less than the speed of light. In **diagram-B**, the light-sphere and the correlate lightcone below it get to expand from t_1 to t_2, describing a new radius for an enlarged light-sphere given as ct_2, and an enlarged lightcone of diameter d_1 given by **2(ct₂)**.

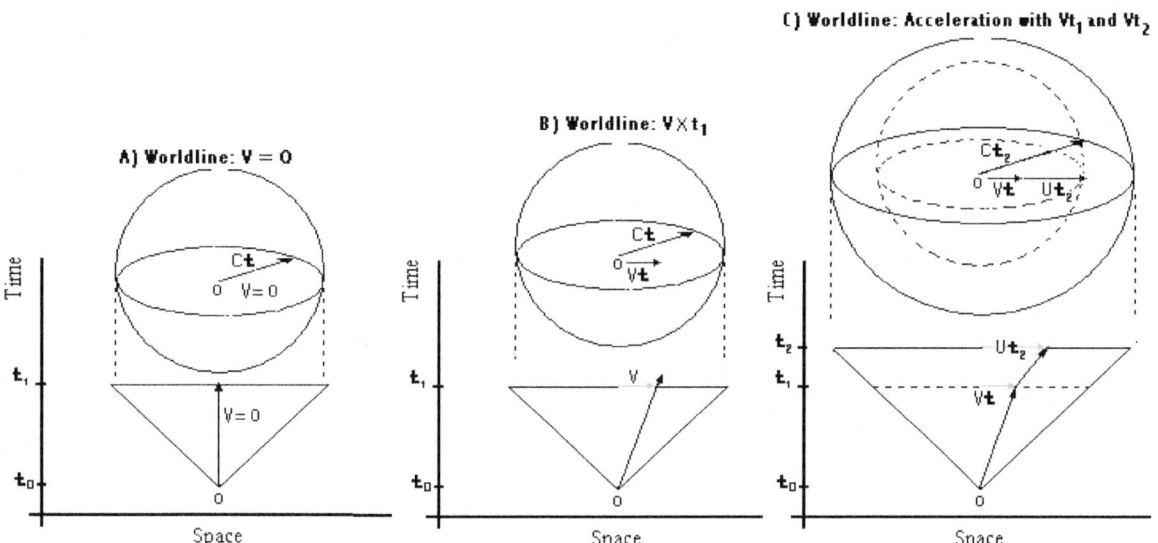

Fig. 2.01: Spacetime diagrams illustrating worldlines.

Notes: In **diagram-A**, an object of zero velocity will describe a simple vertical worldline: i.e., the line and arrow from t_0 to t_1 originating in **O**. This is within the bounds of the lightcone also attached to **O**. The lightcone relates to a light-sphere depicted above the Minkowski diagram, with its diameter given as the top horizontal line component. **Diagram-B** depicts the worldline from t_0 to t_1 which belongs to an object in uniform unchanging velocity. Note that the velocity vector within the light-sphere above the lightcone is included as a clarifying grey vector line at the horizontal of lightcone at t_1. **Diagram-C** illustrates a change in velocity, or acceleration, of the object at t_1, with a new adjoining worldline from t_1 to t_2. To complete our description of Minkowski spacetime, we must develop worldlines. Worldlines account for the paths or the series of time-positional sets that putative bodies, stationary or in motion, describe in the course of spacetime. In short, the worldline constitutes the spacetime 'path', journey, or history of a putative material body; always confined to an embedding lightcone. While bodies describe worldlines, the spacetime history of the light radiation is given by the lightcone. The worldline is illustrated in **Fig. 2.01-A** to **C**.

Per **Fig. 2.01-A**, we assume that, while the light radiation begins its journey from origin-**O** (i.e., t_0), a co-located stationary body at origin-**O** begins its own journey, with an independent velocity, $v = 0$. By t_1 the light radiation will describe a sphere of radius $c \times t$ centred on origin-**O**: (see **Fig. 2.01-A top**). The radius squared will give us the diameter **d** which describes the 'width' of the top of the associated lightcone formed by time t_1: (see **Fig. 2.01-A bottom**). On the other hand, our material body at velocity $v = 0$ will describe a vertical line, with its arrowhead indicating its 'direction' in spacetime: This is the worldline of the material body originating at **O** and drawn from **O** to the top of the lightcone, and which bisect that lightcone into two equal parts. Thus, in Minkowski diagrams, all objects of velocity $v = 0$ describe purely vertical worldlines.

In **Fig. 2.01-B** we describe a similar setup as found in **A**, but for a body with a non-zero uniform velocity-v travelling in a particular direction. In **Fig. 2.01-B top**, by time t_0 to t_1 (= t) this body will have travelled a distance $v \times t$, described as a line-and-arrow within the light-sphere. But in the associated Minkowski spacetime (**Fig. 2.01-B bottom)** we depict the line as a worldline, set at an angle from the vertical proportionate to velocity-v.

In Minkowski diagrams, all bodies that travel at a non-zero velocity always describe worldlines at an angle from the vertical: The greater the velocity the greater the angle from the vertical. If the object could travel at the velocity of light, its worldline would align with one of the sides of the lightcone at 45-degrees. At velocities greater that light, the worldline would have an angle greater than the speed of light at 45-degrees: it would be located outside of the lightcone: in a region called "elsewhere".

In **Fig. 2.01-C** we describe the worldline of an accelerating body: For brevity and simplicity, we assume that the change in velocity is a simple two-step change: Having originally travelled at velocity v, at t_1, the body undergoes an instant velocity-change to velocity **u**. It continues in **u** from t_1 to t_2. Both the old velocity and the new velocity are illustrated as different but co-joined vectors within the light-sphere in **Fig. 2.01-C Top**. However, in the Minkowski spacetime in **Fig. 2.01-C bottom**, the lightcone has expanded beyond the horizontal dashed line (i.e., the horizontal from t_1) and the new worldline given by **u** has a greater angle than the old worldline described by **v**: The greater angle of **u** implies that velocity-**u** is greater than velocity-**v**. i.e., acceleration.

187

*

With the generic Minkowski spacetime model in hand, note that it possesses three implicit features and two key deficits. <u>First</u>, all of spacetime is modelled in the implicit *retrospective mode*: i.e., generic spacetime will model *only* what has already transpired and resolved into conclusion in spacetime. That is, generic spacetime implicitly models only *realised* spacetime, comprised *only* of realised lightcones and *realised* worldlines: essentially a block universe who's past and future is pre-complete and pre-resolved, and in which there is no open as-yet not-realised futurity of alternate possibilities. That is, **generic spacetime *only* models the spacetime-past... hence, *only* the OR-form of spacetime... as a static completed and causally neutered state**.

To put the matter in another way, generic spacetime models only depict classical OR-form physical relics, dispossessed of any subsequent causal significance: a 'dead spacetime', so to speak.

One of our endeavours in Book-II will be to render spacetime into a 'living' spacetime by furnishing it an open future and an active spacetime-past: one that can affect the subsequent development of spacetime by decohering its open future. In short, a growing block spacetime universe. This will be accomplished in Part-II.

<u>Second</u>, overlapping with the first deficit, **generic models do not explicitly incorporate the reality of future spacetime possibilities, such as would be required by a growing block universe. In other words, generic models entirely omit AND-form spacetime and potential-futures of as-yet not-happened events**. Thus, the conventional understanding of spacetime exemplified in the Minkowski model has absolutely no idea of what to do with, much less how to fit or integrate, AND-form quantum mechanical waves and attendant phenomena with OR-form classical spacetime.

<u>Third</u>, bereft of a futurity, **generic spacetime models are equally bereft of time as an unfolding process, despite claims to the amalgamation of time with space in 'spacetime'**. Purely OR-form retrospective spacetime, comprised of realised events only, bereft of a future-frontier of possibilities, is spacetime exhausted of just such future possibilities. It cannot depict any process of 'coming into being': it is dispossessed of any explicit description of a process through which future AND-form possibilities decay or 'collapse' into OR-form actualities. That is, **generic spacetime does not incorporate the process of AND-to-OR collapse synonymous with universe-scale wavefunction collapse... synonymous with time itself**. In short, and ironically, *time* is not part of generic models of spacetime: a bizarre feature of what contends to be a model of space*time*.

The central aim of Book-II will be to remedy these deficits in an intermediate model spacetime that does justice to a growing block universe approach that explicitly incorporates the AND-form future as a quantum mechanical spacetime domain, *and* explicitly depicts the process of time itself... *not* as a 'fourth dimension', but as a universe-scale process of AND-to-OR wavefunction collapse: one that perpetually and partially decays an inexhaustible AND-form future domain of spacetime in potentiality into new OR-form events, and into ever-expanding memory constituted as realised spacetime.

2.05: Integrating Quantum Mechanical & Classical-Relativistic Descriptions of Nature: Illustrating Spacetime in Potentiality via the Intermediate Model Spacetime

We must incorporate into models of spacetime hitherto omitted quantum mechanical reality; the ontic objective future. The attempt is depicted in **Fig. 2.02: A to C**. Thus, **Fig. 2.02-A** depicts a highly simplified universe with one lightcone and one worldline: constituting a retrospectively observed spacetime history from time t_0 to t_1. Note that, it does not display any AND-form quantum mechanical realities: The system has no future.

However, **Fig. 2.02-B**, integrates the quantum mechanical AND-form into spacetime as the domain of *spacetime in potentiality*. Also note that the model possesses a **domain called *realised spacetime***; constituted by the standard lightcone typically found in generic models. Realised spacetime constitutes the sum of hitherto past *realised lightcones* and *realised worldlines* within the bounds defined by t_0 to t_1. Thus, the domain of realised spacetime constitutes the sum of realised informational states, whatever their ultimate physical form, retained as the total past-history of spacetime; as abstract memory, as we shall discover in Book-III and Book-IV.

The domain of *spacetime in potentiality* comprises two features: First, an extension of the hitherto realised lightcone into the future, albeit as an as-yet not-realised **future-potential lightcone** from t_1 to potential-t_2. This potential lightcone is depicted as a hatched segment attached to the realised lightcone but projected beyond t_1 to potential-t_2. The potential lightcone could in principle extend to the infinitely removed future.

Instead of just one realised worldline from t_1 to potential t_2, in **Fig. 2.02 B**, we depict a minimum of two distinct **potential worldlines in AND-form** i.e., two AND-form superposed alternative future possibilities that might become realised in the future. For brevity sake, we assume a simple system with just two AND-form future-potential worldlines. Obviously, there could be many more. The two potential worldlines are... {**potential-p** AND **potential-q**)... attached to the endpoint of the single OR-form realised worldline ending at t_1... and embedded from t_1 to future potential-t_2.

Also note a third feature depicted in **Fig. 2.02 B**: The domains of realised spacetime versus spacetime in potentiality are separated by the relative present-continuous ***event horizon***: i.e. the present-continuous horizontal line which separates the domain of realised spacetime (the past) from the domain of spacetime in potentiality (the future). **The event horizon comprises the leading 'surface' of OR-form realised spacetime but does not constitute the whole of spacetime**. Indeed, all OR-form outcomes obtained from the universe-scale AND-to-OR collapse of spacetime in potentiality must deposit on or along the event horizon: Thus, OR-form realisations are restricted to the event horizon.

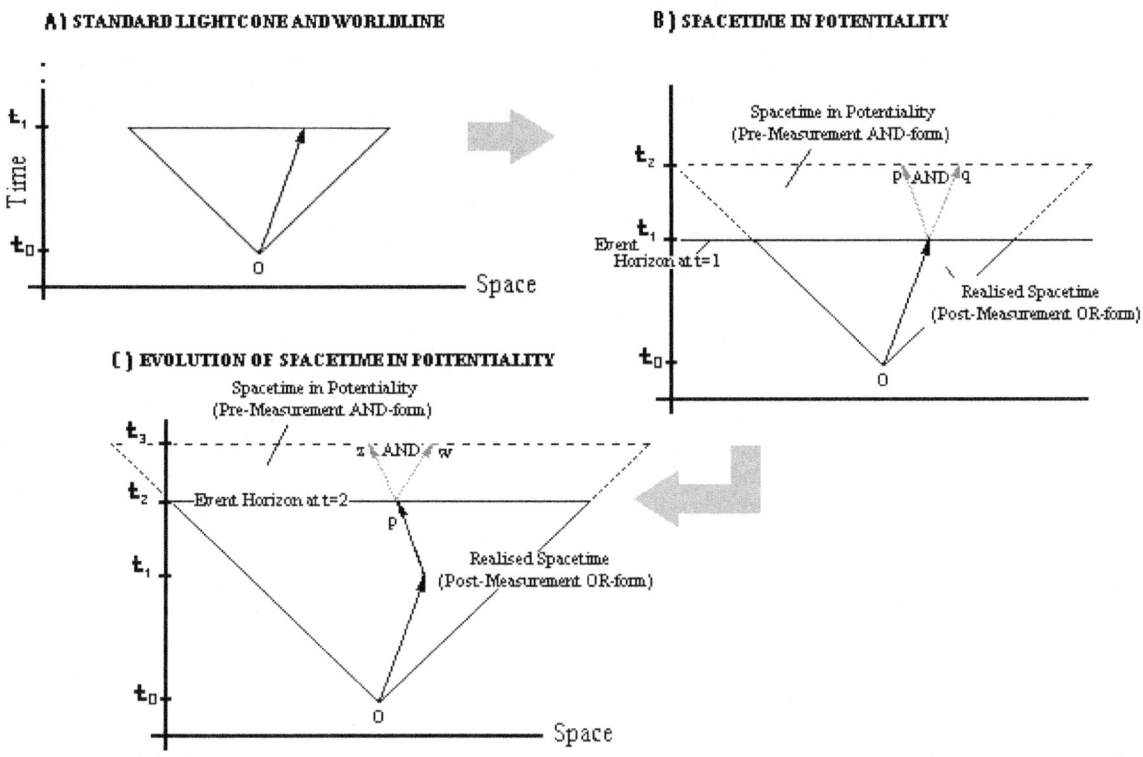

Fig. 2.02: Spacetime unification of quantum and classical-relativistic domains: a growing block model

Notes: **Diagram-A** depicts a generic spacetime diagram for a simplified universe with just one lightcone and one worldline originating at **O**; from t_0 to t_1. Typically, such diagrams have nothing to depict about quantum mechanical realities and how these might relate to and integrate with a growing block spacetime. However, in **diagram-B**, we incorporate just such realities: a simple expedient of extending the lightcone by means of dashed lines into a future-potential lightcone region from t_1 to potential-t_2, separated by a horizontal line called the **event horizon**: the frontier between the **domain of realised spacetime** 'below' it (spacetime past) and the AND-form quantum mechanical **domain of spacetime in potentiality** 'above' it (spacetime future). Realised spacetime contains the basis of the past constituted as the single realised worldline that terminates somewhere at the event horizon. At its endpoint we find attached two quantum mechanical future possibilities or not-yet realised worldlines, or future-potential worldlines: namely, {**p** AND **q**}. These potentially terminate at a future-potential event horizon expected to form at potential t_2. In **diagram-C** the event horizon has advanced to t_2 accompanied by the collapse of {**p** AND **q**} into the OR-form realised worldline-**p**; in turn potentialised with not-yet realised future-potential worldline possibilities: namely, {**z** AND **w**} within the remaining spacetime in potentiality 'above' t_2. In the future, the event horizon is expected to progress from t_2 to potential-t_3: one of the possibilities (either {**z**} or {**w**}) will be realised, further expanding the domain of realised spacetime. The incorporation of the past and future domains constitutes a growing block model of spacetime and universe.

The event horizon is *not* a static: It perpetually progresses 'upward' the spacetime system along the vertical axis: It can never reverse or proceed backwards or 'downwards'. That is, the perpetual formation, reformation and succession of the event horizon is always in conformity to AND-to-OR directionality: the 'arrow of time': (see Book-I **1.9**). Thus, the 'upward' progression of the event horizon is always synonymous with and is **attended by universe-scale AND-to-OR collapse-process and time (or wavefunction collapse)**. Any putative 'downward' regression of the event horizon would constitute causality-violating OR-to-AND de-collapse and time-reversal, and, with it, the 'falling out of evidence' of events back into potentiality, with consequent absurd 'un-science'.

All things being equal, the event horizon progresses at a relative constant rate 'upward' the spacetime model; always entailing general AND-to-OR conversion of spacetime in potentiality and **always accompanied by increase in the entropy of future-potential information** (see Book-I **1.7**). This progression is illustrated in **Fig.2.02-C**; wherein the event horizon is shown progressing from t_1 to t_2; entailing a universe-scale AND-to-OR collapse, or wavefunction collapse from t_1 to t_2, attended by the collapse of spacetime in potentiality (the state {**p** AND **q**}) into the OR-form worldline-**p**; to the erasure of the foregone potentiality, worldline-**q**.

At the new event horizon formed at t_2, we find that the most recent realised worldline, worldline-p and its leading point is itself futurised with two subsequent AND-form future-potential worldlines; both of which superpose within the surviving domain of spacetime in potentiality: namely... {**potential-z** AND **potential-w**}. Again, for brevity, two future-potential worldlines are assumed, even though there could be more than two.

We may reasonably expect the event horizon at t_2 to subsequently progress to **potential-t_3** and into subsequent AND-to-OR collapse of {**potential-z** AND **potential-w**} into one or the other OR-form version of the given worldlines, attended by another event horizon progression... only to be followed by yet another AND-form state of potential worldlines within what remains of spacetime in potentiality. This will also be subject to a further process of AND-to-OR collapse; with yet another progression of the present-continuous event horizon. The process is never-ending and inexhaustible. It could in principle continue to the infinitely removed future, consistent with our growing block universe thesis.

*

Note that **the domain of spacetime in potentiality comprises the perennial universal wavefunction, constituted as an 'instantaneous whole'**; operative for the whole universe, but, in our case, in a highly simplified universe with just one lightcone and one or two future potential worldlines. Subsequent progression of the event horizon from, say, t_2 to t_3 would coincide and comprise the very AND-to-OR collapse of this instantaneous whole... from a pre-measurement AND-form quantum mechanical spacetime state into a post-measurement OR-form realised spacetime state, on or along the event horizon.

Implicit to the above is the view that the quantum mechanical wavefunction for the whole universe, hence the domain of **spacetime in potentiality as a whole... does not 'collapse' comprehensively, wholly, or exhaustively in one single collapse-step**: It does not lead to the depletion or a 'running out' of spacetime in potentiality: The future elaborates to the infinitely removed future. Hence, the collapse of the universes-scale wavefunction cannot convert to realised spacetime wholly or comprehensively, but always marginally or partially.

Since spacetime cannot run out of futures, the universe cannot run out of time, insofar as the ontic character of time *is* the always partial process of AND-to-OR collapse of an otherwise inexhaustible future-domain.

The precursor to the idea of spacetime in potentiality, and the future as quantum mechanical, was first postulated by Alfred North Whitehead[64]. Thus, the idea is not new, although the rational and evidential basis furnished by Alt-R *is* new, insofar as it is critically founded on the firewall principle and other novel features not found in Whitehead's account. For example, our model implies that the universe-scale wavefunction, hence the whole of the domain of spacetime in potentiality, cannot constitute a spatially dynamic state: That is, spacetime in potentiality is not constituted as a quantum wave that supposedly displaces across space, or undergoes undulation in time. Instead, it is a static state: somewhat in similarity to a standing wave, but without dynamic undulation in time; comprised of a vast system of potential future possibilities that spacetime might come to realise. This radical claim for the reality of the static non-displacing universe-scale wavefunction will be fully justified in Part-II (**2.24** to **2.26**) and it prefigures and scaffolds for later de-spatialisation in Book-III and IV: Indeed, if the 'motion' of the quantum wave is not real, what need of the 'space' through which such motion is expected to unfold?

*

The critical feature to the unification of quantum and classical relativistic descriptions of nature and the pertinent spacetime domains is the event horizon itself: the perpetually advancing *relative* present-continuous frontier or 'cut' that divides and segregates realised spacetime 'below' it (the past) from the domain of spatially static quantum mechanical spacetime in potentiality 'above' it (the future). The progression of the event horizon is synonymous with universe-scale AND-to-OR collapse, hence *time*.

Notice how the universe-scale AND-to-OR collapse, identifiable with the progressing event horizon, dynamically and logically integrates both OR-form and AND-form spacetime domains: hence integrates, harmonises and unifies both OR-form and AND-form descriptions of nature at the event horizon, and does so in a causally harmonious, complementary and mutually necessary way, albeit with the AND-to-OR collapse process and time depicted, *not* as a 'fourth dimension', but as a 'living' perpetual AND-to-OR collapse process. So much for the oft claimed incompatibility of quantum descriptions and classical-relativistic descriptions of nature.

2.06: The Intermediate Model Spacetime & the Ontic Meaning of
The Quantum Mechanical Wave: A Summary

From our preliminary analysis of the quantum wave per the analogy of navigational dead reckoning given in **2.02**, we arrived at a view both obvious and hitherto overlooked: i.e., the physical meaning of the quantum mechanical wave as the physical manifestations of the ontic future: the future as an ontic objective reality. The future is not a *bloss* subjective idea, but a physical constitution independent of mind, 'out there', and inherent to the very structure of spacetime and the growing block universe. Indeed, in the opposite block universe, the future would be a pre-resolve complex of definite 'paths' and 'which way' states. In effect, there would be no ontic

---

[64]Whitehead, James, and the Ontology of Quantum Theory: Henry P. Stapp, Lawrence Berkeley National Laboratory, University of California, Berkeley, USA

future. **If the block universe held true, the two-slit experiment could never yield self-interference and the expected interference pattern, given that these could not emerge out of pre-resolved OR-form 'paths', or from a fully resolved pre-constituted future. In really-existing two-slit experiments, we** *do* **obtain self-interference and interference patterns. This can only happen if the future-state of the experiment exists, not as OR-form resolved 'paths', but as an as-yet to happen future possibilities in probabilities, rendered into AND-form superposition.** In short, AND-form logic and the quantum wave, and the wavefunction devised to model it, really *do* pertain to an ontic future: The quantum wave *is* the future-form of nature, phenomenalised as an objective non-illusory existent… *not* in our minds, but 'out there'. And the growing block model is the true model of the universe.

The intermediate model incorporates into generic spacetime models the ontic reality of spacetime futurity as the domain of spacetime in potentiality. Other terms for the same could also be used: e.g., 'quantum mechanical spacetime'; 'AND-form spacetime'; 'spacetime futurity'. Whatever the terms, the domain simply constitutes the ontic future: necessarily structured per AND-form logic and the superposition principle: necessarily quantum mechanical in character; comprised of future events that have not yet transpired, out of which realised events 'come into being' through AND-to-OR wavefunction collapse or time.

Below the future-form of spacetime, below spacetime in potentiality, we find **the past form of spacetime: the domain of realised spacetime**: a domain no more illusory, and in every way as ontic as the ontic future. It is tempting to suppose that the past-form of spacetime, the domain of realised spacetime, must be 'classical', insofar as it is comprised purely of the sum of past states that had decayed into OR-form realisations out of future-potentiality. Yet, the past is retained in abstract-immaterial form at a physical remove from the present-continuous OR-form states of radiation and matter resolved on or along the event horizon. By Book-IV, consistent with abstract memory concepts hitherto intimated in the Preliminary, in Book-I, and presaged in Book-III, we shall discover that **the domain of realised spacetime comprises the abstract memory of the total past. But it is** *not* **constituted as an OR-form corpus.** Indeed, de-spatialisation and temporisation in subsequent books will be key to unravelling the nature of realised spacetime and the form in which abstract memory abides.

*

The relative present-continuous **event horizon, which separates and segregates future-form spacetime in potentiality from mnemonic domain of realised spacetime, does not simply comprise the relative present form of spacetime** but, insofar as it is subject to perpetual progression 'upward' the spacetime system, and insofar as it is the frontier on, along or through which perpetual universe-scale AND-to-OR collapse of spacetime in potentiality into realised spacetime transpires, it must constitute the OR-form 'skin' of spacetime. We will discover that **OR-form outcomes are wholly restricted and confined to the event horizon** and, when relegated into abstract memory within realised spacetime, wherein the past is retained, realised outcomes no longer assume OR-form attributes but transform into abstract form, the basis for which cannot be disclosed until de-spatialisation and temporisation in Book-IV.

Hence, even though critical AND-to-OR collapse and time takes place along or coincides with the progression of the event horizon, and it must appear to manifest as a relative present-continuous state of material and radiational outcomes, the OR-form outcomes so secured cannot remain indefinitely 'materialised': The most recent resolved states of events constituting what we think of as 'matter in space' must pass into an abstract immaterial past-form: into the domain of abstract mnemonic realised spacetime.

The sum of past states embedded and retained within the domain of realised spacetime (succinctly, abstract memory) is expected to play a critical causal role, beyond the role of generic or *mere* **'initial conditions'.** The abstract past in the form of the total domain of realised spacetime, critically decides the manner and direction of, and outcomes obtained from subsequent AND-to-OR collapse, attendant progression of the event horizon, and time. Thus, unlike generic models, realised spacetime is not a dead relic or fossil: It is a 'living' and active causal influence; not merely in terms of the continuity, endurance and perdurance of patterns and identities as memory in time, but as a critical contributor to inertia and gravitation itself, as we shall presage in Part-III, and whose core processes we will completely uncover by Book-IV.

2.07: Prelude to Spatially Static Quantum Mechanical Waves & Static Spacetime in Potentiality

Generic spacetime models presume the ontic reality of spatially dynamic electromagnetic radiation, typically subject to the perennial speed-of-light limit. However, our intermediate model rests on the ontic primacy of the AND-to-OR collapse process over the perennial speed-of-light limit. The process of AND-to-OR collapse and time is evinced and identified with the progression of the relative present continuous event horizon, presupposed to the moment-to-moment resolution of the succession of OR-form events from which putative 'radiation in motion' and the perennial 'speed of light' are inferred or imposed. In other words, it is ontologically primary AND-to-OR collapse and time which manifests the events from which we retrospectively infer 'motion', the 'motion of matter', or the 'motion of radiation' at its putative speed of light. The ontological primacy of AND-to-OR collapse to all things, including the speed-of-light limit, was espoused almost throughout Book-I.

We have already alluded to spatially static quantum mechanical waves in contradistinction to the conventional approach to quantum mechanical waves depicted as spatially dynamic. As can be grasped by reference to **Fig. 2.02**, (see essay **2.05**) **the domain of quantum mechanical spacetime in potentiality is effectively a static state**: a perpetually reducing or AND-to-OR collapsing and grand decohering static state of possibilities. As such, **its AND-to-OR collapse from a static state is ontically presupposed to the very manifestation of the apparent 'motion' of radiation and matter inferred or assumed from the succession of OR-form outcomes obtained from that collapse and manifesting the pertinent state-change.** This succession of partial-perpetual AND-to-OR collapse processes bring into formation the attendant lightcone and the putative speed-of-light limit. Thus, the idea of spatially static quantum

191

waves, synonymous with the static domain of spacetime in potentiality, is inbuilt into our intermediate spacetime structure and its processes.

If electromagnetic quantum mechanical waves are spatially static, as indeed they are per the above and other findings from **2.23**, it must follow that quantum mechanical waves cannot be described as subject to any sort of spatial displacement, much less portrayed as propagating in accordance to a speed-of-light limit… save as a retrospectively imposed pleonastic assumption garnered from the temporal succession of OR-form events that, in the first place, arise from and depend on an ontologically primary AND-to-OR collapse process and on an equally ontologically primary *static* spacetime in potentiality. In short, quantum mechanical radiation cannot move: It also follows that, in its electromagnetic form, it cannot move at 'the speed of light'; consistent with implications from decoupling theory from Book-I.

The explicit proof of spatially static quantum waves, or the static future, will serve to validate the overall structure of the intermediate model spacetime: The conclusion will emerge out of a specific application of John Wheeler's delayed choice approach, applied *against* the notion of motion of the quantum mechanical wave (**2.23**): part of the justification of our intermediate spacetime model.

*

It is the 'upward' progression of the event horizon that coincides with and is the same as the process of AND-to-OR collapse of spacetime in potentiality into the most recent OR-form outcomes. Transformed into abstract form, the outcomes are then relegated and retained within the domain of realised spacetime; generating the expansion of growing block realised spacetime and memory, always accompanied by the irreversible increase in the entropy of future-potential information (Book-I: **1.7**), whatever else might happen to generic entropy... *and* always coinciding with the generation of apparent 'motion of material objects' or the 'motion of light at the speed of light'.

Thus, all things being equal with respect to these findings, and certainly *not* denying time dilation evinced in Einstein's Special Relativity, time is indeed bound to and the same as the 'upward' progression of the event horizon; attendant and synonymous with AND-to-OR collapse. **Thus, the truer account of spacetime must begin, not with propagating quantum mechanical electromagnetic radiation at the 'speed-of-light', but with the ontologically primary AND-to-OR collapse process of the spatially static quantum wave or spacetime in potentiality**, as was argued for in Book-I in the specific context of decoupling. Our deeper understanding of spacetime must therefore begin with the dynamics of the event horizon directly associated with AND-to-OR collapse, itself, grasped as synonymous with time, and **wherein time is no longer a 'fourth dimension'**.

2.08: Mutual Justification of the Intermediate Model of Spacetime, Information-Matter Dualism, and Decoupling Theory

In Book-I Part-IV, Alt-R presented the first foray into decoupling theory, which asserted for the physical decoupling between information and its propagation constant, the speed of light. Decoupling turned out to be synonymous with and obvious from information-radiation and information-matter dualism, grounded in the five distinct proofs of dualism furnished in Book-I (**1.19** to **1.24** and **1.33**): Therein, it was shown that information-matter dualism and decoupling *must* hold universally in the relation between information versus *all* forms of radiation-waves or de Broglie matter-waves.

This dualism, entailing the separation and decoupling of information from quantum mechanical radiation and de Broglie matter, is mirrored in the very structure of the intermediate model of spacetime, as depicted in essay **2.05, Fig. 2.02**: **Information-matter dualism arises from, and it is synonymous with, the segregation of the domain of realised spacetime (the past) physically dissociated from the AND-form quantum mechanical domain of spacetime in potentiality (the future)**.

In turn, decoupling theory, founded on information-matter dualism, is consonant with the implicit decoupling and separation of information per the segregation of realised spacetime from spacetime in potentiality: of the past from the future. Thus, information-matter dualism and decoupling theory justifies, and is in turn justified by, the tripartite structure of our intermediate model spacetime, or the past-future dualism structurally intrinsic therein.

Moreover, **from the view of information-matter dualism furnished by the intermediate model, we must replace the naïve idea of information dissociation from spatially dynamic or moving quantum mechanical radiation and matter, and must succeed this with informational dissociation from a purely spatially static future-form quantum mechanical radiation and matter, constituted as the spatially static domain of spacetime in potentiality, whose partial-perpetual decay or AND-to-OR collapse leads to the effective illusion or the presumption of 'spatial transportation of information across space' by seemingly moving, but *not* actually moving, quantum mechanical waves**.

Of course, even with static quantum waves, the essential logic of information-matter dualism and decoupling theory must abide. Indeed, with spatially static quantum waves, we are furnished a deeper and more exotic form of information-matter dualism and decoupling, with all its radical implications intact; including those that pertain to the dualism of mind and memory vis-à-vis the brain. Yet, if one must insist in the implausible reality of spatially dynamic quantum waves, this would merely culminate into the version of information-matter dualism developed in Book-I, minus significant reference to spacetime structure and process. On the other hand, if one concedes to the reality of spatially static quantum mechanical waves, this must culminate into essentially the same information-matter dualism, albeit furnished by the very structure and process of the growing block intermediate model spacetime.

From the framework of the intermediate model, obviously, the past (constituted as realised spacetime, combined to the most recent OR-form events on or along the event horizon) cannot be carried by the as-yet not-realised future evinced as spacetime in potentiality. It is obviously absurd to suppose that the future itself (i.e. quantum mechanical totality) could 'move' at all,

192

much less 'transport the past across space'. That is, information-matter dualism is essentially per function of the structure of the growing block past-future dualism, seamlessly furnished in the separation of the domains in the model.

Also recall from Book-I that physical systems cannot be purely and only AND-form quantum mechanical; no more than these could be purely OR-form 'classical'. Both the quantum mechanical *and* the classical are physically real, though *not* identifiable or interchangeable per function of AND-OR dualism: a fact also justified and mirrored in our intermediate spacetime model. The model inexorably depicts spacetime as an admixture of the future and past: with AND-form spacetime in potentiality co-joined and admixed with realised spacetime. It is not possible to obtain a purely OR-form spacetime: In other words, generic block spacetime models are untenable. Nor is it possible to describe spacetime in purely AND-form terms; as pure spacetime in potentiality; as pure future. The latter cannot be eliminated by, say a Quantum Zeno effect applied to the whole event horizon and to attendant AND-to-OR collapse processes (time), assuming in the first place that Quantum Zeno effects involving continuous bombardment of a given state could furnish requisite observational inputs of zero-wavelength or infinite frequency: not possible per staple quantum mechanical facts.

*

In Book-I, essay **1.39**, we presented the prime experiment exemplifying decoupling theory: This was Gunter Nimtz's quantum tunnelling experiment, whose peculiar results we justified in terms of information-radiation dualism. Recall that the experiment involved the apparent transmission of Mozart-40 by means of quantum tunnelled microwaves that appear to displace at speeds up to 4.7 times that of light. If we insist, not only in the supposed reality of spatially dynamic radiation, but also in the dubious notion that radiation must constitute a physical 'carrier' and 'transport' of information (in this case, of Mozart-40) all the usual violations of the laws of physics and relativity theory, if not of causality itself, will seemingly arise.

However, in Book-I, the dualism of information vis-à-vis radiation and matter came to the rescue: **Since OR-form information cannot be literally physically carried in or by AND-form quantum mechanical microwaves, or even by quantum tunnelled quantum mechanical microwaves, the information is not physically fated to what happens to that radiation, much less to how fast it is supposedly made to propagate.**

Moreover, the integrity intrinsic to that information will remain intact no matter how fast the 'carrying' microwaves appear to displace across space, as was argued in Book-I. Indeed, and as repeatedly stated, **it is only through the process of ontologically primary AND-to-OR collapse that Mozart-40, dissociated from the 'carrying' radiation in the interim, finally remerges with the 'faster than light' microwave radiation at the point of reception. The speed of that radiation cannot alter or reverse the fundamental ontological primacy of AND-to-OR directionality which, in the first place defines the proper 'arrow of time', and constitutes and *is* time itself, hence justifies causality**, and gets to AND-to-OR resolve the events from which 'faster than light' microwaves are inferred. Our proper focus must shift from the putative superluminal radiation in the Nimtz experiment to the dissociation of that information from radiation and matter per spacetime structure.

However, we must focus on the more central ontologically primary AND-to-OR process of collapse and time, which constitutes the *real* basis for the integrity of information, order and causality in the universe.

Per our intermediate model spacetime, we need only replace our notion of 'spatially dynamic displacing microwaves' at apparent faster-than-light with spatially static microwaves: In so doing, we arrive at the central contention of our intermediate spacetime model, now grasped in terms of decoupling theory recapitulated in terms of spatially static non-moving 'tunnelled' microwaves, in turn subsumed to an equally spatially static non-moving domain of spacetime in potentiality (the future): Indeed, spacetime in potentiality must also include quantum tunnelled forms of microwave radiation as part of its AND-form future potentiality-set. Again, note that the domain of spacetime in potentiality does not 'move across space': a structurally absurd notion per the strictures of the intermediate model, given that, obviously, the future cannot 'move', even at a supposed 4.7 times the speed of light.

Thus, **the logic inbuilt into the intermediate model of spacetime cannot permit the domain of spacetime in potentiality, hence the potentials for quantum tunnelled microwave radiation subsumed to it, to undergo any sort of 'spatial motion', much less at 4.7 times the speed of light**. Consequently, the model cannot permit the treatment of quantum mechanical light radiation, or even quantum tunnelled 'faster than light' microwaves in the Nimtz experiment, as states undergoing *any* sort of physical 'displacement', **precisely because the quantum mechanical domain of spacetime in potentiality must remain purely static, or simply because the future cannot obviously 'move'.**

What then of Mozart-40? What happens to it if, structurally, it cannot be 'stored' in the quantum tunnelled radiation and it cannot be transported by such radiation, given its dissociation and given that the quantum wave cannot in the first place 'move' from one point in space to another, supposedly at faster than light? Where does M-40 go? **Obviously, M-40 physically dissociates and is relegated into the domain of realised spacetime as abstract memory. Eventually, it re-emerges in materialised form at some succeeding event horizon via ontologically primary AND-to-OR collapse**: The 4.7 times speed-of-light 'displacement' of Mozart-40 does not actually occur, since even speed-of-light displacement does not actually transpire, given the spatially static nature of electromagnetic quantum mechanical radiation. In any case, to begin with, per information-radiation dualism, M-40 was never carried by the quantum tunnelled quantum mechanical radiation, because... again... the future cannot carry the past, much less 'transport' it across space, especially given that the future, or spacetime in potentiality, constitutes a static state that cannot 'move'.

*

193

So far, we have secured **four principal justifications of the intermediate spacetime model**: The first is its integration and unification of supposedly incompatible AND-form quantum description (the future) versus OR-form classical-relativistic descriptions of nature (the past): Hence, **justification from primitive unification**.

The second justification derives from implications from **anticipated spatially static quantum waves, which prefigure the static character of spacetime in potentiality**: Given that the future cannot move, but can only decay into successive OR-form events from which we infer or *assume* 'motion', the quantum mechanical waves purported to transport force or information (or memory) do not undergo motion, but are as spatially static as the static domain of spacetime in potentiality (the future) to which they are subsumed. Hence, **justification from static quantum waves**.

The third justification emerges **from the organisation of our spacetime model into its two main domains, demarcated by an 'event horizon' frontier**, through which all three domains fashion for information-matter dualism and decoupling theory as mirror images of past-future duality, and wherein the past, as memory.. as resolved events… *obviously* cannot be carried in or be transported by as-yet not-realised AND-form future, which cannot in any case 'move' such information. Hence, **justification from the mutual mirroring of information-matter dualism and decoupling in growing block spacetime structure**.

A fourth justification of our intermediate spacetime model will emerge out of causality itself, intertwined with the principle of conservation of energy-matter to constitute the firewall principle. The causality argument and the firewall principle first introduced in Preliminary and further anticipated in Book-I (**1.31**) was part of the argument for information-matter dualism. The firewall principle, as a ward against 'signals from the future', and against both the violation of causality and of conservation of energy-matter, hence the necessity of a 'cut' (the event horizon) between spacetime future and the spacetime past, necessarily justifies the need and reality of a spacetime in potentiality and, with it, the growing block intermediate spacetime model that incorporates spacetime-futurity with spacetime-past, *and* hence secures the primitive unification of physics. The causality argument will be capitulated in **2.10** and **2.11**.

2.09: The Event Horizon & Special Relativity: The Event Horizon is *not* Absolute Time

At first glance, the event horizon might be misconstrued as synonymous with long-defunct absolute time from Newtonian physics: i.e., the notion of a common present moment in which all things transpire vis-à-vis all observers… and wherein there is only one clock time for all observers and for the whole universe. **The event horizon is *not* absolute time: It constitutes the cut between realised spacetime (the formed past) versus the quantum mechanical AND-form spacetime in potentiality (the future). The event horizon is integral to the firewall principle that keeps the two domains compartmentalised and constitutes a ward against 'signals from the future'**. But a misconstruing of the event horizon with absolute time might be agitated from the over-simple depiction of it as a flat line that bisects the growing block spacetime system, and appears as a space-like absolute common present moment for the whole universe. The event horizon need not be described as a linear line, or flat. It can be 'curvy' or distorted, as we will discover in Part-III: **2.43 to 2.45**.

Insofar as its progression is synonymous with universe-scale AND-to-OR collapse and *time* itself, AND-to-OR collapse and time per the 'upward' progress of the event horizon does not unfold at the same one rate, much less at a same common present moment on or along the event horizon. In other words, the relativity of simultaneity from Special Relativity, pertinent to events obtained from the process of AND-to-OR collapse, abides for observers and frames distributed on or along the event horizon, *ceteris paribus* with respect to time dilation, length contraction and the famous twin's paradox.

Succinctly, **observers distributed to different frames of reference along the event horizon cannot enjoy instantaneous direct *substantive* observations of each other's frames, or of the generated OR-form events associated with these, and must rely on lightcones from other frames to observe events attached to those other frames: the observations of which can only take place in the future, subject to delayed choice time-intervals, and *never* simultaneously or instantaneously, or in an impossible common present moment**. To obtain information, observers in the said frames will need to rely on the resolution of signals through successions of partial perpetual AND-to-OR collapse processes (or time) which resolve pertinent potential lightcones and potential worldlines into their observation-pertinent forms… and do so *in the future*. Even the seeming exception to this from decoupling theory per putative faster than light anomalies in the Nimtz experiments must be rendered consistent with Special Relativity, as was argued in the several harmonising scenarios in Book-I (**1.41**).

Only grand decoherence will appear to imply seeming space-like relations: an effect that almost approaches absolute time, but never obtains it. (See Part-III **2.35**): We will discover that counterfactual causality and grand decoherence cannot furnish instantaneous transactions comprised of *substantive* information. Thus, instantaneous observations of events belonging to 'distant' frames in absolute time must remain impossible, as must absolute time itself. Ultimately, counterfactuals and grand decoherence enforce consistency upon the common future pertinent to all frames within spacetime in potentiality. By example, if a supernova that might have happened on a 50-50 probability *did not* happen at Alpha Centauri, futures within spacetime in potentiality attached to our Earth frame will decohere into future possibilities that erase the future-potential supernova… because it never happened. Thus, in the future potentiality projected four years into spacetime-future, wherein the lightcone from Alpha Centauri is expected to overlap with our Earth frame, the consistent possibilities enforced within it through grand decoherence of the total future will *not* permit the observation of a supernova that did not happen, but will have erased the set of future possibilities involving a supernova. The rest of the universe, indeed the whole of spacetime in potentiality and its future possibilities, will also decohere into consistent futures, wherein the said supernova did not happen and cannot therefore be observed to have transpired by *any* observer in *any* future frame; even in the infinitely removed future.

Note that **the process of grand decoherence of spacetime in potentiality does not involve *any* transaction of substantive information in violation of the putative speed-of-light limit, much less revive absolute time. In an attendant way, the event horizon does not constitute revival of absolute time.**

2.10: The Causality Argument & the Quantum Mechanical "AND"

Clearly, the reality of quantum mechanical AND-form phenomena cannot be denied and must be incorporated into spacetime models. Hence, the incorporation into our intermediate model spacetime of the domain of spacetime in potentiality. However, it is the causality argument that constitutes the decisive justification for an intermediate model spacetime: **Intermediate spacetime *must* be constituted in the growing block form previously described because causality demands it.**

While quantum mechanics and AND-form logic is often misunderstood as leading to the contravention of causality, causality necessitates the ontic reality of AND-form logic, if not quantum mechanics itself, as was anticipated in the Preliminary **0.10** and in Book-I. Insofar as causality necessitates the objective reality of quantum mechanical waves, we *must* incorporate such realities into our spacetime models. In effect, at this juncture, we are reiterating the precursor to the fifth argument for information-matter dualism from Book-I (**1.33**), but now applied to the structure of spacetime, in justification of it *and* of the domain of spacetime in potentiality.

Recall that, in a flight from London to New York, it is not possible to arrive at New York before having first departed from London. **Causality demands that the future cannot transpire before events leading to it have come to pass… otherwise, we end in 'signals from the future'**, or the occurrence of that future before it transpired. We cannot assign the status of an 'event' in spacetime to anything that has not yet transpired into an event, until it has transpired thus: Otherwise, again, we commit 'signals from the future'.

An event may consist of a specific experimental measurement on a particle, or on a specific attribute of a particle. Therein, an event constitutes an outcome from the collapse of that particle's attribute from its AND-form superpose of future possibilities into its post-quantum mechanical OR-form state. **In the pre-event quantum mechanical state, the attribute to be assumed by our particle, whatever that may be, does not yet justify being thought of as an event: i.e., it has not yet happened: it has not yet converted from future *potentia* into realised *actua*. Thus, the naive realist insistence that the particle must always be in 'objective' OR-form state even before it has transpired into such, is to assign to it the status of an event before it has transpired into one, *in contravention of causality*, and with inevitable presumption of 'signals from the future' and subscription to the block model.**

*

For purposes of brevity and didactic requirement, we will treat light radiation *as if* it constituted the carrier of information, even if this is invalid per information-matter dualism, decoupling theory, and per implications from spacetime structure (see **2.08**). Recall that the event horizon acts as a dynamic frontier between realised spacetime and the quantum mechanical spacetime in potentiality. Insofar as causality is critically fated to the AND-to-OR collapse attendant the progression of the event horizon, causality can be modelled in terms of the progression of that event horizon. By reference to **Fig. 2.03-A** to **C**, it is not possible for the event horizon to resolve at t_3 before it has progressed to t_1: with t_3 constituting arrival at New York before departing from London at t_1, or before such a journey unfolded and was realised in spacetime. Indeed, the prospective events must remain as potentialities of the future, AND-form superposed with alternative future possibilities. From this we garner that the principle of causality demands just the sort of growing block quantum relativistic incorporative spacetime model so far developed: one with a futurity-domain; a spacetime in potentiality.

We can apply the same logic to the measurement process involving a particle. This is depicted in **Fig. 2.04-A** to **C:** a repurposed version of the previous depiction. Therein, t_1 is the point of inception of the given experiment. On the other hand, potential-t_2 is the moment at which the particle will be placed into isolation until potential-t_3; comprising an isolation-interval, t_2 to t_3. At potential-t_3 the planned or expected measurement will hopefully take place: Therein, the particle will be brought out of isolation with the help of observational inputs: its attributes will AND-to-OR collapse into a realised or 'objective' OR-form outcome.

Any planned measurement of a particle is going to transpire at some appointed time *in the future*: i.e. it is a potential *within* spacetime in potentiality, above the event horizon at t_1. Thus, the time of measurement is obviously not going to be within the region below the event horizon, or in the past. Indeed, the time of measurement will not constitute a definable resolved OR-form event. Again, we immediately grasp that the consequence from any naive realist demand that the particle remain physically 'objective' (or OR-form) before measurement, before the event horizon has progressed to t_3, and before the completion of the measurement process, would violate causality and must presuppose 'signals from the future'. (The same logic applies to Afshar's experiment: an experiment wherein a one slit outcome was obtained from a two-slit arrangement, and supposedly usurped the complementarity principle, or the assertion that the particle is AND-from before it becomes OR-form at the point of measurement: (see Preliminary **0.16** to **0.18**). Only after t_3 can the particle acquire OR-form 'objective' characteristics. Before then, in the isolation-interval from t_2 to t_3, we cannot treat that interval and the particle associated with it as possessing a set of OR-form realised 'objective' attributes or events, save by travesty of causality, or through 'signals from a future'… tantamount to the demand that our particle assume the characteristics of an event before it had transpired into such: no different from one's arrival at New York before having departed from London.

Naive realist interpretation of quantum mechanics unwittingly commit such implicit travesties against causality and erroneously conceive spacetime in purely 'objective' OR-form terms, or as a block model spacetime system dispossessed of AND-form futures, or without quantum mechanical properties: a form of generic spacetime strangely bereft of the quantum mechanical; bereft of spacetime future; and a physics blind to the mutual complementarity that resides between quantum and classical relativistic descriptions of nature.

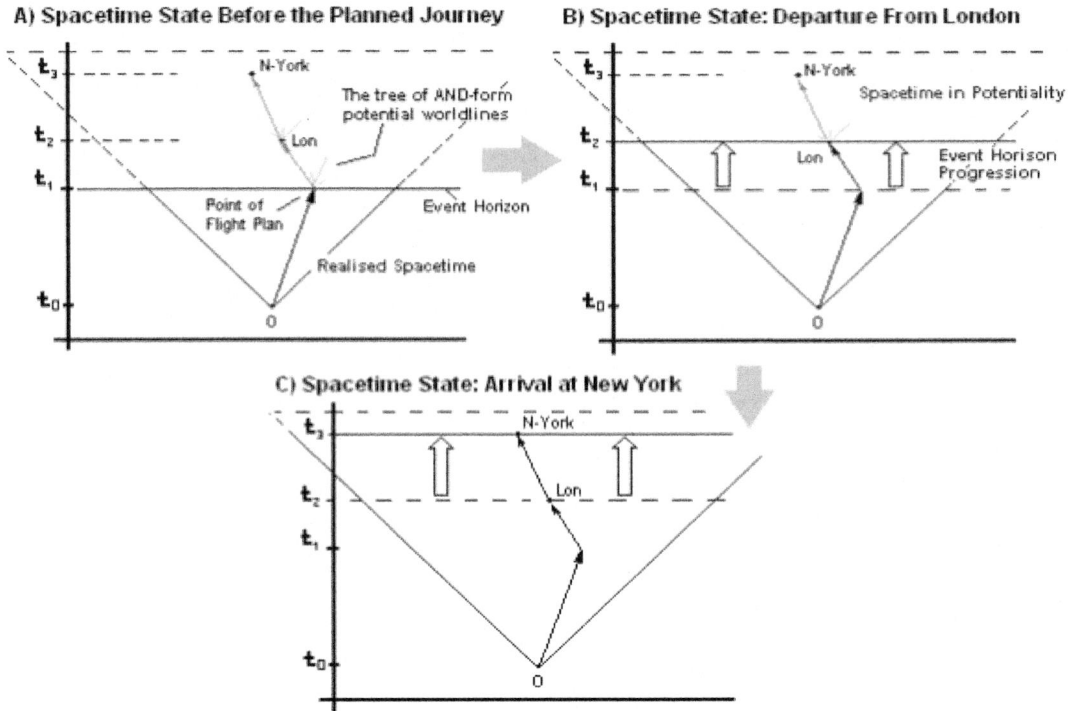

Fig. 2.03: Illustration of the principle of causality using intermediate model spacetime diagrams

Notes: Diagrams **A** through **C** depict the principle of causality from within the intermediate model spacetime. **Diagram-A** depicts a flight from New York to London, incepted at t_1 at the endpoint of a realised worldline originating at **0**, and denoted as the **point of flight plan**. Obviously, one cannot expect to arrive at New York before first departing from London. At the point of inception at t_1, potential events comprised of **departure from London** (denoted as **LON**) and **arrival at New York** (denoted as **N-York**) reside within quantum mechanical spacetime in potentiality above the event horizon. These are not-yet happened futures: The two future-potential events are joined by a potential worldline, also not-yet happened, constituting a future-potential journey from New York to London. Note the possible branch outs to alternative potential worldlines per the intercession of contingencies: One of these might be a future-potential diversion to another airport, for example. This implies that, until the event horizon progresses up the spacetime diagram, the future is not permanently set, and alternative contingent possibilities exist. In **diagram-B**, the event horizon advances from t_1 to t_2: The departure from London is now realised in OR-form terms; it is no longer in spacetime in potentiality, but now part of the past: i.e., the domain of realised spacetime. By **diagram-C**, all alternate future possible worldlines are erased. The event horizon has progressed to t_3. A newly realised worldline now terminates in **arrival at New York**. Note again that it is not possible to arrive at New York before departing from London: Such a feat would entail that the event horizon first starts at t_3 (i.e., at **arrival at New York**) then undergo retrograde time-reversal back to t_2 (**departure from London**): It would also require that we receive 'signals from the future', from a future New York, before having departed from London. The point is that causal order is bound to the progression of the event horizon, which 'cuts' future potentiality from the realised past, or acts as a firewall between future and past, such as to preserve causality. The event horizon cannot occupy just any point on the time-axis, much less regress to a former past-state in the spacetime system.

Consequently, to the naive realist, the AND-form description of the particle must somehow be 'illusory' (see the quote from J. Jeans in **2.2**). Yet, quantum mechanics demands that the attributes of the particle must remain in AND-form until requisite AND-to-OR collapse and time has transpired. By rendering the future into a purely pre-measurement pre-event quantum mechanical state, quantum mechanical reality can be incorporated into spacetime structure and rescue the principle of causality.

   Hence, quantum mechanics is an *essentially* correct description of nature, consummately justified on account of causality; but entirely complementary and mutual vis-à-vis undeniable OR-form classical physics. Ironically, we now find that it is the naive realist view that is actuality incompatible with causality per its rejection of the reality of the quantum mechanical, of the ontic reality of the future, and of the reality of spacetime in potentiality: all of it necessitated by the demands of causality. **Thus, insofar as it possesses a spacetime in potentiality, as is demanded by causality, our growing block quantum-relativistic incorporative intermediate model is entirely justified by causality**.

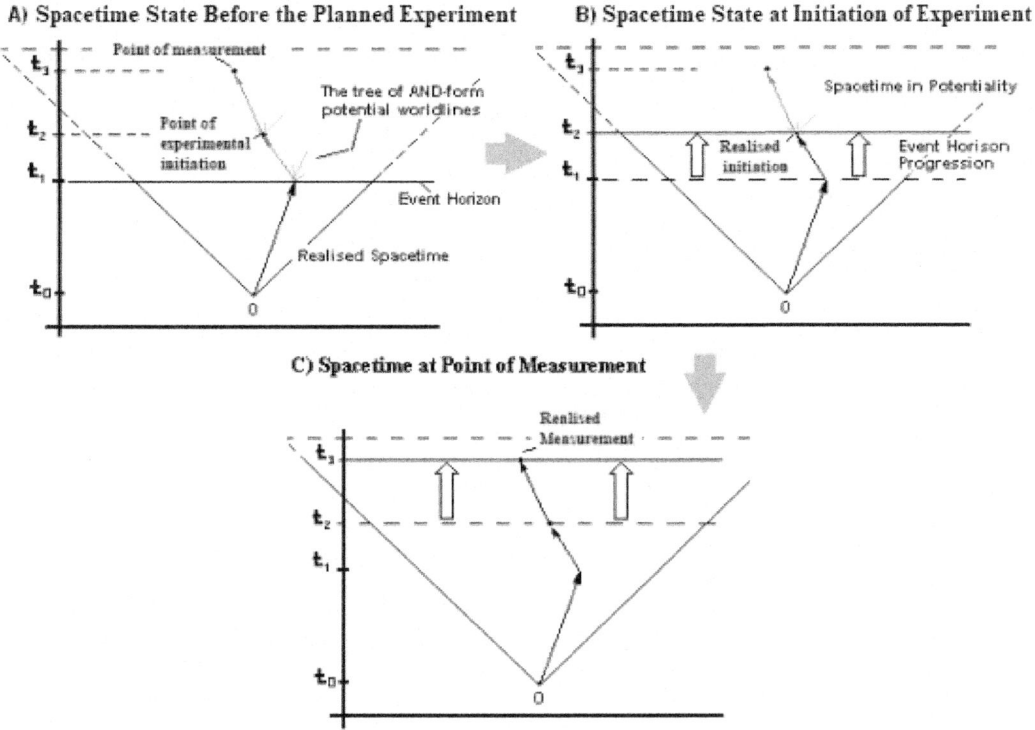

Fig. 2.04: The meaning of measurement in the context of the principle of causality
Notes: The three spacetime diagrams illustrate causality in the context of an experimental measurement carried out on a particle. **Diagram-A** depicts the inception of the experiment when the event horizon is at t_1: Note the denoted **point of experimental initiation**. Also note the denoted final **point of measurement**, which will conclude the experiment. Both reside in the future at potential-t_2 and potential-t_3, respectively, within spacetime in potentiality. These are not-yet realised events. As such, these enjoy no reality, save as pure future-potentialities. Also note **the tree of AND-form potential worldlines** (with most important ones at t_1 and t_2): These are contingencies that spacetime in potentiality must also embed, and may involve, say, the abortion of the experiment per the action of a mischievous laboratory cat (Wheeler's cat). Furthermore, insofar as both the initiation and finalisation of the experiment reside in the future and have not yet happened in OR-form terms, the naïve realist expectation that the particle from t_2 to t_3, ought to have an 'objective' OR-form physical state independent of observation and measurement, is tantamount to treating the futurity of the particle as a collection of resolved OR-form attributes *before* it has transpired into such. The naïve realist expectation is tantamount to 'signals from a future', with attendant violation of causality, per the reception of such signals in proof of the "objective" particle. **Diagram-B** illustrates the progress of the event horizon to t_2, at which point the expected **point of measurement initiation** AND-to-OR collapses into a realised OR-form state: now part of the domain of realised spacetime. **Diagram-C** depicts the resolution of the potential measurement and observation into an OR-form actuality.

2.11: The Principle of Conservation, Causality & the Intermediate Spacetime Model:
Incorporation of the Firewall Principle to Spacetime
In the pertinent flight, it is not possible to arrive at New York without first having departed from London; and it is not possible to treat a particle as 'objective', or as OR-form, before measurement has transpired. The operating principle is the same in both cases: The future cannot transpire into an event before it has been resolved into such, as is demanded by the quantum-relativistic incorporative intermediate model spacetime, in turn justified by the demands of causality. Yet, as was intimated in the Preliminary and reiterated in Book-I, **the principle of conservation is closely linked to, if not enmeshed and inseparably intertwined with, the principle of causality. They are aspects of the same one thing.**

The modern form of the principle of conservation of energy-matter is encapsulated in the famous equation, $E=mc^2$: wherein matter $(...mc^2)$ constitutes congealed energy. The principle states, in one part, that the quantity of energy-matter at the end of a given state-change process must remain unchanged (conserved) with respect to the quantity of energy-matter at the beginning of the state-change process: One cannot create energy-matter out of nothing. Nor can one destroy energy-matter.

Now, it is not usually appreciated that **any violation of the principle of causality would result, or at least necessitate, the tandem violation of the principle of conservation of energy-matter. Hence, the principle of causality and the principle of conservation are intertwined, and together constitute** *the firewall principle*.

As with causality, it follows that **the principle of conservation must also validate and justify AND-form logic, the ontology of synonymous** *spacetime in potentiality*, **and the intermediate spacetime model itself**. To illustrate this, we revisit the flight from London to New York

Thus, let us suppose it were possible to arrive at New York before first departing from London. The proof of such a feat would require that, before departing from London Heathrow, one obtained concrete physical 'signals from the future': i.e., evidence of one's arrival at future New York, Kennedy airport. There are now two selves. But where did the extra energy-matter to constitute the future New York self come from?

What about the plane? The signal received from the future-self at New York must confirm that the same plane must also be at New York, resolved as a future version of the same. Now we have two planes, and the quantity of energy-matter constituting it has also doubled: one in present London and the other at future New York. The extra energy-matter to make up the future plane at New York seems to have been created out of nothing: At the least, this is what 'signals from the future' must require. But it gets worse: **If the future really is resolved before it has transpired and can interact with the present via physical 'signals from the future' (indeed, empirical proof of such a feat requires such an interaction), then one will necessarily be subject to gravitating bodies, not only from the relative present, but also from sources in the future**. We will be subject to gravitational attraction from the plane at London as well as from the plane from future New York. However, we now also have two Moons: the Moon associated with present London and the future-Moon associated with future New York, and never mind the doubling of energy-matter from one Moon into two.

It does not stop there: There exists a clear connection between the Moon and tidal effects on Earth. What will happen to the tidal patterns on Earth per the gravitational influences from both the relative present Moon *and* the future-Moon? Indeed, we must also contend with gravitational influences from two Suns (the relative present *and* future-Sun), with the doubling of energy-matter per two Suns, if not two sets of solar systems...and so on.

The notion of a pre-resolved future, hence the inadvertent naïve realist presumption of a block model universe, viable only if 'signals from the future' are possible (but otherwise beyond falsifiability and credibility) inevitably invites the violation of the principle of conservation or the creation of energy-matter out of nothing.

Nature exhibits AND-form logic, quantum waves, and the phenomenalisation of the ontic future (the quantum mechanical wave) precisely to preserve both causality *and* **the principle of conservation. It follows that our intermediate model spacetime, and its spacetime in potentiality, in service of causality and the preservation of putative conservation principles, is perfectly justified and necessitated by the firewall principle that wards against such 'signals from the future' and attendant violations of the conservation principle.**

2.12: Isolation, Quantum Mechanisation & Futurisation of the Barrier, As Demanded by the Firewall Principle & Spacetime Structure

In Book-I **1.22** we made the apparent contentious assertion that, upon isolation, the barrier in the two-slit experiment *must* quantum mechanise and futurise into an AND-form state. Assuming the absence of memory, the barrier would be relegated to spacetime in potentiality (to the future) into an AND-form superposition of *all* four possible barrier-settings, including 'both slits closed'. Again, in the absence of the memory-retention of initial conditions, combined to the absence of attendant grand decoherence of the future into consistency with the past, the barrier would subsequently AND-to-OR resolve into anyone of its four possibilities in a quantum indeterminate way. Yet, per the dissociation and retention of the initial barrier-setting as abstract memory, upon subsequent AND-to-OR collapse, the initial barrier-setting will be restored to the barrier, even though the barrier had indeed quantum mechanised and futurised into an AND-form state of alternatives upon its isolation and relegation to futurity.

Upon closer inspection, this restoration does not imply that the barrier was retained as an OR-form state throughout the interim of isolation. Indeed, even per facile information-matter dualism, the initial OR-form barrier-setting *must* dissociate from the futurised quantum mechanised de Broglie matter pertaining to the barrier. The initial setting of the barrier will be retained in abstract memory within realised spacetime, but in a form other than OR-form. The restitution of the barrier at the end of isolation simply implies the reality of abstract memory and the decoherence such memory imposes on spacetime in potentiality: one that must AND-to-OR collapse into the restoration of an OR-form of the initial barrier-setting, which was retained as abstract memory. (Here, we begin to see a glimmer of the anticipated recuperation of the conservation principle in its abstract temporised mnemonic form; as the conservation of memory).

It cannot be sufficiently avowed that the barrier cannot constitute an OR-form configuration in the interim of experimental isolation because, **upon isolation, the barrier and the whole system is objectively relegated to the future... to spacetime in potentiality... into an AND-form state. Therein, the barrier** *must* **futurise into an as-yet not-happened state of AND-form possibilities... one open to its alternative settings per function of future contingency, but with the near-certainty of the restoration of the futurised barrier into its original setting per the demands of memory. To treat the barrier otherwise would be tantamount to the claim**

for a pre-resolved barrier, with 'signals from the future', in contravention of the firewall principle, in that the claim that the barrier must be in an OR-form state throughout the interim of isolation requires empirical evidence to that effect, which could only be furnished via 'signals from the future'.

The practical alternative to 'signals from the future' requires intercession to end experimental isolation prematurely, or keep the barrier under near-continuous observation throughout the experiment via observational inputs and feedback that approach the Quantum Zeno effect (see Book-I **1.25**). Yet even this could not constitute absolute seamless-continuous observation, given sundry limits imposed by the impossibility of zero wavelength or infinite frequency inputs, and consequent ineliminability of AND-form states and intervals of isolation. Such AND-form intervals are ineliminable and, from the framework of the intermediate model spacetime, constitute states deferred to the future… deferred into spacetime in potentiality. (See the didactic depiction in **Fig. 2.05**).

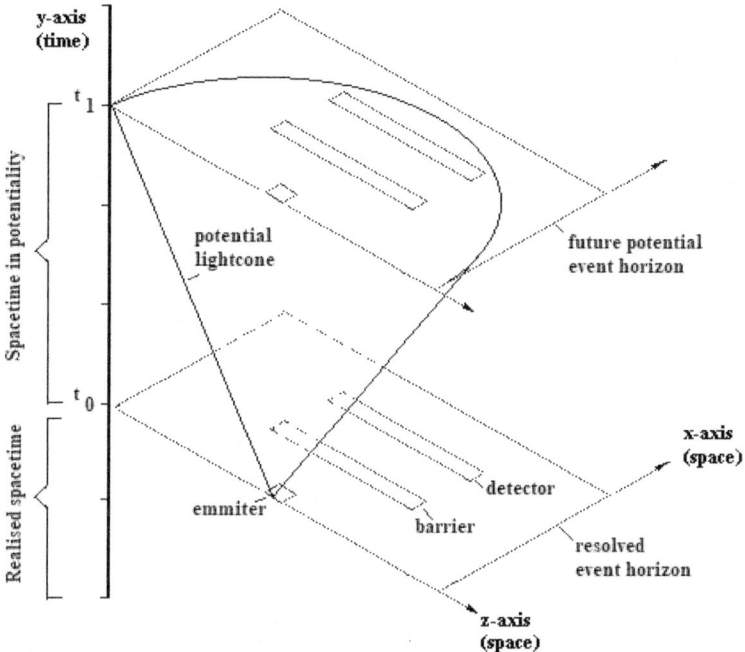

Fig. 2.05: Embedding the two-slit experiment to the intermediate model spacetime: barrier futurisation
Notes: The didactic intermediate spacetime diagram depicts the two-slit experiment from within the context of the as-yet not-happened future: i.e. from within spacetime in potentiality. Note that the various components of the two-slit experiment on or along the two-dimensionally rendered event horizon at t_0 and at future-potential event horizon at t_1, are shown embedded within spacetime in potentiality and, by t_1, are within the reach of the future-potential lightcone: i.e. the semicircle that encapsulates the reach of the light radiation from the emitter point-origin. Also note that the barrier is deferred to the future, to spacetime in potentiality, from t_0 to t_1. This also constitutes the interval of experimental isolation. As such, the barrier is quantum mechanised and futurised into an AND-form state of all its alternate possibilities: i.e. two of "one slit open", one "both slits open", and one "both slits closed". Therein the barrier cannot be treated as an OR-form state, save by the contravention of causality and entwined principles of conservation via 'signals from the future'. However, the dissociation of the initial barrier setup via facile information-matter dualism, or its retention in abstract form within the domain of realised spacetime (below t_0), will restore that initial setting to the future barrier at t_1 at the end of isolation. Yet, there exists future-potential contingencies within spacetime in potentiality (not shown in the depiction) that could change the setting of the barrier or upset the experiment. For example, the laboratory cat might tumble the whole experimental setup to the floor and trip the barrier to "both slits closed", or any other setting: The barrier must remain open to such contingent futures despite abstract memory, which cannot absolutely close off the future only to consistent-only futures and must concede to said consistent future contingencies. Therefore, in the interim of its isolation, the futurised barrier *must* constitute a superposition of AND-form alternative barrier possibilities per ineliminable contingent possibilities inherent to spacetime in potentiality.

*

We have already espoused that abstract memory, as the retained form of initial conditions, is bound to decohere futures within spacetime in potentiality into consistent-only sets: This claim will be proven and enhanced in **2.34**. Consequently, with the end of isolation, the quantum mechanised barrier will resolve back into its initial setting. But must we treat the interim of isolation as a quantum mechanised superpose of alternative barrier-settings, if the initial setting is certain to be restored per the demands of memory? Given memory, why not treat the barrier as a fully resolved state throughout the isolation-interim?

Therein, **the futures within spacetime in potentiality also contain contingent possibilities: For example, the experiment might get struck by lightning… or the mischievous laboratory cat (Wheeler's cat) might tumble the whole contraption to the floor… or some other contingent possibility that might affect and change the setting on the barrier, overriding the decoherence imposed by memory. Such contingencies typically constitute low probability futures. Yet, the ontic future, constituted as spacetime in potentiality, *will include* contingent futures composed of alternative barrier-settings. Therefore, the future-form description of the barrier pertinent to the interim of isolation *must* incorporate such contingencies as part of its AND-form superposition structure, despite the initial barrier setting retained in abstract memory. Hence, the alternative barrier-settings, if nothing else, will be incorporated into the interim of isolation and to the futurity of the barrier, despite implications from memory.**

Thus, even if the initial conditions in memory imposed the restitution of 'both slits open' at the end of isolation, a contingent possibility might override this to, say 'one slit open'. In other words, **abstract memory has no absolute guarantee of closing off spacetime in potentiality in favour of just one consistent-only possibilities. Indeed, memory *cannot* close that future off to contingent alternatives.**

This insight will emerge into full saliency with John Wheeler's delayed choice approach, wherein the alternatives can be suspended in such a way that, given appropriate timing, the experimenter can intercede in a contingent way, playing the role of Wheeler's cat, with remarkable consequences and implications to the structure of the future and of spacetime. Indeed, in Part-II, the implications from delayed choice will be fully exploited to develop the nested futures view of the quantum mechanical wave and of spacetime in potentiality, critical to our Alternative Realist interpretation and thesis.

2.13: The Intermediate Model Spacetime:
Spacetime Structure, New Complementarity & the Sum Over All Potential Worldlines

Wave-particle complementarity, a central feature of generic quantum mechanics, but one that has survived into an improved form in the new complementarity thesis (see Preliminary: **0.14 to 0.17**) follows seamlessly from intermediate model spacetime: The **new complementarity thesis is both clarified and justified by our intermediate spacetime model,** as depicted in the previous rendition in **Fig. 2.02** and in succeeding depictions based on it.

By direct reference to the intermediate model spacetime, the classical resolved **particle facet of the particle-wave duality is identified with the leading OR-form post-measurement event resolved on or along the event horizon**. The domain of realised spacetime, the area 'below' the event horizon, comprises the domain of the sum of past states of the 'particle': i.e., its physical basis as abstract memory, whatever the form such memory turns out to assume: a matter resolved in Book-IV.

A realised worldline within the domain of realised spacetime (one whose leading point terminates at the event horizon) constitutes the history of the sum of realised putative 'positional states' of our particle, or the history of a corpus of related events about the particle across spacetime. From anticipated de-spatialisation in Book-IV, we will need to recapitulate this description *without* space, in terms of pure delayed choice time-interval relations, or as a distribution in time. Through de-spatialisation, the nature of abstract memory will become fully salient (Book-IV). Until then, utilising generic notions of space, we reiterate that the domain of realised spacetime must constitute the total sum of organised past states (the sum of physical memory) retained in anticipated abstract and extant form 'below' the event horizon.

Whereas the 'particle' aspect of the wave-particle duality constitutes its OR-form manifestation and materialisation on or along the event horizon, **the wave-aspect of particle-wave duality and complementarity is to be identified with the quantum mechanical AND-form domain of spacetime in potentiality and futurity 'above' the event horizon.** Hence, particle-wave duality is the empirical manifestation of the very structure of intermediate model spacetime. Thus, wave-particle dualism is the experimental-empirical face of the dualism that resides between spacetime in potentiality versus the event horizon, short of inclusion of abstract memory. Hence, **the new wave-particle complementarity thesis, recapitulated in terms of the intermediate model spacetime, succeeds and is synonymous with the complementarity between AND-form spacetime in potentiality versus OR-form event horizon: a dualism between the AND-form future state of a system and its present-continuous OR-form state, short of abstract memory and the inclusion of mnemonic realised spacetime.**

This same particle-wave dualism and complementarily is part of a set of insights that also supports information-matter dualism and the primitive unification of quantum and classical-relativistic nature: see essay **2.08**.

*

There are other implications from the intermediate spacetime model: In **Fig. 2.06** diagram-A, we present a simple depiction of Richard P. Feynman's path integral formulation, or 'sum over all possible paths', pertinent to the two-slit experiment. Therein, if we now add a second barrier with four open slits, and a third with eight open slits, followed by a fourth with sixteen slits open… and keep adding barriers *ad infinitum*… we will have cut out infinitely many such slits such as to form a quantum mechanical wave containing infinitely many putative potential 'which way' paths from emitter to detector: all in unitary AND-form superposition.

Note that, at this juncture, notional space, hence notional 'paths in space', are treated as valid, despite anticipated de-spatialisation (Book IV) and despite the attendant critique of the notion of motion to be presented in Part-II: **2.23 to 2.26**. We also put aside our doubts

200

about 'which way' paths raised in the critique of the standard evaluation of the two-slit experiment and in the critique of the Afshar controversy (see Preliminary, **0.16** to **0.17**).

In **Fig. 2.06-A**, the most probable 'pathway' out of the manifold of infinite paths is displayed as the shortest-path (the bold-line) from emitter to the detector: As Feynman claimed, this most probable pathway is given by the amplitude of the 'sum over all possible paths', of which there are infinitely many: (Note how issues of renormalisation clearly enter the fray). In **Fig. 2.06-B**, we graft Feynman's 'sum over all possible paths' to our quantum–relativistic incorporative intermediate model spacetime: We characterise the 'paths' as potential worldlines leading on from the endpoint of a realised worldline terminating at the event horizon at t_1. Thus, where the realised worldline meets the event horizon, it is potentialised into an AND-form superpose of manifold future-potential worldlines formed in the fashion of Feynman's 'sum over all paths' constituting a *sum over all future-potential worldlines*.

Our sum over all potential worldlines is comprised of, in principle, infinitely many pre-measurement or pre-event AND-form alternate future-potential worldlines. As is also true in Feynman's conception, the bold worldline illustrated in **Fig. 2.06-B**, embedded within spacetime in potentiality, constitutes the most probable potential worldline, given by the amplitude of the manifold of the least probable worldlines, as a 'sum over all possible worldlines'. This is applicable to the whole of the universe and to the anticipated nested future view of total spacetime in potentiality (see Part-II).

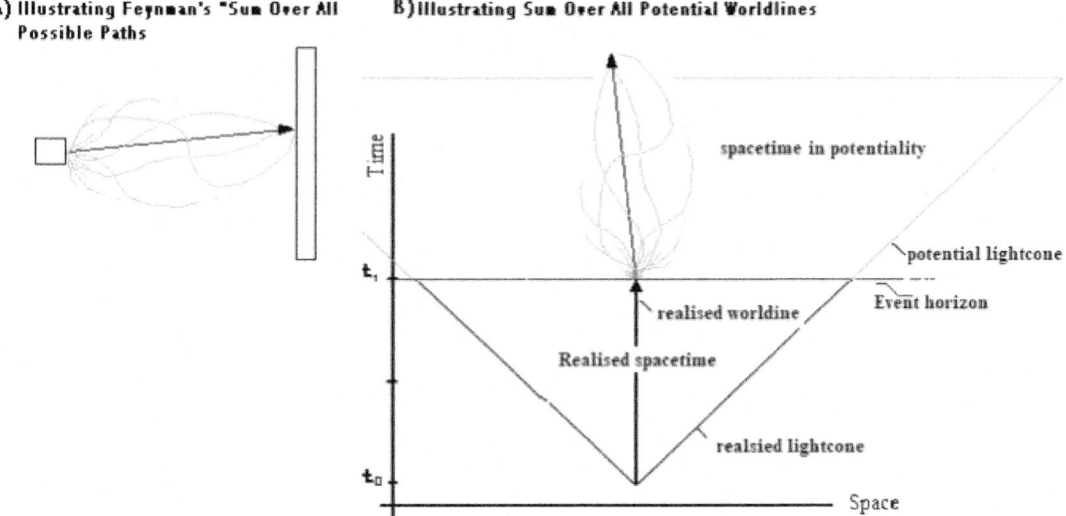

Fig. 2.06: Illustrating the concept of "sum over all potential worldlines"
Notes: Didactic **diagram-A** depicts Feynman's concept of the "sum over all paths" that a particle will assume on its putative journey from emitter to the detector-surface. The sum constitutes a superposition of potentially infinite number of possible paths: the most probable path is rendered in bold: i.e., the sum over all paths: the high amplitude or highest probability path. In a similar vein, in **diagram-B**, in a quantum mechanical AND-form superposition state for future-potential worldlines, the most probable future-potential worldline is also rendered in bold: i.e. "the sum over all potential worldlines". Thus, when an object of zero velocity (a purely vertical worldline within realised spacetime) attains its culmination at event horizon at t_1, to it is attached a future-form quantum mechanical frontier comprised of the "sum over all potential worldlines" within spacetime in potentiality, with the highest probability worldline (bold). The latter constitutes a default spacetime teleology or default future end-goal of spacetime.

Treating **Fig. 2.06-B** as a simplified version of the whole universe, we state that **the evolutionary direction of the whole universe must tend toward a** *universal sum over all future-potential worldlines*, **generated per the amplitude of** *all* **the lesser future potential worldlines constituting that universe-scale AND-form wavefunction and the total domain of spacetime in potentiality**.

This universe-scale 'sum over all potential worldlines', with its most probable worldline, must constitute a strange attractor; funnelling outcomes from subsequent future progressions of the event horizon toward its fulfilment, and, with it, funnelling the general development of the whole universe toward that attractor. **The spacetime attractor, given by the amplitude of the** *sum over all future potential worldlines*, **can be said to govern and bind the developmental history of the universe, so constituting a default teleology; defining the universe's most probable end-state or goal, otherwise placed at the infinitely removed future within spacetime in potentiality**.

However, **the domain of realised space time, the domain of memory, is not going to assume a passive role: The domain of realised spacetime acts as a boundary state, a pertinent set of initial and limiting conditions to which any future-form 'sum over all future-potential worldlines', and its default teleology, must conform**. Any subsequent AND-to-OR collapse that produces OR-form events in conformity with the said default teleology, upon relegation to memory, must feedback as a reinforcing set of boundary conditions; analogous to what happens to a reducing aiming circle or dead reckoning circle in ship-navigation; wherein the circle is perpetually reduced, and ship disposition becomes progressively more certain, or wherein the future is reduced and the scope of future uncertainty is reduced in tandem.

Applied to the context of our intermediate model of spacetime, the 'sum over all potential worldlines' may be perpetually and progressively reduced and, in the process, acquire increasing definition toward a 'crazy' end-goal or conclusion. Yet, it can never attain ultimate or final reduction or absolutely definition, given that such a state is potentialised to the infinitely removed future and subject to interim contingent futures (i.e., Wheeler's cat).

Another useful clarifying analogy is that of a visual image that must perpetually come into sharper and sharper relief, resolution and definition, but can never attain its final absolute relief. The perpetually sharpening and clarifying image stands for both the perpetual reduction of the AND-form 'sum over potential worldlines' *and* represents the process of the expansion and completion of memory and universal history, perpetually relegated to and retained within the mnemonic domain of realised spacetime.

2.14: Clarification of Heisenberg's Uncertainty Principle by Spacetime Structure

According to Heisenberg's uncertainty principle, it is not possible to resolve accurately one attribute (position) without inflicting a simultaneous determinable reduction in the resolution of another non-commutative attribute (momentum). As was argued in Book-I (**1.34**), the account of the uncertainty principle must be anchored to a causality-based naturalistic-intuitive approach furnished by the physical *raison d'etre* of the quantum mechanical wave, combined to the intertwine of causality and the conservation principle that attends the firewall principle, which also justifies the segregation of the growing block spacetime-future from spacetime-past. **The utilisation of the intermediate spacetime model, which employs just such a separation of past versus future, and is also based on the firewall principle, augments the clarification of the uncertainty principle espoused in Book-I.**

First, recall the popular but erroneous account of uncertainty. This proved useful to our causality-based naturalistic intuitive account in Book-I. It began with the correct observation that, to pin-point the position of a particle accurately, it is necessary to confine that particle to as small as possible a 'space', using very short wavelength high frequency observational inputs. If we use long wavelength low frequency inputs instead, the particle is liable to be anywhere within the now-larger area delimited by the longer wavelength, with implied inaccuracy and enlarged uncertainty in position. Therefore, we must shorten the wavelength-input to reduce uncertainty about position.

However, the energy of the input is related to its wavelength or frequency, combined to the Planck's constant. The energy of the inputted radiation increases as its wavelength is shortened or its frequency is increased: The shorter the wavelength the greater the energy of the input. Since accurate resolution of a particle's position requires that we use the shortest possible wavelength of an input, this also entails an input that is highly energetic or high frequency. This will give a big 'kick' to the particle. The kick will imply a large range of possible momentum-values for that particle (i.e. a larger cross-section of values for velocity times the particle mass). Thus, uncertainty about momentum will become enlarged per the wider range of possible momentum-values because of our increased certainty about the particle's position.

Unfortunately, this popular account gives the impression that, despite the effective uncertainty in momentum engendered by short wavelength observational inputs, the particle could yet be treated as having an 'objective' resolved OR-form momentum-value: *not* an AND-form superposition of as-yet not-realised possibilities for such: It leaves open the possibility for the erroneous supposition that a definite 'objective' OR-form momentum-state for the particle might yet abide, despite uncertainty.

Conversely, if we try to resolve the momentum of the particle accurately, we must input long-wavelength radiation to reduce the range of possible momentum-values that the particle might assume, by reducing the 'kick' given to it, given that the longer wavelength input has a lower level of energy and conveys a lower 'kick' to the particle. This will furnish a greater level of accuracy per resolution of momentum. But the long-wavelength input will enlarge the area of confinement of the particle, which could be anywhere within that enlarged area, with consequent greater uncertainty in position, even as we gain greater certainty about momentum.

Again, this popular account allows for the notion that the particle might yet possess a fully resolved OR-form position somewhere within the enlarged area of confinement, instead of an AND-form superpose of not-realised position-possibilities.

Clearly, observational inputs garner either a position-bias or a momentum-bias: Thus, the mutual exclusivity of the position-bias versus momentum-bias of measurement necessarily follows: It is not possible to devise a measuring device or process that simultaneously furnishes high resolution over both position *and* momentum: If we combine both short-wavelength position-bias and long-wavelength momentum-bias, this will merely produce a mixture, without accuracy over any attribute.

In Book-I, it was asserted that, the more accurate resolution of one attribute must defer the other attribute to the future: to *potential*. By deferring it to future-potential, the attribute thus forgone cannot be described as anything but a quantum mechanical set of as-yet not realised AND-form superposition of future possibilities within the future. If two attributes cannot be measured at the same time per structural factors pertaining to different observational wavelength-input requirements, the other attribute *must* be deferred to and remain in its pre-event pre-measurement future-form quantum mechanical state of as-yet not-realised values in AND-form. To treat that future as if it constituted an 'objective' OR-form resolved state... or to assume that a particle's foregone non-commutative attribute in its

202

pre-event pre-measurement future-form could assume a realised OR-form character... would be tantamount to 'signals from the future' and the transpiration of that future before it transpired. If such a proof were experimental realised... if 'signals from the future' could be received... it would most certainly culminate into the violation of causality, the contravention of the conservation principle, and the unravelling of the firewall principle. Thus, Heisenberg's uncertainty is a necessary condition demanded and forced by the intertwine of causality and the conservation principle into the firewall principle, inbuilt to the structure of our intermediate spacetime model. In other words, **Heisenberg's uncertainty principle is demanded and enforced by the structure of growing block intermediate model spacetime, insofar as its structure is compelled by the firewall principle.**

*

In terms of our quantum-relativistic incorporative model of spacetime, we may restate the above firewall principle-based clarification of Heisenberg's uncertainty by using the framework furnished in **Fig. 2.04**. (See **2.12**). Therein, an experimenter makes the decision to plan out a position-measurement on a particle and has set up the appropriate apparatus to apply short-wavelength high energy observational inputs. The plan is initiated from time t_0 to t_1. The experimenter applies the short wavelength input across t_1 to t_2, obtaining an OR-form positional resolution of the particle at t_2. The pertinent event horizon must progress from t_1 to t_2: yielding a highly confined and accurate position-measure, or the potential for such upon subsequent AND-to-OR collapse. Of course, the short wavelength input is *not* appropriate for the accurate resolution of the particle's momentum, and this is structurally deferred to some future-potential resolution at a future-potential event horizon at, say, t_3.

Since the measurement of momentum has been deferred to future-potential-t_3, by t_2, momentum has not yet transpired into an event. From interval t_2 to potential-t_3, momentum must remain a non-event: a pure AND-form potentiality for momentum-values within spacetime in potentiality, *within the future.* **Thus, the momentum state within that interval is** *not* **a concrete OR-form actuality,** *because* **it is deferred to the not-yet happened future.** The momentum of the particle must remain in an AND-form superposition of not-yet realised momenta, proportionate to the 'kick' the particle received at realised-t_1 and within and across spacetime in potentiality up to potential-t_3. As such, the particle's momentum *cannot* constitute an 'objective' single momentum OR-form state. Any view otherwise must claim the transpiration of an event, specifically the resolution of a definite particle momentum-value, before it has transpired into such; with consequent 'signals from the future', the contravention of causality, the usurping of the principle of conservation, and the unravelling of the firewall principle. As to which future-possible momentum values will be assumed at potential-t_3, this can only be assigned a probability.

Conversely, instead of position, the experimenter might measure momentum first, but use a similar time-sequence of measurements described above, but using long-wavelength momentum-biased observational inputs. This would merely invert the above in favour of momentum and to the deference of position to not-yet realised future potentiality, constituted as a broad spread of AND-form position possibilities within spacetime in potentiality, within the future. Again, to insist otherwise would be tantamount to 'signals from the future' in travesty of causality, the violation of the principle of conservation, and the overturning of the firewall principle.

Thus, we apprehend the direct usefulness of our growing block intermediate model spacetime structure to the naturalistic-intuitive account of Heisenberg's uncertainty, first espoused in Book-I: **1.34**.

2.15: Summary & Conclusion to Part-I

The following constitutes a bullet-point summary of the findings in Part-I:

- **AND-form logic and the quantum mechanical wave constitute the ontic future: the future constituted as an ontically real state.** This solves the physical meaning of the quantum mechanical wave and AND-form logic: leading to its incorporation as 'spacetime in potentiality' within intermediate model spacetime: (**2.02** and **2.03**).

- **The intermediate model spacetime furnishes primitive unification of quantum mechanical and classical-relativistic descriptions of nature** via its tripartite structure, undercutting the 'contradiction' said to hold between these: A primitive unification of physics is thus obtained, precursor to a superlative unification that might follow it. (See **2.04** and **2.05**).

- **The intermediate spacetime model,** a growing block model, is comprised of a domain of **realised spacetime** which embeds the sum of past realised lightcones and worldlines; a **spacetime in potentiality**, or the future-form quantum mechanical domain of future-potential lightcones and worldlines in AND-form. Finally, the model has an **event horizon**: the 'cut' between the two domains, comprised of the most recent OR-form realisations AND-to-OR collapsed out of spacetime in potentiality. **The progression of the event horizon 'upward' the spacetime system is synonymous with AND-to-OR collapse or time: no longer a 'fourth dimension'.** The event horizon does not constitute 'absolute time' or the notion of a common present moment, and it remains consistent with the relativity of simultaneity: (See **2.05** to **2.06**).

- **Information-matter dualism emerges seamlessly from the intermediate spacetime structure**: Information-matter dualism is synonymous with AND-OR dualism and wave-particle dualism: both per function of future-past dualism evinced in and as the structure of intermediate model spacetime. Realised OR-form information is obviously not carried in AND-form spacetime in potentiality (the quantum wave), and, obviously, information as it pertains to the past (realised spacetime) is not reducible to or identifiable with the AND-form futurity or spacetime in potentiality: The AND-form

203

future radiation and matter cannot 'carry' or 'transport' realised spacetime, or the past. Hence, information-matter dualism seamlessly emerges from the structure of growing block spacetime. (See **2.08**).

- **The principle of causality demands the reality of AND-form quantum mechanical futurity**: The future cannot transpire before the past: The future must be described as an AND-form superposition of as-yet not-realised pre-event states. Hence, causality ingrained to past-future dualism demands the real existence of a domain of spacetime in potentiality and the ontic reality of the future. It demands the growing block intermediate spacetime. Thus, causality justifies information-matter dualism mirrored in and by spacetime structure, as much as it justifies spacetime structure. (See **2.10**).

- **The principle of conservation and the principle of causality, intertwined and inseparable, constitute the firewall principle**. The future cannot happen before the past and present: Attendant 'signals from the future' would not only end in the subversion of causality but would also engender the creation of energy-matter out of nothing, subverting the principle of conservation. Causality *and* the conservation principle intertwine into the firewall principle. The firewall principle justifies and necessitates the separation of the domain of spacetime in potentiality from the past and culminates into the quantum-relativistic incorporative model of spacetime. Together with causality, the principle of conservation also justifies information-matter dualism mirrored in and by spacetime structure. (See **2.11**).

- **In the interim of isolation, the barrier is objectively deferred to the future and assumes a quantum mechanical AND-form state**: futurised into a not-yet happened AND-form domain of spacetime in potentiality. Indeed, the two-slit experiment can be modelled within spacetime in potentiality. **The barrier must be quantum mechanised into alternate possible settings, if for no other reason that the fact that low-probability contingent future possibilities also reside within spacetime in potentiality, which could alter the barrier-setting**: e.g. our mischievous laboratory cat, or 'Wheeler's cat', could tumble the experiment, switching 'both slits open' into 'one slit open'. Thus, in the interim of isolation, the barrier *must* assume an AND-form superpose of *all* possible barrier possibilities vis-à-vis future-potential contingencies: (See **2.12**). The veracity of this view will follow from the nested futures view in Part-II.

- **The 'sum over all future-potential worldlines' and spacetime teleology emerge from the generalisation to the domain of spacetime in potentiality of Feynman's 'sum over all paths'**, whose sum of amplitudes directs the evolution of spacetime, notwithstanding contingency, but whose final end-goal resides at the infinitely removed future, beyond realisation: (See **2.13**).

- **Heisenberg's uncertainty follows from the structure of intermediate model spacetime, causality and the firewall principle**. When non-commutative attributes are involved, the resolution of one attribute on the event horizon must necessarily defer the resolution of the other attribute to spacetime in potentiality: to the future, which cannot constitute as an 'objective' OR-form state, save by 'signals from the future', the violation of causality, of the principle of conservation, and the firewall principle mirrored in and by spacetime structure: (See **2.14**).

BOOK-II PART-II:
THE DEEPER VIEW OF QUANTUM MECHANICAL WAVES:
NESTED FUTURES, PARTIAL-PERPETUAL COLLAPSE
AND THE SPATIALLY STATIC QUANTUM MECHANICAL WAVE

2.16: Aims of Part-II

Part-II will expand on the growing block intermediate model spacetime approach central to the Alternative Realist philosophy of physics (Alt-R) by unravelling the **structure of spacetime in potentiality as a complex system of nested futures.** The universal wavefunction or quantum mechanical wave, grasped as the total spacetime in potentiality, embeds mutually incompatible and exclusive systems of 'futures-within-futures', niched to the immanent possibilities typical to generic wavefunctions. In short, a wavefunction does not simply contain immanent AND-form possibilities, but subsequent future wavefunctions nested to those possibilities: That is, **a wavefunction also superposes nests of subsequent future wavefunctions.** This nested futures view was always implicit to John Wheeler's delayed choice approach, and Part-II will review and apply delayed choice to render the nested futures structure of spacetime fully salient.

However, conclusions obtained from delayed choice logic and the nested futures view, combined to the critical distinction between *closed-choice* versus *open-choice* quantum mechanical systems (prospective essay, **2.20**), will constitute the basis for **the validation of the spatially static view of the quantum mechanical wave**: This asserts that the wave does not move, and is no more capable of doing so than the ontic future or spacetime in potentiality itself, with which the quantum wave is synonymous, or within which it is embedded and nested. Instead, the quantum mechanical wave constitutes a non-moving non-undulating static state.

We will also obtain proof of the **partial-perpetual model of wavefunction collapse**, which runs counter to the conventional pragmatic apprehension of wavefunction collapse as a one-step all-exhaustive process of AND-to-OR collapse of imminent-only futures. This will be incorporated to the intermediate model spacetime system and the broken-discontinuous partial-perpetual progression of the event horizon. Indeed, it is obvious that the quantum wave, now grasped as a complex nested future state that elaborates to the infinitely removed future, all synonymous with the totality of the ontic futurity constituted as the total spacetime in potentiality *and* instantaneous whole… does not move or undulate. The quantum wave is as spatially static, non-moving, and as non-undulating as the total non-moving and non-undulating spacetime in potentiality with which it is synonymous. And the decay of the future via AND-to-OR collapse is always partial and perpetual. Hence, wavefunction collapse is ultimately always partial and perpetual: *not* all-exhaustive.

The static view of the quantum wave will necessitate a re-framing of both decoupling theory and information-matter dualism; furnishing deeper insights into both; exceeding those obtained in Book-I and hitherto. It will also help establish primers for **the presage to the reformation of the Schrödinger wave-equation** in the light of the nested futures view, the spatially static quantum wave approach, and from the partial-perpetual collapse-model.

Part-II will also elaborate **a taxonomy of nested futures**. The taxonomy will lead to the first foray into the case for non-local and transtemporal **quantum erasure of non-consistent nested futures, obtained through counterfactual causality and, with it, the grand decoherence of the totality of spacetime in potentiality to the infinitely removed future**. Counterfactual causality and grand decoherence will constitute the precursors to the solution to the problem of non-locality generally, and constitute the cornerstone to **the revival of Mach's principle**, foundational to any future **growing block quantum mechanical solution to inertia and the growing block quantum mechanical solution to the permanent magnet**, pertinent to the solutions to problems that pertain to 'vacuum fluctuations' and the zero-point, and to generally overlooked problems associated with particle spin in atomic nuclei.

2.17: Introducing the Nested-Futures View of Spacetime in Potentiality

In Part-I, we discovered that the quantum mechanical wave constitutes the non-subjective ontic future. We integrated this futurity to our intermediate model of spacetime, identifying it as the domain of spacetime in potentiality: the future-form of spacetime (Part-I: **2.05**). However, the domain of **spacetime in potentiality is comprised, not only of immanent quantum mechanical AND-form future possibilities, but constitutes a more exotic and complex system of subsequent futures, embedded and nested to each immanent or preceding future. This nested futures complex encompasses the whole of spacetime in potentiality and elaborates to the infinitely removed future**.

To grasp the nested futures system, consider an approach using Schrödinger's cat thought experiment. Therein, a cat is placed and sealed in a box. A poison vessel is incorporated to the box. The poison vessel is rigged to two quantum spin outcomes. Depending on the spin-outcome from a subsequent AND-to-OR collapse, there is a fifty-fifty chance of the release of the poison into the box. Hence, the cat will die… or live… depending on whether the poison is released or not released.

As a set of as-yet not-happened macro-scale events within spacetime in potentiality, Schrödinger's cat is constituted as the superpose… {**live cat** AND **dead cat**}. But these are immanent possibilities: mere first-round future alternatives that might come into realisation via subsequent AND-to-OR collapse and attendant progression of the event horizon.

A typical description of a quantum mechanical wavefunction with two alternatives supposes that, upon pertinent AND-to-OR collapse, *all* the futures will be exhausted, and one possibility will resolve into OR-form realisation, with the alternative erased. The moment we grasp the quantum mechanical wave as the ontic reality of the future, we are no longer bound to suppose that the first-round immanent AND-form futures could ever constitute or exhaust the total future. Each immanent possibility will have attached to it a nest of subsequent future possibilities, and each nested future will in its turn possess its own subsequent future possibilities further removed, and so on… to the infinitely removed future. Hence, the nested futures view of spacetime in potentiality.

Applying this to Schrödinger's cat, each imminent possibility within {**live cat** AND **dead cat**} will possess subsequent unique nested futures. For example, if AND-to-OR collapse should generate the OR-form of 'live cat', this could nest for a subsequent future, 'happy daughter + miserable physicist'. Another alternative could be 'happy daughter + physicist files for divorce'. There could be many other alternative nested futures: for brevity, we will restrict our approach to just two. Thus, nested to the as-yet not-happened AND-form of 'live cat'…

…AND-form 'live cat': {happy daughter + miserable physicist **AND** happy daughter + physicist files for divorce}

On the other hand, if AND-to-OR collapse generates the OR-form of 'dead cat', this will also nest subsequent futures: such as, 'unhappy daughter + happy physicist'. Another alternative could be 'unhappy daughter + physicist's partner files for divorce'. Again, only two nested futures for brevity…

…AND-form 'dead cat': {unhappy daughter + happy physicist **AND** unhappy daughter + physicist's partner files for divorce}

The above pair of nested future possibilities describe **the divergent category nested futures: Therein, the futures diverge at each AND-to-OR collapse process, and to infinity**. When we embed the above into spacetime in potentiality we obtain the nested futures view: a view that supersedes and corrects the generic characterisation of the quantum wave as containing *only* immanent future possibilities, or one that supposedly undergoes a one-time comprehensive all-exhaustive AND-to-OR collapse. Hence, we obtain a more complex and exotic account of both quantum waves, universal wavefunctions and synonymous spacetime in potentiality… as a niched or embedded complex constituted as a system of futures-within-futures…as wavefunctions-within-wavefunctions.

The nested futures approach to the quantum mechanical wave and to the domain of spacetime in potentiality must imply the overhaul of the Schrödinger wave equation into a system that must incorporate nested futures or nested potentialities, and not just the usual immanent ones. The equation must also accommodate **partial-perpetual collapse** of the quantum wave beyond generic all-exhaustive comprehensive collapse. All of this must obviously apply to spacetime in potentiality. The overhaul must also incorporate **AND-to-OR collapse as time itself**, superseding the 'fourth dimension' view of time.

Furthermore, notwithstanding anticipated de-spatialisation, **the overhauled Schrödinger equation must incorporate the spatially static quantum wave**, which emerges from the nested futures view and from the obvious tautology that the future cannot move or undergo time-undulation: More on this will be elaborated in **2.23**.

2.18: Wheeler's Delayed Choice & the Nested Futures View of Quantum Mechanical Waves

How do we prove the reality of nested futures? It is implicitly to our intermediate model of spacetime. It is also obvious from the observation that the entirety of spacetime in potentiality cannot be reduced to just one immanent set of AND-form future possibilities but must be constituted as a 'tree' and nest of future possibilities that inexhaustibly elaborate to the indefinite future. It was also implicit to Feynman's 'sum over all paths' view of the quantum mechanical state. But **the decisive proof of the nested futures structure of spacetime in potentiality was always implicit to John Wheeler's delayed choice experiment**[65], wherein alternative mutually incompatible futures *do* exist, embedded within a single all-encompassing nesting quantum mechanical wave, albeit in a concealed or implicit way, forming a **primary hedging or 'carrying' wavefunction that nests for two mutually incompatible subsequent wavefunctions attached to the immanent possibilities**, often with nested futures involving more than two nested alternatives.

In the nested futures system, contingency plays a decisive role: Contingent possibilities, typically of low probability of occurrence, can drastically alter the future, or restructure spacetime in potentiality in unexpected ways: analogised by our apocryphal mischievous laboratory cat, or 'Wheeler's cat', who changes the barrier-setting of the two-slit experiment and the future outcome obtained therein by tumbling the apparatus to the floor. We shall discover that **switch-x in Wheeler's delayed choice constitutes the abstract generalised face of contingency; of our apocryphal' Wheeler's cat'.**

The nested futures view is consistent with the expectation from spacetime in potentiality… that any immanent possibility above the event horizon must, upon its AND-to-OR collapse, open the way to subsequent futures and their wavefunctions; all with their own immanent possibilities… constituting a *wavefunctions-within-wavefunctions* approach to both spacetime in potentiality *and* to quantum mechanical waves. Again, as to which nested future wavefunction gets to happen, to the extent that the dictate of memory might be overridden, this is a matter of contingent future possibilities: i.e., Wheeler's cat... or **switch-x**.

*

A simplified version of Wheeler's delayed choice experiment is presented in **Fig. 2.07**. The generic version presumes the operation of spatially dynamic quantum mechanical waves. These are imagined in the usual dubious pseudo-realtime way critiqued in the preceding non-standard appraisal of the two slit-experiment and the Afshar controversy: (see Preliminary: **0.16** and **0.17**). We will later argue for the spatially static view of the quantum wave. At this juncture, we will assume spatially dynamic quantum waves, given that the principal question that confronts us is whether it is possible to experimentally prove the reality of the nested futures view of quantum mechanical waves and, by extension, the nested 'wavefunction within wavefunction' character of spacetime in potentiality.

In **Fig. 2.07**, the simplified delayed choice is comprised of the following components: A 'beam-splitter' at **A** beam-splits a single photon of light radiation into two presumed potential 'which way' paths: {path-ACD **AND** path-ABD}. The mirrors at **B** and **C** are expected to deflect the beam-split quantum mechanical photon toward the detector at **D**. The detector at **D** is specially designed to detect self-interference formed from the convergence of the two split beams.

Also note the critical complex at **X-Y** along **C-D**: This complex is the basis for delayed choice: i.e., **switch-x**.

In the experiment, the beam-splitter accomplishes what the barrier does in the staple two-slit experiment: A single photon will start as a 'light beam', which will be subsequently split by the beam-splitter at **A** into two component beams, in the same sense that a single quanta quantum wave is split into two wavelets per two open barrier slits. One beam will propagate to mirror-**B**, and the other to mirror-**C**. Reflected by the mirrors **B** and **C**, the two split-beams are expected to displace toward and converge at detector-**D**, where they are expected to interfere with each other in the usual constructive and destructive terms and form a generic interference pattern. Thus, the setup for delayed choice is essentially identical to that of the two-slit experiment, save for the beam splitter; save for the inclusion of the delayed choice complex at **X-Y** along path **C-D**.

The delayed choice complex **X-Y** along path **C-D** introduces a critical element to the scheme: one not present in two-slit experiments. The switch at **X** can be suspended between two settings: One setting will allow the beam-split photon from mirror-**C** to pass along path **C-D** to **D**, with consequent self-interference and the formation of the generic interference pattern at detector-**D**. The other setting at **switch-x** will divert the split photon from mirror-**C** toward detector-**Y** along path **X-Y**. In the later possibility, sought self-interference will *not* occur, and the interference pattern will not form at **D**. Of course, **the formation of an interference pattern derives from a wavefunction different from that of the wavefunction without the interference. In other words, the hedging nesting wavefunction nests two distinct wavefunctions, exhibiting a 'wavefunctions within wavefunction' nested futures form. Hence, Wheeler's delayed choice implies *nested futures*.**

---

[65] Mathematical Foundations of Quantum Theory, edited by A. R. Marlow, Academic Press, 1978. P. 39

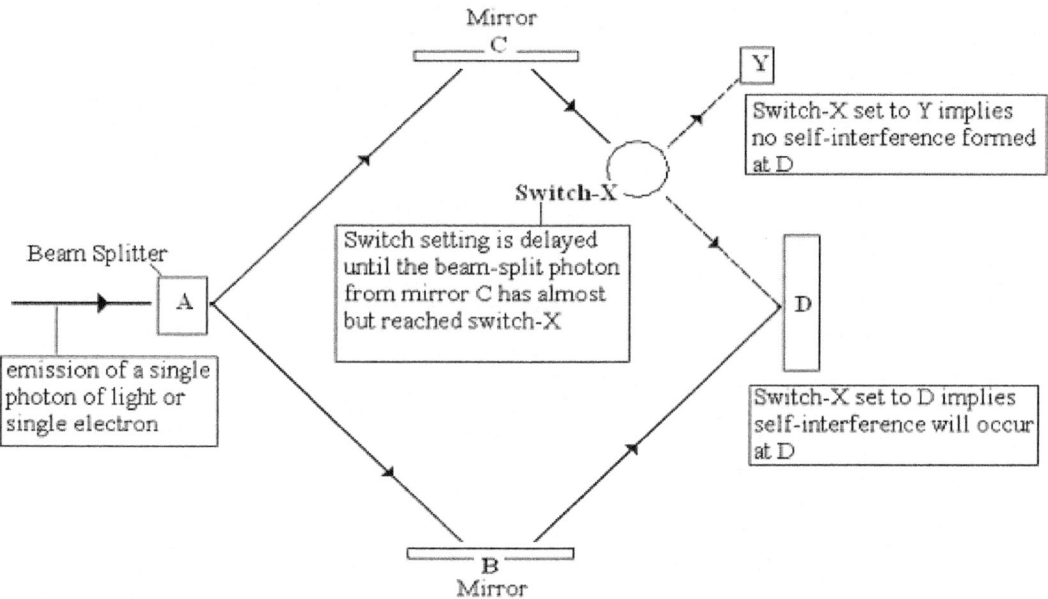

Fig. 2.07: Illustrating the outlay of delayed choice

Notes: The illustration depicts a simplified outlay of John Wheeler's delayed choice experiment. A single incoming photon of light is split by the beam splitter at **A** which works in a similar way on the particle as does the barrier in the double-slit experiment when set to "both slits open". The particle is split into two quantum mechanical wavelets: One heads toward **mirror-C** and the other to **mirror-B**. Both will be deflected by the pertinent mirrors toward **detector-D**. Normally, by the time the wavelets reach **detector-D**, these will self-interfere, forming an interference pattern at **detector-D**. Thus, the experiment is essentially an alternative version of the double-slit experiment, save that the barrier is replaced by the beam splitter. However, the operation of **switch-x** radically alters the experiment. **Switch-x** is suspended at two settings: One setting permits the quantum wavelet from **mirror-C** to reach **detector-D**, forming the expected self-interference and interference pattern. The other diverts the wavelet from **mirror-C** to **detector-Y** so that no self-interference and no interference pattern will form at **detector-D**. However, whether we trip the setting on **switch-x** at the last moment, or suspend it in ambiguity till the last moment, **switch-x** is rendered into an AND-form future-suspension. Thus, the resolution of **switch-x** is *delayed* until after the wavelet from **mirror-C** has passed **mirror-C** and is already on its way to **switch-x**; apparently committed to self-interference at **detector-D**: That is, **switch-x** is set to **Y** at a moment when it ought to be too late for the system to adjust to the alternate possibility implied by **Y**: i.e. no non-interference. It turns out that the quantum mechanical wave hedges and includes within its possibility-set mutually incompatible future wavefunctions that include both wavefunction for self-interference at **D** AND wavefunction for diversion to **Y**, implying the nested futures view of the quantum mechanical wave. The issue is further radicalised when we recast the scheme in terms of spatially static quantum waves within the intermediate model spacetime.

Switch-x need not be committed to either setting until some critical moment or critical delay is obtained, i.e., the critical timing at which **switch-x** may be tripped one way or the other must be at a moment *after* the beam-split photon passes mirrors **B** and **C**, and just *before* the split beam reaches **switch-x**. Thus, the experimenter may *delay the choice* to trip **switch-x** (or choose not to trip it at all) until after the split-beam photon is deflected by mirrors **B** and **C**, and is thereafter committed to self-interference at detector-**D**.

To trip or not to trip switch-x in the allotted time, and the time-limit or constraint vis-à-vis the pertinent delay or decision, clarifies the meaning of *delayed choice*. The consequence from the delay choice complex at **X-Y** and the effective suspension of **switch-x** to an as-yet not-decided setting, is that the nesting hedging quantum wavefunction for the photon from **A** must incorporate or *nest* or *hedge* into its AND-form potentiality the very possibility of **switch-x** being tripped to either possibility. This is encapsulated in... {diversion to **Y**... AND ... self-interference at **D**}. That is, per the delay imposed on the decision over **switch-x** to a moment after the split-photon has passed mirrors **B** and **C,** and has become apparently fated to self-interference at **D**, the system yet remains in a quantum mechanical AND-form suspension: one open to *both* self-interference at **D** *and* diversion to **Y**. **Hence, we obtain a quantum mechanical system that does not just superpose one immanent future possibility-state but hedges for alternate mutually incompatible futures**: It is as if, having allowed the quantum wave to pass through the two slits at the generic two-slit experiment barrier, and fate it to the usual self-interference, we somehow rigged the system to superpose a single-slit outcome…AND vice-versa… per function of the possibility that **switch-x** might be tripped one way or the other.

Hence, we obtain the case for *nested futures*… effectively, a 'wavefunction within wavefunction' system of futures.

*

Suppose we ran an alternative version of the delayed choice experiment: One in which we included the beam-splitter at **A** and the mirrors at **B** and **C**, and the detector at **D**, but removed **switch-x** and, with it, removed the delayed choice complex **X-Y**. What would be the outcome? The AND-form possibility complex for our single beam–split photon after its emission would be an AND-form superposition fated to the formation of an interference pattern at **D**, its wavefunction described as...

…. {ABD AND ACD │ interference pattern formed at D} …

… which is read as… "path-ABD **AND** path-ACD", such that an interference pattern will form at **D**. The outcome would be no more radical than that obtained in generic two-slit experiments with an interference pattern.

Now we reintroduce the delayed choice complex **X-Y** and **switch-x**. Per a delayed choice decision made *after* our split-beam photon has passed mirrors **B** and **C**, we trip **switch-x** to permit our split-photon to get to **D** and end up with the usual interference pattern at **D**: The lay formalism for this has already been expressed above; except that the interference pattern will form if and only if (*iff*) **switch-x** is set to divert the split-photon to **D**. This is given by the wavefunction expressed as…

…. {ABD AND ACD │ interference pattern formed at D iff X set to D} …

Let us call this future **nested wavefunction-(a)**.

On the other hand, if **switch-x** is tripped to divert the split-beam photon from **mirror-C** toward **Y**, we will end up with no interference pattern at **D**. Thus, the pertinent wavefunction will be...

… {photon at D AND photon at Y │ no interference pattern iff X set to Y} …

Let us call this future **nested wavefunction-(b)**.

Now, if we delay our choice to trip **switch-x** until the very last possible moment after the beam-split photon is reflected from **mirror-C** and is apparently fated to self-interference at **D**, but before it has reached **switch-x**, the approaching quantum mechanical wave, or **the initial wavefunction from A, must hedge for *both* nested wavefunction-(a) AND nested wavefunction-(b)**.

However, note that **switch-x** itself must be described as a wavefunction state, distinct from **nested wavefunction-(a)** and **(b)**. Thus, **switch-x** must be described quantum mechanically, simply because its setting is not yet decided until the critical moment. As such, the decision about **switch-x** is deferred to the AND-form future and must constitute as an AND-form superpose of *both* of its possible settings. Indeed, even if **switch-x** is forced into one setting, a later intercession could trip it to the other setting *in the future*. Hence, again, **switch-x** *must* be described as an AND-form future-potential superposition composed of *both* of its settings. Alternatively, we can choose to suspend **switch-x** at a middle setting: i.e., in a state not tripped to either setting, or even to fluctuate haphazardly between one and the other setting; a perpetual Schrödinger's cat device: one that perpetually reset **switch-x** until the last-ditch moment. Again, this implies a future-form **switch-x** rendered as an AND-form superposition for both possible settings. In all cases, **switch-x** must always be described in AND-form superposition terms. Thus, the **wavefunction switch-x** will be comprised of…

switch-x suspended into... {"X tripped to D" AND "X tripped to Y"}

Thus, the primary nesting and hedging quantum wave, or **hedging wavefunction** in the interval *after* mirrors **B** and **C have been passed by the split photons**, but *before* one of these gets to **switch-x**, will contain the possibilities and wavefunctions…

{("X tripped to D" implying nested wavefunction-(a))... AND... ("X tripped to Y" implying nested wavefunction-(b))
│ after mirrors B and C, iff the tripping of switch-x is delayed and suspended}

The **nested wavefunction-(a)** AND **nested wavefunction-(b)** are mutually incompatible alternative futures nested within the **hedging quantum wave**. Up until the very last moment of its approach to **switch-x**, the hedging quantum mechanical wave and wavefunction for the photon split at **A** and deflected by mirrors **B** and **C** will, under the effect of the contingent AND-form suspension of **switch-x**, hedge for…

… {nested wavefunction-(a) ...AND... nested wavefunction-(b)}

Succinctly, in the delayed choice schema, alternative mutually incompatible futures, as 'wavefunctions within wavefunctions', are nested within the initial hedging quantum mechanical wave. The AND-form suspension of whatever constitutes the equivalent of *contingent* switch-x remains critical to this, as is the time-interval throughout which the delay over the choice about the setting of switch-x abides. Hence, delayed choice experiments inadvertently demonstrate the nested futures wavefunction and the nested futures quantum wave.

*

Before **switch-x** is tripped, the hedging quantum wavefunction for the split-beam photon hedges for both mutually incompatible nested future possibilities. When the switch is tripped to one or the other of its settings, the hedging quantum wave (the nested quantum

208

wave) will decohere to one or the other of the mutually incompatible AND-form future complexes. Note that, so long as **switch-x** is effectively AND-form suspended up to the critical moment, the nesting quantum mechanical wave *will* hedge for both nested futures: i.e., it will cohere for both nested future wavefunctions... {**nested wavefunction-(a)** ...AND... **nested wavefunction-(b)**}. But when hitherto suspended contingent **switch-x** is OR-form resolved, the total quantum mechanical system will adjust its future-form content into consistent form by forgoing and erasing some of its future possibilities in favour of the alternative that remains. Yet, to accommodate such an adjustment, **the quantum mechanical system or wave must, in the truest sense, contain embedded and nested alternate mutually incompatible future possibilities, or 'wavefunctions within wavefunctions' in relation to *all* possible delayed choice elements or contingencies that might get to intrude in the perpetual resolution of the system.**

*

In the delayed choice experiment just described, the situation is simple: the system is comprised of just one hedging AND-form delayed choice element and two alternate mutually incompatible AND-form future wavefunction possibility-sets, both of which are rendered precarious to a contingency composed of two AND-form alternative setting-potentials vis-à-vis **switch-x**.

In naturally occurring non-laboratory systems, we can expect the incorporation of manifold potential delayed choice elements or contingencies: i.e., not just one **switch-x** with just two settings, but more than one equivalent of **switch-x** with n-number of settings arrayed along the natural system's equivalent to path-CD featured in our simplified delayed choice experiment. Indeed, the fact that delayed choice supports the nested futures view might not be explicitly obvious in experimental designs with just one delayed choice element **switch-x**, or even when a single split-beam photon is involved. When we elaborate generic delayed choice experiments to include more than one delayed choice element along either path from **A** to **D**, and even include more delayed choice elements that lie between **X** and **Y** itself, the veracity and saliency of the nested futures structure of the quantum mechanical wave, and of the future conceived as spacetime in potentiality, is rendered obvious. Consequently, we must revamp the Schrödinger wave equations according to its demands, and **natural quantum mechanical wavefunctions must be modelled as nested futures systems comprised of *all* mutually incompatible alternate future possibilities and wavefunctions per all possible contingent future delayed choice elements and intrusions.** The wavefunction for a naturally occurring quantum mechanical system is expected to configure into a state of *anticipation* for all and any plausible and possible delayed choice contingencies and the unique futures that attend these: a system of nested futures that elaborate to the infinitely removed future.

Finally, when we generalise this to the whole of the universe-scale wavefunction, hence to the domain of spacetime in potentiality per the intermediate model of spacetime, the domain of spacetime in potentiality is found to be anticipatory of *all* possible manifold future delayed choice contingencies, and must embed, niche or *nest* within its possibility-structure *all* possible pertinent alternate mutually incompatible futures and wavefunctions, elaborated to the infinitely removed future.

In short, the domain of spacetime in potentiality constitutes a complex system of nested futures and contingent delayed choice elements elaborated to the infinitely removed future.

2.19: Delayed Choice & Nested Futures:
Partial-Perpetual Collapse of Quantum Mechanical Waves & Spacetime in Potentiality

How does partial-perpetual collapse of the wavefunction come into the fray in the framework of our nested futures view? Recall that, before **switch-x** is tripped to either of its two possible AND-form suspended settings, the total wavefunction for the split-beam photon will hedge for both mutually incompatible futures implied by **switch-x**. When the switch is subsequently tripped to one or the other of its settings, the hedging quantum wave that passes through it will collapse or decohere to one or the other of the nested future AND-form states. And it is from this fact of partial decoherence and partial erasure that we arrive at the clearest expression of, and the physical-experimental demonstration of, partial-perpetual AND-to-OR collapse of open-choice quantum mechanical waves, or open-choice universe-scale spacetime in potentiality: (By definition, all delayed choice systems are open-choice systems: see prospective essay: **2.20**).

In short, **delayed choice elements, and their consequences to wavefunctions imply the non-exhaustive partial and perpetual collapse and decoherence of wavefunctions. This is generalised to spacetime in potentiality and implies the partial-perpetual collapse and decoherence of spacetime in potentiality**.

Per every contingent intrusion, such as exemplified by **switch-x**, or analogised by our apocryphal Wheeler's cat, the hedging nested futures quantum wave will forgo and erase some of its future possibilities in favour of those that are compatible with the resolved setting at **switch-x**, or with some other contingent delayed choice element. Thus, the quantum mechanical system will behave in a similar way to a partially and perpetually collapsing or reducing navigational dead-reckoning circle for a ship with accumulating certainties about its position and condition, some of which, or even all of which, might constitute critical delayed choice contingencies.

Recall that quantum mechanical systems are only quantum mechanical per the operation and enforcement of isolation. When we refer to 'intrusive' or contingent delayed choice elements, we are speaking of the partial lifting of enforced isolation on the quantum mechanical AND-form system. This 'lifting of isolation' does not constitute the end of isolation. Delayed choice, and with it, the justification for partial-perpetual collapse, implies that there exist open-choice quantum mechanical systems whose isolation is only ever partially lifted or marginally reduced as the event horizon advances into spacetime in potentiality, or as the latter undergoes perpetual increase in future-potential entropy.

209

The isolation-state of the universe-scale quantum wavefunction, synonymous with the domain of spacetime in potentiality, must always be partially lifted and partially-perpetually decayed as the event horizon progresses and spacetime unfolds. Indeed, the isolation-state of the domain of spacetime in potentiality can never be lifted completely, comprehensively, or exhaustively, but only ever partially so, in perpetuity, to the infinitely removed future. **The comprehensive lifting of the isolation of spacetime in potentiality would require its total AND-to-OR decoherence *and* collapse to the infinitely removed future: an impossible feat.**

The same holds for parochial quantum mechanical waves entailed in most laboratory experiments. Due to the finite number of delayed choice elements, the number of partial AND-to-OR collapses undergone by such quantum mechanical systems appear not to involve partial-perpetual collapse *ad infinitum*. Indeed, the reality of partial perpetual collapse and decoherence is obscured when no delayed choice contingents are incorporated into such experiments: a deficiency obvious in generic two-slit experiments and in EPR. Yet, parochial wavefunctions are implicitly embedded within a much grander subsuming quantum mechanical systems and futures, or within the domain of spacetime in potentiality itself: the ultimate open-choice quantum mechanical system. Indeed, even when we eliminate all contingencies and all delayed choice elements, and constitute a pure closed-choice two-slit experiment, or an EPR experiment, or even a Schrödinger's cat experiment, there is always the contingent possibility of the mischievous laboratory cat, or Wheeler's cat, or other unplanned contingencies... such as the possibility of power failure, or even a lightning strike, and many other mundane contingencies that belong to the larger embedding futurity that could affect the embedded closed-choice experiment.

In other words, **what appears to be a laboratory finite-possibility closed-choice quantum wave, such as featured in generic two slit experiments without delayed choice elements (without the deliberate incorporation of contingency, and even in delayed choice experiments with just one delayed choice element), is in truth niched to a larger set: an ultimate open-choice domain spacetime in potentiality itself. Therein, every final resolution of a quantum mechanical system is but a prelude to a larger set of futures which it must niche, necessarily implying a process of incomplete and inexhaustible partial-perpetual collapse, *ad infinitum*... rendering even the most narrow and closed system into an open-choice system.**

2.20: Closed-Choice & Open-Choice Quantum Mechanical Systems:
Implications to the Domain of Spacetime in Potentiality

Nearly all naturally occurring quantum mechanical systems are delayed choice systems. Insofar as these do not incorporate delayed choice elements into their frameworks, **most artificially devised experiments and quantum wavefunctions, including generic two slit experiments, comprise closed-choice quantum mechanical systems**. In the generic two-slit experiment, the state of the barrier is decided from the very beginning, before the pertinent radiation and matter waves have been emitted. The barrier is ordained to exhibit a result in which either both slits are open or only one slit is open. Thereafter, the choices open to the evolution of the system are *closed*. i.e., tautologically, *not open* to any alternative incompatible future, unless there is deliberate inclusion of a delayed choice contingency, or Wheeler's cat.

Since most artificial experiments are closed-choice quantum mechanical systems, or lack explicit delayed choice elements, overreliance on these misleads about the nature of quantum mechanical reality: The near-exclusive utilisation of closed-choice systems in quantum science, and the unquestioned assumption and even lack of an explicit awareness of the fact, serves to conceal the general physical truth of the alternative: namely, **the preponderance in nature of open-choice quantum mechanical systems over closed choice systems**, with the latter invariably embedded to larger open-choice systems and the nested futures quantum mechanical nature.

Naturally occurring quantum mechanical systems in the universe outside of the laboratory are rarely closed-choice systems: **Natural systems will typically have one or more contingent delayed choice elements and consequent nested futures. Most incorporate manifold numbers of contingent delayed choice elements**, not merely in terms of numbers of, but also in terms of serial, parallel and combinatorial structure: (See **Fig. 2.08**). Hence, naturally occurring quantum mechanical phenomena will incorporate alternative mutually incompatible nested future contingencies and wavefunction complexes per function of *all* available delayed choice elements that are arranged in *all* permissible serial, parallel and combinatorial forms. These alternate futures will remain concealed or nested to immanent possibilities but will be no less real and no less open-choice.

Wheeler's delayed choice approach to the two-slit experiment enables us to construct experimental situations that are truer to nature, insofar as these might encapsulate nature's open-choice nested future characteristics, culminating into nested futures form of spacetime in potentiality. Yet, despite appreciation of the reality of delayed choice, science has tended to overlook the distinction between open-choice and closed-choice quantum systems. In nature, closed-choice quantum systems clearly constitute the rarity, if not the eccentricity.

In Part-I, the domain of spacetime in potentiality was described as a futures-state subject to an unending series of perpetual-partial collapses attendant the progression of the event horizon. It so happens that the results obtained from our exploitation of delayed choice and the development of the nested futures approach vindicates the reality of just such perpetual-partial collapse. To reiterate, only in closed-choice quantum mechanical systems is it plausible to conceive of supposed one-step, one-time, all-exhaustive AND-to-OR collapse. But, in open-choice systems, partial-perpetual collapse is necessarily the obvious norm and predominant across nature. **Open choice systems, hence the domain of spacetime in potentiality *en toto*, cannot attain closure or exhaustion. Hence, the growing block spacetime cannot run out of the future, much less run out of time itself.**

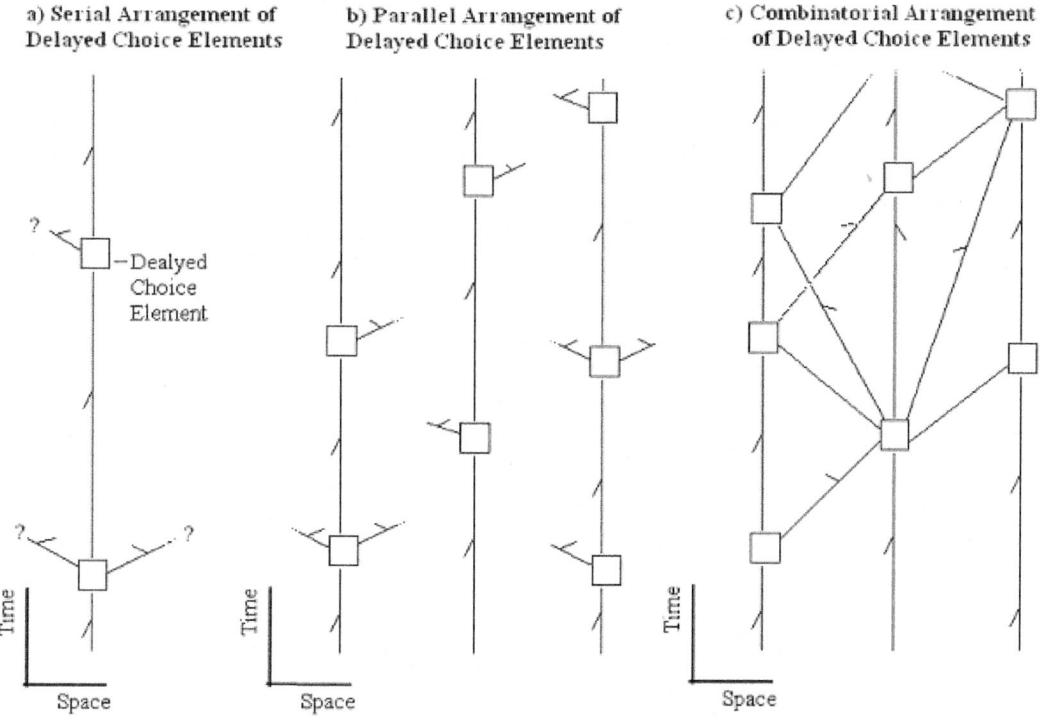

Fig. 2.08: Serial, parallel and combinatorial delayed choice complexes and spacetime "switchboards"
Notes: The series of didactic illustrations clarifies the concepts of serial, parallel and combinatorial potential delayed choice elements and contingencies, pertinent to the open-choice nested futures view of the quantum wave and of spacetime in potentiality. The serial, parallel and combinatorial delayed choice elements are embedded within spacetime in potentiality as AND-form states. In the serial version depicted in **diagram-a**, a single pathway, line or worldline possesses just two serialised delayed choice elements or contingencies (depicted as boxes and equivalent to **switch-x**). Through implicit AND-to-OR wavefunction collapse, each delayed choice element can branch off to alternate mutually incompatible future possibilities implied by the depicted branch and question-marks. In **diagram-b** we have three parallel potential worldlines, each with their own unique delayed choice elements. The parallel future complexes are not cross-linked. Thus, what happens along one worldline does not affect what happens along the others. In **diagram-c**, we depict parallel potential worldlines that *are* cross-linked by potential linking worldlines via delayed choice elements, forming a combinatorial potential spacetime "nested futures switchboard". Therein, outcomes resolved at a given delayed choice "box" along one line or path *will* generate consequences and delimits (decoherence) to the future development and neighbouring parallel worldlines and delayed choice settings. The parallel delayed choice elements make for convergent futures. Indeed convergent futures are definitive of the system of spacetime in potentiality.

2.21: Closed-Choice Systems as Special Embedded States
Within Open-Choice Quantum Mechanical Systems

When we set up any experiment based on conventional closed-choice quantum systems, even when we incorporate into these delayed choice elements, we tend to restrict our consideration and scrutiny of the system to the bounds of the finite delayed choices factors. That is, therein, we typically consider *only* the possible implications from **switch-x**, abstracting out the fact that the finite system so established is part of a larger nested open-choice system, or spacetime in potentiality. This closed-choice notion is implicitly and erroneously applied to John Wheeler's delayed choice experiment when it is scrutinised *after* the switch has been tripped. The closed choice notion is also applied to the two-slit experiment in which the experimental pre-setup of the barrier is almost equivalent to Wheeler's delayed choice experiment *after* **switch-x** has been tripped. **The total number of future complexes available after the passage of the split-beam quantum wave through switch-x will certainly be reduced to, hence *closed off* to, only one possibility complex**. The alternate mutually incompatible futures comprising the original nested future will become irreversibly erased from the

211

total system; no longer part of the pertinent AND-form state. Thus, the quantum system will have become *closed*, save to just one future possibility complex or closed-choice future. **Yet, the system remains part of a larger quantum mechanical nested futures system, despite appearances: The final OR-form outcome from the experiment must constitute a 'tripped switch' in its own right: one that must affect subsequent alternative futures: an *implicit* resolved delayed choice element in its own right: part of a larger switchboard-like order** didactically depicted in **Fig.2.08**: a larger open-choice spacetime in potentiality of inexhaustible nested futures.

To appreciate this, consider that, if the OR-form outcome is comprised of 'no self-interference at detector-**Y**', at the larger spacetime in potentiality, this outcome will relate to a future-form wavefunction that will niche its own subsequent unique future-possibility complexes, at the expense and to the erasure of possibilities that might have applied to the alternative outcome comprised of 'self-interference at detector-**D**'. Put another way, if the outcome is in favour of detector-**Y**, the rest of spacetime in potentiality will need to reconfigure, restructure and decohere its content in such a way as to become consistent with that outcome and erase *all* alternate non-consistent future complexes originally potentialised and attached to the erased possibility comprised of 'self-interference at detector-**D**'. This will ensure that observers in the future do not get to observe any other outcome but the one that had transpired.

It follows that the closed choice system *is* an element in an open-choice system, insofar as its parochial closed-choice outcomes have a decisive contribution to the subsequent evolution of the larger embedding open-choice spacetime in potentiality, and to the subsequent content of spacetime in potentiality in terms of permissible nested future potentials and permitted delayed choice contingencies.

To reiterate, the result obtained from apparent closed-choice quantum systems are merely OR-form resolved delayed choice switches or elements that, in conjunction with other such resolved elements, implicitly decide the direction and unfolding of subsequent possibilities in spacetime in potentiality. Just because we ignore these embedding realities in generic considerations of *apparent* closed choice experiments does not imply their ontic absence or irrelevance. Indeed, we *cannot* ignore the larger open-choice order culminated as spacetime in potentiality.

*

In summary, a system appears to be closed either or both because we scrutinise it after some critical delayed choice element has been tripped, or else fail to consider its final outcome and conclusion as an OR-form 'tripped switch' state in its own right: one that restructures the whole of the larger and embedding open-choice nested spacetime in potentiality. In some final sense, there is no such thing in nature as a closed-choice quantum mechanical system: In some final sense, *all* putative closed-choice systems are, or are partial resolutions of, larger open-choice quantum mechanical systems comprised of mutually incompatible inexhaustible nested futures that elaborate to the infinitely removed future.

2.22: Spacetime in Potentiality of Nested Futures: Structure, Taxonomy & Quantum Erasure

The standard view of the universal wavefunction, typically remiss of the nested futures structure, might incorporate only immanent future worldline possibilities into its AND-form state: see **Fig.2.09-A**: Therein, the immanent potential worldlines {**a** AND **b**} are expected to be embedded within the AND-form domain of spacetime in potentiality, while the single pertinent realised OR-form worldline (**worldline-w**) to which the two future potential worldlines are attached at the event horizon, will be embedded within realised spacetime. However, the nested futures view describes things differently: See **Fig. 2.09-B**: Therein, future-potential worldlines {**a** AND **b**} will each potentialise and nest for their own unique subsequent future-potential worldline possibilities: namely… {**c** AND **d**} AND {**e** AND **f**}, respectively. In principle, the nested futures system so described can elaborate beyond potentials **c**, **d**, **e**, **f** to the infinitely removed future.

Note that, in **Fig. 2.09-B**, the leading point of a realised **worldline-w** constitutes a delayed choice element, a **switch-x**, that leads to either of the two mutually incompatible alternate future complexes {wavefunction-**a** AND wavefunction-**b**} that have yet to happen and are suspended in AND-form within spacetime in potentiality. On the other hand, future-potential worldlines {**a**} and {**b**}, or their end-points, also constitute *potential* delayed choice elements in their own right, whose prospective AND-to-OR collapse promises to trip other equivalent contingencies or **switch-x** equivalents: part of a grand quantum mechanical spacetime switchboard that stretches to the infinitely removed future, as was depicted in **Fig.2.08** in essay **2.20**.

The nested futures view of spacetime in potentiality also implies a modification of Feynman's 'sum over all paths' view of quantum mechanical waves and wavefunctions: This was first presented in Part-I, essay **2.13**. Feynman applied this view to the two-slit experiment, modelled as a series of barriers, up to an infinite number, into which effective infinite numbers of slits are cut. A single particle will thus be potentialised into an AND-form state of infinite number of self-contained 'which way' paths from emitter to detector, with the shortest path constituting the most probable of the lot: i.e. the highest amplitude 'sum over all paths'.

Grasped in terms of spacetime in potentiality, which can be treated as the total interval between emitter (at event horizon) and detector (at the infinitely removed future), but with an intermediate set of infinite numbers of barriers and infinite numbers of open slits, Feynman's infinite 'paths' transforms into infinite future-potential worldlines. Indeed, describing these as potential worldlines is much closer to reality than describing these as 'paths in space'. Our nested futures view implies that each potential worldline in lieu of 'path' must, in its simplest form, split via potential delayed choice contingent elements, which must nest alternate mutually incompatible future-potential nested worldlines at as many points along any potential worldline as one can place delayed choice elements: i.e. potentially infinitely many: (See **Fig: 2.10-A** and **B** for the pertinent illustration and clarification).

Once we modify the 'sum over all paths' view in terms of the nested futures approach, we can generalise its characteristics into a *sum over all potential worldlines* approach to spacetime in potentiality *en toto*: grasped as alternative future-potential

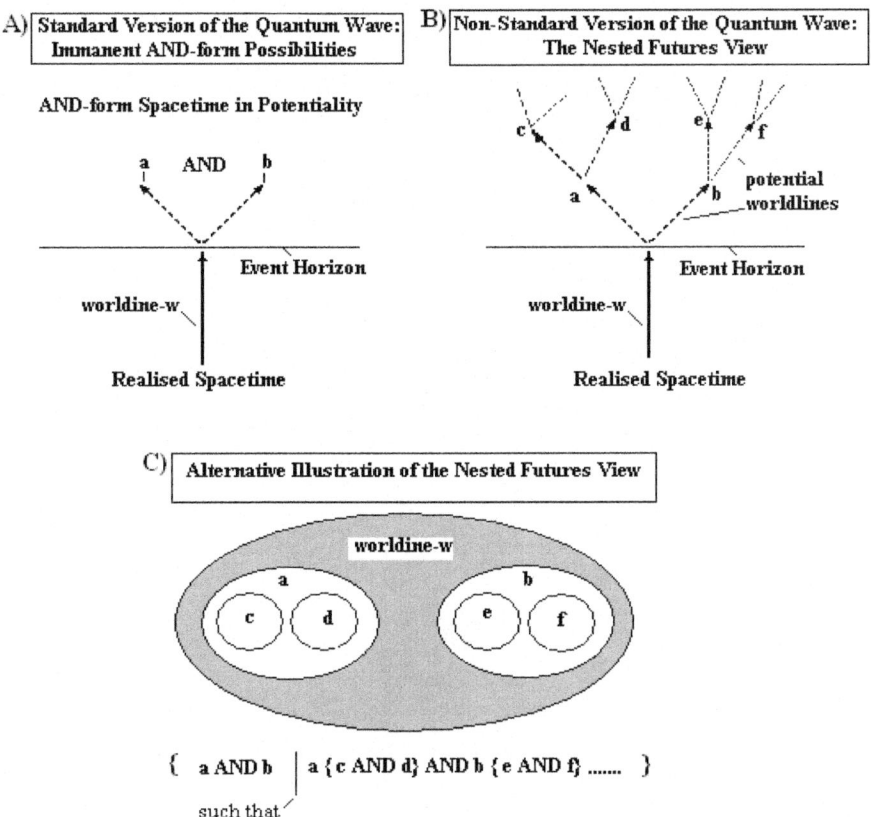

Fig. 2.09: Intermediate model spacetime: standard and nested futures view of potential worldlines
Notes: **Diagram-A** depicts a simplified spacetime version of the standard quantum mechanical wave: For brevity, the space axis and time axis, and the lightcone, are implicit and not shown. Only the event horizon is depicted. The standard quantum mechanical wave, or that aspect of spacetime in potentiality attached to hitherto realised **worldline-w**, is depicted with just two future possibilities: i.e. {**a AND b**}. In **diagram-B** we have a non-standard or nested futures view of the quantum mechanical wave: Therein, subsequent futures {**c AND d**} are nested to **possibility-a**, while a mutually incompatible set of futures comprised of {**e AND f**} are nested to **possibility-b**. Of course, possibilities **c, d, e, f** in turn nests their own subsequent futures, indicated by the dashed lines that extend from these. **Diagram-C** is a Venn diagrammatic portrayal of the nested futures view of **diagram-B**: a top-down version of **diagram-B**.

worldlines in AND-form superposition: (see **Fig: 2.10-C**). Yet, the truer picture is given in **Fig: 2.10-D**: with potential worldlines split into subsequent potential worldlines per delayed choice contingents.

*

We will now elaborate on the taxonomy of nested futures: This is important in its own terms *and* per the critical insight it secures about the process of quantum erasure, counterfactual causality and the larger grand decoherence of spacetime in potentiality: the basis for the revival of Mach's principle, which permits the counterfactually holistic unitary view of spacetime… an 'instantaneous whole' view of the same, and wherein one part of spacetime non-locally and transtemporally counterfactually 'knows' every other part of spacetime; culminating into the first decisive contribution to the quantum mechanical account of inertia, of the permanent magnet, and the first foray into quantum gravity: (More on these later in Part-III).

The nested future taxonomy is illustrated in **Fig. 2.11-A** to **C**. There are three basic categories of nested futures. These are the **divergent nested futures**; the **entangled nested futures**; and the **convergent nested futures**. The *serial, parallel* and *combinatorial* spacetime switchboard system depicted in essay **2.20**, **Fig. 2.08** runs parallel with and implicitly depicts the stated nested future categories.

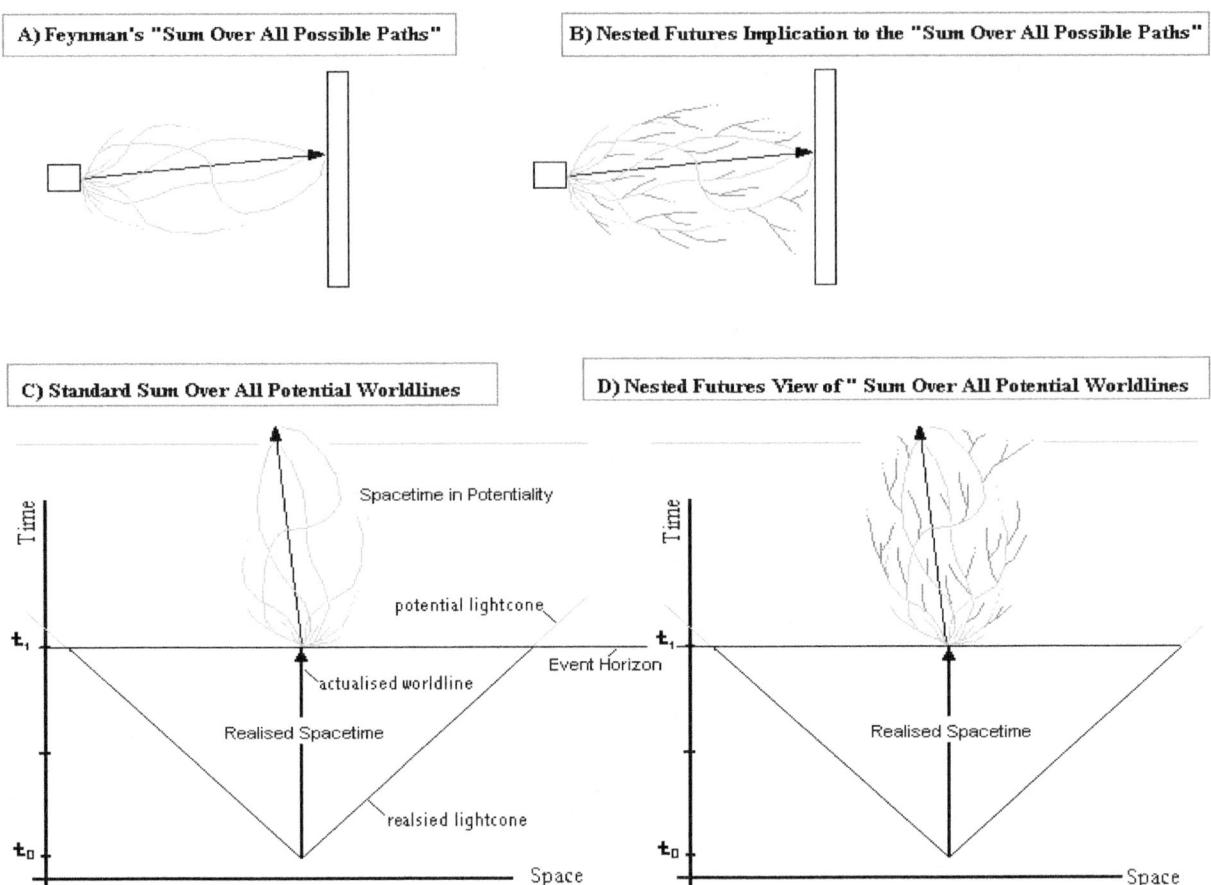

Fig. 2.10: Sum over all potential worldlines according to the nested futures view
Notes: **Diagram-A** depicts the standard version of Feynman's "sum over all possible paths" as it might apply to the two-slit experiment, with the high amplitude path rendered in bold. **Diagram-B** depicts the modification of the standard view in terms of the nested futures view, which turns out to be implicit to Feynman's approach: Consequently, each of the summed paths from emitter to detector possess their own nested alternate future potentials via implicit delayed choice complexes. This must further complicate the summing of possibilities, given that there are potentially an infinite number of possible immanent paths to begin with, without the nested futures. **Diagram-C** depicts a standard "sum over all potential worldlines" structured in a similar way to Feynman's paths but transplanted into spacetime in potentiality within the intermediate model. Spacetime in potentiality from event horizon at t_1 to the infinite future is treated as if it constituted a two-slit experimental milieu composed of an infinite number of barriers, with infinite number of open slits: the same method used by Feynman, but now involving potential worldlines in spacetime-future, with the bold potential worldline depicting the default teleology of spacetime. **Diagram-D** depicts a similar arrangement, but with nested futures explicitly incorporated, wherein each worldline will branch out into alternate mutually incompatible nested futures per implicit delayed choice elements, to the infinitely removed future.

Let us look at the first nested futures category: the divergent nested futures: (see **Fig. 2.11 A**). Within spacetime in potentiality above the event horizon, the endpoint of the single realised **worldline-w** is potentialised into a minimum of two immanent alternate future possibilities: {**a** AND **b**}. But immanent possibility {**a**} nests for subsequent futures that are mutually incompatible with those nested to immanent possibility {**b**}. There are no cross-linkages between the nested futures attached to {**a**} vis-à-vis the nested futures attached to {**b**}. Any subsequent AND-to-OR collapse process and synonymous progression of the event horizon will either realise possibility **a** OR **b**. If **a** is realised, the future complexes nested to **b**, including **b** itself, will be erased from all futures. On the other hand, if **b** is realised, all future complexes nested to **a**, including **a**, will be erased. Since there are no cross-linkages between these two mutually incompatible diverging nested futures, the category must rightly be called *divergent*: i.e., future complexes nested to {**a**} diverge from those attached to {**b**}.

However, **divergent category nested futures are highly idealised. Indeed, these do not exist in nature on account of the larger spacetime open-choice system to which divergent futures subsume,** although EPR entangled couples comes very close to exemplifying it. Even so, the category scaffolds toward the entangled and convergent categories, which *do* exist in nature.

Entangled nested futures are similar to the divergent category, except that the alternate mutually incompatible futures are entangled by means of subsequent cross-links or byways, or alternate routes that bridge future-potential worldlines (as shown in **Fig. 2.11 B**). Thus, divergent future complexes attached to {**a**} are, at key future-potential delayed choice junctions, bridged by potential linkages to future complexes that are nested to {**b**}. Hence, a distant future possibility is ultimately nested to and could be realised through either future-potential worldline: {**a**} *and* {**b**}.

It turns out that the entangled category nested futures constitute a highly idealised depiction of really-existing convergent category nested futures. Note that most of our depictions of nested futures do not explicitly display future-potential lightcones. The illustration given in **Fig. 2.11 C** is the exception: Therein, **the convergent nested futures category emerges out of at least two distinct frames rooted to two distinct realised worldlines: worldline-w and worldline-x, separated by a putative distance; perhaps even lightyears. The nested futures complexes from worldline-w and worldline-x are potentialised to converge at common distant futures, to bring about common potential outcomes for both frames within their overlapping future-potential lightcones.**

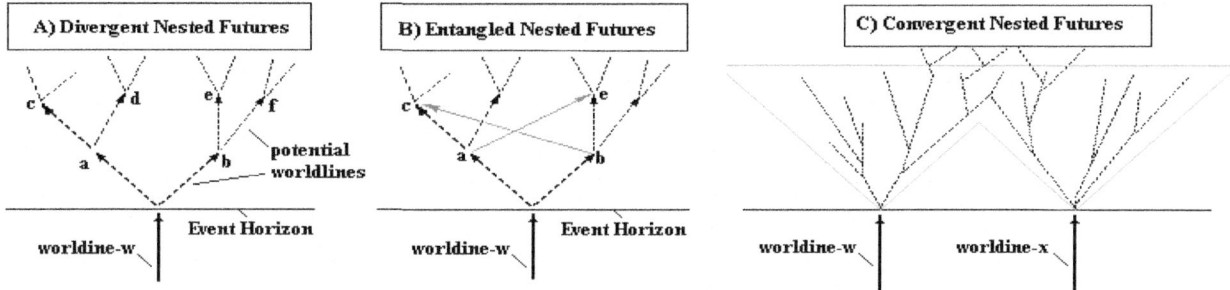

Fig. 2.11: Illustrating the taxonomy of nested futures
Notes: The three diagrams depict the three distinct categorises of nested futures. For brevity, we rely on a simple bifurcation of the future into two nested future possibilities per delayed choice element. **Diagram-A** depicts the *divergent category* nested futures: The endpoint of **worldline-w** is potentialised with immanent possibilities {**a** AND **b**}. These in turn nest mutually incompatible alternate nested futures {**c** AND **d**} and {**e** AND **f**}, respectively. There are no cross-connecting potential worldlines between these: i.e., the structure of spacetime in potentiality is *divergent*. The divergent category is pure idealisation and does not exist in nature. In **diagram-B**, seeming divergent nested futures are cross linked by entangling potential worldlines, constituting the *entangled category* nested future structure of spacetime in potential. Finally, **diagram-C** illustrates the *convergent category* nested future-structure of spacetime in potentiality: Therein, two distant worldlines potentialise for nested future complexes that merge into a common distant future per the overlap of their future-potential lightcones, and are thus convergent in that distant future. This convergence to a common set of future possibilities of otherwise spatially disparate frames of reference proves decisive to solutions to the quantum mechanical non-local account of inertia and other phenomena that ought not to involve 'spooky action at a distance' or non-locality, but do so regardless.

The entangled cross linkages in the convergent category nested futures are not idealisations: they are real. These engender profound consequences that open physics to unexpected advances. Indeed, the convergent category nested futures is critical to the very possibility of non-local counterfactual causality of a form beyond non-locality evinced in generic EPR, and, on that basis, helps found counterfactual causality and unity, and the grand decoherence of spacetime in potentiality, reviving Mach's principle: (More on this in Part-III).

We can now consolidate our nested futures taxonomy with the critical process of the quantum erasure of nested futures. Thus, **Fig.2.12-A** to **F** depict the process of quantum erasure through a series of interrelated diagrams. In **diagram-A**, the leading end of realised worldline {**w**} at the event horizon at t_1 is potentialised into two immanent futures and subsequent nested futures. The immanent pair of possibilities {**a** AND **b**} AND-to-OR collapse in favour of the OR-form of {**b**}. Hence, possibility {**a**} is forgone and completely erased (see **diagram-B**). All nested future complexes formerly potentialised to {**a**} are also completely erased from spacetime in potentiality, as is depicted in **diagram-C**. Consequently, the entropy of future-potential information will increase per the erased nested futures: (see Book-I, essays **1.07** and **1.09** for pertinent entropy theory). Insofar as a nested futures system could in principle elaborate to the infinitely removed future, the consequent quantum erasure of nested futures must ultimately decohere the totality of spacetime in potentiality all the way to the infinitely removed future.

Given the erasure of {**a**} and its nested futures, only those future complexes potentialised to realised outcome {**b**} will enjoy potentiality within what remains of spacetime in potentiality: namely, {**e** AND **f**}; with each of these nesting subsequent nested future complexes {**k** AND **l**} versus {**m** AND **n**}, respectively.

In **diagram-D**, the event horizon progresses to t_3: Synonymous AND-to-OR collapse will convert the quantum mechanical superpose state {**e** AND **f**} into the OR-form realised **worldline-e**.

In **diagram-E**, possibility **f** is foregone and, together with its own unique nested futures comprised of {**m** AND **n**}, it is erased from spacetime in potentiality. We can see that only {**k** AND **l**} remain available for the future evolution of our growing block spacetime. We also find in **diagram-F** that the subsequent erasure to OR-form **k** is potentialised to a new series of nested futures. This process of 'collapse', erasure, and decoherence of spacetime in potentiality will unfold in the partial-perpetual form depicted in **Fig.2.12-A** to **F**: Insofar as spacetime in potentiality elaborates to the infinitely removed future, the process is virtually inexhaustible. As it unfolds, realised spacetime will expand in terms of the accumulation of past memory. Thus, realised spacetime begins with **worldline-w** but progresses and accumulates in content to include the serial of worldlines **w-b-e-k**.

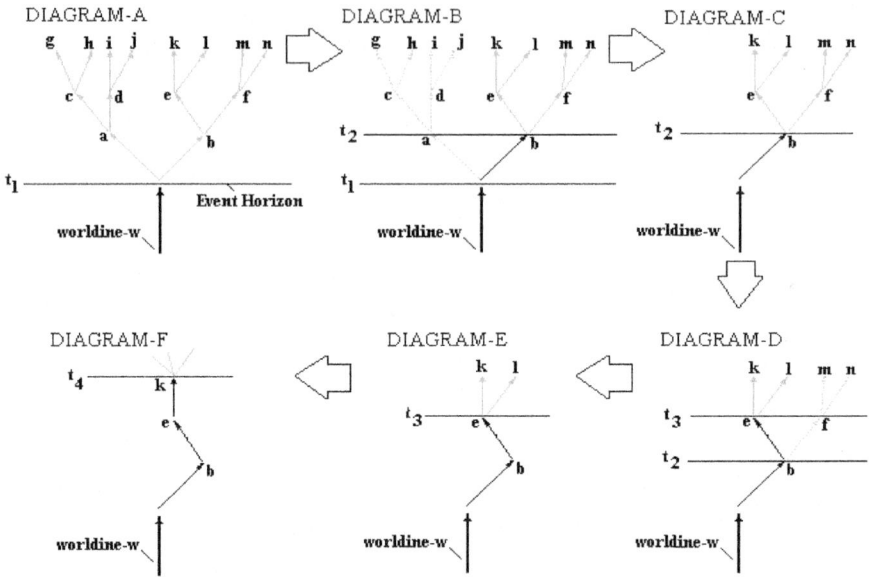

Fig. 2.12: Quantum erasure, partial-perpetual collapse, and the restructuring of spacetime in potentiality
Notes: The sequence of didactic diagrams from **diagram-A** to **F** comprise simplified intermediate model spacetime diagrams and convey a step-by step illustration of the process of quantum erasure pertinent to the nested futures view. The depictions assume the divergent category nested future. Again, we use the simple bifurcation version of nested futures for brevity. In **diagram-A** the event horizon is at t_1, and **worldline-w** is potentialised with immanent future-potential worldlines, {**a** AND **b**}. Each of these hedge subsequent future-potential worldlines complexes, {{**c** AND **d**} versus {**e** AND **f**}}, respectively. In **diagram-B**, subsequent AND-to-OR collapse of spacetime in potentiality advances the event horizon from t_1 to t_2. Hence, the quantum mechanical state {**a** AND **b**} undergoes AND-to-OR collapse to OR-form outcome {**b**}, while {**a**} is irreversibly erased. The OR-form state **worldline-w** now extends to include OR-form {**b**} in an expanded realised worldline system **w-b**, within the domain of realised spacetime below the event horizon at t_2. **Diagram-C** illustrates the consequences of the former outcome: All nested future complexes potentialised to potential {**a**} are irreversibly erased: All subsequent future complexes that further bifurcated from potential {**a**} are also irreversibly erased: The whole domain of spacetime in potentiality must restructure or undergo **grand decoherence**, accordingly. The only future possibilities available for the evolution of spacetime are now nested to {**b**} or to its immanent possibilities, {**e** AND **f**}. In **diagram-D**, the advancement of the event horizon to t_3 collapses {**e** AND **f**} into the OR-form of {**e**}: We now obtain an OR-form worldline system comprised of **w-b-e**. Note that all nested futures complexes nested to {**f**} are irreversibly erased, as is {**f**} itself. The same sort of process is described in diagram-**E** and **F**, albeit with subsequent nested future complexes subject to irreversible erasure, and with spacetime in potentiality undergoing more extensive grand decoherence. Again, the event horizon will advance to t_4: AND-to-OR collapse will take place, generating OR-form outcome {**k**}, and the realised worldline system will expand to **w-b-e-k**… with all remaining nested future alternatives erased.

2.23: Delayed Choice Time-Intervals & Spatially Static Quantum Mechanical Waves: The Retrospective Observational-Inferential Structure Argument

We now turn to the issue of spatially static quantum mechanical waves, opposed to the generic notion that quantum mechanical radiation and de Broglie matter constitute spatially dynamic, time-undulating or moving states. The Alternative Realist case (Alt-R) for spatially static quantum waves is founded on three observations and inferences. To appreciate these, we must first outline three forms or modes of observation and inference, only one of which is viable in our growing block model universe. The first is realtime observation and inference. The second is after-the-fact or retrospective observation and inference. The third is observation and inference before-the-fact, or a-causal observation and inference involving 'signals from the future'.

Realtime seamless-continuous observation and inference would involve the attempt to carry out observations on events and processes as and at the same time as they occur, seamlessly and synchronously vis-à-vis the unfolding physical process... in effective absolute time, despite the case against it per Special Relativity. Unfortunately, key experiments in quantum mechanics, such as the two-slit experiment, and even Wheeler's delayed choice and others, tend to be described in a highly dubious pseudo realtime way, *as if* unfolding per seamless-continuous observation, wherein we imagine the operation of an implicit realtime observation and inference of otherwise empirically and structurally hidden processes subject to full physical isolation. This is exemplified by the dubious and totally pleonastic notion and description of the pseudo-realtime 'moving quantum mechanical wave'... from emitter to detector in the two-slit experiment, in Afshar's experiment, and certainly in Wheeler's experiment.

The second mode of observation and inference about physical processes is the **retrospective observation and inference mode:** involving observations on physical processes sometime *after* these have transpired, reported through intermediary inputs and their return-signals *after* the source-process was resolved and now belongs to the past, or is inferred to have become resolved through past AND-to-OR collapse processes. For example, we always observe the Moon as it was in the past: some 1.5 seconds ago. The constitution of events that make up the observed Moon were AND-to-OR resolved in the past, and arise to us after-the-fact, in retrospective form, *always* as the past, and always subject to the relativity of simultaneity.

Another example: When we measure the attribute of a particle, we must bombard it with observational inputs. The attendant resolution of some attribute of that particle via AND-to-OR collapse will get to us as a constituted past, by means of a return signal *after* the said resolution has transpired, or *in retrospect*. Indeed, in our universe, observations and inferences about physical processes or system are *always* about their past. These get to us after the fact per some delayed choice time-interval. **Hence, all practically possible observations in the physical universe are necessarily and strictly retrospective. Observations are *never* of events and processes 'as they happen', or in realtime: Indeed, realtime observations would constitute assertions of absolute time and the reality of a common present moment in which all things transpire and are simultaneously observed by all.**

In the case of typical didactic descriptions of the two-slit experiment, of delayed choice experiments and other experiments, the ontic retrospective inferences are disguised and pragmatically misrepresented *as if* unfolding in pseudo-realtime. But the didactic value of this is rarely explicitly admitted and tends to engender the presumption of a realtime account of physical processes. Again, the key examples are **the notional spatially displacing quantum mechanical wave in the double-slit experiment, in delayed choice, or across nature: The notion has no empirical or ontic basis and is never observed. The claim for displacing quantum waves cannot be validated in empirical-observational terms: It is a never-corrected assumption imposed on the facts, but only *after* the real facts are garnered *in retrospect*.**

If we shorten the intervening putative distance or space, or, in truth, if we shorten the delayed choice time-interval relation between the observable and the observer across which the feedback mediates, and if we reduce this interval to something approaching, but never obtaining realtime synchronous observation and inference, we might then approximate the relation to something that resembles realtime observation of events and processes as they happen. As we shorten the said putative spatial interval or delayed choice time-interval, the intermediary process, such as the light radiation 'carrying' the information from the observable to the observer, will take much less time to get to the observer. Thus, the observation will occur *almost* at the same time as the process itself. Yet, this must always fall short of any true realtime observation and inference, given that we cannot reduce the said interval between the observable and the observed to zero or simultaneity. Indeed, even if we could reduce it for one discrete event and obtain absolute synchronicity with it, we could not do so simultaneously for myriad other events and observables. Hence, we rightly state that realtime observations and inferences are ultimately unobtainable. This tautology is *not* a trivial fact, as we shall soon discover.

For its part, **the a-causal form of observation and inference of events or observables before they transpire** would involve the reception of 'signals from the future', with reports about events that have not yet occurred, in contravention of the firewall principle from the intertwine of causality and conservation principle formed against just such signals (see **2.11**). Recall the integration of the firewall principle to our intermediate model spacetime structure as part of the event horizon: the 'cut' between the past and the future: (Also see Book-I, **1.17**). Hence, we must discount notional a-causal observation from the list of the three contenders.

*

What is the relevance of any of the modes of observation and inference to the presumed and putative spatially displacing and time undulating quantum mechanical wave? We know that AND-form quantum mechanical waves are real: Generic interference patterns from the two-slit experiment furnish the reality of self-interference; possible only if AND-form logic abides in nature. But how can we know whether such waves are spatially dynamic in the usual pseudo-realtime depictions of them, or spatially static, as Alt-R contends? The answer to the questions is critically bound to the types of observations and inferences that are physically possible. Indeed, experimentally and empirically, we are certain of the physical reality of AND-form quantum waves, but always *after* the completion and conclusion of the physical-experimental process: always *after* the *en masse* particle depositions have formed up to constitute the wave-signature interference pattern... which gets to our senses *after* this corpus has formed... *always* as the past.

217

On the other hand, realtime observations on quantum mechanical waves must remain impossible: **First, one can never reduce the interval between the observable and the observed to zero to affect realtime observation: This is required if the claim about spatially dynamic quantum mechanical waves is to be corroborated. Second, as a matter of staple quantum mechanics fact, the very act of attempting a realtime observation, or even retrospective observation, of the 'moving' or time-undulating quantum mechanical wave, or even of the process of self-interference of that wave, would serve only to wipe out the very same moving, time undulating, and self-interfering quantum wave we sought to observe. Consequently, no self-interference could then occur from which we might garner AND-form logic in any realtime way.**

In short, the act of observation or intercession to observe the wave as it is supposedly undergoing time-undulations, displacement, and self-interference, would wipe out both the sought motion and the AND-form structure expected to generate the interference. This is why when we place detectors at one or both of the two open slits in the two-slit experiment, and then intercede with the use of observational inputs, we destroy the wave-like process and the possibility for self-interference and fail to obtain the signature interference pattern at the detector, *even when both slits are open*. Shorter still, the supposition of the moving and undulating quantum wave in the typical pseudo-realtime way often depicted is only plausible on the strict condition that we deliberately abstain from observing that wave, remove any observational inputs to it, and subsume the interval of the experiment into total isolation.

Hence, only retrospective observations of quantum waves are possible: Realtime and a-causal observations of quantum waves are strictly prohibited and impossible.

Hence, retrospective observations and inferences cannot furnish *any* **basis for the reality of time-undulating and spatially dynamic quantum waves**. In brief, it is obvious that we have never observed quantum waves during their *presumed* 'motion across space', or 'as they move', or 'as they undulate'... and cannot do so for reasons that are now clear: i.e., the impossibility of realtime observation and inference. **The only way to prove the notion of the motion of quantum mechanical waves is either to observe it in realtime, or 'as it moves'... or observe it a-causally, via 'signals from the future'. Both are impossible. Therefore, the notion of the motion of the quantum mechanical wave must also remain observationally impossible... and ontically forlorn.**

It is a true cliché: Eliminate the impossible, and whatever remains... must be the truth: **Succinctly, quantum mechanical waves are spatially static**.

*

We must further augment why retrospective observation and inference cannot furnish the notion of spatially dynamic quantum waves, and why this must force conclusions in favour of spatially static non-undulating wave-states. While the fact of self-interference and AND-form logic is admittedly plausible upon the *assumption* of spatially dynamic quantum waves, the same observation and inference about self-interference and AND-form logic is easily plausible upon the counter-assumption of spatially static quantum waves. Insofar as this is true, it follows that **the ontic reality of retrospective observation of self-interference cannot uniquely compel in favour of spatially dynamic quantum mechanical waves, no more than it could compel** *against* **the possibility of spatially static quantum waves**.

Since, for reasons given, we cannot obtain any realtime observation and verification of purported spatially dynamic quantum wave, all we are left with is a retrospective *assumption* of 'moving quantum waves'. **In other words, we have no evidence for moving and undulating quantum waves; only an** *assumption* **about these**. Indeed, the empirical retrospective observation of self-interference constitutes the only key certainty, but this always involves the ineliminable delayed choice time-interval between the observable and the observation: Again, if we were to exercise the *choice* of ending isolation prematurely, with a view to obtaining observation of the presumed 'moving quantum wave', or 'catch it in the act', we would destroy the 'moving wave' and fail to catch it at all.

Alternatively, if we choose to delay the moment at which we end the isolation of the system and wait for the full formation and manifestation of the interference pattern, and *then* made a retrospective observation of that interference pattern at the detector... even then **we could only be certain of the fact of the delayed choice time-interval of the observation, but obtain** *no* **certainty and** *no* **evidence of and no compelling reason for any 'moving or time-undulating quantum wave'**. The delayed choice time of observational intrusion is *not* the same as, and does not imply the reality of, quantum waves undergoing putative 'spatial motion' or even time-undulation; especially given the fact that *exactly* **the same interference pattern and outcomes can be obtained when we assume** *only* **spatially static quantum waves; subject** *only* **to observations structured in terms of delayed choice time-intervals...** *without* **any presumption of 'motion in space' or of undulation** in the interim of the delayed choice time-interval so-demarcated.

In other words, we can only speak about OR-form events, or corpuses of such, before and up to the onset of isolation, *and* at the end of isolation; all organised and distributed purely temporally. We cannot make any direct realtime claims about happenings in the said time-interval demarcated by the onset and end of isolation: We cannot claim with certainty that the said interval involved any literal moving and undulating quantum mechanical wave. And the fact that we can obtain the same sorts of results from the alternative hypothesis composed purely of static non-undulating waves (subsumed to an obviously purely static and non-undulating spacetime in potentiality; the very ontic future which obviously does not and cannot 'move'), there is no compelling reason why we must assume and describe quantum mechanical waves as 'moving and undulating'.

Once we realise that the retrospective observational-inferential mode can *only* involve recourse to pure delayed choice time-interval relations and assertions between and about successions of OR-form events for observables versus their observations, but never about unobserved non-event AND-form states concealed in and by such time-intervals and attendant isolation-interval, it follows that spatially static quantum waves constitute the inevitable conclusion.

Therefore, **while we might assume and impose 'motion' to the interval residing between the succession of OR-form events, but need not do so, we cannot assert any provable or certain claim about 'motion' or 'undulation' of quantum waves, and, again, *need not assume* moving or undulating waves, when static waves will do just as well, if not better**.

In short, the moving and undulating quantum mechanical wave appears to be *pleonastic*: its assumption is neither compelled by the empirical-experimental evidence nor necessary from it.

2.24: Spatially Static Quantum Mechanical Waves: Structural Argument from Nested Futures

In **2.20**, we discovered how open-choice quantum mechanical systems and spacetime in potentiality hedge for and nest alternate mutually incompatible futures. The discovery from Wheeler's delayed choice, combined with the open-choice nested future structure of spacetime in potentiality, has direct consequences to the question of whether quantum waves are spatially dynamic or static. Indeed, **Wheeler's delayed choice can be configured to attack the notion of the motion of the quantum mechanical wave, forcing conclusions in favour of spatially static wave**.

We employ our previous simplified description of John Wheeler's delayed choice from essay **2.18: Fig. 2.07**. Recall therein that only one nested future will be selected if **switch-x** is set to shunt the split-beam quantum wave to detector-**D**: The other nested future will be selected if **switch-x** is set to shunt the quantum wave toward **Y**. Of course, the setting of **switch-x** is suspended: i.e., it can either be changed to one or the other setting at any moment within the required delayed choice time-interval, or it can be deliberately suspended at a null setting, or even quantum suspended, until a critical moment is reached. In all instances, the spacetime in potentiality description for **switch-x** must constitute as an AND-form superposition with both settings of **switch-x**, whose AND-to-OR resolution will be delayed until the critical moment. Consequently, the ***nesting hedging quantum wave* is forced to back both alternative future complexes, and *must* configure as a nested wave; nesting both subsequent wavefunctions per each stated possibility**.

For the sake of argument, let us assume that the nesting hedging quantum wave approaching **switch-x** is 'displacing across space'… from the beam-splitter… deflected at the mirrors… and with one half of the split-beam heading to **switch-x** and the other to the detector. What are we to say about the mutually incompatible alternate nested futures embedded to the hedging 'moving' wave? Are the embedded and nested quantum waves and wavefunctions also moving? The nested future complexes so hedged are quantum mechanical waves in their own right: different from the hedging wave only in that these constitute nested waves that ought to undergo spatial motion toward **Y** and toward **D** along their intervals **X-Y** and **X-D**, respectively, pending how **switch-x** is subsequently tripped. Indeed, **if we treat the hedging quantum wave as something moving through space, why not treat the nested futures embedded to it as moving across their own intervals? Of course, we cannot do this, given that the nested possibilities have not-yet attained their respective 'paths' and 'spatial intervals' X-Y and X-D, respectively. The nested waves must necessarily be spatially static, even if embedded and nested within a supposedly 'moving' hedging wave. But this realisation brings doubt about the presumption of treating the hedging wave itself as a 'moving wave'**, quite aside implications from the retrospective mode of observation and inference elaborated in **2.23**. In short, the notion of the moving and time-undulating quantum waves constitutes a pure a-empirical and a-evidential assumption and imposition: one not true of nature.

Why cannot we treat the nested quantum waves as moving? Again, as states subsumed to the primary hedging wave yet to reach switch-x, the nested quantum waves and wavefunctions have not yet coincided with their putative pertinent spatial intervals; X-Y and X-D: i.e., these being their prospective 'paths' subsequent to switch-x. Thus, the nested waves cannot be treated as states moving across their spatial intervals, specifically per the fact that these, as subsumed states of the hedging nesting wave, have not yet reached or coincided with *any* part of the paths subsequent to switch-x.

*

Let us further clarify: The hedging quantum wave heading toward **switch-x** subsumes within its content the two mutually incompatible nested future complexes: In one complex, we have the future-potential for… {…the wave has been shunted to and is moving toward detector-**D**}….AND, in the other, we have… {…the wave has been shunted to and is moving toward, detector-**Y**}. That is, the hedging 'moving' quantum wave on approach to **switch-x** nests these other 'moving' quantum waves belonging to the two alternate mutually incompatible nested future possibilities that, as part of their inherent descriptions, are supposedly undergoing motion along **X-D** and **X-Y**, respectively, even though these have not yet passed beyond **switch-x,** and have not physically coincided with their pertinent spatial intervals.

Obviously, the nested future possibility wave, {…the wave has been shunted to and is moving toward detector-**D**} contained in the nesting hedging quantum wave on approach to **switch-x** is not really 'shunted to and moving toward detector-**D**'. It is in truth a purely spatially static state of pure as-yet not-realised possibility, nested as such within the hedging quantum wave on supposed approach to **switch-x**. The same must also hold for the alternate future complex comprised of…. {'the wave has been shunted to and is moving toward detector-**Y'**}: also, a purely spatially static state. To assume otherwise would be tantamount to the claim that the spatial arrangement and intervals of **X-D** and **X-Y** are somehow located in the spatial area before **switch-x,** even though these come after **switch-x**. We would also have to assume that the alternate futures are also moving across **X-D** and **X-Y** as spatially dynamic waves, even though these are putatively 'located' as hedged alternate potentialities within the nesting hedging quantum wave yet to complete its journey along the pertinent intervals before **switch-x**.

The problem can easily be avoided if we appreciate that **the nested future quantum waves are purely static forecast-states (albeit, ontically real futures) that account for both potential events and for the *potential space* (the putative Hilbert space) that**

219

is itself yet to be formed up into any realised spatial interval, and across which spatial displacement has not yet occurred. In short, we must treat the nested waves within the principal hedging quantum wave on approach to switch-x as spatially static waves vis-à-vis their own putative potential spatial intervals, ceteris *paribus*...or *assuming the ontic reality of 'space'*.

Put in terms of the larger system of spacetime in potentiality and grasped from the perspective of the intermediate spacetime model, the future-potential spatial intervals **X-Y** and **X-D** constitute, as-yet not-happened potentials that reside 'above' the event horizon; 'above' the synonymous AND-to-OR collapse and time-processes that have not yet attained to these paths and associated potential events. This implies that these are not realised constituents of spacetime. Therefore, the paths cannot be treated as if they are realised, or as being traversed by *anything*, let alone by 'spatially dynamic quantum waves'.

*

While the nested future quantum waves must be grasped as spatially static, **it also follows that the hedging primary quantum wave on approach to switch-x cannot itself be treated as moving or displacing across 'space' towards switch-x**. Five factors bring the presumption of the 'moving' hedging wave into question:

- <u>First</u>, the fact that **we need an impossible realtime observation of it as a 'moving wave' to claim its reality:** a feat practically and physically impossible per the argument presented in **2.26**.

- <u>Second</u>... that **any attempt at a realtime or other observation of the hedging quantum wave to prove that it is 'moving' toward switch-x, would only serve to destroy the effect we sought to observe i.e., destroyed the moving wave itself**... rendering it impossible to observe and corroborate. (See **2.23**).

- <u>Third</u>... doubt emerges from **the realisation that a spatially static treatment of the quantum mechanical wave is more than adequate to the task of supporting *all* the various empirical outcomes produced**, and that the pleonastic spatially dynamic quantum wave is neither critical nor necessary for the viability of quantum mechanics and physics generally.

- <u>Fourth</u>... doubt from our most recent realisation: **The alternate future quantum waves nested within the hedging primary wave are spatially static vis-à-vis their own potential spatial intervals, X-Y and X-D: *This proves that there really are such things as spatially static quantum mechanical waves*:** i.e., that such static waves are compelled by the structural logic imposed by the delayed choice nested futures view of the quantum wave. **Hence, the ontic reality of spatially static quantum mechanical waves is compelled by the nested futures view, and it must surely apply to the hedging quantum wave itself.**

- <u>Fifth</u>....it follows, **there is no compelling reason to suppose that the hedging quantum wave must be exempt from characterisation as a static quantum mechanical wave**, or treated as any more spatially dynamic than the nested future quantum waves that it nests and hedges. Indeed, what would be the point or scientific utility of treating it so?

*

Our conclusion in favour of spatially static quantum mechanical waves and wavefunctions can be augmented as follows: Let us reiterate that, insofar as the nested quantum waves cannot themselves be treated as 'moving' across their potential as-yet not-realised spatial intervals, and insofar as their pre-AND-to-OR collapse potential spatial intervals do not yet literally exists as realised spatial intervals, the same can and *must* hold true vis-à-vis the hedging nesting and primary quantum wave on approach to **switch-x**. Why? **The hedging and nesting quantum wave is related to a putative spatial interval that itself constitutes a pre-measurement and pre-AND-to-OR collapse state. That is, insofar as physical isolation has been enforced, and no system-intrusive observational inputs are applied, the putative spatial interval leading up to switch-x along which the hedging quantum wave is supposedly 'moving' must also constitute a pre-event state: a 'not-yet happened' interval within spacetime in potentiality that cannot exhibit anything moving across it, *especially* a hedging nesting primary quantum wave.** See essay **2.25**.

So much for spatially dynamic quantum mechanical waves: There are no such things, even assuming the ontic reality of space. It follows that quantum mechanical waves of radiation and de Broglie matter can only be spatially static non-undulating standing states.

Of course, **with anticipated de-spatialisation in Book-IV, we will have removed the whole notion of 'space' from the ontology and lexicon of physics: There is no 'space'. Therefore, there can hardly be anything 'moving' across a thing that does not exist... and the ontic reality of spatially static quantum mechanical waves must automatically and seamlessly follow.** But baring radical de-spatialisation, and assuming the dubious ontic reality of 'space', subversion of the moving and time-undulating quantum wave in preceding and following essays must yet abide, with conclusions equal to the those to be wrought by anticipated de-spatialisation.

2.25: Spacetime Structure Argument for Spatially Static Quantum Mechanical Waves

Realtime observation of a hypothetical spatially displacing quantum wave would require the seamless succession of observations of that wave, at all spatial points associated with its progression, and at all pertinent successions of the event horizon through which the progression and displacement is to be evinced. But, given that realtime seamless-continuous observation of the wave along its putative spatial interval is not physically possible (see Preliminary **0.02** and **0.03** and the three modes of observation in preceding essay, **2.23**) and given that any physically permissible observation entails inabstractable isolation and a delayed choice time-interval between the observed and the observation, *and* an additional delayed choice interval between the wave's incidence at a given moment vis-à-vis the succeeding moment of its retrospective observation, only the retrospective observational mode of the wave's is possible. The closest we can come to realtime seamless-continuous observation might involve the Quantum Zeno effect, but this cannot furnish truly seamless-continuous realtime observation, because it cannot eliminate quantum mechanical intervals of isolation: (see Preliminary **0.13**, and Book-I: **1.25**).

The key aim of this essay is to model retrospective observation and inference processes from within the intermediate model and spacetime in potentiality. The aim is to model our reiterated insights about the retrospective mode in terms of the structure of the intermediate model spacetime, incorporating our previous conclusions, including those from **2.23**, to its processes, and validating these in terms of its framework.

*

What does it mean to introduce a delayed choice time-interval between the moment of experimental initiation and the moment of resolution and observation, assuming the performance of a two-slit experiment? In terms of the intermediate spacetime model, the moment of experimental initiation and systemic isolation that summons the putative non-observable 'moving quantum mechanical wave', followed by the moment of its post-resolution and subsequent observation into an interference pattern, must coincide with successive event horizons: i.e. a first event horizon, resolved at the moment of experimental initiation and isolation, followed by a succeeding event horizon formed at the final moment of retrospective observation at experimental end.

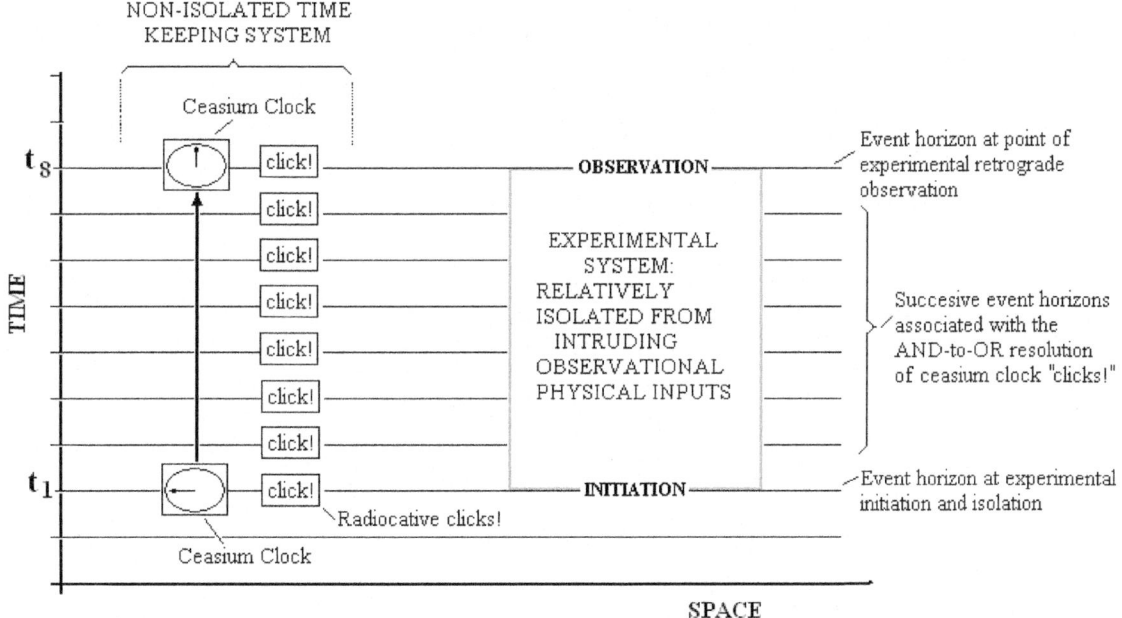

Fig. 2.13: Isolated verses non-isolated physical systems and clarification of delayed choice in experimental initiation and retrospective observation
Notes: The illustration systemically relates an experimental system (the area on the right, subject to experimental isolation via the withdrawal of observational physical inputs) to a relatively non-isolated time-keeping caesium clock system (on the left). Each click of the caesium clock is associated with an AND-to-OR collapse, synonymous time and synonymous event horizon-progression from t_1 to t_8. Yet only two of these event horizons (t_1 and t_8, respectively) and their OR-form outcomes in terms of "clicks!" directly relate to the isolated experimental system: namely, at the point of its initiation at t_1 and at final observation at t_8, respectively. With observation taking place at t_8, experimental isolation ends through the application of physical inputs of observation. The caesium clock system versus the experimental system differs in terms of the number of event horizons (i.e. the number of successive AND-to-OR collapse processes and events). Through isolation, the experimental system is subject to a delayed choice time-interval: one measured by the interval constituted by the "clicks!" from the relatively non-isolated caesium clock time-keeping system. Thus, the illustration clarifies the essential means by which we assign time-measurements, isolation, and the implementation of delayed choice time-intervals to experimental systems and outcomes.

From the first event horizon and thereafter, and with isolation enforced, the system is transformed into an AND-form state of potential as-yet not-realised putative spatial intervals and, by pure *unwarranted assumption*, associated putative 'moving' AND-form quantum mechanical waves. On the other hand, at the subsequent final event horizon pertinent to the subsequent retrospective observation, decisive AND-to-OR wavefunction collapse and measurement will transpire. Thus, by reference to the didactic depiction in **Fig. 2.13**, the said delayed choice time-interval will manifest as a delay in time between the initiation of experimental initiation at t_1 versus the subsequent experimental observation and the end of isolation at **t₈**, at the associated event horizons. The said delayed choice time-interval will correlate with the relatively less isolated neighbouring system comprised of a time-keeping device: say, a caesium clock.

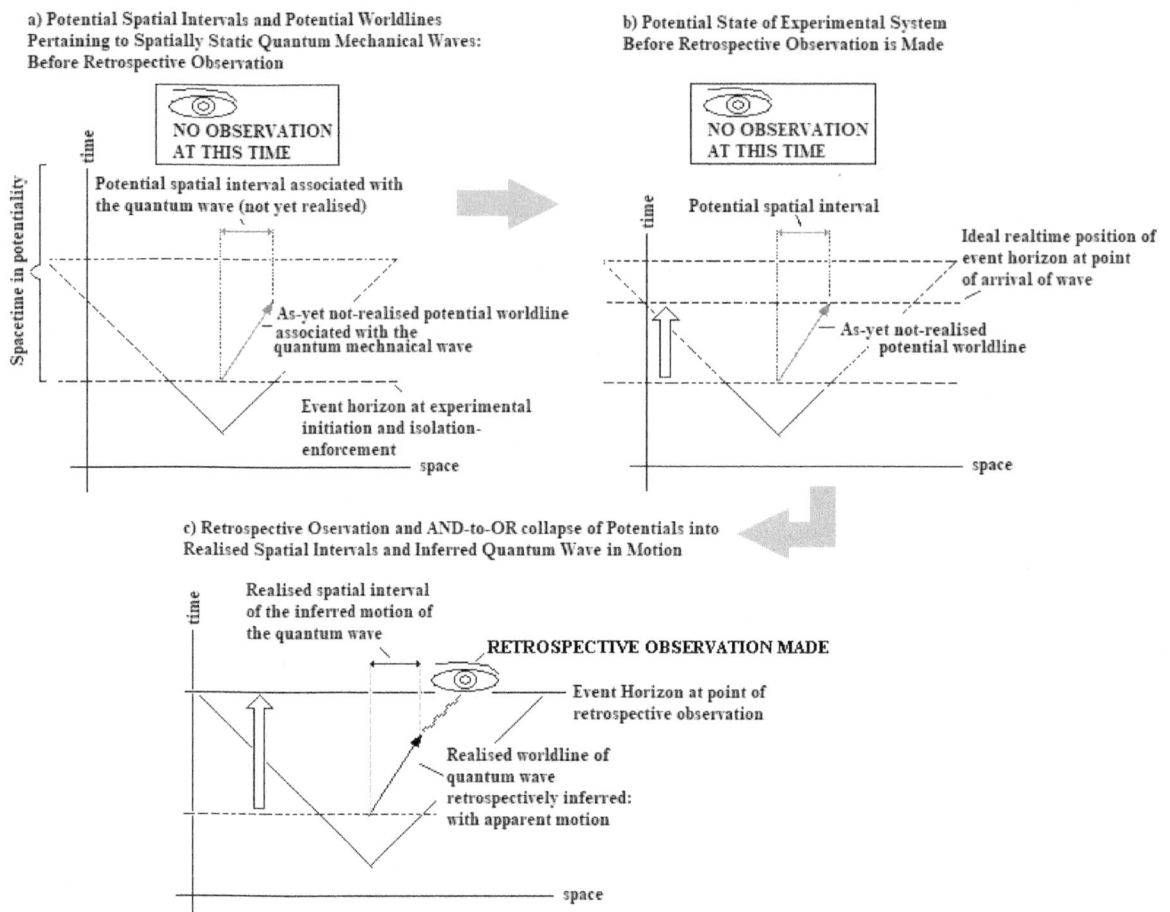

Fig: 2.14: Integrating the concepts of retrospective observation, and implications in favour of static quantum waves in intermediate spacetime models

Notes: Diagrams **a** and **b** depict an experimental system and its embedded quantum mechanical wave before any observation on it is made: The presence of a time-keeping non-isolated system is implicit. The experiment is subject to a delayed choice time-interval relation from the moment of experimental initiation to its final retrospective observation, indicated by the pertinent successive event horizons. In **diagram-a**, the experiment is initiated, and the system is placed into isolation. So long as isolation holds, we can only speak of a potential not-yet realised worldline for a quantum wave associated with a future-potential not-yet realised spatial interval. Moreover, the quantum wave is spatially static, if for no other reason than the fact that its pertinent spatial interval has not-yet formed up for the wave to traverse. In **diagram-b**, the inferred 'arrival' of the quantum wave at the point of its destination (indicated by the second succeeding event horizon) must remain in isolation: it can no more be assigned an OR-form realised event-status than can the associated not-yet realised spatial interval across which the wave is purported to have displaced. Indeed, potential contingencies (i.e. Wheeler's cat) reside between the expected moment of its potential arrival and the expected moment of its final retrospective observation: one that might resolve into an alternative future outcome to that expected. In **diagram-c**, the system is finally brought out of isolation and a retrospective observation is made via intermediating observational inputs and feedback: i.e., the wavy worldline linking the arrival of the quantum wave vis-à-vis the later observer, in turn depicted as an overseeing 'eye' placed at a future-potential moment. Thus, finally, the presumed 'spatial interval' and the worldline pertinent to the quantum wave and experimental system is realised in OR-form terms and falls into observation, but only as the past in total retrospect.

222

QUANTUM MECHANICS AND MIND

The caesium clock-system is subject to a greater frequency of successive, almost continuous AND-to-OR collapse processes, indicated by radioactive caesium decay 'clicks' and attendant retrospective observations of the said clicks, in comparison to the isolated quantum mechanical system. In **Fig.2.13**, the latter is subject to only two main observations: one attending the setup of the system just prior to its isolation, and the other comprised of its final retrospective observational moment: both separated by the said delayed choice time-interval as measured by reference to the neighbouring caesium clock-system and its OR-form series of 'clicks'; with each such click associated with an event horizon progression.

By such means, we forgo any need to interrupt, intercede in, or make a premature observation of the isolated system. Instead, we use the neighbouring less-isolated caesium clock-system to measure the pertinent delayed choice time-interval of the isolation-interval of the experimental system. Thus, while we preserve the never-observed notional *assumed* 'displacing quantum wave' within the isolation-interval, we deprive ourselves of any direct empirical evidence in proof of such a 'spatially dynamic quantum wave'.

The same can be recuperated directly using the intermediate spacetime model: see **Fig.2.14-a to c**. Therein, we not only place the quantum wave, now depicted as a potential worldline, into experimental isolation per the said delayed choice time-interval, but also relegate it to the domain of spacetime in potentiality. That is, the 'spatial interval' across which the 'moving quantum wave' is purported to unfold is relegated to a futurity that has not-yet happened, necessarily in a state of isolation. The said 'spatial interval' must itself become potentialised into an as-yet not-realised spatial interval that, insofar as it has not yet come into being, nothing can be said to move across it, much less a quantum wave.

Following an appropriate time-delay, or delayed choice time-interval relationally to the caesium clock-system, we finally make a measurement on the isolated system: With the end of isolation, its potential spatial interval, together with its associated 'moving quantum wave', must now AND-to-OR collapse into a set of realised OR-form outcomes and be brought out of isolation. Again, this will fail to furnish *any* direct evidence of any putative moving or even time-undulating quantum mechanical wave: (See **Fig. 2.14-b**).

Note in **Fig.2.14-c** that the moment of measurement or observation, and the associated event horizon, will not be the actual moment at which the retrospective observation is obtained: There needs to be physical feedback via the observational input involved in the destruction of the experimental isolation. The feedback rapport must itself unfold and 'travel' across its own purported spatial interval: It will impose its own delayed choice time-interval from experimental endpoint to the final observation. Hence, observation will be further time-delayed until a subsequent progression of the event horizon unfolds and coincides with the moment at which the retrospective observation is finally made.

From experimental initiation to retrospective observation, the whole experimental system must necessarily become relegated to the domain of spacetime in potentiality: But spacetime in potentiality (the future) is not a moving state: the future cannot move. The future constitutes a static state, and the quantum mechanical waves associated with it are necessarily non-undulating and spatially-temporally static standing states.

By integrating our empirical retrospective observational mode to the structure and processes of our intermediate spacetime model, we discover that **quantum mechanical waves are not spatially dynamic: They do not 'move across space', and that they are temporally and spatially static states. Indeed, 'motion' is a false or pleonastic assumption imposed on the succession of OR-form events**, in turn generated from the decay of potentialities out of static spacetime in potentiality. This conclusion for spatially static quantum waves from the structure of intermediate spacetime constitutes the background to a final alternative argument in favour of the same: i.e., the causality argument and the firewall principle for spatially static quantum mechanical waves.

2.26: Spatially Static Quantum Mechanical Waves from the Firewall Principle

In **2.07** (Part-I) it was stated that all apparent motion is merely assumed and imposed upon the succession of events generated by and garnered from successive AND-to-OR collapse-processes (time) synonymous with attendant progressions of the event horizon: It is AND-to-OR collapse and attendant progression of the event horizon itself that constitutes the only dynamic process in spacetime: All else, including the apparent 'motion' of objects, and especially of quantum waves, constitutes nothing more than assumption, imposed or foisted on the *temporal* succession of events generated by the intertwine of AND-to-OR collapse and event horizon progression. Succinctly, assumed 'motion' is apparent only through retrospective observations (**2.23**) of the succession of events constituted as the past, brought about by the process of AND-to-OR collapse and attendant event horizon progression.

As was also argued in **2.25**, the whole of the quantum mechanical AND-form domain of spacetime in potentiality within which the 'moving' quantum wave is embedded, cannot be susceptible to spatial displacement or motion: Spacetime in potentiality, as the ontic reality of the future itself, or of any subsumed possibility within it, cannot undergo motion, simply because the future cannot move... if only because a system of potentialities subsumed therein have not-yet happened, and we cannot imposed upon it the notion of 'motion' or even of 'undulation'. Indeed, to what and to where could the whole or parts of the future displace? To what and to where could the future move? Obviously, the domain of spacetime in potentiality, the future, is static: So also the future-form quantum mechanical waves suspended within it. All of this is consistent with the case for retrospective observation and inference in **2.23**, framed to the structure of our intermediate model spacetime model.

Under what empirical and structural conditions could we prove that quantum mechanical waves are spatially dynamic? **If we could observe things a-causally, before they transpired, and use physical inputs to garner direct observation of the supposed 'moving quantum wave' across its not-yet realised spatial interval** *in the not yet realised future*, **and assume that such a mode of observation did not disturb the as-yet not-realised future so as to destroy the very quantum wave we sought to observe... we** might then hope to observe and empirically confirm the 'moving quantum wave,' and even confirm its time-undulation. **However, this could only be accomplished by usurping the firewall principle and the contravention of both causality and conservation principles; tantamount to 'signals from the future'**... and the inevitable treatment of that future *as if* it constituted a corpus of realised events and happenings before it had transpired and resolved into such.

Of course, the firewall principle cannot be usurped: we cannot obtain 'signals from the future' and violate causality and the conservation principle. **Therefore, we can never observe or prove the 'moving quantum mechanical wave', much less the 'space' through which it is purported to have moved.**

In conclusion, both causality and the principle of conservation, entwined in the firewall principle, demand that quantum mechanical waves, hence the domain of spacetime in potentiality to which such states are subsumed, *must* be treated as spatially static temporally non-undulating states, notwithstanding implications from de-spatialisation that render doubt on the notion of space itself: (Book-IV).

2.27: Presage to the Overhaul of the Schrödinger Wave Equation

The spatially static quantum mechanical wave, combined to the nested futures view argued in **2.24** to **2.26**, together with the case for partial-perpetual wavefunction collapse, forces the anticipated overhaul of the famous Schrödinger equation: the generic wave equation that models the time-evolution of a wavefunction composed of AND-form immanent possibilities, say, for spin, or for putative position along a one-dimensional line; across a two dimensional surface; or within a three-dimensional block. **The Schrödinger equation seeks to model the 'time-evolution' of the wavefunction before AND-to-OR collapse... or before time has happened to it: an implicit and inadvertent contradiction. Indeed, in the equation, AND-to-OR wavefunction collapse is not recognised as constituting time itself. Yet, time grasped as AND-to-OR collapse (or as it attends 'state vector reduction' in the collapse of the Schrödinger wave into a classical OR-form outcome) is *not* incorporated into the equation. In short, the central problem of Schrödinger's equation is in the definition and application of time.**

The following bullet points describe the key overhaul points of the Schrödinger equation, presaged in the light of our discoveries, to incorporate time and AND-to-OR collapse to appropriate models of quantum mechanical waves.

- **Nested futures**: The AND-form potentialities mapped out within the generic wavefunction and formalised in the Schrödinger equation incorporate only immanent possibilities. Yet, from our delayed choice approach **(2.17** to **2.18)** each immanent possibility nests and subsumes subsequent unique sets of nested future complexes and nested future wavefunctions. **The overhaul of the Schrödinger wave equation must accommodate nested futures and the taxonomy of divergent, entangled, and especially convergent nested future forms.**

- **Partial-perpetual wavefunction collapse**: The generic Schrödinger equation does not model for AND-to-OR collapse itself. Yet, when this is incorporated, the equation cannot typically be treated as a closed-choice one-step one-time all-exhausting comprehensive collapse-base; with all its AND-form possibilities completely wiped away. Instead, the nested futures framework, augmented by the intermediate model spacetime, forces a model of this as an inexhaustible process of partial-perpetual process of AND-to-OR collapse-steps through which collapse of immanent possibilities discloses a new nest of future possibilities ready for subsequent AND-to-OR collapse-process... *ad infinitum*. Thus, **the Schrödinger wave must be overhauled into an open-choice nested futures state, subject to partial-perpetual AND-to-OR collapse.** Indeed, we need to replace the possibility set of the Schrödinger wave with the complex nested futures state constituting the whole of spacetime in potentiality, in turn elaborated to the infinitely removed future.

- **Spatially static quantum waves: Spatially static spacetime in potentiality**: The generic Schrödinger wave equation presupposes the ontic reality of spatially dynamic quantum mechanical waves: undulating and displacing across or within a one, two or three dimensional 'space', and the time-evolution of its purported motion... or its time-undulation, if it is a standing wave. We now know from 2.24 to 2.26 that quantum waves are, at best, spatially static, and cannot obviously traverse across any potential as-yet not-realised spatial intervals, even assuming the dubious ontic reality of space itself. It follows that **the Schrödinger equation must be overhauled to accommodate the reality of spatially static quantum mechanical waves, including spatially static non-undulating standing waves incorporative of the nested futures view.** The presaged replacement equation must forgo the assumption and contrivance of the 'moving and undulating wave'. Instead, the replacement equation must assume a time-evolution or decay of static nested future wavefunction block: a time-evolution and decay directly tied to time grasped as partial perpetual AND-to-OR collapse of static immanent future possibilities and attendant grand decoherence of the divergent, entangled, and convergent nested future possibilities elaborated to the infinitely removed future. Note that **the wave undergoing decay is a static set of immanent and nested possibilities...and its time-evolution is the decay and restructuring (decoherence) of its nested future static possibilities: None of this involves a 'motion in time', much less an 'undulation in time' of the quantum mechanical wave.** Per these stipulations, whatever replaces the Schrödinger wave cannot be described in any pseudo-realtime dynamic 'undulating' or 'moving' form, from which time, grasped as AND-to-OR collapse itself, is contradictorily excluded: That is, we must drop the 'time-evolution of the wave', save as a decohering purely static non-undulating spacetime in potentiality.

- **Incorporation into the intermediate model spacetime as spacetime in potentiality**: The presaged overhaul of **the Schrödinger wave equation must integrate into the intermediate model spacetime as spacetime in potentiality itself: wherein the now static wave-equation becomes the whole of the nested future domain of spacetime in potentiality: a static block... wherein**

the progression of the event horizon is grasped synonymous with AND-to-OR wavefunction collapse and time: and wherein the quantum erasure and counterfactual grand decoherence of spacetime in potentiality, or *block-decay,* **replaces the 'time evolution' of the Schrödinger equation**: The replacement equation must pertain to a static pre-time and pre-AND-to-OR collapse future-possibility state, *not* to a dynamic 'moving' or 'time-undulating' one.

These then are the presaged primers according to which an overhaul or replacement of the Schrödinger equation must be realised in the future. Yet, it is by no means a complete account of the changes that would need to be made: **The proposed overhaul neglects anticipated de-spatialisation**, wherein we can no longer model information pertaining to the future constituted as spacetime in potentiality, or as memory constituted as the domain of realised spacetime, as a spatial distribution. Instead, information must be remodelled as distributed purely in time, *without* space: a matter for Book-IV and the Bergson-Whitehead amalgam successor to our intermediate model spacetime. Thus, the replacement to the Schrödinger equation must incorporate the radical implications that arise from de-spatialisation, temporisation, and abstract memory.

That de-spatialisation is obvious from our case for spatially static quantum mechanical waves and from equally spatially static spacetime in potentiality immediately emerges from the observation that, if the quantum mechanical waves are spatially static, then there is not much point or utility in the notional 'space' through which these no longer 'move'.

2.28: Quantum Mechanical Waves, Spacetime in Potentiality, Causality, Irreversibility
And the Ontic Reality of Time

In 1990, the author arrived at what then seemed certain conclusions about the nature of time: The author assumed that inanimate non-conscious states cannot experience time; that time only applies to consciousness; that time is not an objective property of the physical order. Thus, an electron particle, a non-conscious state, cannot discern or 'know' about the things that had happened to it in its past, and must be equally oblivious to things that might happen to it in its future. Hence, time could not possibly exist for the electron... on **the erroneous assumption that the perception of time is necessary to the ontics of tim**e.

This notion of consciousness-imposed and projected time, and time as an illusion, was augmented by a second consideration: According to Special Relativity, any clock attached to a frame of reference that approaches near the speed of light would be observed to slow down, relative to a clock attached to a stationary frame i.e., time dilation. By implication, if a clock could attain the velocity of light, in the clock's frame, time would stop. For observers attached to that clock, a journey from one point in space to another would unfold instantaneously: distances would collapse to zero.

Light itself, and the photon-particles that constitute it, typically propagate at the speed of light. Hence, what for us might appear as a photon particle that took 14 billion years to complete its journey from the putative Big Bang to the present will, from the photon's own frame, constitute a journey completed instantaneously, given that a photon that propagates at the speed of light has no intrinsic tense-sense: no 'experience' of time, given absolute time-dilation. Hence, for the photon, space or distance could not exist either, and a single photon could be everywhere and any-when simultaneously, as viewed from its own frame of reference, in a block universe without time and within a 'future' filly constituted as a realised state.

Of course, time could not be ontically real in such an arrangement. Another crazy consequence would be that all events, including happenings to electrons and to all other particles, would tend ultimately to be mediated by just one photon. Crazier still, it might further be conjectured that electrons and other particles do not possess any sort of intrinsic existence, and that it is the single photon of light that, through its structured simultaneous and instantaneous deflections and refractions throughout its instantaneous history, furnishes the impression of such things.

These ideas are hardly novel and were certainly not original to the author. Yet, they also prove to be erroneous when confronted with our insight into the nature and physical meaning of quantum mechanical waves, and from implications garnered from our growing block intermediate model spacetime.

While it is certainly the case that non-conscious states cannot experience time, to conclude that this must imply that time is not an objective feature of the mind-independent world is dubious: **The perception of time is not a necessary requisite to the ontology of time. The only certain thing we can accord to mind and consciousness is the power to apprehend events, including time itself. Yet this is not the same as the power to create, project or impose events, or even time itself. The apprehension of time does not imply the creation of time, much less its imposition or projection**.

As for the notion that the whole universe could be generated by the journey of a single spatially dynamic photon that completes its own and the whole universe's entire journey and history in a timeless instant... this is brought into question by critical implications from our nested futures view of the quantum mechanical wave constitutive of spacetime in potentiality, and by the spatially static nature of the quantum waves: The nested futures view must rightly treat the quantum mechanical wave-state of the putative lone-photon as a spatially static state, undermining the naïve supposition of the 'spatial motion' of that photon. It is not so much that the distance across which the photon propagates is transformed to zero on account of absolute time dilation, but, as a quantum mechanical state of affairs, hence as a static quantum state within spacetime in potentiality, the lone-photon constitutes both a static non-moving state *and* it has nothing to move across, given that its putative spatial interval has not yet formed up (ignoring de-spatialisation). Thus, spatially static quantum waves upset the generic account of the dynamic lone-photon, or one that supposedly undergoes displacement at the speed of light, or, upon its own-frame absolute time-dilation, gets to realise its journey in no time at all... *and* generates the total history of the whole block universe in one timeless instant.

Moreover, in its spatially static state, the quantum mechanical lone photon is a total futurity of possibilities rendered in AND-form: It is now the whole of spacetime in potentiality elaborated to the infinitely removed future, and, as such, constitutes only the potential history of the universe, cum alternatives and contingencies... all of it structured as a nested futures complex in AND-form superposition state. The implication is that, at best, the lone photon constitutes only a partially realised and materialised history: one constituted as the hitherto accumulated domain of realised spacetime...and that **the lone photon must yet undergo partial-perpetual AND-to-OR collapse in order to realise and complete the total universe out of AND-form future-potentiality. That is, the lone photon must intrinsically and objectively undergo AND-to-OR collapse and time. That is, insofar as time *is* AND-to-OR collapse, and the photon *must* undergo such collapse perpetually, it 'experiences' time... and its experience of time is not dependent on the presence of consciousness, much less projected by perception**.

Indeed, time is not an illusion: The process of AND-to-OR collapse and, with it, the synonymous progression of the event horizon, comprises the universe-scale process synonymous with time itself: time grasped as an ontically real feature of the physical universe, with or without mind and consciousness; with or without the perception and the noetic apprehension of time.

If we could prove that the entire history of the photon was, from its own frame, something completed and exhausted for all time in a single instantaneous moment, thus constituting a block universe, only then could we conclude time as illusory. To put the same in another way, from the photon's frame, if we could decisively prove that that photon possesses no quantum mechanical AND-form characteristics, and, with it, also prove the domain of spacetime in potentiality to be wholly non-existent, hence debunk quantum mechanics and AND-form logic itself... only then could we assert time as an illusion.

However, **the feat above presupposes a block universe of fully resolved futures without the possibility of AND-form logic... one in which the generic two-slit experiment could not generate self-interference or subsequent interference patterns, both of which presuppose AND-form superposition of alternate future possibilities. Yet, generic two-slit experiments *do* produce interference patterns. Therefore, AND-form states *are* ontically real. Therefore, the future is not a resolved state of some block universe. Therefore, we live in a growing block universe in which quantum mechanics is *not* illusory**: Indeed, the reality of quantum mechanics and the ontic reality of time is demanded by the firewall principle (see **2.10** to **2.11**). **Therefore, time is indeed objectively real**, with or without consciousness.

*

There are a multitude of other reasons why we must consider time as ontically real: In Book-I, as a contribution to the growing block universe and as a mirror of it, we discovered that we can no more obviate and invalidate the ontic reality of AND-form logic than invalidate OR-form realities. Consequently, there can be no such thing as a pure post-quantum mechanical OR-form physical system equivalent to a block universe. Nor can there be a pure quantum mechanical system bereft of any OR-form contaminants, equivalent to a pure futurity without events or memory. These considerations are consonant with our growing block intermediate spacetime structure, in that the inevitable admixture of OR-form and AND-form constitutes a mirror of the inevitable admixture and integration of the ontic past with the ontic future; of the domains of realised spacetime *with* spacetime in potentiality.

The ineliminability of OR-form and AND-form ontics, and the intermediate spacetime structure these mirror, is also independently augmented by the new complementarity thesis, first promulgated in Preliminary. Therein, we discovered that, to verify in empirical and experimental terms OR-form realities, we require AND-form quantum mechanical realities, and vice-versa. For example, in the two slit experiment, to infer AND-form logic and the superposition principle, we must garner evidence of interference patterns. But the required interference patterns are formed out of *en masse* OR-form particle depositions: That is, we need the OR-form particles to bring into evidence the reality of AND-form waves: We need both in an observationally simultaneous way. Indeed, this relation is encapsulated in the new complementarity thesis, which was first articulated in the Preliminary essays in relation to the critique of Afshar controversy.

Since the new complementary thesis presupposes AND-form *and* OR-form realities in an observationally simultaneous way, the process of AND-to-OR collapse through which the observational simultaneity of particle and wave is brought into being at the detector and as the interference pattern, necessitates the ontic reality of time, grasped as AND-to-OR collapse itself, with or without consciousness.

*

Our final augmentation of the ontic reality of time is furnished by our new thoughts on the character of entropy and the 'arrow of time', both also mirrored in the structure of our growing block intermediate spacetime model. In Book-I, **1.9** to **1.10**, Alt-R augmented and superseded generic statistical entropy-based views of time and its 'arrow' by making a fundamental link between time, entropy, and AND-to-OR wavefunction collapse: Therein, the perennial 'time's arrow' was shown to be a function of two interrelated things: First is the fundamental process of AND-to-OR collapse, which always proceeds from the quantum mechanical AND-form state (the future) to the post-quantum mechanical OR-form outcome (the present and accumulating past). This AND-to-OR polarity defines the figurative 'arrow of time' as always proceeding from AND-to-OR, never from OR-to-AND. Second, with AND-to-OR collapse, we *always* obtain an increase in the entropy of future-potential information as it pertains to potential information, now grasped as the AND-form nested futures domain of spacetime in potentiality.

Since time must always proceed from AND-to-OR and can never reverse into OR-to-AND, it necessarily and inescapably follows that time always proceeds from a state of lower entropy of future-potential information to a higher level of entropy of future-potential information. These factors, once combined, superseded generic statistics-based entropy-perspectives on time's arrow,

226

even while incorporating and subsuming such statistical entropy… and supersede the conception of time as a 'fourth dimension'. Thus, time's arrow is firmly tied to ontologically primary AND-to-OR collapse and associated entropy.

Of course, AND-to-OR collapse is real. We inhabit a growing block universe. **Time could only be illusory if AND-to-OR collapse itself was illusory. Time and AND-to-OR collapse could only be illusory if AND-form logic did not exist. Yet, AND-form logic is ontically real; we could not otherwise obtain the interference pattern in the two-slit experiment. It follows that the process of AND-to-OR collapse is real. Therefore, time is ontically real**… with or without consciousness and the perception of time.

2.29: The Intermediate Spacetime Model: Irreversibility & the Ontic Reality of Time

The intermediate model of spacetime is divided into two main domains, both separated by the event horizon. The first of these is the 'pre-measurement' pre-event future-form quantum mechanical 'spacetime in potentiality', comprised of all possible future-potential lightcone and nested future-potential worldlines that spacetime might get to evolve into in the future through the process of partial and perpetual AND-to-OR collapse, or time. The other spacetime domain is the post-measurement post-event mnemonic domain of 'realised spacetime', comprised of the sum of memories constituting the ever-accumulating past, generated from perpetual AND-to-OR collapse of spacetime in potentiality into the realised OR-form outcomes and events through successive event horizons, whose abstract basis is then relegated to and adds to the accumulating past.

What has any of this to do with the perennial 'time's arrow', or with the ontic reality of time? **The structure and processes of the intermediate model constitute the growing block model of the universe, in which both the ontic reality of the future *and* the growing accumulation of information and memory require the ontic reality of time and its 'arrow'.**

The domain of AND-form spacetime in potentiality is ontologically primary to the domain of realised spacetime. **It is never possible that AND-form spacetime in potentiality could arise from OR-form realised spacetime, or from OR-to-AND de-collapse and implausible time-reversal. The AND-to-OR directionality which generates realised spacetime from the decay of spacetime in potentiality constitutes the very determiner of the 'arrow of time'.**

With AND-to-OR collapse, an irreversible processes and outcome transpires: The ontologically *a priori* AND-form state of possibilities will not only get to 'collapse' to just one realised outcome on a succeeding event horizon but, per quantum erasure and grand decoherence of the nested futures structure of spacetime in potentiality, the alternative future possibilities embedded within the AND-form state will become irreversibly erased and re-structured into a new constitution of the future. This erasure and decoherence is irreversible, not only in the sense of entailing an irreversible increase in the entropy of future-potential information, but also in the sense of the realisation of a unique and irreversible OR-form outcome.

The formation of a unique and irreversible event, combined to the absolutely unique and *irreversible* re-configuration of spacetime in potentiality to infinity… clearly indicates an irreversible alteration in the potential futures-content *and* the mnemonic event-content of the universe: i.e., a growing block universe. **The 'arrow of time' can therefore be discerned as an intrinsic physical reality that arises in a necessary way from the objective asymmetric processes of spacetime: an asymmetry comprised of two opposed but harmoniously integrated spacetime domains that mirror the primitive unification of classical-relativistic (the past) with the quantum mechanical (the future), and evinces what we call *time*:** Thus, we can no longer dismiss time as a *bloss* subjective illusion imposed by mind, or as a thing that presupposes perception in order to be rendered ontologically real.

In short, time is ontically real, with or without perception, mind, or consciousness.

*

Also, recall from Book-I (see **1.11** to **1.13**) the argument for the time-asymmetry of physical laws. The very viability of physical law is intertwined to AND-to-OR directionality. Physical laws and their affects become manifest through AND-to-OR collapse. But OR-to-AND de-collapse and time-reversal would make all else, including physical law, fall out of evidence, so constituting a bizarre un-science. Hence, physical laws are not viable in *any* time-direction, as convention believes on the basis of its notion of 'time-symmetric physical laws'. Physical laws are evident only in one time-direction: the AND-to-OR direction. Therefore, physical laws are time-asymmetric and viable only in and through AND-to-OR collapse.

The discovery of the time-asymmetry of physical laws retrenches the fact and reality of time and irreversibility, with both as non-illusory. The time-asymmetry of physical law renders time an obvious ontic reality. Without time, without AND-to-OR collapse, there would be no physical law. The ontic reality of physical law is inseparable from the ontic reality of time; its fixed AND-to-OR directionality and irreversibility. Hence, from the ontology of physical law grasped as time-asymmetric, we necessarily arrive at the ontic reality of time itself.

The conclusion from the time-asymmetry of physical laws also augments the independence of time from perception, consciousness, and mind. The latter are ontically real but cannot project physical law: The inference of law by a conscious being requires the necessary objective time-asymmetry of law obtained via irreversible objective non-illusory AND-to-OR collapse. **Our perception of physical law is secured by the ontic reality of time and irreversibility… both of which are presupposed to the very viability of mind and consciousness. Indeed, what absurd fate would await mind and consciousness if AND-to-OR collapse and time did not exist? Mind and consciousness would be rendered impossible.**

*

The above additions to the ontic reality of time, from memory and from physical law, augments the assertions on the same furnished in **2.28**.

227

2.30: Time, Potentiality & the Origin of the Universe

The following is a summary of key points made in Book-I (**1.10**) but now explicitly farmed to the intermediate model of spacetime. The insights garnered in the preceding essays must raise doubts about the usual way we describe time; as something that proceeds from past to future. To the contrary, time proceeds from AND-to-OR… from a pre-measurement pre-event quantum mechanical spacetime in potentiality… that is, *from the future*... to a post-measurement post-event state of realised spacetime… *to the past*. Put another way, **time proceeds from possibility to actuality. Hence, in effect, time proceeds from future to past, *not* from past to future**.

The arrow of time conceived as proceeding from a quantum mechanical spacetime futurity to realised spacetime... from future to past... is not in any sense a trivial tautological spin-off of our findings: it constitutes the basis for resolving problems of fundamental cosmology pertaining to the origin of the universe.

It turns out that, with or without the Big Bang, the error of foisting the origin of the universe to some remote past event (a view which conforms to the dubious characterisation of time as 'flowing from past to future') is displaced in favour of a solution both more exotic *and* closer to common senses, even if it proves to be startling at first glance.

Applicable to a cosmology and cosmogenesis **with or without a Bing Bang, the manifest universe which we perceive as realised spacetime, emerges out of ontologically prior and originary quantum mechanical domain of spacetime in potentiality**, and does so perpetually, as we have discovered in our intermediate model spacetime and the attendant growing block universe approach. It follows that the origin of the manifest universe, succinctly of 'realised spacetime', is *not* a one-time creation-event source-located in the remote past, but one that has emerged out of a 'sea of future-potentiality': a source within which our manifest universe resides as potentiality: one 'located', *not* in the past, but in the *future*...out from which it has decayed and continues to do so in perpetuity.

In short, with or without the Big Bang, the manifest universe emerged out of the future, which is itself beyond the need or want of any explicatory origin-point or a creation-event, given that we cannot derive the future (spacetime in potentiality) from the past (from realised spacetime) or from any temporal origin: Indeed, spacetime in potentiality ontically precedes time.

2.31: Summary & Conclusion to Part-II

The following is a summary of the key discoveries and conclusions obtained in Part-II:

- **John Wheeler's delayed choice implies the nested futures view of the quantum mechanical wave and of spacetime in potentiality**. Per Wheeler's delayed choice, a single delayed choice element implies a wavefunction that hedges for at least two alternate mutually incompatible future wavefunctions. Hence, an n-number of delayed choice elements would imply a complex system of n-number of alternate mutually incompatible nested futures and wavefunctions vis-à-vis a hedging primary wavefunction's immanent futures. This can be linked to Feynman's 'sum over all paths' and modifies the latter in terms of the nested futures perspective. Generalised to our intermediate model spacetime and its spacetime in potentiality, we obtain an inexhaustible nested futures system that elaborates to the infinitely removed future. (See **2.17** to **2.18**).

- **The three-fold taxonomy of nested futures and spacetime in potentiality:** There are three categories of nested futures: the **divergent, entangled,** and **convergent categories**. In the first, nested futures diverge without contact. In the second, nested futures diverge, but are also future-entangled via cross linkages. It is the convergent category that is ontically real. It forms the basis for non-local counterfactual causality and entangles the future of every frame on or along the event horizon to every other frame on the same, while also entangles every nested futures attached to such frames to all other nested futures, defining the totality of spacetime in potentiality to infinity. Counterfactual causality makes for the grand decoherence of the whole of spacetime in potentiality and revives Mach's principle. Both grand decoherence and the revived Mach's principle anticipate the quantum mechanical non-locality solution to inertia and to the permanent magnet: (See **2.22**).

- **Partial-perpetual wavefunction collapse, versus single-step comprehensive collapse.** Spacetime in potentiality cannot be exhaustively converted into OR-form outcomes: The event horizon cannot leap 'upward' the whole of spacetime in potentiality to the infinitely removed future in one step. (See **2.19**). Hence, the end of isolation and wavefunction collapse is ultimately a partial-perpetual process and can never entail an all-consuming collapse of AND-form nested futures to infinity. Partial and perpetual collapse is tacit to Wheeler's delayed choice, rendered explicit by the nested futures approach.

- **Closed-choice quantum mechanical systems**: Most laboratory experimental systems that test for quantum phenomena approximate to closed-choice quantum mechanical systems. Such systems lack obvious, explicit delayed choice elements. Consequently, they appear to test only for immanent possibilities; the collapse of which appear to be all-comprehensive and all-exhaustive, from which nested futures are apparently absent. Closed-choice quantum systems precipitate the bias in favour of the notion of one-step one-time all-consuming comprehensive wavefunction collapse: an error. (See **2.20**).

- **Open-choice quantum mechanical systems**: Almost all natural quantum mechanical systems are open-choice systems: exemplified by John Wheeler's delayed choice experiment; the basis for the nested futures view. Naturally occurring systems (i.e., the whole of spacetime in potentiality) constitute complex open-choice quantum systems per function of vast numbers of delayed choice elements or contingencies, elaborated to the infinitely removed future. Hence, the domain cannot totally or comprehensively decohere or collapse, but could undergo as many partial-perpetual collapses as there are delayed choice elements (potentially an infinite number). Applied to spacetime in potentiality, the attendant event horizon progresses to a new frontier of delayed choice elements and contingencies, but never obtains the infinitely removed future. Hence, partial and perpetual collapse: (See **2.20**).

- **Closed-choice systems are always embedded in open-choice systems**. Apparent closed-choice systems are *always* embedded and nested within decaying open-choice quantum systems. The ultimate open-choice system is spacetime in potentiality itself. Thus, ultimately, there is no such thing as a closed-choice quantum system. Therefore, there can be no such thing as one-time total comprehensive wavefunction collapse: (See **2.21**).

Quantum mechanical waves are spatially static, not dynamic (prelude to de-spatialisation). Quantum waves do not displace across space or undulate in time. This is on account of the following:

- Since **only retrospective observations are physically possible, we can never observe the alleged displacement of the quantum wave by means of an impossible synchronous realtime observation of the 'moving wave'**. The only way to preserve the quantum wave is by refraining from observing it in *any* fashion, and by keeping it in isolation. **Any observation, whether in impossible realtime or retrospective observation, or even via impossible a-causal observation of the wave per 'signals from the future', will destroy the quantum wave and collapse it into an OR-form state**. Thus, we can never directly observe 'moving quantum waves' retrospectively and cannot empirically justify our assumption that these are 'moving across space' or undergoing time-undulation. **The only explicit fact we have about quantum waves is per generic interference patterns furnished in two-slit experiments. But these can equally be formed by spatially static quantum probability waves**: The interference pattern is no certain, necessary, or sufficient basis for assuming the existence of spatially displacing or undulating quantum waves: Both are pleonastic. (See **2.23**).

- Any supposed **spatially displacing quantum wave would need to displace across potential space, *not* across any realised space**. That is, the quantum wave would need to move and undulate across a Hilbert space: But Hilbert space is a future potential as-yet not-realised space within spacetime in potentiality. **How can anything move across a future-potential 'space' that has not yet formed?**

- In john Wheeler's delayed choice approach, a primary hedging wave hedges for, or nests for, mutually incompatible alternative waves and outcomes: Per Alt-R, this leads to the nested futures view of the quantum wave and, by generalisation, a nested futures view of the totality of spacetime in potentiality. But delayed choice can also be used to attack the notion of the motion of the quantum mechanical wave: Therein, we find that **we cannot treat the hedged or nested quantum waves as moving, given that the nested quantum waves have not yet coincided with their putative pertinent spatial intervals when nested in the hedging wave prior to delayed choice switch-x. Yet, what applies to the nested quantum waves must also apply to their hedging wave: also, a static state, given that there is no fundamental reason to suppose that the hedging wave itself is moving, given the viability of spatially static waves, courtesy of the static nested waves.**

- Of course, **spacetime in potentiality itself, to which both the hedging and its nested waves are subsumed, is obviously a static state, given that the future cannot obviously move (where would it 'move'?): Spacetime in potentiality, constitutes a static state that can only AND-to-OR decay into a succession of events… upon which one imposes the 'notion of motion' of anything, including the supposed motion of quantum waves.** (See **2.24** to **2.26**).

Presage to the overhaul of the Schrödinger wave equation (see **2.27**) must follow from the critical insights into the meaning of the quantum wave as the physically phenomenalised future-form of spacetime; from the nested futures perspective derived from Wheeler's delayed choice; and from the spatially static character of quantum mechanical waves from the same. Thus, the Schrödinger wave must be overhauled to accommodate the following truisms:

- **Wavefunctions with nested futures**, as opposed to AND-form wavefunctions with only immanent futures.

- **Partial-perpetual wavefunction collapse**, as opposed to the generic notion of one-time total wavefunction collapse. Generic Schrödinger waves undergo a one-time collapse and state-vector reduction. The overhaul requires the description of partial collapse only, per inexhaustible nested futures in spacetime in potentiality.

- **Spatially static quantum waves,** as opposed to a Schrödinger wave that must undergo a time-evolution with literal physical undulation or spatial displacement. Any future version or replacement of the Schrödinger equation must model a spatially static non-undulating state of nested futures, even for standing waves.

- **Time as AND-to-OR collapse modifies the 'time-evolution' basis of the Schrödinger equation**. The generic Schrödinger equation describes the movement and undulation of the wave in dubious pseudo-realtime terms. Yet, time is AND-to-OR wavefunction collapse itself: Time is the reduction and collapse of the Schrödinger wave state, which must remain as a pure static state of possibilities and probabilities in the interim of isolation, ended by AND-to-OR collapse and time itself. The time-evolution of the Schrödinger wave must be replaced by the decay and decoherence of an otherwise static, non-moving, and non-undulating AND-form nested futures system: ultimately constituting spatially static spacetime in potentiality itself; incorporating the counterfactual causality and grand decoherence of that domain as part of the block-decay process.

Time is *not* a subjective imposition projected by consciousness: Time is ontically real (See **2.28** to **2.29**) and does not need to be perceived to possess ontic reality:

229

- **To invalidate the notion of objective time and to dismiss it as illusory, one would need to obviate and invalidate the reality of AND-form quantum phenomena.** The very existence of quantum phenomena implies the incomplete realisation of spacetime: i.e., the growing block universe...and the process of its completion through AND-to-OR collapse requires, and *is*, time itself.

- **The process of AND-to-OR collapse *is* time itself**... 'coming into being' from AND-form possibility to OR-form actuality: from spacetime in potentiality to a realised spacetime of accumulating events in abstract memory.

- **Time has an 'arrow' or directionality per two main reasons**: There is always an *increase in the entropy of future-potential information* per function of partial-perpetual AND-to-OR collapse of the future and the attendant progress of the event horizon. This supplants and succeeds notions of an 'arrow of time' based on mere statistical generic entropy. The increase in entropy of future-potential information is an absolutely certain disposition, not a statistical or probable one. Lastly, *time, hence causality, always proceeds from AND-to-OR, never from OR-to-AND, and this is corroborated by the time-asymmetry of physical laws* elaborated in Book-I (**1.11** to **1.12**). Time directionality per AND-to-OR is absolute and irreversible. This means that time and causality proceeds, not from past to the future, but from future-potentiality to past-outcome... from future to past... with direct implications to cosmology and to the origin of the universe, with or without the Big Bang: (see **2.30**). Thus, the manifest universe decays out of the future: Therefore, origin does not reside in the remote past, but in the future.

BOOK-II PART-III:
GRAND DECOHERENCE & IMPLICATIONS
FROM THE INTERMEDIATE MODEL SPACETIME TO INERTIA, GRAVITATION AND THE NATURE OF MIND

2.32: Aims of Part-III

As an important part of the Alternative Realist thesis and interpretation (Alt-R), Part-III will elaborate on the quantum erasure of nested futures, on counterfactual causality, hence on **the process of grand decoherence of spacetime in potentiality, affected in the framework of the convergent category of nested futures,** which designs for a form of quantum entanglement and non-locality that supersedes the limited version in EPR, **reviving Mach's principle** and constituting the presage to the quantum mechanical solution to inertia and gravitation. Part-III will also culminate into **the re-evaluation of the character and basis of 'spacetime curvatures' from within the intermediate spacetime model, and as attributes of the event horizon**.

Recall that, in Part-I and II, the event horizon was described as a flat line or surface. In Part-III, we will incorporate time-dilation effects due to gravity to the event horizon, rendering it 'curvy'. One effect from curvy event horizons due to gravity will be the warping it imposes on developing 'light rays' and their lightcones. Curvy event horizons also effectively refract worldlines: generating basic gravitational phenomena. Part-III will contend that the **process of counterfactual causality and attendant grand decoherence of spacetime in potentiality constitutes a critical basis for the process that curves the event horizons, which then affects the refraction of worldlines and the distortion of lightcones, producing gravitational phenomena.** Moreover, Part-III will posit that the basis of counterfactual causality and grand decoherence that affects gravity ultimately resides in realised spacetime: that **gravity emerges as a memory effect which 'tells' spacetime how to decohere, which in turn 'tells the event horizon how to curve'.** Yet the full account of gravity as memory effect will not be obtained until de-spatialisation is achieved in Book-IV.

Recall that the progress of the event horizon 'upward' the intermediate model spacetime does not unfold in a seamless-continuous way, but in a discontinuous-broken form. In Part-III, we will combine broken-discontinuous event horizon progression to event horizon curves and arrive at **the presage to long-elusive quantum gravity, which emerges as a function of the broken-discontinuous (hence, quantised, or *discretised*) progression of the event horizon, or quantised time from attendant AND-to-OR collapse of spacetime in potentiality**.

One obvious implication from the broken-discontinuous progression of the event horizon is the deeper insight it furnishes about face-front information-matter dualism: wherein information-matter dualism is grasped as a function of the very structure and process of spacetime. In Part-III, we will integrate spacetime-derived information-matter dualism to an expansion of decoupling theory itself. Succinctly, Part-III will espouse an addendum to decoupling theory in the light of spatially static quantum mechanical waves and static spacetime in potentiality, and as a function of the very structure and process of the intermediate spacetime. This will also apply to attendant information matter dualism.

Other insights will also be garnered: Part-III will account for the **work function zero-basis and grand decoherence model of the permanent magnet** from the framework of intermediate spacetime, its nested futures structure and other related findings. Yet, **work function zero processing is also implicated in inertia and gravitation,** for essentially the same reasons as found to apply in the permanent magnet. Moreover, the spatially static view of spacetime in potentiality will help eliminate key problems associated with both quantum spin and problems from so-called 'zero-point energy' or vacuum fluctuations. An unexpected turn in such matters will transpire when **we reframe otherwise didactic and pragmatic work, energy, and action concepts basic to modern physics in terms** of intermediate spacetime structure and, in the process, abstract out energy and work concepts from the ontology of physics: **rendering these pleonastic**, with profound implications to the conservation principle of energy-matter, often weaponised and abused by materialism against Cartesian dualism.

Part-III will round up with a **critique of the notion of the 'computational universe' on the basis of implications from quantum erasure of nested futures, counterfactual causality and grand decoherence**: These processes are expected to crash into non-terminating halting problems, but only if mediated by putative computational processes or gears: In such a case, the progression of the event horizon, hence AND-to-OR collapse and time, would 'freeze up'. Thus, the universe cannot be generated or driven by a computational process. Thus, augmenting material covered in Book-I (Part-IV) **with clear implications to the nature of mind**, to the favour of Cartesian revival.

2.33: Non-Local Counterfactual Causality and EPR

In generic spacetime models, time-based relations hold along the vertical y-axis. Putative spatial relations, as relations between events, hold along the horizontal x-axis or z-axis, when two spatial dimensions are added. Permissible spacetime relations can only be obtained within embedding and overlapping lightcones: These comprise 'areas' in which events come into contact with other events, principally via physical signals at the speed of light. Parts of spacetime not within any embedding or overlapping lightcones or outside these reside 'elsewhere': These cannot 'know' of each other by means of substantive signals or causal interactions that abide at the speed of light. Thus, generic spacetime models assert that faster-than-light signal-based substantive physical contiguity is not possible. The models do not permit or espouse instantaneous 'spooky action at a distance'. Yet, this appears to clash with outcomes from Einstein Podolsky Rosen experiments (EPR) that involve instantaneous 'spooky action at a distance', such as demonstrated in the Alan Aspect's Paris experiment of 1981[66] and thereafter.

EPR was explored in the Preliminaries (**0.26**) and in Book-I (**1.47** to **1.49**): it will not be reiterated in detail here. In EPR experiments, the OR-form resolution of particle-spin at one end instantaneously resolves the correlate particle-spin on the other end, in such a way as to conserve spin, albeit without the mediation of signals, and instantaneously across distances that could stretch many lightyears or parsecs. Thus, in EPR, the system begins as an AND-form wavefunction superpose of entangled spin-correlate possibilities, an *instantaneous whole* that can span lightyears. Then, through AND-to-OR collapse, the system quantum indeterminately collapses into a spin-outcome on the one end and an instantaneous opposite spin-outcome on the other end, without regard to distance or any adherence to the speed-of-light limit.

Within EPR, quantum indeterminacy ought to render the transmission of meaningful information impossible, given that there exists no way of obviating scrambling quantum indeterminacy involved in the resolution of the spin-outcome obtained on the first particle and upon the entangled pair, instantaneously. Therefore, quantum noise dominates, and, in this sense, no meaningful information may be transacted, as was also espoused in superlative rigging theory, which utilised EPR quantum noise against strong artificial intelligence: (see Book-I **1.48** to **1.51**).

However, **the quantum indeterminacy-based prohibition against the transaction of meaningful information does not apply to the 'transmission' of the principle of the conservation of spin itself**: This fact formed the basis for the dualism of physical law vis-à-vis putative 'matter and space', as was explored in both the Preliminary (**0.26**, followed by the argument from quantum mechanics against mereological nihilism in Book-I; **1.03**, **1.04**, and the later **1.26 to 1.27**). The essays and arguments clarified that, **since quantum indeterminacy also operates in non-EPR situations, it ought also to prohibit the transaction and formation of meaningful information, or wholes, in 'normal' non-EPR situations and frames that do not involve 'spooky action at a distance', but *do* involve quantum indeterminacy. But no such prohibition applies in such frames, and information is resolved into coherent 'materialised' wholes, regardless**. For example, quantum indeterminacy ought to scramble the formation of a coherent photographic image-pattern or whole, even though the photon depositions involved are *not* resolved via 'spooky action at a distance' but *are* fully subject to ineliminable quantum indeterminacy. The further exploration of these conundrums led to the hypothesis of instantaneous wholes (see Preliminary, **0.26**... and Book-I: **1.03**, **1.04** and **1.26** to **1.27**). It turns out that spacetime in potentiality constitutes the ultimate instantaneous whole in relation to the total event horizon: That is, the totality of spacetime in potentiality must counterfactually decohere into consistent futures vis-à-vis all frames on or along the event horizon, to the infinitely removed future.

In effect, EPR involves counterfactual communications: a counterfactual non-local causality that enforces the principle of conservation of spin on the entangled states and their AND-to-OR outcomes, and maintains consistency in line with the demands of the conservation of spin, regardless of quantum indeterminacy and noise in spin-resolution. It turns out that our intermediate spacetime model, framed in terms of convergent category nested futures, furnishes just such oblique counterfactual non-local causality, but to the *whole* of spacetime in potentiality: wherein the total futurity of the whole physical universe is entangled into a unitary state or instantaneous whole, *not* only along the entire event horizon, but throughout spacetime in potentiality, to the infinitely removed future.

Through counterfactual causality throughout the totality of spacetime in potentiality, one part of the universe meaningfully and instantaneously (or transtemporally) adjusts to any AND-to-OR outcomes obtained in other parts of the universe along the event horizon, even though these could be spatially separated by, say, a million lightyears or parsecs, and despite quantum

---

[66] Experimental Realization of Einstein-Podolsky-Rosen-Bohm Gadenkenexperiment: A New Violation of Bell's Inequalities. A. Aspect, P. Grangier, and G. Roger. Physical review Letters, Vol 49, Iss.2, pp.91-94 (1982)

indeterminacy. Hence, each part of the universe resolved on or along the event horizon 'knows' every other part resolved along the same, and does so through the counterfactual consequences to their converging and common futures embedded within spacetime in potentiality. This recognition revives Mach's principle and underpins the route to the growing block quantum mechanical theory of inertia, gravity, and other things.

2.34: Convergent Category of Nested Futures:
Non-Local Counterfactual Causality & Grand Decoherence

The non-local unity and instantaneous whole of the universe along the event horizon is inextricably linked to and realised through the process of quantum erasure of convergent category nested futures that characterises the totality of spacetime in potentiality. To illustrate, in **Fig. 2.15**, we present a more detailed depiction of the convergent category nested futures.

Therein, incident to the event horizon, we find two OR-form events (or two frames of reference) comprised of the leading moments of **worldline-w** and **worldline-x**, spatially separated from each other by, say a million parsecs: i.e., 3.26 million light years. These events

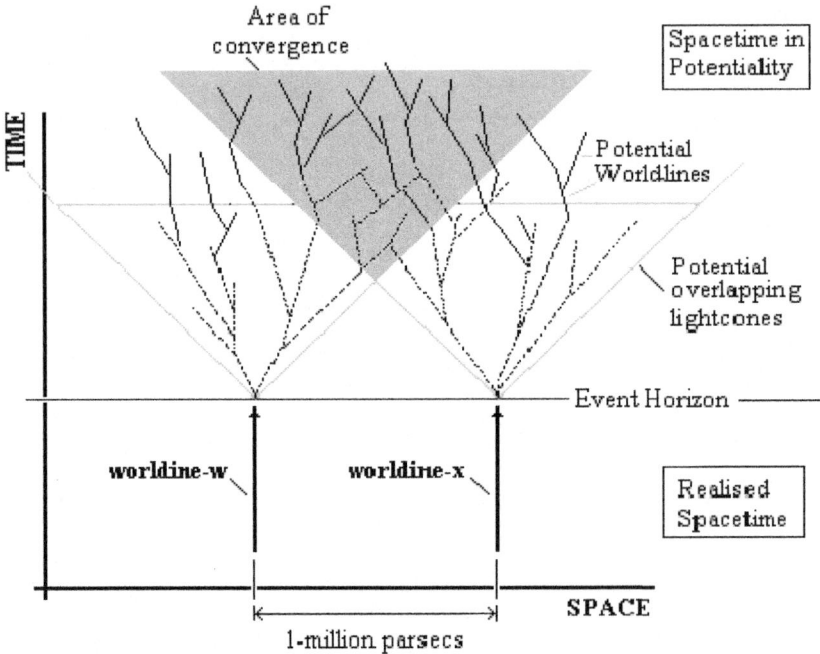

Fig. 2.15: Convergent category nested future
Notes: The didactic depiction is a reprisal of the convergent category nested futures; foundational to counterfactual non-local causality and grand decoherence, beyond the limited version of non-locality evinced in EPR. Worldlines **w** and **x**, whose leading ends terminate at the event horizon, are spatially separated by, say, a million parsecs (3.26 million light years). There is no possibility of any *direct* radiation or matter-based physical connection between these. However, within the domain of spacetime in potentiality above the event horizon, we find two as-yet not-realised overlapping and converging future-potential lightcones: These project from the ends of worldlines **w** and **x**, respectively, into the future: into spacetime in potentiality. The future-potential lightcones quantum mechanically overlap and converge in the potential far-future: i.e., the two events at the termini of **w** and **x** are potentialised to physically 'know' of each other three-millions years in the future. The future possibilities projected from the termini of both worldline **w** and **x** form a complex 'bush' structure of mutually exclusive converging not-yet happened nested future-potential worldlines. When immanent first-level possibilities attached to **worldline-w** AND-to-OR collapse into specific OR-form outcomes, and the alternatives are quantum erased, spacetime in potentiality will undergo appropriate grand decoherence into consistent-only future possibilities, first in the 'area of convergence' and beyond. Hence counterfactual non-local causality or grand decoherence of spacetime in potentiality *en toto*, to the infinitely removed future.

232

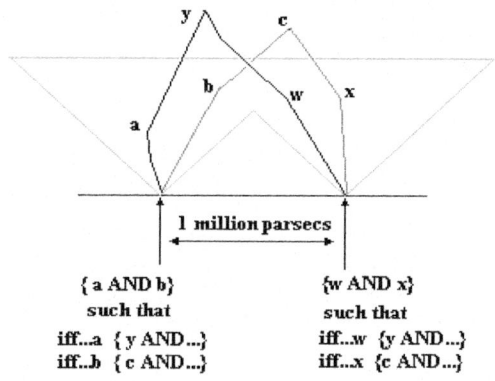

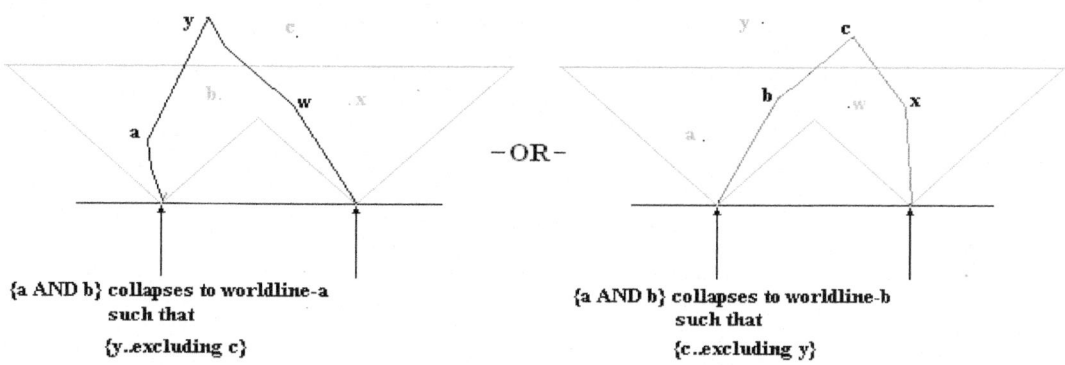

Fig. 2.16: Illustration of counterfactual non-local causality in convergent categories
Notes: **Diagram-A** is a simplified universe constituted as convergent category nested futures. Therein, the ends of two realised worldlines, separated by a million parsecs, are entangled by future potential worldline cross-linkages within the 'area of convergence' formed by the overlap of the potential lightcones. The left worldline is potentialised to futures {**a** AND **b**}: the right to {**w** AND **x**}. But {**a**} on the left and {**w**} on the right converge and nest for the far-future possibility {**y**}: while {**b**} and {**x**} nest for the far-future possibility {**c**}. In **diagram-B**, the left worldline AND-to-OR collapses first because the right frame is subject to time dilation due to gravity, while the left frame is subject to less gravity and events 'unfold faster'. Thus, on the left, potential {**a** AND **b**} collapses into {**a**}, pre-conditional to far-future possibility {**y**}. The collapse on the leading left frame acts as a non-local counterfactual vis-à-vis the trailing worldline on the right: The system on the right cannot now collapse in such a way as to realise far-future {**c**}: This is now erased and untenable. It must collapse in favour of possibility {**w**}, requisite to far-future {**y**}. (The right-depiction shows the alternative development {**c**}). The leading frame biases or erases in favour of specific far-future convergent nested futures, while the trailing frame must decohere into permissible consistent future possibilities per counterfactuals imposed by the leading frame… in a meaningfully but non-directly observable way. Hence, every frame of reference in the universe along the event horizon *does* 'know' of every other frame along the same, in a meaningful counterfactual non-local way, *without* direct physical signalling or contact, via 'spooky action at a distance'. Thus, we secure the case for a counterfactually unitary spacetime in potentiality that inter-decoheres every frame on or along the event horizon to consistency, regardless of 'distance' or the speed-of-light limit.

233

cannot physically and causally 'know' of each by means of any substantive speed-of-light mediation within the lightcones depicted, nor instantaneously on or along the event horizon. However, in terms of the future-form convergent category spacetime in potentiality, the same-said two events and frames of reference *are* potentialised to 'know' about each other at some potential remote future: That is, the two events are expected to physically contact each other, either in terms of the overlap of their attendant future-potential lightcones, or via putative future-potential radiation or matter-mediated signals and worldline contiguity or contact within the overlap of their future lightcones, embedded within spacetime in potentiality, beyond the leading moments of **worldline-w** and **worldline-x** confined to the pertinent event horizon.

From this, subsequent progression of the event horizon and synonymous AND-to-OR collapse of spacetime in potentiality must lead to new OR-form worldline outcomes and extensions that generate critical erasures and transformations of the potentialities within spacetime in potentiality, constituting the counterfactual non-local grand decoherence of the totality of spacetime in potentiality into consistent-only futures. Hence, the two frames attached to the leading moments of **worldline-w** and **worldline-x**, even though spatially separated by millions of light years, across which no direct substantive physical-causal signal or contiguity is possible, are in unity and in counterfactual 'contact' per their common convergent category potential futures.

In **Fig. 2.16-A to C**, in the simplified universe depicted therein, in the far-future of that universe, we find spacetime in potentiality structured into two mutually incompatible alternatives: namely, {**potential-y** AND **potential-c**.}: both nested to the leading moment of two OR-form worldlines separated by a million parsecs, and whose leading moments terminate on or along the depicted event horizon.

In **Fig. 2.16-A**, the AND-form future nested to the left-worldline is future-potentialised to **potential-y** through an intermediary **potential-a**. Note that **potential-y** will also need an additional requisite far-future **potential-w**. But this is nested to and originates from the right-side worldline, via intermediary **potential-w**.

What about the alternate, **potential-c**? Note from **Fig. 2.16-A**, that **potential-c** needs the left-worldline to AND-to-OR collapse into intermediary **potential-b**. Instead, if AND-to-OR collapse realises **potential-a** *but not* **potential-b**, then **potential-c** cannot happen. (For clarification, follow the diagrams in **Fig. 2.16**).

However, **potential-c** also requires **potential-x**, nested to and originating from the right-worldline, but at the expense of **potential w**. In short, the chain of future-potentials nested to both the left-side and right-side worldlines are exclusive: Those that lead to **potential c** exclude the future possibility for **potential-y**. As to the question of which of these potentials will be realised, excluded and erased, this will be fated to the progressions of the event horizon and its attendant AND-to-OR collapse-processes along the event horizon, and to the specific outcomes generated therein to the left or right worldlines: Again, for clarification see **Fig.2.16-A**.

Using a God's eye view, let us assume that the progression of the event horizon and attendant AND-to-OR collapse processes occurred on the left-side first: perhaps because the right-side was embedded in a gravitational field and was subject to time dilation, while the left-side was subject to negligible gravitation and much less time dilation. Therefore, events attached to the left-side happened 'faster', or 'first', constituting the *leading frame of reference* with respect to which all other frames of reference, now *trailing frames of references*, must decohere to consistency. Let us also assume that the left-side leading frame underwent AND-to-OR collapse into **realised-a**… to the erasure of alternative intermediary **potential-b**. Consequently, the right-side trailing frame must now be biased in favour of far-future **potential-y** at the expense of far-future **potential-c**: (See **Fig.2.16-B: Left diagram**). This will constitute oblique, negative, but no less meaningful non-local and non-contiguous counterfactual affectation on the right-side trailing frame by the left side leading frame, even though it is separated from the latter by a million parsecs and cannot be subject to any direct 'speed-of-light' contiguous influence. In any case, the attendant **counterfactual causality is obtained through the restructuring or** *grand decoherence* **of the future, wrought by the attendant counterfactual quantum erasure of non-consistent nested futures from that future. This does not involve any speed-of-light contiguity or 'contact causality' between frames along the event horizon, but yet constitutes abstract elimination and restructuring process that operates via consistency-rules that apply instantaneously across a frame-unifying system of convergent category nested futures, constituted as an instantaneous whole.**

The fact that AND-to-OR collapse on the left-side leading frame has biased the development of the far-future in favour of **potential y** to the erasure of **potential-c** implies that the right-side trailing frame is also forced to evolve toward **potential-y**. Thus, for the right side trailing frame, the future is permanently foreclosed with respect to the now-forgone **potential-c** per the counterfactual affect from the left-side leading frame vis-à-vis their unitary futures.

Since **potential-c** is erased from spacetime on the left-side leading frame, the right-side trailing frame will have its subsequent process of AND-to-OR collapse choices reduced, delimited and *decohered* by an *instantaneous* non-local and non-contiguous counterfactual affect from the left-side leading frame, despite the separation of both by a million parsecs. Consequently, within the wavefunction on the right side trailing frame, intermediate **potential-x** will also be erased: only intermediate **potential-w** will be available for future realisation. The right-side trailing frame will have no choice but to AND-to-OR collapse into intermediate **potential-w**, requisite to far-future **potential-y**.

However, if the left-side constituted the leading frame instead, the evolution and grand decoherence of spacetime in potentiality would assume a different development; one in favour of far-future **potentiality-c**: (See **Fig 2.16 B-right**).

The point throughout is that, **through the counterfactual restructuring and decoherence of the ontic future (of spacetime in potentiality) the evolution and subsequent AND-to-OR collapse of the trailing frame is fated to the outcome obtained at the leading frame, despite a million parsecs of separation.** Decoherence and wavefunction collapse from the leading frame will critically delimit the direction of development open to the trailing frame; restricting and reducing the scope of its future-possible development:

all of it achieved in a *counterfactual*, instantaneous 'faster than light' non-local and non-contiguous way: but in a *meaningful* way, despite putative separation by a million parsecs; despite the speed-of-light limit and lack of direct 'material' contact.

*

We have now discovered that every part of the universe is non-contiguously, instantaneously and *meaningfully*, albeit *counterfactually*, sensitive to and entangled with the AND-to-OR collapse processes and outcomes generated at every other part of the universe on or along the event horizon, despite spatial separation in terms of millions of parsecs, or by any distance up to infinity... and despite the fact that there cannot exists any direct substantive physical signal or contiguity between the said parts in terms of mediating radiation or matter subordinated to realised lightcones or worldlines. **Every part of the universe on and along the event horizon is absolutely entangled to, and unified with, and must critically and counterfactually adjust into mutually consistent common futures with every other part along the same, insofar as these are integrated and subsumed to a common convergent category nested futurity constituted as a unitary spacetime in potentiality.**

Clearly, quantum indeterminacy intrinsic to universe-scale AND-to-OR collapse processes and to the progression of the event horizon, cannot scramble the operation of meaningful non-local counterfactual causality and grand decoherence of spacetime in potentiality, which must decohere into consistent-only futures for all frames of references. Indeed non-local counterfactual causality incorporates the reality of quantum indeterminism: **If, in a leading frame, the local AND-to-OR collapse of immanent possibilities unfolds in a pure quantum indeterminate way, even though the OR-form outcome from it will be secured in a purely random way, its counterfactual implications throughout the convergent nested futures system of spacetime in potentiality vis-à-vis every other frame on or along the event horizon, and to the total future that elaborates to the infinitely removed future common to all frames along the event horizon, will decohere in a deterministic way, insofar as these must be rendered into consistent futures vis-à-vis the otherwise indeterminate outcome generated at the leading frame.** This is the same as found in EPR with respect to spin-correlates, though applicable to the total macro-scale spacetime in potentiality treated as a unitary instantaneous whole: Yet again, **non-local causality will culminate into effective mediation of meaningful information, instantaneously or at effective 'faster than light', despite ineliminable quantum indeterminacy.**

2.35: Counterfactual Causality, Grand Decoherence, the Revival of Mach's Principle, and Inertia

In generic classical-physics terms, if we push a body of mass to a given velocity, it will continue in its new state in perpetuity... so long as nothing disturbs it in its new state. It is only when we attempt to disturb the mass by applying putative 'force' to alter its direction, or to decelerate or accelerate it, that inertia, as resistance to the change implied by that force, emerges. The more massive the body in terms of its nominal mass, the greater its inertia or resistance to any change to its direction or to its velocity.

We must now address the question of what causes a body to continue in perpetuity at given apparent uniform motion when all accelerating or decelerating forces have ceased to operate on it. Succinctly, where does the body get the power to, as it were, 'keep on going' *in perpetuity?* What power-source could be responsible for this, given that none can be physically specified?

We must also address inertia in terms of Mach's principle. If there existed just one particle or unitary body in the universe, it could not have inertia: There would be nothing else in the universe against which it could compare its physical state and thereby 'know' that it is moving in perpetuity in a particular direction, or 'know' of its resistance vis-à-vis any attempt to change its direction or motion; or even 'know' if it is relatively stationary. All of these require the existence of other particles in the universe, and, with it, the 'knowledge' of every other particle or body in the same.

Mach's principle states that the very possibility of inertia must be tied to an effect that permits a body of mass to 'know' that it is moving and resisting vis-à-vis all other bodies in the universe: That is, the local body must 'know' of every other body in the universe in a non-local, instantaneous way. How can this be possible, considering the speed of light limit?

Despite its initial inclusion as one of three basic principles of General Relativity, Mach's principle was dropped from GR [67]. Mach's principle certainly appeared to conflict with the speed-of-light limit. Yet, given that we now have a viable form of non-local counterfactual causality, we can revive Mach's principle. Our case in argument for non-local instantaneous counterfactual non-contiguous causality and attendant grand decoherence within the convergent category spacetime in potentiality, which involves the erasure of non-consistent futures from all frames along the event horizon, revives the non-local basis of inertia espoused by Ernst Mach.

Again, all inertial states treated as distinct frames along the event horizon are non-locally integrated to every other inertial frame from within their common convergent nested category spacetime in potentiality. Therein, the counterfactual grand decoherence of spacetime in potentiality, one wrought by a change of state in the inertial body 'here', will instantly restructure the permissible futures available to inertial bodies 'over there' and anywhere else on or along the event horizon into consistent-only futures: i.e., consistent to the change made 'here'. In short, bodies elsewhere will counterfactually (*not* substantively) 'know' the state of change made to the body 'here'. Conversely, the body 'here' will also know the state of change made to every other frame or body 'elsewhere' on or along the event horizon, through counterfactual instantaneous 'knowledge'... regardless of the distances... *without* **the requirement or mediation by light-signals or any requirement of substantive 'material' contact.**

---

[67] *Subtle is the Lord: the Science and the Life of Albert Einstein* (Oxford University Press, 2005), pp. 287–288.

235

One need only recall **Fig: 2.16** in essay **2.34** to grasp the essential processes behind Mach's principle, and the grand decoherence behind inertia. Of course, this process of common grand decoherence as it pertains to inertia constitutes the new quantum mechanical face of Mach's principle. It also constitutes the first breakthrough and presage into hitherto-elusive quantum mechanical growing block theory of inertia; one that spans the totality of convergent category spacetime in potentiality and the growing block universe.

*

Does inertia, as perpetual motion, require continuous application of force to realise, or is it a work function zero process? **Once a putative 'force' is applied and change is affected, the established direction and motion-state of the body will enforced upon the future-form wavefunction that spans spacetime in potentiality and its nested futures structure only of futures that are consistent with its newly acquired state. This will entail the total erasure of all other non-consistent futures vis-à-vis the newly acquired state. Once spacetime in potentiality is decohered to consistency thus, each and every progression of the event horizon and attendant AND-to-OR wavefunction collapse processes that subsequently attends the body cannot help but realise OR-form outcomes orthodox to the decohered consistent direction and motion-potentials enforced on spacetime in potentiality, given that these are the *only* possibilities now available to the subsequent evolution of the body potentialised within future spacetime, and to the infinitely removed future**.

It is the perpetual counterfactual entrenchment and retrenchment of the direction and motion-state of the body within spacetime in potentiality, constituted as consistent future-potentials, and to which the convergent future-potentials of all other bodies in the universe must counterfactually adjust and abide, that maintains that body in its new established state of direction and motion, and does so in *perpetuity,* **without subsequent need for applied power or force.**

Notice that no mysterious power-source, or any violation of the conservation principle, is necessary to drive our body in its inertial perpetual uniform motion and direction, or decohere its future and all other convergent futures to orthodoxy, per the enforcement of its inertia. And this effect, born as it is out of the quantum erasure, counterfactual causality, and the grand decoherence of spacetime in potentiality, explains the key **quantum mechanical work function zero process responsible for inertia**, without the violation of the principle of conservation, in conformity to the work equation (i.e., **F = ma**)… wherein **F** = zero.

*

Where does resistance to change in the motion and inertial state of the body come from? **When we try to alter the direction and motion-state of the body, we not only impose relevant quantum erasure of non-consistent convergent futures, but also impose demands in similar and consistent terms upon the wavefunctions and futures of every other inertial state in the universe, per the relevant counterfactuals, given the common convergent futures that unite and bring these into an instantaneous whole. Thus, the future-form consequences of our attempt to change the motion-state of an inertial body 'here' will be non-locally and transtemporally 'intercommunicated' as a complex of counterfactuals to every other state or body in the universe on or along the event horizon: a grand decoherence to consistent only futures,** as previously explained via the processes clarified in essay **2.33**. **And it is this that, in part, gives rise to resistance, and, through it, imposes a local 'force', energy or power-cost to our attempt to alter the object's direction and motion**.

Hence, resistance or inertia of a body, is partly per function of the preference of that body, and every other body in their joint convergent category nested futures, for the least change, or no change at all… for the least quantum erasure, or for no quantum erasure and no grand decoherence at all. The greater the demand on a body in terms of the change to its velocity or direction, the greater the counterfactual re-structuring and decoherence of the common convergent future to all bodies along the event horizon and throughout spacetime in potentiality… to infinity. Thus, the greater the resistance to change, or inertia, from the whole of spacetime in potentiality. Similarly, the 'more massive' the body, the greater the implied counterfactual restructuring, decoherence, hence resistance from spacetime in potentiality.

In short, resistance or inertia to change comes from the resistance of spacetime in potentiality to its restructuring attendant that change, consummated as a required force, energy or power to affect that change. Thus, when we try to change the state of the body, F in F = ma can no longer equal zero.

However, non-local counterfactual causality is not the sole basis of inertia: So far, we have only incorporated considerations that apply from within convergent category spacetime in potentiality; the ontic future. What about factors from the mnemonic domain of realised spacetime; the memory-state pertinent to the inertial body? **A complete account of inertia must also incorporate memory effects from realised spacetime and from abstract memory**. This memory-side explanation to inertia must await Book-IV. Yet, it is sufficient to reiterate at this juncture that one of the core mechanisms behind inertia *is* the stated grand decoherence and the quantum erasure of non-consistent futures from within the all-frame unifying and binding spacetime in potentiality, and the resistance of spacetime in potentiality (the future) to any restructuring implied by the locally enforced change… succinctly grasped as resistance to quantum erasure and grand decoherence, or simply as inertia.

*

What about counterfactual causality as it pertains to bodies that undergo non-uniform changes in direction and motion? Indeed, the notion of uniform unchanging motion and direction constitutes, not only the exception but an extreme idealisation: **There is no such thing as a body in uniform motion. The universe is wholly dominated by bodies in states of non-uniform motion: of bodies undergoing continuous acceleration and continuous change in direction, as exemplified by bodies subject to gravitational influence and orbital motion**. Thus, the cup on the table is not at rest… because the Earth-surface upon which it is 'resting' is spinning.

236

The very same non-local counterfactual causality that explains Mach's principle and accounts for idealised and essentially fictitious inertial bodies must also apply to bodies subject to gravitational phenomena. That is, **gravitation is also *in part* generated from the erasure of non-consistent futures via grand decoherence, but of a form that resolves for non-uniform orbital motion and acceleration due to gravity or gravitational acceleration** in lieu of fictitious uniform inertial motion.

As with inertia, counterfactual and quantum erasure-driven gravitational and orbital motion is necessarily a work function zero process: That is, gravitating bodies that induce acceleration and orbital motion on other bodies do not accomplish this by inputting force, energy or power, and consequently do not entail any violation of the principle of conservation, owing to the fact that change is being enforced via work function zero processes, as we had anticipated in the Preliminary **0.33** to **0.34**, and throughout Part-III. Thus, as with inertia, so with gravitation.

However, the complete account of gravitation must also incorporate memory effects from the mnemonic domain of realised spacetime and from abstract memory. This must also await Book-IV. Again, it is sufficient to state at this juncture that the core mechanism behind gravitational phenomena involves non-local counterfactual causality or grand decoherence in the context of a unifying and binding spacetime in potentiality. The counterfactual causality and grand decoherence 'tells space how to curve'... and thus tells 'matter how to move': all on a work function zero basis.

Of course, this fact undermines the notion that one cannot obtain change without the application of energy, work, force or power; oft abused by materialist philosophy against Cartesian causality. But more on this will be said in later essays, suffice to state that Cartesian mind-brain causality is itself a work function zero process.

*

If the state of inertia and gravitational motion is locked into spacetime in potentiality through counterfactual erasure of all of the non-consistent futures via grand decoherence, how is it possible to overcome inertia or even change the gravitational state of a body, if it must perpetually conform to consistent futures? And where does the energy to accomplish this come from?

Clearly, it is possible to overcome inertia *and* change the orbit of a body. Both entail the restructuring of the future through a new series of grand decoherence processes that force that future to a new conformity. This requires seeming energy or force to accomplish. But, again, where does this energy come from?

We infer that the energy with which we do work, while as pleonastic as space itself, is merely the set of contingent possibilities inherent to spacetime in potentiality, which we interpret as available 'free energy'. Such contingencies are pertinent to the overcoming of inertia via applied 'force' and to the change in gravitational acceleration and orbit of, say, satellites.

Thus the answer to how it is possible to overcome inertia or to shift the orbit of a body, was intimated in the contingency role played by **switch-x** in the delayed choice account of nested futures (see **2.18** to **2.21**). The quantum erasure of non-consistent nested futures in the enforcement of gravitation and inertia does not erase future-potential contingent possibilities that might otherwise entail or permit the change in the motion state of the body, or its acceleration about another body, or its orbit: Such contingencies are inbuilt into spacetime in potentiality as alternative future-potentialities (perhaps low-probability ones) just as are all the future-possibilities that keep the inertial state in perpetuity, or affect gravitational acceleration and orbital motion in perpetuity. In short, the future is not and cannot be totally dominated by the structural implications of the counterfactual erasure of non-consistent futures and the projection of the state of inertia and gravitation to the infinite future: The said projection cannot eliminate potential contingent futures (consistent with decohered futures that favour the given inertia or gravitation) but that could alter and override that inertia and gravitation. Generically, we call these contingencies 'free energy'.

To clarify, consider that **open-choice spacetime in potentiality will nest two main sets of mutually incompatible nested future complexes: The first set will consist of those nested future complexes in which the state of inertia or gravitational relation is enforced in perpetuity to the infinitely removed future. On the other hand, the same open-choice spacetime in potentiality will also nest a second set of potential low-probability contingent futures for the "overcoming of inertia", or the "breaking out of orbit". Indeed, the possibility for the quantum erasure of non-consistent nested futures *and* the retention of contingent potentials that might counter that arrangement so imposed, are inherent to spacetime in potentiality per Wheeler's delayed choice approach: (see 2.18). That is, the delayed choice approach naturally justifies the nested futures that incorporate both countering contingent *and* perpetuating future possibilities within spacetime in potentiality, as part of a grand open-choice quantum system.**

2.36: Spatially Static Quantum Waves: The Problem of the Electron's Orbital Spin

Succinctly, **spatially static quantum waves can help solve problems pertaining to certain characteristics that pertain to electron spin about the atomic nucleus**: From a naive approach that envisions electron spin and orbit about the atom in both the pseudo-realtime way and in a way analogous to planetary orbital motion, the electron might be expected to spin about the atom, perhaps near or at the speed of light. Assuming the notion of motion, the electron is typically portrayed as spinning at 1/137 of the speed of light. But recall from **2.23** to **2.26** that **the notion of spatially displacing quantum mechanical waves must be discarded, and, with it, notional pseudo-realtime time-undulating standing waves**: i.e., the basic format for the description of electron spin about the atomic nucleus. Both motion and spin must be replaced by spatially static quantum mechanical waves and wavefunctions, subsumed to the grander spatially static spacetime in potentiality, if for no other reason than the fact that the future encapsulated by quantum mechanical waves and functions cannot obviously 'move' or undulate. Consequently, the electron about the atomic nucleus must no longer be described as a pseudo-realtime standing wave undergoing either motion or time-undulation about its nucleus. Instead, the

237

electron must constitute a purely spatially static time-frozen standing wave-state of not-yet transpired futures, ultimately subsumed within equally static spacetime in potentiality. Thus, **as a static pre-time or pre-AND-to-OR collapse state, the electron cannot literally spin about the atomic nucleus at the speed of light, or at any 'speed', portrayed in the usual false pseudo-realtime way, but must instead be configured as an as-yet not happened static potential-set for future spin possibilities vis-à-vis future-potential observations of it and attendant AND-to-OR collapse**... from which a specific configuration and outcome of 'spin' might be inferred.

Only AND-to-OR partial-perpetual decay, or decoherence of static quantum waves, and the grand decoherence of spacetime in potentiality, together with the perpetual restructuring and erasure of the static nested futures comprising spacetime in potentiality, is ontically real. None of these require literally spinning electrons undergoing pseudo-realtime 'motion' about the atomic nucleus. Indeed, our assertions obviate moving and time-undulating waves, as much as these annul naive notions of literal 'spin'.

2.37: Spatially Static Quantum Waves, Nested Futures, Counterfactuals
And the Quantum Mechanical Theory of the Permanent Magnet

The humble domestic fridge-magnet constitutes an example of the working permanent magnet: a misnomer, given that the pertinent materials must lose their magnetic properties in the long run. Thus, the permanent magnet is more of an idealisation than a reality. Yet, this household object illustrates a key mystery: Any rock climber knows that it takes energy to stay attached to a vertical surface against the pull of gravity. The climber must use energy and power to remain attached to the vertical surface. On the other hand, **the fridge magnet appears to stay put on the vertical surface of the fridge-door in apparent perpetuity, in defiance of gravity, without the apparent use of, and in the absence of, any power-source or input energy or work to perpetuate its attachment**: The initial work expended to magnetise the material and to deploy it on the fridge door is not sufficient to account for the subsequent perpetual attachment of the magnet to its surface.

It is tempting to imagine the violation of the conservation principle secretly at work behind magnetic attraction. But the magnet-surface relation turns out to be a work function zero process. How is this possible? To address this question, we must model the permanent magnet from within spacetime in potentiality and the growing block intermediate spacetime, in terms of nested futures and spatially static quantum waves, *and* incorporate grand decoherence.

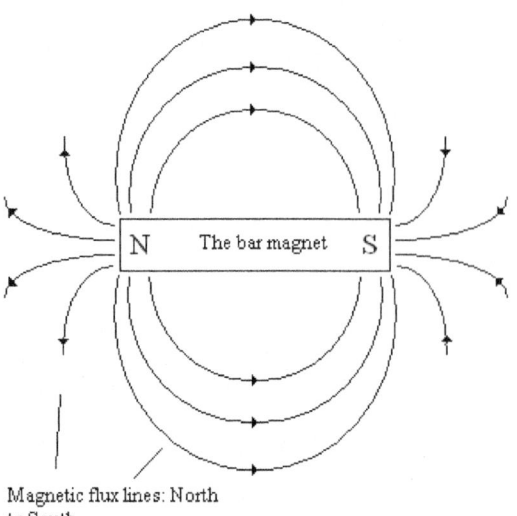

THE GENERIC BAR MAGNET
(THE FRIDGE MAGNET)

Fig. 2.17: The bar magnet
Notes: Diagram–A depicts the magnetic field generated by a bar magnet. Iron fillings will line up with the direction of the magnetic field-lines by means of induction. The fridge magnet is essentially a bar magnet, and the fridge door is substitute to iron filings: the metallic and magnetically susceptible content of the fridge door is subject to induction by the fridge magnet. In consequence, the magnet stays anchored to the fridge door.

The depiction in **Fig.2.17** is that of a generic permanent magnet: a bar magnet. Note the magnetic field lines. Iron fillings will line up with the direction of the magnetic field lines by means of induction. The fridge magnet is essentially a bar magnet, and the surface of the fridge door stands in lieu of iron filings. But the illustration in **Fig, 2.18-A** incorporates the permanent magnet to our intermediate

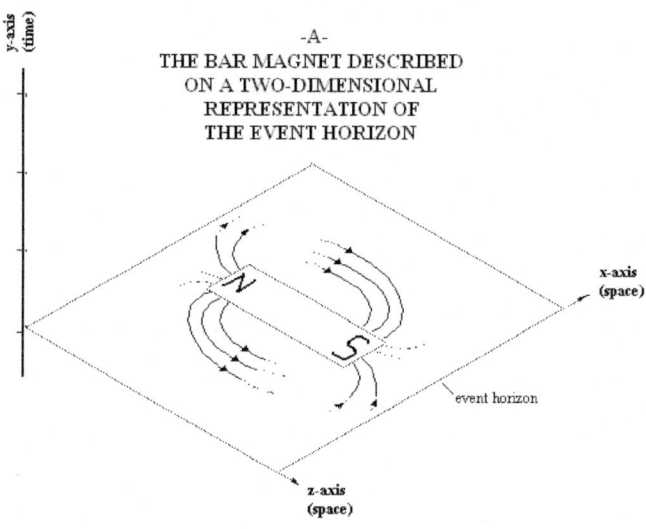

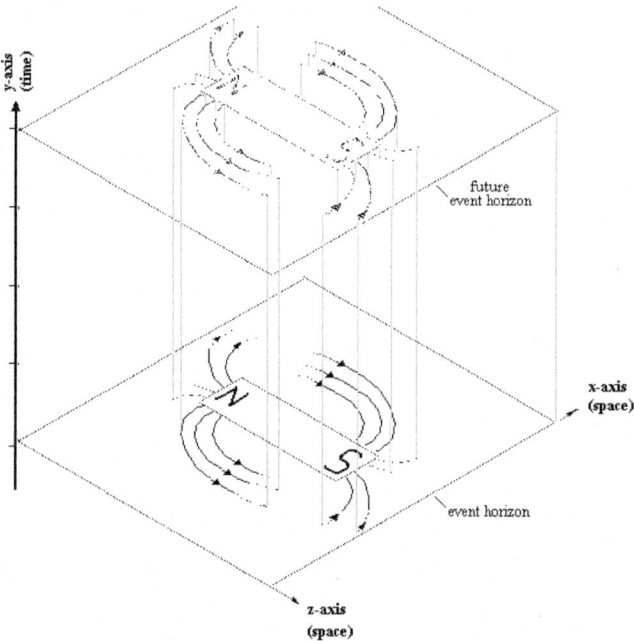

Fig. 2.18: The bar magnet in quantum mechanical spacetime in potentiality
Notes: In **Diagram–B**, the magnet and field are placed on the event horizon. In **Diagram–C** the field is projected into spacetime in potentiality, with respect to which all future-potential iron fillings, fridge doors and electrical fields must conform, owing to the quantum erasure of non-consistent nested futures and grand decoherence of spacetime in potentiality to the infinite future, save contingent futures (Wheeler's cat) that entail the potential removal of the magnet from the pertinent surface. The process of grand decoherence and enforcement is a work function zero process, excluding the initial input: it does not require subsequent energy-inputs to enact and maintain the formed magnet-door relation and attachment.

model spacetime by placing it and its magnetic field lines on and across a two-dimensional depiction of the event horizon. However, **Fig. 2.18-B** extrapolates the magnet and field lines into the whole of future-form quantum mechanical spacetime in potentiality. In doing so, it describes it as a spatially static not-yet realised quantum mechanical future-potential magnetic field: a projection or 'chreode' of potentiality elaborated to the infinitely removed future, at least in principle.

Thus, as a projected chreode into spacetime in potentiality, the magnet and field will decohere and potentialise the whole of spacetime in potentiality to a consistent-only set of possibilities: possibilities orthodox to its field structure and to its established relation to the fridge door. Spacetime in potentiality has no choice but to abide to this orthodoxy, given that all non-consistent alternatives that do not abide or found to be inconsistent with the now-established magnet-surface relation will be quantum erased and eliminated from the future through counterfactuals and through attendant grand decoherence of the totality of spacetime in potentiality, with the latter enforced into said-consistency.

Note that the intermediate model spacetime solution to the mystery of the permanent magnet is essentially the same as the solution to the mystery of inertia, as was elucidated in **2.35**. Now, **subsequent AND-to-OR collapse and time, hence all subsequent event horizon-progressions, cannot help but realise and reconstitute *only* the consistent possibilities projected into spacetime in potentiality, enforced by the quantum erasure of non-consistent futures via grand decoherence. That is, the magnet-door attachment will be recuperated and maintained at each subsequent AND-to-OR collapse, *without* any further energy or power input... on a work function zero basis... baring contingencies or future-potentials that might change the putative position of the magnet vis-à-vis the surface, or even remove it from that surface. i.e., Wheeler's cat.** Such contingencies *will* require energy or power inputs to bring about, in that the removal of the magnet from the surface will entail a positive-sum work function. Indeed, at this juncture, we assume that the pertinent energy-matter content comprising the fridge magnet, the fridge door and the content of the pertinent area, is wholly confined to the event horizon and progresses upward the spacetime system with the progression of the event horizon. As this energy-matter 'moves' with the event horizon, and is unchanged in quantitative terms, no violations of conservation principles will occur, and any energy or work required to remove the fridge magnet and alter its relation will compose contingent alternative possibilities to the system and will not require a 'creation from nothing'.

Again, we infer that the energy with which we do work, while as pleonastic as space itself, is merely the set of contingent possibilities inherent to spacetime in potentiality, which we interpret as available 'free energy'. Thus, our 'free energy' from **2.35** applies to the overcoming of magnetic attachment. Again, such contingencies are inbuilt to spacetime in potentiality. In short, the future is not and cannot be totally dominated by the structural implications of the counterfactual projection of the magnet-surface relation: The said projection cannot eliminate potential contingent futures (consistent with decohered futures that favour the perpetuation of the formed magnet-door relation) that could alter and override the relation, despite its entrenchment to futurity.

2.38: Spatially Static Quantum Waves & Zero-Point Vacuum Fluctuations

We will now account for so-called 'vacuum fluctuations' or 'zero-point energy' in the context of spatially static quantum mechanical waves and the equally static domain of spacetime in potentiality. In generic quantum mechanics, empty space apparently has energy: a function of Heisenberg's energy-time uncertainty[68]. According to the energy-time uncertainty relation, when we relate energy and time as we might putative position and momentum, the shorter the time-interval the more potentially energetic the system becomes: If we could reduce the time interval to zero (although this is impossible), we would obtain an infinite energy potential. However, the time interval might be shortened to but perhaps never obtain the Plank time-interval. **By applying Heisenberg's energy-time uncertainty relation to the vacuum we arrive at the notion of 'vacuum fluctuations' and 'zero-point energy': With it, we arrive at a view of 'empty space' not empty at all: oft described as a 'writhing froth' of fluctuating energy, per the usual pseudo-realtime depictions of it.** Of course, the notion of 'virtual particles', or of very short-lived particles, is essentially synonymous with 'vacuum fluctuations', in that the putative energy fluctuations of the vacuum can be described as a fluctuation of virtual particles. The idea is that, per Heisenberg's energy-time uncertainty, as we observe things at shorter and shorter time-intervals, the so-called 'virtual particle sea' becomes more and more energetic, so constituting the said 'froth'.

The existence of vacuum fluctuations and the zero-point is a confirmed reality. If it were not for vacuum fluctuations, the domestic light bulb could not light up as brightly as it does, even if the difference is not noticeable to human perception. More dramatic evidence for vacuum fluctuations is furnished by the Casimir Effect[69]; experimentally demonstrated in 1997[70]. Therein, as the gap between two metal plates is made smaller, the range of putative virtual particle potential wavelengths and frequencies within that gap will be reduced, and the vacuum energy-content of the gap will also reduce. On the other hand, the range of potential virtual particle wavelengths and frequencies outside the gap will accentuate. Hence, the energy content outside the gap will increase. The generated 'vacuum pressure' from the difference between the reduced energy in the gap and the greater energy outside the gap will push the plates together, implying putative energy and work extracted from the vacuum.

All of this smacks of 'energy out of nothing' in violation of the principle of conservation. Various estimates of the energy content of the vacuum predict levels so high, and its refraction properties so great, that even light ought not to be able to pass across the pertinent

---

[68] "The uncertainty relation between energy and time in non-relativistic quantum mechanics". *J. Phys. (USSR)*.9: 249–254. 1945.

[69] Casimir, H. B. G.; Polder, D. (15 February 1948). "The Influence of Retardation on the London-van der Waals Forces". *Physical Review*. 73 (4): 360–372

[70] *Lamoreaux, S. K. (1997). "Demonstration of the Casimir Force in the 0.6 to 6 μm Range". Physical Review Letters. 78 (1): 5–8.*

'empty space' in which the vacuum flux resides. But the implicit assumption throughout is that the energy state of the vacuum, the 'froth' of energy fluctuations synonymous with it, is something that manifests as a present-continuous realtime writhing and happening, described in the usual suspect pseudo-realtime form. In short, vacuum fluctuations are implicitly assumed to unfold in a spatially and temporally dynamic way.

But what happens when we re-frame zero-point energy and vacuum fluctuations in terms of spatially static quantum waves and equally static spacetime in potentiality which such putative fluctuations inhabit?

*

Recall from our previous accounts of Heisenberg's uncertainty (**2.14**) that when AND-to-OR collapse is resolved in favour of position, we must objectively defer the resolution of momentum to the future; to spacetime in potentiality: Consequently, momentum obtains a broad range of values rendered as as-yet not happened momenta, rendered in AND-form. In a similar vein, when we resolve for smaller and smaller time-intervals, we defer the energy measure contained in such intervals to the future: Consequently, like momentum, the potential vacuum energy of that interval will have an enlarged range of superposed energy values, constituted as a broad range of as-yet not-happened energy fluctuations in spacetime in potentiality. In other words, **the vacuum energies we are speaking of are not realtime happenings, but pure future potentials for such. That is, the flux is a** *static* **AND-form time-frozen complex of not-yet happened possibilities:** *not* **a dynamic pseudo-realtime affective reality.**

Insofar as vacuum fluctuations are not 'real', or *not* **OR-form realised events, but only possibilities relegated to the future and to the domain of spacetime in potentiality, these must also relate to putative 'spatial intervals' that are also not-yet realised spatial intervals.** Again, the 'vacuum fluctuations' are not literal fluctuations at all but constitute pure statics for possibilities for such flux that *might* obtain upon subsequent AND-to-OR collapse of spacetime in potentiality.

Why is the notion of dynamic vacuum fluctuations a problem? **If vacuum fluctuations were 'real', or OR-form, light could not possibly pass through space.** But a partial solution emerges when we utilise the naïve and dubious notion of the 'light ray' and treat it as something that supposedly 'moves' across the vacuum that constitutes the spacetime in potentiality that contains the potentials for zero-point energy fluctuations. The path of our 'light ray' will never be subject to putative vacuum fluctuations. Why? The answer is simple: **The 'ray of light' is not moving across spacetime in potentiality because the light itself is spatially static, as is the spacetime in potentiality through which it is erroneously portrayed to be 'moving'.** The non-existent 'motion' of the spatially static light-ray cannot be subject to any disturbance or refraction by as-yet not-realised vacuum energy fluctuations incident to the spacetime in potentiality through which the light-ray never moves.

Furthermore, our ray-of-light notion presupposes a spatial interval through which it must supposedly 'move'. The putative vacuum fluctuations expected to refract our light-ray are expected to correlate with this as-yet not-realised spatial interval. It is not possible for our 'light ray' to encounter, become perturbed, or refracted, by as-yet not-realised vacuum fluctuations across an as-yet not-realised space. Indeed, **the event horizon simply 'leaps over' the potential putative spatial intervals in which the potential fluctuations reside, collapsing to an OR-form resolution of our 'light ray'** *after* **the said leap, engendering the appearance of the 'motion' of our 'light ray' per the succession of AND-to-OR resolutions of that light ray per event horizon leaps. Thus, the light ray circumvents or leaps over the supposed vacuum froth within that interval, avoiding it completely, never suffering it… although it never actually gets to happen.**

Finally, there is the quantum probability to consider: The shorter the time interval the greater the potential vacuum energy, or the higher the probability of its occurrence. But most light and de Broglie matter is configured as relatively low frequency, relative long-time interval states that do not approach Plank frequency or time-intervals, as is also true of the putative vacuum itself. This fact relates both the radiation, matter, and the space to extremely low probability of vacuum fluctuation, even assuming that our contentions about spacetime in potentiality and spatially static quantum waves are ignored.

The quantum leap of the event horizon over the potential interval of 'space' and its attendant 'vacuum fluctuation' segues with the idea of quantised time, given that the broken-discontinuous form of the progression of the event horizon attends AND-to-OR wavefunction collapse, itself synonymous with time. In other words, time turns out to be as broken-discontinuous or quantised as its related event horizon progressions.

*

If vacuum energy states are not-yet formed energy potentials of spacetime future, how do these get to generate real physical effects, such as the Casimir effect? To grasp how, we subsume the Casimir experiment into the framework of our intermediate spacetime model: We place the two metal plates on the event horizon and spacetime in potentiality. In doing so, we find that, as was true of the permanent magnet, the arrangement of the two plates is projected into the domain of spacetime in potentiality, in principle up to the infinitely removed future. In this way, counterfactuals and grand decoherence will force spacetime in potentiality to abide to consistency and orthodoxy with that projection, as was found true in the same treatment of inertia and in the permanent magnet.

As the plates are brought closer together, the projection into spacetime in potentiality changes: the potential not-yet realised *future* **virtual energy states pertinent to the gap are** *counterfactually* **reduced and decohered vis-à-vis those potential energy states that reside outside of the gap, which increase to a higher potentiality. The probability for the occurrence of such fluctuations is also proportionately decreased within the gap and increased outside it. It is the very structure of our future-form spacetime so-constituted and transformed by the said projection, and** *not* **any putative literal vacuum flux-event therein, that decoheres future-consistency to the plates collapsing onto each other, and evinces such collapse in subsequent progressions and**

realisations of the event horizon per subsequent AND-to-OR collapse sequences… based on 'energy' that appears to come out of nothing. Yet, the 'energy' that brings about and drives this Casimir Effect cannot be described as any 'real' physical existent; no more than the 'energy' that drives inertia, or the workings of the permanent magnet, or even gravitation… no more than can the process of counterfactuals and grand decoherence itself be so-described. As with inertia and the permanent magnet, the Casimir effect, and all other know effects from purported 'vacuum energies', including the lightbulb effect, is brought about by a work function zero process involving the restructuring of future-potential events into consistent-only possibilities. By the elimination of alternative future possibilities per function of the quantum erasure of non-consistent states, subsequent AND-to-OR collapse processes cannot help but realise the Casimir plate configuration into a collapse-configuration… without utilising work or energy… on a work function zero basis. That is, **the Casimir effect is a work function zero effect**.

Thus, the Casimir effect (hence zero-point energy) is an outcome of the structural characteristics imposed on spacetime in potentiality: It does not arise from 'energy relations', but from the quantum erasure of non-consistent futures, pertinent counterfactuals, and spontaneous grand decoherence. Therefore, in this way, 'vacuum fluctuations' do not imply modification of the sum of energy-matter, nor any violation of the principle of conservation.

However, the above is not cognisant of consequences to energy relations that must emerge out of the structure and process of the intermediate model spacetime. This is adumbrated in the next essay.

2.39: The Intermediate Spacetime Model and the Reform of Basic Energy Relations:
The Reformation of the Principle of Conservation

When we frame basic energy relations within intermediate model spacetime, energy concepts and relations turn out to be *pleonastic*, not part of the real and penultimate ontology of physics and physical causality, even if the concepts remain didactically useful and practically pragmatic. The pleonastic status of energy concepts constitutes a key development of the Alternative Realist thesis (Alt-R).

There are two basic energy categories: *potential energy* versus *kinetic energy*. In terms of the famous equation, $E = mc^2$, energy $\mathbf{E}$ is equivalent and interchangeable with the mass $\mathbf{m}$, or matter. But this is potential energy, latent in putative matter. Thus, potential energy is stored in matter, while potential energy released to do work constitutes the kinetic energy, brought into effect as 'action'.

From our intermediate model spacetime, **potential energy, as possibility for energy, must be an attribute of the future… an attribute of spacetime in potentiality**. For its part, **kinetic energy, or work brought into effect, is restricted to the transition between one event horizon and a succeeding one**, obtained through attendant AND-to-OR collapse or time of the future and of the possibility for energy-work therein: This is where $\mathbf{c}^2$ in $\mathbf{E} = \mathbf{mc}^2$ implicitly comes into relevance and relates to time and to AND-to-OR collapse. Restated in terms of change, the potential for change is obviously inherent to and 'contained' as a potentiality that belongs to the ontic future or to spacetime in potentiality. The affected change, or kinetic energy, obviously delimited by the potential for change inherent to the future, is realised through AND-to-OR collapse or time: obtained as a realised event or putative 'action' on a succeeding event horizon, through event horizon transitions synonymous with AND-to-OR collapse.

We justify the above by means of the following. **Even assuming that energy is literally 'stored' in some putative material object as a distribution in and across space…** *and* **conceived of as 'stored' within a point or spatial block… its energy content will only issue out of it in some future moment, through an AND-to-OR collapse that brings it into realisation in that future moment… manifesting it as putative action or 'work' or 'energy'. It follows that, even assuming the naive notion of energy congealed in and as spatial matter, owing to the time-aspect in the manifestation of** *any* **energy relation, the energy so congealed** *must* **reside in the future as a future-potential state, necessarily within spacetime in potentiality**.

It also follows that mass $\mathbf{m}$, portrayed as congealed energy, must *partially* constitute an attribute of spacetime in potentiality, constituted as the future-potential total energy 'store' itself: (This is partial, if only because we have excluded the role of mnemonic realised spacetime in the constitution of mass, or mass as a memory effect. We will exclude this for now, but only for brevity, returning to it fully in Book-IV). It follows that the release of potential energy through its transformation into putative kinetic energy can only come from the decay of that future-potentiality (the energy-pertinent 'mass') in and through AND-to-OR collapse and time… in and through the attendant transition from one event horizon to a succeeding one… in and through the decay of spacetime in potentiality… *from the future*.

*

What about the domain of realised spacetime? Does the energy issued out of spacetime in potentiality via AND-to-OR collapse end up congealed within realised spacetime? No. The domain of realised spacetime serves another purpose: it is the domain in which the past is retained in abstract form: *not* in OR-form, and certainly not as any sort of 'energy', potential or kinetic. Even so, whatever the ultimate basis of abstract memory within the domain of realised spacetime, it constitutes and imposes the critical initial conditions with respect to which both the event horizon and the nested future spacetime in potentiality must abide into conformity, at least in terms of consistency, if nothing else. Implicitly, the putative potential energy stored or potentialised within spacetime in potentiality must abide to the consistency regime imposed upon it by the domain of realised spacetime; by abstract memory acting as the source of initial conditions and consequent grand decoherence of spacetime in potentiality into conformity. Hence, the future potential energy, or simply potential energy potentialised to the future, must be fully consistent with what happened in the past. And it is here that conservation principles enter the fray, albeit in altered form from conventional approaches based on generic $\mathbf{E} = \mathbf{mc}^2$.

Ultimately, the sum of the past, constituted as abstract memory within the domain of realised spacetime, imposes itself as inertia and gravitation, as we shall discover in **2.45** and resolve fully by Book-IV. It is the domain of realised spacetime that tells spacetime (succinctly, the event horizon) how to 'curve', which then tells matter 'how to move', or, rather, restricts the decay of spacetime in potentiality per its imposed consistency regime or 'conservation principle' to the sorts of specific futures and outcomes that we then grasp as 'gravitational phenomena' *and* conserved 'energy-matter'.

Does the domain of realised spacetime in potentiality accomplish the above affect through work or energy? No: The process by which the domain of realised spacetime imposes gravitational phenomena is purely a work function zero process, as we shall discover in **2.45** and fully by Book-IV; consistent with what we have discovered so far about the workings of inertia, gravitation, the permanent magnet, and other phenomena in Book-II.

*

If one seeks to retain basic energy concepts, with all their undeniable didactic and pragmatic value to physics hitherto, one may do so per to the above-described schema, based on the framework furnished by the intermediate spacetime system and its later successor, the Bergson-Whitehead amalgam espoused in Book-IV. However, a loophole emerges directly from the character of AND-to-OR collapse or time, ontically fundamental to and responsible for attendant event horizon transitions and succession, and for supposedly 'drawing out' of future-potential energy from spacetime in potentiality into putative realised kinetic energy, work, or action.

To obtain the manifestation of putative energy relations, we need ontologically prior AND-to-OR collapse or time: No AND-to-OR collapse... no time... and no potential energy drawn out from spacetime in potentiality. Hence, no action or work manifested. Yet, it turns out that ontically fundamental primary AND-to-OR collapse and time is driven by an ineffable gear-less process. That is, **wavefunction collapse does not require an action, an 'energy input', or 'work' to make it happen. Indeed, it is the process of gear-less AND-to-OR collapse that makes putative energy relations happen,** *not* **the other way round. In other words, AND-to-OR collapse or time... is a work function zero process... and so are all its consequences, including those we interpret as 'energy relations' and affects.** This central claim about the ineffable drive of time will be fully disclosed in **2.46** and **2.47**.

Let us reiterate the remarkable implications from the above: It is AND-to-OR collapse that 'draws out' putative future-potential energy within spacetime in potentiality into kinetic energy on or along the event horizon, with attendant events constituted as 'action' and as state-change. Yet, the process that brings about such putative energy-issuance and state-change (namely, AND-to-OR collapse or time) does not itself require energy or work to realise. In other words, *ultimately*, hence *ontologically*, state-change is not brought about by energy, work, or action. Instead, it is gear-less non-energy-driven AND-to-OR collapse and time that brings about change... the realised outcomes of which we then interpret as 'energy-driven change', or upon which we foist assumptions of energy concepts and relations and 'energy-driven change'.

Succinctly, **if AND-to-OR collapse and time is made to happen** *without* **energy or work, then change ultimately need not and** *does not* **require 'work' or energy to happen... and does not ultimately require energy concepts and relations in any ultimate explanatory ontology and framework. Consequently, energy concepts and relations turn out to be as** *pleonastic* **as the notion of 'force', as Ernst Mach once recommended.** Mach was correct, and his appraisal applies fully to energy concepts and relations the moment we appreciate that time and change (and all activity and 'action' pertinent to putative radiation and matter) manifests on a work function zero basis.

However, as with 'force', we can keep and continue to apply energy concepts and relations, given their didactic and pragmatic value, on the strict condition that we appreciate the fact that, *ultimately* and *ontologically*, these are pleonastic and *not* part of the real ontology of physics.

As a final point, energy is usually related to the frequency of radiation or matter. However, frequency is a time-attribute: part of AND-to-OR collapse and time. But time is a work function zero process. Therefore, the frequency relations interpreted as 'energy relations' immediately transform into pure 'frequency of events' brought about by work function zero AND-to-OR collapse, from which the notion of energy can be completely abstracted and rendered pleonastic. In this way, we break the association of energy with frequency, baring legitimate didactic and pragmatic purposes of putative energy-frequency relations.

*

What about the conservation principle? Does this disappear in the face of the pleonastic character of energy concepts? The pleonastic status of energy concepts forces the reform and recuperation of the conservation principle into its alternative abstract form.

Recall that the generic conservation principle refers to the conservation of energy-matter, before and after putative energy-matter transitions, such that the sum after the state-change must equal the sum before state-change. From our intermediate spacetime model, the putative kinetic energy issues out of future-potential energy 'locked' within spacetime in potentiality, but as a quantity less than or equal to that future-potential energy. Hence, no violations of the conservation principle will occur, even though the attendant AND-to-OR collapse is a work function zero process.

Nor will violation occur when we entirely abstract out energy concepts and relations from the ontology of physics and declare these pleonastic, consequent upon the work function zero character of ontologically primary AND-to-OR collapse and time. **That is, grasped as pleonastic notions, when we abstract out energy concepts from time and causality, we leave behind pure informational relations**. No violation of conservation takes place, as the conservation principle is thus transformed into a consistency statement that relates future-possibility to memory and permissible actuality, but does so in abstract-informational terms, *not* in energy-terms, and wherein even frequency transforms into a 'frequency of events in time' as opposed to an energy measure. **In short, the principle of conservation abides, but in pure abstract form, without reference to pleonastic energy, work, or action**.

243

Second, as previously stated, the putative potential energy 'stored' within the ontic future configured as spacetime in potentiality must also abide with a consistency regime imposed upon it by the mnemonic realised spacetime, by abstract memory; by past conditions acting as accumulating initial conditions. Hence, putative future-potential energy must be fully consistent with what happened in the past. And, again, it is here that the now-abstract conservation principle fully enters the fray, different from but consistent with the basic pattern implied by $E = mc^2$. **Thus, when we abstract out energy concepts and relations as pleonastic, what remains is, again, the principle of conservation transformed into a consistency principle with which the future must abide...** *always* **within the permits and delimits of abstract memory.**

With these two approaches to the conservation principle from the pleonastic status of energy relations and concepts, we arrive at a replacement, reform and recuperation of the generic conservation principle into an abstract consistency rule. That is, we abstract pleonastic energy concepts and relations out of the conservation principle to arrive at a pure informational statement about conservation, which must include the following as axiomatic:

- The past must be retained and conserved.
- The future must abide with the conserved past and decohere to consistent-only nested futures with respect to that past.
- Events subsequently brought about via work function zero AND-to-OR collapse and time must issue new events and changes that are consistent with the consistency-demands from the past, *combined to* consequent potentialities and delimits of the future.
- Thus, we arrive at an abstract statement of conservation principles stated in pure informational terms in the context of memory, possibility, and time.

There is no compelling need for energy concepts in our transformed and ephemeralised conservation principle, now grasped as abstract consistency rule tied to the conservation of memory, and which can no longer be used as a basis for materialist assertions about the ontology of nature, much be susceptible to abuse against Cartesian causality.

The conservation principle is thus reconstituted on the basis of the conservation of memory (or the past, which cannot subsequently change) and a consistency principle that enforces conformity of spacetime in potentiality to consistent-only possibilities vis-à-vis that unchangeable past. The conservation principle thus recuperated also works on a work function zero basis, in that the conservation of the past, and the consistency of the future brought about through grand decoherence, is enforced via work function zero process.

The pleonastic status of energy concepts, and the abstraction and transformation of the conservation principle into the conservation of memory and consistency principle, wrought by the integration of these to our intermediate spacetime model and to the fundamental process of AND-to-OR collapse and time, can finally lay bare the full work function zero character entailed in inertia, in the workings of the permanent magnet, in 'vacuum energy' evinced in the Casimir effect, and, finally, in gravitational attraction and acceleration. **However, the pleonastic status of energy, and the abstraction of the conservation principle into consistency principle, must totally unravel the typical tropes raised against prospective Cartesian mind-to-brain causality, and the attendant abuse of the conservation principle on the presumption of the ontology of energy, work or 'force', in ignorance of work function zero processes**: processes evident even in the early history of physics and in generic phenomena. All of these were clarified in Preliminary (Part-III) pre-emptive to the same claims now justified per spacetime structure, the character of time, and from the work function zero profile of time.

*

Finally, **we must bring to bear anticipated de-spatialisation from Book-IV**. This will not only take us to a model of spacetime without 'space' beyond the intermediate model, but completely usurp spatial notions that undergird generic energy concepts and relations... as distributions, instantiations and contiguous-contact relations in and across space. The notion of 'space' supports the notion of 'matter in space', and of congealed energy in matter as a thing distributed and localised in and at point-like particles or spatial blocks, supposedly exhibited as 'mass'. With de-spatialisation, these notions can no longer hold: **There is no space. Therefore, there are no locations at which point-like entities can exist, much less spatial blocks, and even less 'spatially distributed mass' undergoing now-impossible contiguous impact-based interactions. It also follows that we cannot treat energy as something 'locked up' in a location or volume of matter in 'space'**, or supposedly resonating therein via contiguity and the enaction of 'work'.

De-spatialisation necessarily forces the requirement that we reframe now-pleonastic energy relations as relations instantiated and distributed purely in time, *without* **space... and as outcomes attendant non-contiguous work function zero AND-to-OR collapse and time processes.** Thus, our spacetime structure-based reframing of putative energy terms and relations is consistent with implications from anticipated de-spatialisation, and they are essentially correct.

The reader must surely notice the crisis that such a development must bear upon materialism. The pleonastic status of energy terms and relations, and the transformation of the conservation principle into pure abstract principles of consistency, render the materialist assertions on the ontology of physics totally void. Indeed, Cartesian causality and dualism is a necessary consequence of our overhaul and abstraction of energy and work as pleonastic, if not a clear and seamless consequence of prospective de-spatialisation.

2.40: The Intermediate Spacetime Model & Information-Matter Dualism Reprised

Book-I sought to found definitions of physical information suited to the form of information-matter dualism first articulated in the Preliminary (Part-I). In Book-I, one pair of definitions entailed the distinction between AND-form *potential information*, comprised of pure future-form possibilities (the future), versus OR-form states of *realised information*, or post-quantum mechanical realised events (the past). In both the Preliminary and especially in Book-I, the fact that OR-form realised information cannot be carried by or transported in AND-form quantum mechanical radiation and de Broglie matter immediately led to information-radiation and information-matter dualism, entailing the apparent physical dissociation and independent retention of information vis-à-vis 'carrying' radiation, matter, and space.

It turns out that **information-matter dualism is the face-front physical expression, or mirror-image, of intermediate spacetime structure. The reason why OR-form information cannot be carried by AND-form quantum mechanical radiation is per function of the fact that spacetime in potentiality, the future, cannot 'carry' or 'transport' the past, in turn constituted as the domain of realised spacetime, separated from futurity by the OR-form event horizon**. In any case, spacetime in potentiality and its subsumed quantum mechanical waves, are spatially static, such that these cannot 'move' and cannot transport or displace information in and across putative space.

In short, our intermediate spacetime model validates hitherto information-matter dualism, grasped as the mirror-image of tripartite spacetime structure.

From the framework of our intermediate model spacetime, it seems obvious that the basis of retained and dissociated realised information must be the domain of realised spacetime, 'below' the event horizon, inclusive of the most recent OR-from outcomes distributed to the event horizon itself. For its part, the domain of AND-form quantum mechanical *future-potential information* (hence, quantum mechanical radiation and matter) is to be identified with the domain of spacetime in potentiality, 'above' the event horizon. Put another way, realised information retained within the domain of realised spacetime up to the leading event horizon, is physically separate, dissociated from, and removed from quantum mechanical radiation and de Broglie matter relegated to the futurity constituted as spacetime in potentiality. Hence, it is obvious that the past is not physically carried in or by the future. Grasped as a past-future dualism ingrained to the tripartite structure of growing block spacetime, information-matter dualism as an aspect of AND-OR dualism and wave-particle dualism is vindicated in a seamless and natural way on the basis of the intermediate model tripartite spacetime past-future dualism.

To augment this view, let us first address whether the progression of the event horizon 'upward' the spacetime diagram unfolds in a seamless-continuous form, or as something that proceeds in a discontinuous-broken or quantum-discrete form. The question is directly relevant to information-matter dualism attested from the case for broken-discontinuous motion espoused in the Preliminary (**0.1** to **0.4**). Indeed, broken-discontinuous progression of the event horizon 'upward' the spacetime system is synonymous with broken discontinuous motion first espoused therein. Thus, when framed in terms of our intermediate spacetime model and the event horizon, our original question about the possibility of seamless-continuous motion of radiation, matter and of quantum waves 'across space' espoused in the Preliminary is transformed into the synonymous question about the form of the progression assumed by the event horizon 'upward' the time-axis, in the context of spatially static quantum mechanical radiation and matter, and per the subsuming spatially static spacetime in potentiality.

A seamless-continuous hypothesis of the progression of the event horizon would require that the event horizon progress across infinitesimally small time-intervals along the time-axis of spacetime, wherein each point and event horizon constituted along the time-axis would be seamlessly linked to every preceding and succeeding point and associated event horizon: thus, forming an indivisible seamless continuum of such progressions. On the other hand, **a discontinuous-broken hypothesis of the progression of the event horizon would require that the event horizon progress 'upward' the spacetime system in the form of discrete jumps or 'quantum leaps' or transitions, with 'gaps'. Clearly, the broken-discontinuous form of event horizon progression holds true**, for essentially the same reasons as does broken-discontinuous motion in the facile case for information-matter dualism furnished in the Preliminary.

*

Let us further augment our case for information-matter dualism from spacetime structure by incorporating wavelength and frequency relations to intermediate spacetime. Recall that, in the facile case for information-matter dualism presented in the Preliminary, dualism arose from the impossibility of observational intrinsic zero-wavelength states in measurements of putative position and general spatial relations, per foundational solutions to the ultraviolet catastrophe and frequency-wavelength cut-off.

Insofar as putative energy of radiation or matter is related to wavelength and frequency, notwithstanding the pleonastic status of energy relations (see **2.38**), and insofar as energy increases as wavelength is shortened and frequency is increased, **radiation or de Broglie matter-waves that might exhibit intrinsic zero-wavelengths and, on the basis of it, seamless-continuous motion, would be expected to obtain infinite frequency and infinite energy. This is impossible**: Therefore, we have no choice but to conceive of putative 'motion' in terms of discontinuous-broken trajectories. Thus, the seamless-continuous relation of information to radiation, matter and space must remain impossible, as was argued in the Preliminary.

To reframe the above in terms of intermediate spacetime structure, whatever can be physically inferred from the wavelength-based analysis of putative spatial relations pertaining to radiation, matter and 'carried' information, can also be re-stated in terms of frequency grasped in terms of time-intervals… which reside along the time-axis of our intermediate model spacetime, and which attend the progressions of the event horizon. That is, **frequency is fundamentally linked to the broken-discontinuous quantised progression**

245

of the event horizon. **If the event horizon could progress along the time-axis in a seamless-continuous manner, this would require or entail ostensive radiation and de Broglie matter of impossible infinite frequency and infinite (albeit pleonastic) energy**, both in terms of observational inputs *and* in terms of intrinsic frequency of radiation or matter.

Again, **if the event horizon cannot progress 'upward' in the sought seamless-continuous way, then it must do so in the discontinuous-broken quantum-discrete form. This forces information-matter dualism attendant the said broken-discontinuous progression of the event horizon and augments our claim that information-matter dualism is the mirror expression of spacetime structure and process.**

*

Each progression of the event horizon is synonymous with AND-to-OR collapse of the universe-scale wavefunction or spacetime in potentiality. The event horizon 'jumps' to a succeeding one via one AND-to-OR collapse-process. At each jump 'upward', information re-merges along the event horizon to form new OR-form outcomes. Between each such jump of the event horizon, we obtain an interval along the time-axis whose character is purely AND-form: a quantum mechanical interval of isolation subsumed to spacetime in potentiality. The minimum size of this interval is conjectured to approach but perhaps never equal the Planck time-interval.

Let us assume that the specific informational OR-form state formed along the event horizon at t_1 pertains to a particular configuration of the barrier from the double-slit experiment: i.e. the OR-form of 'both slits open'. What happens to this piece of information? Since the event horizon at t_1 must now dissolve and be replaced by a pure quantum mechanical interval until t_2, and since this interval is a pure AND-form state subsumed to spacetime in potentiality and incapable of carrying and transporting the OR-form of 'both slits open', the OR-form state must 'dissolve'. Put another way, **since the event horizon will not progress seamlessly across the interval between t_1 and t_2, the OR-form state 'both slits open' formed on it at t_1 is obviously not going to be carried in any seamless-continuous way across interval t_1 to t_2. Whatever its form of retention as abstract memory, its dissolution at t_1 prior to its re-formation at event horizon t_2 will not imply its erasure: It will be restored to the succeeding event horizon at t_2. Throughout its dissociation, 'both slits open' will be physically retained *within* the domain of realised spacetime.** In this way we gain a deeper grasp of information-matter dualism in terms of tripartite spacetime structure and process: one that requires no 'movement' of the information or of its supposed 'carrying' radiation or matter. We also obtain near-concession to the real existence of abstract memory.

All of this can be applied to the 'transport' of a snowflake pattern, or of a photographic image. Controversially, we can apply the same treatment to the brain itself, with consequent mind-brain dualism per the same deeper information-matter dualism mirrored in and as spacetime structure and process.

2.41: The Intermediate Model Spacetime & Implications to the Nature of Mind and Memory

The dominant materialist consensus on memory-brain and mind-brain relations asserts both to be ultimately reducible, identifiable, and embodied in the hands-on material structures and processes of the brain. However, in the Preliminary (Part-IV) the consensus was shown to be inconsistent with facts and outcomes from memory-side and mind-side experiments and investigations. Indeed, the materialist dogma behind the consensus must wither, not only in the face of facile information-matter dualism espoused in the Preliminary or the deepened account of it in Book-I, but in the face of the same grasped in terms of spacetime structure. To this end, we must now articulate the foundational basis for memory-brain and mind-brain dualism in terms of growing block intermediate spacetime structure.

Consider our previous two succeeding event horizons from **2.40**: with one event horizon originally formed at t_1, followed by a succeeding one at t_2. Again, the interval between t_1 and t_2 will constitute a pure AND-form spatially static quantum mechanical state within spacetime in potentiality: Realised OR-form information cannot be transported, or moved, or carried across this interval per function of AND-OR logic incompatibility first espoused in the Preliminary (Part-I) and deepened in Book-I (mainly Part-II, and in the fifth proof from Part-III). We now apply all of this to the brain within the framework of our intermediate spacetime.

Let us assume that the specific informational OR-form state formed along the event horizon at t_1 pertains to a particular configuration of the brain and its correlate memories, minding and identity structures. Given that the event horizon at t_1 must dissolve into a pure static future-form quantum mechanical de Broglie interval within a static spacetime in potentiality of not-yet happened brain states rendered in AND-form, and since this interval is incapable of embodying or carrying the previous' materialised' pattern of the brain from event horizon at t_1, it follows that the memory, mind and identity from t_1 must dissociate and remain dissociated vis-à-vis the interval t_1-t_2. But it is *not* erased, despite its dissociation vis-à-vis interval t_1-t_2: memory and mind dissociated from the brain at t_1 is no longer 'in the brain' throughout t_1-t_2. The dissolution of information from the event horizon at t_1 entails the quantum mechanisation of the material brain into its future-from de Broglie matter and its nested spacetime in potentiality. As such, this constitutes **the radical dissociation of memory, mind, and identity into a wholly abstract-immaterial non-erasable state *extant* the now quantum mechanised brain, with memory relegated to the domain of realised spacetime. Hence, memory-brain and mind-brain Cartesian dualism must follow per function of the said dissociation per the structure and processes of intermediate spacetime**, otherwise mirrored in information-matter dualism.

Of course, all the above assumes that memory, mind (illusory?), and identity are definable material structures in brains (at least initially or by indulgence) at event horizon-t_1, and accountable as clockable iterational processes carried out by the machinery of brains, without requirement of any 'religious' variables, much less any Cartesian anathema. Even so, the very structure and processes of spacetime, and how information relates to spacetime, will have its way with any initial materialist presumption for the reduction and identification of memory, mind, and identity with brains. **However, from the non-clockability thesis inspired by the intuition of** infinities, but also inspired by the case for the renormalising quantum mechanist, amongst other remarkable hard problem considerations, we need not in the first place presume *any* initial reduction and identification of memory, mind, and identity with brains: (see Preliminary, Part-III **0.35**... and Book-I, Part-V **1.54**).

In conclusion, with the deeper account of information-matter dualism as the mirror image of spacetime structure, if nothing else, we must concede to the Cartesian reality of both memory-brain and mind-brain dualism: We must concede to the dissociation of memory, mind and identity from the brain, even when we assume the veracity of materialism.

2.42: Decoupling Theory: Part-II

Book-I Part-IV presented the first foray into decoupling theory, primarily based on information-matter dualism, of a form that did not directly emerge from the structure of spacetime, and which treated quantum mechanical waves of radiation and matter as spatially dynamic time-undulating and 'moving'. In the peculiar results obtained in Nimtz's quantum tunnelling experiments, which involve the apparent transmission of the musical composition Mozart-40 at 4.7 times the speed of light[71], decoupling theory grappled with how information could be physically transmitted by radiation at apparent faster than light speeds, without subversion of Special Relativity, and without the implied breakdown of causality.

In Book-I (essays, **1.37** to **1.43**), decoupling theory ultimately asserted that, per information-matter dualism, information is not in the first place carried by the presumed spatially displacing quantum mechanical radiation. Indeed, the information is entirely *decoupled* from quantum mechanical radiation and de Broglie matter and retained in dissociated abstract form extant that radiation and matter. Therefore, information cannot be fated to the putative 'speed' of the radiation that, in the first place, was *not* carrying or transporting the information. It followed that the integrity of information is not fated to the quantum tunnelled faster-than-light microwave radiation, no more than it is fated or dependent on the supposed 'carrying' radiation that displaces at the speed of light. It also follows that no violation of principles of causality can arise. Indeed, so long as causality is properly and solely fated, *not* to the 'speed of light', but to the process of ontologically primary AND-to-OR collapse and time itself, which in the first place manifests the putative 'speed of light', and also manifests quantum tunnelled faster-than-light radiation... and insofar as the latter cannot reverse the process of AND-to-OR collapse into OR-to-AND de-collapse or 'time-reversal', there can be no violation of causality by the informationally decoupled quantum tunnelled microwaves entailed in the Nimtz experiments.

All of this was established in Book-I Part-IV. However, we must now reframe original decoupling theory in terms of the now proven spatially static quantum waves and, implicitly, in terms of the nested futures view of spacetime in potentiality and the growing block intermediate spacetime model.

*

The spatially static account of the quantum mechanical wave and of spacetime in potentiality *en toto*, to which such waves are subsumed, leads to the immediate justification of decoupling theory: The question as to how it is possible for quantum waves of radiation or matter to carry and transmit information at apparent faster-than-light is solved, not only per the insight from Book-I to the effect that information is not in the first place carried by quantum mechanical radiation or de Broglie matter due to information matter dualism, but per the fact that quantum mechanical radiation and de Broglie matter, as spatially static states, do not in the first place 'displace', much less accomplish such a feat at faster-than-light, or even at normal 'speed of light'... simply because quantum mechanical waves, light, and tunnelled microwaves, are spatially *static* and do not move.

Thus, we reiterate that **'spatial displacement' is something foisted on the succession of OR-form events generated on or along the event horizon, generated by attendant AND-to-OR wavefunction collapse. That is, we impose the *assumption* of spatial displacement of matter and radiation on the succession of OR-form facts and their event horizons. Yet, quantum mechanical radiation and quantum mechanical de Broglie matter *does not* undergo spatial displacement: These are spatially static, as was established in 2.23 to 2.26**.

Thus, **when we quantum tunnel radiation to an effective speed 4.7 times faster than light** per the experimental arrangement from Nimtz, we naively assume and foist the notion of spatially displacing quantum mechanical microwave radiation, made worse by our typical descriptions of it in the usual pseudo-realtime way, further compounded by our false notion that the radiation is 'carrying' or transporting Mozart-40 across space. **Yet, this is precisely what we must now call into question in the light of spatially static quantum mechanical waves.** We must now grasp the whole process in terms of relations between OR-form resolutions along or on the event horizon, and the dynamics of the event horizon, consistent with spatially static quantum waves.

*

The progression of the event horizon is essentially a broken-discontinuous one: somewhat like an old-school running 35mm film, comprised of otherwise static frames or pictures, from the succession of which we infer or impose the assumption of 'motion'. To reiterate, **the quantum mechanical interval between each event horizon 'frame' implies that the information manifested in or on a preceding frame at t_1 must be subject to dissociation from that frame.** This dissociation holds until the event horizon re-forms at the succeeding frame along the time-axis at t_2, at which point the information previously subject to dissociation from t_1, perhaps with some modification, is restored at t_2; restored to the radiation, matter and 'space' correlate to the new event horizon. Throughout this

---

[71] Enders, A; Nimtz, G. (1992). "On superluminal barrier traversal". *J Phys.* I France 2 (9): 1693-16

process, **there is no literal spatial displacement of quantum waves of radiation or matter across t₁ to t₂, or subsequently, given the spatially static nature of the quantum mechanical waves, even when tunnelled to a supposed 4.7 times the putative speed of light. Thus, such intervals constitute spatially static potentials within the perpetually decohering and decaying domain of spacetime in potentiality,** *not* **'moving' states.**

Hence, when we quantum tunnel spatially static quantum mechanical microwaves to 'speeds' faster than light, we are not in actuality rendering a faster-than-light motion to the pertinent radiation that, in the first place, cannot be subject to any kind of motion. All that we accomplish is a specific alteration in the relation of Mozart-40 within the new corpus of events, within a larger and embedding domain of information restored to a succeeding event horizon, consistent with the various scenarios articulated in Book-I, essay 1.41. The effect is a seeming 'spatial displacement' of Mozart-40 at 'faster than light', when no such displacement has actually or objectively occurred.

Indeed, we must always look at the time-relation between events and observers, not at the presumed 'speed' or 'motion' of the 'carrying' radiation. The various Special Relativity harmonisations offered in Book-I 1.41 remain salient, now grasped in terms of our intermediate model spacetime, comprised *only* of spatially static quantum waves and static spacetime in potentiality. Thus, we advance decoupling theory one step further in the light of spatially static quantum waves and spacetime structure. Further advances to decoupling theory must await subsequent books in this work.

2.43: The Intermediate Spacetime Model: Presage to Gravitation

Generic models of spacetime designed for Special Relativity do not accommodate curvatures of spacetime wrought on by ostensive matter and mass. This is left to General Relativity and pertinent didactic depictions. **The intermediate model spacetime, ostensibly suited to Special Relativity, can easily accommodate spacetime curvatures, albeit according to different premises and forms, as part of the evolution of futures within spacetime in potentiality in conjunction with processes pertaining to the event horizon.**

Any gravitational effect, such as acceleration or change in trajectory of an object per a gravity field, will manifest as a succession of OR-form outcomes on or along successive event horizons and the attendant universe-scale AND-to-OR collapse processes. That is, gravitational phenomena remains critically tied to the process of time; to AND-to-OR collapse, integrated to our intermediate model of spacetime, although time is *not* depicted as a fourth dimension therein. **Given AND-to-OR collapse processes, or time, which attend the broken-discontinuous progression of the event horizon, it should be possible to describe gravitational effects as distortions of event horizons, as pertinent to 'curvy' event horizon.**

Hitherto, we have depicted the event horizon as a flat line or flat surface vis-à-vis the y-axis, with the impression that the event horizon, and attendant AND-to-OR collapse and time, unfolds 'upward' at the same rate on and along it and 'upward' the time-axis. However, some parts of the event horizon may advance 'ahead' while other parts lag; all per function of time dilation from gravitation, recuperated as leading or trailing event horizon progression. This transforms the description of the event horizon from a flat line or surface into something resembling an undulating or 'curvy' line or surface: i.e., a 'curvature of spacetime'.

A caesium clock on the surface of a gravitating body runs slower than a clock placed at some altitude above the same, or placed onboard a high-altitude plane[72]. The clock on the surface is subject to greater acceleration due to gravitation than the clock at altitude. Acceleration ultimately implies attendant time dilation, wherein the on-ground clock will run slower than the clock at altitude further removed from gravitation. For the clarifying didactic depictions, see **Fig. 2.19-A to C**.

In **Fig. 2.19-B**, **clock-a** on the gravitating body-surface runs slower than **clock-b** at altitude. That is, the AND-to-OR collapse rate at the reference frame of **clock-a** runs at a dilated slowed frequency vis-à-vis its higher AND-to-OR collapse frequency at the reference frame of the high-altitude **clock-b**. It follows that, as judged from within an interval of time measured in the frame of **clock-b**, the number and frequency of caesium decays or 'clicks' obtained for **clock-a** subject to higher gravitational acceleration will be measurably less than the number of clicks obtained for **clock-b** subject to lower gravitational acceleration.

We now make a straightforward association between, on the one hand, the number of caesium decay clicks and, on the other hand, the frequency and dilation of the broken discontinuous process of AND-to-OR collapse that resolves the succession of 'clicks' out of spacetime in potentiality into OR-form events. Consequently, the number of AND-to-OR collapses for **clock-b** will be more dilated than those for **clock-a**.

Since both AND-to-OR collapse *and* the progress of the event horizon are synonymous, to state that AND-to-OR collapse (time) is dilated at the frame of **clock-a** is to state that the progression of the event horizon associated with **clock-a** 'upward' the diagram is dilated when compared to the upward progression of the event horizon pertinent to **clock-b**, as is clear in **Fig. 2.19-B**. Since each click of a caesium clock constitutes a definite time-interval, and since more of such intervals will be associated with **clock-b** than with **clock a**, the event horizon at the frame of **clock-b** will have progressed farther up the time-axis than the event horizon at frame of **clock-a**. Consequently, the total event horizon that joins up the frame of **clock-a** with that of **clock-b** is no longer flat: It is a 'curve' or gradient, which descends from **clock-b** to **clock-a**. In **Fig. 2.19-C**, we end with a sufficiently comprehensive didactic geometric description of a gravity field in terms of a graded or 'curvy' event horizon and attendant curvy AND-to-OR collapse and time ... or curvy time... or 'warped' time, furnishing a more intuitive view of perennial 'spacetime curvatures'.

---

[72] *Hafele, J. C.; Keating, R. E. (July 14, 1972). "Around-the-World Atomic Clocks: Predicted Relativistic Time Gains" (PDF). Science.* **177** *(4044): 166–168.*

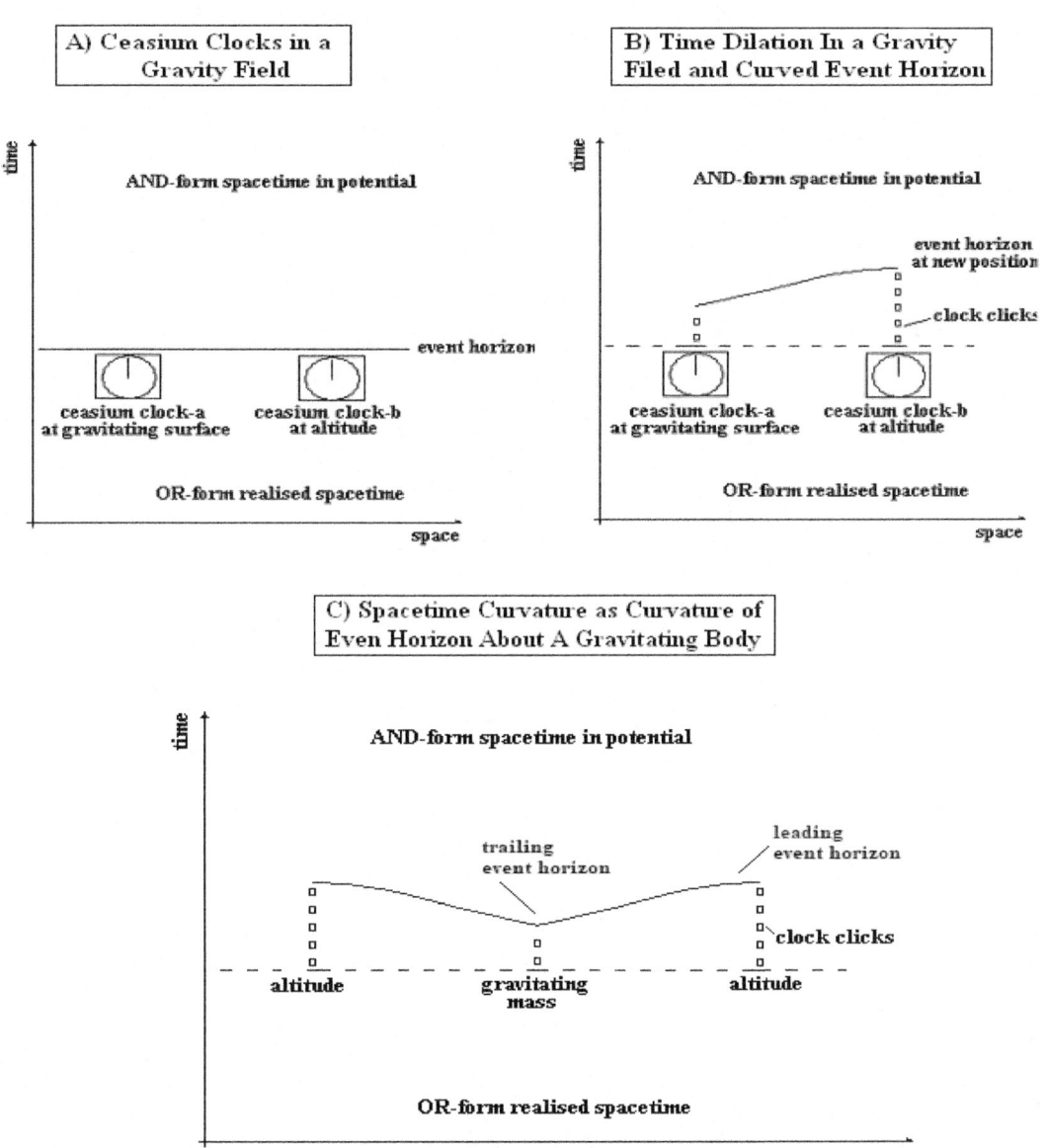

Fig. 2.19: Event horizon, AND-to-OR conversion, and spacetime curvatures
Notes: **Diagram-A** depicts two caesium clocks along a 'flat' event horizon. **Clock-a** is attached to the frame of a gravitating body: **Clock-B** is at altitude above the gravitating body. These clocks are synchronised to run at the same rate before assuming their depicted positions. But in **Diagram-B**, time dilation effects per gravity are incorporated: time dilation at the frame of the gravitating body is accentuated, and the attached caesium clock runs slow. Time dilation due to gravitation is attenuated at altitude: the attached caesium clocks run faster: In **Diagram-C**, three caesium clocks are emplaced, with two at altitude and one clock attached to the gravitating mass: We obtain a differential or gradient for time dilation. i.e., the spacetime curve as the 'curvy' event horizon. Subsequent event horizons will assume the described 'curvy' structure: a 'spacetime curve' depicted as a 'curvy' non-flat event horizon.

To reiterate and augment, the reason why AND-to-OR collapse and time progresses in its attenuated way at the frame of **clock-a** compared to **clock-b** is per gravitation: **The greater the gravitational pull, the more dilated the rate of AND-to-OR collapse or** time: i.e., the attached caesium clock runs slow, as viewed from the fame of clock-b. Conversely, the lesser the local gravity the more accentuated and less dilated the AND-to-OR collapse process: i.e. clocks-b runs comparatively faster, as viewed from the frame of clock-a. Assuming identical twins, but with one of the twins placed at the frame of clock-a, and the other at clock-b, we obtain the infamous twins paradox: The twin at the frame of clock-a will grow older at a much reduced rate than the twin at clock-b. It follows that the part of the event horizon associated with the surface of a gravitating body with higher local gravity will advance upward the spacetime system at a much dilated rate than those parts of the event horizon associated with higher altitudes and lower gravity, describing a gradient for AND-to-OR collapse dilation, or time dilation and the 'warping of time' so-conceived, associated and synonymous with the whole of the gravitational field, as illustrated in Fig. 2.19-C.

The advantage of our intermediate spacetime model approach to spacetime curves is that it incorporates gravity in a non-contrived and seamless way into a model that already incorporates and harmonises both the quantum descriptions and classical-relativistic descriptions of nature. In short, **the model extends the primitive quantum-relativistic unification into a more comprehensive unification that also incorporates gravitation in terms of curvy event horizons.**

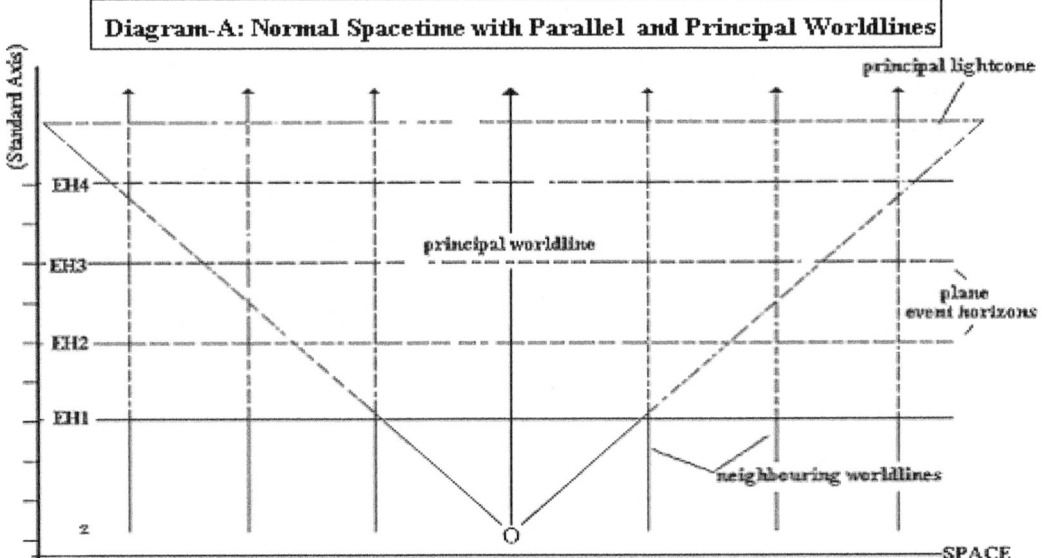

Fig. 2.20: Lightcones an worldlines in a flat event horizon grid
Notes: The didactic diagram illustrates a variant of the intermediate model spacetime, one in which we have added a series of parallel worldlines representing inertial objects vis-à-vis a principal central worldline belonging to a large-mass object; with the latter depicted as attached to a potential lightcone, in turn destined to AND-to-OR collapse and realisation through a succession of 'flat' future-potential event horizons, EH1 to EH5.

The didactic depiction in **Fig.2.20** constitutes a description of successive flat event horizons and attendant AND-to-OR collapse (time) processes related to a principal worldline and lightcone originating at O: all accompanied by parallel worldlines comprising inertial bodies unaffected by the principal worldline.

Therein, only event horizon EH1 is depicted as realised: All worldlines and partial lightcones up to EH1 are also realised. But subsequent event horizons and projected worldlines, and the lightcone beyond EH1, depicted as as-yet not-realised future potential event horizons, potential worldlines and potential lightcones, are all subsumed to spacetime in potentiality.

The moment we incorporate gravitational effects depicted in terms of curvy event horizons, the structure of our didactic spacetime must change. In diagram **Fig.2.21**, a lightcone originating at O will be 'pulled in' per the gradients that attend the succession of curved event horizons; all centred on the principal worldline of the gravitating body originating at O. The 'curvy' event horizon describe gravitation. **The lightcone originating from the gravitating body will be distorted or 'pulled in' (as depicted) from its normal cone-shape into a funnel-form by the succession of curvy event horizons due to gravity.**

In the didactic illustration **Fig. 2.22**, we depict the potential convergence of formerly parallel worldlines due to the succession of curvy event horizons projected into the domain of spacetime in potentiality. **The parallel worldlines will be deflected toward and to convergence with the principal worldline, which constitutes the gravitating body originating at O: That is, what were otherwise parallel worldlines belonging to unaffected inertial bodies will be gravitationally 'refracted' by the broken-discontinuous and discrete succession of curvy event horizons and their attendant AND-to-OR collapse processes; with each such succession entailing a quantum-discrete 'leap' (i.e. a quantum gravitational moment) bringing into OR-form realisation the putative**

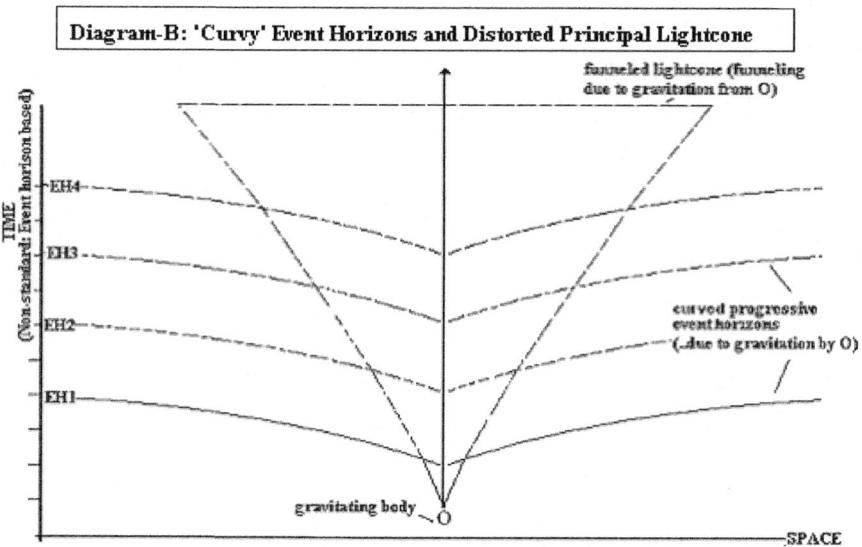

Fig. 2.21: The funnelled lightcone in gravitationally 'curvy' event horizon grid
Notes: In the didactic diagram, we omit the parallel worldlines, but introduce 'curvy' event horizons (the succession of curvy event horizons) per the gravitational field belonging to the central large-mass body represented by the principal worldline. The curvy event horizons will distort the potential lightcone into a distorted 'lightfunnel': i.e., a lightcone affected by gravitation.

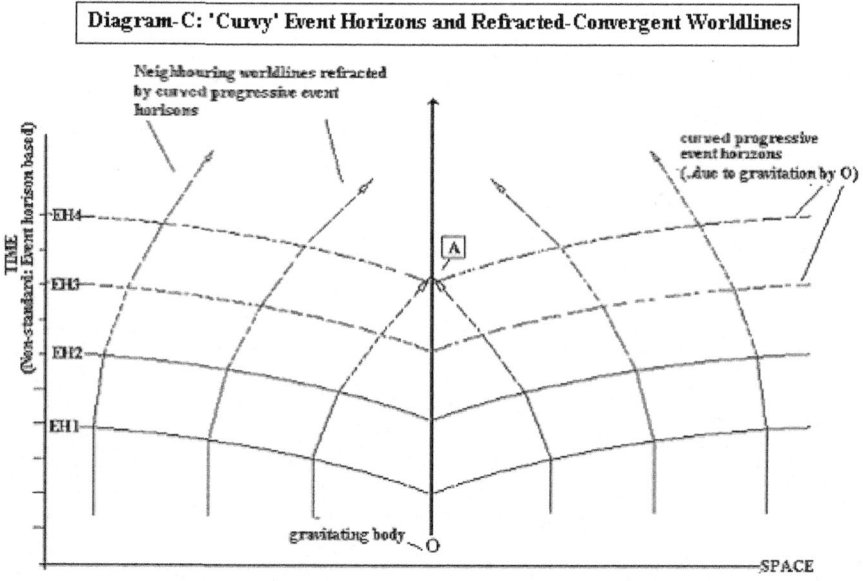

Fig. 2.22: Refracted worldlines in a gravitationally 'curvy' event horizon grid: gravitational attraction
Notes: In **Diagram-C** we omit the 'lightfunnel', now implicit, and reintroduce the original parallel worldlines; the spacetime paths of equidistant low-mass inertial objects. These worldlines must now become non-inertial by normalising vis-à-vis the same 'curvy' event horizons: the otherwise parallel inertial worldlines now curve or change direction, and accelerate into... and converge with the central worldline at point-A: thus, emulating gravitational attraction and in-fall through the said refraction of the worldlines by the succession of 'curvy' event horizons.

251

'gravitational acceleration' and angular change of what were otherwise uniform flat-trajectory inertial bodies. Hence, we obtain gravitational acceleration and synonymous change in the trajectories of such bodies toward future-potential convergence with the gravitating body; with the principal worldline originating at O.

Note that the otherwise parallel inertial bodies, now deflected and refracted into effective acceleration and change in their trajectories, yet remain inertial. That is, their trajectories are not changed by the application of energy or work, but remain forms of perpetual motion, just as is true in inertia, but *with* angular change and acceleration: The change made does not require the application of energy or work to affect: the work function remains zero, even when one ignores the pleonastic status of energy-work concepts established in **2.39**.

2.44: The Intermediate Spacetime Model: A Closer Look at Spacetime Curves

There are marked differences between how intermediate spacetime models gravity via 'curvy' event horizons vis-à-vis lightcones and worldlines, and how didactic depictions from General Relativity accomplish the same. In the intermediate model, we transform

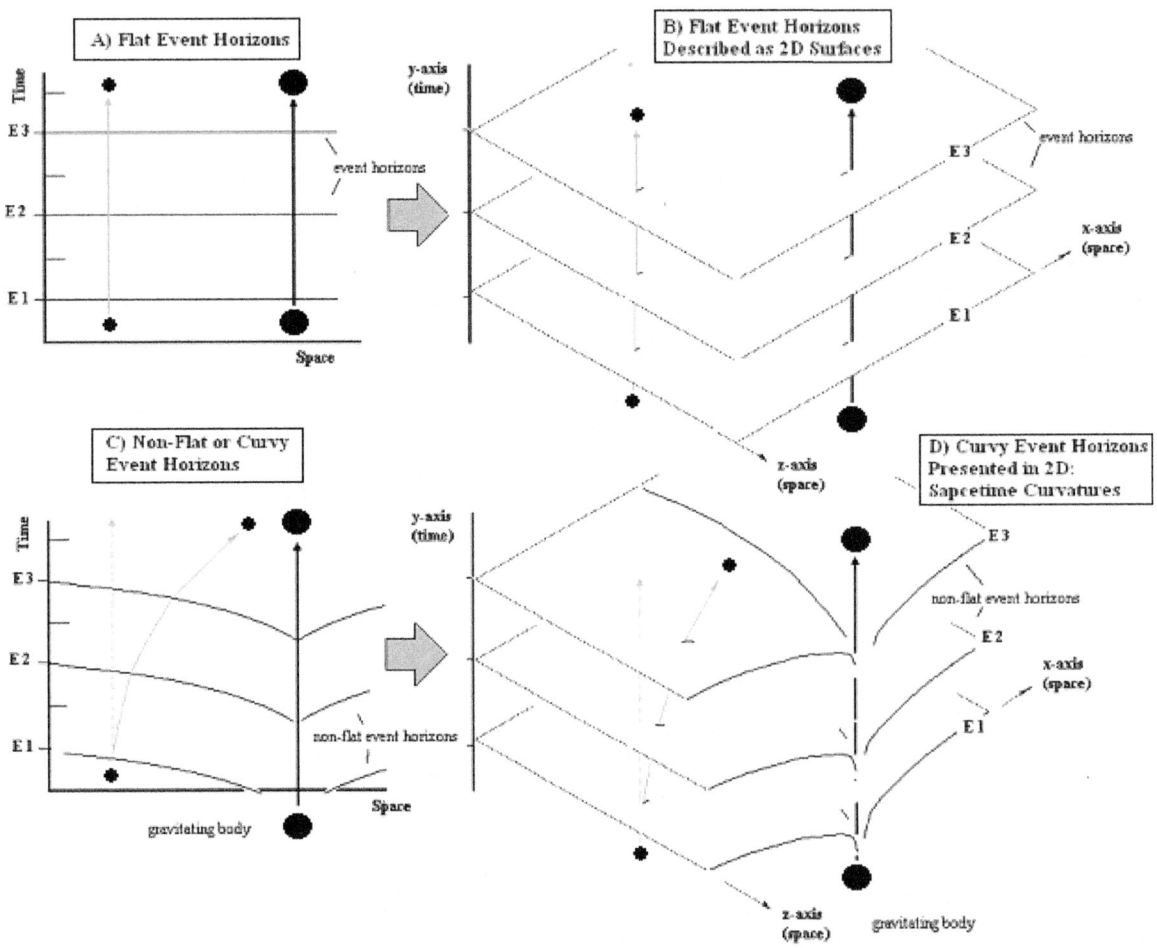

Fig. 2.23: Gravitational spacetime curvatures and gravitational curvy event horizons
Notes: The successive flat event horizons depicted as an edge-on view of the intermediate model of spacetime given in **Diagram-A** assumes the form of two-dimensional surfaces in **Diagram-B**. In both diagrams, the worldlines of two bodies assume a parallel form of vertical worldlines and depict uniform motion. In **Diagram-C**, we incorporate gravitationally 'curvy' event horizons first illustrated in **Fig. 2.20**. Seen edge-on, these are also rendered as successive two dimensional 'surfaces' in **Diagram–D**. Note that the smaller-mass body must now normalise its worldline vis-à-vis the curvy event horizons E1 through to E3, imposed by the gravitating mass. (Note that the curvy event horizons generated by the smaller mass are omitted for the sake of brevity).

stacks of 'flat' successive event horizons depicted in **Fig. 2.23-A** into two dimensional 'surface' representation of the same, as depicted in **Fig. 2.23-B**. Therein, note the two parallel worldlines in both diagrams: each comprising an initial non-accelerating inertial body: The body on the right constitutes the larger-mass central body; the one on the left is the lower-mass peripheral body. Note that the event horizons depicted are 'flat'. The masses are depicted without gravitational effects.

To incorporate gravitational effects, we develop the didactic diagrams depicted in **Fig. 2.23 C** and **D**. These involve stacks of now successive curvy event horizons, the argument for which was given in **2.43**. Notice how the curvy event horizons are centred on and are curved by the larger-mass central body; the succession of even horizon depressions or curves are 'spiked through' by the central mass worldline.

The worldline of the lesser mass peripheral body, whose own curvature of the succession of event horizons has been omitted for brevity, must normalise to the event horizon curve and angle imposed upon it by each succeeding event horizon distorted or curved by the central mass. The peripheral body worldline so refracted must eventually converge with the worldline of the larger-mass body responsible for the curve. In this way, we obtain both gravitational attraction and consequent in-fall of the lower-mass body into the larger-mass gravitating body; a transformation of the peripheral body from a vertical parallel inertial worldline with zero-velocity to a curving and accelerating worldline per function of the stack and succession of curved event horizons: one that yet retains its inertial work function zero characteristics

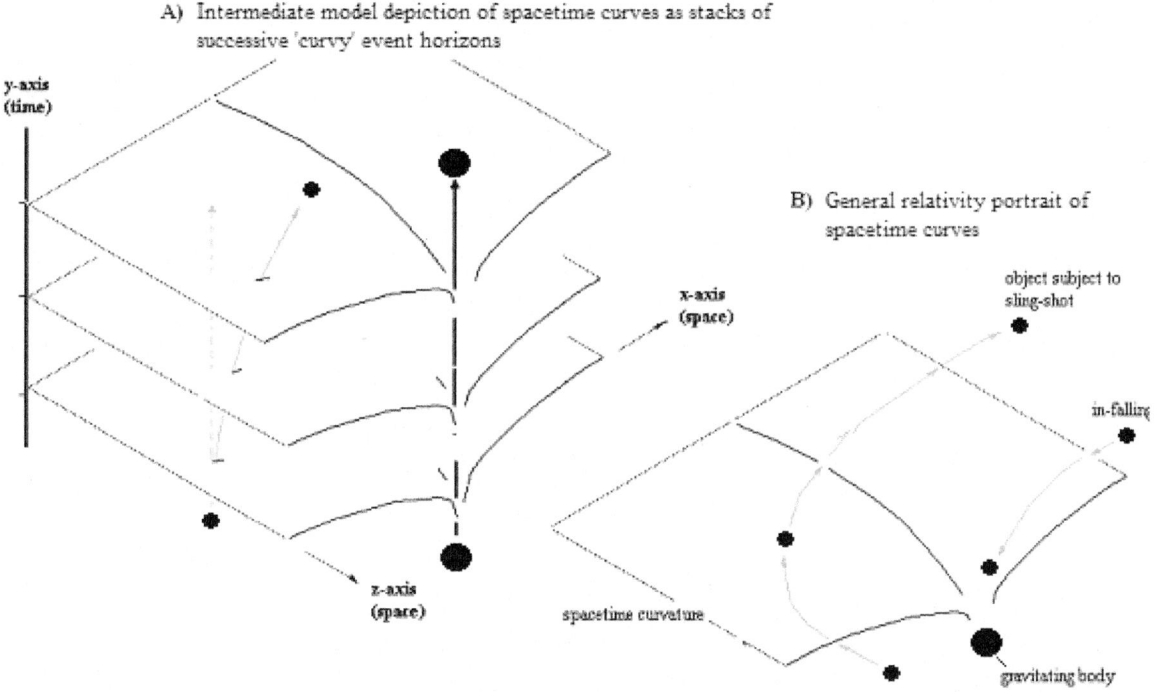

Fig. 2.24: Intermediate model versus General Relativity: graphic of spacetime curvatures
Notes: The illustration contrasts and compares spacetime curvatures and gravitational effects described by our intermediate model quantum–relativistic incorporative spacetime system (see **Diagram-A**) and the vintage depiction of the same per General Relativity (**Diagram-B**). The intermediate model spacetime version proves to be more intuitive than the version from GR. The intermediate model incorporates quantum mechanical realities as a distinct and critical domain of spacetime in its own right (i.e. as the domain of spacetime in potentiality) and depicts spacetime curvatures as a succession of 'curvy' event horizon without recourse to a 'fourth-dimension': It therefore offers a clearer and more intuitive grasp of such curves, linking these directly to the domain of quantum mechanical spacetime in potentiality (i.e. to quantum mechanical realities) insofar as the 'curvature' occurs at event horizon-boundaries that 'cut' realised spacetime from quantum mechanical spacetime in potentiality. Moreover, by incorporating the discontinuous-broken version of the progression and succession of the now-curvy event horizons, the intermediate model also incorporates a quantum-discrete account of that progress, and, with it, quantum-discrete evolution of the proverbial 'curvature of spacetime' synonymous with gravitation: In short, quantum gravity. On the other hand, the model from General Relativity is incapable of incorporating any of these features in any intuitive or natural seamless way.

In truth, the effect is mutual: the lesser-mass body and worldline also imposes its own curve to a same event horizon, and this acts as a counter or resistance to the gravity pull of the central mass as well as a mutually attracting curve. But this is omitted for brevity.

In **Fig.2.24-A** to **B**, we directly contrast how our intermediate quantum-relativistic spacetime model describes spacetime curvatures and gravitational effects versus how didactic General Relativistic (GR) depictions accomplish the same. There are two key advantages of the intermediate model over GR: <u>First</u>, **in GR, an explicit future-form quantum mechanical spacetime domain, a natural and necessary feature of our growing block intermediate model, is** *not* **explicitly modelled or even recognised: i.e., the domain of spacetime in potentiality is absent in GR. It is absent from GR for axiomatic reasons: GR has no place for quantum mechanics. Thus, GR does not incorporate quantum mechanics or AND-form logic and the ontic reality of the future constituted as spacetime in potentiality.** <u>Second</u>, **GR does not explicitly state or recognise time itself as AND-to-OR collapse: Instead, GR assumes time as a 'fourth dimension' inherent to its spacetime curve**. Indeed, the future-less block universe appears almost inbuilt to GR: GR does not incorporate or recognise the growing block universes, and this is in contrast to our Alternative Realist philosophy of physics (Alt-R). All of this is obvious from **Fig. 2.24-B**.

Our contention is that, at least partially, quantum gravitation is to be grasped as, and identified with, the broken-discontinuous quantum-discrete upward progression of curvy event horizons per attendant AND-to-OR wavefunction collapse and time, depicted as a succession of curvy event horizons in **Fig.2.26-A**: **The quantum-discrete progression of spacetime, and attendant AND-to-OR collapse (time) is seamlessly incorporated into our growing block intermediate model, but** *not* **in GR** as is clear in **Fig. 2.24-B**. In these two areas alone, even in its nascent presage form, gravity per curvy event horizons in our intermediate model spacetime is more intuitive, simpler, incorporative of quantum mechanical realities, hence superlative vis-à-vis the more complex block universe GR.

The notion of a 'curvature of spacetime' tends to be problematic to intuition, given its attendant notion of 'four-dimensional spacetime'. When we model spacetime curvatures as stacks of curvy event horizons, and time as a succession of event horizons per synonymous AND-to-OR collapse, itself grasped as time (no longer a 'fourth dimension'), the intuition-problem disappears. We retain the apparent three-dimensionality or stereotemporality of things, but discard the notion of time as a 'fourth dimension'. Instead, time is properly identified as universe-scale perpetual AND-to-OR collapse, or the decay of future potentiality into actuality, with its AND-to-OR polarity constituting its 'arrow', and the impossibility of time-reversal in OR-to-AND de-collapse.

2.45: Spacetime Curves: A First Hypothesis on How Matter Tells the Event Horizon How to Curve

Hitherto, we have described and clarified gravitation in terms of the succession of curvy event horizons and attendant AND-to-OR collapse processes (time), with the curvy event horizon constituting the gradient of time dilation imposed by gravitational mass. But this does not inform us about how gravitational mass tells the event horizon how to curve. Indeed, **the curve itself is not the gravity. The process that causes the curve** *is* **the gravity. With this understanding, we must seek out how the mass makes the event horizon 'curve' and gravitationally affect lightcones and worldlines in the way hitherto described (see 2.44)**. In short, we need a hypothesis about 'how matter tells spacetime how to curve', and how 'the curve tells matter how to move'. The full explanation must await further developments secured in Book-III and Book-IV. Even so, we can provide a first hypothesis of this deeper process in anticipation of the forthcoming full account. Thus, **how matter and mass tell the event horizon how to curve is a work function zero process. How the curve then tells matter how to 'move' is also a work function zero process**. The work function zero character of gravitation derives from essential the same premises found in work function zero inertia and in the workings of the permanent magnet (see **2.35** and **2.37**).

*

Recall from **2.35** how inertia was posited as a function of the quantum erasures, per counterfactuals, of non-consistent futures belonging to the inertial body, and the consequent grand decoherence of spacetime in potentiality, in turn structured as a convergent nested future system. These insights culminated into the revival of Mach's principle and solved the conundrum of how matter 'over here' counterfactually 'knows' the state of matter lightyears away 'over there', instantaneously or non-locally, succinctly through the grand decoherence of their common futures into a common consistent future, according to which their future-potential worldlines and trajectories must conform. We also postulated that inertia so-grasped unfolds according to a work function zero process, consistent with the work function in generic inertia. Almost all of this can be re-purposed to explain, if not how matter tells the event horizon how to curve (a task for Book-IV), then how matter incident to that curve then 'moves'. In this way, we also furnish a partial sketch of hitherto elusive quantum gravity, grasped from within quantum mechanical spacetime in potentiality on entirely different premises to various generic quantum gravity theories.

Whatever is imposing the curve on the stacked succession of event horizons that then distort lightcones and transform otherwise inertial states into future-potential gravitationally in-falling, accelerating and converging worldlines, instigates the attendant quantum erasure of non-consistent nested futures, and, with it, affects the instantaneous grand decoherence of spacetime in potentiality into the said gravitational phenomena; all on a work function zero basis, if only because the process of counterfactual erasure of non-consistent futures, and attendant concordant grand decoherence of spacetime in potentiality, *are* work function zero processes in their own right (see **2.34** to **2.35**). Thus, gravitation is a work function zero process, as was argued in a contentious-seeming way in the Preliminary, Part-III: **0.33** to **0.34**.

The stack of future-potential event horizons will decohere to the same curve-form as the initial curved event horizon, per the same counterfactuals-driven grand decoherence of spacetime in potentiality as attends inertia (see **2.37**). But this requires an initial first-ever curved event horizon. In other words, while our nascent and presaged quantum gravitation explains how the stack of subsequent future

254

potential event horizons assume their attendant curves, it does not tell us what imposes the first curve on the first event horizon, which then grand decoheres consistent futures upon spacetime in potentiality, which then resolve the conformal stack of curvy horizons that then refract potential lightcones and worldlines in the way described.

*

We now know how the curve tells matter how to move: via counterfactuals and grand decoherence of spacetime in potentiality that imposes and projects the same curve upon all subsequent future-potential event horizons, which then refract the attendant worldlines, or 'tell matter how to move'. **But what causes the initial event horizon to curve? To address this question, we must bring into fray the role of realised spacetime, of the abstract past**. But, to address this conundrum, we must first state the generic reason for why the General Relativity account of gravity needs to be doubted: We must clarify why it needs to be replaced by a new theory of gravity hitherto partially disclosed.

Given the quantum mechanical nature of matter, and given its superposition into AND-form 'smear' of positions, to which position does the spacetime curving gravitating mass of the particle relate? Hence, to what position in the superpose of smeared positions does the spacetime curve pertain?

This question is analogous to the conundrum of where potential energy and kinetic energy must relate within the tripartite structure of intermediate spacetime. This was explored and resolved in **2.39**. Therein, we found that, its pleonastic status aside, potential energy must be 'stored' in the future, within spacetime in potentiality. For its part, kinetic energy, or action, was found restricted to the succession of event horizons that attendant AND-to-OR collapse processes (time) that decay potential energy locked up within spacetime in potentiality into pleonastic work, action, and kinetic energy. It followed that the mass equivalent to potential energy espoused in the famous equation $E = mc^2$, must be locked up within spacetime in potentiality, and *not* 'stored' on the event horizon, and even less so within the domain of realised spacetime.

In **2.39,** it was also anticipated that de-spatialisation must further confirm the reform of energy relations per spacetime structure: The obviation of space must inevitably usurp the notion of matter, mass and potential energy as states supposedly 'stored' in spatially distributed localised particle-point or spatial blocks, notwithstanding the pleonastic status of energy that must emerge from the character of ontologically primary AND-to-OR collapse and time.

Again, just as we asked where the potential and kinetic energy must reside, we must also ask where the gravitationally pertinent spacetime curving mass must reside. Is it in spacetime in potentiality? Is it on or along the event horizon? **If, for a given mass, we assume just two alternative future-potential worldlines in AND-form superposition, does the event horizon curve or spacetime curve of that mass attend to one or both of these future potential worldlines?** For the naive realist, wherein the particle has only one real OR-form position, the curve would be expected to relate to the one-and-only OR-form position and attendant worldline. But the reality of interference patterns and self-interference demonstrates that, putative 'positions' *are* superposed into AND-form sets of positions and alternative worldlines. Therefore, **the gravitating mass, hence the spacetime curve of the particle or mass, hence the event horizon curve, would need to relate to *all* the future-possible positions and alternative worldlines of that particle or matter. But this is simply unworkable and constitutes a serious obstacle to the integration and unification of GR (gravity) and quantum mechanics**. In short, GR presupposes a pure OR-form universe and cannot accommodate AND-form realities. Generic physics recognises this problem. It concludes rightly that GR must be fundamentally wrong and admits the incompatibility of General Relativity with generic quantum mechanics.

*

How does our intermediate model spacetime solve the problem at the core of GR? Recall that the future is ontically real and is consolidated as the domain of spacetime in potentiality. Thus, the AND-form quantum mechanical rendition of the putative positions and worldlines of the particle or gravitating matter constitutes the ontically real future-form of that particle or matter: the spacetime in potentiality state of that particle or matter. As such, as a future set of possibilities that have not-yet happened, the particle or matter within spacetime in potentiality *cannot* be characterised as having generated *any* gravity field, and cannot be characterised as having imposed *any* curvy event horizon, or anything else… simply because it has not happened.

In the simplest terms possible, the future has not yet happened. Therefore, the future cannot gravitate. The same holds for the future-form of matter: The state of matter that has not-yet happened cannot gravitate. In this sense, whatever is responsible for the curving of the event horizon does not attend or issue out of the future-potential of that particle or matter, and it cannot issue out of spacetime in potentiality: That is, the gravitating factor, the curving factor, is not produced by the future and does not arise from any potential material 'position', or attend any future-potential worldline within the AND-form superpose of such within spacetime in potentiality. In other words, in our simplified two-possibility system, insofar as it does not issue from the future, the curve does not go to either of the two alternative future-potential worldlines and their leading spatial points.

The second part to the solution comes from the recognition that a particle or matter that has not yet happened, and cannot obviously constitute the generator of any gravitational effect, or of anything else, is the state that must itself become decohered by gravitational effects: That is, matter's future-form possibility-set must decohere in the face of implications from gravity, in conformity to the curve imposed on the initial event horizon upon which that particle is expected to AND-to-OR collapse into an actualised 'position' denoting gravitational influence. **In short, gravity, or the curve, is that which decoheres the future possibility-set of the particle or matter into gravity-consistent form: That is, gravity is that which decoheres spacetime in potentiality into consistent form: In other words, the future is subordinated to and subject to gravity. Hence, it is *not* the future AND-form possibility-set of the matter**

255

that causes the curve, but rather it is the curve that decoheres, narrows and reduces the future-possibility set of the particle or matter to a gravity-consistent set of futures that subsequent AND-to-OR collapse resolve into an OR-form outcomes that we grasp as a gravitational affect: i.e., refracted worldlines and matter infall... to conformity to and in consistency with the demands of ontically *a prior* gravitational grand decoherence... in consistency with the imposed even horizon curve.

To reiterate, it is not the future-form of the particle or matter that causes gravity, but it is gravity that decoheres the AND-form futures of that particle or matter to consistent possibilities permitted within and by gravitationally grand decohered spacetime in potentiality. Hence, we are forced to conclude that, insofar as the gravitational factor does not arise in or issue out from spacetime in potentiality or from the AND-form quantum mechanical state of de Broglie matter, it must arise from an alternative source. This cannot be the event horizon, since it is the event horizon that is being curved by gravity: The event horizon cannot curve itself.

Only the domain of realised spacetime remains as the candidate-source of 'curving power' and of gravitational grand decoherence of spacetime in potentiality to gravity-consistent future-potentials. Thus, we postulate that the curve on the event horizon is imposed by the content of realised spacetime: by hitherto realised history, retained as abstract memory pertaining to putative matter, and by all that has transpired in the growing block universe.

*

The domain of realised spacetime retains the history of the total past as abstract memory, albeit in dissociated form from both the event horizon and from future-form spacetime in potentiality. Hence, we postulate that the curve imposed on the initial event horizon, which then cascades throughout spacetime in potentiality through grand decohering counterfactual erasure of non-consistent futures, is a work function zero abstract memory effect, and ultimately imposed by the totality of the mnemonic domain of realised spacetime. **In short, gravity is a memory effect: Memory tells the event horizon how to curve: Subsequent counterfactuals and grand decoherence then informs future possibilities how to decohere and conform to that curve... or 'how to move'.**

The process by which the total memory constituted as realised spacetime accomplishes this feat will emerge in Book-IV, from anticipated de-spatialisation itself... which will culminate into the supersession of the intermediate model spacetime system by the superlative Bergson-Whitehead amalgam: or spacetime *without* space.

2.46: Is the Universe A Computer?

Is the universe ultimately driven by a computational process? Is the universe a computer? **Succinctly, is it possible for a clockable iterational procedure, an algorithm, to process the counterfactuals-driven quantum erasure of nested futures, hence generate grand decoherence throughout spacetime in potentiality... to the infinitely removed future?** If the answer is yes, then the universe must be a computer, or a simulation projected by some hidden computational system. If the answer is no, then we must overturn centuries of mechanistic-materialistic ontology and ideology.

The answer is obvious: The universe is *not* a computer.

While nature appears iterational and computational, or machine-like as far as what gets to materialise and succeed on or along the event horizon, and given that we tend to foist false notions of contiguity and mechanical causality to the succession of OR-form resolutions on, along and between successive event horizons, neither the AND-to-OR resolution of these (via time itself) nor the broken and discontinuous formation and succession of attendant event horizons, can be generated by any iterational, mechanical, computational or clockable procedure. At the fundamental causality-level behind AND-to-OR collapse itself... at the level of what causes and drives time itself... the universe is anything but iterational, procedural, mechanical, clockable, or computational.

Why?

First, consider the infinities inherent to the system of convergent nested futures that constitutes spacetime in potentiality, elaborated to the infinitely removed future. **The implied infinities involved in the process of counterfactuals-driven quantum erasure and grand decoherence that must perpetually restructure spacetime in potentiality to the infinitely removed future, would force *any* putative mechanical, computational, or clockable process supposedly responsible for it to crash into an insurmountable non-terminating or halting problem.**

Also, given that each immanent possibility within quantum mechanical spacetime in potentiality could potentially branch out into infinitely many alternate and mutually incompatible nested futures (see Feynman's 'sum over all paths' approach to spacetime in potentiality in **2.13**) it is difficult to see how any hidden purely iterational clockable process or algorithmic 'gears'... one that must erase and restructure each of the infinitely many possibility-branches to consistent futures... could accomplish such a task without crashing into insurmountable infinities and non-terminating problems.

It follows that the perpetual counterfactual restructuring of spacetime in potentiality must necessarily unfold by the 'action' of a hidden non-iterational non-computational, hence non-mechanical and non-clockable process: In other words, what makes time possible... what makes AND-to-OR collapse possible... and what grand decoheres spacetime in potentiality to infinity... must constitute an ineffable, non-structured non-clockable process that defies explications in terms of *any* mechanism, or *any* time-bound input-operation-output schema, or 'gears'... or 'computation'.

Insofar as this is true, **it must necessarily follow that the universe is *not* and *cannot* constitute a computer, and that, upon generalisation, the universe is *not* a machine at all.** Hence, we inhabit a 'non-computational universe': one that can apparently materialise machine-like configurations across time and across successive event horizons, while the ultimate process through which it accomplishes such outcomes, and drives time ontologically requisite to such outcomes, is not machine-like at all, and *not* a computation at all.

This remarkable conclusion ought to be obvious independent of the requirements of grand decoherence and the infinities therein: If AND-to-OR wavefunction collapse is time itself, as indeed it is, although AND-to-OR collapse or time can generate clockable outcomes, AND-to-OR collapse itself could not arise from any preceding clockable process. If a hidden mechanism or a computation could furnish AND-to-OR collapse and time, such a 'gizmo' would itself require time to process and unfold. Thus, the 'gizmo' would require its own independent AND-to-OR wavefunction collapse processes to bring about the wavefunction collapse of our own state. Why stop there? Why not assume a whole series of gizmo-driving gizmos… an infinite regress of wavefunction collapsing gizmo-driving hidden gizmos to bring about our own final wavefunction collapse, and, with it, the grand decoherence of our spacetime in potentiality. The consequence of the initial time-generating gizmo, itself at infinite remove, could never ramify to our own moment of wavefunction collapse, even given eternity.

Simply put, any assumption in favour of a clockable mechanism or algorithm as the basis of causality and time must itself crash into an insurmountable non-terminating problem. It necessarily follows that we must abandon mechanism as the ultimate basis or ontology of causality, of time, or of anything else. We must abandon the notion that the universe is a machine or a computer.

Hence, the universe is not a computer.

Insofar as the ultimate causality behind both gravitation and inertia is bound to AND-to-OR collapse and time, and behind attendant grand decoherence to the infinitely removed future, and insofar as time is generated by a non-clockable gear-less process, it follows that gravitation and inertia must also arise from the same gear-less causality behind AND-to-OR collapse and time: the same gear-less time that makes gravity and inertia manifest. This insight augments and is in turn augmented by the work function zero basis of inertia, gravitation and of attendant counterfactual causality and grand decoherence of spacetime in potentiality (see **2.35** to **2.45**). That inertia and gravitation, as with all else, is generated by a work function zero process and, ultimately a non-mechanical and non-clockable a-computational AND-to-OR collapse and time-process, segues harmoniously with the abstraction of pleonastic energy concepts from the ontology of physics, asserted in **2.39**.

Indeed, **since energy or work is pleonastic and *not* part of the real ontology of physics, it follows that phenomena (inertia and gravitation, for example) cannot be outputs obtained from putative energy-work inputs, much less from the mediation of these by some clockable contiguous 'operation' into outputs of gravity and inertia. When we abstract out pleonastic energy and work inputs and outputs from ontology, the schema for mechanical causality or 'gizmo' unravels: Materialism collapses.**

2.47: Gear-less Causality, Time & Mind

At first view, the brain appears to be a finite mechanical-like 'device': It appears to work in an iterational way; a 'computer', it seems. Given the dubious cultural predominance of forlorn materialism, it is tempting to suppose that the mind must be an 'excretion' or output of the brain: a thing purely explicable in terms of the iterational and mechanical processes of the brain. Further galvanised by eliminativist and identity theory, by epiphenomenalism, and by the more recent emergence theories that constitute the materialist contention in philosophy of mind. It seems to appear 'obvious' that mind is nothing but the brain. Yet, the contention breaks down, not merely in the face of information-matter dualism, but from the gear-less and non-clockable process of causality behind the grand decoherence of spacetime in potentiality and behind the AND-to-OR collapse-process of time. Given the fact that all in nature, including time and the perpetual restructuring of spacetime to future-infinity, is driven by a gear-less ineffable work function zero causality, with respect to which even energy-work concepts and relations are rendered pleonastic, it obviously follows that the brain cannot enjoy special exemption from the rest of nature.

In Book-I (**1.44** to **1.55**) we critiqued the notion of mind as mechanism or computation. This was accomplished as part of Rigging and quantum noise-based Superlative Rigging arrayed against the claims of strong artificial intelligence, or the notion that machines could literally think, or, per AI-organic, that we are nothing but machines or computers in our own right. The arguments in Book-I remain completely independent of, but can work in conjunction with, the more fundamental and ontological assertion against materialism from the nature of gear-less causality and time attendant the tripartite structure and processes of intermediate spacetime. **If, per spacetime structure and attendant gear-less AND-to-OR collapse and time, we must concede to a non-mechanical and non-computational universe, it follows that the perpetual resolution of the brain by AND-to-OR collapse and time, and the consequent state-change wrought upon it via successive event horizons, must itself derive and belong to an ultimately non-clockable gear-less causality,** even if the outcomes generated by it can be subsequently clocked as a series of successive OR-form brain-state changes and consequences on and along successive event horizons.

That the mind is non-clockable, and that nous cannot be rendered into a finite configuration of matter or brains, is also clearly implied by the hard problem evinced in the intuition of infinity and the renormalising quantum mechanist: neither of which could be furnished by the finite clockable mechanical brain embodied as a corpus of OR-form resolutions. The intuition into infinities could only be furnished by an ineffable gear-less process that can circumvent halting problems, akin to the same sort of ineffable gear-less process that makes AND-to-OR collapse and time possible.

The conclusion in favour of Cartesian dualism is inevitable when we also incorporate the case against the generic materialist abuse of the conservation principle against Cartesian mind: **The pleonastic nature of energy-work relations and their abstraction from the ontology of physics obtained in 2.39, alone renders the generic materialist abuse of the conservation principle arrayed against Cartesian dualism totally void**, notwithstanding implications that will emerge from anticipated de-spatialisation in Book-IV.

Moreover, even before de-spatialisation, the energy-work based conservation principle is recuperated into its abstract form, based on the conservation of abstract memory (the assertion that memory constitutive of the history of initial conditions cannot be destroyed or altered) combined to consistent futures that must attend the implications from memory, enforced through the process of grand decoherence. None of this requires contiguous causality or energy-work driven causality, and certainly *not* in Cartesian causality in mind-to-brain control. The point is that **recuperated abstract Cartesian causality does not violate the conservation of memory (it does not change the constitution of the past). Nor can Cartesian causality decohere the future into inconsistent possibilities in relation to the conserved past: Cartesian causality can only select future brain-outcomes consistent with the past and with what is permitted in the future through the grand decoherence of that future into consistency by the mnemonic past. It follows that the abuse of the conservation principle against Cartesian causality has lost its viability and utility.**

The full account of how mind exploits the structure of spacetime, or the same in its de-spatialised superlative form, and the implied gear-less nature of AND-to-OR collapse, time, and mind, must await Book-V, as must the exposition about what makes brains, and not pumpkins, the milieu through which nous enacts its will in the milieu of memory, possibility, and time.

2.48: Part-III Summary

The following constitutes a summary of the findings secured in Part-III:

- **Non-local counterfactual causality within the convergent category nested futures leads to the revival of Mach's principle.** Matter 'over here' is non-locally and instantaneously (transtemporally) entangled with matter 'over there'. This is achieved through their common convergent nested futures within spacetime in potentiality, and the joint **grand decoherence** of their common futures. Thus, wavefunction collapse and decoherence 'over here' will produce instantaneous mutually consistent decoherence of and erasure of common inconsistent nested future wavefunctions for matter 'over there', no matter how far 'there' happens to be. Thus, bodies otherwise separated from each other by perhaps millions of lightyears can 'know' of each other instantaneously and transtemporally, in a counterfactual non-local way: a meaningful 'knowing', but only in terms of *what will not transpire* in their common future within spacetime in potentiality. Thus, Mach's principle is revived on the basis of non-local counterfactual causality and grand decoherence of spacetime in potentiality. (**2.33** to **2.34**).

- **Counterfactual non-local causality and the grand decoherence of spacetime in potentiality to consistent-only futures constitutes the basis to the presage to quantum mechanical theory of inertia**: When we interfere with the motion-state and direction of a body 'over here', this will engender counterfactuals and consequent quantum erasure of non-consistent futures within spacetime in potentiality for matter located 'over there' and elsewhere along the event horizon. In this way, matter elsewhere 'knows' the inertial state of matter everywhere and vice-versa, and does so instantaneously, transtemporally, or non-locally. Hence, the resistance to change of the body 'over here' constitutes the resistance (the inertia) of the total spacetime in potentiality to counterfactually mediated quantum erasure of non-consistent convergent futures required by the sought change. Subsequent event horizon progressions and attendant AND-to-OR collapse cannot help but realise the consistent-only future, and accomplish it on a work function zero basis. The energy imposed to the change sought over the motion and direction of the body is per the resistance to that change from spacetime in potentiality. Indeed, this 'energy' comes from contingent future possibilities that underpin the application of putative force or 'work' to overcome inertia. Otherwise, so long as we do not disturb the inertial state of the body, this will demand minimal or no quantum erasure of convergent nested futures within spacetime in potentiality, and no energy or work or force to perpetuate: The result will be a work function zero perpetual uniform motion of the body as an inertial body. (See **2.35**). **Note that inertia is an extreme idealisation and does not really exist in nature**: Inertia gives way to gravitation, which also unfolds on the same basis as ideal inertia.

- **The presage to the quantum mechanical theory of the permanent magnet is also obtained via the quantum erasure of non-consistent nested futures, tandem counterfactual causality, or the grand decoherence of spacetime in potentiality.** The permanent magnet, in its relation to a given surface, projects a future-form spacetime 'chreode' into spacetime in potentiality up to the infinite future, at least in principle. This renders future possibilities within the domain of spacetime in potentiality consistent and conformal to its established relation to a pertinent surface. Subsequent event horizon progressions and attendant AND-to-OR collapse processes cannot help but recapitulate the consistent-only futures: namely the repetition and retention of the magnet's relation to the surface. All of this is accomplished on a work function zero basis, notwithstanding contingent future possibilities that might entail the manual detachment of the magnet from its surface vis-à-vis the resistance to that detachment issued from spacetime in potentiality. So long as we do not disturb the magnet-surface relation, this will demand minimal or no quantum erasure of convergent nested futures within spacetime in potentiality: The result will be a work function zero relation and perpetual attraction of the magnet to its surface without further input of energy. In short, the quantum mechanical solution to the permanent magnet involves the same-form solution as found in inertia and gravity.(**2.37**).

- **Problems pertaining to 'vacuum fluctuation' and electron spin about the atom are normalised per function of spatially static quantum mechanical waves.** Given that spacetime in potentiality, hence quantum mechanical waves, are spatially static, it follows that a spatially static potential vacuum cannot constitute a dynamic fluctuating vacuum. Thus, vacuum fluctuations are obviated by a static potentiality for a vacuum comprised of pure as-yet not-realised potentials for fluctuation.

A 'light ray' supposedly travelling through such a vacuum, itself a spatially static non-displacing state, has nothing to be perturbed or refracted by; explaining why light is *not* normally refracted or obscured by vacuum fluctuations.

Generically, **the electron spin about its atomic nucleus** does not appear to demand *any* energy or power input to realise. In truth, given that it constitutes a spatially static quantum mechanical wave within spacetime in potentiality, the electron is not literally spinning about the atomic nucleus as such, but only has potentiality for such a spin: These have not yet happened. Therefore, these cannot be realistically described as literally spinning in the usual dubious pseudo-realtime way. Given that the spin is not literally happening, the potential spin of the electron about its atomic nucleus is necessarily 'driven' by a work function zero process. You do not need an energy-work input to drive a thing that is not actually happening, much less in the pseudo-realtime way erroneously depicted. (See **2.36**).

Putative **energy from the vacuum** or from the potential sea of virtual particles, such as exemplified by Casimir effect, constitutes a work function zero outcome instigated by the pattern of futures imposed and projected into spacetime in potentiality by two attendant metal plates, and by work function zero counterfactual processes and grand decoherence that also attend inertia, magnetism, if not gravitation itself. (See **2.38**). The metal plates collapse, but the 'force' that makes this happen is per the work function zero grand decoherence of spacetime in potentiality that eliminates all other future configurations other than those futures that entail the collapse of the plates. Subsequent AND-to-OR collapse and event horizon progression cannot help but culminate into the collapse of the plates onto each other, *without* requiring energy or work to accomplish.

- **Reformation and pleonastic status of energy concepts and relations: the reformation and abstraction of the conservation principle (2.39)**: When we reframe energy concepts and relations to the intermediate spacetime model, potential energy is futurised and goes to spacetime in potentiality, while kinetic energy or work goes to the transition of one event horizon to a successor, and into the conversion or issuance of part of the future-potentialised energy within spacetime in potentiality into 'action or work'. However, the event horizon transition and 'energy-work issuance' is realised through gearless AND-to-OR wavefunction collapse or time... which is not itself driven by energy-work, no more than it is made to happen by inputs of energy (see **2.46** and **2.47**). In other words, AND-to-OR collapse and time is a work function zero process. But the ontological primacy of work function zero time requisite the apparent issuance of energy and work compels that state-change ultimately happens on the basis of work function zero time... *without* energy and work. This abstracts energy-work out of the ontology of physics: energy is rendered pleonastic. The consequence is **the recuperation of the conservation principle into abstract form without energy-work. The conservation principle is reconstituted on the basis of the conservation of memory (or the conservation of the past, which cannot subsequently change) and a consistency principle that enforces conformity of spacetime in potentiality to consistent-only possibilities vis-à-vis the unchangeable past.** The conservation principle thus recuperated also works on a work function zero basis, in that the conservation of the past and the consistency of the future through grand decoherence, entails a work function zero process.

- **Superlative form of information-matter dualism, decoupling theory, and mind-brain dualism derive from spacetime structure furnished by the intermediate model of spacetime and from spatially static quantum waves:** (See **2.40**, **2.41** and **2.42**). The dissociation of information from radiation and matter is essentially mirrored in spacetime structure: i.e., the past-future tripartite structure of intermediate model spacetime. The broken-discontinuous progression of the event horizon is synonymous with information-matter dualism, insofar as potential information (the future constituted as spacetime in potentiality) and past-information (from realised spacetime) are always dissociated from the OR-form realisations of resolved radiation and matter on or along the event horizon. The implied dualism also applies to mind and memory vis-à-vis the brain: The want or attempt to reduce mind and consciousness to matter and brains, necessarily on or along the event horizon, cannot circumvent the inevitable dissociation of mind and consciousness from the event horizon, given that memories and minds must necessarily dissociate from the materialised brain OR-form-rendered on the event horizon, per function of the dissociation of realised spacetime (the past and memory) and spacetime in potentiality (the future) from the event horizon and from the materialised brains rendered upon it. Mind-brain dualism inexorably follows.

Decoupling-iii: Quantum mechanical radiation and matter cannot act as spatial transports for information across 'space'...simply because these are spatially static non-displacing wave-states: Quantum mechanical radiation or de Broglie matter cannot move and cannot constitute the displacing carrier for information... because the quantum wave state for these is embedded in a spatially static spacetime in potentiality, and per the fact that quantum mechanical waves are provably spatially static. Hence, purportedly moving quantum waves cannot move at the speed of light, or event at quantum tunnelled 4.7 times faster-than-light in the experiments of Nimtz. To suggest otherwise is tantamount to the claim that the future can move... i.e. that the domain of spacetime in potentiality 'moves'. This is untenable, given that the future, or static spacetime in potentiality, cannot move and could not possibly carry even the most recent resolved events, much less transport the past constituted as, say, Mozart-40. This implies and augments the decoupling of information from quantum mechanical speed of light, matter and 'space' in Book-I, consistent with information-matter dualism and decoupling. (See **2.42**).

- **The quantum-relativistic incorporative spacetime model offers a better way of modelling spacetime curves and gravitation than didactic General Relativity models, while it furnishes critical precursors to prospective quantum gravity.** Per our presage to quantum gravity, spacetime curves are simply features of non-flat 'curvy' event horizons. The intermediate model can help describe the 'curve' and consequent distortion of lightcones and refractions of worldlines per function of stacks of 'curvy event horizons'. This also furnishes a key feature of quantum gravity per its in-built broken and discontinuous quantised formation of the 'stack', the succession of 'curvy' event horizons and the attendant warped AND-to-OR wavefunction collapse of spacetime in potentiality into gravitational outcomes: (See **2.43** to **2.44**).

- **Non-local counterfactual causality and grand decoherence of common futures within spacetime in potentiality constitutes the partial basis for 'how the curve tells matter how to move': a precursor to quantum gravity. Memory effect from the mnemonic domain of realised spacetime generates the gravitational grand decoherence that tells the event horizon how to 'curve'.** The process is essentially the same as holds for inertia and the permanent magnet, grasped in terms of counterfactuals, the erasure of non-consistent convergent nested futures, hence the grand decoherence of a common unifying spacetime in potentiality for the gravitating body *and* the gravitationally influenced body. Per implications from the curve of an initial event horizon, all subsequent same-curve potential event horizons comprising a 'stack' of such, all projected into to the future, cannot help but refract or deflect otherwise inertial bodies and their worldlines into in-falling curving worldlines vis-à-vis gravitating or event horizon-curving bodies, with consequent gravitational acceleration, in-fall, and orbital motion... all on a work function zero basis, or without requirement of pleonastic energy-power inputs to bring into effect. The attempt to alter trajectories or break from orbit will be resisted by the domain of spacetime in potentiality (as is also true in the case of ideal inertia), requiring inputs of pleonastic energy-power to realise, at which point the work function will be non-zero. However, it is mnemonic realised spacetime that constitutes the gravitating 'mass', recuperated as memory. It is the sum of attendant abstract memory that decoheres the future into consistency with that memory and, through gravitational grand decoherence, affects a curvy event horizon. And it is the replication of the same-curvy event horizon throughout spacetime in potentiality via gravitational grand decoherence that then describes future gravitational relations that subsequently AND-to-OR collapse into realised gravitational relations. (See **2.45**).

- **The universe is not a computer**: (See **2.46**). The quantum erasure of nested futures cannot be driven and affected by any hidden iterational, mechanical or clockable 'computational' process. If this were so...

  - **Quantum erasure of nested futures and grand decoherence to the infinite future would crash into an insurmountable non-terminating problem, and the restructuring of spacetime in potentiality could not be affected**. The event horizon and time could not advance, and both inertial and gravitational phenomena could not be realised.

  - Insofar as requisite grand decoherence derives from the process of quantum erasure of nested futures, and insofar as this in turn requires ontologically primary AND-to-OR collapse (time) from which counterfactuals are generated, and insofar as AND-to-OR wavefunction collapse and time must itself be driven by an ineffable gear-less process that defies mechanism, algorithm, 'computation' and clockability, it follows that the universe is not a computational device, much less a simulation generated by some hidden computational process behind time or behind grand decoherence to infinity.

  - If AND-to-OR collapse and time were secretly driven by gears or a clockable process, the gears would itself require its own preceding AND-to-OR wavefunction collapse and gears... and this in turn would require a preceding third wavefunction collapse and gears... and so on, *ad infinitum*. Consequently, no final wavefunction collapse could transpire in our universe. The only way out of this contradiction is ineffable gear-less and non-clockable drive behind wavefunction collapse and time. Consequently, the universe cannot be computationally produced, let alone constitute a computer.

- **Mind and consciousness derive from an ineffable non-structured causal process: the same causal process that attends the non-computational universe and gear-less work function zero grand decoherence, AND-to-OR collapse and time, and gravitation.** Since causality behind all things is non-clockable and non-computational, the causality that resolves mind, consciousness and the apprehension of infinities must also be non-clockable, non-computational and gearless. Hence, the revival of Cartesian dualism. (See **2.47**).

BOOK-II: CONCLUSIONS & INNOVATIONS

The following constitutes a list of the key discoveries and innovations presented in Book-II: Thus, from Part-I...

- AND-form logic and **quantum mechanical waves constitute the ontic reality of the future**, synonymous with the future-form of growing block spacetime and its domain of spacetime in potentiality.

- **Quantum-relativistic incorporative intermediate model spacetime** with a tripartite structure comprised of two domains and a demarcating event horizon: namely, spacetime in potentiality, a realised spacetime, with both partitioned by the event horizon 'cut'. The model constitutes a growing block approach to spacetime, to be enhanced and superseded by Book-IV.

- **Primitive unification of quantum and relativistic descriptions of nature per the intermediate model spacetime**: obtained per the tripartite structure of the growing block intermediate model: the integration of quantum mechanical future-domain spacetime in potentiality (the ontically real future) with the classical-relativistic domain of the past constituted as realised spacetime, constituting a growing block system. The primitive unification of physics is essentially the mutual integration of ontically real future with the abstract past.

- **Time, AND-to-OR collapse, and the progression of the event horizon, are all synonymous**. The progression of the event horizon 'upward' the spacetime system is synonymous with AND-to-OR wavefunction collapse and time itself. Time is not a 'fourth dimension'. Time *is* AND-to-OR universal wavefunction collapse: i.e. the perpetual decay of spacetime in potentiality and the said advancement of the event horizon.

- **The firewall principle**: The future cannot transpire before the past. Thus, causality demands the existence of an as-yet not-happened 'spacetime in potentiality'. Moreover the, **putative principle of conservation and the principle of causality are intertwined and inseparable, constituting the *firewall principle*, which wards against 'signals from the future'**. The firewall principle justifies the growing block universe capitulated in the intermediate model spacetime, and justifies the tripartite structure of intermediate spacetime.

- **Heisenberg's uncertainty follows from the structure of spacetime and causality**. When two attributes of a system are non-commutative, the resolution of one attribute must necessarily be deferred to AND-form spacetime in potentiality, *to the future*: with no OR-form characteristics, unless we invite 'signals from the future'. This is implicit from the very structure of spacetime. To reiterate, the deferred attribute is embedded as an as-yet not realised range of future possibilities within spacetime in potentiality. As such, its uncertainty is objective.

- **The sum over all potential worldlines** is a generalisation of Feynman's 'sum over all paths' approach to the intermediate model spacetime and its domain of spacetime in potentiality (the future), the principal amplitude of which furnishes default teleology to the growing block universe.

Part-II asserted the following discoveries and innovations.

- **John Wheeler's delayed choice necessarily implies the nested futures view of the quantum mechanical wave, of the wavefunction, and of spacetime in potentiality**. Therein, any given quantum mechanical wave and its immanent possibilities in AND-form must also nests subsequent future possibilities, possibility -aves and wavefunctions further removed in the future. In short, a wavefunction does not simply contain immanent possibilities but also subsequent wavefunctions attached to each immanent future. This nested futures structure is implicit to Wheeler' delayed choice, and it is generalised to our intermediate model spacetime; succinctly, to the domain of spacetime in potentiality.

- **The three-fold taxonomy of nested futures:** These are the *divergent*, *entangled*, and *convergent* categories of nested futures. The **convergent category nested futures remain definitive of spacetime in potentiality and is ontically real**: The other categories are either special cases (entangled) or extreme idealisations (divergent).

- **Meaningful non-local counterfactual causality and erasure of non-consistent futures** arises within the framework of convergent category nested futures; culminating into the **grand decoherence of the whole of spacetime in potentiality**; culminating into the **revival of Mach's principle** recuperated in quantum mechanical terms.

- **Partial-perpetual wavefunction collapse and synonymous succession of the event horizon, as opposed to the single one step one-time comprehensive collapse:** Partial-perpetual wavefunction collapse is rendered explicit by the nested futures perspective: Spacetime in potentiality cannot be totally converted into OR-form realised spacetime: The event horizon cannot leap to the infinitely removed future in one step. The following also implicate partial-perpetual collapse (see below):

- **Closed-choice quantum systems**: Most laboratory quantum mechanical systems approximate to closed-choice quantum systems. e.g., the generic two-slit experiment, and the EPR experiment. These lack explicit delayed choice elements and contingencies and appear to allow and test only for immanent possibilities to the exclusion or obviation of nested futures.

- **Open-choice quantum systems**: Ultimately, all systems (the whole of spacetime in potentiality) constitute open-choice nested futures systems per inherent delayed choice elements or contingencies. Therefore, the domain of spacetime in potentiality (the total future) cannot collapse comprehensively (to closure) but must always decay partially and perpetually.

- **Apparent closed-choice systems are always embedded in larger open-choice systems** within spacetime in potentiality. Ultimately, closed-choice systems are idealisations: there are no such systems.

- **Quantum mechanical waves are spatially static**. Quantum waves do not displace across 'space'. Why?
  - **Retrospective after-the-fact observation of putative 'moving' quantum waves**: i.e., the only form of observation of quantum waves possible. Hence, we observe successions of events organised in time, *always* as the past, and never unfolding in 'space' (save by assumption), and we cannot justify the notion of motion on the basis of post-fact inference (except by assumption).
  - **Putative observation of displacing quantum wave in pseudo-realtime is impossible**: The generic pseudo-realtime assumption of what a quantum wave undertakes in time implicitly harks to the notion of 'absolute time', which was debunked in Einstein's Special Relativity: That is, the pseudo-realtime description of the 'moving wave' describes the

whole process as if it is unfolding in a common present moment (absolute time) under continuous realtime observation (not physically possible). Indeed, even the always-retrospective attempt to observe the 'moving wave' cannot help but disturb the quantum wave, destroying the very thing we seek to observe. To reiterate, we never directly observe the quantum wave. Thus, we cannot empirically justify our *assumption* that it is 'moving' or undergoing time-undulation in the usual erroneous pseudo-realtime form... save by pure assumption.

- **A putative spatially displacing quantum mechanical wave would need to displace across a future-potential as-yet not-realised 'space'**, i.e., Hilbert space: Obviously, there can be no displacement across a potentiality 'space' that has not yet formed: There is nothing to move through.

- From Wheeler's delayed choice approach, when we reframe 'moving quantum waves' in the nested futures framework, we crash into the absurdity that the **nested futures embedded in a hedging quantum wave must displace across their own unique as-yet not-realised spatial intervals**, *before* these have obtained their said spatial intervals, before even the hedging wave has reached the delayed choice switch or the intervals beyond. This is impossible: The nested waves are thus static pure possibilities. Yet, the notion of the 'moving hedging wave' is also a pure assumption, as is obvious from the preceding points. Hence, the notion of 'spatially displacing quantum wave' unravels.

- **Anticipated reform or replacement of Schrödinger wave equation**. A future version of the Schrödinger wave equation, or its complete replacement, must incorporate...

  - **wavefunctions with nested futures** that elaborate to infinite future, versus wavefunctions with only immanent futures.

  - **partial-perpetual wavefunction collapse**, as opposed to one-time total wavefunction collapse.

  - **static quantum waves,** as opposed to waves that undergo spatial displacement and/or time-undulation.

  - **the fact that there is no time-evolution or time-undulation of the wave: The time aspect is replaced by the perpetual AND-to-OR collapse or block-decay of a purely static version of the Schrödinger wave** expressed as a spatially static non undulating superpose of immanent *and* nested futures elaborated to infinity. In short, **the Schrödinger wave must transform into the spatially static non-undulating spacetime in potentiality**... simply because the future neither 'moves' nor 'undulates'.

  - **The growing block intermediate spacetime model is effectively the replacement of the Schrödinger equation** and incorporates all the above, short of de-spatialisation; succeeded by the de-spatialised version of the model by Book-IV.

- **Time is not a *bloss* subjective notion but ontically real**. Why?

  - The process of AND-to-OR conversion or wavefunction collapse *is* time itself: i.e., the 'coming into being' of realised events out of the decay of potential future states.

  - Time has an 'arrow' or directionality per facts postulated in Book-I:

  - Increase in the entropy of future-potential information per function of partial perpetual AND-to-OR collapse of spacetime in potentiality and the progression of the event horizon.

  - Time's 'arrow' always proceeds from AND-to-OR, never OR-to-AND de-collapse or time-reversal. This AND-to-OR directionality is the absolute direction of 'coming into being' and *time*.

  - **Spacetime in potentiality incorporates into and constitutes the *monoblock* 'origins' of the universe**, with or without the Big Bang theory. In short, the universe does not emerge out of the remote past: Instead, the universe decays out of the total future... out of spacetime in potentiality incorporated as part of the monoblock, with or without Big Bang.... And time proceeds from potentiality to actuality...from future to past... *not* from past to future.

The following constitutes a summation of discoveries and innovations from Part-III:

- **Non-local counterfactual causality, the elimination of non-consistent futures, hence grand decoherence within convergent nested futures, leads to the revival of Mach's principle and constitutes the first presage into the quantum mechanical theory of idealised inertia**. Wavefunction collapse in one frame on or along the event horizon will imply the alteration of nested future possibilities attached to a distant frame on the same, implying instantaneous mutual decoherence and collapse of their convergent common-futures within spacetime in potentiality into mutual consistency: This revives Mach's principle, leading to inertia per non-local, instantaneous transtemporal counterfactual causality through which matter 'over here' instantaneously *counterfactually* 'knows' the state of matter 'over there', and vice versa, via the decoherence of their common future to mutual consistency. The counterfactual processing of spacetime in potentiality to obtain Mach's principle, and the perpetual motion of matter in its inertial state, is always obtained on a work function zero basis: i.e., inertia *does not require* energy or work inputs to manifest: i.e., grand decoherence does not require energy-work inputs to affect.

- **The quantum mechanical theory of the permanent magnet.** The magnet-surface relation is projected into spacetime in potentiality as a 'chreode', rendering future-form spacetime in potentiality consistent and conformal *only* to the established magnet-surface relation. This is affected to future-infinity. Subsequent event horizon progressions and AND-to-OR collapse cannot help but re-institute the established magnet-surface relation in perpetuity, without requiring energy or work to accomplish. This is achieved through counterfactuals and quantum erasure of inconsistent futures within spacetime in potentiality: a work function zero process, both in the grand decoherence and in the energy-work requirements to keep the magnet attached to its surface in effective perpetuity.

- **Problems of 'vacuum fluctuation' and particle atomic spin are normalised through spatially static quantum waves**. Spacetime in potentiality, hence quantum waves, are spatially static: so is the synonymous 'fluctuating' vacuum: A 'light ray' travelling through it has nothing to be perturbed or refracted by on account of the fact that the vacuum is not a dynamic froth, but a static of low probability possibilities for potential flux. In the case of the electron's orbit, the electron's quantum wave about the atomic nucleus is spatially static: the electron is *not* literally spinning about the atomic nucleus, much less accomplishing this at the pseudo-realtime speed of light, or at any speed.

  Casimir effects are work function zero processes that arise from similar counterfactual-mediated nested futures projections into spacetime in potentiality. The greater potential 'energy' from the vacuum outside the two metal plates 'push' the plates together. But it is the spatially static spacetime in potentiality and its structure that makes this happen, without work or 'energy'... by erasing the non-consisted nested future alternatives from spacetime in potentiality: per an erasure and grand decoherence process that unfolds as a work function zero process that subsequently collapses the plates.

- **Reformation and pleonastic status of energy concepts and relations: the reformation and abstraction of conservation principles**: Potential energy subsumes to spacetime in potentiality and kinetic energy issues from the event horizon transition. But the ontological primacy of work function zero AND-to-OR collapse implies that the energy concepts and relations abstract out of physics per function of the fact that, ultimately, 'action' or state-change is mediated by AND-to-OR collapse itself... by a non-energy based process. The consequent **recuperation of the conservation principle** on the basis of the conservation of memory (or the conservation of the past, which cannot subsequently be changed or erased) combined to a consistency principle that enforces conformity of spacetime in potentiality to consistent-only possibilities vis-à-vis that past, is also enforced on work function zero basis, through work function zero grand decoherence of the total future to future-infinity.

- **A deeper form of information-matter dualism on account of the broken-discontinuous progression of the event horizon and the past-future duality per the tripartite structure of intermediate spacetime.** In short, information-matter dualism reflects spacetime structure and process, with implied information-dissociation of the past (realised spacetime, or memory), which cannot obviously be 'carried' in or by the future (spacetime in potentiality), even less by their event horizon 'cut'.

- **Decoupling theory is augmented by spatially static quantum mechanical waves.** Given that both the quantum wave and its 'space' are static, there is nothing for the wave to displace across, much less 'carry' information across. This implies superlative understanding of decoupling, but from spatially static quantum mechanical radiation, matter, and space in the experiments of Gunter Nimtz, consistent with information-matter dualism. The faster-than-light microwaves in the Nimtz experiments are also spatially static and non-moving and cannot transport Mozart-40 across a space through which they are not moving. We must look to the time-relation between succeeding events and observables formed out of the AND-to-OR decay of spatially static potentialities, and *not* to the putative 'speed' of the alleged M- 40'carrying' radiation.

- **The intermediate spacetime model offers a better didactic take on spacetime curves and gravitation than that of General Relativity, and presages prospective quantum gravity.** Distortions of lightcones and deflections of worldlines are affected by the gradients enforced on stacks of 'curvy' future-potential event horizons projected into spacetime in potentiality. The curve imposed on the event horizons are time dilation gradients. Quantised gravity is grasped as an aspect of the broken-discontinuous quantised organisation and progression of successive stacked 'curvy' event horizons, which are not connected contiguously or seamlessly but are broken-discontinuous: hence quantised.

- **Non-local counterfactual causality and the elimination of non-consistent futures, or grand decoherence, is the partial basis for how matter tells space (the event horizon) how to curve, and how the curve then tells *future* matter and radiation how to 'move'.** But the initial curve imposed on an initial event horizon, which then counterfactually decoheres spacetime in potentiality to conformity to obtain gravitation, is a function of the domain of realised spacetime: i.e., the 'gravitational mass' resides within realised spacetime, not in the future-form descriptions of the matter with which the mass (or **mc²**) is often associated. That mass is from the superpose of the sum of past states of the gravitating body, constituted as abstract memory. That is, gravity is ultimately a memory effect from the mnemonic domain of realised spacetime, of which the event horizon curve is a symptom. It is not the curve that causes gravity, but it is memory that enforces the curve... which then tells future matter and radiation how to happen or 'move'... through the process of work function zero grand decoherence, which then AND-to-OR collapses into OR-form outcomes pertinent to gravitation.

- **The universe is *not* a computer**: The character of counterfactual quantum erasure of nested futures and the processing of spacetime to infinity via grad decoherence cannot be brought about by any clockable iterational or 'computational' process: Such a process would crash into insurmountable infinities and non-terminating problems in the processing and reprocessing of spacetime in potentiality into consistent futures... to the infinitely removed future. Consequently, AND-to-OR wavefunction collapse (time) could not transpire and the attendant 'upward' broken-discontinuous progression of the event horizon could not unfold. The causality responsible for grand decoherence and AND-to-OR collapse and time, is a gear-less ineffable process, *not* a clockable computational one. Thus, the universe is *not* a computer, no more than it is a 'simulation' or 'illusion' concocted by some hidden computation.

- **Cartesian mind-brain dualism is validated by non-clockable causality and its non-computational universe**. Since succeeding brain-states on or along successive event horizons are brought about by AND-to-OR collapse, and since AND-to-OR collapse is driven by an ineffable non-clockable non-computational causality, what resolves the succession of brain states must also constitute a non-clockable non-computational process, ontically extant the brain-states resolved on the event horizons. Cartesian mind-brain dualism necessarily follows. Since AND-to-OR collapse and event horizon-progressions are realised on a work function zero basis in general, the change in the brain-state affected on succeeding event horizon, such as to constitute change in agency and consequence, is also brought about by work function zero process. Thus, **mind-brain causality and control is a work function zero process and consequently *cannot* violate the principle of conservation... of energy-matter... and its recuperation into the conservation of memory and consistent futures...** wherein Cartesian causality cannot be accused of altering the past nor impose non-consistent futures vis-à-vis the conserved past. This saves Cartesian causality and revives mind-brain dualism.

BOOK-III
CRITIQUE OF THE IMPLICIT AMNESIC UNIVERSE THESIS: APAGOGIC MNEMONIC ARGUMENTATION AND SOLUTIONS TO QUANTUM INDETERMINACY

CONTENTS

BOOK-III:
CRITIQUE OF THE IMPLICIT AMNESIC UNIVERSE THESIS:
APAGOGIC MNEMONIC ARGUMENTATION
AND SOLUTIONS TO QUANTUM INDETERMINACY

THE LONG INTRODUCTION TO BOOK-III

3.00: On the Implicit Amnesic Universe Thesis

Generic physics implicitly assumes an amnesic physical order; one without memory. The past is not assigned any stand-alone independent physicality vis-à-vis the present; memory is not placed at an abstract remove from the relative present-continuous state of putative space, matter and radiation resolved on the event horizon, even while the implicit future *is* tacitly accorded reality through the admission of quantum mechanical AND-form descriptions of physical systems. **If there is any concession to memory, it is only as material trace-state; wherein memory or information must be embodied materially; never abstracted or independent of matter**, exemplified by, say, an imprint on a mould; the DNA-based genetic material; or a photographic image-residue of an originary. The originary vis-à-vis the trace is typically assumed erased, while the residual facsimile, as materially embodied trace-state memory, is **fated to** the vagaries of time, and must finally succumb to entropy. Consequently, **memory retention and carry-over, presumed to be never permanent and always temporary, always constituted as a moving present without any independent abstract past, will depend solely on whether material memory trace-states can persist or jump from one material carrying corpus to another, from one event horizon to the next, with final inexorable total erasure on the supposed impossibility of matter-independent or space-independent dissociation or abstraction of information into memory. Hence, the implicit Amnesic Universe thesis**, although it would be just as well to called it the Alzheimer's Universe Thesis.

However, even cursory examination of our growing block quantum-relativistic incorporative intermediate model spacetime from Book-II (and the Bergson-Whitehead amalgam that will succeed it in Book-IV) points to the contrary. The two main domains of tripartite-structured intermediate spacetime are constituted as abstract and dissociated domains vis-à-vis putative 'matter in space' confined to the event horizon. **It is the aim of Book-III to advance the role played by the domain of realised spacetime, grasped as the mnemonic domain of abstract memory; dissociated and extant vis-à-vis event horizon-restricted 'matter in space'**. Indeed, by Book-IV, de-spatialisation alone will establish abstract memory as fundamental to *any* future physics, in the same sense that the ontically real abstract future, conceived as spacetime in potentiality, turns out to be fundamental to the same.

Physics has long hinted at the real possibility of abstract memory, counter to the implicit amnesic universe contention. For example, gravity fields, unlike electromagnetic emissions, cannot be blocked, cancelled, or erased by any known process. If an expanding gravity field can be legitimately thought of as the retention, in whatever form, of all that has happened to its putative generating matter (indeed, what better generic physical candidate for abstract memory, given that the field cannot be blocked, cut, or cancelled by any known means) it follows that such a field *must* also constitute an ineradicable *immaterial* 'memory field', its gravitational effects notwithstanding. Hence, **the gravity field might rightly be treated as an implicit repository of ineradicable historical information or memory**, although gravity fields are never explicitly treated in this fashion, and only their gravity effects are considered; further obscured by their descriptions as 'force fields' or as 'spacetime curves', neither of which readily lend to the idea of an abstract informational repository for memory. Yet, **our own preliminary attempt to model gravity from within the tripartite structure of spacetime in Book-II (2.45) asserted gravity to be a memory effect, issued from the domain of realised spacetime**, and identified gravity with abstract memory, *before* we must reach the same conclusion per de-spatialisation in prospective Book-IV.

The point is, simply, that a realistic basis for abstract memory, dissociated and extant vis-à-vis putative matter, or at least as extant as is the gravity field from matter, and its plausible informational content vis-à-vis its source, is nothing new. It follows that the implicit amnesic universe can be legitimately doubted before subsequent development furnished by our Alternative Realist approach (Alt-R). Book-III will seek to establish the first foray into the **Mnemonic Universe thesis** by means of a series of physics-grounded *apagogic mnemonic arguments*: The pure physics argument, must await Book-IV.

3.01: Apagogic Mnemonic Arguments, Abstract Memory & the Viability of Quantum Mechanics

Apagogical arguments are semantic arguments that furnish the absurdity or impossibility of ultimate implications from a given case or assertion. The series of **apagogic mnemonic arguments in Book-III are posited against the specific anticipated claim that both information and memory have no ontic reality due to information and memory nihilism**. The first task of **Book-III will be to anticipate a series of potentially devastating nihilistic arguments against the ontology of information and against abstract memory**: nihilistic arguments based on a combination of perpetual memory erasure (i.e., the erasure hypothesis) combined to the inherent nihilism of the Manyworlds interpretation of quantum mechanics (MWI), both supplemented by other nihilistic approaches.

The anticipated nihilistic **erasure hypothesis against memory asserts that the past state of a system must perpetually erase from moment to succeeding moment through the operation of a cosmic 'Alzheimer's process'**, so to speak. To this, we will later add the wholly distinct and remarkable *quantum eraser*, developed by Yoo-Ho Kim and associates[73]: i.e., the claim that the past, which ought not to exist at all if the amnesic universe abides, can be directly and mechanically erased through a series of operations carried out in the relative present.

Contrary to both claims, Book-III will argue that the past, or memories, remain ineradicable. This fact is obvious even from gravity fields: *the* primary candidate-phenomena for memory, and empirically vindicating for abstract memory: (see **3.00**).

One of our central contentions in favour of abstract memory consist in the assertion that, **without the dissociated retention of critical information, without the reality of abstract memory retained at a remove from manifest radiation and matter, the basic experimental findings and conclusions of quantum mechanics could not be obtained**. Hence, without abstract memory, quantum mechanics as we know it would be rendered impossible: Hence, the veracity of quantum mechanics and physics forces the necessity and reality of abstract memory.

In our attempt to preserve the finding that, without abstract memory there can be no quantum mechanics, **Book-III will seek to overcome the two principal nihilistic assertions against information and memory through a series of apagogic mnemonic arguments**; expanding on the material against mereological nihilism first presented in Book-I, **1.03** and **1.04**. The project will consist of two main cases: first, the total apagogic character of erasure hypothesis, and second, the falsification of MWI itself, both of which are indispensable to prospective information and memory nihilism.

While MWI designs for complete nihilism of information and memory by virtue of its inherent structure, we will show that the alternative histories into which the root wavefunction supposedly split per Manyworlds must all subsequently culminate into the same global outcome or history: the splits do not constitute alternative histories; exposing MWI to be superfluous.

The nested futures view developed in Book-II will be inadvertently utilised in **the quantum eraser approach of Yoo-Ho Kim**, as will John Wheeler's delayed choice approach, which underpins the quantum eraser. Applying it, we will argue that the putative eraser cannot support the notional erasure of the past, but can only support the erasure of a future as-yet not-happened: one whose resolution has been physically delayed or suspended. To this end, we will develop the **holed past postulate**: **wherein aspects of the past remain in quantum suspension and isolation for potential delayed time-intervals, such as to constitute a 'holed past'. Subsequent intercession into the suspended past finally brings it out of isolation and ends the suspension of its possibilities, belatedly resolving the 'hole' into appropriate OR-form outcomes, while the alternative possibilities are wiped out. But this does not involve erasure of any resolved past**. In short, we cannot erase the past, as Yoo-Ho Kim claimed: By bringing out of isolation the delayed choice suspended past, we merely complete the past *without* erasing memory.

The undeniable conclusion from all this will be that information has real ontology: It cannot be obviated by any anticipated nihilistic hypothesis or procedure, and that attendant abstract memory is also ontically real; comprising the preface to abstract memory directly and seamlessly inexorable from anticipated de-spatialisation in Book-IV.

3.02: The Stochastic-Coherence Paradox, Information & Abstract Memory

The facile consequence of quantum indeterminism, or what we have termed *the stochastic-coherence paradox*, is the question and problem of the emergence of informational coherence, consistency, and pattern-persistence, despite the operation of apparent informationally inimical and scrambling quantum indeterminacy, and, with it, the question of 'impossible' order and structure amid presumed quantum mechanical chaos. Simply put, the stochastic coherence paradox is the problem of order in the context of apparent quantum mechanical indeterminate disorder.

If quantum indeterminism is inimical to the very viability of information, then it ought to be inimical to the formation and retention of information in *any* form, including as abstract memory. Indeed, any solution to the problem of memory must develop out of the solution to the problem posed by quantum indeterminism to the ontology of information and memory. Thus, the solution to the stochastic coherence problem will consolidate our apagogic mnemonic argument in favour of abstract memory and against information and mnemonic nihilism.

Note that the use of the term 'order' in the above and in succeeding material should not necessarily suggest any special arrangement or 'planned' world, much less one consciously designed or created... although it does not necessarily exclude design or plan either. The use of the term applies in the sense that quantum indeterminacy must be problematic to the continuation and carry-over, across time, of *any* initial set of conditions and arrangements nominated as the 'order'. For example, a pack of cards typically gets to us in a generic sequence order. It might as well get to us in a jumbled sequence. Jumbled or otherwise, subject to subsequent random shuffling, the initial sequence is unlikely to survive, whether the initial sequence was jumbled or otherwise. However, what if, despite subsequent shuffling and re-shuffling, the pack retained whatever initial sequence-order or the initial jumble in which it was found? Such a feat is impossible for our pack of cards. Yet, the world experienced and studied by physics, replete with clear persistence of pattern and structure across time, demonstrates the undeniable fact of just such an impossible feat: the retention of default order in the face of ultimate foundational perpetual quantum indeterministic reshuffling, notwithstanding entropy. Hence, the stochastic coherence paradox: the problem of how order and pattern-persistence can abide in the face of ineliminable quantum noise.

Despite quantum indeterminacy and chaos, there really *is* coherency, consistency, and persistence of information across nature; with implied real ontology of memory, insofar as the very persistence of information over time *must* imply memory, even if tacitly.

---

[73] Kim, Yoon-Ho; R. Yu; S. P. Kulik; Y. H. Shih; Marlan Scully(2000). "A Delayed "Choice" Quantum Eraser". *Physical Review Letters*. 84(1):1–5.

QUANTUM MECHANICS AND MIND

Given the reality of just such coherence, consistency, and persistence, it is tempting to conclude that the claim for quantum indeterminacy must be in error: that all things are secretly deterministically orchestrated despite appearing to be indeterminate. But **Book-III will conclusively show that randomness and quantum indeterminacy are ontically and objectively real.**

Book-III will show that, relative to any random indeterminate seeming process, it is possible to imagine and generate very large numbers of unique deterministically operating orchestrating scripts, or 'hidden gears', that secretly orchestrate facile indeterminacy and randomness. Yet, none of the scripts so generated will be so special as to exclude all others so as to constitute the one-and-only script. All the scripts will be equally apt to the task, and there is no structural or qualitative way of deciding which of the scripts get to orchestrate seeming surface indeterminacy and randomness. The only resort left is to randomly select from among the large number of scripts. Moreover, any attempt to establish a deterministic selective sortology from among the scripts will crash into a non-terminating combinatorial explosion to infinity. In short, there is simply no method or process for obviating ultimate randomness and indeterminacy, even onto infinity, even when we imagine or devise secret deterministically orchestrating scripts behind random indeterminate processes. In other words, **there are no 'hidden gears' or variables to orchestrate general randomness or quantum indeterminacy. Thus, quantum indeterminacy is objectively real and ultimately ineliminable**.

The case for the ontic reality and objectivity of indeterminacy ought to constitute the apotheosis of informational nihilism: Since all is random, it ought to follow that all order, pattern, and structure (including the persistence of any initial condition, or even initial jumbles) be rendered illusory, culminating in ultimate cosmic mereological, informational and mnemonic nihilism. One expects that, **when we merge the objectivity of quantum indeterminacy to the erasure hypothesis, in turn co-joined to inherently nihilistic MWI, this ought to permanently secure informational and mnemonic nihilism.** Yet, the expectation will prove to be forlorn: The case against hidden gears and for the objective reality of randomness cannot culminate into nihilism: Book-III will discover that objective randomness pertains to the economics of nature and cannot lend itself to any nihilistic argument against order, information, or memory, save in limited cases, such as in specific applications of EPR.

*

It turns out that **a fundamental harmony exists between randomness or quantum indeterminacy versus order:** This mad seeming insight will constitute the key to the solution to the stochastic coherence paradox. This harmony was always evident, even at the basic level of the bell-curve distribution pattern. Therein, we discover that a random process, such as the roll of several unbiased six-sided die, is always bounded by finite possibilities per the fixed values on the die-facings and the number of dice involved, combined to the range and limit of the sums these can generate. In successive rolls of unbiased dice, randomness creates a state of natural bias in favour of the formation of a generic bell-shaped curve distribution. Indeed, the bell curve made manifest *is* the order, even though it is brought into evidence by random terms. We could even quantum mechanise the die, thus apply true quantum randomness superior to mere mechanical randomness. Yet, the results will be the same.

The idea here is that **no matter how random or quantum indeterminate each roll of the pertinent die, and no matter how causally unconnected these are in sequence or consecutively, the bell-curve distribution (or order) will inexorably emerge: a totally foreordained and fated outcome**. Indeed, the greater the number of sequential and parallel die rolls, or the greater the number of quantum indeterminate terms, the more rapidly will the bell curve distribution get to form up. This is achieved through a three-phase developmental process that transforms a system from an early noise distribution phase into a chaotic distribution phase, and finally into a signal distribution phase, with the foreordained global pattern formed out of *en masse* quantum indeterminate 'chaos'.

In essay **3.26**, and in the four essays following it, we will outline in detail the three-phase developmental process and the key reasons why quantum indeterminacy in general must culminate into a signal distribution in which pattern hitherto concealed must manifest to salience.

*

The development of the bell-curve instances a pre-quantum mechanical illustration of how order manifests *despite* randomness and 'chaos'. If we quantum mechanise the attendant die rolls such as to render the randomness truly quantum indeterminate, it would not make the slightest difference: the foreordained bell-curve pattern *will* manifest. But, to this, we must also incorporate the finding that quantum indeterminacy is not illusory but objectively real... and that there are no hidden gears behind indeterminacy.

The development of the Sierpinski triangle (a 'chaos game' or sieve) involves the operation of the same form of natural odds as found in the development of the bell-curve pattern. It constitutes another way of rendering salient the fact of the harmony between order verses randomness, even when the randomness involved is again quantum mechanised, and even when quantum indeterminacy is rightly judged ontically real and objective, in opposition to 'hidden gears'. As such, the Sierpinski triangle forms another exemplary framework for the solution to the stochastic coherence paradox.

Another example of **the unexpected harmony between order and randomness, one directly native to the world of quantum mechanics, is found in the electronic configuration of the atom itself**: Therein, an electron may assume any number of permissible and possible AND-to-OR resolutions about the atomic nucleus, always in a purely quantum indeterminate way, but always within the AND-form bounded probability regions (i.e., the perennial 'electronic cloud' or electronic 'shells') relative to other electrons in the same. There is a 95% chance that the OR-form of the electron will AND-to-OR resolve within the region specified by the said electronic cloud or probability region. In other words, the objectively indeterminate way an electron comes to realise its OR-form resolution about the nucleus of the atom is subject to strict boundary conditions that do not themselves derive from anything chancy: These boundary conditions are born out of strict mathematical and natural necessity that pertain to the organisation of standing waves about the atomic

271

nucleus, *not* to the subsumed objective randomness and indeterminacy. As such, these abstract boundary states cast the natural odds in terms of their geometry or form, and compel the objective quantum indeterminate succession of electron-resolutions about the atom to collectively render salient the said geometry or form of the electronic configuration of the atom. Thus, as holds in the formation of the bell curve distribution by the accumulation of quantum mechanised die-roll outcomes, and as holds in the development of the Sierpinski triangle by similar means, the greater the number of quantum indeterminate resolutions of the electron about the atom, the more salient the geometry of the electronic configuration of the atom.

In short, structures and boundary conditions, or *bounded probability states,* conceived as 'instantaneous wholes' that derive from strict necessity, get to funnel otherwise a-causal objectively quantum indeterminate terms in such a way that, no matter how objectively random and chancy, these cannot help but collectively reveal the global order and pattern-structure implied by the operating background abstract bounding probability state or 'instantaneous whole'.

In Part-II, and in all the examples so far alluded, **Book-III will posit that order *must* manifest despite objective quantum indeterminate processes, simply because objective quantum indeterminate processes are *always* bound to implicit abstract structures that are themselves born out of necessity, if not out of strict necessity. These subsequently cast the natural odds in their own terms, rendering these salient through quantum indeterminate terms, no matter how objectively quantum indeterminate the terms. This insight is central to the solution to the stochastic coherence problem and paradox.**

*

One consequence of the solution to the problem of quantum indeterminacy will be the demise of the Manyworlds interpretation (MWI). Indeed, MWI emerged as an attempt to wrest a solution to quantum indeterminacy by imagining that *all* the possibilities in the initial AND-form state are realised... albeit as compartmentalised histories and universes: to the effect that, at each round of AND-to-OR collapse, the universe splits into as many histories as there are AND-form possibilities. This appears to forgo the need to explain how order emerges out of indeterminacy by forgoing indeterminacy itself: There is no indeterminacy in MWI, since *all* the possibilities are ordained and fated to become OR-form resolved into histories. And it is this unique feature of MWI that lends it to its anticipated co-joining to memory erasure hypothesis in the dismal promise of total informational and mnemonic nihilism. **The invalidation of MWI will undermine nihilistic memory erasure totally dependent on MWI.** Hence, the anticipated nihilistic argumentation against the ontology of information and memory will fall apart, while the case for the ontology of information and abstract memory will abide... courtesy of the solution to, and the ontic objective reality of, quantum indeterminism.

3.03: Quantum Indeterminism & Abstract Memory

The solution to the stochastic coherence paradox in terms of the ontic objective reality of randomness and quantum indeterminacy in Part-II will incorporate with the welter of apagogic mnemonic arguments in proof of abstract memory furnished in Part-I, while our anticipated solution to the stochastic coherence paradox will independently reform the very conception and definition of physical information first grappled in Book-I, Part-I.

In reforming the conception and definition of information, the solution to the stochastic coherence paradox will also overturn our naive assumptions about memory itself: **It turns out that memory is *not* comprised of retained OR-form information, as was pragmatically supposed throughout the Preliminary, in much of Book-I, and even in Book-II. Indeed, OR-form information is wholly restricted to the event horizon. Memory retained within the domain of realised spacetime cannot assume any OR-form character, no more than does future potentiality within spacetime in potentiality.**

In what form is abstract information in nature retained as memory? It turns out that, in apparent total contradiction to our running expectations, abstract memory retained within realised spacetime is simply the original wavefunction 'cropped' of its former nested futures... and thus distinct from future-form wavefunctions *with* nested futures. Thus, we can distinguish between abstract wavefunction states retained as memories within realised spacetime, but with their nested futures removed (the past) ... versus abstract wavefunction states *with* their nested futures intact, but embedded as future-potentials within the spacetime in potentiality (the future).

The pertinent wavefunctions behave differently: For example, 'un-charged', or nested *wavefunction in potentiality* embedded within spacetime in potentiality possess their nested futures and are subject to counterfactuals-driven grand decoherence (see Book-II, **2.18** and **2.34**). On the other hand, 'charged', or de-nested *wavefunctions in memory*, retained as abstract memory within realised spacetime, are stripped of their nested futures by a preceding process of grand decoherence, and are then relegated to the domain of realised spacetime. Their affectation on subsequent events is weaken per the inverse square law of time. Thus, 'charged' bounded probability states, hence **memories of past states... hence wavefunctions in memory... are distinguished from future-form bounded probability states... from wavefunctions in potentiality**, and relegated to their separate spacetime domains.

3.04: Quantum Indeterminism & Mind-Brain Dualism

The greatest challenge posed to prospective revived Cartesian mind and causality is *not* the ideology of materialism, but quantum indeterminacy: **Quantum indeterminism appears inimical to mind-brain interaction and control.** Conceived as 'chaos', quantum indeterminacy would appear to permanently undermine any prospect of mind's control over its brain: in the same sense that we cannot control or orchestrate spin-outcomes in Einstein Podolsky Rosen experiments (EPR).

Book-III will posit that **Cartesian mind engenders unitary control over the brain by imposing upon the brain confined to the event horizon either instantaneous wholes composed of wavefunctions in memory 'retrieved' from the domain of realised spacetime (the past), or by imposing instantaneous wholes composed of wavefunctions in potentiality, mind-selected out of**

spacetime in potentiality (the future). In both, causality is accomplished through work function zero processes without violation of now-reformed conservation principle (see Book-II **2.39**). Thus imposed, **the wavefunction in memory (or potentiality) will subsequently funnel** *en masse* **quantum indeterminate processes into a now-fated global outcome across the brain, in fulfilment of mind's goals, without contravention by quantum indeterminacy**. In short, our Alt-R solution to quantum indeterminacy and the stochastic coherence paradox will solve the principal problem posed to mind-brain dualism and control by **deflating the problem of quantum indeterminacy versus mind-brain control into a non-problem**, while also undermining quantum mind theories that seek solutions to mind-brain control through consciousness-based orchestration of quantum indeterminacy in brains.

Of course, the detailed account of the unique conditions pertinent to brains and to the whole process of memory retrieval and novel affectation by mind, must await Book-V. Yet, our solutions the problems posed by quantum indeterminism, combined to ancillary solutions against informational and mnemonic nihilism, will scaffold for Cartesian causality and Cartesian revival.

BOOK-III PART-I:
APAGOGIC MNEMONIC ARGUMENTATION AND ABSTRACT MEMORY

3.05: In Search of the Mnemonic Universe Thesis

Memory erasure hypothesis is the first anticipatory devil's advocate nihilistic position against information and memory. The hypothesis asserts that information, the past, is perpetually erased as old events are succeeded by new events, per consequences from the incompatibility of OR-form information versus quantum mechanical AND-form states, combined to the putative information scrambling effects from quantum indeterminism. But, the first apagogic mnemonic argument to challenge nihilism will assert that abstract memory is indispensable to the very possibility of quantum mechanics: that the staple facts of quantum mechanics in the form we know these could not be obtained without abstract memory: hence, **apagogic mnemonic argument 1-a**.

Apagogic mnemonic argument 1-b will incorporate causality and the pertinent tripartite structure of intermediate spacetime developed in Book-II to augment the central assertion from **1-a.** Thereafter, **apagogic mnemonic argument 1-c** will expose memory erasure hypothesis by confronting it with **1-a** and **1-b**. We will then reframe these arguments in terms of the non-standard evaluation of the two-slit experiment from Preliminary, **0.14**. This will culminate into the **apagogic mnemonic argument 1-d**: which will expose the all-or-nothing assumption behind anticipated memory erasure hypothesis.

Thereafter, Part-I will anticipate a more powerful second iteration of memory erasure hypothesis, through the co-joining of the Manyworlds interpretation of quantum mechanics (MWI) with mereological nihilism from Book-I, **1.03**. Thus, **apagogic mnemonic argument 2-a** will hammer out this more potent nihilistic argument against information and memory.

However, subsequent **apagogic mnemonic argument 2-b** will falsify Manyworlds itself. Yet, **2-b** will not be the only argument undermining Manyworlds: A wholly distinct and independent falsification will arise from the solution to the stochastic coherence paradox in **apagogic mnemonic argument 4-a** (to be presented at the end of Part-II). Thus, by falsifying MWI, we will isolate the erasure hypothesis co-joined to it. The latter self-falsifies apagogically without support from MWI.

*

Proceeding directly from **2-b**, **apagogic mnemonic argument 2-c** will show that, even if one were to assume the veracity of Manyworlds and memory erasure, *both* would yet fail on strict semantic grounds: Therein, we will demonstrate that, apagogically, memory erasure hypothesis must undermine its own grounds, *if it is true*. On the other hand, **apagogic mnemonic argument 2-d** will undermine memory erasure co-joined to MWI by directly debunking MWI on the basis of the latter's superfluous status, given that MWI requires that all possibilities attendant any root AND-form wavefunction must be realised into alternative compartmentalised histories. Instead, in **2-d**, we will discover that all such histories must culminate into a same identical global outcome: a same global history, duplicated many times, if not infinitely many times. In this way, MWI will seld-invalidate.

Finally, **apagogic mnemonic argument 3-a** will constitute a critique of Yoo-Ho Kim's quantum eraser: an approach completely distinct from but complementary to memory erasure hypothesis. Kim's quantum eraser approach is a derivative of John Wheeler's delayed choice, but one that claims to physically erase the past. Our aim in **3-a** will be to marshal the nested futures and counterfactuals based grand decoherence from Book-II to cast permanent doubt on the quantum eraser via the **holed past postulate**.

APAGOGIC MNEMONIC ARGUMENTS: SECTION-I: ERASURE & MEMORY

3.06: Anticipating Memory-Erasure Hypothesis

For it to be self-consistent, materialism requires that information must be inseparably carried, transported, and embodied in putative radiation and matter in space. Yet, as was obvious in Book-I and Book-II, long-overlooked evidence leads to the opposite conclusion: Information cannot ultimately be stored in or be transported by radiation and matter: It must dissociate from radiation, matter, and putative 'space', with consequent information-matter dualism *and* attendant abstract memory. If materialism were to accept as true the physical dissociation of information from radiation and matter, materialism would unravel.

The reality that critical physical variables can be retained in abstract dualistic and dissociated form vis-à-vis matter is wholly incompatible with any conceivable materialist monist ontology. Thus, to abide, materialism must make recourse to memory erasure. Hence, anticipatory memory erasure hypothesis is indispensable to the viability of materialism.

Consider memory erasure hypothesis in the context of photography: Therein, per information-matter dualism, de-coupling theory and spacetime structure, our Alternative Realist contention asserts that information must dissociate from the 'transporting' quantum mechanical light radiation, at the point of origin, or at any point along the interim 'space' between the origin and the film-surface. It must subsequently become scrambled by quantum indeterminate resolution-processes at the film-surface, at the point and process of development.

Anticipated memory erasure asserts that, given the dissociation *and* the subsequent quantum scrambling, the image must be totally wiped out from the physical order. That is, while the dissociation of OR-form information from AND-form interims is true, and the scrambling of that information upon quantum indeterminate development ought also to be true, the information and memory ought to be wiped out, and *is* wiped out, without possibility of abstract retention as memory, and without possibility or ontology of information itself.

However, the subsequent temporal re-manifestation of information into the fully developed photographic image on the film surface, and its subsequent perdurance and persistence over time amid matter and space at post-development, could only constitute an improbable miracle, insofar as the otherwise erased information and memory must be reconstituted amidst matter; not just once, but repeatedly over time, despite its alleged total erasure vis-à-vis the AND-form interim that operates from origin to film surface, and by quantum indeterminate processes, subsequently.

Therein lies the problem with anticipated erasure hypothesis, at least in its first iteration: Erasure hypothesis must demand something even more radical and fantastical than the dualism of information and abstraction memory: Anticipated **erasure hypothesis in its first iteration cannot be considered a serious contender against information-matter dualism and abstract memory, given its ultimate reliance on the miraculous**.

If first-iteration erasure hypothesis cannot constitute a serious contender, it yet proves apagogically useful in rendering salient the obvious fact of the persistence of information: a critical ingredient in our first foray into apagogic mnemonic argumentation in support of the ontology of information and the reality of abstract memory.

The first apagogic mnemonic argument will assert that, if memory erasure hypothesis were to hold provably true, it would render quantum mechanics in the form we know it totally impossible and untenable. In other words, deny the reality of information-matter dualism or the dissociation of information, and deny the necessary corollary of its dissociation and retention as abstract memory, then, apagogically, we ought to also deny the very possibility of quantum mechanics.

3.07: Apagogic Mnemonic Argument 1-a:
Abstract Memory & the Viability of Quantum Mechanics

For quantum mechanics and its core conclusions to be tenable, key pertinent informational states, or their ultimate physical basis, *must* be retained and temporally persist in non-erasable form, despite provable dissociation of information from AND-form radiation, matter and space per AND-OR dualism and quantum indeterminism. **The very possibility and viability of quantum mechanics is fated to the dissociation and survival of information as abstract memory. Abstract memory saves information from erasure by apparent informationally inimical AND-form quantum mechanisation *and* subsequent information-scrambling quantum indeterminate processes. The dissociation into abstract memory permits the formation of physical-experimental outcomes from which we can infer the central conclusions of quantum mechanics: conclusions that would otherwise be impossible to obtain if erasure hypothesis held true.** This, at least, is the basic assertion of apagogic mnemonic argument 1-a.

In contrast to abstract memory, **anticipated memory erasure hypothesis must assert the contrary**: It must assert….

- **that the configuration of the barrier (a critical piece of information in the two-slit experiment) must be totally erased upon the enforcement of experimental isolation**. And…

- **that the restoration of the barrier to its original setting at the end of experimental isolation must unfold in a purely quantum indeterminate way**; with no guarantee that another barrier-setting will not resolve in place of the original. And…

- **even if the OR-form barrier setting could be mediated across the AND-form interim without erasure, it is expected to be scrambled into erasure by subsequent quantum indeterminate processes at the end of isolation**.

Thus, according to erasure hypothesis, the critical piece of information pertaining to the original barrier-setting must not and *cannot* be retained as memory: Indeed, **in some final sense, the original piece of information pertaining to the barrier did not really exist. It *never* existed**… *assuming* the erasure hypothesis abides.

Upon the assumption of memory erasure, certain paradoxes follow, even while these permit unusual forms of quantum mechanical outcomes and inferences…of the kind that do not cohere with really-existing quantum mechanics.

*

Throughout **argument 1-a**, and in all subsequent apagogic mnemonic arguments, the reader should keep in mind that, if it could ever be proven to be true, memory erasure would need to apply comprehensively… to *everything*… on an **all-or-nothing basis**: Therein resides its clearest apagogic invalidation: If memory erasure held true, we could not even report the purported fact of memory-erasure itself, given that such a feat would presuppose the retention of *some* pertinent information requisite to our communication of experimental results. In which case, memory erasure could not constitute an all-or-nothing erasure, but only a partial and contradictory erasure… or a convenient and dishonest one, such as to constitute a fallacious form of the liar's paradox.

*

As a piece of information, the status of the barrier must somehow survive the interim of experimental isolation, even if the barrier dissolved into a de Broglie quantum mechanical AND-form state upon the enforcement of experimental isolation, and became subject to quantum-indeterminate sortology at the end of isolation. At the initial point of the experiment, before isolation, the slits making up the barrier may be set to one of following OR-form alternatives:

- **left-slit open**
- **right-slit open**
- **both slits open**...and,
- **both slits closed**: (the setting often ignored, but no less a part of any future possibility-set).

Only **both slits open** can yield self-interference and the formation of characteristic interference patterns at the detector, establishing the reality of AND-form logic, the superposition principle, hence quantum mechanics itself. The question of whether information is erased upon its dissociation from radiation and matter is of direct relevance to the question of whether the status of the barrier survives the interim of isolation in the two-slit experiment. Simply put, if **both slit open** were subject to erasure upon isolation, and if the subsequent state of the barrier at the end of isolation was resolved in a purely quantum indeterminate way against all the other alternative barrier-settings on a 1-in-4 or 0.25 probability basis, we could never be certain of obtaining an outcome in line with *any* critical experimental initial condition or variable in the two-slit experiment or in any other experiment, just as is true in EPR experiments. Thus, we could never be certain of obtaining an interference pattern vis-à-vis an initial **both slits open** conditional. At best, we could only hope that the outcome will correspond to the initial condition we set up at the barrier. But this will transpire on a 0.25 probability basis instead of probability-1… assuming the physical order could retain independent memory of our original barrier-setting to begin with *and* restore this to the experiment outcome: *not possible* if all-or-nothing erasure abides.

Of course, experience informs us that the chance of getting the outcome, 'interference pattern', on the detector vis-à-vis **both slits open** is *always* a matter of probability equal to 1 (absolute certainty), *ceteris paribus*… and that, save by contingent intrusion of 'Wheeler's cat', the setting on the barrier does not change or quantum scramble to an alternative every time we run the experiment.

In short, if memory erasure hypothesis held true, we could never derive certain quantum mechanical facts, such as the staple fact of self-interference under the **both slits open** condition, whose persistence and perdurance is typically taken for granted. Instead, these would be rendered purely uncertain contingents; resolved in a purely probabilistic way; with no possibility of experimental control or certainty over any setting or outcome: just as is true in EPR.

In short, quantum mechanics as we know it would be rendered impossible.

*

How can we be certain that the barrier quantum mechanises upon isolation? Could it not remain OR-form throughout isolation? **Upon experimental physical isolation, the material of the barrier *must* dissolve into quantum mechanical de Broglie AND-form state. Recall that there are powerful causality-based and conservation principle-based reasons why this must happen: part of *the firewall principle* developed in Book-II (see essays 2.10 to 2.13); part of the *raison d'etre* of the quantum mechanical AND-form wave, grasped as the physical phenomenalisation of the future, and the future as ontically real; part of the growing block universe. The barrier must dissolve into an AND-form state upon isolation because, if it did not do so, this would lead to the violation of causality and the conservation principle by the assumption of 'signals from the future'.

Moreover, our intermediate spacetime model, also developed in Book-II, is justified by the same firewall principle. It forced upon us the reality that the barrier must indeed 'dissolve' into a quantum mechanical AND-form state of alternatives upon experimental isolation. **Isolation defers the barrier to the future, to spacetime in potentiality; wherein the same barrier must now become an AND-form state. The same put in another way: since the experimental outcome will happen in the future, it exists only as an AND-form potentiality, both for the initial condition set *and* for the alternatives, co-joined to possible contingent possibilities that might alter the future barrier setup (i.e., Wheeler's cat').**

At this juncture, we must either assume erasure or assume memory retention. In the latter case, the OR-form setting of the barrier, or whatever its final physical basis in memory, must dissociate from that barrier upon the enforcement of experimental isolation and quantum mechanisation. The **both slits open** state must be preserved and retained in abstract dissociated form in the interim of that isolation. It must finally be restored to the barrier in its original setting at the end of isolation per consequent AND-to-OR wavefunction collapse, via *en masse* quantum indeterminate terms, notwithstanding contingent possibilities or Wheeler's cat, who might tumble the experiment into a **left-slit open** set-up, against the assertion of memory.

*

To augment, utilising the tripartite structure of our intermediate spacetime model from Book-II, once isolation is enforced such as to constitute a quantum mechanical AND-form interval between the initiation of isolation at one event horizon (EH-1) and the end of isolation at a subsequent future-potential event horizon (potential EH-2) the barrier must 'dissolve' into a quantum mechanical de Broglie matter state, per function of its deference to the future post EH-1; thus transforming into a future-possibility state... into an

275

AND-form state... which cannot constitute a repository for the initial OR-form of the barrier. It will remain in that state until the end of isolation at the subsequent future-potential event horizon, potential-EH-2.

Across the quantum mechanical interval between the point of isolation and the end of isolation, we will obtain only the spacetime in potentiality-form of the barrier: i.e., the de Broglie AND-form of that barrier... with all its alternative configurations in *presumed* equipotential AND-form superpose per contingent nested futures.

Thus, assuming a generic quantum mechanical de Broglie matter-state for our barrier, described using our lay formalism...

Status of Barrier: {"only slit-one open" **AND** "only slit-two open" **AND** "both slits open" **AND** "both slits closed"}

... even though it was originally set to, say, **both slits open**. The material of the barrier will remain in this quantum mechanical form of alternate equipotentiality until the end of physical isolation, brought about by attendant AND-to-OR quantum indeterminate collapse.

The barrier ought to resolve into any of its four alternative equipotential settings, and not necessarily back into the original setting. **However, as real-world experiment and experience shows, and as is demanded by the very viability of quantum mechanics and science, the real-world resolution of the barrier** *will not* **be a matter of quantum indeterminate 0.25 probability. Notwithstanding contingency or the usurping role of Wheeler's cat, the alternate possibilities other than 'both slits open'** *will not* **be allowed to resolve at the end of isolation, despite undeniable quantum indeterminacy.** *The original setup for our barrier will be restored*, **despite the operation of both AND-form quantum mechanisation** *and* **quantum indeterminacy.**

*

No matter how quantum mechanised the embodying de Broglie matter becomes, and no matter how quantum indeterminate the processes of its development at the end of physical isolation, and as is manifestly true of a developing photographic image *and* true across nature, the barrier must be resolved into its original form, baring Wheeler's cat. Without this certainty, there can be no quantum mechanics as we know it. Hence, **apagogic mnemonic argument 1-a: The very viability of quantum mechanics as we know it necessarily implies the reality of abstract memory of the original setting of the barrier, retained in dissociated form from the interim of isolation, despite the quantum mechanisation of the barrier, and despite subsequent quantum indeterminism. The abstract memory of the barrier must 'leap over' the interim of isolation to the end of isolation, without loss or erasure,** notwithstanding contingent possibilities, or the operation of Wheeler's cat.

Hence, the strict and necessary dependency of quantum mechanics as we know it on abstract memory and its retention extant radiation, matter, and space. Any claim otherwise, or in favour of the assertion of the erasure of memory and the nihilism of information, will culminate into an apagogic contradiction versus really-existing quantum mechanics we know.

3.08: Apagogic Mnemonic Argument 1-b: Causality, Spacetime Structure
And the Dependency of Quantum Mechanics on Abstract Memory

Spacetime structure relates in a mutually justified and mirroring way with the firewall principle, formed from the intertwine of causality with the conservation principle. The tripartite structure of intermediate spacetime, compartmentalised into a mnemonic past separated from a quantum mechanical future by the event horizon, constitutes the firewall principle enacted in spacetime structure. It justifies the ontic reality of abstract memory as part of the system of past-future dualism in a growing block universe (wherein the past cannot obviously be carried by the future)... and from the recognition of the domain of realised spacetime as the proper mnemonic domain: the repository of abstract-form past states and initial conditions, such as the setting of the barrier in the two-slit experiment, the abstract memory of which must play a critical role in the viability of quantum mechanics.

As was indicated in 3.07, the interim AND-form quantum mechanisation of the barrier in the two-slit experiment can be further augmented in terms of the firewall principle inherent to spacetime structure first developed in Book-II. This states that, until it gets to transpire, we cannot describe the future in any definite OR-form terms or as a set of realised events, save by travesty involving 'signals from the future' in violation of causality and the principle of conservation. This is from the assertion that any *detectable* resolved future, observed per the reception of 'signals from that future', would evince a quantity of pleonastic energy-matter above the energy-matter unity of the relative present frame, notwithstanding the pleonastic status of energy-work concepts and relations (see Book-II, **2.39**). This above-unit energy-matter would appear as if created *ex nihilo*... in violation of the principle of conservation.

Also recall from both Book-I and Book-II that the future of a physical system is not an abstract *bloss* subjective state purely in the mind: The future state of a system, as a set of AND-form possibilities, but *not* as a state of concretised events, possesses real ontology as the AND-form quantum mechanical wave itself: Thus, the quantum wave *is* the future, ultimately constituted as spacetime in potentiality. Once this is appreciated, we solve the *raison d'etre* of the quantum mechanical wave and the existence of AND-form logic: We incorporate it to spacetime structure itself, courtesy of the firewall principle.

*

With the end of isolation via AND-to-OR collapse and attendant progression of the event horizon, the material of the barrier will resolve back into an OR-form state; perhaps into the original setup, baring the operation of low probability contingency or Wheeler's cat, the mischievous laboratory cat who overturns the experiment and alters the setup of the barrier into, say, **all slits closed**, instead of the original **both slits open**. This sort of alternative is allowed for by the delayed choice nested futures structure of spacetime in potentiality, justified by the specific insight garnered from John Wheeler's delayed choice approach to the quantum mechanical wave (see Book-II **2.17** and **2.18**). Hence, spacetime structure and the nested future form of spacetime in potentiality, which embeds within it various contingencies that can alter the future *away* from any initial conditions, can only augment the reality that, **upon isolation and relegation to spacetime in potentiality, the barrier is indeed transformed into all alternative AND-form future possibilities: the only basis upon which nested future contingencies might operate and might alter the barrier from both slits open to some other alternative future possibility**.

Upon AND-to-OR collapse and attendant progression of the event horizon and the ending of isolation, the barrier will resolve either into its original barrier setting or it will be replaced by some other setting. The *en masse* quanta will collectively re-materialise the mereology of the barrier, whatever this might turn out to be, with or without the operation of contingency. Yet, to reiterate, **the final outcome for the barrier will *not* be a matter of 0.25 probability per its stated four alternate configuration possibilities, or even per the operation of contingency.** Again, despite quantum indeterminacy, the outcome will be an absolute certain one, culminating into the re-manifestation of the barrier-state originally set up for it, save what contingency or Wheeler's cat might conspire. **The quantum indeterminate processing of the resolution of the barrier will not erase the original setting of that barrier, even if contingency forces it to some other outcome.** As we shall discover, quantum indeterminate processes turn out to be as incapable of erasing the original set-up of the barrier as they are of erasing and prohibiting the correct photographic image from forming up. The question of how this is possible will be resolved in Part-II.

*

With isolation, if the barrier must dissolve into an AND-form quantum mechanical de Broglie matter-wave per the demands of the firewall principle, the critical information constituting the original barrier-setting must be fated to one of two alternate mutually excluding possibilities. First, **for its post-isolation restoration to be possible, notwithstanding Wheeler's cat, the information must be retained as abstract memory, relegated to the domain of realised spacetime**, at a physical remove versus the quantum mechanised and futurised barrier, with the latter now part of spacetime in potentiality. Thus, **the pre-isolation OR-form state of the barrier, or its abstract basis, must survive and persist as abstract memory within the domain of realised spacetime**. At the end of isolation, it will be restored to matter as a resolved OR-form state, baring Wheeler's cat.

If it were at all viable, the second possibility asserts that, **per the erasure hypothesis, once isolation is enforced and the inevitable quantum mechanisation of the barrier is obtained, the pre-isolation OR-form state of the barrier *must* become totally and irreversibly erased from the system, and further erased by the isolation-ending information-scrambling quantum indeterminate process at the end of experiment, constituting the amnesic universe (or succinctly, an Alzheimer's universe) wherein information is neither retainable as memory nor can it 'leap over' quantum mechanisation and quantum scrambling to post-experimental restoration**.

Hence, according to memory erasure hypothesis, the original setting of the barrier ought to be totally erased from the system, regardless of any due consideration to contingency: The resolution of the succeeding post-isolation barrier at the end of isolation would then be a matter of pure information-free quantum indeterminate 0.25 probability: with an outcome that has 0.75 probability of being other than, and therefore inconsistent with, the barrier's originally setting.

However, the issue of consistency could not arise in an information-free amnesic universe: The erasure of information and memory would render the firewall principle itself void, wherein 'signals from the future', and all attendant paradoxes and violations of causality and of the conservation principle would be rendered null, or lose any relevance to reality. Of course, quantum mechanics as we know it, and even experience itself, could not be tenable… *if* erasure hypothesis held true. Hence, the apagogic nature of the anticipated erasure hypothesis.

3.09: Apagogic Mnemonic Argument 1-c: The Consistency Problem of Erasure

If the erasure hypothesis held true, having initially sought to fix the barrier to *both slits open*, it might instead lead to *only slit one open* on a 1-in-4 or 0.25 probability… or to *both slits closed* on the same probability basis. Any consistency between initial experimental conditionals on the one hand versus outcomes on the other, would become impossible to guarantee. The best we could hope for would be a 0.25 probability for a consistent correlation between initial setup and outcome.

Hence, suppose we ran the experiment having initially set the barrier to *only slit one open*, only to end up with the formation of an interference pattern typically obtained from *both slits open*. Alternatively, we might run the experiment, only to obtain a null outcome, since it is also possible, albeit quantum-indeterminately, that the experimental outcome might be consistent with *both slits closed*, even though we originally set it to *both slits open*. Thus, **if memory erasure held true, the experiment might resolve into any of the other possibilities, predominated by a 3-in-4 chance that the conclusion will be *inconsistent* to the initial setup before isolation versus the barrier setup subsequently formed without regard to the initial setup**. Indeed, any accidental consistency between initial barrier setup, subsequent barrier-resolution and post-isolation outcome, will *not involve a temporal* 'leap over' or transfer of abstract memory from the past to the future: There could be *no* memory.

Assuming memory erasure was real, the greater the number of experiments performed over time, the more miraculous would be the result if, as is true in our really-existing universe, we always obtained consistency between setup and outcome, despite the said 1-in-4 probability structure that relates the initial barrier to subsequent barrier and detector-outcome. In a universe where memory erasure operated, for any long-running sequence of two-slit experiments, we would end with around 25% happy consistencies, and 75% inconsistency, whichever barrier setup we initially imposed. We could *not* enjoy any permanent happy streak, as we appear to do in our really-existing universe.

277

Obviously, the universe according to memory erasure is totally inconsistent with the real universe: According to quantum mechanics as we know it, in the real universe, initial conditions persist and specific outcomes in two-slit experiments are *always* consistent and correlate with specific barrier-settings, baring disruptive contingencies, or 'Wheeler's cat'. The setting on the barrier is *not* scrambled or decided on a 1-in-4 quantum indeterminate sortology: it is not subject to inconsistency problems per memory erasure. The preliminary form of memory erasure hypothesis is therefore apagogically false.

It follows that, **abstract memory must be valid, given that quantum mechanics as we know it would not be tenable otherwise, save by the operation of absolute improbabilities and miracles, as espoused by apagogic mnemonic argument 1-c.**

3.10: Reiterating the Non-Standard Evaluation of the Double-Slit Experiment
And the Apagogic Mnemonic Argument

It is speculatively possible to anticipate a form of erasure hypothesis in which some sort of quantum science might yet be possible: albeit different from the form we obtain in our really-existing universe. To appreciate this, we must use the framework of the non-standard evaluation of the two-slit experiment first outlined in the Preliminary; essay **0.14**.

According to the standard evaluation of the experiment, when we set to **both slits open**, this exhibits the famous interference pattern: one that indicates self-interference and demonstrates the reality of wave-like AND-form quantum mechanical states. On the other hand, the *one slit open* condition is typically (and erroneously) purported to exhibit only the particle-facet of quantum mechanical radiation and matter. **The standard evaluation holds that the wave-like facet of quantum mechanical radiation and matter is exclusive only to the *both slits open* condition, while the particle-facet is exclusive only to the *one slit open* condition**: (Preliminary **0.13**). The standard view is also implicit to the Afshar controversy: (Preliminary **0.16**).

By contrast, **our non-standard evaluation asserted that *both* conditionals (*both slits open* and *one slit open*) get to exhibit *both* wave-like *and* particle-like facets (both AND-form *and* OR-form states) in an observationally simultaneous way**, even though wave-particle complementarity (AND-OR complementarity) must yet abide per the *new complementarity thesis* first introduced in Preliminary **0.16** and **0.17**.

New complementarity arose out of the fact that, to discern the generic interference pattern on the detector when both slits are open (to infer the reality of wave-like AND-form phenomena) we need to generate *en masse* OR-form particle-like incidences on the detector-surface that, at a given threshold, collectively manifest the sought wave-like interference pattern. Thus, **both slits open** will generate wave-like *and* particle-like outcomes; both AND-form *and* OR-form outcomes, in an observationally simultaneous way, on the strict condition that a sufficient number or threshold of particle-like incidences accumulate at the detector and do so in a way that simultaneously collectively exhibit the interference pattern that evinces the reality of AND-form self-interference. In short, we need *en masse* OR-form particle-realities to simultaneously manifest the evidence and reality of AND-form wave reality.

Even though the evidence for wave-like AND-form phenomena enjoys greater saliency in the **both slits open** condition, the wave-like AND-form facet of the duality needs the observationally simultaneous manifestation of the particle-like OR-form facet to allow us to infer the reality of self-interference and AND-form logic.

However, **just as it is misleading to state that the *both slits open* exhibits *only* wave-like AND-form phenomena, it is equally misleading to state that the *one slit open* exhibit *only* particle-like phenomena**. Why? The characteristic distribution of OR-form particle incidences formed at the detector when only one slit is left open can be generated by an AND-form wave, just as it does in the **both slits open** setup. That is, the distribution of particles at the detector when one slit is open does not preclude the operation of an AND-from wave that controls for that specific distribution, and wherein the OR-form and AND-form realities are observationally simultaneous, just as they are in the two slit experiment.

The reality of the AND-form state, and its observational simultaneity with the distributed particles it garners under the **one slit open** condition, cannot be obviated, even though no interference pattern is formed. Again, it follows that the **one slit open** condition does not exclude the presence and final evidence of AND-form quantum mechanical waves, even when no interference pattern is formed... no more than can the **both slits open** obviate the reality of OR-form logic and particle-like reality.

In short, **the absence of interference patterns in the single slit experiment cannot imply the absence of AND-form wave-like states, much less the exclusivity of OR-form states, given that the same distribution of OR-form states can be generated by a non-self-interfering AND-form wave when only one slit is open.**

*

To augment, we need to reprise what is happening in the **one slit open** experiment to show that both AND-form *and* OR-form logics are present and observationally simultaneous, in abidance with the new complementarity thesis.

When we pass mechanical waves through a *one slit open* barrier, the intensity of the arriving mechanical wave will be greater at the area of the detector closest to the single open slit, but progressively reduce with respect to areas of the detector more distant from that open slit. But **if we use quantum mechanical probability waves instead of mechanical waves, the probability of the AND-to-OR collapse of that wave into an OR-form particle-like deposition will be greatest at the area of the detector in front of and closest to the single open slit, while its probability of deposition will reduce with respect to those areas of the detector farthest from that open slit. Just as AND-form logic structures the distribution of incident particles at the detector when both slits are open, but does so by function of a self-interference, essentially the same thing happens when only one slit is open, save that there is no possibility of self-interference. In other words, the single-slit condition *does* yield an AND-form outcome implicit to the distribution of *en masse* quanta it generates**: This is the central contention of the non-standard evaluation of the two-slit experiment.

278

If neither barrier setting can exclude AND-form *and* OR-form logics from being observationally simultaneously, what are the implications to complementarity? The question was tackled in the Preliminary essays, **0.16** and **0.17**: In summary, while wave-like and particle-like (AND *and* OR-form logics) *are* observationally simultaneous, this simultaneity always manifests at the post-measurement (post AND-to-OR collapse) and post-isolation stage. In the interim of isolation, the emitted quanta transform into pure wave-like quantum mechanical AND-form states (futurisation), although this holds true under both the single *and* the double slit conditions. As such, in the interim of isolation, the system is constituted as future-form 'which way' states in AND-form superposition: There is futurisation into AND-form possibilities in the interim of isolation.

Thus, **while the wave-particle duality (AND-OR duality) is always observationally simultaneous at the detector at post-isolation, it is *not* simultaneous in the interim or during isolation**. Under isolation, and in both double and single slit conditions and experiments, the system is *always* an AND-form quantum mechanical wave-state. After isolation, or following observation and measurement, the AND-form state collapses into its particle-like OR-form outcomes. And when we run the experiments long enough, we obtain collective or *en masse* formations of OR-form particles that, in their collective, get to manifest what was also true during isolation: namely, the AND-form wavelike character of the system, with or without self-interference, in both **both slits open** *and* in **one slit open** conditionals.

Insofar as the system is not simultaneously wavelike and particle-like in the interim of isolation, and only purely wave-like in that interim, and with its particle-like aspects manifested only at the end of isolation, wave-particle complementarity is preserved: But, it is preserved per the *new complementarity*; one that admits the non-standard evaluation for the observational simultaneity of wave and particle ontics, post-isolation and post-experiment, and the saliency of AND-form logic in one-slit experiments.

*

Let us now apply the non-standard evaluation of the two-slit experiment and its attendant new complementarity thesis to the memory erasure hypothesis. **Assuming memory erasure to be true with respect to the state of the barrier in the double-slit experiment, outcomes obtained at the detector might yet permit conclusions in favour of the reality of quantum mechanics, though only when the experiment is re-framed in terms of the non-standard evaluation, albeit with a different form of quantum mechanics to the form known**, apagogically implying that memory erasure hypothesis cannot hold true, save in a purely fictitious universe.

Recall that, in the non-standard evaluation of the two-slit experiment, the evidence for wave-like AND-form logic could also be inferred from the **one slit open** experiment, even if no self-interference is evident. We now assume the operation of memory erasure: We assume no retention and carry-over of the original configuration of the barrier to experimental conclusion. We also assume that the resolution of the barrier at the end of isolation unfolds in a purely quantum indeterminate way, with no control or guarantee as to how many slits will remain open, closed, or other, given that there are four possible equipotential barrier configurations, hence four possible alternative attendant outcomes at the detector.

An interference pattern might correlate only with the **both slits open** conditional. On the other hand, the two variants of the **one slit open** conditional might constitute the two other outcomes, *without* evidence of any interference pattern, but *with* wave-like AND-form phenomena inherent to the distribution of OR-form outcomes obtained from these, per the non-standard evaluation of the two-slit experiment. Finally, the fourth outcome might be the **both slits closed** state, correlated with a null outcome at the detector. Per our non-standard evaluation, three out of four of the outcomes (not just one) permits the inference of AND-form logic and the founding of quantum mechanics, even if we assume memory erasure of the initial barrier setup and assume the correlation of barrier to the outcome on a pure non-certain miracle-basis.

In summary, assuming that all memory of the original barrier setup is completely wiped out upon the enforcement of isolation; assuming no temporal leap-over of that information is possible; and assuming that the barrier-setting at the end of the isolation is decided upon a quantum indeterminate 0.25 probability; even with all the consistency problems entailed and exposed in the preceding mnemonic argument, it might be possible to obtain evidence of AND-form logic. Thus, it might yet be possible to found quantum mechanics from three out of four possible experimental outcomes per the non-standard evaluation. This would make for a quantum mechanics of a kind, but *not* in the form we know it.

While the non-standard evaluation of the two-slit experiment in the absence of abstract memory and the operation of memory erasure could seemingly allow for the discovery of AND-form logic and the founding of a quantum mechanics of sorts, the sort of process involved could not be of the same form as we observe in our universe. **Even though three out of four outcome-possibilities would allow for the founding of quantum mechanics under conditions of memory nihilism *and* the erasure hypothesis, in the real universe, the barrier-configuration at the start of the experiment is *not* quantum indeterminately scrambled to its alternative configurations, notwithstanding contingency: The resolution of the barrier and outcome is not a matter of quantum indeterminacy of the form observed in, say, EPR.** Therefore, as an apagogic didactic hypothetical case, our nihilistic non-standard evaluation of the two-slit experiment enhances our evolving apagogic case against memory-erasure and supports the real existence of abstract memory, *against* informational and mnemonic nihilism.

3.11: Apagogic Mnemonic Argument 1-d: The All-or-Nothing Problem of Memory Erasure

The preceding critiques of erasure hypothesis are too generous and permit unwarranted licence to nihilism. Why? If we must assume the operation of memory erasure at all, it needs to apply on a comprehensive all-or-nothing basis: applicable at every point and facet of experiment and experience. This is precisely what must be demanded by any self-consistent memory erasure hypothesis, if not by pure information nihilism. It is disingenuous and dishonest to entertain selective memory erasure for some things, but conveniently exclude it for others. Thus, if memory erasure were true, *everything* entailed in and pertaining to *any* experiment and experience would need to be subject to total erasure. Not even reportable experience could survive such erasure. **Assuming all-or-nothing memory erasure, it will be impossible to obtain *any* outcome in proof of quantum mechanical realities, even with the non-standard approach to the two-slit experiment.** Indeed, the three out of four outcomes from which we might infer AND-form logic per the non-standard evaluation would also be erased, together with the experimenters and any reports they might have compiled.

In short memory erasure is either all… or not at all. Self-consistently, it must be *all*. Therefore, memory erasure, at least in its first iteration, is apagogically implausible.

To illustrate, imagine we carry out a two-slit experiment in a universe in which memory erasure applied in the all-or-nothing way, which it must do in order for memory erasure hypothesis to be self-consistent. Suppose therein we obtained the characteristic interference pattern in proof of self-interference and AND-form logic: hence proof of fundamental quantum mechanical reality. But experimental proof is only useful if we can communicate it to our peers.

The evidence we wish to present to our peers might consist of extensive notes that pertain to the fact that we *cannot* assert control over the initial barrier setup per memory erasure: presented as proof of memory erasure. In having accumulated many results from many such experiments, we place all our evidence and documentation in a storage box and await the morning.

When morning comes, we gather our colleagues and open the box… only to find that, like the initial barrier-setting, the original contents have become erased. Even if an accidental consistent correlation should emerge between the experiment carried out yesterday and the content of the box in the morning, it will have nothing to do with any retention or time-transfer of information or memory from past to present. Memory erasure will apply to the material placed in the box as much as it must to the original barrier setup, if not to the whole laboratory, the university… to the universe itself. Indeed, there can be no exceptions to erasure, unless we are willing to commit dishonesty: *All* will be erased: Subsequently, *all* will be decided and sorted on a quantum indeterminate basis, without any recourse to the persistence of information or memory via abstract retention or temporal carry-over.

It follows that the evidence for quantum erasure itself will inevitably evaporate into nothing, given all-or-nothing memory erasure. **Given all-or-nothing memory erasure, the experimental setup, the whole laboratory, the city, the planet, the solar system, the galaxy, indeed, the universe itself, could not be guaranteed to persist in its requisite or initial configuration; or even survive the erasure process; or permit the performance and repeatability of the experiment; much less permit any communication of results in 'proof of memory erasure' itself. And never mind our personal memories, much less our own existence**.

Upon these insights, it necessarily and inescapably follows that memory erasure hypothesis in its first iteration is false, given that, in our really-existing universe, we *do* have control over experimental variables and outcomes; and vital information *does* persist and survive across time; and we *are* thus enabled to secure scientific progress, including the founding of quantum mechanics.

*

Information in nature *is* retained in abstract form vis-à-vis putative materialised radiation, matter, and space. The very possibility for conclusions in favour of quantum mechanical realities demands the reality of abstract memory, if only because quantum mechanical conclusions are critically fated to the very possibility and reality of informational persistence and perdurance over time, notwithstanding Wheeler's cat. Indeed, a central contradiction exists in generic quantum mechanics: On the one hand, as the science of AND-form logic, quantum mechanics cannot furnish the existentiality of OR-form logic and attendant abstract memory. On the other hand, quantum mechanics critically depends on the viability of OR-form logic *and* abstract memory to obtain the resolution of initial conditions and the setup (the apparatus), and depends decisively on the mnemonic retention of said initial conditions and the setup, critical for central conclusions of quantum mechanics *and* for the viability of quantum mechanics. In short, quantum mechanics cannot furnish a decisive claim about the viability of the apparatus upon which it depends, and does not explicitly recognise the apparatus problem as the problem of abstract memory.

One of our goals in Book-III is to make good on this weakness of generic quantum mechanics by incorporating to it what it always lacked: namely, *memory*.

SECTION-II: MEMORY ERASURE HYPOTHESIS CO-JOINS WITH MANYWORLDS

3.12: Apagogic Mnemonic Argument 2-a:
The Manyworlds Interpretation & Revival of Memory Erasure Hypothesis

In its first iteration, anticipated memory erasure hypothesis is not viable. In the first section, we saw that the very possibility of quantum mechanics and science, and our ability to draw out fundamental conclusions about the reality of both AND-form logic and, presumably, quantum indeterminacy, indeed, the very possibility of experience and intelligible communication about *anything*, including about quantum mechanical reality, must remain critically depend on the non-erasure and persistence of critical information. All of this implies the physical retention of informational states vis-à-vis radiation, matter and 'space'. In short, abstract memory is real.

However, **in the following, we will anticipate a revamped superlative variant of memory erasure hypothesis: one co-joined to the Manyworlds interpretation** of quantum mechanics. Together, these will constitute a more profound nihilism and threaten the ontology of information *and* abstract memory.

QUANTUM MECHANICS AND MIND

The Manyworlds interpretation of quantum mechanics (MWI) states that, out of the welter of possibilities inherent to the AND-form superposition, upon AND-to-OR collapse, it is not just one possibility, but *all* the possibilities that get realised. In this way, MWI apparently circumvents the whole problem of quantum indeterminism: the *raison d'etre* for MWI.

Since all the possibilities in the wavefunction get to happen, all will form distinct histories or universes. These possibilities do not happen per quantum indeterminate sortology. Hence, quantum indeterminism is completely obviated. Yet, we can only observe one of the histories and cannot observe the others, even while we misinterpret ours to be a unique uni-history. The reason why we do not get to observe all of the possibilities and histories, save just one, is because, at the point of AND-to-OR collapse, the universe supposedly splits into as many alternate *compartmentalised* histories as there are AND-form possibilities in the original wavefunction. Each alternate history formed must remain cut-off from every other. From our own branch and history, we can have no empirical-physical, observational or experiential access to the other histories. Yet, scientific evidence and utility demands empirical-physical and observational evidence. The very fact that the viability of MWI appears to require the strict impossibility of just such evidence relegates it to the realm of pure conjecture, no matter how mathematically hypertrophic or even internally self-consistent it might otherwise be.

MWI will run into even greater problems when it is framed with respect to fundamental considerations involving the principle of conservation at the multiverse level, and other criticisms: (More on these issues later).

*

Let us briefly reiterate that, in a universe subject to memory erasure and nihilism, but without MWI, the status of the barrier and its resolution would be subject to quantum indeterminate sortology: Whatever configuration we initially set up for the barrier, there ought to be 0.25 probability that the final outcome will coincide with any one of the following four possible barrier settings…**both slits open… only slit-1 open… only slit-2 open… both slits closed**. But, in MWI, any initial barrier setup will be succeeded by *all* the four alternative barrier configurations, not just one. All four barrier settings will form up as alternate histories; with each completely compartmentalised from the others. Thus, if we initially set the barrier to **both slits open**, upon the end of isolation, three histories will be inconsistent with the initial setting on the barrier, while only one history will be resolved into a default consistent outcome (i.e., to **both slits open**) at the end of experiment. But this consistency will not be due to any temporal carry-over of information or memory of the barrier from the initial point of the experiment to a succeeding concluding one. Indeed, the remaining three histories will be wholly inconsistent with respect to the initial setup of the barrier.

Two sets of disturbing implications are obtained from MWI alone:

- **In Manyworlds, there is no memory: Nothing has been retained or needs to be retained, much less carried over across time. Only purely default consistent (sensible) and mostly inconsistent (non-sensible) histories will be generated: all accomplished in brute force exhaustive fashion, without logic, reason, causality, connection, or *any* informational principle. Indeed, the ontology of both memory and information are completely obviated and rendered irrelevant** in a more effective and complete nihilism than the preceding nihilism from forlorn memory erasure hypothesis.

- Moreover, and most critically, in at least one of the histories generated purely by default, and the apparent 'consistency' obtained therein, we will have obtained a default interference pattern contingent to a default **both slits open**, and can thus found quantum mechanics. **That is, in MWI, quantum mechanics will *appear* viable, in at least one history, despite total nihilism with respect to both memory and information itself**, notwithstanding the non-standard evaluation.

Of course, at this juncture, for brevity only, we will avoid the application of our all-out all-or-nothing critique central to **argument 1-d**, and readily applicable to MWI. In truth, the all-or-nothing critique cannot be avoided. But we will pretend that it can. Thus, MWI *appears* to save real experience as we know it: MWI guarantees that we will obtain a small number of pure default 'consistent' histories in which 'order' and science will appear to be possible; albeit, purely by default, and *not* per any retention of abstract information; all on the basis of mereological and informationally nihilistic miracles, courtesy of MWI and its inherent obviation of information and memory (ignoring the all-or-nothing problem from **argument 1-d**). Hence, the seductive and destructive promise of MWI versus *any* notional memory or notional 'real ontology of information'.

*

Apparently, our first foray against anticipated memory erasure hypothesis presented in section-i was totally in vain, insofar as MWI appears to rescue memory erasure and information nihilism completely, even without directly co-joining with memory erasure hypothesis… although memory and information erasure is essentially identical in both MWI and in erasure hypothesis. Therefore MWI and memory erasure hypothesis are indeed co-joined.

Yet, this unfortunate nihilistic result is completely dependent on whether MWI is itself true, notwithstanding the all-or-nothing problem that first arose in **argument 1-d**. (see **3.11**). **If MWI is true, then memory erasure becomes more than plausible; even while it must remain beyond any ultimate and certain proof**, given that, by its very logic, we cannot prove Manyworlds and the reality of alternate histories from within our own history, save by contradiction. Consequently, there will be no way of deciding the issue scientifically, for or against erasure or abstract memory. But, this weakness is not a sufficient basis for an apagogic argument to establish the veracity of abstract memory and the ontology of information.

On the other hand, if MWI can be shown to be false, then hitherto apagogic mnemonic arguments in favour of abstract memory will be automatically validated. In the following, we will seek to show that MWI is indeed provably false.

281

3.13: Apagogic Mnemonic Argument 2-b Invalidates MWI:
Liar's Paradox & the Violation of the Principles of Conservation

The Manyworlds interpretation involves logical and physical problems that are often overlooked. Its viability can only be maintained on a fallacy: Therein, the alternate histories MWI recommends must remain mutually unobservable, rendering the interpretation permanently beyond physical and empirical proof. Indeed, it can only remain 'true' per function of this very condition. But MWI also entails the violation of the principle of conservation: albeit, one conveniently hidden from view.

In this essay, we will enumerate all the self-voiding problems that Manyworlds engenders: Thus...

(a) **The first liar's paradox**: For MWI to be true, it must remain beyond a*ny* physical-empirical proof or test. This culminates into an obvious liar's paradox: **Manyworlds can only be 'true' if, and only if, it cannot be proven to be true in physical-empirical terms.**

A theory or interpretation can be as 'crazy' as one needs it to be. Yet, to be part of science, it must directly or indirectly relate to the physical empirical world. It must act as a resource, or at least as a springboard, to further explorations of nature. Yet, MWI demands that, to remain 'true' it must remain beyond just such physical-empirical grounding. Hence, Manyworlds must remain a non-resource; a non-science... or simply *nonsense*.

(b) **The second liar's paradox**: What if one could prove MWI in physical-empirical terms? Given that, to remain true, MWI must remain beyond physical-empirical proof, any furnished proof would serve to disprove the interpretation. Consequently, MWI would be physically and empirically falsified by its evidential 'proof'. That is, **to be proven false, MWI must be proven true**.

Assuming the simple case of the formation of just two histories, and whichever history we end up in, the empirical-evidential proof of the existence of the other history from within our history would imply that the other history is empirically part of our own history (or vice versa) per the fact that it is *not* compartmentalised at all; not separate from ours; hence part of our empirical-physical order. Never mind consequences to the conservation principle.

(c) **The violation of the principle of conservation**: With or without compartmentalised histories, **MWI would lead to the violation of the principle of conservation, if not anomalies vis-à-vis entropy**. In one extremity, in one AND-to-OR collapse-step involving 'infinite paths', MWI would engender absolute entropy for the multiverse of distinct histories it would otherwise constitute.

*

We will now explore the three noted paradoxes or contradictions of MWI. To this end, let us first envision a simplified scenario involving, say, the spin-state of a particle with just two immanent AND-form spin states: {**spin-up** AND **spin-down**}. Let one of these be rigged to a nuclear detonator. Consequently, if the system resolves into the OR-form outcome of **spin-up**, no detonation will transpire. But if the system resolves into the OR-form outcome of **spin-down**, nuclear detonation will occur.

The Manyworlds interpretation holds that, at the pertinent juncture, the universe will split into two histories: *history-a*, in which we will *not* obtain nuclear detonation... and *history-b*... with nuclear detonation.

Let us assume we ended up in *history-a* and, from within *history-a*, we sought evidence for the existence of the alternate *history-b*, and did so by means of a physical signal from the nuclear detonation engendered by the 'spin-down' outcome in *history-b*. The immediate problem here is that the proof of *history-b* to be obtained from within *history-a* must manifest within *history-a*. Hence, *history-b* could not be truly compartmentalised from our own, against the requirement of MWI. Instead, it must be physically and empirically subsumed to *history-a,* (or vice versa): Hence, part of *history-a*, given its physical-empirical proof from within it. Hence, the second liar's paradox.

Of course, **the evidence for *history-b* would require some residue from the nuclear detonation therein: one that issued through to *history-a*. So now we have two nuclear detonators: one in *history-a* (not detonated) and the another in *history-b*, detonated. In other words, we have just doubled the amount of *knowable* fissile nuclear material** above the original quantity of the fissile material before the system resolved into two histories. Note the use of the term, *knowable*. This arises precisely from the fact that there is a physical signal that ties *history-b* vis-à-vis *history-a*, and allows the former to know about the latter, given the loosened conditions in relation to the first liar's paradox.

Other violations of the principle of conservation will transpire: We have a particle in *history-a* resolved into **spin-up**. But we also get another particle in *history-b* resolved into **spin-down**: We now have two particles instead of the original one particle. The total energy-matter has doubled, given that the particle has doubled. But where is the energy-matter to constitute this to come from?

What about the observers? With MWI, we have doubled the number of observers. Indeed, we have at least doubled the number of universes: i.e., as *history-a* and *history-b*. Again, where is the energy-matter going to come from to constitute these?

Hence, if one must meet the demand for evidence of claims, Manyworlds appears untenable when confronted with the basic conservation principle of energy-matter: This is not going to be obviated by the pleonastic status of energy concepts and relations, nor by the implied abstraction of the principle of conservation espoused in Book-II, **2.39.**

*

If the histories are compartmentalised, the evidence for the above-stated violations of conservation principles will not become evident. However, it does not matter that we cannot obtain evidence of *history-b* from within *history-a*: It does not matter that we will not witness the violation of the principle of conservation per the creation of *history-b*: The violation *will* occur, albeit 'out of sight', at the larger multiverse to which, *conveniently*, we have no physical-empirical access. The said violation could appear to be 'safely ignored' per the framework of MWI, but only on the strict condition that Manyworlds remain permanently beyond scientific evidence and proof. Yet, as stated repeatedly, this only serves to remove MWI from the domain of credible science.

282

Is there any way of rescuing MWI from the violation of the conservation of energy-matter? **Instead of entertaining problematic creation of energy-matter from nothing in violation of the principle of conservation, we might speculate a process involving a sharing out of a given finite energy content between the many histories and universes so formed.**

In our simple scenario with just two histories, we might obtain the doubling of the number of particles by dividing and sharing out a given finite energy-matter content into two equal portions. If the original energy of the particle was, say, $e = h \times v$ (where h is Plank's constant and v the frequency) upon its duplication through formation of *history-a* and *history-b*, the energy of the particle in *history-a* would become immediately reduced to ½ of e, with consequent reduction of its frequency-v. The other half of the original quantity of energy would end up with the particle formed up in *history-b*. **Consequently, even if we could not observe history-b directly, we in *history-a* would observe an unexpected reduction in the total energy-matter per the shared out energy-matter content of the particle under our scrutiny**: a reduction that, short of being interpreted as an inexplicable disappearance or destruction of energy, could be **viewed as an increase in entropy**, consequent upon the Manyworlds bifurcation into *history-a* and *history-b*, and the division of the original energy-matter content into and between the two histories.

From within *history-a*, this could be initially interpreted as a violation of the principle of conservation by function of an unexplained bleed-away of the initial energy, seen as a mysterious destruction of the local energy content. Yet, from the multiverse perspective, this 'destruction' would be due to the increase in the total entropy of energy due to the sharing out of the starting conserved energy content between the two newly formed histories. **The unexpected local 'destruction' of energy, or the equivalent Manyworlds-level increase in the entropy, would constitute just the sort of physical-empirical evidence for the reality of MWI, and of the sort that could prove MWI. However, the specific entropy-observations have never been made**: We do not report any mysterious or unaccountable 'destruction' of energy-matter, or anything remotely approaching events that could be interpreted as evidence of inexplicable drastic increase in entropy, ultimately at the multiverse level. Indeed, such predicted effects ought to be ubiquitous throughout our really-existing universe, given that the creation of histories through MWI is applicable to every instance of every wavefunction collapse-process in nature.

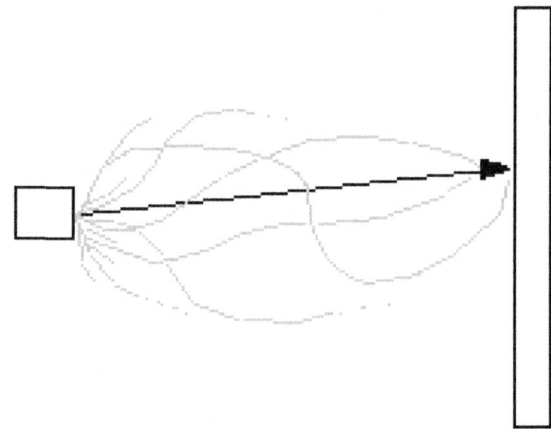

Fig. 3.01 Sum over all paths revisited
Notes: The diagram above depicts Feynman's concept of the "sum over all paths". It is derived from a double slit experiment comprised of a series of barriers (infinitely many) into which many successive slits are cut (also potentially infinitely many) producing manifold potential paths for the particle, potentially infinitely many. Note that the 'sum over all paths' is framed in its generic form without reference to nested futures: The depiction also assumes the onticity of 'space'. The emitter is on the left, and the detector is on the right. Particles released by the emitter will constitute a wavefunction whose AND-form possibilities will comprise all the infinitely many possible paths from emitter to detector, with the bold-line shortest path constituting the most probable path. The constituent probabilities of all the paths must sum to 1. The contribution of each of the remaining infinitely numbered paths from emitter to detector (some of which are depicted in the shaded trajectories) will comprise an increasingly diminishing fraction of the total probability. The fact that it is not practically possible to count all these paths has led to the use of renormalisation. Yet, this does not diminish the fact that there are an infinitely many paths from emitter to detector. Manyworlds will seek to resolve these into alternative universes or histories: an infinite number of histories.

So far, we have assumed that the source of the energy-matter to be shared out between the histories resides in the root pre-bifurcation universe. Instead, **let us suppose the source of that energy resided, not in the histories, but extant each history, at the multiverse level that embeds and subsumes the histories.** Histories would form, split, and materialise, not through the creation of energy-matter out of nothing, and *not* by the sharing out of the energy from a root history, but through the sharing-out of the energy content generalised to the subsuming multiverse level. Therein, from within *history-a* and from any other history, there would be no increase in entropy, nor any sudden apparent or interpreted 'destruction' of the energy. Instead, in the case of a simple MWI bifurcation with two histories, we would obtain an apparent doubling of energy-matter per the doubling of the particles, without any ultimate violation of the principle of conservation, given that the energy would be drawn out from an already-existing energy store at the subsuming multiverse level.

Therein, while there must be increase in entropy of energy, it will occur vis-à-vis the pertinent multiverse level. The universe-scale heat death typically implied by generic entropy would eventually transpire, but at the multiverse level.

While this epicycle might appear to solve the problem that MWI has with the conservation principles, **the whole scheme falls apart the moment we shift focus away from simple AND-form wavefunctions with just two or finite possibilities to wavefunctions with infinite possibilities, even when we loosen the restrictions and imagine an infinite quantity of potential energy available at the subsuming multiverse level**.

There are quantum mechanical AND-form states and universe-scale wavefunctions that embody an infinite number of potential future possibilities. These include the sorts of wavefunctions that give rise to renormalisation in quantum field theory. **Indeed, future AND-form states with infinitely many possibilities might well be the norm in nature**, while those with finite or binary possibilities, such as 'spin-up' AND 'spin-down' in EPR, may well constitute the rarity. Yet, even if AND-form states with infinite possibilities were shown to be uncommon, it would take just one to bring down MWI.

A system with infinite future possibilities must remain implicit to Richard P Feynman's *sum over all paths* approach, first encountered in Book-II, **2.13**. For a basic illustration, see **Fig. 3.01**. The obvious problem posed to MWI by such systems is from the infinitely many histories these ought to generate. Indeed, MWI collapses even if we furnish a multiverse level with a limitless generalised potential energy. **The sharing out of this infinite potential energy across infinitely many histories must imply instant depletion, culminating into a *grand heat death* of the MWI multiverse**. In a single AND-to-OR conversion, analogous to subtracting infinity from infinity, all the infinite histories so formed would suffer absolute entropy of energy and instant heat death, both for each history *and* at the multiverse level. But **the instant heat death of the universe is a thing that we *do not* observe. It follows that MWI cannot be physically true**: Either it must entail the violation of the principle of conservation or produce an instant multiverse state in absolute entropy.

Of course, continuums are bigger than infinities. Surely a continuum-level of energy within the multiverse MWI could satiate the infinitely many histories implied by the *sum over all paths* approach. Unfortunately, within every history, we will find wavefunctions with infinitely many 'sum over all paths' in AND-form, further accentuated by the nested futures approach, which *was* incorporated briefly into the 'sum over all paths' in Book-II, in the framework of the structure of intermediate spacetime (see Book-II, **2.13**). Each infinite sum over all paths from infinitely many histories will constitute a continuum, only to exhaust the putative continuum of energy at the multiverse level: That is, the system effectively constitutes a continuum of future possibilities and putative 'which way paths', with consequent continuum of histories. Again, this would instantly exhaust the continuum of energy at the multiverse level into total heat death or entropy.

*

We must either assume that the conservation principle holds firm, and this also applies when we render energy concepts and relations pleonastic, and wherein conservation is recuperated as the conservation of abstract memory *and* future consistency (see Book-II, **2.39**), or we must accept the possibility of its violation. We must either subscribe to an MWI with an energy-base that is finite, or to one that is infinite. If finite, there ought to be evidence of unaccountable entropy increase. In the case of AND-form wavefunctions with infinite possibilities, there ought to be multiverse-level absolute entropy as well as instant head death per each of the infinite histories. Resort to continuums will be to no avail, given that it leads to the same dismal conclusion.

The only other epicycle left for the viability of MWI is straightforward violation of the principle of conservation (the creation of energy-matter out of nothing) but under the 1st liar's paradox. i.e., without the requirement of any kind of meaningful proof or evidence of Manyworlds or of its violations, and on the strict condition that such proof could never be furnished.

However, **when any given perspective on nature requires that its 'truth' depend on convenient epicycle-based evasions of empirical-physical testability, or when the same perspective entails convenient contrivances comprised of unlimited energy-matter creation out of nothing, and hence the violation of the principle of conservation, even if such violations are conveniently out of sight and need not be observed, we must reject that perspective: we must reject MWI.**

*

In summary, MWI is false because, to be 'true' it would need to be beyond any kind of physical-empirical testability: That is, its 'truth' demands and depends on its scientific falsehood: the First Liar's paradox. On the other hand, MWI is false because any prospect of its physical-empirical 'proof' would constitute the physical-empirical falsification of the very notion of fully compartmentalised and isolated histories; the evidence for which would constitute disproof of MWI, while leading to all manner of problems with respect to the conservation of energy-matter: Hence the second liar's paradox.

Third, and as stated, MWI entails the violation of the conservation principle, even when MWI is placed beyond physical empirical testability, proof, or truth per the first liar's paradox. This alone invalidates MWI.

Fourth, in the case of wavefunctions with infinite paths from the *sum over all paths* approach, the splitting of the universe into infinitely many histories through a sharing out of an infinite quantity of multiverse-level energy, would lead to the instant heat death of our own history as well as all others: a thing we do not observe.

284

It appears that MWI can only be true if it can avoid the heat death problem by contriving a multiverse creation of energy-matter out of nothing. In which case, MWI would require proof of such a feat in terms of empirical evidence of energy-matter creation out of nothing… conveniently beyond observation and proof.

In short, MWI is false either because it evades scientific physical-empirical proof through the travesty of its liar's paradoxes, or, alternatively, because it leads to physical outcomes that we ought to be able to observe daily and everywhere, even when the first liar's paradox holds, and even when infinite paths apply.

*

What then of the revival of memory erasure hypothesis through its anticipated co-joining with MWI? For the erasure hypothesis to have any plausibility, MWI must itself be true. Yet, MWI appears false. It follows that the erasure hypothesis fated to it must also be false. Thus, the ontology of information, the case for information-matter dualism (facile or spacetime-structure based) hence the argument for abstract memory, is saved.

However, three further considerations are needed to complete the picture against MWI: These are…

- **MWI, taken in its own terms, does not in actuality lead to many histories**. Development of pattern in the universe is almost always per function of *en masse* quantum indeterminate outcomes that reveal pattern, exemplified by the abstract electronic configuration of the atom; by a photographic image; or the formation of an interference pattern in the two-slit experiment. In all such cases, the *en masse* quantum resolutions culminate into the same global pattern implied by the initial wavefunction. **That is, if we run the two-slit experiment per MWI, all histories formed from it would culminate into the same global interference pattern, or the same photographic image in every history so formed. And when we apply this insight to the universal wavefunction, all purported histories must again culminate into the same final cosmic pattern-outcome consonant to the starting universal wavefunction**. In short, MWI is superfluous to begin with, and our foray into logic and conservation principle-based criticism of MWI was entertained only for the sake of comprehensiveness.

- **The pleonastic nature of energy concepts and relations**. In Book-II (**2.39**) we showed that, when framed to our intermediate model spacetime structure, and subsumed to gear-less work function zero AND-to-OR wavefunction collapse or time, energy concepts and relations abstract out of the physical order and ontology. That is, energy concepts and relations are pleonastic: not parts of real physical ontology, no matter what their didactic and pragmatic value. But this has consequences to the extent that the conservation principle must be transformed into abstract form in any future physics. Yet **the expected abstraction of conservation principle is not a licence to its violation or to any 'free lunch', given that the conservation of memory and tandem consistent futures (to which MWI is extremely inimical) does not furnish such 'free lunch'. It follows that the abstraction of conservation principles on the basis of pleonastic energy or work relations will be of no use to MWI and its dependency on infinite energy-supply or on the creation of 'free lunch' energy-matter out of nothing… not that this would obviate the inexorable heat death problem…** (*not* observed).

- **Problems from the notion that the real universe is a 'free lunch' proposition**: This view holds that the universe was created out of nothing at the putative Big Bang in a licenced single violation of the conservation principle. Therefore, since this was possible at least once in cosmic history, it ought to be possible perpetually, perhaps in MWI. Unfortunately, as part of the prelude to the pleonastic argument for energy concepts in Book-II, we placed potential energy to the future, to spacetime in potentiality that, as a convergent nested futures structure, elaborates to the infinitely removed future. Therein, in Book-II, essay **2.30**, we argued that **the universe was never a 'free lunch', since the putative potential energy, albeit reframed as 'future-potential energy' locked up within spacetime in potentiality, was directly attached to the Big Bang monoblock from which the universe supposedly emerged: The requisite energy or matter, as a future-potential, was always existent as potentiality and did not require any *ex nihilo* act of creation**. That is, even on the assumption of the ontology of energy, the putative matter that constitutes our universe had AND-to-OR collapsed out of spacetime in potentiality… out of the future… *not* from nothing. But, notwithstanding the pleonastic argument about energy concepts and relations, if potential energy or matter is inherent to and embedded in spacetime in potentiality, it cannot act as a viable source to fuel MWI which, in its infinite sum over all paths formation (or in its continuum form) would exhaust that potential energy-matter, culminating into the instant total heat death of the multiverse and its infinite Manyworld histories.

One must also pay attention to the fact that potential energy and potential matter to make up the histories must be drawn out of the futurity of spacetime in potentiality, but only to the extent or in proportion to event horizon progression. To satiate the required energy and matter to constitute the infinite histories in MWI, the event horizon would need to progress, in one step, to the infinitely removed future in order to issue out the infinite potential energy-matter content therein. A similar feat would be required in a Manyworlds continuum of histories. Both would plunge our own and every other history into a grand instant heat death and maximum entropy. Obviously, this is not happening and has not happened.

However, almost all the above is rendered void if MWI is in the first place pointless on account of the fact that *all* histories generated as Manyworlds, culminate into the same global outcome… into the same interference pattern… into the same photographic image…into the *same* world. This usurps the value of MWI in its claim to circumvent the problem of quantum indeterminism, and certainly undermines its usefulness to any nihilistic case against information and memory.

285

3.14: Preliminary to the Apagogic Mnemonic Argument 2-c:
Non-Arbitrariness & the Meta-Semantic Hypostasis
From the Contingent, the Brute-Force Exhaustive & the Mind-Driven Developmental Process

To complete the case against anticipated MWI-based memory erasure and informational nihilism, we must develop the basis for the forthcoming **apagogic mnemonic argument 2-c**. This requires that we elaborate the case for the reality of the meta-semantic hypostasis: i.e., the existence of an extra order of abstract and extant information upon which nihilistic propositions and assertions must rely in order to constitute as intelligible but wholly apagogic claims, even when posited within an ostensibly nihilistic universe. Indeed, **all nihilistic assertions, in order to be viable and intelligible within a nihilistic universe, must posit the very thing they deny or annihilate, but could only accomplish such apagogy through backchannel to an extant meta-semantic hypostases. But nihilism must constitute as all-or-nothing and, in its inevitable resort to the said backchannel, must fall victim to the same all-or-nothing critique suffered by erasure hypothesis** (see **3.11**).

The historical 'mother of nihilism' was the *impermanence thesis* of Heraclitus. To ancient Heraclitus, *all is change and impermanence*: When you step into the river twice, it is not the same river the second time. On an all-or-nothing basis, the ultimate implication of impermanence is that there can be no ontically real identities, essences, or information (the river)... much less memory and the temporal resonation of such: There can be no possibility for the temporal *endurance* of identities or universals (such as physical laws) nor any possibility for *perdurance* or growing-block time-evolving identities or 'worms in time'… such as the 'river'.

However, the nihilistic universe of Heraclitus is untenable: The nihilistic assertion for impermanence of the river could not itself be viable as information: it could not constitute a communicable idea transmittable across time… unless Heraclitus relied on a backchannel to a semantic hypostasis of terms (including the idea of the 'river') not furnished by his nihilistic impermanence-ridden universe. Indeed, apagogically, Heraclitan nihilism presupposes just such a backchannel to ideas and states that can endure or perdure, and so escape the ostensive impermanence and the mnemonic-informational vacuity of his nihilistic universe.

Essentially the same charge can be made vis-à-vis anticipatory erasure hypothesis, against Manyworlds, and against erasure's co-joining with MWI.

The counter to Heraclitus was the permanence thesis of ancient Parmenides, also implausible: the assertion that *all* is eternal and unchanging, and that time is an illusion. This contradicts time, change, growth, perdurance, decay, and dissolution. Indeed, Plato sought to unify impermanence and permanence through the participation of a meta-semantic timeless hypostasis of principles (permanence) realised in and through time and change (impermanence recuperated as 'appearance'). Our Alt-R case for time-asymmetric physical laws from Book-I (**1.13**) constitutes a contribution to the kind of unification sought by Plato, as was our argument for the ontic reality of time synonymous with AND-to-OR collapse... as is our case against nihilism in favour of the ontic reality of information and abstract memory. Of course, Parmenides designed for a block model of fully OR-form resolved futures. But the growing block universe abides.

*

Throughout **apagogic argument 2-c**, we will assume that MWI is true despite the case against it furnished in **2-b**. We will also assert that the formation of alternative histories in MWI do *not* involve any informational ontology, or any process of carry-over of abstract memory from the past to successive moments. Thus, while we happen to occupy a universe with apparent sensible outcomes, we will treat these as if secured on a nihilistic pure information-free arbitrary process. Apagogically, we will discover that the inherent backchannel fallacy is ineliminable even from MWI-based nihilism, and constitutes inadvertent self-contradictory concession to a meta-semantic hypostasis, counter the professed nihilism.

Thus, let us assume that the initial barrier setup in the two-slit experiment to, say, '**both slits open', appears to persist from initial setup through to experimental conclusion, purely due to the brute-force exhaustive splitting of the barrier, into its four possible alternate histories. Therein, the only history in which default consistency between initial setup and outcome will arise will be resolved purely per the said brute-force exhaustion of the possibilities, *without* the requirement of any temporal resonation of memory from the abstract past, and *without* any reason, cause, order, or principle of consistency, or ontology of information. The default 'consistency' obtained within the single 'sensible history' is expected to be purely illusory: an intrinsically vacuous outcome.**

However, MWI in service to memory-erasure and nihilism must also call into question the very ontology of initial conditions, or information so-conceived: **There is no compelling reason to assume that, in the two-slit experiment at the end of isolation, the barrier must resolve into a barrier: Instead, it could resolve into pumpkin, or a lemon … or into nothing at all**, given that, in the MWI framework co-joined to memory erasure, there is no ontic reality to information or mereological identities, much less their retention as memory, or their temporal resonance across time. **With information nihilism furnished by MWI, we *must* completely dispense with *any* notion of endurance of principle or the perdurance of basic identity, or of initial conditions, and even less the possibility of intellection, mentation and the communication or rapport of ideas… even nihilistic ones.**

However, such reckless Heraclitan-form nihilism involves obvious insurmountable apagogic problems for any nihilistic model, in that it must assert the ultimate denial of the real existence of *any* being upon which the very viability of the nihilistic model must critically depend. We are back to the all-or-nothing problem from **argument 1-d** (see **3.11**), now framed to MWI.

Thus, if we insist on retaining nihilistic MWI, or assert it to be the basis of our universe and existence, the only recourse to secure the viability of nihilistic assertions and ideas will be the existence of and backchannel to a meta-semantic hypostasis: wherein requisite information must exist in some form, regardless of the nihilism inherent to Manyworlds, and wherein we must assume the existence of a backchannel to that hypostases.

286

QUANTUM MECHANICS AND MIND

Such a meta-sematic hypostasis could well exist in the fashion of Platonic abstractions: an ontology that supersedes and stands extant the purported informationally empty Manyworlds system of purely nihilistic vacuous and amnesic histories.

The aim of this essay will be to establish the reality of the meta-semantic hypostasis. To accomplish this, our apagogic preliminary will elaborate three distinct forms of pattern-development that could pertain to the MWI framework, if not to *any* framework.

*

The utilisation of a most simple domain will help lay bare the three forms of development pertinent to the apagogic argument against nihilism: namely, the jigsaw puzzle domain. Any commercially available jigsaw puzzle will suffice. The picture depicted does not matter, although the conformalities between the pieces *do*. In any case, we need only consider one central piece; almost any piece will suffice. We must also include its immediate conformal adjacent or flanking pieces. In principle we could process many hundreds of pieces. But our small sample will capture the essentials pertaining to the assembly of larger puzzle-pictures. We will use the jigsaw puzzle domain to secure truths generalisable to nature and to our putative MWI-driven nihilistic universe.

To this end, the three distinct forms of development we seek to clarify are...

- *the contingent process* of development
- *the brute-force exhaustive process* of development, and…
- *the mind-driven process* of development.

We will model each of these, apply it to Manyworlds, and make the case for the meta-semantic hypostasis, *and* secure the backchannel fallacy.

1. The contingent process of development constitutes a pure randomness-driven process: We may even quantum mechanise the indeterminate sortology involved. Given that our central jigsaw piece, with its four unique jig-jawed sides, is conformal to four other adjacent flanking pieces and their pertinent sides, we assign to each side of our central piece and to the sides inherent to each flanking piece probabilities, using a four-sided die (see **Fig: 3.02**): e.g., 1 for side-a; 2 for side-b; 3 for side-c; 4 for side-d… for each piece involved. We then roll a four-sided die for each, including for the central piece, to find out which sides get to participate in the development of our jigsaw puzzle assembly. Only one assembly can constitute a sensible development due to rules of shape-conformality. Many other ensembles will be non-sensible or non-viable. These will be generated on the basis of randomness and probability: a contingent developmental process enhanced by the quantum mechanisation of the dice involved.

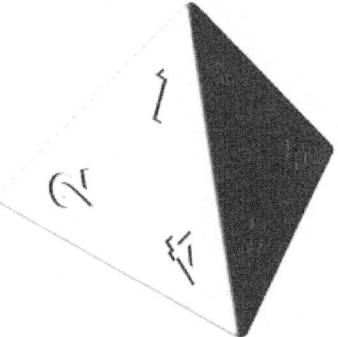

Fig. 3.02: The four-sided die: D4
Notes: Depiction of a four-sided die useful in the contingent development process involved in the thought experiment below.

Whether we resort to classical randomness or to a quantum mechanical sortology in the generation of our jigsaw puzzle ensembles, the selected pieces and sides will either fail or succeed to produce a sensible assembly, but always by default.

The developmental process will *not* be a matter of design or plan, much less involve any ostensive carry-over of critical information or memory. Indeed, given the strong role of contingency and randomness, especially when quantum mechanised dice are used (see **Fig. 3.03**), the final mereology ought to be totally illusory: given that no apparent informational process or ontology was involved.

While the role of contingency and happenstance in the contingent process of development so far depicted is clearly undeniable, the actual jigsaw shapes that allow or disallow for the possibility of default sensible ensembles *will* constitute *a priori* non-contingent ordinances. These are furnished to the domain prior its haphazard indeterminate development. That is, while the developmental sortology is purely contingent, and the developmental outcomes are by default, the ontology of the possibilities furnished are not explainable or furnished by the contingent process of development. Instead, the contingent development process pre-supposes the ontology of a-priori ordinances and possibilities, just as it presupposes the ontology of what constitutes 'sensible and viable' vis-à-vis the nonsense and non-viable… in a similar way that an analogical lightbulb presupposes light radiation, but cannot explain or furnish the ontology of light radiation… nor of itself.

The ordinances whose ontology is presupposed to the contingent process of development, and stands outside and supersedes the random-probabilistic sortology employed, analogises the meta-semantic hypostasis: The meta-semantic ordinances 'burn through' the

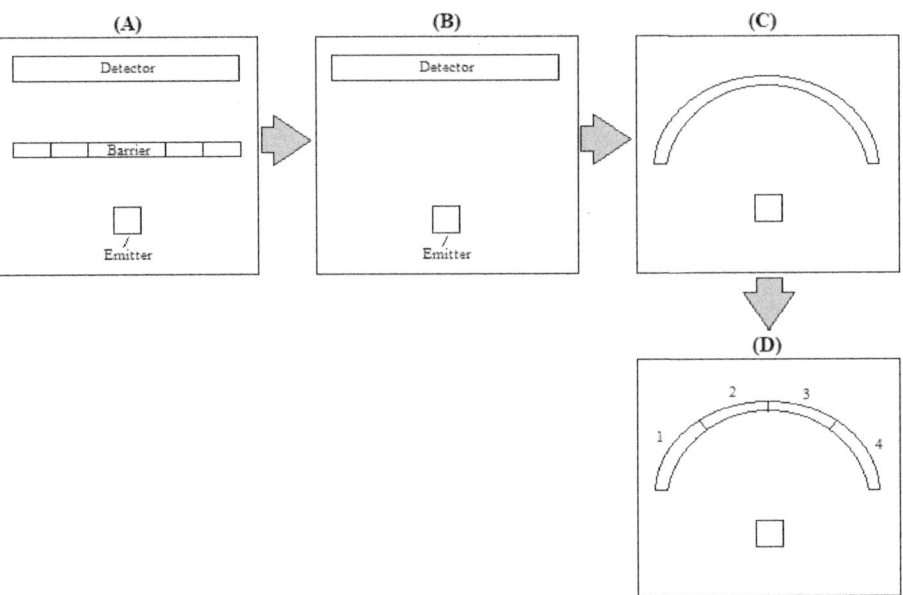

Fig. 3.03: The quantum mechanised die

Notes: The series of depictions illustrate how we can replace our mechanical classical four-sided die with a quantum mechanical die and bring direct quantum indeterminism into the generation of dice values and outcomes in lieu of classical-mechanical process whose randomness could always be doubted. Take the two-slit experimental setup depicted in **(A)**; Remove the barrier in **(B)**; Concave the detector at **(C)** and render it conformal to the arc of the quantum mechanical wavefront issued from the emitter, thus rendering particle-deposition equally probable across the total concave surface. Then partition the detector into four separate equal sectors with 4-sded dice in mind (or as many as there are dice-values or sides) in **(D)**: each lined with a single particle sensitive array; with each in turn wired to an output-reader (not shown) to register the pertinent die-values shown in **(D)**, all quantum indeterminately generated. Hence, we obtain the quantum mechanised four-sided die. Therein, there is a one-in-four equipotential chance that the single emitted particle will deposit in any one of four sectors. This will trigger the appropriate assigned die-value output. Of course, we could partition the concave detector surface into six sectors for the quantum mechanisation if a six-sided die: or ten sectors for the ten-sided die, etc. In this way, we can replace classical mechanical dice with quantum indeterminate quantum mechanised dice, or "true dice", and apply these in the developmental generation of, say, a generic bell curve distribution; or the development of the Sierpinski triangle (later) and, at present, the development of our jigsaw puzzle domain.

subsequent random contingent and even quantum indeterminate developmental sortology to decide what constitutes 'sensible and viable' vis-à-vis the nonsensical and non-survivable outcomes, despite indeterminacy and the default nature of outcome-generation. Also, the fate of the jigsaw puzzle domain (indeed, the probability structure behind its development) are *necessities* imposed on the developing system, even when that system subsequently develops on the basis of a pure contingency-driven process. The said ordinances of the jigsaw puzzle domain constitute the fundamental informational states of the system constituted as the probability structure (and rules of conformality) to which the sensible and random nonsense ensembles are pre-distributed: i.e., these are the *read-only* informational states that delimit and 'burn through' the intermediate process of contingency-based random development.

Now, in our jigsaw puzzle domain, at first glance, there is no obvious extant abstract ontology of ordinances: Yet the *a priori* ordinances are partly etched into the domain as the visceral conformalities of the pieces: These are not ghostly abstractions. Yet, as analogy, these capture the role of the abstract meta-semantic hypostasis that abides vis-à-vis the really-existing universe.

Again, the possibilities for conformality are presupposed to the contingent probabilistic process of their subsequent assemblies. The rules that determine the viability of a possible sensible jigsaw assembly are presupposed to the blind random sortology that governs their combinations and trials, and do not arise from the randomised combinatorial process involved. Beyond analogy, the principles that decide the sensible from the nonsense formations in the really existing universe are truly abstract and include extant, or transtemporal and time asymmetric physical laws and principles… despite the undeniable operation of contingency and happenstance, succinctly exemplified by quantum indeterminate processes.

In short, contingency does not obviate information and order grasped in terms of the *a priori* ordinances: Contingency, happenstance, and 'accident' *presupposes* ordinances as *a priori* information and order in order to be viable. That is, chance and

indeterminacy require necessity and order for 'accident' to produce, *even if by default,* **the coherent pattern-outcomes we observe throughout nature and in the universe.** Our artificial didactic jigsaw puzzle domain captures the essence of this claim.

*

Immediate applications follow. For example, contingency and happenstance play major roles in biological evolution. Yet, it turns out that, even with undeniable and ineliminable contingency, randomness and happenstance, some other factor must enter the fray. **Akin to contingency and happenstance in the development of our jigsaw puzzle domain, in the evolutionary process, living forms issue fort through contingency, randomness, and possibly even quantum indeterminacy. However, their viability is** *not* **a condition decided by the attendant contingency and randomness. It is the** *a priori* **ordinances furnished upon nature that 'decides' what is and what is not a viable biological development or living form, subsequent to the contingency and happenstance that issues the forms into existence, and regardless of the contingency that issued these forth.**

Thus, the *a priori* ordinances are implicit to undeniable evolutionary natural selection. The a-noetic and non-conscious contingent processes in biological evolution certainly issue living forms into existence and generate evolutionary change, apparently *without* the involvement of intelligent direction; i.e., by 'accident'. But it is the *a priori* ordinances inherent to non-blind natural selection that 'decide' the viability and survival of otherwise contingent evolutionary outcomes. In short, natural selection is the junction in and through which abstract ordinances enter the fray and filter evolutionary viable forms from non-viable ones.

None of this is in denial of contingency and randomness in biological evolution. The point is, randomness cannot obviate the role of abstract ordinances in *any* system or domain, including biological evolution, no matter how perfectly both contingency and happenstance prevails. Viability derives from *a priori* ordinances. Indeed, we assert that *all* possible or potential living forms, if not whole ecological possibilities and potential evolutionary pathways, exist as future-potentials and ordinances within the growing block spacetime in potentiality, and did so from the very beginning of the universe, inherent to the putative Big Bang monoblock.

Another application of the contingency-driven developmental process is cosmogenesis in Big Bang itself. The Big Bang monoblock is hypothesised to contain all possible physical constants, laws, and initial conditions, per all possible histories…and even alternative universes, including silly and unworkable ones, consonant to Manyworlds. It is typically speculated that our universe and its unique conditions, constants and laws were blindly selected by pure 'accident': succinctly, a quantum indeterminate leap at the expense of all other possible sensible and nonsense universes. Hence, our universe, its conditions, its constants, its laws, and the fact that it can exhibit stars, worlds and living forms… is claimed to be an outcome of pure 'accidental', or simply a contingent random developmental process, similar to the one envisioned in our jigsaw analogy. **While the selection of the universe into realisation out of the monoblock may well have been a matter of pure contingency and 'accident', the viability of our universe, or what actually 'decides' its viability, is** *not* **a matter of the said contingency or 'accident' that issued it at the putative Big Bang. The viability of our 'accidental' universe is strictly a matter of non-blind (but otherwise non-conscious)** *a priori* **ordinances inherent to the Big Bang monoblock. That is, the universe may be effectively 'accidental', but its very viability, or what 'decides' its viability, is** *not* **accidental, and cannot be explained… or explained away… by resort to the 'accident', no more than a lightbulb could account for the ontology and viability of electromagnetic radiation.**

In any case, the notion of *a priori* ordinances that stand as extant, timeless and transtemporal states, is no more radical than the ontology of abstract physical law, established in our Einstein Podolsky Rosen experimental-treatment of the conservation of spin: an instance of transtemporal law with which otherwise pure contingent quantum indeterminate spin-resolutions must abide, but cannot override. (See Preliminary **0.26**).

Also recall that the generic physics assertions in favour of time-symmetric physical laws, which we disputed and replaced with our time-asymmetric laws in Book-I (**1.11 to 1.13**) must be so constituted as to stand extant time itself. Indeterminacy entailed in the process of AND-to-OR collapse of complex outcomes out of future-potentiality must abide to the 'read only' laws or timeless ordinances that are part of the meta-sematic hypostasis: ordinances which 'burn through' otherwise objectively real contingency, indeterminacy, and happenstance… in exactly the way that the conservation of quantum spin 'burns through' the quantum indeterminate resolution of quantum entangled pairs in EPR experiments.

2. The brute-force exhaustive process of development: The development of our didactic jigsaw puzzle domain is suitable to description in brute-force exhaustive terms. **Instead of selecting just one history or one jigsaw puzzle development,** *all* **possible jigsaw puzzle combinatorial possibilities will be exhaustively realised into a system of Manyworlds puzzle-histories.** Therein, contingency and indeterminism have no role, given that *all* developmental possibilities will be realised on a brute-force exhaustive basis. Indeed, recall that the circumvention of quantum indeterminacy is the *raison d'etre* of Manyworlds. Hence, MWI constitutes a brute-force exhaustive developmental interpretation of quantum mechanics, which we now apply to our jigsaw puzzle domain.

It is tempting to assert nihilism vis-à-vis the total set of developmental jigsaw outcomes, given the blind, unplanned, mindless and brute-force exhaustive default nature of the development process. **The brute-force developmental process may certainly realise all possibilities, hence exhaust these in a mindless way without regard to sensible or nonsense outcomes. But what it cannot accomplish is the principles of viability of those outcomes: Again, this is 'decided' per function of the** *a priori* **ordinances, whose ontology is presupposed to the brute-force process,** *not* **furnished by that process.** Again, the 'lightbulb' cannot explain or furnish the ontology of 'light', its nature, and the laws that attend it, but can only presuppose the ontology of the latter. **This insight exposes the serious limits to any attempt or want to employ MWI as part of an anticipated nihilism in denial of information and memory. If only because such a utility presupposes that the brute-force exhaustive essence of MWI can obviate the ontology and**

289

existentiality of the *a priori* ordinances and the meta-semantic hypostasis from which these derive, and which the nihilistic Manyworlds universe can only presuppose but cannot furnish.

One immediate area of application of the brute-force exhaustive developmental approach is, again, cosmogenesis in the Big Bang. Recall that the Big Bang monoblock is hypothesised to contain the sum of all possible constants, laws and conditions for all possible universes. Whereas the contingency-driven approach to the Big Bang may claim that only one of these universes is brought into realisation. **In the brute-force exhaustive approach (or the application of Manyworlds to the same)** *all* **possible configurations of constants, laws and universes will be issued into realisation, including sensible and nonsense universes.** The typical attendant conclusion asserts that there is nothing special about our universe and history; no more than there is anything special about any other universe or history, even if many of these turn out to be silly and unworkable. **However, as with the lightbulb-light analogy, this cannot obviate the a priori ordinances writ into the Big Bang monoblock: i.e., the meta-semantic ordinances that 'decide' the Manyworlds repertoire of sensible from the nonsense histories and universes in an** *a priori* **fashion, even when all are formed blindly, without selection, exhaustively, or** *without* **divine intercession or intelligent design.**

Indeed, it is obvious that the replacement of contingency with brute-force exhaustive realisation does not explain away or obviate the fact that our universe is viable; and that it can engender life and consciousness; or, that it bothers to exist at all, or that many (if not all of the other universes) are unworkable or silly, even if *all* are brought into realisation by means of a mindless brute-force exhaustive process in Manyworlds.

Again, with the Big Bang, as with all else, **what decides the viability or no-viability of any outcome, history or universe is a non-blind but non-conscious meta-semantic hypostasis that transcends the brute-force exhaustive process that supposedly brought about our universe**... a meta-semantic state that, in some profound way, transcends the putative Big Bang monoblock itself, as much as it exceeds the presumed Manyworlds brute-force exhaustive process of genesis.

3. The mind-driven developmental process: We now come to the third and final process: namely, the mind-driven process of development. Therein, interceding mind supposedly 'decides' and orchestrates the development of our jigsaw puzzle domain. There are three variants of the mind-driven process of development: In the first, interceding mind plays at randomness. In the second, it plays at brute force exhaustion. But it is the third variant that will agitate the usual disputants.

a. Minding by arbitrary fiat: Herein, the interceding mind plays the role of a mindless random selector. **Minding by arbitrary fiat is nothing more than a minded version of contingency-driven development. Therein, the mind selects any combination of the jigsaw pieces for assembly, but does so arbitrarily and without concern to whether these form sensible or nonsense outcomes...** *without* **preference, plan, or design.** As before, what decides the outcome into sensible or nonsense will have nothing to do with the arbitrary will of the interceding mind playing at randomness, but everything to do with the non-blind non-conscious *a priori* ordinances presupposed to the arbitrary orchestrating mind. As was also the case with mindless randomness, minded randomness cannot obviate ordinance, order or information. Again, **it is the meta-semantic hypostasis that 'burns through' minded randomness and 'decides' what constitutes sensible or nonsense outcomes.**

b. Exhaustive minded intercession: Herein, mind simply plays at Manyworlds. Mind consciously generates *all* the jigsaw puzzle ensembles; sensible or otherwise; thus carrying out a minded variant of mindless brute-force exhaustive developmental process. But, just as was found in the same without mind, **minded brute-force exhaustion cannot decide what constitutes sensible outcomes from nonsense ones: these are decided by the non-blind a-noetic a priori ordinances that pertain to a meta-semantic hypostasis presupposed to the mind playing at Manyworlds.**

c. Minding by ordinance: Herein, **mind exercises conscious deliberate selection; discerning correct jigsaw puzzle juxtapositions from nonsense ones. Yet, what decides the correct selection from the incorrect combination ontically precedes the interceding mind, and it is furnished to mind as mind-independent ordinances, upon which mind totally depends, with respect to which mind plays the role of the supplicant.**

Insight into minding by ordinance recuperates the insight from Plato's dialogue, the Euthyphro. Rendered into contemporary form, Euthyphro's dilemma can be encapsulated thus: Is a thing good because God esteems it to be so, or does God esteem it to be good because it is *good in itself*? In the latter, the *good in itself* derives from a higher ordinance above God; one ontologically superior to God. In that case, what need of God?

However, in the former case, God Himself decides what constitutes the good on the basis of pure divine will, fiat, or whim… *without* resort to any higher God-transcendent order of principles…essentially through *divine minding by arbitrary fiat*… or perhaps through a *divine brute-force exhaustive process,* realising all alternatives into Manyworlds of the good or not-so good. In either case, assuming a non-nihilistic universe, what gets to constitute the good, the viable, the sensible, will depend on a set of ordinances that transcend God, even though God participates by either playing dice with the possibilities, or brute-force exhausts the possibilities, all *without* any divine discernment or concern in either case.

On the other hand, if God deliberates over the good, or plays *divine minding by ordinance*… God must again rely on higher principles that transcend it. Yet again, the basis of the good must transcend God. Indeed, ultimately, God will have no choice or decision in the matter of what constitutes the Good: a supplicant to the Good. To insist otherwise is to brook profound nihilism, beyond the egregious form entertained by materialist and secular nihilists.

We simply replace or repurpose the moral concern central to the original Euthyphro's dilemma with concern over the ontology of physical laws, principles, constants, and the pattern-structure of nature, as evinced in biological evolution, in

cosmogenesis, and across nature as a whole, analogised by the ordinances of our didactic jigsaw puzzle domain. When repurposed thus, we arrive at precisely the same conclusion: **What constitutes natural law, principle and ordinance must transcend the putative interceding and 'deciding' God. To insist otherwise constitutes a profoundly nihilistic move**; perhaps worse than nihilistic as resort to pure 'accident' or pure brute-force exhaustive cosmogenesis and evolution. And, as with the latter, God can only presuppose the ontology of the sum of possibilities, captured perfectly by the analogy of the lightbulb to 'light'.

We reiterate that, **omniscient or not, divine or not, mind is always ordinance-dependent. For mind to be viable, it requires access to an ontologically primary or independent meta-semantic hypostasis composed of abstract ordinances that are ontologically *a priori* to mind participating in the intellection and manifestation of order, though not itself the ultimate ontological source of order**, just as the lightbulb is not the ultimate ontology source of light. The set of ordinances, and the hypostasis from which these issue forth, are both non-blind and non-conscious; always *a priori*, even to an omniscient supermind, or God misconceived thus. Thus, **order is *not* ultimately mind dependent. But mind is *always* order-dependent**: an order of intelligibles furnished to mind, but not conjured or 'created' by mind. The very viability of mind and consciousness, whether merely human, or else super-conflated to God, remains wholly dependent on mind-transcending and God transcendent meta-semantic hypostases.

*

Note the two immediate applications: From a creationist approach, it might be postulated that God selected our universe out of the welter of other possibilities writ the Big Bang monoblock. But did the putative God select our universe by arbitrary fiat? If so, the pertinent ordinances came into operation by default: i.e., the ordinances that make for our sensible universe were *not* decided by God, but, at best, belonged to a meta-semantic hypostasis co-eternal with, but ontologically transcending God playing at randomness. Thus, that our universe is viable, and makes for interesting developments, including life and consciousness… exceeds and supersedes the power of any putative God that may have engaged in arbitrary selection of the universe from the Big Bang monoblock menu.

Did the putative God realise all the Manyworld universes writ in the monoblock through a brute-force exhaustive means? If so, the viability or otherwise of the formed universes will be per the meta-semantic ordinances, *not* per a God playing at 'multiple choice', but one that 'ticks *all* the choices'.

However, if we assert that God conjured the ordinances *ex nihilo*, without resort to any meta-semantic hypostasis, dismal and profound nihilism will follow.

Thus, did the putative God create the monoblock and it's *a priori* ordinances, or were these co-eternal with God? If the ordinances were co-eternal with God, then God did not create them. In that case, what need of God? If God created the ordinances *ex nihilo*… necessarily on the basis of arbitrary fiat… in that case, there could be no real informational ontology, no meta-semantic hypostasis; and only dismal and profound information nihilism… unless the arbitrary whim of God was subject to meta-semantic ordinances that transcended God, or had semantically 'burned through' mere divine whim. In that case, it was because of the order issued from the meta-semantic hypostasis beyond God that cosmos emerged, despite the supposed arbitrary fiat of God.

The same sorts of problems abide in the matter of the formation of life: Did the putative God arbitrarily select the basis for what constitutes viable lifeforms, or did God rely on a pre-existent set of ordinances beyond God? The former furnishes total nihilism. The latter obviates the need for God, now reduced to the status of a midwife to order, at best.

Subscription to mind, even in the form of God, as the *ex-nihilo* arbiter of the constants and laws of the universe, of potential evolutionary principles and possibilities, or even of moral principles… but *without* concession to a meta-semantic hypostasis above and transcending God, is tantamount to placing arbitrary divine whim over and above order and principle, degenerating into total dismal nihilism.

*

There is no denying the role of contingency and happenstance in the development of the universe and in the formation of life. But we cannot use this fact to deny the ontology of ordinance, of order, and of the meta-semantic ontology of information. In its turn, the MWI-pertinent brute-force development approach to the constants, laws, and patterns of nature, including the evolution of living forms, cannot be utilised to assert the arbitrariness of those same-said ordinances, or support the case for informational or mereological nihilism. The constants and laws of nature, its patterns, or even critical experimental informational states, such as the barrier setup in the two-slit experiment, cannot be asserted as arbitrary by resort to brute-force exhaustive development.

Ironically, the ultimate way to assert for the arbitrariness of constants, laws and patterns of nature is by resort to God, at least in the form misconceived as omnipotent supermind… wherein the said constants and laws are allegedly brought into existence by God, *ex nihilo*… purely on the basis of absolute divine whim or fiat…by a God that stands above, ontologically 'creates' (from out of divine whim) the meta-semantic order or principle: culminating into a most profound nihilistic turn that outcompetes even generic secular materialistic forms of nihilism.

*

In conclusion, any argument for information nihilism that involves resort to mindless contingency and randomness, or to MWI style brute-force exhaustion of possibilities, must fail. Neither contingency nor brute-force exhaustion of possibilities can obviate final resort to meta-semantics. Ironically, the ultimate method to deny the real existence of the meta-semantic hypostasis involves resort to God, of a form misconceived as super-conflated omnipotent mind.

All of the above constitutes contributory foundational material to a *non-Arbitrariness* and *meaningful universe thesis,* based on the plausibility of the meta-semantic hypostasis, involving semantic 'burn-through' of ordinance and principle vis-à-vis contingency-driven, brute-force exhaustive driven, and even mind-driven developmental processes. Although these matters require a whole book in their own right, the material furnished hitherto remain sufficient for our present purpose.

The universe is *peculiarly* non-arbitrary and meaningful, and remains so despite undeniable mindless quantum indeterminacy or prospective brute-force exhaustive processes: It would remain meaningful despite the vagaries and mischiefs of any consciousness, or even of a putative 'God'. It also follows that nihilistic assertions for a supposed meaningless universe are void. Indeed, the following essay in direct critique of Manyworlds will constitute the *coup de grace* against information and memory nihilism, and it is key to the anticipated thesis on non-arbitrariness and the meaningful universe.

3.15: Apagogic Mnemonic Argument 2-c: The Meta-Semantic Critique, Manyworlds-based Nihilism & Memory Erasure

In really-existing two-slit experiments, consistency tends to hold between the initial condition and subsequent outcomes. Thus, if we enforce the **one slit open** condition on the barrier, we are not going to obtain the interference pattern-outcome expected from the **both slits open**, notwithstanding Wheeler's cat. Conversely, we will not obtain a single slit distribution from a **both slits open** condition, notwithstanding Afshar (**0.16** to **0.17**). Is the observed consistency brought about by the carry-over of information in memory, or is it secured by a pure a-causal and a-informational brute-force exhaustive processes per Manyworlds? Two illustrations (**Fig. 3.04** and **Fig. 3.05**) help address these questions in the light of the preceding meta-semantic critique of contingency-based, brute-force exhaustive based, and mind-driven development-based approaches (**3.14**). Therein, we found that these processes cannot obviate basic informational ontology or the requisite meta-semantic hypostasis.

Recall from **3.14** that the brute-force exhaustive developmental approach is essentially synonymous with the form of developmental process inherent to Manyworlds. To this end, **Fig. 3.04** depicts the formation of Manyworlds brute-force exhausted alternative histories in relation to an initial root-history composed of a two-slit experiment with **both slits open**.

For reasons argued in essay **3.06**, upon isolation, the barrier must futurise and dissolve into an AND-form superposition-state of alternative futures comprised of...

> ... {both slits open...AND...**only left slit open**...AND...**only right slit open**...AND **both slits closed**
> **| 0.25 probability of occurrence of any barrier setup}**

With subsequent measurement and attendant AND-to-OR collapse, the above wavefunction is normally expected to resolve into just one OR-form outcome, with the elimination of all alternatives, on the basis of 1-in-4 probability, baring memory. Yet, **according to MWI, the root will generate *all* four future possibilities into distinct histories, on the basis of brute-force exhaustive development. Consequently, the history with the consistent outcome will be guaranteed to occur, but only on a pure brute-force default basis...** *not* **because of any informational resonance of the initial barrier-setting vis-à-vis the outcome. As a consequence of the same brute-force exhaustive process, three out of the four histories will be nonsense or non-consistent histories, wherein the barrier will subsequently resolve into an entirely inconsistent form vis-à-vis the root setup.** Only Universe-A, will be 'sensible' or consistent: (see **Fig 3.04** for clarification).

Hence, with Manyworlds, it is possible to generate a universe with *apparent consistency* and *apparent information*, and even the *apparent retention* of initial conditions, but with no ontically real information or abstract memory... with no *real* consistency. If so, what need of the notional temporal carry-over of information across time, or of the retention of such information as memory, let alone the supposition that information is ontologically real?

Thus, through the brute-force exhaustive approach furnished by Manyworlds, we restore memory erasure espoused in **3.06**, albeit on a more effective form of nihilism. Indeed, both nihilisms are effectively co-joined, although MWI can circumvent the quantum indeterminism central to erasure hypothesis.

How do we escape this ultimate nihilistic *cul de sac*? To solve the crisis, we must go beyond consistency issues pertaining to the retention of initial conditions and consistent outcomes vis-à-vis the barrier: We must investigate consistency issues that pertain to the barrier-detector correlations. These are depicted in **Fig. 3.05**.

Per its inherently nihilistic brute force processing of histories as mere default outcomes, **Manyworlds *seemingly* obviates causality *and* information: Its framework cannot furnish a meta-semantic hypostasis, much less any backchannel to such a hypostases, much less support abstract memory, so as to secure consistency between the barrier vis-à-vis detector-outcomes** in 16 distinct histories: see **Fig. 3.05**. Indeed, **any claim about 'consistency', even to support the assertion that many of the histories are 'inconsistent' and nonsensical, *must* require backchannel to a meta-semantic hypostasis, given that the information pertinent to such a judgement, and its ontological basis, *cannot be furnished* by the face-front processes inherent to Manyworlds.**

How will the denizens attached to non-consistent nonsense histories in both **Fig-3.04** and **Fig. 3.05** discern inconsistency and nonsense in their barrier-detector correlations, if full-fledged nihilism must reign absolute? In the fifteen inconsistent and nonsense histories depicted in **Fig. 3.05**, there exists no intrinsic information furnishable by the brute-force exhaustive *modus operandi* of Manyworlds, upon which attendant denizens of these histories could judge the barrier-detector correlations and histories to be 'inconsistent'. It follows that the said denizens could never discern 'inconsistency', much less grasp or communicate its opposite 'consistency'. Indeed, it will remain totally impossible to discern, intellect and communicate about *anything* in *any* Manyworlds generated 'non-consistent' history; not merely because of the 'inconsistency', but simply because there is no possibility for the requisite

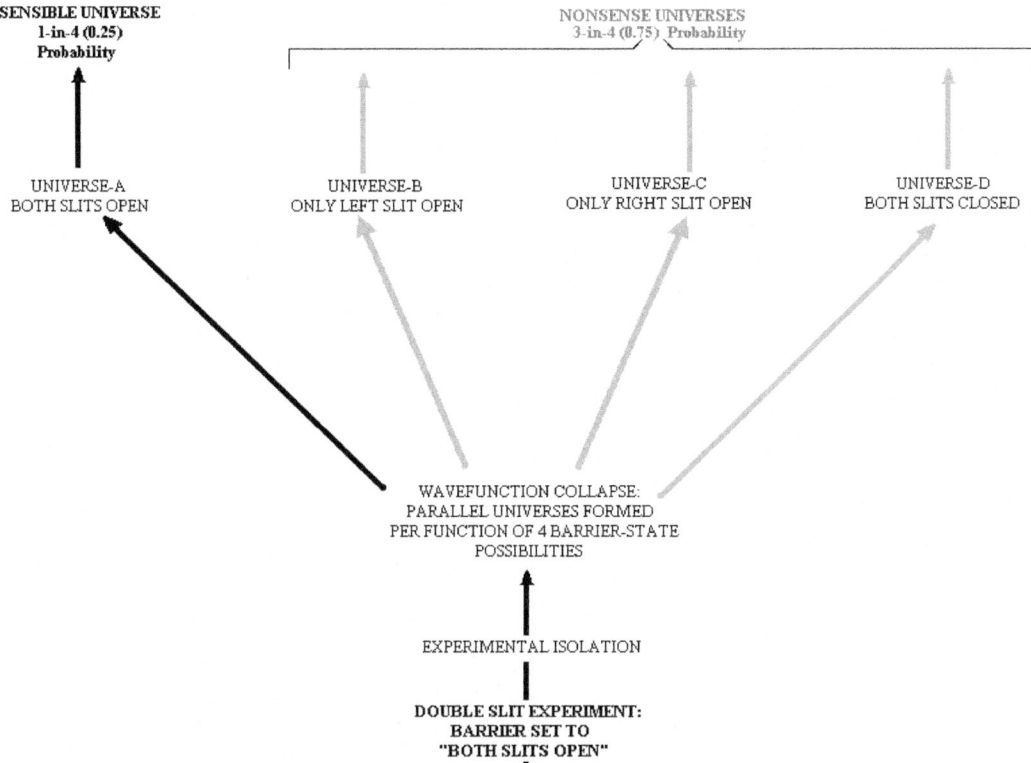

Fig. 3.04: Manyworlds interpretation revisited: sensible-consistent and inconsistent-nonsense histories
Notes: The illustration depicts the Manyworlds account of the two-slit experiment and the formation of many histories, wherein the ontic reality of information and memory is seemingly rendered superfluous. A two-slit experiment is set up at the root history: The barrier is set to **both slits open**, followed by isolation and consequent quantum mechanisation. With quantum mechanisation, per futurisation and the firewall principle, the barrier m*ust* 'dissolve' into a future-form quantum wavefunction comprised of all the four possible barrier-settings as future alternatives. Per Manyworlds, all the barrier settings will be realised as distinct compartmentalised histories. Note that only **Universe-A** will constitute a sensible history, wherein the barrier at the end of isolation is the same as the barrier at the initial setup. All the other histories will be inconsistent nonsense histories. Yet, the only sensible history, **Universe-A**, will be realised purely by default: by an a-causal brute-force exhaustion of possibilities, without memory carry-over and, *without* information, *without* the principle of consistency: The outcome happens, not because consistency demands, but purely by the exhaustion of possibilities.

information or memory from within Manyworlds and its histories: Manyworlds cannot furnish any basis or ontology for such a possibility from within itself. Therefore, the denizens of the 'inconsistent' histories could not judge their attendant barrier-detector correlations and histories to be 'inconsistent', or discern anything else for that matter.

Similarly, in the single Manyworlds history with *apparent* 'consistency' between barrier and detector-outcome, given that there exists no intrinsic information furnished by the brute-force process inherent to MWI upon which the attendant denizens could judge the barrier-detector relation to be 'consistent', and given that they are not provisioned with *any* backchannel to a meta-semantic hypostasis by the MWI that created their 'consistent' history, the denizens could not discern 'consistency', much less grasp or communicate the difference between 'consistency and inconsistency', or discern about *anything*. Hence, the denizens in the one 'consistent' history could not judge their attendant barrier-detector outcome and correlation to be 'consistent', or discern, intellect or communicate about *anything*, because MWI in itself cannot furnish any basis for this.

But we must reintroduce the all-or-nothing rule. The above assumes that the denizens, and we who are looking in from without, could agree on the basic identity-constituents involved in the nonsense *and* sensible histories generated: such as the identity "barrier"; such as the identity "detector"; such as 'both slits open', etc. By what right do we suppose that these survive and abide in the nihilistic a-mnemonic and a-informational Manyworlds daz? How can we be sure that these identities do not, say, transform into pumpkins instead, or into lemons instead,… or into amorphous 'nothings'? The answer is obvious: Given the nihilistic nature, process and content of Manyworlds *in itself*…or *without* the possibility of backchannel to a meta-semantic domain… there can be no assurance of the

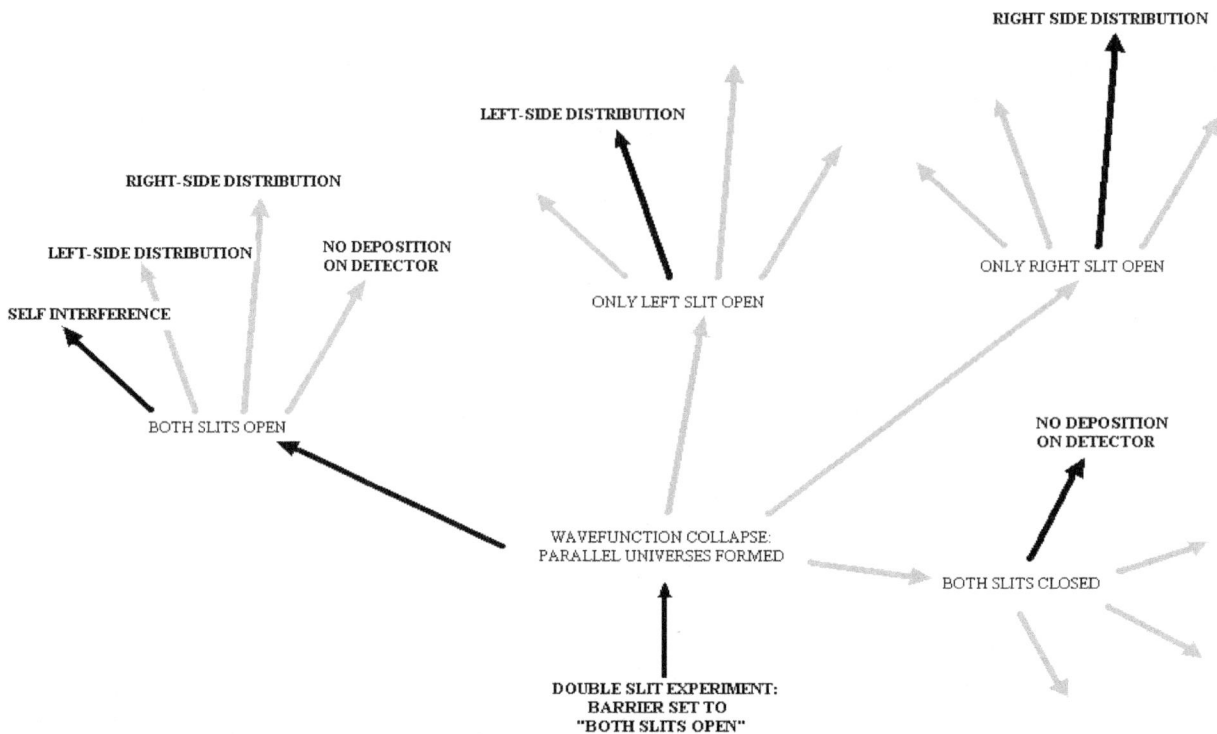

Fig. 3.05: The apagogic invalidation of Manyworlds-based informational and mnemonic nihilism

Notes: The Manyworlds process of brute force exhaustive a-causal, a-informational and a-mnemonic development process, insofar as it is inherently nihilistic, cannot furnish the consistency or otherwise of the four possible barrier-to-detector correlations and histories pertaining to the development of the two-slit experiment. Hence, when the root wavefunction generates ***both slits open*** as one of four Manyworlds histories, this must further develop into four subsequent histories, wherein ***both slits open*** correlates with ***self-interference*** … with ***left-side distribution*** … with ***right-side distribution*** … and with ***no deposition on detector***, respectively. Only one of these will constitute a seeming consistent history, wherein ***both slits open*** correlates with ***self-interference***: The others will be nonsense and inconsistent histories, wherein ***both slits open*** will correlate with, ***left-side distribution***; with ***right-side distribution***, and with ***no deposition on detector***, respectively. In total, only one out of sixteen default histories will constitute the 'sensible history' in which 'consistency' abides (highlighted by continuous black arrows) with the rest constituting 'inconsistent' nonsense-verses. Notwithstanding a universal Alzheimer's effect, wherein all information and memory must wash out even from the one-and-only sensible 'consistent' history, Manyworlds-based information and memory nihilism must assume its own intelligibility as a non-subjective claim *and* objective state of the universe: This ought to be impossible in a nihilistic ontology, even for denizens in the one sensible history, given that information and memory cannot temporally carry-over, and it cannot have any ontologically real status *from within* the totally nihilistic Manyworlds scheme*, even in the one sensible history*. Thus, any claim about the state of any history, and even a Manyworlds-based assertion for information and memory nihilism from within *any* history, including from within the one sensible history, must presuppose backchannel to a meta-semantic hypostasis composed of *a priori* ordinances and ideas about 'consistency' and 'inconsistency', beyond what Manyworlds *in itself* could furnish, and which *cannot* be furnished by Manyworlds at all, owing to its inherent nihilism per function of its brute force exhaustive a-causal, a-mnemonic and a-informational modus operandi. The fact that we *can* furnish such judgements in our own universe implies either a Manyworlds *with* a backchannel to a meta-semantic hypostasis, or a single unique history without Manyworlds, wherein information is inherent and objective to the system. Both options undermine nihilism *and* the utility of Manyworlds to prospective informational and mnemonic nihilism.

pertinent identities, or of *any* identity,… or of *anything*. Indeed, the all-or-nothing approach asserts that we can only be assured of the total absence of information *if* we rely on Manyworlds *in itself*, or conceive Manyworlds *without* backchannel to any meta-semantic domain. In other words, by the deliberate exclusion of the all-or-nothing problem of nihilism, we had indulged in over-generosity to nihilism, but can now appreciate that *all* nihilisms, including Manyworlds co-joined to memory erasure, are truly apagogically superfluous and false.

QUANTUM MECHANICS AND MIND

How is it possible for *us* to judge about barrier-detector outcomes and consistency issues in *our* history, if it is nothing but the product of an a-causal, a-informational and a-mnemonic nihilistic developmental processes, without even the possibility of a backchannel to a 'religious' meta-semantic hypostasis, even ignoring the all-or-nothing problem? Clearly, we could not rely on any intrinsic information ontology, transfer, or memory resonance to scaffold for such judgements, given that, *in itself*, neither nihilistic memory erasure nor nihilistic Manyworlds could furnish such miracles. Yet, given that we *can* and *do* judge and transact about barrier-detector-outcome-consistencies in *our* universe, it follows that, at a minimum, we *indeed* rely on a backchannel to a meta-semantic hypostasis, despite and extant erasure and Manyworlds, *assuming* either erasure or Manyworlds held true, vis-à-vis to our history and universe.

Hence, assuming the all-or-nothing argument was not sufficient to the task... the *coup de grace* against MWI-based nihilism is that Manyworlds must impale itself on one fork or the other: Either MWI is true, and the only explanation for our consistency-assertions is a backchannel to an ontologically extra meta-semantic hypostasis... in which case the whole nihilistic Manyworlds enterprise against information and memory is rendered void... or we do *not* inhabit a Manyworlds universe. In which case, to secure our judgements about consistencies in our universe, we likely rely on an integrated ontology of universe *and* information (on memory, possibility and time vis-à-vis time-asymmetric law and principle) and the retention of information as abstract memory temporally resonated ...*without* erasure. Consequently, mnemonic, and informational nihilism conceived in terms of Manyworlds must fall apart, as surely as had the preceding memory erasure hypothesis.

*

We assert that, in our really-existing universe, both intrinsic information *and* abstract memory, *and* the meta-semantic hypostasis, abide. There really is a meat-semantic hypostasis extant the world, whether the latter is conceived as a single unique history or as a parallel universe of many histories: in either possibility, we enjoy a backchannel to that hypostasis. However, the unique history or parallel histories really do possess inherent information, wherein the inherent information is retained as abstract memory, and wherein its temporal resonation abide.

To grasps why both the meta-semantic hypostasis *and* intrinsic information must abide, we must clarify the Kantian error. Immanuel Kant claimed that it is the categories of the mind that furnish to experience its intelligibility. Therein, the Kantian categories would be similar to Platonic forms, in that mind intellects these... or they are furnished upon the mind, which then imposes these upon experience to render the world intelligible and sensible... and render experience as we know it possible. Our own backchannel to the meta-semantic hypostasis fulfils the same role as Kantian categories vis-à-vis the world, albeit recuperated in the framework of Manyworlds. Yet, Kant also claimed uncertainty about the world *in itself*. To summarise, Kant claimed that, though the mind imposes the categories upon experience, the world *in itself* might be totally different from the categories so imposed. Indeed, one may as well deny the world *in itself*. But the move renders Kantian philosophy vulnerable to the charge of hidebound idealism.

However, for Kantian categories to work, or for our meta-sematic hypostasis to work, the 'world in itself'... or the histories espoused in Manyworlds... must enjoy conformality to the categories imposed on them, in the same sense that two jigsaw pieces must conformally and seamlessly fit together, even though their forms might otherwise be totally distinct. That is, the 'world in itself'... hence the histories of a putative MWI... must conformally 'fit' with the categories of Kant, as surely as the categories must fit vis-à-vis the world. But the necessity of 'fit' necessarily implies that the 'world in itself' possess inherent content conformal to the abstract categories and forms imposed upon it. That is, the 'world in itself'... the object attended by the subject... possesses an informational character all its own, albeit conformal to the abstractions furnished upon mind and upon the world through backchannel to a meta-semantic hypostasis, even as the latter transcends both mind and world. Otherwise, the meta-semantic hypostasis could not furnish the intellection and observation of 'consistency' if the world according to Manyworlds did not possess an appropriately conformal stand-alone content *in itself*: one conformal to 'consistency'. Hence, the world *does* possess content *in itself*. Thus, we assert that *both* intrinsic information *and* memory, *together with* the transcending meta-semantic hypostasis, abide. Hence, mnemonic, and informational nihilism per Manyworlds must wither, as surely as had nihilistic assertions from previous memory erasure hypothesis.

*

To augment our *coup de grace*, consider that, *in itself*, the Manyworlds universe would constitute, not merely an amnesic universe, but an Alzheimer's universe, wherein the communication of *any* message, including messages with nihilistic assertions, could not resonate from one moment to any succeeding moment, even if a default 'sensible' and 'consistent' history should form up *and* include an intact nihilistic 'message'. Even in the single 'consistent' and sensible history depicted in **Fig. 3.05**, an Alzheimer's effect would certainly reign, *unless* by resort to a meta-semantic hypostasis that might otherwise support the ontology of the nihilistic message by means of semantic burn through.

On the other hand, if the meta-semantic hypostasis does not exist, and even less the possibility of 'semantic burn through', then the intellection and communication of any fact, observable, idea, or even a 'messages with a nihilistic assertion', will be rendered impossible, even in default 'consistent' histories, **The fact that, in *our* universe, we can and *do* intellect facts, observables, *and* communicate absurd apagogic nihilistic messages and assertions, *and* can transact about consistency and non-consistency of barrier-detector correlates, necessarily implies that Manyworlds-based nihilism and memory erasure are false, and that any nihilistic assertion must culminate into an apagogic liar's paradox. Consequently, information and abstract memory are real... not possible to obviate by resort to a brute-force exhaustion of possibilities via Manyworlds, any more than by resort to memory erasure hypothesis consequent upon AND-OR dualism and quantum indeterminacy.**

295

As is obvious from our apagogic mnemonic **argument 2-c,** in our universe... *in the real universe...* we can and *do* communicate even absurd apagogic messages. We also possess peculiar metalative notions or intuitions about 'consistency', which cannot be dismissed by resort to any arbitrary brute-force exhaustive default correlations of MWI, much less by resort to memory and information-erasure; and even less by resort to the more vintage variant of nihilism per causality-free 'constant conjunctions' of David Hume[74], or to the older Heraclitan notion of absolute impermanence. At the minimum, we must concede to a backchannel to, and the existence of, an abstract meta-semantic hypostasis... *and*, with it, the ontology and resonation of information across time.

3.16: Anticipating the Falsification of Manyworlds from
The Solution to Quantum Indeterminism & the Stochastic Coherence Paradox

Historically, the Manyworlds interpretation[75] emerged out of the want to circumvent the problem of quantum indeterminacy. At face value, it appears implausible that any kind of pattern and structure could ever arise out of AND-to-OR collapse processes whose outcomes are quantum indeterminately resolved. The solution from MWI was to suppose, not just one AND-form possibility, but *all* possibilities that constitute the wavefunction form up into compartmentalised alternative realised histories. This appeared to forgo the need for quantum indeterminacy, at least at the Manyworlds or multiverse level, since outcomes from AND-to-OR collapse are no longer ultimately matters of indeterminate probabilistic sortology but of inexorable all-exhaustive brute-force development.

However, it is not a trivial note that *some* possibilities inherent to the AND-form wavefunction enjoy higher amplitudes or higher probabilities of occurrence than others. If all the possibilities are destined to be generated as distinct histories, why does nature bother with amplitudes of occurrence, wherein some possibilities are favoured more than others? Indeed, all the possibilities ought to be equi potentialised and equi-probable, or set to equal amplitudes. Manyworlds cannot address this puzzle, although it can certainly ignore it and survive it.

There is no need to reiterate the arguments so far made against Manyworlds from preceding apagogic mnemonic case. In the following, and in Part-II, we will finally resolve the stochastic coherence paradox; the problem of quantum indeterminism. Therein, we will find that the solution to the stochastic coherence paradox entails the objectivity and ontic reality of randomness and of quantum indeterminism: *not* its circumvention by MWI or by any other interpretation. Moreover, our solution to quantum indeterminacy in Part-II will serve to bring about certain insights into the nature of both information and abstract memory, *and* constitute the final argument against Manyworlds, which, as stated, was historically postulated as solution to the problem posed by quantum indeterminism.

If Manyworlds is to be falsified by our anticipated solution to quantum indeterminism and the stochastic coherence paradox, so is information nihilism, or any nihilistic memory erasure hypothesis, or attendant postulate for an amnesic universe, all of which ostensibly require the veracity of Manyworlds and the presumption of quantum indeterminism as a 'problem' that needs a Manyworlds 'solution'.

3.17: Apagogic Mnemonic Argument 2-d:
The Superfluency & the Invalidation of the Manyworlds Interpretation

Does wavefunction collapse generate only one unique history, with all the alternative possibilities erased, or does wavefunction collapse resolve *all* the AND-form possibilities into multiple alternative histories, as espoused by the Manyworlds interpretation? The **Alternative Realist thesis (Alt-R) posits the former; the single unique history model... against Manyworlds.** To demonstrate the invalidity of MWI we require a simple model of the universe: one in which we can run the brute-force exhaustive process of development central to Manyworlds. The chaos game based on the Sierpinski triangle offers just such a model.

The Sierpinski triangle, as a bounded probability state that regulates seed-orbits otherwise generated indeterminately, serves as a useful model for a wavefunction with three AND-form possibilities (i.e., the tendency of the seed to gravitate toward any one of the three vertexes of the Sierpinski triangle, expressed as future-potentials implicitly embedded within spacetime in potentiality). On the other hand, the orbit of the seed is equivalent to successive AND-to-OR wavefunction collapse-processes, with the successive orbits of the seed constituting OR-form outcomes resolved on a 0.33 probability via AND-to-OR collapse.

As stated, we could replace the seed-orbit generator with a quantum mechanised dice, thus guaranteeing that the seed-orbit is truly quantum indeterminate: useful when reasonable doubt might exist about the authenticity of randomness furnished by the use of classical mechanical dice. However, the incorporation of quantum mechanised die presupposes that the quantum indeterminate generation of outcomes do not involve hidden gears or hidden variables that orchestrate the whole process in a determinate but hidden way. Later, in essay **3.32**, we shall discover that there are no hidden gears behind quantum indeterminacy, and that randomness is ontically and objectively real. Therefore, quantum mechanised dice are authentically indeterminate.

Normally, a quantum mechanised Sierpinski triangle would emulate the development of a single unique-history universe, wherein the quantum mechanical resolution of each seed-orbit constitutes a string of unique outcomes that form a single history, culminating into the expected fractal pattern unique to the Sierpinski triangle. But how do we apply the Sierpinski triangle approach to a Manyworlds framework? To accomplish this, **we posit that all the three possible seed-orbits will AND-to-OR collapse into**

---

[74] Hume, David, (1748) An Enquiry Concerning the Human Understanding, Essay VII

[75]; Wheeler, J. A.; DeWitt, B. S.; Cooper, L. N.; Van Vechten, D.; Graham, N. (1973). DeWitt, Bryce; Graham, R. Neill (eds.). *The Many-Worlds Interpretation of Quantum Mechanics*. Princeton Series in Physics. Princeton, NJ: Princeton University Press. Everett, Hugh

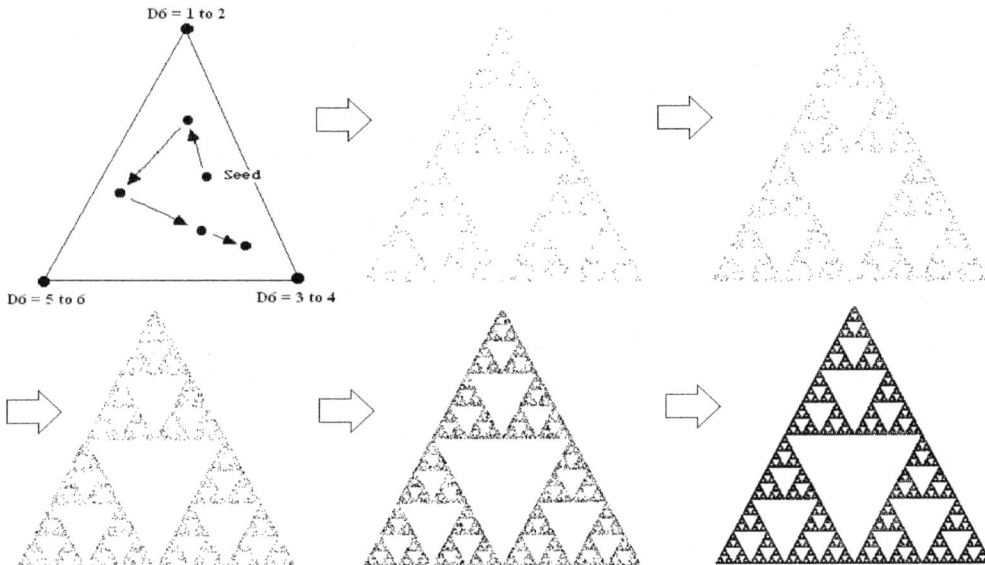

Fig.3.06 Sierpinski triangle, and physical information as minimal abstract bounded probability state
Notes: The illustrations depict the "chaos game" involving the Sierpinski Triangle, pertinent to the development of a single unique history through a quantum indeterminate seed-orbit process. The chaos game is played by describing an equilateral triangle whose three vertexes are labelled "1 to 2", "3 to 4" and "5 to 6", respectively, as a reflection of possible outcomes from the roll of a quantum mechanised six-sided die, D6. We pick a point within or just outside the triangle: the **seed**. We roll a D6 and, pending the outcome, move the seed halfway toward the pertinent vertex. We mark this new position and roll another D6, repeating the process. After a hundred or so trials, we obtain a fractal pattern as shown above (bottom, far right) despite intrinsically random and probabilistic processes of resolution of the seed-orbits. With quantum mechanised dice, the process need not be restricted to classical random-probabilistic processes.

realisation into distinct histories. The seed-orbit generation will no longer be a matter of probability but of brute-force exhaustive certainty for all three possible seed-orbits and histories.

Thus, per MWI, *all* the possibilities will get to be realised, constituting a first round exhaustive Manyworlds with three alternative distinct histories. In one history, the seed-orbit will develop toward vertex-**A**: In the accompanying history, it will orbit toward vertex-**B**. In the third, toward vertex-**C**.

By the second round, **each of the three histories will resolve into a second set of histories: i.e., in their turn, each will niche a set of successions composed of three seed-orbits and histories, making for nine histories at that second round**. The nine histories at the 2nd round are also expected to form distinct histories linked to preceding histories.

What happens at the Nth round? The nine histories from the second round will be succeeded and extended into 27 histories at the third round. Thereafter, 81 histories. Thereafter, 243 histories...and so on. The defining fractal pattern characteristic of the Sierpinski triangle will begin to emerge at around the 100th seed-orbit. At which point, we will obtain the critical contradiction which undermines MWI. At the 100th round or above, **the supposed 'alternative' histories will prove *not* to constitute alternatives at all, but *identical* global histories**. (See **Fig. 3.07**).

What is the point of envisioning so many histories if, to obviate quantum indeterminism, or to foist anticipatory informational and mnemonic nihilism, we accomplish a vast number of redundant *identical* global histories, whatever the unique pathways and seed-orbits that produced these? We will end up with a colossal number of identical Sierpinski-verses, notwithstanding all the hidden violations of the principle of the conservation of energy-matter, pondered in **3.13**.

Could we obtain any different a conclusion if, in place of the Sierpinski triangle, we substitute the Manyworlds development of a complex photographic image? None. Even if the universe split into as many histories as the deposition-paths and possibilities inherent to the complex photographic wavefunction, these would all culminate into the same global outcome: the same photographic image. **And what if we substituted in place of the photographic image the complex wavefunction of the universe: Hugh Everett's very own universal wavefunction itself?** Again, we would end up with an unimaginable number of supposedly 'alternative' developmental histories...all culminating into a huge number of *same* universes.

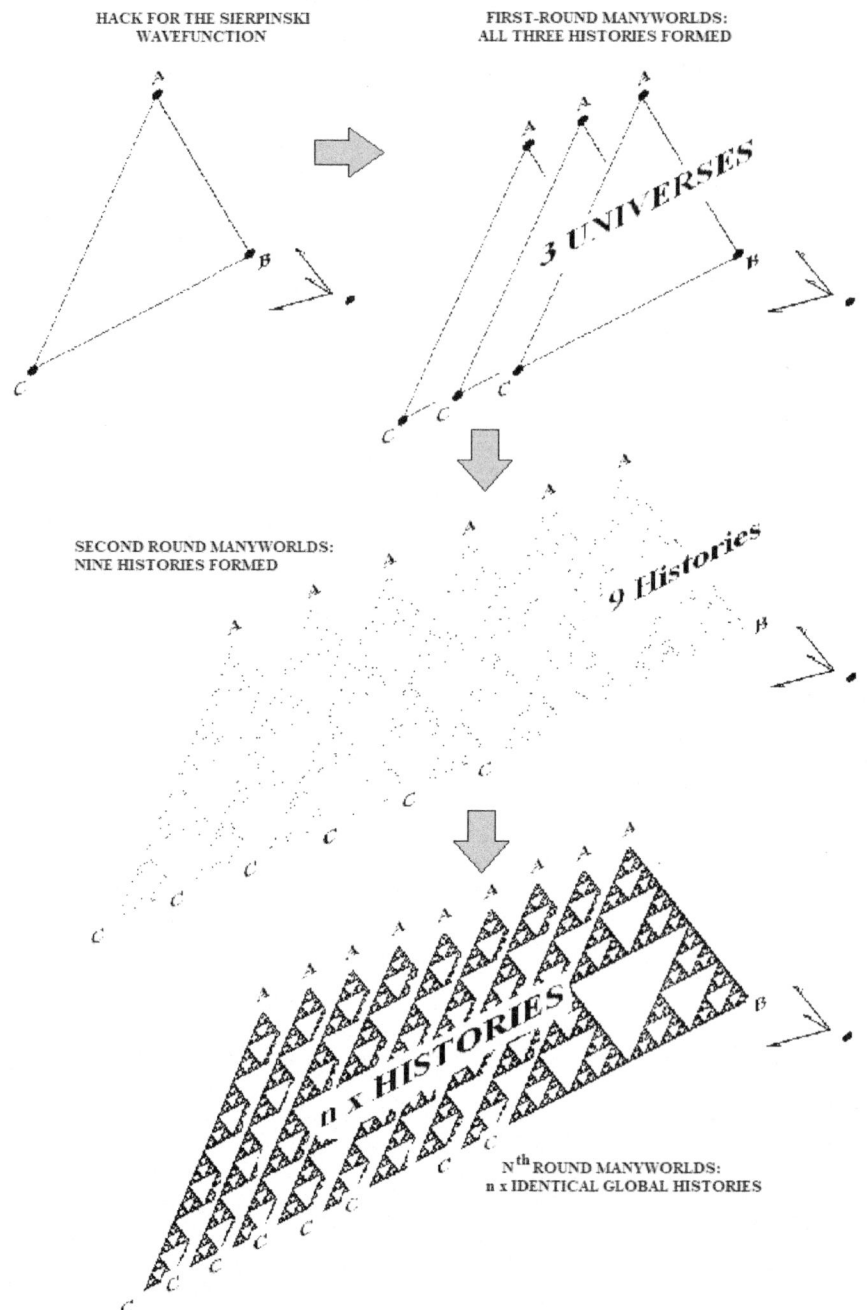

HACK FOR THE SIERPINSKI WAVEFUNCTION

FIRST-ROUND MANYWORLDS: ALL THREE HISTORIES FORMED

3 UNIVERSES

SECOND ROUND MANYWORLDS: NINE HISTORIES FORMED

9 Histories

n x HISTORIES

N^th ROUND MANYWORLDS: n x IDENTICAL GLOBAL HISTORIES

Fig. 3.07: Manyworlds brute-force exhaustive development of a quantum mechanised Sierpinski triangle
Notes: The fallacy of Manyworlds illustrated using the quantum mechanised Sierpinski triangle: By the Nth round, the vast number of supposed 'alternative' non-identical histories culminate into a large number of identical global histories constituted as the definitive same-fractal pattern of the Sierpinski triangle itself. The same would be obtained from the universal wavefunction: What is the point of Manyworlds, if it must culminate into vast numbers of same or duplicate histories?

Note that **our approach has exposed the false notion that an alternate quantum indeterminate outcome, generated from the welter of AND-form possibilities inherent to a root wavefunction, *must* imply wholly different universes or histories, as espoused in MWI. Yet, it is the global state and outcome that finally matters, *not* the 'uniqueness' or distinction of any subordinate**

constituent discrete event, seed-orbit, or pathway, all of which must culminate into the same final global pattern-outcome through the materialisation of a same-identity 'instantaneous whole' or root complex wavefunction. It is *not* the formation-path or behaviour of any 'pixel' that matters, but what the 'pixels' get to form *en masse*. When we fetishise the significance of quantum discrete outcomes, only to ignore the grander pattern and whole manifested by the collective of such, we miss the fallacy inherent in Manyworlds.

What, then, is the point of Manyworlds?

*

Recall that our search for abstract memory revolved on the fate of the barrier as a critical piece of information; a concern fundamental to the viability of generic quantum mechanics as we know it (see **3.7** and **3.8**). That the barrier is resolved at the end of isolation through a quantum indeterminate process should no longer be of concern. As is true of the Sierpinski triangle, it is the abstract bounded state or complex wavefunction, or the *instantaneous whole* pertaining to the barrier as a whole (and, of course, *its* retention by the physical system as abstract memory) that regulates and *fates* otherwise quantum indeterminate outcomes into fated global culminations, as exemplified in photographic development *and* in the pertinent restoration of the barrier to its original setting, notwithstanding contingency or 'Wheeler's cat'.

It is the retention of the abstract boundary state pertaining to the barrier, and its resonation in time from the past, itself embedded within realised spacetime vis-à-vis the event horizon, that matters. The quantum indeterminate resolution of the discrete terms that are otherwise collectively and globally fated to re-manifest the barrier in its original setting will matter, but much less than we usually suppose, even if proven to be objectively indeterminate.

Also note that, even if Manyworlds was somehow correct and not superfluous, and it culminated into many histories, it would all culminate into the restitution of the same memory: the same barrier-setting with the same slits open; hardly any basis for a nihilism of information or abstract memory, or, with it, the supposition of the amnesic universe, or of the Alzheimer's effect, or espousal for a totally meaningless universe.

Hence, through the self-invalidation of Manyworlds we obtain the solution to the problem of memory, culminating into a vindication of abstract memory and against the nihilism of information; almost completing our series of apagogic mnemonic arguments in favour of abstract memory.

3.18: From Manyworlds to Little Manyworlds

A more limited version of Manyworlds may well be plausible, even though it must again entail violations of the principle of conservation, generate various liar's paradoxes, and remain beyond any form of empirical provability. In any case, Little Manyworlds (LMW) could not obviate the ontology of information or invalidate abstract memory, as we shall discover in this essay. That is, **LMW might be plausible, but it is useless to any nihilistic framework arrayed against the ontology of information or memory.** To illustrate LMW, consider the consequences that would obtain from a nuclear detonator fated to the resolution of a single-particle quantum spin. Thus…

… {**spin-up** AND **spin down** | OR-form **spin-up** leads to nuclear detonation}

At a succeeding moment, if the particle spin AND-to-OR collapsed into the OR-form **spin down**, no detonation will occur. But, if particle spin resolves into OR-form of **spin up** then, in addition to the destruction of a target city, it will produce a permanent and irreversible alteration of the structure of possibilities available to spacetime in potentiality, and affect the very structure of decoherence and erasure of nested futures to the infinite future, fated on a single quantum indeterminate incident.

What if, instead of a single unique history, we envision Manyworlds in which both spin states were resolved into existence; forming two truly distinct histories, precisely on account of the fact that each spin-outcome must exhibit profoundly different global consequences and histories that can never culminate into a same final global outcome? Hence, it is conceivable that, under LMW-pertinent circumstances, a single or a collection of quantum indeterminate consequences might coincide to produce profoundly different global outcomes and histories, and not a same final global outcome and history, argued in **3.17**. Thus, a limited form of Manyworlds might be plausible, baring violations of the principle of conservation and all the other stated problems.

However, note that seemingly viable LMW cannot be utilised to circumvent the problem of quantum indeterminism : LMW is no more capable of obviating indeterminacy than is forlorn MWI, as we shall discover in Part-II. Moreover, **while MWI offered the anticipatory promise to obviate the ontology of information and memory, LMW is incapable of being utilised for such a nihilistic purpose.**

LMW can only apply to limited types of AND-form states. And the brute-force exhaustion of possibilities operating in LMW cannot obviate the ontology of information, in the sense that the nuclear detonators will be real, and so will the consequences, whatever these are, despite brute force exhaustion of the possibilities, notwithstanding the meta-semantic hypostasis that renders these so: see **3.17**.

Finally, LMW cannot come to the aid of forlorn memory erasure and memory nihilism: It cannot seek to accomplish for the erasure hypothesis what MWI promised, even though the later failed (see **3.17** and **3.18**). Thus, the specific identity of the nuclear detonator will resonate into both histories, while the only loose-end or a-causal and a-mnemonic variable will be constituted as the bomb-triggering quantum spin.

299

Hence, **the Alternative Realist assertion to the effect that nature possesses memory, and that abstract memory is of a dissociated form that can resonate across time, even while it is retained extant the 'matter' confined to the event horizon, abides.** Hence, our case against the implicit *amnesic universe* is almost complete, short of **apagogic mnemonic argument 3-a** and beyond.

SECTION-III: CRITIQUE OF THE QUANTUM ERASER DELAYED CHOICE APPROACH

3.19: Apagogic Mnemonic Argument 3-a:
Critique of Yoon-Ho Kim's Quantum Eraser & Consolidation of the Case for Abstract Memory

Before we proceed, it is necessary to make clear distinctions between following similar-sounding terms. Thus…

- **Quantum erasure of nested futures**, extrapolated from John Wheeler's delayed choice approach in Book-II, refers to the process of the restructuring of future-form quantum mechanical spacetime in potentiality: a domain modelled as a grand convergent category nested future. Quantum erasure of nested futures unfolds when converging nested future possibilities are non-consistent in relation to newly resolved events on or along the event horizon. Despite the lack of direct connection between recently resolved events along the event horizon, which may be separated by parsecs, these remain counterfactually entangled through their common convergent futures through spacetime in potentiality. This counterfactual binding is responsible for both the erasure of non-consistent futures *and* attendant grand decoherence of spacetime in potentiality. The quantum erasure of nested futures, hence grand decoherence, revives Mach's principle pertinent to inertia, amongst other things. (See Book-II: **2.34** to **2.37**).

- **Memory erasure hypothesis**. The hypothesis is an anticipation that claims that there is no carry-over of abstract information across time from one moment to succeeding moments, and no integration of the present and future to the past per any binding logic or causality, or in any informationally meaningful way. When erasure hypothesis co-joined with Manyworlds (see, **3.15** to **3.18**), it constituted the basis for an ultimate nihilistic argument against memory and against the ontology of information.

- **Quantum Eraser**: Proposed by Yoon-Ho Kim *et al* and inspired by John Wheeler's delayed choice. It employs a similar process to delayed choice, but claims to be a method by means of which the past, hence memory, could be erased. This is contrary to Alt-R's case for abstract memory, and contrary to the conservation of memory.

Yoon-Ho Kim's delayed choice quantum eraser alleges the possibility of the physical erasure of the past. As such, the quantum eraser constitutes an additional challenge against abstract memory, portraying it as a thing erasable: a paradoxical claim.

The quantum eraser approach predates and does not recognise or incorporate the nested futures view of the quantum mechanical wave or wavefunction. Nor does it recognise the critical insight that the quantum mechanical wave constitutes the physical manifestation and phenomenalisation of the future, encapsulated as the domain of spacetime in potentiality, as espoused in the growing block intermediate model of spacetime developed in Book-II. When we correct the background to Yoon-Ho Kim's quantum eraser in terms of such findings, its claim to erase the past will be cast into doubt.

It turns out that the quantum eraser can only ever operate on the as-yet not-realised future, or else on those aspects of the past that remain quantum suspended into isolation and have not yet resolved into OR-form outcomes. As such, isolated untampered pockets or 'holes' embedded within realised spacetime , within the past, constitute quantum suspended as-yet not-realised delayed choice states: the 'holed past'. Hence, the *holed past theory of the quantum eraser*, espoused in Alt-R

In our holed past approach, the future, or certain aspects of it, under certain circumstances, can be delayed from undergoing AND-to-OR resolution, while the rest of spacetime advances 'upward' the spacetime system and accumulates as the resolved past retained as a growing block abstract memory complex within realised spacetime.

If parts of the past remain in isolation and quantum AND-form suspension, relegated to probable AND-to-OR resolution at some future moment, then these constitute 'holes in the past', or 'bubbles' of suspended possibilities embedded within realised spacetime, even though these belong to the future and are simultaneously embedded within spacetime in potentiality, constituting inter-protrusions of the domains. Just such 'holes' are precisely the states subject to AND-to-OR collapse in Yoon-Ho Kim's quantum eraser.

Therefore, the past, in any *resolved* form, is *not* and cannot be erased in Yoon-Ho Kim's quantum eraser, which can only erase the future suspended as a bubble within the holed past. Therefore, using Wheeler's delayed choice, **the quantum eraser experiment sets up a 'holed past' situation; resolves that holed past into a realised event; erases the quantum potential pertaining to the alternative possibility co-suspended in that 'hole', and, on that basis, erroneously claims to have erased the past**.

Fig. 3.08 recapitulates the logic of Wheeler's delayed choice experiment originally presented in Book-II: We will not be recapitulating it in detail here. The reader is encouraged to reprise the illustration and its attached notes.

The critical factor in the illustration is **switch-x**. A beam of light composed of just one particle is split up at the beam-splitter at **A** into two component beams. Described in the generic but erroneous pseudo-realtime terms, and involving the invalidated notion of spatially dynamic and displacing quantum waves (see Book-II, **2.23**) one half of the split beam 'displaces' toward mirror **C**; the other towards mirror **B**. The experimenter keeps **switch-x** in suspension (deliberately or by default, given that its setting will be decided at a last-ditch moment *in the future* and, from a future spacetime view, it is necessarily AND-form superposed into its two future-possible settings) until after each split-beam is deflected from mirrors **C** and **B**, respectively.

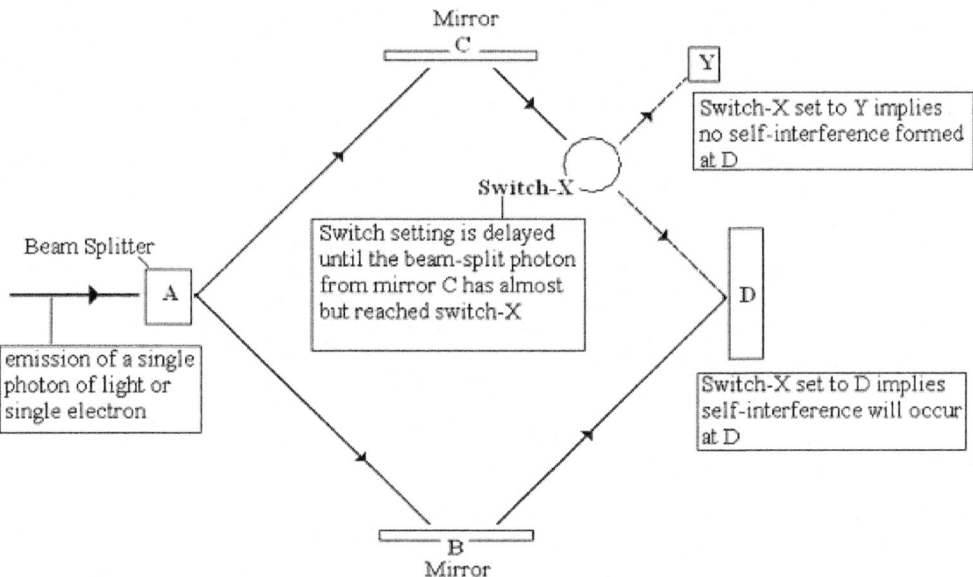

Mirror
C

Switch-X

Switch-X set to Y implies
no self-interference formed
at D

Beam Splitter

A

Switch setting is delayed
until the beam-split photon
from mirror C has almost
but reached switch-X

D

emission of a single
photon of light or
single electron

Switch-X set to D implies
self-interference will occur
at D

B
Mirror

Fig. 3.08: Illustrating John Wheeler's delayed choice
Notes: The illustration depicts the basic outlay of John Wheeler's delayed choice experiment. An incoming photon of light is split by a beam-splitter at **A** into two quantum mechanical wavelets or split-beams. One heads toward **mirror-C** and the other to **mirror-B**. The split-beams will be deflected by the mirrors toward **detector-D** to self-interfere, and to form an interference pattern at **detector-D**. The experiment is essentially a two-slit experiment, save that the barrier is replaced by the beam splitter. However, **switch-x** is *effectively* AND-form suspended at its two settings: One setting permits the split-beam from **mirror-C** to reach **detector-D** and to evince interference, while the other is deflected per **mirror-C** to **detector-Y**, and no interference pattern forms at **detector-D**. Thus, the throw of **switch-x** is *delayed* until after the beam has passed **mirror-C** and well on its way to **switch-x**, when it ought to be too late for the quantum mechanical system to adjust to the alternate possibility implied by **switch-x** diversion to **Y**. It turns out that the system hedges within its possibility-set mutually incompatible alternate future outcomes that include subsidiary wavefunctions for both prospective self-interference at **D** AND diversion to **Y**. This tacitly implies the nested futures view of the quantum mechanical wave.

Of course, there can be no direct realtime observation of any of this, given that such an attempt would destroy the purported 'displacing waves' and the quantum mechanical split-beams. Thus, without resort to direct realtime observation, the delay over the choice vis-à-vis the last ditch setting on **switch-x** will remain in force until a critical moment in time is reached after the split-beams have supposedly passed mirrors **C** and **B** and are seemingly locked in to form up the prospective self-interference and interference pattern at detector **D**.

Recall that, in the framework of Alt-R, the delayed choice experiment is important because it demonstrates that the primary or hedging quantum wave of the split beams embeds within it alternate mutually incompatible nested futures by function of the quantum suspension imposed on **switch-x**. In other words, the nesting hedging wave and wavefunction contains and nests within its AND-form possibilities two mutually incompatible and distinct nested wavefunctions that the system may assume per the suspension of **switch-x** and per the delayed choice of the moment at which we trip **switch-x**.

As a generalisation to all quantum mechanical waves, but with the conditional exception of closed-choice quantum systems (see Book-II **2.20** and **2.21**) wavefunctions will nest as many mutually incompatible alternative futures and contingent possibilities as there are delayed choice elements inherent to a given system. Consequently, as a description of the future, the whole of spacetime in potentiality is essentially a *convergent* nested future system: a nested future quantum switchboard that elaborates to the infinitely removed future. This insight was the basis for the nested futures view of the quantum wave and wavefunction, and of the convergent structure of spacetime in potentiality; both developed in Book-II (see **2.17**).

*

How does Yoon-Ho Kim's quantum eraser of the past, of memory, claim to work, and how does it relate to the delayed choice process described above?

Imagine that a very distant star in the Andromeda galaxy exploded into a supernova, and that this event transpired some two million years ago in the past. For convenience, we ignore the Alt-R case against spatially dynamic quantum waves from Book-II and assume that these displace across space in the naive way assumed and taken for granted. For brevity, we will also assume that just one photon of light from the said supernova has travelled towards the Earth, taking two-million years to get to Earth. Midway along its journey, our single photon had to negotiate the gravity field of an interceding supermassive object. The gravity field of that object had the effect of lensing the path of our lone photon about the supermass. Generically, we assume that our single photon must have taken either the left-curved path or the right-curved path about the supermass. But, in its quantum mechanical AND-form state, our photon is configured into an AND-form superposes for *both* paths. This scenario is equivalent to the split-beam photon featured in delayed choice and the two mirrors and associated paths. Indeed, deflection by **mirror-C** and **B** are equivalent to the potential supermass-imposed left-curve and right curved paths in our scenario.

Some two million years after the supernova event, astronomers on Earth train their instruments toward the Andromeda galaxy. At just that moment, our lone photon completes its two-million-year journey and is observed on Earth. The question of which type of instrumentation our astronomers will use to observe the lone photon, hence the critical decision made at the delayed-choice switch-point, will play the same role as suspended **switch-x** featured in Wheeler's delayed choice. The type of observational instrument chosen will have selected the one possible past-history of our lone photon vis-à-vis its other possible past-history, hitherto subject to a one million yearlong suspension as a not-yet resolved AND-form superposition. Our observation of the lone photon will resolve one past at the expense of the other per the specific kind of instrumentation used. In other words, the delayed choice over the form of observational instrumentation used, and the decision to choose that instrument at the stated critical moment, will *apparently* bring about the erasure of some aspect of our photon's history. Hence, the quantum eraser of the past. Hence, the erasure of memory of Yoon-Ho Kim *et al.*

*

We need to take a closer look at what transpires in the choice made about the observational apparatus vis-à-vis the lone photon from Andromeda, and how this translates into a purported quantum eraser of memory.

At the pertinent critical moment, we either choose a binocular means of observation (i.e., a dual lens and dual chamber system) or a monocular means of observation (i.e., a single lens, single chamber telescopic apparatus): These constitute the two settings pertinent to **switch-x**. If we choose the binocular apparatus, we will obtain a fifty-fifty chance that our photon will end up in one or the other of the two chambers. Indeed, it is the binocular observation that is of utmost significance to the claims of Yoon-Ho Kim's quantum eraser, given that it is the only mode upon which we can assume a purported 'which way' path for our lone photon's navigation about the supermass one million years ago.

A key delayed choice element (again. the equivalent to **switch-x**) is constituted as the quantum suspension of the astronomer's decision over which instrument (monocular or binocular) to use. Thus, the delayed choice suspension is between… {**binocular** AND **monocular**}

As previously stated, the passage of the lone photon about the interceding supermass one million years ago is equivalent to the pair of mirrors **C** and **D** in the delayed choice experiment (again, see **Fig. 3.08**) described in the usual erroneous pseudo-realtime terms involving dubious notions of 'spatially displacing quantum waves'.

Let us assume the selection of the binocular apparatus per **switch-x**. Thus, **switch-x** is now tripped to the binocular setting just before the arrival of our lone photon. There is a 0.5 probability that the photon will end up in one or the other of the binocular chambers. It is certain that it will not end in both: If our lone photon ends up in the left-side chamber, this will apparently resolve for a past-history of our lone photon that took the left-curved path about the supermass one million years ago: the OR-form of the left curved path about the supermass. If the photon ends up in the right chamber, then the photon must have taken the right-curved path about the supermass one million years ago: i.e., the OR-form of the right-curved path.

It seems exceedingly disturbing that a binocular observation made at the present (i.e., one million years after the presumed passage of the photon about the interceding supermass) will get to determine which curved path the photon *actually* took about the supermass *in that past*… and erase the other alternative past. Put another way, *the present gets to determine which past actually happened*; thus inverting causal relations and principles of consistency that integrate the present to the past... wherein we assume that the past determines, or at least delimits to consistency, the content of the present... and wherein the present does not decide what happened in the past, much less create that past.

Far more disturbing is the notion that, having resolved the past through the present, the same means can be used to assert the claim for the erasure of the alternative past: The quantum eraser appears to imply that there is no fixed or stable past, and that it is resolved by the intercession of the present, and, more profoundly... the past can be erased.

By implication, the quantum eraser appears to undermine our claim that information is something retained by the natural order as permanent abstract memory upon which the recuperated new conservation principle can be assumed: i.e., the conservation of the past as abstract memory, and tandem consistent futures. Thus, in the scenario depicted, depending on which chamber the photon ended up in per quantum probability, the observation will favour one past history at the expense and to the erasure of the other; even if the latter was the one that actually transpired in the past.

As far as the actual past-path assumed by the photon is concerned, we will be unable to claim any universally absolute fixity about its history. Instead, the past-history of the photon will arise, albeit in a quantum indeterminate way, only *after* we collapse its quantum mechanical state by means of our present-moment binocular observation of the photon, carried out a million years *after* the photon supposedly crossed paths with the said interceding supermass.

It is logically consistent to suppose that, if sufficiently distant alien observers could subsequently subject our own history to a quantum eraser process, in doing so, they could erase our entire history, or at least our observation and judgement about the history of the lone photon and our creation of its history through the erasure of the memory of its alternative. They could do

this as easily as we have erased the history of our lone photon from one million years ago, although neither party could control what gets erased, since this is a matter of quantum indeterminate resolution.

Reframing the implications in terms of our intermediate model spacetime from Book-II, according to the quantum eraser, **instead of the past delimiting and honing the development of both the event horizon and future-form spacetime in potentiality through the grand decoherence and elimination of non-consistent futures (all of which supposes a fixed or conserved set of past events permanently retained as the abstract past, thus recuperating the principle of conservation in new terms) Yoon-Ho Kim's quantum eraser, operating from the event horizon, can get to select, erase and replace past history and memory we thought was permanently locked-in within the domain of realised spacetime;** implying that the mnemonic domain of realised spacetime has no fixed or final content or status, much less one retained as memory at a remove from present-continuous matter, radiation and 'space' on or along the event horizon. **Consequently, our claim for abstract memory ought to fall apart as readily as the whole corpus of nested futures, grand decoherence, the revival of Mach's principle**, and the whole gamut of solutions to physical mysteries, including gravity itself, as espoused in Book-II.

The implications from Yoon-Ho Kim's quantum eraser are doubly shocking, if only because Yoo-Ho Kim and associates have carried out eraser experiments and have produced results.

*

One cannot deny the operational-experimental reality of the quantum eraser. However, it erases only future-form states of spacetime in potentiality 'holed up' within realised spacetime: it does *not* erase any *realised* past states embedded within realised spacetime.

The quantum eraser is understandably framed in the absence of key insights into the ultimate nature of the quantum mechanical wave as the phenomenalisation of the future; as an aspect of future-form of spacetime, and the nested futures view of the latter. Once we incorporate these insights into the quantum eraser, these change its character; undercutting its claim to erase the past.

The erasure of the past is no more possible than is the viability of quantum mechanics in the absence of abstract memory (see **apagogic mnemonic argument 1-a**, essay **3.07**). It all comes down to whether, throughout its journey and in its negotiation about the supermass, our lone photon remained wholly in physical isolation or was instead AND-to-OR resolved into an actualised 'path'. If it remained in isolation and in AND-form superposition for both potential paths about the supermass, and consequently did *not* engender counterfactual consequences and grand decoherence to the larger spacetime in potentiality at that point, with respect to which *all* convergent mutual futures must restructure into consistency, the subsequent observation of the photon on Earth will not generate the radical implications Yoon-Ho Kim and associates claim. **If the photon never interacted with anything and remained in AND-form suspension vis-à-vis the supermass and throughout its journey to Earth, then its OR-form resolution was deferred to a potential first-time future observation on Earth, which transpired one million years hence. In that case, the photon's state throughout its journey, or throughout its suspension, constituted a 'hole in the past': part of the past that remained in a pure non-resolved future form quantum suspension for one million years...** and as a protrusion of the future from spacetime in potentiality into and within realised spacetime; made possible per the said suspension and deference of the photon's resolution to the future. **Hence, the subsequent binocular observation and resolution of the photon can hardly constitute the 'erasure of the past', or of memory, which had never formed into a resolved past or into memory to begin with, but was merely deferred to the future and quantum suspended thus.**

On the other hand, if the lone photon had been brought out of isolation by events about the supermass, and this had occurred one million years ago, and the photon's path was AND-to-OR resolved into a specific OR-form path about that supermass, the past of that photon, and hence the path it took about the supermass, would have been resolved prior to its subsequent binocular observation on Earth. In that case, two things abide: First, the objective reality of memory, supported by our various apagogic mnemonic arguments against memory erasure hypothesis and nihilistic MWI, including the case for memory as critically necessary for the viability of quantum mechanics, would abide, as would the tripartite structure of spacetime developed in Book-II. Thus, as far as the later Earth observation of the photon's past is concerned, the critical factor that *must* be retained by the physical order, and which must then help delimit its observation on Earth consistent to that past, is the actualised past-path of the photon about the supermass; retained as memory; necessitating the temporal resonation of that information from the past from one-million years ago to its present-day confirmatory binocular observation one million years hence.

But this requires the tandem operation of the second factor, which must also abide: The second factor consists in counterfactual causality and grand decoherence, based on the nested futures view hammered out in Book-II. This *must* apply to the lone photon: **If the photon's past-path had indeed undergone AND-to-OR collapse into a definite OR-form path about the supermass one-million years ago, the future-potential observation of it on Earth one-million years hence will be delimited to a decohered set of consistent-only possibilities: succinctly, consistent with the realised OR-form path of that photon about the supermass one million years ago, even if a binocular observation of it is carried out one million years in the future hence**: If the photon traversed the supermass from the left, and had resolved into this one-million years ago, the photon will arrive in the left-chamber one million years later... because the process of grand decoherence affected one million years ago will force this outcome into this consistent-only future possibility. If the photon had OR-form resolved into the right-path, it would arrive in the right chamber of the binocular one million years hence, again because grand decohere affected one million years ago will have forced it into the requisite consistent-only possibility one million years later.

303

The future-potential Earth observation of our photon to a consistent-only possibility with respect to what *actually* transpired in the past will be enforced throughout spacetime in potentiality through counterfactual erasure of non-consistent convergent nested futures. These must encompass and delimit the potential observation of the lone photon on Earth one-million years in the future into consistency with the actual path taken by our photon one million years ago.

The resolved past-path of our photon will constitute a counterfactual; eliminating from the convergent category nested futures elaborated one million years into the future *all* future observational possibilities that are inconsistent with the actualised path... enforced on the photon's final binocular observation on Earth. **The alternative path was erased one million years ago and will *not* be erased by any action from a binocular observation affected in the present-day observation one-million years hence. Thus, the future binocular observation cannot erase any realised past. Instead, the future is forced into consistency with that past, assuming it actually transpired and was not quantum suspended into AND-form potentiality, or into a 'holed past'.**

*

In summary, the lone photon either remained in physical-observational isolation and AND-form suspension as a 'hole in the past' throughout its one million year 'journey', with no resolution of any OR-form actualised path about the supermass, and no consequences in terms of counterfactuals and grand decoherence vis-à-vis future-form spacetime in potentiality... or the lone photon *was* physically interfered with about the supermass and *did* AND-to-OR resolve into an actualised OR-form path. If the latter, the resolved path generated a counterfactual that restructured future spacetime in potentiality in a manner fully consistent with what had transpired. **If the photon remained in isolation, the photon had no resolved past or memory to subsequently erase one-million years hence. Therefore, no erasure of the past or memory could transpire one million years hence on Earth. But if the photon *had* resolved into a definite path one-million years ago, then that path is all that we will be permitted for observation one-million years hence: The alternative path will not be erased at the moment of our observation of the photon on Earth, since it was already erased one million years ago, per pertinent AND-to-OR collapse of the lone photon about the then supermass, and by the process of consistency-enforcing grand decoherence of spacetime in potentiality.**

In short, the apagogic mnemonic arguments from Part-I are saved: abstract memory is real, and the past cannot be erased: Its content had been decided irreversibly and absolutely for *all* future observers by counterfactuals and by grand decoherence in the context of the convergent category nested future domain of spacetime in potentiality.

*

Finally, consider what would happen if the quantum eraser could erase some past event. How would one go about proving that this erased past had actually happened and that it had been subsequently erased by an operation carried out in the present? Indeed, Yoon-Ho Kim's quantum eraser engenders its own lira's paradox.

The *first liar's paradox* of quantum eraser asserts that the erasure of the past may be true, but only on the strict condition that we are not able to produce *any* evidence or even cognisance of that erased past, much less of the erasing process itself, in any meaningful empirical-experimental terms. In short, quantum erasure of memory could only hold 'true' if, and only if, it was scientifically or evidentially false, or beyond any proof.

On the other hand, **if the quantum eraser could furnish proof of the very erasure of the past by producing empirical experimental evidence of its very erasure, this would culminate into the *second liar's paradox* of the quantum eraser: The proof of erasure must consist of some trace of the very past it erased**: a thing that must be incorporated into the evidence of the eraser process. At which point, the very evidence and presence of that trace will have demonstrated the failure of the purported erasure.

One possible anticipated way out would be for Yoon-Ho Kim's quantum eraser of memory to attempt marriage with Manyworlds: essentially identical to the one between memory erasure hypothesis and MWI. In the case of our lone photon, the MWI based quantum eraser scheme would entail the splitting of the universe into two histories. **Upon subsequent binocular measurement of our lone photon a million years hence, the photon would instantly resolve into two distinct histories entailing two distinct realised curved paths. But, in such a scheme, the binocular observation could not constitute a quantum eraser: since none of the past possibilities could be erased by the binocular interaction, and both would be retained, albeit in a Manyworlds framework. The quantum eraser would then amount to a pseudo-quantum eraser.**

In a second variation of the same, our photon is brought out of isolation and resolved into a definite OR-form path at the supermass. **At the point of our photon's OR-form resolution one million years ago, the universe will have split into two distinct histories. In each history, following a million-year delay, subsequent binocular observations would transpire, and the universe would then split into four histories. In one history, the right-path will have happened, but it will be pseudo-erased in favour of the left path. In the other history, the inverse will happen involving the left curve path. The quantum eraser involved in either will be a pseudo-eraser, simply because, at the Manyworlds level, erasure could *not* really happen, even if the pertinent observers believe it to be happening.** Of course, the observers could never furnish proof of their claims, but only make equiplausible non-provable postulates about the eraser effect, unless they had access to the Manyworlds level, only to grasp that all possible histories were fully realised, and the eraser never actually worked.

However, **MWI itself is shown to be invalid (see 3.17)** aside its own problems with information and memory (see **3.15**). **Therefore, any anticipated merger between Yoon-Ho Kim's quantum eraser of memory *and* MWI must remain forlorn. In any case, the combination of quantum eraser with MWI leads to the pseudo-eraser. The pseudo-eraser is not a real eraser,** as was argued above, and it defeats the whole point of Yoon-Ho Kim's quantum eraser of memory.

*

304

Throughout, the basic process that justifies the holed past theory is furnished by counterfactual causality and grand decoherence, both per function of the convergent category nested futures. When these are applied to Yoon-Ho Kim's quantum eraser, we discover that the past, once formed, *cannot* be erased: Counterfactuals always eliminate alternatives foregone as well as non-consistent futures for all frames along the event horizon and futures potentialised within spacetime in potentiality, to the infinitely removed future. Hence the grand decoherence of spacetime in potentiality into consistency with what happened in the past, and the delimitation of possibilities open to the future that must deny future observers *any* future outcome that is inconsistent with what had transpired in the past. Thus, if the past was not suspended and had been resolved into a definite outcome, Yoon-Ho Kim's eraser could not erase it, since only an outcome consistent with what had transpired is permissible, per grand decoherence. On the other hand, if the past had not transpired and had remained suspended as a holed past, the action from the 'eraser' will indeed resolve the past in and through the present, *not* by erasing any realised past, but only by erasing suspended delayed futures holed up in the past. The outcome thus resolved will be consistent with all that had transpired across the universe, as guaranteed by grand decoherence.

The retention of the past so-implied must again recommend that abstract memory cannot be erased: Abstract memory abides. It follows that the apagogic mnemonic arguments so far articulated must prevail. Hence, abstract memory is real. There really is memory in nature, locked within the domain of realised spacetime; always extant with respect to the manifest radiation, matter and 'space' restricted to the event horizon; always resonating across time, but *not* across space, as we shall discover in the advent of superlative de-spatialisation and temporisation in Book-IV.

3.20: Summary of Book-III, Part-I

The following constitutes a bullet-point summary of the key points and discoveries of Book-III, Part-I:

- **Memory erasure hypothesis**. The case for information-matter dualism from Book-I is inverted: Instead of dissociation and retention of information as memory in abstract form, both AND-form logic *and* quantum indeterminacy is posited against information-matter dualism and especially against the possibility of abstract memory. (See **3.06**). According to erasure hypothesis, both AND-form logic and quantum indeterminacy ought to be inimical to the possibility for abstract memory. Experimental isolation and futurisation of the two-slit experiment must imply that the barrier assume an AND-form configuration inimical to any specific OR-form setup of the barrier before isolation, which must subsequently become erased, *not* retained as abstract memory. Information and memory erasure must also occur per *apparent* information-scrambling quantum indeterminacy during subsequent re-formation of the barrier at the end of isolation.

- **Apagogic mnemonic argument 1-a asserts that the viability of quantum mechanics critically depends on the reality of abstract memory**. If memory erasure hypothesis held true, and the original barrier-setting was erased, and its fate at the end of isolation was a matter of quantum indeterminate sortology, quantum mechanics in the form we know it would become impossible. If erasure hypothesis held, we could never control experimental conditions and outcomes, or establish consistent correlations therein. Our determination of the barrier setup would be as forlorn as the want to deterministically orchestrate spin-outcomes in EPR-type experiments. For quantum mechanics as we know it to be possible, the information pertaining to the original barrier setup *must* survive both the interim of isolation *and* the quantum indeterminate process at the end of isolation, and subsequently restore the barrier to its original setting. Thus, the information pertaining to the original barrier-setting must dissociate from the system and remain in that dissociated abstract form until its restoration to the barrier at the end of isolation. Hence, abstract memory must abide, and memory erasure hypothesis must be false: the very viability of quantum mechanics attests to this (See **3.07**).

- **Apagogic mnemonic argument 1-b: The dependency of quantum mechanics on abstract memory justified by the firewall principle and the structure of spacetime**. The argument utilises both causality and the structure of spacetime to further justify the claim that, upon isolation, per causality and the conservation principle entwined into the firewall principle, any system (such as the barrier in the two-slit experiment) whose resolution and content-formation is deferred to the future, *must* dissolve into an AND-form superposition state of future alternative possibilities. Any contrary claim to the effect that the barrier continues as a materialised OR-form state even under isolation and deference to the future, requires the reception of 'signals from the future' in proof of the claim, with tandem violation of causality and principles of conservation in contravention of the firewall principle and of spacetime-structure itself. Therefore, upon isolation, the barrier *must* transform into an AND-form superposition of alternatives, notwithstanding low probability contingencies (Wheeler's cat), while the abstract basis of the OR-form information pertaining to the barrier setup must dissociate and be retained in abstract form, if the very viability of quantum mechanics is to abide. If memories were erased, quantum mechanics as we know it would become untenable, as stated in **argument 1-a**, despite implications from isolation and quantum indeterminacy: (See **3.08**).

- **Apagogic mnemonic argument 1-c exposes the consistency problem of memory erasure hypothesis**. For quantum mechanics as we know it to be possible, there must exist strong consistency between the barrier configuration we setup before isolation versus the same setting (*and* the detector outcomes) at the end of isolation: In effect, this is **argument 1-a** by alternative means, but restated in terms of consistency. The resolution of the barrier must be a matter of certainty, otherwise there would only be a 0.25 probability that the outcome at the end of isolation be resolve into the consistent outcome in relation to the original barrier setup. Indeed, there would be a 0.75 probability that the original barrier setup be replaced with one of the other three inconsistent settings, with no certain control of the barrier setting at the end of isolation. According to quantum mechanics as we know it, in the real world, specific outcomes in two-slit experiments are *always* consistent with any initial barrier setup and the continuity of that setup, baring low probability contingencies (i.e., Wheeler's cat). Hence, the setting on the barrier is *not* quantum indeterminately scrambled: It is *not* subject to the inconsistency problems implied by memory erasure. The barrier setting *must* survive the interim of isolation and can only do so if the setting dissociates into an abstract extant information-state retained 'elsewhere' or within mnemonic realised spacetime. It follows that the postulate for abstract memory must be true. (See **3.09**).

- **Non-standard evaluation of the two-slit experiment reprised.** The non-standard evaluation of the two-slit experiment (Preliminary, **0.14**) asserts that the reality of AND-form logic is evident in all experimental outcomes, including in single-slit experiments: *not* just in experiments with both slits open. The basic argument is that the specific outcome in single-slit experiments can also be produced by an AND-form quantum mechanical probability wave and, as such, also constitutes evidence of AND-form-logic, even though the interference pattern will be absent. Indeed, in all experiments, AND-form and OR-form logics are evident in an observationally simultaneous way... both when both slits are open *and* when only one slit is open, as was argued in the *new complementarity thesis* in the Preliminary (**0.16** and **0.17**). Thus, it becomes possible to envision a fictitious universe wherein memory is erased, and in which the initial barrier setup does not resonate to the end of isolation, and the barrier resolved at the end of isolation is resolved in a purely quantum indeterminate way, and from which AND-form logic could yet be inferred from 3 out of 4 outcomes, not just from **both slits open**: (See **3.10**). Only **both slits closed** barrier-state furnishes a null outcome. Consequently, even with erasure, we could resolve at least three outcomes from which we might infer AND-form logic and basic quantum mechanical knowledge. The contribution of the non-standard appraisal to information and memory nihilism is as a preliminary to **apagogic mnemonic argument 1-d**: (below).

- **Apagogic mnemonic argument 1-d: The all-or-nothing critique of memory erasure hypothesis**. If we assume the operation of memory erasure, it ought to operate on an all-or-nothing basis: *all* must be erased. It is disingenuous and dishonest to entertain selective erasure of some things (say, of the initial barrier setup) but conveniently exclude others: i.e., the rest of the apparatus; the laboratory; the data collected; the scientists; their memories; or even the nihilistic assertion and message that reads, '...there is no information or memory in nature', as a piece of communicable information in its own right. All would be erased or impossible to manifest. Not even reportable experience should survive memory erasure, rendering impossible *the* proof of quantum mechanical realities, even when we resort to the non-standard evaluation (summarised above). Even the default consistencies between initial conditions and outcomes, and even the assertion of memory erasure itself, must evaporate into nothing, as demanded by the all-or-nothing approach. But this runs counter to real world experience in which it *is* possible to make communicable nihilistic assertions, run experiments with high levels of consistency despite indeterminacy, and discuss meaningful results... all consequent upon the reality of information and memory. Thus, memory erasure must be false, and abstract memory must abide. (See **3.11**).

- **Apagogic mnemonic argument 2-a revives forlorn memory erasure hypothesis by co-joining it with the Manyworlds interpretation**. According to Manyworlds, outcomes from any AND-to-OR wavefunction collapse are not subject to quantum indeterminate resolution. Instead, *all* the possibilities get to form up into separate histories. This ensures that, in some histories, we obtain a series of events that are *apparently* sensible and consistent, but only by default... without resort to or need of an ontology of information, causality, or principles of consistency... and minus any temporal resonation of information as memory. If the **both slits open** setup survives from isolation to the experimental conclusion, this is not because of any retention or transfer of the barrier setup, but purely a matter of brute-force exhaustion of *all* possibilities of the root wavefunction into all possible outcomes, consistent or otherwise vis-à-vis the initial barrier setup: Thus, in MWI, there is no memory as such, and ultimate memory nihilism fully abides. But this also comes with information nihilism. The only difference between memory erasure hypothesis and Manyworlds is the inclusion of quantum indeterminism in the former versus its exclusion in Manyworlds-based memory and information nihilism. Thus, memory erasure hypothesis appears revived in Manyworlds, and the former is effectively co-joined to the later in at least its central claim of memory erasure and ultimate information nihilism. (See **3.12**).

- **Apagogic mnemonic argument 2-b: Manyworlds interpretation, liar's paradoxes, and the violation of the principle of conservation.** The Manyworlds interpretation cannot escape from semantic contradictions, nor can it escape from the violation of the principle of conservation, even if this is apparently hidden from sight. For MWI to be 'true', it must not furnish empirical-experimental evidence of the other histories. Hence, the *first liar's paradox*. i.e., For MWI to be 'true', it cannot be proven to be true, and must remain false. However, if physical evidence for other histories could be obtained by observers within any history, the other histories could no longer be considered compartmentalised, but rendered part of just one history, with consequent large-scale violations of the principle of conservation of energy-matter. Hence, MWI would be rendered false by its very 'evidence', given the dependency of its 'truth' on the strict impossibility of its proof: Hence the *second liar's paradox*. It gets worse: With or without physical proof of other histories, MWI would produce dramatic violations of the conservation of energy-matter: or, given the pleonastic status of energy concepts and relations espoused in Book-II **2.39,** the principle recapitulated as the conservation of memory and the assertion of attendant consistent futures would also be violated. Ignoring the pleonastics of energy, the energy-matter to constitute many histories must come from somewhere. One must either contrive a source for this, conveniently beyond proof or demonstration, or discard the principle of conservation entirely via creation of energy-matter *ex nihilo*. Even given an infinite energy supply at the Manyworlds multiverse level, the whole scheme must break down when confronted with wavefunctions with infinite possibilities, such as those that inspired 'the sum over all paths' approach in quantum field theory. These would generate infinite numbers of histories, while the histories thus generated would suffer instant absolute heat death, even with an infinite energy source. Since the absolute heat death of our universe has not happened, we can safely conclude that MWI is false...unless the requisite energy-matter is being created out of nothing. But this would render the principle of conservation of energy-matter useless even when the latter is finally recuperated into a pure informational abstract form: (For full argument, see **3.13**).

- **The three forms of developmental process, the meta-semantic order, and semantic burn-through**. To enhance our case against Manyworlds, and as a preliminary to the mnemonic argument 2-c, we described three distinct developmental processes pertaining to the formation of *any* pattern.

   The **contingent development process** involves random processing of pattern via quantum indeterminacy. From a given set of combination possibilities, only one combination will be randomly selected into possibility: the alternative combinations will be foregone. The combination that gets to form will be by chance or probability, or in a purely contingent indeterminate way, into a sensible or non-sensible combination: all *without* recourse to logic, causality, memory, or *any* ontology of information. This is essentially the universe espoused in memory erasure hypothesis (**3.06**). How, then do denizens of the universe get to intellect sensible from non-sensible combinations, if neither the memory nor the information requisite to this is intrinsic to a nihilistic universe founded on the contingent developmental process? (i.e., The all-or-nothing problem, again). Indeed, the intellection of pattern would be wholly anomalous in such a nihilistic universe. If the universe is an outcome furnished by purely contingent process of development, hence inherently nihilistic, the only basis for the said intellection would be by way of backchannel to ordinances furnished by a meta-semantic hypostases, and their 'burn-through' into the nihilistic contingency-driven indeterminate universe… a 'burn through' from an extant ontology and hypostases.

   The **brute-force exhaustive process of development** has direct affinity with the same process entailed in Manyworlds: It appears to circumvent contingency and happenstance, not by resort to the indeterminacy, but by generating *all* the possible developmental combinations of terms, sensible or otherwise... thus circumventing any role for inherent logic, causality, information resonation or memory. However, we must yet again account for the reality of pattern-judgement in our universe and the possibility and recognition of sensible from nonsense combinations and patterns, whose intellection ought to be impossible in a nihilistic universe generated through a-mnemonic nihilistic brute-force exhaustion. Again, if our universe is the outcome of a nihilistic brute-force exhaustive process, what decides the very possibility of intellection and observability of the sensible and consistent will be ordinances furnished to the *in-itself* nihilistic brute-force exhaustive system from outside of that system... via semantic 'burn-through' from a meta-semantic hypostasis extant the in-itself nihilistic system.

   The third process, the **mind-driven process of development**, comes in three forms: In *minding by arbitrary fiat*, mind plays randomness, similar to randomly selecting cards from a pack… by the non-attentive fashioning of combinations into arbitrary outcomes. Yet, regardless of the non-attentiveness of mind, what finally decides whether the outcome is sensible or otherwise will be the ordinances that stand outside the interceding mind and system: ordinances that cannot be obviated by mind playing at randomness. The ordinances will 'burn through' into the formed outcomes from a meta-semantic corpus.

   On the other hand, in *exhaustive minded intercession*, mind acts in lieu of mindless brute-force exhaustion of terms. It forms *all* combinations into *all* the outcomes: Therein, mind emulates the brute-force exhaustive mode claimed in Manyworlds by consciously but non-selectively realising *all* the developmental possibilities or histories of the system. But, from among these, what defines and decides sensible from non-sensible outcomes will be the meta-semantic ordinances extant the interceding mind and system. The ordinances will 'burn through' from a meta-semantic hypostasis extant both the system and the mind playing at brute-force exhaustive processing.

   Finally, in *minding by ordinance*, the interceding mind guides development through attentive selective direct pattern recognition, circumventing both contingency-like as well as brute-force exhaustive-like modes of development (whether minded or mindless) … directly and consciously 'deciding' what constitutes sensible from non-sensible outcomes by consciously abiding and applying principles of consistency and order in the light of the ordinances. That is, mind permits the ordinances to decide the outcome: mind surrenders to the ordinances and becomes a conscious conduit for these. But the ordinances are furnished to mind from a meta-semantic hypostasis beyond mind. The ordinances are not created *ex nihilo* by mind or by whim. Such a feat would constitute wholly nihilistic arbitrary outcomes and relations, necessarily devoid of the ontology of information, and subject to the all-or-nothing critique featured in **argument 1-d**. Thus, as was esposed in Plato's dialogue, the Euthyphro, ordinances cannot be arbitrarily generated by mind or by fiat: Order is not mind-dependent or 'mind-created'. Rather, mind is order or information-dependent: dependent on the ontological primacy of information, whose ontology makes intellect and agency possible, but which intellect and agency has not the power to conjure *ex-nihilo*, consistent with conclusions from Plato's Euthyphro.

   Hence, in general summary, what constitutes consistent and non-consistent, sensible versus nonsense, ordered versus disordered... hence, the ontology of information in cosmogenesis, in evolutionary processes, and in the humdrum processes of nature... supersedes and 'burns through' developmental processes involving both mindless and minded contingency and brute-force exhaustion, as much as burns through selective and attentive minded development, even when the latter is super-conflated to God, or wherein God is misconceived as mind. In this way, we secure the demise of information nihilism and memory nihilism: presage to the meaningful universe: (See **3.14**).

- **Apagogic mnemonic argument 2-c against Manyworlds and the meta-semantic Alzheimer's problem.** In a two-slit experiment subject to the brute-force exhaustive process definitive of a-mnemonic nihilistic Manyworlds, only one history will resolve into default 'consistency' between an initial **both slits open** condition and a subsequent interference pattern. The other universes will be inconsistent in their barrier-outcome correlations. This single consistent history will not be due to intrinsic informational logic of the in-itself nihilistic brute-force exhaustive Manyworlds, much less per the resonance of memory therein: neither of which apply in Manyworlds. In that case, first, how could denizens in non-consistent histories know that their barrier-outcome correlations are inconsistent? They cannot: there is no intrinsic information in the in-itself nihilistic Manyworlds to furnish such knowledge, given that the outcomes are brute-force generated *without* recourse to logic, causality, memory, 'consistency', or information. Likewise, and assuming that we happen to reside in the one-and-only consistent history, how do *we* know that our correlation is 'consistent'. We ought not to, simply because the in-itself nihilistic Manyworlds could not furnish any such information. Yet, we *do* make judgements about correlations and garner meaningful observations of conditions and outcomes. W can even form and exchange messages composed of nihilistic assertions. How is this possible in a nihilistic universe courtesy of MWI? Indeed, we come full circle: back to the all-or-nothing argument posited in **argument 1-d**. The critical information that render possible judgements about barrier-outcome consistency possible, or even nihilistic messages possible, cannot be furnished from within the in-itself nihilistic and void Manyworlds, but, at a minimum, would require the real existence of a meta-semantic hypostasis of ordinances, and a backchannel to it... from which principles of 'consistency' and other ideas are furnished to observers within the in-itself nihilistic Manyworlds and its histories. Apagogically, this undermines the very purpose of MWI as basis for an anticipatory nihilism of information and memory. Thus, MWI must be false in the sense that, even if it were true, it could not obviate resort to information and memory which it cannot in-itself furnish. It follows that the want to co-join MWI to memory erasure and nihilism must fail. Hence, both the ontology of information *and* the reality of memory must abide, even in Manyworlds: (See **3.15**).

- **Apagogic mnemonic argument 2-d falsifies MWI**. The above presupposes that Manyworlds is possible, viable, or even true. But MWI fails in its own terms: To illustrate, we take the Sierpinski triangle as a simple model of a simulated universe and treat its seed-orbit as a process generated in the brute-force exhaustive fashion furnished by Manyworlds. This produces Manyworlds histories of the Sierpinski universe... all of which culminate into a huge number of *identical* global histories, all embodying the *same* fractal pattern generic to the developed Sierpinski triangle. The same would happen to any real quantum wavefunction: even the universe-scale wavefunction. That is, even if Manyworlds held true, the culminating global outcomes from it would constitute a vast series of *same or identical global histories of the same universal wavefunction... not* divergent alternative histories. This can hardly constitute a basis for the promised obviation of either information or memory. What is the point of MWI-based nihilism if it cannot in the first place obviate information and sensible correlations between initial conditions and outcomes, and if it cannot generate a pure information-free and memory-void default histories, in the way it promised to accomplish in **argument 2-a**? Thus, the utilisation of MWI in any anticipatory informational-mnemonic nihilism is permanently rendered void. In its own terms, Manyworlds is superficial and self-falsifying: (See **3.17**).

- **Little Manyworlds (LMW)**: While MWI is falsified, it is plausible that a quantum spin-tripped bomb, triggered by one of two possible OR-form spin-outcomes, could generate alterations in the structure and content of a given universe-scale wavefunction that, in the context of the nested futures view and the intermediate model spacetime, could decisively alter the whole history of the universe into at least two truly diverging non-identical histories: i.e., a bombed-out city with a profoundly different historical evolution versus an intact city with a normal history. While one can use such an LMW scheme to argue for a speculative bifurcation of histories similar to the one featured in Manyworlds, LMW cannot lend itself to a nihilism of information or the obviation of abstract memory, given that all the previous and subsequent apagogic mnemonic arguments are not undermined by LMW, while LMW depends on the retention of the tripped bomb in abstract memory. Moreover, LMW must also entail violations of the principle of conservation of energy-matter, especially in the context of 'sum over infinite paths'. Thus, LMW must be called into doubt, despite its comparative plausibility and non-nihilism vis-à-vis information and memory, counter to Manyworlds: (See **3.18**).

- **Apagogic mnemonic argument 3-a**: **The 'holed past theory' casts doubt on Yoon-Ho Kim's delayed choice quantum eraser**, (see **3.19**). It turns out that the past embedded within realised spacetime is not always totally or wholly resolved: Some parts of the past remain in quantum suspension and isolation, owing to the fact that the rest of the physical universe did not interact with or help collapse the wavefunction pertaining to that suspended non-resolved past. A 'holed past' is thus formed wherein spacetime in potentiality (the future) protrudes into realised spacetime. Subsequent observation on that holed past (say, a million years hence) will then resolve it into an OR-form past outcome, but *not* by erasing that alternative past, nor by replacing it with an alternative past. Memory cannot be erased per Yoon-Ho Kim's methods. Recall from Book-II that counterfactuals always eliminate non-consistent futures from all frames along the event horizon, in relation to any OR-form outcome realised on or along that event horizon and throughout spacetime in potentiality, up to the infinitely removed future. This generates grand decoherence of spacetime in potentiality, rendering the total future consistent to what happened, and thus restructuring possibilities open to the future, and, in doing so, denying future observers any future-outcome inconsistent with what had transpired in the past. (This is the new form of the principle of conservation) Thus, if the past was not suspended and was resolved into a definite outcome, Yoon-Ho Kim's eraser could not erase it, since only an outcome consistent with what had transpired is permissible in *any* future or future observation. On the other hand, if the past was not resolved and was AND-form suspended into a 'hole in the past', the action from the present will indeed AND-to-OR collapse the 'hole', but *not* by erasing any *realised* past or memory state, but erasing a suspended potentiality-state that never got to happen. The outcome thus resolved must yet be consistent with all that has transpired in the rest of the universe, courtesy of grand decoherence. The retention of the past so implied shows that abstract memory abides.

Thus, Part-I secures abstract memory against anticipated nihilistic a-informational and a-mnemonic erasure and MWI-based arguments, and against the implicit amnesic universe postulate, prefiguring the conditions and contention for a non-arbitrary meaningful universe.

BOOK-III PART-II:
SOLUTION TO THE PROBLEM OF QUANTUM INDETERMINISM AND CONSOLIDATION OF THE CASE FOR ABSTRACT MEMORY

3.21: Aims of Part-II

The goals of Part-II include the further advancement of the definition of information. But the primary goal will be the solution to the problem of quantum indeterminacy at the core of the stochastic coherence paradox, and of how order and pattern can arise, and temporally persist or perdure, despite the apparent 'chaos' of quantum indeterminacy. On the heels of the solution to quantum indeterminism, and further advancement in the classification of information per the growing block spacetime structure, the completion of decoupling theory will follow. Part-II will also culminate into the last apagogic mnemonic argument, so completing the goal set out in Part-I

Part-II will evolve the definition of physical information to a form in diametrically contrast to the near-exclusive OR-form characterisation of it featured throughout the Preliminary and into Book-II. Yet, Book-II conceded that, in the framework of the growing block intermediate spacetime model, OR-form information must be restricted and confined to the event horizon, while the ultimate form of physical information must assume an abstract form, either in its future-form version distributed within the domain of spacetime in potentiality, or in its memory form retained within mnemonic realised spacetime. Part-II will resort to the Sierpinski triangle as the didactic demonstrator of **the abstract character of physical information: a probability structure that funnels and aggrandises discrete OR-form events into manifest and fated wholes: indeed, into 'instantaneous wholes' synonymous with the quantum wavefunction. Thus inverting the key working assumption held about the nature of information throughout the previous material**.

Even with the anticipated overhaul of the definition of information, **information-matter dualism will abide**, as will abstract memory: the latter destined to be fully secured on the basis of de-spatialisation in Book-IV.

Once we grasp the general form of information as a predominantly abstract state, we can then integrate to it the read only category of information from Book-I. To this end, Part-II will show that read-only categories of information (such as physical laws) are states effectively embedded within spacetime in potentiality, (within the future) even while they transcend spacetime in potentiality as ordinances of the meta-semantic hypostasis from the infinitely removed future. Recall that the case for the remarkable meta-semantic hypostasis was developed in **apagogic mnemonic arguments 1-a** and **2-c**, in Part-I (see **3.7** and **3.14**). We will reiterate and firmly establish that **read-only categories, hence physical laws, are fundamentally dualistic and extant the OR-form materialisations that laws help manifest on or along the event horizon**: The electronic configuration of the atom, which regulates the distribution of the electrons about the atom, exemplified by d_{z2} orbital pattern, illustrates how read-only abstract informational states structure and fate OR-form realisations, and, in the process, render the operation of physical law salient. Yet, the d_{z2} orbital pattern is dualistic and abstract vis-à-vis the series and succession of OR-form electron-resolutions about the atom.

*

The abstract approach to information will scaffold for two key solutions to the problem of quantum indeterminacy: a core problem of quantum mechanics subsumed to the stochastic coherence paradox. Once we clarify read-only information, with the primacy of abstract states entailed in the configuration of natural odds in the regulation of quantum indeterminate processes, and from the observation that quantum indeterminate processes always *unfold* within such abstract states , with the latter born of necessity and law, and *not* from randomness, we will obtain the first component to the solution to quantum indeterminism: namely, the three-phase developmental process.

The first of the three phases is the **below-threshold noise phase**. Therein, it is found that an insufficient sample of quantum indeterminate terms cannot exhibit *any* pattern. The global outcome will appear perfectly scrambled. Indeed, our prejudices about randomness and chance… as 'chaos'… derive from the below-threshold developmental phase. The pattern implied by the background abstract structure and instantaneous whole is not at all obvious from the noise-distribution of quantum depositions formed therein.

We will discover that the second developmental phase is the **transitional chaotic phase**: Therein, sufficient numbers of quantum indeterminate terms obtain an emerging pattern: albeit a volatile, unstable, easily perturbed one in the face of subsequent quantum indeterminate additions. Hence, a chaotic distribution of quantum indeterminate terms.

The third phase is the **noise-reduction pattern-saliency phase.** Beyond a certain threshold, quantum indeterminate terms bring developing systems out of their transitional chaotic phase into full noise-reduction pattern-saliency phase; wherein the hitherto hidden abstract state and instantaneous whole finally becomes fully salient or 'materialised'. The addition of more indeterminate terms can have no perturbing effects on the emergent pattern, but only serve to sharpen it: i.e., randomness becomes a source of noise-reduction and a friend to information. Therein, the noise-reduction role of randomness and quantum indeterminacy becomes obvious, and it becomes clear that neither can be viewed as inimical to pattern and order, or viewed as synonymous with 'chaos'.

Thus, **quantum indeterminacy in the context of the three-phase developmental process will be understood as a noise-reducing aspect of the economics of nature. With it, the stochastic coherence paradox, or the question of how order, pattern and its persistence can be attained despite quantum indeterminate 'noise', will be partially solved**.

Our three-phase process of development will be applied to the quantum mechanised version of the Sierpinski triangle, and to the **dz2** orbital (or any other orbital) governing the electronic configuration of the atom. It will be applied to the development of a photographic image formed out of quantum indeterminate depositions. It will be applied to the drifting snowflake. Thereafter, it will be generalised to all pattern and form in the universe. Thus, we will grasp that the natural odds always exist in the context of abstract information, rendered as an abstract bounded probability state, and that, through the three-phase developmental process, *some* abstract forms can 'materialise' the natural odds into coherent global patterns and structures.

However, the solution to quantum indeterminism from the three-phase developmental approach cannot prove presupposed quantum indeterminacy to be objectively and ontically real, but merely assumes the reality of indeterminism. **The proof of the objective and ontic reality of randomness and quantum indeterminacy will constitute the second component to the solution to quantum indeterminacy and of the resolution of the stochastic coherence paradox**. Therein, we will show that there are ultimately no hidden gears behind quantum indeterminism: We will show that the attempt to found a secret deterministic orchestration of surface quantum indeterminate processes can only generate unfathomable numbers of deterministic orchestrating scripts or 'hidden gears', but cannot generate a one-and-only exclusive or special deterministic script, and cannot secure any fundamental qualitative or structural basis for deciding which of the huge number of equally apt scripts must get to orchestrate apparent indeterminacy. Consequently, one must again resort to random selection from among equally apt scripts to 'solve' the problem. Yet, the resort hardly constitutes the solution to or obviation of randomness and indeterminacy. Indeed, the finding will demonstrate the ineliminability, hence the objectivity and ontic reality, of randomness and quantum indeterminacy.

With the two key component solutions to the problem of quantum indeterminism, Part-II will culminate into the **four-fold classification of physical information**: almost completing the goal initiated in Book-I. The first will be OR-form information, always confined to the event horizon, which constitutes **read-write Lagrange OR-form information**. It is Lagrange because it constitutes the combination of two things. i.e., conditionals from read-only future-form **wavefunctions in potentiality** from the future (the second of our four-fold information types)… and read-write **wavefunctions in memory** from the past (the third of the four definitions of information). The fourth in the list is read-write **contingent wavefunctions in potentiality**, analogised by 'Wheeler's cat': the apocryphal mischievous laboratory cat that overturns the physicist's experiment and changes conditions and outcomes.

When wavefunctions in memory from the future and wavefunctions in memory from the past combine with contingencies, the combined system then AND-to-OR collapses to generate read-write Lagrange OR-form information on or along the event horizon: i.e., putative 'matter in space'.

*

Book-II will but complete the classification of information *and* justify abstract memory through an **addendum to decoupling theory**. In the experiments of Gunter Nimtz first described in Book-I (**1.38** and **1.39**) the anomaly from the quantum tunnelled displacement of Mozart-40 at a purported 4.7 times the speed of light will be resolved by treating the **putative displacement of M-40 at 'faster than light' as a process involving abstract memory interrelated and structured in time, *without* resort to either 'motion' or 'space'**. While the critique of motion and space is not central to Book-III, the notion of the motion of radiation and matter was called into permanent doubt in Book-II (**2.24** to **2.26**). Obliquely, Book-III will warrant the addendum to decoupling theory on the basis of the obviation of motion and space by presaging de-spatialisation, wherein space itself is explicitly obviated, and can no longer constitute a fundamental part of physical lexicon or ontology.

Finally, **apagogic mnemonic argument 4-a** and **4-b** will constitute the last set of such arguments in Book-III, albeit secured from the solution to quantum indeterminism and the objectivity of randomness: Both falsify the erasure hypothesis and Manyworlds in a supplementary way to **argument 2-d (3.17), 1-a** and **1-b (3.7** and **3.8)** from Part-I. Thus, if quantum indeterminism is not inimical to information, as will be show, then erasure hypothesis in supposition of the utility of quantum indeterminism to nihilism, is immediately rendered void. And if the role of Manyworlds was to obviate quantum indeterminism, the reality of its objective ontic reality must

310

necessarily obviate the utility of Manyworlds, if not Manyworlds itself. Thus, nihilism from erasure *and* MWI totally unravels, and the ontology of information and memory is secured.

3.22: Information as Minimal Abstract State:
Wavefunction in Memory Versus Wavefunction in Future-Potentiality

In information-matter dualism espoused in the Preliminary (essays **0.1** to **0.4**) and further developed in Book-I and Book-II, we sought the history and provisional working definitions of information in terms of AND-form and OR-form dualism. We intuited our AND-OR categorisation as AND-OR (wave-particle) dualism. We subsequently incorporated both into the tripartite structure and dual domain growing block intermediate spacetime model in Book-II. But **we seek a consolidation of the definition of information that spans the whole gamut of physical states and processes in the wake of our various apagogic mnemonic arguments, and to perfect our overall case for abstract memory**, short of noetic information vis-à-vis mind and consciousness. The task will solve the problem posed by quantum indeterminism and the problem and paradox of how nature can manifest the persistence of coherent patterns and wholes, in the face of quantum indeterminate scrambling and randomness. The task will also address the question of whether quantum indeterminate processes are secretly orchestrated by hidden deterministic procedures or 'hidden gears'.

*

Consider information in the context of a photographic image. Is information to be identified with the set of coordinate addresses of each photon 'pixel' that composes it? Is there a simple overarching way of accounting for information pertaining to the photographic image, without resort to endless lines of pixel-addresses; and of a form that transcends the OR-form myopia? Is there a form of information that can act as a pure abstract basis for the image without the said complexity, such that, once it is subject to a specific type of development processes, the abstraction can manifest and re-manifest the full and complete photographic image? **Our previous interlude into the Sierpinski triangle in the context of the criticism of Manyworlds (see 3.17) suggests that an abstract minimal complexity conception of information *is* possible: It is key to the solution to quantum indeterminism**. To appreciate this, we must first reiterate the developmental process that pertains to the Sierpinski triangle (see **Fig 3.09**). Recall that the Sierpinski triangle is resolved via an equilateral triangle with three vertexes: each labelled "1 to 2", "3 to 4" and "5 to 6", respectively. This is to reflect outcomes obtainable by the roll of a six-sided die (D6) which could be easily quantum mechanised. We designate a point just outside the triangle as the 'seed'. The die roll outcome indicates the pertinent vertex: the new position of the seed will be halfway toward that vertex from the initial seed-position. We repeat the process for each successive 'seed orbit'; marking each as it develops. After a hundred or so seed-orbits we obtain the remarkable fractal pattern depicted in **Fig. 3.09 (right)**. This global pattern-outcome will 'materialise' despite the random and probabilistic process of resolution of each seed-orbit and constituent pixel.

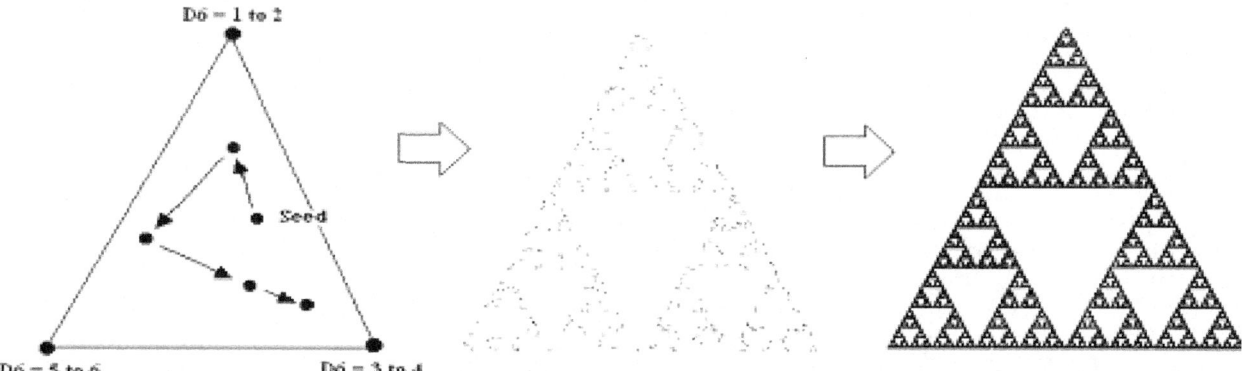

Fig. 3.09 Sierpinski triangle and physical information as minimal abstract state
Notes: The depiction is of the chaos game within the Sierpinski Triangle. The attendant equilateral triangle has three vertexes labelled "1 to 2", "3 to 4" and "5 to 6", respectively, This reflects the possible outcomes from a six-sided die, D6: see left-most diagram. We select a point just outside the triangle: i.e., the **seed.** We roll a D6 and, pending the outcome, move the seed halfway toward the pertinent vertex. We mark this new position and roll another D6, repeating the process. After a hundred or so orbits we obtain a fractal pattern as shown above (bottom, far right): we obtain this despite intrinsic random and probabilistic processes of resolution of the pixels. The process is analogous to a quantum wavefunction with three AND-form possibilities, resolved repeatedly and successively to constitute the succession of seed-orbits; outlining a single unique-history or universe that culminates into the fully developed fractal pattern universe: the right diagram.

How does any of this clarify the sought alternative definition of information beyond the OR-form myopia? The Sierpinski triangle exemplifies two types of information: the **maximal materialised state** versus the **minimal abstract state**. The maximal materialised state constitutes the fully developed or 'materialised' Sierpinski pattern composed of 'triangles within triangles': It is the OR-form of the Sierpinski triangle. We tend to think of any sort of information, including physical information, in just such OR-form maximal fully

311

'materialised' terms. The transmission of the fully developed Sierpinski triangle, or of any information portrayed in this maximal way, supposedly requires that we compose endless lines of code pertaining to every pixel, term, or position that constitutes the whole pattern, both before transmission and at completion of transmission. Indeed, this is how we transmit information using generic information and communications technologies. The Sierpinski triangle could be transmitted in just this way, in its fully developed maximal material form.

In contrast, the minimal abstract state of information pertaining to the transmission of our Sierpinski triangle would be stripped down to the indispensable: It would include the pertinent equilateral triangle in its pre-developed form, including its three vertices, and including the simple rules of construction that regulate the seed-orbit. This is all we would need to package into the transmission: constituting information as a minimal content bounded probability state. Upon reception, the Sierpinski triangle would be re-constructed into full pattern-saliency by the enaction of the simple rules governing the seed-orbit.

Consider what could happen to the efficiency of information and communications technology if, instead of storing, processing, and transmitting maximal states of information, we found some formal way of stripping *any* salient maximal state to its minimal abstract state, as can easily be done for our Sierpinski triangle, but cannot be so easily accomplished for other complex patterns. At the point of reception, rules of construction would come into operation, involving random-probabilistic sortology culminated into the maximal state.

What is lacking in modern information technology is a formal general process for deconstructing and stripping down *any* **complex pattern to its base minimal abstract state. If such a method were found, it would fashion for an unprecedented revolution in the efficiency of information storage, transmission, and processing**.

Is it possible to achieve such a 'magic bullet'? Nature has already accomplished what we can only speculate: It is our contention that nature retains and processes information in this radical minimal abstract state, of a form illustrated by the Sierpinski triangle 'chaos' game. **Nature handles information principally in the minimal abstract form and potentialises it in this form, both within spacetime in potentiality (the future) and subsequently retains it in this minimal form within realised spacetime (the past) ... 'materialising' these into OR-form maximal materialised states on and along the event horizon**. The implication is that the grander portion of nature's totality is configured in this minimal abstract form; namely as its abstract past and its potential future. Thus, in the growing block universe espoused in Alt-R, abstract memory retained within the mnemonic domain of realised spacetime constitutes a modified variant of this minimal abstract structure state: The superpose from such memories participate in the manifestation or 'materialisation' of events into maximal materialised states vis-à-vis the event horizon. Such 'materialisations' always involve quantum indeterminate sortology: i.e., the physical equivalent of the random-probabilistic construction and re-construction analogised in the Sierpinski triangle in **Fig. 3.09** but generalised across nature.

As with memory, future-form information, or potential information, is also constituted in minimal abstract form. The key difference between minimal structure potential information versus minimal structure memory is two-fold: First, potential information belongs to the future-form of as-yet not-realised spacetime in potentiality. For its part, the past, or memory, is retained within mnemonic realised spacetime. Second, **minimal abstract states of future-potential information are 'un-charged': they are comprised of immanent futures that also nest for complexes of nested futures elaborated to the infinitely removed future. This contrasts with minimal abstract states of memory, with their former nested futures 'cropped', and no longer possessing their former nested futures, but can yet act as 'instantaneous wholes' from the past, transforming into re-materialised mereologies on the event horizon.** In either case, 'materialisation' is via AND-to-OR collapse.

*

It seems we have contradicted notions about information hitherto developed in preceding material. Is not the minimal abstract state the same as generic quantum waves and wavefunctions? It is, but with the added nuance that **some wavefunctions belong to the past and constitute abstract memory; as 'charged' wavefunctions in memory 'cropped' of their former nested futures. These minimal abstract states constitute** *wavefunctions in memory*.

On the other hand, **not-yet realised 'un-charged' or non-cropped, wavefunctions belong to and constitute the ontic future, with both their immanent possibilities** *and* **nested futures fully intact:** These constitute wavefunctions of the future, or *wavefunctions in future-potentiality*.

We have arrived at a peculiar conclusion seemingly against the grain from the Preliminary into Book-II: We started with the assumption that real information must be OR-form; that AND-form potentialities do not contain, carry, or act as repositories of information; and that OR-form information *must* dissociate from AND-form quantum mechanical states, with implied inexorable information-matter dualism. **It now turns out that information** *is* **predominantly AND-form**, *save that information in memory is comprised of AND-form states 'cropped' of their nested futures, while future-potential informational states are AND-form states with nested futures*.

A given wavefunction in potentiality (the future) transforms into a wavefunction in memory (into the past) upon AND-to-OR collapse and attendant progression of the event horizon (time). Consequently, it is 'cropped' of its nested futures and materialised into an OR-form corpus at the event horizon. Thereafter, the cropped or 'charged' wavefunction is relegated and abstracted into the domain of realised spacetime. Therein, it is retained as abstract memory (as minimal abstract AND-form state *without* nested futures) extant the event horizon and extant moments of 'materialised' OR-form radiation and matter. In short, none of this obviates information-matter dualism espoused throughout the Preliminary and in Book-I, no more than it invalidates abstract memory: **Information-matter dualism is simply per function of the fact that abstract memory conceived as** *wavefunction in memory* **is retained within realised spacetime: That is, memory does** *not* **involve the retention of OR-form complexes. Therefore, the almost pure OR-form approach**

312

to information, and especially to memory, espoused from the Preliminary through to Book-II, proved useful and didactic. But it must now be superseded by the above scheme. Yet, this eventuality was anticipated at several points in the preceding books.

*

Henceforth, the reader must keep in mind that the 'charged' minimal abstract states (i.e., quantum waves and wavefunctions cropped of their nested futures) are interchangeable and synonymous with wavefunctions in memory retained within realised spacetime. On the other hand, 'un-charged' minimal abstract states (i.e., quantum mechanical waves and wavefunctions *with* their nested futures intact) are synonymous with wavefunctions in future-potentiality embedded within spacetime in potentiality.

Together, **wavefunctions in memory and wavefunctions in future-potentiality constitute a more nuanced non-trivial account of the quantum wavefunction, albeit differentiated into a memory-form variant within realised spacetime and into its future-form variant within spacetime in potentiality: Both constitute stripped-down minimal abstract configurations of information whose first-time AND-to-OR decay out of the future, or subsequent re-constructions out of the past, must proceed in the way, analogued by the developmental process evinced by the quantum mechanised Sierpinski triangle, but subject to 'materialisation' via quantum indeterminacy**...consistent with the overall case for information-matter dualism, wherein OR-form information is restricted to the event horizon and is effectively dissociated from AND-form configurations of radiation and matter.

3.23: 'Read Only' Bounded Probability States: The Electronic Configuration of the Atom

In our first forays into a working definition of information, **Book-I differentiated read-only from read-write categories. We must now integrate these distinctions into our minimal abstract state-based approach**; conceiving information, not just as a minimal abstract state pertaining to future potentiality or past memory, but also as read-only *and* read-write states.

In this brief essay, we will explore nature's clearest and simplest example of minimal abstract state in read-only form: namely, the electronic configuration of the simple atom, which describes the probability regions or complex wavefunctions within which electrons are likely to manifest per subsequent AND-to-OR collapse.

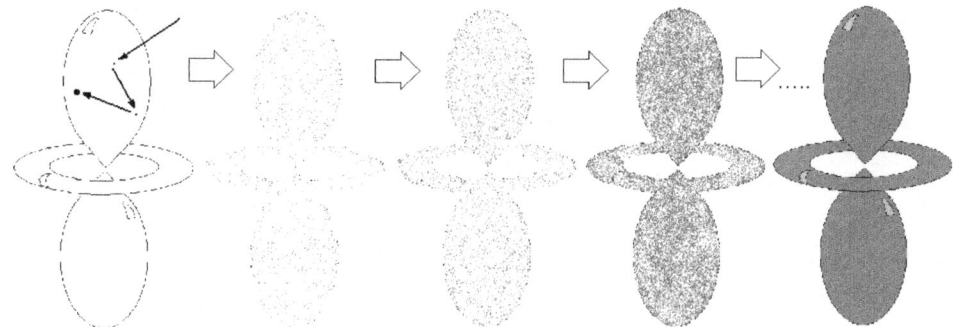

Fig 3.10: Bounded probability states, the electronic configuration of the atom: exemplification of read-only information
Notes: The illustration depicts the atomic orbitals for electrons for a simplest atom. The illustration on the far left shows the so-called **dz2** orbital, grasped as three basic regions within which an electron might be found: i.e., the torus region about the nucleus, and the two 'teardrop' shaped regions above and below the nucleus. The shown regions, torus, and teardrop-like, are formed out of standing AND-form quantum mechanical waves within purported three-dimensional space about the nucleus. These will permit 'seed orbits' from successive measurements of electron position about the atom. There will be a quantum indeterminate 95 percent chance that an electron will assume a position within these regions. Thus, if we made a succession of positional measurements of the electrons, and marked these positions within the bounded probability regions, the effect would be essentially the same as found in the previously explored quantum mechanical development of the Sierpinski Triangle: a coherently 'filled in' region that reflects and manifests an otherwise abstract-immaterial bounded probability state for where the electrons are permitted to resolve. Of course, the example illustrates the retention by the physical order of read-only future-potential AND-form states as these pertains to the atomic electronic configuration. The question must arise as to whether it is also possible for nature to retain 'charged' bounded probability states that belong to the **past**: namely, wavefunctions in memory and read-write states, at a remove from both the domain of spacetime in potentiality as well as from manifest states of radiation, matter, and 'space' on and along the event horizon.

The electronic configuration of the atom is illustrated in in **Fig. 3.10**: This is the **dz2** orbital. It is not the only possible form of the electronic configuration, but the same principle applies to all four alternatives: (for brevity, these are not shown). **Orbitals of electrons about the atomic nucleus can be grasped as regions of quantum mechanical standing waves within which an electron might be**

313

found upon putative 'position' measurements. Hence, in the illustrated case of the **d$_z$2** orbital, the torus-shaped region about the central nucleus, combined to the two teardrop-shaped regions above and below the nucleus, constitute forms made up of standing waves.

How are such regions formed? **Only at certain radii about the nucleus, and only at certain frequencies, can AND-form quantum mechanical waves form up into viable standing waves. This cannot happen at any radii. (This can be restated in pure temporised form**[76]**), 'Hence, the d$_z$2 orbital probability structure is not itself an arbitrary or probabilistic outcome: it emerges out of strict necessity**, even though it constitutes the bounded limits within which otherwise quantum indeterminate random resolution of electrons may arise, in the same sense that the configuration of the Sierpinski triangle bounds and funnels seed-orbits: (see **3.22**). **The d$_z$2 constitutes natures consummate example of a read-only state.**

There is a quantum indeterminate 95% chance that an electron will assume a putative position within the depicted probability regions or 'electronic clouds'. To reiterate, the regions formed into the **d$_z$2** configure a bounded probability state that delimits and funnels quantum indeterminate resolution of electrons about the nucleus: Hence, **if we make successive positional determinations of a given electron within any of the 'electronic clouds', and mark these down onto a map of the system as these developed... in the same way we mark the seed-orbits in the developing Sierpinski triangle... the cumulative funnelled effect will constitute a 'filled in' maximal materialised state, rendering explicit or 'materialised' the 'read only' minimal abstract state behind it: this being the 'instantaneous whole' abstract of the d$_z$2 orbital itself.**

*

Any future-form wavefunction within spacetime in potentiality will constitute the pre-measurement isolation-state of a given system: a superpose of as-yet not-happened future possibilities available to it. Applied directly to the electronic configuration of the atom, the complex wavefunction now formed up as the **d$_z$2** orbital, also within spacetime in potentiality, constitutes the pre-measurement read-only instantaneous whole isolation-state for the superpose of future electron position-possibilities within spacetime in potentiality. Over time, this AND-to-OR 'collapses' into a succession of resolved OR-form positional states for the electron. Yet, the **d$_z$2** orbital minimal abstract read-only state ultimately delimits the possible and probable succession of OR-form positions of the electron into those specified by the abstract state. Indeed, for **every successive AND-to-OR resolution of the electron, as 'law', the d$_z$2 minimal abstract state is re-imposed from the future; from spacetime in potentiality**. In other words, the **d$_z$2** orbital permanently and ineradicably pre-exists, albeit as a future-form potentiality and abstract law, *within* spacetime in potentiality... and enforced to the infinitely removed future.

Insofar as it pre-exists the OR-form materialisation of the electrons, and is *not* erased by any successive OR-form resolutions of electron on the event horizon (and since the resolution of the electron cannot obviate the future, much less the whole of spacetime in potentiality in which the **d$_z$2** system ineliminably resides) the read-only **d$_z$2** orbital state is necessarily retained at a physical remove or dissociated form from the event horizon... from the instances of OR-form materialisation of the electron. That is, **the d$_z$2 orbital ultimately subsist as a fully dissociated abstract-immaterial state and law within spacetime in potentiality, vis-à-vis the materialised world of electrons restricted to the event horizon, or from 'matter' confined to the event horizon. This confirms our case for information-matter dualism, even if from the vantage point of read-only future potential information... from the vantage of spacetime in potentiality and wavefunctions in future-potentiality vis-à-vis the event horizon... vis-à-vis 'matter in space'.**

3.24: Presage to the Solution to Quantum Indeterminism:
The Relation of Abstract Necessity and Order to Subordinate Quantum Indeterminacy

The stochastic coherence paradox can be restated thus: **How could random probabilistic sortology, or 'chance' and 'noise' in the resolution of electron-positions about the atom, or in the formation of a photographic image out of quantum indeterminately deposited 'pixels', produce and *reproduce* coherent structures, patterns, and corpuses of information, despite apparent pattern-inimical and pattern-scrambling quantum indeterminism?** Essentially the same question may be posed in the context of the emergence of all natural patterns and cosmic structures, including living forms, given the inherent role of contingency and happenstance in cosmology and in the bio-evolutionary process, some of which may be decisively quantum mechanical in character, before natural selection takes hold of the indeterminately resolved outcomes and sieves these for survival or otherwise.

The supposition that randomness and quantum indeterminacy must be illusionary is tempting... as tempting as the want to sweep the problem away by resort to the Abrahamic God, or even to a universe-scale quantum consciousness super-conflated to God. More prosaic circumventions of indeterminacy have also been entertained: For example, James Jeans' navigation-based analogy to sweep away the notion of possibilities in AND-form: see Preliminary (**0.10**) and Book-II (**2.02** and **2.03**). Another is David Bohm's hidden variable and implicate order approach that utilises pilot waves[77]. We must not forget Manyworlds itself, which seeks to obviate quantum indeterminism by having *all* the possibilities in the AND-form superpose realised. Another minor but popular contrivance involves resort to scale: wherein quantum indeterminacy is asserted to apply only to the very small, licencing the notion that we can

---

[76] It is possible to recapitulate radii in pure de-spatialised temporised form by dropping the wavelength and framing the system in terms of the wave's potential moments of undulations, or 'frequency': only certain frequencies will yield the temporal equivalent of the standing wave.

[77] Bohm, David (1952). "A Suggested Interpretation of the Quantum Theory in Terms of 'Hidden Variables' I". *Physical Review*.85 (2): 166–179.

ignore it at the larger macro-scale, especially vis-à-vis the brain. Yet, scale is irrelevant, as we shall find out. We do not need to resort to *any* of the above contrivances to obviate or circumvent quantum indeterminism.

The indeterminate probabilistic deposition of quanta across nature can be treated in the same way as the quantum mechanised indeterminate seed-orbit in the Sierpinski triangle. The process thus analogised can be generalised to the rest of nature and can help solve the problem of how nature furnishes coherent corpuses and mereologies of maximal materialised form through otherwise intrinsic quantum indeterminate processes. Moreover, the approach exemplified in the Sierpinski triangle helps hint at the solution to the problem of the persistence of identities and patterns, despite the presumed ravages of quantum indeterminism: A rose will remain the rose over time: It is not a rose one moment and a pumpkin the next, despite the view that such continuity of identity ought to be impossible, given ineliminable quantum indeterminism *and* from nihilism from forlorn memory erasure hypothesis (**3.6** to **3.11**), as was made atrocious by its later co-joining to equally forlorn Manyworlds (**3.12** to **3.18**).

In the Sierpinski triangle, in the **dz2** orbitals and the electronic configuration of the atom, *and* in the developing photographic image… and **throughout nature, it is always the case that quantum indeterminate processes unfold within some minimal abstract state. This does not abide just at the micro-scale but also and especially at the macro-scale, as evident in the complex wavefunction in photography, and in the universal wavefunction at the universe-scale.** Such abstract states are ubiquitous throughout nature *at all scales*, and are infused into spacetime future and spacetime past. They are always at a dissociated remove from 'matter', radiation, and space, or OR-form outcomes that are confined to the event horizon: They are constituted as either wavefunctions in future potentiality or as wavefunctions in memory within the two pertinent spacetime domains. Quantum indeterminate processes of resolution, always incident at the event horizon, cannot help but generate *en masse* outcomes that exemplify organised pattern and attendant continuity of identity implicated by the said abstractions. The conclusion is that, we need not resort to Manyworlds-based circumventions, or invent contrivances in the form of hidden gears, much less resort to quantum consciousness, or even to the Abrahamic God, to solve the supposed 'problem' of quantum indeterminism, much less resort to notional micro-scale scale-restriction of quantum mechanics in order to evade the 'problem' at the macro-scale.

Thus, in the **dz2** orbitals, successive quantum indeterminate resolutions of electrons will render salient the hidden abstract pattern encapsulated in the abstract form of **dz2**. But, if we consider minimal abstract states belonging to the macro-scale, such as the outcome obtained in the two-slit experiment when both slits are open via cumulating *en masse* indeterminate terms, or by *en masse* deposition of photons in the formation of a macro-scale photographic image, **the culmination of indeterminate depositions in all such cases render explicit the macro-scale minimal abstract instantaneous whole behind the formed up electron shells, behind the interference pattern involved in the two-slit experiment, and behind the complex photographic image. In other words, the basic relation between minimal abstract states and the *en masse* indeterminate quantum resolutions that bring these into saliency, is the same, *whatever the scale*. i.e., The relation is scale-invariant and *not* confined only to the 'very small', but equally applicable to the larger-scale universe.** That is, quantum indeterminacy is not circumvented by resort to scale.

*

Another key to the solution to the problem of indeterminacy and the stochastic coherence paradox is **the supremacy of necessity over subordinate indeterminacy, given that quantum indeterminate processes *always* unfold within minimal abstract bounded probability states that are either functions of read-only natural law and necessity or partake of law to form viable read-write abstractions: These are *not* products of contingency or happenstance, as was readily exemplified by dz2 orbitals (3.23). Consequently, while order *does* arise through quantum randomness and contingency, the minimal abstract states that get to funnel these into *en masse* global outcomes are products of necessity and law… whose origins ultimately reside in a meta-semantic hypostasis,** as was posited in **3.15**. No matter how truly indeterminate the resolution of positions and photographic photon depositions that collectively make up a given pattern, these *always* produce and reproduce the pattern implicated and enforced by the minimal abstract state behind it.

In short, while the constituent quantum terms are certainly indeterminate, the global outcome is almost always ordained, fated, and determined, baring contingency, or 'Wheeler's cat'.

The electronic configuration of the atom exemplified in the **dz2** orbital, and the non-haphazard way in which the standing waves for electron shells are constructed about the nucleus of the atom, amply demonstrates the necessitarian read-only and lawful nature of certain minimal abstract states and outcomes: to the effect that there are indeed specific fixed patterns and ordinances in nature that derive out of law and necessity, *not* from indeterminacy… while the former funnel and regulate indeterminacy and chance into delimits and into fated culminating patterns. **Hence, determinacy *and* indeterminacy harmoniously co-join *without* contradiction, so long as indeterminacy is subordinated to and subject to necessitarian read-only abstract order.**

*

Consider the distribution that a student involved in a school statistics project might obtain from the serial roll of three unbiased six-sided dice. The distribution curve obtained for the sums from these will form a generic bell curve for probabilities of values ranging from 3 to 18. In the first few rolls, the anticipated bell curve distribution will not be at all obvious. It will take a threshold number of die rolls and outcomes to render explicit the implicated bell curve distribution. Over time, as more such random samples are generated, the sharper and more well-defined the bell curve will become.

We could perform this simple school statistics experiment using quantum mechanised die, and obtain the same results.

315

Given the usual prejudices we have about randomness and indeterminacy, one might be forgiven for assuming that, as the number of random outcomes generated increases over time, the bell curve ought to become obscured and occulted by 'noise'. The 'secret' to how the bell curve becomes sharper and more well-defined over time despite randomness or quantum indeterminacy, and not more occluded or haphazard, is obvious: The randomised outcomes from the three six-side dice are *always* manifested *within* the delimits of an abstract minimal structure ordinance, one comprised of a fixed boundary state: a domain and range of 15 possible outcomes, fated per fixed relations or sums obtainable from the sum of die-rolls ranging from 3 to 18... fated to fixed probabilities of outcome. Obviously, given the six sides per each dice, given the finite number of dice, and given the fixed outcomes possible from consecutive rolls, if we generate a large enough sample over a sufficient time, we are bound to obtain the appropriate and expected bell curve distribution, simply because the situation cannot help but realise this outcome, even with quantum mechanised dice. The bell curve will peak at the sum of 10, simply because the odds are naturally biased (at just above 12% probability) to sums equal to 10 vis-à-vis other sum of lower probability.

However, the less the number of rolls of dice, the less the number of sample-sums generated, and the less salient the bell curve distribution obtained, given that there will be less opportunity for the system to culminate or realise the long-run inherent higher probability: 12% amplitude in favour of 10. On the other hand, the larger the number of random sample-sums, the sharper and more salient the bell curve distribution centred on outcome 10.

The same logic can be generalised to probability distribution of successive electron resolutions in the d_z2 orbital... in the probability distribution of electrons in the developing interference pattern in the two-slit experiment... in the probability distribution of photons in a developing photographic image... and, indeed, in the development of the whole of nature, including the development of the universe scale wavefunction itself that bestrides the totality of spacetime in potentiality to the infinitely removed future.

Thus, **insofar as it manifests within minimal abstract bounded probability states, quantum indeterminacy has the long-run astonishing-seeming, but otherwise unremarkable effect of increasing the precision and saliency of pattern. Indeed, randomness and indeterminacy reduce noise over the long run, especially when generated *en masse*,** as we can ascertain even from our school statistics project.

In short, beyond the minimal threshold state, randomness is *not* noise: a conclusion that runs against the grain of prejudgement. By prejudgement, we tend to take randomness to be synonymous with noise. Indeed, one reason why Darwinian Evolution appears to 'fail' in the minds of a number of people is due to the presumption that randomness, or 'accidents', must constitute noise, and that nature cannot generate any signal out of noise, much less evolve into complex living forms or even whole ecologies. The presumptive 'rationalists' are no different: they treat randomness as noise and posit it as an obviation of pattern and purpose in nature, with a view to obviate God and religion. For an obviation of both forms of confusion and contradiction, recall the conclusions reached in essay **3.15** in Part-I.

3.25: Presage to the Solution to Quantum Indeterminism:
Obscuration of the Problem of Indeterminacy from Limiting Cases in Quantum Mechanics
Over-reliance on limiting cases in quantum mechanics tend to obscure the harmonious relationship that resides between indeterminacy versus all-scale order, pattern, and information. The Einstein Podolsky Rosen experiment (EPR) as well as the Schrödinger's Cat thought experiment, exemplify just such limiting cases.

In EPR, the spin from one of the entangled particle pairs may resolve either into an OR-form of **spin-up** or into **spin-down**, with an opposite corollary-spin resolved for the other particle; with 0.5 probability of one or the other spin-pair outcome resolved. In a similar vein, in the imaginary Schrödinger's Cat experiment (or the nuclear detonator featured in our previous Little Manyworlds in **3.18**) the fate of the cat will be rigged to an EPR outcome, with a future-form wavefunction constituting a bounded probability state comprised of the **live cat** AND **dead cat**... destined to degrade into the complex *en masse* OR-form of the one outcome or the other.

In EPR, there will only be one quantum incident; not a mass of such. Hence, with the single exemption of the law of the conservation of spin and the read-only abstraction pertinent to that principle, the single-run EPR scenario is not complex enough to infer any abstract ordering process that might otherwise funnel *en masse* long-run quantum indeterminate outcomes into large-scale determinate meaningful global materialisations. Relying on such limiting cases, one will be tempted to assert the operation of *a-causality* throughout nature; assert causal nihilism... and perhaps even assert information and memory nihilism itself. Consequently, the matter of whether we obtain a live cat, or a dead cat rigged to EPR will *not* be grasped as a matter of significant causality or temporal carry-over of critical information across time, or foster the inference of any temporal endurance and perdurance.

Clearly, only sufficiently complex cases that require *en masse* quantum indeterminate incidences, or a long-running series of successive single incidences accumulated into *en masse* outcomes, such as featured in the two-slit experiment interference pattern and presupposed in the evolution of **d_z2**, can permit correct generalisations about the physical relation of randomness to information, to order, and to large-scale pattern.

Once corrected for the scotoma imposed from the overuse of limiting cases, quantum indeterminacy and randomness turn out *not* to be inimical to information and order. Therein, quantum indeterminacy ceases to be a 'problem' that one either must dismiss as an artefact, (the inference from Jean James: **2.02**), or a thing that belies secret underdetermining 'hidden gears', as David Bohm proposed; or a problem that must be circumvented via Manyworlds... or obviated through the intercession of consciousness or the Quantum God... or a thing that must be swept under the 'scale' rug.

316

QUANTUM MECHANICS AND MIND

It turns out that, quantum indeterminism is nature's powerful economising process; transforming abstract forms of future-potentiality and past-memory into maximal 'materialised' states of information on the event horizon... into resolved complex photographic images; into drifting snowflakes; into whole galaxies... and accomplishing all of this *without* having to specify endless lines and 'addresses' of 'code' for every pixel, particle or constituent, or resort to secret fine-structure orchestration, direction or control.

Quantum indeterminate processes cannot help but unfold and realise the global pattern implied by the funnelling background abstraction. The more such indeterminate incidences get to unfold, the sharper and more salient the pattern rendered, and less the 'noise', even though the randomness and indeterminacy remain objective and ontically real, as we shall discover in **3.32**.

Yet, according to generic prejudgements, randomness and indeterminacy are erroneously perceived *as* nothing but noise: which then leads to the problem of how coherent pattern could at all be produced by noise, and, even more remarkable, *reproduced* from *noise*... and wherein seriously misleading limiting-cases, such as Schrödinger's Cat or EPR, are brought into the play to perpetuate confusion.

3.26: The Solution to Quantum Indeterminacy Part-I:
The Three-Phase Developmental Solution to the Problem of Quantum Indeterminacy

Given the centrality of quantum indeterminism across nature, **it might appear that indeterminism has triumphed against determinism. Yet, the presumed eternal battle between these prove to be part of a long-running false dispute. The harmony between determinism and indeterminism is inherent to our developmental solution to quantum indeterminacy.**

The Alternative Realist (Alt-R) solution to the stochastic coherence problem from the three-phase developmental approach is based on two complementary postulates. The proof of the first postulate must await **3.32**.

Thus...

- **1st Postulate: Randomness and quantum indeterminism remain ontologically and objectively real: there are no hidden gears or any secret orchestration of quantum indeterminate terms.** However, ...

- **2nd Postulate: The long-run sequential and *en masse* quantum indeterminate outcomes nearly *always* culminate into the production or reproduction of pre-ordained, fated, and fully determined global outcomes and patterns** per function of minimal abstract states that funnel otherwise quantum indeterminate outcomes into pre-fated global formations. The said abstract states come in two main forms:
  - 'charged' minimal abstract states, or **wavefunctions in memory**, relegated to the domain of realised spacetime.
  - 'un-charged' minimal abstract states, or **wavefunctions in future-potentiality**, inherent to the futurity... to the domain of spacetime in potentiality.

This essay is concerned with the second postulate. To prove the second postulate, we will detail a simple statistics project involving the roll of six-sided quantum mechanised dice, and the development of the pertinent bell curve distribution viewed from the framework of our the three-phase developmental process. The implications will later be generalised to the three-phase development of the Sierpinski triangle; to the development of $dz2$ orbitals; to the development of the interference pattern in the two-slit experiment; to the development of a photographic image or an electron micrograph... and, thereafter, generalised to the rest of nature.

*

The statistics project involves the simple roll of three six-sided dice. To avoid the charge that classical die-rolls are not truly random, we furnish a quantum mechanised dice-shaker device: see **Fig. 3.11**. Our student records the sums thus produced; generating an **n** number of quantum mechanised three-die rolls; a large tally of such, culminated graphically into a generic bell curve distribution; with the most probable sum-outcomes given by the peak (the amplitude) of the distribution curve.

Therein, the sum-outcome of 10 constitutes the peak: with a probability just above 12%. The less probable sum-outcomes are found along the flanking declines of the bell curve, with the least probable at the extreme ends of bell curve: (See **Fig 3.12**).

However, for the sake of our three-phase developmental approach, for each developmental phase constituting it, we graph the quantum mechanical die-outcomes generated, *not* by plotting the frequency of the accumulated outcomes *after* large tallies are generated. Instead, we seek to illustrate the developmental process *in realtime*... as the samples are generated and garnered marginally; so constituting a developmental realtime evolution of the process of accumulation of terms and the culmination into pattern. This will be accomplished through a succession of sample graphs, with each graph incorporating the results obtained and accumulated in the preceding graph, while adding the new outcomes to the succeeding one. Simply put, we plot on the first of such graphs the outcome generated by the very first die roll. Then, on a second graph, we plot the outcome from both the first *and* the second roll. Then, on a third graph, we plot the outcome from the first, second and third rolls... and so on. Through a succession of such graphs, we inspect the *time-evolution development* of the distribution-curve, almost in realtime, or as it forms.

Our approach would require a substantial number of graphs that could easily exceed the number of pages constituting Book-III. Therefore, we will restrict our examination to a few key samples-graphs; more than sufficient to illustrate the three-phase developmental process per quantum indeterminacy and the role of minimal abstract states therein.

What would we discover in the application of the above-described process?

We would find three basic developmental phases, including a final phase that exhibits the fully salient bell curve distribution. Thus...

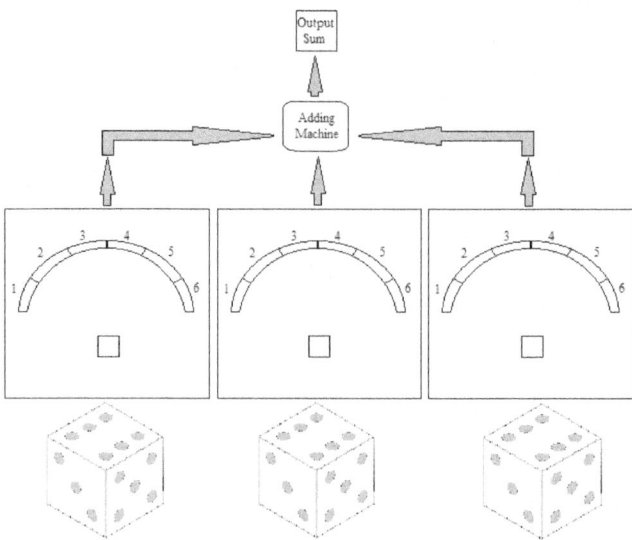

Fig. 3.11: Quantum mechanised 3D6 for the generation of a bell curve distribution: range 3 to18
Notes: The depiction of quantum mechanised six-sided 'dice-shaker' for the generation of a bell curve distribution. Each of the three D6 die involved are replaced by a modified two-slit experiment setup from which the barrier is removed and whose detector is rendered concaved conformally to the wave-front arc of a single photon emission; with each concave detector divided into six equal arcs for the equal probability of photon deposition, forming a 1 to 6 range of possibilities per die. The outcomes from each quantum mechanised die are inputted to an adding machine, whose output can be any value from 3 to 18. Using the clear advantage of such "true dice", whose randomness cannot be disputed owing to the quantum indeterminate character of the process, a school student uses the quantum dice-shaker to generate a realtime tally of quantum indeterminate sum-outcomes from which the student can generate the generic bell curve distribution.

*

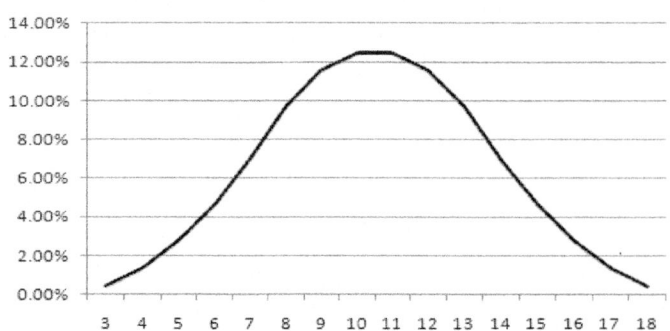

Fig. 3.12: Bell curve distribution obtained from rolling three quantum mechanised six-sided die
Notes: Our school student's statistics project involves the simple roll of a three-quantum mechanised six-sided dice using the quantum 'dice shaker' device illustrated in **Fig. 3.11**. The student tallies the sums from repeat 3D6 rolls. These then express the expected bell curve distribution (depicted) with the most probable sum-outcome given by the peak or amplitude of the curve. The sum-outcome of 10 is the peak. It has a probability just above 12%. i.e., greater than the probability of the other sum-outcomes. The less probable sum-outcomes are distributed to the flanks of the curve: the least probable are 3 and 18.

Below-threshold noise-phase: This is the phase of development in which the quantum indeterminate samples generated do not exhibit any definitive frequency-distribution or collective tendency into a bell curve form, or *any* form. Collectively, **the samples appear to be perfectly scrambled: a haphazard distribution of outcomes. In other words, a *noise distribution*: (see Fig. 3.13)**. If we stopped generating any new samples while within this phase, we could not subsequently infer or predict *any* final pattern-outcome, or *any* pattern. **At this below-threshold sample level, there really is no general pattern to discern: Seemingly, there is no structure or discernible information in the system**. The randomness obtained truly constitutes 'noise', and we readily presume randomness to be synonymous with nothing but noise.

318

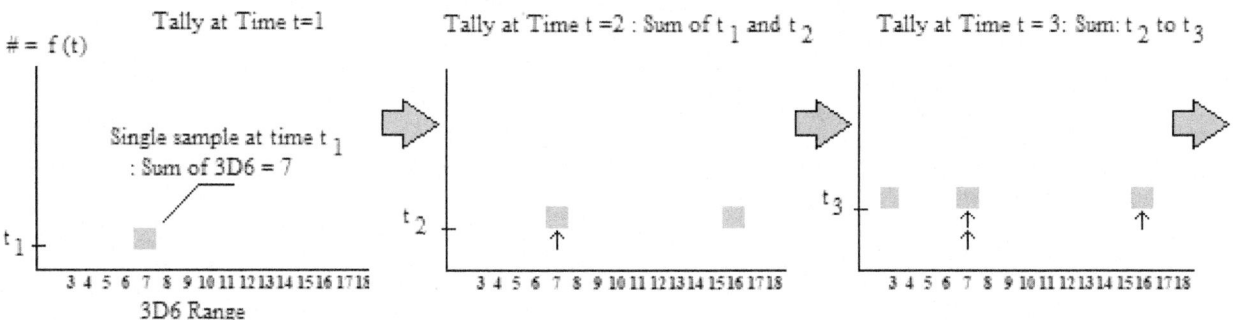

Fig.3.13: Below-threshold noise-phase as a succession of realtime accumulation of random samples
Notes: The succession of graphs depict the accumulation of random samples from a quantum mechanised 3D6 system. The number of samples is a function of time. At t_1, only outcome 7 is generated. At t_2, we have 7 *and* the new outcome of 16. At t_3, we get a new outcome 3 in addition to the 7 and 16 from the preceding rounds. The accumulation of few samples yields no evidence of pattern: a *noise distribution* that appears to conform to our prejudices about randomness; as a thing that embodies no pattern or information, but mere noise.

The common 'understanding' of randomness and indeterminacy is steeped in this below-threshold developmental phase and framework: i.e., randomness as noise. In the context of quantum indeterminacy furnished by our quantum mechanical dice-shaker device, from this limited threshold, we are bound to pose the dubious question of how 'quantum noise' could ever generate the coherent highly structured world we live in, at *any* scale, much less permit the persistence, endurance or perdurance of identities and mereologies over time. We might then seek contrived 'solutions' to overcome this 'problem', ranging from Manyworlds to the secretly orchestrating Quantum God. However, there is another phase of development, assuming we let run the experiment long enough to generate additional samples above the below-threshold level, above the noise distribution attendant that phase:

Transitional chaotic phase: Once an above-threshold number of quantum indeterminate random sum-outcomes are generated, we begin to distil the emergence of pattern, but not yet the saliency of the bell curve distribution. Instead, within this phase, **we obtain a distorted unstable and ever-changing distribution** (see **Fig. 3.14**). Each new accumulation of quantum indeterminate sum-outcomes can change the distribution curve in unstable and *chaotic* ways. Thus, **the phase constitutes the developmental transitional chaotic phase. One might also call it 'the chaotic phase', with its evolving *chaotic distribution*.** If we stop the generation of any further random sample sum-outcomes, we will obtain something better than the below-threshold noise distribution, but we will fall short of the fully salient bell curve structure-pattern. Even so, we begin to appreciate that something is different: that randomness within this transitional chaotic phase, despite ultimate quantum indeterminism, instability and volatility, no longer assumes the characteristics of mere noise. Instead, **as we let run this developmental phase to its conclusion, the 'noise' in the system will begin to reduce and dampened. It will do so with each successive marginal sample-accumulation: i.e., randomness will begin to *reduce* noise**, albeit chaotically at first, without any seeming stable structure.

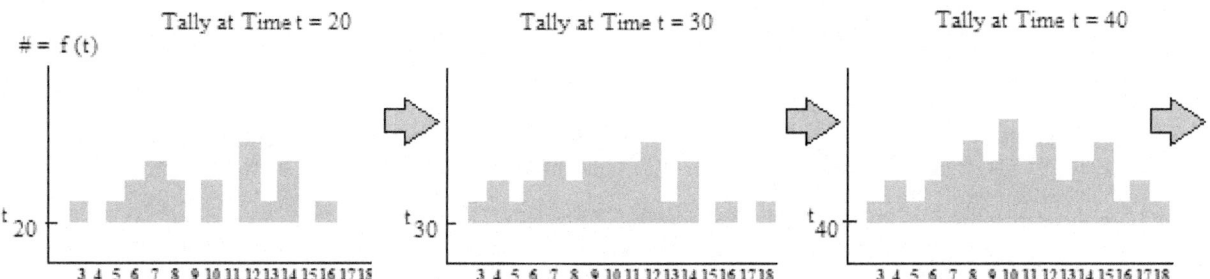

Fig. 3.14: Transitional chaotic phase and volatile structure from realtime random samples
Notes: In the transitional chaotic phase, the accumulation from quantum mechanised 3D6 random samples will render salient unstable volatile pattern, which could shift from one to a succeeding form chaotically: In the above, increments of just 10 quantum indeterminate samples render huge differences in global pattern-outcomes: i.e., see the depiction for the differences for 20 samples versus 30 samples, versus 40 samples, respectively. While what is observed is no longer a noise distribution, it is certainly a *chaotic distribution*. The transitional chaotic phase captures another prejudgement we have about random processes: randomness as perturber and disrupter: a thing we must seek to expunge from all systems, "…or there be chaos".

Finally, with sufficient random indeterminate samples accumulated, we enter a final developmental phase:

Noise-reduction pattern-saliency phase: When the next threshold is reached, the further accumulation of quantum indeterminate sum-outcomes will bring the system out of its transitional chaotic phase into its full *noise-reduction pattern-saliency phase*; **wherein the anticipated bell curve distribution finally becomes fully salient: We finally obtain a** *signal distribution*: (See **Fig. 3.15**).

If we continued to add more such quantum indeterminate samples, these will have no perturbing or disrupting effect on the emergent pattern in the way it did in the transitional chaotic phase. Subsequent samples could not be dismissed as constituting a noise distribution of the sort that attended the earlier below-threshold phase. Thus, each additional random indeterminate sample will merely contribute to the further and higher definition of the now fully developed bell curve distribution.

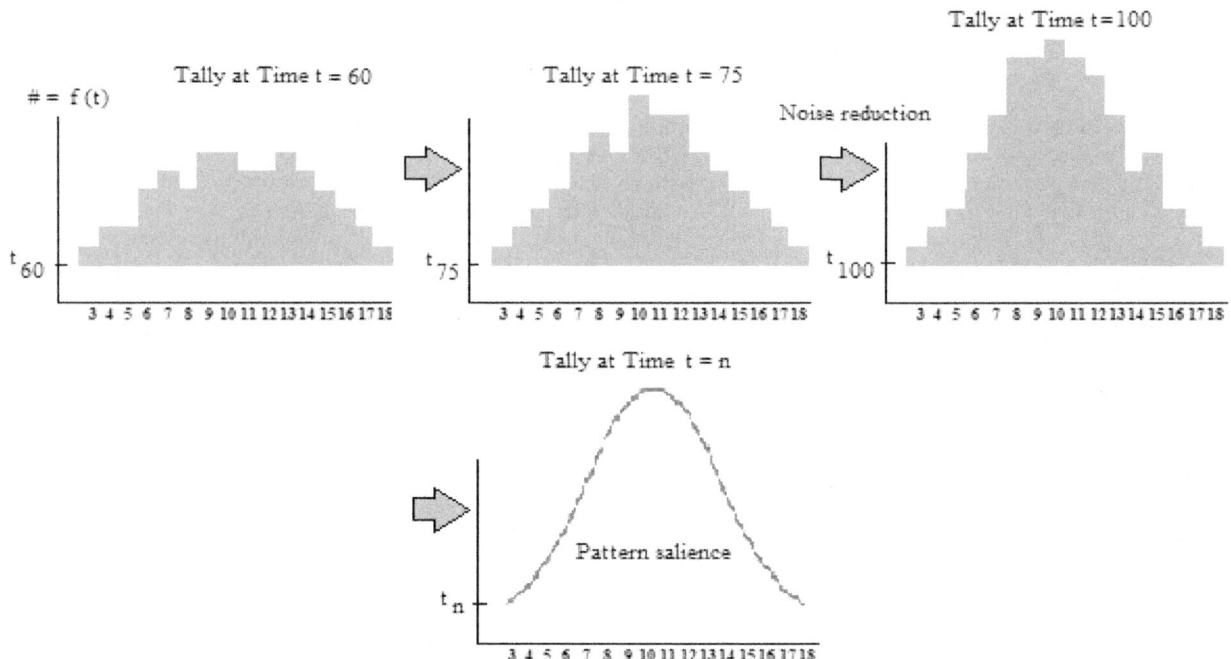

Fig. 3.15: Noise-reduction pattern-saliency: the accumulation of samples signal pattern-emergence
Notes: Finally, with sufficient samples accumulated, we transit from chaotic (t = 60) to noise-reduction and pattern-saliency: The bell curve gains saliency by t = 75 and has formed up substantially by t =100. Subsequent quantum indeterminately generated random samples cannot perturb or disrupt the now-salient bell curve distribution but can only increase its definition and resolution into a *signal distribution*: In this way, randomness is rendered evident as a signal-magnifier and can only bring forth pattern-salience at ever sharper and higher definitions (t = **n**, where **n** may equal several hundred samples). At which point, one will search in vain for the notion that 'randomness means chaos', or the claim for the supposed 'impossibility of pattern and order from randomness'.

Thus, in this developmental phase, quantum indeterminate randomness assumes, not merely the role of a noise-reducer, but acts as a *signal-magnifier*: The bell curve, hitherto abstract, now constitutes a full signal: a maximal 'materialised' state magnified by the over-sample of quantum indeterminate terms.

At the noise-reduction pattern-saliency phase, the noise-reducing role of randomness is rendered salient, and we can generalise it to nature as a whole. With it, questions such as, "… How can quantum noise generate a coherent highly structured world, or allow for the continuity of identity and pattern over time?"… begin to appear naive: Beyond a certain threshold, and certainly beyond the transitional chaotic phase of development, **randomness, quantum indeterminacy and chance transform into noise-reducing signal-magnifiers**. Within this phase, randomness can no longer be misconstrued as any sort of scrambler of information, even in an otherwise chaotic system. In the noise-reduction pattern-saliency phase, **randomness fully transforms into a** *de-scrambler*: **a friend to the emergence and saliency of information, order, and pattern**.

Insofar as any system might remain within its below-threshold noise-phase of development, randomness and indeterminism will constitute noise: it will furnish a mere *noise distribution*. Judgements about natural systems inspired by below-threshold developmental phase, or when relying on misleading low-complexity limiting case systems such as EPR and Schrödinger's Cat (see **3.25**), will tend to culminate into the dubious judgement that all randomness must constitute noise or the absence of information. How, then does order and pattern in nature constitute and re-constitute itself out of such noise?

320

QUANTUM MECHANICS AND MIND

Once a system graduates into the transitional chaotic phase of development, randomness is no longer noise. At worst, randomness transforms into a perturbing factor: fashioning for a *chaotic distribution*. Though no longer characterised as informationally vacuous, the system in development exhibits the time-evolution of unstable and volatile structure. As is amply demonstrated in chaology, such systems are sensitive and unstable vis-à-vis further random inputs or with respect to the slightest change in initial or attendant and evolving conditions: i.e., randomness becomes a system-perturber and disturber. And if our view of randomness is inspired by such systems, we will falsely generalise randomness to be inimical to stable order and pattern: a source of haphazard volatility that we and nature must eliminate or control.

However, once a system transits into the noise-reduction pattern-saliency phase of development, we find that randomness assumes its obvious undeniable role as noise-reducer and signal-amplifier. It engenders a *signal distribution*: It transforms into a friend to pattern saliency and information: no longer an adversary or a problem to dismiss or obviate (David Bohm's hidden variables) or circumvent (Hugh Everett's Manyworlds).

*

One additional insight into how the transition from noise to pattern-saliency unfolds is to think in terms of the **marginal bandwidth of randomness.** Let us examine what happens to our system at every subsequent 10-sample addition of quantum indeterminate terms: At every ten samples we accumulate, we check the evolution of the global pattern to see if we obtain anything worthy of comment. In truth, in the development of the bell curve distribution from quantum mechanised 3D6, we should mark every 15th sample (i.e. the domain and range attendant the die-sum possibilities). But, for brevity, we assume observation at every 10 indeterminate samples.

We ignore the below-threshold phase, as this makes for mere noise distribution. Yet, **as the system graduates into the transitional chaotic phase, every addition of ten samples will furnish wildly different pattern-outcomes.** Thus, starting at the 20th sample threshold, the next ten samples will graduate the system to the 30th sample-threshold: i.e., one-third bandwidth of the new total per dividing the total samples by the range = 10. The next ten will total 40 samples and the bandwidth of randomness will reduce from the previous one-third to one-quarter. The next ten samples will constitute a marginal bandwidth of randomness of one-fifth... and so on. Note that, **as the bandwidth narrows and reduces per every marginal addition of ten samples, the chaotic volatility of pattern-development will be proportionately reduced, given that it can only transpire within an ever-reducing and narrowing marginal bandwidth of randomness.**

In simpler terms, an addition of ten samples to a previous twenty will furnish significant pattern-difference. But the addition of ten to a previous total of 60 will make less of a pattern difference.

As we graduate into the noise-reduction pattern-saliency phase, the marginal bandwidth of randomness will narrow to a very small fraction of the total: one-tenth by the 100th sample, and narrower still as we generate more samples. The addition of 10 samples thereafter will involve terms no less objectively quantum indeterminate. But these cannot perturb the now-salient pattern in any chaotic manner. **Indeed, if we continued to increase the samples indefinitely, the marginal bandwidth of randomness would tend to the infinitesimal, and, consequently, a higher definition of the distribution curve and near-total noise reduction would obtain,** even if with diminishing marginal returns per every ten samples.

The marginal bandwidth of randomness, as an approach, and as clarifier of why randomness cannot help but culminate into higher definition and pattern-saliency within the rubric of implicit minimal abstract states, must apply to all pattern-development in nature: It applies to our bell curve distribution using quantum mechanised dice: It applies to the quantum mechanised Sierpinski triangle: It applies to the formation of the interference pattern in the two-slit experiment, wherein quantum indeterminacy is native and does not require any 'device' to furnish it: It will apply in photography and in electron micrography, to which quantum indeterminism is also native. It will apply to the drifting and perduring snowflake; to the formation and continuity of galaxies; to the formation and development of living forms; and to the global operation and pattern-saliency evinced in brains.

*

Throughout the described three phases of development, the emergence of the bell curve distribution is fully preordained: a matter of systemic fate, *not* happenstance. The fate of the system is not per function of the quantum indeterminate randomness itself, if only because all truly random processes must *always* unfold within abstract ordinances constituted as minimal abstract states, such as quantum mechanical wavefunctions. Notice that the abstraction operating in our statistic project is comprised of the three six-sided die (quantum mechanised via the device depicted in **Fig. 3.11**) co-joined to the domain and range of possible sum-outcomes that can be generated (i.e., 3 to 18) as permitted by the possible combinations of fixed die-values, and the inherent marginal bandwidth of randomness. Clearly, while the specific outcomes are quantum indeterminate and matters of happenstance, the said minimal abstract boundary conditions are *not* matters of happenstance: These are the fixed non-fungible read-only *ordinances* of the system, prior to and attendant its subsequent development.

The quantum indeterminate random constituents will *always* be funnelled to a pre-ordained and fated culmination per the pertinent minimal abstract bounding states, no matter how random, unpredictable, or probabilistic the samples obtained. The longer the process of development (or, alternatively, the greater the number of consecutive and parallel quantum indeterminately generated samples) the more rapidly the system will fulfil its final ordained and fated end-state: that being the specific bell-curve distribution pattern in the quantum mechanised simple statistics project. In other words, **the development of the system is *necessary* and *deterministic* per the background bounded probability minimal abstract state, despite the quantum indeterminate randomness entailed in the three**

321

phase developmental process. The ordinances of the system will culminate objective quantum indeterminate randomness to an ultimate noise-reducer and signal-magnifier role.

But where do nature's ordinances, the minimal abstract bounding states, come from? Any serious attempted answer must incorporate the repurposed Euthyphro's dilemma to the examination of physical order and information: see **3.14**. We ned not further berate the implied point.

*

It now appears that determinism is the winner. However, before we reach such a premature conclusion, **the quantum indeterminate random samples that finally get to reduce noise and bring about the pattern-saliency of the bell-curve distribution, are truly objectively and ontically random**: This assertion will be proven in **3.32**. The quantum indeterminate terms remain objectively random *throughout* the three-phase development process **and do not cease to be random in *any* phase**, even at the noise-reduction and pattern saliency phase, or even when functioning as noise-reducers and signal-magnifiers. The ordinances constituting **the implicit bounded probability state that fates the development to a totally deterministic necessitarian culmination cannot eliminate the fact and operation of objective randomness and indeterminacy**.

Thus, determinism is not the winner, given the ineliminability of randomness. But then neither is indeterminism the winner, given the ineliminability of development-ordaining terms and minimal abstract states: states that derive from law and necessity, not from the objective randomness or happenstance. Again, any serious attempted to address the origin of the ordinances must incorporate the Euthyphro's dilemma repurposed to the examination of physical order and information: see **3.14**. Again, we shall not berate the implications.

*

Randomness abides, despite ordinances and pertinent minimal abstract states. But, beyond the transitional chaotic phase of development, randomness must play the role of noise-reducer and signal-magnifier. **It follows that the historic debate between determinism and indeterminism was totally superfluous and false**. We must transcend the strictures of this false debate by appreciating the reality of *both* necessity *and* objective randomness, afforded from the vantage-point of our three-phase developmental approach, which incorporates the best of both views without disqualifying either, but without succumbing to one or the other... incorporating both into mutuality and harmony.

In summary…

- **Randomness and quantum indeterminacy will always unfold within a necessity-born structure of ordinances composed of minimal abstract bounded probability states.**

- **Quantum indeterminate resolutions are *always* collectively *fated* to produce perfectly pre-determined global outcomes.**

- **This, and the harmonious co-joining of indeterminacy *and* determinacy, unfolds through three phases of development**:
  - **The *below-threshold noise phase***; wherein randomness constitutes what most think of as 'noise'... succeeded by…
  - **The *transitional chaotic phase***; wherein randomness acts as perturbation factor vis-à-vis an unstable but emerging mereology... finally, culminating into…
  - **The *noise-reduction pattern saliency phase***: wherein structure and information is finally evinced; wherein randomness clearly reveals itself to be critical in noise-reduction, signal-magnifying and pattern-sharpening: *not* an enemy, but an aid and friend to the emergence and consolidation of coherent structure, pattern, and information in nature.

3.27: The Three-Phase Developmental Process & the Quantum Mechanised Sierpinski Triangle

We will now apply our three-phase developmental approach to the Sierpinski triangle. Previously, the Sierpinski triangle served to illustrate the operation of indeterminism that, within the framework of specifiable ordinances and abstract states, explicitly produced a totally fated global pattern-outcome. We need not repeat the full construction rules for the Sierpinski triangle: Here, we apply a simplified version of the quantum mechanised dice-shaker depicted in **Fig. 3.11**, but involving only a single quantum mechanised D6, with a single concave detector segregated into three arcs to output the sequences of seed-orbits in the sought quantum indeterminate way.

The manner in which our three-phase developmental process applies to the Sierpinski triangle is presented in the sequence of pertinent illustrations in **Fig. 3.16**: Therein, in phase (A), a certain threshold of accumulated quantum mechanically generated seed orbits obtain nothing more than a truly random collection: a *noise distribution* of seed-orbits: a haphazard randomness of the sort that meets our typical prejudgements about randomness: This is the below-threshold noise phase of development.

Once we enter the transitional chaotic phase of development in (B), we grasp that our otherwise random and haphazardly distributed seed-orbits begin to coalesce into some sort of pattern: ever shifting, volatile and chaotic though it may be: a *chaotic distribution*, with some areas developing more rapidly than others: i.e., equivalent to 'developmental sites' found in photography; perhaps similar to 'hot spots' found in radioactive material.

This transitional chaotic phase eventually gives way to the emergence of stable or salient manifest pattern in (C). Therein, successive quantum indeterminate resolutions of the seed-orbit will act as noise reducers: bringing about the *signal distribution* of pattern saliency, no matter how truly random or quantum indeterminate the seed-orbits.

322

In the noise-reduction pattern-saliency phase, **the marginal bandwidth of randomness reduces toward insignificance, and we can no longer deny that randomness is truly a noise reducer: a friend to information, pattern and order: *not* an enemy, as is erroneously judged per the usual aspertions we entertain about randomness**. In the context of the Sierpinski triangle, we recognise that subsequent quantum indeterminately generated seed-orbits will only serve to sharpen the pattern-structure, increasing its definition, as implied by the embedding minimal abstract bounded probability state.

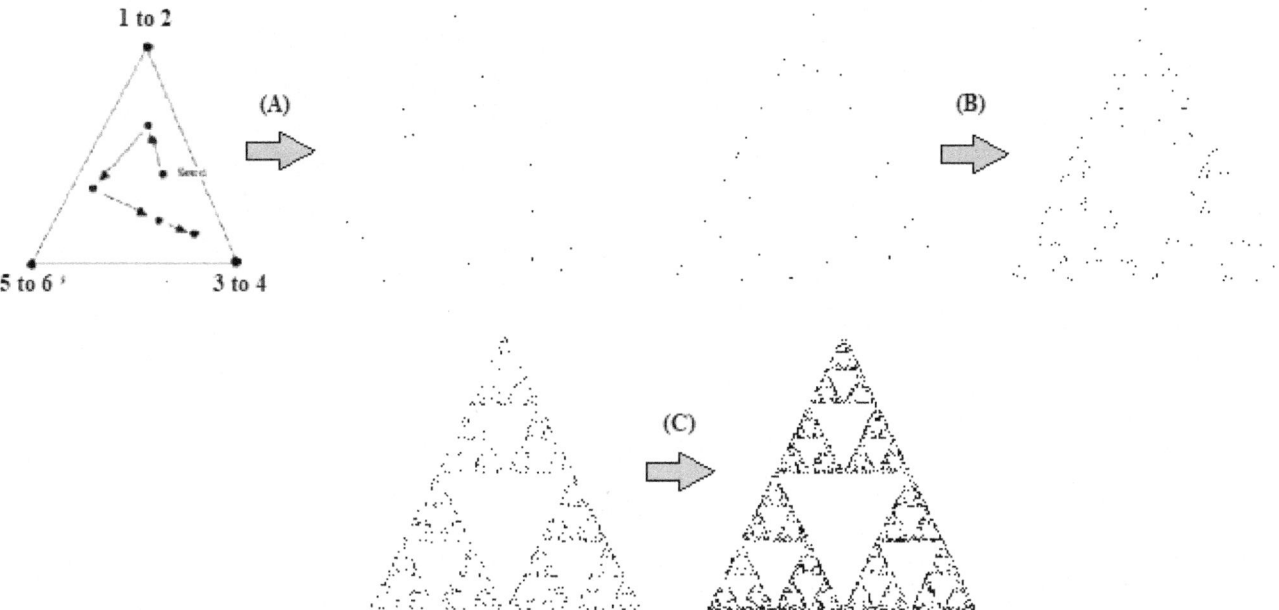

Fig. 3.16: The three-phase developmental of the Sierpinski triangle
Notes: The three-phase development of the Sierpinski triangle: **(A)** depicts the below threshold phase: no pattern-structure is evident: the development conforms to prejudices about randomness as noise distribution. But **(B)** depicts the transitional chaotic phase, wherein structure begins to form out of random elements, but it is volatile and unstable. Eventually, this culminates into the barely recognisable form of the Sierpinski triangle. By **(C)** we arrive at the noise-reduction pattern-saliency phase, wherein subsequent random seed-orbits, even if quantum mechanically generated, culminate into the salient pattern-structure typical to our Sierpinski triangle. Subsequent random indeterminate terms cannot obviate or over-ride this fate.

3.28: The Three-Phase Developmental Process Applied to d$z2$ Orbitals

We now apply our three-phase developmental approach to the electronic configuration of the atom, exemplified by the **d$z2$** orbital. Generic physics typically concerns itself with the calculation of a single instant of electron orbital resolution about the atom. In our case, **we seek the cumulative global pattern arising from the accumulation of successive electron position-resolutions about the atomic nucleus, and what the accumulation must reveal through *en masse* distribution**. Here, we do not need a quantum mechanical dice-shaker device: The quantum indeterminism is native to electron position-resolution about the atom, expected to transpire typically within the pertinent electronic shells or clouds: These are regions formed out of necessity and law, where the wavefunctions for the electrons at certain frequencies can form viable standing waves, though only at certain radii.

Assume an experiment in which substantial numbers of successive electron position-resolutions are made; with each OR-form resolution marked on a map of the **d$z2$** system. We can use these to case for our three-phase developmental approach. As with the preceding quantum mechanised bell curve distribution and the quantum mechanised Sierpinski triangle, we first obtain the expected below-threshold noise phase, wherein successive electron resolutions will not collectively yield recognisable pattern: a *noise distribution*: effectively, no information. Thus, the implicit background ontology of the minimal abstract state, or the bounded probability structure and wavefunction in potentiality of the **d$z2$** orbitals, will not yet be salient: (See left-most depiction in **Fig 3.17**). However, as quantum indeterminate electron position-resolutions accumulate, our mapping will time-evolve into the transitional chaotic phase of development: As with the Sierpinski triangle before it, the transitional phase will be characterised as an unstable and shifting structure: a *chaotic distribution*. Eventually, the accumulation of subsequent quantum indeterminate electron-position resolutions and attendant narrowing of the marginal bandwidth of randomness, will institute noise-reduction: (See next four depictions from left to right in **Fig 3.17**). Hence, the system will enter the noise-reduction pattern saliency phase. Therein, through the *en masse* accumulation of successive electron resolutions, the framing minimal abstract state of ordinances constituting the **d$z2$** orbital will become fully salient: the *signal-distribution* of electron positions will be obtained. Order out of seeming quantum indeterminate 'chaos' will be at hand.

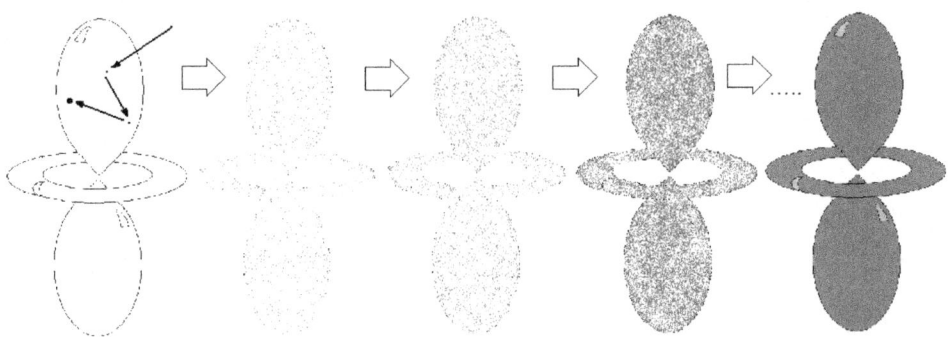

Fig. 3.17: The d$_z2$ shell system: application of the three-phase developmental process
Notes: The didactic series of diagrams depict the three-phase developmental evolution of the accumulation of the successive electron-position resolutions within the **d$_z2$** system. Assuming an analogical and didactic similitude between seed-orbit generation in a Sierpinski triangle and the turnover in electron position-resolutions about the atom, we follow through the three-phase process into the culminating noise-reduction pattern-saliency phase, wherein the full minimal abstract structure of the **d$_z2$** system is rendered maximal and salient.

3.29: Three-Phase Development in Photography & in Interference Patterns in Two-Slit Experiments

What held true in the formation of the quantum mechanised bell curve distribution, in the quantum mechanised Sierpinski triangle, and in the succession of electron-position resolutions in **d$_z2$** orbitals, must also hold true in the development of a photographic image: Therein, the three-phase developmental approach to quantum indeterminism must also abide. But there are three key differences: First, **photography is not restricted to one-at-a-time sequences and successions of quantum indeterminate photon depositions**, but could involve *en masse* consecutive depositions. The second difference is that **what holds true for *en masse* photon depositions can also be extended to *en masse* depositions of de Broglie matter,** such as in the formation of an image out of electrons (matter) in electron micrography: essentially the same as photography, but with electrons. The third difference is that **photography often involves a complex macro-scale wavefunction as its background minimal abstract 'instantaneous whole' state**: one that funnels *en masse* quantum indeterminate photons into a pre-fated global scale-outcome. In this way, the process of photography is kindred to the formation of large-scale patterns and processes across nature, further aided by the fact that what holds true for radiation (photons) also applies to de Broglie matter (electrons, etc, including corpuses of material patterns).

   Thus, photography scaffolds for the generalisation and universalisation to the whole of nature of the *all-scale* three-phase developmental process and its attendant solution to quantum indeterminism (and of the stochastic coherence paradox). As such, the three-phase developmental process abides in the formation and drift of a snowflake; in the formation and operation of living cells and living forms; in the development and activity of the brain as a whole; in the formation of stars, worlds and galaxies… in the formation of the whole universe conceived in the fashion of a cosmos-spanning complex universal wavefunction (i.e., spacetime in potentiality *en toto*).

   The development of a photographic image upon an appropriate chemical or digital medium will first evolve through a below threshold noise phase (See the top illustration in **Fig 3.18**). The few photon depositions will appear to be completely random, without form, without coherent structure, pattern or information: a *noise distribution*, consistent with our usual prejudices about randomness and 'accident'. But as indeterminate photon depositions accumulate on the film-surface, 'development sites' will form up. The final picture or image will not be obvious from such sites, which will emerge chaotically, constituting a *chaotic distribution*. Yet, structure, albeit unstable, will begin to emerge despite indeterminate photon depositions. This is the transitional chaotic phase in which additional indeterminate photon-depositions might generate perturbing effects to the global distribution and attendant pattern. Even so, the *en masse* quantum indeterminate photon-depositions will be issuing some sort of structured outcome.

   As more photon depositions accumulate and the marginal bandwidth of randomness narrows, the background minimal abstract bounded probability state will fate these into a pre-ordained globally deterministic outcome; into a *signal distribution:* see **Fig. 3.18**: right and bottom-left depictions.

   That the photographic film is fated to a final non-random non-arbitrary pre-determined outcome, both despite *and because of* quantum indeterminate photon depositions, becomes obvious when the process finally transits the transitional chaotic phase into the noise-reduction pattern-saliency phase, wherein subsequent quantum indeterminate depositions of photons, subject to a narrowing marginal bandwidth of randomness, act as noise reducers and signal-magnifiers; and wherein randomness and indeterminacy

BELOW THRESHOLD TO TRANSITIONAL PHASE

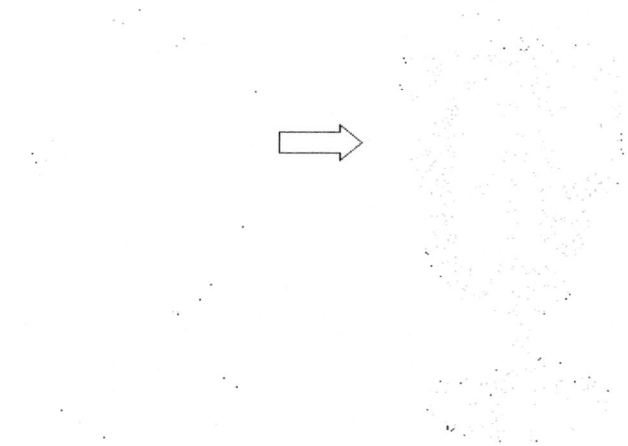

TRANSITIONAL PHASE TO PATTERN SALIENCY NOISE REDUCTION

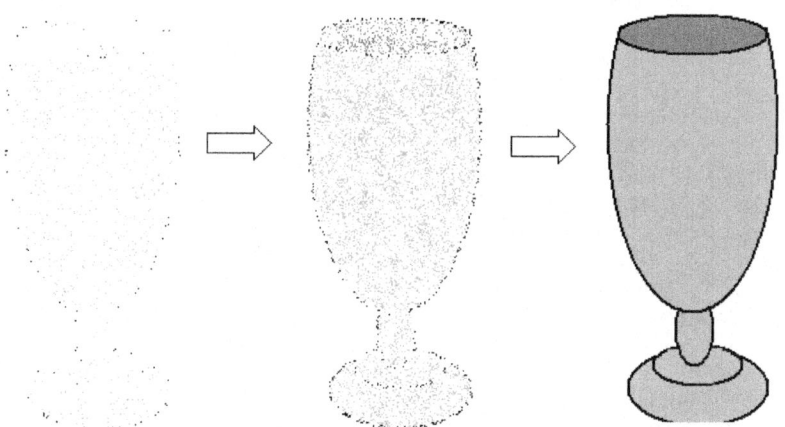

Fig. 3.18: Photography and the three-phase development process
Notes: As with all other systems, in photography the three-phase developmental process unites quantum indeterminism with determinism, as *en masse* quantum indeterminate photons are funnelled into a pre-determined global outcome by the background minimal abstract state; in this case, a complex wavefunction pertaining to the photographic image of a vase. The same process holds in the formation of electron micrographic images and in the formation of interference patterns in the two-slit experiment, and even in the distribution obtained when only one slit is open.

are finally appreciated as friends of information, as these help constitute the *signal distribution* constitutive of the photographic image into full high definition and high resolution saliency. (See **Fig. 3.18** bottom-set depictions: centre and right).

*

What holds true in photography holds in essentially the same way to quantum indeterminate electron depositions (to de Broglie matter) in the two-slit experiment. Therein, the three-phase developmental process finally culminates into the recognisable interference pattern when both slits are open. But it is also possible to take electron micrographs, or photographs with electrons. In other words, our developmental approach applies to complexes constituted by matter.

In its below threshold noise phase, we will only find random electron-depositions on the double slit experiment detector and in a developing electron micrograph, with no obvious saliency of either the interference pattern or of the micrographic image: a *noise distribution* for electron depositions. This will then transit into the transitional chaotic phase, wherein the formation of developmental sites will begin to form out of *en masse* electron depositions; accompanied by the emergence of unstable pattern: a *chaotic distribution* of electron depositions.

Finally, once enough electron depositions are resolved on the detector or electron micrograph, the system will graduate into the nose-reduction pattern-saliency phase of development, wherein the expected interference pattern, or electron micrograph image, will be rendered fully salient, despite fundamental quantum indeterminacy. Hence, the *signal distribution* of electron depositions.

In both the two-slit experiment with electrons *and* in the formation of images in electron micrographs, quantum indeterminate electrons will deposit on an appropriate film-surface *en masse* and develop according to the three-phase development process as found in photography. Thus, the minimal structure state pertinent to the **both slits open** barrier configuration, or pertinent to the source-origin of the electron micrograph image, will draw the attendant large-scale pattern into a pre-ordained, fated and wholly determined global outcome, despite the operation of inherent inelinimable quantum indeterminacy.

We are now able to generalise and universalise our oft repeated finding to the rest of nature and universe in the framework of the growing block intermediate spacetime model developed in Book-II.

3.30: Generalisation & Universalisation of the Three-Phase Development Process
To the Intermediate Model Tripartite Spacetime Structure

The generalisation and universalisation of our three-phase developmental process to nature takes the **below-threshold noise phase, the transitional chaotic phase, and the noise-reduction pattern-saliency phase, and the inherent perpetual narrowing of the marginal bandwidth of randomness implicit to all three phases, and frames these to a time-evolution format that incorporates the tripartite structure of the intermediate model spacetime**. Our specific candidate consists in the time-evolution of a drifting snowflake, although it and attendant conclusions apply to all patterns undergoing growing block time-evolution.

- **The process begins with the purely immaterial minimal abstract state** on the basis of which our snowflake will develop into maximal materialisation and saliency on or along a succession of event horizons. **This minimal abstract state resonates across time**; from one moment to the next, from one event horizon to the next, in the fashion of a complex wavefunction… typically as the **"snowflake in memory" (i.e., a wavefunction in memory) retained within the domain of realised spacetime** (the past) and projected onto the event horizon thus. Alternatively, the minimal abstract state for the snowflake will be constituted as a **"snowflake in potentiality" (i.e., a wavefunction in future-potentiality) embedded within the domain of spacetime in potentiality** (the future). It will emanate to and impose itself upon successive event horizons, albeit from the future. Upon its emanation to a succeeding moment and attendant event horizon, the *en masse* quantum indeterminate resolution of constituent terms via the process of AND-to-OR collapse, wavefunction collapse… or time… is destined to bring the potentiality of the snowflake into maximal materialised saliency.

- In **the ascent,** at first, our snowflake will be **at its below threshold noise phase** of development: wherein the quantum indeterminate resolutions of its constituent terms per the minimal abstract state, whether projected from the past or emanated from the future upon the event horizon, will be truly random; without form or information: a *noise distribution.*

- Eventually, we will arrive at the **transitional chaotic phase** of development: The snowflake pattern will begin to coalesce and 'materialise'; a *chaotic distribution*: The rules of chaology abide, as it does in all systems that are sensitive to initial or cumulating conditions. Yet, with subsequent accumulation of quantum indeterminate terms and the narrowing of the marginal bandwidth of randomness, the perturbation and chaos will dampen. The pertinent ordinances of the minimal abstract wavefunctions in memory (or future potentiality) attendant our snowflake will funnel the *en masse* quantum indeterminate terms into the 'materialised' snowflake on a succeeding event horizon.

- Finally, our snowflake will attain the **noise-reduction pattern-saliency phase** of development. Therein, quantum indeterminate resolutions will no longer constitute 'noise', nor terms of perturbation and volatility. Instead, quantum indeterminate events will begin to affect noise-reduction, sharpness, and high resolution of pattern. At this apex of development, quantum indeterminacy will begin to act as a noise-reducing and information-resolving signal-magnifying 'force', even while it retains its fundamental random and indeterministic character, forming the *signal distribution* through which our snowflake will fully manifest and 'materialise' on the succeeding event horizon.

- In the **descent process**, beyond the said developmental apex attendant the signal distribution, and inferred from the frequency based nature or the group-frequency nature of things, we conjecture the reversal of the process described above: i.e., the transition of our snowflake from its highest amplitude fully materialised maximal structure state, back to its minimal structure abstract wavefunction. The reversal *does not* involve any time-reversal of AND-to-OR directionality. Yet, from maximal pattern saliency, we will go back to the transitional phase, and then back to the below threshold phase. Within this reversal of the order of phases, hitherto resolved constituent terms will dissolve back to superpositions of possibility: That is, the snowflake will open up to new future possibilities: it will open up to spacetime in potentiality. Thus, we arrive back at the minimal abstract structure state attendant our snowflake, but with any modifications to it now incorporated into realised spacetime as part of growing abstract memory; constituted as a new wavefunction in memory. This will decohere future-potential possibilities for the snowflake into futures consistent with the accumulated evolution of the snowflake. Any change implied will manifest in the future.

- Thereafter, time-evolution of our snowflake will repeat the three-phase process of development: from point-1 through to point-4; culminating back into a new succeeding maximal materialisation and saliency, with change now rendered salient. The whole process will repeat in this described way again and again, constituting and supporting perdurance… until contingent conditions obtain to permanently block the subsequent formation or reformation of our snowflake (i.e., nothing lasts forever).

The frequency at which this developmental ascent and descent occurs is likely bound to the group or average frequency of the constituents making up the snowflake's complex wavefunction in memory cum future-potentiality. Hence, the process of coming into 'material' being, and then falling out of it, and the frequency-cycles that attend it, will be reflected in the structure of the quantum wave wavefunction, which ultimately resonates, *not* across 'space', but across time… forcing the domain of spacetime in potentiality (the future) to conformity through grand decoherence. **The successive formations and re-formations of our snowflake in the fashion just described, will constitute oscillations between moments of maximum saliency and maximum abstraction… as our snowflake journeys across time, 'materialising' and dematerialising back into abstract memory-cum-potentiality** via successive event horizons.

This oscillation between ascent into 'materialisation' and descent into abstraction was implicit to the broken-discontinuous relation of information to putative radiation, matter, and time, espoused in our early foray into information-matter dualism in Preliminary **0.01** to **0.04** and throughout Book-I, Part-II: The said oscillation *is* the broken-discontinuous relation, merely mirrored in the tripartite structure of intermediate model spacetime.

*

What holds true in the time-evolution of our snowflake also attends the time-evolution and the perpetual 'coming into being' of, say, a cloverleaf; of the 'motion' of the billiard ball; or of the development and time-evolution of any pattern… up to and including the time-evolution of a solar system of worlds; the constellations of stars; of whole galaxies… and, indeed, of the brain itself.

Note again that **the false dichotomy between determinacy and indeterminacy, its specific rendition in the context of quantum indeterminism, is almost solved, although we have yet to address whether quantum indeterminacy is truly objectively and ontically real, or else secretly orchestrated by hidden gears, and only seemingly indeterminate.** Otherwise, we have solved the problem of quantum indeterminacy by grasping quantum indeterminate process as always unfolding within abstract minimal structure states of the future (spacetime in potentiality, or wavefunctions in future-potentiality) or the past (realised spacetime, or wavefunctions in memory), but often as admixtures involving both, constituting a Lagrange. This abstraction or 'instantaneous whole', born out of necessity in its read-only aspects, and not itself a matter of indeterminate scrambling, funnels and fates singular or *en masse* quantum indeterminate terms into entirely fated and pre-determined global outcomes across a below threshold noise phase, into a volatile transitional chaotic phase, and finally into a noise-reduction pattern-saliency phase. In the last two phases, per function of the narrowing of the marginal bandwidth of randomness, quantum indeterminacy begins to act as a noise-reducing signal-magnifier and can no longer be characterised as inimical to pattern-structure, or deemed destructive of persistence and perdurance of identities across time. Hence, the false dichotomy between determinacy and indeterminacy is broken and resolved into an integrated mutuality and harmony, *without* contradiction.

3.31: Presage to the Ontic Reality & Objectivity of Quantum Indeterminism: Primacy of Form over Natural Odds & the Hidden Variables Problem

En masse quantum indeterminate outcomes collectively manifest the fated global implications from any natural odds cast into abstract form. This is accomplished via the three-phase developmental process espoused in **3.26**. Yet, the structures and patterns of the world arise, not merely from the implications of the natural odds, but from the same rendered into specific arrangements, shapes, or forms in putative spacetime. Some of these are read-only: such as the d_{z2} electron orbitals; such as the enforcement of the conservation of spin in EPR. Others are read-write: such as the abstraction pertaining to a photographic image and other complex wavefunctions in memory or future-potentiality. Thus, when we make the erroneous assertion that nature is wholly a product of 'chance', or of quantum indeterminism and the natural odds, this fails to clarify that the same odds are typically rendered into specific abstract forms and mereologies… into instantaneous wholes (Book-I **1.03**, and **1.26** to **1.27**) constituted as memory or potentiality… admixture of necessity and law when read-only, or compounded abstractions when read-write. Anticipated de-spatialisation in Book-IV will not alter this truism but will force a pure temporised distribution and 'shaping' of the odds, *without* space.

To reiterate, with or without de-spatialisation, the natural odds can only ever operate in the context of necessity and form. **Indeed, it is not possible to arrive at forms from 'pure' natural odds in the absence of a spatial-geometric or ultimately temporal frameworks: Simply, there *are* no 'pure' natural odds in the absence of a context that can casts these as distributions in time or 'space'.** Restated in Aristotelian terms, we cannot have substance without form: No form, and there is nothing for the odds to fulfil or bring into materialised saliency. **Therefore, natural odds are *always* subordinated to the ontological primary forms; to the reality of *information*** grasped as minimal abstract bounded probability states… or as wavefunctions in potentiality or memory, embedded within the futurity of spacetime in potentiality or retained within mnemonic realised spacetime.

For example: The standing waves that constitute d_{z2} orbital can only form up viably and in a sustainable way at certain radii (the pure temporal equivalent of radii will be clarified in Book-IV). If the radii are slightly less or more than that required, the standing waves for potential electron positions cannot form. If we tried to cast the natural odds inherent to quantum mechanical waves for electron position about the atom by forcing the odds into non-viable radii, these and attendant natural odds will not be viable.

The point is also this: Not all forms to which natural odds are cast can lead to coherent stable outcomes and complexes. Only some spatio-temporal forms and arrangements of the natural odds can fate otherwise quantum indeterminate outcomes into coherent global developmental end-states through the three-phase developmental process. In short, the undeniable role of natural odds cannot imply the dominance of odds over form, much less lead to the irrelevancy of form. On the contrary: form dominates and structures the odds. **Hence, the natural odds matter *only* when cast into very specific forms**, such as those of the **d$_z$2** orbital: one of five patterns capable of sustaining structurally coherent phenomena over time in the context of the electronic configuration of a simple atom.

Of course, the forms that structure and dominate over the odds range from atomic electronic clouds to galaxies, and to forms that pertain to living systems. **Hence, form is ontologically primary vis-à-vis ontologically subsidiary natural odds. Hence, to dismiss the world as merely the outcome of averages per the odds, but ignore the fact that it is *form that casts and shapes the odds, without itself being cast by the odds*, is to ignore the existence of undeniable non-arbitrary abstract spatial and finally temporal ordinances (ultimately emanating from a meta-semantic hypostasis) writ within spacetime in potentiality, materialised on the event horizon, and relegated to and retained as memory within mnemonic realised spacetime**.

Yet again, any serious attempted to address the origin and ontology of the forms must incorporate Euthyphro's dilemma repurposed to the examination of physical order and information: see **3.14**. We need not reiterate or berate the implications here.

Fundamentally comprised of read-only categories of information, as was espoused in Book-I, such forms constitute the timeless Platonic states (for example, the physical laws) emanating from a meta-semantic hypostasis that transcend *and* abide across time and change, even while these might fall in and out of time, or transform in due course, as Feynman speculated in a dialogue with Sir Fred Hoyle:

..it's interesting that in many other sciences there's a historical question, like in geology - the question how did the Earth evolve into the present condition? In biology - how did the various species evolve to get to be the way they are? But the one field that hasn't admitted any evolutionary question - is physics.

Given that the natural odds cannot imply the rule of chaos or 'happenstance' in nature despite inherent quantum indeterminism and randomness, much less imply the absence of form and necessity per the ontological primacy of form vis-à-vis the natural odds, and given that the natural odds and quantum indeterminate outcomes cannot produce coherent and sustained complexities when cast just in any way, and given that indeterminate processes are strictly form-dependent, it follows that the primacy of form constitutes an important truth attendant solutions to the stochastic coherence paradox and the 'problem' of quantum indeterminism: In summary, the insight asserts the following:

- **The ontological primacy of form (abstract information) over subordinated natural odds and quantum indeterminate processes.**

- **The subordinating form casts the natural odds in *its* terms (not the other way round) and subsequently funnels quantum indeterminate terms into manifest or materialised pattern in the likeness of ontically primary abstract form.**

*

While our three-phase developmental process, combined with the primacy of form vis-à-vis the natural odds, constitute critical components of the overall solution to the stochastic-coherence paradox, the complete solution will require two further developments: These are…

- **The solution to the cause of AND-to-OR wavefunction collapse, synonymous with time itself**: We need a solution to the 'drive of time'. Our three-phase developmental approach entails and presupposes AND-to-OR collapse that generates quantum indeterminate terms, but does not furnish an account of ontologically primary AND-to-OR collapse and time pertinent to the generation of quantum indeterminate terms. Whatever the explanation for wavefunction collapse or time, it will prove decisive to the question of whether quantum indeterminacy is ontically real.

- **The question of whether randomness itself (hence quantum indeterminism itself) is ontologically and objectively real, and *not* an appearance underscored by a wholly deterministic orchestrator-process or 'hidden gears'**. That is, we need to demonstrate that quantum indeterminism is not 'illusory', but objectively and ontically real.

Hence, if we can prove the objective reality of quantum indeterminacy, we can then establish the final veracity of the three-phase developmental process that presupposes objective indeterminacy.

On the other hand, if wavefunction collapse is mediated by hidden variables or 'hidden gears', then we might conclude that quantum indeterminacy is merely facile and illusory, and that the hidden nature of it is that of a fully scripted deterministic orchestration process. In such a case, we would be forced to completely reframe our three-phase developmental approach.

*

By Book-V we shall discover that wavefunction collapse, hence the perpetual process of 'coming into being', or, simply, *time*, is *not* generated by any hidden variables or gears, although this much was secured in Book-II 2.46 and 2.47. Thus, the process of deep causality, hence the process of causality in general at the heart of wavefunction collapse and time, must itself be essentially structure-less; with attendant quantum indeterminacy also rendered non-illusory, given that it cannot then be characterised as orchestrated by any definable process or gears, if it is ultimately generated by a time-process *without* 'gears'.

However, there exists a second and more direct method of proving the ontological reality and objectivity of quantum indeterminacy. Independent of the primary problem of causality behind wavefunction collapse and the question of the 'drive of time', we can address the objectivity of randomness decisively by generating *putative* deterministic orchestrating scripts (i.e., 'hidden gears') behind apparent quantum indeterminate random processes. This is accomplished in the next essay.

3.32: Solution to Quantum Indeterminism, Part-II:
The Ontic Objective Reality of Randomness & Quantum Indeterminacy

If minimal abstract states can generate entirely predictable pre-ordained outcomes out of apparent indeterminate terms, then there appears to be no need to hypothesise for any hidden gears to deterministically orchestrate 'surface' indeterminate terms. But we seek completeness: We seek independent proof of the ontic objectivity of randomness and quantum indeterminacy.

There exists a simple method for accomplishing just what we seek: **We will again resort to our old friend, the Sierpinski triangle. The orbiting seed will again be generated by means of the quantum mechanised dice-shaker, but with one critical difference: The quantum indeterminism therein will be treated *as if* governed by a secret set of hidden gears or hidden variables, constituted as deterministic orchestrating scripts. The scripts will produce the appearance and *illusion* of quantum indeterminism, but *without* its reality or ontology**.

Thus, in the process of successfully developing our method, we will discover that our scripting will run into insurmountable problems that cannot be solved by resort to any definable orchestrating mechanics. That is, **we will discover that the very attempt to deterministically script quantum indeterminism must self-unravel: The resort to unambiguous deterministic orchestrating scripts will compel for the very opposite: namely, inescapable resort to indeterminism and randomness. In this way and generalising our discovery to natural processes involving quantum indeterminacy, we will conclusively prove that quantum indeterminacy *is* objectively real, and that notional hidden gears... or the supposed illusion of randomness... is forlorn.**

*

Let us recall how the orbit of the seed in the Sierpinski triangle is generated via quantum mechanised die in relation to the three vertexes of the pertinent equilateral triangle: To summarise how the triangle develops, a point outside the triangle is pin-pointed and constitutes the first seed. The quantum mechanised die is rolled to determine to which vertex the subsequent seed orbit will tend. The succeeding position of the seed will be placed halfway to the stated vertex. Thereafter, the die will be rolled again, and the next seed orbit will be determined in essentially the same way.

The process is repeated for a hundred seed-orbits or more, until the characteristic fractal pattern of the Sierpinski triangle emerges.

Assuming the operation of our quantum mechanised dice-shaker (see **Fig. 3.11** in essay **3.26**), this will generate the succession of seed orbits attendant the three-phase development of the Sierpinski triangle through AND-to-OR collapse processes (see **3.27**). Throughout the experiment, we will also assume that our deterministic orchestrating scripts secretly directs the operation of our dice-shaker, and that what appears indeterminate at the surface-output will be secretly determined: That is, we will assume that quantum indeterminism and randomness are illusory.

To this end, and to generate our orchestrating scripts, first, we must replace the system of denotation of the Sierpinski triangle vertexes with a simpler one: with the vertexes labelled **A** instead of 1, 2; **B** instead of 3, 4; and **C** in place of 5, 6. (see **Fig. 3.19**).

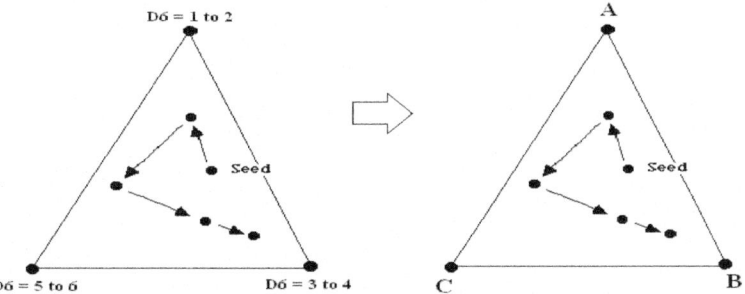

Fig. 3.19: Relabelling the vertexes of the Sierpinski triangle
Notes: In the variant of the Sierpinski triangle to be used for the present purpose, we relabel the vertexes with letters A, B and C, while the seed-orbit is generated by a quantum indeterminate process per quantum mechanised die shaker.

Given the three vertexes (**A**, **B** and **C**), **we can generate 'block-scripts' that deterministically orchestrate the orbit of the seed to a maximum of three orbits**. The total number of such block-scripts might be given by the simple equation...

$$n \times \left[2^{\frac{n \times (n-1)}{2}} \right]$$

Given the three vertexes, it follows that **n** equals 3. Applying the equation above, we can generate 24 fundamental unique block-scripts, each deterministically orchestrating from one orbit to up to three successive seed-orbits. These block-scripts are depicted as network-sets in the sequence of illustrations below, with the first set of block-script exclusively orchestrating for seed-orbits that tend toward vertex-**A**, or 'begin' with **A**. Thus...

.... BLOCK-SCRIPTS BEGINNING WITH "A"

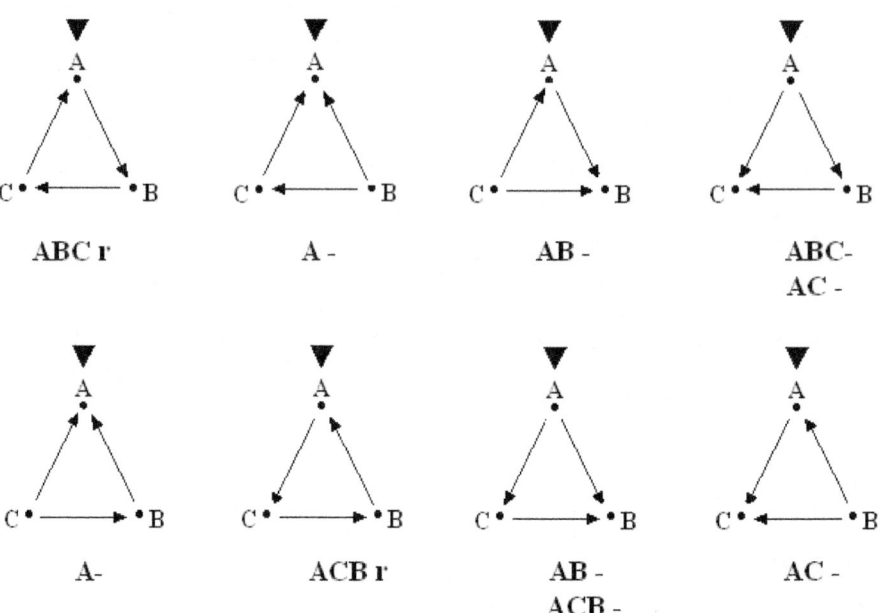

The remaining 16 scripts are given below as block-scripts beginning with **B** (below) and those beginning with **C** (also below); with each unique block-script accounting for a minimum of one to a maximum of three seed-orbits.

.... BLOCK-SCRIPTS BEGINNING WITH "B"

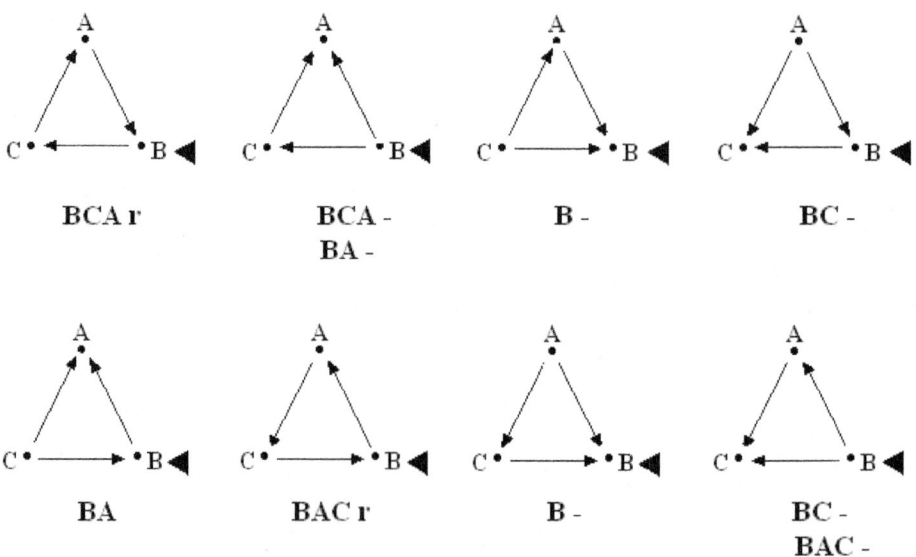

…. BLOCK-SCRIPTS BEGINNING WITH "C"

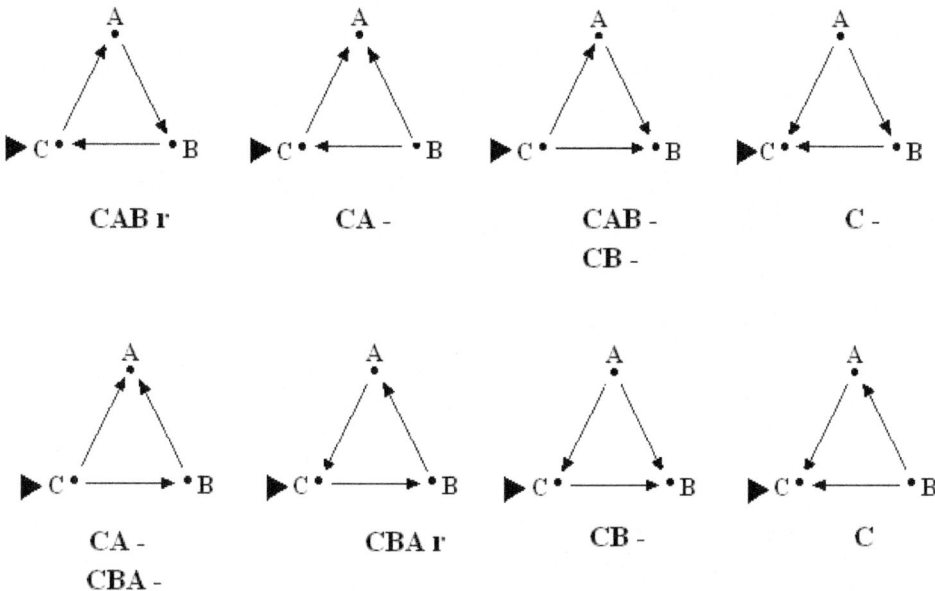

Note that each block-script has a sequence of letters below it, describing the orbit of the seed pertinent to that block: e.g., **BCA r** or **BC -**.Now, when a block-script ends with… "-", the orbit stops and it is presumably taken over by another block-script. When any given block-script ends with "**r**", it is recurrent and may repeat itself for as many seed-orbits as permitted by the script: but it is by no means certain that it will do so indefinitely, since another block-script could 'intervene' and take over the subsequent orchestration of the seed-orbit, as we shall soon discover.

Also note that certain block-scripts allow for, or embed, more than one script, simultaneously. Such constitute the *equiplausible block-scripts*. To identify these, look for a block-script with two sets of seed-orbit plots. For example, a block-script may start at, say **A**, but will then plot for seed-orbit **AB** *and* **ACB**, respectively; thus, generating two distinct simultaneous orbits of the seed, hence two distinct sub-scripts embedded in one block-script generator. Which of these sub-scripts does the actual seed-orbit take? Even at this juncture, such **instances of equiplausible blocks demonstrate that total and unambiguous determinism is impossible to script, in the sense that some the scripts are not closed-off to alternatives or restricted only to single conduction paths. Consequently, in such instances, it is inevitable that resort to an indeterminate sortology and randomness will emerge, even with the presumption or want of unambiguous deterministic scripting**.

Of course, we could ignore such problems and simply assume, purely for the sake of argument, that every possible block-scrip deterministically directs the orbit of the seed, *without* ambiguity. This is obviously not true, but the reader will discover that our indulgence and error will not make the slightest difference to the culminating conclusion.

Thus, certain unambiguous block-scripts allow only for one sequence of seed orbits: such as…

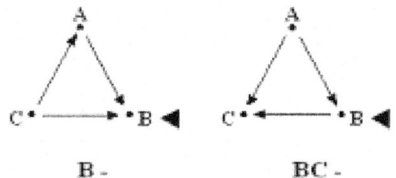

While other block-scripts (namely, the equiplausible block-scripts) can script for more than one sequence. For example…

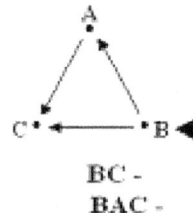

BC -

BAC -

Equiplausible block-scripts imply that the fundamental 24 block-scripts given by **n** = 3 can embed more sequences than the base 24 scripts. To account for these, we might add… **n x (n-1)** to the original simple equation. Thus, when **n** = 3, an extra six block-scripts will be added: When n = 4, twelve extra block-scripts will be added. And so on.

So modified, our simple equation will now look like…

$$ n \times \left[2^{\frac{n \times (n-1)}{2}} \right] + n \times (n-1) $$

Thus, the total number of fundamental block-scripts for the Sierpinski triangle at n = 3 will be… 24 + 6 = 30… *if* we choose to incorporate equiplausible scripts to the fray.

*

Now, each of the 30 block-scripts from the **1st order combinatorial script-sum** from n=3 (or the original 24 base block-scripts, if we choose to ignore the modified equation) are in no way sufficient to manifest the fractal pattern peculiar to our Sierpinski triangle, given that each script orchestrates anywhere from just one seed-orbit to a maximum of three. We need over a hundred or so seed-orbits to render salient the fractal pattern sought. Obviously, we need longer scripts.

How do we obtain these? We do so by generating a **2nd order combinatorial script-sum**: We treat the 30 fundamental block-scripts from the 1st order combinations (or the 24 blocks from the simpler equation) as the starting input for **n**. We take **n** to be equal to 30. In doing so, we construct a new combinatorial network comprised of 30 nodes; more than the original three; with each node representing one of the 30 scripts from the 1st order combinatorial generated when **n** = 3

As was done for the three vertexes, by drawing linking arrows between the 30 block-scripts, and by systematically changing the direction of the arrows that join these, we end up generating larger and longer 2nd order scripts: the majority of which will possess a sufficient number of seed-orbits suited to the task of deterministically scripting over a hundred seed-orbits; sufficient to generate the expected fractal pattern of the Sierpinski triangle.

The total number of 2nd order scripts generated when **n** = 30, using the same formula as had applied to the 1st order combinations, constitutes an extremely large number: The number is so large that it is simply not worth stating. We will simply call it **VLN2nd** (i.e., 2nd order Very Large Number).

If we were to treat the total number of 2nd order scripts as a new input for a **3rd order combinatorial script-sum**, using the same method as before, the number generated (i.e., **VLN3rd**) would be similarly meaninglessly large, if not more so. The good news is that we can garner the final implications and conclusions without resort to unfathomable quantities.

*

At the 2nd order, the combination of up to a maximum of 30 block-scripts (from the 1st order) can orchestrate from a minimum of 30-length orbits up to a maximum of 90-length orbits. Each of these will comprise unique combinations of the 30 block-scripts furnished by the 1st order script-sum.

Each unique 2nd order script will be *presumably* deterministic in how it orchestrates the sequences of seed orbits so generated, ignoring equiplausible scripts and their sub-scripts, or treating these as if they constituted non-ambiguous unique script, since, ideally, there ought to be no indeterminacy or chance involved in the generation of sequences and successions of seed-orbits.

In the development of our Sierpinski triangle, if the observer was kept ignorant of the operation of the purely deterministic 2nd order scripts, what might appear to constitute quantum indeterminate succession of seed-orbits on the surface-level will arise from fully deterministically orchestrating processes at the hidden or essential level. This is the very point of generating hidden orchestrating strips or 'hidden gears'. **The question must arise: How could we ever be certain that apparent quantum indeterminate processes are truly indeterminate and random, and not secretly deterministically orchestrated by hidden processes of nature?** We could pose this question about *any* quantum mechanical system, ranging from EPR to the two-slit experiment; to quantum indeterminism in photography, to the drifting snowflake; indeed, to the whole universe, notwithstanding the conclusion independently secured in the three-phase developmental approach (**3.26**).

*

Now, each 2nd order 90-length seed-orbit script will be unique and deterministic. Yet, no script will be so special, exclusive or distinct as to obviate or eliminate the *equipotentiality* of all other unique 2nd order 90-length scripts; They are all equally capable of

332

generating the same fractal pattern peculiar to the Sierpinski triangle. **There will be no single *one-and-only* script that excludes all other equally valid scripts**. Indeed, given the unfathomable size of **VLN2nd**, the equiplausibility of any given 90-length script vis-à-vis other same-length scripts within **VLN2nd** is expected to be great. **This fact poses a serious problem to the want to arrive at an unambiguous deterministic orchestrating hidden gears to obviate quantum indeterminism**.

Indeed, out of the very large number of unique deterministic 90-length scripts from within **VLN2nd**, how could we ever decide unambiguously for the existence of a "one and only" special or exclusive script? Given that there cannot exist a script so special as to be the only possible script to the exclusion of all the others, it follows that **there is no distinct structural or qualitative basis for any unambiguous decision or preference for one or any other equiplausible script**.

*

Given the absence of any special quality inherent to any of the 90-length scripts within **VLN2nd**, **perhaps we could resort to some random means of selection**. That is, we take all the unique equally apt but non-special 90-length scripts from VLN2nd and simply 'roll' for the one that gets to operate and orchestrate. We could even devise a quantum mechanised die-shaker whose concave detector is segregated into as many arcs as there are 90-length scripts within **VLN2nd**.

The script so selected will certainly orchestrate the orbit of the seed in the expected fully deterministic way (ignoring implications from equiplausible scripts). **But this will not do: Our aim was to eliminate any resort to random indeterminate sortology** in favour of full unambiguous determinism. Resort to a random selector to accomplish what we could not do so on a deterministic basis constitutes the failure of our project.

*

Is there an other method we might try? **We might try to settle the matter by delegating the problem to a 3rd order combinatorial generator and script-sum**: We apply the same equation and procedure as before. But this time, we take **n** to be equal to **VLN2nd**. Thus, we generate a 3rd order script-set, with unique scripts equal to the even larger combinatorial number, **VLN3rd**.

To clarify, let us assemble sample-scripts from the 2nd order 90-length unique scripts. Each of the 90-length scripts in our sample can be denoted as Ă, Č, Ĕ,... Nth, where Nth means "any additional 90-length 2nd order script", and any numbers of such from within the 2nd order combinatorial range. We can now combine these to form new larger sample-scripts, each up to 90-orbits long, that now belong to the 3rd order script-sums, **VLN 3rd**.

Below is a list of assembles of these, each term a 90-length script, all starting with Ă. Thus…

Ă, Č, Ĕ,...Nth

Ă, Ĕ, Č,...Nth

Ă,...Nth, Č, Ĕ

Ă,... Nth, Ĕ, Č

The idea is that we overcome the resort to random sortology wrought by the 2nd order script-sum by simply forcing and subsuming the 2nd order scripts to larger combinations belonging to the 3rd order script-sum. But the want to circumvent resort to random sortology will fail: We cannot circumvent the resort to random sortology in the 2nd order script-set by hiding the problem in the 3rd order set. Indeed, the move will only magnify the problem, turning a 2nd order ambiguity into a very much larger 3rd order ambiguity.

For example, script Č is embedded within each of the listed 3rd order deterministic script-samples that start with Ă (see above list). The embedding of Č appears to entirely circumvent the need to 'roll for' the selection of Č as the operating script from within the 2nd order set. Consequently, it might now appear that we have successfully delegated the selection of Č to some 3rd order deterministic script-set without resort to any random sortology. But this move will not work.

Let us take a closer look: Script Č is now part of a larger deterministic 3rd order script: say, script Ă, Nth, Č, Ĕ. (i.e., third script from the list above). As such, the activation of Č will be automatic; part of a *presumably* deterministic iteration of the 3rd order script, Ă, Nth, Č, Ĕ. Yet, this move now raises the matter of which of the four unique 3rd order scripts from the list above is the special and exclusive script. Given that all four are equipotential to the task of successfully generating the pattern of our Sierpinski triangle, we have merely crashed into the same problem we ran into at the 2nd order level, but now magnified to the 3rd order level.

It is obvious that, just as was true in the 2nd order set, **the 3rd order script-set does not furnish any special or exclusive "one and only" final script. While it circumvents resort to random selection at the 2nd order level, it achieves this by creating an even bigger version of the need for random sortology at the 3rd order level. Hence, we are back to either resort to random sortology from amongst our new 3rd order scripts, or we must again repeat the same method of delegating and abnegating the problem to a 4th order script-set…. Perhaps later followed by a 5th order set… Then a 6th set… and so on**... without ever solving the fundamental problem of the want to circumvent resort to random selection; without ever ridding the system of the inevitable re-emergence of randomness and indeterminacy, despite the presumption of unambiguously deterministic orchestrating scripts or 'hidden gears'.

*

Our inability to overcome ambiguity, despite resort to hard-core determinism, must compel resort to random and probabilistic sortology: We are compelled to 'decide' by means of the utilisation of random means the unique deterministic script that will get to operate and orchestrate 'apparent' surface indeterminism.

333

In short, in the attempt to found a pure deterministic solution to 'surface quantum indeterminate randomness', we could not eliminate final acquiescence to randomness and indeterminate sortology. That is, despite the presumption of a purely deterministic 'hidden gears', we could not eliminate indeterminism; We transformed presumed 'surface randomness' into essential ineliminable objective randomness.

*

Any self-consistent hidden gears and 'implicate order' approach to the development of our quantum mechanised Sierpinski triangle, wherein the seed is treated as an evolving quantum object, must assert, not merely the operation of a hidden deterministic script, but the operation of a single *'one and only'* uniplausible and unipotential exclusive script. If and only if such a script existed, and all other scripts of appropriate length were found incapable of generating the expected fractal pattern, could one assert in favour of a hidden implicate order or hidden gears behind quantum indeterminism. **In the attempt to generate such a script, we end up, not with a *one-and-only* script, but a combinatorial explosion of equiplausible and equipotential scripts: each equally capable of generating the pattern of the Sierpinski triangle. There exists no special or exclusive script to exclude any one or all of the others**.

In the absence of any exclusive script, we can only resort to random probabilistic selection from among the vast welter of equiplausible and equipotential scripts. Any resort to a delegation and abnegation of the problem to some N^{th} order combinatorial domain will fail to solve against the re-emergence of randomness: Indeed, the attempt will 'explode to infinity'. **The forlorn search for that special uniplausible unipotential exclusive deterministic script might well be obtained if we take the said combinatorial explosion to infinity. But such a move would only crash into an insidious halting or non-terminating problem.**

In short, we cannot eliminate randomness, even when we presume hidden deterministic processes behind 'surface' randomness: Ultimately, randomness and indeterminacy abide.

*

How does any of this directly relate to quantum indeterminism and the question of whether it is ontologically objectively real? In the above material, we imagined a quantum indeterminate process that governed the orbit of the seed in the development of the Sierpinski triangle. Our want to under-determine the 'surface' quantum indeterminate orbits by means of a final uniplausible exclusive deterministic script utterly failed. We failed to eliminate the final role and resort to random selection and randomness, even with the full presumption of determinism. Hence, the quantum indeterminism involved in the development of our quantum mechanised Sierpinski triangle turned out *not* to be facile: The orbit of the seed is *essentially* quantum indeterminate, simply because there is no possibility of a final deterministic hidden script to eliminate the intrusion of quantum randomness and indeterminism.

What if, instead of the Sierpinski triangle, we applied the same sort of scripting approach to the successions of electron position-resolutions in d$z2$ electron orbital? Our want to eliminate quantum indeterminism in the resolution of electron position about the atom, and in all successive resolutions of the same, would totally fail… for the same reasons as found in the case of the quantum mechanised Sierpinski triangle: **We will not be able to garner a *one-and-only* exclusive electron position-orchestrating script. Therefore, the indeterminism in electron-position resolution cannot be secretly under-determined by resort to purely deterministic hidden gears or variables.** The want to found such hidden gears or variables will degenerate into a combinatorial explosion and an unstoppable halting problem, with no solution.

The same problem will hold in the generation of the interference pattern in the two-slit experiment: The want to deterministically orchestrate the sequence of quantum indeterminate depositions that render manifest the interference pattern by means of some exclusive hidden deterministic script will also fail… for the same reasons: There can be no uniplausible *one-and-only* deterministic script to orchestrate the three-phase development of the said interference pattern. Objective quantum indeterminism will abide, as was shown to do so in our quantum mechanised Sierpinski triangle.

If we sought a similar script to deterministically orchestrate the *en masse* photon-depositions in a developing photographic film, this will end in the same failure, The same failure will arise in the context of electron micrography. There can be no uniplausible *one-and-only* deterministic script to orchestrate the three-phase development of our photographic image or of our electron micrograph. In the resolution of the pertinent photons and electrons, objective quantum indeterminism will abide.

Indeed, we can generalise our core finding to all of the myriad processes of nature that *ultimately* entail quantum indeterminate terms… such as in the drift of a snowflake over time; in the formation of radioactive hotspots; in the perpetuation of Jupiter's Red Spot; in the formation of whole galaxies…and, in the long run, in the evolution and formation of living forms. All these ultimately entail *en masse* objective quantum indeterminate terms without resort to hidden scripts, variables, or gears.

Hence, there *are no* hidden gears; no hidden variables; no Bohmian 'implicate order', supposedly secretly orchestrating 'illusory' quantum indeterminism in purely deterministic ways. In short, ultimately, randomness and quantum indeterminism remain objectively and ontically real.

*

Randomness in its quantum indeterministic form is objectively real. Moreover, we have now resolved a critical aspect of the cause of wavefunction collapse: **Insofar as the ultimate process behind quantum indeterminate resolution is kindred with the process behind AND-to-OR wavefunction collapse, and insofar as quantum indeterminate processes are not driven by hidden gears, it follows that there are no hidden gears behind wavefunction collapse either: That is, there exists no definable 'drive of time'.** Later, in Book-V, we shall develop a wholly independent means for validating this claim.

334

Thus, resort to hidden variables, gears, or implicate order, is shown to be wholly forlorn. This finding, as it pertains to quantum indeterminism specifically, through its co-joining with our three-phase developmental approach to quantum indeterminism, completes our solution to the stochastic-coherence problem, as thoroughly as it solves against hidden variables or gears. **Thus, we arrive at the comprehensive solution to the stochastic-coherence problem of how order, coherence and pattern could arise out of quantum indeterminate 'chaos',** *and* **we secure the viability and basis for the continuity of form and identity over time, despite ineliminable objective quantum indeterminism.**

The stochastic-coherence paradox is now solved: Quantum indeterminism is objectively real, and the three-phase developmental process obviates the fallacy of presuming enmity between randomness and pattern, or between indeterminism and determinism, notwithstanding the fact that minimal abstract ordinances exemplify how form dominates over and 'shapes' natural odds (see **3.24**).

3.33: Definition of Information Part-I:
Integrating Read-Only Physical Law to the Intermediate Model Spacetime

This essay constitutes one step in the completion of the definition of information initiated in Book-I; a pursuit that was augmented by the various apagogic mnemonic arguments for abstract memory, *and* by the two principal solutions to the problem of quantum indeterminacy and the stochastic coherence paradox (see **3.26** and **3.32**).

In material preceding Book-III, it was stated that **the initial working definition that treated information in purely OR-form terms must give way to a more complex abstract definition, reflective of the tripartite structure of intermediate model spacetime, culminating into information in its three principal forms per spacetime.** This will survive into the successor to the intermediate spacetime in Book-IV: namely, the Bergson-Whitehead amalgam (i.e., spacetime *without* space).

Information is certainly constituted as OR-form 'materialised' states. But **OR-form information is confined to the event horizon**. Spatially static abstract *wavefunctions in future-potentiality* remains confined to spacetime in potentiality (the ontic future), comprising the form of information to which we may also subsume contingent future potentialities (or Wheeler's cat). On the other hand, spatially static abstract *wavefunctions in memory* (the past) remain confined to the domain of mnemonic realised spacetime, thus constituting the other key form of information. A Lagrange of these also constitutes a key form of information, subsumed to the preceding two: a combination of future-potentiality and past-memory.

Another category is noetic information as it pertains to mind, consciousness and though: It does not directly concern us at this juncture.

Future-form abstract wavefunctions in future-potentiality certainly possess read-only characteristics and variants (i.e., physical law and principle) as well as read-write variants (see Book-I, **1.05**). The original assertion from information-matter dualism to the effect that OR-form information cannot be retained in or transported by AND-form quantum mechanical waves survives our tripartite definition of information, given that the resolved OR-form present, confined to the event horizon, cannot obviously be carried or transported in or by future-form spacetime in potentiality. It is also clear that resolved information pertaining to the past cannot obviously be carried in or by the future either, but must also exist in dissociated form from the future, even while the latter must decohere to consistency with the past per implications from abstract memory. Thus, **information-matter dualism abides, but only as part and as mirrored reflection of the tripartite structure of our growing-block spacetime.** Information-matter dualism also survives into the successor Bergson-Whitehead amalgam in Book-IV… or spacetime *without* space.

Recall that the *domain of realised spacetime* constitutes the domain in which all that has transpired in the universe is retained as abstract memory: part of the growing-block universe of accumulating wavefunctions in memory. **Imposed on past events and on subsequent abstract memory, physical laws obviously do not originate from the domain of realised spacetime or from the past.** Even so, the domain has an obvious critical role to play in read-write information (such as in the retention of the barrier setup in the two-slit experiment). It also has a most profound role in gravitation grasped as a memory effect, as was adumbrated in Book-II **2.45**, and will be explicitly espoused in Book-IV.

Also recall that the second of the tripartite domains is the *event horizon,* on and along which OR-form outcomes are 'materialised' into radiation and 'matter in space'. We typically misconstrue the event horizon and its 'materialisations' for the whole of spacetime and for the whole of physical existence: (i.e., materialism). In truth, the event horizon constitutes the mere relative present-continuous 'skin': a 'moving' boundary or 'cut' between the abstract total future and an abstract total past; both of which are structurally dissociated and dualistic with respect to the event horizon, or with respect to 'matter in space': hence, information-matter dualism.

Finally, 'above' the event horizon, we have the third domain of our tripartite growing-block universe: i.e., the quantum mechanical AND-form *spacetime in potentiality*, composed of nested future pre-event potentials for putative radiation, matter, and 'space' that have not yet come into OR-form realisation. The domain of spacetime in potentiality elaborates to the infinitely removed future. **It is within or through spacetime in potentiality that we 'locate' and depict the operation of read-only physical laws; completing our thesis espousing the dualism of physical laws with respect to OR-form radiation, matter and 'space' confined to the event horizon.**

*

If a series of Einstein Podolsky Rosen experiments (EPR) involving quantum entangled couples were run, within the early below-threshold noise phase, with few random terms thus generated, there would be no discernible indication of the expected pre-ordained distribution constituted as 50% **spin-up** and **spin down**: Hence, the EPR version of *noise-distribution.*

With the onset of the transitional chaotic phase, unpredictable short-run deviations might arise in which one spin-correlate temporarily outnumbers the other: a *chaotic distribution*. However, with the advent of the noise-reduction pattern-saliency phase, the chaotic distribution would graduate to a *signal distribution*; exhibiting the expected 50-50 distribution of spin-correlates; tending towards perfection with each subsequent addition of the quantum indeterminate resolution of entangled spin-correlates, or as the marginal bandwidth of randomness reduced toward zero. The longer and greater the number of successive EPR trials, the greater the saliency of the pre-ordained distribution, with more samples culminating into a higher definition of that distribution. From this, now-proven objective quantum indeterminate events *are* subordinated to an EPR-specific minimal abstract bounded probability state that fates the distribution of spin-outcomes and entangled couples into a near-perfect 50-50 distribution, **constituting an EPR read-only category of abstract information: one that incorporates the law of the conservation of spin: a future-form wavefunction in potentiality embedded within the domain of spacetime in potentiality.** The quantum spin-outcomes and couples, in the long-run, merely materialise the law of conservation of spin within the read-only EPR-specific abstract-form information.

Notice that the law of the conservation of spin is resolved into saliency through objectively real quantum indeterminate processes, given that, as previously shown in **3.32**, we cannot foist any hidden gears or hidden variables to quantum indeterminacy. This conclusion contrasts with staple generic inferences often drawn from EPR: to the effect that 'meaningful information' cannot be transacted in or as quantum indeterminate entangled or quantum coupled states. Clearly, one cannot control spin outcomes in quantum entangled couples, and one cannot utilise such forlorn control to affect faster-than-light or instantaneous Morse code-like signalling process (see Book-I 1.**50** and 1.**51** for a 'challenge' to this view based on superlative rigging, which does not obviate quantum indeterminacy in EPR).

We reiterate that, in EPR, ultimate information *is* the minimal abstract bounded probability state itself: the abstract wavefunction for quantum entangled spin-states. Inherent to this wavefunction in potentiality embedded to that futurity is the principle of conservation of spin (a law of nature). Hence, the dualism of natural law vis-à-vis matter, time and putative space: **The fact that it is impossible to transact meaningful information using EPR 'action at a distance' obscures the fact that the bounded probability state orchestrating the succession of spin-outcome correlates and the inherent law of the conservation of spin, *is* the critical piece of information.** As law, it does not itself need to be transacted, transmitted or transported by any putative particles; much less across 'space'. **Indeed, the principle of conservation of spin cannot be described as even residing in space, or, even in time; although its implications and its evidence thus must decay out of future-potentiality through AND-to-OR collapse or time. The law of the conservation of spin must be ascribed the status of an instantaneous whole** that spans putative space in a non-local way, and spans time itself in a transtemporal way… to the infinitely removed future.

Physical law as the read-only aspect of physical information, as exemplified by the principle of the conservation of spin in EPR, framed to the intermediate model spacetime, transcends the matter, the radiation, and the 'space' it otherwise subordinates, while it also transcends time itself. The physical principle is ultimately *not* material or 'spatial'; *not* carried by, *not* transported in, *nor* even epiphenomenal or emergent with respect to the 'stuff' or 'spatial medium' it governs and structures. And, although physical law requires AND-to-OR collapse and time to manifest in the time-asymmetric way espoused in Book-I (**1.12**), physical law is necessarily extant time, even as it participates in time via AND-to-OR collapse. Thus, physical laws are Platonic-Cartesian dualistic abstractions vis-à-vis matter, radiation, and 'space', if not versus time itself, despite their time-asymmetry per AND-to-OR polarity or 'time's arrow'.

*

If read-only physical laws are not located in or originate from putative space or from amidst manifest matter and radiation restricted to the event horizon, and if laws are dualistic-irreducible with respect to these latter states, then where are the physical laws 'located', and from where do they originate?

We need only consider the electronic configuration of the simple atom (i.e., the read-only d_z2 orbital) as an example of the operation of natural law in our growing block spacetime, and as part of the method for 'locating' and sourcing physical law. Successive electron position-resolutions within the read-only d_z2 orbital will be made manifested through successive AND-to-OR wavefunction collapse-processes. Over the long-run, through the three-phase developmental process, successive OR-form position-resolutions of the electron will accumulate and render salient the information-proper: i.e. the law-laden d_z2 pattern. We must now explicitly describe this process from within the intermediate model spacetime. Therein, we will model how **structures born out of necessity and law (the physical laws, such as the conservation of spin in EPR or the law-laden d_z2 itself) constitute the 'read-only' abstract states that *emanate* from the domain of spacetime in potentiality, and ultimately originate from the infinitely removed future, or from timelike infinity**.

To this end, we consult **Fig. 3.20**, of which diagram-A comprises a standard depiction of successive event horizons within the intermediate model spacetime from Book-II. Recall that each event horizon is an OR-form resolution of spacetime: a relative present continuous 'frame': in this case comprised of EH-1, EH-2 and EH-3: with EH-3 constituting the leading event horizon. Each EH is intermediated by a quantum mechanical AND-form interval. The intervals design for a broken-discontinuous succession of the three event horizons and the OR-form resolutions deposited on these, reflected in and exhibited as information-matter dualism.

Diagram-B in **Fig. 3.20** expands the structure by adding a z-axis; depicting each event horizons as a two dimensional 'surface': In this way, 'three-dimensionalising' our intermediate spacetime system. Thus, the system lends itself to the spacetime depiction of our d_z2 orbitals: That is, we can use our intermediate spacetime model to depict abstract physical law and necessity from within it. This is what we accomplish in **Fig. 3.21**.

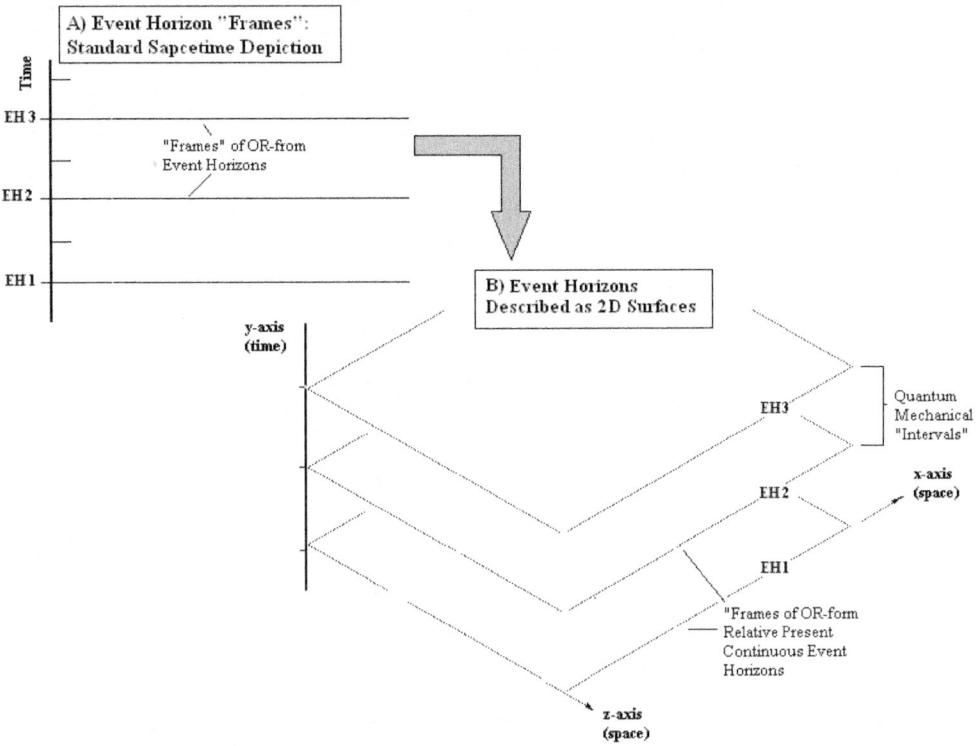

Fig. 3.20: Illustrating intermediate model spacetime tripartite structure

Notes: Information-matter dualism grasped from within intermediate spacetime: We move from a wavelength-based approach to a pure frequency-based approach in the relation of information to radiation and matter. **Diagram-A** is the standard 'edge-on' version of intermediate spacetime composed of a series of event horizons, EH1 through to EH3. In **Diagram-B** the same event horizons are depicted as two-dimensional surfaces. Note the quantum mechanical AND-form 'intervals' between the event horizons. We will discover that physical laws are imposed from and through the future, or from spacetime in potentiality, and made manifest via AND-to-OR collapse and time.

In **Fig. 3.21**, diagram-A constitutes a variant of our three-dimensionalised version of intermediate spacetime, but with differences. Therein, EH1 and EH2 are portrayed in dashed form, while EH3 is depicted in salient form. The dashed form of EH1 and EH2 emphasises that these are event horizons belonging to the resolved past, within realised spacetime (below EH3). However, EH3 is depicted in salient form to express the fact that it is the most recent 'leading surface' of relative present continuous OR-form outcomes. Indeed, it is to EH3 that leading OR-form resolutions are restricted and confined.

Also note that, in **Fig. 3.21** diagram B, we illustrate the didactic 'location' of physical law exemplified as **dz2**. Indeed, we portray **dz2** as an emanating abstraction from spacetime in potentiality; emanated upon EH3. It is now obvious that the projection of **dz2**, and **dz2** itself is ultimately sourced from and located at the infinitely removed future: EH-infinity.

The domain of spacetime in potentiality, together with the physical laws imprinted and imbued throughout it, exist in a non-reducible dissociated always-extant form vis-à-vis the 'materialised' discrete OR-form electron position-resolutions that constitute 'matter in space' in the context of **dz2** orbitals. Hence, the dualism of physical law vis-à-vis materialised OR-form resolutions and 'matter'.

The domain of spacetime in potentiality, as the ontic future, is obviously always extant and dualistic in relation to all the preceding OR-form resolutions of the same rendered on or along past-EH1 and past-EH2 that reside within realised spacetime. The future is also dualistic in relation to the leading EH3 itself. In other words, physical law and necessity, epitomised by the system of **dz2** infused into spacetime in potentiality, is dissociated from both the past constituted as realised spacetime (and abstract memory) *and* from the 'materialised' OR-form present restricted to the event horizon at EH3: This is consistent with our espousal for the Platonic-Cartesian character of physical law in the Preliminary (**0.26**), in Book-I (**1.13**), and elsewhere.

If physical laws imbue and delimit the totality of spacetime in potentiality, and consequently operate on nature from that future, *and* emanate from the infinitely removed future itself, then law and necessity must also transcend the very same spacetime in potentiality through which it imposes itself on the leading event horizon. That is, given that the 'location' of physical law immanently resides in the future, and the future constituted as the total spacetime in potentiality elaborates to the infinitely

removed future, this 'locates' physical laws at the infinitely removed future, at timelike infinity. It follows that physical laws are 'located' effectively outside of the total future constituted as spacetime in potentiality… outside the whole of spacetime… despite imbuing the totality of spacetime in potentiality by virtue of their 'location' at the infinitely removed future.

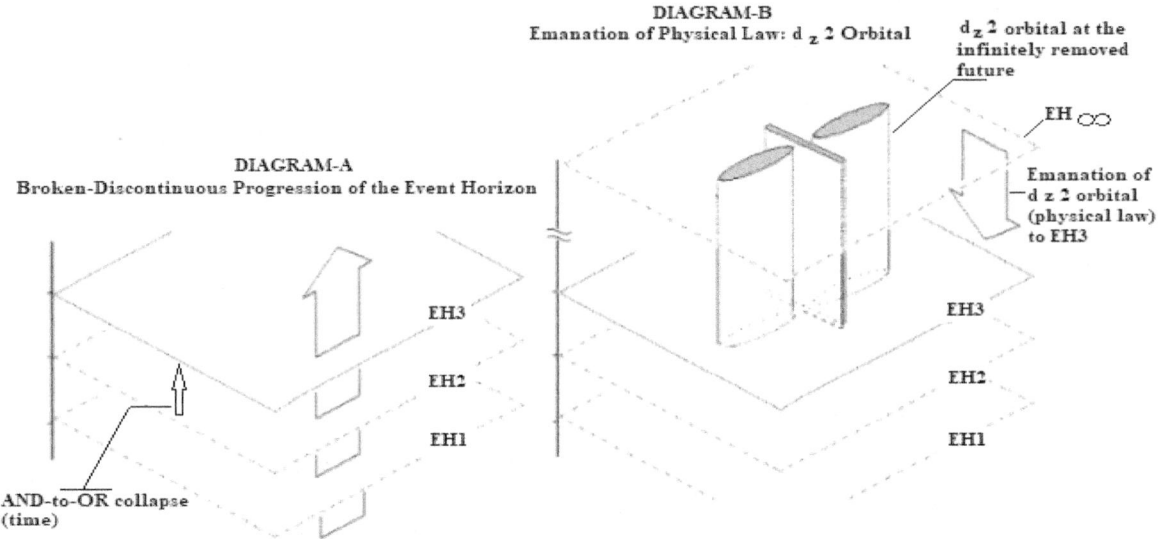

Fig. 3.21: The 'location' of physical laws within spacetime; dualism of physical law

Notes: **Diagram-A** is a variant of the three-dimensionalised version of intermediate spacetime with two-dimensional 'frames': i.e., EH1 through to EH3, formed through AND-to-OR collapse or time. Therein, the dashed form of EH1 and EH2 emphasises that these are event horizon belonging to the past, now relegated to realised spacetime, no longer manifested in 'materialised' form. But EH3 is depicted in salient form as the most recent or leading event horizon, 'materialised' into OR-form resolution. In **Diagram-B**, we can finally specify the 'location' of physical laws exemplified by the lawful nature of read-only **dz2** orbitals. Obviously, **dz2** does not arise from any past state within realised spacetime below EH3 or even from EH3, but it is imbued to the whole domain of quantum mechanical spacetime in potentiality, stretching to final potential event horizon, **EH∞** at the infinitely removed future. The 'arrow' related to **dz2** and the direction of that arow indicates that physical laws as emanated 'downward' from, or *emanate* from, the infinitely removed future. Consequently, natural laws are dualistic and irreducible with respect to manifest radiation, matter and space confined to EH3, but they are also dualistic and irreducible vis-à-vis realised spacetime and memory. The law operates on EH3 from outside of manifest radiation, matter, and putative space… from a meta-semantic extra-ontology or hypostasis 'located' at the infinitely removed future… effectively *extant* the whole tripartite system of intermediate spacetime. The Platonic-Cartesian status of physical law will survive into the successor model to the intermediate spacetime: the Bergson-Whitehead amalgam espoused in Book-IV.

Hence, while it is tempting to posit that physical laws must originate from the future constituted as the total spacetime in potentiality, physical law and necessity does not ultimately originate from spacetime in potentiality itself: law is imposed on it from 'outside', from the infinitely removed future. Hence, spacetime in potentiality assumes the constitution of laws and necessity deriving from an abstract meta-semantic hypostasis (see apagogic argument 2-c: essay **3.15**) that ultimately transcends even spacetime in potentiality, as much as it transcends the whole of the tripartite growing-block system. That is, **physical law and necessity is not so much infused from spacetime in potentiality, but it is spacetime in potentiality that is *infused with* physical law and necessity, whose origin resides in a meta-semantic ineffable ontology that transcends the whole of tripartite spacetime; transcends its futurity; and even transcends mind and consciousness itself. Thus, physical laws are ultra-abstract Platonic-Cartesian forms that emanate from the meta-semantic hypostasis 'located' at the infinitely removed future.**

*

In conclusion, physical laws, hence read-only informational states, must justifiably be regarded as abstract *Platonic-Cartesian physical laws* that are extant and ultimately dualistic-irreducible with respect to both space and time and matter, and must even transcend the total futurity constituted as spacetime in potentiality, in full vindication of the nascent dualistic view of physical laws anticipated in the Preliminary essay **0.26**, and in Book-I, essay **1.13**. As further augmentation, we must also add the argument from the meta-semantic hypostasis uncovered in **3.14**. Indeed, we must combine our meta-semantic argument with the argument for the primacy of form over natural odds in **3.31**: both enhance our overall case for the dualism of abstract-immaterial physical law vis-à-vis the whole of spacetime and the 'material world' wholly confined to the event horizon 'skin'.

3.34: Definition of Information Part-II:
Integrating Read-Write States & Abstract Memory to the Intermediate Model Spacetime

In the previous essay, we modelled physical law as imbued throughout spacetime in potentiality, emanating from the infinitely removed future, and ultimately ontically transcending the whole tripartite structure growing block spacetime. However, as demanded by the series of apagogic mnemonic arguments from Part-I and from related arguments elsewhere, nature must also retain the past in the form of abstract memory. **Memory must also get to influence new events, albeit in a consistent way with respect to read-only Platonic Cartesian physical laws emanating, delimiting and super-superimposing co-jointly with future potentialities cum abstract memory**.

Memories comprise read-write states of information, exemplified by the retention of the barrier-setting in the two-slit experiment. Indeed, as the mnemonic arguments in Part-I assert, without abstract memory projected from realised spacetime, quantum mechanics and physics as we know it would be rendered impossible (see essay **3.07**). This fact, in combination with others that render all nihilistic arguments against the ontology of information and abstract memory apagogic and self-voiding, render abstract memory both ontically real *and* dualistic-irreducible to the 'material world' confined to the event horizon. This certainly goes against the grain of generic materialist notions about memory, which only concede to memory as matter-trace state and cannot concede to memory as immaterial abstractions retained extant a 'material world' restricted to the event horizon. But what holds true in the case of emanated read-only laws must also hold true in the case of read-write abstract memories retained within realised spacetime.

Unlike potentials and physical laws that emanate from the future, abstract memories are projected onto the event horizon from the past: This is ultimately accomplished via grand decoherence of the future into the consistency of the event horizon with the accumulating past. The projected memories then funnel objective quantum indeterminate events into coherent 'materialised' outcomes consistent with the demands of both memory and emanated physical laws.

Also, **whereas future-form potential information, both read-only and read-write, constitute *uncharged* wavefunctions in potentiality (*with* nested futures), the read-write memory states projected from realised spacetime have had their nested futures 'cropped' or erased, constituting read-write *charged* wavefunctions in memory**.

To model read-write memory, we apply the same treatment to the barrier in the two-slit experiment that we previously applied to the emanation of read-only **dz2** orbitals and physical law. But this time, the influence is from realised spacetime (the past) 'below' the event horizon, not from future spacetime in potentiality 'above' it.

*

How do read-write memories get to structure and funnel objective quantum indeterminate processes into global materialisation on the event horizon? Memories in the form of charged wavefunction in memory operate from 'below' the event horizon and upon it: **Memories *project* 'upward' upon the event horizon, and, through this, *illuminate* pertinent objective quantum indeterminate processes therein, per the three-phase developmental process espoused in 3.26into *en masse* OR-form outcomes that materialise or re-materialise the memory so-projected**.

The process of illumination by memory is depicted in **Fig. 3.22**: Therein, diagrams A through to C portray the memory state of the barrier from the two-slit experiment with **both slits open**. Diagram-A depicts the configuration of the barrier as an OR-form materialised state on or along event horizon EH1 at experimental initiation. Upon subsequent isolation, diagram-B depicts the dissolution of EH1 barrier-state into an abstract charged wavefunction in memory specific to **both slits open**. Note the quantum mechanical suspension or 'interval' between the past EH1 and expected succeeding potential-EH. This is per function of the broken-discontinuous 'upward' progression of the event horizon, first described in Book-II. From the interval of experimental isolation, all observational inputs are withdrawn, and the internal content of the two-slit experiment, including the barrier itself, are quantum mechanised into a read-write wavefunction in memory.

Diagram-C depicts the end-point of experimental isolation and the tandem conversion of potential-EH into realised EH2 through attendant AND-to-OR wavefunction collapse of the abstract memory for **both slits open**. This must resolve at EH2 through *en masse* objective quantum indeterminate terms via the three-phase developmental process, even though the wavefunction in memory constituting the barrier is not itself subject to randomised sorting or scrambling, as was shown in the key **mnemonic arguments 1-a and 1-b** (see **3.07** and **3.08**).

What is the critical role of counterfactuals and the quantum erasure of nested futures espoused in Book-II to abstract memory? **The background operation of counterfactuals and the erasure of non-consistent nested futures via grand decoherence at the start of the experiment at EH1 will have erased all non-consistent alternative nested future-potential barrier settings from within spacetime in potentiality, save contingent possibilities that might yet alter the outcome**, epitomised by **switch-x** in Wheeler's delayed choice and analogised by the mischievous *Wheeler's cat* who might leap onto the apparatus and overturn it; altering the settings and changing the course of both the experiment *and* future spacetime, counter to what memory otherwise demands.

The moment the experimenter sets up the barrier as **both slits open**, the act will counterfactually erase all alternative non-consistent barrier settings from spacetime in potentiality: **Only those future-possible states that are consistent with *both slits open* will remain within future spacetime in potentiality for future realisation, save what 'Wheel's cat' might incur otherwise. In short, it is the**

339

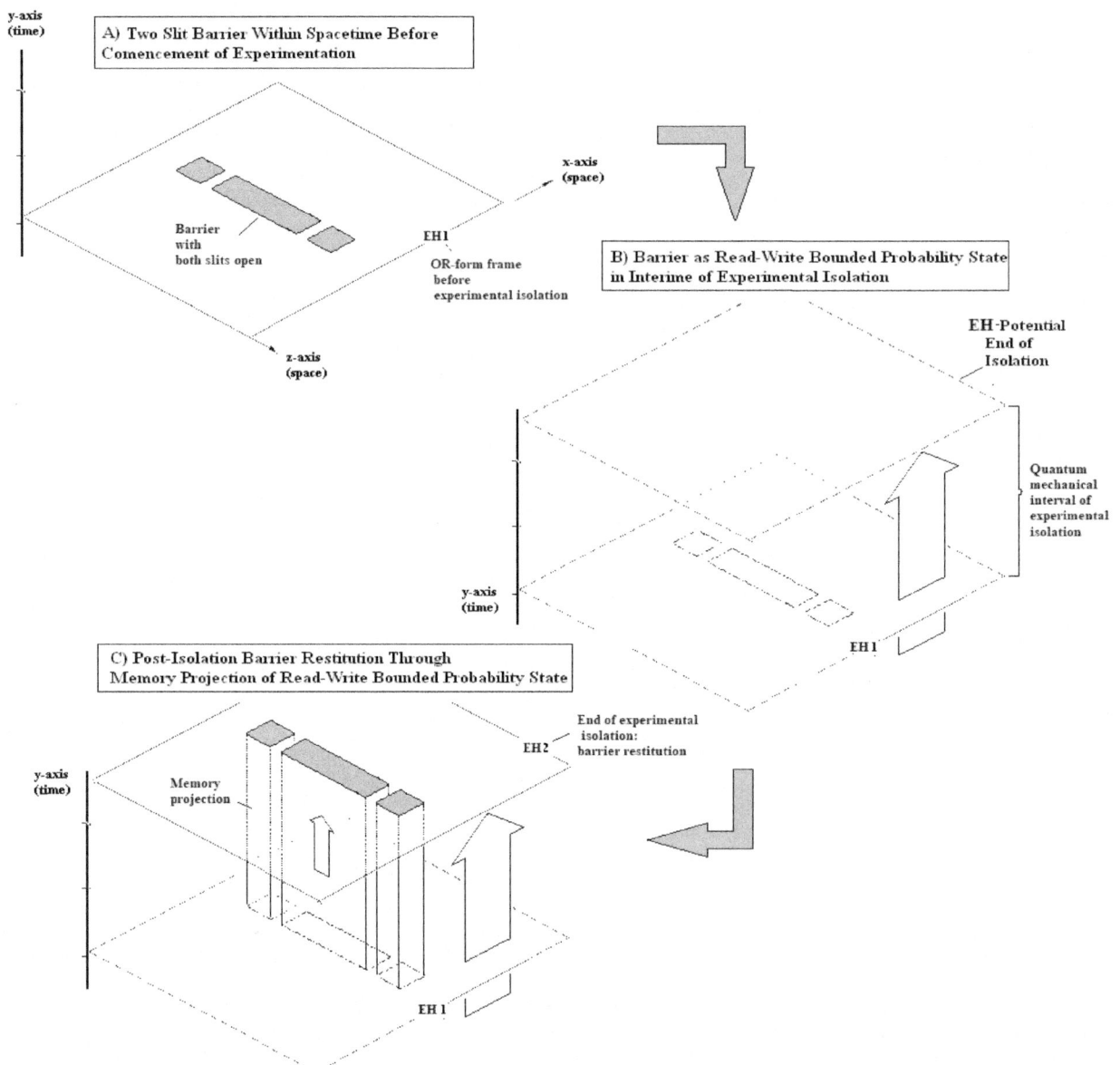

Fig. 3.22. The 'location' of memories within spacetime: dualism of abstract physical memory
Notes: Diagrams **A** through **C** portray the spacetime projection of abstract memory specific to the 'both slits open' barrier setup from the two-slit experiment. **Diagram-A** depicts the barrier as an OR-form state on or along event horizon EH1. **Diagram-B** depicts the dissolution of EH1 and barrier into an abstract 'charged' state, cropped of its nested futures, but retained in and consigned to realised spacetime: *not* erased. Note the quantum mechanical 'interval' between past EH1 and future-potential-EH per the broken-discontinuous progression of the event horizon. **Diagram-C** depicts the end of experimental isolation and AND-to-OR collapse of potential-EH into realised EH2, and the resolution of the barrier into its original 'both slits open' configuration. By EH-1, all other alternative barrier-settings were erased per counterfactuals and the quantum erasure of nested futures (i.e., grand decoherence). The 'two slits open state' from past EH was projected onto the newly formed EH2: Quantum indeterminate terms were funnelled through the three-phase developmental process back into the materialised recuperation of 'both slits open'.

process of grand decoherence that projects memory onto the event horizon 'canvas' and 'illuminates' it thus. Thus, at EH2, only future developments consistent and conformal to the initial **both slits open** (retained as abstract memory within realised spacetime) will transpire into materialisation through *en masse* objective quantum indeterminate processes. Indeed, this idea was espoused in the augmenting **apagogic mnemonic argument 3-a**, (see **3.19**) and in pertinent parts of Book-II that had hammered out the role of counterfactuals, quantum erasure of non-consistent futures, hence grand decoherence (see **2.34** and **2.35**).

*

The same form of projection of abstract memory from the past, from realised spacetime, will apply in the case of the drifting snowflake, in photography, and in electron micrography. Therein, the complex wavefunction involved in the formation of the image is necessarily a wavefunction in memory, given that we can never take a photograph of the future, but *only* of the past... necessarily retained within realised spacetime in abstract form. Thus, the complex wavefunction in memory of the vase from essay **3.29** (**Fig. 3.18**) projects onto the photographic film on the event horizon. The instantaneous whole that constitutes the wavefunction in memory pertinent to the expected image then funnel *en masse* objectively quantum indeterminate terms through the expected three-phase process described in **3.26** into a pre-ordained globally fated outcome: i.e., the vase-image itself.

The same projection process of memory will apply to any 'material' structure or complex that had formed up in the past and must now perdure through time through successive AND-to-OR collapse processes, courtesy of its retention as abstract memory within the domain of realised spacetime.

However, we must combine memory projection and emanation: we must grasp the whole process as a super-superposition of the total past *and* of the total future, with appropriate decoherence, and with their joint implications amalgamated upon the event horizon.

*

The same process of memory projection must inescapably apply to the time-evolution of the brain: involving the projection of a superposed past brain-state at EH1 (an instantaneous whole and a complex wavefunction in memory) onto the most recent resolution at EH2, with obvious implications about how memory is related to the 'material' of the brain, **wherein memory is grasped as abstract and necessarily extant and dualistic-dissociated vis-à-vis the materialised brain, with the latter restricted to the event horizon**. Hence, the implications from our growing block projection and illumination theory of abstract memory is that memory is *not* in the OR-form material brain, confined as this is to the event horizon 'skin'. Instead, memory is extant the brain: in dissociated and abstract form vis-à-vis the brain, within and projected from the abstract past: 'illuminating' the brain from within mnemonic realised spacetime... even when we assume an initial reduction and identification of memory with brains.

3.35: The Definition of Information Part-III: The Four-Fold Classification of Physical Information

In Book-I, we posited two basic categories of physical information, and, by extension, two basic categories of minimal abstract bounded probability states: These were the *read-only* and *read-write* states, which in turn come in two variants: namely, *future-potential,* and *past-realised* forms. We now expand this categorisation into a more comprehensive form, co-joining it with our findings secured in Part-I and Part-II; integrating these to our new understanding of abstract memory; and, short of implications from explicit superlative de-spatialisation, completing the mission to found the definition of information started in Book-I.

The first category of information is comprised of read-only future-form minimal abstract states: states constituted as *wavefunctions in future-potentiality*. These incorporate abstract physical principles and laws, such as the conservation of quantum spin in EPR, as well as fundamental structures such as the **d_{z2} orbital**. These emanate from the infinitely removed future and imbue spacetime in potentiality through to the event horizon, thus securing the lawfulness of events resolved on the event horizon. While the counterfactuals-based quantum erasure of nested futures through grand decoherence cannot eliminate read-only future-form abstract states of physical laws and fundamental structures, these are nonetheless integral to the system of convergent category nested futures that define the whole of spacetime in potentiality to the infinitely removed future.

Second, we also have read-write mnemonic charged *wavefunctions in memory*. These reside 'below' the event horizon and pertain to the stock of past-realised events; to memory retained in dissociated abstract form vis-à-vis 'materialisations' on or along the event horizon. These are 'charged' (i.e., previously resolved and utilised) states that appear to have the same AND-form structure as generic future-form states within spacetime in potentiality, but *without* nested futures.

It turns out that most quantum waves and wavefunctions are wavefunctions in memory: For example, the complex wavefunction pertinent to a photographic image *necessarily* constitutes a wavefunction in memory, insofar we cannot take a photograph of the future (*obviously*), and the film is always exposed to information from the past. The same applies to the barrier in the two-slit experiment.

In Book-IV, we will grasp that the influence of wavefunctions in memory on the event horizon fade per function of the inverse square of time. The influence of wavefunctions in memory remain constrained by emanating physical laws embedded as read-only future-form wavefunctions in potentiality within spacetime in potentiality. Thus read-write memories cannot over-ride read-only laws and structures: For example, the wavefunction in memory pertinent to **both slits open** cannot obviate physical laws emanated and enforced from and through spacetime in potentiality.

*

All things being equal, within the domain of future-form spacetime in potentiality, there will exist a third category of information: namely, *'read write' contingent wavefunctions in future-potentiality*. These are contingent possibilities that constitute future-potential overrides of, say, the given barrier setting. Contingent wavefunctions in potentiality are analogised by 'Wheeler's cat'. For example, the experimenter might decide to re-set the barrier to the **left slit open**, thus replacing the original **both slits open** state.

341

Another example: the experimenter might interrupt and terminate the experiment before its conclusion to change the barrier setting to **left-slit open**. Another example: in Wheeler's delayed choice framework, the scientist my delay the final decision of the setting on the delayed choice **switch-x**. Observers who might happen to observe our activity in some remote future will note that the barrier was indeed set to **both slits open** before it was changed by contingency: Thus the memory of the initial condition will survive, even if it was subsequently overridden by an alternative then-potential contingent event, or by the mischievous Wheeler's cat itself.

The point of contingent wavefunctions in future-potentiality is this: The future is not closed off to alternate outcomes from those imposed by wavefunctions in memory. From within spacetime in potentiality, the contingent possibility for change exists as nests of potential overrides, epitomised by the delayed-choice **switch-x**, and analogised by Wheeler's cat. Of course, such contingencies cannot erase memory, no more than they could undermine the erasure of non-consistent futures, or over-ride consistent futures: Indeed, contingent futures are part of the repertoire of consistent futures.

*

We now have our fourfold categorisation of information, comprised of...

- **Read-only wavefunctions in potentiality**, confined to spacetime in potentiality (the future)
- **Read-write wavefunctions in memory**, retained in abstract form within realised spacetime and...
- **Read-write contingent wavefunctions in future-potentiality**, analogised by Wheeler's cat and by the delayed choice element, **switch-x...** also within spacetime in potentiality.

The 'materialised' outcomes or events obtained with every partial-perpetual AND-to-OR collapse (i.e., time) and attendant progression of the event horizon, resolves Lagrange OR-form outcomes from the super-superposition of all three forms of information listed above, whose super-superposition is compelled by the tripartite structure and process of growing block intermediate spacetime. Thus, to the above list, we must also add...

- **Read-Write Lagrange OR-form information**... confined to the event horizon.

There is a fifth category of information: namely, noetic information pertaining to mind and consciousness, but this exceeds the bounds of the subject in hand.

3.36: Definition of Information Part-IV:
Addendum to Decoupling Theory: Nimtz Experiments & Wavefunctions in Memory

Recall from Book-I how Gunter Nimtz utilised quantum tunnelling procedures to displace microwaves at apparent 4.7 times the speed of light[78] and had transmitted Mozart-40 at that speed (Book-I: **1.38** and **1.39**). In both Book-I and Book-II, we presented solutions in-progress to this anomaly: a graduated evolution of decoupling theory. Thus...

- **Decoupling-i:** Book-I treated AND-form quantum mechanical waves as states that undergo displacement across putative space, but called into question the notion that AND-form states can act as 'carriers' of information. This was done on the provisional assumption that information must be purely OR-form, and on the correct judgement that OR-form information cannot be carried by quantum mechanical AND-form states, as demanded by the first iterations of information-matter dualism. Thus, OR-form information must dissociate from its AND-form 'carrier'. From information-matter dualism, **it followed that, even if the quantum wave 'carrying' M-40 was quantum tunnelled to faster than light by Nimtz, M40** *could not* **in the first place be carried by any quantum mechanical wave, much less by waves tunnelled to 4.7 times faster than light**. Thus, as a piece of information, M40 (and, with it, causality) could not be fated to the apparent violation of the speed of light limit, given that M40 was objectively dissociated from the quantum tunnelled radiation, and was never presented in or carried by it to begin with, given information-matter dualism. Book-I also espoused the **ontological primacy of AND-to-OR collapse and directionality to time and causality: Thus, the apparent faster than light transmission of M40 could not bring about time-reversal into OR-to-AND de-collapse. The integrity of causality and M40 was found to be per function of non-reversible ontologically primary AND-to-OR directionality and the attendant 'arrow of time'; both independent of and decoupled from the speed-of-light limit as well as from accelerated faster-than-light microwaves.**

- **Decoupling-ii** Book-II (**2.42**) arrived at a more profound variant of decoupling theory: By application of Wheeler's delayed choice, the **quantum mechanical radiation or matter, hitherto treated as if spatially dynamic and displacing, turned out to be purely spatially static: i.e., a non-moving and non-undulating state** (Book-II: **2.23** to **2.26**). Consequently, the OR-form resolution of events at any succeeding event horizons is a function, *not* of the supposed 'motion' of a 'carrying' quantum mechanical radiation or matter, but the organisation of temporally successive events into pure delayed choice time-interval relations... upon which we impose the assumption of 'motion'. The space through which the quantum wave is purported to displace can *only* be treated as a pre-event potentiality or possibility embedded within spacetime in potentiality: i.e., within the not-yet happened future: constituting a not-yet realised 'space'. Obviously, nothing can displace through a 'space' that has not yet formed up. The 'space' presumed from the succession of OR-form events, misinterpreted as evidence

---

[78] *Enders, A.; Nimtz, G. (1992). "On superluminal barrier traversal". J. Phys. I France. 2 (9): 1693–1698.*

of 'motion', is at best a post-measurement post-quantum mechanical space; *not* a directly observable space: it cannot be observed or proven in realtime. **Thus, the whole notion of the 'motion' of quantum mechanical waves was brought into question, including that of the 'motion' of quantum tunnelled faster-than-light radiation and matter. Indeed, it ought to be obvious that, insofar as quantum waves constitute the ontic future, hence constitute spacetime in potentiality, the future cannot 'move', and the post hoc interpretation of 'motion', even quantum tunnelled 'faster-than-light motion', is foisted upon the decay of successive events out of a non-moving and non-undulating static spacetime in potentiality or futurity.** It follows that, even when we tunnel radiation to apparent 4.7 times the speed of light, neither the radiation nor the physical information (M-40) is 'carried' by the non-moving spatially static nested futures-state residing within non-moving spacetime in potentiality, simply because neither move. The apparent superluminal character of radiation in Gunter Nimtz' quantum tunnelling experiment, and the supposition of *any* 'motion' of radiation and matter, is a false inference and an imposed assumption. Instead, we must look at pure time-interval relations between the critical events involved in the experiments. Of course, this also calls the whole notion of 'space' into question: Since quantum waves are spatially static non-moving and non-undulating states, what need of the 'space' across which these cannot and need not 'move'?... a matter to be fully addressed in Book-IV.

Throughout Book-I and Book-II, we portrayed information almost exclusively in OR-form terms. However, in Book-III, we discovered that OR-form information is strictly confined to the event horizon and is not the exclusive and dominant form of information. Beyond its 'materialisation' on the event horizon, OR-form information must 'dissolve' into its dominant minimal abstract state... into *wavefunction in memory*, retained within the domain of realised spacetime. In short, **the retained form of information is not OR-form but abstract wavefunctions in memory, projected onto the event horizon** (see 3.34). Based on this finding, we can now fashion for a third graduation of decoupling theory.

*

It turns out that, **in nature, information, including M40, is retained within realised spacetime as a complex wavefunction in memory; an abstract instantaneous whole.** This is also dissociated from the 'material world' confined to the event horizon. Thus, the state that resonates in the Nimtz experiments constitutes a read-write wavefunction in memory of M40 in abstract form, dissociated into and retained within mnemonic realised spacetime. Hence, M40 is certainly *not* carried by future-form wavefunctions in potentiality embedded within the non-moving total futurity or static spacetime in potentiality.

However, wavefunctions in memory do not require transportation or displacement across 'space', even at or approaching the speed of light, much less at 'faster-than-light': Indeed, **wavefunctions in memory are as spatially static and non-moving as wavefunctions in potentiality. It follows that the inference of purported 'spatial displacement' of the memory state of Mozart 40, even at 4.7 times the 'speed of light' in the Nimtz experiment, is an imposed post-hoc assumption.**

Only a pure delayed choice time-interval that relates one event horizon (pertinent to the 'transmission' of M40) to a succeeding one (pertinent to the reception of M40) is relevant. This pure temporal approach will be seamlessly compelled per function of anticipated de-spatialisation in Book-IV. As such, we reiterate that there is no literal 'spatial displacement' of the pertinent wavefunction in memory for M40: Instead, as per projection and illumination theory espoused in **3.34**, M40 is projected onto the event horizon 'canvas', and 'illuminates' that canvas through grand decoherence and subsequent three-phase developmental process that funnel pertinent quantum indeterminate terms into global outcomes: namely, realised M40. This process is an abstract resonation instantiated purely temporally and non-contiguously, *not* via 'transportation' through a contiguous spatial resonation, but via grand decoherence..

In the Nimtz experiment, what transpires does not involve the motion of anything at *any* speed. Instead, an initial super-superposition between the wavefunction in memory pertaining to Mozart-40 *and* spacetime in potentiality generates the grand decoherence of the totality of spacetime in potentiality and forces the latter into consistency vis-a-vis the memory of M40. The rest is a matter of the funnelling of *en masse* quantum indeterminate terms through the three-phase developmental process into the 'materialisation' of M40 on or through successive event horizons, supposedly 'transmitted' (but not so) at 'faster than light'... or otherwise at the normal 'speed of light'(but, in truth, not so).

But how do we account for the faster than light effect? How does this affect the delayed choice time-interval relation involved? In answer, **we need only reiterate the various scenarios and the maincone approach furnished in Book-I (1.41): These harmonise decoupling with Special Relativity: They are consistent with the impositions from grand decoherence from the totality of mnemonic realised spacetime *and* the wavefunction in memory therein, constituted as M40. The rendition of spacetime in potentiality into consistent form cannot help but AND-to-OR decay into the 'materialised' recollection of M40, whatever the erroneous supposition about the motion and 'speed of transmission', even at '4.7 times speed-of-light',** *ceteris paribus* per contingency or Wheeler's cat.

In short, it is the process of the projection of wavefunctions in memory, and the attendant illumination process grasped in and as temporal resonation and relation, co-joined to grand decoherence of spacetime in potentiality into consistency, and *not* the naïve putative spatial transport of M40 at some 'speed', that comprises the critical *modus operandi* in decoupling.

*

Let us clarify and augment these claims and their implications: Consider that, to see 'objects in space', or even to listen to a recording of M40, is to see or apprehend the light generated or reflected from pertinent past events. Generically, the light, supposedly travelling at a fixed speed, takes time to travel to us across purported 'space'. Even with the assumption of space and of a speed-of-light limit, the peculiar result is that the 'objects' we see are transformed into other than 'objects in space': they are transformed into events of the past organised and interrelated to each other and to the retrospective observers in purely temporal terms; in terms of delayed choice time relations. We reiterate that this conclusion is unavoidable, *even when we assume space,* unless one asserts dubious notions of absolute time undermined in Einstein's Special Relativity [79] .

Moreover, appraisal of the physical order as a state constituted by events interrelated purely temporally is both necessary and sufficient, while the appraisal of the same as 'objects in space' is neither necessary nor sufficient: This is because, without time, events cannot arise, and purported 'spatial relations' between events could not be established or observed, and even putative 'motion', much less the putative 'space' within which it supposedly unfolds, would be rendered untenable.

A physics without time would lead to a non-physics, even if such a physics could somehow furnish 'space'. That is, we cannot remove time from the lexicon of physics, although we could remove space *and* keep a functioning physics, so rendering 'space' into a mere pleonastic term. In that case, what need 'motion' or even motion at the 'speed of light', or at 4.7 fold 'faster' than light?

Yet, even if we choose to ignore the consequences of anticipated de-spatialisation from Book-IV, even *with* the assumption of forlorn 'space', what we *always* observe are *not* objects spatially related to each other and to us in terms of co-ordinates and distances: Instead, the 'objects' *always* constitute corpuses of *past* events related to each other and to us in terms of time-intervals; always organised in and by *time*. Indeed, it is *never* the case that one object is 'closer' than some other more 'distant' object: Instead, what we observe is one corpus of events (say, the vase on the table) defined as being farther removed in the past vis-à-vis another past corpus of events (say, the tip of one's nose) vis-à-vis a more recent past. Thus, what we observe, and what any 'point' in the universe 'sees' and is exposed to, is *always* a system of structured past states temporally interrelated and integrated, ultimately *without* 'space'… *even when we assume the false notion of 'space'.*

The pertinence of all this to the Nimtz experiment is that the observation of M40 from its transmission to its reception, constitutes a pure time-relation between M40 and its observed replay, *without* any spatial carry-over of M40 at the speed of light or even at faster than light: a transition and relation brought about via AND-to-OR collapse. Thus, only a pure time relation applies: one that cannot alter the integrity of M40, nor reverse or upset the 'arrow of time' vested purely in AND-to-OR collapse-directionality… even when the time relation is, post-hoc, interpreted as a 'faster than light' affect.

Again, in order to accommodate the anomaly of faster-than-light processes in the Nimtz experiment and in the reception of M40, we simply rehash the several scenarios for the harmonisation of decoupling with Special Relativity; especially the maincone theory… all reframed in terms of the super-superimposition of memory and futurity, and a futurity compelled into harmony via grand decoherence, and spontaneous AND-to-OR collapse; and the *en masse* quantum indeterminate terms into the re-materialisation of M40.

3.37: Apagogic Mnemonic Argument 4-a:
Abstract Memory from the Solution to Quantum Indeterminism & the Stochastic Coherence Paradox

According to the Manyworlds interpretation of quantum mechanics (MWI), for a given wavefunction composed of AND-form possibilities, *all* the possibilities will be realised into compartmentalised distinct histories. In this way, Manyworlds appears to forgo the issue of quantum indeterminacy by circumventing the need to explain how just one outcome or unique history gets to form. Yet, our Alternative Realist (Alt-R) solution to both quantum indeterminacy and the stochastic coherence paradox in Part-II obviated resort to Manyworlds. In doing so, it also constituted the finial apagogic mnemonic argument against informational-mnemonic nihilism, consolidating Part-I and II.

The nihilistic assertions anticipated from Manyworlds must unravel per function of the following conclusions:

- **Quantum indeterminism is objectively real and ineliminable,** (3.32) rendering the utility of Manyworlds superfluous, given its professed claim to obviate and 'solve' the now non-problem of quantum indeterminism.

- **Manyworlds does not lead to alternative histories, but only to the same global outcome in all its purported histories,** hence into many *identical* histories (3.17).

- **Manyworlds cannot obviate memory, given that the root wavefunction it exhausts constitutes a wavefunction in memory** (3.35), and constitutes abstract memory... wherein all the Manyworlds it produces turn out to be nothing more than manifold recollections of that same root memory.

First, randomness and quantum indeterminacy are provably objectively real; *not* illusory; *not* artefact; *not* born of ignorance. Even when fully deterministic orchestrating scripts are hypothesised, these apagogically fail to eliminate ultimate resort to random selection and sortology from among such scripts: (see **3.32**). That is, we cannot ultimately eliminate resort to randomness and indeterminacy, even when we presume deterministic orchestrators or 'hidden gears'. Therefore, quantum indeterminacy is objectively real: There are no hidden gears, no hidden variables, and no 'pilot waves' that resolve outcomes from wavefunction collapse.

Second, from our three-phase developmental approach, objectively real *en masse* quantum indeterminate outcomes are *always* collectively funnelled into determinately fated coherent global pattern-outcomes that are either born out of necessity and law, or else from read-write wavefunctions in memory or potentiality, or by contingent potentialities analogised by 'Wheeler's cat' (**3.35**). From our three-phase developmental approach, it is obvious that, beyond a determinable threshold, objective quantum

---

[79] Einstein, Albert: "Ether and the Theory of Relativity" (1920), *Sidelights on Relativity* (Methuen, London, 1922)

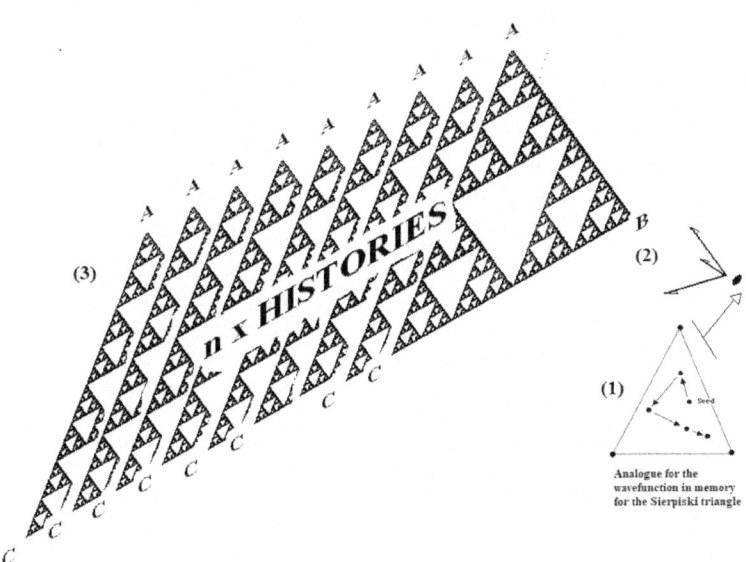

Fig. 3.23: Identical global histories at the nth seed-orbit invalidates Manyworlds

Notes: In this modified version of a previous illustration, **(1)** depicts the analogue for the root wavefunction in memory pertaining to the Sierpinski triangle, retained within realised spacetime; although this could instead constitute the complex root 'wavefunction in memory' for a photographic image, or pertain to the 'both slits open' setup, or pertain to any other formation acting as an initial set of conditions retained as abstract memory within realised spacetime. In most instances, Manyworlds inadvertently seeks many histories from an implicit root 'wavefunction in memory'. Therefore, Manyworlds presupposes the ontic reality of memory, at least per the starting wavefunction, and obviously cannot presume its erasure. In **(2)** either a presumed Manyworlds-based brute force exhaustive process resolves *all* possible 100th seed-orbit and beyond (as shown) or we implicitly assume a single unique history for the development of our Sierpinski triangle. In the first case, in **(3)**, what we obtain is brute force exhaustive culmination into *identical* global histories of the same-pattern Sierpinski triangle... *not* into different pattens and histories, as might be expected from generic Manyworlds; all will constitute the recapitulation of the root wavefunction in memory; all will constitute the recollection of memory, *not* its obviation or erasure. However, if we assume a single unique history for the development of our Sierpinski triangle, through non-illusory quantum indeterminism funnelled per the three-phase developmental process into the expected global pattern, we again obtain the ontic reality of memory, given that, in the development of the expected pattern, what is being reconstituted and recalled *is* memory. In short, Manyworlds must fail in its anticipated nihilism against memory and information, simply because Manyworlds apagogically presupposes memory.

indeterminate terms can only serve to reduce noise *and* sharpen the saliency of 'materialised' information. In other words, quantum indeterminacy is not a 'problem' that requires a solution by resort to contrivances, or especially MWI. This fact renders void the utility of MWI to informational and mnemonic nihilism, simply because of its lack of utility and superfluous standing vis-à-vis quantum indeterminacy. In short, **MWI cannot lend itself to any form of nihilism against both information or abstract memory, simply because its *raison d'etre*, or the assumption that quantum indeterminacy is a problem that needs MWI to overcome it, is nullified, rendering MWI itself nullified.**

To augment the case against MWI from our own Alt-R solution to quantum indeterminacy, we again make use of our old friend, the Sierpinski triangle: We treat the bounded probability structure and construction rules for the Sierpinski triangle as a root wavefunction in memory, and we utilise a quantum mechanical dice-shaker to generate the seed-orbits. Of course, in the Manyworlds treatment of the triangle, we assume the brute-force exhaustive developmental process for *all* possible seed-orbits, forgoing quantum indeterminacy. Therein, the mass of near-countless histories must converge to a *same* culminating global outcome: i.e., the same Sierpinski fractal pattern in *all* histories, as shown in the modified reprise-illustration in **Fig. 3.23**. **What, then, is the point or value of MWI and of its purported 'solution' to or obviation of quantum indeterminacy, when all histories culminate into identical histories, and into the same culminating history that would be obtained without Manyworlds, assuming we need to resort to MWI in the first place?** Thus, we could easily assume a uni-history with the same global culmination, but without the redundant identical histories... *without* Manyworlds. Why not simply submit to one unique history with ontically real objective quantum indeterminacy, *without* resort to Manyworlds and without totally superfluous same-global outcomes and histories? Indeed, this was the critique of Manyworlds furnished in essay **3.17**.

Of course, in place of the Sierpinski triangle, we might choose the complex root wavefunction for a two-slit experiment with **both slits open**. If we apply the Manyworlds treatment to the two-slit experiment, we will *again* obtain the same global outcome (the same global interference pattern) in *all* the Manyworlds histories: i.e., a *same history* reduplicated potentially into infinitely many. But why not abide with one history, with one outcome, without the reduplicated waste, *and* retain objective quantum indeterminacy? Again, MWI proves superfluous.

Recall that, according to our Little Manyworlds (LMW) replacement of MWI in essay **3.18**, there *are* systems wherein a single quantum indeterminate incident could generate critically different global outcomes, such as in Schrödinger's cat; such as in the quantum spin-tripped nuclear detonator. In principle, both artificial and naturally occurring variants of LMW-pertinent conditions could produce true non-identical histories, assuming we overlook the problematic relation of both MWI and LMW to the principle of conservation, in both its modern pleonastic energy-based form and in its recuperated abstract form as conservation of memory cum consistent futures. Yet, **even if LMW were viable, we could not utilise it to obviate the objectivity of quantum indeterminism in every instance, nor annul the conclusions drawn from our three-phase developmental approach that asserts quantum indeterminacy is *not* a 'problem' that requires a solution.** Thus, both the objectivity of quantum indeterminacy and the three-phase developmental approach render both Manyworlds and our Little Manyworlds superfluous and unnecessary as obviators of indeterminacy and mnemonic information, even if it should turn out that LMW might enjoy *some* plausibility.

Of course, the obvious problem for both LMW an MWI is the implication from the root wavefunction: In many instances across nature and in experiment, the root wavefunction constitutes a wavefunction in memory. Per the new conservation principle pertinent to memory, **Manyworlds, or even LMW, could not erase that memory, and cannot help but develop that root wavefunction in memory into identical Manyworld culminations that constitute the reconstitution of that memory, as must LMW in a more limited form. In other words, it is obvious that MWI presupposes memory, constituted as its root wavefunction... and it will recuperated that memory in *all* its purported histories, baring Wheeler's cat. As such, Manyworlds ceases to be useful to mnemonic nihilism, or no more useful than informational nihilism on its own.**

In short, MWI cannot obviate memory, or argue for its nihilism, even ignoring the other apagogic arguments in Book-III, given its reliance upon really-existing memory at the root wavefunction level, as the root wavefunction itself.

This concludes the apagogic argument 4-a against Manyworlds.

3.38: Apagogic Mnemonic Argument 4-b:
The Final Apagogic Case Against Memory Erasure Hypothesis

Now that we can grasp the superfluency of Manyworlds, we come full circle to the original contention for memory erasure. Recall that, in the anticipated memory erasure hypothesis **(3.06)** quantum indeterminism, supposedly inimical to information and scrambling of it, ought to obviate the very possibility of the re-materialisation of information at successive event horizons: That is, the **both slits open** setup ought not to be possible to reconstitute at the end of experimental isolation, and it ought to be scrambled by quantum indeterminacy to some other barrier setting, or some other identity (a pumpkin?) or into an amorphous white noise… or even into nothing. **The erasure hypothesis for the nihilism of information and memory assumed that randomness and quantum indeterminacy could only constitute noise, and culminate into noise. This was shown to be erroneous in essay 3.26.**

Thus, from our three-phase developmental approach involving objective quantum indeterminacy, abstract information, specifically in the form of a root wavefunction in memory (in this case, the abstract state for **both slits open**) will always fate *en masse* terms into globally coherent outcomes, maintaining both the persistence and perdurance of identities and states over time, including that of **both slit open** itself. Per our three-phase developmental process, quantum indeterminacy is a friend to information and memory. As such, quantum indeterminacy cannot scramble or erase information or memory, but can only re-materialise it in the noise-reduction pattern-saliency phase of development.

Thus, the root wavefunctions in memory pertinent to **both slits open** is *not* subject to quantum indeterminate scrambling or erasure. Indeed, **the memory, or the information pertaining to *both slits open*, is vested in a wavefunction in memory of initial conditions. This wavefunction in memory is *not* erased, neither by the experimental isolation process nor by subsequent quantum indeterminate processes at the end of isolation… and *both slits open*, together with the consistent outcome on the detector, will be brought into recuperation, baring the mischievous 'Wheeler's cat'.**

Thus, the presumption of memory erasure by the action of quantum indeterminacy is false, and the basic assumptions upon which the original anticipatory memory erasure hypothesis was constructed must fall apart.

Hence, both the ontology of information *and* abstract memory abide. It also follows that the assumption of the implicit amnesic universe is invalidated. The mnemonic universe is at hand.

This completes Alt-Rs' apagogic mnemonic arguments in favour of the real ontology of information and abstract memory.

3.39: Implications to Mind, Memory & Brains

Materialism asserts that memories are stored in brains in the form of 'material trace-states', and that mind and consciousness is produced by, and reduced to, the events and processes of brains. These assumptions cannot abide when confronted by the key findings in Book-III.

QUANTUM MECHANICS AND MIND

For brevity, we will refrain from reference to noetic information first espoused in Book-I, essay: **1.54**. Even so, using our growing block universe intermediate spacetime model, the materialist contention necessitates that the pertinent information be defined solely in terms of OR-form states that are restricted to the event horizon pertinent to brains. But what holds true across nature must equally hold true for the brain: In the tripartite structure of intermediate spacetime, we discover that physical information in general is *not* constituted as 'material trace-states' on the event horizon, but principally as minimal abstract wavefunctions in memory, dissociated from the event horizon and retained within the domain of realised spacetime. Again, **memory in general is always 'located' extant the event horizon, 'below' the event horizon, dualistic and irreducible with respect to the event horizon. Obviously, memories supposedly reduced to and pertinent to brains are not going to be exempt from this truism. Therefore, radical implications from noetic information aside, the actual form that memory pertinent to brains *must* assume is as abstract wavefunctions in memory, extant and dissociated vis-à-vis the OR-form brain restricted to the event horizon.**

All of this inexorably leads to mind-brain dualism and Cartesian revival: **Mind must 'reach out' to the dissociated abstract domain of memory below the event horizon... into the complex of wavefunctions in memory therein... in order to select, reconstitute and recollect these into OR-form expressions at successive event horizons via 'materialised' brains. In this way, mind must become extant the brain or *extant* the event horizon. Hence, mind-brain dualism is the inexorable conclusion**.

Also, given anticipated de-spatialisation, the mind must reach out into the past constituted as abstract memory *across time... not* across space. Mind must temporally relate to both brains vis-à-vis the event horizon *and* to the abstract memory extant and dissociated within the domain of memory. A mind that can relate across time to memory removed in time, and in a purely non-spatial non-contiguous way, is *necessarily* extant both the brain on the event horizon *and* the memory dissociated from it residing in the abstract past. **That is, mind must transcend both the brain and the abstract system of past-in-memory in order to recollect such memory**.

*

One other critical point to consider is the relation of objective quantum indeterminism to Cartesian mind-brain control: How does mind overcome quantum indeterminism in order to affect control over its brain? Mind gets to express its memories in and through brains through the three-phase developmental process and *en masse* objective quantum indeterminism. Although the full account of this must await Book-V, ontically real objective quantum indeterminacy *and* the three-phase developmental process helps underscore how both first-time novel expressions of the mind *and* recollective expressions of memory, or declarations of memory, get to form within brains, without crashing into presumed information-scrambling quantum indeterminacy. Thus**, our Alt-R based solution to quantum indeterminacy completely circumvents and overcomes the only serious threat to Cartesian mind-brain control, by showing that quantum indeterminacy could never obviate such a control, given that mind operates and controls its brain, not by orchestrating its quantum indeterminate terms, but by imposing a complex wavefunction on the brain. The imposed complex wavefunctions then AND-to-OR collapses and funnel *en masse* objective quantum indeterminate processes in brains to an inexorable global outcome sought by the mind**.

Briefly, Cartesian mind can reach into and 'draw out' brain-pertinent wavefunctions in memory from realised spacetime, or even draw out potentialities from the future. Mind then imposes these on the brain. The brain then spontaneously develops the selected wavefunction in memory, or the selected future potentiality, in the same way that a photographic image 'materialises' through and out of *en masse* quantum indeterminate terms through the three-phase developmental process, culminating into a noise-reduction pattern-saliency signal-distribution: i.e., the expression or declaration of memory or will.

Also, in Book-V, memory-retrieval, and noetic imposition of attendant wavefunctions on the event horizon and on the brain therein, will be shown to unfold **through a work function zero process that exploits the work function zero character of AND-to-OR collapse or time itself.**

The great advantage of our summarised approach to mind-brain memory-expression is that it completely circumvents the need for mind to act in lieu of hidden gears: **Our Alt-R explanation obviates the need for mind to deterministically orchestrate *en masse* quantum indeterminate terms into the required macro-scale sought outcomes of brain activity**, so rendering some quantum mind based theoretic speculations unnecessary.

3.40: Summary of Book-III, Part-II

The following findings from Part-II relate to the solution to the problem of quantum indeterminism and the attendant stochastic coherence paradox. The solution brings our pursuit of the definition of information to near completion. Indeed, the solution to quantum indeterminism required that we overturn our didactic working definition of information as exclusively OR-form, and expand it to a new level of abstraction. Yet, the core of the original OR-form based information-matter dualism abides, in that OR-form states cannot be retained in or carried by AND-form states.

The following is the summary of our expanded definition of information furnished in Part-II: Thus...

- **Minimal abstract states versus maximal materialised states of information, and implications to information-matter dualism (3.22)**: *Minimal abstract states* constitute fundamental information, but not in OR-form terms: These are stripped down abstract ordinances from which OR-form 'materialisations' of information are generated. Some are read-only (e.g., physical laws) while others are read-write future-potentialities in spacetime in potentiality or past memory-states from realised spacetime. The *en masse* AND-to-OR resolutions of quantum indeterminate terms into fully 'materialised' structures on the event horizon constitute *maximal materialised states of information*, which are *not* carried by, but resolve out of AND-to-OR collapse of AND-form minimal abstract-states. However, whereas minimal abstract states within spacetime in potentiality constitute wavefunctions with nested futures, abstract memory retained within realised spacetime constitute past wavefunctions; or wavefunctions with their former nested futures permanently 'cropped' or erased. Both types of minimal abstract states are AND-form. Insofar as memory is retained 'below' the event horizon within realised spacetime, it is always physically dissociated and dualistic in relation to its event horizon-restricted 'materialised' OR-form states: In short, information matter dualism yet abides. Since the same memory is also dissociated from future-form abstractions within spacetime in potentiality, it follows that the abstract past is *not* retained in, carried, or transported by the abstract future: Again, information matter dualism abides.

- **Wavefunctions in future-potentiality** and **wavefunctions in memory** (3.22): Of course, minimal abstract states within spacetime in potentiality (the future) constitute complex *instantaneous wholes* (first pondered in Book-I **1.03**), formed out of or constituted as *wavefunctions in future potentiality*: These possess nested futures that elaborate to the infinitely removed future, and, through AND-to-OR collapse, decay into global OR-form outcomes; materialising through *en masse* quantum indeterminate terms on the event horizon. On the other hand, abstract states within realised spacetime (the past) constitute instantaneous wholes formed out of *wavefunctions in memory*: their former nested futures 'cropped' or erased. Again, these collapse or decay into OR-form outcomes on the event horizon. There is an interplay or superposition of wavefunctions in memory with wavefunctions in potentiality: a *super-superposition* culminating into the grand decoherence of spacetime in potentiality based on counterfactuals and the erasure of non-consistent futures (see Book-II **2.34**) decohering into a *wavefunction Lagrange* that unites memory and potentiality, but which then AND-to-OR collapses into *en masse* quantum indeterminate terms on the event horizon through a three-phase developmental process (see below)... into 'materialised' OR-form manifestations of the original abstract instantaneous whole.

- **The three-phase developmental process** (3.26) constitutes the developmental solution to quantum indeterminism (one of the two component solutions) entailed in the collapse of minimal abstract states grasped as wavefunctions in memory or wavefunctions in potentiality. To appreciate the solution to the problem posed by quantum indeterminism in the context of the three-phase developmental approach, it is necessary to forgo highly misleading views on quantum indeterminism based on limiting cases: such as EPR or the Schrödinger's cat (**3.25**). Instead, we must look to situations in which quantum indeterminacy operates *en masse* in the framework of larger instantaneous wholes formed out of complex wavefunctions in memory or potentiality, such as evinced in photography. The three-phases of development are...
  - **Below-threshold noise phase**, in which the quantum indeterminate terms generated do not exhibit any definitive distribution, signal or pattern. Collectively, the samples appear perfectly scrambled into a haphazard distribution of outcomes: a *noise distribution*. Our common prejudgements about randomness and chance derive from views garnered from this below-threshold developmental phase.
  - **Transitional chaotic phase**: When a sufficient or above-threshold number of random or quantum indeterminate terms arise, we observe the emergence of pattern; albeit distorted and unstable: not yet formed into full stable saliency. Each new succession or addition of quantum indeterminate terms will change the nascent pattern in unstable and *chaotic* ways. Hence, an evolving *chaotic distribution*. However, as the *marginal bandwidth of randomness* narrows (as the ratio between the accumulated samples versus the subsequent samples enlarges, and subsequent additions reduce to a perpetually smaller fraction of the accumulated total) this gives way to stability.
  - **Noise-reduction pattern-saliency phase**: The further addition of quantum indeterminate terms will bring the system out of its transitional chaotic phase into its full *noise reduction or pattern-saliency phase*; wherein hitherto hidden abstract form and pattern, or 'instantaneous whole', hence the information initially concealed as invisible minimal abstract state of wavefunctions in potentiality or memory, finally obtains saliency into *signal distribution*. The attendant marginal bandwidth of randomness will be so narrow that the addition of subsequent random terms will have no obvious perturbing effects on the developed pattern, nor engender a noise distribution. As the marginal bandwidth of randomness closes to and approaches the infinitesimally small, additional randomness can only further sharpen and define the now fully developed and fully 'materialised' form. Thus, in this developmental phase, randomness assumes, not merely the role of a noise reducer, but acts as a *signal magnifier:* a de-scrambler of noise: a friend to the emergence and saliency of information, order, and pattern. The noise-reduction role of randomness and quantum indeterminacy becomes obvious and can be generalised to nature as whole. The questions, "...How can quantum noise and 'accident' generate a coherent and highly structured world?" ...is rendered void.

The above three-fold process of development can be applied to the quantum mechanised Sierpinski triangle using quantum mechanised dice (3.27); to the naturally occurring quantum mechanised **dz2 electron orbitals** (or any other electronic configuration) (3.28); to the development of a photographic image out of *en masse* indeterminately resolved photon depositions (3.29); to the time-evolution of the drifting snowflake; to biological evolution and to cosmic evolution... indeed, to all pattern and form in the universe, of any complexity and scale... including the brain.

- **The ontological primacy of form over subordinate natural odds** (3.24). Given the fact of quantum indeterminacy, it is tempting to dismiss the world as merely the outcome of averages obtained from the odds (i.e., the universe as 'accident') and ignore the fact that it is *form that structures and renders the odds, while form is not itself conjured by the odds*. For example, in the electronic configuration of the atom, the standing waves that constitute **dz2** orbitals can only form up viably at certain radii, but *not* at any radii. That is, the form of **dz2** is born out of necessity: The form structures the odds consistent to its abstract pattern, *not* the other way round. That is, the odds do not make the form, but it is the form that casts and structures the odds that, in this instance, then govern otherwise quantum indeterminate resolutions of electrons about the atom. Thus, the generic want to resort to the odds (or to happenstance and to 'accident') to explain away the manifest and undeniable fact of order in nature ignores the existence of undeniable non-arbitrary form, such as exemplified by **dz2**: an order comprised of *ordinances* that fate otherwise *en masse* quantum indeterminate terms into wholly fated global outcomes, harmonising determinism and indeterminism. In **3.14** and **3.15**, we discovered how this implied the real ontology of a meta-semantic hypostasis comprised of timeless read-only Platonic-Cartesian states (such as **dz2** and the conservation of spin) that are extant time and change, even if these prove to be time-asymmetric (see Book-I: **1.12**). The assertion of the primacy of form over the odds recapitulates the argument for 'instantaneous wholes' first presented in Book-I (essay **1.03**) as part of the argument against mereological nihilism, and subsequently against generic emergence theory of mind. Of course, the ultimate origin or ontology of ordinances and forms constitutes a great mystery, the appreciation of which can only be enhanced by Euthyphro's dilemma: (for clarification, see essay **3.14**).

- **The ontological reality and objectivity of randomness and quantum indeterminism** (**3.32**). The quantum indeterminate terms funnelled into determinate global outcomes through the three-phase developmental process are provably objectively indeterminate: Indeterminacy and randomness are ontically real. i.e., there are no hidden gears behind non-illusory quantum indeterminism and randomness. Indeed, the want to found hidden deterministic orchestration of apparent facile quantum indeterminate processes must fail, insofar as this cannot obviate the final resort to random selection and indeterminate sortology, and will crash into an insurmountable non-terminating combinatorial explosion in the attempt to circumvent resort to random selection. The want to found deterministic orchestrating scripts or 'hidden gears' can certainly generate very large numbers of deterministic orchestrating scripts, all apt to the secret orchestration of surface quantum indeterminacy. But the method cannot generate a one-and-only exclusive script. The equi-possibility ratio of any one deterministic script over all other alternative equally apt scripts cannot be subject to an unambiguous selection and decision that favours just one script. We cannot resort to either a qualitative basis or a structural deterministic one to obtain qualitative or structural selection from among the vast number of equally apt scripts. We must instead resort to random selection out of the very large number of equally apt scripts. Thus, resort to randomness re-emerges, even when we imagine the possibility of hidden deterministically orchestrating scripts or 'hidden gears', *and* compute these. The subsequent attempt to break out of this dead-end crashes into an insurmountable combinatorial explosion of scripts, and attendant halting problems. In short, randomness and quantum indeterminacy are ineliminable, inexorable, and finally ontically and objectively real: Randomness and indeterminacy are not illusory, but *real*. Hence, there are no hidden gears.

- **The four-fold classification of physical information** (**3.35**). The original classifications of information in Book-I, reconsidered in the light of the three-phase development process (**3.26**), combined to the ontic reality and objectivity of quantum indeterminism (**3.32**), integrated into the tripartite structure of growing block intermediate model spacetime developed in Book-II, culminates into a new four-fold classification of physical information. These are…
  - **read-only future-form <u>wavefunctions in future-potentiality</u>** embedded within spacetime in potentiality (the growing block future, with nested futures), imposed on the event horizon; and which contain emanating physical laws and structure-patterns that arise from necessity and ordinance. These are dualistic-irreducible to, hence extant, the 'material world' restricted to the event horizon. The read-only future-forms emanate from a meta-semantic ontology of ordinances that imbue spacetime in potentiality: the meta-semantic hypostasis is 'located' at the infinitely removed future and is comprised of Platonic-Cartesian physical laws and forms (see **3.33**).
  - **read-write past-form <u>wavefunctions in memory</u>** embedded within realised spacetime (memories of the growing block past). These are projected and enforced on the event horizon from realised spacetime and are comprised of past wavefunctions that had 'collapsed', with their former nested futures permanently 'cropped' or erased. These are the abstract memories sought in the apagogic arguments in Part-I. Retained within and constituting the domain of realised spacetime, wavefunctions in memory are extant and dualistic-irreducible with respect to the 'material world' restricted to the event horizon.
  - **read-write future-form <u>contingent wavefunctions in future-potentiality</u>** (**3.34**); also nested within spacetime in potentiality. These comprise nested futures and contingent potential overrides of the assertions of memory upon the event horizon: e.g., analogised by *Wheeler's cat* (the metaphorical stand-in for **switch-x** featured in delayed choice experiments) who overturns the two-slit experiment and alters the original setting of the barrier therein, thus overriding the stipulations of memory upon the future, without erasing that memory. This exemplifies contingent potentialities within spacetime in potentiality. The nested futures perspective from Book-II (Part-II) best encapsulates the physics of this form of potential information, especially in its convergent category nested futures framework.

349

- ○ **read-write** <u>Lagrange OR-form information</u> **at the event horizon (3.35)** constituting the maximal 'materialised' state of information formed via the AND-to-OR collapse of the super-superpose formed out of wavefunctions in memory combined to wavefunctions in potentiality (and contingency): so forming a Lagrange complex wavefunction that then collapses into OR-form 'materialised' being on the event horizon, appearing as 'materialised form in space'.

- **Addendum to decoupling-ii (3.36)**. The putative quantum tunnelled 'displacement' of Mozart-40 at purported 4.7 times faster-than-light in the experiments of Gunter Nimtz is brought to further sensibility by our expansion of the definition of information beyond OR-form Lagrange state confined to the event horizon. The contribution of Book-III to decoupling theory involves grasping that the critical form of memory as it pertains to M40 is constituted as an abstract wavefunction in memory retained within realised spacetime; extant the event horizon-restricted 'material world'. It turns out that w*avefunctions in memory are as spatially static and as non-undulating as wavefunctions in potentiality,* as was asserted in Book-II (**2.23** to **2.26**). Hence, the 'transmission' of M40, or the wavefunction in memory pertinent to M40, cannot entail a spatial displacement of that wavefunction in memory, either at the 'speed of light', and even at 4.7 times 'faster-than-light'. Instead of motion, we must look to characterise the 'transmission' and subsequent re-constitution of M40 in pure delayed choice time-interval terms: In any case, this approach will be forced upon us by de-spatialisation in Book-IV. Yet, this requires the incorporation of 'faster-than-light' effects from Nimtz experiments: (See scenarios and *maincone theory*: Book-I, **1.41**). Indeed, therein, maincone theory turns out to be memory theory in alternative form. *The memory of M40, constituted as a wavefunction in memory, subjects the totality of spacetime in potentiality, including the part of it within the maincone, to a future-form state consistent with past M40.* In this way M40 projects onto any subsequent event horizon, regardless of the presumption of a 'speed' to light, or even 'faster-than-light'. *Thereafter, the event horizon cannot help but reconstitute M40,* realised through the expected three-phase development process from **3.26**... *but without any 'motion' of M40, much less at the purported faster-than-light.*

- **Apagogic mnemonic argument 4-a (3.37)** constitutes the penultimate mnemonic argument for the ontic reality of abstract memory, thus completing the mission of Part-II. It is based on the solution to quantum indeterminism and the stochastic coherence paradox secured in Part-II which directly usurps the Manyworlds interpretation (MWI). *While MWI purports to obviate quantum indeterminism by realising all possibilities in the AND-form wavefunction, all the histories so formed must yet culminate into the same global outcome. This obviates the claimed usefulness of MWI. In any case,* <u>the root wavefunction involved is invariably a wavefunction in memory,</u> *given that most wavefunctions in nature constitute memories. Even if subject to putative brute-force exhaustive processing,* <u>the root wavefunction in memory can only culminate into both the utilisation (not the obviation) of that memory,</u> *and* <u>the materialised recollection of that memory in the form of identical global culminations and histories: a feat that could be achieved without Manyworlds.</u> *It follows that Manyworlds cannot be used to erase the root wavefunction in memory, much less constitute a basis for mereological and general informational and mnemonic nihilism.*

- **Apagogic mnemonic argument 4-b (3.38)**: The last of the apagogic arguments in support of abstract memory. The original memory erasure hypothesis (**3.06**) presumed that both the interval of isolation *and* subsequent quantum indeterminism must erase information and memory. Yet, <u>isolation does not erase the root wavefunction in memory pertaining to, say, the 'both slits open' condition.</u> This is retained in abstract form within realised spacetime. Finally, <u>the quantum indeterminate resolution of the system at the end of isolation reconstitutes the 'both slits open' condition by re-developing the root wavefunction in memory that pertains to it through the three-phase developmental process that funnels otherwise objective quantum indeterminate terms into the recollection and 're-materialisation' of 'both slits open' setup.</u> Thus, the initial assumptions in the original erasure hypothesis, to the effect that both isolation and especially quantum indeterminacy must wash out and scramble information and memory, are shown to be false.

- **Cartesian memory and mind (3.39)**: Given the tripartite structure of growing block spacetime, we discovered that information is retained throughout nature at a remove from 'materialised' Lagrange OR-form states restricted to the event horizon. In nature, memory is not constituted as 'material trace-states' on the event horizon, but as abstract wavefunctions in memory within the domain of realised spacetime, dissociated from the event horizon. Memory pertinent to brains, treated as if physical memory, is not going to be exempt from this. <u>Memories pertinent to brains ultimately constitute as wavefunctions in memory dissociated and extant from the brain,</u> as is true of all memory in nature. This implies the dualism of memory in relation to the brain: Hence, recollection requires a Cartesian mind that can 'reach out' to memories across time to the temporally dissociated past within realised spacetime, and re-impose these on the brain at successive event horizons. This implies tandem mind-brain dissociation. Also, <u>the ontic reality of objective quantum indeterminism, co-joined to the three-phase developmental process, harmonises the Cartesian mind with 'scrambling' quantum indeterminism, obviating quantum mind theories that posit mind and consciousness as secret orchestrators of quantum indeterminism in brain</u>s. Mind draws out wavefunctions in memory from realised spacetime and imposes these on the brain at succeeding event horizons. The brain then spontaneously develops the mind-imposed wavefunction in memory in the same way that a complex wavefunction pertaining to a photographic image spontaneously 'materialises' out of *en masse* quantum indeterminate terms, through our three-phase development, *without* secret orchestration by hidden gears or by mind, wherein objective quantum indeterminism is no more an enemy of ordered brain activity than it is in enmity with the formation of an ordered photographic image.

BOOK-III: CONCLUSIONS & INNOVATIONS

Book-III set itself three main goals: First, an advance toward the completion of the definition of information, and second, the solution to the problem of quantum indeterminacy and randomness. The latter segued into the definition of information. A third goal consisted in the apagogic semantic-physical argumentation for the reality of abstract memory, *against* information nihilism and memory nihilism. This also made the case in favour of the mnemonic universe in contrast and in opposition to the belief in the implicit amnesic universe. All three goals proved indispensable to our Alternative Realist philosophy of physics and quantum mechanics, even though the apagogic vindications of abstract memory must fall short of a more blatant physical proof of abstract memory, which must await de-spatialisation and temporisation in Book-IV.

Starting with quantum indeterminacy, **Book-III offered two independent but mutual solutions that show quantum indeterminism to be ontically real *and* objective...** to the effect that there are no hidden gears that secretly deterministically orchestrate quantum indeterminism. It also established **that the three-phase developmental approach to randomness and quantum indeterminacy unravelled our prejudices about indeterminism as 'chaos' or 'noise'**. The latter showed that objective randomness and indeterminism are *not* in enmity with information, memory, order and pattern; and that, on the strict provision that we grasp information as either wavefunctions in memory or as wavefunctions in potentiality, and refrain from defining information exclusively in OR-form terms, abstract instantaneous wholes constituting memory and future-potentiality cannot help but funnel *en masse* objective quantum indeterminate aggregates into ordained and fated outcomes and mereologies on the event horizon.

Throughout, the tripartite structure of intermediate spacetime and growing-block universe (developed in Book-II) constituted the unifying framework, scaffolding for the interplay and integration of abstract information grasped as abstract memory or future-potentiality; scaffolding for the AND-to-OR collapse process (time) which 'materialises' these upon pertinent event horizons; scaffolding for the three-phase developmental process that fates ontically objective quantum indeterminate terms into coherent 'materialisations' of the said abstractions on the event horizon, either out of realised spacetime or out of spacetime in potentiality.

In a series of apagogic mnemonic arguments, Book-III also sought to render the case for abstract memory beyond its mere supposition or common-sense claim. The series of apagogic mnemonic arguments showed that anticipated hypothetical erasure of information and memory (such as the erasure of the initial conditions pertinent to the barrier setup in the two-slit experiment) otherwise expected from the conjunction of both AND-form quantum mechanisation during isolation (which cannot carry or transport OR-form information) *and* from the presumed information-scrambling per quantum indeterminacy at the end of isolation, ought to erase information and memory, or prohibit any possibility of the resonance, endurance and perdurance of past initial conditions and events through to the future.

Of course, we can now appreciate that **quantum indeterminacy is not only objective, but it does not at all scramble information: It is a friend of information per the three-phase developmental process**, and fates quantum indeterminate terms into signal distributions. In most of our apagogic mnemonic arguments, we did not entertain this truism. Instead, we supposed the 'reality' of information-scrambling by quantum indeterminacy... and attendant amnesic universe thesis... and the attendant inexorable-seeming informational and memory nihilism, all made worse by the co-joining of memory erasure hypothesis with the Manyworlds interpretation.

But memory erasure hypothesis cannot utilise the now-false notion of information-scrambling quantum indeterminacy, given that quantum indeterminacy ultimately proves to be a noise-reducing signal-amplifier, *not* an eraser of information or memory. For its part, Manyworlds cannot utilise its brute-force exhaustion mode of operation to obviate information and memory, given that...

- Quantum indeterminism is objectively real and ineliminable, rendering the utility of Manyworlds superfluous per its professed claim to obviate and 'solve' the problem of quantum indeterminism... (quantum indeterminism is not a 'problem' that needs a solution, as the three-phase developmental approach amply demonstrates).

- Manyworlds does not lead to alternative histories, but only to the replication of a same global outcome in all its purported histories, hence into a mass (or even an infinite number) of same histories.

- Manyworlds cannot obviate memory, given that the root wavefunction which its exhausts brute force into identical histories *is* a wavefunction in memory, and all the identical histories that Manyworlds generates constitute same-recollections of that same memory.

Hence, Book-III solved the problem of information and memory nihilism by showing that abstract memory is indispensable to the very possibility of quantum mechanics as we know it, and by demonstrating that both memory erasure *and* MWI, which might be employed in anticipatory information-memory nihilism, are fundamentally erroneous.

*

There are **two main innovations furnished by Book-III: The first is the apagogic proof of abstract memory**, short of direct proof from anticipated de-spatialisation (promised in Book-IV). **The second innovation is the non-illusory ontic reality and objectivity of randomness and quantum indeterminism, co-joined to the three-phase developmental process**, with the stochastic coherence paradox rendered void... wherein both culminate into the penultimate completion of the definition of information first started in Book-I, resolved into a basic four-fold definition set in **3.35.**

The **derivative innovations** of Book-III were...

- **Falsification of MWI**
- **Harmonisation of randomness and indeterminism vis-a-vis determinism and order.**
- **Apagogic argument for an ontic meta-semantic hypostasis**… one exceeding and subsuming the tripartite growing-block intermediate spacetime structure, and transcending and even subsuming mind and consciousness, beyond the caricature God conceived as super-conflated mind.
- **The non-arbitrariness thesis** against general informational nihilism, given the apagogic argument for the meta-semantic hypostasis, and implied anticipatory **meaningful universe thesis**.
- **Emanation theory of physical laws from the infinitely removed future and from the meta-semantic hypostasis**: succinctly, Platonic-Cartesian physical law.
- **Projection theory of abstract memory from realised spacetime** upon the leading event horizon, through grand decoherence via the super-superpose of memory and potentiality.

Meta-semantics prove to be fundamental, both against nihilism in general (including informational and memory nihilism) and as the springboard for an anticipatory meaningful universe thesis, against the grain of dominant secular-nihilistic *and* theo-nihilistic ideology: an issue that requires due treatment in an entirely separate set of books, but intimated throughout the Preliminary and in the five books of the present corpus.

Of course, **all of the innovations stated above prove critical to Cartesian revival**: Indeed, abstract memory alone, especially when grasped in terms of intermediate spacetime structure, with the 'materialised' brain restricted only to the event horizon, and with both memory and future-potentiality pertinent to brains now rendered extant the brain, forces a conception of memory-retrieval and recollection by mind that must reside extant the event horizon and its brain.

BOOK-IV:
THE BERGSON-WHITEHEAD AMALGAM:
MEMORY, GRAVITATION AND CONSOLIDATIONS

CONTENTS

BOOK-IV:
THE BERGSON-WHITEHEAD AMALGAM:
MEMORY, GRAVITATION AND CONSOLIDATIONS

GENERAL INTRODUCTION

The core aim of Book-IV will be the **supersession of the intermediate spacetime model from Book-II by the superlative Bergson-Whitehead amalgam**: 'spacetime' *without* space. The intermediate model in Book-II asserted the reality of abstract memory, but only obliquely: identifying it with the domain of realised spacetime. On the other hand, the series of apagogic mnemonic arguments furnished in Book-III espoused the reality of abstract memory through an admixture of physical *and* semantic apagogic argumentation. In Book-IV, **the anticipated Bergson-Whitehead amalgam will furnish the case for abstract memory directly and seamlessly. It will do so through the several proofs of de-spatialisation and temporisation.**

De-spatialisation removes space from the lexicon and ontology of physics and asserts its pleonastic status. It permanently undermines notional 'matter in space' culminating into **the unravelling of contiguous impact causality**: De-spatialisation also compels for a purely temporised physics, constituted as abstract memory *and* future possibility, organised and distributed into delayed choice time-interval relations, all *without* 'space'. Consequently, the generic claim that memory and information constitutive of world and brains must be comprised of material trace-states, supposedly carried and embodied in and as spatio-material states, or as distributions and extensions in 'space', is permanently usurped.

Our abstract memory theory is almost identical to the form of memory Henri Bergson had conceived in the early twentieth century, espoused in his book, *Matter & Memory*. Bit Bergson did not question the ontics of space or avow de-spatialisation.

The attack on absolute time by Einstein ought to have culminated into de-spatialisation. Unfortunately, history took a convoluted turn: Einstein *et al* amalgamated space with time into 'spacetime'. Our Alternative Realist case for de-spatialisation and temporisation will begin with the implicit **synonymity that resides between space and defunct 'absolute time'**; the latter undermined by Albert Einstein in 1905. Absolute time is essentially the idea that all events and objects transpire in or within a unitary common present moment of total simultaneity. The synonymity between space and absolute time, and the first demise of both, will emerge out of **the problem of contiguity and extensionality through simultaneity**. (See 4.**01**).

Beyond the problem of extensionality through simultaneity, Book-IV will present additional reasons in support of de-spatialisation. Following an interlude into the economics of science and the consequent pleonastic status of space, Part-I (**4.04**) will delve into implications from empirical-observational strictures of quantum mechanics and physics vis-à-vis the empirical-evidential requirement to furnish proof of the motion of radiation and matter and the attendant space: **The evidential validation of motion and its attendant space requires seamless continuous observations of the radiation or the de Broglie matter across its putative spatial interval. But this is prohibited by quantum mechanics, notwithstanding even the Quantum Zeno effect.** Indeed, according to quantum mechanics, continuous observation is no more possible than continuous energy output from a radiating black body. The output from a black body is always and necessarily non-continuous, broken, discrete, or *quantised*. The same must inevitably hold with respect to the observation of radiation and de Broglie matter 'undergoing motion in space': Permissible observations could only be discontinuous and discrete, or broken-discontinuous, as foreshadowed in the Preliminary (**0.03**). Hence, the notion of the motion of radiation and matter constitutes an imposed *assumption*, as is the assumption of the 'space' across which either is purported to move: (see **4.02**).

Part-I will also argue that, **all existing empirical claims about the nature of radiation and matter can be saved even when re-framed in terms of observations and events organised and interrelated purely in time...** *without* **resort to space.** Thus, not only is the ontology of space beyond empirical proof, but it is ultimately unnecessary and superfluous, or pleonastic to physics: a foisted and imposed concept; ultimately *not* necessary to physics.

To this, Part-I will add the independent problem posed by Hilbert space: the generic quantum mechanical portrayal of space. Hilbert space assumes the ontic reality of space. It is through Hilbert space that radiation and de Broglie matter must undergo its purported motion. But, **at best, Hilbert space constitutes a future-potential space that has not yet come into formation... because it belongs to an as-yet not-realised future. But if it has not yet come into formation, how can anything move through it?** (see **4.05**). Indeed, insofar as the quantum mechanical wave *is* the future phenomenalised, and quantum mechanics is finally grasped as the physics of futures that attends the growing block model, and since any attempt at portraying a thing as 'moving across Hilbert space' is tantamount to describing it as displacing in and through a future that has not yet transpired, any forlorn empirical-observational proof of motion and space thus would culminate into the violation of the principle of causality and the entwined principle of conservation of energy-matter, as was asserted in the firewall principle from Book-II: **2.11**, insofar as any attempt to gain direct information about the future in proof of radiation or matter 'moving' through that future (Hilbert space) must constitute 'signals from the future', with said attendant usurping of the firewall principle. The insight can only contribute to de-spatialisation: (also see **4.06**).

Finally, Part-I will complete *John Wheeler's delayed choice approach to the question of space*. In Book-II, delayed choice asserted **the quantum wave as a spatially static, non-time-undulating non-displacing state. In which case, what need of the space across which motion or undulation does not need to transpire?** In Book-II, this conclusion was augmented and proved consistent with the nested futures view (**2.17** to **2.18**). In the delayed choice experiment, we found that the static character of the nested future waves nested to the hedging wave must also apply to the 'moving' hedging wave itself supposedly 'carrying' the nested future waves. But the hedging wave was treated as 'moving', and its status as such was questioned obliquely. In the meta-delayed choice approach in Part-I (**4.07**), we will embed the original delayed choice system, with its nested futures and 'carrying' hedging wave, into a subsuming preceding delayed choice system: In this way, we will nest delayed choice systems into larger embedding delayed choice systems. From this, the static nature of the original hedging wave will be rendered fully salient and undeniable: We will no longer be able to sustain any notion that the hedging wave, or *any* quantum wave, might be moving or time-undulating in a meta-delayed choice system elaborated to the indefinite past, or at least to the Big Bang. Hence, both the notion of the moving wave *and* space will unravel. Thus, in Book-IV, **the meta-delayed choice approach will directly attack and unravel space itself**.

Consequent de-spatialisation will compel for the comprehensive reframing of almost all physical information constitutive of the growing block past: the latter's reformation into a temporised abstract memory-system organised into pure delayed choice time-interval relations. This pure temporisation of the past, *without space*, will complete abstract memory theory espoused in Book-III, consonant to Bergson's view of memory and time. A brief integration and consolidation of de-spatialisation *and* abstract memory must lead to **the reformation of Einstein's Special Relativity, recuperated *without* space** (see **4.08**) recast purely in terms of delayed choice time-interval relations. The 'speed of light', together with uniform and non-uniform motion, will also be recuperated into their pure temporal forms and successions.

*

De-spatialisation will also compel the reformation of the 'field' in physics into its de-spatialised purely temporal form (4.09). Given de-spatialisation, and as was intimated in our intermediate model spacetime treatment of the magnetic field in   Book-II **2.37**, magnetic fields and quantum field, require that we no longer apprehend these as 'areas', 'spreads' or as spatial *extensa*, but, at least in their quantum mechanical forms, as projections in potential time that encapsulate future AND-form potentials for possible aligned iron filings, and as *instantaneous wholes* organised into potential delayed choice time-interval complexes: all projected into a futurity constituted as the Whitehead facet of the Bergson-Whitehead amalgam (formerly, spacetime in potentiality). With this last, Book-IV will arrive at the long-anticipated superlative **Bergson-Whitehead amalgam: the replacement to the intermediate spacetime model, and successor to spacetime in general.** (see **4.10**) Therein, the Whitehead facet will replace the domain of spacetime in potentiality; the Bergson facet will succeed the domain of realised spacetime; and the event horizon will abide in its usual form as the 'cut' between the domains. The primitive unification of quantum and classical-relativistic physics furnished by the intermediate model in Book-II will survive into the Bergson-Whitehead amalgam. But the Bergson-Whitehead amalgam will prove superior because it can accommodate de-spatialisation and temporisation explicitly, and, by doing so, secure the indispensable and critical basis to presage solutions to inertia and gravitation, and to other enduring mysteries.

*

In previous material and in Book-III, we moved beyond the notion of time as a 'fourth dimension': we finally apprehended time as synonymous with AND-to-OR wavefunction collapse: the process of 'coming into being'… synonymous with the partial-perpetual and broken-discontinuous progression of the event horizon into the future... into 'spacetime in potentiality'… into the now Whitehead facet of the Bergson-Whitehead amalgam. With de-spatialisation co-joined to the Bergson-Whitehead amalgam, we will take the next step: **the replacement of 'three dimensions of space' with *stereotemporal time*: i.e., with three-dimensional time: (see 4.10)**.

Book-IV will also integrate the new definitions of physical information that had emergent in Book-II and Book-III into the superlative Bergson-Whitehead amalgam: such as, *wavefunctions in memory* versus *wavefunctions in future-potentiality*. It will also integrate attendant solutions to the problem of quantum indeterminism into the processes of our growing block Bergson-Whitehead amalgam; including the three-phase developmental process co-joined to the ontic objectivity of quantum indeterminism… all recapitulated into the de-spatialised temporised form of the Bergson-Whitehead amalgam.

*

Book-IV will pay due homage to North Whitehead and Henri Bergson: North Whitehead had presaged the growing block approach and the quantum mechanical wave-description as the description and physics of futures. But Whitehead did not develop any causality-based argument to compel the necessity for his view, as we had done as part of the firewall principle in Book-II. For his part, Henri Bergson presaged abstract memory as information structured and distributed in time; consonant with our own dualistic account of information-matter relations. But Bergson never challenged the false notion of 'space'. With de-spatialisation, almost all relations of nature are immediately recuperated into relations of information and memory in time, notwithstanding the growing block futures… all of it recuperated *without* space. Yet, the very name we have given to the replacement to spacetime, namely, the superlative *Bergson-Whitehead amalgam*, stands as open acknowledgement of both Whitehead and Bergson, in recognition of their overlooked contributions, and in affirmation and incorporation of their essential intuitions and ideas, albeit with key modifications and additions from Alt-R. **Book-IV will consolidate the said homage in a *new process philosophy* that unifies the approaches of Whitehead *and* Bergson, but within the framework of innovations and advances furnished by Alt-R (see 4.12)**.

*

With de-spatialisation and the development of the pure temporised growing block Bergson-Whitehead amalgam, **Book-IV will reiterate the reform of basic energy relations from Book-II, but without resort to space, and per the structure of the superlative Bergson-Whitehead amalgam**. In a similar form as espoused in Book-III **2.39**, putative 'potential energy' will be shown to belong to the Whitehead facet; the future (formerly, spacetime in potentiality) while 'kinetic energy' will be shown to relate to the transition from

358

a preceding to a succeeding event horizon, entailing the conversion of potentiality and potential energy into kinetic actuality. But, as was shown in Book-II, **we will again abstract out and supersede energy concepts as we tie these to ontologically primary and subsuming process of non-energy and non-gears driven AND-to-OR collapse, which is synonymous with time itself. In so doing, we will again render energy concepts as pleonastic as space itself**.

*

If space is pleonastic, then physics cannot speak of 'position', let alone espouse about 'uncertainty in position' in Heisenberg's uncertainty. **Book-IV will save Heisenberg's uncertainty by displacing it with the well know time-based variant of the same: namely, Heisenberg's energy-time uncertainty**, notwithstanding the pleonastics of energy concepts. Energy-time uncertainty will prove to be in perfect accord with de-spatialisation and temporisation: integrating energy-time uncertainty and now-pleonastic energy relations to the de-spatialised, and to the purely temporised tripartite structure of the Bergson-Whitehead amalgam. The integration of pleonastic energy in energy-time uncertainty will emerge from the abstraction of energy concepts from the frequency and its attending 'probability of futures', none of which require energy concepts. Yet, the essence of energy-time uncertainty will abide.

This will be followed by **the integration of memory and entropy relations to the de-spatialised and temporised processes of the Bergson-Whitehead amalgam; culminating into the perfection of the concepts of** *entropy of future potential information* **and the** *entropy of memory* first espoused in Book-I (**1.07** and **1.08**). To this end we will reiterate the non-statistical objective basis of entropy of future-potential information and the 'arrow of time' now synonymous with the AND-to-OR polarity of wavefunction collapse, co-joined to the impossibility of OR-to-AND inverse-collapse attendant the progression of the event horizon in the decay of the Whitehead future into the Bergsonian past. The implied accumulation of abstract memory through the expansion of the Bergson facet (formerly, 'realised spacetime') will be show to be consistent with entropy of information: as had been presaged in Book-I. Thus, while the past accumulates and total memory increases, it is progressively weakened over time by the *inverse square law of time*, so constituting the entropy of memory; part of a more fundamental form of entropy compared to generic statistics-based entropy.

*

Given de-spatialisation and the tandem development of the Bergson-Whitehead amalgam, combined to the attendant reform of the physical field, of energy concepts, of entropy, and of Heisenberg's uncertainty, **Part-II will seek to model physical interactions and forces** *without* **resort to space: recuperated purely in terms of delayed choice time intervals... as the temporal super-superpose of memory and future-possibility**. Therein, Whiteheadian expression must necessarily constitute expressions *only* about future potentials of as-yet-not-realised events; all structured into potential delayed choice time-relations. On the other hand, the Bergsonian form assumed by events and relations that had transpired to form the past must also be remodelled as realised past events incorporated into attendant realised delayed choice time-interval relations and temporal distributions; all retained as *wavefunctions in memory* within the Bergson facet of the Bergson-Whitehead amalgam. Insofar as this is to be accomplished without space, physics must transit out of contiguous or contact-based notion of 'interactions of matter in space' into a physics recapitulated purely in terms of the non-contiguous temporal resonance of the past upon the event horizon. This will culminate into **a physics of memory: one that completely transcends materialistic 'matter-trace', contiguity or impact-causality based approaches to physical interaction and to information and memory: a presage to Bergsonian mnemonic mechanics** that will take us beyond defunct spatial impact and contact causality, superseding 'matter in space' with *memory in time*. To this end, our first aim will be to establish **the five basic postulates of mnemonic mechanics**. These will include the inverse square law of time: Given that 'distance' is superseded by time, the weakening of influence 'across space' is recuperated into the weakening of the same across time.

Key certainties aside, the five postulates, together with nascent mnemonic mechanics, will be proposed as prospective and conjectural. It is likely that the approach will be replete with errors. Yet, the need for a pure temporal physics of memory and information is inevitable, given the veracity of de-spatialisation. **De-spatialisation compels that** *all* **physical interactions, including those that attend inertia and especially gravitation,** *must* **be recuperated and modelled as non-contiguous resonations in time... as outcomes engendered by the non-contiguous summation and superpose of abstract memory,** *without* **resort to contiguity or space**. Our crude presage to mnemonic mechanics, though it might prove erroneous and lacking in key details, cannot obviate the final need for just such a pure temporal and de-spatialised mnemonic physics.

With this core certainty in mind, Part-II will elaborate upon the several conjectures of Bergsonian mechanics: Utilising the concept of wavefunctions in memory, it will assert that the superpose formed out of past wavefunctions in memory always involve constructive interference and fashion for the non-erasure of memory that, once formed, *must* remain ineradicable and permanently retained within the Bergson facet, notwithstanding the attenuation of memory per the inverse square of time and by primacy-recentcy rules.

Galvanised by nascent conjectural mnemonic mechanics, Book-IV will segue into the mnemonic presage to inertia and gravitation; all framed in terms of de-spatialised and temporised mnemonic mechanics and the growing block Bergson-Whitehead amalgam.

Admittedly conjectural, the necessity for a mnemonic physics of inertia and gravitation is inexorable, given de-spatialisation and temporisation, and the necessity to model almost all interactions in nature in terms of the superpose of de-spatialised and temporised Bergsonian past. Yet, this move will lead to unexpected possibilities about ever-elusive dark matter, purportedly responsible for excess gravitational influence in the universe. It will also lead to implications to dark energy, purportedly responsible for the accelerated expansion of the universe. Therein, we will conjecture that **both dark matter and dark energy are natural facets of Bergsonian mnemonic mechanics in a de-spatialised universe formed out of information rendered into pure time-relations... to the effect that there is no literal 'dark matter nor 'dark energy' as such**.

359

The issue of dark matter and energy will involve assertions about singularities and supermassive objects in the context of Bergsonian gravitation, in support of asymptotic 'black holes'. Therein infinities or singularities are rendered impossible in a de-spatialised purely temporised mnemonic physics of inertia, gravity and 'mass'. In any case singularities cannot form in a de-spatialised universe: It is not possible for mass to collapse to a point, given that there is no such thing as a spatial point in a de-spatialised universe; and given that attendant point-singularities have no plausibility in a purely temporised order.

Despite its conjectural content, Part-II will establish the certainty that the manifest universe is *not* constituted as contiguous 'matter in space'. Instead, the manifest universe, as the superpose of the past *en toto*, is constituted as memories in time; essentially consonant with Henri Bergson's view. Hence, the basis of all things manifest is not 'matter in space', but *memory in time*. The world is not made of 'matter'. Instead, it is made up of memories… notwithstanding future potentialities. This central insight, together with the very processes of our putative mnemonic mechanics, will complete our long-drawn foray into both information-matter dualism and decoupling theory. Thus, **we will come full-circle and validate, at least in *essential* terms, our initial explorations into information-matter relations, decoupling theory, and the then-nascent abstract memory**.

*

Part-IV will culminate into the revival of Cartesian mind-brain dualism: Cartesian dualism asserted a dualism between extended brains, or *res extensa*, versus non-extended mind, or *res cogita*. This led to a myriad of purported 'unsolvable problems', the general abandonment of Cartesian dualism, and the ascendancy of dismal materialist philosophy of mind. It turns out that the problem of Cartesian dualism had hitherto partially rested on **Descartes' minor error: namely, the notion of *res extensa*… or, simply, his implicit subscription to the ontic reality of 'space'. With de-spatialisation, we usurp *res extensa* and correct Descartes' minor error**. With attendant temporisation, we inexorably conclude that mind must interrelate to world and brains in a purely temporised non-contiguous way.

Hence, world and brains constitute a corpus of information and memory distributed in time, *without* space. It follows that a 'where' of conscious awareness, cognition and mind, (even if dismissed 'illusory') or the demand for a spatio-material 'location' for 'illusory' consciousness, thought and mind, cannot be secured in a purely temporal universe, world and brains.

Purely temporal abstract memory has no spatial address… simply because there is no space. Instead, we are forced to seek the temporal *moment* of mind, thought and consciousness vis-à-vis temporised abstract memory; a noetic moment that temporally succeeds and supersedes the world and brains formed out of memory in time. **It follows that the noetic moment must necessarily temporally succeed and supersedes the totality of the events constituting the temporal world and brain**. Hence, the witnessing nous must come *after* the events of the world and brain… *not in* space, *not as* the brain, *not from* the brain… but temporally *after* the brain. **This necessarily compels Cartesian mind-brain dualism, now recuperated in pure temporal terms, *without res extensa*…** consistent with findings from the experiments of Benjamin Libet and the peculiar a-synchronicity therein between consciousness and brain activity, among other things: (see Preliminary: **0.44**).

BOOK-IV PART-I:
DE-SPATIALISATION AND THE BERGSON-WHITEHEAD AMALGAM

4.00: Aims of Part-I

Part-I has two principal aims: The first consists in the falsification of space as a pleonastic notion, or de-spatialisation. The second is the further development of the growing block model: namely, the Bergson-Whitehead amalgam, or spacetime *without* space. Part-I will show that space is synonymous with the notion of absolute time; the notion that all events arise in a common present moment, or simultaneity. Absolute time was falsified by Einstein in 1905. If the stated synonymity is true, then the demise of absolute time in 1905 ought to have implied the tandem demise of space. Unfortunately, pleonastic space was retained and incorporated into 'spacetime'.

The key to the unravelling of absolute time was the relativity of simultaneity, integral to Einstein's Special Relativity. But **the first key step to the unravelling of space will emerge from the physical impossibility of contiguity and extensionality per the impossibility of said contiguity through simultaneity in time, specifically the impossibility of contiguity of a string of position-ordinates along a putative line (4.01)**. The impossibility of contiguity through simultaneity prohibits the physical possibility of the one-dimensional line or 'distance-interval', 'path' or 'trajectory', and, by generalisation, the physical impossibility of the two-dimensional 'surface-space', followed by the equal untenability of the three-dimensional 'block-space' per the same.

If two putative positions on a line cannot be contiguously and seamlessly linked, simply because they belong to different moments in time, and are not in simultaneity or in a common present moment of 'absolute time', and cannot thus form a putative extended line of contiguous ordinates, then the possibility of extensionality or 'line in space' is rendered untenable. By generalisation, the same must apply to the two-dimensional surface-space and to the three-dimensional block-space; also rendered impossible. Of course, the impossibility of contiguity through simultaneity also exhibits the relativity of simultaneity, furnished in Einstein's Special Relativity. Indeed, the impossibility of contiguity through simultaneity (of space) *is* the relativity of simultaneity (the impossibility of absolute time): the former is merely the latter in alternative form.

Part-I will also argue that **time is ontologically indispensable, while space is entirely pleonastic or disposable**. Without time there can be no inference or possibility of putative 'motion', which requires the *temporal* succession of a minimum of two events.

360

Without time, nothing would happen... not even events interpreted as 'objects in space'. But we can reframe the said objects and succession of events as events organised *only* in time, *without* resort to space. Thus, a physics without time is not possible; but a physics *without* space is perfectly feasible, if not ultimately necessary.

The third argument for de-spatialisation will show that **quantum mechanics prohibits the continuous observation of events, even assuming the Quantum Zeno effect. Hence, we cannot prove seamless-continuous motion. Consequently, we cannot prove extensionality or the 'space' across which purported motion supposedly transpires.** Insofar as such an observation is prohibited by the very structure of quantum mechanics, the oft supposition that the quantum wave is a time-undulating spatially dynamic state, one that supposedly undergoes 'motion', constitutes a pure assumption about the quantum wave, about motion, and, succinctly, about the ontics of space.

A fourth argument will assert that **the invalidation of space must emerge naturally from the critique of Hilbert space.** Even if we want to treat the quantum mechanical wave as moving, it could only move through a Hilbert space. But, at best, Hilbert space is a potentiality for space that has not yet come into realisation... through which nothing could move, much less any quantum wave.

However, **the fifth and final de-spatialisation argument will spring from Wheeler's delayed choice approach to both motion and space**, respectively.

In Book-II (Part-II), John Wheeler's delayed choice approach culminated into the spatially static view of the quantum wave as a non-moving non-undulating state. Therein, nested waves carried by a hedging wave, were shown to be spatially static. This alone invalidated space in the sense that, as far as the now-static nested waves are concerned, and if there is no motion to begin with, what need of requisite space? Yet, Part-I will take this conclusion to its next stage: It will treat the nesting or *hedging wave* as static and nested in its own right; belonging to a preceding meta-delayed choice system of preceding static hedging quantum waves that elaborate back to the putative Big Bang or to the indefinite past. Strangely, only just after the putative Big Bang might one entertain notional 'space' in relation to which motion might have transpired, rendering space of little use to physics in general, or pleonastic: eliminating it from the ontology of physics entirely.

With de-spatialisation in hand, the reform of Einstein's Special Relativity (SR) will follow. This will require replacements to the velocity-terms and of the 'speed of light'; all reconstituted in terms of pure delayed choice time-interval relations, *without* 'space'. The reform of SR will segue into **Apagogic Mnemonic Argument 5-a: the final mnemonic argument: A physical world conceived as a system of information organised into delayed choice time relations is a system of abstract memory, interrelated and resonating across time... without contiguity, contact or impact. Hence, *without* space. And wherein all 'contiguous spatial relations between objects' is transformed into non-contiguous pure temporal relations and distributions of events. Hence, de-spatialisation will constitute direct seamless proof of abstract memory,** complementary to findings in Book-III.

De-spatialisation will then force the reform of the physical field, exemplified by the magnetic field, into a pure time-distributed state: an *instantaneous whole* projected in stereotemporal time. These reforms or recuperations, combined with the preceding case for de-spatialisation and temporisation, will culminate into **the succession of the intermediate model spacetime from Book-II with the Bergson-Whitehead amalgam: i.e., spacetime *without* space.**

The tripartite structure from the intermediate spacetime model will survive; so will the primitive unification secured in Book-II, as will the contribution to prospective quantum gravity from grand decoherence. But the domain of 'spacetime in potentiality' will be succeeded by the Whitehead facet of the amalgam; while the domain of 'realised spacetime' will be superseded by the mnemonic Bergson facet of the amalgam. The event horizon will be retained in its given form, save insofar as it must relate to a Whitehead facet and a Bergson facet through pure delayed choice time-interval relations and expressions.

The nested futures perspective, counterfactual causality and grand decoherence, and the revival of Mach's principle furnished in Book-II, will also survive into the Bergson-Whitehead amalgam, but *without* space, as will the basic mechanisms that pertain to gravitation, albeit reframed to the Whitehead facet.

Obviously, **with de-spatialisation, the usual three dimensions must be replaced by stereotemporality**, wherein the system of events are integrated into pure delayed choice time-interval relata, albeit distributed in three axes of time. In other words, time-relations will be shown to be three dimensional or stereotemporal. In Book-II, time was shown *not* to be a 'fourth dimension': It was argued that time *is* the process of AND-to-OR wavefunction collapse itself, tandem to the progression of the event horizon. Hence, we will show that AND-to-OR collapse unfolds stereotemporally in a de-spatialised three dimensional temporal order.

Penultimately, Part-I will elaborate on the *new process philosophy,* wrought on by de-spatialisation, by 'spacetime' *without* space, and by stereotemporality. Therein, we will pay tribute to both Alfred North Whitehead and Henri Bergson. In the context of de-spatialisation and the superlative Bergson-Whitehead amalgam, *process* will be identified with the AND-to-OR collapse of the universe-scale wavefunction (the Whitehead facet of future potentiality) combined to the tandem progression of the event horizon; accompanied by consequent formations of new memories into a perduring state within expanding memory in a de-spatialised growing block universe, wherein time and the process of coming into being constitute the central dynamic. Definitions and problems of *endurance* and *perdurance*, key terms in *process*, will be subsequently expanded on and perfected in our new process philosophy.

Part-I will culminate into the consequent **replacement and succession of generic notions of 'atomos in space' with 'durata in time'**. Therein, the spatial concept of 'particles' as points, or else as spatial blocks, must transform into durations *without* extensionality. Hence, our view will evolve from forlorn spatial *atomos* to ontic temporal *durata*... accompanied by the unravelling of contiguity and contact causality: their succession and replacement by quantised time-transitions, or the non-contiguous 'quantum leap in time' … all to the demise of materialism, which must also unravel in the face of the demise of pleonastic space and of forlorn contiguous causality.

4.01: Preliminary Case for De-Spatialisation:
Impossibility of Contiguity Through Simultaneity & the Synonymity of Space with Absolute Time

In Special Relativity, relying on the relativity of simultaneity in conjunction with the speed of light (see **Fig. 4.01**) Albert Einstein falsified the notion of absolute time: i.e., the notion that there is a common present moment in which all events or objects occur together at the same time, or in simultaneity. Ironically, it turns out that putative space is ultimately synonymous with absolute time. Thus, the falsification of absolute time in 1905 should have precipitated the demise of space and its removal from the lexicon of physics.

When we observe the Earth's Moon, we see it as it was around 1.5 seconds ago in the past: We see it as a corpus of events that did not happen at the same time as the more immediate events about us upon the Earth. On the assumption of space, combined to the speed of light, it takes putative event-bearing light radiation some 1.5 seconds to cross the purported space between the Moon and the Earth. Per the same reasoning, when we observe the Sun, we see it as it was some nine minutes ago in the past: It takes nine minutes for light from the Sun to traverse the purported intervening space between the Sun and the Earth. The Sun as we see it at a given moment constitutes a corpus of events that did not happen at the same time as events immediate to us on Earth.

Again, when we observe the closest star system to our own Solar System (namely, the binary star system with Alpha Centauri) we see it as it was some four years and three months ago in the past: It takes that much time for light to cross the putative space between Alpha Centauri and our Solar System. Clearly, Alpha Centauri as we see it at any given moment is not a corpus of events that arose at the same time as events that are immediate to us on Earth.

Finally, the most distant parts of the visible universe belong to the most remote moments in time: perhaps some fourteen billion years ago in the past. Presumably it took light time to cross the purported expanding space that resides between the most remote corpus of past events of the early universe vis-à-vis present-time events upon the Earth.

All the above relations constitute implicit memory states: The relations constitute the presence and resonance of the past with respect to our own relative present frame of reference on Earth. Yet, what holds true for the Moon, the Sun, to Alpha Centauri and to the most 'distant' events of our universe, must also hold true with respect to the corpus of events that are seemingly immediate to us on Earth.

The following objects are present before the author: a keyboard on which one is typing; a cup of tea to the right; and, beyond both, the screen upon which writing is being displayed. Even with the assumption of space, all these objects and events are mediated to the author ultimately by means of quantum mechanical light radiation, which supposedly propagates across space at a fixed speed.

Per the limits of human perception, the author grasps the said objects *as if* these were transpiring simultaneously with the author, at the same time; at a same moment, or within an all-encompassing common present moment of simultaneity: in short, in absolute time.

However, if the purported spatial distances between the author and the said objects were as vast as those between the Moon and the Earth, the Sun and the Earth, or Alpha Centauri and the Solar System, the author's want to foist the notion that the said events were "happening at the same time" in some all-encompassing common present moment of simultaneity, or in absolute time, would be rendered doubtful. The putative distances between the author and the keyboard, the cup of tea and the computer screen, are so short (and the putative speed of light so swift) and the author's human perceptions so inadequate to the actuality of the condition, that the objects appear to arise within an all-encompassing common present moment... in simultaneity... or in absolute time. And it is upon such common experience, garnered through inadequate senses and perceptual limits, that the notion of absolute time, and, with it, the synonymous notion of 'space' (implicit) experientially arose.

The peculiar characteristic of consciousness also has a role to play here; but we will leave that matter to the very end of Book-IV: essay **4.41**.

*

When we describe objects as sitting on a 'common surface', a two-dimensional space, or else within a common three-dimensional block, what exactly do we mean? Do these purported objects sit on a common surface or inhabit a common three-dimensional block *at the same time*? Does time operate as a common present moment across the surface or throughout the block in which the objects sit, such as to fashion real contiguity of ordinates and co-ordinates into a system in total simultaneity, hence a 'surface'? But if the objects do not sit on a common surface or within a common block in any temporally simultaneous way then, the objects could not be said to sit on *any* 'surface' or within *any* spatial block. Indeed, *when* is their common two-dimensional surface or three-dimensional block-ness?

Nor could mere reference to the objects themselves as means to discern their shared spatial surface or shared inhabited block-space deliver what is sought, save by the imposition of a mere post-hoc assumption for such a 'space' upon the objects, followed by treating it all *as if* in proof of space. Per the relativity of simultaneity and the nature of light furnished in generic relativity theory (and even when, ironically, 'space' is assumed per the notion of the 'speed of light') reference to the objects will *always* constitute reference to different past moments and past events: *never* to their 'places'... *always* to *whens*: never to *wheres*.

*

Let us examine the issue more closely. Assume that there is an approachable object placed at some putative 'distance' from us: It could be a gnome at the end of the garden, or, if one must be more dramatic, the remains of the Apollo landing module from the first Moon shot. We might hope to prove this claim beyond reliance on passive observation and assumption: **We might seek to prove the 'place' of the object viscerally by approaching it and getting to it... by specifying its place and space in a hands-on way. But, as we approach the object, we notice that the initial past events that had constituted it and its presumed 'place' have been succeeded**

362

and replaced by newer events. These also belong to the past, albeit to the more recent past. But the original 'place' is now gone, and we can never prove its onticity in the visceral sense sought: it is gone with the original past events associated with it. As we approach 'closer' to our object (again, *assuming* space, place, and motion) the observational time-delay per function of the light radiation from the object vis-à-vis our senses will be shortened: The past events that had constituted these will be succeeded by even newer events: and the 'place' associated with *these* will also be gone.

We continue our journey until we obtain almost-simultaneity with the object. Yet, the object, the garden gnome, the old Moon Lander, continue to belong to what is now the very recent past, albeit at almost-simultaneity, though *not in* simultaneity. This shows that, viscerally, **we never get to any putative 'place' or location, unless we reduce ourselves to a putative dimension-less point and obtain simultaneity with some other similar point that belongs to the object... though *not* to the totality of its constitutive events. If we could become such a point and obtain simultaneity with the point of destination, then we could argue for 'position' and 'place' of sorts.** But such a simultaneity with a single point is not possible in practical-physical terms. **Even if we could obtain such simultaneity, we could only report the point as an event in perfect simultaneity *in time... not* in a 'place'.** At the very least, there will be ambiguity to the want to portray the point-destination as a 'place'. Yet, no ambiguity can arise in the want to posit it as a moment or event distributed in time. Yet, note that there could be no possibility of imposing 'place' to any other point aside the destination-point, now in seeming simultaneity with us. These others are past events, or *whens* distributed in time, not places, locations or *wheres*.

Viscerally, we *always* register events related to us in time... as the past, as the very recent past, or as the remote past... but *never* in simultaneity. The original 'place' we attribute to the object at the beginning of our journey never really existed: it could not be obtained physically, evidentially, or subsequently. Of course, we could assert a notional 'location' to the object at the start of our journey and do so even at the end of our journey, when we obtain near-simultaneity with it. But only on *assumption, not proof*. In real-physical terms, despite our assumptions, we are restricted to truth-claims purely about events organised into time-relations... to truth-claims about the primacy of events in the past vis-à-vis the recentcy of other events... but *never* about 'locations', 'positions' or 'places'; all of which could be abstracted and discarded without causing any diminution to physics. The only meaningful physical informational claims we can fashion about nature, even when we assume and assert rapports based on the presumption of 'space' and the 'speed of light', will always be comprises of events and intervals organised into pure time-relations apprehended in retrospect (see **2.23**). The same holds true with respect to the interval between the Earth and the Moon; the Earth and the Sun; between Alpha Century and our Solar System, and into our relations to the most remote ancient events of the universe.

*

Why is space and absolute time synonymous? The key is simultaneity: its absolute requirement for contiguity to be possible, and, consequently, for extensionality and space to be possible: **In order to prove the ontic reality of space in a one dimensional line, in two dimensions constituting a plane or surface, or in three dimensions comprising a block-space, one must prove the simultaneity of the ordinates and co-ordinates along or throughout the line, the surface or the block, respectively,** with consequent seamless-continuous contiguity of the infinitesimal number of ordinates and co-ordinates.

Let us star with an examination of the one-dimensional line or distance-interval or 'path'. We can then extrapolate our conclusions to the two and three dimensions.

A line of any length will do. The line is made up of an infinitesimal number of positions or points or ordinates, seamlessly or contiguously inter-connected. But in order to assert this as a true line, and a true distance-interval or 'path', the observer and the ordinate points must obtain simultaneity (hence contiguity) with each preceding and succeeding ordinate of the line... so as to assert that all such points arise in a common present moment... so as to constitute a line-contiguity, a line-extensionality, or 'space'. This cannot be achieved.

Consider just two ordinates along the putative line. Do these connect and join to form an extended line? Are they contiguous? Treated as distinct events, as demanded by the relativity of simultaneity, if these happen at different times, then they are detached and discontinuous; incapable of forming a real line or extensionality. This is because time-relations are not contiguous: time-relations do not support contiguity or spatial connection between two neighbouring ordinate points. If we introduce a third middle point to join the two points into a seamless unbroken line or extension, as an event, this middle point must also happen at a different time to the preceding and succeeding points. Thus, by adding a middle point, we merely complicate the original problem and fail: The middle point is also an event in time and fails to join the flanking events into a same-time seamless-continuous line.

Consequently, we can obtain a string of putative 'ordinates' or points... as many as one might desire... to form 'a string of disjointed dots'. But these can never constitute a true line or extension, given that the 'dots' comprise different moments from different times in interrelation to each other and to observers. Each point along the line with respect to every other is subject to the relativity of simultaneity (see **Fig. 4.01** for clarification). Even for a point or observer at the centre of the line, the simultaneity achieved with respect to the rappor about the two flanking points or events cannot obviate that these belong to past moments or *whens*; *not* to *wheres*. The latter cannot be established physically or ontically, since such a feat would require contiguity of the points in and through simultaneity, such that all the said points and the middle observer arise in a common present moment, or in absolute time But, such a feat would collapse into a singularity and unravel the very line and extensionality sough, as we shall later discover.

Without simultaneity, there can be no contiguity of ordinates. Without contiguity, there can be no putative 'seamless line of ordinates'. Since we cannot rid the situation of the relativity of simultaneity, we can only assert for the real existence of a series

and successions of pure *whens*; all distributed at different moments in time vis-à-vis the observe. However, there can be *no* simultaneity for all the points or observers into some impossible state of absolute time... or the synonymous 'space'.

The same can be argued with respect to co-ordinates vis-à-vis other co-ordinates in a putative two-dimensional surface-space. And the same can be extended to a three-dimensional block-space. In all cases, it remains impossible to assert for the onticity of contiguity

Fig. 4.01: The relativity of simultaneity for stationary observers

<u>A) Initial Setup at t = 0</u>

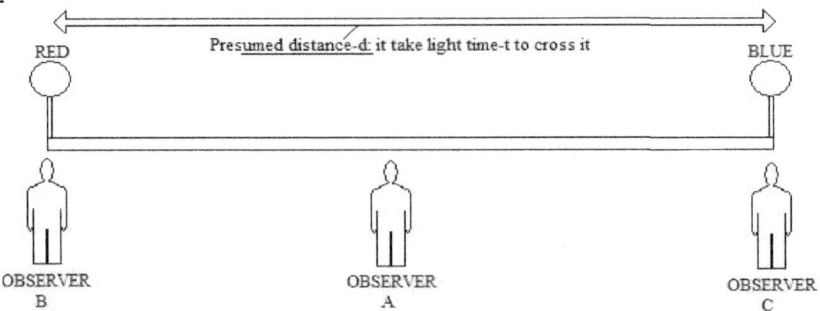

Notes: In the initial setup, two flashbulbs are separated by putative distance-d which can be traversed by light in time-t. Note the three observers: Observer-A is equidistant from the red and blue flashbulbs, both of which are synchronised to flash at a pre-set moment in the future.

<u>B) Observers B and C: what they see by time-t</u>

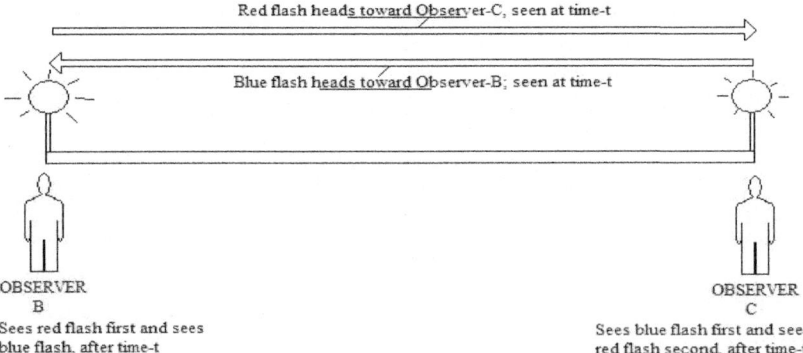

Notes: It takes time-t for light-flash from one bulb to reach the other bulb. Observer-B sees the red flash and then sees the blue flash at the later time-t. Observer-C sees the blue flash first, then the red flash at the later time-t: The order of events is relative. On the other hand, observer-A sees the red and blue flashes simultaneously. The differences illustrate the relativity of simultaneity: see Diagram-C, below for observer-A...

<u>C) Observer A: simultaneity</u>

Light from the red flash and the blue flash meet at the mid-point, located half of distance-d, at half of time-t: both travel at the purported 'speed of light'

RED   BLUE

OBSERVER
A
Observer-A sees red flash and blue flash
at the same time: simultaneity achieved

D) Moving observer: initial setup

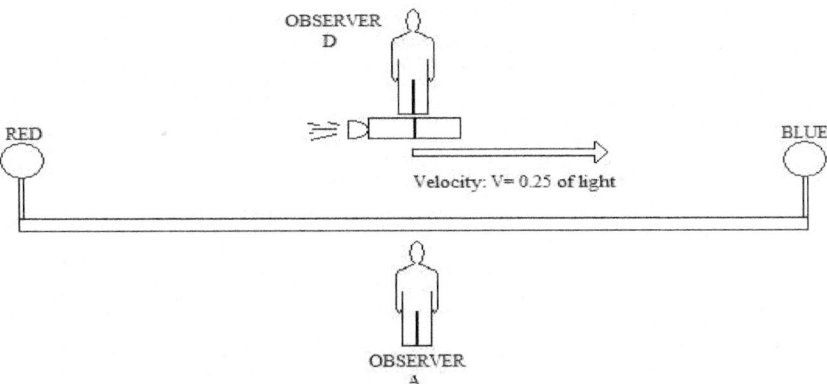

Notes: Moving observer-D coincides with observer-A at the mid-point at time = 0, just as the flashbulbs are triggered. Note that observer-D has a putative uniform velocity of 0.25 of light toward the right.

E) Relativity of simultaneity due to motion

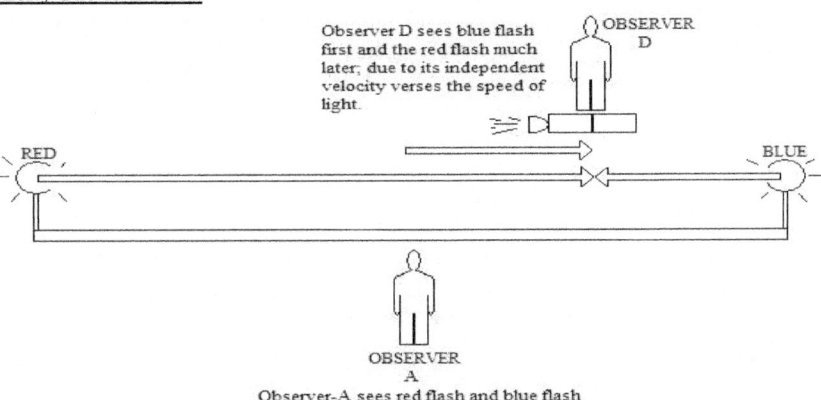

Notes: Observer-A will see the bulb flashes occur simultaneously, as usual. But the moving Observer-D will see the blue flash at a quarter of time-t due to his motion-v, and will see the red flash at three quarters of time-t. Hence, the relativity of simultaneity due to putative motion.

and extensionality, given the relativity of simultaneity and the impossibility of furnishing simultaneity (and hence contiguity) to all ordinates and co-ordinates along the purported line or two-dimensional surface, or throughout a three-dimensional block-space,. **The impossibility of contiguity through simultaneity inexorably culminates into the impossibility of spatial lines, surfaces, and blocks. The impossibility of contiguity through simultaneity necessarily implies the demise of space**.

Of course, we can always pretend 'space', but only when the time-intervals that interrelate attendant events and observers are so short as to be treated effectively *as if* 'arising in a common present moment', in seeming simultaneity, or in apparent absolute time. Hence, the naïve perceptional and experiential basis for the insipid idea of 'space'.

*

However, there is another peculiarity that we must mention: Any want to assert that lines formed out of contiguous ordinates are ontically and physically real, such as to justify the notion of one-dimensional line-space, requires proof of contiguity. But contiguity requires the simultaneity of each ordinate with respect to every other ordinate along that line. However, if this is true, then **the contiguous points must not only arise at the same moment but must coincide at a same ordinate. Thus, the presumption of contiguity through simultaneity in a common present moment, such as to justify a one-dimensional line-space, collapses that line to a single point-singularity,** *without* **the possibility of extensionality into** *any* **sort of line… so destroying the very possibility of the line-interval and of 'space', so-conceived.**

The same can be extrapolated to the simultaneity of co-ordinate points or positions throughout a putative two-dimensional surface-space. For the same reasons as had applied to the line, this must also collapse into a point-singularity, bringing about the demise of the surface-space purported. The same confronts the three-dimensional block of co-ordinate point-positions: This must also immediately collapse into a point-singularity. Thus, the 'block-space' renders itself impossible through the very contiguity through three-dimensional simultaneity that it requires.

Apagogically, 'space' *absolutely* requires the contiguity of point co-ordinate positions, rendered into total simultaneity: i.e., space requires absolute time with which it is synonymous. Yet, based on this indispensable requisite, space renders itself untenable through its own collapse into singularity... and destroys the very extensionality or 'space' thus sought.

In short, apagogically, space unravels itself. Hence, we are left to ponder what it is that the physical universe consists in: Does it consist in contiguous relations extended in one, two and three-dimensional states in absolute time, so constituting the possibility for 'objects in space'? Our Alternative Realist contention asserts that there is no contiguity; that there is no extensionality... and that such a feat requires absolute time... and leads to apagogic self-collapsing space.

Future-potentials aside, the physical universe consist in past events comprised of pure *whens*... of *memory*... organised into stereotemporal time, structured according to primacy-recentcy principles... all without contiguity through simultaneity; all without absolute time... *without space*.

*

With the downfall of absolute time in 1905, Einstein ought to have discarded 'space' synonymous with absolute time: Space should have gone the way of the old eather. An alternative Special Relativity theory, and a different take on succeeding General Relativity, might have been founded on a purely temporised physics, *without* 'space'. In such a physics, the 'speed of light' would have been replaced with a pure temporal equivalent: the *delayed choice time-interval standard*; structuring and integrating past events vis-à-vis other past events vis-à-vis relative present frames of events and observers... *without space*.

Instead, Einstein *et al* kept 'space' and amalgamated it with time into "spacetime"; a distorting schema.

Throughout much of Book-I, Book-II, and Book-III, we abided with the notion of space and of spacetime, and even sought to develop it further into the growing block quantum-relativistic incorporative intermediate model spacetime. Yet, throughout, we also warned that space is not ultimately viable; and that spacetime will need to be drastically modified into a form *without* space. It is time to dispense with space completely and transform our old intermediate spacetime model from Book-II into a superlative model *without* 'space': namely, the *Bergson-Whitehead amalgam*.

4.02: De-Spatialisation from the Analysis of the Motion of Light Radiation

Against our assertion that space is no more ontically real than synonymous absolute time, we anticipate the following counter:

When you observe the Moon, you observe it as it was 1.5 seconds ago in the past. This does not imply that space does not exist. Succinctly, because of its speed-limit, light takes time to cross the intervening space between the Moon and the Earth. Consequently, you observe the Moon, not in some approximate common present moment, but as it was in the past, precisely because of the effect generated by the relation of the speed of light to a really-existing spatial interval. Therefore, space is real. It is the relation of the speed of light to space that furnishes the relativity of simultaneity: You cannot use the relativity of simultaneity as an argument against the ontology of space and the speed of light that, in the first place, renders the relativity of simultaneity possible and viable. Lastly, Special Relativity incorporates space with time into spacetime. Hence, space is inherent to the very success, both of generic Special Relativity and succeeding General Relativity. In these theories, absolute time is certainly dethroned, even while the concept of space is retained.

This is contradictory: Space *is* synonymous with absolute time, given the impossibility of contiguity and extensionality through simultaneity (**4.01**), and we cannot keep space in a physics from which absolute time has been removed. That aside, on what physical empirical terms do we judge light itself to be "travelling through space" and requiring space to bring about the relativity of simultaneity? **Does the relativity of simultaneity and the attendant temporal distribution of events emerge out of the relation of the speed of light vis-à-vis space, or does the *notion* of a 'speed of light' and of its 'space' constitute an assumptional interpretation foisted on what are otherwise pure temporised distributions of events and observations that render viable the relativity of simultaneity, but do not furnish either 'space' or the spatial conception of the 'speed of light', save as an assumptional pleonastic foisting?**

*

We begin by treating the putative event-bearing light radiation from the Moon to the Earth as a form of object: a "light-ray" undergoing purported perpetual change in position along an assumed spatial line. The spatial line or 'path' embedded in a two dimensional slice of surface-space joins the Moon to the Earth. The light-ray is destined to disclose to our observers on Earth its content; namely, the corpus of past events of the Moon as it was 1.5 seconds ago.

Note that, we cannot observe our putative light-ray undergoing its purported motion across space in realtime, unless we travelled with it: something we cannot do per basic constrictions from Special Relativity: i.e., we cannot travel at the speed of light. Therefore, we cannot confirm the purported motion of our light ray in any pseudo-realtime empirical way. Indeed, to reiterate a conclusion from Book-II (**2.23**), realtime observational inferences are physically impossible. Only retrospective observational inferences are permitted.

We certainly could not watch the light-ray undergo its purported motion continuously, even if we could travel with it at the purported speed of light. The reason for this is simple: notwithstanding the Quantum Zeno effect which might furnish almost near-continuous observation of a system, but fall short of doing so, and given that continuous observation would require infinite frequency and energy inputs and return-signals to and from the 'continuously observed light-ray', this is simply impossible to achieve: Contiguity, hence

366

'space', could never be established or proven in this way: Quantum mechanics prohibits the possibility of seamless-continuous observation of *anything*, for the same reason that it prohibits continuous-seamless infinite frequency and infinite energy inputs and outputs to and from *any* system... including vis-à-vis the 'moving light-ray' supposedly displacing at the alleged speed of light.

Our other recourse is to forgo forlorn realtime observation of the light-ray and restrict ourselves to the retrospective observational inferential mode from **2.23** (as we must); admit to the relativity of simultaneity, and place a string of equidistant detection devices between the Moon and the Earth. Our putative light-ray will be expected to pass through these as it displaces from the Moon to the Earth across the said spatial line or path.

We may place as many detection and observation devices as deemed necessary, perhaps at shorter and shorter equidistant intervals and strings along the Moon-Earth path. But we cannot place seamlessly and continuously linked devices along the putative path.

We have three choices about where to place our human observers: We can place some observers on Earth. We can place as many as is required directly at each of the said devices strung along the Moon-Earth path. And we can place other observers somewhat sidereal to the Earth-Moon mid-point about the path of our light-ray. We can arrange radio communication between all the observers and devices involved.

Each device will be triggered when our light-ray passes through; with each trigger transmitting a radio signal. The signal must also travel at the speed of light. The signal will report the passage of the said light-ray to all observers, whether they are attached to the devices or established at other positions.

Clearly, each device vis-à-vis every other along the purported path of our light-ray cannot be treated as occurring in a common present moment of simultaneity or in absolute time: This is simply because there is no absolute time or unifying simultaneous moment. The platforms are interrelated to each other in terms of the relativity of simultaneity. Therefore, we must yet again ask the *when* of each of our devices and their trigger-events per the passing light-ray along the putative Moon-Earth path.

Notice again that we are not inquiring about the *where* of each device and trigger-event. Reference to the devices and their tripped radio signals and observers, as events, can only be referenced as past moments or past events organised in time and subject to the relativity of simultaneity. The interrelations between the devices and their attached observers, to each other and to observers on Earth, are to be treated in the same way. There is no *where* in such a system: It consists purely of a collection of time-organised older versus more recent *when*s, subject to the relativity of simultaneity. Also, while the composition of events will be conserved, the order of the events will be different to different observers: see previous **Fig. 4.01**: diagrams B and C.

Recall that, we rely on the said trigger-events at each device strung along the Moon and the Earth to confirm the putative 'spatial motion' of our light-ray along a 'path' that strings all the devices together… in proof of that path. On that account, each trigger that our light-ray induces, transmitted to all observers including those on the Earth, ought to prove the reality of the motion *and* of the space or path along which our light-ray supposedly moves, presumably at the speed of light. But, since we cannot assert any *wheres* to these, but can only assert relative *when*s for each, and can only treat each as unique past events distributed in time (i.e., from the oldest to the most recent trigger incidences vis-à-vis and relative to any given device and observer) *and* subject to the relativity of simultaneity, it follows that we cannot assert a 'space-path', much less a string of contiguous *wheres* in simultaneity, such as to constitute a spatial extension along and across which the light-ray purportedly travels. Indeed, such an assertion for contiguity through simultaneity remains impossible to assume, much less demonstrate per reliance on pertinent empirically evidence (the detectors and observations just depicted) given that the trigger-events, devices, and observers are comprised exclusively of *when*s.

It follows that the only claim we can make about our light-ray must consist of a minimum of two temporally succeeding events constitutive of it: These are distributions in time, *not* distributions along or comprising a spatial line or path, much less supposedly undergoing spatial motion at the speed of light. Indeed, **it is not the putative 'speed' and 'motion' of the light-ray across putative space that generates the delayed choice time-interval between the Moon and the Earth, and thus furnishes the relativity of simultaneity. On the contrary, it is the delayed choice time-interval relation between the event (the Moon) and its rapport 1.5 seconds later on Earth (the observation), that garners the impression of an *"effect generated by the relation of the speed of light"* to a (supposedly) *"real spatial interval"*. In other words, the *assumption* of a signal-bearing 'moving' radiation is foisted on the factity of events distributed and relationally delayed in time vis-à-vis the rapport of the pertinent events. Post hoc, the imposition of the assumption is then presented *as if* it constituted 'evidence' of a spatially moving signal-bearing light-ray, supposedly moving at the speed of light, and as 'proof' of the space that supposedly constitutes the path of the light-ray.**

Attaching observers directly to each device will not make the slightest difference: If the observers communicate to each other by radio signals to confirm the passage of our light-ray per each device, for reasons given, they cannot relate this sort of information to each other in terms of places, positions or 'space', save by pure *assumption* foisted on what are otherwise exclusive pure time-relations and *when*-distributions that relate to a presumed 'speed of light' across putative 'space'. The passage of our light-ray at each device, as events structured and related in time, imply that the related reports, and the 'motion of the light-ray' foisted on these, and the foisted 'space' across which the ray purportedly unfolds, are nothing but organised successions of older events vis-à-vis more recent ones (relative), organised in pure time and subject to the relativity of simultaneity. Again, the 'speed of light' is 'inferred' from a time structured series of *when*s, to which only a time-delay attribute can be accorded, not a superfluous spatial or contiguous one.

*

There is no space. Physics must discard space in the same way it discarded both absolute time *and* the eather. It also follows that there is a need to reform Special Relativity, perhaps also General Relativity, into alternate pure temporised forms. In short, we need a replacement-model for spacetime: We need a spacetime *without* 'space'.

4.03: A Brief on Scientific Economics Against Space: The Pleonastic Character of Space

Our second counter against space is based on the straight-forward economics of science, and the need to dispense with concepts that are not ultimately needed or otherwise superfluous to the scientific understanding: i.e., concepts that, no matter how didactic and pragmatically useful, prove to be ultimately pleonastic, and very possibly limiting of science, if not liabilities to future advancement.

Aside the fact that the practicalities of observation falsify both the moving 'light-ray' at its purported 'speed of light' *and* its space, neither can be physically proven, much less proven as necessary or indispensable. And **insofar as we can easily account for observed phenomena in terms of events expressed and structured purely in temporal terms, *without* space... or as structured and distributed *whens* without resort to *wheres*... the economics of science recommends that we completely discard the concept of space; removing it from both the lexicon and ontology of ultimate physics and science.** That is, space is neither supportable per the *ultimate* ontology and observational procedures of science, nor is it ultimately needed as a significant term, no matter how convenient it might be for daily purposes (such as giving directions to the local gas station) or in naive physics suited to the description of the trajectories of artillery shells. Indeed, while we can certainly discard space from the lexicon of physics, we cannot discard time: Thus, space is perfectly discardable from physics, and it is thus pleonastic. But time is indispensable to physics, and therefore fundamental.

It is very easy to imagine a physics without space: Indeed, we accomplished it in the previous essays. More will follow; each more complete against the notion of 'space' than the preceding account. On the other hand, we *cannot* conceive any physics (even a limited perception-based Galilean and Newtonian physics) without resort to time. How could observations be made and reported if neither the reported events nor their observations never get to transpire? This requires time as an indispensable ontic. Space on its own, without time, is useless, given that, in a physics of space alone, nothing could happen, much less any facts upon which we might want to foist the notion of 'space'. (One wonders why anyone bothered with block model precepts in generic apagogic 'spacetime' models). On the other hand, time alone will furnish all we need into a viable physics, *without* space.

The only factor favouring space is its mere assumption; an insidious one born out of limited perception-based experience, foisted upon the whole of physics and universe thereafter. Thus, **time is truly fundamental to and ineliminable from physics. On the other hand, as we can now begin to appreciate, space turns out to be ultimately superfluous, pleonastic and disposable**: a fiction that we can and *must* discard if we want to advance physics and natural science to its successor iteration.

4.04: De-Spatialisation from the Empirical-Observational Stricture of Quantum Mechanics: Critique of the Moving Quantum Mechanical Light Radiation

Quantum physics prohibits any direct continuous observation of the quantum mechanical wave and of the light radiation that purportedly displaces at the speed of light. Insofar as such observation is prohibited, the supposition that the quantum wave is a spatially dynamic state, oft described as undergoing 'motion' in the dubious pseudo-realtime form, is at best a didactically useful assumption about the quantum mechanical wave, but *only* an assumption: a disprovable one.

In Book-II, using John Wheeler's delayed choice, we argued that the quantum mechanical wave constitutes a spatially static and non-moving non-undulating nested futures state (see Book-II, **2.24** to **2.26**): This will be recapitulated in Part-I as part of the anticipated meta-delayed choice approach against space. But the discovery in Book-II constitutes an implicit independent attack on the notion of space, insofar as a spatially static quantum mechanical wave does not undergo motion; in which case, what need of requisite space? Apply this to our thought experiment from **4.02**, in which our quantum mechanical 'light-ray' is purported to displace across the intervening space between the Moon and the Earth, and to which observational strictures from quantum mechanics *and* the spatially static quantum waves must apply, one immediately grasps the implausibility of the "moving light ray" *and* of the 'space' across which it is purported to displace.

However, let us return in detail to the matter of the impossibility of the direct continuous observation and corroboration of spatially displacing quantum mechanical waves and 'light rays': **The attempt to observe the moving quantum wave or 'light-ray' necessitates that we apply observational inputs to it. But these have the effect of destroying the very thing we seek to observe... namely, the purported time-undulating and moving quantum wave itself.** Subjecting it to AND-to-OR collapse and destruction leaves behind a collection of OR-form quantum discrete events or *whens* that are only ever retrospectively observed... observed *after* the wave has undergone AND-to-OR collapse... observed as past events or 'whens', *not* 'wheres'. The quantum mechanical light-ray and radiation before AND-to-OR collapse can never be observed simultaneously or even in realtime.

The strictures of quantum mechanics demand that **the notion of the purported moving quantum mechanical wave could be maintained, but *only* on the strict condition that we refrain from observing the quantum wave, and keep it in its state of physical and observational isolation... by refraining from bombarding it with any wavefunction-disturbing observational inputs** that might bring it out of isolation and transform it into a post-facto purely retrospectively observed collection and *temporal succession* of OR-form outcomes: outcomes that are *always* reported as having transpired in the past per the retrospective observational stricture. Indeed, the restriction of all observation to the retrospective, and the impossibility of realtime or pseudo-realtime observation, was fully developed in Book-II, essay **2.24**.

368

If we make a minimum of two temporally successive observations of our purportedly moving quantum wave, or at moments across a given time interval within which the wave is isolated, we can only *assume* the notion of its motion within the defined isolation-interval... as something that we *presume* to have transpired but cannot demonstrate or prove to have actually transpired, forming the two required succession of past events from which we can then draw an inference, but can only be evidentially certain of the time-interval interspersing the two OR-form events, *not* of any spatial interval: Again, at best, we can only *assume* the never-observed 'motion of the quantum wave' purported to have transpired between the two temporally successive events, *without* ever proving this to be true in empirical-evidential terms.

*

To exhaust all possibilities and be certain of our claims, we might want to reduce the intervals between the temporal succession of events and their observations to shorter and shorter time-intervals or frequencies. We can do this to approach, but never obtain, continuous observation of the quantum wave or 'light-ray'. Notwithstanding the Quantum Zeno effect, so long as some time interval of isolation between *any* succession of observations must abide, and given that we cannot make any direct observation regarding such isolation-intervals per the very stricture imposed by quantum mechanics itself, it again follows that we can only *assume* a purported 'motion of the radiation' across the now much-shortened but never-eliminated time-intervals. Of course, what we get to observe, but *only* in the retrospective mode, will constitute short time-interval successions of OR-form events that belong to the past, or are distributed in time, without any proof or evidence for 'space'. The observations always involve an empiricism of past events or *whens*, and processes distributed in time: in this instance, distributed per short delayed choice time-interval relations.

But suppose we were to reduce the intervals that distribute our retrospective observations to the infinitesimally small, and hence to infinitesimally short time-intervals? Such a feat is strictly necessary, if we seek to overcome the problem of inherent isolation and the non-observability of the said intervals, and achieve direct continuous observation of our 'moving' quantum wave or 'light-ray' within and across the said interval. First, the frequency of events and their observations per time would shoot to infinity, culminating into the ultraviolet catastrophe and impossibility. Nature abhors an ultraviolet catastrophe: Her evidence against it forced the development of quantum mechanics, and, with it, the abandonment of the fictitious then-suspect notion of 'continuous radiation'. It also implicitly undermined the tandem possibility of *any* observation of *any* physical systems, or of the quantum wave itself, through continuous radiation inputs and outputs... or continuous observations even the vaunted Quantum Zeno effect cannot furnish.

Quantum mechanics favoured quantum-discrete or broken-discontinuous radiation and matter, with tandem co-requisite grainy and broken-discontinuous retrospective-mode observational structure, *always* of events belonging to the past... always involving the temporal succession of discrete events whose time-intervals might approach Plank time, but might never obtain it. Consequently, we cannot get rid of intervals of isolation pertaining to forlorn spatially dynamic quantum waves and 'light-rays'. Therefore, per quantum mechanics, we cannot furnish empirical proof of the notion of the motion of quantum waves and 'light-rays'. Hence, we cannot secure continuous observation of our light-ray along its purported spatial path, no more than obtain or furnish continuous radiation and de Broglie matter.

*

As stated, we can only assume the notion of the motion of light radiation across an interval, on the strict condition that we never observe it or disturb its AND-form quantum mechanical state in that interval. But the notion of the motion of light radiation is just that, a *notion...* an *assumption*. Its attendant 'space' is also an unprovable assumption that cannot be supported empirically or observationally.

However, the main point is, the notion that the moving light radiation in relation to its 'space' generates the distribution of events in time, *and* forms the basis of the relativity of simultaneity... and that this supposedly necessitates the ontic reality of space, or of the need to incorporate it with time into 'spacetime'... is falsified.

There is no empirical-experimental and quantum mechanical necessity for the ontics of either motion or space, notwithstanding generic quantum mechanics flirtation with 'moving quantum waves'. It follows that, **since all we can meaningfully report about are events distributed in time, it is the distribution of events in time, and the delayed choice time-intervals that intersperse these, that generates the relativity of simultaneity, and otherwise permits the imposition of the false notions of space and of the 'moving quantum mechanical light radiation at the speed of light'. Both space and the 'speed of light' prove to be ultimately pleonastic.**

*

In our alternative counterfactual history, by 1927, quantum mechanics independently attacked the empirical-observational viability of space in terms similar to those just described, usurping the notion of the motion of radiation and matter, and permanently undermining the presumed onticity of 'space'; opening the way to a pure temporal quantum mechanics and physics. Unfortunately, history did not transpire this way: it ought to have done so.

4.05: The Problem of Hilbert Space:
Relevance of the Principles of Causality & the Conservation Principle:
The Critique of the Motion of Quantum Mechanical Light Radiation

A central concept in generic quantum mechanics is Hilbert space. It is essentially AND-form potential space; oft demarcated or defined by the quantum wavefunction itself, or by whatever observationally isolated 'well' or 'area' that the wavefunction spans. The concept of Hilbert space assumes the reality of space, even if as a future potentiality. Thus, the wavefunction of a particle in isolation

369

and within its area of confinement constitutes a superpose of all its potential positions, while the said area constitutes its potential space, or Hilbert space.

For brevity, we will put aside the nested futures description of the wavefunction and the static view of the quantum wave developed in Book-II, Part-II. Even without the nested futures approach, the notion of a Hilbert space poses problems vis-à-vis the notion of the motion of quantum mechanical radiation: If we assert the Hilbert space of the wavefunction to constitute a future-form state, we must give up the notion of the motion of the wave across its future-form Hilbert space... if only because its future-form space has not yet formed up. Insofar as its space has not yet come into being, nothing can move through it.

*

Generic quantum mechanics does not utilise an explicit argument from causality intertwined with the conservation principle of energy-matter to justify the ontology and objectivity of AND-form logic and the *raison d'etre* of quantum wave and wavefunction. It was our Alt-R approach that compensated for this remiss: (see Book-I, **1.17**, and **1.31** to **1.32**, and in Book-II: **2.10** through to **2.12**). Causality intertwined with the principle of conservation of energy-matter, hence the *firewall principle*, compels the acceptance of the quantum wave as the literal objective manifestation of the future, and the wavefunction as the mathematical attempt to model that objective future. Hence, the future constitutes non-resolved potentiality, necessarily rendered in AND-form... unless we treat it as a resolved set of outcomes *before* it had transpired into such, as perhaps Einstein *et al* would have inadvertently preferred:

> "Like the moon has a definite position" Einstein said to me last winter, "whether or not we look at the moon, the same also holds for the atomic objects, as there is no sharp distinction possible between these and macroscopic objects. Observation cannot *create* an element of reality like a position, there must be something contained in the complete description of physical reality which corresponds to the *possibility* of observing a position, already before the observation has been actually made."[80]

As we can see in the quote above, in Einstein's block universe, there is no concession to an as-yet not-realised objective future: All things including the Moon and the particle must be rendered in OR-form. The Moon as observed is certainly OR-from. But what about the Moon just above the 1.5 second delayed choice time-interval in its future?

Ironically, generic quantum mechanics did not explicitly furnish a growing block universe or the ontic reality of the future, even though both are inexorable from the very content and *raison d'etre* of quantum mechanics. Hence, we will assume, and have assumed, the contrary position to that of Einstein: a growing block universe with an as-yet not resolved ontic future domain; a domain necessarily AND-form quantum mechanical in character. It becomes obvious that the Moon (or a particle) five minutes or two seconds from now can only be described as an AND-form future potentiality state, not as an OR-form resolved state. If we treat either *as if* these were resolved into concrete events, and empirically proved this to be so, the future would have transpired before it actually happened, in violation of causality, through 'signals from the future'.

If such 'signals from the future' were possible, where would the extra energy-matter to make up the Moon (or the particle) five minutes from now come from? Hence, the violation of the principle of conservation of energy-matter would also follow in an Einsteinian block universe.

It is because generic quantum mechanics made no explicit assertion about the intertwine of causality and the conservation principle, and rendered no explicitly combination of these to resolve the ontology of the AND-form wavefunction, that it missed out in grasping Hilbert space as a potential not-yet-realised future-form state. Once we grasp the *raison d'etre* of AND-form logic and, with it, conclude that the future is ontically real *and* phenomenalised as the quantum mechanical wave, the nature of Hilbert space as future-form not-yet realised 'space' follows seamlessly.

*

Consider the generic two-slit experiment. The entire apparatus (the active point of the emitter, the barrier, and the surface of the detector) is placed into physical isolation. Any intrusion by the application of physical inputs at any moment in time during the experiment will be tantamount to the inclusion of decohering or even collapsing delayed choice switch-throws: (See John Wheeler's Delayed Choice in the subsequent account or in the original account in Book-II (**2.18**)).

By way of isolation, the whole 'region' demarcated by the two-slit experimental apparatus is rendered a quantum mechanical potential; a not-yet realised Hilbert space. Into this Hilbert space we release single quanta of 'particles' of radiation or matter via the emitter, one at a time. In the usual dubious pseudo-realtime depiction of the whole process, the quantum mechanical wave and associated wavefunction of the emitted particle purportedly displaces through the as-yet not-realised Hilbert space pertaining to the isolated experimental area. Baring interruption via physical-observational inputs, baring the introduction of any delayed choice switch-throws, we can only be certain of the time-interval interspersing the activity of the emitter vis-à-vis the subsequent 'click' we might set up to obtain for each quantum particle deposited at the detector. In other words, we can be certain of a delayed choice time-interval relation that integrates the older *primacy* state composed of the past emission-event at the emitter versus the succeeding *recentcy* quantum incidence resolved on the detector; both *also* observed retrospectively per the rules first elucidated in Book-II (see **2.23**).

---

[80] Letter from Pauli to Niels Bohr, February 15, 1955: Charles P.; Meyenn, Karl von, eds. (1994). *Writings on physics and philosophy by Wolfgang Pauli*. Springer-Verlag. p.43

QUANTUM MECHANICS AND MIND

What, then, of the space and the alleged motion of the quantum wave through and within the confines of the experiment? How is this and the generic pseudo-realtime depiction of the purported motion (the thing we can never observe directly and cannot rightly assume as a given) to be realised across the said Hilbert space... across a 'space' that has not yet come into formation? The answer is obvious: The notion of the motion of the quantum mechanical wave across such an as-yet-not realised Hilbert space is entirely implausible. Thus, per the same account, and in *any* other context, **the notion of a propagating light radiation through space, whether from the Moon to the Earth, or from emitter to detector, or from anything to anything else, is implausible, insofar as it must ultimately displace across a space... a Hilbert space... that has not yet come into realisation. In which case nothing can move through it, or be depicted to do so**; especially as regards any dubious pseudo-realtime depiction pervasive in the various description of experiments in modern physics.

*

We assert that, even assuming the notion of 'space', when a physical system is placed in isolation, or when the quantum wave of a 'particle' is not being subject to system-disturbing physical inputs of observation, within a given time-interval, the system thus isolated must be immediately relegated to the future: i.e., to the future-form domain of spacetime in potentiality, or to the Whitehead facet of our prospective Bergson-Whitehead amalgam. In that state, the quantum wave constitutes a purely static state of nested futures: a delayed choice quantum switchboard of nested futures: (see Book-II, Part-II). As such, the quantum wave and wavefunction must elaborate to the infinite future in and as a pure spatially static state.

In its superlative form, Hilbert space in its totality constitutes a complex *static* state of delayed choice nested futures switchboard: one constitutive of the whole of spacetime in potentiality, even when we assume the notion of 'space' (see Book-II Part-II). Again, assuming space, the domain of spacetime in potentiality, hence Hilbert space', cannot permit the literal 'motion of the light-ray', or of the motion of *any* quantum mechanical wave, including de Broglie matter-waves.

All we now need to do is transplant the Moon-Earth relation and its 'moving light-ray' from 4.02 into the domain of spacetime in potentiality: a static domain of complex nested-future states, synonymous with Hilbert space... through which nothing can move, simply because it has not yet formed up into a purported 'space'.

*

From within the framework of the intermediate spacetime model developed in Book-II, and its successor in the anticipated superlative Bergson-Whitehead amalgam, the purported travel of our light-ray from the Moon to the Earth requires a minimum of two things: First, it requires the broken-discontinuous advancement of the event horizon from a prior past-primacy at event horizon-1 to a successive new recentcy at event horizon-2. It is through this that the apparent 'motion' of our light-ray becomes manifest. Second, the purported motion of our light-ray requires that the broken-discontinuous advancement of the event horizon resolve a succession of pertinent OR-form events: namely, the emission of our light ray from the Moon at even-horizon-1 (or from an emitter in the two-slit experiment) succeeded by its reception on Earth at event horizon-2 (or its deposition on the detector in the two-slit experiment). All we can be empirically certain of is the time-interval (implicitly, the Hilbert 'space') demarcated by event horizon-1 and event horizon-2. We will *not* be nor need be concerned with any purported 'spatial interval' alleged to reside between the Moon and the Earth, or between the succeeding event horizons, simply because we cannot empirically justify such a spatial interval, nor the supposed 'motion' along it of *anything*, including our 'light ray'... save by post-hoc assumption and per the post-event *as if* foisting of such a pleonastic notion.

In any case, the interiority that resides between the event horizons (ultimately a pure time-interval) could only comprise a spatially static quantum mechanical state: an as-yet not-realised Hilbert space through which nothing could move... and thus renders the notion of space superfluous and entirely dispensable. **The only sorts of intervals delineated by the processes and the progressions of the event horizon are time-intervals, not spatial ones: That is, 'Hilbert space' is in truth a Hilbert time-interval potential, not a 'space'. Successive progressions of the event horizon may give the impression of an "object moving" across a space, or of "displacing radiation at the speed of light". Yet, these are entirely dispensable pleonastic impressions.**

Hence, yet again, it is not the 'motion of light' at the 'speed of light' that furnishes the relativity of simultaneity and the attendant time-distribution of events and moments, but it is the temporal distribution of events, combined to the inherent relativity of simultaneity from the de-spatialised temporality of nature, that furnishes the *impression* of the 'speed of light' and of its 'space'... both of which turn out to be pleonastic.

4.06: Integrations:
Modes of Observational Inference, the Firewall Principle, the Intermediate Model Spacetime,
The Character of Time, & the Pleonastic Status of Space

As interim to our delayed choice argument in proof of the pleonastic status of space, we must make a summary reprise of pertinent background material from the preceding books. We must include such material as the *retrospective mode of observation and inference*; or the ontology of AND-form logic and the ontic reality of the future, and the nature of time; or the tripartite structure of the growing block intermediate model and the firewall principle.

We have already incorporated the consequences from the retrospective observational and inferential mode, the firewall principle, and many of the stated concerns... albeit in passing. Even as we seek to supersede the intermediate model of spacetime from Book-II, its tripartite structure, especially its future-form 'spacetime in potentiality', garners obvious implications that cohere seamlessly with the firewall principle, with the above-stated restrictions from quantum mechanical reality to the notion of the motion of the 'light-ray', and to implications from the retrospective observational-inferential mode vis-à-vis the time-distribution of information and the question of space… all of which segue to our central assertion for the impossibility of contiguity through simultaneity: more than sufficient to render space pleonastic, even before we develop our delayed choice and meta-delayed choice case for de-spatialisation.

*

Let us start with the retrospective observational and inferential mode: the only viable form of relation that frames of reference (noetic or a-noetic) are permitted to have vis-à-vis realised events and future-potentialities: (See Book-II, **2.23**). Recall that **we cannot garner information about the physical order in realtime observational mode, since this would require that we obtain simultaneity (or absolute time) with the event under scrutiny, as well as simultaneity with *all* events or ordinates** along the pertinent line-interval, across the totality of the co-ordinates that render two-dimensional surface-space, and throughout the totality of a three-dimensional block-space.

Nor can we garner information about the physical order in any a-causal *a priori* way, via 'signals from the future', *before* the events that constitute the observables have transpired, given the operation of the firewall principle from the intertwine of causality and the conservation principle (Preliminary: **0.15** and Book-II: **2.23**). Hence, **we can only rely on the retrospective observational-inferential mode vis-à-vis observables of nature. It follows that, insofar as realised events and outcomes are concerned, we can only obtain information about the temporally distributed past:** All we are permitted to observe *is* the past, whether of the most recent past or the most remote past, notwithstanding as-yet not-realised futurity. It necessarily follows that we are 'surrounded', *not* by 'objects distributed in space' but by past events distributed and organised in time, notwithstanding future potentiality. That is, **by the imposition of the retrospective observational-inferential mode alone, and even if we erroneously ignored the impossibility of contiguity and extensionality through simultaneity, we can only observe resolved *whens*: distributed according to primacy-recency rules, *without* 'space'… constituted as *memory in stereotemporal time*.**

*

Recall that, as part of the *raison d'etre* of the AND-form quantum mechanical wave, we espoused that the wave constitutes the ontic objective state of the future component of the growing block universe: (Preliminary: **0.10**; Book-I: **1.07** and **1.09**… and Book-II: **2.03**). From this emerged **the firewall principle: the intertwine of causality and conservation laws, which ward against 'signals from the future'**. The firewall principle contributed to the tripartite structure of intermediate spacetime, in that **it forced the ontic reality of the future, compelling the incorporation to intermediate spacetime a futurity**, or 'spacetime in potentiality'.

Time was grasped, not as a 'fourth dimension' but recuperated as synonymous with AND-to-OR wavefunction collapse itself: Succinctly, time *is* the collapse of AND-form futures comprising spacetime in potentiality into leading or recency OR-form events realised on the event horizon.

Essentially, **spacetime in potentiality constitutes the total 'Hilbert space' and engenders the same consequences to our 'light-ray' espoused in the previous essay (4.05), only augmented by implications from the firewall principle**: Indeed, Hilbert space (more appropriately, Hilbert time) and spacetime in potentiality are essentially synonymous. Recall in the previous essay that, since Hilbert space constitutes a future possibility for space within spacetime in potentiality, but one that has not yet formed up, it followed that nothing could move through it. From the spacetime in potentiality treatment of the same, the light-ray is expected to undergo its motion *in the future:* It has not yet done so, and, as such, it must exist as a static state of future possibility within spacetime in potentiality: a non-moving 'light-ray'. Hence, the light-ray is no more capable of 'motion' than is the whole of static non-undulating spacetime in potentiality or futurity.

To generalise, light radiation is quantum mechanical: its future potential, erroneously interpreted as 'motion', constitutes a static potentiality within static non-undulating spacetime in potentiality. **The domain of spacetime in potentiality cannot move, because… obviously, the future cannot move (Indeed, where could the future move?). It follows that the electromagnetic light radiation embedded within that future is no more capable of moving than is the future in which it is embedded**.

It follows that the series of observations we garner about events reported via light radiation and electromagnetic processes do not entail inherent motion of that light radiation along a 'path' supposedly formed out of contiguous positions through simultaneity, or absolute time, or 'space'. Indeed, the reports issued by light constitute nothing more than the succession of OR-form events, always subject to the relativity of simultaneity, that have decayed out of non-moving static spacetime in potentiality. *This* cannot be otherwise: **To presume that the light radiation is 'moving' within space is to presume that it is moving through spacetime in potentiality: the future. But physical proof of this would entail the observation of such motion via 'signals from the future'**, at the expense of causality and the contravention of the conservation principle. In short, **the firewall principle prohibits 'signals from the future'. Therefore, the sought physical proof of the 'motion of light' at the supposed 'speed of light', cannot be furnished,** and, for reasons given, the notion of 'moving light' is rendered totally pleonastic.

*

The expected replacement of spacetime in potentiality with the Whitehead facet, part of the anticipated Bergson-Whitehead amalgam, will not gainsay any of the above: First, the demise of space is integral to our prospective Bergson-Whitehead amalgam, which constitutes spacetime *without* space. Second, the Whitehead facet, as the total future, does not move: It is as static as the former spacetime in potentiality that it is destined to replace. Indeed, all the arguments above, including those from the firewall principle, abide in the anticipated amalgam.

Moreover, as 'spacetime' without space, the Bergson-Whitehead amalgam is perfectly consistent with the notion of a non-moving non-undulating static quantum mechanical light radiation, or de-spatialised light radiation.

372

QUANTUM MECHANICS AND MIND

*

Time is not a 'fourth dimension'. Time is AND-to-OR wavefunction collapse. As such, all putative claims require that critical signature-events furnished in supposed 'proof' of motion must decay out of the future through AND-to-OR collapse into OR-form exhibits in proof of that motion. Notice that, even if one insists in the veracity of the motion of light across its putative spatial interval, the minimum required two successive OR-form events to assert such a forlorn notion will have decayed out of a purely non-moving static futurity, which undercuts the ontology of motion and space per function of its static nature alone.

Moreover, the very character of AND-to-OR collapse (or time) usurps the possibility of contiguity through simultaneity by simply undercutting the presumed contiguity between 'strings' of 'points' or 'positions' associated with the succession of generated OR-form events. **Thus, AND-to-OR collapse and time generates a discontinuous string of events (of *whens*) but cannot furnish any contiguity between such events**. In other words, time is quantised and leaps over the presumed 'points of contiguity' without realising these, from one OR-form event to a temporally succeeding OR-form event, *without* integrating contiguity. This is consonant with our original claim from information-matter dualism furnished in the Preliminary and in Book-I: the case for the broken and discontinuous form in opposition to the seamless-continuous form of association of information with putative radiation, matter and space: In its own right, information-matter dualism is also destructive of contiguity through simultaney. This should be of no surprise, given that information-matter dualism constitutes merely the mirror reflection of the deeper tripartite structure of intermediate spacetime and its successor, the Bergson-Whitehead amalgam.

Finally, **given the demise of space, our old suspicions regarding generic pseudo-realtime descriptions of quantum mechanical waves in the various staple experiments, and notional presumptions about 'paths' and 'which way' information claims first deconstructed in the Preliminary, are fully justified**. The demise of space, and all other contributory findings garnered from the inexorability of the retrospective observational mode, from the ontology of the quantum mechanical wave as the ontic objective future, from the firewall principle, and from the static character of the future, especially *en toto* and of quantum mechanical radiation and matter potentialised within that future... all consolidated in the tripartite structure of growing block spacetime *and* in its de-spatialised successor... undermine pseudo-realtime depictions of 'displacing quantum waves' and of moving light radiation at the supposed 'speed of light', dependent as these are on contiguity through simultaney, the tandem implicit and erroneous assumption of absolute time, and the synonymous error in the supposition of 'motion in space'... and of forlorn the ontics of space.

4.07: John Wheeler's Delayed Choice & the Problem of Space

In Book-II we discovered and developed the nested futures view of the quantum mechanical wave. It was extrapolated to the intermediate spacetime model and to spacetime in potentiality: A nested quantum mechanical wave hedges for and nests mutually incompatible divergent, entangled, and convergent futures (Book-II, **2.22**). Indeed, the nested futures wavefunction constitutes a wavefunction of alternative nested wavefunctions, and not simply or only of discrete or immanent possibilities. This discovery had direct consequences to the question of whether quantum waves are spatially dynamic or static. The generic assumption of and presupposition for a dynamic time-undulating and spatially moving quantum wave suffered in the light of both the nested futures view and from the character of intermediate spacetime: This is obvious from the assertion that 'spacetime in potentiality', or the future, cannot 'move', and neither does the nested futures quantum wave constitutive of that future. What remained was concession to spatially static quantum mechanical waves.

In this essay, we will complete the case for de-spatialisation by expanding our application of Wheeler's delayed choice approach beyond the assertion for non-undulating spatially static quantum mechanical waves. We will obtain a superlative meta-delayed choice approach that directly attacks the very notion of space itself, independent of spatially static waves, and independent from the case for the impossibility of contiguity through simultaney furnished in **4.01**, and from the other critiques of space and of the notion of the motion of light (see **4.02** through to **4.06**). To this end we must first reiterate the case for spatially static quantum waves furnished in Book-II. As stated, this involved the utilisation of John Wheeler's delayed choice approach: (see **Fig. 4.02**). Therein, we envision a split-beam nesting and **hedging quantum mechanical wave** that spans the interval from **A**, to **mirror-C** and **mirror-B**, through to just before **switch-x**. This quantum wave nests or embeds (i.e., hedges) two distinct mutually incompatible futures by function of the delayed choice element **switch-x**, which is placed along path **C to D**. Indeed, **switch-x** is effectively AND-form suspended between two alternate settings. One setting can shunt one leg of the split beam to **Y**, while the other setting can allow the same split beam to **detector-D**. Only one of these nested futures will be selected for realisation when **switch-x** is tripped.

As stated, **switch-x** is suspended between two possibilities, either because the setting is decided at the last-ditch moment, in which case the switch-possibilities must be treated as AND-form suspended futures, or perhaps an EPR type device is attached to randomly select which switch-setting will get to transpire at the required moment: again, **switch-x** must be treated as an AND-form suspension of futures.

The 'delay' in delayed choice pertains to the timing of the impeding switch-trip, which abides up to the very last moment of approach of the pertinent leg of the split-beam to **switch-x**. There is a time-window of opportunity in which **switch-x** is suspended and might be tripped into one setting or the other: The future has not yet happened: It follows that **switch-x** has not yet tripped. Therefore, again, it *must* be described as a future-state in AND-form suspension: constituted as...{**toward-Y** AND **toward-D**}. It is this characteristic of the default quantum suspension of **switch-x** that fashions the split-beam wave into a *nesting wave*, generalised to the total future grasped as 'spacetime in potentiality', to the infinite future.

What is being nested therein are two alternative mutually incompatible future wavefunctions, pending the AND-to-OR resolution of **switch-x** to one or the other setting, with both alternative wavefunctions in AND-form nested future suspension owing to the ambiguity of the setting on **switch-x**. The wavefunction that describes the split-beam if **switch-x** is tripped to **toward-Y** constitutes a different wavefunction and outcome vis-à-vis the other nested wavefunction per **switch-x** tripped **toward-D**. Hence, the two distinct wavefunctions are contained in (or nested to) the primary hedging quantum wave and wavefunction describing the split-beam from **A**, to **mirrors-C** and **B**, through to just before **switch-x**.

Notice that we have described our nesting and hedging quantum wave *as if* it constituted a 'moving wave'… as if we could follow its travels in realtime from **A** to **switch-x**. However we treat quantum waves, clearly, in the framework of delayed choice, the hedging quantum wave must hedge two mutually incompatible divergent wavefunctions, per function of the default suspension of **switch-x**.

If we assume the *nesting hedging quantum wave* on its approach to switch-x as spatially dynamic, could the mutually incompatible nested future possibilities hedged to it *also* be described as spatially dynamic? Could these be described as undergoing literal spatial displacement toward-Y and toward-D along their respective X-Y and X-D paths, even though their nesting hedging wave has not even obtained switch-x? Or are these nested futures exceptions: to be treated as spatially static states?

Before **switch-x** is tripped, the nested future quantum waves could not yet have coincided with any part of their *supposed* spatial paths: namely, **X-Y** and **X-D**. Thus, **the nested waves cannot be treated as states displacing across X-Y and X-D, respectively, precisely because these have not yet obtained or coincided with any part of their putative paths. Yet, both states, as nested futures, *are* contained 'inside' the hedging supposedly 'moving' quantum wave from A through to switch-x.**

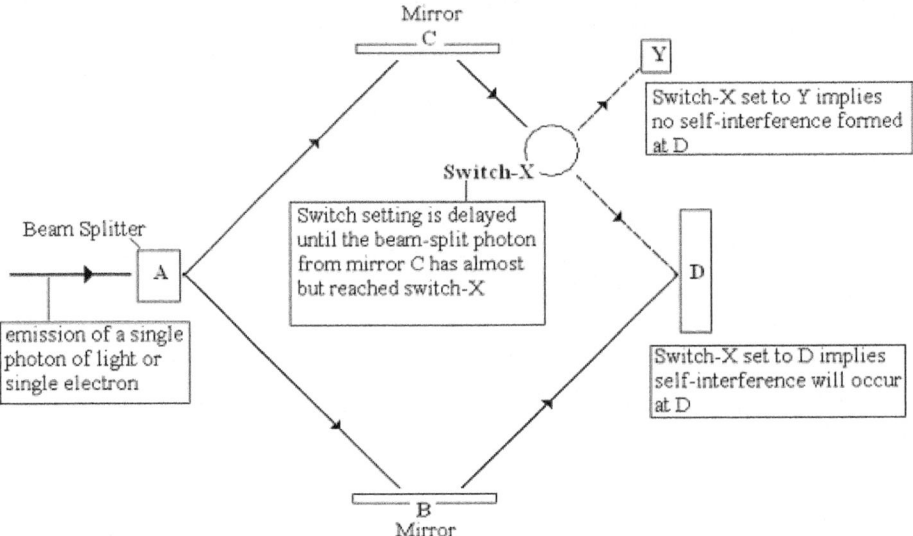

Fig. 4.02: Illustrating the outlay of delayed choice
Notes: The diagram illustrates the basic outlay of John Wheeler's delayed choice experiment. A single incoming photon of light is 'split' by the beam splitter at **A** which works in a similar way on particles as does the barrier in the double-slit experiment at 'both slits open'. The particle is split by **A** into two quantum mechanical wavelets of possibility: One wavelet heads toward **mirror-C** and the other to **mirror-B**. The wavelets will be deflected by the pertinent mirrors toward **detector-D**. Normally, once the wavelets reach **detector-D**, they will self-interfere to form an interference pattern at **detector-D**. Thus, the experiment is essentially a version of the double-slit experiment, save that the barrier is replaced by the beam splitter. However: The operation of **switch-x** radically alters the experiment. **Switch-x** is effectively suspended at two settings until the last moment: One permits the quantum wavelet from **mirror-C** to reach **detector-D** and form the expected self-interference. The other setting diverts the wavelet from **mirror-C** to **detector-Y**: thus, no self-interference is formed at **detector-D**. Now: the setting of **switch-x** is suspended: Thus, the setting is *delayed* until after the wavelet from **mirror-C** has passed **mirror-C** and is already on its way to and has almost reached **switch-x**: That is, suspended **switch-x** is only set to **Y** when it ought to be too late for the quantum mechanical system to adjust to the alternate possibility implied by **Y**, if only because the split beams have already passed the pertinent mirrors. It turns out that the quantum mechanical wave hedges, or *nests*, within its possibility-set the mutually incompatible alternate futures. i.e., the wavefunctions for both self-interference at **D** AND diversion to **Y**; implying the nested futures view of the quantum wave, further radicalised when re-cast in terms of the spatially static quantum wave in the framework of intermediate spacetime and its de-spatialised successor, the Bergson-Whitehead amalgam.

*

Let us clarify and augment: The nesting or hedging quantum wave, purportedly moving toward **switch-x**, hedges the two mutually incompatible nested futures: One nested wave constitutes the future-form of…

{… "the wave has been shunted to, and is moving toward, detector-**D,** along **X-D**": such that, other leg goes along BD} …

….**AND**…the other nested wave constitutes...

{… "the wave has been shunted to, and is moving toward, detector-**Y,** along **X-Y**": such that, other leg goes along BD}.

That is, although it has not yet coincided with putative paths **X-D** and **X-Y**, the hedging quantum wave on approach to **switch-x,** must yet nest and hedge the said two alternate mutually incompatible futures expected to coincide with their own said paths *in the future,* even though neither these nor their hedging wave has yet passed beyond **switch-x.**

How do we solve this contradiction?

Now, the subsumed nested future wave comprised of...

{"…the wave has been shunted to, and is moving toward, detector-**D,** along **X-D**"…} …

is in the hedging quantum wave on approach to **switch-x**. Yet, in truth, it is *not* literally "shunted to and moving toward detector-**D** along **X-D**": Instead, it and its path **X-D** constitute purely static as-yet not-realised *future* potentialities embedded within the future, precisely because the future has not yet transpired.

The same must also hold for the alternative nested future comprised of...

{… "the wave has been shunted to, and is moving toward, detector-**Y,** along **X-Y**"…}.

This is also a spatially static pure future-potentiality: neither moving nor undulating... embedded as a static future state within the supposedly 'moving' hedging wave. Again, to assume otherwise (to assume that the mutually incompatible nested waves must be dynamic and moving across **X-Y** and **X-D**, even though these have not yet reached their pertinent 'paths') would be to assume that putative spatial intervals **X-D** and **X-Y** are somehow 'located' before **switch-x,** even though both come after **switch-x,** both in terms of spatial distribution and in terms of potential temporal sequence and order.

The problem can easily be avoided if, first, we appreciate nested future quantum waves for what they are: static non-moving and non-undulating forecast and possibility-states embedded within spacetime in potentiality; the future. Even assuming that as-yet not happened intervals **X-Y** and **X-D** are spatial, at best, these could only constitute paths within *potential space* (i.e., Hilbert space): Hence, they are *not* realised spatial intervals but purely future-potential ones. Obviously, nothing can move within, across, or through such as-yet not-realised space-intervals, and this can only reinforce the static quantum wave account of nested future quantum waves and wavefunctions garnered from the nested futures delayed choice approach.

To put it in terms of the larger system of spacetime in potentiality from the intermediate spacetime model, the intervals **X-Y** and **X-D** are potential spatial intervals that reside 'above' the event horizon... 'above' the synonymous AND-to-OR collapse and time processes that have not yet attained or realised these. Indeed, as repeatedly stated, the paths are not-yet realised constituents of future-form spacetime in potentiality. Therefore, the paths **X-Y** and **X-D** cannot be treated *as if* they are being traversed by *anything*, let alone by 'spatially dynamic quantum waves'. Thus, whatever assumption we may garner about the hedging wave, the futures that it hedges are certainly static, non-undulating and non-moving quantum mechanical waves.

To augment, if we can envision immanent possibilities in AND-form superposition quantum waves as not-yet-realised events, or as pre-event static potentials, it is not an impossible demand that we envision the futures and wavefunctions nested to those immanent possibilities as not-yet happened possibilities in their own right, in AND form… as nested waves… and as *also* static. And because these have not yet transpired, we can only treat them as static potentia, whatever we might erroneously attribute to the hedging wave. i.e., its supposed motion in space.

*

Throughout the above, the nesting hedging quantum wave was treated *as if* it was spatially dynamic; as something moving and displacing across space along its path from **A** through to **switch-x**. But four things must bring this assumption into question: First, per Book-II, essay **2.23,** *any* realtime observation of this "moving hedging wave" remains physically impossible, given the exclusivity of the retrospective mode of observation and inference, recapitulated, and summarised in the preceding essay, **4.06.**

Second, per generic quantum mechanics, any attempt at *any* observation of the hedging quantum wave will only serve to collapse and destroy that "moving quantum wave"; destroying the very thing we sought to confirm through observation; denying it to observation; rendering the sought observation and proof impossible to obtain. (See **4.04** and **4.05**).

Third, a pure spatially static treatment of the hedging quantum wave will be more than adequate to furnish *all* the various known and expected empirical outcomes produced by the delayed choice experiment. Therefore, the presumption that our nesting wave on approach to **switch-x** *must* be a spatially dynamic quantum wave is neither indispensable nor necessary for the viability of the science, of nature, or of observation and experience. Again, it is a matter of the economics of science, but arrayed against the notion of motion and space. (See **4.03**).

375

Fourth, **when we considered that the nested futures within the hedging quantum wave are spatially static, this shows that there *are* such things as a spatially static quantum mechanical waves. The finding can be generalised to *all* quantum mechanical waves and wavefunctions. Even on the assumption of the reality of 'space', if the reality of static quantum waves remains unavoidable in one area, there is simply no reason to treat the hedging quantum wave as specially exempt. Hence, there is no compelling reason, or *any* reason, to suppose that the nesting hedging wave itself is any less spatially static than the static nested waves and wavefunctions that it embeds.**

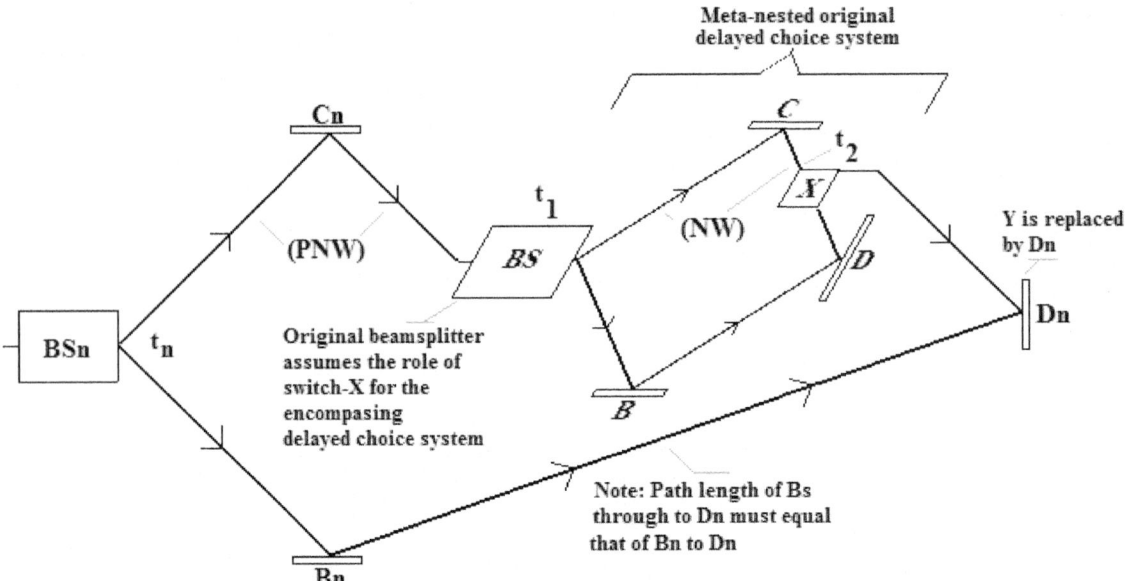

Fig. 4.03: Meta-delayed choice: the original delayed choice nested into a larger delayed choice system
Notes: In the meta-delayed choice system, the original delayed choice experiment, together with its purportedly moving nesting wave, together with its original components (i.e., the beam splitter **BS**, the mirrors **B** and **C**, **switch-x** and detector-**D** –but with **Y** omitted) is nested to a preceding encompassing delayed choice system. This latter system is comprised of its own beam splitter **BSn**; mirrors **Bn and Cn**; and detector **Dn**. Only **switch-Xn** is omitted: It is substituted with the original beam splitter **Bn** from which the original delayed choice experiment setup now extends as a niched system. In the now-subsumed original delayed choice experiment, **Y** is omitted entirely. Instead, **Switch-X**'s alternate 'path' now leads to the new detector **Dn**: part of the larger delayed choice system. Recall how, in the original delayed choice experiment –now the embedded system- the nesting wave **NW** along **Bs** through to mirror-**C** and just before **switch-x** (i.e., from t1 to t2) was presumed to be spatially dynamic and moving, even though the mutually incompatible futures that it nested could not be so characterised and were understood to be spatially static. Applying the same logic as was found to operate in the original delayed choice experiment, insofar as the original hedging nesting wave **NW** is now a nested split-beam wave in its own right (i.e., embedded as one of the mutually incompatible nested futures within the precursor nesting wave **PNW** along **BSn** to **BS**) consequently, the original nesting wave **NW** along **BS** and up to **switch-x** must now be grasped as a spatially static state. On the other hand, the preceding encompassing nesting wave **PNW** from **BSn** up to **BS** (i.e., from tn on approach to its subsequent split at t1) must be considered the *actual* spatially dynamic moving wave. In short, from the meta-delayed choice perspective, the original nesting wave from **BS** up to **switch-x** turned out to be spatially static. Indeed, there is no single stand-alone delayed choice system to which a moving nesting wave status could be attributed: All delayed choice systems are embedded within larger precursor embedding delayed choice systems that elaborate back to the indefinite past or to the Big Bang moment: at which resides the very first and last purported nesting 'moving wave', and, with it, the very first and last instance of 'space'.

There is a fifth and final reason why we must assert that the hedging wave (and *all* quantum mechanical waves) must constitute spatially static states. This is rendered obvious when we make the hedging wave *itself* the subject of an earlier or preceding hedging quantum wave: one belonging to an earlier or preceding delayed choice system. In other words, **it is possible to treat our hedging wave as itself one of two preceding nested futures hedged to a *preceding* hedging wave, subsumed to a larger 'meta-delayed choice' system.** Such a system is portrayed in **Fig. 4.03**.

As can be seen in **Fig. 4.03**, the delayed choice experiment, together with its purportedly moving hedging wave **NW**, together with its original experimental components (i.e., the beam splitter **BS**; mirrors **B** and **C**; **switch-x**, and detector-**D**... but with **Y** and path X-**Y** omitted) is nested to the larger preceding and subsuming meta-delayed choice system. This latter subsuming system is comprised of its own beam-splitter **BSn**; mirrors **Bn** and **Cn**; and detector **Dn**. Only its **switch-x** is missing: This is replaced by our original whole delayed choice experiment complex, now integrated to the larger meta-delayed choice system at beam-splitter **Bn** and at the modified **switch-x-to-Dn**.

Again, Y and X-Y path is omitted entirely. Instead, **switch-x**'s alternate 'path' now leads to the new detector **Dn**: part of the larger embedding meta-delayed choice system. Thus, our delayed choice system is portrayed as fully nested within an embedding meta-delayed choice system.

Applying the same logic as was found to operate in our original delayed choice experiment, insofar as the original hedging nesting wave **NW** is now itself nested to an embedding meta-delayed choice system, and merely constitutes one of two mutually incompatible nested futures within a hedging *meta-nesting wave* from **PNW** to **BS**, it turns out that the original hedging wave **NW** along **BS** and up to **switch-x** must now be grasped as a spatially static state, nested and embedded to the meta-nesting wave **PNW**.

Consequently, **PNW** along **BSn** up to **BS** (i.e., from **tn** on approach to its subsequent split at **t1**) must itself be treated as the s*eeming* spatially dynamic quantum wave, while **NW** must now be rendered static. In short, from the meta-delayed choice perspective, the original *hedging* quantum wave **NW** turns out to be a spatially static state, nested to a preceding meta-nesting wave **PNW**.

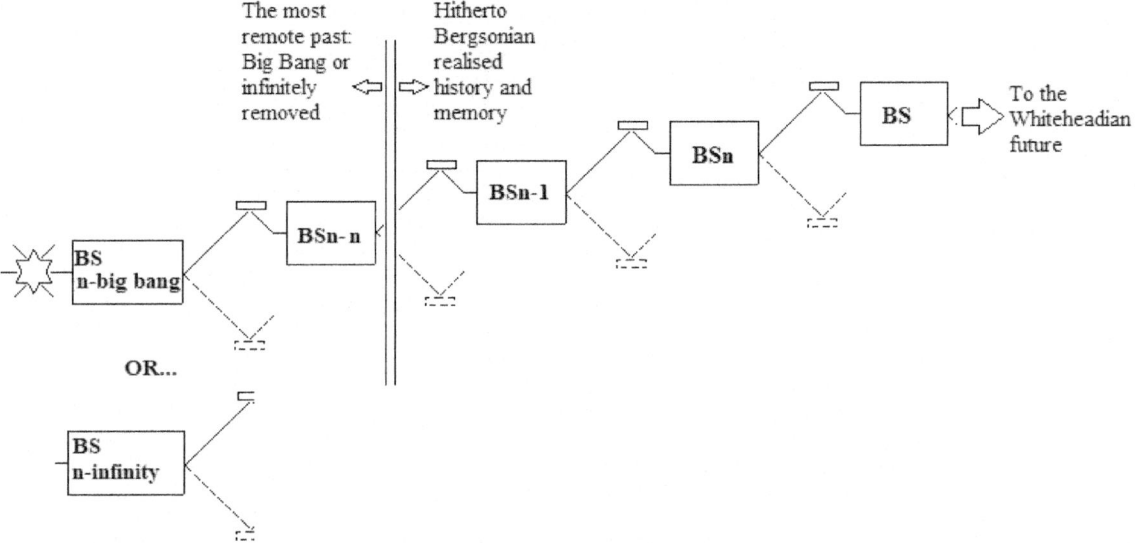

Fig. 4.04 Penultimate and ultimate meta-delayed choice system
Notes: The original delayed choice system from **BS**, together with its purportedly 'moving nesting wave' **NW**, turns out to be subsumed to the meta-delayed choice system that extends from **BSn** and its nesting wave **PNW**. Thus, **NW** -nested to **PNW**-turned out to be spatially static. But before we presume **PNW** itself to be spatially dynamic, we find that the delayed choice system from **BSn** is itself subsumed to a larger preceding meta-delayed choice system: one that extends from **BSn-1**. Thus, **PNW** must also have been spatially static: it was a static state nested to an even earlier precursor nesting wave from **BSn-1**. However, the delayed choice system from **BSn-1** also turns out to be subsumed to an even larger multiple successive preceding meta-delayed choice systems: a penultimate system characterised as a meta-delayed choice systems nested within larger meta-delayed choice systems: elaborating back to the most remote past ending at **BSn-n**. This penultimate meta-delayed choice system is in turn subsumed to an ultimate meta-delayed choice system extending out to just after the Big Bang itself, or, to one that elaborates to the indefinite infinitely removed past. Hence, to maintain the idea of the motion of the spatially dynamic quantum waves in 'space', we must insist that both the purported 'motion' and the 'space' along which it happened were plausible *only* at a moment just after the Big Bang, or else at the indefinite and infinitely removed past. Neither of these is empirically demonstrable, provable, or eligible to experience. There is no 'space'. There is only time. There is no 'matter in space'. There is only memory and possibility in time.

However, before we erroneously conclude that **PNW** must be spatially dynamic, consider the penultimate meta-delayed choice system depicted in **Fig. 4.04**. Therein, we garner that *all* delayed choice systems and their hedging quantum waves (including **PNW** itself) are embedded within larger preceding and subsuming meta-delayed choice systems that elaborate to the infinitely remote past, or at least to the remote Big Bang moment itself; presumably the very first and last instance of a 'moving quantum wave', and the first ever prime meta-delayed choice system. Therein, **any given presumed moving hedging wave is rendered into a static possibility state; one embedded and nested to a preceding hedging wave. But the preceding hedging wave cannot itself be considered moving: it must also be static: it must itself be embedded and nested to an even earlier hedging wave. This in turn cannot be** considered the moving wave... and so on, until we either reach the interval just after the Big Bang moment or obtain the infinitely remote past.

This last thought, illustrated in **Fig. 4.04**, demonstrates the spuriousness of the notion of the motion of the quantum wave: the absurdity of motion in general, and, ultimately, the absurdity of the notion of the 'space' in which forlorn motion purportedly transpires. To maintain the plausibility of the spatially dynamic quantum wave, of motion in general, hence the 'space' in which all motion supposedly transpires, one must entertain a first and last brief moment in which 'space and motion' might have been possible: namely, at the moment just after the Big Bang itself, or at the beginning of time at the infinitely remote past. In either case, **the first and last instance of both motion and space would belong to a remote and inaccessible past, beyond** *any* **possibility of direct observability, demonstrability, or any viable empirically provability and science. But what possible usefulness could such a brief interval of non-provable 'space and motion' bequeath to the rest of cosmic history or subsequent physics, save as an ultimately false and unnecessary spurious pleonastic term?**

It is finally obvious that 'space' is indeed wholly pleonastic: not part of the real ontology of physics, and ultimately superfluous to science, no matter how pragmatically useful both motion and space might appear to be to the formation of, say, artillery trajectory tables and other a naive but useful forms of physics.

4.08: Toward the Reformation of Special Relativity:
From Spatial to Realised & Potential Delayed Choice Time Interval Relations

With de-spatialisation, all things must transform into potential or realised information and memory, organised and distributed in time. But what is to be the replacement for 'distance'? How should we recuperate into pure temporal form such things as uniform and non-uniform motion and acceleration, or even the vaunted 'speed of light'? To address these issues, we must first assume the false notions of distance and space, but with a view to superseding these at the opportune moment: Thus, first, we take an object located at some purported 'fixed distance' from a given observational frame of reference. For example, the Moon roughly assumes something analogous to an object at a stable unchanging putative distance from observers on Earth, *ceteris paribus*. We mount a laser reflector on the Moon and use it to bounce laser signals to and from the Earth; clocking the time taken for the beam to reach the Earth and back to the Moon: a thing historically accomplished by both NASA and the Soviets [81]. Knowing the purported 'speed of light' and the time taken for the beams to bounce to and from Moon and the Earth, we can calculate the 'distance'.

However, the moment we move beyond purported 'space and distance', as we *must*, given de-spatialisation in **4.01** through to **4.07**, what we obtain is an expression of the relation between the Earth and the Moon (or between any 'objects' or events, or reference frames) expressed purely in terms of uniform or non-uniform delayed choice time-interval relations. The typical delayed choice time-interval relation that applies to the Moon-Earth is around 1.5 seconds, *ceteris paribus*: That is, in order to see the Moon 'now' from a given relative reference frame on Earth, or for the tides on Earth to be affected by events constituting the Moon 'now' (and this would transpire even in the absence of conscious observers on Earth) will require a delay in the time required equal to 1.5 seconds. At the end of the 1.5 second interval, as observers, we may choose to look or not look at the Moon, although the tides will do what they will regardless even if we were not present on the Earth.

The' choice' in delayed choice does not espouse any subjectivist or idealist component: All it asserts is the *termination of suspension* (the 'throw of **switch-x**') often brought about by non-conscious causal processes. The subsequent throw of the proverbial switch often constitutes an intercession by some physical system or variable, such as the interaction of the gravitational effects of the Moon with the waters of the Earth that then generate tidal effects, even in the absence of conscious beings... or the training of a telescope toward the Moon by some observer on Earth, with key events on the Moon witnessed retrospectively *before* the attendant observer (or a pure software-driven non-conscious detector) get to witness, or get affected by them. Given the retrospective mode-based relation of *any* conscious observers or a-noetic devices vis-à-vis past events organised into delayed choice time-interval relata, the term 'choice' hardly supports idealism or exclusive subjectivity.

Of course, the 'speed of light' is implicit to the delayed choice time-interval relation that resides between the Moon and the Earth, or between any two corpuses of events subject to the relativity of simultaneity, whether or not these are undergoing purported 'motion'. If the Moon was 'closer' to the Earth, the delayed choice time-interval relation between the Moon and the Earth would be shorter than the approximate 1.5 seconds. In anticipation of what is to come, the replacement to the 'speed of light' reframed in terms of pure delayed choice time relata will be the **delayed choice time-interval standard.**

In the framework of the intermediate model spacetime, the delayed choice time-interval relation between the Moon and the Earth, or between any corpus of events in the universe, is an AND-to-OR collapsed realised delayed choice time-interval relation; one generated through the attendant broken-discontinuous progression of the event horizon 'upward' the system; a progression ultimately generated by gear-less causality and time (see Book-II **2.46** and even **2.47**). In the case of the Moon-Earth relation, in its idealised form, we obtain a delayed choice time-interval equal to 1.5 seconds, demarcated by, say, a preceding event horizon at EH-1 and a succeeding event horizon at event horizon EH-2.

---

[81] Chapront, J.; Chapront-Touzé, M.; Francou, G. (1999). "Determination of the lunar orbital and rotational parameters and of the ecliptic reference system orientation from LLR measurements and IERS data". *Astronomy and Astrophysics*. 343: 624–633

Thus, let us first re-conceptualise and recuperate uniform motion as a de-spatialised pure temporised process. To this end, we assume that an object is approaching an observer (or some reference object) at a purported 'uniform motion', or 'constant velocity'. How is this to be recuperated in pure delayed choice time-interval form, given that we can no longer use 'space' or 'distance', or even attendant 'motion'? We do so by grasping it as a uniform change in the delayed choice time-interval relation residing between the object and observer: The time-relation will change and shorten from moment to succeeding moment in a uniform way: say 0.05 seconds of reduction to the delayed choice time-interval standard at the first succeeding moment, followed by another 0.05 second reduction at the second succeeding moment... and so on. **From moment to succeeding moment, the delayed choice time-interval relation between the two objects will be reduced by the same fixed *uniform* interval of time; and the "approaching object" ... approaching from the past or from a former state of relative primacy into an evolving state of relative recency. The object is destined to realise a state of *convergence to simultaneity* vis-à-vis the other reference-point or 'object', or observer.**

Once the object passes the moment of convergent simultaneity, the delayed choice time-interval will change versus the said standard. This will again transpire in a uniform way from moment to succeeding moment, in a fashion that now increases the delayed choice time-interval relation, but away from a state of simultaneity, with the 'moving' object receding to the past... receding from recentcy to primacy.

Notice that 'movement' as such is replaced and recuperated as either a uniform time-relation reduction approaching convergence to simultaneity, or uniform time-rarefaction away from simultaneity toward recession into the past.

In the background, both within the context of the intermediate spacetime system *and* within its successor amalgam, the moment-to-moment uniform change in the delayed choice time-interval relation will be generated through, and in tandem with, the broken and -discontinuous quantised progression of the event horizon; with the events (i.e., the 'objects') at first relationally converging toward a common relative present-continuous moment of recentcy and simultaneity, and, thereafter, diverging and receding toward past-primacy, in which their delayed choice time-interval relations will become accentuated or rarefacted over time.

Thus, if the Moon were to fall toward the Earth with 'uniform motion', from moment to succeeding moment, there would be a uniform reduction in the delayed choice time-interval relation that resides between the Moon and the Earth as the Moon shifted toward greater and greater recentcy relation vis-à-vis the Earth, until convergence to simultaneity occurred at an approximate common present moment: one in which the Moon and the Earth merged into a state of near-simultaneity. Assuming no collision occurred, thereafter, from moment to succeeding moment, there would be a divergence in, and a rarefaction of, the delayed choice time-interval relation between the Moon and the Earth, as the Moon receded into past-primacy per the uniform rarefaction of its delayed choice time-interval relation vis-à-vis the Earth.

*

How would we recuperate acceleration in terms of events organised in time, given that we cannot use space or distance, or any attendant notion of motion? In this case, we obtain a non-uniform change and reduction in the delayed choice time-interval relation between the Moon and the Earth: The delayed choice time interval might reduce by 0.05 seconds against the delayed choice time-interval standard (the replacement to the 'speed of light') at the first succeeding moment... followed by 0.06 seconds of reduction at the second succeeding moment; followed by a reduction by 0.07 seconds at the third moment... and so on. Hence, **the delayed choice time-interval between an approaching 'accelerating' object and the destination-frame will reduce in a temporally non-uniform way, converging toward a state of simultaneity. As the object 'passes' the pertinent reference moment of convergence to simultaneity and 'accelerates away', the delayed choice time-interval defining their time-relation interrelation will again change, but in a non-uniform way... from their near-simultaneity to the object's non-uniform time-recession into past-primacy.**

Within the context of our intermediate model spacetime from Book-II, but true also with respect to the Bergson-Whitehead amalgam successor, this moment-to-moment non-uniform change in the delayed choice time-interval relation between the object and the destination frame will be generated through the broken-discontinuous progression of the event horizon; say, from EH-1 to EH2... through to successive EH-n... with the events (the 'object' and destination-frame) temporally converging in non-uniform fashion to a common moment of simultaneity, and, thereafter, diverging and receding... in terms of non-uniform time-interval change... into the past, toward an ever-accentuating primacy.

Thus, if the Moon were to 'fall' toward the Earth with a 'non-uniform motion', from moment to succeeding moment, there would be a non-uniform reduction in the delayed choice time-interval relation between the Moon and the Earth, until convergence occurred at an approximate common present moment of simultaneity. Thereafter, and baring collision, there would follow a non-uniform recession in the delayed choice time-interval relation between the Moon and the Earth, as the Moon receded into past-primacy vis-à-vis the Earth and its observers.

In summary, the idea of uniform and non-uniform 'motion in space' is entirely replaced by uniform and non-uniform change in the delayed choice time-interval relation between observables and between frames: Either corpuses of past events constituted as the 'objects' are converging from primacy to recentcy toward a common state of simultaneity... or they are receding 'away' into past-primacy, per relations expressed *purely* in terms of time and in terms of the relation of the past to the relative present vis-à-vis any given frame of reference... through relative uniform or non-uniform change in the delayed choice time-interval relation between stated corpuses of events, against a delayed choice time-interval standard: i.e., the replacement to the 'speed of light'.

Thus, we have replaced 'motion' with *pure* uniform or non-uniform change in delayed choice time-interval relations, *without* 'space' and *without* 'motion'.

*

Fig. 4.05: Basic equations of Special Relativity recapitulated in terms of c^i : Special Relativity without space

A) Relativistic time dilation equation reformed

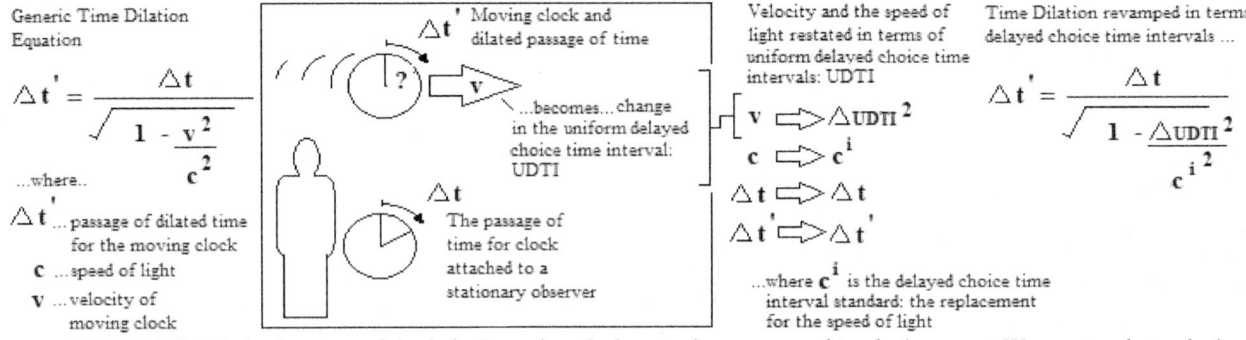

Notes: In Special Relativity (see central depiction) moving clocks run slow compared to clocks at rest. We assume that only 1 second has elapsed per the clock attached to a stationary observer. A moving clock passes the stationary observer at velocity $v = 0.95c$ (c = the speed of light … = 1). By how much does the moving clock run slow? Per the generic equation on the left, it will have slowed by **delta-t':**(Λ-t'). Plugging in the values for elapsed time per the stationary clock **delta-t**, velocity v, and the speed of light c into the generic time dilation equation, **delta-t'** will equal 0.0975 seconds: i.e., for every second elapsed at the stationary clock, only 0.0975 seconds will have elapsed at the moving clock. Given de-spatialisation, and the need to express time dilation in terms of delayed choice time-intervals, on the equation on the right, we substitute in place of v, **delta-UDTI**, (the uniform delayed choice time-interval) and replace c with c^i: the replacement for the speed of light (which still equals 1), where **delta-UDTI** equals 0.95 of c^i. We obtain the same result for time dilation, but *without* 'space'.

B) Relativistic length contraction reformed

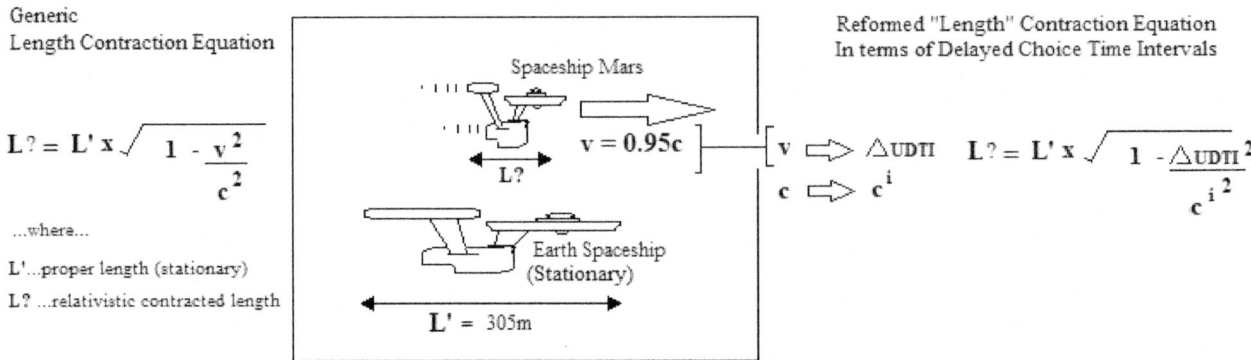

Notes: In the above, 'Earth Spaceship' (proper length L' = 305m) is at rest in stationary position, while a similar 'Spaceship Mars' flies by at velocity: $v = 0.95$ of the speed of light (see central depiction). What is the relativistic contracted length (i.e., **L?**) of Spaceship Mars? In the generic equation on the left, we plug in all the said values. We find **L?** contracted from 305m to 29.73m: contraction of 90% of proper length. Given de-spatialisation, length contraction no longer applies: Per equation shown on the right, apparent length contraction will be due to the delayed choice time-interval difference between events evinced by the reflected light from the leading and tail-end of Spaceship Mars, respectively, and, per the operation of the relativity of simultaneity, observed and judged by the crew of stationary Earth Spaceship. (Relativistic aberration, and the distortion of shapes are also functions of the delayed choice time-interval differences between pertinent events from different parts (moments) of an object. For brevity, aberration and distortion is not depicted). With de-spatialisation, we substitute **UDTI** in place of v (and c^i in place of c) and obtain the same result as found in generic "length contraction", without 'space' or 'length', per pure delayed choice time-relations delimited by c^i

380

C) Lorentz velocity transformation reformed

Lorentz Velocity
Transformation: S to S'

$$u'_x = \frac{u_x - v}{1 - \dfrac{u_x \times v}{c^2}}$$

....where...

v ...velocity of rocket Sr

u_x ...velocity of rocket Sr'

u'_x ...the sum of velocities u_x and **v**

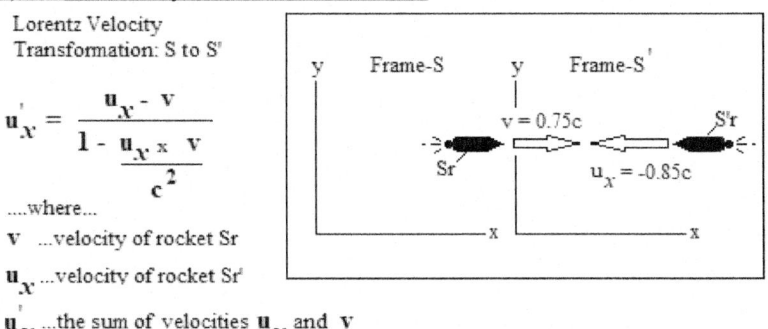

Velocity Transform S to S' Reformed
In terms of Delayed Choice Time Intervals

$$\triangle UDCTI^{u'_x} = \frac{\triangle UDCTI^{u_x} - \triangle UDCTI}{1 - \dfrac{\triangle UDCTI^{u_x} \times \triangle UDCTI}{c^{i2}}}$$

v ⇨ $\triangle UDCTI$ (...of Sr)

u_x ⇨ $\triangle UDCTI^{u_x}$ (... of S'r)

u'_x ⇨ $\triangle UDCTI^{u'_x}$ (...addition of Sr and S'r)

Notes: From the central depiction, viewed from a stationary observer attached to frame **S**, in the moving frame **S'**, a rocket **S'r**, with a velocity U_x -0.85c in the negative direction (right to left) is travelling toward another rocket Sr with its own independent velocity **v** in the positive (left to right direction. What is the velocity of **Sr** as viewed from **S'r**? Using the generic equation on the left, by treating **Sr** as attached to frame **S'**, and by plugging in the appropriate values where these are denoted, we obtain **U'x** (the sum of the velocities as viewed from **S'r**) to be equal to 0.9771c, *not* the simple sum, 1.5c. Given de-spatialisation, we must express this form of relativistic 'velocity transformation' in terms of the transformed sum of converging uniform delayed choice time-interval relata. The equation on the right is the Lorentz S to S' transformation recuperated per pure delayed choice time interval-based expression, *without* space.

D) Inverse Lorentz velocity transformation

Inverse Velocity
Transformation: S' to S

$$u_x = \frac{u'_x + v}{1 + \dfrac{u'_x \times v}{c^2}}$$

u_x ... sum velocity from both rocket stages

v ...velocity of the 2nd stage

u'_x ...velocity of the 1st stage

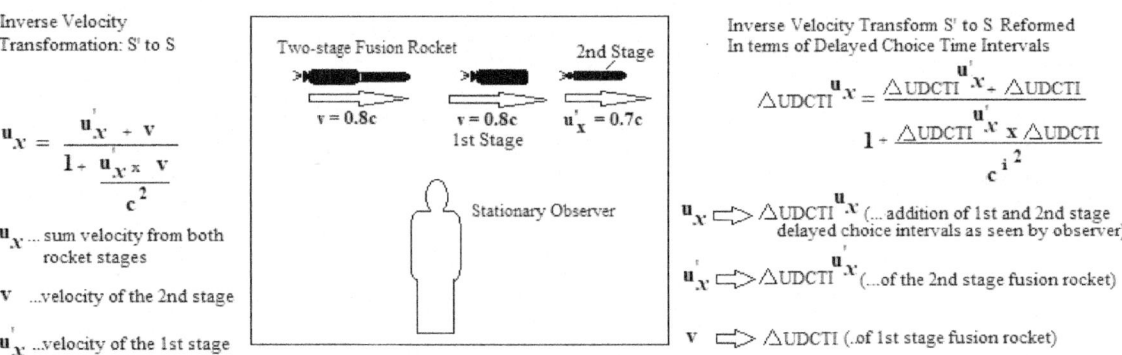

Inverse Velocity Transform S' to S Reformed
In terms of Delayed Choice Time Intervals

$$\triangle UDCTI^{u_x} = \frac{\triangle UDCTI^{u'_x} + \triangle UDCTI}{1 + \dfrac{\triangle UDCTI^{u'_x} \times \triangle UDCTI}{c^{i2}}}$$

u_x ⇨ $\triangle UDCTI^{u_x}$ (... addition of 1st and 2nd stage delayed choice intervals as seen by observer)

u'_x ⇨ $\triangle UDCTI^{u'_x}$ (...of the 2nd stage fusion rocket)

v ⇨ $\triangle UDCTI$ (.of 1st stage fusion rocket)

Notes: The inverse Lorentz velocity transformation (S' to S) has the same form as the transformation from S to S', but the subtractions become additions. In the central depiction, a two-stage fusion rocket obtains velocity-v equal to 0.8 of light by expending fuel from its 1st stage. Thereafter, its second 2nd stage detaches and ignites: It is independently capable of achieving velocity-**ux'** equal to 0.7 of light. In classical velocity addition, one would expect the velocity of the second stage to be the sum of its independent velocity plus the already acquired velocity of 0.8c (i.e., **ux** ought to sum to 1.5c). Instead, the actual sum is Lorentz transformed to a value of **ux** = 0.962c, less than the speed of light. However, given de-spatialisation and the demise of 'velocity', and given the need to reframe the latter in terms of uniform changes in the delayed choice time-interval relations (i.e., delta-UDTI) we recuperate the inverse transformation purely in terms of the uniform change in the delayed choice time relation. Thus, we obtain the same result in terms of delayed choice time relations: right equations.

Given de-spatialisation, and given that we can no longer presume the 'motion' of light-rays or quantum mechanical waves across forlorn 'space', the 'speed of light' must also transform into a de-spatialised pure temporal form: namely, *the delayed choice time-interval standard*. First, we use a familiar scheme and mark two points separated by a known 'distance'. We then time the passage of light across the presumptive spatial interval and treat light as if it were a 'moving light ray' or displacing quantum wave of radiation. We find that the time taken was 1 second. Given demands from de-spatialisation, we replace both space and distance with a pure delayed choice time-interval relation, as was done in the case of uniform and non-uniform 'motion': The two marked points are not points distributed per 'distance' or in space, but now constitute two distinct *moments* in time, constituting time-differentiated frames of reference. To obtain a temporal standard in lieu of the generic 'speed of light', we establish a delayed choice time-interval standard of 1 second between our two temporal events and frames. We then use this 1 second delayed choice time-interval standard as the constant and divisor in any calculation of any other delayed choice time-interval relation between events or between frames of reference, whether these are cast in uniform or non-uniform terms and temporal relations.

Hence, we replace the symbol for the speed of light **c** with the symbol for the **delayed choice time-interval standard**: succinctly, c^i. Thus, c^i = 1 second per second.

What about relativistic time dilation? How do we recuperate in terms of delayed choice time-relations and our standard c^i the observation in Special Relativity that clocks run slow when attached to 'moving' frames versus clocks attached to relative 'stationary' frames? First, we make the initial false assumption that space and motion are real. We then apply both to a scenario entailing a 'rest frame' (observers on the Earth) versus a 'moving frame' (star travellers on their way to Alpha Centauri). Both frames will have caesium clocks attached. The star traveller's 'moving frame' will be treated as 'moving' at 0.95 of the 'speed of light' i.e., in our terms, 0.95 of c^i. We will assume that there exists near-continuous communication between the frames, mediated by the exchange of near-continuous laser pulse emissions synchronised to 'clicks' on pertinent attached caesium clocks. From the signals received and compared, it will be calculated that the caesium clock attached to the traveller's moving frame will run 90% slower than caesium clocks attached to the relative non-moving Earth-frame. This is time dilation: i.e., "Moving clocks run slow".

The assertion, "0.95 of the speed of light", attributed to the 'motion' of the star traveller's frame can be replaced by an equivalent *rate of change in the delayed choice time-interval relation: i.e., between the Earth- frame versus the star traveller's frame.* i.e., 0.95 of c^i. This rate of change in the delayed choice time-interval relation between the frames can be calculated based on the near-continuous laser mediated signal-exchanges between the frames… mediated per the delayed choice time-interval standard c^i = 1 second per second: our replacement of the 'speed of light'.

Without resort to either the spatial 'speed of light' or 'motion', and by relying on the delayed choice time-interval standard, and relying on both uniform and non-uniform changes in the delayed choice time-relations between frames, almost all of the conclusions from generic Special Relativity, including time dilation, will follow. See **Fig. 4.05** for expressions and substitutions of the basic equations of Special Relativity, starting with the reformed equation for time dilation in **Fig. 4.05 (A)**.

In the recuperated Special Relativity depicted throughout **Fig. 4.05**, we obtain the Lorentz transformations for time, for time dilation and transformations from the addition of uniform and non-uniform delayed choice time-interval relations. We no longer need the Lorentz transform for length or distance, given de-spatialisation, but this is included in **Fig. 4.05 (B)** as an honorific. Indeed, putative 'length contraction' could be judged as the result of differences in the delayed choice time-intervals between events that happen at the leading end of a 'moving' object versus events happening at the trailing end of that object. See **Fig. 4.05 (B)**.

Consider the relative 'velocity' of two mutually approaching frames: each independently 'moving' toward a common moment of convergence to near-simultaneity, with each frame at velocity slightly less than the purported speed of light, such that the velocity of one frame judged from the other would be expected to obtain a velocity-sum much greater than that of light, but is instead Lorentz transformed to a sum less than the generic speed of light in generic Special Relativity: (See **Fig. 4.05 (C)**).

However, in our delayed choice time-relation terms, the notion of 'velocity' is replaced by the *uniform change in the delayed choice time-interval relation* as judged from one frame vis-à-vis the other. Moreover, the 'speed of light' will be replaced by our *delayed choice time-interval standard*. By substituting these new terms in place of the old terms in the generic equations, we anticipate the replacement of velocity transformations in Special Relativity with a same-logic, but as pure temporised Lorentz transformations, expressed purely in terms of delayed choice time-interval relata and the standard c^i. See **Fig. 4.05 (D)**. What is transformed therein is the relative change in the uniform delayed choice time-interval of the one frame vis-à-vis the other.

*

What about the famous Twins Paradox, wherein the traveller ages at a reduced rate compared to a twin on stationary Earth? If the star traveller's frame is an 'accelerating' frame, and if we now recuperate 'acceleration' in terms of non-uniform change in the delayed choice time-interval relation between the 'rest' frame on Earth and the accelerating frame of the star travellers, and apply our c^i = 1 second delayed choice time-interval standard in lieu of the 'speed of light', we find that the twins paradox of Special Relativity can be recuperated in our purely temporised terms, *without* 'space'.

*

We now need to integrate recuperated de-spatialised Special Relativity to the structures and processes of the intermediate model spacetime from Book-II: Essentially the same will abide in the Bergson-Whitehead amalgam successor to spacetime.

Recall the tripartite structure of intermediate spacetime, comprised of the growing block 'domain of realised spacetime', or the past retained as abstract memory from past collapse-processes. Also recall the 'event horizon' at which AND-to-OR collapse and universe-scale wavefunction collapse (and time) actually get to happen. Finally, recall the domain of AND-form 'spacetime in potentiality'; the future, out of which all things decay into realisation via AND-to-OR collapse on or along the event horizon. It is in this context that 'motion' and 'acceleration', and, with these, time dilation and the twin's paradox of Special Relativity, must AND-to-OR decay into signature-events out of future-potentiality… all now grasped purely in terms of delayed choice time-intervals and the delayed choice time-interval standard c^i, given the demise of 'space'.

Thus, **all realised uniform and non-uniform delayed choice time-interval relations of the past, together with all relativistic relations pertaining to time dilations and other transforms, are confined to the domain of realised spacetime** (or memory) residing 'below' the event horizon: always observed in the retrospective observational-inferential mode from frames attached to the event horizon; *always* subject to the relativity of simultaneity.

On the other hand, **all future-potential not-yet realised uniform and non-uniform delayed choice time-interval relations, and their future-potential relativistic time dilation and transforms, are embedded within spacetime in potentiality: the future *en toto*.**

It is the broken-discontinuous progression of the event horizon 'upward' the system, accompanied by synonymous process of AND-to-OR collapse of spacetime in potentiality (or the universe-scale wavefunction collapse) that gets to decay the said future-potential AND-form potentialities for uniform and non-uniform relativistic delayed choice time-interval relations and transformations into realised OR-form delayed choice time-interval relations and transforms.

Succinctly, the AND-to-OR collapse (time) that unfolds synonymously with the progression of the event horizon will involve the decay of the AND-form superpose of future potential states for changes in delayed choice time-interval relations, whether uniform or non-uniform, into actualised recent and past rates of change, expressed as successions of actualised OR-form delayed choice time interval relations via the event horizon. Moreover, realised, and potential time dilations, Lorentz-like transformations of converging rates of change unfolding as delayed choice time-interval relations, with both past-realised and future-potential twin paradoxes... will be implicit to their respective and appropriate spacetime domains, recapitulated in pure temporal terms, *without* 'space', per c^i.

Of course, the delayed choice time-interval standard c^i attends the realised lightcone within realised spacetime, or c^i **in actua** (later, the Bergson facet) as well as the potential lightcone within spacetime in potentiality (later, the Whitehead facet) attached to the realised lightcone, or c^i **in potentia**.

However, the reader is reminded that the intermediate model spacetime must be superseded by the Bergson-Whitehead amalgam, as de-spatialisation demands, and which seamlessly accommodates relativistic relations recuperated in pure delayed choice time-interval based terms, *without* space. Prospective Bergson-Whitehead amalgam will constitute a pure temporised 'spacetime', *without* space.

4.09: From 'Fields in Space' to Projections in Time

If the notion of spatial intervals and distances must be abandoned in favour of past-memory and future-potential events distributed in pure time... into networks of realised *and* potential stereotemporal delayed choice time-interval relations... then key consequences follow. To start with, the notion of **'fields' in physics must be de-spatialised into pure temporised forms, expressed in terms of pure delayed choice time-interval relations.**

From the putative gravitational field of Newton to the classical electromagnetic and magnetic fields of Maxwell, to fields in quantum field theory, the field was hitherto modelled as a region or extension in space. Examples are legion: the unitary arrangement of iron fillings in a magnetic field; the unitary arrangement of OR-form particle incidences by a collapsed quantum mechanical wave treated as a projection across space, manifesting such things as the interference pattern in the two-slit experiment, or the photographic image on a film. All such fields must now be re-conceptualised in terms of temporal projections and distribution in and across pure time: Spatial fields must be recuperated into projections from the past temporally insinuated upon the event horizon as 'instantaneous wholes' (see Book-I **1.03,** and **1.26** to **1.27**), and into the Whitehead future as future-potential decohered projections and as potential instantaneous wholes.

The field observed is the manifest or 'materialised' form, built up out of composite AND-to-OR collapsed discrete OR-form events, each of which belong to a different moment in time per the relativity of simultaneity and per de-spatialisation; temporally organised per delayed choice time-interval relata. For example, **take any interference pattern formed at the detector in the two-slit experiment; select just two quantum depositions from among the mass of depositions. Treat these as discrete events belonging to two different past-moments (i.e., *not* in simultaneity) as part of a larger distributions of events organised in time: What applies to the two incidences and their temporal interrelation applies to every other discrete OR-form event belonging to any *en masse* deposition that render salient the whole interference pattern: rendered salient as a distribution of events in time, *not* in space.** Hence, the interference pattern transforms from a notional 'materialisation in space' into a pure de-spatialised purely temporised distribution and organisation of events or moments, deriving from the decay and collapse of an abstract instantaneous whole (again see Book-I **1.03. 1.26** and **1.27**) constituting the complex quantum mechanical wave. **Hence, the underlying field behind the subsequent materialised interference pattern must be treated either as an instantaneous whole embedded in the Whitehead future (formerly spacetime in potentiality of as-yet not happened events) or as an instantaneous whole projected from the abstract Bergsonian past upon the event horizon... from the former domain of realised spacetime.** This instantaneous whole must subsequently undergo AND-to-OR collapse, bringing into materialisation the instantaneous whole into a stereotemporal complex of events internally subject to the relativity of simultaneity, yet manifesting a complex whole, such as the interference pattern... such as a photographic image. **Obviously, this instantaneous whole, whether as abstract memory or future potentiality, is *not* a spatial projection or outlay, but a wholly temporal projection rendered across the growing block universe constituted as the Bergson-Whitehead amalgam.**

Indeed, Book-III **3.33** and **3.34** furnished two forms of projections: i.e., emanation from the future versus projection from the past... essentially, of instantaneous wholes. For its part, Book-II modelled the permanent magnet and magnetic field in similar terms (see **2.37**). We simply recapitulate these into de-spatialised purely temporal forms, formerly emanated or projected as abstractions within the intermediate model, but now emanated or projected within the de-spatialised Bergson-Whitehead amalgam: (see forthcoming: **4.10**).

Therefore, the same treatment as had applied to the materialised interference pattern above must also apply to the magnetic field rendered salient by quantum constituents or even iron filings. The latter must also be treated as events and outcomes that belong to different past-moments: distributed and interrelated in time, *not* in space. Consequently, the magnetic field, even in its materialised form, constitutes a distribution in time and a projection thus. But the form of the magnetic field that is projected purely across time is an abstract instantaneous whole that must funnel subsequent AND-to-OR generated OR-form events (the iron fillings) into *en masse* time-distributed and stereotemporal coherent wholes that, though their constituent events, are inter-relationally subject to the relativity of simultaneity, even while in abidance to the precursor time-projected abstract instantaneous whole.

Obviously, the field must also be projected beyond the event horizon into and across the quantum mechanical future: into spacetime in potentiality… into the Whitehead facet of the forthcoming Bergson-Whitehead amalgam in **4.10**. Consequently, this decoheres into a future-form AND-form potentiality state… into the decohered quantum wavefunction that structures all relevant not-yet realised future-possible events, but all organised into as-yet not realised potential delayed choice time-interval relata therein.

However, we must also incorporate to this field the nested futures structure forced upon us by John Wheeler's delayed choice approach… with futures nested to the immanent potentials and to the futures beyond these; organised by their own internal potential delayed choice time-interval relata, and structured according to the divergent, entangled, and convergent categories enumerated in preceding Book-II (**2.22**). The basis for these claims were furnished in Book-II in relation to the intermediate model spacetime, its nested futures description, and the attendant presage of the quantum mechanical and growing block model of the magnetic field; rendered in terms of the nested futures view, in terms of grand decoherence, and per the erasure of non-consistent nested futures (see Book-II **2.37**). We retain these breakthroughs from Book-II, but recuperate these into de-spatialised pure temporal forms, into a system of stereotemporal delayed choice time-interval relations and complexes, constituting in these new terms the physical field as a projection in time and as an instantaneous whole: exhibiting potential and realised events or moments organised into time-relations and distributions, *without* space.

*

The following affords a means by which we can arrive at a post-space physics of the purely temporal and stereotemporal physical field. Taking the humble magnetic field and its corpus of organised iron filings as our example, we first assume the forlorn spatial approach: i.e., the conception of magnetic field as a projection in and across space that organises each iron filling related to every other by a network of putative positions and distance-intervals. But we now change this: The iron fillings are no longer objects arranged in space but constitute distinct non-simultaneous past events; with each iron filling over-simplified and reduced and treated as single simple OR-form events in time, *without* space. We replace the 'distance relations' between the iron fillings in the field, and the same to any observer, with delayed choice time-interval relations subject to c^i: i.e., the delayed choice time-interval standard in lieu of the 'speed of light': (see **4.08**). Consequently, the magnetic field that organises these events and their web of delayed choice time-interval inter-relations can no longer be treated as a projection across 'space' but immediately transformed into a projection across time; organising the said non-simultaneous past events (the iron fillings) into a purely temporal formation, in accord with the relativity of simultaneity.

We must also consider the same magnetic field projected from the past, or from mnemonic realised spacetime, or its de-spatialised successor in the Bergson facet... projected into the future, into spacetime in potentiality, or its replacement Whitehead facet, constituted as future-potentials for 'iron filling' events networked into a web of alternate future-potential delayed choice time-interval relations vis-à-vis other future-potential iron-filings and future-potential observers... all organised as an instantaneous whole projected into the future… into spacetime in potentiality, or into the Whitehead facet.

Finally, the comprehensive account of the magnetic field cannot comprise of just the combination of its past form in conjunction with its future-potential form: The comprehensive magnetic field must incorporate the key process of partial-perpetual AND-to-OR wavefunction collapse, synonymous with the broken-discontinuous progression of the event horizon... with 'coming into being', or *time* itself... which manifests the magnetic field into its 'materialised' form.

Thus, we obtain the three-fold description of the magnetic field and of any other field: a pure stereotemporal time-projection in terms of its past; in terms of a grand decohered nested future and instantaneous whole; and in terms of 'materialising' process of AND-to-OR collapse and time attendant the progression of event horizons... all integrated into a comprehensive description of the de-spatialised purely temporised magnetic field in the framework of the anticipated Bergson-Whitehead amalgam.

This model applies universally to *all* physical fields: whether these are magnetic, electromagnetic, or from quantum field theory, and even to gravitational fields: All physical fields must be recuperated into abstract purely temporal instantaneous wholes, projected across past-realised and future-potential time, *without* space.

Hence, we leave behind 'fields in space' and arrive at fields and instantaneous wholes in de-spatialised stereotemporal form.

4.10: From the Intermediate Model Spacetime to the De-Spatialised Bergson-Whitehead Amalgam

Our three-fold description of the physical field as a pure distribution and projection in time constitutes a good segue into the reform of the intermediate model spacetime into the Bergson-Whitehead amalgam: into spacetime *without* space. Without alteration, the Bergson-Whitehead amalgam accomplishes and accommodates what the intermediate model furnished: such as the growing block primitive unification of supposedly irreconcilable classical-relativistic and quantum mechanical physics: i.e., the unification and harmonisation of 'particle-wave' OR-AND logics and physics. The Bergson-Whitehead amalgam retains all the other developments furnished in Book-II and by the other books: such as the nested futures view of the quantum wave (Book-II **2.17**); counterfactual causality and grand decoherence (**2.34**), and the revived Mach's principle (**2.35**); the quantum mechanical 'spacetime' solution to the permanent magnet (see **2.37**) modified in the previous essay into its pure temporal field rendition; and the process of how mass (the past) tells spacetime in potentiality (the future) how to decohere into the events constitutive of gravitational phenomena (**2.43** to **2.45**). But the Bergson-Whitehead amalgam incorporates these as projections, distributions, and processes in time, *without* space.

Recall that the intermediate model spacetime from Book-II had three components. **In the Bergson-Whitehead amalgam, the domain of spacetime in potentiality (the future) is de-spatialised and replaced by the Whitehead facet, and the domain of realised spacetime is succeeded by the Bergson facet: the de-spatialised domain of abstract memory. Finally, the relative present and continuous event horizon survives into the Bergson-Whitehead amalgam and plays exactly the same function as before, albeit without 'spatial' characteristic,** constituting the 'cut' between past-future, or the Bergson facet versus the Whitehead facet of the amalgam. As with the intermediate model, along the event horizon, OR-form events will arise into being through AND-to-OR collapse and time, but these will be structured into temporal distributions per delayed choice time-intervals relations, subject to the relativity of simultaneity... all *without* space.

Essentially, we retain the same structure and relations in the Bergson-Whitehead facet that we had obtained in the intermediate model, but *without* space: namely, a past-future growing block system. But the de-spatialisation of spacetime is not a trivial tweak: It will prove decisive in casting light on the nature of inertia and gravitation, beyond the partial form furnished by the intermediate spacetime model in Book-II.

*

At the very dawn of quantum mechanics, Albert North Whitehead was the first to realise the significance of the AND-form quantum wave as the future-form configuration of the then 'spacetime'[82]. Unfortunately, Whitehead's growing block insight was never incorporated into historical quantum mechanics as an explicit assertion about the ontics of the quantum wave and the ontic reality of the future. What Whitehead lacked was a causality-based argument, combined to de-spatialisation itself.

Our Alternative Realist approach (Alt-R) adds the missing ingredients. Indeed, in Book-I, and throughout succeeding material, the key Alt-R addition to Whitehead's original insight was the firewall principle: the intertwine of causality with the principle of conservation epitomised by $E = mc^2$. **The future as an objective physical phenomenon, and the wavefunction itself as a mathematical model of that objective future, emerged from the demands of the firewall principle that ward against 'signals from the future': the first contribution from Alt-R to Whitehead's growing block thesis,** reinforced by the structural approach that asserts that any system placed in physical-experimental isolation-state must necessarily constitute a system deferred to the future. As such, such a system cannot assume any resolved OR-form configuration: it can *only* configure as an AND-form potentiality for future possibilities that have not yet transpired. An isolated system... a system deferred to the future *and* constituting the future... cannot be treated in any other way. This is *not* because we merely imagine it or prefer it to be so. **To insist, as Einstein and others had inadvertently done, that a system implicitly deferred to the future must yet exist as an "objective" OR-form state, would entail the violation of causality and of the principle of conservation.** Indeed, any physical proof to the effect that an isolated system... the future... ought to configure as an "objective" resolved state requires nothing less than the reception of 'signals from the future', from before that system transpired into resolved events: Hence, the violation of causality.

In tandem, such a violation through 'signals from the future' would evince that future as a resolved complex of putative energy and matter state...as an above-unity energy-matter content, with attendant violation of the generic form of the principle of conservation of energy-matter, always intertwined with the principle of causality. The pleonastic status of energy concepts does not alter or annul this assertion, but merely recuperates the principle of conservation into its abstract informational form; as the intertwine between the conservation of memory vis-à-vis consistent futures: a post-energy recouperation (see Book-II **2.39**).

As was stated in Book-I, and against the expectations of Einstein *et al*, the entwining of causality with the conservation principle demands the necessity and objective reality of a non-resolved potential AND-form domain for isolated systems whose resolution is deferred to the future. Hence, the firewall principle renders the future ontically real, phenomenalised as the quantum mechanical wave. Hence, the *raison d'être* for AND-form logic and the superposition principle: (see Book-II: **2.10 and 2.11**).

The second key contribution from Alt-R to North Whitehead's original insight is de-spatialisation itself and the removal of pleonastic space from the ontology and lexicon of physics, with consequences to how we must model the ontic future (see **4.01** to **4.07**). Thus, Whitehead's quantum mechanical future-form idea is incorporated, first as spacetime in potentiality in the intermediate model, and now as the Whitehead facet of the Bergson-Whitehead amalgam; as the de-spatialised reiteration of the growing block future. Within the Whitehead facet, there is no space; not even a 'potential space': There is only a Whiteheadian temporal distribution of future possibilities arranged into potential delayed choice time-interval relations.

The other contributions by Alt-R to Whitehead's central insight consist of such things as the utilisation of John Wheeler's delayed choice approach developed in Book-II, which fashions for the nested futures perspective of the wavefunction and of the ontic future, and grants the incorporation of a nested futures view to the physics of futures through the processes of grand decoherence: all now incorporated into the Whitehead facet of the Bergson-Whitehead amalgam.

---

[82] Whitehead, James, and the Ontology of Quantum Theory: Henry P. Stapp, Lawrence Berkeley National Laboratory, University of California, Berkeley, USA

385

As a fourth consideration, Whitehead's insight into the ontic reality of the future is fully consonant with the most basic contention of the growing block model: The universe must be modelled as a growing block universe, simply because, according to Alt-R, quantum mechanics is untenable in a block universe. **Thus, according to the fourth contribution, in a block universe, the future is constituted as a fully resolved OR-form relic: But, in a block universe, the generic two-slit experiment could not generate quantum interference, much less generate the known interference pattern at the detector. In a block universe, the emitted particle is rendered purely OR-form, not only at one or the other open slit, but seamlessly OR-form before and subsequent to the open slits, and all the way to the detector. Therein, seamless resolved OR-form 'which way paths' would not only constitute the norm, but there could be no possibility of either constructive or destructive self-interference of attendant particles. Consequently, no interference pattern at the detector could form, and quantum mechanics and science in** *any* **form would become untenable. Yet, experiment and experience show otherwise. Thus,** *ipso facto,* **we live in a growing block universe... and the future is both ontically real** *and* **rendered in AND-form. Hence, Whitehead was correct.**

*

North Whitehead was not the only historically obscure contributor to the understanding of nature. It was Henri Bergson who, in the late nineteenth century, in the context of his ruminations about consciousness, matter and memory, pondered the idea of information and memory as a distribution in time [83], in contradistinction to the generic notion of 'objects distributed in space', or constituted as 'material memory-trace'. Unfortunately, Bergson never developed a path to de-spatialisation. Instead, he accepted 'space' as a given: Doubt about space was never part of his later forlorn debate with Albert Einstein[84].

Bergson's idea of memory as information in and across time would have followed automatically, directly, and ineluctably from de-spatialisation and attendant temporisation. Of course, the same can be said of North Whitehead: Whitehead did not get rid of space either but sought to found his ideas about 'becoming' on the notion of 'motion': a spatial idea (but more on this later in **4.12**). Thus, a potential revolutionary moment was lost to scientific history, and significant parts of Alt-R could well have been achieved over a century ago.

Our own contribution to the validation of the basic kernel of Bergson's idea of memory as information organised in time... albeit, without 'space'... first involved a renewal of the critique of absolute time (see **4.01**) which showed that Einstein's discarding of absolute time was tantamount to de-spatialisation and a prelude to a pure temporal physics, given that de-spatialisation immediately recuperates all things into a pure temporal distribution... into Bergsonian memory. Unfortunately, the truism between 'space' and 'absolute time' was not grasped in 1905. Consequently, it remained hidden in plain sight for more than a century; obscured by 'spacetime'.

The immediate consequence of de-spatialisation is that the mnemonic domain of realised spacetime from our short-lived intermediate model from Book-II must now be wholly replaced by the de-spatialised Bergson facet of the Bergson-Whitehead amalgam; now constituted as Bergsonian memory... *without* **space**. Thus, the Bergson facet of the amalgam is truly the domain of abstract memory... of the totality of cosmic history organised into memories structured into pure temporal and non-contiguous primacy and recentcy distributions: a grand mnemonic of all realised events and relations that have transpired since the putative Big Bang, if not since the indefinite past... amalgamated into a unitary holonomic stereotemporal domain within the growing block whole. In short, Henri Bergson's view is vindicated. This explains why his name is incorporated into the Bergson-Whitehead amalgam.

*

De-spatialisation, and the consequent replacement of the spurious notion of 'objects in a common present moment' with a system comprised of past and future-potential events purely distributed temporally, all subject to the relativity of simultaneity, and *without* **'space', necessarily undercuts materialism**. De-spatialisation usurps the conception of causality as a contiguity-based process involving impact and contact, sundering the notion of "the interaction of matter in space", simply because there are no 'locations' for matter, and no spatial points-of-contact for the materialist notion of 'contact-interaction'. Real interactions are interactions in time, and are non-contiguous, and unfold according to the processes of the Bergson-Whitehead amalgam. Thus, relations between memory in time within the Bergsonian facet vis-à-vis the decohered nested futures within the Whitehead facet, processed through grand decoherence and AND-to-OR wavefunction collapse (time), replaces the false notion of contiguous causality via spatial contact and impact... thus supplanting materialism.

The demise of materialism is augmented by the discovery that time, hence the drive behind AND-to-OR wavefunction collapse, cannot be reduced to any contiguous causal variable in its own right, much less mediated by a now impossible contact or impact causality within a purely temporised framework and processes of the Bergson-Whitehead amalgam.

Furthermore, there are no 'hidden gears' behind AND-to-OR wavefunction collapse: no mechanism or definable iterational operational structure behind 'coming into being' in fundamental causality, vested in and as the process of time. This was established throughout the main work, especially in Book-II **2.46** and **2.47**, and it will be comprehensively vetted for in Book-V. In other words,

---

[83] *Matter and Memory*, 1990 (*Matière et Mémoire*, 1896), translators N.M. Paul and W.S. Palmer. Zone Books

[84] Canales J.,*The Physicist and the Philosopher: Einstein, Bergson and the Debate That Changed Our Understanding of Time*, Princeton, Princeton Press, 2015.

ultimate causality is structure-less. As such, it is even *extant* the Bergson-Whitehead amalgam, as much as it is extant the AND-to-OR collapse and time that it renders possible, and through which it operates.

Ultimate causality operates from 'outside' of the growing block past and future… from *outside* the Bergson-Whitehead amalgam, but through gear-less AND-to-OR collapse and time. As such, causality is *not* reducible to 'matter', much less to its 'spatially mediated contact', or to contiguous interactions. Hence, wither materialism.

4.11: The Demise of Space & Apagogic Mnemonic Argument-5a:
The Completion of the Mnemonic Argument for Abstract Memory

The demise of space and its implication with respect to how information is structured in nature undercuts generic presumptions and prejudices about memory as 'material trace-state', supposedly contained in and transported by 'matter in space'. De-spatialisation secures the view long-intimated from the more limited vantage afforded by information-matter dualism, and further augmented by the tripartite structure of the former intermediate 'spacetime', and by its Bergson-Whitehead amalgam successor. De-spatialisation compels conclusions in favour of abstract memory, sealing and superseding the series of mnemonic apagogic arguments for abstract memory furnished in Book-III. Thus, **apagogic mnemonic argument-5a asserts that what de-spatialisation compels… namely, the reframing of information into a stereotemporal system distributed purely in time, and organised into delayed choice time relata…** is abstract memory, retained extant of and in dissociated form from *any* 'materialised' recentcy on or along the event horizon 'skin', with the latter inadvertently construed as 'matter in space'.

Recall that the original mnemonic arguments for abstract memory in Book-III emerged out of the anticipated nihilistic argument against information and memory based on the Manyworlds interpretation (MWI) and other contentions. These appeared to question the very ontology of information and, in doing so, appeared to nullify the plausibility of abstract memory itself.

The nihilistic arguments were successfully overcome in Book-III through the various apagogic mnemonic arguments. These were no less significant in their casing for abstract memory than the now direct case from de-spatialisation and temporisation. Yet, by relying far more on physics and less on semantic arguments, de-spatialisation directly and seamlessly obtains abstract memory as information distributed in time, in the form Henri Bergson had intuited long ago. De-spatialisation constitutes a stark undeniable and 'higher' argument for abstract memory, given that contemporary culture accords a higher status to physical arguments than semantic ones.

*

What applies to the whole of nature must necessarily apply to the brain. Throughout nature, with de-spatialisation, we discover that the manner of how information is retained and how it is modified by other information is not a process unfolding at 'location' along contiguous lines, or across contiguous surface-spaces, or in contiguous three-dimensional block-spaces, given the impossibility of contiguity and extensionality through simultaneity (see **4.01**). The same necessarily applies to the relation of memories vis-à-vis the brain, mind and consciousness vis-à-vis the same, *and* the inevitable dissociated temporal distribution of both vis-à-vis the same. Hence, memory-brain and mind-brain dualism, or Cartesian dualism, follows ineluctably from de-spatialisation.

The notion that our memories have 'location' in brains, and are 'stored' in brains, or restricted and reduced to that part of the event horizon pertinent to brains, is immediately undercut upon the clear recognition that *all* information, including memories pertinent to brains, are interrelated and distributed purely in time, *without* space… in dissociated form within the Bergson facet vis-à-vis the event horizon…extant the manifest brain otherwise seemingly restricted to the event horizon.

This also leads to radical implications pertaining to the process of the retrieval of memories, putting aside the question of the locus of the mind itself in a de-spatialised temporised universe, even if the latter could somehow be portrayed as 'brain-writ'. **Even if the mind were nothing but the brain… even if the retrieval process of memory was totally confined to the brain… given de-spatialisation and temporisation, this could not involve a retrieval of memories from any 'spatial location' or corpus restricted to the event horizon or 'in the brain'.** Indeed, even, the general non-mental process of how the past gets to resonate and influence the present can no longer be modelled as a contiguous process unfolding 'across space': noetic memory retrieval is not going to be any different.

Given de-spatialisation and temporisation, the retrieval of memory requires a process that must 'reach out', *not* across space, but *across time*… into memories partly retained within the Bergson-facet of the Bergson-Whitehead amalgam… to a past moment in time, *not* to a 'location in space' or in brains. The minded control of the brain, and the expressions of the mind through the 'keyboard' brain… expressions that might involve declarations of memory… realised upon the brain or upon the event horizon… must involve a minded imposition-process on how the attendant Whitehead facet… the future… gets to decohere and develop into a pertinent global outcome… into subsequent brain activity consistent with the memory thus recollected from the Bergson facet and expressed in the generated brain activity. This process must also be extant the brain, given that the Whitehead facet is obviously not reducible to or identifiable with the event horizon, nor with the materialised brain confined to it. The active mind must reach out into the future and become extant the brain, assuming it was initially confined to the brain. **Hence, de-spatialisation and temporisation compel for the 'anathema' of Cartesian dualism, against the grain of culturally sanctioned now-defunct materialism.**

4.12: Time, Stereotemporality & New Process Philosophy

Generic ideas in physics assert space to be three dimensional, amalgamated to a 'fourth dimensional time'. In Book-II, we asserted that time is not a fourth dimension (see Book-II **2.03**): Time is the process of the transformation of future possibility into actuality. Thus, time is the broken-discontinuous decay of the future-form quantum mechanical superpose of nested possibilities within the Whitehead facet (formerly 'spacetime in potentiality) into OR-form actualities. Thus, time is synonymous with AND-to-OR wavefunction collapse, rendered tandem with the broken-discontinuous progression of the event horizon 'upward' the growing block system; and the collapse of the future into memories retained subsequently in the Bergson-facet; formerly 'realised spacetime'.

With the Bergson-Whitehead amalgam, almost all of the features of the intermediate model survive intact, but space completely disappears. If there is no such thing as space, where do we place the idea of the 'three-dimensional' universe? We addressed this conundrum in the preceding material, in the concept of stereotemporality. While time is correctly identifiable with AND-to-OR collapse of the future into OR-form actua, the sense of the purported three-dimensionality of things abides, but not as 'three-dimensional space', simply because there is no space. Instead, we obtain *stereotemporality*, or the three-dimensionality of time. Thus, **the three-dimensions survive, but only as three-dimensional time; as the stereotemporal pattern of delayed choice time-interval relations residing between successive events;** resolved through AND-to-OR collapse and time, and through the tandem progression of the event horizon into stereotemporal or three-dimensional memory... into the stereotemporal Bergson-facet.

On the other hand, potential events within the Whitehead-facet of the amalgam (the future) constitute the total AND-form superpose of future possibilities, including nested ones, *also* organised according to potential stereotemporal delayed choice time-interval relations. Of course, each future possibility within the Whitehead facet will exhibit a complex of potential stereotemporal nested futures.

In these two ways, **'three-dimensional space' is replaced by the three-dimensional time; replaced by the growing block stereotemporality of past-memory and future-potentiality.**

*

Whereas Alfred North Whitehead spoke of the centrality of 'motion' to emphasise the reality of 'becoming' and posited this against the notion of 'static material objects in space', or against purported notions of a "stand-still" reality, his insights and contributions to process philosophy remained incomplete: Whitehead did not get rid of 'space'.

Recall from previous material (see **4.08**) that putative uniform motion must be entirely supplanted by the uniform transformation in delayed choice time-interval relations that reside between events and reference frames. For its part, putative non-uniform motion (acceleration) was also be replaced by the non-uniform transformation of the said delayed choice time-interval relations. In either case **we arrive at the centrality of the process of 'coming into being' in a non-standstill universe, consonant with Whitehead's intuition, but with the process of change now grasped purely in terms of transformations in time-relations,** *without* space... *without* **'motion'.** The only static or 'stand still' that remains is the static Whitehead facet that elaborates to the infinitely removed future: part of the Bergson-Whitehead amalgam. Yet, even this static 'stand still' futurity must perpetually grand decohere and decay per function of accumulating and perduring Bergsonian memory *and* AND-to-OR time.

Moreover, since there is no space, there can be no synonymous 'absolute time'; there can be no objects in any sort of common present moment. Thus, there can be no 'static' or 'stand still' state of relations against which Whitehead argued, save as Alt-R's 'instantaneous wholes' of the static futures rendered into future-potential delayed choice nested futures, elaborated to the infinite future, combined to the conserved past. **Thus, as Whitehead had intuited, there can be no timeless 'slice' of space with 'objects sitting on it 'in the fashion of a 'stand-still' state, given that there is no such thing as a 'slice' of absolute time (synonymous with 'slice of space') to constitute such a forlorn stasis of 'material objects in space'.**

Given that there is only time, we can only obtain perpetual unfolding actuality drawn out from the future: Therefore, the essential point made by Whitehead abides, and we can now scaffold for a truly coherent new *process philosophy* to succeed and supersede the one originally developed by Whitehead *et al.*

*

Reframed in terms of the processes of the growing block Bergson-Whitehead amalgam, and in tandem with and similar to Whitehead's view against the notion of 'stand-still' reality, we must also investigate Henri Bergson's insight into what one might call *inseparability*, a view also held by Whitehead, albeit recuperated in de-spatialised purely temporal terms in Alt-R. In short, **we cannot separate manifest events from their total past (the Bergson facet) any more than we could separate these from their total future (the Whitehead facet). The past, the realised present, together with the future, are rendered into a growing block unity. But this is not a spatial unity, given de-spatialisation.**

Essentially similar to how Bergson had intuited it, it turns out that the most recent events resolved on the event horizon do not constitute isolated 'bits' vis-à-vis other bits upon that horizon, no more than these could be considered isolated in relation to the total Bergsonian past or to the total Whiteheadian future. To begin with, any event resolved on the event horizon is always present to and inseparable from the whole of the Bergson-facet and from the whole of stereotemporally organised history and memory. However, the consequence of the total past upon any formation on or along the event horizon will be realised in and through the AND-to-OR collapse of the Whiteheadian future and the latter's attendant grand decoherence into consistent nested futures in line with the demands of the conserved but perpetually growing past. Thus, any event that has come into being as a state of recentcy is always intertwined with the whole abstract past, and helps decohere the Whitehead future into consistency. In short, **through the grand decoherence of the future, the whole past, in organised form, and in abidance with the inverse square of time, is present to *any* new formation or event recently formed on the event horizon, just as the totality of nested futures within the Whitehead facet of the total futurity is counterfactually present to the same** (see counterfactual causality and grand decoherence: Book-II, **2.35**).

We will presage that both inertia and gravitation are aspects of this 'presence of the past' imposed by the Bergson facet of abstract memory, co-joined to the process of counterfactual causality and grand decoherence. Therefore, **'process' is the tandem advancement**

388

of the event horizon in partial-perpetual and broken-discontinuous form, into the Whitehead facet of future possibilities... or the perpetual decay of the latter into the former through AND-to-OR collapse and time... always accompanied by the embedding of newly realised events (the recentcy) *inseparable* vis-à-vis the abstract holonome of Bergsonian memory and history, with consequent expansion or growth of Bergson history and memory; all accompanied by attendant Whiteheadian grand decoherence and revived Mach's principle... almost completing *process philosophy* recuperated as the growing block universe.

In summary, our Alternative Realist framing of *process* incorporates and unifies key insights from both North Whitehead and Henri Bergson, paying homage to but superseding their critical contributions and insights. The central dynamic involved in process is *not* 'motion', as Whitehead grappled, by which he meant 'coming into being': Upon de-spatialisation, it becomes obvious that the only dynamic is the process of 'coming into being', or pure time: That is, perpetual coming into being, hence *time*, replaces 'motion'.

*

The conundrum of *endurance* and *perdurance* is inherent to process and can be reconciled in and through the Alt-R recuperation of process. While timeless read-only principles and abstractions *endure* without change through emanation, read-write identities and states *perdure* as part of the growing block process, as does the whole growing block universe.

Perdurance **involves the unity of two seeming contradictions: namely, the retention and continuation of identity (the definition of a thing) combined to and in contrast with the change and transformation of that thing, although this no longer pertains to 'object in space' but to a formation in time, constituted as projection in stereotemporal time, or as a 'worm' of memory-cum-potentiality spanning the Bergson-Whitehead amalgam, caught in a state of perpetual coming into being, time or** *process.* Therein, the most recent formations of the perduring 'worm' reside on or along the event horizon, while the perpetually accumulating growing 'worm' resides within the Bergson facet as the totality of memory pertinent to its total past, whose influence upon the event horizon is subject to the inverse square of time. The 'worm' does not terminate at the event horizon: It is projected into future potentiality within the Whitehead facet, in turn grand decohered into a 'future-potential worm' and into consistent nested futures *and* plausible future-potential contingencies (Wheeler's cat) that might deflect the 'worm' to alternative future developments, or even transform it into a superseding identity, or even bring about its demise through destructive contingencies.

Change is facilitated by the AND-to-OR decay of the future (time), but this is often change that retains the identity of the worm hitherto accumulated, if only because the worm that has transpired remains safely retained and conserved within the Bergson facet, and insofar as the change that decays out of the Whiteheadian future is often a futurity grand decohered into *consistency* with the worm, unless destructive contingency intervenes (i.e., Wheeler's cat). Thus, we harmonise the seeming contradiction between endurance and perdurance, and resolve the conundrum of identity and change, albeit within a pure temporised growing block universe and *new process philosophy*... all *without* space.

4.13: Mach's Principle in the Context of De-Spatialisation
And the Bergson Facet: Stereotemporal Memory

Book-II revived Mach's principle based on the process of counterfactuals-driven grand decoherence of the convergent nested future structure of 'spacetime in potentiality' (see Book-II **2.35**), now succeeded by the Whitehead facet. The revival of Mach's principle constituted the critical part of the presage to the growing block quantum mechanical theory of inertia, which must involve the certain 'non-locality' of the whole growing block universe. Therein, all frames along the event horizon share common converging potential futures embedded within the total Whitehead facet futurity; a convergence and entangled unity elaborated to the infinitely removed future. Hence, a change 'here' will counterfactually and instantaneously restructure the totality of mutual future possibilities attached to frames elsewhere: It does so by eliminating non-consistent potential futures from all frames that are convergent in the future, even if the attending frames are separated by a 'million lightyears'.

The grand decoherence involved in the elimination of said non-consistent futures up to and beyond 'one million years' into that future permits matter 'here' to counterfactually 'know' what matter is *not* doing elsewhere, and the implication of this to their convergent futures; so configuring its future course... or inertia... accordingly or consistently, in response to the unifying grand decohered common future, and into what is permitted to happen in their common futures up to timelike infinity. Hence, the revival of Mach's principle.

While the account furnished in Book-II abides, it was originally rendered in spatial terms, without de-spatialisation. Also, it was almost exclusively rendered in terms of a futures perspective; mostly revolving around 'spacetime in potentiality', now succeeded by the Whitehead facet. With de-spatialisation, Mach's principle must be recuperated in purely temporal terms from within the superlative Bergson-Whitehead amalgam. Hence, we must now incorporate the specific and total effect of the Bergson facet, of abstract memory, upon the grand decoherence of the future, in a de-spatialised purely temporal form.

*

In Book-IV, intermediate spacetime has given way to the de-spatialised Bergson-Whitehead amalgam. Former 'spacetime in potentiality' has re-emerged as the temporal Whitehead facet. Since all of this entails de-spatialisation, it follows that the instantaneous affect per Mach's principle, attained through the grand decoherence process described above, must now transpire within the a-spatial Whitehead facet, within the system of convergent nested futures therein, subject to the pertinent grand decoherence that generates inertia.

How does the Bergson facet and the total realised history of the universe relate to inertia? Reverting briefly to the notion of defunct 'space', the inertial object 'here' appears to correlate to the rest of the visible universe and its matter content: It does so in an

389

instantaneous and 'non-local' way. But, with de-spatialisation, the vast 'space' of 'distant objects' immediately transform into a complex de-spatialised purely temporised complex past, organised in terms of primacy and recentcy; with the most recent past and the most primal past superposed into a whole. It follows that **the mystery of inertia does not transpire across a 'space of objects', but in a system of memory and cosmic history organised temporally. In short, inertia is an instantaneous reaction to the total past, albeit realised through the modification of the common future for all matter via the process of consistency-imposing grand decoherence of that common future, as espoused in Book-II, now recuperated into a de-spatialised purely temporised form**.

It seems absurd to suppose that mass 'here' is instantaneously and transtemporally responding to some state of the remote past 'over there'. Yet, it does just that: Indeed, the 'matter-states' of the remote past *and* of the most recent past had generated common counterfactuals that then decohered the Whitehead facet of future possibilities into mutual consistency with those past states. In other words, it is the grand decoherence implied by 'matter' from different moments in cosmic history, affected through the total Bergson facet, that AND-to-OR collapses the consequences of inertia, and accomplishes the consequences of inertia in a transtemporal or instantaneous way. In short, **the totality of grand decoherence is per function of the totality of Bergsonian history and memory generated since at least the putative Big Bang. It follows that, naturally, the matter 'here' is *certainly* a response and an adjustment to that total past-history affected by the total history of grand decoherence imposed by matter at other past-times. The inertia is a consequence from the total history of grand decoherence (the history of matter *everywhen*, affected per function of the total restructuring of a common unitary convergent potential futurity for *all* matter from different times, but achieved in a de-spatialised purely temporal way**, superseding classical inertia conceived as an instantaneous space-like non-local effect, recuperated into an instantaneous transtemporal growing block process.

Thus, the total history of the Bergson facet, acting as an instantaneous whole upon the total future, decoheres the totality of Whitehead facet futurity into consistency, such that matter 'here' transtemporally and instantaneously obtain future outcomes consistent to other matter and other events in memory retained and imposed by the total Bergson facet: One outcome of this is inertia.

4.14: From '*Atomos* in the void' to '*Durata* in time'

The minimum implications from de-spatialisation to quantum mechanics consist in the obviation of the spatial wavelength attribute of particles in favour of, and with total replacement by, the time-based frequency-attribute. In short, the general immediate implication to quantum mechanics specifically, and to physics generally, consists in recapitulation purely in terms of frequency, or purely in terms of time.

It is a true peculiarity that entities typically imagined in spatial terms...as 'points in space', or as spatial blocks, 'spaceballs' or *atomos*... should possess a temporal attribute: a *frequency*. Even when we ignore de-spatialisation, it is certainly odd that a supposed 'point in space' should possess *frequency*. Frequency is a time attribute: a claim about duration: the description of a 'particle' as a duration. Whereas wavelength is consonant with entities presumed to possess *extensa*, even when these are modelled as point-like states. However, given de-spatialisation, space is pleonastic: There is no 'space'. Hence, the notion of a point-like particle, or a 'point in space'...or an extensionality... is obviated into a form that supersedes 'point' and 'extension'. With de-spatialisation, we can no longer portray putative matter, particles… the *atomos*… as spatial states: We cannot characterise these as possessing 'wavelength'

There is no 'space'. It follows that there are no point-like 'particles' with 'locations', no more than there are extended 'bits' of matter, or 'spaceballs'. Hence, there is no onticity to 'wavelength'. Hence, *wavelength* is as pleonastic as tandem forlorn 'space'.

Given de-spatialisation and the tandem pure temporisation of the physical order, what happens to extended 'matter', to 'particles? These by no means disappear. Instead, they are superseded. In a purely temporised physics and ontology, only time is real. Hence, only the frequency-attribute of the said entities enjoy onticity. Hence, point-like particles and 'extended matter' transform into pure durations… into *durata*. Thus, the 'particles', the 'bits', constitute intervals or moments in time, framed in terms of 'frequency': grasped as *durata*. Hence, the old *atomos in the void* transform into and is succeeded and totally replaced by *durata in time*.

This is no trivial consequence: it obviously usurps materialism. It is not merely the implied replacement of 'stuff' with *durata*, but the usurping of any possibility of contiguity or contact-causality in a universe composed of durata that are non-contiguously and only temporally inter-related. Thus, 'particle interaction', insofar as it can certainly be modelled as the superposition of durata and modelled in terms almost *only* of frequency, does not entail spatial contact, much less the conveyance of 'force' or action through collision, contact, or impact at some pleonastic spatial point or 'location'. It also follows that causality can no longer be modelled as the transfer of influence across 'space'. Instead, it must consists in the pure superposition of possibilities and probabilities rendered in terms of frequency and *durata*; resolved through gear-less AND-to-OR collapse and time, out of Whiteheadian potentialities into OR-form actualities, and, thereafter, into Bergsonian abstract memory.

Hence, the fabled '*atomos* in the void' of ancient Democritus must unravel; superseded by *durata* in a stereotemporal daz; mediated through causality *without* contiguity or without spatial contact; mediated through gear-less ineffable AND-to-OR collapse and time.

4.15: Summary of Part-I

The following constitutes a bullet-point summary of key findings in proof of de-spatialisation and implications from Part-I:

- **The problem of contiguity through simultaneity and consequent impossibility of extensionality or space: The synonymity of space with 'absolute time' abandoned in 1905:** Pleonastic space is inseparable from the notion of a 'common present moment' in which all things supposedly arise i.e., absolute time. Per the relativity of simultaneity and other findings, absolute time was invalidated by Einstein. There is no common present moment such as to constitute a collection of *wheres*, or such as to constitute a contiguity through simultaneity into extensionality out of seamlessly linked ordinates along a line, or a contiguity of co-ordinates through simultaneity to form a surface-space, or the same to form a three-dimensions block-space: These are all impossible states of absolute time... or 'space'. Einstein *et al* did not notice the synonymity of space with absolute time, and did not furnish criticism of contiguity though simultaneity, or the falsification of space that would have followed it. Instead, Einstein *et al* incorporated space and time into 'spacetime': (See **4.01**).

- **The relativity of simultaneity does not arise from the relation of the speed of light to ontic space (essay 4.02). Instead, the relativity of simultaneity emerges out of pure time relations residing between successive events and observers**, the distribution of which requires reference only to the said time-relations and intervals, which emerge out of AND-to-OR time, *without* space. Indeed, the notion of a 'speed of light' and of its space is foisted on AND-to-OR resolved time-relations formed between the succession of events and their observers: The 'speed of light' can be recuperated and replaced with the *delayed choice time-interval standard*: all outcomes would be the same, *without* space. The 'speed of light' *and* space are pleonastic impositions foisted on pure temporised events cast into pure time relations. Yet, the 'speed of light can be recuperated, while 'space' cannot. Space must be discarded.

- **Time is ontologically indispensable; space is entirely dispensable and pleonastic: (essay 4.03):** Space cannot be conceived without time. Without time, nothing would happen... not even events interpreted as 'objects in space' supposedly undergoing 'motion in space'. States purported to be 'objects in space', and motion itself, can be expressed purely in terms of delayed choice time-interval relations between events, frames, and observers. Thus, time proves both necessary and exclusively sufficient, *without* space. Space proves to be both unnecessary and superfluous: Thus, space is dispensable and pleonastic.

- **Quantum mechanics prohibits the possibility of the continuous observation of events, and especially of the moving quantum mechanical wave of light. The attempt to observe the wave would wipe out that wave. Thus, we cannot empirically prove 'motion', much less the motion of light at the 'speed of light'. Thus, we cannot empirically prove the reality of 'space' attendant such motion**: (essay **4.04**): Insofar as such observation is prohibited by the empirical stricture of quantum mechanics, the depiction of the quantum wave as undergoing 'motion' in notional pseudo-realtime constitutes an *assumption* about the wave. Indeed, to prove the purported motion of the wave (and hence its motion in space) it must be observed continuously, without any gaps, without non-observable intervals or physical-experimental isolation: with all spatial ordinates revealed into perfect seamless contiguity through simultaneity. Quantum mechanics cannot permit such seamless and continuous motion because it cannot permit seamless-continuous observation of *any* system by means of *any* return-signals (or radiation outputs), necessarily set to infinite frequency, even with the Quantum Zeno effect. Hence, both the proof of the motion of the light wave *and* its attendant space would culminate into the equivalent of the infamous ultraviolet catastrophe, even if we choose to disregard the fact that any attempt to observe the quantum wave of light radiation during its purported motion would undo the very thing we want to observe: namely, the putative 'moving quantum wave' or 'light ray'.

- **De-spatialisation from the critique of Hilbert space: (4.05):** Even if we ignore the above, and treat the quantum mechanical wave *as if* moving, it could only move through a Hilbert space: i.e., quantum mechanical *potentiality* for space that has not yet come into realisation: That this is so emerges from the demands of causality intertwined with the principle of conservation (the firewall principle) both of which justify the AND-form quantum mechanical wave and its Hilbert space as the phenomenalised future (see Book-I, **1.31** and **1.32**), augmented by spacetime structure (see below). But if Hilbert space is putative future-space that has not yet come into formation, then, obviously, nothing can move through it: A space through which nothing could move is as good as no space at all. In terms of intermediate spacetime structure, the 'moving quantum mechanical light radiation' at the speed of light, and its Hilbert space, is tantamount to the absurd notion of the moving spacetime in potentiality with which both light and 'space' must be identified; with the future-domain of the growing block universe constituted as its futurity. Obviously, the future cannot move, although its decay into temporally successive events and outcomes could be misinterpreted as 'light undergoing motion in space'. (**4.06**). The same conclusions survive into the Bergson-Whitehead amalgam.

- **Wheeler's meta-delayed choice and the final case for de-spatialisation (4.07):** In Book-II (**2.18**), John Wheeler's delayed choice was presented as the obvious basis for the nested futures view of future-from spacetime. It culminated into the spatially static non-undulating and non-moving quantum wave. This alone obviated 'space' per the invalidation of the notion of the 'moving quantum wave' (Book-II: **2.23** to **2.26**): Therein, nested quantum waves cannot be characterised as moving across their pertinent intervals, even while their 'carrying' hedging wave might be misportrayed as moving, given that the nested possibilities therein could not yet have coincided with their respective putative future spatial intervals, and have yet to transpired into 'spatial intervals'. Hence, the nested waves must be static (not moving). By inference, so must the hedging wave carrying the spatially static nested waves: In **4.07**, we took the matter to the next stage and treated the hedging wave itself as a nested future in its own right: one subsumed to a preceding hedging wave; part of a meta-delayed choice system of hedging waves that elaborate to the putative Big Bang or to the remote indefinite past. We discovered that only the first ever 'carrying' nesting wave immediately after the putative Big Bang, or at the indefinite remote past, might be *assumed* to have been 'moving'. At which point, both the notion of the motion of the wave *and* its pertinent 'space' is rendered superfluous and fully pleonastic to physics. Hence, the delayed choice approach forces the reality of spatially static quantum waves *and* the usurping of space by the meta-delayed choice approach, constituting the complete argument for pleonastic space. There is no space. Only time is real.

- **Reforming Special Relativity per de-spatialisation: (essay 4.08):** De-spatialisation requires the reformation of the basic terms of Einstein's Special Relativity. The 'velocity' component is replaced by the *uniform change in the delayed choice time interval relation*. Acceleration is recuperated as *the non-uniform change in the delayed choice time-interval relation*. The 'speed of light' is replaced by the *delayed choice time-interval standard*. Time dilation, length contraction and velocity transformation equations are modified accordingly in terms of the relative delayed choice time-interval relations and standard vis-à-vis events, frames, and observers. Hence we obtain a pure temporal SR, all *without* space.

- **From 'fields in space' to projections and distributions in time via the AND-to-OR decay of instantaneous wholes: (4.09)** With the demise of space, we can no longer consider 'physical fields', such as magnetic fields, quantum fields, or even the quantum wave and wavefunction itself, as 'outlays in space'. There is no space. Therefore, a putative magnetic field and its events must be recuperated as bounded projections in and across time, constituted as instantaneous wholes, cohering both potential and realised events (the iron fillings in a magnetic field) organised purely according to delayed choice time-interval relations, *without* space.

- **The Bergson-Whitehead amalgam succeeds the intermediate model spacetime: (4.10):** The Bergson-Whitehead amalgam, or 'spacetime' without space, is born out of de-spatialisation and temporisation. It replaces the preceding intermediate spacetime model from Book-II. The tripartite structure from the intermediate model survives, albeit without space. The primitive unification of classical and quantum relativistic physics also abides. Thus, 'spacetime in potentiality' is replaced by the *Whitehead facet* of the amalgam, comprised of potential events and nested futures cast into potential delayed choice time-interval relations. The domain of 'realised spacetime' is replaced by the mnemonic *Bergson facet* of the amalgam: the sum of abstract memory that preserve the abstract basis of past realised events, to which de-spatialisation and temporisation is directly pertinent, given that realised information organised in and distributes purely temporally *is* abstract Bergsonian memory. The event horizon, the 'cut' between the said two facets of the amalgam also survives intact. The Bergson-Whitehead amalgam opens the way to the presaged mnemonic solution to gravitation and inertia.

- **De-spatialisation and the final apagogic mnemonic argument-5a: (4.11)** De-spatialisation compels the reframing of information purely in terms of delayed choice time-interval relations *without* 'space': as a pure non-contiguous distribution in time. But information so conceived *is* abstract memory. Therefore, de-spatialisation constitutes the seamless physical proof of abstract memory, or Bergsonian memory... of memory that transcends the dubious notion of 'material trace state in space', recuperated into superlative abstract memory conceived as a 'distribution in time'... in the form Henri Bergson had intuited. This seals the case for abstract memory and completes the series of apagogic argumentations started in Book-III.

- **Stereotemporal time, or three-dimensional time: (4.12):** With de-spatialisation, three-dimensional space transforms into stereotemporality, while the old 'fourth dimension' (or time) is completely replaced by AND-to-OR wavefunction collapse. There is no space. All events are interrelated and integrated in terms of pure delayed choice time-interval relata, and these are three dimensional or *stereotemporal*. Time is not a fourth dimension. Instead, time is identified with the process of stereotemporal AND-to-OR wavefunction collapse and with tandem event horizon progression, and the realisation of events into pure stereotemporal distributions.

- **New Process Philosophy, de-spatialisation, stereotemporality and the Bergson-Whitehead amalgam: (4.12):** In his contribution to process philosophy, Alfred North Whitehead spoke of the centrality of 'motion' in challenge to the notion of static material objects in space, or "stand-still" reality. However, motion is a spatial concept. With de-spatialisation, motion disappears and transforms into pure temporal *becoming*, usurping 'stand-still' reality: This is truer to the essence of Whitehead's intuition. As Whitehead suggested, the central process is *becoming*, but now recuperated within a de-spatialised stereotemporal order constituted as the growing block Bergson-Whitehead amalgam, facilitated through AND-to-OR collapse and time, wherein only time is ontologically real. For his part, in addition to the conception of memory as temporal (perfectly secured by de-spatialisation) Henri Bergson emphasised *inseparability* (as did Whitehead) in opposition to the false notion of discrete and isolated 'bits' of 'matter' distributed in space. With de-spatialisation, the inseparability of everything from everything else becomes automatically obvious: Any event on or along the event horizon is temporally inseparable from the stereotemporal past in memory (the whole of mnemonic Bergson facet) as well as inseparable from the future *en toto* and *its* future-potentials (the Whitehead facet). A 'bit' cannot be separated from its abstract past, no more than from its potential futurity, both of which it shares with other 'bits' along the event horizon. Indeed, it is through grand decoherence of the Whitehead facet that the Bergsonian past enforces Mach's principle: i.e., inadvertent Bergsonian inseparability as it pertains to inertia. In the framework of the Bergson-Whitehead amalgam, *process* is comprised of the combination of time, or the AND-to-OR collapse of the universe-scale wavefunction; the tandem progression of the event horizon; the formation of new memories and attendant expansion of the Bergson facet; combined to the perpetual grand decoherence to consistency and decay of future potentials out of the Whitehead facet. All of this secures the perdurance of formations and identities as 'worms' retained within the Bergson facet, grand decohering the future within the Whitehead facet to consistency, or to a future-potential 'worm': all *without* space; and wherein the process of 'coming into being' and time constitute the central dynamic or *process*.

- **Mach's principle augmented by de-spatialisation**: **(4.13)**: How does de-spatialisation modify Mach's principle based on grand decoherence of convergent nested futures developed in Book-II? The matter-state 'here' constitutes a response and adjustment to the total Bergsonian past-history and consequent grand decoherence of the total Whitehead facet, which then reduces the future for matter here and elsewhere to consistent futures... with which matter here and elsewhere must abide, and do so in their AND-to-OR decay into their successive states. Hence, inertia. Thus, inertia is obtained through the total history of grand decoherence affected in a de-spatialised instantaneous transtemporal way. Book-II focused on grand decoherence from a futures perspective. From a Bergson facet approach, we must include the sum total of the history of grand decoherence, necessarily in de-spatialised and temporised form.

- *Durata* **succeeds** *atomos*: **(4.14)**: From the materialist view, all things are made of 'matter in space': i.e. putative spatial point-like or extended entities said to possess both wavelength and frequency, owing to their wave-like natures. However, with remarkable de-spatialisation, the spatial 'wavelength' falls away, leaving only frequency. The putative 'particles' can no longer be rendered as point-like entities or as extensions, given the impossibility of spatial points, 'location' or extension per function of de-spatialisation. Instead, on the exclusivity of the temporal frequency attribute, the entities of nature recuperated into durations, or into temporal *durata*. Hence, the spatial *atomos* of ancient Democritus unravels and is superseded by the pure temporal *durata*, realised or potential... cast in stereotemporal growing block daz encapsulated as the Bergson-Whitehead amalgam.

BOOK-IV PART-II:
REFORMING ENERGY-MATTER RELATIONS IN THE CONTEXT OF DE-SPATIALISATION & THE BERGSON-WHITEHEAD AMALGAM

4.16: Aims of Part-II

Much of Part-II will consist in recapitulations of previous insights and findings in terms of the de-spatialisation, the consequent temporisation of physics, and the superlative Bergson-Whitehead amalgam, culminating into conjectures about matter-creation.

The character of the quantum mechanical wave and of the wavefunction must change per implications from de-spatialisation and temporisation. Indeed, the reformation of the physical field into its de-spatialised temporised form in **4.09** prefigures the de-spatialised and temporised wavefunction. But we must also recapitulate quantum indeterminism and the three-phase developmental process in their de-spatialised and temporised form, attendant the AND-to-OR collapse of quantum mechanical waves within the growing block Bergson-Whitehead amalgam.

This will follow into **the recuperation of new entropy and the 'arrow of time' from Book-I Part-I, rendered in terms of the de-spatialised temporised Bergson-Whitehead amalgam**. Thus, the entropy of future-potential information, and the entropy of memory, and the AND-to-OR directionality constitutive of the 'arrow of time', and the role of the inverse-square of time, will be recuperated in de-spatialised and temporised form within the Bergson-Whitehead amalgam.

Recall the reformation of energy concepts and relations started in Book-II (**2.39**). Hence, recall the pleonastic character of these, framed to the intermediate model spacetime in spatial terms. Part-II will advance **the reform and pleonastics of energy concepts and relations into their de-spatialised temporised Bergson-Whitehead forms**. Therein, as a reiteration, *potential energy* will be shown to issues out of the Whitehead facet... while kinetic energy will be reiterated as part of future-potential energy manifested via the transition and succession of event horizons per AND-to-OR collapse and time, recuperated as de-spatialised temporised kinetic energy. The de-spatialisation of potential energy was obvious even in Book-II. But matter and mass are oft conceived as congealed potential energy; succinctly, congealed in and as a spatial point or a spatial block particle or 'nugget' of matter. Superlative de-spatialisation and the transformation of *atomos* into pure temporal *durata* necessarily implies that the mass associated with matter must itself be de-spatialised and temporised into the recapitulation of mass as Bergsonian memory: culminating into **mass as a memory effect**. We will also discover that matter itself must be issued out of the potentiality of the Whitehead facet... out of the future: Thus, forming a critical insight-point into matter creation at the putative Big Bang and thereafter.

The pleonastic status of potential energy and kinetic energy... and, ultimately, of matter itself, generically misconceived as spatially 'congealed energy', but recuperated as 'potential matter' within the Whitehead facet, and as mass per function of the memory effect of the Bergson facet... must come to the fore from the character of time and the gear-less nature of time's drive: Thus, we will reiterate that the process of AND-to-OR collapse and time is *not* ultimately an energy or work-driven process, but a gearless work function zero process. Given the ontological primacy of AND-to-OR collapse and time vis-à-vis putative 'energy relations' brought into being by time, and given that what engenders work and change is time itself, not the other way round. It must necessarily follow that energy concepts and relations must abstract out of the ontology of physics, as was originally espoused in

Book-II **2.39**... but now recuperated within the framework of the Bergson-Whitehead amalgam in explicit de-spatialised and temporised form, with remarkable addendums and implications about the nature of mass and gravity.

Indeed, **de-spatialisation itself cannot help but render void generic notions of contiguous energy and action: Remarkable de-spatialisation robs presumed spatial contiguity requisite to the supposition of 'physical contact' and to impact-causality inherent and indispensable to materialist notions of 'work and action'.** Energy and 'action' in a de-spatialised purely temporal physics must involve nothing more than pure transitions in time, via AND-to-OR collapse (itself a gear-less and work function zero process), *without* contiguity, *without* contact or impact... *without* the requirement of 'energy or work'. That is, only time itself engages in 'work' or change. But time itself does not require energy or work to transpire, and nor does the state-change that time makes happen.

What about the creation of matter? In Part-II, we will reiterate that, **at the putative Big Bang, putative matter was issued from or 'created' out of Whiteheadian potential matter... out of future-potential matter inherent to the Whitehead facet**. Hence, matter was *not* created 'out of nothing', notwithstanding the pleonastic status of the space and 'energy'. But **Part-II will also conjecture for post-Big Bang 'matter creation' by combining Heisenberg's energy-time uncertainty to grand decoherence, with both co-joined to profound time-dilation effects on the surfaces or at low altitudes vis-à-vis asymptotic 'black holes'.**

4.17: The Quantum Wave, Quantum Indeterminism, & the Bergson-Whitehead Amalgam

In **4.09**, the notion of the physical field was recuperated in terms of stereotemporal delayed choice time-interval relations that attend AND-to-OR collapsing instantaneous wholes , or abstract memories from within the Bergson facet, or future-potentialities from within the Whitehead facet. We can accomplish essentially the same for quantum waves and attendant wavefunctions: We can integrate these and the tandem three-phase developmental solution, co-joined to an anti-hidden gears approach to quantum indeterminism (Book-III-Part-II: **3.26** and **3.31** to **3.32**) to the framework of our de-spatialised Bergson-Whitehead amalgam from **4.10**.

To this end, consider the self-interfering quantum mechanical wave from generic two-slit experiments: The self-interfering quantum wave is destined to exhibit an interference pattern at the detector-surface. However, **given de-spatialisation, the detector can no longer be characterised as a 'spatial surface'**. Consequently, each putative point co-ordinate constituting the detector 'surface' pertaining to each potential particle deposition vis-à-vis every other potential deposition-point must in truth constitute a complex of future-potential delayed choice time-interval interrelations-based distributions between future-potential incidences.

Upon AND-to-OR collapse and attendant event-horizon progression, these potentials will resolve into OR-form realised delayed choice time-distributed depositions. Thus, **what resides between the 'particles' (succinctly, the *durata)* on the detector 'surface' are pure delayed choice time-interval relations: *not* 'spatial intervals'... and the detector 'surface' must itself be recuperated into a sets of temporal distribution potential or actua: no longer a 'surface', but itself a temporal projection and distribution of events. It also follows that the quantum wave and wavefunction attendant the detector 'surface' does *not* constitute a region or area in 'space', but an instantaneous whole and temporal projection of potential durata in potential delayed choice time-interval relata across time**, consistent with the temporisation and de-spatialisation of physical fields in **4.09**.

The relativity of simultaneity, or potentialised relativity of simultaneity, applies throughout, given that the future particle-incidences or *durata* on the detector, destined to collectively form up into the 'materialised' interference pattern, are not going to constitute simultaneous incidences manifesting in an impossible absolute time. Thus, yet again, delayed choice time-interval relations must supersede and replace the notion of particle 'positional relations'. **This also implies that the whole wavefunction no longer constitutes the superpose of positional possibilities in 'space': Instead, it must comprise the domain of potential temporal incidences: their potential minimum to maximum (highest amplitude) potential delayed choice temporal interrelations, rendered into unity or into an 'instantaneous whole' within the Whitehead facet.** And it is within this de-spatialised temporised quantum wave and wavefunction that we must re-envision the unfolding of quantum indeterminate processes.

*

In the three-phase developmental process (see Book-III **3.26**) non-illusory objectively real quantum indeterminate incidences were shown to accumulate over time towards the highest quantum probability locus (the highest amplitude) of the wavefunction. Generically, we erroneously treated the quantum mechanical wave *and* the attendant three-phase developmental process *as if* it unfolded in a 'region in space'. **Given de-spatialisation, there is no space: (see 4.01 to 4.07). Therefore, we must treat the whole three-phase developmental process as unfolding and transpiring, *not* in space, but across a stereotemporal domain of potential events and potential delayed choice time-interval *relata***, with the abstract wavefunction behind the three-phase development process projected within the Bergson-Whitehead amalgam as an instantaneous whole (see Book-I **1.03**), but subsequently 'materialised' through AND-to-OR decay out of the Whitehead facet into a set of temporal events and relations subject to the relativity of simultaneity.

We begin with **the below-threshold phase** of the three-phase development process, but applied to the two-slit experiment: Within this phase, the expected interference pattern that will form out of quantum indeterminate incidences is not yet obvious: only a *noise distribution* of temporally distributed quantum discrete durata resolutions are salient (see Book-III **3.26**). But the phase is succeeded by **the transitional chaotic phase** of development. Therein, pattern begins to emerge out of the larger collective of temporally distributed indeterminate quantum durata resolutions, but only as a *chaotic distribution*: Pattern therein remains precarious and volatile. Finally, with more quantum indeterminate depositions, development leads into **the noise-reduction pattern-saliency phase**; wherein the anticipated *signal distribution* will fully manifest the expected stereotemporal interference pattern, despite and *because of* objective quantum indeterminacy.

QUANTUM MECHANICS AND MIND

All three developmental phases unfold across pure time into realised stereotemporal distributions... from out of the AND-to-OR decay of an instantaneous whole wavefunction embedded within the stereotemporal Whitehead facet. With this recognition, **we integrate our three-phase developmental solution to quantum indeterminism and objective randomness from Book-III into the de-spatialised and temporised framework of the Bergson-Whitehead amalgam.**

*

The above de-spatialised and temporised three-phase developmental approach to quantum indeterminacy specifically pertains to *wavefunctions in future potentiality,* now cast within the Whitehead facet of the Bergson-Whitehead amalgam. Therein, the various nested futures categories abide. Yet, **we must also account for the same three-phase development process in de-spatialised and temporised form as it applies to instantaneous wholes constituted as *wavefunctions in memory* from the Bergson facet** of the Bergson-Whitehead amalgam; (formerly, 'realised spacetime'). The wavefunction in memory, also an instantaneous whole, must also undergo AND-to-OR decay **in a process of 'memory resonation' and perdurance unfolding within a growing block superlative Bergson-Whitehead amalgam,** as explicated in new process philosophy (see Part-I, **4.12**).

Recall from **4.10** that wavefunctions in memory are subsumed within the Bergson facet according to primacy-recentcy rules. Whereas wavefunctions in potentiality within the Whitehead facet are encapsulated as a system of nested futures (ultimately convergent). For their part, wavefunctions in memory constitute former future wavefunctions, but with their nested futures 'cropped': hence, *charged* (i.e., no longer nested) as opposed to *uncharged* (i.e., nested).

In a similar way to wavefunction in potentiality, **wavefunctions in memory within the Bergson facet must be described, *not* as distributions or projections in 'space', but as projections across time**... as pure temporal resonations from the Bergsonian past. Therefore, **the quantum indeterminate incidences pertinent to the AND-to-OR processing of wavefunctions in memory** *cannot* **unfold in 'space' but must also manifest into discrete moments... into *durata*... across time**... organised into pure delayed choice stereotemporised time-interval relations, subject to the relativity of simultaneity. Thus, **the three-phase developmental process that applies to objective quantum indeterminism must also generally apply to the AND-to-OR collapse of Bergsonian wavefunctions in memory,** albeit now grasped as a de-spatialised pure temporal three-phase development process.

4.18: Entropy Relations & the Bergson-Whitehead Amalgam

Book-I posited the *entropy of future-potential information* and *entropy of memory*. It did so in contradistinction to generic forms of entropy (see Book-I: **1.07** and **1.08**, respectively). In this essay, **we will integrate the growing block universe view of entropy to our de-spatialised and temporised Bergson-Whitehead amalgam.**

Recall from Book-I that generic entropy consists in the transformation and dissipation of energy into relatively less useful and scattered forms, less readily available for subsequent work. A hot cup of coffee loses its heat energy to its surround. This scattered heat energy would require an even greater input of energy to recollect in order to restore it to the coffee, or to use it for other useful work. All systems tend toward the increase in entropy, even when treated as open systems, although only closed systems completely run down. Moreover, all things appear to tend from higher complexity to lower complexity, from a lower state of information entropy to a higher state of information entropy, unless putative energy is inputted to recover, sustain, or expand that complexity, or unless the entropy of energy perpetually increases to sustain that complexity: a feat typically observed in living forms.

Generic entropy is often purported to define the 'arrow of time': a direction from higher complexity to lower; or from lower entropy of energy to higher entropy of energy, as the price paid for sustained complexity increases, as it does in living forms. The tendency enjoys higher probability of occurrence; a statistical tendency toward increased disorder, with any increase in order a lower probability outcome.

The reversal in entropy is often characterised *as if* it could constitute 'time reversal'. However, in Book-I, we clarified the ontologically primary basis for the 'arrow of time': Time's 'arrow' is a function of the increase in the entropy of future-potential information attendant AND-to-OR wavefunction collapse (time); and the 'arrow of time' always proceeds unequivocally from potentiality to actuality... from AND-to-OR... never from OR-to-AND: (See **1.07** and **1.09**). That is, time always proceeds from AND-form potentiality to OR-form actuality. Reframed to our replacement to 'spacetime', time always proceeds from the decay of the Whitehead facet (the future) into the accumulation of memory in the growing block Bergson facet. That is, time proceeds from ontic future-potentiality to past-actuality. **Whether or not generic forms of entropy increased or decreased, either possibility must be evinced in events brought about by ontologically primary AND-to-OR collapse or time, per function of 'time's arrow' configured into irreversible AND-to-OR directionality. Decrease in generic entropy could not constitute 'time-reversal' or any OR-to-AND inversion of the direction of wavefunction collapse. This insight culminated into the non-statistical objective basis for 'time's arrow', based on ontologically primary AND-to-OR directionality and irreversibility,** *not* **on any mere statistical generic entropy tendency.**

*

How do we integrate new entropy to the de-spatialised Whitehead facet of the amalgam? Note that the *entropy of future-potential information* is a feature of the future-form Whitehead facet of the Bergson-Whitehead amalgam. The de-spatialised Whitehead facet comprises the total future-potential information of the universe: the sum of both immanent future possibilities vis-à-vis the event horizon, as well as the potential of complexes of nested futures that elaborate to the infinitely removed future. As the Whitehead facet perpetually decays, the future-potential information loss increases. Thus, the entropy of future potential information perpetually and *irreversibly* increases, given the impossibility of OR-to-AND de-collapse or time-reversal to render it otherwise. This is not a statistical tendency: nor is it a matter of probability. The increase in future-potential entropy of information is a matter of absolute certainty attending perpetual AND-to-OR collapse and time.

Recall that, **given de-spatialisation, the nested future potentialities within the Whitehead facet, hence the future potential information therein, must be grasped as potential future events organised purely into potential delayed choice time-interval relations, *without* 'space'. It follows that the increase in the entropy of future potential information per AND-to-OR collapse constitutes the irreversible loss of alternative sets of potential events and attendant potential delayed choice time-interval relations embedded within the de-spatialised temporised Whitehead facet.**

*

We now turn to the integration of entropy considerations vis-à-vis the event horizon, recuperated as part of a de-spatialised and temporised Bergson-Whitehead amalgam. Before any progression of the event horizon transpires, the total potential information embedded within the Whitehead facet is rendered into a relative high level of potential complexity and high potential content: Thus, the entropy of potential information is at its relative lowest.

However, as the event horizon progresses from, say, EH-1 to a successor EH-2, the immanent AND-form future potentialities will decay, undergoing AND-to-OR collapse into resolved OR-form actua or durata on or along the new event horizon. Hence, the entropy of potential information will increase proportionate to the alternate immanent possibilities erased, given that AND-to-OR collapse will reduce these to just one OR-form realised outcome.

This much was anticipated in Book-I. But we must now integrate it into the Bergson-Whitehead amalgam. Therein we must also incorporate the role of grand decoherence and counterfactual erasure of non-consistent nested futures as part of the entropy of future potential information.

As immanent possibilities are realised or erased through the progression of EH-1 to EH-2, within the convergent category nested futures that permeates throughout the Whitehead facet, counterfactual processes first described in Book-II (Part-III) will erase non-consistent nested futures. This decay will degrade the total potential information embedded within the Whitehead facet, above the mere elimination of immanent alternative possibilities, thus contribute to the increase in the entropy of future-potential information, and, with it, increase the *entropy of future-potential delayed choice time-interval relations* that attend the de-spatialised erased nested futures.

*

We now turn to the *entropy of abstract memory*, also anticipated in Book-I (**1.08**), and integrate it to the de-spatialised and temporised Bergson-Whitehead amalgam. Abstract memory is explicit to the Bergson-Whitehead amalgam, especially to its Bergson facet. Therein, memory is rendered abstract per its pure temporisation per function of de-spatialisation (**4.12**). Hence, only time is indispensable and ontologically real: All events are states of past-primacy and past-recentcy structured into delayed choice time-interval relations. Memory is the totality of the mnemonic Bergson facet of the Bergson-Whitehead amalgam. Memory is the past retained as a series of charged wavefunctions in memory (i.e., with their former nested futures eliminated) necessarily structured and distributed in accordance to delayed choice time-interval relations within the Bergson facet. Therein resides the basis of the entropy of memory.

The accumulation of memory within the Bergson facet ought to imply a perpetual reduction in the entropy of realised information: It ought to imply an increase in realised information and historic complexity vested as the totality of the growing block Bergson facet. **However, this expectation is counteracted by the *inverse square law of time*.** Recall from Book-I that the most remote past (the primacy) has the weakest influence on the realisation formed on or along the event horizon. On the other hand, the most recent past (the recentcy) has the greater influence in the formation of the said present outcomes. (The anticipated Bergsonian mnemonic mechanics that models all of this will be developed in Part-III (**4.25** to **4.28**)). Indeed, Book-I anticipated that the diminution of the power of the past over what forms at the event horizon is *not* due to 'distance', but entirely due to time: succinctly, the inverse square law of time. The power of past events vis-à-vis the evolving present is perpetually reduced by the operation of the inverse square of time. In short, through the inverse square law of time operating upon Bergsonian memory, we obtain the entropy of memory. **The entropy of memory is grasped as the perpetual weakening of Bergsonian memory per the inverse square law of time.**

*

The ultimate character of AND-to-OR collapse and time, in tandem with de-spatialisation and temporisation, generates immediate consequences to entropy-considerations. Given that generic entropy notions are steeped in the notion of 'particle interactions in space', wherein the purported interactions and 'collisions' between particles affect a spatial diffusion, scattering and dissipation of work and energy, with consequent increase in generic entropy of energy and information… these are obviously exhibited in and through the succession of event horizon-progressions combined to the ontologically primary gear-less causal process responsible for AND-to-OR wavefunction collapse and time.

However, gear-less AND-to-OR collapse and time constitutes the true face of causality behind the realised diffusion, scattering and dissipation pertinent to generic entropy of energy; often tandem with generic entropy of information. Recall that AND-to-OR collapse and time is a work function zero process. Hence, **the impetus for the scattering and entropy of energy is *not* a feature of, nor carried by, the 'particles' themselves, but by the ontologically primary gear-less work function zero AND-to-OR collapse and time process**… which cannot itself be grasped as an outcome from contiguous particle, variable or energy inputs or interactions.

However, the very assumption that the mediation of work and energy, and attendant increase in generic entropy, is furnished by contiguous contact causality, or that it transpires as a function of a diffusion in space, is usurped by de-spatialisation. **There is no space.**

396

It follows that putative energy or work cannot be mediated from particle to particle through 'contact in space', much less diffuse contiguously across a non-existent 'space'. Entropy cannot therefore be mediated thus. Indeed, it cannot be sufficiently avowed that a change-of-state does not unfold spatially, but purely temporally: What brings about this transition is the discontinuous process and successions of AND-to-OR collapse and time (itself brought about *without* 'gears' (Book-II **2.46** and **2.47**)… and *without* spatial contiguity (see Part-I **4.01** to **4.07**)). **Thus, entropy is a pure temporal process, mediated through gear-less AND-to-OR collapse and time, without contiguity, contact, impact, or 'spatial interaction'.**

What happens when we incorporate the pleonastic status of work and energy (see Book-II 2.39 and the next essay: 4.19) into our superlative entropy considerations? Simply, the generic entropy of energy totally abstracts out of the ontology of physics: All that is left is the pure temporised growing block form of the entropy of information: namely, the entropy of Whiteheadian future-potential information accentuated by grand decoherence, coupled to the entropy of Bergsonian memory per function of the inverse square of time.

*

What would the impact of prospective matter creation be on entropy of potential information and entropy of memory? The question of 'matter creation' is not obviated by the pleonastic status of energy, as we shall discover in subsequent essays. In the forthcoming essays, we will reconsider putative matter creation at the Big Bang and thereafter, all within the framework of the de-spatialised and temporised Bergson-Whitehead amalgam. Matter creation will emerge from the postulate that both putative pleonastic potential energy *and* related potential matter are future-potentialised within the Whitehead facet, and were so at the 'beginning of time'… and that matter issues out of future-potential matter. Hence, with or without the Big Bang, it turns out that matter is *not* created 'out of nothing': the universe is *not* a 'free lunch'; a reiteration of similar assertions espoused in Book-II. Hence, conjectured creation of matter out of the potentiality of the Whitehead facet attached to the Big Bang monoblock, or even post-Big Bang, or even as continuous matter creation in a revived Steady State cosmology, does not undercut or contradict entropy, generic or otherwise.

Matter creation at the Big Bang or subsequent to it must also entail the tandem increase in the entropy of future potential information inherent to the Whitehead facet, whose decay and entropy must always attend the 'creation' of matter out of future potentiality. Put another way, **conjectured matter creation at the Big Bang subsequently always entails the tandem decay and the increase in *the entropy of future-potential matter*. Matter-creation does not reverse or annul entropy**, notwithstanding the entropy of memory per the inverse square of time that must also attend matter-creation.

4.19: Kinetic Energy & Potential Energy in the Context of the Bergson-Whitehead Amalgam

This essay reiterates conclusions first drawn in Book-II **2.39,** but recuperates these per de-spatialisation, temporisation and the growing block Bergson-Whitehead amalgam. Thus, starting with elementary concepts, recall that *potential energy* is the stored, latent, possible or potential capacity of a system to do work on some other system. Analogically, potential energy is the 'battery' of stored energy or power. On the other hand, *kinetic energy* is the actualised work done, or work in the process of being done, per the conversion of potential energy to the said actualised work. Also recall the relation between energy and matter exemplified by $E = mc^2$. Therein, energy is inter-convertible with matter, and matter is congealed energy. The equation also relates that the sum of energy-matter is unchanging: i.e. the conservation principle in its generic modern form. On the other hand, energy E relates to the kinetic energy doing the work, issued from matter… issued from mc^2.

In a similar way to what was achieved in Book-II in the context of 'spacetime in potentiality', we will subsume potential energy to the Whitehead facet of the Bergson-Whitehead amalgam. Thus, as was espoused in **2.39**… **potential energy does not issue from 'matter' or from any state restricted to the event horizon, but issues out of the ontic future… out of the Whitehead facet**. Indeed, since kinetic energy is expected to actuate at some future moment, it follows that its potential must 'reside' at that possible future moment. Therefore, it will issue out from that future moment, or from the ontic future… *from* the Whitehead facet.

Moreover, the want to locate potential energy with or in matter presupposes 'space', it presupposes that energy is 'stored in' a spatial point, area, or 'location': or in 'matter in space'. However, per 4.01 to 4.07, there is no space. Therefore, there is no 'location' at which purported potential energy is stored. Hence, de-spatialisation independently 'delocalises' and temporises energy. Instead, again, potential energy is necessarily a function of time, not of place, given that only time enjoys claim to physical ontology, given the pleonastic status of space. Indeed, the link between mass m and the 'speed of light' c in $E = mc^2$ implicitly forces the 'relocation' of potential energy from 'matter in space' to future-potential time… to the Whiteheadian future. It accomplishes this in two senses: First, insofar as the 'speed of light' is in truth the *delayed choice time-interval standard* c^i, and refers to the ultimate bounds governing the passage of time, this implies kinetic energy as that which issues in and through the process of AND-to-OR collapse and time… out of the Whitehead facet… out of the future… albeit within the bounds of a delayed choice time-interval standard, Second, given the ontological primacy and exclusivity of time per de-spatialisation, without time, there can be no action; no kinetic energy, much less any presumed potential 'store' of it, released or actuated in and through time.

*

Potential energy *appears to be* an attribute of the AND-form Whitehead facet of the Bergson-Whitehead amalgam: the possibility and potentiality for work 'stored' within the future. This view might give the false impression of 'stuff' stored inside the future, and the future treated as if it constituted a vessel or 'battery'. But the future potentiality is *not* a vessel: It is *not* full of 'stuff'.

397

Instead, it is configured as an abstract future possibility state. In short, potential energy, as not-yet transpired future-possible energy, constitutes abstract future possibility for putative 'doable work'.

Where do we fit kinetic energy? **Kinetic energy is evinced in the transition between one event horizon and a succeeding one: Kinetic energy is the actuation of future-potential energy from the *immanent* Whitehead potentiality that resides between the last realised event horizon (realised EH-1) and the next future-potential event horizon (potential EH-2): a time-interval demarcated by the said event horizons.**

Again, de-spatialisation renders the spatial framework of kinetic energy void, undermining the notion of it as a spatial 'point of action' or impact, and by compelling physics to drop the wavelength measure of energy and kinetic energy in favour of a purely frequency-based form... with the consequent transformation of forlorn 'point of action' into a *durata*. As such, radical de-spatialised kinetic energy is identifiable as the "frequency" (or the group frequency) of a given state, manifested at the transition of one event horizon into a succeeding one. Indeed, it is not a coincidence that a key measure of energy in physics is the frequency of radiation and de Broglie matter, attended by the Plank constant. Thus, the frequency is the potential time-interval that resides between successive event horizons within the Bergson-Whitehead amalgam. Hence, frequency, or *durata* in 'action', through the transition of the event horizon from EH-1 to EH-2, constitutes the conversion of immanent Whitehead potential energy into 'work done'... into kinetic energy proportionate to the frequency via AND-to-OR collapse and time.

What remains of potential energy after the realisation of kinetic energy or work issued from the immanent Whitehead facet resides beyond the immanent possibilities: beyond potential EH-2... within what remains of the Whitehead facet elaborated to the infinite future.

*

How do energy concepts relate to the Bergson facet of our Bergson-Whitehead amalgam? From the above, **clearly, kinetic energy is not and cannot be an attribute of the mnemonic Bergson facet**, given that putative kinetic energy is restricted to the quantum transition of one event horizon into a succeeding one, while the future potential energy that yields it resides within the Whitehead facet... within the future... *not* within the past. Consequently, the Bergson facet of the amalgam cannot be treated as a 'store of energy'.

Moreover, Bergsonian mnemonic processes that are critical and contributory to inertia, gravitation and other phenomena cannot be grasped as energy-driven process. That is, the physical influence of the Bergson facet, as a memory effect, or the effect of the past-in-memory upon the present and future, such as to affect inertia, gravitation, etc, does not involve the employ of putative 'work' or 'action'. Instead, inertia and gravitation (hence their root mnemonic Bergsonian processes) proceed and affect change through the grand decoherence of the Whitehead facet to consistency with memory, through a pure work function zero process. Indeed, the principal effect of the Bergson facet upon the Whitehead facet *is* grand decoherence per function of the superposition of wavefunctions in memory...itself a work function zero process that does not involve or utilise action, or 'energy' or 'work' to unfold... even on the dubious assumption that energy concepts are ontically real, not pleonastic.

*

Throughout the essay, we have treated energy concepts as if ontologically real. Energy concepts and relations have pragmatic value. But the presumption of their real existence is not sound, as was admitted and explored as early as Book-II, **2.39. Ultimately, potential and kinetic energy concepts and relations are pleonastic with respect to the Bergson-Whitehead amalgam: The de-spatialisation of energy also contributes to the pleonastic status of energy concepts and relations. It does so by breaking the erroneous notion of energy and work as 'located in space' or spatially actuated through contiguous impact causality, undermining our pragmatic notions of 'matter-interaction' or 'force-bearing particles'**: These are transformed into alternative temporal forms, *without* contiguous contact-causality, and per the total abstraction of energy concepts and relations out of the ontology of physics.

How do we arrive at the pleonastic status of energy concepts? As was argued in Book-II, the issue is wholly dependent on the ontological primacy of time vis-à-vis energy relations, and the ultimate character of time as a gearless work function zero process.

Recall that time is not a 'fourth dimension'. Time is synonymous with AND-to-OR wavefunction collapse attendant the progression of the event horizon. Also recall that neither quantum indeterminate processes nor the process of wavefunction collapse and time itself, are driven by any hidden inputs, variables, or *gears* (see Book-I **1.35**, and Book-II: **2.46** and **2.47**). This clearly implies that there is no 'kick', 'variable-input'... much less *any* definable operation... that might drive AND-to-OR collapse or time... or make time happen.

The secret to apparent 'kinetic energy' made manifest through AND-to-OR collapse and time *and* attendant event horizon transitions, resides in the ultimate *gearless* and structure-less process of causality that drives time. Unless time happens, *apparent* 'kinetic energy' cannot manifest. Thus, time has ontological primacy over what it manifests; namely, 'kinetic energy' and 'action'. But kinetic energy or work is *not* required to bring about AND-to-OR collapse and time. Nor is energy or work required to bring about the tandem transition of the event horizon and attendant state-change, both of which depend on ontologically primary AND-to-OR collapse. Thus, time and wavefunction collapse does not happen because there is a 'power' or energy input giving a 'kick' to time from behind the scenes.

In short, **it is not energy that drives time, but it is gear-less work function zero time that manifest *putative* energy relations. We reiterate that, if time does not happen, then notions of 'kinetic energy', 'action', 'work', 'power', 'force'... could not be imagined or foisted**. Since time does not happen per any push-factor from within the Bergsonian past, nor per any pull-factor from within Whiteheadian future, and since the drive of AND-to-OR wavefunction collapse or time (i.e., ultimate causality) is gear-less, structure-less and non-definable, it necessarily follows that the ultimate basis of *apparent* 'kinetic energy', 'action', 'work', 'power', 'force' is the very same structure-less and gear-less work function zero non-energy based and non-work based causality responsible for the process of AND-to-OR collapse and time.

398

The consequences are as follows: **If we submit that the ontologically primary time-process is a work function zero process, not requiring any energy-work input to realise… or, if we submit that state-change must generally happen as a result of work function zero gear-less time… and if we simply subtract all energy concepts out of the lexicon of the Bergson-Whitehead amalgam… or if we drop 'potential energy' from the Whitehead facet and simply treat it as abstract possibility… or if we drop 'kinetic energy' out of event horizon transition process, and treat it as nothing but realised event and durata… all we are left with is gearless work function zero time and, in the ultimate ontology, pure possibilities and realisation obtained in and through gera-less time, *without* 'energy' or 'action'. In short, energy concepts are rendered purely pleonastic: unnecessary to the lexicon of physics, and equally unnecessary to its ontology.**

The only hope for re-asserting the ontics of energy requires that we demonstrate that AND-to-OR wavefunction collapse and time are energy-driven, or require energy inputs to realise, or that energy concepts and relations are not ontologically dependent on ontically primary time, but that time is dependent on ontological primary energy and work. But this is obviously forlorn: The ontological primacy of gear-less work function zero AND-to-OR collapse and time vis-à-vis subsumed merely putative and ultimately pleonastic energy concepts and relations, abides.

*

We now incorporate the Bergson facet of abstract memory to pleonastic energy. To this end, we reiterate that the past is actively present to the event horizon and its transition processes. The past generates inertia and gravitation. But the active presence of the past must not be misconstrued as a 'push', or an 'energy' or 'power input' upon the event horizon: Nor can the influence from the Bergson facet upon the event horizon be characterised as 'causing' wavefunction collapse and time, or as furnishing the 'energy' for AND-to-OR collapse. It also follows that gravitational phenomena brought about by the Bergsonian past, affected through attendant work function zero grand decoherence *and* AND-to-OR collapse and time, must also proceed on a work function zero basis. In short, gravity is not an energy driven process, and the mnemonic processes within the Bergson facet (the superposition of wavefunctions in memory within the Bergson facet) are also *not* energy driven. This is obvious, if for no other reason than the fact that energy concepts are wholly pleonastic, and gravitation and the mnemonic processes of the Bergson facet are not going to be exempted from this finding.

From its depiction as a 'force field' to its depiction as a 'curvature of spacetime', gravitation has served both as definitive of physics *and* as its central mystery: We know that gravity is *not* a 'force field' projected by matter across space. Indeed, in General Relativity (GR), gravity is grasped as the curvature of spacetime, wherein matter tells spacetime how to curve, while the curve tells matter how to move: a circular causality that fails to disclose how this comes about.

When we drop the notion of space… thus drop notions of 'motion in space' … what remains of GR is the curvature of time engendered by 'matter'. This curvature of time is expected to tell 'matter' how to develop in and through the succession of event horizon transitions in a de-spatialised purely temporised universe, while matter tells time how to warp. In our terms, and as was elaborated in Book-II (**2.45**), this 'warping' is a concaving of successive event horizons, which then tell collective quantum indeterminate processes and their delayed choice time-interval relations in what fashion, pattern and decohered delimits they must collectively AND-to-OR resolve. The outcomes from this then manifest putative "orbital motion and acceleration of matter", or gravitation.

However, de-spatialisation compels that we drop the notion of a 'material body in space' altogether: from which follows that it is not 'matter' spatially conceived that is generating the gravity or the curvy event horizons. We must reframe matter as a complex of realised events rendered into stereotemporal delayed choice time-interval relations, relegated to the Bergson facet as abstract memory. Thus, redefine the 'material body' in the radical way just portrayed, and it is no longer a 'material body in space': Instead, matter recuperates into information and memory in time, retained as abstract memory within the Bergson facet.

In short, the 'matter' generating the gravity is replaced by the Bergson facet and its memories. It is not 'matter' that tells the event horizon how to curve, but it is the Bergson facet of abstract memory that tells the event horizon how to curve: a pure temporal and abstract process that also unfolds on a work function zero basis.

Consequently, **if gravitational phenomena emerge out of attendant processes of the superposition of abstract memory within the Bergson facet, and if this memory-superposition process is *not* energy-driven *and* constitutes a work function zero process, implying that the OR-form outcomes of gravity are obtained through work function zero AND-to-OR collapse and time, gravitation and the attendant warping of time manifested as curvy event horizons cannot be driven by any contiguous energy-based process. That is, gravitation must also constitute a non-contiguous work function zero process.**

The implications to mind-brain causality are also obvious, insofar as the admission to non-contiguous work function zero processes in gravity, in the drive of time, all combined to the pleonastic status of energy concepts and relations, could only undercut the most banal materialist contentions against Cartesian mind-brain dualism.

4.20: The Bergson-Whitehead Amalgam & Big Bang Matter-Creation: Eliminating the 'Free Lunch' Thesis

According to one contention, putative energy-matter was created apparently out of nothing, as a 'free lunch', at the moment of the putative Big Bang: the only tolerated instance of the violation of the principle of conservation of energy-matter in contemporary science. We will now couple our de-spatialised Bergson-Whitehead amalgam approach to the Big Bang and resolve the problem of matter creation in the light of our reform and pleonastic approach to energy concepts and relations.

We will assume that the Big Bang contention is true. Clearly, per entropy considerations alone, observable matter could not be eternal. It must have been created in the distant past: perhaps in a Big Bang, although a revamped Steady State approach could also

399

account for new matter-creation in continuous form instead of in one instant or Big bang torrent. However, even if Big Bang turns out not to be true, as a sound basis for general matter creation, the essence of the following is liable to abide.

According to the Big Bang model, the pre-Big Bang monoblock constitutes the sum of possible potential configurations of initial conditions and possible physical constants and laws. Per function of a 'quantum leap', or a quantum tunnelling process, or a process that involved Heisenberg's uncertainty principle… one set of initial conditions, constants, and laws (or all of these, on the presupposition of Manyworlds) came into being in a great torrent or 'bang'. Whatever the merits or demerits of using the concept of the Big Bang monoblock, we can augment it by adding the following postulates and assertions:

- **First postulate:** Time's 'arrow' proceeds from AND-to-OR: That is, **time proceeds from future-potentiality to the past**, as do now-pleonastic energy-matter relations and outcomes. Thus, the Whitehead facet ontologically precedes the process of coming into being and time, and it constitutes a crucial part of the Big Bang monoblock: That is, the Whitehead facet constitutes the potentiality for time and future events, including for the Big Bang event itself.

- **Second postulate:** The Whitehead facet (the ontic future) ontologically precedes the beginning of time at the Big Bang, at which the first-ever AND-to-OR collapse transpired... in the sense that the Whitehead facet was presupposed to the Big Bang moment, as is the monoblock, in a necessary and indispensable way as part of the Big Bang monoblock. Indeed, **we must incorporate the Whitehead facet into a new Big Bang monoblock**: Its first-time decay constituted the first-ever progression of the event horizon; the first ever temporal transition and state-change; the first ever elapse of time; the first ever event, or durata and the first ever formation of abstract memory per the Bergson facet.

- **Third postulate:** Insofar as now-pleonastic 'potential energy' is an attribute of the Whitehead facet, and insofar as kinetic energy or 'work done' apparently issues out of the decay of the Whitehead future now rendered integral to the new monoblock, **it follows that pleonastic potential energy was already in existence, albeit as a pure future-potentiality of the Whiteheadian futurity, incorporated into the new monoblock.**

- **Fourth postulate:** Insofar as future-potential energy was integral to the Whitehead facet, and potential matter is but a variant of future-potential energy (and with both integral to the new monoblock), it necessarily follows that **energy or 'matter' was *not* created 'out of nothing' at the Big Bang. Instead, it was always present to the Big Bang monoblock, albeit as 'future-potential energy', or as 'future-potential matter' within the incorporated Whitehead facet. Future-potential matter issued out of the Big Bang monoblock… out of the Whitehead future… without violating the principle of conservation, *without* 'creation from nothing'.** Thus, we eliminate generic 'free lunch' thesis as unnecessary and superfluous.

Recall that, in the relation exemplified by $E = mc^2$, energy is inter-convertible with matter; and the sum of energy-matter is unchanging: i.e., the conservation principle in modern form. But we 're-located' the source of now-pleonastic potential energy (hence that which is convertible to matter) to the Whiteheadian future, as an *apparent* aspect of the Whitehead facet of the Bergson-Whitehead amalgam (see **4.19**).

Recall that now-pleonastic kinetic energy issues out of future-potential energy inherent to the Whitehead facet, in turn integral to the Big Bang new monoblock. If we wish to treat 'matter' as 'congealed energy' then, like energy, such matter must also issue out of the Whitehead future-potentiality inherent to the Big Bang new monoblock. Through the process of AND-to-OR collapse and time, matter issued out from the Whiteheadian complex of future-potentiality. Thus, we reiterate the **fourth postulate: Matter was created out of the Whitehead facet; out of the future. It emerged from the decay of the future. It did not emerge 'out of nothing'** and was not created *ex nihilo*: Matter creation was not a 'free lunch'. Time… the process of gear-less AND-to-OR wavefunction collapse of the Whiteheadian future... *revealed* matter into OR-form actuality by issuing it out of the futurity in which it pre-existed as potential matter.

*

If 'matter' emerged out of the Whiteheadian future at the Big Bang, is it possible for subsequent additional matter to arise from what remains of the Whitehead facet since the putative Big Bang? Can more matter 'fizz out' from the futurity comprised of the Whitehead facet? **If Big Bang-like conditions could be instantiated post-Big Bang, these could well comprise sites for further issuance of future-potential matter into realised matter out of the Whiteheadian future, without violation of the principle of conservation of energy-matter... given that any new or additional matter creation would constitute the mere conversion of existing Whiteheadian future-potential matter into actualised matter, but not *ex nihilo*.**

We are perfectly free to conjecture a cosmology with matter creation at the Big Bang only. We are also free to conjecture matter creation at the Big Bang *and* subsequently, as a hybrid or combination of Big Bang and Steady State. Finally, we are at liberty to conjecture matter creation as an ongoing process, *without* the Big Bang, as part of a new Steady State theory. None of these conjectures entail the violation of the principle of conservation of energy-matter, given that the matter thus 'created' could not emerge *ex nihilo*, but must always issue out of the future-potentiality for matter, pre-existent within the futurity of the Whitehead facet.

*

What are the special conditions for matter-creation subsequent to or without the Big Bang, and 'where' could these be occurring? A key ingredient to matter-creation must entail a variant of Heisenberg's uncertainty: i.e., the energy-time uncertainty (see **4.21** below), of a form critically related to future-potential matter locked within the Whitehead facet. As we shall discover, the process of grand decoherence also has a decisive role to play in conjectured matter creation.

The conjecture is that an amplified energy-time uncertainty due to extreme time dilation at low altitudes or on the surfaces of asymptotic black holes transform extremely improbable matter creation event-possibilities within the Whitehead facet to accentuated probabilities. With accentuated probability, future-potential matter within the Whitehead facet will enjoy a much greater chance of AND-to-OR collapse into realised matter. Without appropriate extreme time dilation, the probability of the sought matter 'fizz-out' is expected to remain extremely low, or so improbable as to be effectively impossible.

Another way of stating the same is to imagine that the probability for future-potential matter to AND-to-OR collapse into realisation across extremely short time-intervals is normally so low that one may as well declare that it can never happen. However, when we time dilate such intervals, as is expected to happen at low altitudes or on the surfaces of asymptotic black holes, what was potentially extremely improbable or fleeting is time-stretched into acquiring a significant 'cross-section' in terms of relative duration: transforming into a 'bigger' or longer-lasting 'target', so to speak. This then increases the probability that incoming radiation or matter might coincide with such accentuated potentiality, with a now-increased probability of attendant AND-to-OR collapse of future-potential matter into realised matter. For its part, upon such 'creation', grand decoherence will inform and adjust the totality of Whiteheadian futurity and universe into consistency with the existence of the new matter so created.

As suggested, this might happen at, near, or on the surfaces of supermassive objects or asymptotic black holes typically located at the centre of galaxies and in active galaxies, such as Seyfert, B-L-Lac.. and in quasars.

*

We must now recast the above speculation about matter-creation in terms of the pleonastic status of energy concepts, per function of the work function zero process of AND-to-OR collapse or time, and per function of de-spatialisation and its usurping of the very possibility of contiguous contact-based causality and work-energy relations so understood.

While we have characterised the creation of matter as a 'drawing out' of realised matter out of future-potential matter embedded within the Whitehead facet, and while we have linked this to future-potential energy *also* inherent to the Whiteheadian futurity, from conclusions garnered in essay **4.19**, we must reiterate that **energy concepts and relations are pleonastic: Only pure information organised in time is real, and putative future-potential matter, like future-potential energy before it, is ultimately a specific state of future possibility in potentiality,** *not* **a 'stuff', nor a 'congealed energy', neither of which is ontically plausible in superlative de-spatialised and purely temporised universe driven by gear-less work function zero time-processes, wherein energy concepts and relations are rendered pleonastic.** Indeed, what we think of as 'matter' consist of stable and lasting *durata* resolved into being through the conjectured special conditions pertaining to 'matter creation', and wherein the durata thus formed is retained as abstract memory within the Bergson facet, wherein it is 'conserved' as memory, with respect to which convergent nested future-possibilities *must* grand decohere into consistent futures. Hence, the recuperation of the principle of conservation in abstract-mnemonic form first intimated in Book-II **2.39**.

Moreover, **given the dependency of matter creation on ontologically primary AND-to-OR collapse and time, and given that time itself is not an energy-driven process... and given that energy concepts can be wholly abstracted out of the ontology of physics per function of the gear-less work function zero time… it follows that time-driven matter creation must itself constitute a work function zero process.** Indeed, the process of matter creation thus speculated involves pure possibilities and special conditions for heightened probabilities for events or durata from which 'energy' is totally abstracted and rendered pleonastic. Ultimately, 'matter creation' does not involve or need any work, action or 'energy inputs', or issuance of such, to realise... unless fundamental AND-to-OR collapse and time itself requires energy-work inputs to realise… which it does not. **Thus, 'matter creation' is simply the heightened probability for the AND-to-OR collapse and formation of stable and perduring** *durata*, **which we then interpret as 'matter'.**

4.21: Energy-Time Uncertainty,
The Bergson-Whitehead Amalgam & Conjectural Matter-Creation

Recall that, as one measures the *putative* position of a particle more accurately, one obtains a calculable uncertainty in particle momentum (i.e., mass-times-velocity). Conversely, as one measures particle-momentum accurately, one obtains calculable uncertainty in purported position. Of course, position is a spatial term, as is velocity. Given de-spatialisation, we need to reframe both the position and velocity-component of momentum in Heisenberg's uncertainty into pure temporal terms, *without* 'position' or 'velocity'. The short cut to this goal is the energy-time uncertainty variant to Heisenberg's relation, which also furnishes a critical component to the mechanism of putative matter-creation espoused in **4.20**.

*

Recall that, to measure putative position accurately, we need to bombard a particle with a short-wavelength high-frequency observational input. Conversely, to measure particle-momentum accurately, we need to bombard the particle with a long-wavelength low-frequency physical input. The resolution of both position *and* momentum is not possible to obtain by applying both requisite short-wavelength high-frequency inputs for position *and* long-wavelength low-frequency inputs for momentum: This would merely give us a mixture of tendencies, with neither position nor momentum resolved accurately. We must either use short-wavelength and high-frequency inputs for positional accuracy or long-wavelength low-frequency inputs for momentum-accuracy.

When we use exclusively short-wavelength high-frequency inputs to obtain accurate positional measures, we necessarily defer the resolution of momentum to an AND-form future moment within the Whitehead facet of possible as-yet not realised momenta. Consequently, there will be no 'objective' momentum for our particle (i.e., no OR-form resolution of momentum) but only a deferred

AND-form superposition or 'spread' of as-yet not-realised future momentum possibilities within the Whitehead facet. Thus, we will obtain an accurately resolved position-measure, but a wide range of *non-resolved* future-potential values for momentum. Thus, uncertainty in momentum.

On the other hand, if we apply exclusively long-wavelength low-frequency inputs to measure momentum accurately, we must defer the resolution of putative position to an AND-form potentiality within the futurity of the Whitehead facet. Consequently, there will be no 'objective' position (i.e., no OR-form position) but only a wide 'spread' of as-yet not-realised AND-form future positional possibilities within the Whitehead facet. Consequently, we will obtain an accurate resolved momentum-measure, but at the expense of the AND-form spread of future-potential *non-resolved* positions. Thus, uncertainty in position.

Thus, we obtain Heisenberg's uncertainty principle, now grasped in the novel future-form approach just summarised, per the demands of the firewall principle (Book-II **2.10** to **2.11**), in turn furnished by the intertwine of causality and the principle of conservation.

Recall that, having resolved position accurately, if we could obtain information about now-deferred future momentum, this could only be furnished by 'signals from the future' in a block model universe. The same, but conversely, would apply to deferred potential future position. Thus, the deference of one attribute to the future implies its suspension into non-resolved AND-form futurity consonant with the growing block model of the universe, unless we could obtain 'signals from the future' and demonstrate otherwise. If we could accomplish 'signals from the future', it would culminate in the violation of both causality and, insofar as such a signal would constitute evidence of additional putative energy-matter above known unity, the creation of energy-matter *apparently* out of nothing, in violation of the principle of conservation of energy-matter.

*

Of course, 'position' and the 'velocity' component of a particle's momentum are spatial terms. **Given de-spatialisation, there is no space**: there are only events organised in and by time. Purported 'matter in space' is in truth constituted by discrete events, or *durata*, organised purely in terms of delayed choice time-interval relations, *without* space. **Thus, there is no 'position' for a particle**: There is only its OR-form incidence as duration, temporally inter-related to other realised incidences, or to durata, to observers or to frames of reference... all interrelated and integrated via delayed choice time-intervals.

Similarly, due to de-spatialisation, there is no 'motion'. Hence there is no 'velocity' component to momentum. There is only a uniform or non-uniform change in the delayed choice time-interval relation of a given particle or durata vis-à-vis other particle or durata, observer or frame. Thus, 'momentum' in our reconceived 'spacetime' *without* space must be recapitulated in terms of uniform (or else non-uniform) realised or potential changes in the relative delayed choice time-interval relations, realised through the transition of event horizon EH-1 to potential EH-2, or by attendant AND-to-OR collapse and time.

What about the mass of the particle as might relate to its momentum? Recall that, generically, momentum is mass times the velocity, and both need reframing in terms of pure delayed choice time-interval transformations in a de-spatialised order. To this end, there are two distinct notions of mass: The first is the *manifest mass* of a purported body, evinced as inertia; evinced as gravitational mass. But, insofar as there is no space, we are forced to explain both inertia, gravitation, and attended mass, as the superpose of past wavefunction in memory retained within the Bergson facet of the Bergson-Whitehead amalgam, subject to the inverse square law of time, and critical to the subsequent or attendant grand decoherence of the Whitehead facet (the future) rendered consistent with the implications of 'mass' re-conceived as abstract memory. Hence, the memory theory of inertial and gravitational mass.

To clarify, **insofar as there is no space, mass cannot have a 'location', no more than we could obtain a 'location' for energy; potential or kinetic. Thus, mass is *not* a physical attribute of 'place' or of a point, or of a block of 'matter in space'**. Insofar as mass is attendant putative 'matter', and 'matter' is in its turn a naive place-holder for the complex corpus of stable durata organised into delayed choice time-interval relations, it follows that, as is true of putative potential energy and kinetic energy (albeit per function of the futurity constituted as Whitehead facet), **the inertial and gravitational mass of a 'body' is brought into manifestation in and through time as an effect of the total past of a body constituted as the Bergson facet of the Bergson-Whitehead amalgam. Again, inertial and gravitational 'mass' is recuperated as a Bergsonian memory effect**. This is to be elucidated in Part-III.

In addition to mass reconceived as Bergsonian memory effect, the second form of mass might be called *Platonic mass*. As such, its basis is totally mysterious; perhaps beyond any physics-based solution. Per pure speculation and conjecture, formal mass might be derivable from the application of number theory to physics. For example, an electron is barely above 1/1836 the mass of a proton, but this has nothing to do with the active presence of its past, nor its inertial and 'gravitational mass' as a memory effect. Like all fundamental laws and constants imbued into the Whitehead futurity and emanating from timelike infinity, this aspect of 'mass' is formal or Platonic, *not* ultimately phenomenal in origins... even though it structures phenomena that AND-to-OR collapse out of futurity, and is thus phenomenalised, as is all physical law. This formal mass is furnished upon, yet remains extant of, memory, futurity and time. Formal mass is extant the growing block superlative Bergson-Whitehead amalgam, even while it informs and delimits the amalgam.

The Platonic characteristics of nature was furnished in our EPR-based approach to the ontology of physical law, exemplified by the principle of conservation of quantum spin. This argued for the dualism of physical law vis-à-vis memory, future-possibility and time: (see Preliminary **0.26**, and Book-I **1.13**). The approach argued that a higher ontology (in effect, a Platonic-Cartesian ontology, or a meta-semantic hypostasis) must constitute the basis of otherwise time-asymmetric and time-extant physical laws and natural constants and principles that attend nature: This must constitute the basis for the formal Platonic masses of particles. Also, the emanative theory for physical law (and, implicitly, of formal mass) was argued in Book-III **3.33**. Therein, it was argued that physical laws and constants (and presumably formal mass) are emanated from and imbued to Whiteheadian potentiality from timelike infinity, or the infinitely removed future.

*

Given that there is no space, the 'wavelength' attribute of the particle must be dispensed with and succeeded by a pure frequency based approach, as part of the purely temporal approach to Heisenberg's uncertainty relation. Consequently, the 'position' attribute sough must also abstract out as pleonastic to physics. **Conveniently, the 'alternative' to the spatial approach to Heisenberg's uncertainty is the long-established *energy-time uncertainty relation*.** Of course, in generic physics, energy-time uncertainty often references the 'wavelength' of the particle. Again, this disappears per the demands of de-spatialisation and the ontic exclusivity of a purely temporal frequency-based physics.

There is no space. There is only time. Hence, Heisenberg's energy-time uncertainty perfectly fits our need to obviate space and wavelength-based approaches to Heisenberg's uncertainty. In doing so, we obviate 'position' or 'velocity', both of which are spatial notions: In an energy-time uncertainty relation, we need not refer to 'velocity'. It also follows that we need not refer to any 'momentum' attribute, which requires 'velocity', while the 'mass' remains implicit either as part of the background memory effect or as formal Platonic mass, or as a combination both.

In Heisenberg's energy-time uncertainty, the shorter the potential time-interval, the greater the uncertainty in energy, and greater the potential AND-form spread of energy-level possibilities potentialised to that time-interval. Conversely, the longer the time-interval, the smaller the range of energy levels potentialised to the time-interval. In Alternative Realist terms, when we reduce the time-measure to smaller and smaller intervals, we necessarily defer the potential energy of the system to a spectrum of greater future potential energy-level possibilities within the Whitehead facet and futurity. As before, this 'deference to the future' is per the demands of the firewall principle from the intertwine of causality and the conservation principle. **The range of future-potential energy levels and ranges rendered in AND-form do not constitute OR-form realised or materialised energy. These are *not* actions or 'work' in OR-form, but only constitute future-potential as-yet not realised (hence, *not at all* realised) ranges for putative energy, action, or work.**

In terms of the Bergson-Whitehead amalgam (and even in terms of the intermediate model from Book-II) the pertinent time-interval that defines the time-variable in the energy-time uncertainty relation is the demarcation defined by, say, the event horizon at EH-1 versus future-potential succeeding EH-2, with the interval recuperated as frequency. If the time-interval is very short, and the 'gap' between EH-1 and potential EH-2 is small, the range in potential as-yet not-realised future energy will be greater. If the time-interval between EH-1 and potential EH-2 is long (lower frequency), the range of future-potential as-yet not-realised energies will be lower or reduced.

From our recapitulation of putative kinetic and potential energy in terms of both the intermediate spacetime model in Book-II (**2.39**) and the same in terms of the Bergson-Whitehead amalgam successor (**4.19**), recall that potential energy is not stored in 'matter in space', given that there is no space. Therefore, there is no 'location' from which potential energy can issue out as kinetic energy, work, or action. Instead, potential energy is potentialised to the future: formerly, 'spacetime in potentiality', which is now the de-spatialised Whitehead facet of the Bergson-Whitehead amalgam. Thus, the uncertainty in energy, as potential not-yet realised energy that must attend short time-intervals, is necessarily a future-potential of the Whitehead facet.

*

What about future-potential matter? Recall from **4.20** the invalidation of the notion that the universe is a 'free lunch': Putting aside the pleonastic status of energy concepts and relations for the moment, the putative energy-matter that makes up the universe had issued out of the Whitehead facet incorporated into the new monoblock at the Big Bang. This future-potential energy-matter existed as a future potentiality: it pre-existed its 'creation' at the Big Bang. It was never created *ex nihilo*.

With the Big Bang, the potential energy-matter was somehow realised out of the Whitehead facet through AND-to-OR collapse.

*

In our speculations about subsequent post-Big Bang matter-creation in 4.20, we considered the prospect of matter creation at or near asymptotic black holes. Therein, we posited the above Heisenberg energy-time uncertainty as part of the mechanism of matter-creation. Galvanised by the reiterations and clarifications above, we will now posit a more detailed speculation about how ongoing matter-creation might well unfold, post-Big Bang.

Recall from the energy-time uncertainty relation that, as shorter and shorter time-intervals are considered, the greater the range of (and uncertainty in) future-potential energy within the attendant futurity or Whitehead facet. A similar notion must hold for future potential matter. **The shorter and shorter the time-interval, the greater the range of as-yet not-realised future-potential matter within the attendant Whitehead facet.** Indeed, since matter is thought to be 'congealed energy' or interchangeable with energy, it is elementary to equate future-potential matter as akin to future-potential energy. **Thus, the uncertainty in energy vis-à-vis time is effectively uncertainty in future-potential matter vis-à-vis time**, putting the pleonastic status of energy aside, at least for now.

However, matter-creation based on energy-time uncertainty encounters two problems:

- First, future-potential matter per the energy-time uncertainty does not constitute actualised or realised energy, no more than it could constitute actualised or realised matter. As such, as future-potential matter not-yet existent, potential matter is *virtually* not existent. It is merely the possibility for matter, *not* its actuality.

403

- The second problem is imposed by the short time-intervals involved. The short time-interval *reduces* the probability that future-potential matter might AND-to-OR collapse into realised matter. Why? The shorter the time-interval in which potential events constitutive of energy-matter might transpire, the less likely that some physical input will coincide with that interval and attend its AND-to-OR collapse into realised matter. That is, the 'cross-section' for matter-creation, grasped in terms of time... is simply too temporally short or 'small'. **Thus, while the range of future-potential matter vis-à-vis very short time-intervals involved might well be very large, the probability that the rest of the universe will interact with it, and, through grand decoherence, participate in its AND-to-OR collapse into realised OR-form stable durata or 'matter', will be almost close to the zero, owing to its very-short time-interval or short temporal 'cross-section'.**

We must reframe what we have just stated in terms of nested futures counterfactuals, grand decoherence, and the revived Mach's principle from Book-II (**2.35**): These are perfectly relevant to the subject in hand, as we shall shortly discover.

Ignoring de-spatialisation for the moment, recall that matter elsewhere in the universe 'knows' about matter here, and vice versa, via grand decoherence. This is 'non-local' and instantaneously counterfactually mediated, because matter here and elsewhere, though not directly integrated via the event horizon or by putative 'mediating radiation and matter' limited to the 'speed of light', is yet integrated through a system of mutually convergent nested futures elaborated within their common Whitehead facet to the infinitely removed future.

The nested futures perspective and its convergent category structure survive de-spatialisation and into the Whitehead facet, and it is recuperated into the revived Mach's principle, but *without* 'space', as is the counterfactual causality and the grand decoherence that attends our revived Mach's principle.

But how does counterfactual causality and grand decoherence relate to matter creation? Unless our very short time-interval and its broad range of future-potential matter-states is resolved into an OR-form actuality, matter-creation will not transpire, or will remain so negligible as to be effectively irrelevant. Consequently, with no matter-creation in effect, no grand decoherence of nested futures into consistency with that matter will transpire within the Whitehead facet to reflect the creation of matter.

If the attendant time-interval in the energy-time uncertainty relation remains so short as to reduce the probability of matter creation almost to zero, switching from short-time to long time-interval states will obviously not solve the problem: The long time-interval state will certainly present a greater temporal 'target cross-section', and will constitute a much higher probability that some extant physical input might get to coincide with the 'bubble' of future possibility for new matter. Yet, the range or levels of future-potential matter therein will approach almost zero as the time-interval is increased.

We need an additional factor to co-join with the two basic requirements of energy-time uncertainty *and* grand decoherence to affect meaningful matter-creation.

*

What happens when we combine short time-interval energy-time uncertainty relations with profound time-dilation effects, of the form that might occur at low altitudes or on approach to the Schwarzschild radii of supermassive objects or asymptotic 'black holes'? For observers attached to the frame of the short time-interval energy-time relation at the said altitude or on the surface of the asymptotic black hole, there will be no apparent time dilation, and the future-potential energy or matter within that short time-interval will be subject to the expected extreme low probability of matter-creation.

However, from the frame of an observer at a 'distance', making observations on the energy-time relation frame at the surface of the supermass, profound time-dilation will be observed to apply. Because of profound time-dilation, the otherwise short time-interval of the relation will be dilated to a relative long time-interval duration, even while the probability and distribution for potential matter within that interval will remain the same. In short, relativistically, the temporal cross-section of the otherwise short time-interval Heisenberg energy-time relation will be dramatically increased, at least as viewed by observers outside and at 'distance' from the system. Hence, its disposition, size, or profile as a 'target' to incoming physical inputs will increase accordingly. Given its dilated and enlarged cross-section, there will now emerge a significant probability that some physical input emanating from another part of the universe might get to coincide with the now profoundly time-dilated time interval cross-section and its higher probability for-future potential matter. Profound time dilation will transform what was otherwise a 'small cross-section target' into a 'large cross-section target', so increasing the likelihood that something from outside might coincide with it. Indeed, given the nature of the supermass and its tendency to 'vacuum in' both radiation and matter inputs from the rest of the universe, the probability of such a coincidence between infalling inputs and the profoundly time-dilated short time energy-time uncertainty cross-section will become very significant, with a higher probability of AND-to-OR collapse, with attendant higher probability of the issuance of significant quantities of potential-matter into realised new matter out of the future-potentiality of the Whitehead facet.

The higher probability and possibility of matter-creation per the now higher probability of coincidence of incoming physical inputs to the accentuated temporal cross-section of future-potential matter, must necessarily constitute a greater possibility of counterfactuals vis-à-vis the rest of the Whitehead facet. The Whitehead facet must grand decohere into consistency, even to the infinitely removed future. That is, per grand decoherence, the Whiteheadian future will be rendered consistent with the possibility and subsequent higher probability of AND-to-OR collapse of, and final OR-form actuality of created matter from out of its future-potentiality, so informing all other matter in the universe that new matter has come into being; with the new matter counterfactually 'knowing' about all the other matter in the universe, and vice-versa, by means of the grand decoherence that attends our revived Mach's principle: see Book-II **2.35**.

Hence, we obtain an almost-certain mechanism for new matter-creation: one possibly useful to the Big Bang, but especially to subsequent new matter-creation, post-Big-Bang; and perhaps especially significant to prospective revival of Steady State cosmology.

In summary, the mechanism of matter creation proceeds thus:

- The shorter the time-interval pertaining to the energy-time uncertainty relation, the smaller the temporal 'cross-section' or target profile for future-potential matter. Thus, the less possibility that any input might coincide with that energy-time 'bubble' for matter-creation to occur: In temporal terms, the target is too fleeting: it has a 'peek a boo' character and will very likely be 'missed' by any incoming radiation or matter. Therefore, the possibility for matter-creation will remain so improbable as to be effectively non-existent.

- **When an otherwise short time-interval energy-time uncertainty relation 'bubble' is placed at low altitude or on the surface of an asymptotic black hole, profound time-dilation effects that abide therein will transform our otherwise 'small target' into an accentuated 'large target' temporal cross-section, increasing the probability that incoming inputs might coincide** with the 'bubble' for potential matter.

- The supermassive object will also increase the likelihood of such coincidence; predicated by its very nature to draw in more radiation and matter-in-fall from elsewhere, so increasing the likelihood of the said coincidence.

- **The probability of coincidence between incoming inputs and the now larger time-dilated temporal cross-section of the energy-time uncertainty 'bubble' will increase the probability of AND-to-OR collapse of future-potential matter within the Whitehead facet into realised or new matter.**

- **Per general grand decoherence, the rest of the universe will be counterfactually informed of the existence of this new matter: All other matter in the universe will have their mutual futures within the Whitehead facet decohere into consistent convergent future possibilities... consistent with the fact and future possibilities imposed by the new matter.** The new matter will in turn be counterfactually informed about all matter elsewhere; **its future decohering to mutual consistency and, through it, acquiring its own inertia per our revived Mach's principle**: (Book-II, **2.35**).

- **Other mechanisms will take over and eject the newly created matter from the supermassive object:** Observations of matter-streams in Seyfert and BL-Lac galaxies and in quasi-stellar objects conform to this expectation. But this presupposes asymptotic black holes that cannot collapse into singularities (see below for a solution against such singularities). It is expected that, as the matter is created, the supermass will become more massive; the time-dilation more protracted; which then must escalate the rate of matter creation...until a critical threshold is reached, and the matter thus created is either explosively shod off from the supermass or it is ejected via matter-streams of the form observed in active galaxies and quasars.

Note that none of this involves the creation of matter 'out of nothing', given that the future-potential matter pre-exists as future-potentiality within the Whitehead facet, as it did at the putative new monoblock at the Big Bang. Thus, we are not entertaining a 'free lunch' thesis or the violation of the generic conservation principle.

Supermassive objects do reside at the centre of galaxies. In active galaxies, such as B-L-Lac and Seyfert galaxies, and in quasi-stellar objects: Therein, prodigious energies, matter streams and ejecta are observed. Hence, we conjecture that future-potential matter is being converted into actualised new matter in such sites, forced to 'fizz out' of the Whitehead facet of the Bergson-Whitehead amalgam through the mechanism of matter-creation presaged above.

*

However, there is another problem we need to solve before matter-creation so-described could be rendered plausible. **The said mechanism for matter creation requires that supermassive objects do not collapse into singularities.** Otherwise, the created matter would fall into the singularity and become lost without trace. **The idea here is that supermassive objects are always asymptotic: not real. That is, there are no singularities.** Thus, real supermassive objects might well collapse and approach the temporal equivalent of the Schwarzschild radii, but they never obtain it. At critical altitudes or on the surfaces of such asymptotic 'black holes', the sought profoundly time-dilated energy-time relation bubbles will come into formation: increasing the probability of matter-creation out of future-potential matter within the Whitehead facet. Consider two attendant assertions:

- **The prospect that supermassive objects might not collapse into singularities is explicit in de-spatialisation: Indeed, singularities are only conceptually plausible on the supposition of 'space'. There is no space. Hence, notional 'collapse' of matter to a spatial point-singularity is implausible and forlorn.** In short de-spatialisation, as ultimate renormalisation, gets rid of infinity-singularities, and renders all supermassive objects into asymptotic black holes, or non-singularities. Thus black holes at the centre of galaxies, in BL-Lack and Seyfert active galaxies and in quasars, are all *asymptotic* black holes.

- Also, in our de-spatialised Bergson-Whitehead amalgam, **Bergsonian memory, hence the Bergson facet of the amalgam, constitutes the basis for gravitational mass, as a memory-effect born out of the superposition of the total past (see Part-III). But the attendant mnemonic superpose process cannot engender a singularity or sum to infinity,** notwithstanding obvious limits from de-spatialisation against the very notion or possibility of a spatial point or 'singularity'. Again, the implication is that all inferred and observable supermassive objects or black holes are asymptotic.

What would a singularity look like from the Bergson-facet approach to gravitational mass? Therein, such a strange state could not obtain 'infinite mass', given that it would require the superposition of an infinite number of Bergsonian wavefunctions in memory, but even then it will be delimited by a Lorentz-like transformation of the sum of memories away from infinities and singularities.

Also, assuming the false notion of 'space', a singularity is essentially equivalent to a frame of reference that is receding from us, not merely at the putative 'speed of light', but at infinity: equivalent to a wavefunction in memory now receded to an infinitely removed past-primacy. However, the power of memory to affect the present is reduced by the inverse square of time. Hence, the power of a wavefunction in memory receded to the infinitely removed past would become attenuated to zero. What sort of singularity could *that* produce? Or, succinctly, what sort of gravity could it produce? The answers are obvious and tautological.

In any case**, it cannot be sufficiently avowed that, in a purely temporised de-spatialised universe, there are no singularities. It follows that 'black holes can only ever be asymptotic**... perpetually collapsing to their temporal equivalent of the Schwarzschild radius, but never obtaining that radius. It is more than conjecture to state that, **while supermassive objects or asymptotic 'black holes' are plausible and seemingly evident, point-collapsed singularities to 'infinite mass' must remain impossible, per de-spatialisation and other reasons. In that case, our matter-creation mechanism gains a much greater viability, if not certainty**, to the extent that newly created matter need not disappear into a singularity-sink.

*

Finally, we return to the issue of the pleonastic status of energy, and what it implies to 'matter creation'. **Surely, if energy concepts abstract out of nature such as to become not part of the real ontology of physics, the same ought to apply to matter, given that matter is often portrayed as merely 'congealed energy'. How does 'matter creation' fare in the face of pleonastic energy?**

When putative matter is created, what issues out of the Whitehead facet is the formation of a permanent, or at least stable and enduring *durata*... whose basic measure is its frequency (a time-attribute). The consequent grand decoherence that attends the creation informs and reforms the total mutual convergent futurity of the universe to the factity of the newly created durata, with respect to which the rest of the universe must abide into counterfactual consistency, and into permissible consistent Whiteheadian futures.

To reiterate, **stable *durata* (or 'matter', if you will) possesses a basic frequency-attribute, which is a time-attribute that could be foisted a notional 'energy', but need not be so foisted, given the pleonastic status of energy. Therein resides the harmony of 'matter creation' vis-à-vis pleonastic energy**. Thus, the creation of new durata can be characterised as an issuance of otherwise extremely low-probability future potential stable durata out of the Whitehead facet, but it need not be characterised as an 'energy issuance', save for didactic and pragmatic purposes that have no bearing on the ultimate ontology of nature, or on the process of work function zero AND-to-OR time that finally makes such 'matter creation' happen.

In summary, the Whitehead facet possesses a future-potentiality for stable durata whose otherwise extremely low probability of realisation is accentuated per the mechanisms described above, backed by the impossibility of singularities in a de-spatialised and temporised universe.

Such new durata, or 'new matter', need not be portrayed in energy terms, and must ultimately be characterised as pure possibilities and realisations in an otherwise abstract, de-spatialised purely temporised and ephemeralised informational universe.

4.22: The Matter-Creating Hyper-Spinning Flywheel & Recapitulation of the Critique of Manyworlds

How convenient that we have no direct access to supermassive objects, nor to naturally occurring profound time-dilating phenomena to test the above contention for matter-creation. Yet, there may be a way of generating profound time dilation using small hyper-spinning flywheels, and we might even induce matter-creation therein.

A solid flywheel of an appropriate size (the smaller the better), constructed out of appropriate materials (perhaps mainly carbon nano-tubular), magnetically levitated, suspended, and encased in a vacuum chamber, could be made to spin at near the 'speed of light'. Such a feat would require huge quantities of putative energy; plausible if the acceleration of the flywheel is built-up over time. We could accelerate the flywheel using lasers, in the fashion that hypothetical 'lightcraft' might be so accelerated. Presumably, the energy could come from your future local thorium salt reactor, or Eric Lerner's prospective focus fusion device.

If we plug out a cavity through the material near the circumference of the flywheel, we then create an internal surface or region in which profound time dilation effects will abide to whatever extent we accelerate the flywheel to near the speed of light. A caesium clock, or other radioactive material that decays at predictable frequency may be attached to the flywheel near the cavity and can be used to indicate the time dilation therein, compared to a similar clock attached outside the flywheel.

Hence, in this way, we might achieve the sort of profound time dilation in a similar way to what attends the surface of supermassive objects or near their Schwarzschild radii. Within the cavity, the temporal cross-section of any 'bubble' of energy-time uncertainty and future potential matter could accentuate into a large cross section or 'larger target'.

What would happen if we shone coherent radiation into the said cavity, discharged from a gun positioned next to the flywheel? The laser-shot would need to be synchronised to coincide with the cavity and must be discharged in pulse-form with appropriate frequency. In this way, we create the physical inputs that might coincide with now time-dilated accentuated cross-section energy-time uncertainty and future-potential matter 'bubbles', and so increase the probability of new matter creation.

Such matter creation would not be free, given the vast quantities of putative energy required to spin the flywheel to near the speed of light. Moreover, the process could not involve 'creation from nothing' in violation of the principle of conservation, given that the said matter pre-exists in future-potential form within the Whitehead facet: See **4.20**.

*

If matter creation is possible, what are the implications to the Manyworlds interpretation of quantum mechanics? In Book-III **3.13** and in other parts of Book-III, Manyworlds was portrayed as a violator of the conservation principle. If each possibility encapsulated in a wavefunction was OR-form realised into a history, and given infinite possibilities entailed in certain wavefunctions, where would the energy-matter to constitute these histories come from? Could these issue from future-potential energy and matter in the Whitehead facet?

Ignoring the pleonastic status of energy concepts and treating energy *as if* it was ontically real, it appears that MWI might get a new lease of life, if the future-potential matter within the Whitehead facet is inexhaustible and infinite. However, this cannot avail the superfluent character of MWI, given that, all its alternative histories culminate into the same identical global outcome (see Book-III **3.17**) among other criticisms that render MWI implausible as any basis for information and mnemonic nihilism (see Book-III, Part-I).

In any case, **even if the future-potential energy or matter inherent to the Whitehead facet is inexhaustible and infinite, matter creation yet requires very special conditions obtainable only at supermassive objects: Obviously, profound time-dilation does not abide across the whole universe. Moreover, at such sites, it is not certain that we could create infinite quantities of new matter to constitute infinite Manyworlds histories. Indeed, at best, our case for matter creation at such sites could only support the Little Manyworlds (LMW) thesis:** (see Book-III **3.18**).

When we bring back pleonastics and recuperate 'potential matter' into potential stable durata, the same possibilities, problems and limits stated above must yet abide vis-à-vis Manyworlds, but without reference to any energy-based conservation principles.

4.23: Summary of Part-II

The following constitutes the bullet-point summarisation of key findings furnished in Part-II:

- **Wavefunctions and quantum indeterminism recapitulated in the framework of the de-spatialised and temporised Bergson-Whitehead amalgam (essay 4.17):** From Book-III, minimal structure informational states of future-potentiality or past actuality constitute wavefunctions in future-potentiality and wavefunctions in memory. But *de-spatialisation implies that both wavefunctions in memory and wavefunctions in future-potentiality (subsumed to the Bergson facet and Whitehead facet, respectively) can no longer be conceived as projections and emanations in or across space, but only in time.* It also follows that objective quantum indeterminate terms resolved on the event horizon are not resolved across 'space' or on 'detector surfaces', but are funnelled into coherent pattern-bearing global outcomes per the operation of the three-phase developmental process (Book-III **3.26**) rendered into stereotemporal time, *without* 'space'. Thus, interrelations between quantum indeterminate terms and observers are always composed of delayed choice time-interval relata: *not* 'distances' or spaces.

- **Entropy relations recuperated in the framework of the de-spatialised and temporised Bergson-Whitehead amalgam (essay 4.18):** a reprise of material first articulated in Book-I, Part-I, but incorporated into the Bergson-Whitehead amalgam:

  - The 'arrow of time' proceeds, not according to a statistical tendency for generic entropy of energy, but per the objective polarity of AND-to-OR collapse synonymous with time. *Time proceeds from AND-to-OR... from future-potentiality and the decay of the Whitehead facet into recency states on or along successive event horizons. Time and its 'arrow' can never proceed from OR-to-AND,* or time reversal. The mere statistical reversal in the tendency of generic entropy cannot imply time-reversal, given that such outcomes require ontologically primary AND-to-OR collapse in order to manifest, and cannot furnish OR-to-AND de-collapse and time-reversal.

  - The *entropy of future-potential information* supersedes generic statistical-based entropy of information. Future-potential information constitutes the total potential for events and delayed choice time-relations (including nested futures) within the Whitehead facet of the Bergson-Whitehead amalgam. Whenever AND-to-OR collapse or time unfolds, and whenever counterfactual-driven grand decoherence of nested futures occurs, *the total future-potential information of the universe within the Whitehead facet proceeds from a lower state of entropy (higher potential information) to a higher state of entropy (to lowered potential information).*

  - While AND-to-OR collapse and time increases the memory-content of the Bergson facet of the Bergson-Whitehead amalgam, this cannot constitute an inversion in the tendency for increased in the entropy of information. *While the memory-content of the universe increases and the Bergson facet thus expands, the total accumulating memory is subject to the inverse square of time. Thus, memory weakens per the inverse square of time, so constituting the entropy of memory per the inverse square of time.*

- **Reform and pleonastic status of kinetic and potential energy concepts and relations recuperated in the context of the de-spatialised temporised Bergson-Whitehead amalgam (essay 4.19):** A reprise of the pleonastic arguments from Book-II but reframed to the Bergson-Whitehead amalgam. On the false assumption of 'matter in space' both kinetic and potential energy concepts presume kinetic energy to be instantiated as 'action' or 'work' between 'bits of matter', while potential energy is treated as 'congealed' into 'matter in space', but issued out as kinetic energy or work. But *there is no space. Hence, both energy and 'matter' must be recapitulated purely in terms of time. It turns out that potential energy issues out of the Whitehead facet (the future) as future-potential energy. Kinetic energy is that part of future-potential energy that manifest on the event horizon through the AND-to-OR decay of the Whitehead facet*, or when event horizon EH-1 is succeeded by EH-2. Note that the transition from EH-1 to EH-2, attendant *AND-to-OR collapse or time, is not ultimately an energy or power-driven process, but arises by the 'action' of a gear-less causality with a work function zero profile* (see Book-I **1.35**, and Book-II: **2.46** and **2.47**). *Thus, what appears to be the conversion of Whiteheadian future-potential energy into realised kinetic energy is dependent on ontologically primary gear-less work function zero AND-to-OR time process: a non-energy-based causality. Thus, at the primary ontology, we obtain pure information physics, wherein energy considerations as basis of 'action', causality and time are rendered pleonastic.*

- **There is no 'free lunch': Realised matter collapses out of pre-existent future-potential matter... out of the Whitehead facet at the Big Bang, and likely subsequently (essay 4.20).** *If future-potential energy resides in the Whitehead facet as an as-yet not realised energy, so could future-potential matter.* At the Big Bang, both the Whitehead facet (the future) *and* the future-potential matter within it, were potentialised to the Big Bang new monoblock. *Whatever the cause, AND-form future potential matter within the Whitehead facet was mass-converted into OR-form realised matter, but not 'out of nothing'*, and without the violation of generic conservation of energy-matter, even putting aside the pleonastic status of energy and matter.

- **Heisenberg's energy-time uncertainty and presage to matter-creation mechanism subsequent to the Big Bang (essay 4.21):** With de-spatialisation, we must abandon position-momentum uncertainty in favour of purely temporal energy-time uncertainty. Even so, a matter-creation mechanism is conjectured from the combination of energy-time uncertainty with profound time dilation on the surfaces or near the Schwarzschild radii of asymptotic 'black holes', with critical participation from grand decoherence. The mechanism proceeds thus: Typically, the shorter the time-interval of the energy-time uncertainty the smaller the 'cross-section' of the uncertainty relation grasped in temporal terms, and the less possibility for any radiation or matter input to coincide with that energy-time 'bubble' to attend matter-creation. However:
  - When the otherwise short time-interval energy-time uncertainty bubble is placed at low altitude or on the surface of an asymptotic black hole, profound time-dilation effects transform the small cross-section target into an accentuated dilated 'large target'; increasing the likelihood that incoming radiation or matter-inputs might coincide with that 'bubble'.
  - The supermassive object increases the likelihood of this by drawing in more radiation and matter-infall.
  - This will increase the probability of AND-to-OR transformation of future-potential matter locked within the said 'bubble' out of the Whitehead facet... into realised or 'created' matter.
  - Per general grand decoherence, the rest of the universe will be counterfactually informed of the existence of this new matter: All other matter in the universe will find their convergent futures within the Whitehead facet decohere and erase into consistent-only futures: consistent with the factity of the new matter. The new matter in turn will be counterfactually informed about all matter elsewhere, with its own Whiteheadian future grand decohering to consistency and, through this, it will acquire inertia per the revived Mach's principle (or grand decoherence).
  - Other mechanisms will take over and eject the newly created matter from the supermassive object: Observations of matter-streams in Seyfert, BL-Lac and quasi-stellar objects conform to this expectation.

- But **all of this presupposes asymptotic black holes that cannot collapse to singularities**. Thankfully, de-spatialisation obviates point-singularities of infinite mass. There is no space, and no spatial point to which matter could collapse to infinity. A purely temporal universe obviates spatial infinite singularities. Moreover, the fact that gravitational mass is a memory effect from the Bergson facet also obviates singularities: the superpose of memories cannot generate such singularities. Thus, all 'black holes' are necessarily asymptotic... or *not* singularities. Therefore, matter-creation is rendered plausible and likely.

- **Flywheels for matter-creation (essay 4.22):** *We conjecture that it might be possible to emulate both profound time-dilation effects requisite to matter creation and, possibly, matter creation itself.* This requires a small solid flywheel, with a cavity, accelerated to spin at very close to the speed of light, with requisite time dilation effects. We input radiation into the cavity to help induce matter creation. This would not be a 'free lunch', but an extremely energy-inefficient and costly process, pleonastics aside.

BOOK-IV PART-III:
PRELUDE TO THE PHYSICS OF MEMORY:
INERTIA, GRAVITATION, AND OTHER THINGS

4.24: Aims of Part-III

With the succession of intermediate model spacetime by the Bergson-Whitehead amalgam, with de-spatialisation and the temporisation of physics, Part-III will fulfil several goals. With de-spatialisation and temporisation, 'matter in space' gives way to the *durata* of 'information distributed in time', or Bergsonian memory, with the consequence that fundamental interactions in physics transform into memory-addition processes. Hence, the necessity for a mnemonic mechanics as a first goal that, even if erroneous in specific details in our presage, it *must* yet be developed and *must* supersede notions of 'matter in space' and 'spatial interactions'.

Presaged mnemonic mechanics also promises to furnish memory-based gravitation from within the growing block Bergson-Whitehead amalgam, with presage to quantum gravitation implied.

Five postulates will be stated as key to the presage to mnemonic mechanics and gravitation theory: These are, the **independent retention postulate** of past states or wavefunctions in memory; the **superpose of past states across time**, involving the addition of primacy and recentcy memories from within the Bergson facet of the amalgam; the weakening of memory through **the inverse square law of time**; and the **exclusivity of constructive interference in the addition of wavefunctions in memory**. The final postulate will be the **inapplicability of grand decoherence to Bergsonian memory**. Hence, the conservation of memory so-implied.

Beyond the brief sketch of mnemonic mechanics attending the presage to inertia, Part-III will presage nascent **Bergsonian mnemonic mechanics for gravitation**, wherein the wavefunction in memory attached to one body, operating across time and subject to the inverse square of time, superposes with the culminating superpose from the stock of memories that attends another body, bringing about the gravitational attraction of the latter to the former. This nascent mnemonic mechanics for gravitation will constitute counterfactuals to the Whitehead facet, generating the grand decoherence of the nested possibilities within the Whitehead facet into consistency. Through grand decoherence, the background mnemonic mechanics inscribes into the Whitehead facet' the **future-form work function zero 'paths' of orbital motion, acceleration and attraction**: Such paths are work-function zero paths in the sense that apparent 'motion' AND-to-OR resolved along these requires no putative energy, work, 'force' or power-inputs in order to affect, while any subsequent attempt to change these to some other path *will* require putative energy-power inputs, with attendant non-zero work functions. If no attempt to alter that future-form path is made, the 'motion' of the body along the said future-form path will remain a work function zero state, with 'acceleration' and 'orbital motion' along such work function zero paths realised in perpetuity, *without* energy inputs: *without* violation of the conservation principle, generic or mnemonic.

Insofar as all of this is realised through universe-scale AND-to-OR collapse of Whiteheadian possibilities through attendant broken discontinuous progressions of the event horizon, **per the quantised or discrete nature of the attendant time-process, quantised gravity is implied. Hence, quantum gravity is presaged**.

Finally, we will discover that nascent Bergsonian mnemonic mechanics and gravity implies a **limit against the formation of singularities**, aside obvious implications from de-spatialisation against spatial points in general and against attendant singularities, with an implied **alternative explanation for both 'dark matter' and 'dark energy'**.

4.25: Conjectures on Bergsonian Mnemonic Mechanics:
Five Postulates of Bergsonian Mnemonic Mechanics

Grasped as a de-spatialised pure temporal process, **mnemonic mechanics seeks to describe how an earlier state from the past (primacy) superposes with a succeeding state from the past (recentcy) and, in so doing, forms the quantum probability structure of a culminating superpose**, and, **through grand decoherence, compels the Whitehead facet to consistency with the memories and with their culminating superpose**.

Bergsonian mnemonic mechanics is compelled by de-spatialisation and temporisation: Both necessitate that fundamental interactions take place across time, without contiguity or contact, and involve the superpositions of memories in primacy with memories in recentcy. What we naively conceive as 'material interactions in space' are ultimately temporal resonations and superpositions of different moments and memories from the Bergsonian past, culminating at a given reference frame or succeeding event horizon into new events. This process is accompanied by tandem grand decoherence of the Whitehead facet of the Bergson-Whitehead amalgam. Co-joined mnemonic mechanics *and* grand decoherence almost complete the theory of causality, but for the additional critical role of distinct AND-to-OR collapse and time *and* its distinct drive.

Insofar as mnemonic mechanics involve some variant of wave-addition involving wavefunctions in memory, we must first clarify key features true of general wave-addition and superposition. Superpositions of generic quantum mechanical waves clearly apply to the superposition of futures embedded within the Whitehead facet, but with de-spatialisation and temporisation now included, and with the nested futures structure *and* grand decoherence also incorporated. Indeed, Book-II (Part-II, **2.18** to **2.22**... and Part-III **2.23** to **2.35**) accounts for the superposition of quantum waves and futures, enhanced by the incorporation of nested futures and grand decoherence, but in the framework of now-superseded 'intermediate spacetime', *without* explicit de-spatialisation and temporisation. At this juncture, we simply recuperate the findings from Book-II into their de-spatialised and temporised forms via the Bergson-Whitehead amalgam.

To arrive at Bergsonian mnemonic mechanics, we must elaborate a wave-addition process that specifically applies to the superpose of wavefunctions in memory, and that recognises that wavefunctions in memory are bereft of their former nested futures, and consequently *not* subject to grand decoherence: with no possibility of the elimination or erasure of past states. Bergsonian mechanics is expected to involve similar features evinced in the addition of generic quantum mechanical waves, but recapitulated in pure temporal form, save for the other differences stated above and clarified below. At the most basic level, the superpose of memories should involve the addition of a minimum of two states: a *primacy* and a *recentcy* wavefunction in memory, relative to a given future-potential event horizon.

While the addition of generic quantum waves typically entail that the original two waves completely disappear into the new superpose they form, and, at the level of grand decoherence, involve the erasure of possibilities through destructive interference *and* the elimination of non-consistent futures... in the addition of memories, the primacy and recentcy wavefunctions will *not* disappear into the superpose formed. Instead, together with their new superpose, the primacy and the recentcy memories will be independently retained within the Bergson facet of the amalgam and survive to contribute to subsequent rounds of memory superposition, even if in time-attenuated form. Hence, in summary...

- The non-erasure of superposed wavefunctions in memory implies the **independent retention postulate**: the first postulate of mnemonic mechanics. In short, **wavefunctions in memory that participate in Bergsonian superposition are never erased by that superposition, nor by their subsequent AND-to-OR resolution into 'materialised' outcomes**. The addition of primacy to recentcy memories within the Bergson facet involve the addition of wavefunctions in memory across time: i.e., the superposition of different moments of the past; culminating into a new superpose at a succeeding event horizon.

- Recall that we have permanently called into question 'space' via de-spatialisation: Thus, the second postulate of mnemonic mechanics: **the temporal superpose of past states... a superpose across time**, *without* 'space'... *without* interactions involving spatial contiguity, or contact or 'impact'.

- In the Bergson facet, primacy and recentcy memories from different moments of the past are subject to the inverse square law of time: The **amplitudes and other attributes of involved wavefunctions in memory are reduced or modified according to their time-relation vis-à-vis their culminating new superpose** at future-potential event horizon. The more removed in the past a wavefunction in memory is, the weaker its subsequent contribution to the formation of any new superpose on or along a given event horizon. Thus, the third postulate of mnemonic mechanics: the **postulate for the weakening of memory per function of the inverse square of time**. The inverse square law is non-relativistic under normal conditions, but presumably relativistic under extreme conditions.

- Moreover, **there can be no destructive interference or erasure of past states per the superpose of wavefunctions in memory**. In the superposition of memory, only constructive interference is permitted, given that the past cannot be erased. This also implies that the effect of a past state on the development of events on a succeeding event horizon must always constitute a non-zero influence, regardless of how removed it is in time, since it cannot be cancelled out through destructive interference or otherwise erased. Moreover, the influence of the mnemonic past can never be blocked. Thus, the fourth postulate of mnemonic mechanics asserts the **exclusivity of constructive interference in the superposition of wavefunctions in memory**. The fourth postulate also explains why gravitational influence, which is a memory effect, always produces constructive effects and can never be interfered with, blocked, or cancelled: i.e., gravity cannot be cancelled. Yet, the exclusivity of constructive interference in memory-superposition is counterbalanced by the third postulate: the weakening of memory per the inverse square of time.

  ○ To clarify, **destructive interference is always possible when two alternative wavefunctions in future potentiality are made to superpose**. For example, the wave interference and interference pattern obtained in the two-slit experiment is per the superpose and mutual cancellation of futures, *not* of memories. (The memory attends only the configuration of the barrier).

  ○ Moreover, **if a wavefunction in memory superposes with wavefunctions in future potentiality, destructive interference could occur**, but at the expense and elimination of the whole or part of the wavefunction in future potentiality (e.g. grand decoherence) while the wavefunction in memory will remain intact. Indeed, counterfactuals driven erasure of nested futures *is* the destructive interference inflicted by the past upon the future, to the cancellation and elimination of futures per function of non-consistency with the past-in-memory: i.e., again, grand decoherence.

  ○ Finally, **when one wavefunction in memory superposes with another wavefunction in memory, the culminating superpose is essentially constructive**, or a Lagrange of the first two, with the retention of all three within the Bergson facet as part of a memory-accumulating growing block mnemonic universe.

- The quantum erasure of nested futures through grand decoherence cannot apply to Bergsonian memory and cannot modify or erase the past, or affect its destructive interference. Memory states constitute pure un-nested wavefunctions in memory, in similitude to generic wavefunctions, but *without* nested futures. This is why **counterfactuals and the erasure of nested futures cannot apply to Bergsonian memory... given that memories do not possess alternate mutually incompatible futures to erase, or upon which counterfactuals could operate, or upon which consistency rules can operate and eliminate**. Moreover, the past can never be cancelled out or erased, despite claims based on the delayed choice quantum eraser experiments of Yoo-Ho Kim *et al*: (see *Holed Past Thesis* from Book-III, **3.39**, for the key insights). The past must always remain ineradicable, even when it is highly attenuated per the inverse square of time. Thus, we arrive at the fifth postulate of mnemonic mechanics: **the inapplicability of grand decoherence to Bergsonian memory**. This can also be expressed as **the compartmentalisation postulate**: i.e., the processes specific to the Whitehead facet (the future) cannot apply to the Bergson facet (the past).

Hence, the five principal postulates of conjectural Bergsonian mnemonic mechanics.

4.26: Nascent Bergsonian Mnemonic Mechanics: Conjecture, Certainty & Core Validation

The use of the term 'conjecture' in the title and in subsequent material should not be misleading: **The action of the past on the present is physically real and undeniable: It unfolds across time, *not* across 'space': This is a matter of absolute certainty, *not* conjecture. It is axiomatic to any fundamental physics of the future, and emerges inexorably from de-spatialisation and**

temporisation, which compels that we recuperate nature as a distribution of information in time, subject to the inverse square of time and to grand decoherence. **Hence, Bergsonian mnemonics is inexorable**, even if the nascent model furnished in Part-III turns out to be incomplete or erroneous in key parts. Our nascent theory presages mnemonic mechanics and is expected to remain intact in at least its five postulates. These were espoused in 4.25:

- The **independent retention postulate for past states**, as wavefunctions in memory within the Bergson facet.
- The **superpose of past states across time** per de-spatialisation and temporisation espoused in Part-I.
- The **inverse square of time**, wherein the more remote past in memory is the weaker memory.
- The **exclusivity of constructive interference in the superposition of memories**, which can never produce internal destructive interference between memories.
- The **inapplicability of grand decoherence to memory**: Grand decoherence applies only to Whiteheadian futurity.

To augment the inexorability of mnemonic mechanics, consider the Moon's affects upon the Earth: specifically, the tidal effects consequent upon the Earth by the influence of the Moon's gravitation. If we model this on the notion of 'forces and interactions of matter in and across space', we will observe and calculate the tidal effects of the Moon upon the Earth per function of gravitational 'force' across 'space' or, per General Relativity, as the implication of a spacetime curvature engendered by the Moon vis-à-vis the Earth, projected across the space between these. But, **even when we assume all of this to transpire 'across space', the tidal effects brought on from the Moon upon the Earth (and other effects) are effects from the past: namely the Moon as it was 1.5 seconds ago in the past. In short, the Moon's gravitational effect upon the Earth is a memory effect from 1.5 seconds ago, from the past. Therefore, we really do need a physics of mnemonic mechanics to account for this… especially because de-spatialisation forces this, and attendant temporisation compels that we model the Moon's gravity effect as past-memory effect enacted across time, *without* contiguity or space.** Indeed, the Moon's tidal effects upon the Earth are *not* brought about by any contiguous material interactions of the Moon with the Earth 'across space', even when we assume space and contiguous causality. Again, there is no space; no more than there is any synonymous 'common present moment' in which the Moon and the Earth arise in simultaneity. De-spatialisation demands that the Moon is *not* an 'object in space': Instead, the Moon is a corpus of past events integrated with the past events constitutive of the Earth, ultimately according to a complex set of stereotemporal delayed choice time-interval relations.

In relation to the Earth, the Moon is a complex corpus of non-nested wavefunctions in memory, embedded within the Bergson facet of the amalgam that attend the Moon. The Moon's influence upon the Earth, including its tidal effects, are generated by the superposition of the Moon's complex *primacy* memory states, subject to a typical 1.5 second delayed choice time-interval relation   vis-à-vis the Earth; one subject to the inverse square of time per function of 1.5 seconds. These are then added to the *recency* wavefunctions in memory that constitute the Earth. The past states from the Moon and the Earth superpose and, through subsequent AND-to-OR collapse, culminate into a new corpus of events at the event horizon: a corpus of events in which the said mutual gravitation and tidal effects due to the Moon are manifested upon the Earth.

Moreover, the orbital 'motion' (and any 'motion': or, succinctly, the non-uniform change in the delayed choice time-interval relation) per the gravitation of the Moon about the Earth, and universally between all 'objects', can be explained in essentially the same way: Therein, putative 'orbital motion' arises from the superposition of primacy verses recency wavefunctions in memory belonging to the Moon and the Earth. Thus, and as we shall find, 'gravitational mass' and gravitational effects can be understood as consequent upon the total superpose of Bergsonian primacy-recency wavefunctions in memory belonging to a 'body' or to gravitating 'bodies'; producing the warping or 'curvature' of time, furnished through 'curvy event horizons' that attend gravitation: (see Book-II **2.43** to **2.45**).

In further clarification, we simply take the gravitational phenomena we normally characterise as 'unfolding in and across space' and recapitulate these as a culmination of the superpose of past states unfolding in pure time. The attendant past states within the Bergson facet are abstract non-nested wavefunctions in memory. These are extant successive event horizons upon which their culminating outcomes are realised. The memories are *not* extant in terms of 'space' but in the sense they are placed at a remove in terms of time: a temporal remove, not a spatial one. That is, memories are not in event horizon-confined 'matter'. Instead, they are abstractions at a temporal remove within the Bergson facet, at remove from any OR-form events realised on the event horizon, often grasped as 'matter'.

*

All the above claims are certain; *not* speculative; *nor* conjectural. They, and the five postulates underpinning these, are certain… on account of equally certain de-spatialisation secured in Part-I and from the Bergson-Whitehead amalgam that attends it. **Thus, we are certain about the key postulate for the superpose of past states, simply because there is no 'space'… and almost all culminations of nature necessarily arise across time… in part, from the addition of past states or memories: It follows that, the Whiteheadian futurity aside, the manifest universe really is made up of abstract memories, *not* 'matter'.**

We can also state unequivocal certainty about the reality of the inverse square law of time entailed in the superpose of past states, from the fact that a spatial variant of this has been known since Newton, and can be recuperated purely in terms of time, and we are compelled to do so per de-spatialisation.

The conjectural aspect pertains only to the details involved in the subsequent presage to mnemonic mechanics and attendant gravity theory as memory effect: For example, we will later attempt to sketch a form of wave-addition likely to underpin the superposition of primacy and recency memories. But we may have missed out on some crucial detail without which the schema cannot work. Moreover,

411

the inverse square law of time must be subject to modification in the face of relativistic conditions: We can make guesses about such matters, but our guesses are likely to entail errors.

With the admission of the real possibility for errors, we must yet develop the schema for mnemonic mechanics as a useful didactic primer, if not as a useful foil leading to a truer theory of mnemonic mechanics; a physics of memory that *must* emerge, if only because de-spatialisation and temporisation permit no other possibility.

4.27: Conjectures on Bergsonian Mnemonic Mechanics: Superposition of Memories

We first describe the simple addition and superposition of two coinciding mechanical waves, such as water waves or sound waves. Assuming sinusoidal or 'snake-like' waves, each will have a 'height' and 'depth' above and below an imaginary horizontal line or 'zero level' along which each undulates: hence, the amplitude of the waves. When the waves overlap or coincide, coinciding amplitudes of the two waves add. If the peaks of the waves coincide, this leads to constructive interference and reinforcement, with the amplitudes adding to a sum greater than the original component amplitudes. If the peak of one wave coincides with the trough of the other wave, this leads to destructive interference and mutual cancellation. Constructive interference obtains a more powerful wave: a higher sum amplitude. Destructive interference obtains a less powerful wave, or even mutual cancellation of the amplitudes to zero.

The addition of the two mechanical waves leads to a superpose wave: a third wave. The original two waves are subsumed to and replaced by the superpose so formed. But the original waves are *not* erased: Indeed, two approaching water waves may pass through each other and emerge intact at opposing ends. Also note that, in the case of mechanical waves, the amplitudes (the vertical 'heights') can contribute to the energy or power of the wave. Thus, the superpose gives the sum of the energy or power from both waves.

However, in quantum mechanical waves, the amplitude, the height of the wave, is the measure of the probability of a particle being resolved at a putative 'position' in pleonastic space, while the energy of the wave resides in its frequency. Of course, a wave is a spread of probability amplitudes, with the greatest probability given by the peak of the wave: the highest probability of resolution upon subsequent AND-to-OR collapse. The spread of amplitudes embodied by the wave are component amplitudes that, in total, sum to 1 and cannot exceed 1. With de-spatialisation and temporisation, the probability amplitudes are recuperated into the probability set for alternative future-possible delayed choice time-interval relations for future-potential events vis-à-vis detectors or other frames of reference comprising noetic or a-noetic affectable states.

In the case of constructive interference, the probability amplitudes of the coinciding quantum waves are added together, generally increasing the probability of particle-resolution at the pertinent point. However, **the addition of coinciding component probability amplitudes of one quantum wave with those of the coinciding amplitudes of a second wave can never obtain or exceed the probability value of 1, and will always sum to less than 1… wherein 1 constitutes absolute certainty. This limit applies, not only to the two peak coinciding amplitudes but to *all* amplitudes along the span of one wave coinciding with all the amplitudes along the span of the second wave.** This is unlike what happens in the addition of amplitudes in mechanical waves, wherein amplitude additions are not delimited or transformed to a limiting value. Indeed, **we can coincide and superpose as many quantum waves as we wish: The final superpose from these can never exceed amplitude of probability sum of 1. That is, quantum wave-additions can never produce a probability singularity, or even a pleonastic 'positional' singularity or 'point-position'.** This will become pertinent when we recapitulate mass as a memory effect: a function of the superpose of a stack of wavefunctions in memory within the Bergson facet attendant a 'body'. The implication is that such a sum, even of an infinite number of wavefunctions in memory, could never superpose to a probability singularity, or to a point mass-singularity or 'black hole'. This is augmented by de-spatialisation, given that de-spatialisation disallows the formation of a mass-point in space and prohibits singularities so conceived, simply because there is no space, and there is no 'spatial point' at which to form a 'point-position' let alone a 'point-mass singularity': (see **4.32** below).

With the above-stated differences, the mechanics of wave addition and superposition hold true in both mechanical waves and in the superposition of quantum waves, save for the fact that, when quantum waves are superposed, their amplitudes never exceed 1, and, except for the addition of wavefunctions in memory, quantum waves effectively disappear into, and are effectively erased in the new superpose formed. This is especially true in the grand decoherence entailed in the superposition of wavefunctions in future-potentiality.

*

From the perspective of Alternative Realism, generic quantum wave-addition and superposition is very relevant to the Whitehead facet of the Bergson-Whitehead amalgam and the addition of wavefunctions in potentiality, although we must also incorporate the spatially static character of such waves, their nested futures structure, and counterfactuals-driven processes of grand decoherence espoused in Book-II: This much was admitted in the reform of the Schrodinger equation in (see Book-II, **2.27**).

There is a difference between the addition of generic quantum waves and wavefunctions in future-potentiality versus the addition of wavefunctions in memory, and these differences are of import to prospective mnemonic mechanics. To grasp this, in wave-addition and the superpose formed from a minimum of two wavefunctions in memory within the Bergson facet, the original two waves superpose to form a culminating third wave. Yet, the original two wavefunctions in memory, plus the new superpose, are retained within the Bergson facet, and the original two do not disappear into the new superpose, although they will be subject to the effects from the inverse square of time. This is consistent with the retention and non-eradicability of memory generally espoused throughout this work.

All the key differences between the superposition of future possibilities versus memories are summarised thus:

- **Wavefunctions in memory are retained within the Bergson facet, while wavefunctions in future potentiality are embedded within the future-from Whitehead facet** of the Bergson-Whitehead amalgam.

412

- **Wavefunctions in memory are un-nested because of past grand decoherence and AND-to-OR collapse**. Whiteheadian wavefunctions in future-potentiality are nested into convergent category nested futures (see Book-II **2.34**).

- **When wavefunctions in future-potentiality superpose, these disappear into the superpose they form and are irreversibly erased via destructive interference, if by nothing else than the process of grand decoherence and the elimination of non-consistent futures** (Book-II, **2.34** and **2.35**).

- **But wavefunctions in memory sum together to form a culminating superpose *without* disappearing or erasure: They and their new culmination are independently retained within the Bergson facet**. Indeed, the total Bergsonian past is internally consistent, while the totality of Whiteheadian future must grand decohere into consistency, both internally *and* in relation to the total Bergsonian past. **That is, wavefunctions in memory can never be erased, but are always retained intact within the Bergsonian past**, albeit subject to weakening per the inverse square of time.

- Finally, **superposing wavefunctions in memory from within the Bergson facet are comprised of <u>primacy states</u>** (i.e., the oldest wavefunction in memory from an earlier moment in cosmic history relative to a culmination on the event horizon) **versus <u>recentcy states</u>** of more recent or most recent wavefunction in memory formed from the superpose of the total past, relative to the same culmination point on the event horizon.

4.28: Constructive Superposition of Memory, Non-Erasure & the Inverse Square of Time

The illustration in **Fig. 4.06** depicts the growing block Bergson-Whitehead amalgam. Therein, above the event horizon, we find the potential lightcone projected into the ontic futurity constituted as the Whitehead facet. Below the event horizon, adjoined to the vertex of the potential lightcone within the Bergson-facet, the inverted cone serves the purpose of rendering salient the inverse square law of time: a feature essential to prospective mnemonic mechanics.

To reiterate, the potential lightcone within the Whitehead facet constitutes the abstract for the 'speed of light' limit, here recuperated as a static future-form cone of as-yet not realised future-potential delayed choice time-interval relations, delimited by the *delayed choice time-interval standard* in lieu of the old 'speed of light' in the framework of a de-spatialisation universe: (see **4.08**).

As stated, **the inverted cone within the Bergson facet constitutes the rendition of the inverse square law of time and the attenuation of wavefunctions in memory retained within that cone, within the Bergson facet**.

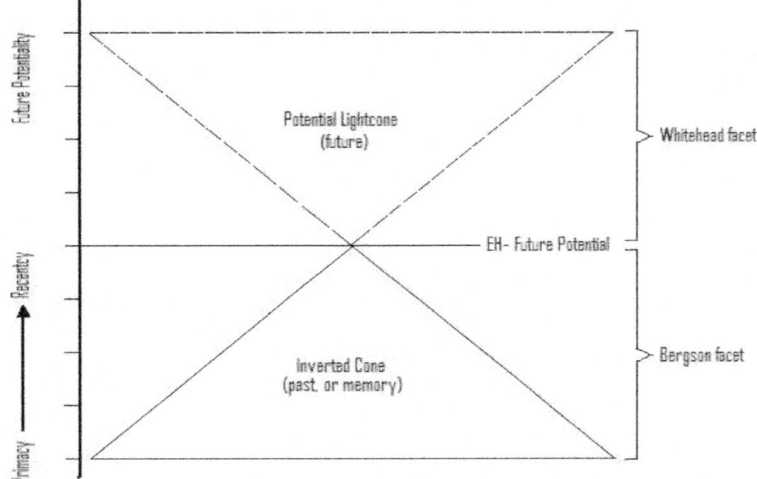

Fig. 4.06: Future potential cones and inverted cone of the Bergson-Whitehead amalgam
Note: Depiction of the potential lightcone within the future Whitehead facet of the Bergson-Whitehead amalgam and the inverted lightcone within the Bergson facet of the same. The potential cone within the Whitehead facet is in lieu of the speed of light limit, now a static opening cone of potential delayed choice time interval relations, as required by de-spatialisation. The inverted cone is the abstract for the inverse square law of time vis-à-vis the recentcy to primacy memory within its confines. The inverted cone is essential for putative mnemonic mechanics and important to the development of putative gravitational theory and inertia, both requiring anticipated Bergsonian mnemonic mechanics. Note that the vertexes of both the potential and the inverted cones coincide at a point on the future potential event horizon that divides the Whitehead facet from the Bergson facet of the amalgam.

In **Fig. 4.07** we illustrate the operation of the inverse square law of time on memory by means of the inverted cone. The inverted cone is projected into the Bergson facet, but it must *not* be misconstrued with displacing radiation: **The inverted cone is a conceptual construct** that we impose on the Bergson facet to readily intuit the operation and delimits of the inverse square law of time.

The inverse square of time obviously operates on memory embedded within the delimit of the inverted cone. As any given memory state is succeeded by a new one, it is relegated into the Bergsonian past... into *primacy* within the inverted cone. Consequently, **memory weakens per the inverse square law of time, or weakens as the inverted cone widens: Hence, we obtain the weakening and the entropy of memory**: (see Part-II **4.18**).

The inverted cone's projection into the Bergsonian past also models the expansion and perpetual weakening of a gravity field attendant a 'body' of origin. This is per the attendant 'stretching' of what we normally think of as the 'wavelength' of the wavefunction in memory, also depicted in **Fig. 4.07**. Per de-spatialisation, what is being stretched is the delayed choice time-interval relation-set encompassed by the wavefunction in memory. Indeed, **the weakening of memory, hence the entropy of memory, combined with the weakening of gravitational power over time, are essentially interchangeable**.

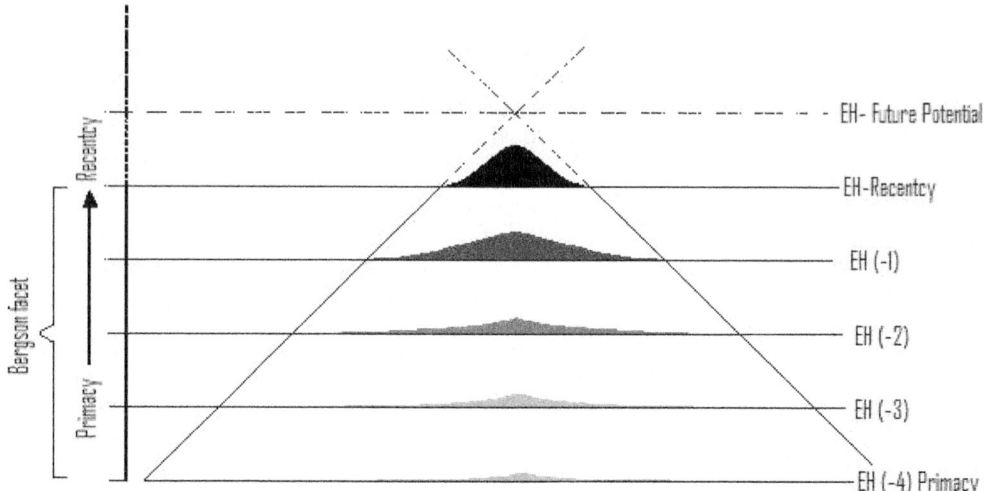

Fig. 4.07: Inverted cone and the inverse square law of time in mnemonic mechanics
Note: Didactic depiction of the operation of the inverted cone and attendant inverse square law of time on memory. The inverted cone is an abstract that prescribes and describes the operation of the inverse square law of time. Its static widening into the past is essentially synonymous with the generic expansion of a gravity field attached to a body. Of course, memory states constitute the un-nested wavefunctions in memory from preceding moments of the past. Wavefunctions in memory are didactically exhibited 'along' past event horizons... from EH-recency to EH (-1), through to EH (-4) and beyond... elaborating to the most remote past-primacy, even to an infinitely removed past-primacy. Note how the un-nested wavefunction in memory coinciding with the preceding event horizons are coded from black (at recency) to light grey at EH (-4) primacy. Note how the amplitude or 'height' of memory at EH (-1) will be reduce to half of the amplitude of EH-recency, and what we would normally think of as the wavelength (but, with de-spatialisation, now reframed as the delayed choice time interval relation set) has 'stretched' in proportion to the widening inverted cone: This has also implications to gravitation: the generic 'expansion' of a gravity field is essentially synonymous with this 'stretch' of memory into the past per the widening inverted cone. The same reduction and stretching applies through EH (-1) through to EH (-4), and to the remote past-primacy. Note how the memory states grow weaker and weaker the more relegated to past-primacy they are vis-à-vis the relative present. *Ceteris paribus*, these will grow weaker according to the inverse square law of time described by the inverted cone.

In **Fig. 4.08**, we exhibit successions of wavefunctions in memory undergoing time-evolution from EH-recency to EH (-1), through to EH (-4). Through four successive AND-to-OR collapse process and the attendant progression of the event horizon, the memories are 'pushed down' to EH (-4) ... and, eventually, below EH (-4). Each such wavefunction in memory belongs to a distinct past event horizon; shaded from black (at recency) to attenuating shades of grey by EH (-4) primacy. Note how **the amplitude or height of each depicted wavefunction in memory reduces by half compared to its amplitude at the event horizon immediately 'above' it. This reduction is per function of the inverse square of time, per function of the inverted cone**: The total probabilities of the pertinent wavefunction in memory must be stretched over a 'wider' time-interval 'area' as the inverted cone widens, while the amplitude must also decline in proportion, although **the total sum of the probabilities, or the sum of range of amplitudes demarcated and constituted by the whole of each stretched wavefunction in memory, must always sum to 1**.

To reiterate, **the relegated and weakening wavefunctions in memory should not be thought of as having any wavelength attribute, given de-spatialisation**. Per de-spatialisation the spatial wavelength concept must be replaced with the pure temporised delayed choice time-interval relation. Hence, per the inverted cone, a wavefunction in memory will be 'stretched' in terms of its delayed

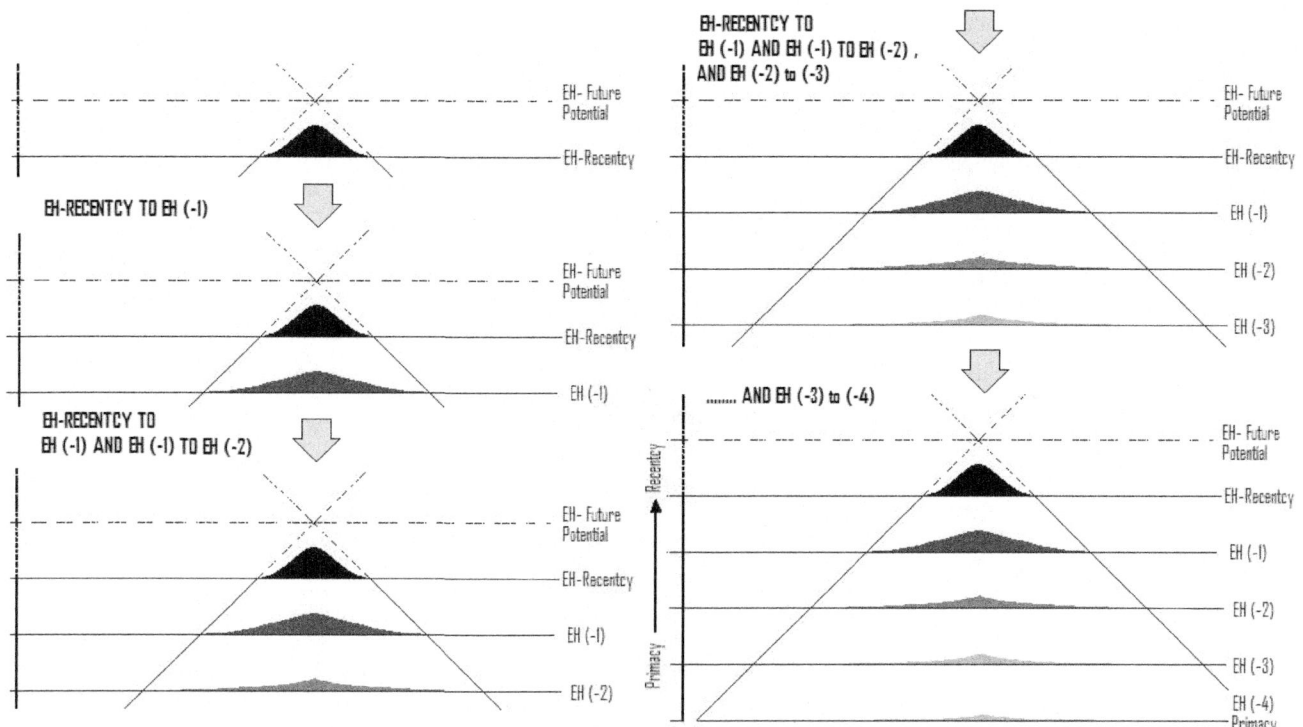

Fig. 4.08: Time-evolution of memory and the inverse square of time
Note: The didactic illustration depicts the operation of the inverse square law of time on memory through the process of discrete quantised time-evolution of the Bergson facet. A new memory state at EH-recency (top left diagram, black) progresses to a state in which the original EH-recency is relegated to EH (-1) and its amplitude is calculably reduced, with its domain of delayed choice time interval relations 'stretched' per the widening of the inverted cone (second diagram from top left). The old EH-recency is replaced with a new EH-recency through the progression of the event horizon and tandem AND-to-OR collapse of Whitehead future into a new set of events in memory. With successive AND-to-OR collapse and time, memory is further relegated to the past and further weakened per reduced amplitude and increased 'stretch': This attenuation of memory is carried through EH (-3) to EH (-4); with the last depiction equivalent to the full diagram shown in **Fig. 4.07**.

choice time-interval scope in proportion to the 'widening' of the inverted cone, or per function of the inverse square of time. As stated, this also engenders implications to gravitation: wherein the generic 'expansion' of a gravity field is essentially synonymous with the said 'stretching' of wavefunctions in memory, as these are relegated into past-primacy per the inverted cone, in turn synonymous with the inverse square of time. Indeed, to reiterate again, **the weakening of memory, hence the entropy of memory, combined with the weakening of gravitational power over time, are essentially interchangeable.**

It must be stated that **the attenuation of memory per the inverse square of time or the inverted cone does not require energy or work to accomplish. The process is a work function zero process.** Consequently, the component of inertia and gravitation attended by mnemonic mechanics based on the inverted cone approach is also furnished through a work function zero process, as will be argued in the subsequent essays.

4.29: Conjectures in Mnemonic Mechanics & the Theory of Inertia

In Newtonian physics, inertia is described thus: *An object will proceed in its current state of uniform motion, and will do so in perpetuity, so long as there are no countervailing forces acting upon it to render otherwise. The body will resist the change implied by any countervailing force and, in doing so, the body will display inertia or resistance.* In short, inertia is the resistance of the body to change.

How do we account for inertia using our nascent mnemonic mechanics? In **Fig. 4.09** we depict a conjectural mnemonic mechanic sketch of how inertia might arise: The explanation has three components: The first component to inertia consist in the **Bergsonian component**: wherein the past states of the body, its stack of wavefunctions in memory from EH-Recentcy through to EH (-4), though subject to diminution in amplitude and other attributes per the inverse square of time, will collectively superpose, albeit without destructive interference. The superposition of the stack of memories will form a **potential culminating superpose** at future potential-EH. With the subsequent progression of the event horizon, this culminating superpose will undergo gearless work function zero AND-to-OR 'collapse' to yield an OR-form outcome, depicted as a worldline arrow in conformity to the higher probability direction implied by the culminating superpose. About this culminating superpose, *en masse* AND-to-OR generated OR-form quanta will tend to cluster. That is, the 'body' will 'materialise' into its inertial state. So long as no acting countervailing 'force' is incident on the culminating superpose, each successive culminating superpose will resemble the preceding one: The AND-to-OR resolution of each culminating superpose will regenerate the same 'vector' or 'worldline in perpetuity, as 'perpetual uniform motion' per work function zero.

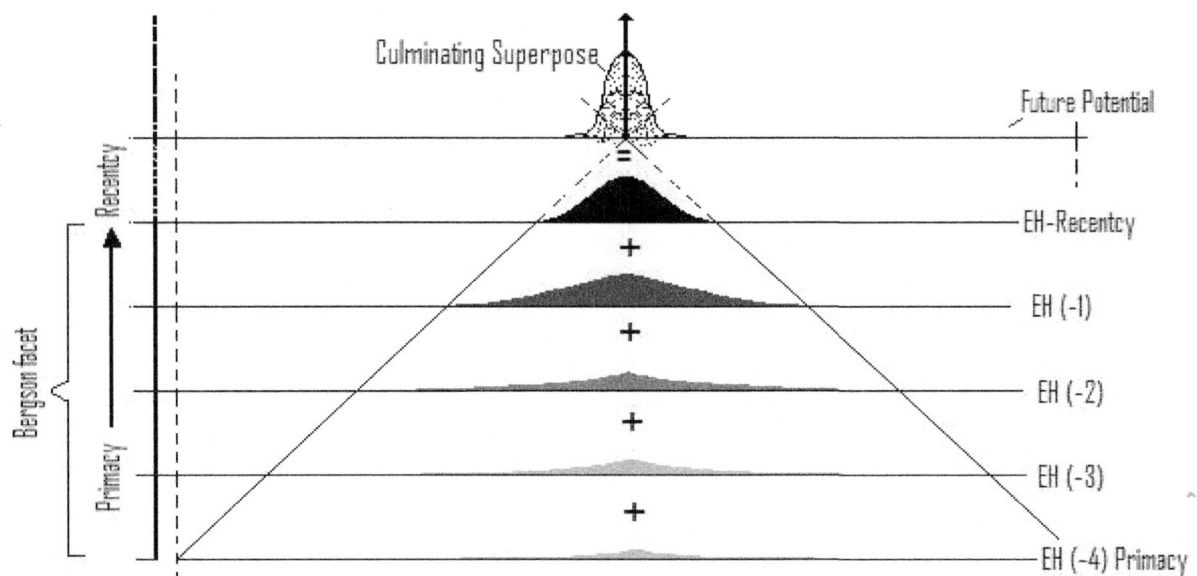

Fig. 4.09: Inertia from the culminating superposition of memories
Note: Inertia is defined thus: "An object will proceed in its current state of uniform motion in perpetuity for as long as there are no culminating forces acting on it to do otherwise. The body will resist such forces and display inertia". In our own conjecture on inertia, we first clarify the **Bergsonian component**: Past states of the body from EH-Recentcy through to EH(-4) and older, notwithstanding the inverse square of time, will superpose in a similar way to how quantum waves superpose -albeit without destructive interference- to form a **culminating superpose** at the Future Potential EH. With subsequent progression of the event horizon, this culminating superpose will undergo AND-to-OR conversion. The second component to inertia (not depicted) is its **Whitehead component**. This entails counterfactual causality and the quantum erasure of nested futures throughout the Whitehead facet: Counterfactuals and erasures will scope and delimit the nested future possibilities open to the future development of the body and, insofar as these unfold in a convergent category of nested futures that ties the body to every other body, what happens to the body will instantaneously delimit the future scope of possibilities for every other body in the universe, and vice versa: hence the revived Mach's Principle first articulated in Book-II. Note that, so long as no acting countervailing force is present, each successive culminating superpose will resemble the preceding superpose: the AND-to-OR resolution of each culminating superpose will generate the same 'vector' or worldline for the body in perpetuity. However, a third component to inertia is the **force component**: The putative force acting on the body to change its state will act on the described culminating superpose at future-potential EH: It will do so as an independent quantum wave. Unlike that of memories, the addition to the culminating superpose of an 'acting force quantum wave' may entail both constructive *and* destructive interference: (The acting force is not depicted above). In effect, the acting force is attempting to overcome the implication of the Bergsonian component –i.e., the superpose of past states from EH-Recentcy through to EH (-4) and older. Its attempt to overcome implies a new superpose (also not shown) according to which our affected body must subsequently AND-to-OR resolve. The source of resistance, or inertia, vis-à-vis the acting force is from Bergsonian memory *and* from the grand decohered Whitehead facet. All other bodies in the universe will be informed of the change per Mach's principle per subsequent grand decoherence. In the absence of the countervailing force, the body will continue in perpetuity according to its given state, vector or worldline.

The second component to inertia (not shown in the depiction) is the **Whitehead component**: This entails the operation of counterfactual causality and the instantaneous grand decoherence of the whole of the Whitehead facet into consistency *with* the culminating superpose and, subsequently, into consistency with the outcomes generated by the AND-to-OR collapse of the culminating superpose. This memory-induced grand decoherence of the Whitehead facet was first expounded in Book-II Part-II (essays **2.34** to **2.35** and **2.45**) and is exemplified by the operation of Mach's principle synonymous with the instantaneous grand decoherence of the total Whitehead facet. Therein, counterfactuals-based grand decoherence and consequent erasure of non-consistent nested futures, forced by the culminating superpose at future potential-EH, will instantaneously decohere the totality of nested future possibilities into consistency, but not only the nested futures belonging to the 'body' pertinent to the culminating superpose, but of the whole convergent category future that encompasses the whole Whitehead facet'... to the infinite future. Therein, every other 'body' in the universe along the event horizon whose convergent futures are integrated to the Whitehead facet will have their convergent futures instantaneously grand decohered into mutual consistency. In this way, the body 'here', through the formation and subsequent collapse of its culminating superpose, will instantaneously affect and reduce and render into consistency the convergent future scope of possibilities open to every other body in the universe (i.e., revived Mach's principle), through the erasure of their mutual non-consistent futures. This will furnish inertia upon the other bodies. Conversely the grand decoherence generated by all bodies elsewhere will furnish the same decoherence to the body 'here', informing it of its inertia vis-à-vis the rest of the universe, by affecting its future scope and modifying the structure of its culminating superpose: (see Book-II **2.35** for full account).

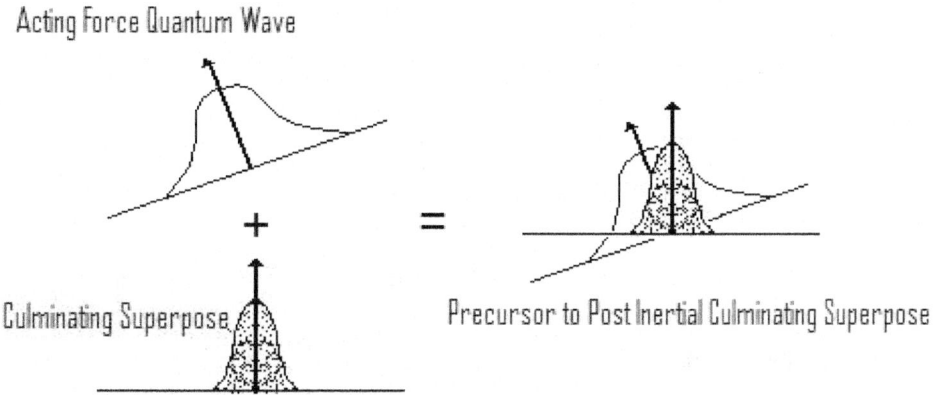

Fig. 4.10: Inertia: acting force quantum wave and culminating superpose.
Note: On the left side are two wavefunctions: the **culminating superpose** at EH-potential (bottom left) with its inherent highest probability worldline, plus the countervailing **acting force quantum wave** (top left) with its own implied worldline at 20 degrees to a same 'plane' potential future event horizon. On the right diagram we have the superpose of the two wavefunctions: the **precursor to the post inertial culminating superpose**. The implication is that the acting force quantum wave will increase the amplitude of AND-to-OR collapse probability of the original culminating superpose in an asymmetrical way to the 'left'. Conversely, the culminating superpose furnished by Bergsonian memory will in turn have its own indigenous probability, tending its potential worldline directly ahead, so constituting the inertia or resistance to the 20-degree left tilt implied by the acting force wave. The two wavefunctions will be superposed to form the **post inertial culminating superpose** (not shown but easily inferred as the union of both). Upon subsequent AND-to-OR collapse, a new worldline between the two tendencies will form, manifesting both change and the resistance to it. The post inertial culminating superpose (again, not shown) will be subsequently incorporated into the Bergson facet as a new memory-state of the putative body, as its new recentcy. When it in turn superposes with the preceding older memories, a new culmination will form; one different from the original and constituting reduced resistance or inertia. Any subsequent acting force quantum wave will then need to act against this new culmination. The process will repeat until inertia is totally overcome. Note that this is *not* a work function zero process, as implied by the acting force quantum wave: i.e., putative 'applied work'.

The third component to inertia consists in the **force component**, depicted in **Fig. 4.10**. A putative 'force' will superpose with the culminating superpose at future potential-EH. The acting 'force' constitutes another quantum wave: i.e., the **acting force quantum mechanical wave** , possibly from non-gravitational sources. Unlike the simple superposition of Bergsonian memories, the combination of the culminating superpose with the acting force quantum wave will be subject to both constructive *and* destructive interference. The acting force is in effect attempting to overcome the implications of the culminating superpose from by the stack of memories from potential-EH to, say, EH (-4) embedded within the Bergson facet (we are assuming the profile depicted in **Fig. 4.09**). This generates a new **post-inertia superpose** (not shown in **Fig. 4.10**, which depicts only the precursor combination: the third diagram) according to which the body, its worldline, and the amplitude controlling the subsequent cluster of OR-form outcomes 're-materialising' the attendant 'body', must adjust and unfold.

417

Also note that, **notwithstanding grand decoherence, the resistance or inertia is inherent to the interference between the acting force quantum wave and what it would imply vis-à-vis the culminating superpose constitutive of what memory seeks to retain and retrench within the Whitehead facet.**

On the other hand, it could also be argued that the acting force quantum mechanical wave' is seeking to grand decohere and restructure the totality of the Whitehead future according to its own impositions. The Whitehead facet can be said to prefer the original configuration of futures, and *it* seeks to 'resist' the implied change. This resistance will be overcome through the persistence of the 'acting force quantum mechanical wave', even against the resistance generated by the Whitehead facet and, effectively, from the whole universe.

Which of these components constitute the real source of resistance and inertia? Is it Bergsonian memory or the Whitehead totality? We posit that *both* are complementary in the gestalt sense:

- If we focus on the Whitehead facet as the sole bearer of resistance and inertia, this could only have come about per the grand decoherence of the Whitehead facet per function of the superpose of Bergsonian memories.

- If we treat the Bergson facet as the sole bearer of inertia and resistance, it could only generate its implications by forcing the grand decoherence of the total Whitehead facet to resistance against the 'acting force quantum mechanical wave'.

- Hence, one approach cannot exclude the other. The one must include the other. Both are mutually complementary or gestalt.

Subsequently, upon AND-to-OR collapse, the new state implied by the acting force quantum mechanical wave will form a new post-inertia culminating superpose. This will pass into the Bergson facet as a new recentcy memory state… also subject to relegation to the past and to diminution per the inverse square of time.

Here is the key: So long as the acting force quantum mechanical wave temporally persists into the next potential event horizon, it will impose another round of superposition and countervailing implications to a subsequent now-slightly modified culminating superpose… modified by the incorporation of the implications of the previous round of acting force quantum wave, or by the incorporation of the latter's implications into the most recent Bergsonian memory state of the body. Another round will incorporate another memory into recentcy, but with an accentuated modification and reduced resistance. **The new superpose incorporating the recent acting force quantum wave will constitute a possibility state of reduced resistance, relegated into memory. Thus, the acting force quantum mechanical wave, or a series of such, will 'overcome' the inertia per the original culminating superpose by generating a new stack of increasingly modified wavefunctions in memory. These will superpose into a culminating superpose with much-reduced resistance vis-à-vis the persisting acting force quantum wave… until a threshold is reached and we attain the post-inertial culminating superpose and the overcoming of inertia… the overcoming of the preferred implications of past memory. Once the acting force ceases to operate, a new inertial state will have been achieved.**

Notwithstanding contingent acting force quantum waves, mnemonic mechanics of inertia is a work function zero process:

- It is gear-less work function zero AND-to-OR collapse and time that affects the post-inertial culminating superpose.

- It is with subsequent gear-less work function zero AND-to-OR collapse that the memory content of the Bergson facet, hence inertia due to memory, expands.

- It is per the gear-less work function zero AND-to-OR collapse process that the grand decohered Whitehead facet furnishes the affectation of Mach's principle.

*

The superposition of memories does not transpire across space, given de-spatialisation: these transpire across time. The coincidence of wavefunctions in memory and the subsequent superpose these form do not require 'energy' inputs to realise. Neither Bergsonian mnemonic processes nor Whiteheadian grand decoherence is affected via 'force-bearing' particles or by 'work' or 'action'. Only the consequences of these processes at the event horizon, realised into OR-form outcomes through work function zero AND-to-OR collapse and time, *may* be interpreted as 'work' and 'action'. But the energy concepts and relations thus foisted on time are ultimately pleonastic: (see **4.19**). Time is not driven by energy or work inputs… and neither is the inertia that time brings into realisation.

We reiterate that any contiguous-mechanical action-mediated or work-mediated grand decoherence process of the Whitehead facet to the infinitely removed future would crash into an insurmountable halting problem: Grand decoherence itself constitutes a gear-less work function zero process, notwithstanding distinct gear-less work function zero AND-to-OR collapse and time; both ontologically primary to *any* phenomena, including inertia.

Of course, the same work function zero contention must be asserted for gravitation, given that this also arises out of mnemonic mechanics modelled as a work function zero process.

4.30: Conjectural Bergsonian Mnemonic Mechanics of Gravitation:
From Force Fields and Space-Curvatures to Memory in Time

The illustrations in **Fig. 4.11** through to **Fig. 4.13** depict conjectural mnemonic mechanics and memory superposition responsible for gravitational attraction and acceleration, perpetrated by **Body-A** upon **Body-S**. The complementary process that operates from **Body-S** to **Body-A** will have the same profile, but in converse. For brevity, the latter is not depicted, but it is implicit. The simple relations depicted constitute the cornerstone to the conjectural replacement to General Relativity via elusive quantum gravity, insofar as gravitation is depicted as arising from the superpose of pertinent Bergsonian wavefunctions in memory, or from the superpose of

418

past-memory attendant the participating 'bodies'. This superpose of memory is affected, *not* across 'space', but purely non-contiguously and temporally, given de-spatialisation from Part-I, **4.01** through to **4.07**. Moreover, the pertinent wavefunctions in memory from the Bergson facet are subject to the inverse square law of time, per function of the inverted cone approach first introduced in the presage to inertia (see **4.29**), now to be incorporated into our description of gravity.

Insofar as there is no space, and only time is ontologically real, it necessarily follows that physical relations oft portrayed as if unfolding across 'space' are ultimately non-contiguous temporal relations instantiated across time; structured in and by pure delayed choice time-interval relata. For example, the mutual 'attraction' between the Earth and the Moon, or between any ostensive 'bodies', is an outcome engendered from the temporal relation between past events; between information and memory organised in time. Even when we assume the reality of space, the gravitational effect of the Moon upon the Earth is obviously an affect from a past state of the Moon… from 1.5 seconds ago, supposedly communicated to the Earth 'across space'. Yet, it invariably constitutes a temporal effect of the past state of the Moon engendering tidal effects *and* gravitational attraction upon the relative present Earth per function of a 1.5 second time-delay. Upon de-spatialisation, the assumed spatial Moon-Earth interval abstracts out, but the temporal relation of the Moon upon the Earth abides. Hence, the conception of gravity as a memory effect of the past upon a relative present frame is simply inexorable: It follows that the requirement for a mnemonic mechanics as the basis for gravity and quantum gravity is also inexorable, and the following mnemonic mechanics-based conjecture is an attempt to model gravity in temporal and mnemonic terms.

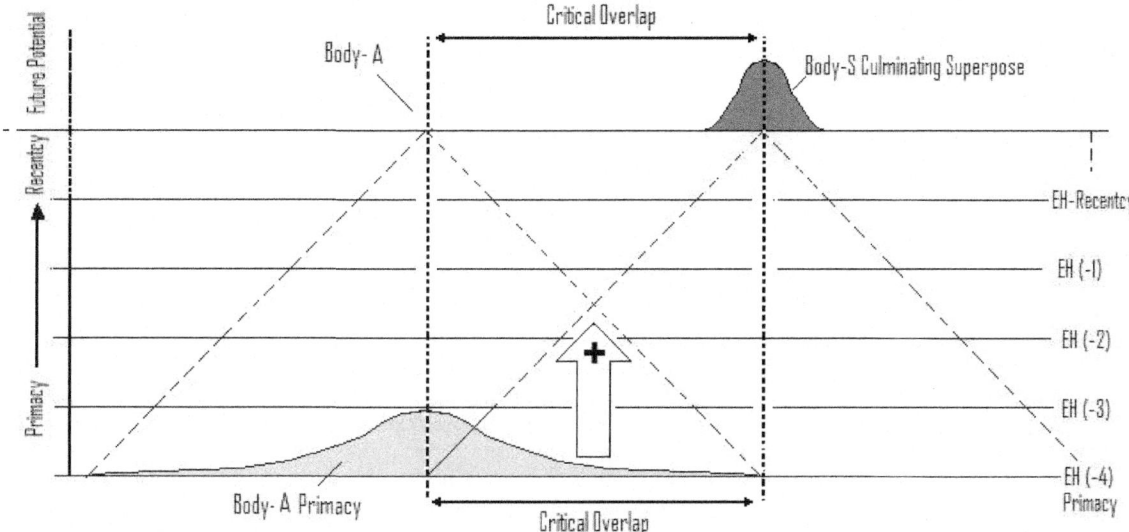

Fig. 4.11: Simple gravitational mnemonic mechanics: Body-A to Body-S
Note: A didactic depiction of the process of simple mnemonic mechanics through the mnemonic superpose responsible for gravitational attraction and acceleration perpetrated by **Body-A** upon **Body-S**. Given de-spatialisation, only time is ontically real. It follows that putative relations in and across 'space' are purely relations in and across time, structured per pure delayed choice time-interval relata. Thus, the mutual attraction between the Earth and the Moon is a product of a temporal relation: an abstract memory relation. Hence, we depict two independent inverted cones attached to both **Body-A** and **Body-S**, respectively. These project into the Bergson facet. Note how the inverted cones overlap: This **critical overlap** between the vertical bold hatched lines selects for the respective parts involved in the superposition of wavefunctions in memory. This will engender gravitation. Also note that only two pertinent wavefunctions in memory are depicted: i.e., the dark grey **Body-S culminating superpose** as the culmination of all the wavefunctions in memory attached to Body-S, from its **EH-recentcy** all the way to **EH(-4) primacy** (implicit), superposed with the wavefunction in memory belonging to **Body-A**: i.e., **Body-A primacy**, (depicted in light grey), which constitutes the oldest memory pertaining to **Body-A** from its EH(-4) past-primacy, subject to the inverse square law of time. (The sizes of the pertinent wavefunctions in memory are exaggerated for purposes of didactic clarity). It is the superpose and addition of **Body-A** primacy to **Body-S** culminating superpose, or those parts belonging to these that reside in the 'critical overlap', that engenders gravitational attraction of **Body-S** to **Body-A**: Hence the past of **Body-A** from EH(-4) is present to and superposes with the immanent future of **Body-S**: a 'presence of the past' of **Body-A** that modifies the future possibility open to **Body-S**, tilting its subsequent AND-to-OR resolution toward Body-A: i.e., gravitational attraction and acceleration. While the presence of the past from **Body-S** upon itself constitutes the basis of its inertia, the presence of the past from **Body-A** upon **Body-S**, and the superpose of that past vis-à-vis the immanent future of **Body-S**, makes for gravitational attraction of **Body-A to Body-S**.

In the didactic depiction furnished in **Fig. 4.11**, we find two distinct inverted cones depicting the inverse square of time, attached to both **Body-A** and **Body-S**, respectively. These project into the Bergson facet of the amalgam per rules elucidated in **4.29**. Note how the two inverted cones overlap in the same fashion that lightcones from different origins overlap in generic spacetime models, but in inverse, and within the Bergson facet. The area of overlap, or the **critical overlap** depicted in vertical bold hatched lines in **Fig 4.11**, selects for the relevant parts of the wavefunctions in memory from **Body-A** *and* **Body-S**: These get to superpose and, in doing so, engender gravitational attraction.

Again, per de-spatialisation, the superpose of the pertinent wavefunctions in memory involved in the critical overlap will not transpire 'across space' but across time: The formation of the superpose from the overlap is non-contiguously temporally mediated, and does not involve spatial contact, or any 'particle exchange'… or any 'force exchange'… or any energy-work based 'action'. Consequently, gravity from mnemonic superposition is a work function zero process.

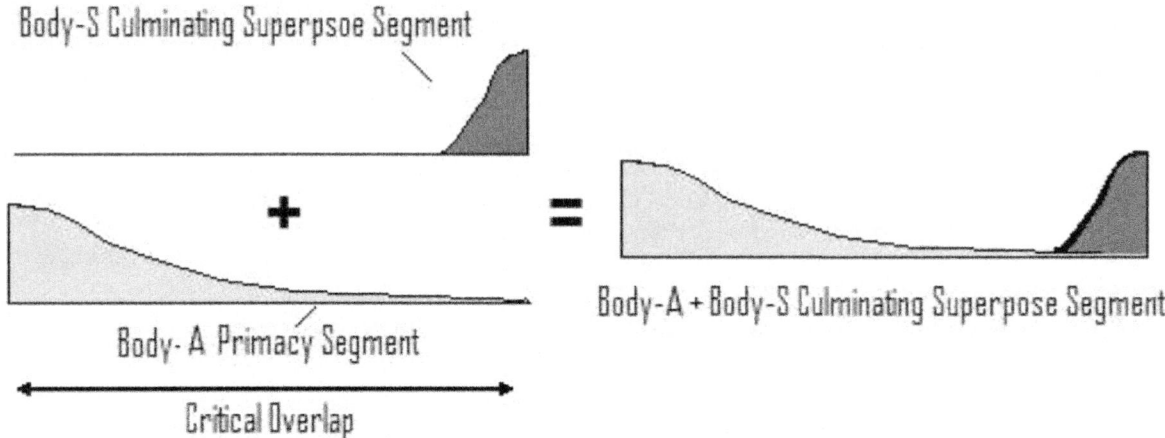

Fig. 4.12. A Closer look at Body-A to Body-S superpose
Notes: In this depiction, we segment out those parts of Body-A primacy and Body-S culminating superpose that fall within the critical overlap (see **Fig. 4.11**). Having sorted out the segments as shown on the left diagram-set, we now superpose the mnemonic **Body-A primacy segment** from EH (-4) to **Body-S culminating superpose segment** per usual constructive addition. (There is no destructive interference; only constructive interference is permissible). From it, on the right-depiction, we obtain the new asymmetric superpose **Body-A + Body-S culminating superpose segment** formed across the **critical overlap**. Note how the new superpose exhibits the vestiges of the original two memory segments (the two light and dark grey areas). Note also the black-shaded area, which constitutes the modified amplitude distribution of prospective *en masse* quantum indeterminate AND-to-OR outcomes, constituting a probability bias to the left. The black-shaded area is critical: the *asymmetric* modification of amplitude and probability distribution will structure subsequent 'motion' of Body-S vis-à-vis Body-A in such a way as to engender worldline consequences for gravitational 'attraction' and 'acceleration' toward the 'left', towards Body-A (See **Fig. 4.13**).

In **Fig.4.12**, in the left-hand pair of diagrams, note that the two key components involved in the superpose of gravity-generating memories are depicted as the dark-grey **Body-S culminating superpose** (the culmination of all the wavefunctions in memory attached to **Body-S**) and the light grey **Body-A primacy** (the bottom-left depiction) composed of the oldest memory pertaining to **Body-A** from EH (-4) past-primacy, attenuated by the inverse square of time. (Note that the depicted sizes of both wavefunctions in memory are deliberately exaggerated for purposes of didactic clarification and are not meant to facilitate calculation).

From **Fig.4.12**, **it is the superpose of Body-A primacy with Body-S culminating superpose within the 'critical overlap' that engenders gravitational attraction**: Thus, **Body-A** past-primacy from EH (-4) is present to Body-S culminating superpose. The pure temporal 'presence of the past' from **Body-A** upon **Body-S**, and the superpose of that past upon the immanent future of Body-S, engenders the gravitational attraction of **Body-A** to **Body-S** via the **Body-A + Body-S culminating superpose segment**.

In **Fig. 4.13**, we take a closer look at **Body-A + Body-S culminating superpose segment**. The process of mnemonic mechanics will superpose **Body-A primacy segment** from past-primacy at EH (-4) to that segment of **Body-S culminating superpose** with which it overlaps within the critical overlap shown in **Fig.4.11**, thus forming a new superpose: the **Body-A + Body-S culminating superpose segment**. We discover that the new **Body-A + Body-S culminating superpose** forms an asymmetric wavefunction for gravitation; (the top depiction in **Fig. 4.13**). The asymmetry refers to a 'tilt' in the probability structure and the implied direction of the worldline (depicted as a black arrow): That is, the left portion of the culminating superpose from **Body-S** is distorted and biased toward future AND-to-OR collapse and the *en masse* cluster of OR-form 'materialised' quanta to the left; towards **Body-A**. The said bias is per function of the black-shaded area which constitutes that part of **Body-A** superpose which modifies the possibility structure of **Body-S** culminating superpose, thus tilting the normal vertical worldline toward the left… toward **Body-S**.

Again, **the said 'tilt' is depicted as a thin black shaded area in the asymmetric bounded probability state for gravitation:** top diagram in **Fig. 4.13: The 'tilt' constitutes the modification of the amplitude of probability that will subsequently structure the 'motion' of Body-S vis-à-vis Body-A, and 'turn' the future-potential worldline attached to Body-S to the 'left': Hence, the gravitational attraction and acceleration of Body-S toward Body-A, as highlighted by the 'left tilting' potential worldline** in the top diagram.

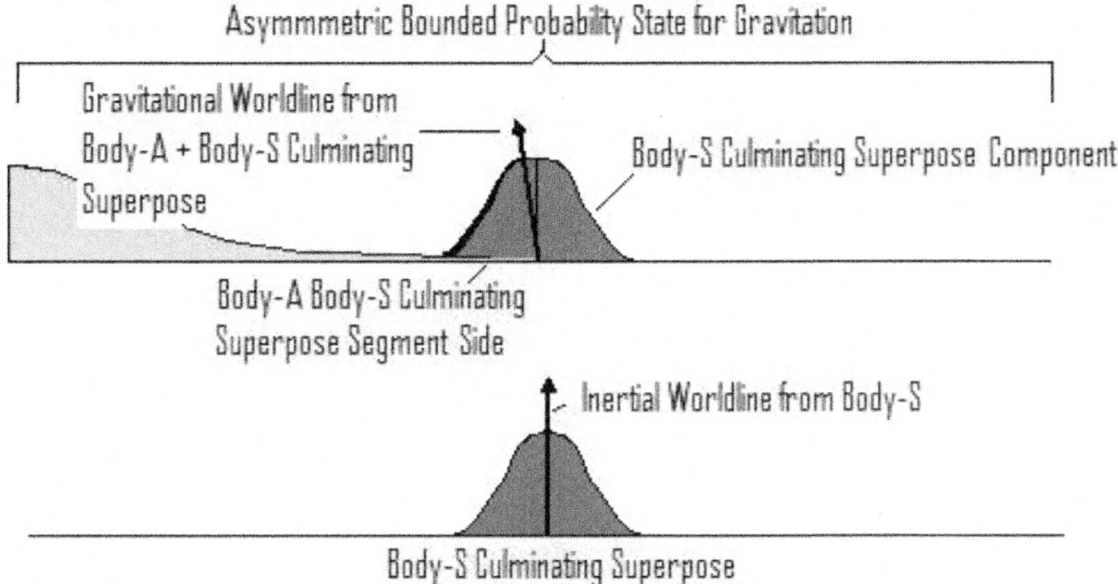

Fig. 4.13 Worldline consequences of Body-A to Body-S mnemonic superpose.
Notes: A didactic depiction of two distinct worldline consequences generated by mnemonic mechanics, from inertia (bottom diagram) and from Body-A to Body-S gravitation in **asymmetric probability modification and gravitation** (top diagram). If Body-S never fell under the gravitational influence of Body-A, its culminating superpose at a given future-potential event horizon would favour the vertical non-modified inertial worldline outcome (bottom diagram): Therein, upon AND-to-OR collapse of **Body-S culminating superpose** and tandem event horizon progression, the collection of putative 'particles' related to Body-S would resolve about the said vertical inertial worldline: The worldline is 'perfectly vertical' because the amplitude distribution of possibilities for Body-S culminating superpose is symmetrical. On the other hand, in the top depiction, the probability amplitude distribution is *asymmetrical* per the black shaded area, formed by the superpose of the right half of **Body-S culminating superpose** with **Body-A + Body-S culminating superpose segment side**: (Also see **Fig. 4.12**). The probability amplitude distribution in the asymmetric superpose is slightly higher on the left-side (indicated by the black-shaded area, toward Body-A). Consequently, the probability distribution of subsequent AND-to-OR resolutions will be biased toward the 'left': i.e., 'toward' gravitationally attracting Body-A. Thus, upon subsequent AND-to-OR conversion and event horizon formation, our asymmetric bounded probability state for gravitation will generate a worldline tilted to the left; toward Body-A (see top diagram). Hence, its cluster of OR-form constituents will assume *en masse* resolution toward Body-A, or 'move towards body-A'. The angle pertinent to the tilt will constitute ''gravitational acceleration' of Body-S toward Body-A. In effect gravitational attraction unfolds because the relative present state of Body-S is influenced by the past-memory state of Body-A: i.e., gravitation arises from the 'presence of the past' grasped as a relation of information and memory across time. In short, gravity is an abstract memory affect and effect, wrought as a work function zero profile process.

Of course, this is expected to be counteracted by the inertial symmetric worldline that attends the superpose of Bergsonian memories attached to **Body-S** (bottom depiction in **Fig. 4.13**), in a similar way to how the inertial worldline and culminating superpose resists the 'acting force quantum wave' in inertia (see **4.29**).

Consider that, with memories attendant only **Body-S** we would obtain only the symmetric distribution of amplitudes constitutive of **Body-S culminating superpose,** implying a perfectly 'vertical' potential inertial worldline, again as depicted in the bottom illustration in **Fig. 4.13**. Upon AND-to-OR collapse, putative OR-form events would collectively cluster in favour of this higher probability amplitude, rendering salient the vertical inertial worldline and the inertial state of Body-S, with no gravitation.

In contrast, in the asymmetric probability state for gravitation (i.e., top diagram in **Fig. 4.13**) subsequent AND-to-OR collapse cluster OR-form outcomes in favour of the now-higher probability amplitude 'left-tilt' worldline, manifesting gravitational attraction or 'orbit' of **Body-S toward Body-A** per the modification of the probability structure depicted as the black shaded area (again, see top diagram in **Fig.4.13**). Hence, we obtain gravitational attraction, motion *and* acceleration, courtesy of Bergsonian mnemonic mechanics.

What about the inverse relation, **Body-S to Body-A**? This would constitute the same form as depicted in the top illustration shown in **Fig. 4.13**, but in converse. Both relations are simultaneous and constitute mutual gravitational attraction.

Note that **none of the above involves contiguity or contact, or any 'particle exchange'… or any 'force', or any energy-based 'action'. Consequently, gravity from mnemonic superpositions unfold on a work function zero basis. Consequently, gravity is a work function zero process, rendered salient by the process of AND-to-OR collapse or time, itself a gear-less work function zero process**.

Finally, consequent upon gravitation by mnemonic superposition and mnemonic mechanics, the 'curve' we are destined to obtain is not a 'spacetime curve', given de-spatialisation, but a 'curvy event horizon': (see Book-II **2.43** to **2.45**). Yet, the curvy event horizon does not 'tell matter how to move'. Instead, the curve is a time dilation curve affected by background mnemonic mechanics in conjunction with grand decoherence. In short 'spacetime curves' are completely circumvented, even as these survive in vestigial form as affected 'curvy event horizons', but *not* as gravity-affecting factors. Hence, **we supersede General Relativity and arrive at gravity recuperated as an abstract memory effect**… *assuming* our mnemonic mechanics conjecture is foundationally and essentially correct.

Again, the reader is reminded that the details are conjectural and will likely be corrected and superseded. Even so, what cannot change is the inexorability of some form of growing block de-spatialised, non-contiguous and temporised mnemonics as the basis of gravitation, or gravity as a memory effect.

4.31: Gravitational Grand Decoherence:
Whiteheadian Implications from Conjectural Bergsonian Mechanics for Gravitation:
Work Function Zero Orbital "Paths", Counterfactuals & Quantum Gravity

In Book-II we developed the presage to the Whiteheadian perspective about gravitation, centred on grand decoherence of convergent futures into consistency with the past, complementary to the Bergsonian memory-based approach developed in the preceding essays: **4.29** to **4.30**. Here, **we will seek to recapitulate the Whiteheadian approach and incorporate it to our Bergsonian format, so forming a unified growing block account of inertia and gravity, thus presaging quantum gravity**.

First, we must recapitulate how grand decoherence, hence Mach's principle, operates within the Whitehead facet. From Book-II, both counterfactual causality *and* the quantum erasure of non-consistent convergent futures, hence grand decoherence, helped explain the operation of the permanent magnet (see Book-II, **2.37**). These processes enable the permanent magnet to stay put on its vertical surface (the fridge door), against gravitation, without any further input of force or power to maintain it on that surface.

When we place our magnet on the fridge door, this has the effect of eliminating all non-consistent futures from the system of nested futures pertaining to the door-magnet relation within the Whitehead facet. Hence, this is projected to the infinite future to the exclusion and erasure of all non-consistent possibilities, reinforcing the established magnet-surface relation on a near-permanent footing upon all futures within the Whitehead facet. However, the futurity of the magnet-surface relation will include future-potential contingent possibilities: often, low probability future possibilities that might remove the magnet from its surface, epitomised by Wheeler's cat: the mischievous pet who overturns laboratory equipment, altering expected experimental outcomes. Otherwise, the magnet-surface relation will perpetuate in the same form from event horizon to succeeding event horizon, reinstated at each attendant AND-to-OR collapse, simply because the Whiteheadian future has decohered to *almost* eliminate any other future alternative.

The initial input of pleonastic putative energy-work to place the magnet on the fridge-door aside, thereafter, the magnet-surface relation will be sustained in effective perpetuity *without* subsequent energy-power inputs… on a work function zero basis. The magnet will remain attached to the door and follow an 'orbit' within the Whitehead futurity, thus retaining its attachment in perpetuity, effectively for free, or without subsequent work… *without* the violation of the conservation principle.

Essentially the same form of process abides in gravity, but for orbiting bodies in mutual gravitational coupling.

However, returning to the role of contingency, or to 'Wheeler's cat', recall from Book-II that, when we seek to detach the magnet from the fridge door (via a non-zero work function) the detachment will imply a new conformal restructuring of the Whiteheadian system of nested futures to the infinite future, counter to the established door-magnet relation. To the extent that the detachment of the magnet will imply a more profound counterfactual re-restructuring of the system of nested futures, the attempt will meet 'resistance' from the established magnet-door relation and from the total conformal Whitehead facet itself. As such, the removal will require putative energy-work, one proportionate to that resistance: a non-zero work function and 'free energy': (For the basis of the latter, see Book-II **2.35**). A similar form of resistance will arise in the attempt to shift the orbit of an object from high to low altitude or to a new orbit, or accelerate or decelerate it when under gravitational influence, or under the regime of otherwise work function zero 'paths', orbits, or the work function zero 'perpetual acceleration' of bodies when subject to gravity. (The same applies in inertia and in the attempt to alter or overcome inertia).

Indeed, we must also not forget the role of memory: The resistance is also wrought by the background mnemonic mechanics that generates the attendant grand decoherence of the Whitehead futurity to conformity with established orbits or paths and their consequent reinforcement on the Whitehead future (see previous essays **4.29** and **4.30**). And it is this that we must now further elaborate on: Thus, **we seek to unify the findings and key discoveries pertaining to grand decoherence as attends gravitation developed in Book-II** with conjectural mnemonic mechanics for gravity from essay 4.29 to 4.30, thus furnishing a full growing block Bergson-Whitehead conjecture on gravitation; prelude to quantum gravity.

*

Recall that, as was true of the former 'spacetime in potentiality' which it replaced, the de-spatialised Whitehead facet constitutes a nested futures system that exhibits, not just immanent possibilities and potential worldlines, but the convergent nested futures that are elaborated to the infinitely removed future. Hence, in the context of Bergsonian mechanics for gravitation depicted in the preceding series of essays and illustrations (see **4.30**), the resolution of any culminating superpose and worldline from **Body-A to Body-S** (and inversely, from **Body-S to Body-A**) must restructure all subsequent convergent nested futures constituting the Whitehead facet into consistency… to the infinitely removed future: This is **gravitational grand decoherence**.

By reference to essay **4.30**, what is the counterfactual implication from the **Body-A-to-Body-S** memory-relation vis-à-vis the Whitehead facet and to the future evolution of the said bodies? **The Whitehead facet pre-empts the said Body-A-to-Body-S relation in that, attendant Bergsonian memory states of that relation will act as counterfactuals to the shared convergent futures pertinent to the bodies within the Whitehead facet. Hence, the Whitehead facet as a whole will undergo gravitational grand decoherence into consistent-only future possibilities (i.e., consistent with the Body-A to Body-S relations in memory), eliminating almost all non-consistent futures within the Whitehead facet to the infinite future, thus inscribing a work function zero 'orbital path' within that future,** even if this entails the possibility of low probability contingent possibilities that might otherwise change the **Body-A to Body-S** relation (i.e. Wheeler's cat).

What are the energy-power requirements for all this? What does the attendant work function for gravitation look like? **Beyond the pleonastic energy-power that may be required to bring Body-A and Body-S into their initial relation (such as when we place a weather satellite into first-time orbit about the Earth) no further power will be required to perpetuate the 'orbital path', now inscribed into consistency-rendered Whitehead facet by gravitational grand decoherence,** such that all future realisations of the two-body relation will adhere to that future-form 'orbital path' so inscribed. All subsequent AND-to-OR collapse processes and progression of the event horizon (the progress of time) will resolve Body-A and Body-S into their ordained potential mutual 'orbital paths' gravitationally grand decohered and inscribed into the Whitehead facet.

The enforcement of consistency through gravitational grand decoherence itself will not involve contiguity or contact, or any 'particle exchange'… or any force or energy-based 'action'. Consequently, gravity per attendant gravitational grand decoherence will be a work function zero process, with orbital paths, acceleration due to gravity, and mutual gravity rendered salient by the process of perpetual AND-to-OR collapse or time, itself a gear-less work function zero process.

*

How does 'matter' tell space how to curve, or the curve tell matter how to 'move'? There is no 'space'. Therefore, there is no 'space curve'. **Instead, the collective Bergsonian memory informs the future how to grand decohere into consistency, in lieu of the expected 'curve'. The gravitational decoherence of the Whitehead facet then tells** *en masse* **AND-to-OR resolution process to cluster up, or 'tells matter how to move'… or, given de-spatialisation, generates all the non-uniform transformations in delayed choice time-interval relations that attend the gravitational relation between Body-A and Body-S.** Again, this will not involve contiguity or contact, or any 'particle exchange'… or any 'force' or energy-based 'action': It is a work function zero process.

*

What applies to inertia also applies to gravity, different only in the details. In inertia, a body will continue in its given state in perpetuity unless acted upon by an 'acting force quantum wave': (see **4.29** and **Fig. 4.10**). The resistance to the implied change will arise from the Bergsonian memory in conjunction or gestalt with the grand decoherence and consistent futures vis-à-vis the Whitehead facet. However, in the absence of an 'acting force quantum wave' to affect change, the body will continue in its given state in perpetuity: That is, the mnemonic superpose from the stack of wavefunctions in memory pertinent to the inertial body, stacked within its inverted cone within the Bergson facet, will enforce and reinforce the same-form culminating superpose, with exactly the same-form probability distribution of *en masse* quantum indeterminate outcomes realised at every future-potential culmination, with the same unmodified vertical worldline affected in perpetuity through the perpetual of successive AND-to-OR collapses and event horizon progressions. That is, the body will continue along its same-form worldline without any further application of force, in perpetuity, per function of a work function zero process, enforced by the accompanying **inertial grand decoherence** of the total Whitehead facet to appropriate consistency.

Of course, inertia so described (and even in its classical form) constitutes an extreme idealisation: Indeed, it is a didactic fiction… since all putative pleonastic 'motion' in the universe is ultimately non-inertial and non-uniform, and what appears to be inertial is nothing more than an approximation subsumed to a larger non-uniform non-inertial framework.

The key differences between grand decoherence or Mach's principle as applies to gravity in contrast to the same as applies to inertia are three-fold:

- **Gravitational grand decoherence** applies to gravitating bodies, and always entails the enforcement or reinforcement of non-uniform change, 'acceleration', or the non-uniform transformation in the delayed choice time-relation (realised or future potential) that resides between two mutually gravitating bodies, and between all bodies on or along the event horizon: This always entails 'acceleration' along future-potential work function zero 'orbital paths' inscribed into the Whitehead facet.

423

- **Mach's principle and <u>inertial grand decoherence</u> as applies to highly idealised inertia always entail realised or future potential *uniform* change ('uniform velocity') in the delayed choice time-interval relations between successive states of a body to itself (to its past) and to all other bodies in the universe.**

- **Inertia is a useful didactic idealisation:** a fiction, *ceteris paribus*. It is not possible to abstract the mnemonic influences from many body relations, or from and between all bodies along the event horizon. Thus, idealised didactic inertial mnemonic mechanics must give way to the mnemonic mechanics for non-uniformly integrated bodies: i.e., gravity.

*

Finally, **we must clarify implications to prospective quantum gravity**. Consider that, per any mutually influencing bodies, their potential work function zero pathways or orbits, decohered and formed within the Whitehead facet, must be brought into 'materialisation' through a series or successions of AND-to-OR wavefunction collapse-processes and attendant event horizon progressions. The gravitational phenomena described above must always entail a series of broken-discontinuous event horizon progressions and synonymous process of AND-to-OR collapse. Thus, the succession of event horizons that bring into realisation the said gravitational attraction, acceleration and orbit do not unfold in seamless-continuous form, but always in a broken-discontinuous form… in a discontinuous quantised fashion. Hence, quantised gravity.

Of course, there is no 'space'; and there is no 'motion' as such: There is only the non-uniform transformation in the delayed choice time-interval relations that reside between 'bodies', otherwise interpreted as pleonastic 'motion'; as 'acceleration'; as mutual attraction, and as 'orbits'. The point is that **the succession of non-uniform changes in the delayed choice time-interval relations between bodies have a discrete character: The orbital 'paths', or 'acceleration curves', are not seamless-continuous affectations, but are *always* broken into successive 'bits' or discrete moments… or *gravitational durata*. In other words, gravitational phenomena are quantised. Thus, we arrive at quantum gravity**, necessarily incorporative of and arising from Bergsonian mnemonic mechanics co-joined to Whiteheadian grand decoherence… all of it unfolding on a non-contiguous temporal basis via fundamental work function zero AND-to-OR collapse and time.

*

Does not the work function zero profile of inertia and gravitation usurps conservation principles? Recall that energy relations and terms are pleonastic and subsumed to the ontologically primary process of time, or AND-to-OR collapse; itself *not* an energy-driven process, no more than it is driven by hidden gears.

Moreover, and per de-spatialisation, gearless work function-zero time, and, per the pleonastics of energy, the principle of conservation is recuperated into abstract form (see Book-II **2.39**), based on the intertwine of two things: The conservation of memory, wherein the constitution of events that have come into realisation in the past, cannot be annulled, erased or altered. That is, **Bergsonian memory cannot be destroyed or erased, replacing the notion that 'energy or matter cannot be destroyed'**. Grand decoherence of the Whitehead facet per function of the conserved memory-content of the Bergson facet, will reflect the conservation of memory into the Whitehead facet, as the latter's consistent futures. Hence, the second factor in the intertwine is the principle of consistent futures: That is, **the convergent futures that bind all 'matter' or all memory will be rendered into consistent forms per function of grand decoherence, courtesy of the conserved and ineradicable past. So long as any natural process does not alter the content of the past or bring about inconsistent futures, it cannot violate the recuperated now-superlative abstract form of the principle of conservation.** Thus, gravity and inertia accounted for per the growing block mnemonic mechanics and grand decoherence can add to the past, but cannot erase or destroy the past, and it cannot introduce inconsistent futures. In this way, neither gravity nor inertia can undermine the newly constituted and transformed conservation principle of memory and consistency… and neither can Cartesian mind.

4.32: Bergsonian Mnemonic Mechanics & De-Spatialisation: Conjectures on Singularities & "Black Holes"

Our foray into prospective Bergsonian mnemonic mechanics-based gravitation appears to imply certain restrictions against the possibility of 'singularities' or of bodies self-collapsing to infinity, limiting these to asymptotic forms of supermass or 'black holes'. The same rubric behind the critique of singularities also engenders implications to the nature of putative dark matter and dark energy.

First, in the case of singularities or 'black holes' that one might want to obtain from the superposition of wavefunctions in memory attached to a 'body', even a stack of infinite number of temporally non-diminished wavefunctions in memory could not sum to a singularity, or self-collapse the body to infinity. To put it another way, **since the mass of the object is the sum of its memories, the rules that govern and delimit the summation of memories cannot obtain 'infinite mass'**… even when we might imagine the participation of a stack formed of an infinite number of superposing wavefunctions in memory and ignore effects from the inverse square of time.

Also, from the addition of wavefunctions in memory pertinent to the Bergsonian past of a gravitating body, **the sum of coinciding amplitudes of the pertinent superposing wavefunctions in memory can never sum to a probability equal to 1 (or effectively, to a singularity) in *any* culminating superpose, even if an infinite number of memories were superposed.**

Moreover, even when we superpose an infinite number of memories, **the frequency of the culminating superpose formed at the succeeding event horizon can never reduce to a 'zero time interval' or infinite frequency, or to a single amplitude at a probability equal to 1… or into an effective singularity.** Such outcomes are impossible to obtain per function of the limits that govern the superposition of wavefunctions in general and in the Bergsonian superposition of memories specifically.

424

The implication from the sum of memories is this: **Subsequent AND-to-OR collapse that manifests inertia or gravitational acceleration and attraction per function of Bergsonian mnemonic superposition described in 4.29, 4.30 and 4.31, can never impart a 'self-gravitation' or a collapse to infinity via an 'acceleration due to gravity' equivalent to the 'speed of light', much less at 'faster than light'. In short, singularities are not possible in a de-spatialised temporised universe in which de-spatialised temporised 'mass' is recuperated as the sum of Bergsonian memories.**

Of course, given de-spatialisation, such things as 'acceleration due to gravity', or the 'collapse of a gravitating body toward its own centre', or even the forlorn notion of a dimension-less spatial point of infinite mass (a singularity) must be recapitulated in pure temporal terms. Given de-spatialisation, gravitational mass is necessarily temporised into the superpose of wavefunctions in memory within the inverted cone embedded within the Bergson facet, so constituting 'the mass of the body' in its purely temporised de-spatialised mnemonic form. As such, **the superposition of memory cannot affect a zero delayed choice time-interval relation (a singularity) … no more than it could affect to both the self-gravitator *and* to the affected body a convergent infinite frequency attribute (a singularity)** per processes described in **4.30**.

Moreover, **from pure de-spatialisation itself, there is no spatial point to which mass could collapse, given that there is no space and there are no 'points in space'. Hence, from de-spatialisation directly, there can be no point-singularity. And since remarkable de-spatialisation compels for a purely temporised universe comprised of *durata* organised into delayed choice time relations, these cannot constitute into infinitesimally small durations or infinitesimally small time-relations of infinite frequency or 'singularity'.**

Thus, in a de-spatialised purely temporised universe, self-gravitating bodies cannot induce 'faster than light collapse' into a singularity and cannot induce a faster than light in-fall to another body under its gravitational influence, much less affect a situation from which 'even light cannot escape'.

It follows that all purported 'black holes' are at best asymptotic: That is, objects may collapse onto themselves in perpetuity, but can never obtain infinity or singularity, assuming key details and postulates involved in our nascent mnemonic mechanics are correct, and assuming de-spatialisation itself is correct: (We *do* assume the latter to be true and certain).

We also assume that, at extremes, the relationship between gravitational power and mass (succinctly, abstract memory) transform in a similar fashion to how the addition of velocities are transformed in Special Relativity: to a value always less than the putative 'speed of light', or, in temporal terms, to less than the delayed choice time-interval standard, no matter how many 'velocity components' are added, or no matter how many wavefunctions in memory are superposed.

If the normal or presumed proportionality residing between gravitational power and mass must breakdown in the way just suggested, it conjecturally follows that the much vaunted *universal gravitational constant* G from Newton, a constant retained in Einstein's General Relativity, might well turn out to be variable; *not* a constant... if only because the mass (the sum of Bergson facet wavefunctions in memory) is no longer constant but is itself transformed… in that the superposition of even an infinite number of wavefunctions in memory must culminate into a finite sum and state below a certain limit... and that this must also apply to the superposition of a large but finite number of wavefunctions in memory. This is analogised by Lorentz transformation in the addition of velocities, even when involving the addition of an infinite number of component velocities: This must always sum to a value less than the generic 'speed of light'.

One obvious implication is that, **our finding against singularities segues into and supports conjectural matter creation process espoused in 4.21**, given that the putative realised matter drawn out from Whiteheadian future-potential matter at asymptotic black holes need not disappear into a now-impossible singularity.

The other implication is to the Big Bang itself: **A central problem of Big Bang was the Big Bang singularity, or the notion that the universe emerged out of a singularity.** To overcome implications from the Big Bang singularity, contrivances were devised, such as Alan Guth's cosmic inflation theory first proposed in 1981[85]. **But both de-spatialisation *and* the summation process of Bergsonian wavefunctions in memory obviate singularities *and* totally obviate the 'necessity' for any Bing Bang singularity**; perhaps even obviating inflation theory itself.

The invalidation of singularities could save the Big Bang theory from its present-day hypertrophy, although the same could also undermine the viability of Big Bang and revive Steady State theories based on clouded and continuous matter creation at asymptotic 'black holes' throughout the universe (again, see **4.21**). It is also possible that our ponderings culminate into a hybrid cosmology and cosmogenesis: the fusion of Big Bang *and* Steady State.

4.33: Conjectures on Dark Matter

Observations in the 1970s showed that the orbits of stars about the centre of the Milkyway galaxy and other galaxies follow paths that imply the presence of much greater gravitational mass than observationally inferred. The orbital speeds of even the farthest stars about the galaxy were found to be too fast to be held by the know mass of the galaxy or the mass at its core[86]. Furthermore, other

---

[85] *Guth, Alan H. (1981). "Inflationary universe: A possible solution to the horizon and flatness problems". Physical Review D. 23 (2): 347–356*

[86] Rubin, V.; Thonnard, W.K. Jr.; Ford, N. (1980). "Rotational Properties of 21 Sc Galaxies with a Large Range of Luminosities and Radii from NGC 4605 (*R* = 4kpc) to UGC 2885 (*R* = 122kpc)". The *Astrophysical Journal*. 238: 471

observations showed strange 'dark haloes' around galaxies. It is as if galaxies are embedded in a 'cosmic gel' of hitherto elusive matter. This elusive matter allegedly comprises over ninety percent of the total matter of the universe. Its gravitational presence apparently ubiquitous, even while it is itself elusive and unknown: Hence, 'dark matter'.

Moreover, in the late 1990s, upon observation of the radiation from exploding stars within distant galaxies, the fugue of these galaxies was observed to be anomalous: The fabled expansion of the universe exceeded the Hubble prediction of constant expansion into accelerating expansion[87]. Some sort of mysterious repulsing energy was inferred, and its nature remains elusive to this day: Hence, 'dark energy'.

The implications from the impossibility of singularities in Bergsonian mnemonic mechanics and per our de-spatialised temporal universe espoused in 4.32 must raise questions about generic assumptions about dark matter and dark energy. As a first insight, per de-spatialisation, if a supermass cannot collapse to infinity and singularity, the addition of further mass, even tenfold the original supermass, will not make the slightest difference: no singularity will form. But what will happen to the gravitational effects of this tenfold addition vis-à-vis the rest of the universe? It is our contention that the key puzzle to 'dark matter' and 'dark energy' resides in the implied modification of gravitational power, which can no longer be considered proportionate to the mass in a normal way: That is, either the mass must transform, or the gravitational power must transform, or the gravitational constant must transform, given the prohibition against singularities from de-spatialisation and from the superposition of wavefunctions in memory.

*

In view of such considerations, let us imagine the gravitational outcome from the addition of two wavefunctions in memory, but each belonging to two distinct gravitating supermasses or frames: similar to the schema depicted in **Fig 4.11**, essay **4.30**. Therein, wavefunctions in memory, treated as past-primacy within the inverted cone belonging to gravitating supermass at **frame-A**, is superposed to the recentcy wavefunction in memory belonging to the other supermass at **frame-B**, forming a new culminating superpose, with consequent change and gravitational attraction of **frame-B** to **frame-A**. Given de-spatialisation, this effect needs to be recuperated as an evolving change in the delayed choice time-interval relation between **frame-B** and **frame-A**, manifested through a successions of AND-to-OR collapse-processes (time) and tandem successive event horizons.

The same, but in inverted form, will apply to the mnemonic relation; **frame-B** to **frame-A**, hence, mutual gravitational attraction.

Insofar as the culminating superpose formed from the addition of **frame-A**'s past to **frame-B**'s recentcy can be treated approximately equivalent to, or at least analogous to, the generic 'addition of velocities' and Lorentz transforms in Special Relativity (see Part-I, **4.08**) involving the velocity (i.e., gravity) from **frame-A** added to that of **frame-B**, we can use such Lorentz velocity transformation in a didactic way to garner what might happen in the addition of memories belonging to interacting supermasses.

To reiterate, in the superposition of memories, when we coincide the pertinent wavefunctions in memory, the broad stroke of component amplitudes that compose the primacy wave from **frame-A**, summed to the broad stroke of amplitudes composing the recentcy from **frame-B**, cannot individually sum to amplitudes equal to 1 or to absolute certainty. Instead, the addition of the component amplitude from one wave to the coinciding component amplitude of the other wave must transform to a sum-amplitude less than 1.

The implication is that, in the superpose of the wavefunctions in memory within and between the inverted cones of two distinct supermassive objects, the relation within and between the supermasses both before and after their convergence to a single combined supermass, can never equal to or exceed the 'speed of light', or obtain or exceed the delayed choice time interval standard: the replacement to the speed of light. In short, the gravity from **frame-A** to **frame-B** must always be less than the speed of light, or less than the delayed choice time-interval standard, even when these frames subsequently converge to form a single combined supermass.

The same will hold in the inverse relation, **frame-B** to **frame-A**.

Shorter still, gravitational additions... per the addition of 'masses' recuperated as functions of memory superpositions... transform in a similar way implied by the Lorentz transformations in the addition of velocities in Special Relativity... to a value and gravitational power 'less that the speed of light', or less than the delayed choice time-interval standard... even if the component gravities per each body and frame were equivalent to, say, 0.75c and 0.85c, respectively... and with a consequent sum less than the 'speed of light': i.e., *not* the expected sum into 1.5c: hence *not* 1.5c (See case C and D in **Fig. 4.05**: Part-I, essay **4.08**, for the analogical relation).

It is our contention that, if we were to treat mass-convergence in Bergsonian mnemonic mechanics terms, in terms of the addition and superposition of wavefunctions in memory of converging supermassive objects, we would obtain essentially the same form of outcome as found in the didactic and analogous Lorentz transformation of velocity: We would find the transformation of gravities and gravitational masses to finite limits below the equivalent to the 'speed of light'. Hence, the new combined supermassive object's frame, formed from the convergence of two distinct preceding supermasses from **frame-A** and **frame-B**, could never obtain or collapse beyond their expected new temporal equivalent of the Schwarzschild radius or the 'speed of light'.

*

How does any of this relate to putative dark matter?

If two supermasses of gravities equal to 0.9c each should converge to simultaneity and add together to give a gravitational mass-value less that 1c (and much less than the naive sum of 1.8c) where has the rest of the total gravitational mass gone? A quantity of gravitational mass equivalent to just above 0.8c has apparently gone missing.

---

[87] Huterer, D.; Turner, M. (1999). "Prospects for probing the dark energy via supernova distance measurements". *Physical Review D*. 60 (8): 081301

Using the didactic analogy of Lorentz transformation-like relations for gravitational mass and in the addition of gravities, our contention is that **gravitational mass and excess 0.8c equivalent of associated gravity will be transformed out of the local, only to be 'displaced', transferred or re-emerge at 'distance', at the larger scale. That is, the locally transformed 'disappeared mass' and gravity will reveal itself at 'distance' from its moment or point of transformation. Erroneously, this 'displaced' mass and gravity will be interpreted as 'dark matter'.** Of course, all of this must be recuperated in de-spatialised temporised physical terms.

To illustrate in some detail, we again assume two supermassive objects, each of gravity equivalent to 0.9c. Naiveté would expect that the convergence of these would form a new combined supermass of 1.8c gravity. But, applying our didactic Lorentz transform-like approach, the converging supermasses would form a new supermass with a gravity less than the putative 'speed of light': i.e., these would sum to 0.994475c with just over 0.8c equivalent of gravitational mass now 'missing'.

However, *across* the putative 'distance-d' over which the total gravities unfold, the inverse square law will abide... *not* just at 1d or at the point of the convergence of the two original supermasses, but at 2d and through to 4d, with the distinct gravities from each supermass modified per the inverse square law before they are summed and Lorentz transformed at d2, d3 and d4, respectively. Thus, when we add the gravities from one supermass to the other at distance 2d, given that each starting 0.9c equivalent of gravity will be reduced to a lower gravity per the inverse square law, the transformed sum at 2d will be less that 1c, and certainly below 0.994475c at d1. The sum at 2d will also be transformed to a value below the naive sum. The same will apply at 3d, but with the transformation even less accentuated. However, by d4, the summing gravities will be almost the same as the value one would obtain from the naive sum from 0.9c + 0.9c of the original supermasses, but with each component gravity reduced by the inverse square law by d4.

The consequences are that observers at d1 will insist that the gravity of the new combined supermass *is* the transformed 0.994475c. But observers at d4 will infer from their gravity measures at their d4 local that the combined supermass at d1 must be equal to the original 0.9c plus 0.9c...i.e., the reported 0.994475c, *and* the extra 'missing' 0.8c worth of gravitating mass. If they did not know any better and relied on the observed gravity effects of the converged supermass on nearby stars, and inferred from these that the converged supermass is 0.994475c, observers at d4 will speculate that the extra 0.8c equivalent of mass inferred from d4 local is 'missing' from d1. They might further speculate that the missing 0.8c equivalent of mass must be constituted as a mysterious 'dark matter'.

Thus, our contention is that there is no 'dark matter', but merely the transformation of gravitational power below the 'speed of light' limit, and its 'displacement' to the periphery.

*

To obtain a mnemonic mechanic account of the same, we would need to replace the didactic and merely analogical Lorentz like transformation of the said gravities and masses with a transformation that emerges seamlessly out of the mnemonic mechanic superpose of the Bergsonian memories and their amplitudes attendant the converging supermasses. Recall that these can never sum to any singularity, and the 'transformation of mass' from out of the sum of wavefunctions in memory belonging to the converging supermasses must delimit the converged-value and gravitational consequences to below infinity, and certainly below the delayed choice time interval standard in lieu of the 'speed of light'.

Recall that the summation and superpose of wavefunctions in memory within an inverted cone, even if composed of a stack of infinite number of wavefunctions in memory, cannot sum to an amplitude equal to or exceeding 1, or to the equivalent of a spatial point and infinite frequency expected to attend a culminating amplitude of 1, or singularity. Consequently, we contend that, in the addition of wavefunction in memory attendant two supermasses, the 'transformation of mass' from out of the superpose of all the wavefunctions in memory, constituting the basis of both now-converged supermasses, *must* delimit their converged culminating mass and gravity to below infinity, and certainly below the delayed choice time-interval standard in lieu of the 'speed of light'.

Of course, **given de-spatialisation, we must replace all 'distances' and 'motion' attendant the above-depiction of gravitational transformation with purely temporised de-spatialised delayed choice time-intervals and their non-uniform alterations.** Consequently, our 'distant' observers, will obtain similar conclusions to the effect that there appears to be inexplicable 'extra mass' in the system or region observed, or 'dark matter'. Yet, again, in truth, **there is no such thing as dark matter. There is only the transformation of gravitational mass and gravitational power... to values below some upper finite limit, similar to what is obtained in Lorentz velocity additions in Special Relativity, but of a form inherent to the process of the superposition of wavefunctions in memory in prospective Bergsonian mnemonic mechanics for gravitation.**

4.34: Conjectures on Dark Energy & the Mnemonic Approach to the Expansion of the Universe

In the late 1990s, upon observations made on exploding stars within distant galaxies, the recession of galaxies was observed to be anomalous, and the putative expansion of the universe no longer a constant, but accelerating. The cause was attributed to a mysterious repulsing factor: its nature elusive to this day. Hence, 'dark energy' was born.

In this essay, we offer several distinct conjectures about 'dark energy'. Yet, the ultimate answer may well comprise a combination of all our conjectures: The first is the *new tired light conjecture*. The second is the *Bergsonian pseudo-repulsion,* or *time rarefaction conjecture*... the temporal replacement to 'spatial expansion'.

In generic terms, gravitational fields normally have a drag effect on electromagnetic light radiation. Radiation leaving the surface of a gravitating body will be red-shifted by the gravity field: it will become 'tired light' and 'lose energy' [88]. Normally, tired light could not constitute an explanation of the Hubble-expansion red-shift and the expanding universe. Instead, the generic explanation for the observed Hubble red-shift is given by the putative expansion of the universe per the Big bang contention.

However, given our postulates about transformed gravitation mass and power in **4.32 and 4.33**, the profile of a gravitational field will not assume a generic form. To appreciate this, consider what would happen to the red-shift of electromagnetic radiation moving away from our new combined supermassive object formed out of the convergence of two initially distinct supermasses, as described in in essay **4.33**. Therein, the hidden gravitational mass (hidden per transformation process to a new supermass with a gravitational power less than the 'speed of light') must yet reveal itself at 'distance', seemingly as elusive 'dark matter': in truth, 'displaced' or transformed gravitational mass and gravity. Radiation traveling away from the converged supermasses and subject to the transformed gravitational profile therein, will be red-shifted to a degree greater than normally expect per normal gravitational profile.

Put another way, **given our conjecture for the transformation of gravitational power, and given the prohibition against singularities, and given the possible affects that this might have on the behaviour of radiation in the cosmic environment, we obtain a new 'tired light' conjecture: one that might be responsible for some of the** *apparent* **accelerated fugue of the universe: now conjectured as an interpretational artefact from 'new tired light' per the new mnemonic gravity-transforming mechanism that forces light to lose frequency and 'energy' in the way just described.**

Thus, the expansion of the universe might actually be proceeding normally, if at all, but for our new tired light affect, which gives the false impression of an accelerated expansion. **The implication is that 'dark energy' is ultimately illusory and non-existent.**

*

Our second conjecture on dark energy is the **Bergsonian pseudo-repulsion conjecture, although this has several alternative variants: such as the time-rarefaction version**. Therein, the total transformation that regulates the *en toto* superpose of Bergsonian wavefunctions in memory within the totality of the Bergson facet at the universe-scale, and since the Big Bang, or since the indefinite past, reaches a limit. In an infinitely old universe, this would be tantamount to the superposing an infinite number of wavefunctions in memory, but *never* obtaining a singularity. Hence, therein, to prevent gravitational infinities and to obviate singularities in probability amplitudes, an alternative balancing act comes into play: one that subsumes local transformation and regulation of memory and gravitation into a generalised cosmic scale version of the same: manifested as 'repulsion'. Hence, the observed accelerated fugue of the universe.

What we have just conjectured is the mnemonic mechanic equivalent of Olbers' paradox from 1823. Therein, in an infinite universe, Olbers claimed that every point in the sky would have a star or some other luminous celestial object. Consequently, in time, the total night sky would become as bright as the surface of the sun…. unless something reduces the frequency of light incoming from the night sky…or unless the universe is expanding such as to reduce the frequency of the radiation… or unless the universe is *not* infinite but finite, and Big Bang thus constitutes the solution to the Olbers' paradox.

However, in our mnemonic mechanic take on Olbers' paradox, the issue is no longer light radiation from celestial objects, but the background mnemonic mechanical superposes, or the total superpose of total cosmic memory within a whole universe-worth of 'inverted cones' that elaborate to the indefinite past. We posit that this undergoes transformation and regulation against infinities, with possible consequent 'repulsion' or pseudo-repulsion. That is, unless we concede to some dampening and regulating principle against infinities and singularities therein, we could not explain our normal universe. We need not even contemplate an infinite or infinitely old universe, but one sufficiently old such as to constitute a mnemonic variant of Olbers' paradox. Therein, we must yet search for the required regulation against infinities and singularities inherent to our conjectured universe-scale mnemonic mechanics.

Therein, **the accelerated expansion of the universe would not be a consequence of any literal 'dark energy', but a spill-over from nature's attempt to regulate against the formation of singularities from the total superpose of the Bergson memory, mass, gravity and quantum probability: a regulation achieved through a** *seeming* **'push' of all things against all, at least at larger scales, and more so at the cosmic scales,** *without* **requiring any inexplicable work or 'dark energy': all achieved on a work function zero basis. Hence, in such a scheme, 'dark energy' would be rendered superfluous.**

Given de-spatialisation, there is no space. Therefore, what manner of 'expansion' or seeming 'repulsion' (accelerated or otherwise) are we speaking of? Given de-spatialisation and the pure temporisation of physics, what we obtain in Bergsonian pseudo-repulsion is the *time-rarefaction of the universe*, which we naively interpret as a 'repulsive' effect, or as 'dark energy', and as 'accelerated expansion'. One immediate consequence of time-rarefaction would be new tired light, recuperated as, *not* a loss or reduction of frequency or 'energy': what we think of as a 'loss of energy' is in truth the transformation of it to *more* time, with the consequent reduction in the frequency of light interpreted as 'loss of energy' *and* red-shift. It is this that we will adumbrate below.

*

What form does the 'expansion' of the universe assume in a de-spatialised purely temporised physics? The 'expansion' must be recuperated in pure temporal terms, without 'space'. Since distance-measures must now be superseded by pure delayed choice time interval relata, and the generic 'speed of light' must be succeeded by the *delayed choice time-interval standard*, **the observed expansion**

---

[88] Zwicky, F. (1929), "On the Red Shift of Spectral Lines through Interstellar Space", *Proceedings of the National Academy of Sciences*, 15 (10): 773–779

of the universe must involve a rarefaction in the delayed choice time-interval relation that resides between any corpus of events that tend to past-primacy (formerly the 'most distant' objects in the universe) vis-à-vis any relative present-continuous frame of events or relative recentcy. That is, the 'expansion' of the universe is a rarefaction in the time-relation between primacy and recentcy, *not* an 'expansion' in the pleonastic 'distance relation', and with consequent reduction in the frequency of light and 'tired light' per function of time-rarefaction. The said time-rarefaction is apparently non-uniform and accentuating, assuming the accelerated expansion, or *any* expansion of the universe, has merit.

What could be the cause of cosmic time-rarefaction, uniform or otherwise? Convention supposes a Big Bang 'push', often erroneously popularised as a cosmic 'explosion' from which the universe emerged. The Big Bang theory, steeped in spatial terms, assumes that, the further back in the past we extrapolate, the 'smaller', hotter and denser the universe must have been... until one extrapolates to a past-moment at which the universe is so confined as to constitute a high energy domain, wherein the sorts of high energy physics explored in particle accelerators begin to unfold: and wherein we obtain a thing as good as an 'explosion' or a 'hot Big Bang'. Further extrapolation into the past tends to conclude in a Big Bang singularity from which the universe *somehow* emerged... the consequences of which we now infer as the 'expansion of the universe'.

Yet, problems emerge: The first is the notion of a Big Bang singularity, impossible in a de-spatialised purely temporised universe, and prohibited in any prospective growing block Bergson-Whitehead amalgam-based physics and mnemonic mechanics. The problem is not so much the notion of a sudden emergence of the universe, which need not be a 'free lunch', and could readily issue out of the Whiteheadian future-potentiality attached to a putative Big Bang monoblock (see Part-II 4.20). The problem resides in the forlorn spatial notion of the 'singularity' and the Big Bang singularity itself, circumvented in **4.32** and **4.33**: **Simply put, there are no singularities from which the universe could emerge, as standard Big Bang presumes. If the Big bang theory has any merit, it must be recuperated without a singularity. This is demanded by de-spatialisation and temporisation, if not by the character of our presaged mnemonic mechanics.**

However, in addition to the Big Bang, or even instead of it, and as might be espoused in a revamped Steady State cosmology, we also conjectured continuous matter-creation, obtainable under special conditions, with new matter issuing out of future-potential matter inherent to the Whitehead facet (see Part-II: **4.20 to 4.21**). The implication is that the 'expansion' or, succinctly, the *time-rarefaction* of the universe, may be due to some other principle, and *not* necessarily due to a Big Bang 'push', aside the question of whether or not the Big Bang really happened. (An alternative explanation of expansion need not obviate Big Bang itself).

*

What principle could the time-rarefaction of the universe be due to? But the answer was previously furnished: The universe scale superposition of the total stack of wavefunctions in memory within the totality of the inverted cones constituting the universe-scale Bergson facet, must reaches a limit at which so much memory is superposed that, as a regulatory feature to prevent the formation or possibility of probability-singularities, mass-singularities and 'faster than light' gravitation, or to obviate the mnemonic equivalent of Olbers' paradox, we obtain 'spill-over' into an alternative balancing and regulating act: namely, seeming 'repulsion' or pseudo-repulsion... recapitulated in pure temporal terms as the 'temporal stretch' or *time rarefaction of the universe*... without 'dark energy'.

Alternatively, and to augment the point that time-rarefaction of the universe could transpire without the Big Bang, the Bergson facet and the totality of memory or cosmic history superposed therein, could elaborate to the infinitely remote past: so constituting a universe *without* a Big Bang origin; one infinitely or indefinitely old. Therein, the said time-rarefaction of the universe must yet transpire, and for the same reasons as previously adumbrated: namely, to circumvent the formation of infinities and singularities in probability amplitudes, gravitation masses and affects from the superpose of indefinite numbers of wavefunctions in memory.

To re-integrate the new tired light conjecture to our time-rarefaction conjecture in lieu of 'dark energy', we simply impose on the culminating superpose from the whole-universe Bergson facet an incident electromagnetic radiation-potential residing within the Whitehead facet: an electromagnetic radiation-potential that must undergo grand decoherence to consistency with what the total memory of the universe demands. **Our conjecture holds that time-rarefaction from the total mnemonic superpose from the whole-universe defined by the total inverted cone, must grand decohere and reduce the frequency-range of the said potential radiation,** *against* **the formation therein of probability amplitudes that equal to 1, so as to prevent infinities and singularities. Consequently, the radiation will 'lose energy', or is reduced in frequency. Hence, it is 'red-shifted' to 'tired light'.**

To prohibit infinities, the set of possibilities and amplitudes that comprise the electromagnetic radiation in its summation with the culminating superpose from the total superpose of Bergsonian memory, will be attenuated. Given its origin from a 'distant' object, the reduction in the frequency of the radiation will give the impression of a recession and 'spatial expansion' of the universe, even if in excess of the actual level of recession and expansion, assuming any recession or expansion abides... so furnishing the impression of an expansion-accelerating 'dark energy'.

The question now arises: Will the radiation lose its frequency and become red-shifted in concordance to the Hubble constant and expectation, or will its red-shift and frequency-loss be so exaggerated as to give the false impression of an accelerated expansion due to 'dark energy'? Tentatively, our contention is that **the resulting 'tired light' will give the false impression of an expanding or accelerating universe above and beyond the actual level of expansion, assuming that the universe is 'expanding' at all (a spatial concept) even while the time-rarefaction of the universe would certainly abide.**

429

However, we must reiterate one more factor to complete our conjecture. If time-rarefaction per the stated causes leads to the frequency-loss of radiation from 'distant objects', where does the 'lost energy' go? The pleonastic status of energy concepts cannot gainsay this question. Our answer is simple, even assuming the dubious onticity of energy and work: The loss in frequency of light is compensated by and leads to the said time-rarefaction. That is, **just as time is flexible in Special Relativity and regulates against the violation of the putative 'speed of light', both the frequency of the red-sifted radiation, *and* the age of the universe, are flexible so as to prohibit infinities and singularities. In short, the pleonastic 'energy', proportionate to the lost frequency, or to 'lost energy', is *not* destroyed: Instead, it is exchanged into the time-rarefaction of the total cosmic milieu comprised of memory in time. Shorter still, the 'lost energy' is manifested as a non-illusory temporally older universe. But is this not equivalent to a spatially expanding universe? Only if space was ontically real. But space is pleonastic. There is only time. A time-rarefacted universe is *not* a 'spatially expanding universe'. A time-rarefacted universe is, simply a temporally older universe... wherein 'energy' lost in tired light is exchanged for an older universe**.

4.35: Summary of Part-III

The following constitutes a summary of the key findings and conclusions of Part-III:

- **De-spatialisation renders abstract memory and anticipatory mnemonic mechanics necessary to any future physics**: Given de-spatialisation (see Part-I **4.01** to **4.08**), there is no space. Only time is real. Thus, processes and interactions of nature do not transpire between objects or 'particles' in space, but are non-contiguous relations between Bergsonian memories and Whiteheadian futures realised across pure time. Hence attendant, gravitational relations between bodies are temporal relations; wherein the influence of the Moon upon the Earth is per function of the past-memory state of the Moon superposing with the relative present culminating superpose of the Earth... which then grand decoheres the convergent Whitehead facet to consistent possibilities: i.e. gravitation. The same applies to inertia per the superpose of a body's own memories.

- **The five postulates of mnemonic mechanics (4.25):** Mnemonic mechanics, born out of de-spatialisation and temporisation, is based on five main postulates. These are...
  - **The independent retention postulate** of past states, of wavefunctions in memory, independently of and extant subsequent OR-form 'materialisations' upon the event horizon. Wavefunctions in memory are former futures, but with their nested futures cropped. Memories are retained within the Bergson facet of the Bergson-Whitehead amalgam.
  - **Wavefunctions in memory superpose across time**, the non-contiguous addition and superposition of primacy to recentcy wavefunctions in memory residing within the Bergson facet transpire, *not* across space, but across time.
  - **The inverse square law of time:** Wavefunctions in memory are subject to the inverse square of time, modelled as the inverted cone within the Bergson facet: Memory amplitudes and other attributes are reduced proportionate to the extent memory is temporally removed vis-à-vis a given future-potential event horizon.
  - **The exclusivity of constructive interference in the addition of memory (4.28):** Wavefunctions in memory can never be blocked or be subject to destructive interference when superposed with older wavefunctions in memory. This does not apply to the superpose of Bergsonian memories with Whiteheadian wavefunctions in potentiality, although it is the latter that become destructively interfered with, primarily via grand decoherence of the total Whitehead facet.
  - **The inapplicability of grand decoherence to, and the non-erasure of, Bergsonian wavefunctions in memory:** Unlike wavefunctions in future-potentiality, wavefunctions in memory possess no nested futures. Hence, the processes of grand decoherence and the erasure of nested futures (Book-II) do not apply to wavefunctions in memory in Bergsonian mnemonic mechanics. Thus, wavefunctions in memory are consequently ineradicable: i.e., Memory is ineradicable.

- **Bergsonian mnemonic mechanics of gravitation (4.30):** With de-spatialisation, the mutual gravitational influence of the Moon versus Earth cannot be mediated across space, much less as a 'force' or as a 'curvature of space'. There is no space: The Moon's gravity vis-à-vis the Earth does not emanate from a 'material body in space' at a presumed 'distance' from the Earth, but it is constituted as the past state of the Moon, rendered into abstract memory, separated from the Earth by approximately 1.5 seconds in time. The influence of the Moon upon the Earth is a non-contiguous purely temporal one, per a delayed choice time-interval relation approximating to 1.5 seconds, and weakened per the inverse square of the time; finally superposing with the relative present frame of the Earth. It is the Bergsonian wavefunction in memory constituting the Moon from 1.5 seconds ago, subject to the inverse square of time, that superposes with the culminating superpose from the Bergsonian memories attendant the Earth. This then 'tilts' the potential worldlines of both the Moon and the Earth toward each other: That is, *en masse* AND-to-OR collapsed 'particles', resolve into realised OR-form clusters in conformity to the 'tilt' enforced by the mnemonic superpose, leading to moments of mutual 'gravitational attraction' and orbital motion via the successions of event horizon-formations. In conclusion, gravity is a Bergsonian memory effect, while 'gravitational mass' is per function of the total memory within the pertinent Bergson facet attending the 'body' and retained within the Bergson facet. The attendant grand decoherence of the Whitehead facet will 'inscribe' to it work function zero 'orbital paths' and pertinent gravitational relations: the said conformity of the Whitehead facet to Bergsonian memory, according to which gravitational phenomena unfold.

- **Work function zero 'orbital paths', Whiteheadian grand decoherence, and quantum gravity (4.31):** While Bergsonian mnemonic mechanics gets to superpose wavefunctions in memory to generate culminating superposes and gravitational effects, attendant processes must also transpire within the Whitehead facet of the Bergson-Whitehead growing block universe, culminating into a growing block account of gravitation. The key is **gravitational grand decoherence** and the counterfactuals based erasure of non-consistent futures within the total Whitehead facet. Bergsonian memory-affected counterfactuals generate the grand decoherence of the total Whitehead facet to consistency with those counterfactuals, so inscribing within the Whitehead facet the work function zero future-potential 'orbital paths', subsequently followed by the pertinent bodies or their attendant *en masse* quantum indeterminate AND-to-OR collapse processes. The inscribed 'paths' constitute work-function zero paths in the sense that, any subsequent attempt to change these to some other contingent path (i.e., any attempt to 'push' bodies to a lower or higher orbit, or to change their apparent 'motion-state' under gravity) will require putative pleonastic energy or work, with attendant non-zero work function relations. But in the absence of 'applied force quantum mechanical waves', the 'paths' inscribed onto the Whiteheadian future will constitute 'paths' that do not require any subsequent energy or power inputs to realise. In short, we obtain acceleration, perpetual motion, and orbital coupling under gravity, with attendant change in 'angular velocity'... *without* the requirement of energy or work, but *without* the violation of the principles of conservation, which transform into abstract form per function of de-spatialisation and the growing block universe: wherein conservation attends the immutability and unchangeability of the accumulated past, and wherein the future must be rendered consistent to the past per attendant grand decoherence. Insofar as all the above are realised through work function zero processes of AND-to-OR collapse and attendant broken-discontinuous progression of the event horizon, we arrive at the presage to discrete quantised gravity, or simply *quantum gravity*.

- **Bergsonian mnemonic mechanics and inertia (4.29):** With de-spatialisation and pure temporisation of physics furnished in Part-I, inertia arises from the dual consequence of the summation of wavefunctions in memory via Bergsonian mnemonic mechanics *and* attendant **inertial grand decoherence** of the Whitehead facet to consistency, both of which entrench and retrench the tendency of the 'body' to remain at its existing state and resist the change to that state... or *inertia*. Mach's principle abides through the grand decoherence of the total Whitehead facet; forcing mutual consistency vis-à-vis bodies elsewhere, through which all bodies acquire their inertia. Of course, inertia is ultimately an idealisation: Only non-inertial non-uniform gravity-based relations truly exist, characterised by perpetual non-uniform delayed choice time-interval interrelations and changes.

- **Conjectured implications from Bergsonian mnemonic mechanics to singularities (4.32):** In a de-spatialised purely temporised universe, singularities and black holes cannot exist, although asymptotic nominal supermassive objects could and *may* exist, on the condition that these do not collapse to infinity and never obtain their putative Schwarzschild radii (or its temporal delayed choice time-interval equivalent). Thus, when masses (succinctly wavefunctions in memory) are added to each other or converge to form new masses or supermasses (or when their gravities are added) the sums transform in analogous way to the Lorentz transform of velocity in Special Relativity: wherein the sums always culminate into outcomes that are less than the 'speed of light' or its replacement, the *delayed choice time-interval standard*. Of course, the nominal addition or multiplication of masses in the context of gravitation consist ultimately of the mnemonic mechanical superposition of wavefunctions in memory, also pertinent to the converging supermass 'bodies': Thus, both inertial and gravitational mass are but the sum and superpose of Bergsonian wavefunctions in memory. However, when wavefunctions in memory add, and as is true in the superposition of all wavefunctions, their coinciding probability amplitudes never obtain or exceed the probability of 1. This prevents the bodies and masses so combined (hence the wavefunctions in memory so superposed) from obtaining infinity or singularity. Consequently, no singularity can form out of the superpositions of wavefunctions in memory. Indeed, even if we superposed an infinite number of wavefunctions in memory, this could never constitute infinite mass, much less singularities in the probabilities of how the Whiteheadian future is permitted to evolve. Hence, 'black holes' are prohibited. The finding segues and supports conjectured matter-creation detailed in Part-II **4.21**, given that the putative realised matter drawn out from Whiteheadian future-potential matter will not fall into and disappear into an impossible singularity.

- **Conjectured implications from Bergsonian mnemonic mechanics to 'dark matter' (4.33):** When two supermasses, say, each of 0.9c gravity, are added together, the gravitational mass of the combine will transform to a value less than 1c, *not* to the naïve sum of 1.8c. For observers attached to the new supermass, around 0.8c worth of gravitational mass will go 'missing', even though it is merely transformed or hidden. However, at distance from the now converged supermasses, the gravity additions emanating from the two original supermasses that had converged will sum almost normally and will reveal the hidden 0.8c equivalent of gravitational mass, showing it to be merely locally transformed and 'displaced' to distance. Observers at distance will detect the gravitational effects of that transformed and hidden mass at that distance, but interpret it as a mysterious 'dark matter' . Of course, in all such instances, it is the pertinent wavefunctions in memory associated with the supermasses that are being superposed and transformed... to outcomes that exhibit coinciding amplitude-additions that always sum to less than 1, generating a transform of implied 'mass' and gravitational power: a culminating superpose that is less that what would normally be implied by a simple summation and superposition. In other words, there is no 'dark matter', but only the transformation and delimit of gravitational mass and power to a limit always below the delayed choice time

431

interval standard of 1 (the 'speed of light') as demanded by Bergsonian mnemonic mechanics and the inbuilt transformation and renormalisation against infinities and singularities entailed in the superposition of memories; all of it recapitulated in temporal terms, as demanded by de-spatialisation.

- **Conjectured implications from Bergsonian mnemonic mechanics to 'dark energy' (4.34):** Our conjectures on dark energy and the supposed accelerated expansion of the universe come in several forms: These are, *new tired light* and *Bergsonian pseudo-repulsion*. As gravitational mass and gravity field-profiles transform to reveal the hidden mass otherwise transformed out of the locus at origin, and it is 'displaced' to the periphery, radiation moving away from that origin will be subject to a redshift above the normal expected level, contributing to 'tired light'. Hence, for observers unaware of gravitational transformation and the delimit against singularities, the extra redshifted light will be misinterpreted as consequent of an accelerated expansion or recession of the origin and of the cosmic environment, supposedly due to a repulsive 'dark energy'. In the case of conjectured Bergsonian pseudo-repulsion, the total wavefunctions in memory constituting the hitherto resolved events of the whole universe, or of the total Bergson facet, will sum up and transform in such a way that, to delimit probability distributions and discrete amplitude sums to values less than 1, and to prevent singularities, a counter balancing 'repulsive effect' will come into being. The universe will apparently 'expand' at an accelerated rate, perhaps in just the accelerated way observed, conveying a greater level of redshift upon observed radiation into 'tired light'; although the expansion cannot be in terms of 'space and distance' but, in a de-spatialised temporised universe, it must be recuperated as an 'expansion in time' (see below). In short, there is no such thing as 'dark energy'.

- **De-spatialisation, temporisation, and the time-rarefaction of the universe (4.34):** The implications of de-spatialisation require that we completely replace 'spatial expansion', if not even the pseudo-repulsion conjecture (above), with *cosmic time-rarefaction*, with or without a Big Bang. Therein, the superposition of the total sum of wavefunctions in memory constituting the total Bergson facet of the universe up to the Big Bang, or elaborated to the infinitely remote past, transform in such a way that, in order to delimit probability distributions and discrete amplitude sums to values less than 1, and to prevent gravity and mass singularities, the universe undergoes cosmic time-rarefaction in a uniform or non-uniform way... with consequent frequency loss and pleonastic 'energy loss' in the electromagnetic radiation 'from distant objects'... again enmeshing with 'tired light' conjectures... but, ultimately, *without* dark matter or dark energy. The 'lost energy' from reduced electromagnetic frequency is transformed, not into spatial expansion, but into temporal rarefaction: i.e., the universe becomes temporally older, albeit in a non-illusory way. That is, 'energy' is transformed into time, and into an older universe.

BOOK-IV PART-IV:
DE-SPATIALISATION, THE BERGSON-WHITEHEAD AMALGAM AND MIND-BRAIN RELATIONS

4.36: Aims of Part-IV

In Part-IV, de-spatialisation, attendant temporisation, combined to the Bergson-Whitehead amalgam approach to the growing block universe, will furnish a **final account of information-matter dualism** first started in the Preliminary and in Book-I. Therein, information-matter dualism will be espoused as the mirror image of the temporal structure of the Bergson-Whitehead amalgam. This will also apply to related decoupling theory provoked by the experiments of Gunter Nimtz. Our preceding want to reform physical fields, pattern and form into their de-spatialised temporised variants in **4.09** turn out to be fully consistent with information-matter dualism and with implications from decoupling theory. In Part-IV, these and the Bergson-Whitehead amalgam, will fashion for **a new cymatics**, and proposals on how cymatics could be further developed in a de-spatialised temporised growing block paradigm of the life-sciences.

However, our driving goal was always the discovery of solutions to the mind-brain conundrum. Thus, on the heels of insights and conclusions garnered thus far, **the demise of materialism is finally secured through de-spatialisation and temporisation, the consequent usurping of contiguity and contact causality, and per the structure and ephemeralised processes of the superlative Bergson-Whitehead amalgam, furnishing the final ingredient to the anticipated revival of Cartesian dualism** and, with it, the demise of academic philosophy of mind; namely, the system of materialist apologetics contrived against Cartesian dualism. Hence, Part-IV will sketch the key features pertinent to the revival of mind-brain dualism, including the correction of **Descartes' minor error**: his supposition of *res extensa*, or 'space'.

Revived Cartesian dualism will require that we remodel brain and memory, and the attendant noetic processes in relation to both, in pure de-spatialised and temporised terms, in the framework of the growing block Bergson-Whitehead amalgam. Indeed, radical de-spatialisation and the demise of *res extensa* compels the pure temporisation of memory-brain and mind-brain relations, accomplished in two modes: First, the **passive mode** relates the **noetic moment** to the system of past events (memories) retained in the Bergson facet. Therein, **the noetic moment arises temporally *after* (and *not* from, or not from the same moment) as the Bergsonian past: The noetic moment cannot therefore be reduced to that Bergsonian moment of the past, nor to any past state of the brain.**

The second is the active mode: Therein, the mind engenders causal agency, not by acting on any realised brain of the Bergsonian past, but by selecting a brain-pertinent future-potential possibility within the Whitehead facet. In this instance the mind is *not* acting on a 'place', but instead acts on an abstract future-potentiality of the brain; one residing within the Whitehead facet. That is, the mind 'reaches into the future', and draws out from that future-potentiality what it seeks to realise in the future brain. What it selects from that future is subsequently spontaneously AND-to-OR resolved into 'materialised' brain activity. The minds' 'action' on the Whiteheadian future does not involve contiguous causality, since it is not enacted or mediated via either space or through contact causality: The process is work function zero and involves the equivalent of the throw of suspended delayed choice **switch-x**, which selects the specific complex wavefunction in future-potentiality that will AND-to-OR resolve into realised brain activity at some future-potential event horizon. Yet**, in the active mode, a noetic moment of mind, as the selective causal moment of agency, temporally *precedes* the sought future potentiality for brain and world. Hence, mind is not 'locatable' or reducible to (and cannot arise from) the future potential Whiteheadian brain and world that it seeks to realise, given that the latter has yet to transpire. On the other hand, the noetic moment of minded agency, insofar as it can project into the future, cannot be confined to or arise from the most recent resolved brain-state confined to the event horizon: That is, it cannot arise from the 'materialised' brain.**

Given the above-stated temporal relation and extant status of nous to both Bergsonian memory *and* Whiteheadian potentiality, and its implied extant status vis-à-vis the 'materialised brain confined to the event horizon, Cartesian dualism ineluctably follows.

Part-IV will culminate into **noumnemonae**, or **'memory in consciousness'**: It turns out that the noetic moment, the moment of consciousness in its passive mode, and the moment of mind in its active mode, possesses no 'parts': This will become especially obvious from radical de-spatialisation and the consequent depiction of nous as an indivisible noetic moment. Therefore, the noetic moment cannot be subject to an internal relativity of simultaneity. Indeed, the noetic moment does not constitute a temporal distribution of events, no more than it could consist of a spatial distribution of such. Instead, the noetic moment, as a noetic monad, stands 'outside' of the Bergsonian time-distributed events: It subsumes the time-distributed events into consciousness. That is, information and memory in time arises to and 'in' nous, but *not from* nous. **Onto itself, nous constitutes a single 'point': but *not* a 'point in space'. Nous is constituted as a temporal monad. As such, nous constitutes a radical form of the 'common present moment': a state of noetic 'absolute time', but *without* space.** This contrasts with the temporal Bergsonian world and brains in abstract memory, effectively comprised of 'parts'… wherein the 'parts' are constituted as events temporally scattered and distributed into past-primacy and past-recentcy, necessarily subject to the relativity of simultaneity.

Hence, we end with two distinct types of memory and attendant dualisms: namely, **noumnemonae, constituted as 'part-less' noetic moment, or as a temporal noetic monad: an absolute time-like state constituting 'memory in consciousness'**, or consciousness-memory. This is in contrast to **mnemonae**, subject to the relativity of simultaneity, comprised of Bergsonian world and brains, both subsequently arising to and *into* nous. This division of memory engenders radical consequences unpalatable to contemporary culture, insights that must await Book-V and beyond.

4.37: De-Spatialisation, the Bergson-Whitehead Amalgam
And the Completion of Information-Matter Dualism

In the next two essays, we complete information-matter dualism and decoupling theory, with both recuperated per function of the demands from the de-spatialised and temporised Bergson-Whitehead amalgam.

Information-matter dualism and decoupling theory are intertwined (see Book-II **2.08**). Also, both segue into the mnemonic mechanics-based understanding of form and morphology.

But first, we recollect the various iterations of information-matter dualism hitherto developed: These are:

- **Facile information-matter dualism** from the Preliminary (see **0.01** to **0.04**): This assumed the onticity of information carrying radiation and matter in 'space'. Even so, on a largely pre-quantum mechanical basis, we discovered that information must dissociate from radiation, matter, and space per function of the impossibility of seamless-continuous motion in the relation of radiation and matter to space. This compelled for the broken-discontinuous form of the motion of radiation and matter vis-à-vis space. As such, it implied ineliminable intervals across and along which any purported 'carried information' cannot coincide. Hence, the information 'carried' *must* dissociate from such ineliminable intervals, and 'leap over' the intervals… until subsequent reconstitution to radiation, matter and space transpires.

- **Quantum logic-based information-matter dualism**: In Book-I, facile information-matter dualism evolved out of the key distinction between classical OR-form versus quantum mechanical AND-form logic and attendant AND-OR (wave-particle) duality. Therein, we assumed the erroneous view that realised information must be constituted purely as an OR-form configuration, and that, insofar as a quantum mechanical AND-form radiation, matter and attendant intervals cannot be eliminated from systems, and insofar as we cannot realise a pure OR-form state despite the promise of the Quantum Zeno effect, information grasped in pure OR-form terms must indeed dissociate from ineliminable AND-form radiation, matter and attendant spatial intervals. It followed that OR-form states (presumed definitive of real information at that juncture) could not be structurally or logically retained in and by AND-form radiation, matter, or 'space'. Hence, information-matter dualism was recapitulated in terms of wave-particle AND-OR dualism.

433

- **Causality-based information-matter dualism**: In Book-I and Book-II, information-matter dualism was further augmented by the demands of causality intertwined with the principle of conservation: i.e., *the firewall principle*. (Book-II, **2.10** to **2.11**). Book-I compelled that we cannot eliminate the AND-form characteristic of a system, given that this would be tantamount to eliminating the system's total future (see Book-I, **1.17** and **1.20**) directly culminating into the reality of information-matter dualism per function of the ineliminability of AND-form logic, recapitulated as the ineliminability of the future *at any and all scales*, forcing the realisation that AND-form states or quantum mechanical waves (the future) cannot be restricted to the micro-scale alone, but must equally abide at the macro-scale and at the universe-scale. To remind, the AND-form future is ineliminable from *any* system, unless we assert for a block universe in which the future is 'objective' or fully OR-form pre-resolved. But such an assertion, otherwise undermined by the very viability of the generic two-slit experiment, would render wave-interference impossible to obtain, and furnish empirical-physical proof of a form that requires the reception of 'signals from the future'… in violation of causality. The proof would also evince above-unity pleonastic energy-matter attendant the received 'signal from the future', with tandem violation of the principle of conservation. In conclusion, we cannot eliminate the future-form description of any systems at *any* scale, unless we violate putative causality *and* the principle of conservation. Thus, we are forced to accept that, whatever the scale, a system must always constitute as an admixture of AND-form *and* OR-form attributes. Consequently, we must concede to inescapable all-scale information-matter dualism that attends such an admixture.

- **Intermediate spacetime structure-based information-matter dualism**: By Book-II (essay **2.40**) we had arrived at the quantum-relativistic incorporative intermediate model spacetime. Information-matter dualism turned out to be a mirror image of ineliminable AND-OR wave-particle dualism inbuilt into intermediate spacetime: Since AND-form states constitute the ontic reality of the future from which 'signals from the future' cannot be received per the firewall principle, the recuperation of spacetime as the growing block intermediate model spacetime system must also recuperate information-matter dualism as past-future dualism. In short, information-matter dualism turned out to be the facile reflection of the tripartite spacetime structure divided into *realised spacetime* (the past) and *spacetime in potentiality* (the future), both demarcated by the *event horizon* 'cut'. Naturally, OR-form information (the presumed form of the past) cannot be carried in or as the AND-form quantum wave or interval (the future), simply because the past is not carried in or mediated by the ontic future: i.e., events that have transpired are not reducible to future-potential events that have not-yet transpired. Hence, the dissociation of OR-form information from AND-form potentiality is simply the dissociation of *realised spacetime* (the past) from *spacetime in potentiality* (the future): confirming that OR-form information is truly *not* carried by AND-form radiation or de Broglie matter, as was inferred in facile information-matter dualism and in dualism based on the AND-OR wave-particle duality. Moreover, per a specific application of John Wheeler's delayed choice, it was shown that AND-form waves embedded within a static spacetime in potentiality (the future) cannot constitute a 'moving' or displacing state: Obviously, the future cannot 'move'. Therefore, the quantum wave (the future) cannot displace or 'transport' OR-form (past) information across space. The conclusion drawn was that resolved OR-form information could not be carried in or by any AND-form quantum mechanical radiation and matter-potentials embedded within static spacetime in potentiality (the future). Hence, spacetime structure-based information–matter dualism emerged as a deeper alternative basis for information-matter dualism.

Note that, in all the variants and evolutions of information-matter dualism recalled above, the notion of 'space' was presumed. Yet, by Book-III and Book-IV, space turned out to be completely pleonastic, and even OR-form events had to be recuperated in terms of pure delayed choice time-interval distributions, *without* 'space'. Moreover, the provisional OR-form based definition of information we had hitherto relied on had to be replaced and succeeded, given the case for abstract memory in Book-III (**3.22**). Therein, we had arrived at the minimum complexity form of information: the abstract structure-state and *bounded probability state* approach or definition of information, culminating into a more superlative view of information: one in which OR-form information remains relevant, but only with respect to and confined to the event horizon. However, realised information itself (or information in memory) turned out to be the abstract Bergsonian *wavefunction in memory,* retained within 'realised spacetime', itself succeeded by the mnemonic Bergson facet of the Bergson-Whitehead amalgam. Succinctly, by implication of the Bergson-Whitehead amalgam, we arrived at three distinct forms of information:

- Whiteheadian future-potential information in the form of nested AND-form states, or **wavefunctions in future potentiality**. These articulate for future-potential events, durata, and their future-potential delayed choice time-interval relations; all embedded within the Whitehead facet (the de-spatialised replacement of 'spacetime in potentiality'); structured as a universe-scale convergent category nested future; subject to 'non-local' or transtemporal grand decoherence (i.e., the basis for the revival of Mach's principle).

- States comprised of the most recent OR-form 'materialisations' exhibited on and along the present-continuous event horizon: namely, **the recentcy**. Event horizon recentcy states conform to our naïve conceptions of 'matter in space' (that is, to "pure OR-form information") with which we then tend to confuse the whole of physical reality and the whole of naturalism; and with respect to which we erroneously seek to reduce the totality of real existence and ontology, causality, and even mind itself, via forlorn closed-causality schemas and materialism.

434

- Bergsonian mnemonic information is comprised of non-nested **wavefunctions in memory** (the past in abstract memory) retained 'below' the event horizon within 'realised spacetime', now succeeded by the Bergson facet. Memories superpose upon successive event horizons through a process of conjectured mnemonic mechanics (see Book-IV: Part-III: **4.25** to **4.28**). The culminations of the superpose then undergo AND-to-OR collapse and, through the three-phase developmental process (Book-III **3.26**), are resolved into objective *en masse* quantum indeterminate OR-form outcomes at the event horizon... and into resolved inertial, gravitational, and other outcomes recuperated in their de-spatialised purely temporised forms (Book-IV: **4.29** to **4.30**).

With the three types of information in hand, we can now furnish **the final iteration of information-matter dualism**. First, **Bergsonian mnemonic information constituted into wavefunctions in memory (or the total Bergson facet) does not physically reduce to the OR-form 'materialised' recentcy states confined to the event horizon. Hence, immediate information-matter dualism**; one similar to both the facile and to the AND-OR based iterations of it, but justified per the relation of the Bergson facet to the event horizon. We see just such information-matter dualism implicit in the presage to Bergsonian mnemonic mechanics of inertia (see **4.29**) and gravitation (see **4.30**)... a dualism between the abstract-immaterial memories of the Bergson facet versus the 'materialised' inertial and gravitational outcomes realised on successive event horizons. The same must clearly apply to the relation of abstract Bergsonian memories vis-à-vis the 'materialised' brain.

Second, **Bergsonian mnemonic information, together with outcomes garnered upon the event horizon, do not physically reduce to Whiteheadian future-potential information**. From the structure of the Bergson-Whitehead amalgam, and from the three types of information given above, it is clear that neither the Bergsonian past-primacy, nor the event horizon of OR-form recentcy, can reduced to, be 'carried in', or be 'transported by'... or 'moved by'... *any* static nested AND-form future-potentials embedded within the Whiteheadian facet, no more than by the total Whitehead facet itself. Both Bergsonian memories *and* event horizon realisations must remain fully dissociated from the Whiteheadian future, even though both are present to that future through counterfactuals and consistency-enforcing grand decoherence.

Finally, **given de-spatialisation, it would be absurd to treat past-primacy Bergsonian information as 'carried in', within, or 'spatially transported' by, ultimately static Whiteheadian future-form quantum mechanical radiation and matter: These are static non-moving and non-undulating states,** and remain so even on the false assumption of 'space'. Ultimately, the whole notion of space presupposed to the idea of 'moving' materio-spatially transported information is rendered pleonastic and false (see Part-I, **4.01 to 4.07**). The supposition that information could be reduced to radiation and matter presupposes space; or presupposes the 'location' of information in space and in spatially displacing radiation or matter. **Debunk space and it forces the conclusion that information *must* relate to the past, to the present and to the future, in a purely temporised non-contiguous form... in terms of abstract realised or potential delayed choice time-interval relations and distributions. Thus, information relates to a past, to a relative present frame, or to a future-potential frame,** *always* in temporally *extant* and dissociated form: a Cartesian-like dualistic conclusion about information, with ramifications to how mind, thought and consciousness must also relate to memory, to futurity, and to emergent events from AND-to-OR collapse and time... if not to the brain itself.

Hence, we arrive at our final iteration and completion of information-matter dualism: grasped in terms of our de-spatialised and purely temporised Bergson-Whitehead amalgam replacement to spacetime.

4.38: The Bergson-Whitehead Amalgam & the Completion of Decoupling Theory

Decoupling theory was developed in response to the anomalous experimental results obtained by Gunter Nimtz (Book-I, Part-IV). Using microwaves, Nimtz quantum tunnelled Mozart-40 at an apparent 4.7 times the speed of light, in apparent violation of the speed of light limit. Decoupling theory solves this violation by exploiting information-matter dualism: thus *decoupling* information from the apparent 'faster than light' radiation, from radiation and matter at any speed, and from the speed of light itself.

The first decoupling theory from Book-I comprised a synthesis of the *maincone approach* (Book-I: **1.41**: especially scenario-4). This exploited nascent information-matter dualism. The maincone approach asserted that a maincone (an embedding larger or preceding lightcone originating at the formation and setup of the apparatus, or even originating at the original recording of M40) will embed any subsequent lightcones and events subsequently attached to the Nimtz's experiment. The approach asserted that it is the maincone that contains the critical information, *not* the embedded lightcones: Insofar as critical information does not arise from the lesser lightcones subsumed to the maincone, and insofar as it is the information attached to the maincone that decides the content of subsequent outcomes, and does so without violation of the speed-of-light limit, we avoid the apparent violation of the speed of light limit that attend Nimtz experiments.

However, the maincone approach needed information-matter dualism in tandem: Information-matter dualism was used to show how information gets around within *any* system, quantum tunnelled or otherwise, in dissociated and clouded form within the maincone that subsumes the subsequent corpus of relations and events. Recall that physical information (originally presumed to be purely OR-form) was expected to dissociate from quantum mechanised AND-form matter and radiation and from *any* lightcone, as required per nascent information-matter dualism. Hence, M40 must also dissociate from the quantum tunnelled 'carrying' microwave radiation in the Nimtz experiment.

435

Given that M40 was attached to the maincone in any case, M40 can be said to be de-localised or 'clouded' throughout the circumscription delimited by the maincone. Hence, M40 is retained in dualistic dissociated form *throughout* the system. Thus, in the Nimtz experiment, per nascent information-matter dualism, M40 must also 'pull out' from both its point of origin as well as from the quantum tunnelled 'carrying' microwave radiation. That is, M40 must *decouple* from the 'carrying' tunnelled microwave radiation into its de-localised clouded form, until restored to that radiation at the moment of its destination and final observation by Nimtz.

*

By Book-II we had arrived at the growing block intermediate spacetime system. Our first use of John Wheeler's delayed choice approach, combined with implications from the structure of intermediate spacetime, forced us to modify nascent decoupling theory from Book-I. In Book-II, using delayed choice, we obtained decisive conclusions in favour of spatially static quantum mechanical radiation and matter waves (Book-II **2.24** to **2.26**... also recuperated above in Part-I, **4.07**). It turned out that quantum mechanical radiation, even if quantum tunnelled by Nimtz to an apparent 4.7 times 'faster than light', belonged to the domain of future-form 'spacetime in potentiality': As such, the quantum tunnelled radiation constituted what we then posed to be a pure spatially static nested futures non-moving state. In which case the quantum tunnelled radiation cannot 'move' as such, even at a purported normal speed of light, let alone at 4.7 times 'faster than light'. In short, the 'movement' of the carrying radiation, at the speed of light, or slower, or faster, constitutes an imposed or foisted assumption or artefact: Radiation does not actually move or displace at all: it is static, insofar as the futurity to which it belongs is also *obviously* static: i.e., the future does not move. It also followed that M40 is not being 'carried' or 'transported' by such static quantum waves, and it is not itself moving or displacing across space at *any* 'speed', presaging de-spatialisation, later secured in **4.01** to **4.07**.

Thus, even within the region demarcated by a potential maincone, and independent of all its subsumed or embedded potential lightcones (including the quantum tunnelled lightcone) Mozart-40, ultimately retained within realised spacetime, and not transported by the non-moving non-transporting spatially static quantum waves within spacetime in potentiality, must remain dissociated from... hence *decoupled* from... both the event horizon *and* from the future-potential moments of its 'materialisation' via subsequent AND-to-OR collapse of the 'quantum tunnelled' static state.

In short, growing block intermediate spacetime structure delineated in Book-II, combined with the case for spatially static quantum mechanical waves, compelled the decoupling of M40 from the quantum tunnelled 'faster-than-light' radiation and from *any* ultimately non-moving static AND-form 'speed of light' radiation, or even from 'slower' de Broglie matter.

*

We now need to reframe decoupling from Book-II in terms of the structure of our superlative de-spatialised and purely temporised Bergson-Whitehead amalgam: the successor to the intermediate model. We must include the evolved definition of information superlative to our myopic pure OR-form view of it entertained throughout much of the Preliminary and through to Book-III. Thus, information (and M40 itself) must be grasped as a complex abstract wavefunction in memory, retained in de-spatialised and purely temporised form within the mnemonic Bergson facet of the Bergson-Whitehead amalgam. As such, M40 is necessarily dissociated from both the event horizon *and* its OR-form recentcy states, as well as from the AND-form nested future wavefunctions in potentiality within the Whitehead facet. **The maincone from decoupling theory in Book-I and II is superseded by the Bergsonian *inverted cone* (see 4.28), while the maincone itself transforms into the zone of Whiteheadian potential that has grand decohered into consistency per the memory of M40 retained within the Bergson facet. Thus, the original notion that information was attached to and distributed throughout the maincone is the mirror-image of the superlative view that information *is* the wavefunction in memory pertinent to M40 retained within the Bergson facet; retained in purely temporised form within its inverted cone.**

Per de-spatialisation, we remove 'space' itself from the fray of decoupling theory. Thus, the decoupling of M40 must transpire in a purely temporised way. **Thus, there can be no transmission of M-40 'across space', whether normally at the 'speed of light' or quantum tunnelled radiation at 'faster than light'. Thus, M40 within the Bergsonian inverted cone superposes itself upon the event horizon through a purely non-contiguous temporal mnemonic mechanics and grand decoherence, with or without quantum tunnelling by Nimtz.**

Retained within the Bergson facet as wavefunction in memory, dissociated from both the event horizon *and* from the static Whiteheadian future, M40 may re-manifests at a future moment, but without requiring 'spatial displacement'. **Of course, M40 from within the Bergson facet constitutes the critical counterfactual that grand decoheres the Whitehead facet, or that area of it constitutive of the potential maincone and subsidiary potential lightcones, into consistency. Thus, the Whitehead facet *is* potentialised into a future recollection-potential of M40...** *not* **through the 'spatial motion' of M40 via 'transporting radiation', but through the very grand decoherence of the Whitehead facet into consistency and near-inevitability, baring 'Wheeler's cat', even as M40 is dissociated and decoupled from the Whitehead facet or from the AND-form radiation definitive of the consistency-rendered potential maincone.**

4.39: Mnemonic Mechanics, Morphology, Morphogenetics
And New Cymatics: Preface to a Radical Biology

Mnemonic mechanics presaged in Part-III sought to sketch an explanation for both inertia and gravitation in the light of superlative de-spatialisation, attendant temporisation, and the development of the Bergson-Whitehead amalgam replacement to intermediate spacetime. We must now consider the retention and transfer of more complex mereology and morphology beyond the simple forms exemplified by inertia and gravitation, such as the resonance of pattern and morphology evinced in living forms. These must also be

436

recuperated as expressions and distributions in time, *without* space, to the negation of generic notions about living forms and biological processes misconceived as 'spatial distributions', supposedly mediated and affected through contiguous causality.

Morphology and morphogenetic forms evinced in biota must ultimately be 'transmitted' and retained across time, from the past to the future, through a non-contiguous causality espoused and presaged in mnemonic mechanics: This idea has been presaged by Rupert Sheldrake in his *formative causation* and *morphic resonance*. De-spatialisation, the concomitant temporisation of information and form, and the revamping of both information-matter relations *and* decoupling in terms of the superlative Bergson-Whitehead amalgam, necessarily implies that the critical information pertaining to *all* pattern-structures, including those that attend living forms, cannot be exclusively reduced to pleonastic spatial materio-genetic and DNA-based concomitants, or to the pure OR-form recentcy states confined to the event horizon comprising 'materialised' living nature. Indeed, the living form of the organism, its processes, and its embedding environment, must be recuperated in terms of abstract memory combined to future-potentiality in the framework of the temporised growing block Bergson-Whitehead amalgam, *without* 'space'. In short, **de-spatialisation and temporisation implies that all pattern-structure and form, including biological form, process, development, and even evolution, cannot ultimately be rendered in terms of 'distributions in space' or per contiguous processes in space, but must be totally recuperated purely as, if not *only* as, distributions and non-contiguous processes unfolding in time, all within the framework of a de-spatialised growing block universe: within the Bergson-Whitehead amalgam**. Indeed, what applies to de-spatialised 'physical fields' (see **4.09**) necessarily apply to living form and process; the pure temporisation of form into the growing block framework.

*

With the above goal in mind, we might start with the most essential biological factors: the DNA-based information system and the genome, and the manner through which it integrates to the cell, to the whole organism, and to the embedding environment... typically treated in spatial distributive and contiguous terms, as is the genome itself. Yet, the said DNA-based system of information must ultimately be grasped as embedded in a de-spatialised purely temporised system of Bergsonian abstract memories projected in and across de-spatialised pure time, with attendant Whiteheadian future-potentialities and their pertinent grand decoherence; all *without* contiguity, without 'spatial transportation', without 'spatial distribution', *within* the growing block Bergson-Whitehead amalgam. Thus, the DNA and the DNA-based genome can no longer ultimately be characterised as a spatial distribution of process or pattern-outlays, or as 'matter in space', but must be treated as a distribution in time: as perduring 'worms' and projection within the Bergson facet, subject to mnemonic superpose processes; integrated to general AND-to-OR wavefunction collapse or time-process; the very processes that must perdure and perpetually re-materialises the DNA and genome on successive event horizons, as it does the whole cell, the whole organism, and its whole growing block domain of memory and future-possibility, including its environ.

The genome must be present as a Bergsonian state of memory in dissociated form from any subsequent materialisation of that genome upon subsequent event horizons, with much of the genome constituted as abstract memory, embedded to a grander domain of Bergsonian memory that constitutes the organism as a whole, its environment, and its history as a whole... if not its whole evolutionary history... subject to the inverse square of time. Therein, the interaction between genes, the cellular environment, the whole organism, the embedding environment, and its evolutionary history, must be grasped as a remarkable non-contiguous and non-spatial process of grand mnemonic superpose, co-joined to the perpetual grand decoherence of Whiteheadian system of future-potentials for subsequent consistent-only future-form, process, development, and even evolutionary potential.

The prospective recuperation of life sciences in these de-spatialised purely temporised growing block terms must engender radical implications to genetics, to molecular genetics, to molecular mechanics, to morphogenetics, to embryology and development... if not to the process of biological evolution itself: incorporating and transcending ultimately pleonastic materio-spatial presumptions about living nature, process, and evolution. This is inevitable, simply because of de-spatialisation and temporisation, to which *all* processes of nature subsume, including biological ones, *forces* the recuperation of life sciences into a pure temporal Bergson-Whitehead framework and its attendant mnemonic processes and grand decoherence.

*

A potentially useful preliminary to the stated goal for a growing block purely temporised life-science paradigm might start with a reconstituted cymatics[89]. **Generic cymatics is the mechanical study of form and pattern obtained in and through vibration;** with pattern typically generated upon a flat metal plate or other surface, subject to vibration. Much-neglected cymatics successfully demonstrates how complex pattern and wholes can seemingly resolve per function of confined vibration, and how complex pattern and form can be sustained thus, wherein the particles constituting the material formed into pattern (often sand, and even toothpaste) are informed by the whole, and *not* the other way round, against what is supposed in reductionist life-sciences and even in subsequent forlorn emergence theory. The particles are random constituents: the whole informs and forms these into pattern, in essentially the same way that the form of the photographic image-pertinent complex quantum mechanical wave and wavefunction ultimately informs and forms the *en masse* quantum indeterminate photons into a coherent whole photographic image.

It is tempting to conjecture that constituents of biology, ranging from atomic elements to macro-molecules, could also be informed and formed by sustained complex 'vibrations' constituted as complex wavefunctions, implying that the secret to living form, its development, and even its evolution, might lie more in the implicit cymatics of organising wholes exemplified or analogised as the

---

[89] Jenny, Hans (July 2001). *Cymatics: A Study of Wave Phenomena & Vibration* (3red.). Macromedia Press. ISBN 978-1-888138-07-8.

437

generic cymatic confined vibration, or as complex wavefunction. This is also consonant to our own Alternative Realist approach to instantaneous wholes (see Book-I **1.03** and **1.26** to **1.27**) and the reform of the physical field per de-spatialisation and temporisation (**4.09**). We can immediately grasp that our solutions to quantum indeterminacy in Book-III and the three-phase development process in the context of abstract minimal states or complex wavefunction, easily segue into the vision intuited in cymatics.

But, to render cymatics useful to our goal, we must de-spatialise and temporise it, such that the 'vibrations-set' upon a metal plate surface, and the plate or the surface itself, are no longer 'distributions in space', but transformed into 'distributions of information in time', organised per delayed choice time-interval relata... concordant to a similar reform of physical fields compelled by the demise of space in essay **4.09**. In short, **we de-spatialise and temporise the 'plate' or 'surface' entailed in generic cymatics,** in the same way we had de-spatialised the 'detector-surface' in the two-slit experiment, and had de-spatialised the 'surface' of a photographic film.

Having de-spatialised and temporised the 'plate' or surface, **we must also quantum mechanise or futurise cymatics by incorporating it to the growing block Whitehead facet**: To this end, we replace the usual and often silly candidates for pattern demonstrators (i.e., 'toothpaste', sand, etc) with quantum mechanical AND-form versions of particle, atomic, molecular, and biomolecular states pertinent to living forms, now transformed into vast complexes of temporal *durata*: i.e., the temporal replacement to spatial *atomos*. In this form, our bio-pertinent durata will constitute a vast complex of Whiteheadian future-potentialities, and, as evolutionary potentials, form up into a system of nested futures that elaborate to the infinite future.

However, in the growing block universe, it is the Bergson facet that attends the grand decoherence of the Whitehead facet of bio-potential futures into consistency, into development, and ultimately into evolutionary paths. The Whiteheadian future of cymatic potentials for life must unfold into consistency per attendant complex wavefunction in memory and the mnemonic superposition of such: i.e., the perduring 'worm' of the living form projected across time, subject to mnemonic superpose; subject to the inverse square of time; and engendering grand decoherence of the Whiteheadian future-potentiality into consistency, which then undergoes global AND-to-OR collapse and perpetual 'materialisation' into realised perduring living pattern via the succession of event horizons.

That is, **we incorporate abstract memory to cymatics and, in the process, supersede cymatics into a new form; solving the key problem of how the 'vibration' pertinent to generic cymatics is sustained and perpetuated in the growing block temporised version of cymatics**... pertinent to the entrenchment and retrenchment of morphological form (as is also true in rudimentary forms in inertia and gravity), and the narrowing of genetic, biochemical, developmental and evolutionary potentials *grand decohered* into consistent futures within the Whitehead facet.

Finally, we consolidate our new cymatics by **incorporating the three-phase developmental process:** Thus, the time-evolution of our growing block new cymatics of life will entail the perpetual unfolding of *en masse* objectively quantum indeterminate processes (Book-III **3.32**) pertinent to the basic bio-constituents of life, according to the developmental regime espoused in the three-phase developmental process on Book-III, **3.26**. These unfold and are funnelled per abstract instantaneous wholes (first espoused in Book-I **1.03**) imposed from the combination or super-superpose of Bergsonian memory *and* the grand decohered Whiteheadian future. Of course, throughout, the ineluctable operation of information-matter dualism and decoupling must abide, as it does in other areas, even if as surface features of the deeper purely temporised growing block universe.

Such a new cymatics, as presage to an anticipated superlative future science of life and to a **growing block quantum biology** of living form, would constitute the basis for revolutionary biology, with inevitable implications to the character of evolutionary processes, insofar as the latter must concede to the existence of abstract memory rendered in pure time, as was intuited by Bergson and the contemporary Rupert Sheldrake, and to genetic and morphogenetic mnemonic temporised processes of non-contiguous information transfer, superseding and exceed the materialistic assumptions of spatial biology inadvertently restricted to the event horizon.

Indeed, what we anticipate for future life science is but the application of *process*, espoused in essay **4.12**.

We are destined to return to this topic in Book-V, within the context of the 'brain in the box' approach espoused as part of the anticipated fine-structure theory of the mind-brain.

4.40: De-spatialisation, the Bergson-Whitehead Amalgam
And the Demise of Academic Philosophy of Mind

Much of Western philosophy constitutes two-millennium worth of footnotes against Plato and against philosophy, or against abstract reality. This is supplemented by three-centuries worth of footnotes against Rene Descartes, or against abstract mind, consolidated into academic philosophy of mind. Modern philosophy of mind constitutes almost nothing but materialist apologetics and polemics against Cartesian dualism; almost totally constituted as forlorn and unworkable materialist contrivances arrayed against dualism.

However, **materialism is forlorn, notwithstanding implications from de-spatialisation and the temporisation of physics. As materialism withers, so must generic academic philosophy of mind**.

The materialist 'philosophy' of mind suffers from a recuperation problem. Much of science can be recuperated even in the face of de-spatialisation and temporisation, in the face of gear-less non-contiguous causality, and even when confronted by the breakdown of closed causality and the growing block framework. For example, the much-abused conservation principle transforms into the conservation and retention of memory combined to consistency-rules vis-à-vis the Whiteheadian future; with the latter obtained via grand decoherence (see Book-II **2.39** for the first full recuperation of the conservation principle per the growing block framework). This culminates into a recuperated form of the conservation principle that cannot be abused by materialist philosophy against Cartesian mind: a recuperated principle of conservation, permissive of Cartesian causality, instantly unravelling the whole enterprise of academic philosophy of mind critically dependent as it is on the materialist supposition and abuse of the conservation principle and other things.

QUANTUM MECHANICS AND MIND

However, given pure temporisation, the notional axiomatic status of 'space' and 'motion' in physics cannot be recuperated. Nor can particle contiguous contact causality, as this is succeeded by the 'quantum leap in time' approach (see Book V **5.30** for the full rendition). Nor can the various materialist contentions about mind-brain relations espoused in and definitive of academic philosophy of mind survive or recuperate in the face of de-spatialisation and temporisation.

*

Causality turns out to be structureless, undefinable and *extant*: i.e., dualistic, and irreducible to purported 'matter in space'. Causality is insinuated to the drive behind time. Without AND-to-OR wavefunction collapse, or time, nothing can transpire. Therefore, time is ontologically primary to the possibility of events from AND-to-OR wavefunction collapse... through which purported 'matter in space' is resolved into being at successive event horizons.

The case against closed causality (the case against the IOO, or input-operation-output schemas) entails that the drive of time, or what causes time, cannot be rendered as a definable clockable mechanism or 'operation' that 'gears' physical inputs (realised events) into outputs of future events, much less collapse and resolve future-potential AND-form states of the Whitehead facet into OR-form *actua*. In short, there are no hidden gears driving wavefunction collapse and time: Any putative gears will crash into insurmountable infinities and halting problems (see Book-II **2.46** to **2.47**...and as implied in Book-III, **3.31** and **3.32**, wherein the search for orchestrating gears vis-à-vis quantum indeterminism prove both unnecessary and untenable, and the resolution of attendant AND-to-OR collapse and time via gears must also crash into infinities and halting problems).

Insofar as causality and the 'drive of time' cannot be reduced to definable or clockable 'gears', it follows that, as with inertia, gravity, and much else, including mind-brain relations (or, parsimoniously, the peculiar causal processes specific to and coincident with brains, which conducts state-change upon brains and generates supposedly 'illusory' agency and nous) cannot be reduced to or encapsulated by definable closed IOO schema-based clockable process, much less be rendered as such into spatially locatable and identifiable hands-on structures and processes of the brain... a conclusion independently justified by the demise of space, which denies the very ontology of 'locations' and 'contiguous interactions', and undercuts the presumption of brains and minds as 'distributions in space', as much as it usurps the very plausibility of mechanistic contact causality in *any* account of causality, including in mind-brain causality, even when nous is declared 'illusory' or presumed reduced to 'nothing but the brain'.

Indeed, it would be very odd if causality in general is found to be gear-less and non-contiguous, dualistic, irreducible, and undefinable, but somehow causality vis-à-vis brains is reducible to and identifiable with... and as... brain-structures and process, or that causality in brains requires contact-mediated 'materio-mental gears', so to speak.

Materialism also collapses when confronted with the Bergson-Whitehead amalgam and, with it, the de-spatialisation and the pure temporisation of physics therein. Facile materialism asserts that 'all is matter in space'. De-spatialisation renders this notion false and void: Nature is comprised of future potentials *and* mnemonic abstractions rendered into pure temporal distributions. Nature is not made up of 'matter in space': Instead, it is made up of memories, potentialities, and of *durata* in lieu of now-defunct *atomos*. There are no *wheres* in nature: there are only *whens*. There is only Bergsonian memory in time, and, beyond it, the frontier of Whiteheadian future possibilities. Obviously, the brain is not going to be exempted from this growing block framework: **The brain cannot *ultimately* be treated as a 'material object in space'**, or as a purported spatio-material state with spatial co-ordinate addresses at which supposed material trace-memories are stored in the form of Karl Lashley's engrams (Preliminary **0.38**), or in the form of Karl Pribram's materio holographic states (Preliminary **0.39**), from where they are supposedly retrieved by an analogue Waldo arm-like mechanism or its neuro-algorithmic equivalent, itself supposedly embedded in and as the material of the brain.

If we must consider the brain, *not* as material corpus of spatial memory-addresses, but as an arrangement of *whens* in time, linked to a preceding system of past *whens* in abstract Bergsonian memories, such as to constitute a perduring 'worm'... combined to a world similarly constituted as abstract *whens* to which the brain must relate... and all of this necessarily embedded within the non-spatial purely temporised Bergson facet... it necessarily follows that the want to find in-brain event horizon-restricted spatial 'address' for both memory and nous, or else reduce both to identity with hands-on spatial structures confined to the event horizon... or as epiphenomenal or emergent upon such presumed spatio-material brains... must remain as hopelessly untenable as forlorn academic philosophy of mind.

All of this explains why the experiments of Rose, Harding *et al* (see Preliminary: **0.40**) were unable to validate the materialist contention of even easy-problem procedural memories, and why the materialist holographic-holonomic contention of the same espoused by Karl Pribram *et al* was also invalidated in the same inadvertent crucial experiments.

It turns out that a true holographic and holonomic theory of memory requires the hologram or holonome of information to be distributed, *not* in 'space' or in 'matter', but purely temporally...in stereotemporal abstract form... *without* contiguity or 'space'. Thus, **a holographic and holonomic approach to memory and mind is viable on the strict condition that it is modelled *without* space, in pure temporal terms, within the growing block framework of the Bergson-Whitehead amalgam. In this way we obtain a non-materialist dualistic holonomic theory of memory and brains, beyond Karl Pribram's materialism-strapped version, but friendly to Cartesian dualism.**

*

Consider the fate of any materialist contention on the nature of mind, memory, brain, and world, or of any materialist hypothesis from academic philosophy of mind (from the Churchland's Thesis through to John Searle's radical-seeming materialist ideas) pitted against a de-spatialised temporised growing block system. The moment we try to recuperate these in terms of the de-spatialised temporised Bergson-Whitehead amalgam, all unravel.

Why?

If nothing else, the non-recuperability and demise of materialism in the face of de-spatialisation and other things instantly leads to the demise of the contrivances of academic philosophy of mind, steeped as these are on forlorn materialism.

*

From our growing block approach, it is obviously that we cannot reduce the whole of world, brains, memory, and nous to the Whitehead facet of the Bergson-Whitehead amalgam: **At best, the abstract future constitutes the not-yet-realised nested system of possibilities for potential world and brains. Clearly, events that have not yet transpired do not constitute memories, much less prove eligible to characterisation as 'stored in brains'.**

If we cannot reduce and identify world, brains, and memories with the Whitehead facet futurity, perhaps we might attempt to reduce these to the OR-form outcomes restricted to the event horizon. While the event horizon is the growing block component that comes closest to our false notions of 'space', and thus constitutes the most recent OR-form states garnered from perpetual AND-to-OR collapse and time, and could be erroneously misconstrued as a 'distribution of matter in space', it turns out that **the event horizon is *not* a spatial plenum on or along which we could outlay the material brain, much less its purported material memories, and even less its retrieval and depository mechanism, and even less treat it as the locus of 'illusory' mind.**

There is no 'space': The event horizon is not a 'space', but it is the relative present continuous boundary or 'cut' that resides between the purely de-spatialised and exclusively temporised Whiteheadian future versus the abstract Bergsonian past. This implies that one part or point on the event horizon as it pertains to brains can only inter-relate to any other part on the same event horizon and brain in a non-contiguous fashion, indirectly, and always in full abidance with the relativity of simultaneity... through a succession of delayed choice time-interval relations that are realised in and through gear-less AND-to-OR collapse or time. Succinctly, there is no contiguous causality proceeding on or along the event horizon, from one part of the realised brain to another, as one imagines in generic notions of 'displacing material transmissions, impacts and interactions across space'. Indeed, de-spatialisation forbids material, mechanical and contiguous memory-formation and retrieval of the form presumed in both brain science and in academic philosophy of mind. Instead, **the delayed choice time-interval relations between events realised on or along successive event horizons are *not* 'horizontal' outlays along any event horizon-formation, but are ineluctably 'vertical' *temporal* relations between succeeding events, and, so conceived, constitute relations organised purely in time, and as relations with abstract memories in time and abstract future potentialities in time... *not* contiguous material relations in 'space'.**

Moreover, the deeper gear-less, undefinable, and structure-less causal process that generates AND-to-OR collapse (time) *and* attends the succession of event horizons and brain state-changes, constitutes a causal process extant the event horizons. Hence, time is extant and dualistic-irreducible vis-à-vis the brains manifested in and through the succession of event horizons. **It follows that the 'illusory' minding entailed in the formation and retrieval of memories itself, fated to whatever drives AND-to-OR collapse and time, and wholly dependent on AND-to-OR collapse and time to realise the expressions of 'illusory' witness and agency in and through events formed on the event horizon, cannot obviously reside on or along the event horizon.**

What about the Bergson facet? Could this constitute the locus of mind and consciousness? Insofar as the Bergson facet is indeed the abstract non-material repository of wavefunctions in memory, the Bergson facet is most certainly the partial memory-repository pertinent to a 'materialised' brain otherwise confined to the succession of event horizons; with memory dissociated from the event horizons and brains, and integrated to the succession of event horizons and brains via non-contiguous temporal relations. It necessarily follows that brain states, or their abstract background wavefunctions in memory, are most certainly within the non-spatial Bergson facet of the amalgam, at least partially. Thus, memory partially resides within the Bergsonian facet and presumably constitutes a mnemonic perduring 'worm' therein.

*

Does the 'illusory' mind reside in the Bergson facet? The answer is no. Bergsonian mnemonic mechanics cannot account for noetic superlative processes, such as moral and aesthetic intuitions, or even the more fundamental apprehension of infinity, or even the apprehension of infinitive, 'to be'. The abstract bases for these cannot arise from or reduce to the world, even when that world is finally grasped as a system of abstract memory organised in time. Indeed, such superlatives supersede Bergsonian memory: They are not in Bergsonian memory: nor is the nous that can apprehend them.

However, there is a deeper reason why nous cannot be reduced to the Bergson facet. The relations within the Bergson facet are relations subject to the relativity of simultaneity. Insofar as relations between brain states are also relations between Bergsonian memory states, these are also necessarily subject to the relativity of simultaneity. Why is this significant? **It is impossible to establish unitary binding of time-disparate time-distributed Bergsonian brain states subject to the relativity of simultaneity within the Bergson facet, and even less so from brains confined to the event horizon. Hence, the minding of the unity of brains cannot be 'located' in the Bergsonian 'worm' that attends the brain, no more than it could be 'located' on the event horizon brain.**

What would the expected unitary binding require? It would require nothing less than a noetic version of absolute time... a noetic version of the 'common present moment' in which the said relations of brain and world arise. That is, the sought unitary binding of the brain (hence mind and consciousness) would require a condition that approaches absolute time, which cannot be furnished by the brain or world, both constituted as distributions in time, and subject to the relativity of simultaneity....*unless the states of the brain subject to and arranged per the relativity of simultaneity arise to and within some other radical ontic: an ontic different from the Bergsonian*

440

brain: one *not* subject to the relativity of simultaneity: one that can impose on the temporally disparate events of the brain and world a mysterious retrospective unity peculiar to the form of unity exhibited by nous.

The temporal brain and world, subject to the relativity of simultaneity arise to nous, which constitutes a peculiar state of absolute time: a unity conceived as a 'common present moment in consciousness'; a temporal noetic moment... a temporal noetic monad that yet preserves within it a world and brains subject to the relativity of simultaneity. Again, there is nothing in the Bergson facet that can generate such a thing. It follows that the radical *something* that exhibits the sought feat (namely, mind and consciousness) must be a thing very different from the Bergsonian domain of memory: necessarily extant from, irreducible to, and dualistic vis-à-vis the Bergsonian domain of memory in time.

Hence, Cartesian mind-brain dualism follows ineluctably from the irreducibility of nous to Whiteheadian futurity; its irreducibility to the event horizon; its irreducibility to the 'materialised' brain therein; and its irreducibility to Bergsonian memory. Hence, the demise of the materialist enterprise against Cartesian mind. Hence, the demise of academic philosophy of mind incapable of recuperation per the new realities that engender the demise of the essential materialism foundational to academic philosophy of mind.

4.41: Descartes' Minor Error: Time, Mind & Noumnemonae

We now posit what might be the final footnote against Descartes, albeit in allegiance to his dualism. *Descartes' minor error* consisted in a dualism stated in terms of abstract non-extended mind versus extended brains; a dualism between *res cogitas* versus *res extensa*. How could non-extended *res cogita* have any causal influence over spatially extended brains, or *res extensa*? The loaded question presupposed the operation of contiguous impact causality between the mind and its brain, supposedly mediated by 'particle interaction' within a closed causality input-operation-output schema (IOO): an approach that presupposes the ontics of 'space' and the viability of IOO-schema closed-causality *and* contiguous causality.

The worn-out challenge against Cartesian duality remains obviously oblivious to prospective de-spatialisation and attendant temporisation, and of the long-overlooked existence of work function zero processes in nature that unfold without the violation of the much-abused principle of conservation; itself recuperated into a form emancipated from materialist misuse and abuse against the ontology of nous.

Descartes' assumption of *res extensa*... of the ontology of extensionality or 'space'... constituted his minor error. As such, it licensed and gave seeming purchase to subsequent materialist critique against dualism. But while the correction of Descartes' minor error rescues dualism, it is fatal to materialism and to subsequent academic philosophy of mind. As stated repeatedly, materialism, and academic philosophy of mind cannot abide without forlorn 'space': Materialism, and the notion of *res extensa*... and, with it, the supposed implausibility of causality between abstract non-extended nous versus 'extended brains'... must collapse... as must materialism itself... as must academic philosophy of mind founded on forlorn materialism.

Simply put, the brain is not a *res extensa*, given de-spatialisation and temporisation. Therefore, the materialist purchase against dualism collapses, as surely as it must itself wither before de-spatialisation and the temporisation of physics and nous (see **4.40**).

We cannot spatially 'locate' either memory or mind in the brain; not in principle, nor in practice... simply because the very notion of 'spatial location' and *res extensa* is bunk. **We can only attempt to specify the temporal relation and moment of mind, consciousness and thought i.e., the *when* of mind and consciousness vis-à-vis the events of the Bergsonian brain and world, in turn grasped as memory in time, *without* 'res extensa'.**

*

Even so, could not mind and thought yet assume identity with, or be treated epiphenomenal to, or as an emergent culmination of, Bergsonian brains and world, even in the absence of space, and even in the face of the unavoidable reframing of brain and world purely in Bergsonian temporal form? The answer is an emphatic *no*.

Succinctly, we must relate the *moment* of unitary and undivided mind, thought and consciousness as a state *temporally removed* from the brain and world. Thus, **brains and world in Bergsonian memory must temporally and subsequently arise to mind and consciousness, while mind and consciousness cannot arise from Bergsonian world in memory. Moreover, mind and consciousness must *always* temporally succeed or come *after* Bergsonian brain and world, and can never be synchronous with the later or reduce to the latter via forlorn synchronicity.** It is from the vantage of temporal dissociation through temporal succession of the noetic moment that mind and consciousness 'looks into the past'... looks into the brain and world held in Bergsonian memory. And, thereafter, it is from just such a succeeding noetic moment that mind imposes unitary apprehension and binding upon the time-disparate Bergsonian brain and world... or incorporates that brain and world into its unitary-form noetic moment or 'common present moment in consciousness'.

*

What is the temporal relation of unitary nous to the past Bergsonian brain and world? We must go further and also address the question of the temporal relation of nous to the Whiteheadian future potentiality of brain and world? Given de-spatialisation (i.e., no *res extensa*) we are permitted to relate the moment of mind and consciousness to memory and potentiality *only* in pure temporal terms. Thus...

- In the **passive mode**, we can relate nous in its witnessing form, as consciousness, as a temporally succeeding moment vis-à-vis the temporally preceding complex of past-events constituted as the Bergsonian domain of world and brains in memory. Therein, **consciousness constitute the critical moment that comes temporally *after* the world and brain have**

transpired. Therefore, nous cannot arise at the same time as world and brains, much less be physically reducible to or 'locatable' in, or as, world and brains. Hence, temporally successive and extant nous vis-à-vis temporally preceding brains and world, or mind-brain dualism as a function of the temporal relation and dissociation of nous vis-a-vis brains and world.

- The more radical possibility is the **active mode**. It is through the active mode that nous, in its causal mind-form, engenders selective agency over the menu of Whiteheadian future-form possibilities pertinent to the future of the brain and world: a process to be fully modelled in Book-V. The noetic selective process over Whiteheadian futures certainly employs a work function zero process and operates without the violation of the much-abused conservation principle, without undoing or erasing memory, and without bringing about futures that are inconsistent with the past. In its active mode, in its selective power, the noetic moment is critically related to, but not physically reduced to, a future-possible brain and world held in Whiteheadian potentiality. Therein, active mode mind temporally *precedes* the future Whiteheadian potentiality for brains and world, and does not arise with or as, nor from, the subsequent AND-to-OR realised brain activity and world. The temporally preceding mind is obviously temporally dualistic and irreducible vis-à-vis both the Whiteheadian future-potentiality for brains and world *and* from the latter's subsequent OR-form realisation. **In its state of time-precedence, active mode mind accomplishes the equivalent of the throw of the suspended delayed choice switch-x, deciding the form in which the Whiteheadian brain-in-potentiality will AND-to-OR decay out of potentiality in concordance to mind's intents. Succinctly, when mind 'throws' its delayed choice switch, it does not collapse the wavefunction attendant the brain, nor does it orchestrate *en masse* quantum indeterminate terms into determinate coherent brain activity. Instead, mind selects the complex wavefunction or 'instantaneous whole' imposed on the future brain, drawn out of the welter of future-potential nested alternatives. Note again that the mind is *not* the subsequent 'brain activity': nor can it arise from instigated brain activity. Active mode mind temporally precedes its brain activity.**

Thus, the mind precedes the brain activity it instigates: The brain activity will arise because of mind, but the mind does not arise from brain activity: In other words, the mind is not the brain. Indeed, outcomes obtained in the experiments of Benjamin Libet *et al* (see Preliminary **0.44**) are consistent with expectations from the stated passive and active mind-brain modes and the inherent dualism therein: a dualism per function of the Bergson-Whitehead amalgam and the de-spatialised temporised approach to the world, to brains, and to mind-brain relations therein. Indeed, any putative reduction and physical identification of mind with the brain would require mind's exact temporal synchronicity with the Bergsonian brain and related world, contradicting the findings from the experiments of Libet.

Note that the key fault in generic materialist interpretations of the results of the Benjamin Libet's experiment consists in misreading the undeniable a-synchronicity observed between the moment of consciousness (supposedly volitional) and brain activity. Ironically, Libet's a-synchronicity turns out to be a consequence of the above-described temporal dualism of mind vis-à-vis the brain and world, per function of the structure and processes of the Bergson-Whitehead amalgam and attendant articulation of brains, world and mind in their de-spatialised and temporised forms.

In the Libet experiment, in its active mode, mind arises *before* future-potential brain activity. In the interregnum before the formation of the succeeding event horizon, mind engages in its peculiar active mode. It is through the subsequent outcomes manifested on the event horizon that mind declares its purported conscious volition, or 'wiggles its finger' to that effect.

In the same experiment, in its passive mode, mind and consciousness arise *after* brain activity is resolved at a succeeding event horizon, wherein the passive nous assumes the role of a pure passive *witness* vis-à-vis the Bergsonian brain and world, and wherein passive mode consciousness apprehends that it *had* voluntarily 'wiggled its finger'. But in neither mode is nous 'locatable' or synchronised as an event manifested in or as the brain, much less as the latter's activity. Therefore, mind and consciousness must always elude *all* physical or temporal identification or synchronisation with *any* event of the brains and world. Therefore, mind cannot arise from the brain, no more than it could do so as an epiphenomenon of brains, and even less as an emergent aspect of the same.

*

When we consider the dualistic extant relation of the noetic moment of unitary mind and consciousness vis-à-vis Bergsonian memory in time, and when we combine this to the dualistic relation of the same to its Whiteheadian future-potential brain and world, this designs for comprehensive dualism. Yet, this is *not* a spatial dualism between *res cogitas* versus *res extensa*, but a temporal dualism of the noetic moment vis-à-vis the temporised past and future: a pure time-based temporal dualism of the mind-brain. And, when we relate all of this to the experimental results obtained by Libet *et al*, it becomes obvious that the noetic moment of consciousness, volitional or not... 'illusory' or not... *cannot* be produced by non-conscious brain-activity or by its 'readiness potentials'.

*

Let us conclude our case for mind-brain dualism by returning to the *binding problem*. We must address the question of how unitary coordination of the brain is possible, and seek to address how, if at all, the brain supposedly generates the unitary sense of the common present moment of consciousness, or the *witness as singular durata*.

We must recall *ad nauseum* that the brain is not an 'object in space': It is not a complex of 'matter' distributed in some impossible common present moment of absolute time. Instead, the temporal brain, like all else in nature, is subject to both external and internal relativity of simultaneity: The brain is a system of disparate past-events organised into relative primacy-recentcy relations: a complex of time-organised Bergsonian system of wavefunctions in memory: a perduring 'worm' within the Bergson facet. And it is to such a brain that we must relate unitary mind and consciousness and the binding problem evinced in the unitary coherence of the brain.

To reiterate, nous is absolute time-like, albeit without any spatial extension. In its passive mode, nous is the recipient of retrospective information about the brain and world held in Bergsonian memory: Nous is the retrospective unitary witness of brain and world, constituted as a noetic moment or durata. Hence, nous is different and distinct from the brain-world by the fact that the brain and world is subject to the relativity of simultaneity, but nous is internally temporally singular, absolute time-like, or unitary and monadic.

Thus, while the brain, world and universe have 'parts' constituted as different events from different moments in time, all subject to the relativity of simultaneity... in the case of the noetic moment of mind and consciousness, **one 'part' of the witnessing moment is *not* subject to the relativity of simultaneity vis-à-vis some other 'part' of the same. Indeed, witnessing nous has no 'parts', given that it is not constituted as *res extensa*, nor as a temporal distribution of disparate events. Indeed, in its noetic moment, the witness is equivalent to a spatial 'point' with no internal structure: Succinctly, nous is constituted as a temporal monad; as a singular noetic *durata*,** with absolute time-like characteristics.

One's perception of the environment as a visual, tactile, and olfactory complex of past events (as Bergsonian memories in time) arises to one's witnessing consciousness retrospectively and as a unity, even though the origins of the sense-data belong to a past distribution of disparate events inter-relationally subject to the relativity of simultaneity. Yet, **past events subject to the relativity of simultaneity converge to a succeeding noetic moment or monad... arise *into* unitary-singular experience and apprehension... at which the memories, otherwise internally and demonstrably subject to the relativity of simultaneity, and lacking in any inherent binding or unity in the sense of an encompassing instantaneous whole, are brought into a common present moment of consciousness, one constituted as a noetic absolute time-like instantaneous monad and unity**.

Let us relate our argument to the experimental findings of Penfield and Cameron *et al*. Therein, even though the complex events that constitute the brain are distributed into imperceptible delayed choice time-interval relations, this does not obviate the fact that absolute time does not apply to the brain. Brain-states are *not* rendered into unity in some impossible common present moment, or in a 'space' imposed from within the brain. Thus, **brain states, and even less the whole brain, remain incapable of constituting unitary perception from the vantage of any of its 'parts', given the relativity of simultaneity prohibitive of such a feat. Any attempt to found a brain-based unitary mind and consciousness must fail, precisely because absolute time does not apply to the brain and cannot be foisted on mind and consciousness by that brain**. This explains why Penfield *et al* could not find any brain-based basis for unitary experience and nous, much less any basis for the unitary orchestration of brains (see Preliminary: **0.46**). It also explains why the notorious Edward Cameron, with his destructive methodology, could irreversibly disrupt the contingent structures of the brain so as to permanently destroy the expression (hence effective recollection) of biographic memories and personal identity, but could not destroy the ever-elusive unitary "I": i.e., the unitary content-less self, with all its peculiar absolute time-like characteristics, and which resides beyond the capacity of the brain to furnish... and must therefore temporally reside beyond the brain itself: (See Preliminary: **0.46** through to **0.48**, including anomalous results from split-brain patients).

Hence, again, mind-brain dualism ineluctably follows.

*

In the Preliminary, and culminating into Book-I, we began with the peculiar claim that information and memory remain dissociable from matter, radiation, and space. We then superseded this view in our subsequent findings to the effect that memory is information organised in and by time, *without* 'space'. This discovery was implicit to the spacetime model articulated in Book-II, but it was rendered explicit by de-spatialisation, temporisation, and the attendant Bergson-Whitehead amalgam successor to 'spacetime'.

The brain and world belong to the Bergsonian mnemonic past distributed in time, subject to the relativity of simultaneity, and bereft of absolute time-like characteristics. Yet, while the world is retrospectively grasped by nous as non-conscious past states subject to the relativity of simultaneity, these are temporally convergent and integrated to a state of unitary consciousness that *does* exhibiting internal absolute time-like characteristics. From this follows an additional form of memory to which Bergsonian memory of world and brain must be subsumed; a form of memory encapsulated into consciousness with absolute time-like characteristics: This is **noumnemonae**, or *memory in consciousness*, or simply *consciousness-memory*... incorporative of, yet distinct and different from, Bergsonian a-noetic mnemonic content.

Hence, we arrive at a dualism rendered in terms mnemonic categories: Thus:

- **Noumnemonae: the noetic moment characterised as an absolute time-like state** or 'the present moment in consciousness': It subsumes 'memory in consciousness', distinct from Bergsonian non-conscious memory constitutive of the non-mental world and brain.

- **Mnemonae, subject to the relativity of simultaneity, *is* the Bergsonian world and brain in temporised form... hence constituted as abstract memory. It is non-mental or without consciousness (a-noetic) but later arises into mind and consciousness** as 'noumnemonae'... as memory (or world) in consciousness.

Noumnemonae always emerges *after* the formation of Bergsonian memory... *after* the formation of *mnemonae*... in and through the subsummation of the latter to consciousness. Thus, noumnemonae is temporally extant to and dualistic vis-à-vis subsumed mnemonae.

Again, mind-brain dualism follows.

443

4.42: Summary of Part-IV

The following constitutes Part-IV in summary.

The final form and completion of information-matter dualism (4.37) emerges out of the following:

- Whiteheadian future-potential information of nested AND-form states, or *wavefunctions in potentiality* embedded within the Whitehead facet of the amalgam.

- Event horizon recentcy states, comprised of the most recent OR-form 'materialisations' on and along the event horizon. These conform to our naïve conceptions of 'matter in space': i.e., pure OR-form types of information with which we tend to confuse the whole of physical reality.

- Bergsonian mnemonic information, comprised of wavefunction-like states with nested futures cropped: i.e., *wavefunctions in memory*: This is the retention of the past in abstract form 'below' the event horizon. Memories are subject to mnemonic mechanics and primacy-recentcy rules, including the inverse square of time. Memories underpin inertial and gravitational phenomena.

- De-spatialisation forces the recapitulation of information as memory in pure temporal form: Information as a distribution in time. It is just such a distribution in time that is dissociated from the event horizon and from the Whitehead facet, culminating into a final-form information-matter dualism per the de-spatialised temporised tripartite structure of the Bergson-Whitehead amalgam.

Bergsonian memory is dissociated from the event horizon *and* from the Whitehead facet of future wavefunctions in potentiality. This furnishes the form of information-matter dualism evinced in the Preliminary, and in Book-I and Book-II, albeit recuperated in terms of the superlative Bergson-Whitehead amalgam, and from the de-spatialisation and temporisation that undermines the 'space' across which information is purportedly 'carried' by putative radiation and matter. Thus, we arrive at the final form and completion information-matter dualism.

Completion of decoupling theory provoked by outcomes from the experiments of G. Nimtz (4.38): De-spatialisation, the Bergson Whitehead amalgam, and the culmination of realised information into identity with Bergsonian wavefunctions in memory, demands the recapitulation and completion of decoupling theory from Book-I and Book-II. To this end, information, instanced as Mozart-40, is grasped as a complex wavefunctions in memory, retained within the Bergsonian facet, and necessarily dissociated or decoupled from both the event horizon *and* from the AND-form wave belonging to the Whitehead facet. The explanatory maincone approach from Book-II is superseded by the Bergsonian inverted cone, within which M40 is distributed in de-spatialised temporised form. Moreover, insofar as there is no space, there is no 'transmission of M40 across space', whether at the speed of light or by quantum tunnelled faster than light radiation in the experiments of Nimtz, or at any 'speed'. Hence, de-spatialisation, temporisation and the Bergson-Whitehead amalgam fashion for decoupling theory in its final form.

New Cymatics and the presage to radical biology (4.39): The demise of space and prospective Bergsonian mnemonic mechanics, combined to final-form information-matter dualism and decoupling, and per Whiteheadian grand decoherence, anticipates a purely temporised form of cymatics: the science of form per function of vibration: a science that could incorporate a new life-science and a growing block quantum biology. New cymatics is anticipated thus:

- **Temporisation of the plate**: To render cymatics useful to our goal, we must de-spatialise and temporise the 'plate' or 'surface' entailed in generic cymatics.

- **Incorporation of the ontic future: the Whitehead facet.** We quantum mechanise or futurise cymatics by incorporating it to the growing block Whitehead facet, and by recuperating the usual spatial particle, atomic, molecular, and biomolecular states pertinent to living forms into vast complexes of *durata*: i.e., the temporal replacement to spatial *atomos*. Thus, constituting a vast complex of Whiteheadian nested futures for genetic, molecular cellular, developmental, and even evolutionary future potentials.

- **Incorporation of the abstract past: the Bergson facet.** We incorporate abstract memory or the Bergsonian 'worm' to cymatics and, in the process, supersede cymatics into its new form, solving how the 'vibration' pertinent to generic cymatics is partly sustained and perpetuated in the growing block universe per function of the abstract Bergsonian past and its attendant grand decoherence imposed upon the Whitehead facet, rendering the latter to consistency and entelechy.

- **Incorporation of the three-phase developmental process** from Book-III: Finally, the time-evolution of our growing block new cymatics entails the perpetual unfolding of ultimate *en masse* objectively quantum indeterminate processes (Book III **3.32**) pertinent the said bio-constituents, according to the regime espoused in the three-phase developmental process (Book III **3.26**)… with quantum indeterminate terms always unfolding and funnelled per the abstract instantaneous whole attendant genetic, morphogenetic, developmental, and evolutionary potentialities, in turn enforced by the combination of Bergsonian memory with consistency-rendered grand decohered Whitehead futures.

Revival of Cartesian dualism and the demise of academic philosophy of mind (4.40): The revival of Cartesian mind-brain dualism arises ineluctably from de-spatialisation, temporisation, and from the Bergson-Whitehead amalgam. Any future model of the mind and brain must forgo the notion of spatial 'memory-addresses', while mind and consciousness (even if somehow proven 'illusory') can no longer be construed as states generated by brains. All of this permanently undermines academic philosophy of mind, which is comprised almost totally of materialist apologetics arrayed against Cartesian dualism. As such, the philosophy of mind is tied to the fate of materialism. The latter's demise in the face of de-spatialisation, temporisation and the growing block Bergson-Whitehead amalgam implies the tandem demise of academic philosophy of mind. In detail...

- The breakdown of input-operation-output (IOO) schemas undercut the materialist want of closed causality requisite to the materialist case against mind-brain dualism. Closed causality requires the definability of the 'operation' term in IOO schema causality. But *the case against hidden gears behind wavefunction collapse and the non-clockable ineffable drive behind time, permanently undercuts the possibility of any definable 'operation' to causality, both generally and in brains, the presumption of which would crash into insurmountable infinities and halting problem*s. It follows that one of the central postulates against Cartesian mind-brain dualism is undercut.

- Independent of the demise of IOO schemas and the unravelling of 'hidden gears', *de-spatialisation and temporisation dissolves the want to reduce memories into spatial distributions in brains, while attendant temporisation forces the temporal distribution of key aspects of memories that now extend beyond the materialised brain confined to the event horizon, into abstractions temporally dissociated into and retained within the Bergson facet*... wherein memories can no longer constitute *wheres* within the brain, and the notion that the brain itself constitutes a 'distribution in space' is undercut. Indeed, there is no space. Therefore, there are no 'locations' in the brain at which purported memories are addressed or 'stored'. De-spatialisation and temporisation supersedes the materialistic memory theories of both Karl Lashley's engrams and Karl Pribram's holonomic theory, recuperating the latter in pure temporal form within the Bergson-facet.

- *Dualism emerges from the tripartite structure of the Bergson-Whitehead amalgam, which unravels any attempt to reduce memory to brains.* The resolved brain is obviously *not* reducible or physically identifiable with the Whitehead facet of futures: the future has not yet happened. Nor can the brain (and memories) be reduced and restricted to the event horizon or rendered wholly 'material' or spatial upon it, given the de-spatialised purely temporised 'vertical' organisation of events across succeeding event horizons, all subject to the relativity of simultaneity. In any case, memories reside within the Bergson facet 'below' the event horizon and are obviously not reducible or confinable to the event horizon. It follows that the manifest brain in laboratory experiments, including in the experiments of Libet, does not constitute the full ontology of the brain itself, since most of it is rendered as an abstract 'worm' within the Bergson facet, dissociated from the recentcy of OR-form states scrutinised in experiment. Hence, Cartesian memory-brain dualism.

- *Insofar as memory is dissociated from the manifest brain restricted to the event horizon, so is the process and agency of its retrieval and recollection*, which must somehow reach 'outside' of the brain and 'below' it into the Bergson facet, where key aspects of memory reside. The retrieval and recollection process must also be recapitulated as a pure temporal process, obviating the forlorn hope for spatio-material retrieval and recollection supposedly constituted as brain structure confined to the event horizon. In other words, insofar as mind and consciousness is deeply implicated vis-à-vis the process of memory retrieval and recollection, it follows that mind and consciousness, even if 'illusory', must be as dissociated and dualistic in relation to the brain as is the dualistically dissociated purely temporal process of retrieval and recollective power. Hence, the revival of Cartesian dualism and the demise of materialist academic philosophy of mind.

- *Insofar as the principle of conservation is recapitulated into a de-spatialised and temporised form that permits Cartesian causality, the principal materialist tenet that licences academic philosophy of mind to contrive forlorn materialistic theories against dualism on the supposed veracity and want to save the principle of conservation immediately unravels, as does almost the whole of philosophy of mind.* In Book-II **2.39**, the principle of conservation is recuperated into the conservation and non-eradicability of memory, co-joined to consistent futures via the grand decoherence of the future... of the Whitehead facet... per function of the conserved Bergson facet. Mind-brain Cartesian causality cannot alter the content of the past or usurp the conservation of memory. Nor could mind-brain causality select for or bring about inconsistent futures vis-à-vis the conserved past. In short, recuperated conservation principle is perfectly consistent with and abides in the face of Cartesian dualism. Hence, wither academic philosophy of mind critically dependent on the trope and abuse of the conservation principle against Cartesian dualism.

Descartes' Minor Error (4.41) is Rene Descartes' forlorn commitment to *res extensa*, or space, fashioning a dualism of *res cogitas* versus *res extensa*. But how could non-extended immaterial mind have any causal influence and power over extended material brains? The question unravels the moment we acknowledge de-spatialisation and the demise of *res extensa*; combined to the reality of work function zero processes in nature through which change and transformation is brought about, *without* the violation of the principle of conservation: (Also see Preliminaries, **0.32** to **0.33**). Simply put, the brain is not *res extensa*, given de-spatialisation, and this obviates problems of Cartesian causality and saves mind-brain dualism. The consequences of unravelling *res extensa* are...

- *Memory cannot have a 'location' in the brain, simply because there is no space, and, therefore, no 'location' for memory.* Abstract memory is retained in the Bergson facet and temporally relates to brains, albeit in temporally dissociated form vis-à-vis manifest brains.

- *Mind cannot be 'produced' by the brain (even as an 'illusion'), simply because we cannot spatially 'locate' mind (or its 'illusion') in the brain,* neither in principle nor in practice. Even if we insist that nous is illusory, the illusion remain dualistically temporally dissociated from brains, and cannot arise from brains.

- *We can only state the moment of mind, consciousness and thought (i.e., the 'when' of mind consciousness and thought, or the noetic moment) vis-à-vis the non-spatial and purely temporal Bergsonian brain and world.*

- *The moment we rid of 'res extensa', we rid of the central problem posed by 'res extensa' in Cartesian dualism,* given that, in general, in a de-spatialised and temporised universe, causality is *never* mediated contiguously or across 'space': Mind-brain causality is not going to be the exemption.

Passive and active modes in mind-brain relations (4.40 and **4.41):** Given that there is no space… no *res extensa*… we can only relate the noetic moment of consciousness and mind to memory and world in pure temporised terms, in two modes:

- In the **passive mode**, we relate nous as a temporally succeeding noetic moment and witness versus the past events retained in the Bergsonian domain. Consciousness, or the witnessing moment, arises temporally *after* the world and brain have transpired. Therefore, the past Bergsonian brain and world must temporally arise to a succeeding moment of nous that emerges temporally *after* world and brain have transpired, while consciousness passively witnesses the memory of the events constituting that world and brain in retrospect, *without* temporally reducing to the events and memories of that world and brain.

- Is the **active mode**, the nous as mind engenders its causal agency upon brain and world through the decay or wavefunction collapse of the potential future, in the process of coming into being out of AND-to-OR collapse (time) from out of the Whitehead facet. The key details about mind-brain causality must await Book-IV. In anticipation, in the active mode, the mind is critically related to a future-possible brain and world rendered as an AND-form future-potentiality. Temporally, the active mind *precedes* that future potentiality for brain and world. Therefore, it is *not* 'locatable' or reducible to that future potential brain and world, which, in any case, has not yet transpired. In its relation to its potential future brain-state, mind accomplishes the equivalent of the throw of a suspended delayed choice switch-x, deciding the way future-potentiality decays out of the Whiteheadian future into the sought realised brain-state and world… so manifesting the will of mind over its brain and world. This is accomplished on a work function zero basis ubiquitous throughout nature; hence, without violation of the much-abused conservation principle.

The noetic moment (4.41): The temporal relation of consciousness and mind to the Bergsonian world, brain and memory, as well as to future-potential Whiteheadian brain and world, is constituted as a noetic moment, distinguished into passive and active modes of *durata*. Yet, the noetic moment cannot be registered as any event of the Bergsonian brain, no more than it could be specified as a Whiteheadian future-potential event. Indeed, the noetic moment defies identification even with the Bergsonian past, or with any moment therein, since the latter arises retrospectively to and within the noetic moment, even if not from that noetic moment. Thus, mind and consciousness, constituted as the noetic moment, must always remain *extant* the brain and world (extant the Bergson-Whitehead amalgam itself) by virtue of arising before, after, and *from between* successive states of the brain and world, but never at the same time as, and *never as*, the brain and world. A new Cartesian dualism without *res extensa* necessarily and ineluctably follows.

Noumnemonae, or *memory in consciousness* (4.41): While the combination of world and brains is constituted as Bergsonian memory distributed in time, the succeeding noetic moment that apprehends the brain and world constitutes a remarkable monadic absolute time-like moment. The noetic moment of mind and consciousness is not internally divided into different components or past events organised in time: One 'part' of the noetic moment is not related to another 'part' of the same per the relativity of simultaneity. To itself and in itself, mind and consciousness constitutes the equivalent of a single 'spatial point': a state of absolute time, but *without* space… even while the world and brains in Bergsonian memory that arises to the noetic moment of absolute time-like nous *is* comprised of 'parts', or of events scattered in time *and* subject to the relativity of simultaneity. Thus, world and brains in memory, but *not* in absolute time, arises in and into a noetic moment constituted as a temporal monadic 'point' or durata *in* absolute time, constituted as a common present moment in consciousness. Hence, this culminates into two types of memory and dualism:

- *Noumnemonae:* the noetic moment characterised as an absolute time-like state: with the world and brain arising in and to it to become 'memory in consciousness', or 'consciousness-memory', distinct from the time-distributed non-conscious Bergsonian memory constitutive of world and brains.

- *Mnemonae,* subject to the relativity of simultaneity, constituting the Bergsonian world and brain in abstract memory, without consciousness, but later arising to and within the noetic moment of retrospectively witnessing consciousness.

BOOK-IV: DISCOVERIES AND CONCLUSIONS

Two key conclusions were secured in Part-I: namely…

- **De-spatialisation: space as pleonastic, and the removal of space from the ontology of physics:** This was based on…
  - **…the impossibility of spatial contiguity and extensionality through simultaneity** in the face of the relativity of simultaneity, and the consequent rendition of physics as a pure distribution of events in time, or temporisation: (**4.01**).

○ **Space is synonymous with absolute time.** Therefore, the demise of absolute time in the face of the relativity of simultaneity (Special Relativity, in 1905) should have led to de-spatialisation and the comprehensive temporisation of physics. Instead, space was incorporated into 'spacetime': (**4.01 and 4.02**... from the analysis of light).

○ De-spatialisation also arose from the fact that, while **time is ontologically indispensable, space is entirely dispensable**: Without time, not even the succession of events falsely interpreted as 'the motion of an object in space' could transpire. On the other hand, we can recuperate the same events temporally... as organised purely in terms of delayed choice time interval relations, *without* space, rendering the notion of space entirely superfluous: (**4.03**).

○ **Quantum mechanics prohibits continuous observation of events. But such continuous observation is necessary to establish the veracity of space**: (**4.04**). Consequently, we cannot prove continuous motion. Therefore, we cannot prove the 'space' presupposed to it. The naive supposition that the quantum wave is a spatially dynamic state undergoing 'motion' is an assumption about the quantum wave and, succinctly, about the ontology of space: Neither can be proven empirically or even in principle, according to basic quantum mechanics... if only because the attempt to observe the seamless-continuous motion of the quantum wave will destroy the effect we want to observe, while observables constitute successions of resolved past events. Hence, the notion of the motion of the wave is an *assumption* at best. In any case, even with a Quantum Zeno approach to observation, we cannot eliminate the AND-form intervals of isolation inherent to the purported motion: intervals about which we can only *assume* the motion of the quantum wave, but cannot prove it empirically or observationally.

○ De-spatialisation also emerged from a critique of Hilbert space: **At best, Hilbert space constitutes a future-potentiality for space** that has not yet come into realisation. Obviously, **nothing can move through a 'space' that has not yet come into realisation**: (**4.05**).

○ Our final and conclusive argument arose from **John Wheeler's delayed choice approach to, and in invalidation of motion and space, respectively (4.07)**. In Book-II (**2.23**), the delayed choice approach established the quantum wave as spatially static: a non-moving and non-time undulating state: i.e., no motion. Therefore, no need for 'space' presupposed to such 'motion'. However, in Book-IV **4.07** we took our method to the next level by treating the hedging nesting wave from the delayed choice experiment as itself nested to a preceding meta-delayed choice hedging wave: an arrangement that can be repeated and elaborate to the indefinite past, or at least to the putative Big Bang. From this meta-delayed choice approach, we discovered that only just after the putative Big Bang might one entertain a notional non-provable 'space' through which a notional first-ever hedging nesting wave might have undergone 'motion', but *not* thereafter: with both motion and space rendered permanently superfluous and pleonastic thus.

Certain consequences followed from de-spatialisation: The **reform of Special Relativity (4.08)** purely in terms of delayed choice time interval states, *without* 'space'... and the replacement of 'spacetime'. Information rendered purely temporal constitutes the seamless case for abstract memory, completing the central goal of Book-III per **Apagogic Mnemonic Argument 5-a: (4.11)**. De-spatialisation also forced the reform of the physical field (such as the magnetic field or quantum fields) into temporised or time-distributed forms. Hence, the 'field in space' was recuperated in to a pure temporised form, *without* space... culminating into **the reform of the physical field (4.09)**. Of course, de-spatialisation also culminated into the replacement of the intermediate model spacetime from Book-II: Thus...

• **The Bergson-Whitehead amalgam**: i.e., spacetime *without* space: (**4.10**). The tripartite structure from the intermediate model survived into the amalgam, as did the primitive unification of physics from Book-II (**2.05**). Thus, spacetime in potentiality was replaced by the Whitehead facet: Realised spacetime was replaced by the mnemonic Bergson facet: The event horizon was retained as the 'cut' between the two stated facets, but recapitulated in purely temporal form, *without* space. Certain attendant and additional consequences followed:

○ The **stereotemporality of time: or three-dimensionality of time (4.12)**, in which all events are interrelated and integrated per pure three-dimensionalised or stereo-temporal delayed choice time interval relata: From Book-II, time is not a 'fourth dimension': Instead, time is the process of AND-to-OR wavefunction collapse of the Whitehead facet. Thus AND-to-OR collapse (hence time) unfolds stereotemporally into stereotemporally distributed realised events.

○ **New Process Philosophy (4.12)** emerged from both de-spatialisation, temporisation and stereotemporality. Our new process philosophy vindicated and honoured the contributions of Alfred North Whitehead and Henri Bergson, but corrected for minor omissions: In the context of de-spatialisation and the Bergson-Whitehead amalgam, *process*, and the evidencing of both *endurance* (of physical law or timeless principle) and *perdurance* (the accumulating growing block identity undergoing continuity *and* change) turns out to be fated to the very AND-to-OR collapse-process (i.e., time) of the universe scale wavefunction (the total Whitehead facet), with tandem progressions of the event horizon, accompanied by the consequent formations of new memories; leading to the expansion of the Bergson facet (perdurance and the growing block of accumulating memories: i.e., the Bergsonian 'worm') per the perpetual decay of future potentials out of the Whitehead facet... all *without* space; and wherein time and the process of 'coming into being' constitute the central dynamic.

Key achievements of Part-II: With de-spatialisation from Part-I, Part-II recapitulated the reform of energy concepts first initiated in Book II, but within the framework of the Bergson-Whitehead amalgam. This dovetailed the following...

- *Wavefunctions in future-potentiality* and ***wavefunctions in memory* as two key forms of abstract information**: Per the framework of the non-spatial purely temporal Bergson-Whitehead amalgam, the Whitehead facet is comprised of *wavefunctions in future-potentiality*, while the Bergson facet is constituted by *wavefunctions in memory* (the past comprised as abstract or temporised memory). With de-spatialisation, neither facet could be portrayed as a spatial distribution, but only as temporal and stereotemporal distributions.

- **Recapitulation of the reform of entropy in the framework of the Bergson-Whitehead amalgam (4.18)**. Given the centrality of time in the ontology of wavefunctions and their decay, Part-II augmented the reform of entropy relations presaged in Book-I; recapitulating that the 'arrow of time' proceeds from AND-to-OR: i.e., from future potentiality (the decay of the Whitehead facet) to recentcy states via the succession of event horizons. Time can never proceed from OR-to-AND: i.e., there can be no such thing as 'time reversal'. The critical implication was that any 'statistical reversal' in the tendency of generic entropy to increase could not constitute an objective basis for 'time-reversal', given that the reversal in the statistical tendency could only be brought about by ontologically primary AND-to-OR collapse, *not* through time-reversal in OR-to-AND 'de-collapse'. This key insight led to the following:
  - **Entropy of future-potential information** (or the validation of the original thesis presaged in Book-I): Whenever the process of AND-to-OR collapse and time unfolds, the total potential information of the universe proceeds from a lower state of future-potential entropy (higher Whiteheadian future-potential information) to a higher state of future-potential informational entropy (lowered or reduced Whiteheadian future-potential information).
  - **Entropy of memory:** On the other hand, while AND-to-OR collapse increases the memory content of the Bergson facet of the amalgam, the total increase in Bergsonian memory is subject to the inverse square of time. Consequently, memory accumulates, but weakens or fades per the inverse square of time, constituting the increase in the *entropy of memory* per function of the inverse square of time.

- **Recapitulation of the reform of kinetic and potential energy relations in the context of the Bergson-Whitehead amalgam: (4.19)**. Completing a reform started in Book-II 2.39… There is no space. Consequently, potential energy does not issue out into kinetic energy from spatial positions or locations. Thus, energy itself has no 'location' or 'spatial address'. Instead, potential energy necessarily issues out of the Whitehead facet... from future-potential energy. On the other hand, kinetic energy is that part of future-potential energy that manifest in and through the transition of one event horizon to a successor event horizon, entailing the conversion of Whiteheadian future-potential energy into realised kinetic energy (putative work or action) via attendant AND-to-OR collapse, via time itself.
  - **Ontology of pure information physics, and energy as pleonastic.** Kinetic energy is that part of future-potential energy that manifest in and through the transition of one event horizon to a successor event horizon. This is brought about by ontologically primary AND-to-OR collapse or time. But time is not itself brought about by *any* energy-work inputs or hidden gears. Thus, AND-to-OR collapse and time arises from, and is driven by, a gear-less causality characterised by a work function zero profile. Thus, putative energy relations are brought about by non-energy work function zero gearless time. Thus, it is not 'energy' or work that produces change or 'action'. Instead, it is work function zero time that produces events and outcomes that we then interpret as 'energy' or 'work', supposedly causative of change. Change is brought about by work function zero time: i.e., the real face of causality. Hence, at the ultimate ontology-level, we obtain pure information physics, wherein 'energy' considerations totally abstract out of the ontology of physics and are grasped as pleonastic. i.e., superfluous to and not part of the real ontology of physics.

- **The universe is not a free lunch (4.20)**: The moment we re-define pleonastic potential energy as *not* stored in or issuing out of 'matter in space' but out of the Whiteheadian future through a gearless pure-information work function zero time-process, it follows that 'matter', as supposed 'congealed energy', must be treated the same way: i.e., as future-potential matter, synonymous with future-potential energy. Thus, we posit that, at the Big Bang, 'matter' collapsed out of future-potential matter potentialised within the Whitehead facet, in turn part of the Big bang monoblock. Consequently, the universe was *not* a 'free lunch', in that matter was *not* created 'out of nothing', and certainly *not* by any violation of the principle of conservation. But this conjecture required the application of Heisenberg energy-time uncertainty relation, given that, with the demise of 'space', we needed a replacement to Heisenberg's position-momentum uncertainty with a pure temporised version of the same, *without* space… notwithstanding the pleonastic status of energy.

- **Heisenberg's uncertainty as exclusive energy-time uncertainty (4.21)**: Given that space is bunk, both position and the velocity component of momentum in generic uncertainty relations can easily be replaced by a pure energy-time uncertainty relation. Conveniently, the relation is well known and established in generic quantum mechanics and physics… notwithstanding the pleonastic status of energy.

- **Post Big Bang collapse of future-potential matter into realised matter (4.21)**: We posited a Big Bang and post-Big Bang 'matter creation mechanism', which could generate realised matter out of Whiteheadian future-potential matter, all without violation of the conservation principle, given the pre-existence of that matter as future-potential matter within the Whitehead facet, and that it is not created out of nothing thus. As time-intervals are shortened, potential energy or matter-possibilities increase per Heisenberg's energy-time uncertainty. But these are extreme short-time interval potentialities. Therefore, the probability that the future-potential matter within the time-interval might issue out and generate counterfactuals and grand decoherence to the Whitehead facet remains extremely improbable. However, when we incorporate profound time dilation, such as found on or around asymptotic supermassive 'black holes', the short time-interval is profoundly dilated, transformed into a larger temporal cross-section for incoming radiation and matter, which drastically increases the probabilities for the AND-to-OR collapse of future-potential matter into realised matter out of the Whitehead facet, and the generation of counterfactuals pertinent to grand decoherence through which new matter acquires inertia or 'mass'. Hence, conjectural post Big-Bang matter-creation at supermassive objects or asymptotic 'black holes'. We also conjectured a **matter creation flywheel approach (4.22)** that could test the feasibility of the scheme.

In Part-III, de-spatialisation posited the clear assertion that 'interactions' in physics are ultimately between informational states resonated and superposed across time, *without* spatial contiguity or contact... as non-contiguous 'interactions' between memories, *not* between 'matter in space'. Hence, the necessity for a presage to mnemonic mechanics to supersede notions of 'matter in space', and to presage memory-based gravitation (a precondition to future quantum gravity) in the framework of the de-spatialised temporised Bergson Whitehead amalgam. As such, mnemonic mechanics was shown to require five postulates (**4.25**) also essential for prospective quantum gravity:

- The **independent retention postulate:** the retention of past states as Bergsonian wavefunctions in memory.

- The postulate for the **superpose of past states across time**, involving the addition of primacy and recentcy wavefunctions in memories from within the Bergson facet of the Bergson-Whitehead amalgam.

- The postulate for **the inverse square law of time,** which affects the weakening of wavefunctions in memory by time, *not* per pleonastic distance.

- The postulate for **constructive interference in the mutual superpose and addition of wavefunctions in memory**. In other words, unlike in the addition of generic quantum waves, there can be no destructive interference from the superposition of wavefunctions in memory with other wavefunctions in memory... although destructive interference can arise in the superposition of memories with Whiteheadian wavefunctions in potentiality, albeit to the destruction or cancellation of Whiteheadian potentials, such as evinced in grand decoherence.

- Finally, the postulate for the **inapplicability of grand decoherence to Bergsonian wavefunctions in memory**. That is, memory cannot be erased or destroyed (replacement of the old conservation principle). Hence, grand decoherence can only apply to the Whitehead facet of non-consistent futures and to wavefunctions in future-potentiality.

All five postulates were framed to the Bergson-Whitehead amalgam. Following a brief interlude into the similarities and differences between the superposition of mechanical and quantum mechanical waves, and of wavefunctions in memory, we hammered out the following:

- **The Bergsonian mnemonic mechanics of inertia (4.28 and 4.29)**, or inertia per the addition of wavefunctions in memories pertaining to just one 'body'; wherein the past memory-states associated with that body entrenches and retrenches the form in which the body is AND-to-OR resolved across successive event horizons, with tandem grand decoherence of the Whitehead facet to consistency with implications from memory, constituting the Whiteheadian aspect to resistance of that body to change.

- **The Bergsonian mnemonic mechanics of gravitation (4.29, 4.30 and 4.31)** in which the wavefunction in memory attached to one 'body', across time and subject to the inverse square of time, superposes with the culminating superpose from the stack of memories generated by and attached to the other 'body'; bringing about the gravitational attraction of the latter to the former; with implications to the Whitehead facet per grand decoherence into consistency, inscribing into the Whitehead facet **work function zero orbital "paths"** which the bodies subsequently resolve into via AND-to-OR collapse and time, and in and through successive event horizons. Thus, gravitation is a memory effect. Thus, we obtain the presage to quantum gravitation per the broken-discontinuous (i.e., quantised) manner in which the work function zero orbital paths and mutual gravity are AND-to-OR resolved from event horizon to succeeding event horizon; all realised on a work function zero basis, *without* the violation of the conservation principle.

- Finally, we discovered that plausible implications from Bergsonian mnemonic mechanics implies two things...
  - **The limit against the formation of singularities (4.32)** per the logic that the superposition of wavefunctions in memory are incapable of generating sums of probability amplitudes equal to 1... incapable of generating singularities... although supermassive objects, as asymptotic 'black holes', approaching but never obtaining their Schwarzschild radii, are likely possible and appear to be evident.
  - **The impossibility of singularities is also independently affirmed by de-spatialisation and temporisation**: There is no space. Therefore, there is no spatial point to which mass could collapse. Therefore, in a de-spatialised and temporised order, singularities or 'black holes' are not permissible, unless constituted as asymptotic supermasses.

449

○ **Alternative explanations for both dark matter and 'dark energy'** must emerge (**4.33** to **4.34**) without the requirement of any literal dark matter or dark energy.

Part-IV tied up a few loose ends, but principally culminated into the revival of Cartesian mind-brain dualism:

- **Completion of information-matter dualism per the Bergson-Whitehead amalgam and de-spatialisation (4.37):** Information is not 'distributed in space' but is configured as immaterial abstractions distributed in time within the Bergson facet, dissociated from the 'materialised' OR-form states confined to the event horizon, *and* dissociated from *wavefunctions in potentiality* (i.e., quantum mechanical radiation and de Broglie matter) confined to the Whitehead facet.

- **Completion of decoupling theory (4.38):** Given de-spatialisation, information is not transported across space at *any* speed, whether quantum tunnelled to faster than light, at the speed of light, or 'slower'. Thus, wavefunctions in memory, such as Nimtz' Mozart-40, retained within the Bergson facet, are not 'transported' by Whiteheadian wavefunctions in potentiality (i.e., quantum mechanical radiation and de Broglie matter) quantum tunnelled or not. Thus, M40 in the Nimtz experiment is decoupled from the quantum tunnelled radiation and from any radiation... and 'clouded' to the inverted cone within the Bergson facet, which succeeds and grand decoheres the Whiteheadian maincone into forms consistent with the future recollection of M40.

- **New Cymatic of form and pattern as a pure distribution in time, inspired by the reform of the physical field per radical de-spatialisation (4.39).** We take obscure cymatic process of 'form through vibration' and quantum mechanise the implicit complex vibration entailed; de-spatialise and temporise the process, and recapitulate it to the Bergson-Whitehead amalgam. Form and pattern transform into a pure time-distributed projection perpetually coming into being through AND-to-OR collapse and time out of the Whitehead facet and its potentiality for such form and pattern, constituting a perduring Bergsonian accumulation and projection of form and pattern: the perduring 'worm'. Consequences and implications follow with respect to life sciences, morphology, morphogenetics and development... if not to the evolutionary process.

Revived Cartesian mind-brain dualism (4.40 to 4.42): De-spatialisation exposes the false notion of *res extensa*, correcting Rene Descartes' minor error. The pure time-based description of the brain so-compelled usurps the materialist mind-brain and memory-brain reduction, undercutting anti-Cartesian academic philosophy of mind. Attendant de-spatialised and temporised brain per the Bergson-Whitehead amalgam revives mind-brain dualism as a function of the time-relation and a-synchronicity of the noetic moment vis-à-vis brain and world: wherein the noetic moment cannot arise at the same 'place' as the brain, and is shown to be temporally extant that brain. The noetic moment turns out to be a temporal monad, without extension, with absolute time-like characteristics... versus the Bergsonian world and brain subject to the relativity of simultaneity. The world and brain as memory arises in and to the noetic moment, but *not from* the noetic moment... while the latter furnishes unitary binding vis-à-vis the brain. The obvious revived Cartesian dualism culminates into radical conceptions of memory that surpass even the Bergsonian wavefunction in memory approach: namely, **noumnemonae**, or *memory in consciousness*. The full account of attendant mind-world theory must await Book-V.

BOOK-V
THE ALTERNATIVE REALIST INTERPRETATION
AND MIND-WORLD THEORIES

CONTENTS

Page

BOOK-V
THE ALTERNATIVE REALIST INTERPRETATION
AND MIND-WORLD THEORIES

GENERAL INTRODUCTION

Book-V has four main aims: As preliminary to mind-world theory, Part-I will seek to condense the developments garnered from all preceding Books into **a comprehensive summary of the Alternative Realist thesis and interpretation** (Alt-R). Part-I will also furnish a **critique of key interpretations of quantum mechanics**: from the Copenhagen interpretation through to David Bohm's hidden variables. These will be contrasted with Alt-R. Part-II will continue this goal into a **full critique of 'consciousness causes wavefunction collapse'**, against the notion that consciousness secretly directs apparent quantum indeterminism.

Quantum idealism is probably the most popular interpretation of quantum mechanics outside of the physics community: It often colours and distorts popular understandings of quantum mechanics. Yet, Alt-R will undermine quantum idealism. However, it is not idealism but materialism that is ubiquitous throughout the physics community, if not throughout Western culture and in 'educated' sections across much of the world. Part-III will condense **the position against materialism from Alt-R**. The case against materialism proves to be as decisive as the case against quantum idealism. Thus, Part-I through to Part-III will condense Alt-R and show that idealism *and* materialism are both provably dubious. Perennial object-subject relations, often undermined in both idealism and materialism, will survive the challenge from Alt-R, given that Alt-R constitutes a *realist* contention about quantum mechanics, physics, mind, and ontology.

Per the second aim, in Part-IV, V and VI, material from Alt-R, in combination with the case against idealism and materialism, will develop **the general, intermediate, and fine-structure mind-world theories**, respectively, culminating into the **revival of Cartesian dualism.** Part-IV will develop the general mind-world theory, starting with the four-fold definition of mind and consciousness: It will relate the said definition of nous to a de-spatialised temporised physics of memory, possibility and time (i.e., the growing block Bergson-Whitehead amalgam) with Cartesian dualism arising from memory and possibility subject to the relativity of simultaneity, versus the absolute time-like character of mind and consciousness. The theory will furnish **an inversion of the object-subject relation while preserving that relation, leading into general solutions to the binding problem and to the unitary orchestration of brains**, culminating into an unexpected **Anti-Copernican revolution and its inversion into the *en-worlded mind*.**

Per the third aim, in Part-V, the intermediate mind-world theory will lay bare the Bergson-Whitehead amalgam-basis for the **developmentally reversible critical milieu**: it will furnish the de-spatialised growing block definition of the brain. It is the critical milieu that renders the superlative Bergson-Whitehead amalgam and the brain susceptible to control and direction by extant abstract mind. It is the brain-based critical milieu that projects into the Whitehead facet a **Whitehead equalised possibility plateau**. This consists of possibilities that the future brain could be made to assume, so constituting the projection of the critical milieu within the Whitehead facet, essential to mind-brain control, to habit formation, and to memory expression. The mind selects from within the equalised possibilities in the plateau its preferred future possibility for its brain. It accomplishes this selection through a radical **ephemeralised delayed choice switch-x trip; so realising Cartesian mind-brain control**. The process will be accounted for as part of the **ephemeralised 'punch card' theory**, which will espouse that the above-noted mind-brain control unfolds on a work function zero basis; from the 'gaps in time' that reside between the succession of pertinent event horizons.

Finally, per the fourth aim, in Part-IV , Book-V will espouse the fine-structure account of mind-world theory. Therein, the search for the critical milieu in brains will be sought in terms of enabling constitutive fine-structure and cellular components and organisation. From it, **four fine-structure conjectures will be postulated as the likely form assumed by the critical milieu in brains; the basis of the equalised possibility plateau projected into the Whitehead facet**. Also, more mundane non-exotic conjectures will be proposed for how the brain could further delimit the generally over-conflated 'problem' of randomness and quantum indeterminism in brains, thus circumventing various approaches in generic quantum mind theory and their forlorn attempt to circumvent or otherwise orchestrate quantum indeterminism in brains.

BOOK-V PART-I:
PRELIMINARY TO MIND-WORLD THEORY:
INTERPRETATIONS OF QUANTUM MECHANICS
AND THE ALTERNATIVE REALIST THESIS

5.00: Aims of Part-I

What follows is **a comprehensive summary of the Alternative Realist interpretation and thesis of quantum mechanics and physics** (Alt-R), congealed from the Preliminary through to Book-IV. The summary will dovetail into a critique of key generic interpretations of quantum mechanics, in contradistinction to the interpretation from Alt-R.

The summary will start with **the four assertions at the core of Alt-R**: First, that the AND-form superposition expression of a system is the objective future-form expression of that system (i.e., the ontic objective reality of the future). Second, that the OR-form expression of a given system is its resolved state generated out of its AND-form future. Third, that the process of AND-to-OR wavefunction collapse through which resolved OR-form events are generated out of AND-form futures constitutes time itself; no longer a 'fourth dimension'... and the synonymity of time and wavefunction collapse. And fourth... the growing block model of the universe.

From these, all the other key findings of Alt-R will follow: including the now-defunct **intermediate model spacetime** and the **nested futures approach** therein; with tandem counterfactual causality and **grand decoherence**; the **de-spatialisation** and the temporisation of physics... all of which was detailed in Book-II through to Book- IV... and congealed into the de-spatialised and temporised **successor to the intermediate model spacetime: the growing block Bergson-Whitehead amalgam**, incorporating the case for **abstract memory** secured directly from de-spatialisation and temporisation... incorporating the **solution to quantum indeterminism** and **the case against hidden gears** behind wavefunction collapse, recapitulated from Book-III.

Part-I will also briefly recapitulate the Bergson-Whitehead amalgam-based mnemonic and grand decoherence-based solutions to inertia, gravitation, and the permanent magnet.

Finally, Part-I will culminate into the **comparative critique of the hegemonic Copenhagen interpretation**, followed by the **critique of the popular Manyworlds** interpretation; followed by the **critique of Hidden Variables** or 'pilot theory'. Wigner's 'consciousness causes collapse' will receive its own full treatment and critique in Part-II.

5.01: The Alternative Realist Thesis: A New Philosophy of Physics in Summary

The **Alternative Realist interpretation of quantum mechanics and attendant philosophy of physics (Alt-R) constitutes the critical prelude to the solution to nous and to the mind-brain conundrum**. In this essay, we condense the core of Alt-R into comprehensive summary.

We reiterate that the central problem of quantum mechanics is **AND-OR dualism**, generically known as wave-particle dualism. All other problems, such as the measurement problem and the apparatus problem, derive from AND-OR dualism. Thus, ignoring remarkable de-spatialisation for the moment, and assuming the reality of just two 'positions', a particle is expected to be found at either position-a OR position-b; *not* at both. However, quantum mechanics discovered that, before measurement and before the resolution of putative particle-position, it assumes *both* position-a AND position-b: a superposition encapsulating both possibilities. Without explanation, generic quantum mechanics asserted that **a particle must constitute both OR-form *and* AND-form attributes, encapsulated as particle-wave duality and complementarity**... wherein, OR-form logic constitutes the classical-relativistic 'particle' aspect of the duality, while AND-form logic pertains to the wave and superposition aspect and principle of the duality.

Expressed as AND-OR dualism and logics, our lay formalism, founded on the AND *and* OR distinctions, or AND-OR duality, helped furnish ready clarification and simplification of the core problem and logic of quantum mechanical nature.

Is nature OR-form logic (particle) or AND-form logic (wave)? How can nature be OR-form *and* AND-form at the same instant? The hitherto inability to resolve these questions despite the great success of quantum mechanics lies behind Richard Feynman's well known assertion to the effect that no one understands quantum mechanics. Yet, the Alt-R interpretation asserts that quantum mechanics is perfectly understandable and AND-OR dualism perfectly sensible. To appreciate this, we must reiterate the *raison d'etre* of AND-form logic and recapitulate the question why nature assumes such logic. Thus...

- **The AND-form expression of a system is its objective future-form expression:** The future is *not* a *bloss* subjective abstraction in our minds. Nor is the wavefunction a mere pragmatic mathematical expression. Instead, the future is an ontic thing 'out there', independent of our subjective imaginings, concepts, and epistemology. Therefore, the mathematical wavefunction is a legitimate response to a phenomenally objective ontically real future. The principle of causality intertwined with the principle of the conservation of energy-matter (putative) into the *firewall principle* demands the ontic reality of the AND-form rendered future.

- **The growing block universe approach:** In the block model universe, the future is fully pre-resolved into definite OR-form 'paths' and 'which way' states. *If the block universe held true, the two-slit experiment could never yield self-interference or generate scale-amplified interference patterns. Interference patterns could not emerge out of pre-resolved OR-form 'paths'.* Yet, in really existing two-slit experiments, we *do* obtain self-interference *and* interference patterns. This could only happen if the future-state of the experiment objectively exists as an as-yet to-happen superpose of futures rendered into potentiality. Hence, AND-form logic, the quantum wave and the wavefunction really *do* pertain to an ontic future; the quantum wave *is* the objective future phenomenalised as a real-existent. Hence, our universe is a growing block system, with an ontic past-future domain, espoused in Book II and IV.

- **The OR-form expression of a system is its resolved state**, brought out of the ontically real AND-form superpose of future possibilities (the wavefunction) into a single realised OR-form outcome; the 'classical' outcome.

- **The process of AND-to-OR collapse, or wavefunction collapse, constitutes time itself**. Thus, time *is* wavefunction and quantum wave collapse. Time is *not* a 'fourth dimension'. This requires that we modify our block models of 'spacetime' to the growing block models. Time is ontologically primary to the process of measurement and observation: Therefore, no time, no measurement. *It is not measurement or observation that causes wavefunction collapse and time. Instead, it is AND-to-OR collapse and time that brings about measurement and observation... of the OR-form actua decayed out of AND-form futures attendant the process of measurement.* Consequently, the mechanical process of observation and measurement cannot be the causal factor behind wavefunction collapse and time, given the ontological primacy of AND-to-OR collapse and time vis-à-vis measurement.

458

QUANTUM MECHANICS AND MIND

The four tenets stated above constitute the indispensable core of Alt-R. To grasp these is to obtain core understanding of quantum mechanics and the *raison d'etre* for the existence of quantum mechanical waves and AND-form logic: i.e., the superposition principle and the ontic reality of the future. Yet, the tenets fall short of furnishing the deeper insight as might be garnered from the intertwine of causality and the principle of conservation into *the firewall principle*, both of which justify the existence of AND-form logic. We must first elaborate on the specific nature of the AND-form wave-state: The way Alt-R models this is critically distinct from how generic quantum mechanics and generic interpretations accomplish it. The latter tends to attribute pseudo-realtime descriptions of AND-form waves and experiments, even when these are subject to strict isolation. It also assumes the possibility of 'which way' paths and states. Alt-R rejects such assumptions and many others. Thus, according to Alt-R…

- **Any system or interval from which observational or perturbing inputs are removed is a system rendered into observational and experimental isolation. Any system or interval subject to such isolation is immediately rendered a future possibility state: an AND-form state.** The possibility of obtaining information from a system requires that we bring it out of isolation by applying to it observational-perturbing inputs. This, and attendant 'measurement', is deferred to the future… to a future-possible moment, wherein measurement itself constitutes a future possibility state; itself slated to transpire in the future.

- Under conditions of enforced isolation, **the AND-form quantum mechanical wave will *remain* in AND-form, and cannot be attributed any resolved OR-form 'which way' spatial trajectory: It is not possible for the quantum mechanical wave in isolation, or in experimental intervals in which isolation abides, to assume *any* OR-form attributes, even if subsequent observation might entail a foisted inference of a putative 'which way' path to that interval.** The AND-form character of the isolated system will abide even in single-slit experiments that supposedly emphasis the particle aspect of the duality: The single-slit arrangement cannot obviate AND-form logic, even though self-interference and the formation of interference patterns is not permitted therein: (See the non-standard interpretation of the two-slit experiment: Preliminary: **0.14** and **0.15**).

- In the single slit experiment, even when the approaching quantum wave is restricted to a single slit or small hole, this does not imply the manifestation of an OR-form particle-like state at that pertinent point, unless active detectors are attached therein and participate in the AND-to-OR collapse of the quantum wave. Yet, subsequently, if observational inputs are withdrawn, the particle will revert to its AND-form state. This is because the AND-form wave-state applies to *any* interval of isolation, while all isolation implies the deference of the system into its future-form state, even in single slit experiments or arrangements. The AND-form wave will abide throughout isolation-intervals... even in single slit or single hole experiments with or without active detectors, save when subject to active detectors that reduce isolation.

- Thus, **whether designed to amplify the wave-aspect (via the two slit experiment) or deny the possibility of self-interference (one slit), the AND-form of the wave abides if isolation abides.** (See the non-standard appraisal of the two-slit experiment, and the Afshar controversy: Preliminary: **0.16** and **0.17**). The Afshar controversy claimed that one could obtain single-slit outcomes from double-slit conditions, or a definite 'which way' path from an arrangement that supposedly denies that possibility and only emphasises the wave aspect. In this way, Afshar called the wave-particle AND-OR complementarity into question. However, even with Afshar's experiment, under isolation, only the AND-form state must abide. In any case, even a single slit experiment must unfold under strict isolation and must exhibit AND-form logic therein. Hence, the Afshar experiment cannot in truth assert for an OR-form particle 'which way' path in the interim of isolation. A similar assumption about 'which way' paths is repeated in delayed choice experiments and in the more radical quantum eraser experiments of Yoon Ho-Kim (see Book-III: *Holed Past Theory*: **3.19**); again, premised on the dubious assumption that we could obtain OR-form 'which way' information from systems under isolation, or even erase such information tantamount to the erasure of the past and of memory. In any case, with de-spatialisation, the impossibility of this becomes fully salient: Isolation or not, there are no 'which way' paths, simply because there is no space across which such paths could be inscribed.

- **Even on the false assumption of 'space', the AND-form quantum mechanical wave does not actually 'move' through the experimental system. The quantum mechanical wave is, at best, *always* spatially static.** The *assumption* of the 'motion' of the quantum mechanical wave in the interim of its isolation is just that: an *assumption*. There is not a shred of evidence to support that assumption. When we attempt to furnish the evidence, the act of the observation destroys the putative 'moving AND-form quantum wave' observationally sought: From the moment of the initiation of the two-slit experiment (or even of the delayed choice experiment, or even of the quantum eraser experiment) to the moment of its conclusion, and throughout the relevant intervals of isolation, we can only describe radiation or de Broglie matter as a non-moving static quantum mechanical wave; one that superposes all possible 'which moments' of future-potential interruptions and observations… with potential alternative outcomes and contingencies nested to each such 'which moment'. Indeed, **the static quantum mechanical wave is inherent to the nested futures account of the wave**, as was explicitly developed in Book-II, Part-II.

- Thus, the 'motion' of the quantum wave, or of anything…is a thing we foist upon the temporal succession of events generated out of the AND-to-OR collapse of an otherwise spatially static non-undulating AND-form future; Hence, the error of interpreting the really-existing time-interval between two temporally succeeding events as a 'spatial interval'. But the interval is nothing more than a temporal interval. The ontologically real future obviously does not move, although its decay into at least two temporally succeeding OR-form events permits our naive foisting upon that succession the notion of 'motion'. Therefore, the quantum mechanical wave is truly a non-moving non-undulating spatially static state. (See delayed choice experiment Book-II **2.23 to 2.26**... and in Book-IV **4.07**, respectively).

- Of course, **given de-spatialisation, space is not ontologically real, but pleonastic. It follows that the notion of the motion of anything, including that of the quantum mechanical wave, is not tenable**. There is no 'space' through which any 'moving' quantum mechanical wave could displace, or be teased or tricked into describing an OR-form 'which way' path under two-slit conditions, or in Afshar's experiment, much less become erased in Yoon Ho-Kim's quantum eraser. For the comprehensive case for de-spatialisation and pure temporisation of physics, see Book-IV **4.01** to **4.07**: De-spatialisation will be summarised in **5.02**.

From the above view about the structure and nature of the AND-form quantum mechanical wave, combined to the four indispensable tenets of Alt-R, follow other key tenets of Alt-R. Thus...

- **Time is irreversible and 'flows' from AND-to-OR, never from OR-to-AND**. Recall from Book-I (**1.09** to **1.12**) that time, or AND-to-OR collapse, decays AND-form futures into OR-form outcomes. The 'arrow of time' proceeds from AND-to-OR. Ultimate entropy is the entropy of the future: i.e., the irreversible entropy and decay of AND-form futures into OR-form outcomes, grasped as loss or degradation of future-potential information formerly locked up in the future. The elimination of futures constitutes the *entropy of future-potential information* which supplants conceptions of 'time's arrow' based on generic statistical accounts of entropy: The very process of AND-to-OR collapse is ontologically primary and presupposed to any generic entropy-measures or statistical effects pertaining to energy or information. It is time that resolves generic entropy. It is *not* generic entropy that resolves or defines time, much less it's 'arrow.' While generic statistical entropy could be reversed, such a reversal could only come into evidence through AND-to-OR collapse. Thus, the reversal of generic statistical entropy could not imply the reversal of the ontologically primary AND-to-OR time-process into OR-to-AND de-collapse, or time-reversal. The proper reversal of time would require the reversal of AND-to-OR into OR-to-AND. Science could not proceed in a universe where OR-to-AND time-reversal took place, if only because we need to resolve our outcomes, observations, and our communications about these, into accumulations of OR-form evidence, and then into accumulating memory. But an instance of OR-to-AND time-reversal would serve to "un-make" the sought accumulation, rendering the phenomenological and experimental facts of science 'un-evinced'; transforming science into an absurd 'un-science', or nonsense; rendering sophont experience untenable.

- **New Complementarity thesis**. Before measurement, a system is purely AND-form; constituting a static state of future possibilities projected across a given potential time-interval, which may subsequently AND-to-OR decay into an OR-form actua. Per our case for the **New Complementarity thesis** (see Preliminary **0.15** to **0.17**) generic complementarity abides, but with the added proviso that, post-experiment, *both waves and particles are empirically observationally simultaneous*: Hence, the evidence for AND-form wave-interference in the two-slit experiment is obtained per the interference pattern at the detector, rendered salient by the *en masse* OR-form 'particle' depositions that are observationally simultaneous with the interference pattern so-render. We need the OR-form depositions to render salient the AND-form reality. Thus, the OR-form reality is empirically-observationally simultaneous with the AND-form reality it reveals at post-experiment. The single-slit experiment is not exempt, in that the *en masse* distribution of 'particle' depositions therein must also empirically simultaneous evidence the preceding AND-form wave at post-experiment, even if with the absence of an interference pattern. In both forms of the experiment, AND-OR realities are observationally and empirically simultaneously evident. Hence, the non-standard appraisal of the two-slit experiment, which helped solve the Afshar and other controversies.

- **New Complementarity thesis gains saliency when we grasp the *raison d'etre* for AND-form logic as the ontic future**. Thus, any system at *any* scale, including the macro-scale, must first exists as a suspension of future possibilities rendered in AND-form: part of an ontically real and objective futures-domain. This ontically real future then decays into OR-form outcomes via the process of AND-to-OR wavefunction collapse or time, producing a growing block admixture of what remains of the AND-form future, combined to what now constitutes the OR-form present, combined to the accumulation of the growing block past-in-memory. Generic complementarity abides in the sense that the quantum mechanical description of the system is the ontically real future. Generic complementarity also abides in the sense of the saliency of the growing block universe... whose past-future division perfectly captures new complementarity, integrated by AND-to-OR process of wavefunction collapse and time. This deeper complementarity was first expressed in our case for the intermediate spacetime model (Book-II: **2.13**); finally consolidated into its de-spatialised temporised form in the Bergson-Whitehead amalgam (Book-IV: **4.10**).

So far, we have asserted that the AND-form state of the system constitutes the ontically real future. The following two tenets pertaining to the **firewall principle** explain directly and explicitly why this *must* be so. To understand this is to understand quantum mechanics and why AND-form logic exists in nature. Thus...

- **AND-OR dualism and complementarity is demanded by causality**. According to Albert Einstein's Naïve Realist view, an object must always remain in an OR-form state independent of any observation of it and independent of any interaction with it from the larger environ... even when subject to isolation and consequently deferred to the future: The Moon must remain the Moon, even if we do not look at it. However, quantum mechanics shows that a system must be rendered AND-form before any subsequent observation of it. But generic quantum mechanics never explained why this must be the case. Once we grasp the AND-form state of a system as its objective ontic future, we grasp that causality itself demands AND-form logic. Indeed, the future cannot be described as a concrete OR-form event before it gets to transpire into such. To treat the future event before it had transpired into an OR-form state *as if* it constituted an OR-form state (as is inadvertently demanded by the Naïve Realist's conception of 'objectivity') is tantamount to treating it as if it transpired *before* it had transpired. This would be tantamount to the reception of 'signals from that future', necessarily in violation of causality. Hence, AND-form logic, hence quantum mechanics, *saves* causality by imposing a growing block 'cut', frontier, or 'firewall'... an *event horizon*... between the as-yet not realised future and the accumulating past. Also, objectivity is saved inasmuch as it is understood in proper terms, free of the erroneous assumptions of Naïve Realism: Objectivity holds true because the AND-form future *is* objectively *and* ontically real.

- **AND-OR dualism and complementarity is demanded by the putative principle of the conservation of energy-matter**. If it were possible to observe the future before it transpired, this would be tantamount to the reception of 'signals from that future'. If it were possible to treat your future arrival at New York as a resolved OR-form event, *and* prove it in empirical terms through the reception of physical signals by your present-self in London (via a phone call), the result would be your duplication. Where is the required extra energy-matter for your 'future self' at New York going to come from? If we could treat the future of a putative particle two picoseconds from now *as if* it were an OR-form resolved event before it transpired into such... *and* obtained physical signals from that future in proof... we would gain physical evidence for the duplication of our particle. Where would the extra energy-matter for the second future particle come from? If it could decay out of future-potential matter using our matter-creation mechanism conjectured in Book-IV **4.21**, even this would require the ontic reality of the growing block AND-form future, one constituted as an as-yet not-happened potentiality for future matter, but *not* as a resolved OR-form state belonging to an implausible block model universe. In short, the said duplication would be tantamount to the violation of the principle of conservation of the sum of putative energy-matter, given that the signal received from the pre-resolved future would entail evidence of an over-unity quantity of energy-matter. (This criticism is not usurped by the pleonastic status of energy espoused in Book-II and Book-IV: **2.39** and **4.10,** respectively). Thus, AND-form logic, hence quantum mechanics, *saves* the putative principle of the conservation of energy-matter, even if 'energy' so-conserved proves to be finally pleonastic.

- **The 'principle of causality' and the 'principle of conservation of energy-matter' intertwine into** *the firewall principle*. Thus, AND-form logic arises from the intertwine of causality and the conservation of energy-matter, and the growing block 'cut' and partition of the future from the past via the event horizon: a sensible feature of nature, and a solid basis for the existence and ontic reality of AND-form logic therein, and, with it, the justification of the growing block universe. Hence, quantum mechanics is perfectly harmonious with legitimate common sense tense-sense and generic notions of a future-past divide. Thus, Alt-R constitutes the return of common sense to physics, but without Naïve Realism. It was originally North Whitehead that intuited that quantum mechanical waves must constitute the phenomenalisation of the future. Yet, Whitehead never posited a firewall principle upon which the onticity of the future is justified.

One of the key culminations of Alt-R consists in the primitive unification of quantum and classical nature. This falls short of the integration of General Relativity with quantum mechanics; further limited by its inability to explain the putative particle zoo. Even so, Alt-R unifies the classical-relativistic (OR-form) and quantum (AND-form) worlds in and through the growing block intermediate model spacetime (Book-II), and, finally, in the de-spatialised temporised Bergson-Whitehead amalgam (Book-IV). Thus...

Primitive unification of AND-form quantum and OR-form classical physics: Generic models of spacetime treated the future as a fully pre-resolved collection of OR-form worldlines and lightcones: a block universe. Our justification of AND-form logic from the intertwine of causality and the conservation principle demands that we partition spacetime into two main domains, such as to constitute a growing block universe: The domain of *spacetime in potentiality* (the future in AND-form) and, 'below' it, the domain of *realised spacetime* (the past in retained form). This implies a distinction between realised parts of any given lightcone and worldline attendant realised spacetime versus future-potential AND-form parts of the same lightcone and worldline embedded within AND-form spacetime in potentiality. Developed in Book-II, the intermediate spacetime model asserted the following:

- **Spacetime in potentiality**... or the total future, rendered purely as a static, non-moving and non-undulating AND-form of possibilities; comprised of all possible future-potential lightcones and alternative future-potential worldlines, rendered into equal or unequal probabilities. Spacetime in potentiality sits 'above' realised spacetime, demarcated by the event horizon (see below).

- **The event horizon.** This is the 'cut' comprised of the most recent OR-form events that have decayed out of static non-moving and non-undulating AND-form spacetime in potentiality: The event horizon is typically depicted as a horizontal line or surface between spacetime in potentiality 'above' it and realised spacetime 'below' it. It is not necessarily flat: it can be 'curvy'; a component expression for gravitation (See Book-II: **2.44**). The event horizon is not absolute time: It accords with the relativity of simultaneity.

- **Time and synonymous AND-to-OR collapse (or wavefunction collapse) attends the broken-discontinuous advancement of the event horizon** into the domain of spacetime in potentiality, with subsequent decay of that future into resolved OR-form events along the vertical time axis of a spacetime diagram. In this approach, wavefunction collapse (or time) is not a one-time exhaustive comprehensive decay of the total future. Instead, it is always a partial-perpetual decay of an imminent part of a larger future potentiality elaborated to the infinite future.

- **Time is no longer a 'fourth dimension'**. Instead, time is synonymous with the AND-to-OR partial-perpetual wavefunction collapse attendant the broken-discontinuous (quantised) progression of the event horizon. This implies that **time is quantised**: Hence, time is not infinitely divisible, and it is *not* seamless-continuous. This fact must have bearing on prospective quantum gravity, in that quantised time implies quantised gravitational processes, if only because gravitational effects must unfold through quantised time as broken-discontinuous successions of gravitational moments. Of course, time always proceeds from AND-to-OR, *never* from OR-to-AND. Thus, time-reversal is impossible.

- **Memory**: The domain of realised spacetime below the event horizon constitutes the implicit ontic reality of abstract memory in nature, although the intermediate spacetime model cannot tell us in what form this memory assumes, it is yet memory effectively 'outside' of putative matter confined to the event horizon. With prospective de-spatialisation in Book-IV, the reality of abstract memory becomes obvious: explicitly incorporated into the successor to intermediate spacetime: namely, the Bergson-Whitehead amalgam.

- **Nested futures**: Implications from the application of John Wheeler's delayed choice to spacetime in potentiality forces a reconsideration of how the future is structured and phenomenalised: (See Book-II **2.17** to **2.18**). The future is wholly non-moving and static: The future does not merely consist of immanent possibilities in AND-form. Instead, each immanent possibility subsumes and nests additional AND-form futures, and even nests alternative complex wavefunctions. These in turn nest their own subsequent potential future possibilities further removed into the future. Nested futures can be categorised into the *divergent, entangled,* or *convergent* categories. The convergent category predominates within spacetime in potentiality and, through a counterfactuals driven process of grand decoherence, makes possible the 'non-local' or transtemporal unity of the universe; instantaneously binding each frame of reference on the event horizon to every other frame of reference on the same, through their common convergent nested futures, and to the infinitely removed future.

- **Grand decoherence and the revival of Mach's principle**: Our frame of reference on Earth is attended by a system of nested futures that elaborate to the infinitely removed future, and includes future-potential observations of events four years in the future pertaining to Alpha Centauri 'located' some four lightyears from Earth. This can be represented as overlapping future lightcones from the Earth frame and Centauri frame, respectively, with both projected into spacetime in potentiality. If the stars at Alpha Centauri should turn supernova now, and do so on a 50-50 probability basis, on Earth, we will not know of this until some four years have transpired, on the reasoning that, per the assumption of 'space', it takes light four years to get to us from Alpha Centauri. Yet, the nested futures attached to our frame must incorporate future events and future-potential observations of Alpha Centauri four years hence, and must decohere to the elimination of non-consistent futures therein. This *grand decoherence* will permit only future-possible observations and events that are consistent with the fact and reality of the said supernova at Alpha Centauri. In other words, even without direct observation or direct physical signals, the wavefunction of nested future possibilities attached to our Earth frame and elaborated into four years into the future, must counterfactually 'know' what is *not* happening at Alpha Centauri. Indeed, through the same convergent futures and grand decoherence, the Earth frame will counterfactually 'know' about every other frame in the universe that lies on or along the event horizon. The sensitivity of our frame and its nested futures to every other frame and future-convergent nested futures in the universe, including that of Alpha Centauri, constitutes *grand decoherence*; a universe-scale macro-scale decoherence. That is, inconsistent possibilities that cannot happen in the future… because they have been eliminated by the actual turnout of events elsewhere in the universe… must imply the counterfactual decoherence of every other frame of reference into mutually consistent futures: (See Book-II **2.35**). In this way, matter here can 'know' what matter elsewhere is *not* doing. In this way matter 'here' adjusts to such counterfactuals through the tandem decoherence of its own future possibilities into consistent futures: i.e. consistent with the future-content of all other frames. The future possibilities attached to matter 'here' must adjust (decohere) according to what happened and did not happen to matter elsewhere in the universe, along the event horizon. In addition, matter elsewhere must necessarily adjust to the implied futures based on what has happened here. This explains inertia, which requires matter here to instantaneously 'know' about matter elsewhere, and vice versa. That is, counterfactual causality and grand decoherence in the context of convergent nested futures-based growing block spacetime model revives Mach's principle and designs for the long-elusive growing block quantum mechanical explanation of inertia: Succinctly, inertia is per function of grand decoherence.

- **Counterfactual causality and grand decoherence, and the all-scale reality of AND-form logic**. It turns out that, against the want to confine the AND-form superposition principle and quantum physics to the 'very small', macro-scale objects *are* subject to the superposition principle *and* to AND-form logic. This is furnished by the convergent category nested futures, and through the grand decoherence of convergent futures into consistent futures… through the total quantum mechanical future that converges, unites, and binds every frame of reference to and with every other frame into an 'instantaneous whole' (see Book-I **1.03**, **1.26** and **1.27**) otherwise known as the 'universal wavefunction'. Indeed, the demand for AND-form logic from the intertwine of causality and the generic conservation of energy-matter necessitates that the future (the AND-form state of the system) cannot be confined only to micro-scale or to the very small: it must apply equally to the very large, as does causality and the conservation principle intertwined. Thus, per grand decoherence, from the perspective of the convergent category nested futures, the universe as a whole self-interferes: Hence, billiard balls 'here' instantaneously decohere and interfere with billiard balls elsewhere, and vice-versa, into consistent futures. Hence, quantum interference at the grandest scale demonstrates the all-scale pertinence of quantum mechanics, beyond the mere micro-scale.

QUANTUM MECHANICS AND MIND

5.02: Alt-R, De-Spatialisation & the Bergson-Whitehead Amalgam

The intermediate model spacetime was revolutionary enough. Yet, it scaffolded for conclusions even more radical. The decisive key was de-spatialisation and temporisation, furnished in Book-IV (**4.01** to **4.07**). Hence, **there is no space. Only time is real. Almost all things are events organised or distributed in time according to primacy-recency rules; expressed in terms of delayed choice time-interval relations in lieu of 'spatial intervals'**... culminating into the replacement and succession of spacetime with the superlative Bergson-Whitehead amalgam: 'spacetime' *without* 'space'. Thus...

The Bergson-Whitehead amalgam.

De-spatialisation emerged out of the following key considerations:

- **The problem of contiguity through simultaneity and the impossibility of extensionality or' space'... and the synonymity of space with absolute time, the latter of which was abandoned in 1905:** (Book-IV **4.01**). Pleonastic space is inseparably tied to the notion of the 'common present moment' in which all things supposedly arise i.e., absolute time. Per the relativity of simultaneity and other considerations, absolute time was invalidated by Einstein in 1905. There is no common present moment such as to constitute a collection of *wheres*... or such as to constitute a contiguity through simultaneity of seamlessly linked ordinates along a line... or of co-ordinates through simultaneity to constitute a surface-space... or the same in three-dimensions such as to constitute a block space: All of these, or 'space' in one, two or three dimensions, would constitute impossible states in absolute time. Hence, the synonymity of space with absolute time... and the demise of space with the demise of absolute time. Unfortunately, Einstein *et al* did not grasp the synonymity of space with absolute time and did not develop a criticism of contiguity though simultaneity. Instead, space and time were incorporated into 'spacetime'.

- **The relativity of simultaneity does not arise from the speed of light in relation to space (essay 4.02). Instead, it emerges out of pure time relations that temporally intersperse the temporal succession of events and observers**, the organisation of which requires reference only to time-relations and intervals that emerge out of ontologically primary AND-to-OR collapse or time, *without* space. Indeed, the notion of the 'speed of light' and of its space is foisted on AND-to-OR resolved time-interval relations that intersperse successive events and observers: The 'speed of light' and space are pleonastic imposition foisted on the succession of events: The speed of light is replaceable by *the delayed choice time-interval standard* and relata (**4.08**).

- **Time is ontologically indispensable; space is entirely dispensable and pleonastic: (essay 4.03):** Putative space cannot be conceived without time. Without time, nothing would transpire... not even events supposedly interpreted as 'objects in space', or undergoing presumed 'motion' in space. Insofar as things purported to be 'objects in space', or in 'motion', can be recuperated and expressed purely in terms of delayed choice time-interval relations between temporally succeeding events, frames and observers, it follows that time proves to be both necessary and exclusively sufficient, minus 'space'. Insofar as the succession of events can be expressed purely in terms of delayed choice time-interval relations *without* space, space proves to be unnecessary and superfluous: Thus, per the 'economics of physics', space is shown to be entirely pleonastic.

- **Quantum mechanics prohibits the possibility of the continuous observation of events, of motion, or especially of the 'motion of quantum waves'. It follows that we cannot empirically prove 'motion', much less the motion of light at the 'speed of light', and even less the 'motion' of the quantum wave itself. Therefore, we cannot empirically prove the reality of 'space' attendant such motion:** (essay 4.04). Insofar as such observation is prohibited by the empirical stricture of quantum mechanics, the oft depiction of the quantum wave as spatially dynamic and undergoing 'motion' per notional pseudo-realtime descriptions, constitutes an *assumption* about the quantum wave. The assumption is wholly disprovable: Indeed, to prove the purported motion of the wave (and hence its motion in space) it must be observed continuously, without any gaps or non-observable intervals, or any moments of physical-experimental isolation: with all attendant spatial ordinates revealed in perfect seamless contiguity through simultaneity. Quantum mechanics cannot permit such seamless-continuous motion because it cannot permit seamless-continuous observation of *any* system by means of observational return-signals (or radiation outputs) that must necessarily be set to infinite frequency, notwithstanding the Quantum Zeno effect. Consequently, both the proof of motion *and* of attendant space would necessarily culminate into the equivalent of ultraviolet catastrophe, even if we choose to disregard the fact that any attempt to observe the quantum wave of light radiation during its purported motion would destroy the very thing we sought to observe.

- **De-spatialisation from the critique of Hilbert space:** (essay 4.06): Even if we ignore the argument for static quantum waves from Book-II, and treat the quantum mechanical wave as moving, it could only move through a Hilbert space: i.e., quantum mechanical space: a *future-potentiality* for space that has not yet come into realisation; imposed by the demands of causality intertwined with the principle of conservation (the firewall principle), both of which justify the AND-form quantum mechanical wave and wavefunction as the ontically real future (see Book-I, **1.31** and **1.32**). If Hilbert space is putative space that has not yet come into realisation, then, obviously, nothing could move through a space that has not yet come into realisation... tantamount to the assertion that there is no space.

Finally, Alt-R presented **the delayed-choice based over-kill of both motion and space**: Here, we will not furnish the full details of the approach but will simply state the conclusions. For the fuller account, see attached references.

John Wheeler's delayed choice approach can be tweaked to interrogate the notion of the motion of the quantum mechanical wave *and* attack the notion of the space across which the wave is purported to be moving.

- **Wheeler's delayed choice and spatially static quantum mechanical waves obviates 'space'**: Quantum mechanical waves are shown to be spatially static. This obviates the space across which quantum waves are purported to move. (See below).
- **Wheeler's meta-delayed choice and final de-spatialisation** (Book-IV, **4.07**): In Book-II (**2.18**), John Wheeler's delayed choice approach was presented as the obvious basis for the nested futures view of future-from spacetime. It culminated into the spatially static non-undulating and non-moving view of the quantum wave:(Book-II: **2.23** to **2.26**). This alone obviated 'space' per function of the invalidation of the notion of the' moving quantum wave'. Therein, pertinent nested quantum waves cannot be characterised as moving across their pertinent intervals, even if their 'carrying' nesting, or hedging wave might be so depicted, given that the nested future possibilities could not yet have coincided with their respective putative spatial intervals. Hence, the nested future waves must be static (not moving). By inference, so must the nesting hedging wave carrying the spatially static nested future waves: The hedging wave must also constitute a static quantum wave. In **4.07**, we took delayed choice to the next stage and treated the hedging wave as a nested future in its own right; one subsumed to a preceding nesting hedging wave; part of a meta-delayed choice system of preceding nesting hedging waves elaborated to the putative Big Bang or to the remote indefinite past. We discovered that only the first ever 'carrying' nesting or hedging wave immediately after the putative Big Bang might be *assumed* to have been 'moving'. At which point, both the notion of the motion of the quantum wave *and* of its 'space' was rendered superfluous and pleonastic. Hence, delayed choice nested futures forced the reality of the spatially static quantum waves, combined to the meta-delayed choice approach that directly usurped space itself, so constituting the final argument for de-spatialisation and the pleonastic status of space. Conclusion? There is no space. There is only time.

De-spatialisation unravels the notion of 'matter in space' and succeeds it with physics conceived as an abstract system of events organised in and by pure time: a physics grasped as a purely temporised and stereotemporised events distributed in time... *without* space. Thus, purely temporised delay choice time-interval relations replace 'spatial intervals', and 'motion' is grasped as uniform or non-uniform change in relative delayed choice time-interval relations that bestride event, generated through the succession of AND-to-OR collapse of realised events and time-relations out of static future-potentiality, including the 'motion of light' itself. Of course, the 'speed of light' must itself be replaced by the *delayed choice time-interval standard* (see Book IV **4.08**) given that electromagnetic radiation in ultimate quantum mechanical form cannot 'move', simply because there is no 'space' across which it could move, much less do so at a 'speed'.

*

The implications from de-spatialisation to our spacetime model culminated into the **Bergson-Whitehead amalgam**: the successor to spacetime, or spacetime *without* space (Book-IV: **4.10**). The de-spatialisation and temporisation of the physical order in Book-IV imposed the radical overhaul of the growing block intermediate model spacetime from Book-II. De-spatialisation took physics beyond 'objects in space': It replaced 'matter in space' with purely temporally distributed events, rendered *without* 'space'. De-spatialisation also brought into question the notion of contiguous or impact causality between 'particles in space', rendered impossible by the obviation of the 'spatial point-of-contact', thus undermining the notion of the 'spatial particle', transforming spatial *atomos* into temporal *durata* (see Book-IV **4.14**), undermining materialism: (See Part-II for the comprehensive account of the decomposition of materialism).

With space removed and the growing block universe recuperated into events distributed in realised or future-potential time, this led to memory in the form envisioned by Henri Bergson... culminating into Bergsonian abstract memory, or pure temporal memory, and into the Bergson facet of the Bergson-Whitehead amalgam.

Hence, the Bergson-Whitehead amalgam replaced the intermediate model spacetime per the following:

- **The Bergson-facet** replaced 'realised spacetime'. The Bergson facet constitutes the abstract basis of all events that have come into being, retained as the total growing cosmic history, and as the repository of all *wavefunctions in memory* i.e., wavefunctions of the past cropped of their former nested futures. The system of memories within the Bergson facet elaborate to either the infinitely remote past, or at least to the Big Bang moment in that past. Memories therein are organised and radicalised into pure temporal relations per the demands of de-spatialisation.
- **The Whitehead facet** replaced 'spacetime in potentiality', also de-spatialised and temporised into future-potential convergent nested futures and attendant future-potential delayed choice time-intervals and events. The Whitehead facet is the repository of *wavefunctions in future-potentiality* that exhibit nested futures and elaborate to the infinitely removed future. The Whitehead facet, constituted as an instantaneous whole, is subject to perpetual grand decoherence, furnishing Mach's principle and inertia.
- **The event horizon** remains intact as the leading realisation of the newest OR-form events that had decayed out of the Whiteheadian future. Thus, OR-form states ('matter'... if you will) remain confined to the event horizon. The Bergson and Whitehead facets are purely abstract and immaterial, dissociated from and extant with respect to the event horizon and its 'materialised' outcomes.

Certain developments followed on from the Bergson-Whitehead amalgam. These are noted below:

- **Physical information is structured either as *wavefunctions in memory* or as *wavefunctions in future-potentiality*.** The conundrum of physical information is thus resolved. As memory, information is comprised of past *wavefunctions in memory* cropped of their former nested futures. Wavefunctions in memory are relegated into and are retained within the Bergson facet, so constituting abstract memory at physical remove from attendant OR-form 'matter' confined to the event horizon. On the other hand, information exists as future-potential information within the Whitehead facet, in the form of *wavefunctions in* *future-potentiality*. These latter are like generic quantum mechanical waves and possibility functions, save that they are static, non-undulating, *and* incorporate nested futures. As for **OR-form information** from AND-to-OR decay, this is resolved 'matter' confined to the event horizon.

- **The problem of gravitation.** Hitherto, serious incompatibility between AND-form quantum mechanical and OR-form classical-relativistic physics was instance in the problem of gravitation. If we assume point-like or billiard ball-like masses in space, *and* assume the reality of space, OR-form matter would be expected to produce the usual notional gravitational field, either in the form of a force field or as a curvature of spacetime. The said gravity would be centred at and emanate from the OR-form point-like or billiard ball-like mass. Yet, any attempt to describe a gravity for matter in its AND-form state must fail, given that the sought 'force field' or the 'spacetime curvature' could not be centred and emanate from all of the AND-form positions of the said matter. De-spatialisation can only make this worse by undermining the whole notion of point-like entities and billiard ball-like objects from which fields supposedly emanate or are attended by spacetime curves, by undermining position, place, and space, and by derailing the want to locate a point or centre about which the putative gravity field could form. To reiterate, de-spatialisation also forces us to abandon *any* spatial description of the field: it removes 'space' from the 'spacetime curve'. Moreover, the future-form state of a particle's position constitutes an expression of what has not yet happened to it. Therefore, as a thing that has not yet happened, obviously, it cannot generate a gravity field. In short, gravity fields cannot be associated with future-form AND-states at all, but only with realised states, or with the past. Therein lies the part-solution to gravity and the circumvention of the contradiction said to reside between classical physics and quantum mechanics in their specific relation to gravity: namely, *gravity is a memory effect of the Bergson facet* upon both the event horizon *and* the possibility-structure and grand decoherence of the Whitehead facet.

- **Solution to gravity and addendum to the primitive unification of classical and quantum physics.** Gravity does not emanate from AND-form future possibilities or from the Whitehead facet pertaining to a putative mass or matter resolved on the event horizon. Indeed, gravity is that which decoheres the probability structure of the wavefunctions in potentiality and the Whiteheadian future. The latter then AND-to-OR collapses to generate pertinent OR-form outcomes. The comparison of the OR-form outcomes to a preceding series of outcomes describes a state-change over time: a state-change that we describe as 'gravitational acceleration', or as 'orbital motion'. It is Bergsonian memory (the retained history of the past) that ultimately grand decoheres to consistency the nested future possibility complex embedded within the Whitehead facet, biasing its most immanent wavefunction to AND-to-OR outcomes that we then read as gravitational phenomena. Of course, this solution goes beyond the primitive unification of physics furnished by both the intermediate model and by the succeeding Bergson-Whitehead amalgam: We arrive at an almost full unification of quantum and classical physics through a growing block mnemonic mechanical and quantum mechanical description of gravity. (Book-IV, Part-III: **4.24** to **4.31**)

Consider the **tidal effects of the Moon upon the Earth brought about by the Moon's gravity. This is exemplary of gravity as a Bergsonian memory effect**. Even assuming the false notion of 'space', from the frame of the Earth, the Moon constitutes a corpus of events relegated to a past: a past that had transpired around 1.5 seconds ago: an influence differentiated from that of the Earth, but engendering tidal effects upon Earth, *not* from across space, but purely across time… *from the past*. It follows that the influence of the Moon upon the Earth, or of anything upon anything else, is necessarily the physical influence of the past… the physical influence of memory. Thus, notwithstanding attendant counterfactual causality, grand decoherence, and the revived Mach's principle, gravity and inertia are effects generated from the superpose of the Bergsonian past, resonated as memory *across* time, *not* as a 'force' or 'curve' across pleonastic space.

We need only incorporate de-spatialisation and pure temporisation to the Moon-Earth relation to arrive at the direct and seamless assertion that gravity is indeed a memory effect: gravity is the resonation of the past across pure time, without space or contiguity.

It is now obvious that gravity is *not* a thing that arises from the future: It does *not* emerge out from the Whitehead facet. Instead, it arises from a past (of the Moon) that restructures the future (i.e., the future-potentiality of tidal effects on Earth) according to the total superpose of that past upon the possibility-state of the Earth. The Bergson facet and its memories impose a consistency-enforcing *gravitational grand decoherence* and restructuring of the Whitehead facet. Succinctly, gravity involves the superposition of all the *wavefunctions in memory* that belong to different past moments of a body retained within the Bergson facet (Book-IV **4.27**). Indeed, the gravitational mass *is* the total Bergsonian memory superpose… or so it is argued in Book-IV Part-III.

The said superposition of wavefunctions in memory (the generation of 'gravitational mass') proceeds according to a work function zero process. That is, the superposition of the past with itself does not require energy inputs or work to bring about.

The wavefunctions in memory involved in the superposition of memories are subject to the *inverse square law of time* (**4.28**). The process of mnemonic mechanics generates counterfactuals with respect to the Whitehead facet, enforcing consistency to that future *en toto* (**4.31**… and Book-II, **2.43** and **2.45**). In short, *memory tells the future how to grand decohere and restructure, while the decohered future tells the OR-form outcomes generated from subsequent AND-to-OR collapse and time how to form up*. This replaces the adage that 'matter tells space how to curve… and the curve tells matter how to move'. The 'motion' brought about is a non-uniform change in delayed choice time-interval relation of attendant events in the Moon-Earth relation. It is the associated quantised nature of time and the broken-discontinuous moment-to-moment realisation of the non-uniform change in the delayed choice time-relation between the Moon and the Earth that then fashions for quantised gravity… or *quantum gravity*.

5.03: Preface to the Critique of Generic Interpretations of Quantum Mechanics

In this and following essays, we will summarise the key interpretations of quantum mechanics. Our task will include the Copenhagen interpretation and Manyworlds, both of which will be contrasted with the Alternative Realist interpretation (Alt-R), founded on the philosophy of physics recapitulated in the preceding essays: (**5.01** to **5.02**).

All interpretations of quantum mechanics address eight critical areas of concern, articulated as eight key questions, such as… is the wavefunction ontologically real? Does wavefunction collapse really occur? Is quantum indeterminism real? Are there hidden gears or variables behind seeming quantum indeterminism and wavefunction collapse? The eight-fold questions of quantum interpretation and philosophy will be detailed in the forthcoming essay (**5.04**).

In the forthcoming material, **we will contrast the position of Alt-R in terms of the standard eight-fold questions of quantum interpretation. We will also critique the generic interpretations in the light of Alt-R's own premises and conclusions. Obviously, the point of these essays is to espouse Alt-R vis-à-vis key generic interpretations of quantum mechanics.**

*

Unlike Alt-R, generic interpretations do not infer new phenomena, nor furnish novel advancements, much less presage solutions to known phenomena whose explanations have remained elusive. New material from Alt-R include de-spatialisation (see Book IV Part-I), with all its implications; the nested futures view of the wavefunction, with attendant counterfactual causality and grand decoherence (Book-II: **2.18** to **2.21**). Alt-R also posits the revival of Mach's principle through grand decoherence (Book-II: **2.25**) and the presage to the long-elusive quantum mechanical description of inertia and the permanent magnet (Book-II: **2.35** and **2.37**). Alt-R also presages solutions to gravity that circumvent the dichotomy between General Relativity and quantum mechanics, positing gravitation as a memory effect; one co-joined to grand decoherence; unfolding in pure temporal form in a growing block universe; presages to quantum gravity itself (Book-IV, **2.43** to **2.45**).

These and other findings emerged from the new insights into the quantum wave and AND-form logic, with both grasped as indices of the phenomenalisation of an ontically real future, and the attendant growing block reform of 'spacetime', in turn justified by the intertwine of causality and conservation law in the firewall principle (Book-II, **2.21**).

Unlike Alt-R, generic interpretations do not make a case even for a primitive unification of quantum and classical-relativistic physics. The unification Alt-R asserts culminates into the growing block Bergson-Whitehead amalgam (Book-IV, **4.10**) with attendant contributions to the solution to inertia and gravity from the processes of the de-spatialised temporised Bergson-Whitehead amalgam and the mnemonic mechanics that attends it: (Book-IV **4.29** to **4.30**).

As stated, **generic interpretations do not espouse de-spatialisation (Book-IV Part-I) or even recognise the problem of space, much less recognise AND-to-OR wavefunction collapse as synonymous with time itself; no longer a 'fourth dimension'. Generic interpretations have nothing to say about the ontological primacy of time, much less the ontic status of abstract memory beyond forlorn materialistic 'trance memory'** (Book-III Part-I and Book-IV Part-I), or of the reframing of basic relations in nature into pure delayed choice time-interval relations, *without* space (See Book-IV **4.08**). These and other findings illustrate the superlative character of Alt-R in contrast to the various generic quantum interpretations. Thus, Alt-R, or its future variations, improvements, and perfections, is surely destined to constitute the final and decisive interpretation of quantum mechanics and philosophy of physics: **the preface to quantum mechanics-ii.**

*

Two key conundrums at the centre of quantum interpretation are… **the cause of wavefunction collapse**, combined to the related but distinct issue of **whether quantum indeterminism is objectively real**. From discoveries made about the basis of quantum indeterminism (Book-III Part-II) the position of Alt-R on these issues is that there is no clockable tick-tock click-clock 'hidden gears' or 'hidden variables' behind wavefunction collapse (or the drive of time itself), and that quantum indeterminism is no more secretly orchestrated and determined by hidden gears or variables than is attendant wavefunction collapse. Therefore, wavefunction collapse… hence time itself… are 'spontaneous', and quantum indeterminism is non-illusory and objectively real.

Now, **quantum interpretations that espouse hidden gears or hidden variables constitute default materialistic interpretations**: they are consonant with the want to find a closed input-operation-output schema (IOO-schema) or a definable operational basis for the cause of wavefunction collapse and the drive of time, combined to the want of a similar IOO-schema in the orchestration behind surface quantum indeterminate processes. Indeed, any future project to revive materialism through the interpretations of quantum mechanics would require that it revamp IOO-schema *and* revive 'space'. Only with space is the notion of contiguity, impact, and contact causality viable, given that there needs to be a 'location' or spatial point-of-contact between 'spaceballs' such as to facilitate the materialist concept of causality, or contiguously mediate causality, energy, work, or 'force' between supposed energy-bearing or force-conveying spatial entities. **Without space… or in any purely temporised physics asserted by Alt-R… tenets indispensable to materialism are rendered untenable. No space, no 'location'. Therefore, no contiguity. Therefore, no impact causality, much less closed-causality IOO schemas for energy-force bearing or conveying 'spaceballs'.** Nevertheless, the idea of the 'particle' survives within Alt-R, but recuperated into its temporal form, retaining *only* its 'frequency' attribute. Hence, *durata* replaces *atomos* (Book-IV, **4.14**).

Since time itself is quantised, so are the durata issued out of AND-to-OR collapse. The attendant AND-to-OR collapse process equates to a 'packet of energy' proportionate to the frequency of the durata: a pure time measure and time-description of 'particles' and of 'energy and work'. **Yet, even as frequency survives Alt-R, energy and work wholly disappear as these are rendered pleonastic**, (see Book-II **2.39** and Book-IV **4.19**)

466

Insofar as Alt-R espouses a structure-less and ineffable causality as the drive of time; insofar as it rejects IOO-schemas and closed causality, and, per de-spatialisation and temporisation, rejects contiguous impact causality and rejects the secret orchestration of quantum determinism, it follows that **Alt-R permanently usurps the dubious equivocation of physicalism and naturalism with materialism, or with contiguous impact causality and IOO-schemas, 'gears', closed causality, and closed universe.** Consequently, a recuperated naturalism, physicalism and science abide… but materialism withers.

*

Not all interpretations espouse hidden gears: **Not all interpretations of quantum mechanics are materialist in form. Quantum interpretations that do not espouse hidden gears could be deemed default non-materialistic interpretations. However, one would need to scrutinise these on a per-case basis to judge correctly.** For example, Manyworlds (MWI) does not espouse hidden gears in the orchestration of quantum indeterminism. Yet, upon scrutiny, MWI turns out to be a materialistic interpretation per function of its modus operandi; one involving the a-causal brute-force exhaustion of possibilities into histories. On the other hand, 'consciousness causes collapse', and the related 'Manyminds' interpretation, constitute consummate non-materialistic interpretations. Of course, 'consciousness causes collapse' is comorbid with idealism, which espouses the ontological primacy of nous and seeks to undermine object-subject relations by collapsing the object-world to subject-nous: the inverse of materialisms' attempt to collapse subject to object.

Alt-R constitutes a realist *and* non-materialistic interpretation of quantum mechanics and of general physics. Alt-R usurps hidden gears *and* IOO-schema closed causality. It undermines contiguous or impact causality through de-spatialisation and temporisation, transforming *atomos* to *durata*. Yet, Alt-R *also* casts permanent doubt on 'consciousness causes collapse', as we will discover in Part-II. Therefore, **Alt-R is as anti-idealistic as it is anti-materialistic, and seeks to preserve the object-subject relation**.

Furthermore, Alt-R is shamelessly Cartesian in its support of the object-subject divide, as we shall discover in Part-III and thereafter.

5.04: Eight Questions of Quantum Mechanics & the Claims of Alt-R

Below is a brief list of the eight central questions that attend the various interpretations of quantum mechanics.
All interpretations seek to address these questions in unique ways. In logical sequence, the questions are…

- Is the wavefunction ontologically real?
- Is wavefunction collapse real?
- Is quantum indeterminism objectively real?
- Does hidden gears apply to quantum indeterminate resolutions and wavefunction collapse?
- Does wavefunction collapse lead to unique history or to many histories?
- Is there a role for the observer in wavefunction collapse and in quantum indeterminism?
- Locality or non-locality? Is there such thing as instantaneous 'action at a distance'?
- Is there a universal wavefunction?

The attached table constitutes a summary of the key assertions of the Alternative Realist interpretation of quantum mechanics vis-à-vis the eight questions of quantum mechanics: A similar tabulation will apply to generic interpretations of quantum mechanics.

ALT-R INTERPRETATION OF QUANTUM MECHANICS							
W-function real?	**W-function collapse real?**	**Indeterminism real?**	**Hidden gears?**	**Observer role?**	**Unique history?**	**Non-locality?**	**Universal W-function?**
YES (it is the ontically real future)	YES (it is time itself)	YES (objectively real)	NO	decoheres but does not collapse	YES (but w/ Little MW)	YES …but as counterfactuals driven grand decoherence and transtemporality	YES Whitehead facet to the infinitely remote future

In summary, our Alt-R interpretation claims the following:

- **The wavefunction is real. It constitutes the growing block future**: The wavefunction is the inadvertent formal mathematical response to an ontically real future: i.e., "spacetime in potentiality", later succeeded by the Whitehead facet of the superlative de-spatialised and temporised Bergson-Whitehead amalgam. The wavefunction approach gains completion with the addition of the *nested futures* approach implied by John Wheeler's delayed choice, with attendant counterfactual causality, *grand decoherence*, and attendant 'non-local' transtemporal binding of each frame of reference to every other in the universe. The ontic future is justified per demands from the *firewall principle* from the intertwine of causality and the conservation of energy and matter. Also see the presage to the overhaul of the Schrödinger equation in Book-II **2.27**.

- **Wavefunction collapse is real. It constitutes time, synonymous with AND-to-OR collapse**. According to Alt-R, wavefunction AND-to-OR collapse is time itself, i.e., the process of the conversion of ontically real AND-form futures into OR-form actua. Hence, *time is not a 'fourth dimension'*: it needs to be modelled as partial-perpetual wavefunction collapse of the Whitehead facet... of the future. Moreover, wavefunction collapse does not constitute a comprehensive complete exhaustion of the total future: The future is a convergent category nested futures system that elaborates to the infinitely remote future, and it is always subject to partial-perpetual AND-to-OR collapse, *not* to single-step comprehensive collapse. Moreover, *time proceeds from AND-to-OR... never from OR-to-AND*. Thus, there can be no time-reversal.

- **Quantum indeterminism is real**: (see below for 'no hidden gears').

- **There are no hidden gears behind wavefunction collapse or time**: According to Alt-R, wavefunction collapse (time) is not procured by means of hidden clockable tick-tock click-clock 'gears'. Any purported gears that cause time to happen would crash into insurmountable infinities and halting problems, given that a clockable process could not systematically sort and process infinities attendant nested future counterfactuals and grand decoherence to time-like infinity without crashing into said halting problems. Indeed, time itself would fail to transpire; 'freezing up', so to speak. It follows that wavefunction collapse and time is made to happen by an ineffable structure-less causal process; one that cannot be subject to or suffer any non-terminating problem; one that cannot be identified with or reduced to physical inputs, outputs, or any putative definable clockable 'operation', 'gears', algorithm, or computation. Furthermore, time and wavefunction collapse is itself ontologically primary to any putative or seeming mechanical tick-tock click-clock process: In other words... no time, no mechanical process... and no manifest input-operation-output schema or gears. Thus, it is not a hidden clockable and time-dependent mechanism that 'computes' or generates time. Instead, it is time and its structure-less ineffable drive that gets to AND-to-OR collapse the succession of events into seeming clockable 'computational' and 'mechanical' unfolding: (Book-II **2.46** to **2.47**).

- **No hidden gears behind quantum indeterminism:** Similarly, hidden gears could not deterministically and secretly orchestrate quantum indeterminate processes. While hypothetical deterministic orchestrating scripts for quantum indeterminism could be imagined, we obtain a very large number of such; all equal to the task, and with none exclusively special. The question of which script from among the very large number of such get to secretly orchestrate seeming randomness can only be decided by resort to a random sortology: (see Book-III **3.31**, and forthcoming **5.08** and **5.16**). Thus, randomness and indeterminism cannot be obviated, even when we imagine the operation of secret or hidden deterministically orchestrating scripts. Therefore, insofar as randomness cannot be obviated, it follows that quantum indeterminism is also objectively real. This is not a disaster. Indeed, quantum indeterminacy is not inimical to information and pattern: it does not prevent the formation of pattern. The information *is* the complex wavefunction itself, often retained as a wavefunction in memory within the Bergson facet. The abstract wavefunction complex, as an 'instantaneous whole', can be 'materialised' by *en masse* objective quantum indeterminate terms according to the three-phase developmental process, without resort to any secret orchestration of terms: (See Book-III **3.26** and **3.31** for solutions to quantum indeterminacy).

- **Observer and consciousness assume only a moderate role**: According to Alt-R, the act of observation (the instantiation of observational inputs and outputs vis-à-vis any system) can decohere a system to a reduced set of future possibilities. However, the conversion of the now-reduced AND-form state into an OR-form outcome (i.e., AND-to-OR wavefunction collapse, hence time itself) cannot be engendered by any tick-tock click-clock mechanism or hidden gears attendant the mechanics of observation and measurement. The mere input of observationals, followed by the subsequent reception of return-signal outputs, is *not* capable of acting in lieu of the ineffable structure-less process of causality that brings about wavefunction collapse and time, and the outputs pertinent to observation with time. But what are we to say about 'consciousness causes collapse'? While the observer, hence nous, appears to decide the schedule and nature of the observation and measurement to be carried out, and can certainly bring about decoherence through imposing a specific setup through the mechanical act of observation itself, consciousness cannot bring about time: i.e., it cannot bring about wavefunction collapse. This is because time is ontologically primary to mind and consciousness as much as it is to 'mechanism': In other words...no time, no consciousness. Thus, it is not consciousness that brings about time. Instead, time and synonymous AND-to-OR collapse brings about the retrospective passive consciousness of events, and generates the resolution of outcomes sought by active mind. Thus, nous cannot be substituted in lieu of hidden gears in the generation of wavefunction collapse and time. Consciousness and the 'observer' can no more bring about time than it could secretly orchestrate quantum indeterminism by means of a secret consciousness-driven will into OR-form outcomes and observables. The issue of 'consciousness causes collapse' will be comprehensively usurped in Part-II.

- **Unique history**: According to Alt-R, time and wavefunction collapse, or AND-to-OR collapse, generates a unique history, *not* many histories. The generation of unique histories is attended by unique counterfactuals-driven quantum erasure of nested futures, or grand decoherence, elaborated to the infinitely removed future within the Whitehead facet of the superlative Bergson-Whitehead amalgam. This assertion is further augmented by the case against the Manyworlds interpretation (see Book-III **3.12** through to **3.18**; and forthcoming **5.07**: See further references to MWI below). However, there is "Little Manyworlds" or LMW to consider (**3.18**). LMW might abide where a single quantum incidence is capable of bringing about

468

the drastic alteration of the much larger universal wavefunction or the Whitehead facet as a whole, exemplified by the single quantum incident-triggered poison vessel in the famous Schrödinger's cat thought experiment. Obviously, a single quantum incident-triggered nuclear detonator would drastically alter the probability and possibility structure of any complex wavefunctions that attends the future of a city. The universe could conceivably split into two distinct compartmentalised cities at just such a LMW junction.

- **Non-locality is real, but is advanced into counterfactual unity and transtemporality through grand decoherence**: On the dubious assumption of space, the sort of 'non-locality' made evident in the Einstein Podolsky Rosen experiment, and in subsequent Paris experiments of the 1980s, and in others since, turns out to be real, according to Alt-R. However, we need an overhaul of spacetime itself to gain a better grasp of putative non-locality. Succinctly, we find that profound non-locality does indeed attend the process of grand decoherence and the counterfactual erasure of non-consistent nested futures throughout "spacetime in potentiality" or the Whitehead-facet of the Bergson-Whitehead amalgam. Therein, non-locality manifests as grand decoherence (Book-II **2.34** to **2.35**) and the counterfactual binding of each frame of reference with and to every other frame on and along the event horizon, if not in all future-potential event horizons embedded to the convergent category that bestrides the totality of the Whitehead facet. Consequently, a supernova at Alpha Century might not be directly observable from Earth's frame for some four years. Yet, insofar as the futures that attend Earth's frame converge with futures that attend the Centauri frame (in that, their potential lightcones and potential future worldlines overlap and enmesh some four years from now within the futurity of the Whitehead facet) what happens at Alpha Centauri 'now' will instantaneously decohere the nested futures that attend our Earth-frame. In Earth's frame, this decoherence will allow only the potentialisation of consistent futures vis-à-vis the supernova at Alpha Centauri: It will erase those futures that are inconsistent with the supernova fact, leading to *grand decoherence*. Such *grand decoherence revives Mach's principle; rendering possible the quantum mechanical explanation of inertia*, wherein 'matter' here counterfactually, instantaneously, and transtemporally 'knows' what matter is *not* doing elsewhere, and vice-versa; with their convergent future-form wavefunction decohering to consistent-only possibilities, while erasing non-consistent futures. Of course, the moment we de-spatialise and temporise physics, 'locality' in its spatial form can no longer abide. Consequently, counterfactual binding and grand decoherence must be recapitulated in pure transtemporal terms: wherein one frame is counterfactually coupled and instantaneously responsive to every other frame in a *transtemporal* way i.e., without taking time, or instantaneously. However, transtemporality (or de-spatialised 'non-locality') also applies in a profound way to the enforcement of physical law. In EPR paradox, it is not possible to transact meaningful information from one of the pairs of entangled particles to the other by controlling how one of the couples is resolved. This is due to objective quantum indeterminism. However, the conservation of spin constitutes a piece of information in its own right; a fundamental one. The principle is enforced on the entangled couple in a *transtemporal* way... without taking any time, whatever the indeterminate spin-outcomes generated. Hence, 'law enforcement' is from 'outside' of time. In short, we obtain the transtemporality of abstract physical law: (see Preliminary **0.24**, Book-I **1.11** to **1.13**, Book-III **3.33**). This confirms that laws are indeed Platonic-Cartesian abstractions. As such, laws are enforced on the physical order in a transtemporal Cartesian way. The issue in EPR was never ultimately 'faster than light' information-conveyance, but one of enforcement of law and principle from outside of 'spacetime' or, more succinctly, from outside of the system of de-spatialised and temporised memory, possibility, and time: a system recuperated as the Bergson-Whitehead amalgam.

- **The universal wavefunction is real**. Otherwise referred to throughout our work as the 'universe-scale wavefunction', the universal wavefunction is constituted by the whole of "spacetime in potentiality" to the infinitely removed future. With the de-spatialisation and the obviation of 'space', the universal wavefunction constitutes the whole of the de-spatialised temporised Whitehead facet. Alt-R also asserts that the universal wavefunction is an instantaneous whole structured as a convergent category nested future system elaborated to the infinite future; one subject to transtemporal counterfactual causality (the replacement of 'non-locality') and unitary binding via grand decoherence.

We have not included other findings, such as *the primitive unification of quantum mechanical and classical-relativistic physics*; first, as the intermediate model spacetime in Book-II, then, with de-spatialisation, as the temporised Bergson-Whitehead amalgam in Book-IV. We must again briefly mention *mnemonic mechanics* and other inferences and implications, including the reform of and *pleonastics of energy relations*. In a de-spatialised universe posited by Alt-R, there are no 'locations' for objects and particles. The energy associated with particles and with matter is therefore not stored in space or in 'matter located in space'. Instead, the putative potential energy must issue out of the abstract future: from the Whitehead facet (See Book-II **2.39** and Book-IV **4.19**). Moreover, since AND-to-OR collapse and time does not itself require a work or energy-input to bring about (since the drive of time is a work function zero process) it follows that the energy relations brought into realisation by time emerge through a *non-energy driven work function zero causality that generates time*. Given that time has ontological primacy over energy relations, in the obvious sense that, no time, then no transformation of potential energy to putative kinetic energy via AND-to-OR collapse, hence no putative energy relations without time, the non-energy-driven causality behind time must necessarily imply that the universe is *not* an energy or work-driven order. The ontic primacy of time over energy relations implies that putative energy relations are ultimately *not* parts of real existence or ontology: they are wholly pleonastic, given that state-change is brought about by non-energy driven time itself. We will explore further the demise of energy relations in the context of ontologically primary time when we hammer out the critique of materialism in Part-III.

5.05: Critique of the Copenhagen Interpretation: Contrast with the Alternative Realist Interpretation

We start with **the Copenhagen interpretation**, formulated by Niels Bohr and Werner Heisenberg in 1927. The pertinent tabulation given is identical to the one presented for Alt-R, but specific to the Copenhagen interpretation. Note the second row comprised of answers to the eight questions. This should help differentiate between the Copenhagen interpretation and Alt-R, even where the interpretations agree. The agreements are coincidental and not necessarily due to fundamental accord. Note that some of the answers are given in bold: This is to highlight critical differences between the two interpretations.

Note that, in the pertinent table, the Copenhagen interpretation states that the wavefunction is not real, but then states that wavefunction collapse is real. The contradiction arises from the fact that **the Copenhagen interpretation does not recognise the wavefunction as the ontic future. Nor does the interpretation explicitly recognise AND-to-OR wavefunction collapse as time itself. This is per the tendency throughout physics to regard time as a block universe 'fourth dimension', or as just 'what the clock reads'; not as thing identical to wavefunction collapse itself. Thus, in the Copenhagen interpretation, the state of the particle before measurement and collapse is considered 'meaningless'; merely a mathematical abstraction of convenience... with only post-collapse observations asserted to be real or meaningful.**

Clearly, the state of the particle before measurement is *not* a classical OR-form state. It is purely an AND-form state of futurity. To expect it to be OR-form per the demands of Naïve Realism (i.e., Einstein *et al*) is certainly futile and 'meaningless', and the supposition invites violations of causality and the usurping of the conservation of energy-matter. This was explained in Book-II (**2.10** and **2.11**) and in previous summary notes (**5.01**). However, the AND-form state, hence the superpose of possible states of a particle grasped as its future potentialities, *are* physically meaningful, as Alt-R asserts: These constitute the unrealised *objective* ontic future of the particle (or of the *durata*) rendered in AND-form. That future, consummated first as "spacetime in potentiality", is finally superseded by the de-spatialised temporised Whitehead facet of the Bergson-Whitehead amalgam (Book-IV **4.10**).

As can be seen in the pertinent table, **the Copenhagen interpretation coincidentally agrees with the reality of quantum indeterminism. Yet, it does not demonstrate its objective reality, but relies on the probability distribution of possibilities in the AND-form wavefunction.** In contrast to the Copenhagen interpretation, Alt-R seeks to prove the objective reality of quantum indeterminism (see Book-III **3.26**, and Part-I **5.01**).

Moreover, **the Copenhagen interpretation assigns the cause of wavefunction collapse to the mechanical interaction of the experimenter or observer with the system under scrutiny. However, if wavefunction collapse is time itself, as Alt-R asserts, then this later claim is tantamount to asserting that the experimenter's mechanical interaction with the system, hence the mechanical process of observation and measurement itself causes time,** as opposed to the ontological primary time vis-à-vis the act and process of measurement, and which renders the time-subordinated and time-dependent processes of experimental mechanical interaction possible. **By emphasising the observer's interaction as causal to measurement and time, the Copenhagen interpretation renders itself vulnerable to the "consciousness causes collapse" interpretation and to idealism: It could be argued that it leads to idealism.** The latter substitutes consciousness itself, beyond the mere mechanical intercession with the experimental system, as directly causal of wavefunction collapse and time, if not as the secret orchestrator of indeterminism itself: Perhaps this vulnerability underpins the charge sometimes made against the Copenhagen interpretation as 'subjective'.

COPENHAGEN INTERPRETATION OF QUANTUM MECHANICS							
W-function real?	**W-function collapse real?**	**Indeterminism real?**	**Hidden gears?**	**Observer role?**	**Unique history?**	**Non-locality?**	**Universal W-function?**
NO …does not recognise ontological reality of the future	**YES** …but no recognition as synonymous with time; contradictory	**YES** …but no proof of objectivity of randomness, nor proof of absence of gears in W-function collapse	**NO**	**YES** …observer interaction causes W-function collapse	YES	Locality	**NO** …does not recognise ontological reality of the future

However, according to Alt-R, time is ontologically primary to experimental interaction and measurement: No time, and no interaction and measurement can occur (mechanical or otherwise). It is time or wavefunction collapse that brings about the putative mechanical interaction and generates the outcomes from the decay of future possibilities to subsequent OR-form observables. It is *not* the observer, much less the experimenter's mechanical interaction, which brings about time and wavefunction collapse. Yet, the observer and the interaction can certainly decohere the set of future possibilities; a role enhanced by the nested futures view and by the operation of counterfactual causality and grand decoherence. The observer and his mechanical interaction can also specify whether we will obtain a single-slit outcome or a double-slit outcome in a pertinent experiment: a profound impact on the repertoire of possibilities and on how the future is subsequently re-structured. However, this could not obviate the primacy of wavefunction collapse and time over the observer and over the observer's mechanical intercession, and over the experimental setup. Again, it is ontically primary wavefunction collapse and time that renders these possible: the latter do not bring about wavefunction collapse or time.

5.06: Definition of the Apparatus: The Copenhagen Interpretation versus Alt-R Solution to the Apparatus Problem

One area of concern not directly stated in the tabulated summary of the Copenhagen interpretation, is the apparatus problem. The Copenhagen interpretation cannot define the apparatus. Consequently, its attempt to grapple with measurement is incomplete. In fairness, the interpretation is not the only one unable to solve this problem.

Is the apparatus OR-form or is it AND-form quantum mechanised? According to Alt-R and its growing block model, the future abide at all scales, as especially testified by the ubiquity of time at all scales…hence the ubiquity of synonymous AND-to-OR collapse at all scales. Finally, the ubiquity of time at all scales is justified by grand decoherence (Book-II **2.34** and **2.35**) and the Whitehead facet of the growing block Bergson-Whitehead amalgam. Consequently, **the apparatus itself must collapse out of the AND-form Whitehead facet (out of the future) for the experiment and apparatus to resolve into a complex OR-form outcome,** as must everything else in the universe. In other words, in one important respect, the apparatus is certainly AND-form and quantum mechanical in character. Succinctly, the apparatus has a future state, and the apparatus must decay out of its initial future AND-form state into a culminating OR-form state, according to Alt-R.

More comprehensively, **the basis of the apparatus is a combination of two things. The first is the nested futures facet of the setup, with attendant counterfactual causality and the quantum erasure of non-consistent nested futures** (Book-II **2.18** and **2.22**) that this latter implies. **The second is the relegation of the setup, once it is formed and configured, and once it is finalised into a culminated set of outcomes, into Bergsonian memory. Bergsonian abstract memory, retained within the Bergson facet, will project itself onto the Whitehead facet; decohering the latter into possibilities consistent and conformal with the apparatus and its setup, reducing the scope of subsequent AND-to-OR resolution of experimental outcomes into the recapitulation and temporal recouperation of the apparatus in its initial setup**: (see Book-IV **4.31**).

In short, what is lacking in the Copenhagen interpretation (and in other interpretations) is abstract memory, and the role played by Bergsonian memory in conjunction with future possibility-states in the definition and solution of the apparatus problem. **Abstract memory is absent from the Copenhagen interpretation and from its attempt to define the apparatus, because it implicitly subscribes to the a*mnesic universe thesis*,** as do all other interpretation of quantum mechanics: (see Book-III **3.00** to **3.03**).

Yet, as was discovered in Book-III 3.07, quantum mechanics would be rendered impossible *without* the implicit role and ontic reality of abstract memory and, ultimately, of Bergsonian memory: (The basis of the latter was established as a direct result of radical de-spatialisation in Book IV **4.10**). In Book-III we discovered that, upon isolation, and even with the futurisation and quantum mechanisation of the two-slit barrier setup per function of that isolation, and the relegation of it to the AND-form future, the initial setup of the barrier must yet survive in retained form as abstract memory, and in a form independently and extant its now futurised AND-form state. Otherwise, if memory is not ontologically real, and the experimental setup is not retained by the natural order as abstract memory, the barrier-setting could subsequently AND-to-OR resolve into any of its alternative configurations, including 'all slits closed', if not into a pumpkin instead… or into nothing at all. Thus, an amnesic universe would produce vast inconsistencies, and it would render physics and quantum mechanics as we know it untenable and impossible.

Hence, even if implicitly so, memory is indispensable to the very viability of generic quantum mechanics. Of course, this is memory retained in an abstract-immaterial form *extant* the system: a fact obvious from the various forms of information-matter dualism from the Preliminary through to Book-IV. It is also compelled by the de-spatialisation of physics, which forces the recuperation of the apparatus and the barrier configuration into a *retained* temporal distribution of information… as abstract memory... with 'materialised' OR-form facets of the apparatus restricted and confined only to the event horizon.

Notwithstanding implausible 'space', the only recourse against abstract memory would be the informational and mnemonic nihilistic approach comprised of the total erasure of informational and memory, augmented by its combination with inherently nihilistic Manyworlds interpretation. In Book-III **3.12** to **3.18**, we showed that such an informationally nihilistic approach must fail, if for no other reason than the fact that MWI is not itself viable.

The Alt-R definition of the apparatus, based on the indispensability of abstract Bergsonian memory to the very viability of generic quantum mechanics, is a conclusion beyond the purview of the Copenhagen interpretation, and beyond all generic interpretations. This is because the Copenhagen interpretation recognises neither the ontic reality of the future nor the equally ontic status of a non-eradicable past; one constituted as abstract memory.

Of course, Alt-R does not furnish direct acquaintance with the causal process of generation of OR-form outcomes from AND-to-OR collapse. It is time and wavefunction collapse that mediate the now consistency-rendered future into expected recuperation of the apparatus and of conformal outcomes, furnished by counterfactuals and the grand decoherence imposed by the apparatus retained in abstract memory. However, since AND-to-OR wavefunction collapse and time cannot itself be reduced to an IOO-schema, and since time is not explicable in terms of forlorn hidden gears (see Book-I **1.35** and Book-II **2.46** to **2.47**), it clearly follows that it is *not* the apparatus itself that makes time happen. Time or wavefunction collapse is generated by a structure-less ineffable causality that operates on a work function zero basis: This causal process cannot be identified with or be affected by the apparatus. In short, **'measurement' and the operation of the apparatus, insofar as it is brought about by ontologically primary time and wavefunction collapse, is not reducible to or accountable in terms of the mechanics and configuration of the apparatus and measurement.**

5.07: Critique of the Manyworlds Interpretation of Quantum Mechanics

Developed by Hugh Everett in 1957 and later popularised by Bryce DeWitt, the Manyworlds interpretation[90] (MWI) was scrutinised in detail in Book-III in the context of its anticipated forlorn union with information erasure hypothesis. It was also critiqued in its own terms (Book-III **3.12** to **3.18**). We will render only a summary recapitulation of our Alt-R-based critique of MWI in relation to the eight questions of quantum mechanics.

As is true of all other interpretations, MWI subscribes to the *amnesic universe thesis*: Indeed, per its very structure and *modus operandi*, memory and informational nihilism is built-in to Manyworlds. It affords no place or postulate for memory in nature, even though quantum mechanics in the form we know it would be untenable without abstract memory (see Book-III **3.07** to **3.08**). Nor does MWI form a basis for any form of unification of physics equivalent to the growing block unification of quantum and classical-relativistic descriptions, such as furnished by the intermediate spacetime, and by the later Bergson-Whitehead amalgam.

The principal novelty of MWI is that it seeks to obviate the problem of quantum indeterminism by postulating that *all* the possibilities inherent to the AND-form wavefunction get OR-form realised... supposedly into separate compartmentalised histories conveniently beyond provability, and never mind violations of the conservation principle that the many histories would entail.

In short, MWI seeks a solution to the problem of quantum indeterminism by asserting for *many* histories instead of one unique history. Thus, **it asserts that quantum indeterminism is not real, and hidden gears is rendered unnecessary,** insofar as the want of a secret deterministic orchestration of apparent quantum indeterminate outcomes is rendered void.

However, MWI is contradictory in two senses: **It claims that the wavefunction is real (indeed, it postulates the universal wavefunction (see table)). Having postulated the reality of the wavefunction, it then denies the reality of wavefunction collapse.** From our Alt-R perspective, this contradiction is tantamount to accepting the ontological reality of the future (the wavefunction) but denying the process of time synonymous with AND-to-OR wavefunction collapse, from which the future decays into actuality (see notes in table at relevant columns). All of this emerges from a fact typical to all interpretations: simply, **MWI does not recognise wavefunction collapse as time itself, or even AND-form logic as an expression of the ontic future, part of the growing block system.**

MANYWORLDS INTERPRETATION OF QUANTUM MECHANICS							
W-function real?	W-function collapse real?	Indeterminism real?	Hidden gears?	Observer role?	Unique history?	Non-locality?	Universal W-function?
YES …but does not recognise ontological reality of the future	**NO** … and not recognised as synonymous with time	**NO** Deterministic but exhaustive: all possibilities are realised as distinct histories	**NO**	**NO** …observer not needed, given all possibilities realised	NO …many histories per each possibility in w-function	Locality	YES (proposed by Everett)

However, in Book-III, MWI was further faulted per the following findings:

- **Quantum indeterminism is not a problem that requires a solution:** Therefore, the resort to MWI, or any other contrivance to 'solve' quantum indeterminism, is superfluous and unnecessary. This Alt-R-based charge against MWI is one that can be arrayed against any other interpretation that treats quantum indeterminism as a 'problem' in need of a solution. From the solution to the stochastic coherence problem from Book-III (see **3.26** to **3.32**) we discovered that both **randomness and quantum indeterminacy are objectively real**, with no secret deterministic orchestration or hidden gears. Thus...
  - When we take any random or indeterminate developmental process and imagine fully deterministic secret orchestrating scripts for it, we end up with a very large number of such scripts: each equally apt to resolve the whole process of development deterministically, but none exclusive or special as to constitute the actual script.
  - The only way to select from between the very large number of equally apt scripts is to resort either to arbitrary selection by conscious agency (no different from resort to blind randomness) ... or by resort to a sortology process involving blind random selection. In short, we cannot obviate resort to randomness and indeterminacy even when we form deterministically orchestrating scripts and push the generation of such to infinity. (See Book-III **3.32**.)
  - Therefore, quantum indeterminate processes are truly objectively real, and cannot be obviated, even when we imagine or assume the operation of hidden orchestrating script or 'gears'.

---

[90] *Everett, Hugh; Wheeler, J. A.; DeWitt, B. S.; Cooper, L. N.; Van Vechten, D.; Graham, N. (1973). DeWitt, Bryce; Graham, R. Neill (eds.). The Many-Worlds Interpretation of Quantum Mechanics. Princeton Series in Physics. Princeton, NJ: Princeton University Press. p. v*

- **Objective quantum indeterminate processes unfold according to a three-phase developmental process.** (Book-III **3.26**): These are...
  - The **below threshold phase**, which approximates to our prejudices about randomness: A small collection of objectively indeterminate quantum depositions on a photographic film, or in the detector in the two-slit experiment, cannot comprise any sort of recognisable pattern or information. Thus, randomness and quantum indeterminacy at this threshold conforms to our notions of the lack of information in randomness, i.e., 'noise'.
  - In the **chaotic-transitional phase**, despite objective quantum indeterminacy, beyond a threshold of quanta, unstable and transitional pattern begins to emerge, with the smallest number of subsequent inputs generating unpredictable short-run pattern-outcomes and shifting developmental sites. We find that some sort of pattern and structure is emerging, but we are uncertain about the developmental endpoint due to the said volatility.
  - Development is concluded in the **noise-reduction pattern-saliency phase**, wherein the further addition of objective quantum indeterminate terms now leads to the saliency of stable pattern and the reduction in 'noise', despite the ineliminable reality of objective quantum indeterminacy. Pattern saliency could involve, say, the formation of a stabilised distribution curve or 'bell-curve'; one obtained from sums from consecutive or parallel rolled three six-side quantum mechanised dice. It does not matter that the die rolls, governed by objective quantum indeterminate processes, are objectively random: These cannot help but be funnelled into pre-fated culminating pattern-outcomes (the bell curve itself) per the ordinances inherent to the bounded probability state that governs the three six-sided quantum mechanised die-roll system.
  - Indeed, **the bounded probability state 'instantaneous whole' that govern systems *is* the information; as is the complex wavefunction in memory or potentiality. This fates quantum indeterminate processes in photography (or in anything else) to an ordained fated image-outcome**, even if the photon depositions involved are objectively quantum indeterminate.
  - **Randomness and indeterminacy constitute nature's economic solution to pattern-development and saliency.** Nature would be far more complex and cumbersome if it had to secretly assign the deposition-addresses per each pixel or quanta vis-à-vis every other such in order to render salient complex pattern, whether of a photographic image, a drifting snowflake, or a whole galaxy.

  In short, randomness and quantum indeterminacy do not constitute a 'problem' in need of a 'solution', or in the formation, viability and perdurance of pattern and information. Succinctly, randomness and quantum indeterminacy are not inimical to information and pattern, but necessary economising factors to the possibility of both. Quantum indeterminacy is not a problem that needs a 'solution'. Therefore, there is no need for an MWI contrivance, any more than there is need for Bohmian hidden variables, and even less a need for resort to a back-time retro-causal process claimed in the intriguing *transactional interpretation* of quantum mechanics[91].

  However, MWI is invalid for another reason. Recall that MWI admits to a universal wavefunction. What if we treated the Sierpinski triangle as a universal possibility function in lieu of the actual universal wavefunction? What if we treated the development of a quantum mechanised Sierpinski triangle as an analogous simplified universe, and used it to test MWI? We did just that in Book-III **3.17**. Here we summarise and recapitulate our findings. Thus...

  - At each orbit-point in the development of the Sierpinski triangle, we obtain, not just one seed-orbit, but the realisation of all three possible orbits per the modus operandi inherent to MWI, treated as three (3) distinct 'histories'. At the second orbital resolution, all three 'histories' branch into three subsequent histories, now making for nine (9) distinct histories. At the third orbit-point we obtain 27 distinct histories...and so forth.
  - As we proceeded from the below-threshold phase of development through to the chaotic-transitional phase, and finally into the noise-reduction pattern-saliency phase, the very large number of histories thus generated will *not* constitute globally distinct histories and universes. Indeed, all will converge to the same developmentally salient Sierpinski triangle pattern. Thus:

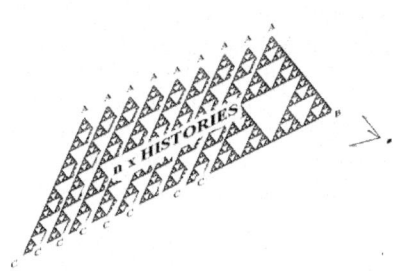

---

[91] Cramer, John (July 2009). "Transactional Interpretation of Quantum Mechanics". *Reviews of Modern Physics*. 58 (3): 795–798.

- **MWI fails because its' manyworlds all culminate into the same final global history. We need only substitute the actual universal wavefunction of MWI in place of our simple quantum mechanised Sierpinski universe to appreciate the same conclusion.** If all the 'unique histories' lead to the same global historical culmination, what is the point of MWI? Instead, we ought to recover the idea of just one unique history and circumvent resort to MWI. We would also circumvent the question of where all that extra energy-matter is going to come from to constitute the now defunct identical histories and universes.

- However, in those instances where a Schrödinger's cat-like scenario applies, some aspect of MWI might become plausible: A nuclear detonator emplaced in a city, its trigger sensitive to a single quantum outcome, will obviously alter the structure of the larger universal wavefunction pertinent to that city. In such instances, the final global outcomes will *not* culminate into a same global outcome or history, obviously. From this, **Little Manyworlds** (LMW) may well follow as a plausible alternative to an otherwise self-invalidating MWI (See Book-III **3.18**). However, even LMW cannot avoid the question of where all the extra energy-matter is going to come from to constitute the alternative universes and histories of LMW. In any case, LMW cannot obviate quantum indeterminism: it actually depends on it… while MWI cannot obviate quantum indeterminism because quantum indeterminism is *not* a 'problem' that requires any solution or circumvention.

5.08: Critique of Bohm's Hidden Variables Interpretation of Quantum Mechanics

We now come to David Bohm's hidden variables or 'hidden gears' interpretation. The point of contention in **Bohm's interpretation lies in the notion of hidden gears that might orchestrate and obviate quantum indeterminism** by means of a 'pilot wave'. **Pilot wave theory assumes implicit well-defined guiding OR-form trajectories for particles inbuilt into the wavefunction. The particles follow these paths deterministically, but the outcome appears indeterminate to the observer that has no access to the hidden variables of the pilot wave or to the initial state of the particle at the emitter**. This circumvents apparent quantum indeterminism. We are uncertain of the initial condition of the particle, and about the attendant well-defined path it will take within the pilot wave that it will otherwise assume. But this is due to ignorance, not due to inherent uncertainty. The manipulation of the two slit experiment simply modifies, permits, or eliminates some of the well-defined trajectories of the pilot wave, depending on whether one or both slits are open, and on whether there is a detector attached to one of the slits, as well as points of wave-cancellation that produce generic interference patterns. Since it is not possible to know the initial state or position of the particle, its final deposition on the detector appears random, but it is ultimately secretly deterministic or orchestrated.

BOHM'S HIDDEN VARIABLES INTERPRETATION OF QUANTUM MECHANICS							
W-function real?	W-function collapse real?	Indeterminism real?	Hidden gears?	Observer role?	Unique history?	Non-locality?	Universal W-function?
YES …but does not recognise ontological reality of the future	**Phenomeno-logical** … but no recognition as time itself	**NO** Deterministic, orchestrated by means of the pilot wave	**YES**	NO …observer not needed,	YES	Locality	YES

Our own attempt to formulate orchestrating scripts that guide otherwise seemingly indeterminate outcomes plays the same role as in-built guiding paths in Bohm's pilot wave. However, pilot waves turn out to be as unnecessary as MWI: **Both randomness and quantum indeterminacy are objectively real**, with no secret orchestration or hidden gears required, and the well-defined paths of a pilot wave prove unnecessary. As with MWI, the very venture of Bohmian hidden variables rests on the supposition that quantum indeterminism is a 'problem' that needs a deterministic solution. Again.

- When we take any random or indeterminate developmental process and imagine fully deterministic orchestrating scripts for it, we end up with a very large number of such scripts: each equally apt to resolve the whole process of development deterministically, but none exclusive or special as to constitute the one-and-only actual orchestrating script.

- The only way to select from between these very large numbers of scripts is by resort either to arbitrary selection (if done by conscious agency, and therefore no different from resort to randomness) or by resort to blind process of random selection. In short, we cannot obviate the final resort to randomness and indeterminacy, even when we form deterministically orchestrating scripts and push the generation of such scripts to infinity. (See Book-III **3.32**).

- Therefore, *quantum indeterminate processes are truly objectively indeterminate,* even when we imagine or assume the operation of some hidden orchestrating script or gears, or pilot wave, behind it all.

The above insight is embedded into the three-phase developmental process espoused in Book-III **3.26**. Thus…

- **Quantum indeterminacy is not a 'problem' that requires a secret deterministic solution. Randomness and indeterminacy constitute nature's economic solution to pattern development and saliency**. Nature would be far more complex if it had to secretly assign the deposition-addresses' of each pixel or quanta vis-à-vis every other in the formation of any complex pattern, whether via orchestrating scripts or via 'pilot waves', whether culminating into a photographic image, a drifting snowflake, or a whole galaxy. In short, neither secret scripts nor pilot waves are necessary to a 'problem' that does not exist. Therefore, the 'solutions' are unnecessary.

*

What about hidden gears and variables with respect to wavefunction collapse and time? This is a distinct area of concern, given that the issue is one of what orchestrates time itself, and not just what secures the outcome from time. Bohm's hidden variables and pilot wave theory did not deal directly with the causality of wavefunction collapse and time. As with other interpretations, **in Bohm's approach, time is taken for granted; wavefunction collapse is never explicitly grasped as time itself,** while time is implicitly treated as a 'fourth dimension', and the notion is not question.

Any schema for the secret mechanical orchestration of wavefunction collapse and time would crash into infinities and an ineluctable halting problem. Consequently, time could never happen. (See Book-II **2.46** to **2.47**). Thus...

- **Any hidden gears for the mechanical processing or computation of wavefunction collapse and time would need to eliminate all foregone immanent possibilities, of which there may be infinitely many**, given that some generic quantum mechanical systems have infinite degrees of freedom. A step-by-step clockable procedure or gears for eliminating each of these possibilities could never halt, and wavefunction collapse or time could not transpire: time would 'freeze up'.

- **Any hidden gears for the mechanical processing or computation of wavefunction collapse and time would need to eliminate the nested future attached to the immanent possibilities... to the infinitely removed future**. Again, any step-by-step clockable procedure or gears for the erasure of each nested possibility, even if finite at every potential future, yet infinite once it tends to the infinitely removed future, could never halt, much less obtain its conclusion-point. Hence, wavefunction collapse and time could not transpire.

- **Any hidden gears for the mechanical processing or computation of wavefunction collapse and time would need to mechanically process *grand decoherence* per the counterfactuals-driven quantum erasure of all non-consistent nested futures in the context of convergent category nested futures ...to the infinitely removed future.** Again, any step-by-step procedure, operating on an input-operation-output schema, would crash into a non-terminating problem, given the infinities involved. Consequently, wavefunction collapse and time would 'freeze up' and could not transpire.

Hence, whatever is responsible for wavefunction collapse and time constitutes a non-clockable structure-less ineffable causality, manifesting 'spontaneous' wavefunction collapse and time. In any case, a secret orchestrator of time must itself require time and a 'clock speed': a Chinese doll of 'gears within gears, within gears...', regressed to infinity. Since, time is ontologically primary to any process, including mechanical input-operation-output schemas, it follows that the structure-less ineffable causality that makes time happen is also ontologically primary to and transcends process and outcome. **It is absurd to envision gears as causative of time and wavefunction collapse when, in the first place, these require the ontic primacy of time and wavefunction collapse to obtain**.

5.09: Brief Summary of Part-I

Alt-R consists of the following assertions, axioms, and postulates:

- **The quantum mechanical wave constitutes the ontic objective reality of the future**: The wavefunction is the mathematical attempt to grapple with that ontic reality.

- **The quantum mechanical wave is a static state of nested futures**, but not just of immanent futures: It is a system of nested futures and constitutes the Whitehead facet: the de-spatialised replacement of 'spacetime in potentiality'.

- **Grand decoherence**: The future is subject to grand decoherence through counterfactual causality. This involves the erasure of non-consistent nested futures and it designs for the **revival of Mach's principle** foundational to the *quantum mechanical basis for inertia*. The *quantum mechanical basis for the permanent magnet* also follows suit, as do contributions to quantum gravity.

- **AND-to-OR wavefunction collapse is time itself: time is not a 'fourth dimension'**. Time is wavefunction collapse, and wavefunction collapse is time. Time's arrow proceeds from AND-to-OR, never from OR-to-AND, with the *entropy of future-potential information* attending it, rendering it superior to the theory of time obtained from generic statistical entropy.

- **De-spatialisation**: There is no space. Only time is real. IOO-schemas and impact causality cannot abide, given that 'position' or 'spatial point-of-contact' has no ontic reality...and, per function of related and independent reasons, **there are no hidden gears or hidden variables behind wavefunction collapse** or time.

- **The Bergson-Whitehead amalgam succeeds 'spacetime'**: The intermediate model was comprised of a 'spacetime in potentiality' consisting of nested futures; a 'realised spacetime' of past states; with both domains separated by an 'event horizon' comprised of the most recent OR-form outcomes from perpetual AND-to-OR collapse and time. This intermediate

475

model was de-spatialised, temporised and replaced by the superlative growing block Bergson-Whitehead amalgam, with a Whitehead facet (the future) and a Bergson facet (the past in abstract memory) separated by an unmodified event horizon.

ALT-R INTERPRETATION OF QUANTUM MECHANICS

W-function real?	W-function collapse real?	Indeterminism real?	Hidden gears?	Observer role?	Unique history?	Non-locality?	Universal W-function?
YES (it is the ontically real future)	YES (it is time itself)	YES (objectively real)	NO	decoheres but does not collapse	YES (but w/ Little MW)	YES ...but as counterfactuals driven grand decoherence and transtemporality	YES Whitehead facet to the infinitely remote future

COPENHAGEN INTERPRETATION OF QUANTUM MECHANICS

W-function real?	W-function collapse real?	Indeterminism real?	Hidden gears?	Observer role?	Unique history?	Non-locality?	Universal W-function?
NO ...does not recognise ontological reality of the future	YES ...but no recognition as synonymous with time; contradictory	YES ...but no proof of objectivity of randomness, nor proof of absence of gears in W-function collapse	NO	YES ...observer interaction causes W-function collapse	YES	Locality	NO ...does not recognise ontological reality of the future

MANYWORLDS INTERPRETATION OF QUANTUM MECHANICS

W-function real?	W-function collapse real?	Indeterminism real?	Hidden gears?	Observer role?	Unique history?	Non-locality?	Universal W-function?
YES ...but does not recognise ontological reality of the future	NO ... and not recognised as synonymous with time	NO Deterministic but exhaustive: all possibilities are realised as distinct histories	NO	NO ...observer not needed, given all possibilities realised	NO ...many histories per each possibility in w-function	Locality	YES (proposed by Everett)

BOHM'S HIDDEN VARIABLES INTERPRETATION OF QUANTUM MECHANICS

W-function real?	W-function collapse real?	Indeterminism real?	Hidden gears?	Observer role?	Unique history?	Non-locality?	Universal W-function?
YES ...but does not recognise ontological reality of the future	Phenomenological ... but no recognition as time itself	NO Deterministic, orchestrated by means of the pilot wave	YES	NO ...observer not needed,	YES	Locality	YES

- **Nature has abstract memory**: De-spatialisation implies that all events are organised temporally into delayed choice time interval relata, all according to primacy and recentcy rules. That is, information is time-distributed, not space-distributed. This *is* abstract memory, ultimately identified with the Bergson facet itself: a system of temporally distributed *wavefunctions in memory* (with nested futures cropped) in contrast to Whiteheadian *wavefunctions in future potentiality* with nested futures intact, but counterfactually erased via grand decoherence.

- **Gravitation is an abstract memory effect:** an effect of the Bergson facet upon the Whitehead facet secured via grand decoherence.

We then contrasted and compared our Alt-R interpretation with key generic interpretations of quantum mechanics, thus outlining the superlative nature of Alt-R. This is summarised in the reiteration of pertinent tables furnished above.

As can be garnered from the compiled tabulations of the interpretations, even when the various interpretations appear to agree in some of the eight-fold areas of concern with Alt-R…

- **…generic interpretations fail to recognise the wavefunction as the default mathematical attempt to model the ontic future, or assert the ontic reality of that future** (i.e., Spacetime in potentiality, or the de-spatialised Whitehead facet). Generic interpretations cannot scaffold for the growing block universe requisite to the nested futures approach and to grand decoherence, to abstract memory, and to mnemonic mechanics, or to quantum mechanical growing block solutions to the permanent magnet, of inertia, and of gravity and quantum gravity. These reside in the purview of Alt-R.

- **None of the interpretations assert that the quantum wave or AND-to-OR wavefunction collapse constitutes time itself**, as does Alt-R, whatever they might otherwise claim about the ontics of wavefunction collapse. Thus, all interpretations garner time as a 'fourth dimension', and implicitly assume the ontics of the forlorn block model universe.

- **None of the interpretations offer solutions to quantum indeterminism, or to the stochastic coherence paradox**, whatever their positions on its ontics. Alt-R offers solutions to the effect that quantum indeterminism is objectively real (**3.32**); that the development of any pattern follows a three-phase development process (**3.26**); both co-joined to render indeterminism a friend to pattern and structure, *not* a problem that needs a 'solution'.

- **None of the interpretations resolve the question of hidden gears**. Only Alt-R solves *against* hidden gears, in both quantum indeterminism (**3.26**) *and* in wavefunction collapse (Book-II **2.46 and 2.47**). The Alt-R solution leads to the collapse of materialist input-operation-output schema, causality, and contiguous impact causality. Alt-R asserts for work function zero gear-less causality as the drive of wavefunction collapse or time.

- **As we shall see, only Alt-R offers decisive solutions against 'consciousness causes collapse'** (Part-III) and resolves the observer or apparatus problem (see **5.06** above), while recuperating 'non-locality' into grand decoherence, and espousing transtemporality in relation to the ontic reality of the instantaneous whole (i.e., 'universal wavefunction'), in turn recuperated as the total nested futurity of the Whitehead facet.

Moreover, only Alt-R offers solutions to mind-world relations and conundrums, without degeneration into materialism or idealism; scaffolding for a viable growing block theory of mind-world, and casing the revival of Cartesian dualism, as we shall see.

BOOK-V PART-II:
THE CASE AGAINST "CONSCIOUSNESS CAUSES COLLAPSE"

5.10: Aims of Part-II:
Eugene Wigner's 'Consciousness Causes Wavefunction Collapse'

'Consciousness causes collapse' emerged out of the groundwork to John von Neumann's *The Mathematical Foundations of Quantum Mechanics,* which opened the path to the speculation that consciousness might constitute the causal basis for AND-to-OR wavefunction collapse *and* consciousness-orchestrated quantum indeterminism. The pertinent interpretation was developed and then later abandoned by Eugene Wigner[92]. But since then, the want to posit a full idealist tilt to Wigner's interpretation and espouse the thesis of a consciousness-created universe, wherein consciousness supposedly constitutes the primary ontology, while all else is an illusion conjured by nous… wherein the object-world collapses into the subject-nous… has run its course in popular culture and in the New Age arena. **Our aim in Part-II will be to cast permanent doubt on the notion of 'consciousness causes collapse', but without succour to materialism. Indeed, the demise of 'consciousness causes collapse' will constitute an indispensable basis for the revival of Cartesian dualism *and* contribute to the demise of materialism.**

*

---

[92] *Wigner, Eugene P. (1961). "Remarks on the Mind-Body Question". In Good, I. J. (ed.). The Scientist Speculates: An Anthology of Partly-Baked Ideas. London: Heinemann*

Eugene Wigner's original contention was that consciousness must be critically different from any measuring apparatus and from the system subject to measurement, and effectively different from the physical universe. Neither the apparatus nor the system measured are causally capable of bringing about wavefunction collapse. The wave-collapsing causality appears to reside outside of the apparatus and the physical-observational inputs entailed. While this contention is true enough, therein lies the vulnerability of quantum mechanics generally and the Copenhagen interpretation specifically to subjectivism and idealism: The inability to account for wavefunction collapse per the causal power of the apparatus and of physical-observational inputs in the Copenhagen interpretation opens the door to quantum idealism. Upon this vulnerability, in lieu of apparatus or non-mental mechanisms, **Wigner claimed that only consciousness could cause wavefunction collapse.** This intriguing idea can be generalised to the universe as a whole; wherein consciousness presumably causes the collapse of universe-scale wavefunction itself.

Moreover, according to Wigner's original contention, consciousness does not simply cause wavefunction collapse, but orchestrates apparent quantum indeterminate processes, circumventing the 'problem' of quantum indeterminism. Both purported wavefunction collapse by consciousness *and* the alleged circumvention of quantum indeterminism by the same, forms the basis of general quantum idealism and the want to collapse object-world to subject-nous.

When we apply 'consciousness causes collapse' to the universe-scale wavefunction, and, with it, arrive at the notion of requisite universe-scale consciousness that supposedly brings about such collapse, we come full circle: we end in Bishop Berkeley's classical idealism, now quantum mechanised into quantum idealism and quantum theology; wherein God is reconceived as the super-conflated universe-scale wavefunction-collapsing Nous.

The late John Wheeler's *participatory anthropic principle* is another thread to quantum idealism and theology. Wheeler conjectured that consciousness must play a central role in shaping the character of physical laws, constants, and conditions[93]. It could only do so by means of a retro-causal orchestration (from the present, or essentially from the future) of the past evolution of the universe into conditions suited to the emergence of life and of the retro-causal consciousness itself. **If we accept Wigner's now-abandoned contention that consciousness and mind are the only things capable of causing wavefunction collapse and the orchestration of quantum indeterminism, the requisite universal wavefunction collapse that occurred at the purported Big Bang might well have been brought about by a retro-causal Grand Consciousness; one operating from the future; albeit, a form of nous beyond mere human consciousness; within the purview of God, itself now re-conceived as a universe-scale wavefunction-collapsing Nous: one presumably operating from the infinitely removed future.**

However, we reiterate that Wigner later abandoned 'consciousness causes collapse'.

*

Our two aims in the following material will be to unravel foundational quantum idealism exhibited in Wigner's thesis *and* cast doubt on inadvertent and attendant quantum theology. Thus, **Part-II will show that consciousness does not collapse the wavefunction; and that consciousness is *not* the drive of time, and it is certainly *not* the primary ontology.**

To this end, first, we will detail Wigner's thought experiment or *Wigner's Friend*, in its own terms, and follow it with an account of why Wigner abandoned his notion that consciousness is fundamentally different from other physical states, systems and apparatus. We will argue that, while Wigner was correct to abandon 'consciousness causes collapse', his reasons for doing so were questionable. **The correct reasons for why we must abandon 'consciousness causes collapse' will show that consciousness is indeed provably different from the world and its apparatus: a difference that fashions for Cartesian mind-world dualism, but one that cannot support 'consciousness causes wavefunction collapse', or quantum idealism.** We will then make an independent case for why nous is neither necessary nor sufficient in the orchestration of quantum indeterminacy, basing our case on the three-phase developmental process, combined to the scripting approach, both previously developed in Book-III, Part-II.

In a follow-on part, we will show why consciousness is *not* the causal drive of AND-to-OR wavefunction collapse and time. This conclusion is arrived at from the fact that any future-potential state of consciousness must itself need a collapsor, while time-bound consciousness cannot resolve infinities and halting problems inherent to the process of grand decoherence and in the collapse-process requisite to the transformation of future possibilities into OR-form outcomes. Hence, something other than consciousness must be responsible for AND-to-OR collapse: a causality whose ontology must transcend mind and consciousness as much as it transcends 'matter in space', as it must also transcend even nous super-conflated to God.

We will then seal the case against 'consciousness causes collapse' and quantum idealism by investigating intriguing claims that statistically significant effects are generated by the interaction of consciousness vis-à-vis physical systems, especially in the Quantum Zeno effect.

5.11: Wigner's Friend

How did Wigner arrive at the intriguing notion that consciousness is different from the apparatus and measured system, let alone responsible for the collapse of the AND-form state of the apparatus *and* the co-joined measured system? **Wigner took the well-known Schrödinger's cat experiment and developed a variant of it: namely, Wigner's Friend.** The Schrödinger's cat experiment involves a cat placed inside a box containing a poison vessel. The latter is rigged to a quantum superposition state with two AND-form

---

[93] Wheeler, John A. (1990). "Information, physics, quantum: The search for links". In Zurek, Wojciech Hubert (ed.). *Complexity, Entropy, and the Physics of Information*. Redwood City, California: Addison-Wesley.

possibilities. This AND-form state is comprised of, say, a particle suspended into "spin up" AND "spin down". The vessel releases its poison when the AND-form state collapses into the OR-form "spin-up" through subsequent AND-to-OR collapse. However, if "spin down" is obtained, the vessel will not release its poison: the cat will live.

The box with cat and poison remains sealed off from the laboratory. After a certain time, the experimenter opens the box to find either a live cat or a dead cat, pending the spin-outcome obtained upon wavefunction collapse, in turn brought on upon the opening of the box.

In his variant of the same, Wigner departs the laboratory while his friend proceeds with the experiment. Later, Wigner returns to the laboratory and obtains the results of the experiment form his friend. By placing the cat in the box and sealing it, Wigner and friend have placed the cat and system into a state of physical-experimental isolation. With isolation, the content of the box is apparently transformed into a superposition state comprised of both "live cat" AND "dead cat". However, by departing the laboratory itself, Wigner presumably places the whole laboratory into its own state of physical-experimental isolation. If this is true then the contents of the laboratory, including his friend therein, must transform into an AND-form superposition state comprised of, say...

… {"live cat + happy friend" **AND** "dead cat + unhappy friend"}

Upon his return, Wigner 'pops' the above wavefunction 'bubble': The laboratory resolves into either "live cat and happy friend" **OR** "dead cat and unhappy friend".

If a non-conscious apparatus is placed in lieu of Wigner's friend, such as a respiration-reading mechanism linked to the cat, what would happen? Upon Wigner's departure and isolation, and in the interim until Wigner's return, the joint superposition comprised of cat in the box *and* respiration-reader would quantum mechanise into an AND-form superposition containing two alternative possibilities: live cat AND dead cat, co-joined to all the attendant AND-form states of the respiration device. However, according to Wigner, this cannot be true for an experiment involving the presence of his friend's consciousness. **According to Wigner, unlike the non-conscious measuring apparatus in its state of isolation, the consciousness of Wigner's friend (and consciousness generally) cannot itself be rendered into a superposition state; or so Wigner claimed. Thus, his friend's consciousness must always be rendered into one resolved OR-form state or another.**

To put Wigner's assertion in our terms, he asserted that consciousness is always in an OR-form state, never in any AND-form state. **Consequently, Wigner's departure and the attendant relegation of the laboratory system of Schrödinger's cat plus friend into physical-experimental isolation, could never render it into a full AND-form superposition state, owing to the presence of the consciousness of Wigner's friend. The very presence of his friend in the laboratory supposedly pulls the system out of (or, in the first place, prevents it from entering into) a state of complete isolation and quantum mechanisation into the expected AND-form superposition state of futures. For this reason, according to Wigner's original contention, consciousness must be special and different from any non-conscious apparatus and arrangement.**

Thus, so long as Wigner's friend is inside the laboratory, per his consciousness, the laboratory will remain in a resolved state comprised only of OR-form outcomes. The box with the cat, placed in isolation and separated from the consciousness of Wigner's friend, *might* internally remain or transform into an AND-form superposition: assuming the cat's consciousness had no contribution to make. However, Wigner's friend's consciousness, always in OR-form, will resolve the part of the laboratory to which he has observational-experiential access into a (presumed) continuous OR-form corpus. The friend's consciousness would also resolve the cat in the box, if he chose to look inside the box before Wigner's return. This again assumes the cat's own consciousness had no independent wavefunction-collapsing contribution to make.

In short, according to Wigner, consciousness is different from the usual apparatus insofar as it is never itself in an AND-form state: Its very presence will OR-form resolve (and keep resolved) the wavefunction that, left to a mere non-conscious non-mental measuring apparatus, would completely dissolve into an AND-form wavefunction-smear of future possibilities.

*

The cat's consciousness does not need to be essentially different from the friend's consciousness in its purported power to affect wavefunction collapse. Just as Wigner's friend resolves the wavefunction for the laboratory (but not the internal state of the box) **the cat's consciousness might resolve the state inside that box, including that of the poison vessel itself**, before Wigner's friend opened the box. Just as Wigner's own consciousness does not resolve the consciousness of his friend plus laboratory, the consciousness of Wigner's friend does not resolve the state of the cat's consciousness and the box-internal. If consciousness of any form is always in OR-form, and thus different from all other non-conscious systems and apparatus liable to AND-form quantum mechanisation upon isolation, then the friend's consciousness is not required to resolve the cat's consciousness. The state of the poison vessel will presumably depend on the cat's consciousness within the purview of the physically isolated box.

What happens when we replace Wigner's friend with the ultimate super-conflated consciousness: namely, God? His consciousness, just like Wigner's friend's mortal consciousness, cannot be characterised as subject to an AND-form superposition state. Presumably, God's Mind must always be in OR-form: this is what makes consciousness special and different, according to Wigner.

Moreover, presumably, God is omnipresent: there are no observational or experimentally isolated parts to the universe as far as God is concerned. As such, God supposedly resolves and supports the whole universe in which you, the reader, Wigner's friend, the cat, and Wigner himself, exist. Of course, if such a God-consciousness existed, there ought to be no evidence of any AND-form quantum mechanical phenomena at all (since all will be fully OR-form resolved), and quantum mechanics or AND-

form logic could not abide in such a universe. Yet, the fact that AND-form reality is ineliminable, and that there can be no such thing as a pure OR-form system or fully resolved block universe, casts doubt on the notion of a quantum God-consciousness, or of a universe based on 'consciousness causes collapse'.

However, if the cat was replaced by a mere non-conscious apparatus designed to detect the released poison (or lack thereof) then the apparatus, together with the poison vessel, would quantum mechanise into a joint AND-form superposition state. If a non-conscious apparatus were to replace Wigner's friend, the whole laboratory (the box included) would become immediately quantum mechanised into an AND-form superposition of possibilities… only to collapse per Wigner's consciousness and *his* return to the laboratory.

Of course, if the latter God-consciousness did not exist, or if consciousness as a possibility did not exist, and if Wigner was correct, the universe would presumably remain in permanent quantum AND-form suspension, given that AND-to-OR wavefunction collapse (time) could never transpire in a universe without consciousness… assuming Wigner's core contention was correct.

*

In summary, according to Wigner, consciousness matters, although Wigner has since abandoned this position.

Recall that Wigner claimed that consciousness could never be in a quantum superposition AND-form state: It must always be in OR-form: *This*, according to Wigner's original contention, is what makes consciousness critically different from non-conscious states and from any non-conscious apparatus; all liable to quantum mechanisation into AND-form expression. Moreover, Wigner contended that consciousness is essential to bring about wavefunction collapse *and* decide the outcome of that collapse in a selective way. Thus, according to Wigner, consciousness is different from mere non-conscious apparatus in that it can bring about wavefunction collapse as well as select and determine the end-result of that collapse. Or so Wigner contended, until he changed his view and abandoned 'consciousness causes collapse'.

However, all of the above presupposes the ontological primacy of consciousness vis-à-vis wavefunction collapse and time, inherent to the notion that consciousness causes the collapse of the wavefunction… or causes time itself. **What if we demonstrated that time or AND-to-OR wavefunction collapse is ontologically primary vis-à-vis consciousness? What if consciousness must itself be resolved out of a noetic AND-form future-potentiality via AND-to-OR wavefunction collapse, in parallel to the AND-to-OR collapse of the world out of its Whiteheadian futurity?**

5.12: Wigner Abandons "Consciousness Causes Collapse"

The disposition in Wigner's approach to solipsism was one of the reasons he abandoned 'consciousness causes collapse': Whose consciousness is primary? Is it Wigner's consciousness? Is it that of his friend? Is it the cat's consciousness? Is it God's consciousness? There appears no clear way to resolve this conundrum within the framework Wigner had developed.

While Wigner never resorted to God as a solution, there are no inbuilt limits in 'consciousness causes collapse' to prohibit it. Insofar as the interpretation assigns to consciousness the principal causality over wavefunction collapse (and time), it cannot escape solipsism, nor escape the incorporation of God misconceived as consciousness, consequent upon the said solipsism.

The matter of what causes the collapse of the wavefunction must segue into the question of whether wavefunction collapse occurs per non-mental hidden gears or variables, or per some other principle; neither mental nor non-mental. The argument against 'consciousness causes collapse' is better addressed from the argument against hidden gears furnished in Book-II and Book-III. Therein, the solution to the causality behind wavefunction collapse and time, which served to usurp the materialist notion of non-mental hidden gears, formed the implicit inadvertent basis for the critique of 'consciousness causes collapse', as we shall soon discover.

*

There is a view commonly held within and outside of the physics community that one should not confuse and apply what happens at the micro-scale to and with the macro-scale. It is also asserted that macro-scale experiments cannot be treated as if quantum mechanical in character, much less exhibit AND-form superposition characteristics. This is on the general assertion and consensus that AND-form characteristics and descriptions *only* apply to the micro-scale, and that quantum mechanics (hence AND-form expressions of nature) must remain confined only to the 'very small'. For example, billiard balls do not undertake a superposition of paths from cue to hole; suggesting that AND-form logic must be confined to the very small and cannot abide at the scale of Schrödinger's' cat in the box'. But there is a reason why billiard balls do not appear to behave like quantum objects, and it has nothing to do with the supposed absence of quantum reality from the macro-scale, much less the confinement of it to the very small, as we shall discover. Aside the solipsism problem, unfortunately, **Wigner agreed with the restriction of quantum mechanics to the micro-scale. Consequently, he abandoned 'consciousness causes collapse': Wigner's Friend experiment is a macro-scale state to which AND-form logic cannot supposedly apply. However, the premise that AND-form logic is restricted to the very small constitutes an** *error.* **Quantum phenomena and AND-form logic** *do* **apply to the macro-scale,** even if often in subtle ways. The following is a bullet-point summary of why this is so:

- **Temporisation of scale per de-spatialisation:** De-spatialisation implies the overturning of spatial notions of 'scale'. The 'very small' is recapitulated into very short time-interval physics. The 'very large' is recapitulated into long time-interval physics, notwithstanding the event-density approach to scale (see below). Obviously, quantum mechanics is not confined to the very short time-interval domain, and we can easily obtain quantum phenomena in a two-slit experiment using low frequency long time-interval microwaves. A pure time-based approach to scale implies that AND-form logic cannot be restricted only to high frequency (short duration) gamma-ray phenomena, but must equally abide at the low frequency long

480

duration end. That is, scale in terms of de-spatialised purely temporised physics does not confine AND-form potentiality states only to short-time high frequency frameworks.

- **The absence of self-interference in the single-slit experiments (and at the 'macro-scale') does not constitute the absence of AND-form logic or of the quantum wave of possibility from the macro-scale**. This is demanded by the non-standard evaluation of the two-slit experiment (see Preliminary, **0.14**), which shows that the distribution pattern obtained in the single slit condition is producible by the AND-form quantum wave, and therefore implies its presence therein, *not* its absence. What we call the 'macro-world' approximates to conditions that attend single-slit experiments in which AND-form logic abides, despite the absence of salient self-interference or its obscuration. Thus, the typical or seeming absence of self-interference in macro-scale phenomena (including billiard balls) does not imply the absence of AND-form logic from the macro-scale, no more than it does in single-slit experiments.

- **Scale considerations must be addressed in terms of the event-density of quantum incidences in a framework involving the temporal distribution of *durata*, as demanded by de-spatialisation and temporisation**. When this is done, we find that we cannot obviate the operation and fact of AND-form logic from the macro-scale: For example, in the two-slit experiment, pertinent interference patterns formed at the detector are comprised of *en masse* quantum incidences or *durata*. Therein, one cannot obtain evidence of self-interference by way of small number below-threshold corpus of OR-form incidences or at the 'microstate': One needs *en masse* incidences (macro-scale) at an appropriate threshold-quantity and density to bring into saliency the fact of self-interference supposedly confined to the micro-scale. That is, the evidence for how nature operates at the 'very small' needs quantum mechanical effects amplified to the 'very large', wherein quantum mechanical phenomena are rendered salient and evident at the macro-scale, such as in the formation of interference pattern… and can no longer be deemed confined 'only to the very small'.

- **AND-form logic and the quantum wave constitute the phenomenalisation of the growing block ontic future: This is an all-scale principle**. All systems have objective AND-form futures, whatever their scale. Futures are not confined to the 'very small'. Thus, macro-scale systems also have futures in the form of 'spacetime in potentiality', or its replacement by the superlative de-spatialised and temporised Whitehead facet. In short, small or large, we cannot obviate the future from any system, much less from the whole growing block universe constituted as the Bergson-Whitehead amalgam. The dubious assertion that quantum phenomena are confined only to the very small is tantamount to claiming that the future possibility state is confined only to the very small, or that conditions that attend the growing block universe are confined *only* to the micro-scale, while only block universe conditions attend the macro-scale; or that time abides only at the small scale, or that there is no time or AND-to-OR collapse at the macro-scale. No! The growing block AND-form logic, the superposition principle and the quantum wave of possibilities, *and* the ontic future itself, apply at *all* scales. Therefore, quantum mechanics and AND-form logic applies at all scales. Thus, the all-scale pertinence of quantum mechanics.

- **AND-form logic is demanded by the firewall principle at all scales: The intertwine of the principles of causality with the conservation principle, which abides at all scales**. To suggest otherwise, or to posit that the firewall principle abides only at the micro-scale, would be tantamount to the claim that 'signals from the future' must abide at the very large, insofar as a non-quantum mechanical macro-scale must constitute as a block universe, with all its futures wholly resolved into OR-form outcomes, to timelike infinity. But its proof would require the reception of 'signals from the future' and the evidence of the transpiration of the future before it transpired (in violation of causality) *and* the creation of above-unity energy-matter to constitute that OR-from future as an energy-matter corpus (in violation of conservation principles). Since such fantastic outcomes do not obtain in normal nature at the macro-scale, we must conclude that the future at the macro-scale must be constituted as an AND-form potentiality for not-yet transpired macro-scale futures. Consequently, the macro-scale *does* possess an ontic AND-form facet: that is, quantum mechanics *does* abide at the macro-scale.

- **Grand decoherence of the total Whitehead facet, of the future constitutes quantum cross-frame self-interference… or self-interference at the macro-scale, and the operation of AND-form logic and quantum mechanics at that scale.** The nested futures view and the inexorable reality of grand decoherence, and, with it, revived Mach's principle and attendant 'non-local' transtemporal solutions to inertia, to the permanent magnet, and to much else besides, implies macro-scale universe level self-interference of one frame along the event horizon vis-à-vis every other frame on the same… to the infinitely removed future. This entails that billiard balls 'here' decohere in concert with all other billiard balls elsewhere in the universe, no matter how mutually remote they appear to be, courtesy of grand decoherence operating at the grandest possible scale, not merely at the macro-scale. But grand decoherence *is* quantum self-interference and mutual interference. It seems that billiard balls *do* behave like quantum objects, at least when grasped from the grand decohering Whitehead facet framework. It also follows that quantum phenomena do indeed apply to the very large, and do apply to Schrodinger's cat, and *must* apply to Wigner's friend…and that quantum mechanics is *not* confined to 'the very small'.

*

With the bulleted summary-points above, we now adumbrate further.

The **reason why macro-scale billiard balls do not *appear* to behave like micro-scale particles has nothing to do with the absence of AND-form logic from the macro-scale**. The moment-to-moment time-evolution of a billiard ball is treated as equivalent, in at least its global effect, to a series of repeat single-slit experimental conditions and outcomes, wherein no manifest interference pattern will appear amid the *en masse* quantum incidences. Recall that our non-standard evaluation of the double slit experiment furnished in the Preliminary (**0.14**) showed that AND-form logic is *not* absent from the single-slit experimental system and outcome. For essentially similar reasons, AND-form logic cannot be absent from billiard balls that approximate to global outcomes from single slit conditions. Indeed, from Book-I **1.20**, if nothing else, AND-form 'gaps' reside ineliminably between each resolved succession of the billiard ball's constitutive *en masse* quanta along their collective 'path'. Recall from Preliminary **0.01** to **0.04**, that this 'path' is *not* ultimately a seamless-continuous state, but a broken-discontinuous one, with ineliminable quantum mechanical AND-form 'gaps'. Therefore, given the ineliminability of such 'gaps', AND-form logic and the superposition principle is present across the macro-world of billiard balls, cats, laboratories, people, moons and galaxies, even if its obvious saliency circumvents limited human perception.

Moreover, the seeming fact that self-interference is absent from the putative macro-scale, and that this must mean the absence of AND-form logic and of quantum mechanical phenomena from that scale, is belied by the fact that one could perform two-slit experiments with 1000m wavelength radiation. This would entail a very large apparatus... *and* obtain the evidence for interference in the *en masse* (large-scale) conglomeration of quantum incidences across the surface of a much-enlarged detector. But what we easily forget is that **most two-slit experiments *are* macro-scale laboratory outlays; *not* restricted to gamma ray frequency time-intervals. As such, generic two-slit experiments *are* macro-scale, and so are their self-interference patterns... and so is the AND-form logic that manifests at the macro-scale, evinced in, as, and by, the attendant interference pattern.**

*

As indicated in the previous bulleted list, just as we must ultimately reform scale per de-spatialisation and express it in pure temporal form, we must also reform scale in terms of the quantity and density of quantum incidences. This is clearly indicated by our three-phase development approach vis-à-vis quantum indeterminism: (See Book-III **3.26**). For example, in the two-slit experiment, a single quantum incidence, or even a very small number of such at the below threshold phase of development (i.e. the event-density version of the 'micro-scale') could not produce evidence for self-interference or even of the superposition principle. **Unless *en masse* quantum incidences of sufficient number (or of sufficient event-density) are resolved so as to obtain the interference pattern in the two slit experiment, the empirical reality of AND-form logic and the superposition principle could not be obtained.** That is, we need an amplification grasped in terms of a threshold number of quanta at scale (and the recuperation of scale in terms of event-density thus) to infer the empirical reality of AND-form logic, and superposition principle operating at, but no longer confined to, the very small. But this can only be obtained at the macro-scale, never at the micro-scale, given that we only have access to the macro-scale. To admit this is to admit that AND-form logic and the superposition principle *must* manifest at the macro-scale: We must ramp up the very small into evidence at the very large (the interference pattern itself), but, in doing so, fail to confine quantum mechanics to the very small. **Indeed, if quantum phenomena were strictly confined to the micro-scale, then one could never amplify self-interference and attendant interference patterns into evidence at the laboratory macro-scale. Quantum mechanics would be rendered untenable.**

*

Ultimately, **the presence of AND-form logic at the macro-scale is a reality forced upon us by the principle of causality intertwined with the principles of conservation; both of which apply at *all* scales and cannot be confined to the 'very small'. Hence, *the firewall principle*:** (see Book-I, **1.31** to **1.32**).

Recall the *raison d'etre* of AND-form logic: It springs from the fact that, as-yet not realised future events belonging to the Whitehead facet must exist in pure AND-form, *regardless of their scale*. This includes billiard balls, cars, people,... planets. We cannot entertain the transpiration of the future before it has transpired, and we cannot entertain the reception of 'signals from that future' before that future has happened. To entertain otherwise is to invite violations of the principle of causality and tandem violations of the conservation principle, at all scales... unless one deems fit to restrict both causality and the conservation principle only to the micro-scale.

Signals from the future would constitute obvious violations of causality. The same signals would entail evidence of above-unity energy-matter, now required to constitute future events as pre-formed and 'materialised' OR-form affairs. To avoid such violations, AND-form logic, and quantum mechanics with it, *must* apply at all scales, and cannot be confined just to the very small, given that, at the macro-scale, we do not observe signals from the future, nor suffer violations of either causality or of the principle of conservation.

To augment, consider the development of the Moon as a macro-scale state five minutes from now. Insofar as it is five minutes in the future, it has not yet transpired. Shall we presume to describe it as an OR-form resolved state-of-affairs five minutes from now, or should we describe it as an AND-form state comprised of *possible* future variations of the Moon that might transpire five minutes from now? Indeed, how should we treat of the Moon two minutes from now, or even two seconds from now?

Restating the same questions, but in a form pertinent to the Wigner's friend experiment, how shall we treat the laboratory and Wigner's friend five minutes from now? Shall we describe the laboratory and friend five minutes from now as comprised of OR-form fully resolved states of affairs? Should we instead describe it as an AND-form state of possible future laboratory, friend, and cat possibilities? Indeed, how shall we treat the laboratory, etc, *et al* two minutes from now... or two seconds from now? What about two nanoseconds from now?

There is a clear reason why nature exhibits AND-form logic. The reason has absolutely nothing to do with 'scale'. At *all* scales, nature will exhibit AND-form phenomena simply because this is what the firewall principle from causality intertwined with the conservation principle demands. If the future remains not-yet resolved (or until it is finally transformed into an OR-form complex of events) we must treat it as an as-yet not happened AND-form superposition of future possibilities. It does not matter if we are scrutinising the future state of a 'micro-scale' lone electron, or the future sate of a 'macro-scale' billiard ball, or the future-state of

482

the 'macro-scale' Schrödinger's cat, or Wigner's friend, or the laboratory as a whole. That is, we must treat the future state of the lone electron, the billiard ball, Schrödinger's cat, Wigner's friend experiment, people, cars, and the Moon... as AND-form quantum mechanical superpositions of future possibilities, no matter what their scale... at *all* scales.

To treat the macro-scale laboratory and Wigner's friend five minutes from now, or even a few milliseconds from now, or a few nanoseconds from now, as a non-quantum mechanical OR-form state is tantamount to treating the laboratory in the future as a resolved state of affairs *before* it had transpired into such: Tantamount to 'signals from the future' in violation of causality and the principle of conservation.

Nature prohibits such violations by always expressing the future (at *any* scale) in non-resolved AND-form superposition form. Hence, the *raison d'etre* for AND-form logic in nature and the existence of quantum mechanics itself as the scientific attempt to grapple with that ontic future. Indeed, this fact, consolidated as the firewall principle, is incorporated into the very structure of the intermediate model spacetime from Book-II and into its successor; the Bergson-Whitehead amalgam from Book-IV.

The Whitehead facet abides at all scales. Hence, attendant AND-form logic must also abides at all scales, including the macro-scale Wigner's friend experiment. Hence, again, quantum mechanics is not restricted to the 'very small' but abides as an all-scale condition.

*

The clearest and most decisive case for the relevance of AND-form logic and the superposition principle at the macro-scale and at the level of Schrödinger's cat and Wigner's friend experiment, is *grand decoherence*. Grand decoherence arose from the following:

- **The nested futures view:** (see Book-II **2.17** to **2.22**). The Whitehead facet of the growing block Bergson-Whitehead amalgam (formerly, 'spacetime in potentiality') constitutes a nested futures system; one ultimately organised according to the convergent category nested futures.

- **The counterfactual quantum erasure of nested futures**: In the convergent category, a supernova explosion of one of the stars of Centauri will restructure the Whiteheadian nested future possibilities belonging to our Earth frame up to some four years into our future, and even to the infinitely removed future, even though we cannot obtain direct rapport of the supernova at the present moment. Our own frame of nested futures will instantaneously undergo counterfactual quantum erasure, with those futures not consistent with the fact of the supernova erased from our frame and from all frames. For example, if the supernova occurred, there is no possibility of observing it in four years' time as having *not* occurred. The possibility of 'no supernova' will be subsequently erased from any branch of our nested futures and from all other futures, even if we have no direct rapport of that supernova on Earth and must await four or so years to observe it, so rendering the Whitehead facet grand decohered into consistency with the said counterfactual.

- **Grand decoherence**: The quantum erasure of nested futures pertaining to the said supernova will thus decohere, not merely our own system of nested future attached to *our* frame, but *all* futures belonging to *all* frames along the event horizon, to the infinitely removed future. Yet, this is a macro-scale quantum interference phenomenon. The whole of the universe along the event horizon and into the Whiteheadian future will transtemporally or 'instantaneously' decohere into consistent futures with respect to the supernova event at Centauri. This is because all future-potential lightcones within the Whitehead facet attached to all frames along *any* event horizon are potentialised to contact, overlap or converge into a common future, and into full convergence at the infinitely removed future.

- Grand decoherence revives and supports two things, both directly pertinent to the question of whether AND-form logic and the superposition principle apply to the macro-scale, to billiard balls, to cats, to people and to Wigner's friend:
  - **The revival of Mach's principle.** (See Book-II **2.34** to **2.35**): Inertia requires that matter 'here' must know about matter 'elsewhere', and vice versa. How does matter here 'know' about matter elsewhere? How does matter elsewhere know about matter here? This is through grand decoherence, described above: Just as our Earth frame decoheres into consistency with the fact of the supernova at Centauri, the future possibilities for matter here must decohere into consistent futures with the futures for matter 'elsewhere'... and vice versa. In this way, matter here acquires inertia, as does matter elsewhere, via the 'non-local' or transtemporal binding, or through Mach's principle facilitated through grand decoherence. Hence, the revival of Mach's principle on the basis of macro-scale counterfactuals, consistent futures, and grand decoherence. Hence, the pertinence of AND-form logic at the macro-scale and at the universe-scale, wherein inertia *is* the mutual quantum interference and self-interference of matter here with matter elsewhere.
  - **The quantum mechanical account of the permanent magnet.** (Book-II **2.37**). What is it that makes the fridge magnet stay put on the fridge door, even against the pull of gravity? There appears no power-input involved in this, thus constituting another example of a phenomenon involving a work function zero profile. Once we place the magnet on the door, all Whiteheadian nested futures that are non-consistent with the now-established fridge-door relation are quantum erased from all futures. This is via grand decoherence. The only nested future possibilities that remain are those that conform to the now-established fridge-door relation, plus other contingent futures. Consequently, subsequent process of AND-to-OR collapse and time cannot help but reproduce and retrench the now-established door-magnet relation, with no further power requirements or inputs to achieve this: all on a work function zero basis.
  - **Grand decoherence is the universe-scale and macro-scale equivalent of generic quantum wave interference and self-interference**. As such, quantum self-interference and interference cannot be confined only to the 'very small'. We

reiterate that grand decoherence makes for both macro-scale and universe-scale wave-interference and self-interference. All billiard balls in the universe are part of a grand convergent nested futures wavefunction: the Whitehead facet itself. All billiard balls are essentially one object to the extent that they constitute the terms of a grand future-form wavefunction. Hence, whatever else these do, all billiard balls cross-interfere and self-interfere at the grand scale through grand decoherence, so garnering inertia... so reviving Mach's principle and the quantum mechanical explanation of inertia, magnetism, and much else.

In short, with grand decoherence, AND-form logic, the superposition principle, hence quantum mechanics itself, fully abides at the macro-scale, as it most certainly abides at the universe-scale. The quantum erasure of non-consistent futures through counterfactuals constitutes the gross reality of the self-interference' of the physical universe, with itself, at the macro and universe-scales: It implies the self-interference of billiard balls with each other instantaneously or transtemporally across vast 'distances', with their own and common futures. It implies the self-interference of the brain with its own future, and presumably with other brains, as much as with the rest of the universe: (This does not mean 'telepathy': It simply implies the elimination of mutually non-consistent futures).

Thus, the notion that quantum mechanics must be confined to the very small is false. To abandon consciousness causes collapse on such a false premise was an inadvertent error. Hence, the quantum superposition principle and AND-form logic abides in Wigner's friend experiment... and Wigner committed inadvertent error in capitulating to the false notion that quantum mechanics abides only per the very small.

5.13: Four Preliminary Objections to Quantum Idealism:
The Alternative Realist Interpretation Against 'Consciousness Causes Collapse'

We have our own reasons for doubting 'consciousness causes wavefunction collapse'. These are bulleted below and subsequently detailed:

- Consciousness and mind, or nous in general, and in its future-potential form as potential consciousness, must also be rendered into AND-form superpose of future noetic possibilities: It cannot be exempt from such a rendition.

- If future-potential consciousness must also be rendered in AND-form, what causes the collapse of potential consciousness into realised consciousness? It cannot self-collapse.

- Bergsonian abstract memory is the basis for the retention of the apparatus and of the measured system: (see **5.06**). The apparatus and measuring system are not retained by consciousness, nor need consciousness to be retained or order to abide.

- Consistent with the idea of future-form potential nous, consciousness is indeed different from the measuring apparatus and from observational inputs and observables, though not in the way originally posited and later abandoned by Wigner.

We now elaborate on each point in detail: Thus...

- **Nous in its future-potential form, as potential consciousness, must also be rendered into AND-form: it cannot be exempt from such a rendition.**

Implicitly, the state of the laboratory five minutes from now, or two milliseconds from now, constitutes the future. This potentiality includes the future possibility of Wigner's and his friend's as-yet not realised laboratory-associated consciousness states, both rendered into an AND-form superposition of as-yet not realised noetic possibilities. This complex will also include the future-potential consciousness state comprised of the... "consciousness of the live cat" AND "last conscious moment of the dying cat" ... five minutes from now.

The moment-to-moment partial-perpetual event horizon progression, hence AND-to-OR collapse and time, of the laboratory-cum-Wigner's friend, until Wigner's return... will unfold independently of Wigner's presence or absence. In that interim, **the laboratories' perpetual AND-to-OR collapse will have nothing to do with any causal power of Wigner's friend's consciousness, nor with Wigner's consciousness upon his return; nor with the cat's consciousness throughout; nor with Bishop Berkeley's quantum mechanised God-consciousness throughout**. In general, the collapse of wavefunctions occur, not by the action of consciousness (no more than by the action of a 'material' input per some hidden gears) but by an ineffable structure-less causality (see Book II **2.46** to **2.47,** and Part-I). This deeper causality (comprised of neither non-mental gears, nor by consciousness, and superseding both 'matter' and consciousness) is responsible for the collapse even of consciousness itself out of its own AND-form future-potentiality.

We can restate the same in terms of the intermediate spacetime model (Book-II) as well as in terms of its superlative successor; the Bergson-Whitehead amalgam. The laboratory and associated future-possible states of consciousness remain embedded within 'spacetime in potentiality'... within the Whitehead facet of the Bergson-Whitehead amalgam. The five-minute interval comprising Wigner's departure and return to the laboratory, together with attendant states of consciousness, will decay out of the future through a series of partial-perpetual wavefunction collapse-processes or synonymous progressions of the event horizon; accompanied by the quantum erasure of non-consistent nested futures (grand decoherence) inherent to the said five-minute interval and beyond. The said collapses and attendant event horizon progressions will occur independently of the conscious beings involved, insofar as consciousness is not causally responsible for the progression of the event horizon or of wavefunction collapse and time. That is, consciousness is not the drive of time.

In the above, **we cannot treat any state of future-possible consciousness as non-quantum mechanical: the future of consciousness can only constitute as AND-form potentials for alternative states of consciousness. To state otherwise... to assert**

that Wigner is directly acquainted with his consciousness state five minutes from now... would be tantamount to 'signals from the future and the violation of at least the principle of causality, and certainly the violation of actual experience: Neither we nor Wagner can experience our future-states of consciousness: These have not yet transpired.

What about retro-causality entertained in the transactional interpretation of quantum mechanics? Therein, consciousness states (of the cat, of Wigner's friend, of Wigner... of Yahweh) would need to be treated as resolved in the future. These would then reach back to the 'now' in order to resolve events out of immanent potentiality to those outcomes consistent with that future-consciousness and its future observations and experience.

However, the transactional interpretation of quantum mechanics is not only beyond proof or falsifiability, but it is neither necessary nor sufficient as an explanatory framework. In any case, if one's consciousness five minutes from now is resolved, why not ten minutes from now as well? Why not at each and every moment , obtaining one's moment of death, if not at the infinitely removed future? One need only examine their own conscious experience to realise that the retro-causality claim is not viable.

What about a possible employment of the Yoon-Ho Kim's delayed choice quantum eraser? In this instance, a supposed resolved future consciousness five minutes from now quantum erases alternative past states attached to the present and from subsequent nested futures that elaborate up to the five-minute mark. However, the delayed choice quantum eraser option is foreclosed. It is foreclosed by abstract memory theory and its culminating *holed past theory*, (see Book-III **3.19**), all based on the nested futures reframing of spacetime and its successor in the Bergson-Whitehead amalgam (see Book-IV **4.17**).

- **If future-potential consciousness must also be rendered in AND-form, what causes the collapse of potential consciousness into actualised consciousness? This cannot be through self-collapse.**

Was it the cat's consciousness that collapsed the wavefunction-collapsing consciousness of Wigner's friend? Was it Wigner's consciousness that collapsed the wavefunction-collapsing consciousness of the cat? Or was it Berkeley's God all along? What collapsed the universe-scale wavefunction-collapsing God-consciousness into *its* consciousness state? The interrogation degenerates into an infinite regress of non-provables: an inevitable consequence of the notion of 'consciousness causes collapse'.

From Alt-R, **we posit that the collapse of future-potential consciousness into realised consciousness out of its own distinct parallel AND-form noetic future-potentiality is accomplished by a deeper superseding ineffable causality that transcends nous and even supersedes God-consciousness, at least as thoroughly as it transcends a-noetic 'gears', 'matter in space', or the system of memory, possibility and time capitulated as the growing block Bergson-Whitehead amalgam.**

- **Bergsonian abstract memory is the basis for the mnemonic growing block retention of the apparatus and of the measured system and its observables (see 5.06). The apparatus and measuring system are *not* retained by or have need of consciousness.**

Recall the various mnemonic arguments for abstract memory from Book-III. These culminated into how information pertaining to the configuration of the barrier in the two-slit experiment is preserved and retained in the physical order as abstract memory: Book-III, **3.07** to **3.08**. Therein, the barrier setup was retained even when the very same experimental system was subject to physical isolation, and, upon its deference to the future, into inevitable AND-form quantum mechanisation. If quantum mechanisation meant the erasure of the barrier setup, quantum mechanics and science as we know it would be rendered impossible. This implied that, even with quantum mechanisation in the interim of isolation, the barrier established before isolation *must* be retained in abstract-immaterial form: as Bergsonian wavefunction in memory. Any attempt to circumvent the case for abstract memory *and* keep generic quantum mechanics, *and* accomplish this by resort to various forms of nihilism co-joined to Manyworlds, will simply fail, if for no other reason than the fact that Manyworlds is itself void: (see **5.07** and Book-III **3.16 to 3.17**).

By Book-IV, we had arrived at de-spatialisation (Book-IV, Part-I). This compelled that we drop space from 'spacetime' and redefine the physical order into information distributed in time. The temporal distribution of information is directly synonymous with abstract memory intimated in Book-III, almost synonymous with what Henri Bergson had envisioned about the nature of memory. **It is within the mnemonic Bergson facet that the box-cat system, the laboratory, Wigner's friend, and Wigner himself, are retained... as complexes of wavefunctions in memory.** The complex of Bergsonian wavefunctions in memory pertinent to the apparatus and laboratory also generate counterfactuals, engendering the grand decoherence and erasure of non-consistent nested futures within the Whitehead facet; so rendering all possible and permissible futures therein fully consistent with the said memory states of the apparatus, the cat, Wigner's friend, the laboratory, and with Wigner himself. **Through grand decoherence, the Whiteheadian future conforms to the memories (the apparatus) and all other participants of the Wigner's friend experiment.** Thus, the Whitehead facet is potentialised into a system of possible nested futures in which the cyanide vessel attached to the box-cat system is a 'read-write' future possibility state; potentialised to an attendant future-potential spin-resolution; with the alternative spin states part of the same Whiteheadian nested future system of alternative potential future outcomes.

Note again that the apparatus and laboratory, with the other participants, are retained in their original configuration within the Bergson facet (see **5.06** and the mnemonic solution to the apparatus problem). In this way, these survive the process of isolation, and are 're-materialised' at succeeding event horizons through the three-phase developmental process (see Book-III **3.26**) ... into outcomes conformal to and consistent with the original setup. Only the possible spin outcomes, combined to possible consequences to the cyanide vessel and to the cat, are rendered into open quantum mechanical states.

Note that the dissociation and retention of the laboratory system, as abstract memory, within the Bergson facet, will have nothing to do with consciousness, save insofar as Wigner and friend will be involved in physically setting up the experiment and obtaining the results. Moreover, the memory of the pertinent system will be carried independently of consciousness within the Bergson facet. This carry-over (*not* across space, and certainly not carried by displacing 'matter' or radiation) happens across the whole of nature and it is automatic: It is *not* dependent on the absence or presence of nous. Succinctly, the retention and carry-over of the apparatus, the cat-and-box, and of the physical order as a whole, does not require consciousness.

- **Consistent with future-potential consciousness, consciousness *is* different from the measuring apparatus and from observational inputs and observables, though not in the way posited and later abandoned by Wigner.**

Wavefunction collapse is AND-to-OR collapse and time. It transforms the Whiteheadian state of AND-form future possibilities into OR-form realised 'materialised' apparatus and systems, as well as into tandem and attendant realised states of consciousness. **Consciousness must also decay out of AND-form future-potentials, albeit specific to consciousness. As stated, this is accomplished by an ineffable structure-less causality that transcends consciousness as thoroughly as it transcends non-mental mechanism and 'matter in space'. Insofar as consciousness is not responsible for wavefunction collapse and time, but utterly dependent on time, consciousness is no different from non-conscious apparatus and states, at least in terms of ultimate causality as it relates to the drive behind wavefunction collapse and time**: i.e. neither consciousness nor non-mental gears constitute the drive of time.

However, consciousness *is* profoundly different form non-conscious apparatus and world. The list of profound differences remains manifold and could come under the rubric of David Chalmer's *hard problems of consciousness*[94]. For example, unlike non-conscious apparatus and 'material' states, consciousness can apprehend infinities; it can discern processes that can halt verses process that cannot. This segues into the critical fact that consciousness inheres to itself the sole privilege of the *renormalising quantum mechanist*. In short, one key difference between the non-conscious apparatus and the conscious being is that the latter can carry out renormalisation (see **5.14** below). If the only difference between consciousness and non-conscious apparatus was per function the fact that the former can carry our renormalisation and the latter cannot, this would be enough to establish a decisive and profound difference: one that does not require wavefunction collapsing consciousness, nor requiring consciousness as the cause of time itself.

However, one attribute that is not generally recognised, yet clearly part of the hard problem, was arrived at at the end of Book-IV (see **4.41**. Also see **5.15** in this book): This is the absolute time-like characteristic of consciousness, manifested in it noetic moment: its unitary sense of the common present moment. We shall have more to say about this in Part-III.

5.14: The Renormalising Quantum Mechanist:
A First Differentiation of Mind & Consciousness from Mere Apparatus

Renormalisation was applied in quantum filed theory and in quantum electrodynamics to resolve situations that generate infinities. It has direct relevance to the nature of consciousness vis-à-vis apparatus and world. Quantum field theory relies on extremely short lived (or, on the erroneous assumption of space, 'short distance') virtual particle interactions. The system of particle interactions enjoy many 'degrees of freedom' and key details about this will be clarified in Part-III, **5.30**. Indeed, the degrees of freedom available to a system explode to infinity. It then becomes necessary to use a mathematical 'trick': The physicist must incorporate a constant or a limiting factor to reduce the said degrees of freedom to a non-infinite manageable range. **In brief, the renormalisation ultimately involves subtracting infinity from infinity to obtain meaningful and manageable calculations**.

For a long time, renormalisation within quantum field theory was viewed with doubt by many, including Freeman Dyson, Paul Dirac, and Richard Feynman. Yet, renormalisation permitted successful calculations and predictions that would otherwise be impossible to obtain. By the 1980s, renormalisation had become widely accepted.

Both our Bergson-Whitehead amalgam approach and de-spatialisation entail implications to renormalisation. Instead of thinking about particle interactions as unfolding 'in space' via contiguous contact causality, or in terms of purported 'short distance interaction particles', the same interactions must now be recuperated purely in terms of short-lived delayed choice time-interval relations, without resort to any spatial intervals and, obviously, without contiguity, contact, or impact, given that there is no space in which such could transpire. (More on this in Part-III).

In terms of the Bergson-Whitehead amalgam successor to 'spacetime', and insofar as the progression of the event horizon is shown to be broken-discontinuous (as opposed to seamless-continuous) it follows that, unlike purported spatial intervals, time-intervals cannot be infinitesimally small: That is, time proves to be grainy: Time is quantised. Thus, where particle interactions are concerned, natural and manageable time-based limits and intervals to renormalisation could be developed if interactions are expressed purely in terms of delayed choice time-intervals. These cannot ever be infinitesimally small and will likely cut-off on approach to the Plank time interval. In short, the move to a pure temporal approach in science and quantum science, combined to the fact that time is not divisible to infinity, could put a cap on the tendency for infinities in quantum field theory, circumventing generic renormalisation, or at least removing the 'trick' aspect from it.

Even if our pure temporal approach might not eliminate the resort to renormalisation, it might help solve hitherto unrecognised causality-related problems exhibited in the resolution of systems with infinite degrees of freedom. Calculation aside, if nature itself had

---

[94] *Chalmers, David (1995). "Facing up to the problem of consciousness". Journal of Consciousness Studies. 2 (3): 200–219.*

to process the infinities, she would crash into insurmountable halting problems and causality, and state-change could never transpire. This hitherto obscured problem arises as an artefact from the assumptions of spatial particle-mediated contact causality. The assumption of spatial and contact-based causality must crash into insurmountable halting problems in the face of the said infinities. But the same, now expressed in a de-spatialised pure temporal terms could entail processes that temporally 'leap over' the infinities... from one event horizon to a succeeding one... *without* 'contact': a 'quantum leap in time' that circumvents the interaction-possibilities subject to infinite degrees of freedom that reside in the leaped-over interval; a 'leap in time' that circumvents the infinities within the interval and obviates the possibility of halting problems in causality and time. More will be said on these two possibilities in Part-III, **5.30**.

However, the matter of the validity of renormalisation in standard quantum field theory is not our specific concern in this essay. **Our purpose is to use renormalisation as evidence to the effect that consciousness, though in no way responsible for bringing about wavefunction collapse, and no different from any other non-conscious apparatus in at least that respect,** *is* **profoundly different from non-conscious apparatus and non-mental world. Renormalisation, trick or otherwise, is the unique and exclusive purview of nous**, whether one accepts it as a legitimate procedure, or seeks to circumvents it by means of our 'quantum leap in time' approach, but which does not obviate the implication to consciousness from the renormalising quantum mechanist.

Even if one could argue successfully that consciousness is an 'illusion', and accomplish it without crashing into apagogic absurdity, one will be hard pressed to deny the specific powers that attend 'illusory nous', evinced in and through renormalisation. **Indeed, the reality of renormalisation (trick or not) clearly implies the profound existence of a really-existing** *woo-woo* **factor in the midst of the physical order**: one profoundly different from mere apparatus or other existents of the world.

If we assert that renormalisation is not a 'trick', we are confronted with the fact that the 'illusory' consciousness belonging to the renormalising quantum mechanist must possess a power vis-à-vis infinities that, on the typical presumption of materialism and the assertion that the physicists is nothing but a finite mechanical and computational brain, should not be possible for the physicist to possess. The renormalising quantum mechanist must rely on some special process; one that could not be furnished by the finite, iterational, mechanical or computational brain.

Only a non-iterational non-computational process (inherently without structure of *any* kind... hence non-definable and ineffable... hence non-material) could endow the physicist with the power to recognise halting problems, handle infinities, subtract infinities from infinities, and 'trick' renormalisation. Thus, a deeper causal process (a non-definable process that resolves consciousness itself into being, but itself transcends nous) conveys to that consciousness a unique power that a mere apparatus or co-joined system could never possess or convey. Thus, the renormalising quantum mechanist (hence consciousness) *is* very different from the apparatus and from the generic physical universe, even when the latter is temporised per de-spatialisation, and we begin to envision the operation of a 'quantum leap in time' that circumvents interval-infinities, tandem with AND-to-OR collapse and time, itself driven by a gear-less ineffable causality... the same causality that renders nous and renormalisation possible.

The real existence of the renormalising quantum mechanist tells us that nature's ultimate attributes, insofar as nature can produce beings that can grasp and manage infinities, defies any categorisation in terms of definable structure, mechanism or 'gears'. **As we had posited in Book-II 2.47 and Book-III 3.32, non-definability is characteristic of the structure-less causality responsible for fundamental and ontologically primary wavefunction collapse and time. This deep ineffable causality is also responsible for manifesting thought, consciousness,** and all the other hard problem categories adumbrated by Chalmers *et al*: **a list that must include the renormalising quantum mechanist.**

*

What if renormalisation proved to be false: perhaps an artefact arising from the assumptions of space, corrected by de-spatialisation and the anticipated 'quantum leap in time' approach? This could not gainsay the implications from renormalisation. Insofar as the 'trick' of renormalisation requires a being that can grasp and handle infinities, we must explain how such a being is possible, regardless of whether or not renormalisation is a contrivance born out of the assumption of space *and* contiguous causality.

Clearly, we do not derive our knowledge of infinity from perception or sense-data. Infinity is not an empirical state apprehended by means of the senses. The Big Bang suggests that the universe is a finite number of years old, or, assuming the dubious notion of 'space', the universe is a finite size. Even if Big Bang was proven false and the universe turned out to be indefinitely old and vast, we could not prove this empirically or experientially: The proof would require an infinite number of observations of states elaborating to the unattainable remote past: a non-terminating incompletable goal.

Where do we get our idea of infinity, the very basis for the possibility of the renormalising 'trick'? If not from the empirical world, if not from the senses, it ought to be by some action of the brain. Yet, the brain is a finite system. No matter how complex, the brain cannot enact and output the idea of infinity. Indeed, if we gave the brain the simple task of counting just the whole integers, and even gave it infinite 'memory capacity' per notions of generic memory from information technology, the brain would require eternity to reach its conclusion. How does the brain know the truth that there are an infinite number of whole integers, or even irrational numbers...and never mind the continuum? It cannot. It cannot accomplish such a feat by means of the finite brain or of any putative tick-tock click-clock mechanism therein. The power belongs to a *something* related to brains, but neither reducible to nor generated by the brain.

On this basis, we arrive at *the* profound difference of nous vis-à-vis apparatus and the non-conscious world and nature, if not versus the brain: a profound difference derived from the fact that, 'trick' or not, renormalisation is possible only upon our ability to grasp and handle infinities and halting problems, even though the infinities so apprehended are not obtained from

sense data; nor from any finite apparatus in sense-data; nor from the brain as merely organic finite apparatus. In short, Wigner was mistaken in subsequently re-thinking that consciousness is no different from mere apparatus. Consciousness remains *profoundly different* from apparatus and world.

5.15: Absolute Time-Like Characteristics of Mind & Consciousness: The Second Differentiation of Consciousness from the Apparatus

We now come to the most radical characteristic that renders mind and consciousness profoundly different from mere apparatus, and from laboratory and world. It was presaged at the end of Book-IV (**4.41**): This is **the absolute time-like characteristics of mind and consciousness in its noetic sense and witness of the 'common present moment'**.

Recall from Book-IV (**4.01**) that absolute time, hence the notion of the 'common present moment' in which events supposedly transpire, constituting a common 'surface' or block upon and within which objects supposedly occupy 'locations', is perfectly synonymous with the notion of space. The invalidation of absolute time in Special Relativity, and the necessary correlate that all of nature is subject to the relativity of simultaneity, ought to have led to the demise of space as early as 1905, or as late as 1927, given the independent implications against 'motion' and 'space' from then-nascent quantum mechanics; (see Book-IV: **4.04** to **4.05**). Our belated de-spatialisation furnished in Book-IV (Part-I) asserted the exclusive ontological primacy of time, *without* space. It espoused nature as a pure temporal order formed out of events organised into delayed choice time-interval relata, with the Bergson facet of the Bergson-Whitehead amalgam constituting the domain of memory of events in their time-relations.

The invalidation of absolute time necessitates that all things, including the cat-box system, the laboratory, and both Wigner and his friend, must be subject to the relativity of simultaneity, and must be retained as complexes of abstract wavefunctions in memory within the Bergson facet of the Bergson-Whitehead amalgam, *without* space.

The distinctive characteristic of the 'material world', its apparatus, its systems, is that, even ignoring de-spatialisation, it is incapable of exhibiting *any* absolute time-like characteristics. However, the absolute time-like characteristic of nous, the visceral sense of the *common present moment* in the noetic moment through which nous apprehends the world, resides at the core of the unity of experience and monadic nous. That is, **consciousness *in itself*, as the apprehending state or witness, is not internally divided into parts. It is consequently not internally subject to relative time-relations and the relativity of simultaneity. To itself, the moment of witnessing consciousness is a singular state in absolute time: Nous is absolute time-like, *albeit* without space.** This remains true even if the world it apprehends *is* subject to the relativity of simultaneity; a world in which absolute time is impossible, and in and from which the sense of the common present moment and of monadic witnessing nous cannot be generated… and from which the noetic moment in absolute time cannot arise. Hence, in truth, the noetic moment is extant and dualistic-irreducible with respect to its associated brain and world, if only because the brain and world cannot generate *any* form of absolute time-like state or noetic moment.

This much was arrived at in our presaging case for *noumnemonae*. (See Book-IV **4.41**). **Thus, the brain, wholly internally subject to the relativity of simultaneity, cannot generate the visceral sense of the common present moment or unitary undivided experience. Hence, the brain cannot generate monadic nous or witnessing consciousness.** The dualism evinced therein is between mind and consciousness characterised by definitive absolute time-like characteristics versus world and brains subject to the relativity of simultaneity. Indeed, the perennial object-subject relation can be reframed as a relation between subject-nous that has absolute time-like characteristics vis-à-vis object-world and brains, with both strictly subject to the relativity of simultaneity.

However, nous exhibits absolute time-like characteristics, not because it is spatial, but precisely because it is not. The closest spatial analogue for monadic absolute time-like nous ought to the 'spatial point', but one necessarily without space or 'parts': a temporal noetic monad. As such, mind and consciousness can exhibit absolute time-like characteristics of a form that would otherwise attend an impossible point-like state.

We shall have more to say on the absolute time-like characteristics of consciousness and other things in Part-IV, in the *general theory of mind-world*.

*

The cat, the box; the laboratory, Wigner's friend, Wigner himself, their de-spatialised brains structured into pure temporal forms… are all retained within the Bergsonian past. These are all co-joined to a Whiteheadian system of nested futures, in turn subject to potential time-relations that are also structured per potential relativity of simultaneity, without any possibility for realised or potential absolute time, notwithstanding the grand decoherence of the Whitehead facet and the erasure of non-consistent nested futures.

However, **the consciousness of Wigner, his friend, and certainly the consciousness of the cat, constitute visceral individuated unique noetic moments: part of their inherent quality and state of experiencing. The consciousness involved cannot emerge from the experienced system of memory, possibility, and time, all of which are subject to the relativity of simultaneity, precisely because the pertinent consciousness of Wigner, his friend, and that of the cat possess unitarity absolute time-like characteristics that cannot be garnered by the apparatus, the laboratory and world, much less by attendant brains. Hence, consciousness and nous *is* profoundly different from apparatus and world, against Wigner's claim to the contrary.**

Is consciousness an illusion? There is no shortage of proponents who assert this apagogic claim, including Daniel Dennett, *et al*. We immediately run into the problem of how the 'material world' and universe, wholly subject to the relativity of simultaneity, could ever generate the 'illusory consciousness' that can conceptualise the notion of absolute time *and* experience it viscerally as an integral part of its 'illusory' self-reference, of its self-awareness, and in the retrospective apprehension of the world. Thus, the 'illusion' ought not to exist, *even as an illusion*. To assert consciousness as an illusion could only be accomplished on the false premise that we need the former material world and brains to generate or produce that 'illusion'. However, a world and brains strictly subject to the relativity of simultaneity, and without absolute time-like characteristics, could not produce even the abstract concept of absolute time, much less the 'illusory' sense of the witnessed 'common present moment'. The fact that this latter can exist, even as a supposed 'illusion', puts the lie on the notion that it is anything other than real.

The absolute time-like characteristics of 'illusory' consciousness forces admission to the real existence of nous in relation to and distinction from a non-conscious non-mental world subject to the relativity of simultaneity, notwithstanding the Cartesian dualism that must arise from the differentiation between consciousness in absolute time versus brain and world subject to the relativity of simultaneity.

5.16 Critique of 'Consciousness Causes Collapse'
From the Alternative Realist Solution to Quantum Indeterminacy

Some interpretations of quantum mechanics, such as 'hidden variables', look for non-mental definable causes responsible for both wavefunction collapse and the orchestration of quantum indeterminism: These seek a set of fully definable variables; part of a complete closed causality input-operation-output schema to mediate the inputs of possibility into determinable outputs of realisation in the generation of state-change in wavefunction collapse *and* in the orchestration of apparent quantum indeterminism. However, in Book-II and III, it was asserted that the process of both AND-to-OR wavefunction collapse (time) as well as the resolution of quantum indeterminate outcomes derive from a structure-less gear-less causality; one elusive to description as any form of clockable mechanism, 'gears', or computational process. (See Book-II, **2.46** and **2.47**, respectively; Book-III **3.26** and **3.32**, and what follows below).

However, the interpretation under scrutiny is Wigner's 'consciousness causes wavefunction collapse'; wherein consciousness is inadvertently asserted as the drive of time, even if Wigner did not recognise the synonymity of wavefunction collapse with time.

In 'consciousness causes collapse', consciousness replaces the said materialist non-mental hidden variables or 'gears'. Our general aim in Part-II is to call into question the central notion that consciousness wills wavefunction collapse and drives time itself, and, in the process, orchestrates tandem quantum indeterminate processes into determinate outcomes, obviating quantum indeterminism.

In this essay, we will focus on quantum indeterminism and reiterate the case for the objectivity of randomness furnished in Book-III. *We will then place consciousness in lieu of purported hidden gears and see if we can obtain conclusions fundamentally different from forlorn materialist non-mental gears.*

*

The want of a solution to quantum indeterminism arose from the unquestioned notion that it constitutes a problem, one solvable, if not by resort to Manyworlds, then by resort to either non-conscious hidden gears or consciousness in lieu of gears. However, in previous material, we discovered the following:

- **Randomness, especially in its quantum indeterminate form, is objectively real**: The want of a solution to the 'problem' of quantum indeterminacy must entertain indeterminism as a facile illusion, one undercut by a deeper hidden deterministic process (e.g., Bohm's 'pilot waves': see **5.08**). This requires proof that quantum indeterminism is not objectively real. However, it *is* objectively real, as was cased in Book-III **3.32**. *Neither resort to a materialistic hidden gears nor to consciousness in lieu of gears can obviate the objectivity of randomness and quantum indeterminacy.*

- **Quantum indeterminism is not a problem that requires a solution. The notion that it is a problem is invalidated by our three-phase developmental approach** in Book-III **3.26**. Therein, it turned out that quantum indeterminism is not inimical to coherence, pattern, or even to the retention, continuation and perdurance of pattern. Indeed, randomness and quantum indeterminism constitute the economising power vis-à-vis information and complexity: friends to the formation, perpetuation and retention of pattern and order. Hence, consciousness is no more necessary to the 'solution' to the non-problem of quantum indeterminism and order in nature than is resort to non-mental hidden gears, pilot waves, or even Manyworlds. *We do not require a 'solution' to a non-problem.*

The proof that randomness is objectively real, combined to the fact that the three-phase developmental approach obviates the notion that randomness is a problem in need of a 'solution', was fully developed in Book-III. Also, in the context of mind-brain issues, it turns out that the secret to how Cartesian mind resolves the brain according to *its* design does not require any mind-driven orchestration and obviation of in-brain indeterminacy and randomness, simply because quantum indeterminism and randomness in the brain is no more a problem that requires a consciousness-based solution than does indeterminism extant the brain, such as evinced in photographic image development. More on this and how mind affects control of its brain will be given in Part-II to Part-VI.

*

In Book-III, the proof that randomness is objectively real arose from the attempt to find hidden deterministic scripts that might orchestrate the development of any random-seeming process. Therein, we utilised the Sierpinski Triangle as our framework and model of a developmental process involving quantum indeterminacy via quantum mechanised six-sided dice. Therein, each subsequent orbit and evolution of the seed was treated as a quantum object: with each indeterminate resolution of that orbit resolved through a process of AND-to-OR collapse, in quantum indeterminate fashion. (See **Fig. 5.01**: (A)).

In the Sierpinski triangle, there are three possible directions toward which the orbit of the seed can evolve. Thus, using the appropriate formula from Book-III **3.32** (not depicted here) the 1ˢᵗ order combinatorial possibilities obtained when n = 3 summed to a

489

maximum of 34 block-scripts (see **Fig. 5.01: (B)**); each script constituting the deterministic orchestration of up to 3 seed-orbits in any sequence. But 34 block-scripts are inadequate for the purposes of fully developing the Sierpinski triangle. At least 100 orbits of the seed are required in order that the sought fractal pattern of the Sierpinski triangle acquire developmental saliency. In Book-III **3.32** we solved this problem by re-combining the 1st order 34 block-scripts to form larger deterministic scripts of up to 102 seed-orbit lengths: We obtained a very large number (VLN) of these. Each unique 102 orbit-length 2nd order script was found to deterministically orchestrate successive 'quantum indeterminate' seed-orbits into the sought pattern-culmination.

<u>(A) Developing the Sierpinski Triangle</u>

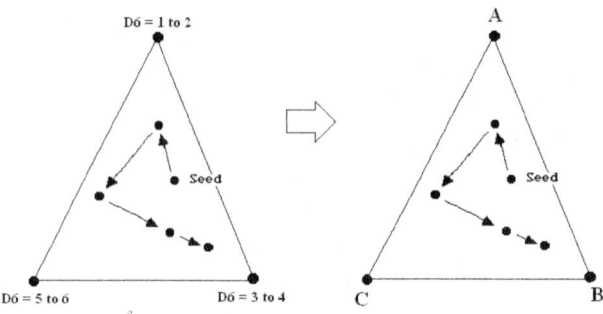

<u>(B) Fundamental Block-Scripts when n=3</u>

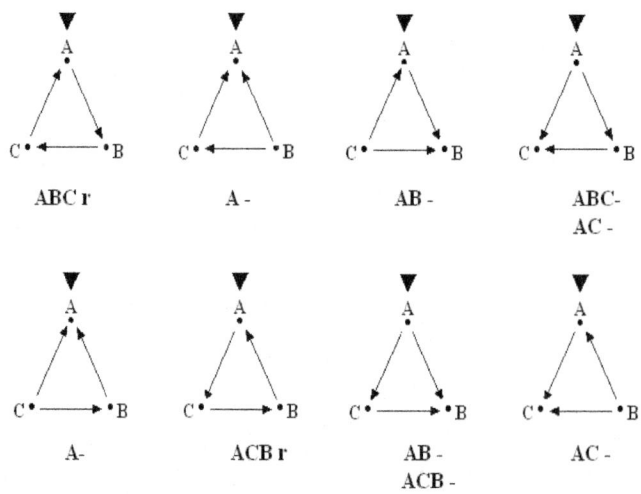

Fig. 5.01: Orbit of the seed in the Sierpinski triangle and fundamental block-scripts reprised
Notes: In (A) we reprise the diagrammatic description of the development of the Sierpinski triangle as was featured in Book-III: an equilateral triangle whose vertexes are specified by the letters **A, B, C** respectively. For every seed orbit, we roll a six-sided die. When we get a 1 or 2, we script this as **A**. If 3 or 4, we script it as **B**. And when 5 or 6, we script this as **C**. From the three vertexes (**A, B** and **C**), we can generate block-scripts for the orchestration of the orbit of the seed of up to 3-orbit lengths: a total of 24 fundamental unique block-scripts. As was shown in Book-III, this can be extended to 34 scripts per certain redundancies. Figure (B) shows all block-scripts starting with "A", with each unique block-script accounting for one, two, or a maximum of three seed-orbits. Only the block-scripts beginning with **A** are given: Those beginning with **B** and **C** have the same form but are not illustrated here.

Thus, surface indeterminism in the quantum mechanised Sierpinski triangle appeared to be orchestrated by entirely deterministic 102-orbits long scripts: i.e., the analogue to hidden gears that supposedly operates behind apparent quantum indeterminism in nature and may be considered synonymous to essentials espoused in Bohmian pilot waves. Yet, with closer inspection, we discovered that we did not generate just one deterministic 102 orbit script, but a VLN of such scripts; each equally apt to deterministically orchestrate apparent quantum indeterminate orbits of the seed to full pattern-development. We could not find a one-and-only exclusive or special 2nd order script to obviate any of the other equally apt scripts from fulfilling the role of a secret deterministic orchestrator.

How do we select a one script out of the VLN of equally apt scripts? In Book-III, it was argued that the script could only be decided by resort to a random selection process or sortology from amongst the said VLN of scripts. That is, we could not eliminate resort to randomness and indeterminacy, even vis-à-vis otherwise deterministic scripts. We could not arrive at a point at which we obviated and eliminated resort to a randomness-based sortology, such as to prove randomness and quantum indeterminism to be non-objective and illusory, or establish the unambiguous final case for hidden determinism to obviate all randomness, or rescue input-operation-output (IOO) causality schemas thus.

Nor did resort to a 3rd order combinatorial domain solve our problem. By combining each of the scripts constituting the 2nd order domain, we generated an even larger VLN set of deterministic orchestrating scripts. However, this merely produced an even larger number of unique non-special and non-exclusive orchestrating scripts. Again, we had to resort to an arbitrary random selection sortology to decide the orchestrating script. The only resort left was to carry on into a 4th order combinatorial generation; followed by a 5th; then by an nth order combinatorial set. In principle, this could be carried to infinity, culminating into a combinatorial explosion and an insurmountable halting problem: both of which constitute clear indicators that the search for hidden variables or gears, and the elimination of randomness, is a wholly forlorn goal. This conclusion abides across nature and in the generation and development of *all* patterns out of either single or *en masse* quantum indeterminate processes, such as in the context of the electronic configuration of the atom; the deposition of photons involved in the formation and development of a photographic image or, similarly, of electrons in electron micrography; in the drift of a snowflake; and in unfolding brain-activity. In all of these, **the general conclusion is that there are no hidden gears behind quantum indeterminism, and that quantum indeterminism is a non-illusory objective feature of reality, despite the want to imagine it otherwise in Bohmian approaches, for example**.

*

In what sense would consciousness be better at developing the Sierpinski triangle, or of any other pattern that involves quantum indeterminate terms, than forlorn non-mental a-noetic hidden gears? Given that there is no special one-and-only exclusive script to under-determine apparent surface quantum indeterminism, we might place mind and consciousness in lieu of non-conscious orchestrating script. In this way, consciousness would become the decider, if not of the seed-orbit from one moment to the next, then in the selection of the orchestrating script from among the VLN of such: supposedly circumventing resort to *any* form of random sortology and selection. Thus, obviating randomness and quantum indeterminacy.

Which one-and-only script from the VLN of unique equally apt scripts would consciousness select? Will consciousness select it on the basis of a qualitative principle inherent to a candidate script? Given that all appropriate-length scripts from *any* combinatorial order are equally unique and equally apt to the task, with none so special as to exclude any other, and given that there exists no structurally or qualitatively superior script, it follows that there is no requirement for a qualitative judge, selector, or decider-mind: a role that would be uniquely suited to mind and consciousness *if* confronted with issues of qualitative sortology. Hence, *the qualitative superficiality problem of consciousness-directed circumvention of quantum indeterminism*.

Given the qualitative equality of the scripts, combined to their efficient equality, the selection by consciousness from among these would constitute, not a qualitative decision, but a completely arbitrary decision, as good as resort to a mindless random selector. One may as well resort to a non-conscious random arbiter in lieu of consciousness instead of consciousness in lieu of non-conscious random arbiter. Hence, the *arbitrary selection problem in consciousness-mediated circumvention of quantum indeterminism*. In that case, what would be the point of *any* resort to nous in lieu of hidden gears, if it is no better than an a-noetic non-mental random sortology vis-à-vis quantum indeterminacy and randomness, or if the resort to arbitrary selective consciousness is effectively as random and haphazard as non-mental haphazard indeterminism? What need of nous in the development of *any* system, whether this be the development of interference patterns in the two-slit experiment; the development of a photographic image, or even in the unfolding of quantum indeterminate events inherent to the activity of the brain… or in the resolution of outcomes in the Schrödinger's cat or Wigner's friend experiment?

Again, the resort to consciousness turns out to be superfluous and unnecessary: It can have no critical, special or decisive role in obviating quantum indeterminism, even assuming the latter constituted a 'problem' in need of a solution. It necessarily follows that consciousness is not involved in the determination or orchestration of outcomes obtained from wavefunction collapse. It follows that physical systems (including the brain) are objectively developmentally quantum indeterminate, with or without the presence of consciousness. Hence, consciousness in lieu of equally forlorn non-mental hidden gears is as forlorn as a-noetic hidden gears. This is true even in the context of brains, calling into question quantum mind theories that espouse consciousness as the orchestrator and obviator of indeterminacy in brains.

*

We will now seal our case against consciousness-obviated quantum indeterminism. The three-phase developmental approach from Book-III **3.26** demonstrated that objective quantum indeterminism and randomness is not the 'problem' it is purported to be: It does not require a 'solution', whether one based on material hidden gears or on consciousness in lieu of hidden gears. To that end, recall that there are three distinct developmental phases for any system subject to quantum indeterminism. This applies to a developing

491

photographic film; to the formation of an interference pattern in the two-slit experiment; in the drift of a snowflake; and, of course, in unfolding brain activity. These are…

- The **below-threshold phase**
- The **transitional-chaotic phase**
- The **noise-reduction pattern-saliency phase**

Fig. 5.02: The three developmental phases illustrated

(A)

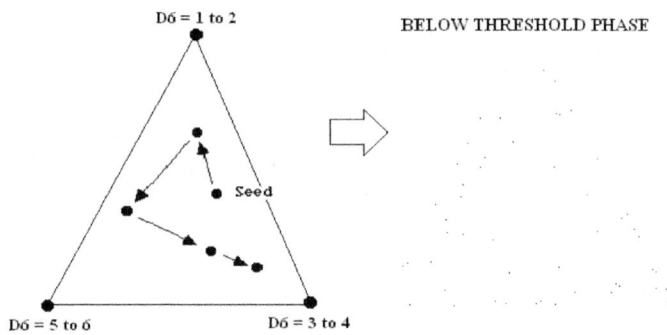

(B)

(C)

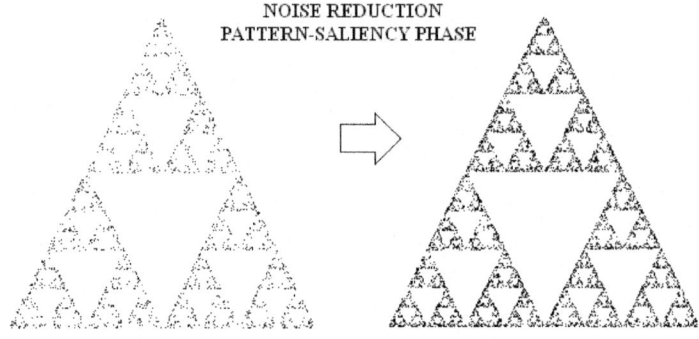

Recall that, in Book-III **3.27**, the development of the quantum mechanised Sierpinski triangle unfolded according to the said three phases noted above: This is repeat-illustrated in **Fig. 5.02: A to C**. Also recall that randomness and quantum indeterminacy remain objectively real: i.e., there are no non-mental hidden gears, and no consciousness-mediated orchestration in lieu of hidden gears. Also recall from Book-III that, below a certain threshold of accumulated orbits, the distribution of seeds appeared to be random and haphazard; effectively constituting noise: **(see Fig. 5.02 (A))**. This is the ***below-threshold phase***, **and it corresponds to our usual prejudgements about randomness as 'chaos' or noise, devoid of information.**

When we obtained sufficient numbers of seed-orbits and enter the *transitional chaotic phase*: (see Fig. 5.02 (B)) we found that quantum indeterminate orbits began to coalesce into volatile unstable patterns: This unstable transitional phase eventually gave way to the emergence of stable and highly resolved pattern.

Throughout the three phases of development, successive quantum indeterminate resolutions of the seed-orbit served to sharpen emerging pattern, despite their objectively indeterminate characteristics, given the forlorn search for hidden deterministic orchestrating scripts recapitulated above. **In the *noise-reduction pattern-saliency phase* (See Fig. 5.02 (C)) objective quantum indeterminacy was not at all eliminated: It remained present and ineliminable. Yet, full order emerged: i.e., the highly structured fractal pattern definitive of the Sierpinski triangle.**

Recall that the quantum indeterminate orbits of the seed throughout the three phases of development are objectively indeterminate: There are no hidden gears. Nor is there any need for an orchestrating consciousness in lieu of gears. Instead, accumulating objective quantum indeterminate processes act as the very means for noise-reduction and the crystallisation of pattern. We can no longer deny the role of randomness as noise-reducer: a friend to information and pattern. It is certainly *not* an enemy of it, as typical prejudgement would have it. The same conclusion can be generalised to the development of a photographic image; to the drift of a snowflake; to the emergence of interference patterns in the two-slit experiment...in *all* process of nature that manifest pattern, including the evolutionary process of life... and including the formation of global coherent activity in brains.

Thus, the attempt to establish hidden gears behind randomness and quantum indeterminacy turns out to be wholly unnecessary; and resort to consciousness in lieu of mechanical hidden gears turns out to be equally unnecessary...simply because randomness or indeterminacy is not only objectively real, but a friend to the emergence and perpetuation of pattern and information across nature. Therefore, randomness and indeterminacy is not a 'problem' that requires any 'solution', either by resort to non-mental hidden gears or by resort to nous in lieu of gears.

*

Let us embed these insights into 'spacetime', or to its successor, the Bergson-Whitehead amalgam. The abstract background possibility function that funnels objectively real quantum indeterminate processes pertinent to *any* system (such as that of the developing photographic film, the formation of interference patterns; the drift of the snowflake; and in unfolding brain activity) comes in one of two types. These either attend Whiteheadian wavefunctions in potentiality (*with* nested futures) or Bergsonian wavefunctions in memory (i.e., past wavefunctions cropped of their former nested futures). These functions, or instantaneous wholes, fate objectively real purely indeterminate quantum outcomes into pre-ordained global pattern-structures. **The instantaneous wholes, present throughout the attendant three phases of development, constitute the developmental guides to pre-ordained fated pattern-structures. Hence, we do not require resort to hidden gears or to consciousness in lieu of gears... or even to God misconceived as consciousness... to obviate the non-problem of quantum indeterminacy. In short, consciousness has no necessary or sufficient role to play in *any* 'solution' to the non-problem of objectively real and order-friendly quantum indeterminacy.**

*

In summary, three reasons invalidate the need for a consciousness-orchestrated quantum indeterminacy. These are...

- **The qualitative superficiality problem of consciousness-mediated quantum indeterminism**: In any imagined set of deterministically orchestrating scripts posited to under-determine quantum indeterminate processes, there is no qualitatively superior script. Consequently, the need for consciousness as the qualitative recogniser, decider and selector from among the scripts is fully obviated: There exists no basis for any qualitative judgement or selection powers of consciousness in *any* scheme devised for the obviation of randomness.

- **The arbitrary selection problem of consciousness-mediated quantum indeterminism:** Given that there is no qualitatively superior exclusive script, robed of any need of its qualitative selection powers, consciousness would be forced to resort to a wholly arbitrary selection from among the VLN of equally apt scripts: This would be as good as resort to a purely mindless random selective process. There is no point in replacing a mindless random selector with a merely conscious arbitrary selector. A conscious arbitrary selector would be as 'bad', or as arbitrary, and as effectively random, as a non-conscious random selector. Thus, the want of a consciousness-directed quantum indeterminacy is necessarily superfluous or pleonastic. But what if there are no scripts, but mere alternative possibilities at each moment of evolution? Again, a conscious arbitrary selector from among these mere alternatives would be as arbitrary, and as effectively random as a non-conscious random selector. Thus, again, the want of a consciousness-directed quantum indeterminacy is necessarily superfluous.

- **Consciousness has no necessary or sufficient role to play in any 'solution' to the non-problem of objectively real randomness and quantum indeterminacy.** From the three-phase developmental approach (combined to the fact that

493

objective randomness and quantum indeterminacy do *not* constitute enemies of coherence and pattern) it is obvious that we do not need to 'solve' the non-problem of randomness and quantum indeterminacy, much less resort to consciousness in lieu of equally forlorn a-noetic materialistic hidden gears to found such a 'solution'.

5.17: Why Consciousness Cannot Circumvent Quantum Indeterminism: Augmentation from MWI

The following constitutes an independent addendum to why consciousness cannot circumvent provably objective quantum indeterminism. It marshals some of the material used in the critique of the Manyworlds interpretation in Book-III. Recall that Manyworlds posited a hypothetical solution to quantum indeterminism by asserting that *all* the possibilities in the AND-form wavefunction are resolved into distinct compartmentalised histories. However, when we applied this to the quantum mechanised Sierpinski triangle (a stand-in for the universe from which we generalised to the really-existing universe) we discovered that, by the 100th seed-orbit and attendant 'histories', we had produced a vast number of identical Manyworlds global outcomes: identical to the Sierpinski triangle in its fully developed form.

Notional unique histories, though seemingly plausible at the 'below threshold phase' at which unique alternative seed-orbits can seemingly be treated *as if* differentiated histories, completely break down as the many 'histories' culminate into a vast redundant number of *identical non-unique global histories,* as they must, once the noise reduction pattern-saliency phase is obtained. This can be generalised to the universe-scale wavefunction, and the developmental process that attends the whole Whitehead facet of the Bergson Whitehead amalgam, with exactly the same conclusion. All histories attendant the real universal wavefunction merely produce a huge number of culminating identical global histories. What is the point of Manyworlds, if the histories it produces are finally globally non-unique and merely identical?

This finding can be generalised to the development of the bell curve distribution for the probability of outcomes and frequencies generated by a number of quantum mechanised dice (see Book-III, **3.14**). The Manyworlds treatment of this would also culminate into a mass of identical distribution curves. Indeed, the same conclusion can be generalised to the development of a photographic image, culminating into a mass of identical photographic images distributed into Manyworlds. It can be generalised to the unfolding of brain activity, culminating into a mass of identical brain patterns rendered into Manyworlds 'alternatives'. In all cases, Manyworlds constitutes a superficial thesis by virtue of the fact that it must ultimately generate a mass of identical global culminations and histories.

Of course, MWI also fails to circumvent the non-problem of quantum indeterminism, which turns out to be objectively real in any case: a 'problem' that does *not* require a 'solution' (See Book-III **3.32** and **5.16** above). If quantum indeterminism is not a problem that needs a solution, it follows that MWI must remain wholly superfluous, given its principal claim to fame as a 'solution' to quantum indeterminism.

Now, what happens if we substitute in place of the non-mental brute-force process of Manyworlds a consciousness-driven process for the resolution of seed-orbits in our quantum mechanised Sierpinski triangle universe and, by generalisation, a God-consciousness driven process for the resolution of the universal wavefunction? Speculatively, two variations are possible:

- The universe-scale consciousness will resolve just one seed-orbit at a time to generate only one history… at the expense of all the other orbits, histories, and worlds.

- The universe-scale consciousness will resolve *all* the seed-orbit possibilities exhaustively… into as many alternative histories and worlds as is possible.

In the first, we crash into….

- the **arbitrary selection problem of consciousness-mediated quantum indeterminism…**
- in tandem with the **qualitative superficiality problem of consciousness-orchestrated outcomes** (see **5.16,** above for both).

Recall that, in the arbitrary selection problem, consciousness determines each outcome of the seed-orbit based on arbitrary whim; in a manner no different from that generated by a pure a-noetic random selector; with the outcomes finally culminating into the same global end-point, regardless of the orchestration of quantum indeterminism by consciousness.

What of the second scenario, wherein consciousness (again operating at the universe-scale) gets to realise *all* the three seed-orbit possibilities in the model Sierpinski triangle universe into supposed distinct histories and for all subsequent seed-orbits? **Instead of arbitrarily selecting one unique history, consciousness would exhaustively realise *all* of these. (See Fig. 5.03 A to B). In doing so, consciousness would generate a very large number of *identical* non-unique global outcomes and histories, as does putative Manyworlds *without* consciousness.**

Let us now entertain a third scenario: What would happen if consciousness simply applied all of the very large number of hidden deterministic scripts, from any combinatorial order, to generate all the supposedly unique histories of the Manyworlds universe? Again, we would end up with a very large number of non-unique fully identical global outcomes. Hence, the resort to consciousness would be rendered superfluous: reduced to a brute-force exhaustive process; *effectively*, a mindless one, even in the presence of or when mediated by consciousness, with the same absurd very large number of identical global histories and culminations per a-noetic Manyworlds.

Again, what is the point of including consciousness as the causal agent and decider in MWI, much less in non-Manyworlds scenarios, if, in *all* cases, the role of consciousness is rendered unnecessary and superfluous? Consciousness cannot prove itself indispensable and necessary to the resolution of *any* developmental process, even in the Manyworlds context.

If a thing is neither necessary nor indispensable to a given 'problem' (and with quantum indeterminacy now recognised as a non-problem) its presumed role must remain wholly superfluous. In either case, the notion that consciousness is required to overcome and 'solve' quantum indeterminacy completely unravels.

In conclusion, **where quantum indeterminism is concerned, idealism based on consciousness orchestrated indeterminism in lieu of non-mental hidden gears turns out to be as superfluous as any materialism-based a-noetic orchestration per hidden gears**.

Fig. 5.03: Consciousness-driven quantum resolution of seed-orbits and Manyworlds
(A) The Sierpinski triangle and the choices for consciousness

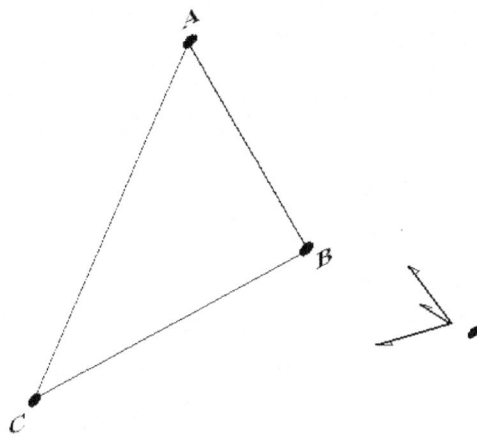

(B) Consciousness generates an uncountable number of identical non-unique global histories

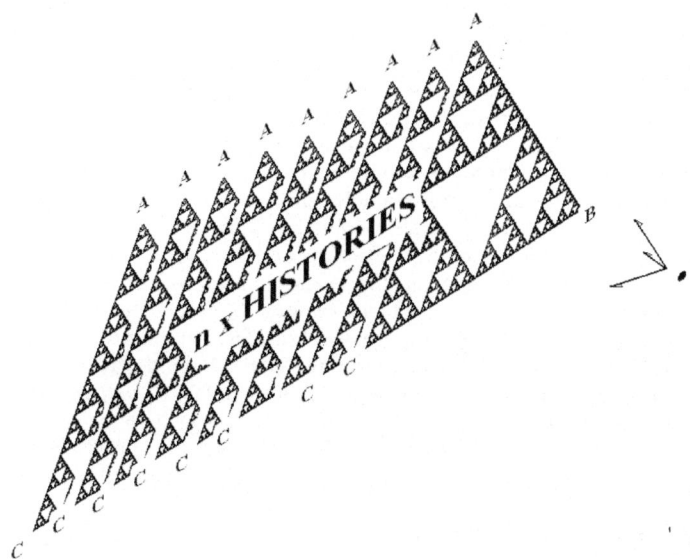

5.18: Alt-R Critique of Consciousness-Mediated Wavefunction Collapse

The essential argument presented below is that wavefunction collapse, or *time*, is *not* driven or caused by consciousness or nous. The foundation to this argument is almost identical to the one made against a-noetic hidden gears vis-à-vis wavefunction collapse espoused in Book-I **1.35** and in Book-II, **2.46** and **2.47**.

Time is ontologically primary to the very possibility of consciousness. Without time, events will not unfold out of AND-form future-potentiality into OR-form actualities, and neither will realised consciousness collapse out of its own peculiar AND-form noetic future-potentiality. In short, consciousness presupposes the ontological primacy of wavefunction collapse or time. To expect that a state totally dependent on time should somehow make time possible is dubious and contradictory.

Also, from the case for the retrospective observational mode furnished in Book-II **2.23**, and completely consistent with the ontological primacy of AND-to-OR collapse and time vis-à-vis consciousness, the attained witnessing consciousness of events is always retrospective with respect to the events that had transpired *before* the later witnessing noetic moment was resolved. That is, **consciousness as the witnessing nous is always retrospective consciousness of the past. The past comes to consciousness pre-resolved (by preceding AND-to-OR collapse processes): the results of any experiment, or whatever else is retrospectively experienced by consciousness, are resolved *before* these gets to consciousness.**

Thus, summarised in two bulleted points…

- It is AND-to-OR collapse that furnishes consciousness. It is not consciousness that furnishes AND-to-OR collapse

- The world of outcomes, experiments and measurement is always apprehended by consciousness *retrospectively*: The world and experiment temporally precedes consciousness. Consciousness does not resolve what it apprehends retrospectively: consciousness does not resolve the past.

Later, we will consider active mode mind, which is not conscious and actively selects future possibilities for the brain from the Whitehead facet via the 'gaps in time'. Yet, active mode mind must itself collapse out of its own noetic AND-form potentiality, as must the passive conscious witness, as must nous in general. It cannot be sufficiently avowed that mind and consciousness, or nous constitutes a future possibility and subsequent actuality, and requires its own explanation as a caused outcome. Thus, in our third bullet point, we assert…

- Nous must itself AND-to-OR collapse out of future possibility, so undermining the naïve notion that nous instigates the process of AND-to-OR collapse and time, while neglecting what causes nous itself.

However, we must now overkill 'consciousness causes collapse', exposing insurmountable problems inherent to the hypothesis that first emerged out of the attempt to found an a-noetic materialistic hidden gears solution to the generation of AND-to-OR collapse and time: problems that consciousness can no more solve than could a-noetic materialistic hidden gears.

*

The fact that neither consciousness nor a-noetic hidden gears can causally drive wavefunction collapse arose out of implications from the superlative version of the quantum mechanical wavefunction: namely, the nested futures model and attendant counterfactuals driven quantum erasure of nested futures: all furnished in Book II (see **2.17** and **2.34**). These are recuperated to the Bergson-Whitehead amalgam furnished in Book-IV **4.10**. Therein, **any putative non-mental gears for processing the counterfactuals-driven quantum erasure of nested futures and general grand decoherence must crash into insurmountable infinities and halting problems.** This is because the nested futures system is elaborated to the infinitely removed future. Consequently it entails inbuilt infinities. In the face of said infinities, the attendant process of grand decoherence *cannot* be mediated by any clockable iterational procedure without crashing into an insurmountable non-terminating problem... failing to bring about final AND-to-OR collapse and time. Consequently, time would 'freeze up'… but *only* if we insist that the process of grand decoherence *must* be driven by a clockable tick-tock click-clock 'gears', or some hidden input-operation-output schema, conveniently hidden from view or rendered 'implicate'.

The inevitable collision between putative a-noetic gears versus insurmountable halting problems in grand decoherence implies that there can be no iterational or even clockable 'hidden gears' behind wavefunction collapse and time, and even less so in the total process of grand decoherence that attends to the Whitehead facet *en toto*, to the infinite future.

When this finding is co-joined to the related but distinct forlorn attempt to found implicit orchestrating deterministic scripts belying quantum indeterminism, the case against a-noetic materialistic-mechanistic hidden gears, and against materialistic input-operation-output (IOO) schema for closed causality that attend it, attains completion. Hence, **there are no hidden gears behind wavefunction collapse or time.** Time is driven by a non-definable non-clockable ineffable process that defies any form of description, and the finding permanently usurps materialist philosophy at the point of core critical causality.

*

We will explore in better detail how infinities arise in the presumption of hidden gears behind grand decoherence and AND-to-OR wavefunction collapse later in this essay. But how does any of this segue into the critique of 'consciousness causes collapse'? **While it is true that mind and consciousness can apprehend infinities and renormalise, and this feat might be construed to aid the dubious idea that consciousness could cause wavefunction collapse… a process in which infinities and halting problems are inherent… it turns out that consciousness cannot apprehend infinities *in time*. That is, the apprehension of infinities and halting problems by mind and consciousness do not equal on actual process of resolution of infinities and halting problems in and through time.**

496

QUANTUM MECHANICS AND MIND

It is one thing to assert that there are an infinite number of whole integers, it is quite another to count the series of whole integers to infinity *and* manifest such a feat in and through the process of *time,* or in any clockable time-interval. In any case, even given such an impossible feat, the idea of infinity is given to nous as ready-made: it is not realised by nous.

It is one thing for the renormalising quantum mechanist to abstract infinity from infinity in order to render quantum field equations workable, it is quite another proposition to suppose that the renormalising physicist literally subtracts infinity from infinity or carries out that subtraction in and through clockable time. In any case such an impossible feat must yet presuppose the idea of infinity furnished to the mind of the renormalising quantum mechanist as ready-made and pre-existent.

From **5.14**, the fact that consciousness can handle infinities and circumvent halting problems certainly implies that nous constitutes profoundly different ontics vis-à-vis mere 'measuring apparatus' and world, if not vis-à-vis the brain itself, compelling the revival of Cartesian dualism. Yet, despite this profound ontological difference, consciousness cannot resolve infinities (never mind continuums) in and through time, and must presuppose the idea of infinity as a thing pre-furnished to it.

And when we add the total dependence of consciousness on ontologically primary time… when we realise that it is time that resolves consciousness into realisation, and, with it, resolves the apprehension of infinities into realisation.. and that it is not consciousness that resolves time… the case against consciousness as 'collapsor' gains near-completion.

It is ontologically primary time that AND-to-OR collapses potential consciousness into realised consciousness *and* into the apprehension of infinity, *and* into the recognition of halting problems furnished to it through AND-to-OR collapse. **That is, a mind-transcending and consciousness-transcending process and ontology is responsible for collapsing and realising both time and nous, *and for* furnishing the idea of infinities and renormalisation attendant nous.**

There is a deeper causality and grander ontology at work; one *without* definable structure. This deeper ineffable non-clockable causality *and* ontology is complicit to the resolution of wavefunction collapse and time, *even in the absence of consciousness*… as much as it is complicit in the resolution of nous itself. Hence, an ineffable consciousness-transcending non-mental and non-conscious causality is responsible for our ability to grasp infinities, if not continuums, without crashing into halting problems. Hence, we are able to handle infinities without having to count to infinity in and through time; without the need to subtract infinities from infinities in a time-expressed clockable way. Indeed, this deeper structure-less causality and ontology furnishes mind and consciousness the answer and understanding of infinities and other ideas ready-made and issued to nous… from 'outside' of memory, possibility, time… *and* from outside of nous.

To summarise in bullet points the essential ingredients to our argument:

- Consciousness or nous in general is incapable of resolving infinities in and through time
- Consciousness or nous is incapable of furnishing the very ontology of the idea of 'infinity' requisite to renormalisation and the recognition of halting problems
- Consciousness or nous is itself dependent on being furnished the idea of infinity by a consciousness-superseding ontology.
- Consciousness and nous, as non-illusory ontics, must emerge out of this subsuming meta-noetic ineffable causality and ontology: an ontology that renders memory, possibility time, ideas… *and* nous… possible and existent.

In differentiating consciousness from the ineffable ontology that makes memory, possibility time and nous itself possible, and by refusing to identify mind and consciousness *as if* it constituted the grander ineffable ontology, we dethrone consciousness from its claim to primary reality or ontology. In doing so, we permanently usurp the notion that consciousness is causative of wavefunction collapse or time, given that it is 'something else'; a *something* that supersedes into ineffability, a *something* responsible for driving time *and* existentialising nous. **Therefore, consciousness or nous is not the driver of AND-to-OR wavefunction collapse or time; but that 'something else'** *is* **the drive of time** *and* **the resolver of time-supplicant consciousness.**

*

While it might be salutary to declare all as 'Being', and even assert nous as Being itself, clearly, the attempt to posit the growing block universe as the totality of Being… the attempt to assert memory, possibility and time as fully exhaustive of and definitive of Being… reaches obvious limits, if for no other reason than the fact that **the system of memory, possibility and time cannot account or explain its own ontology, and requires a subsuming and superseding ontology or Being within which memory, possibility and time reside, and from which these gain existentiality. The same fate awaits the attempt to posit nous and consciousness as exhaustive or definitive of Being, or as primary ontology that supposedly brings about AND-to-OR wavefunction collapse and time. Nous must itself subsume to a grander Being that exceeds nous, and from which nous acquires its existentiality.**

The ontology of ideas is fated to the same. Thus, the idea of infinity is furnished by the nous-transcending Being onto nous ready-made. To assert otherwise is to crash into Euthyphro's dilemma and run headlong into ultimate nihilism, wherein nous supposedly creates the idea of infinity *ex nihilo*, but in doing so it must either concede to a convenient backchannel to an ontology of ideas that co-exist independently form nous and not created by nous… or else crash into apagogic total nihilism, wherein the ideas 'created' cannot exist at all, and, consequently, nothing ought to exist. (see Book-III **3.14** to **3.15** for the non-arbitrariness thesis and related explorations of Euthyphro's dilemma, the backchannel fallacy, and the problem of nihilism).

*

In the following we recapitulate in critical detail how an a-noetic counterfactuals-driven quantum erasure of nested futures must unfold, and how attendant halting problems arise and invalidate the notion of materialistic hidden gears, and how the conclusions garnered from the case against materialistic hidden gears must also usurp 'consciousness causes collapse' at the critical juncture of the processing of infinities.

497

Fig. 5.04: The nested futures categories reprised

(A) Divergent category nested futures

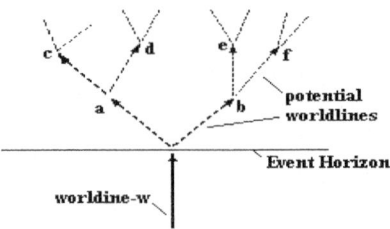

(B) Entangled category nested futures

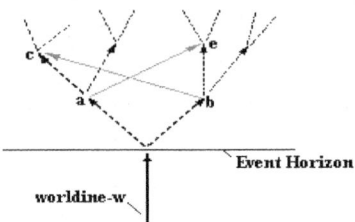

(C) Convergent category nested futures

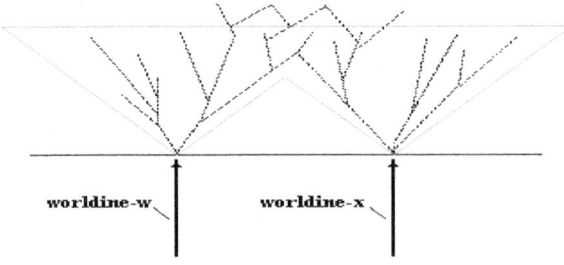

Recall that the nested futures view of the wavefunction, embedded in the Whitehead facet, arose out of John Wheeler's delayed choice approach (Book-II **2.18**). Also recall that three nested futures categories exist: i.e., the *divergent category*; the *entangled category*; and the preponderant *convergent category*... depicted in the repeat illustrations in **Fig. 5.04**.

In **Fig. 5.04**, diagram (A) depicts the divergent category: an explosion of divergent nested futures that elaborate into the future. Diagram (B) depicts the entangled category: similar to the divergent category, but *with* cross-linking possibilities. Both are idealisations. Indeed, the divergent category is effectively fictitious. The entangled category is real but subsumed to the convergent category. Thus, it is the convergent category in (C) that enjoys both real-existence *and* remains central to our stated goal.

Within the divergent category depicted in (A), a minimum of two immanent alternate futures nests a minimum of two subsequent sets of mutually incompatible and exclusive future possibilities. This elaborates and explodes to infinity at the infinitely removed future. Wavefunctions in potentiality always undergo partial-perpetual collapse, accompanied by attendant progression of the event horizon. It is never the case that the whole of the Whiteheadian future to the infinitely removed future undergoes comprehensive collapse, even though the latter *does* undergo grand decoherence and consistency-enforcement to infinity with each such partial-perpetual collapse.

In (A), in the divergent category with two initial possibilities in AND-form, the first partial AND-to-OR collapse decides which of the two will transpire, with all the future nested possibilities attached to the foregone possibility totally erased... erased together with *all* its nested possibilities to the infinitely removed future. This will involve the erasure of essentially an infinite number of nested futures attached to the forgone immanent possibility. At first glance, this does not appear to run into any halting problem. It appears as

498

if a single wavefunction collapse-step is sufficient to handle the infinities therein. The said erasure of the forgone immanent possibility and all its divergent nested futures to timelike infinity will appear to unfold in a single step: involving just one broken-discontinuous AND-to-OR transition, from a previous event horizon to a subsequent one. Thus, the possibility of a halting problem appears superficially circumvented, although this is not ultimately true even in the framework of the divergent category.

Could we not generalise this to *all* quantum erasure of nested futures in the other nested future categories? The divergent category of nested futures is an extreme idealisation: The real Whitehead facet turns out to be constituted by a vast convergent category system of nested futures. The counterfactual restructuring of this latter cannot be simplified into a single step quantum erasure of nested futures. Consequently, the expected halting problem cannot be circumvented if one expects the process of grand decoherence to be furnished by a clockable definable process.

Let us imagine we have two distinct frames of wavefunctions projecting into the future, with the one wavefunction frame separated from the other by, say, a million parsecs: i.e.,3.26 million years. The frames cannot have direct physical contact or integration. Even so, as was posited in Book-II **2.34** and in subsequent essays, the said frames are potentialised to, overlap and convergence in some 3.26 million years' time… *in the future,* (see depiction (**C**) in **Fig. 5.04** for overlapping and converging lightcones). The frames will converge absolutely in the infinite future. Hence, at the very least, their respective potential lightcones are expected to overlap at the said potential future date.

Fig. 5.05: The convergent category and the quantum erasure of nested futures by counterfactuals

(A) Convergent nested futures: before AND-to-OR wavefunction collapse

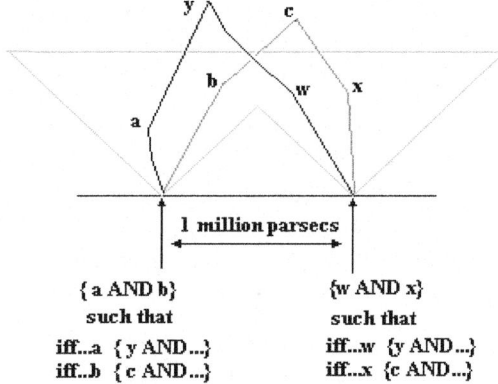

(B) Convergent nested futures: after AND-to-OR wavefunction collapse.

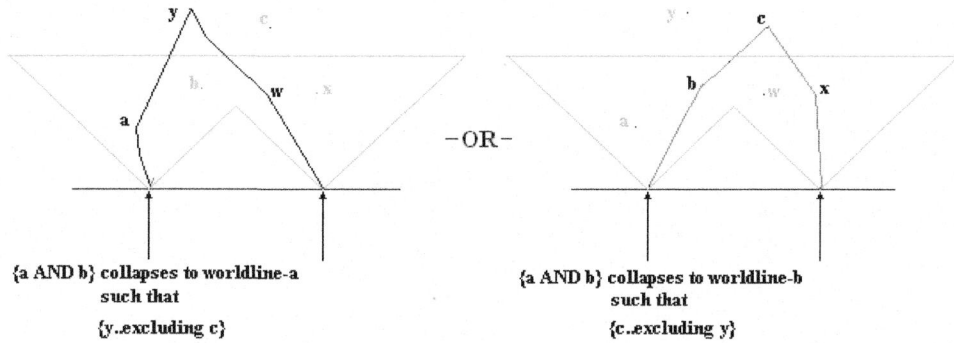

A highly simplified version of this is also presented in the diagram **Fig. 5.05 (A)**. Therein, each of the said wavefunction frames are initially treated as distinct divergent category nested futures. But their potential lightcones overlap to form a convergent category system of nested futures. Per **Fig. 5.05 (B-Left)**, if immanent possibility-**a** gets to happen, then, on the right-side frame, only possibility-**w** will be permitted to happen. Consequently, both frames are fated to realise and converge to the nested future comprised of possibility-**y**.

Of course, the alternative nested possibilities (immanent possibility-**b** on the left-side, combined to possibility-**x** on the right frame, with alternate convergent possibility-**c** in their mutual remote future) will all be erased by the counterfactual generated per the realisation of possibility-**a** on the left frame.

On the assumption of forlorn 'space', this decoherence and reduction of futures will appear to be an instantaneous 'non-local' effect, and involve 'action at a distance' With de-spatialisation and temporisation, the attendant grand decoherence will be grasped as a transtemporal process that operates, scans, selects, erases, and restructures the total system of nested futures, ultimately to the infinite future, in a wholly non-clockable, gear-less, transtemporal and instantaneous fashion.

On the other hand, **Fig. 5.05 (B-Right)** constitutes the inverse that favours convergence to possibility-**c**.

Although the infinitely removed future is not illustrated in the diagrams, nor are all the subsequent nested possibilities attached to possibility-**y** or possibility-**c**, it is yet obvious that, even the simplest convergent category counterfactuals-driven relation must generate uncountable quantum erasure-implications elaborated to the infinitely removed future. **Insofar as the whole of the Whitehead facet must be restructured or grand decohered into consistent futures, and this decoherence must elaborate to the infinitely removed future, any presumed tick-tock click-clock 'gears' that must scan, select and eliminate the infinite number of non-consistent nested futures, and achieve the grand decoherence of the total Whitehead facet through an iterational clockable procedure or 'gears', must crash into an insurmountable halting problem… and fail.**

Again, any presumed tick-tock click-clock time-bound or clockable 'gears' expected to handle all of this will crash into an insurmountable halting problem. **Any time-bound step-by step sorting and erasure procedure would require that time simply 'freeze up' until all the non-consistent nested futures are scanned, sorted, and eliminated to the infinite future: a process that could never obtain completion or conclusion. The obvious inadequacy of such a putative time-bound process (obvious from direct experience, insofar as time obviously *does not* 'freeze up' in the really-existing universe) necessarily implies that the actual process that generates and handles the quantum erasure of nested futures *and* grand decoherence to the infinite future, *and* brings about time itself, *must* constitute an a-temporal, non-iterational, non-definable and non-clockable ineffable process; part of an equally non-definable and ineffable ontology that subsumes, but cannot be subsumed to, memory, possibility, nous and time.**

In short, the process of 'coming into being' in AND-to-OR collapse, hence *time* itself, inseparable from the very viability of the quantum erasure of nested futures and of grand decoherence, must be brought about by a process that cannot in principle be describable as any sort of 'gears' or variables, hidden or otherwise. Hence, hidden gears is forlorn; And so is IOO-causality central to materialism; and so is materialist philosophy absolutely dependent on the notion of a definable 'operation'; the middle term in IOO schema.

*

What if we substituted consciousness in lieu of putative non-mental hidden gears? Could consciousness resolve the required erasure of nested futures to the infinitely removed future *and* obtain attendant wavefunction collapse and time? Again, the answer is an emphatic no.

Again, recall that consciousness is profoundly different from the rest of nature because of the fact that, amongst other things, it can apprehend infinities and halting problems. However, as previously argued, consciousness cannot resolve infinities and halting problems in any literal operational time-bound and time-expressible way. Nous cannot count to infinity in any clockable way, even though it can indeed apprehend the idea of infinity pre-furnished to it as a ready-made idea. Mind and consciousness cannot practically resolve any halting problem operationally or in any time-bound time-expressible way, since such an attempt could never halt or obtain completion or conclusion.

In short, the same problem that abides in the presumption of necessarily time-bound mechanical hidden gears in the processing of grand decoherence and wavefunction collapse must also abide vis-à-vis nous in lieu of materialistic hidden gears. And it is ironic that a critique of materialism per a-noetic hidden gears and time should lend itself to an equally decisive critique of idealism and 'consciousness causes collapse'.

The critical factor involved here is the inherent inseparability and ontic dependency of consciousness vis-a-vis time: As asserted, the very possibility for nous, expressed in and through time, *requires* time. No time, no consciousness. Indeed, mind and consciousness must itself AND-to-OR collapse out of its own noetic AND-form future-potentiality, and, even precluding noetic potentiality, at the very least, nous could only acquire *retrospective* consciousness of events that have themselves decayed out of prior AND-to-OR collapse and time from the Whitehead facet of futures.

The ontological dependency of nous on time does not hold both ways: Thus, AND-to-OR collapse and time *can* occur without consciousness: hence the infinities inherent to grand decoherence of the Whitehead facet to the infinitely removed future *can* decohere without consciousness. But consciousness cannot happen without AND-to-OR time and attendant grand decoherence of the Whitehead facet to the infinite future, if for no other reason than the fact that there would be nothing to be conscious of if the temporally prior process of time did not generate the events that then attended consciousness, retrospectively.

Without consciousness, the universe could yet obtain AND-to-OR collapse and time *and* the grand decoherence of the Whitehead facet to infinity, without crashing into halting problems. How do we know this? We know this from the assertion

500

that time is not consciousness-dependent, while consciousness *is* time-dependent. Consciousness could not itself come into actuality without ontologically primary and consciousness-independent time-process and its attendant grand decoherence. That is, consciousness could not be realised without the non-clockable ineffable causality and ontology behind and transcending both time and grand decoherence; one that is also resolves nous *and* transcends nous.

*

We reiterate that there is no necessary, sufficient, or compelling reason to suppose that we cannot have time (wavefunction collapse) without consciousness. Time and synonymous wavefunction collapse remain ontologically primary and requisite to the very possibility of consciousness. Time and wavefunction collapse can bring consciousness into being. Yet, consciousness, totally dependent on time, totally dependent on primary wavefunction collapse, cannot bring about time and wavefunction collapse.

To reiterate again, consciousness does not arrive at the apprehension of *a priori* infinities furnished to nous by counting its terms in and through a clockable process. Nor does consciousness recognises halting problems in and through such a clockable process. Instead, mind and consciousness is, so to speak, 'furnished' this sort of irrational and transcendental information from a sidereal ontology, from 'outside' of time: from a domain that transcends memory, futurity, time *and* nous.

Because of its time-bound and time-manifested nature, if mind and consciousness were placed in lieu of non-mental materialistic hidden gears, it would confront the same nested future infinities that it could not temporally scan, sort, or process, much less erase to infinity. Consciousness in general, the cat's consciousness, Wigner's friend's consciousness, and Wigner's own consciousness... would require eternity to accomplish such a feat, even within the limited purview of Wigner's friend scenario. Nous would be subject to the same halting problem and non-viability that a forlorn a-noetic mechanistic gears must face when placed in its stead, since both nous and a-noetic gears are time-dependent and time-bound.

*

Clearly, we need a form of causality beyond both time-bound consciousness and beyond a-noetic materialistic mechanism: one that renders both possible, yet transcends both ontologically, *without* identification with either mind or mechanism: a form of causality that transcends both 'matter' and nous, but brings into being 'matter', wavefunction collapse and time, *and* nous… *and* furnished to nous ideas of infinity.

5.19: Augmentation: Why the Universe is not a Computer

We will now reiterate doubt vis-à-vis the popular view that the universe is a computer. Succinctly, this view asserts that a hidden Turing machine drives wavefunction collapse and *time*. Its demise will augment the above case against 'consciousness causes collapse'. The essay is an opportune reiteration in summary form of material from Book-II **2.46** and **2.47** and Book-III **3.32**. To this end, let us apply the putative universe-scale Turing machine to simulate wavefunction collapse, taking into consideration our findings against hidden gears, and even against 'consciousness causes collapse' in lieu of a-noetic gears, as was done in the preceding essay.

Let us assume that a hidden Turing machine drives the quantum erasure process that erases non-consistent nested futures vis-à-vis consistent ones, hence affects grand decoherence of the total Whitehead facet to the infinite future. Assuming a simple universe structured according to the divergent category nested futures, how will our universal Turing machine carry out the scanning, selection, and elimination of the infinite number of non-consistent nested futures? Previously, we stated that, on a superficial level, the divergent category of nested futures appears to escape the halting problem (see **5.18**). We will now show that the erasure of nested futures in the divergent category is no more capable of escaping the halting problem than it is in entangled and convergent nested future categories.

The simplest divergent category has just two alternate immanent futures, each embedding subsequent diverging nested futures. A single step wavefunction collapse of the immanent possibilities and attendant erasure of just one of the two possibilities, with the immediate erasure of all the nested futures attached to it to the infinite future, ought to be obtained without crashing into any halting problem.

A useful analogy to the above is the deletion of a single file on one's personal computer: one that niches within it many other subfiles. The complete deletion of it and its contents ought to be possible to obtain in a single step. On the surface, it appears that a separate and discrete deletion of the subsumed content within the file is not required, and we can imagine an infinite number of such subfiles therein. One need only delete the root file, and, with one deletion-step, erase them all.

Although the deletion of the root file and its subfiles involves merely a few clicks at the end-user level, at the machine level (at the actual computational level) there *is* a scanning, selection and removal process whose clock-time partly and critically depends on the size of both the root file and the number and size of its embedded subfiles. If the size and content happen to be very large, the erasure will require more clock-time to complete. In short, the instant-seeming deletion of a single file and all its contents is not in truth achieved in an instant or even in a single step. Instead, it is brought on by an underlying iterative process that always requires clock-time to complete.

Keep in mind that, even the much-vaunted quantum computational process, though superior to digital processing, would require clock-time. These could not circumvent halting problems if applied to the resolution of infinities. If the above erasure was carried out by quantum computation, it would yet require clock-time.

Our divergent category nested future is equivalent to a single 'root file' (i.e., immanent future possibility) that contains, not just two sub-files (i.e., a minimum of two nested futures) but an infinitely branched and elaborated system of divergent nested futures that stretch to future-infinity. If its attendant quantum erasure of nested futures was to be enacted by a hidden universe-scale Turing machine, or

501

even by a universe-scale quantum computational process, the pertinent scanning, sorting and removal of non-consistent nested possibilities would need to elaborate to infinity. The putative computation could not obtain any final end-point or conclusion. Consequently, our universe-scale Turing machine (or our quantum computational process) would crash into an insurmountable halting problem, given that the attendant wavefunction collapse (and time) could not be realised until the total grand decoherence and attendant quantum erasure of nested futures was obtained to future-infinity. Again, time would 'freeze up', so to speak.

If wavefunction collapse *must* be driven by a hidden Turing machine-like computational procedure, or even by a quantum computational process, the computation of wavefunction collapse could never obtain completion. Even partial-perpetual wavefunction collapse of just the immanent possibilities would fail to unfold, given the inseparability from the AND-to-OR collapse process of the process of grand decoherence, to future-infinity.

However, as we know from real world experience, wavefunction collapse *does* get to unfold. Time does *not* 'freeze up'. Thus, time and wavefunction collapse cannot be realised by means of a universe-scale computational procedure, quantum computational or otherwise. **It follows that quantum erasure and associated wavefunction collapse (hence time itself), even in the context of an idealised divergent category state, must be driven by a non-computational, other-than-mechanical, and wholly non-clockable ineffable process.**

*

What happens when the same universe-scale Turing machine is applied to convergent category nested futures? Since the convergent category subsumes the entangled category, we need not make a separate examination of the latter. Clearly, if a computational universe cannot resolve halting problems in the context of the simpler divergent category, it is destined to equal failure when confronted by grand decoherence and wavefunction collapse in the convergent category context. Again, it follows that the process responsible for the quantum erasure of convergent non-consistent futures *and* wavefunction collapse (time) cannot be computational or mechanical. **It follows that our really-exiting universe is ultimately *not* driven by a computational process. Therefore, the universe is *not* a computer.**

*

In a similar vein, consider that any purported computational process behind and driving time must itself unfold in time or in a clockable way, given that all computation (even quantum computation) is time-bound and time-dependent. **Time and wavefunction collapse remain ontologically primary to the very possibility of computation itself. Computation needs ontologically primary time. But computation is not a fundamental requisite to the possibility and generation of AND-to-OR collapse and time. In short, no time, no computation.**

If we posited the hypothetical assumption that wavefunction collapse and time is a thing produced by a hidden computational process, the process must itself be clockable. It must itself require its own distinct wavefunction collapse in order to unfold. But this in turn must also derive from an even more fundamental computational process, with its own clock time, and so one. The whole scheme degenerates into an infinite regress, with the ultimate root-computational process infinitely removed from the one that attends ours. As each computation leading from the infinitely removed root-computation must itself require its own clock-time and its own attendant wavefunction collapse, it would take eternity for the infinitely removed root-computation to culminate into the final computation that then generated *our* AND-to-OR collapse and time... and time would 'freeze up' yet again.

We must again conclude that wavefunction collapse and time cannot be driven by a computational process.

Again, the universe is not a computer, and, consequently, even less a 'simulation'.

*

What would happen if we substituted mind and consciousness in lieu of our hypothetical universe-scale Turing machine or quantum computation? As already argued, this sort of thing is beyond the province of mind and consciousness, even when super-conflated to God. The power to apprehend infinities and halting problems is furnished to time-bound time-dependent consciousness from 'outside' of the process of time. Nous does not and cannot apprehend infinities and halting problems in and through time, even though it expresses and communicates its understanding of infinities and halting problems in and through time, but must itself unfold in and through AND-to-OR time, collapsing out of its own noetic future-potentiality. Given its time dependent nature, consciousness has not the power to resolve infinities in and through time, no more than could time-dependent computation.

In short, wavefunction collapse is beyond the province of mind and consciousness as much as it is beyond putative a-noetic universe-scale Turing machines, or even of a quantum computational variants of the same; all of which are purely and strictly time-dependent, time bound, and subject to the ontic primacy of AND-to-OR collapse, without which neither could happen. The same applies to nous. Hence, the notion that consciousness collapses the wavefunction (or drives time itself) is again undermined; albeit from the perspective afforded by the forlorn hypothesis of the computational universe.

5.20: On Purported Statistically Significant Effects of Consciousness:
The Quantum Zeno Effect & the Zeno-Memory Conjecture

What if experiments could show that consciousness can generate statistically significant effects *and* measurably skewer what would otherwise constitute a normal distribution of quantum probability? While there are myriad non-quantum mechanical

502

experiments that claim mind-world effects, only few incorporate quantum mechanics[95]. The latter experiments, whatever one might think of these, may be treated *as if* they furnish proof of 'consciousness causes collapse', or even of the full orchestration of quantum indeterminacy by consciousness, if not garner support for the supposed ontic primacy of consciousness.

Our aim in this essay is not to dispute the claims of such experiments. Instead, **we will argue that, even if the pertinent experiments turned out to be true in their claims about statistically significant effects from consciousness, this could not constitute proof of either 'consciousness causes collapse' or of the supposed primacy of consciousness.** Thus, for the sake of argument, we will assume the experimental claims are true, but show that, regardless, these cannot be used to foist quantum idealism.

Note that the Alternative Realist case for the revival of Cartesian dualism accepts the ontology of mind and consciousness vis-à-vis world and brains, but it does not require consciousness to bring about wavefunction collapse or time, and, as we discovered in the previous essays, nous *is* time-bound and totally dependent on the ontological primacy of AND-to-OR collapse and time. Consequently, any experiment that could prove mind-world effects and attendant statistically significant outcomes could not furnish proof of the more radical notion that mind drives time itself. Nor could such experiments demonstrate the ontological primacy of nous vis-à-vis time and universe.

*

Three factors work against the use of statistically significant effects in favour of 'consciousness causes collapse'. As was shown in essay **5.18**, consciousness is not required as a secret orchestrator of quantum indeterminism. Hence, **even if consciousness could marginally modify the probability distribution of experimental outcomes, this could not constitute proof of the secret deterministic orchestration of surface quantum indeterminism by nous**. Indeed, mind affects the world (the brain) by imposing on the brains an analogous 'punch card': a complex wavefunction. More on this will be said later, save that mind-brain control does not require mind to collapse its imposed complex wavefunction, much less orchestrate the attendant mass of quantum indeterminate discrete outcomes garnered from that collapse, let alone engender statistically significant effects therein.

The second problem, assuming the claims about the statistically significant effects from consciousness are empirically valid, and we would not find it objectionable if shown to be true, is that **the statistically significant effects are *marginal*: The effects are *not* absolute. Succinctly, the effects are *not* binary. For mind and consciousness to constitute the primary ontology and generate the 'illusion' we take to be the world, it must have an unambiguous, absolute certain yes-or-no relation to its conjured world. Its effects on its conjured world cannot be comprised of a marginal slightly "yes", or a fudged "no".**

Moreover, **any statistically significant effects from consciousness must necessarily admit the reality of a mind-independent objective world, one bound to a normal statistical disposition in the absence of nous. Indeed, the fact that the purported effects of consciousness are marginal against the dominant grain and inertia of normal-tending distribution indicates the existence of a mind-independent world**, as much as it might also prove the reality of a world-independent causal mind. This latter is the more profound conclusion to draw from any vetted statistically significant effects from consciousness. But none of it requires 'consciousness causes collapse', much less a Berkeleyan ontological primacy of consciousness.

Only in the context of brains does mind and consciousness come close to binary causality: one that goes beyond mere statistically significant marginal effects. When one wants tea, attendant processes pertinent to one's will get to transpire. When one wants coffee, different attendant brain-processes will unfold. One's wanting tea or coffee is not a matter of statistically significant effect that *might* lead to tea slightly more often than coffee per some volatile statistical marginal influence. In this sense, the mind's control of the brain is unambiguous and binary. Mind-brain control is not always unambiguous, if for no other reason than the fact that there are autonomous brain processes that can over-ride the mind's will, even to the point of forcing one to breath water.

We speculate that statistically significant effects, if these are happening at all, may be the result of **mind-brain bleed-over**. Using the brain's unique conditions, mind imposes on the brain a 'punch card' of complex wavefunctions concordant to its will. However, the imposed wavefunction state might not be wholly confined to the brain. It may project beyond the brain to the future, albeit weakly, through latent grand decoherence of that total future, thus skewering the probability of outcomes in futures 'outside' of the brain. None of this requires the orchestration of quantum indeterminism, much less that mind bring about wavefunction collapse and time, and even less the ontological primacy of consciousness.

*

The Quantum Zeno effect[96] is an empirically verified mystery. It was posited by Henry Stapp as a possible avenue to 'consciousness causes collapse'[97]. The effect shows that the 'continuous' observation of an unstable particle can forestall its decay, skewering its normal probability and tendency to just such decay. But is this another statistically significant effect from consciousness? **If one assumes that observation involves consciousness in some ontologically primary and necessary way, then it might be argued that, in the Quantum Zeno effect, we have evidence of significant influence of consciousness, and that consciousness is orchestrating apparent quantum indeterminate decay-possibilities or delaying their probability of occurrence in a statistically significant way.**

[95] Maier, M. A; Dechamps, M. C; Pflitsch, M. *Intentional Observer Effects on Quantum Randomness: A Bayesian Analysis Reveals Evidence Against Micro-Psychokinesis; Frontiers, Psychology, Vol. 9; 2018*

[96] Sudarshan, E. C. G.; Misra, B. (1977). "The Zeno's paradox in quantum theory". *Journal of Mathematical Physics*. 18 (4): 756–763.

[97] Henry Stapp, *Mindful Universe: Quantum Mechanics and the Participating Observer* (2007)

The observation process in the Quantum Zeno effect is certainly high frequency, but it is never continuous, given that neither inputs of observations, nor outputs engendered from these, can ever be seamless-continuous, as we discovered even in the Preliminary (**0.02 to 0.04**). There can be no seamless-continuous radiation or matter-streams of inputs or outputs. This restriction is central to quantum mechanics: Radiation and matter must always be discontinuous, or we end in ultraviolet catastrophes and infinite power sources.

Recall that AND-form logic constitutes the objective reality of the future. The future was consolidated as spacetime in potentiality in Book-II and as the de-spatialised temporised Whitehead facet of the Bergson-Whitehead amalgam in Book-IV. Also recall from material in Book-I (Part-II) that there can be no pure AND-form quantum mechanical state of affairs. Even before we developed 'spacetime' and its successor in the Bergson-Whitehead amalgam, in Book-I, we discovered that nature is always an admixture of both OR-form and AND-form states. No system can ever be purely and only the future (AND-form). The corollary is also true in that there can be no pure OR-form states either... in the sense that there is no such thing as a system without future AND-form possibilities, much less absent of AND-form intervals that reside between inputs, outputs and events and observations. By Book-II, we realised that this inexorable AND-OR admixture is required by the very past-future structure of the growing block universe.

From all of this, it necessarily follows that the Quantum Zeno effect cannot entail any 'continuous observation' of any impossible seamless-continuous OR-form radioactive unstable state, much less obviate AND-form logic from that radioactive state. That is, the quantum Zeno effect cannot remove or obviate the future.

Indeed, it is not compelling to conclude that it is consciousness that is delaying particle decay in the Quantum Zeno effect: The assertion is an *assumption*, given that it is clearly the case that consciousness is not needed in the orchestration of quantum indeterminism (see **5.16** and **5.17**), nor even in bringing about time and synonymous AND-to-OR wavefunction collapse (see **5.18** and **5.19**). Even if consciousness were somehow necessary for the observed decay-delay, **the same argument used against the primacy of consciousness from the purported statistically significant effects of nous could easily be applied against the notion of consciousness-mediated Quantum Zeno effects. Indeed, the decay-delay is not an absolute binary effect: it only constitutes a significant reduction in the probability of decay.**

*

Our own Alt-R explanatory conjecture is the ***Zeno memory conjecture***. In the absence of continuous observation, the stock of Bergsonian wavefunctions in memory associated with the given radioactive unstable particle retained within the Bergson facet, will not convey any countervailing probability effects on any subsequent development of the particle on any subsequent event horizon. The particle decay will occur at a normal probability.

However, high frequency inputs and interactions vis-à-vis the particle (with or without involvement of consciousness or observation) will generate a Bergsonian past comprised of wavefunctions in memory states mostly of non-decay. The more often the attendant AND-to-OR resolution resolves the particle to its non-decayed state, the more such non-decay memory states will be retained within the Bergson facet. Consequently, through mnemonic mechanics (see Book-IV, Part-III) these will get to superpose and impose non-decay outcomes upon successive event horizons, so reducing the probability of radioactive decay.

Of course, this is conjecture. However, any future experimental proof of this, with probabilities for decay calculated per implications from prospective Bergsonian mnemonic mechanics, would constitute independent confirmation of both Bergsonian memory *and* the Zeno memory conjecture.

The key point is obvious: The ready attribution of the Quantum Zeno effect to consciousness, and the presumption of some form of quantum idealism or noetic causality behind it, is not finally compelling.

5.21: Final Atheism & Transcending Atheism

The case against 'consciousness causes collapse' must culminate into Final Atheism; the permanent invalidation of the notion of God, but *only* insofar as God is misconceived as super-conflated nous. There is no necessary or compelling reason to suppose that God, or ultimate ontology, or Being, must be nous. And the serious limits of nous, in its inability to bring about wavefunction collapse and time, can only augment against the dubious notion of Being misconceived as nous.

In any revamped Berkeleyan or quantum idealistic theology, 'God causes wavefunction collapse' would act in lieu of 'consciousness causes collapse', with God conceived as super-conflated nous. God-consciousness would presumably drive the collapse of the universe scale wavefunction, if not that of the multiverse wavefunction. The notion that a God-consciousness collapsed the universe into existence, *ex nihilo,* from the Big Bang monoblock, and out of whatever remained of its Whiteheadian futurity thereafter... must unravel, given the comprehensive demise of 'consciousness causes collapse' obtained in **5.18** to **5.19**, and of the demise of the forlorn notion that nous constitutes the ultimate ontology. God misconceived as nous would suffer from the same **qualitative superficiality problem** and the same **arbitrary selection problem** as mere nous: (also see the Euthyphro's dilemma in Book-III **3.14**). There is also the *non-arbitrariness thesis* to consider, which challenges God-as-nous on the basis of its profound nihilistic implications, incompatible with our really-existing universe: Again, see Book-III **3.14** and **3.15**.

However, one could always conceive God as an abstract a-noetic ontology that transcends nous as thoroughly as it transcends memory, possibility and time; the latter unified into the growing block Bergson-Whitehead amalgam. This view gains succour from the non clockable and ineffable gear-less causality that drives AND-to-OR time and renders all things possible and existent, including nous.

However, one could equally assume alternative atheism on the basis of neutral monism or neutral ontology, garnered from the same ineffable causality and ontology; one transcending even the putative God-consciousness, but only if we conflate God into nous, and on the premise that, since nous cannot constitute primary ontology or Being, God does not exist.

504

Either way, we arrive at an essentially non-nihilistic irreligious ontology and worldview.

Clearly, even when we assume mind-transcending a-noetic Being as God, such a non-mind purely existential God could not furnish succour to any Abrahamic or Judeo-Christian-Islamic corpus, or any other religious corpus based on the forlorn conflation of God into nous. The conflation of God into nous accounts for the Abrahamic penchant for 'intelligent design', creationism, and the inability to come to terms with evolutionary theory, among other things. The most moderate position one might hope to obtain from the Abrahamic fold is deism. Yet, even in deism, God is conflated into nous.

Obviously, a non-mind God could not support any system of piety, propitiation and absolution, or any religious authority, myth or system entailing a narcissistic-exploitative relation between an all-demanding God-as-Mind versus an all-supplicant co-dependent wretched humanity. It is also convenient that the a-noetic conception of Being sits well with standard-fare cosmology, cosmogenesis and evolutionary theory.

Any future theism that reconceives God as non-mind Being in the fashion postulated above, must supersede religion, so constituting a non-religious secular-philosophic theism. God as non-mind belongs to an entirely irreligious secular philosophy in the likeness of Platonism or Neoplatonism, or even Hindu esotericism emancipated from the conflation of *Atma* (consciousness) with *Brahman* (the ineffable).

5.22: Delayed Choice Time-Intervals: An Alternative Basis for the Perennial Object-Subject Relation

The successor to the intermediate model spacetime was the Bergson-Whitehead amalgam: Therein 'space' was invalidated based on five independent arguments (see Book-IV Part-I, and summary in Part-I **5.02**). Spatial and distance-relations were replaced with purely temporised delayed choice time-interval relations. With it, we obtained the replacement to the speed of light: namely, the *delayed choice time-interval standard*: the 1 second-worth of delayed time-interval between an event and its subsequent rapport or observation.

With respect to the Moon-Earth relation, at the end of 1.5 seconds, we may choose to look or not look at the Moon. If we do so, we will see it as it was 1.5 seconds ago. The same applies to the Sun vis-à-vis the Earth: a delayed choice time-interval relation of around nine minutes. For Alpha Centauri vis-à-vis the Sun, a delayed choice time-interval of some four years and three months abides.

Recall from Book-IV **4.08** that what we think of as a 'moving object' approaching and receding from an observer at 'constant velocity' or 'uniform motion' is recuperated as the moment-to-moment uniform reduction in the delayed choice time-interval relation between the 'object' and its observer, leading to near-simultaneity between object and observer. Once the object 'passes' the observer, an increase in the delayed choice time-interval relation between observer and 'object' will transpire as the latter recedes to the past or 'moves away'. The implication is that there is no 'motion' as such, but only a succession of events and changes in delayed choice time relations between events, observers and consequences, or between relative frames of reference; furnishing the impression of 'motion'.

On the other hand, what we think of as an object 'accelerating toward' and approaching our observer, and then receding from that observer, can no longer be characterised as undergoing non-uniform motion in space. Acceleration is recapitulated as the moment-to-moment evolving *non-uniform* reduction to near-simultaneity between the object and the observer; followed by the non-uniform recession in subsequent delayed choice time-interval relations between the same. Again, there is no 'space' or 'motion': there is only uniform or non-uniform transformation in the delayed choice time-interval relation between events, observers, and consequences.

Finally, we must not misread the 'choice' involved in the idea of delayed choice time-intervals. **The 'choice' does not convey any special ontological status to mind or consciousness**: Given a certain delayed choice time-interval standard and minimum between an observer and an observed corpus of events, the observer may choose to observe or choose not to observe that corpus of events. The choice to observe must be made at the end of the elapsed time, equal to that time-interval. For example, if you want to observe the Moon as it is now, you must wait 1.5 seconds. You have no choice about the 1.5 second delay. This implies objectivity: a stand-alone status to the Moon, independent of the observer or subject. It also implies the stand-alone objective status of the time-interval restriction in the Moon-observer relation, about which the observer has no subjective choice. Thus, the 'choice' aspect has no special power or idealistic connotation over the independently existing objective delayed choice time-interval relation in *any* interrelation between frames and events. **The time-interval is objectively set, and it is both choice-independent and subject-independent. Nor does consciousness have any paranormal or mental power to modify the set time-interval relation, unless one can prove otherwise.**

In any case, the argument against consciousness-driven wavefunction collapse and grand decoherence, as well as against consciousness-orchestrated indeterminacy, was concluded in **5.18** and **5.16**. These unravelled the want or notion of the consciousness created universe and of any possibility of consciousness-created time-relations and events.

The corpus of events that constitute the Moon get to us as a 'package' comprised of pre-constituted and resolved past-events. This is per their retention in the Bergson facet and as abstract memory, which must include all pertinent past events vis-à-vis the Moon, integrated to the stated 1.5 second delayed choice time-interval limit. The pertinent memories are retained as a succession of wavefunctions in memory within the Bergson facet, subject to the inverse square of time: (see Book IV **4.10** to **4.11**). And **insofar as consciousness cannot alter the delayed choice time-interval structure between it and corpuses of events furnished to it from the Bergsonian past-in-memory, the prosaic object-subject relation must abide and prevails against the want to reduce subject to object (materialism) or reduce object to subject (idealism), and against forlorn 'consciousness causes collapse'.**

Hence, prosaic object-subject relations must be recuperated in terms of information in time, *not* as functions of spatial distribution of events, given de-spatialisation, and the demise of space, with consequent temporisation.

505

Insofar as resolved information gets to manifest in and to witnessing consciousness as memory, but *always* as the past, the 'object' incident to the observing subject must manifest to that subject as an already decided or AND-to-OR resolved corpus of past events; its content resolved *before* it has subsequently manifested to consciousness. It follows that consciousness, intentionality, will and thought can have no power over the pre-constitution of past events or over the memory constituting these. Mind and consciousness, always subject to memory, always subject to events presented to it as the pre-resolved past-in-memory, and always the witness to events in retrospect (never in real-time, and never through 'signals from the future') necessarily implies that the nous-independent universe composed of memory, possibility and time, is *not* made up of Eddington's 'mind stuff'.

Simply put, we do not live in a consciousness-created universe.

*

How can we be certain that consciousness does not implicitly create the very constitution of past events before these arise to the attention of nous?

There are two forms of delayed choice time-interval relations. The first type is comprised of Whiteheadian future-form potential time intervals: the superposition of as-yet not realised future-possible time intervals and events rendered in AND-form within the Whitehead facet of the Bergson-Whitehead amalgam. The second type of delayed choice time-interval pertains to the *realised* past, or between past-realised events. These are embedded within the Bergson facet of the Bergson-Whitehead amalgam.

The borderline between the Whitehead and the Bergson facets is constituted as the event horizon, comprised of the most recent resolution of OR-form events, the successions of which get to mark realised delayed choice time-interval relations. Potential delayed choice time intervals and events from the Whitehead facet are AND-to-OR collapsed into OR-form events and time relations on or along successive event horizon through attendant partial-perpetual wavefunction collapse or time. The outcomes generated are subsequently relegated to the Bergson facet as de-nested wavefunctions in memory, with their relative time-relations preserved.

Is it possible for mind and consciousness to pre-empt the collapse of the Whitehead facet into the Bergsonian, and accomplish this in such a way as to generate realised events by pure will? With the special exemption of brains, whose control does not involve 'consciousness causes collapse', nor does it involve consciousness-orchestrated quantum indeterminacy, the answer is an emphatic no. **Such a feat would require that consciousness somehow bring about both wavefunction collapse *and* control the quantum indeterminate processes leading to the resolution of delayed choice time-interval relations and attendant event-outcomes. From the preceding material, both are simply impossible, if not totally unnecessary.**

Thus, the object-subject relation, recuperated per unalterable delayed choice time-interval relations that distribute and organise information and memory in lieu of 'objects in space' must remain ontically independent of nous: The events of the world are objectively real. The object-subject relation is an ineliminable attribute of the universe. It is *not* an illusion supposedly generated by nous. The 'object' comprising the corpus of realised events is indeed resolved prior to its 'subject' conscious witness, *always* as a pre-constituted corpus of events in Bergsonian memory. Also, the resolution of the events and their delayed choice time-interval relations prior to their emergence to consciousness, are obtained by means of ontologically primary AND-to-OR wavefunction collapse and quantum indeterminate resolution, neither of which require nous as the causal driver or orchestrator, as was argued in **5.16** and **5.18**.

*

The final case for the onticity of the object-subject relation comes from the role of counterfactuals in the grand decoherence of the Whitehead facet to the infinite future. Although the delayed choice time-interval separating Earth from the star system Alpha Centauri approximates to four years and three months, **a supernova explosion of one of the suns in the Centauri binary cannot be directly observed until the choice is made to observe it on Earth *after* four years and three months have transpired. The supernova event of some four years past will not be altered by the whim, will, or want of consciousness on Earth,** notwithstanding Yoon-Ho-Kim's quantum eraser (see Holed Past theory; Book-III **3.19**). **The observation finally made at the end of the four or so years will be consistent with what happened over four years ago, as is demanded by the memory of it retained in the Bergson facet of the Bergson-Whitehead amalgam, and by the process of grand decoherence that must compel the future into consistency, and restrict consciousness on Earth to the observation of consistent-only outcomes. Therein resides the objectivity of the supernova event vis-à-vis subjective observers on Earth: Therein resides the onticity of the object-subject relation, courtesy of grand decoherence in the enforcement of a stand-alone potential object-world with which subject-nous must abide per consistency rules.**

As was argued previously, the moment one of the stars in the Centauri binary exploded, it generated a set of counterfactuals vis-à-vis the Whitehead facet and its convergent category nested futures. These counterfactuals contributed to the quantum erasure of non-consistent nested futures vis-à-vis the counterfact of the supernova event. Attached to Earth was a potential lightcone frame comprised of a set of nested futures that projected into the Whitehead facet and to the infinite future. Before the supernova transpired at 'far away' Centauri binary system, the nested futures attached to Earth had superposed *all* the future possibilities: The superpose included the to-be-observed 'distant' supernova event AND the future-possible observation of the non-exploded star in the Centauri stellar binary. However, once the supernova occurred, and although any direct observation of it could not be made by observers on Earth, and although the observers had to wait some four years and three months to observe the event, the nested futures attached to Earth, and the noetic future possibilities attending it, became subject to the quantum erasure of non-consistent nested futures. Only nested futures consistent with and containing the future-potential events and noetic observations of the supernova remained within the Whitehead facet and in the future-potentiality of the Earth-frame, as well as in future-potential nous pertinent to that frame and event. That is, **events in the universe elsewhere decohered future-potentials of nous to consistency with those elsewhere events, and nous grand decohered**

506

into consistency with the universe, *not* the other way round. Thus subject-nous was forced into consistency by the grand decohered world-object. Obviously, idealism cannot possess validity in such a framework.

The implications to consciousness and to object-subject relations are as follows: Recall that consciousness must also collapse out of its own AND-form future potentiality: it must do so into consistency with what it is subsequently permitted to witness and affect per function of grand decohered and enforced consistent futures: futures rendered consistent to a pre-resolved Bergsonian past that had retained the supernova explosion at Centauri star system. Hence, future-potential consciousness will be decohered into and restricted only to future observations and consciousness-states that must confirm or be consistent with the fact of the supernova event at Centauri binary, independent of whether consciousness on Earth chooses to observe or not observe that event four years hence.

Again, the content of the 'object' (in this instance, the supernova belonging to a stand-alone mnemonic reality retained in Bergsonian memory, and consequently counterfactually imposed upon the Whiteheadian future) is a process and configuration that unfolds *entirely independent* of subject-nous and of the consciousness subsequently making (or not making) the observation of the supernova at Alpha Centauri four years hence. Moreover, *without volition*, and no matter the presumed will or intent, the future-potentiality of mind and consciousness must decohere to said consistent futures: That is, the subject, or its future scope, is delimited to the demands of the object, and in this sense, object-world enjoys critical primacy over subject-nous.

The choice to observe is real enough, and the mind and consciousness carrying it out is certainly a non-illusory ontic, and, as stated in **5.14** and **5.15**, nous enjoys profound difference vis-à-vis the supernova event, the apparatus of observation, and vis-à-vis the manifest world and universe of memory, possibility, and time. Yet, the object-subject relation is real, even with all the provisos, including remarkable de-spatialisation, temporisation, and consequent revival of Cartesian dualism.

However, our conclusion does not require regression back to dismal forlorn materialism, or to the forlorn want to reduce mind and consciousness to 'matter' and brains, no more than it requires any acquiescence to idealism. Indeed, we can now grasp that Cartesian dualism is real: clearly necessitated by the viability of object-subject relations, rendered so by grand decoherence, if nothing else.

5.23: Comprehensive Summary of Part-II: The Case Against 'Consciousness Causes Collapse'

The two tables attached contrast 'consciousness causes collapse' versus the Alternative Realist interpretation of quantum mechanics: Also note the bullet point comparisons of the two philosophies of quantum mechanics and physics. Thus…

"CONSCIOUSNESS CAUSES COLLAPSE" INTERPRETATION OF QUANTUM MECHANICS							
W-function real?	W-function collapse real?	Indeterminism real?	Hidden gears?	Observer role?	Unique history?	Non-locality?	Universal W-function?
YES	YES	YES	NO	**CAUSAL** Brings about W-function collapse	YES	YES	YES

ALT-R INTERPRETATION OF QUANTUM MECHANICS							
W-function real?	W-function collapse real?	Indeterminism real?	Hidden gears?	Observer role?	Unique history?	Non-locality?	Universal W-function?
YES (it is the ontically real future)	YES (it is time itself)	YES (objectively real)	NO	decoheres but does not collapse	YES (but w/ Little MW)	YES …but as Counterfactuals driven grand decoherence and transtemporality	YES Whitehead facet to timelike infinity

- While both interpretations agree that the AND-form wavefunctions and the universal wavefunction are real, **'consciousness causes collapse' does not explicitly assert that the quantum wave and wavefunction constitutes the ontic reality of the future**, nor identify it as the Whitehead facet of the pertinent growing block amalgam, as does Alt-R per function of the firewall principle from the intertwine of causality with the principle of conservation (**5.01**, and Book-II **2.10** and **2.11**).
- While both interpretations agree on the reality of wavefunction collapse, **only Alt-R identifies wavefunction collapse as synonymous with time itself**, and *not* a 'fourth dimension'.

- Both interpretations agree that quantum indeterminism is real, but **only Alt-R furnishes the case for the objective reality of quantum indeterminism against hidden gears** (**5.01** and Book-III **3.32**).

- While 'consciousness causes collapse' seeks to place consciousness in lieu of hidden gears as the drive behind wavefunction collapse and time, **Alt-R asserts that consciousness is wholly time-dependent, and that time is ontologically primary to the very possibility of consciousness, not the other way round**, as inadvertently espoused by 'consciousness causes collapse': In other words, no time…or no wavefunction collapse,… then no consciousness.

- Consequently, **'consciousness causes collapse' assigns causality to the observer (to consciousness) while Alt-R does not**. However, Alt-R admits that the choices made by the observer about the procedures and mechanisms of observation have the power to decohere the future into consistent outcomes with initial conditions via consequent grand decoherence.

- Both interpretations agree about unique history, although Alt-R speculates for Little Manyworlds (LMW) in lieu of forlorn Manyworlds: (see **5.07**, and Book-III **3.18**). But **'consciousness causes collapse' is vulnerable to a *many-minds interpretation*, wherein the mind and consciousness itself splits into as many supposed histories out of the same root AND-form wavefunction. In Alt-R, the equivalent of many-minds, namely LMW, is speculatively plausible, but it does not emerge from an ontological primacy of nous or its purported wavefunction collapsing time-driving powers.**

- Both interpretations admit 'non-locality'. However, **'consciousness causes collapse' does not challenge the ontology of 'space'. De-spatialisation is unique to Alt-R**. In Alt-R, 'non-locality' gives way to the temporisation of physics *and* to *transtemporality*, exemplified by the transtemporal enforcement of physical law (Preliminary **0.26**; Book-I **1.12** and **1.13,** and Book-II **3.33**; and by grand decoherence itself (Book-II **2.34**), culminating into Mach's principle (Book-II **2.35**), without absolute time or synonymous 'space'.

The following constitutes a comprehensive summary of matters explored in Part-II:

Eugene Wigner was the first to articulate 'consciousness causes collapse'. He later abandoned it.

Wigner held consciousness to be critically different from the measuring apparatus, and did so for two main reasons:

- **Consciousness is always in OR-form**, never AND-form, in contradistinction to non-conscious a-noetic systems (e.g., the apparatus). Therefore, consciousness cannot be subject to AND-to-OR (i.e., wave-particle) dualism and wavefunction collapse, according to Wigner's thesis. **(5.11)**.

- **An always-OR-form consciousnesses will AND-to-OR collapse, and keep in that state of OR-form collapse, any AND-form wavefunction.** Hence….

○ **Consciousness is placed in lieu of non-mental materialistic hidden gears**… as that which brings about wavefunction collapse (and, implicitly, brings about time), *and* assumes the role of the secret orchestrator of quantum indeterminacy...and…

○ **Quantum theology follows almost seamlessly**: One can replace mere consciousness with God-consciousness as collapsor of the universal wavefunction: a claim made by John Wheeler, but not by Wigner.

However, from **5.12**, **Wigner abandoned 'consciousness causes collapse, for the following reasons:**

○ **The disposition towards solipsism**: Whose consciousness is primary? Wigner's consciousness? That of his friend? The cat's consciousness? God's consciousness?

○ **Quantum mechanics and AND-form logic are restricted to the micro-world** and cannot apply to the macro-world of the Schrödinger's Cat and the Wigner's Friend experiment.

What follows is a summary criticisms of Wigner's reasons for abandoning his interpretation (**5.12**). We agree with Wigner's abandonment of 'consciousness causes collapse', but not with his reasons for doing so:

- **Quantum mechanics cannot be restricted to the very small, because causality entwined with the principle of conservation (the firewall principle) demands that AND-form logic must abide at all scales.**

○ The assumption that the macro-scale is always OR-form (that it has no AND-form characteristics) is tantamount to treating all futures attendant the macro-scale as if OR-form to the infinitely removed future (i.e., the block model universe)… tantamount to 'signals from the future'.

○ **The reception of 'signals from the future' would be in violation of causality at the macro-scale,** unless the macro-scale future is constituted as an AND-form state partitioned from the resolved domain of the growing block universe.

○ **The violation of the conservation of energy-matter would also arise in tandem,** given that putative energy-matter would be needed to constitute any futures proven to be OR-form pre-resolved via 'signals from the future'. Where is all the extra energy-matter going to come from?

○ **Grand Decoherence**: The reality of counterfactuals-driven erasure of non-consistent nested futures unfolds at all scales, and constitutes the macro-scale operation of AND-form logic and the superposition principle, if not the self-interference of the universe with itself, and of matter 'here' with matter everywhere else (revived Mach's principle and new inertia theory). Hence, quantum mechanics *is* pertinent to the 'macro-scale' Whitehead facet, and it is necessarily and ineluctably relevant to the macro-scale Schrodinger's cat and Wigner's friend scenarios.

There are better reasons why Wigner's thesis must be abandoned (see **5.13**):

- **Consciousness in its future-potential form, as future potential nous, must also be rendered into an AND-form state of noetic future potentiality, and must also AND-to-OR collapse out of its noetic potentiality. Hence nous is not wholly or purely OR-form, as Wigner had contended.** This is evident from the fact that the reader's consciousness two minutes from now is *not* a resolved state, but constitutes AND-form complex of future-potential alternative consciousness-states.

- **If potential consciousness must be rendered in AND-form, what causes the collapse of potential nous into actualised mind and consciousness?** It cannot self-collapse.

- **Bergsonian memory constitutes the proper basis for the retention, as Bergsonian wavefunction in memory, of the apparatus and of any resolved past events. This retention is not accomplished by, nor has any ultimate need of, mind and consciousness.** Even in the absence of nous, past states of the apparatus and of the world will be independently retained as ontically real states within the Bergson facet of the growing bock Bergson-Whitehead amalgam (see **5.02** and Book-III **3.07**, **3.08** and **3.34**).

However, **consciousness *is* different from the measuring apparatus and world,** but *not* per reasons espoused by Wigner:
- ○ **Consciousness can apprehend and apply infinities to halting problems and in renormalisation,** making possible the renormalising quantum mechanist: This is exclusive to nous and cannot be furnished by mechanism or 'gears'.
- ○ Unlike the apparatus and world subject to the relativity of simultaneity, **consciousness has absolute time-like characteristics**: Nous is constituted as a noetic moment or temporal noetic monad, and has no internal parts separated either in 'space' or scattered and distributed in time. Hence, consciousness is not internally subject to the relativity of simultaneity, no more than could a 'point in space' be internally subject to the same.

Decisive invalidation of 'consciousness causes collapse' necessarily followed. The first set of objections attended the quantum indeterminism issue (see **5.16** to **5.17**). Thus:

- **Consciousness cannot circumvent quantum indeterminism and cannot act in lieu of any orchestrating non-mental materialistic 'hidden gears'.** This is due to the following:
  - ○ **The qualitative superficiality problem**: In the set of generatable 'secret' orchestrating scripts that supposedly under-determine surface quantum indeterminate processes, there exists no one qualitatively superior script for nous to select. *This robs consciousness of any qualitative selection function from among the very large number of equally apt non-exclusive scripts.*
  - ○ **The arbitrary selection problem.** Given that there exists no one qualitatively superior script, the selection by consciousness from amongst the very large number of equally apt scripts will be tantamount to arbitrary selection. *There is no point in replacing a mindless arbitrary random selector with a merely conscious arbitrary selector; one as good as a random selector. This could not solve the problem of indeterminacy or arbitrariness.*
  - ○ **Consciousness has no necessary or sufficient role to play in *any* 'solution' to the non-problem of objective quantum indeterminism.** Per the three-phase development approach posited in Book-III **3.26**, quantum indeterminacy is shown not to constitute a problem that needs a 'solution'. Resort to consciousness to solve this non-problem is as superfluous as resort to non-mental hidden gears, 'pilot waves' or even Manyworlds, to solve the same non-problem.

- **Augmentation from the invalidation of Manyworlds (MWI): Book-III, 3.17: (**see **5.07):** Originally, MWI posited a solution to quantum indeterminism by asserting that *all* the possibilities in the AND-form state resolve into distinct unique histories. When we applied this to an imaginary Sierpinski triangle universe, we ended up with a vast number of redundant *identical* globally culminating histories. The same would apply to the real universe. Hence the pointlessness of MWI as a solution to the 'problem' of quantum indeterminacy. When we substituted consciousness in lieu of the brute force process generation of Manyworlds histories, this culminated into two speculative possibilities:
  - **Consciousness resolves just one possibility and unique history** by resort to an arbitrary selection from among the possibilities: no different from the same furnished by an a-noetic random selector, culminating into the same *arbitrary selection problem* from the critique of 'consciousness causes collapse'… or...
  - **Consciousness resolves *all* the possibilities inherent to the universal wavefunction**: A superfluous role for nous, given that this could be equally achieved by an a-noetic dumb brute-force exhaustive process, and would yet culminate into globally identical 'alternative' histories, succumbing to an equivalent of the *qualitative superficiality problem* from the critique of 'consciousness causes collapse'.

We then contrasted a-noetic hidden gears with nous in lieu of hidden gears and discovered that neither is capable of causing wavefunction collapse and time. Both face insurmountable halting problems: (see **5.18**):

- **Single-step quantum erasure in an idealised divergent category nested future within the Whitehead facet does not appear to run into any halting problems…but only at first glance**: Further scrutiny showed otherwise: hidden gears therein would also crash into an insurmountable halting problem. The same would apply to nous in lieu of materialistic hidden gears.

- **In the convergent category nested futures, quantum erasure of nested futures and grand decoherence would most certainly run into inescapable halting problems**. Time-bound nous in lieu of gears to mediate the scanning and selection processes must generate all counterfactuals and the erasure of infinite numbers of non-consistent futures to timelike infinity, *and* bring about attendant AND-to-OR collapse or time, and thus also crash into insurmountable halting problems.

- Consciousness cannot bring about wavefunction collapse and time because of the following:
  - **Consciousness can apprehend infinities and halting problems, but it cannot resolve these in and through time in a clockable way:** It is one thing to assert, 'there are an infinite number of whole integers', and express this in time, and quite another to count the series of whole integers to infinity in time. It is one thing for the quantum mechanist to abstract infinity from infinity in order to obtain renormalisation, and quite another to suppose that renormalising involves an operational subtraction of infinity from infinity in and through time… ignoring how time-bound nous collapse its noetic moment of insight into infinity out of its own noetic AND-form future-potentials.
  - **Consciousness apprehends infinity through an ineffable non-definable causal process and ontology beyond nous itself, with no 'clock time'… from 'outside' of time.** A deeper ineffable causality and ontology is at work, one also complicit to wavefunction collapse and time, even in in the absence of nous. It constitutes a form of causality that cannot be identified or reduced to any form of consciousness. i.e., **Consciousness is not the primary causality or ontology.**
  - **Inseparability of consciousness from ontologically primary time, or the time-boundedness of consciousness.** No time, no consciousness. No wavefunction collapse, no nous. The universe can fashion AND-to-OR wavefunction collapse and time *without* consciousness. But it cannot fashion consciousness without time. Indeed, consciousness must itself collapse into realisation out of its own AND-form noetic future-potentiality. Hence, we cannot substitute consciousness in lieu of forlorn hidden gears *and* declare consciousness the primary ontology or God, supposedly driving the very time process upon which it itself depends for its own generation into noetic actua… unless one can demonstrate that nous self-collapses out of its own noetic future-potentiality: a non-falsifiable contention.
  - **Substituting time-bound nous in lieu of hidden gears requires consciousness to resolve the quantum erasure of nested futures and grand decoherence of the total Whitehead facet to the infinite future… through a time-based process… thus crashing into an insurmountable halting problem.** If time-dependent consciousness (even God-consciousness) acted in lieu of forlorn non-conscious hidden gears, it would confront infinities that it could not scan, sort, process, or erase in and through time, and even less up to the infinite future. Consciousness would need an eternity to accomplish such a feat, and it would face the same insurmountable halting problem as does a-noetic materialistic 'gears' in its stead.
  - **Causality and ultimate ontology behind AND-to-OR collapse, time *and* nous… transcends memory, possibility, time *and* nous.** We must concede to a form of causality and ontology *beyond* nous and beyond the growing block system; one that resolves for both, while remaining not capable of identification or reduction to either in ontological terms. Ultimate causality and ontology transcends time, 'matter' *and* nous, while it furnishes time and 'matter' their existentiality, *and* furnishes nous *and* its intuition into infinities.

From the critique above, other implications followed:

- **The universe is not a computer; a reiteration (5.19):** If we substitute in place of hidden gears an essentially identical Universal Turing machine to process the quantum erasure of nested futures and counterfactuals to future-infinity, it will run into the same insurmountable halting problems. Indeed, computation is entirely time-bound or dependent on AND-to-OR collapse: Thus, no time, no computation. The notion that computation could generate AND-to-OR collapse and time is an erroneous inversion. Whatever is making time happen, *and* the universe happen, is *not* a computation: The universe is *not* a 'simulation'. These claims were prefigured in Book-I **1.55**; in Book-II **2.46** and **2.47**, and in Book-III **3.32**.

- **Purported statistically significant effects from consciousness on experimental outcomes cannot imply 'consciousness causes collapse', much less the ontic primacy of nous: (5.20).** Per the claim that the presence of consciousness generates statistically significant effects on experimental outcomes, aside the fact that consciousness is superfluous in both the orchestration of quantum indeterminism and in wavefunction collapse (time itself), the claim, otherwise possibly true, suffers certain problems:
  - **The claimed statistically significant effects are *marginal*:** These are *not* unambiguous: They are *not* binary. For mind and consciousness to constitute the primary ontology *and* bring about wavefunction collapse, nous must have an unambiguous absolutely certain yes-or-no relation to the outcomes of AND-to-OR collapse… not a slight "yes", followed by a fudged "no". Yet, the claimed statistical effects from consciousness are just such slights and fudged outcomes.
  - **The closest that consciousness comes to an unambiguous binary mind-world causality (one that goes beyond mere statistically significant marginal effects) is in its relation to the brain.** When one wants tea, attendant processes pertinent to one's brain *will* transpire. When one wants coffee, different attendant brain-processes will transpire. One's want of tea or coffee is not a matter of statistically significant outcomes that *might* lead to tea slightly more often than coffee.

510

○ **That purported statistical effects of consciousness are marginal runs against the grain of disposition toward normal probability distributions. The latter constitutes consummate proof of a mind-independent objective world**. Of course, statistically significant effects from consciousness upon the world would also constitute independent proof of a world-independent causal mind: enhancing the object-subject relation, without furnishing a 'primacy of consciousness'.

Moreover,...

● **The Quantum Zeno effect does not constitute proof of 'consciousness causes collapse' (5.20)**. Apparent 'continuous' observation of an unstable radioactive particle can forestall its decay, skewering its normal probability away from its typical probability towards decay. It might appear that, here, we have evidence that consciousness can deterministically orchestrate indeterminate radioactive decay-possibilities in a statistically significant way. Yet...

○ **The observation process in the Quantum Zeno is certainly high frequency. However, it is not a 'continuous' observation, and cannot be so**. From Book-I Part-I, there can be no pure AND-form quantum mechanical states of affairs: Nature at all scales is *always* an admixture of both OR-form *and* AND-form: a requirement of the growing block universe. There can be no pure OR-form states either, such as to constitute a block universe. It follows that there is no 'continuous observation' as such, even with the Quantum Zeno. We must always obtain irreducible AND-form intervals in observation, even when subject to the Quantum Zeno.

○ **It is neither necessary nor sufficient (nor even compelling) to conclude that it is consciousness that is delaying radioactive decay**, given that consciousness is not necessary to the orchestration of quantum indeterminism, nor in the causality behind AND-to-OR collapse and time. Even if consciousness were somehow necessary for the observed delay in the decay, the same argument previously posed against statistically significant outcomes from consciousness would hold against consciousness-mediated Quantum Zeno: i.e., the decay-delay is not an absolute certainty or a binary effect.

○ **The Zeno Memory conjecture (5.20)**. From Alt-R, it is the stock of Bergsonian memories brought about by high frequency observation of the particle that reduces the probability of its decay to below normal probability, with or without consciousness. The accumulation of wavefunctions in memory of past non-decayed resolutions will superpose and generate a probability bias toward non-decay. The more frequent the inputs of 'observation' and attendant AND-to-OR resolution of the particle into non-decayed states, the more their attendant memories will superpose such as to bias successive event horizons to promote a higher probability in favour of non-decay. Consequently, the ready attribution of the Quantum Zeno effect to a consciousness effect is rendered null.

All of the above culminated into **Final Atheism, *and* the Transcendence of Atheism,** (see **5.21**). Thus...

● **Our Alt-R objections against 'consciousness causes collapse' applies against 'God causes collapse'**... unless one opts to re-conceive God as consciousness-transcendent non-mind Being... synonymous with the ineffable process of gear-less and structure-less causality that generates time itself. However, God as trans-noetic Being is not absolutely compelling: i.e., one could equally assume pure atheism within the same framework, but *only* on the assumption that any viable God requires that it be a mind, and the fact that Being is not mind must imply atheism. But the atheistic move is not compelling either, given that there is no reason to conceive of Being as mind. Yet, the theistic possibility for non-mind Being could not support an Abrahamic God, or any similar religious system, although it could fashion for an irreligious secular-esoteric philosophy, perhaps in the likeness of Platonism, Neoplatonism, or Hindu esoterics emancipated from its idealistic dispositions.

Finally,...

● **Object-subject relations are real** (see **5.22**). As stated, neither AND-to-OR collapse nor quantum indeterminacy require nous as driver or causal primary ontology. Moreover, the relation of witnessing nous vis-à-vis the world is always rendered in retrospect: the object-world arises to subject nous temporally pre-constituted, and the object-world does not temporally arise from subject-nous. But the clearest case for the reality and independence of the object-world from subject-nous is furnished by grand decoherence. Therein, a supernova explosion of one of the suns in the Centauri binary cannot be directly observed until the choice is made to observe it on Earth *after* four years and three months have transpired. The observation finally made will be consistent with what happened over four years ago, as is demanded by its retention as Bergsonian memory, and the grand decoherence of future possibilities to consistency with that memory. Therein resides the objectivity of the supernova event vis-à-vis subjective observers on Earth: the onticity of the object-subject relation, courtesy of grand decoherence and enforcement of a stand-alone object-world with which subject-nous must subsequently abide. Subject-nous is forced into consistency with the world-object. The content of the 'object' (i.e., the supernova belonging to a stand-alone reality retained in Bergsonian memory, and consequently counterfactually imposed upon the Whiteheadian future) is a process and configuration that unfolds *entirely independent* of subject-nous and of the consciousness subsequently making (or not making) the observation of the supernova four years hence: That is, the subject, or its future scope, is delimited to the demands of the object, and in this sense has clear primacy over the subject, *without* requiring the subject to collapse to the object. Nous subsequently collapses out of its own AND-form state into consistent consciousness: consistent with the fact of the supernova.

BOOK-V PART-III:
THE DEMATERIALISATION, OR TEMPORISATION, OF NATURALISM

5.24: Aims of Part-III

It is more than reasonable to suppose that the physics community specifically, and the science community in general, espouse materialism. In the physics-specific context, historically, materialism gained saliency with Galilean mechanics. It thus constituted a commitment to presuppositional IOO-schemas, or input-operation-output schemas, for closed causality and closed ontology, with no possibility of extra-causality outside of the closed loop IOO-schema vis-à-vis contiguously interacting *atomos*, from which all phenomena, including 'illusory' nous, supposedly derive.

Of course, in such a daz, there can be *no* possibility for Cartesian minds, or abstract ontology or 'metaphysics'… perhaps dismissed as childish, 'religious', irrational, or considered scientific anathema.

However, materialism does not constitute the indispensable concomitant to physics, to science, or even to quantum mechanics. Indeed, naturalism and physicalism can survive well and do even better without materialism, as should be obvious from all of the preceding material, and as it will become even more obvious in the explicit dematerialisation of physics in the following essays comprising Part-III. Indeed, de-spatialisation alone, and the consequent pure temporisation of physics, completely unravels materialism, emancipating physics and science from its crippling pall.

While materialism was arguably inevitable from Galilean and even Newtonian physics despite the abstract nature of physical law espoused in both, by 1905, with the demise of absolute time, the notion of space synonymous with absolute time should have gone the way of the aether. At the very least, with the advent of quantum mechanics in 1927, and the empirical-observational stricture to the effect that one cannot empirically verify the notion of the motion of radiation and matter in space, should have put to doubt the onticity of both motion and space, and compel for the pure temporisation of physics by at least the 1930s. Unfortunately, 'space' yet abides as an anachronism; an obstruction to quantum gravity and to a truer account of nature and its ontology. Indeed, **it is primarily the notion of 'space' that foists materialism upon physics. Hence, de-spatialisation necessarily constitutes the dematerialisation of physics**.

Of course, the search for IOO-schema causality in physics, another key factor that foists materialism upon physics, is ultimately synonymous with the forlorn search for hidden gears or variables behind wavefunction collapse and time. The fact that hidden gears constitutes a problematic issue in physics ought to have raised doubts about materialism, at least since the emergence of quantum interpretations as an area of legitimate concern.

*

The de-spatialisation of physics immediately undermines the notion of contiguous causality between particles. There is no space. Therefore, there can be no 'position' or 'location' at which purported 'impact' can take place between supposed spatial block-like or point-particles, much less allow causality, energy, or work to be mediated through spatial contact. Indeed, the assumption of particle spatial contact and impact (hence the assumption of contiguity and space) has led to infinities, and has forced renormalisation approaches in quantum field theory and elsewhere. The historical success of renormalisation has served to obscure the reality that the ultimate form of the problem rests in the *assumption* of space and of contact and impact causality between 'particles in space', all consistent with materialistic closed-causality IOO-schemas.

With de-spatialisation, it emerges that **only time is real, and a pure delayed choice time interval-based distributions of temporal successions of observations and events, or *durata,* emerge to saliency as a key feature of a now-dematerialised physics**. What mediates putative input to output and affects state-change in a pure temporised physics is the 'quantum leap in time'; a version of generic quantum transitions or the 'quantum leap', but grasped as a pure time-interval transition, proportionate to the pertinent frequency of the 'particles'. Therein, nature simply leaps over (not in space, but in time) the infinite degrees of freedom and the infinities in need of renormalisation presumed to hold within the pertinent time-interval in which 'impact' and force-mediation is erroneously presupposed to take place, thus circumventing the need to realise any 'particle contact' or contiguous impact erroneously presumed to transpire within that interval. **In short, causality is not mediated by impact or contact in 'space' but participates in a 'leap in time' over the infinities that attend the concern of renormalisation: a quantum transition involving pure time, *without* motion, *without* contiguity or contact… *without* 'spatial mediation'.**

The notion of 'quanta of energy', hence of 'particles', albeit in de-spatialised form… now recuperated into *durata*… succeed and supersede spatial *atoms*, wherein only frequency abides, and notions of wavelength or 'space' fall to the wayside.

Of course, de-spatialisation also brings about the demise of the 'operation' facet of IOO-schema causality. **Since only time is real, and since time is wavefunction collapse itself, it follows that we must address the problem of causality, and what mediates it, in terms of what causes AND-to-OR wavefunction collapse and time, *and* the attendant 'quantum leap in time' that circumvents the need for contiguous spatial causality**… and which obviates materialism thus.

*

While physics and science can survive and prosper in the midst of de-spatialisation and consequent temporisation, materialist ontology *cannot* survive the attendant dematerialisation of physics. The central factors in the demise of materialism also include…

- **The demise of IOO (input-operation-output) causality furnished by the invalidation of 'hidden gears' in the orchestration of quantum indeterminism** (Book-III **3.32**).

- **The demise of IOO causality furnished by the invalidation of hidden gears in the generation of wavefunction collapse and time** (Book II **2.46** and **2.47**, and this book: **5.18** and **5.19**).

- Both usurp the middle 'operation' term in IOO schemas, as well as usurp notions of closed causality *and* closed existence that attend it, often abused against any prospective extra-ontology, and weaponised especially against Cartesian mind.

- **Finally, de-spatialisation renders impossible the presupposition of point-like particles in space; of 'matter in space'; and of contiguous causality that supposedly transpires between spatial *atomos*.** It even undermines the belief in the spatial distribution of memories in brains, and, with it, the ready supposition of an in-brain 'locus' that supposedly produces 'illusory' consciousness. De-spatialisation forces the pure temporisation of physics (and of the brain) into a system composed of temporally organised *durata*, rendered into a pure stereotemporal order, *without* contiguity or space (Book-IV **4.14**).

As can be grasped from Part-II and the case against 'consciousness causes collapse', **the demise of materialism does not imply the validation of idealism: The demise of 'consciousness causes collapse' puts to rest prospective idealism. Indeed, materialism and idealism are *both false*: both share the same essential corpus of errors.** More generally, the attempt to collapse subject-nous to object-world is as forlorn as the want to collapse object-world to subject-nous. Indeed, mind, thought, and the growing block system of memory, possibility and time derive from a transcendent ontology that cannot be reduced to memory, possibility, time, *nor* even to nous.

*

In Part-III, the consolidation of the case against materialism will begin with the summary reiteration of **information-matter dualism** from Preliminary through to Book-IV. Of course, the **de-spatialisation of physics** solely unravels materialism by leading directly to the temporisation of information; recuperating information-matter dualism in radical de-spatialised and temporised terms.

Part-III will also restate and further develop **the case against IOO-based contiguous impact causality per implications from renormalisation**: The materialist assumption on the nature of causality generates the infinities that subsequently require renormalisation. Consequently, if materialism held true, putative materialist causality and particle-interaction would crash into insurmountable infinities and halting problems: a fate exposed by the background to renormalisation, compelling for the non-contiguous 'quantum leap in time' approach to state-change and particle 'interaction'; circumventing the interim of infinities residing between the input and the output.

The recapitulation of the case against hidden gears from both the solution to the problem of quantum indeterminism *and* from the basis of wavefunction collapse and time, will only serve to augment the invalidation of IOO causality independently of the implications from renormalisation, retrenching our 'quantum leap in time' approach to state-change, and segueing into the **reiteration of the case *against* the notion that the universe is a computer**.

Part-III will also reiterate and augment **the case for the transtemporality and dualism of physical law garnered from the Einstein Podolsky Rosen experiment** (for preceding material, see Preliminary **0.26**, Book-I **1.13** and Book-III **3.33**), revamped in the context of de-spatialisation and the Bergson-Whitehead amalgam... demonstrating that physical laws are indeed transtemporal and belong to an ontology extant the growing block system of memory, futurity and time... undercutting the closed ontology espoused in materialism, and usurping materialism thus.

The transtemporality of physical laws will be enhanced by implications from Bell's inequalities, which we will introduce for the first time in Part-III.

Finally, Part-III will reprise **the pleonastics of energy concepts, garnered from both de-spatialisation *and* from the framework of the temporised Bergson-Whitehead amalgam**, prefigured in Book-II **2.39** and Book-IV **4.19**. The conservation of energy encapsulated in $E = mc^2$ survives the pleonastics of energy, but it is recuperated into a new abstract form composed of the conservation of memory co-joined to the principle of consistent futures, while it must also remain subject to the ontologically primacy of wavefunction collapse and time.

It is time that, in the first place, brings about the now-pleonastic energy relations and the recuperated conservation principle. Yet, time or wavefunction collapse is not energy-driven but a gear-less work function zero process... implying the transcendence of the physics of energy-matter by pure information physics, completing the de-materialisation of physics and naturalism.

5.25: Information-Matter Dualism & the Demise of Materialism

Why must we repeat the claims from information-matter dualism and the problematic relation of information to space, if superlative de-spatialisation and pure temporisation abide? Information-matter dualism presaged de-spatialisation and temporisation: As such, its reality demonstrated that, even when we utilise the false notion of space, the problematic relation between space and information undercuts the viability of materialism. Therefore, **the recapitulation of information-matter relations, from its simplest to its more developed forms, despite and presaging subsequent de-spatialisation and temporisation, can only serve to augment the goal of the dematerialisation of naturalism**.

The simple-sufficient case for information-matter dualism (Preliminary **0.01** to **0.04**) attacked the generic belief in the notion of a seamless-continuous relation of physical information with 'space': Rudimentary positional information pertaining to, say, a particle in motion, proved *not* to be seamlessly or continuously related to space: its purported trajectory turned out *not* to be comprised of an infinite number of seamlessly linked or contiguous successive positions constituting an unbroken projection, this being necessary for the assertion of any seamless relationship between information and 'space', and indispensable to the viability of materialism.

On closer inspection, based on wavelength and frequency considerations, the purported trajectory of the particle in its forlorn seamless-continuous form proved to consist of a series of broken-discontinuous succession of a finite number of 'positions', each separated by a quantum mechanical interval to which one cannot assert *any* concrete OR-form position or attribute. Hence, information pertaining to purported 'position' was found to be broken and discontinuous vis-à-vis space, even on the assumption of space.

The same conclusions followed when we considered a 'spatially stationary particle' undergoing only time-evolution, without relative motion. We again obtained the broken-discontinuous succession of OR-form moments for the particle, albeit in time; *not* the expected seamless-continuous infinite number of moments, or 'infinite frequency'. This was but a preface to temporisation and a presage to *durata*. In short, we discovered that physical information is *not* seamless-continuous, *even* with respect to time. Of course, with the independent demise of space, 'wavelength' must be dropped, and *only* frequency abides. Hence, *atomos* is succeeded and supplanted by *durata*.

*

The facile reality of information-matter dualism cannot be denied. To do so would imply impossible outcomes, such as ultraviolet catastrophes. The latter was one of the central problems that garnered the development of quantum mechanics. It is a generic fact of quantum mechanics that continuous observations of any system cannot be obtained, notwithstanding the Quantum Zeno effect, which cannot furnish continuous observation. Therefore, any series of observations, of *any* system… large or small... whether short or long time-interval… hence, any observation and extraction of physical information…must always configure into a broken-discontinuous format. This fact alone necessarily compels facile information-matter dualism and demonstrates the implicit complementarity between quantum mechanics and information-matter dualism. From the framework of our nascent information-matter dualism, **it is as if information pertaining to radiation and matter 'pulls out' of space and matter. Following a quantum mechanical interval, the information comes back into 'space and matter'**, *ceteris paribus* with respect to the deeper and superior insights about the relationship between information, matter, radiation and putative space, garnered from later de-spatialisation. This effective 'pull out' and dissociation of information from presumed space appears certain, even when we assume forlorn pleonastic space, with obvious inimical implications to materialism, insofar as the latter, in its naïve presumption of 'stuff in space', cannot accommodate such a dissociation, much less the attendant retention of information extant the space, radiation and matter.

Thus, information-matter dualism, even in the form conceived in the Preliminaries and in Book-I, posited the informational dissociation of a photographic image from its 'carrying' photons in photography, and from 'carrying' electrons in electron micrography, with the same form of 'pull out' in the moment-to-moment dissociation of information entailed in the drifting snowflake, and, of course, in brains. Even when we insist in the materialist supposition of subject-to-object collapse, wherein both mind and memory are presume reduced to the hands-on structures and processes of the brain, the very reality of the dissociation of information from that 'space' unravels materialism and materialist contentions about the brain.

*

Book-I recuperated facile information-matter dualism in terms of AND-OR dualism (i.e., wave-particle dualism) inherent to quantum mechanics. But Book-I assumed the onticity of space. Again, **information-matter dualism emerged from the ineliminability of AND-form and OR-form contaminants vis-à-vis any and all physical systems, whatever their scale**, with obvious implications to brains. While generic wave-particle dualism does not appear to explicitly imply information-matter dualism, the discovery and conclusion that it is impossible to eliminate OR-form and AND-form contaminants (or wave-particle dualism) from any physical system *does* imply information-matter dualism. Thus, Book-I asserted that one cannot obtain a pure OR-form resolved system (a block universe configuration) no matter how much we saturate that system with observational inputs. Thus, the AND-form characteristics of that system cannot be totally eliminated. The vaunted Quantum Zeno effect (previously explored in Part-II **5.20**) cannot furnish 'continuous observation', as this would conduce conditions similar to those that would attend the notorious ultraviolet catastrophe.

As was obvious even from the simple-sufficient case for information-matter dualism espoused in the Preliminary, the attempt to totally resolve any system in pure OR-form terms, both in its motion or in its time-evolution, and accomplish this in a seamless-continuous form, requires that we input to it zero-wavelength (infinite frequency) observational inputs in order to garner the required infinite number of successive seamless-contiguous 'positions' or moments. Of course, zero-wavelength inputs imply infinite frequency inputs, putting us back into the realm of impossible ultraviolet catastrophes. Consequently, it might be possible to resolve a physical system up to its minimum Plank wavelength or Plank frequency, or Plank time, but we might not be able to eliminate AND-form intervals equal to Plank time-intervals.

It is also the case that we cannot eliminate any system of its OR-form contaminants. The inevitable admixture of AND-form and OR-form in all systems, no matter what their scale, necessarily implies broken-discontinuous trajectory and discontinuous time-evolution, and the ineliminability of AND-form intervals that intersperse equally ineliminable OR-form states from that system. This necessarily implies the dissociation of information from 'space' and time. Insofar as this is true, information-matter dualism must follow as a truism of AND-OR dualism, and so must the attendant demise of materialism.

*

Our case was further augmented by the information-scrambling role played by quantum indeterminism, which was later shown to be both non-illusory *and* objectively real.

How do identities and patterns survive, perdure and endure in time in the face of quantum indeterminism? When a system is resolved into its succeeding OR-form moment, the information from its preceding moment somehow jumps over to the succeeding moment (*ceteris paribus*) across a quantum indeterminately scrambled AND-form time-interval. The quantum indeterminism therein ought to

514

scramble the information and prohibit *any* carry-over, and so render impossible memory-retention *and* its temporal resonation. Yet, we obtain the opposite, *despite* quantum indeterminism.

Of course, the problem posed by quantum indeterminism has since been solved per the objectivity of randomness and from the three-phase developmental approach, both furnished in Book-III: **3.32** and **3.26**, respectively. These solutions do not gainsay information-matter dualism. Indeed, **information-matter dualism was given a new boost from the growing block framework of the Bergson-Whitehead amalgam-replacement of spacetime. Therein, we obtained the ultimate version of information-matter dualism in terms of the dualism of Bergsonian** *wavefunctions in memory* **(the past) versus Whiteheadian** *wavefunctions in potentiality* **(the future). Indeed, information-matter dualism is nothing more that the surface expression of the structure of the growing block Bergson-Whitehead amalgam: a future-past dualism interchangeable with AND-OR (wave-particle) dualism.**

The 'dissociation' of information evinced in previous iterations of information-matter dualism from the Preliminary through to Book-II simply constituted its retention as memory within the Bergson facet of the Bergson-Whitehead amalgam (or within the preceding 'realised spacetime' from Book-II). As such, information as Bergsonian memory is *not* retained in 'matter' or 'space'... or in any OR-form resolution or state confined to the event horizon. In this way, information-matter dualism re-emerged, albeit in a recuperated radically de-materialised form, involving de-spatialisation and temporisation definitive of the Bergson-Whitehead amalgam, or 'spacetime' without space. Obviously, the brain *cannot* be exempt from such a past-future temporised dualism.

Recall that the tripartite structure of the Bergson-Whitehead amalgam consisted of the relative present-continuous *event horizon*. The event horizon separates the Bergson facet of memory (the accumulating history of the growing block universe; formerly, the domain of 'realised spacetime' (i.e., the past)) from the future-potential Whitehead facet... formerly the domain of 'spacetime in potentiality' (i.e., the future). Information was shown to dissociate from the moment-to-moment dissolution, reformation, and succession of the event horizon. **The dissociated information would be relegated to the Bergson facet, only to be restored or reinforced repeatedly to and upon each succeeding event horizon, so forming a projection in time (i.e., perdurance).** The Bergson facet was inferred as the accumulation of the past: i.e., abstract memory, necessarily configured in dissociated form from both the event horizon *and* from the Whitehead facet of future-potentiality. Therein, obviously, information and memory could not be retained within the domain of the future of quantum mechanical waves i.e., within the Whitehead facet. The domain is nothing more than the totality of not-yet realised future-potential complexes of nested futures, all rendered in AND-form. Nor is information retained in the 'materialised' OR-form states manifested on or along the perpetually dissolving and reforming event horizon. Hence, it was inferred that information could only be retained within, and re-projected from, the Bergson facet... from dissociated memory vis-à-vis the event horizon. Hence, per the said dissociation, we obtained the final superlative evolution of information-matter dualism: information-matter dualism as temporised past-future dualism from the tripartite structure of the Bergson-Whitehead amalgam.

*

Materialism inadvertently asserts that all information (including the brain... and all memory stored therein) must be reducible to and identifiable with OR-form states, which we now grasp as wholly confined to the event horizon. In short, materialism is the implicit claim that all of physics and reality must be reduced to the event horizon. But this is not tenable per the structure of the growing block Bergson-Whitehead amalgam. Informational dissociation into Bergsonian memory proved to be an explicit feature of the Bergson-Whitehead amalgam if not of the preceding model in Book-II. The model forced us to describe all physical systems, whether these are particles, snowflakes, or galaxies, in terms of the growing block tripartite structure of the Bergson-Whitehead amalgam. Obviously, brains and memory cannot be exempt from this structure, and materialism must unravel while attendant information-matter dualism, as the mirror projection of the tripartite structure of information organised in time, prevails. Therefore, Cartesian-like dualism must also necessarily follow as a truism across general nature, especially vis-à-vis brains.

5.26: De-Spatialisation Implies the Explicit Demise of Materialism

There is no space. It follows that all and sundry must be recuperated into temporal interrelations and distributions, *without* **space. In a pure temporised physics that attends de-spatialisation, there is no 'place' for a billiard ball-like 'matter', much less for point-like 'stuff'.** Of course, the quantum of energy, purely a function of frequency (hence, purely a function of time) appears to abide, although even energy is finally rendered pleonastic. Thus, 'particles' so-conceived remain plausible, but are recuperated into their temporal form... into *durata*... given that we can no longer model these as 'points in space', nor as blocks therein, simply because there are no such 'points' or blocks in a de-spatialised purely temporised universe: Time has no 'points' or 'locations', and transform such notional entities into temporal relations and into *durata*.

One might hope that the survival of quantised energy per the survival of frequency in a de-spatialised purely temporised universe, notwithstanding the pleonastics of either, might yet preserve a modicum of materialism. However, the fact that the very process of ontologically primary wavefunction collapse and time (the very process that manifests putative energy relations and the conservation principle) is brought about by an ineffable structure-less causal process that unfolds as a work function zero processes, cannot avail materialism. Indeed, **if we cannot model billiard balls or point-like 'stuff' in a pure time-based physics, how could we ever hope to model spatial interactions, contacts and impacts between such entities, much less the 'energies' supposedly mediated by and between these? How can we preserve any aspect of contiguity, contact or impact causality in a de-spatialised purely temporised daz? We cannot: The interactions must unfold through pure non-contiguous work function zero time... without contact or impact.**

515

Indeed, even in generic quantum mechanics and physics prior to de-spatialisation and temporisation, when we 'hit' one particle with another, we were superposing their AND-form possibility states into a single culminating joint possibility superpose. With de-spatialisation, there is no 'wavelength' component to the said superpose: only a frequency component, itself exhibiting potential delayed choice time-relation possibilities, embedded in and related to a larger system, itself rendered in the same terms. **The state-change transition of the superposed possibility states into outputs must involve AND-to-OR collapse, itself *not* brought about by means of contact causality, and driven by an ineffable gear-less process that can neither be clocked nor structurally or operationally defined, entailing the aforementioned 'quantum leap in time', with *no* possibility of any form of spatial contact or impact between 'particles' or in the state-change so-instigated. But when we introduce de-spatialisation and temporisation, what was obvious even before de-spatialisation and temporisation is immediately rendered into truism.**

5.27: De-Spatialisation Co-Joins Information-Matter Dualism

By Book-IV, information-matter dualism had obtained its superlative form: The model of spacetime from Book-II was superseded by the Bergson-Whitehead amalgam; a development forced on us by de-spatialisation and temporisation. Yet, Book-III had furnished a series of preceding mnemonic apagogic arguments in favour of abstract memory, before the same form of memory explicitly emerged directly and seamlessly from de-spatialisation and temporisation in Book-IV.

From the mnemonic arguments in Book-III, we discovered that quantum mechanics as we know it, and the kind of universe it describes, could not be tenable without the implicit reality of abstract memory, combined to the attendant retention of critical physical information extant putative radiation, matter and 'space', as required by preceding information-matter dualism. Indeed, abstract memory and information-matter dualism turned out to be aspects of the same thing: the latter a surface-signature of abstract memory retained in dissociated form within the tripartite Bergson-Whitehead amalgam.

Also recall that, in Book-III, our assertion about the total dependency of quantum mechanics on the reality of abstract memory survived every anticipated nihilistic attempt to obviate it. The nihilistic arguments were based on information erasure hypothesis and its plausible co-joining with Manyworlds into ultimate-form mnemonic, mereological and informational nihilism.

In Book-III, the case for the indispensability of abstract memory to the very viability of quantum mechanics, as a solution to the apparatus problem recapitulated in **5.06,** also emerged out of the re-examination of the two-slit experiment. Therein, we imagined what would happen to the OR-form 'both slits open' configuration of the barrier upon its physical-experimental isolation and quantum mechanisation: i.e., its relegation into the future, hence its dissolution into and AND-form futurity of alternative possibilities. Yet, with the end of isolation, the barrier was restored to its original setting, *not* to any of the other possibilities: *not* into a pumpkin instead… *nor* into nothing instead. From this, we garnered the provisional postulate that critical information pertaining to the apparatus-barrier ('both slits open' state) had abstracted out of the system upon its isolation and quantum mechanisation (i.e., the dissociation of information per information-matter dualism) and was retained intact elsewhere (i.e., as abstract memory). Hence, again information-matter dualism was shown to be the mirror reflection of abstract memory.

The nihilistic extreme entertained in Book-III was to imagine the total informational erasure and wipe-out of the barrier-setting upon physical-experimental isolation. The resolution of any consistent outcome at the end of isolation would then become a matter of the improbable and the miraculous… and one could assert that that, in nature, there is no memory… *and* no information… culminating into the nihilism of memory and information in physics. Indeed, in Book-III **3.12**, by combining the above erasure hypothesis to Manyworlds in relation to the barrier in the two-slit experiment, it was argued that the universe split up into four unique histories: each history exhibiting one of the four possible barrier settings. In only one history did we obtain default continuity and consistency between the resolved outcome versus past initial conditions attendant the barrier. Such a default consistency did not appear to involve *any* temporal informational carry-over, or memory, much less require the existence or ontology of information itself. Hence, ultimate mnemonic and informational nihilism.

However, **a series of apagogic mnemonic arguments in Book-III rescued abstract memory from the clutches of absolute nihilism. Therein, full arguments were presented for why MWI and mereological nihilism are non-viable; why both must fail; and why MWI itself is provably false:** (see Book-III **3.13** to **3.17** and **3.38**).

The apagogic arguments from Book-III will not be reiterated here but can be examined per the stated essay references.

Yet, the final blow to nihilism emerged in Book-IV, from de-spatialisation. De-spatialisation automatically lead to abstract memory in a wholly independent and seamless way in comparison to the apagogic mnemonic arguments from Book-III. **De-spatialisation implied that *all* information in nature must necessarily be rendered into pure temporised distribution-form, seamlessly culminating into the retention of the past, *not* as 'material trace-state in space', but as a set of non-contiguous purely temporal relations within the past and between the past-future.** This led to the validation of the central ideas of Henri Bergson's theory on memory; ideas synonymous with our own abstract memory theory from both Book-III and IV. Indeed, remove space, and all we get *is* abstract or temporised memory. **Remove space, and what previously appeared to be 'objects in space' immediately transform into *durata* organised purely in and by time… into the non-contiguous resonation of information and abstract memory, *not* across space, but across time, from past to present: almost exactly the way Henri Bergson had intuited.**

*

Other achievements in Book-III perfected our understanding of information and memory; supplementing and completing the case from various mnemonic arguments. From the solution to the stochastic coherence problem and quantum indeterminism, (Book-III **3.26** and **3.32**) a new understanding of information emerged: In its form as future-potential information, information manifests as uncharged bounded probability state: i.e., as wavefunction in future-potential*y;* similar to the generic wavefunction, but exhibiting nested futures that elaborate to the infinitely removed future.

On the other hand, in its charged bounded probability state form, information manifests as wavefunction in memory, comprising wavefunctions of the past, with their former nested future complexes cropped, erased or removed.

Hence, the principal difference between information in potentiality (the future) and information in memory (the past) is that information in potentiality possesses nested futures, whereas information in memory is cropped of its former nested futures. Yet, both are modelled as wavefunctions. This forced the move beyond the provisional myopic classification of information as part of a dualism between OR-form versus AND-form states (See Book-III **3.35**): It turned out that information is *not* retained as memory in OR-form terms: Memory is *not* comprised of OR-form information. Yet, OR-form expressions of information remain valid, albeit confined *only* to the event horizon.

The key point was that wavefunctions in memory (the past) cannot obviously be carried by, or contained in, or reduce to, or physically identify with, wavefunctions in potentiality (the future). Consequently, **information-matter dualism was recuperated as the dualism between wavefunctions in memory versus wavefunctions in potentiality, with both the future and the past dissociated from the event horizon *and* its OR-form resolutions, including the OR-form 'materialised brain'.** Hence, re-emerged information-matter dualism, but as the dualism of wavefunctions in memory versus wavefunctions in potentiality, or past-future dualism.

*

Applying all of these findings to the apagogic mnemonic arguments from Book-III in the context of the two-slit experiment, we discovered that, upon isolation, our setup of the barrier passes into and is retained as a state of abstract wavefunction in memory within the mnemonic Bergson facet of the Bergson-Whitehead amalgam. As such, it is necessarily dissociated from the event horizon. The **memory generates counterfactuals that erase from within the Whitehead facet those future barrier-possibilities that are *not* consistent with the initial barrier-setup, retained as the said wavefunction in memory**. Of course, the attendant wavefunction in memory (the original barrier setup) now belongs to the Bergsonian past, and it is temporally and non-contiguously carried over to subsequent experimental resolution, hence to the succeeding event horizon, as the abstract-immaterial past.

The wavefunction in memory pertinent to the barrier setup lacks any sort of nested futures and lacks the alternative barrier settings that would otherwise attend it. With the alternatives shorn off, subsequent AND-to-OR collapse and time, hence the end of isolation realised at the attending event horizon, cannot help but re-manifest the initial barrier setup, and no other possibility. (Notice how this constitutes the recuperation of the principle of conservation, but as the conservation of memory, co-joined to consistent futures).

Thus, if the barrier was 'both slits open' in the past, with only consistent futures enforced per grand decoherence, 'both slits open' could not help but subsequently 're-materialise' and perdure. The same sort of process of memory-recollection turns out to apply in the temporal carry-over of information pertaining to the drifting snowflake; in the carry-over and retention of identities and image-patterns in photography; in the continuation of forms and identities across nature, in general and in living nature. Finally, it applies in the retention and carry-over of memories pertinent to brains, with memories retained within the abstract Bergson facet, *not* in the materialised brain confined to the event horizon. The same dualism must also hold true for mind and consciousness involved in the recollection of pertinent memories: a recollection, *not* from the brain confined to the event horizon, but through the temporal reach of nous into the Bergson facet of the Bergson-Whitehead amalgam: a feat accomplished dualistically, and *without* contiguity or contact causality.

Recall that the spur to the development of the Bergson-Whitehead amalgam was de-spatialisation itself. Information, in whatever form, especially as the past, can no longer be contained or distributed in any sort of 'space', much less as 'spatial matter trace'. It followed that the wavefunction in memory (*any* wavefunction) cannot be a projection or distribution in 'space'. Wavefunctions in memory are organised within the Bergson facet, interrelated to each other and to the OR-form 'materialised' event horizon purely in terms of delayed choice time-interval relata. **With de-spatialisation, information pertaining to the resolved past, or even of the as-yet-to-happen Whiteheadian future, cannot have a 'location' or 'place'.** How such location-less information is retrieved and put to use by nature (i.e., through a process of perpetual wavefunction collapse, or time) which must itself defy encapsulation in terms of any hidden gears, will necessarily defy materialist explanation. This is because materialism strictly presupposes spatial-based notions of retrieval and transfer of information, and cannot accommodate the temporal forms of the same.

It also follows that de-spatialised memory retrieval must be affected in and across time, affected by nous, in turn related to brain and world in pure temporal terms. This totally undercuts the notion that nous could be generated by the brain, even as an 'illusion'. Indeed, *where* exactly is 'illusory' mind and consciousness supposed to be generated in the brain, given de-spatialisation and the pure temporisation of the brain and world? Instead, **the locus of nous must be sought as a process that transpires purely through time; modelled as a non-contiguous de-spatialised time-relation vis-à-vis memory, futurity, and brains**: one that can resonate into the Bergsonian past as surely as it resonates into the Whiteheadian future: a processes necessarily extant the event horizon *and* the 'materialised' brain confined to it. Hence, heralding cartesian revival.

5.28: The Nature of Causality & the Demise of Materialism

The materialist notion of impact causality is instantly captured in the billiard ball analogy. In order to make one billiard ball 'go' we must knock it with another. The latter must first be subject to an applied force or energy in order to affect the former via impact. An account of all the physical inputs in terms of combined forces and billiard balls would constitute a conserved total. The output would

517

be totally predicated from the inputs: a causally closed system. Hence, billiard ball-like impact causality abides to an input-operation-output scheme (IOO causality), with all inputs operations and outputs defined, manifested and unfolding in 'space'.

Now, if time was infinitely devisable, and if we could no longer pin down the final point of contiguity or 'impact' that made our billiard ball 'go', the scheme would collapse to infinity in the same way that Zeno's arrow did. However, time is *not* infinitely devisable. Time is discontinuous or quantised.

Even at its most rudimentary, information-matter dualism demanded that each moment at which an OR-form state manifests must be flanked by a preceding and succeeding AND-form quantum mechanical time-interval, the minimum of which might approach Plank time. Such intervals must be devoid of any OR-form information. It is across and from such intervals that causality operates and collapses the AND-form state into the succeeding OR-form outcome. Indeed, it cannot be sufficiently avowed that the form of causality that brings about the requisite wavefunction collapse (time) is very different from what materialism supposes, as we found out in **5.18**.

Given that materialism consists in the implausible reduction and identification of all critical variables with OR-form states inadvertently restricted to the event horizon, materialism must somehow prove that causality itself must also be reduced and identified with such OR-form traces resolved upon the event horizon. Yet, in Book-II and IV, we discovered that critical information resides in abstract form in two main domains: as wavefunctions in memory 'below' the event horizon within the Bergson facet, and as future potentials configured as wavefunctions in potentiality 'above' the event horizon within the Whitehead facet. Of course, OR-form outcomes reside on and are confined to the event horizon 'skin' that separates the pertinent domains of memory and future possibility.

From the above schema, **the putative 'motor' of time and wavefunction collapse cannot reduce to or identify with any event horizon or its OR-form outcomes.** Instead, OR-form outcomes must emerge out of the process of time, not the other way around. The drive of time must be 'located' between succeeding event horizons, *not* in or on any event horizons, nor in or as its OR-form resolutions. Thus, **causality itself cannot be reduced to states on or along the event horizon (i.e., to 'matter'), much less to their putative contacts and 'applied forces' supposedly evinced or hypothetically mediated along the event horizon.** It follows that the sort of generic billiard ball impact causality and attendant input-operation-output schema causality sought by materialism must unravel, if one seeks to rescue materialism by reducing all reality and causality to the event horizon 'skin' and its OR-form terms.

*

What then brings about wavefunction collapse? What is the true basis of causality? We have already addressed this question and can only reiterate the answer in summary form. Two facets to causality relate to and bring about wavefunction collapse. The first is the resolution of quantum indeterminate outcomes from wavefunction collapse. The second is wavefunction collapse itself. The presumption of putative billiard ball causality cannot obviate the requirement that we must model causality in terms of what causes wavefunction collapse and time, and in terms of what causes the resolution of quantum indeterminate outcomes that, *en masse*, furnish the impression of 'colliding billiard balls'.

Now, in previous material, we took to task the notion of hidden variables or gears purported to be causative of wavefunction collapse. We attacked this notion as thoroughly as we attacked the dubious notion of 'consciousness causes wavefunction collapse' (Part-II). For the materialist conception of causality to hold true, the hypothesis of hidden gears would also need to hold true, both in wavefunction collapse *and* in the orchestration of its outcomes. **Only hidden gears could provide a form of causality that might rescue materialist impact causality and IOO closed causality, both in the orchestration of quantum indeterminism *and* in the generation of wavefunction collapse or time. Clearly, materialist causality is false by virtue of the fact that hidden gears is untenable.**

We know that quantum indeterminism is objectively true. Yet, **quantum indeterminate processes are *not* orchestrated by any secret deterministic operations or gears**: (See Book-III **3.32**). Therein, in any system purported to be governed by surface indeterminism, a very large number of hidden deterministic scripts could be computed, with each equally apt to orchestrate apparent quantum indeterminism. But we could only select a script by resort to random and arbitrary sortology, and could not obviate randomness. This is problematic to materialism. Materialism needs a definable operation to mediate inputs into outputs and support its IOO-schema. Hence, **the objectivity of quantum indeterminism directly attacks the middle 'operation' term in IOO: It usurps materialism.**

Of course, resort to consciousness in lieu of forlorn a-noetic middle-term "O" in IOO must equally fail: (See Part-I **5.08** and Part-II **5.18**).

It follows that the resolution of one moment of 'interaction' between billiard balls, followed by a succession of outputs and outcomes from the same, cannot ultimately be accounted for in terms of a contiguous gears-driven mediation and orchestration, no more than through mediation by nous.

As for wavefunction collapse and time itself, we discovered that **any attempt to furnish a hidden gears account of the process of counterfactuals-driven quantum erasure of nested futures to infinity, *and* bring about wavefunction collapse and time, must crash into an insurmountable halting problem. In other words, a hidden gear to facilitate wavefunction collapse and time cannot be obtained, even in principle**. In short, the drive of time is *not* constituted as any form of contiguously operating mechanism: Time is *not* made to happen by some hidden tick-tock click-clock gizmo. In the context of a Whiteheadian future (formerly, 'spacetime in potentiality') constituted by the convergent category nested futures, just one counterfactual will imply the total restructuring of that system to the infinitely removed future: (Part-I **5.08** and Part-II **5.18**). A putative hidden gears to process this would need to orchestrate the erasure of *all* of the infinite number of non-consistent nested future possibilities, across the whole convergent nested futures system, to infinity. If it is to accomplish such a feat through a step-by-step 'definable operation' or 'gears', with each step clockable, the putative procedure could not attain completion to the sought outcome. Consequently, state-change through wavefunction collapse and time could not transpire; and the new event horizon could not succeed the preceding event horizon… and no new OR-form events could decay out of the AND-form Whiteheadian future. Obviously, no new Bergsonian memories would form. In other words, time would 'freeze up'.

The actual process that brings about the quantum erasure of non-consistent nested futures to infinity, and hence facilitates the process of attendant AND-to-OR time itself, cannot be any process that has a 'clock time'. Thus, **the 'operation' that drives wavefunction collapse and time must have a structure-less ineffable form. Hence, causality must have an a-materialistic a-mechanical ineffable form.**

The implications are clear, given the total dependency of materialism on the want of a definable operation to supposedly mediate 'spaceballs' into state-change of inputs into outputs: The process of time that brings this and the purported 'collisions' about has no definable operational attribute and cannot be incorporated or reduced to a materialist input-operation-output schema. Indeed, any attempt to fully account for causality per the demands of materialism, per impact causality, per hidden gears, or per the IOO-schema, must run into **a 'black box' whose mysterious processes, responsible for wavefunction collapse and time, defies any and all possibility of definition and specification as gears, mechanism, as materialistic gizmo.** Thus, wither materialism.

5.29: Breaking the Link Between Materialist Causality & the Conservation of Energy in Time: The Abstraction & Dematerialisation of the Conservation Principle

The materialist conception of causality is often enmeshed with the conservation principle of work or energy, often to the point of indistinguishability, notwithstanding work function zero processes in nature. The identification of impact causality and IOO schemes *as if* these are interchangeable with or justified by the conservation principle, is an over-simplification and error. To appreciate this, recall that the process of wavefunction collapse is synonymous with time itself, and while it is true that the putative (ultimately pleonastic) energy-matter inputs from one moment appear to carry over to the next moment following wavefunction collapse and state-change… and the quantities and sums remain unchanged… hence, the principle of conservation abides… this indicates nothing about what causes ontologically primary wavefunction collapse and time, or causality, or 'action'. Without wavefunction collapse and time, there could be no nominal energy-matter relations, conserved or otherwise, simply because pertinent events could not transpire from which we then infer said relations and principles. It is through ontologically primary work function zero wavefunction collapse and time that the said equality of inputs to outputs is realised in state-change.

Recall that **the process that brings about wavefunction collapse and time, and appears to preserve the principle of conservation of energy-matter, defies rendition in terms of *any* mechanism or gears, or even in terms of energy-matter relations or work (i.e., wavefunction collapse and time has a work function zero profile).** Wavefunction collapse and time is *not* brought about by an input or 'impact', much less mediated by a mechanism or definable 'operation', whether or not the attendant sum of putative and ultimately pleonastic energy is conserved after wavefunction collapse and time has transpired, and the putative inputs have been succeeded by the outputs in a form that is consistent with the principle of conservation.

As has already been conclusively shown, the process of wavefunction collapse is per function of a structure-less and undefinable causality that 'works' at a remove from (hence from 'outside' of) the specifiable inputs, outputs and the attendant conservation principle preceding and following wavefunction collapse and time. This implies that wavefunction collapse and time is a work function zero process. It also implies a decoupling between, on the one hand gear-less causality attendant the drive of time and, on the other hand, the principle of conservation ontologically subordinate to AND-to-OR collapse and time. Therefore, the materialist want to render materialist IOO causality indistinguishable from the principle of conservation is in error.

*

Within the framework of the Bergson-Whitehead amalgam and its attendant de-spatialisation and pure temporisation of physics, what form does the conservation principle of energy-matter assume? We must incorporate to the answer to this the pleonastic status of energy concepts and relations, or the abstraction of energy concepts and relations out of the ontology of physics, as was argued for in Book-IV **4.19**: to be recapitulated in **5.35**. The pleonastic status of energy concepts and relations emerged out of the very character of time: from the fact that wavefunction collapse and time that succeeds inputs into outputs constitutes a work function zero process, robbing causality of the notion of 'work', or the notion that 'all in nature is energy-driven'. What exactly is it that is 'conserved' in such a daz? In truth, the framework answer to this question was furnished in Book-II **2.39**.

To address this, we need only recall the structure of the growing block Bergson-Whitehead processes in brief, but co-joined to the admission that energy concepts are indeed pleonastic. **In the abstract structure of our growing block universe, the principle of conservation transforms into a post-energy principle of the conservation of memory… intertwined with the enforcement of consistent futures via the tandem process of grand decoherence, in line with the past-in-memory. In short, the principle of conservation transforms into the conservation of memory co-joined to the principle of consistent futures vis-à-vis memory, the past, or the set of 'initial conditions' retained in abstract memory.** This is *not* in the sense that memory remains unchanged. Indeed, memory accumulates and grows, despite attenuation due to the inverse square law of time. The conservation of memory… the conservation of all things that have transpired hitherto… is affected through the grand decoherence of the total Whitehead facet into consistency with the total past.

In a very real sense, the conservation of memory is equivalent to the conservation of 'initial conditions'… except that we must recuperate the retention of initial conditions as non-contiguous abstract information, temporally organised into abstract memory. To this, we must also incorporate the perpetually growing or accumulating past, also retained as abstract memory within the Bergson facet.

Grand decoherence renders all futures consistent with the accumulating past and, in doing so, project the conservation of the abstract past onto the total future. This understanding of conservation principles is totally abstract, informational, *ephemeral*... and dematerialised... especially when we appreciate that energy concepts and relations so dear to the philosophy of materialism are rendered wholly pleonastic and ultimately abstracted out of physical ontology: (See Book-II **2.39** and Book-IV **4.19**).

*

Notice that conservation principles prevail in nature, albeit in alternative abstract post-energy form: emancipated from pleonastic work or energy relations and concepts, so constituting a dematerialised form that supersedes generic $E = mc^2$... even if the latter is retained in future physics for purely pragmatic purposes, with full recognition of the pleonastic character of energy-matter concepts.

Now, consider the inadvertent abuse of the principle of conservation inherent to the materialist critique of Cartesian causality and mind-brain dualism: **We grasp that the conservation of memory, combined to consistent future through grand decoherence, cannot support the generic accusation against mind-brain dualism as a violator of the conservation principle, given that Cartesian mind-brain causality could not render the Whiteheadian future inconsistent with the Bergsonian past, nor modify or usurp the content of Bergsonian memory. Nor could mind-brain dualism usurp energy-work relations, given their pleonastic nature, and given that mind-brain causality is a work function zero process, in the same way that general causality vested in AND-to-OR collapse and time is also a work function zero process**.

Thus, the abstraction of the conservation principles into ephemeralised dematerialised form necessarily renders both materialist philosophy *and* the long-running materialist critique of mind-brain causality, forlorn and void.

5.30: Demise of Materialism: Renormalisation & the Unravelling of Impact Causality

The implications from renormalisation were covered in **5.14**: Therein, we discovered that quantum field theory describes contiguous particle interactions with so many degrees of freedom that the system explodes to infinity. It then becomes necessary to use a mathematical 'trick' that essentially involves the subtraction of infinity from infinity.

The explosion to infinities in particle interactions is related to the fact that interactions tend to be modelled as unfolding in contiguous fashion: via contact between point-like or billiard ball-like 'particles in space', forming vertexes of contact. **Our contention is that the tendency of particle interactions to exhibit infinities is an artefact of the forlorn *assumption* of space and of spatial contiguity, contact and impact causality, combined to the notion of spatial 'particles'.**

However, as we discovered in Book-IV Part-I, with de-spatialisation, we can no longer describe putative matter as composed of spatial billiard ball-like entities, much less as point-like particles. Hence, with de-spatialisation, there can be no contact or impact causality between particles: Indeed, *where* would such contact or impact occur? There are no *wheres*, given that there is no space or attendant 'positions' or 'points-of-contact' at which contiguity could transpire.

The infinities attendant the purported particle-impacts can be circumvented the moment we accept the implication from de-spatialisation and move to a pure time-based schema for physical processes. From input to output, real 'interactions' do not unfold 'across space' and will not involve any form of physical-spatial contact. This much is obvious even in generic physics, wherein 'interaction' necessarily involve the superposition of two quantum mechanical waves pertaining to distinct particles: But this is a summation and 'interaction' of possibilities, *not* spaces... thereafter subject to *non-contiguous* AND-to-OR collapse and time-transition into subsequent state-change. Thus, per de-spatialisation and temporisation, through a 'quantum leap' realised across time, *not* across space, **wavefunction collapse time-transitions will 'leap over' the infinities, circumventing *ontological renormalisation,* in the sense that physical reality does not renormalise in the way that a physicist renormalises, even if renormalisation must be carried out by the physicist for purposes of rendering calculations viable. Nature renormalises by simply temporally leaping over the real exiting potential infinities via a AND-to-OR time-transition** and consequent state-change.

*

However, even if we refuse to accept the implications from de-spatialisation against impact causality, critical implications that emerge from renormalisation must yet unravel materialism. First, as previously stated, trick or not, renormalisation itself implies radical assertions about the ultimate nature of the renormalising quantum mechanist: i.e., the power to intuit infinities and recognise halting problems: (see Part-II **5.14** and elsewhere). We will not reiterate the full argument made therein: part of the critique of 'consciousness causes collapse'. On a strict materialist assertion that the physicist must itself constitute nothing but the finite brain (one that supposedly operates on a closed contiguous causality and the input-operation-output schema) the renormalising quantum mechanist's handling and managing of infinities ought not to be possible, and renormalisation itself ought to be untenable thus. When we combine this insight with the ontological primacy of time, and realise that, without time, there can be no consciousness; and that, whatever causes time must also bring about nous *and* the physicist's renormalising power, a consciousness-transcending non-definable causal process must be complicit to the resolution of both nous and the noetics of renormalisation. The consciousness-transcending ontology must furnish to the physicists the power to recognise infinities and halting problems, abstract infinities from infinities, and renormalise in quantum field theory.

However, **historically, the physics community never took the renormalisation problem beyond debate about the veracity of calculation (i.e., Is it or is it not a 'trick'?) and did not apprehend the significance of the renormalising quantum mechanist as a basis for delineating nous as a truly different ontics vis-à-vis the scrutinised world, apparatus, system, or even the brain. The community also missed out on the direct falsifying implications that arise from renormalisation *against* spatial contiguous impact causality.**

520

*

The Feynman diagrams in **Fig. 5.06** should help facilitate the conclusion why IOO-causality must breakdown in the face of the said infinities in renormalisation. In the diagram, two particles approach in presumed spacetime: (we could instead use the superlative Bergson-Whitehead amalgam, but it would culminate into the same conclusion). Note that the particles do not directly interact or contact. Instead, the interaction is facilitated by means of 'virtual particle' exchange: i.e., the wavy line between one particle and the other. The term, 'virtual particle', is responsible for many confusions: we will hereby refer to such particles as *very short-lived interaction particles*.

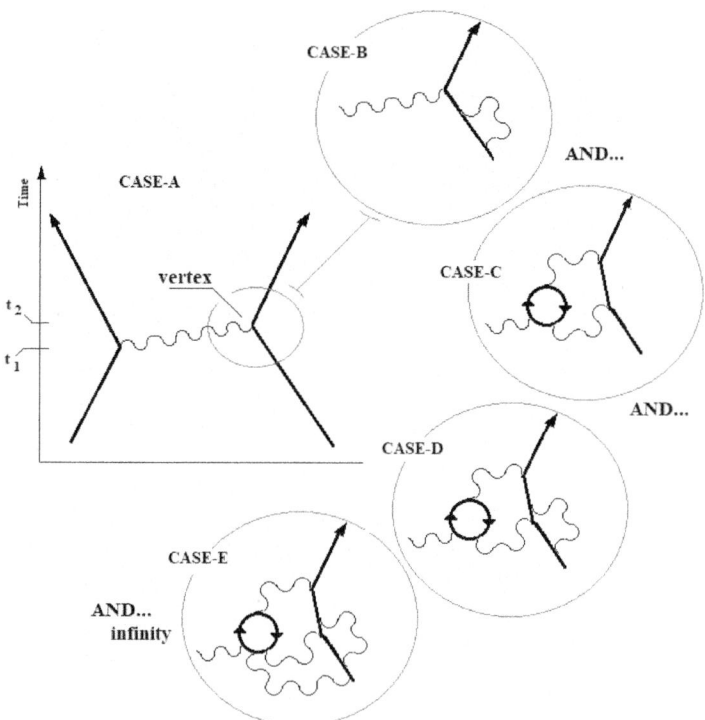

Fig. 5.06: Feynman diagrams, infinities and renormalisation
Notes: The illustration depicts infinities in quantum field theory using Feynman diagrams. In Case-A, two electrons, approach and rebound, with the effect on the right-side particle brought about by a simple interaction via a "very short-lived particle" (VSLP) originally emitted by the left-side particle, forming an interaction vertex. However, this is not the only possibility: Case-B through to Case-E constitute alternative cases. Before the interaction is realised via AND-to-OR collapse, these cases are in superposition AND-form, with some more probable than others (Case-A is the post probable). However, there are an infinite number of alternatives, not just Case-A to E. Moreover, there are additional manifold vertexes per each case (at each point of potential interaction). These also 'blow up into infinity'. Where this presents a problem for calculation, it has been solved through renormalisation: a procedure essentially involving the subtraction of infinity from infinity. However, if physical causality were itself mediated by such means, or by means consistent with spatial contiguity and contact per input-operation-output schemas, due to the said infinities, the process of causal mediation could never halt or reach its conclusion: there would be no transition from t_1 to t_2: the sought interaction could not transpire. The solution resides in a combination of de-spatialisation (which obviates the need for particle contiguous contact) and the "quantum leap in time": De-spatialised purely temporal nature simply 'leaps over' the infinities between t_1 to t_2 forgoing contiguity and circumventing any prospective halting problem.

To reiterate in detail, in **Fig. 5.06 (Case-A)**, two distinct particles and their worldlines are on approach. At t_1, the left-side particle emits a very short-lived interaction particle and loses momentum. Consequently, its worldline clocks to the left, away from its original heading. The said very short-lived interaction particle carries the momentum lost by the left-particle. It heads toward and interacts with the approaching particle on the right side. The momentum lost by the left-particle is thus transferred to the right-particle by t_2. The latter now clocks its direction to the right.

The short-lived interaction particle disappears into the right-side particle. The two particles previously converging must now fly apart in opposing directions, as indicated by their worldlines *after* the interaction process from t_1 and t_2.

All of this seems in full accord with what materialism expects from its 'spaceballs' conception of the universe, albeit transferred to the realm of interactions between elementary 'particles in space'. However, the stated 'infinite degrees of freedom' that then forces the need for renormalisation emerges at the vertexes i.e., at the points at which the short-lived interaction particle is emitted and subsequently absorbed: see Case-B to Case-E… of which there are infinitely many.

One may be forgiven for thinking that the said vertexes are connected seamlessly; that a true physical contiguity or contact *must* be taking place between the short-lived interaction particle and their emitting and absorbing 'non-virtual' particles. This is not true: If we could zoom in on the vertexes, from Case-B to Case-E, we will find alternative possibilities for our interaction vertex: a complex web of finer interactions involving more short-lived interaction particles and even some weird 'backwards in time' loops. Yet, these alternatives in turn exhibit their own myriad interaction vertexes. Presumably, if we opened *these* up for inspection, we would find another complex web of interactions. Indeed, we could carry this to infinity without ever obtaining any final bedrock of contact or impact at *any* vertex of interaction.

Note that **none of this implies the demise of particle physics or 'particles' as such. We can keep the particle zoo and the empirical facts that attend these by reframing it all in terms of pure time or frequency measures for attendant quanta (or *durata*) rendered into a pure de-spatialised and wholly temporised framework. We need not retain the idea of 'space', much less insist that the particles are point-like or spatial entities undergoing spatial contact. We need not model causality itself in such pleonastic terms.**

From the independent invalidation of hidden gears and IOO-causality furnished by the analysis and solution to the process of causality behind both quantum indeterminism and wavefunction collapse and time (also see **5.16** and **5.18**), we now appreciate that a structure-less ineffable causal process is responsible for both. Therefore, the fact that we must invariably fail to reduce particle interactions to some final seamless contact-causality bedrock, or to an attendant closed-causality IOO-schema, comes as no surprise.

The *actual* mediation of particle interactions must be by the said structure-less causal process inherent to AND-to-OR collapse and time-transitions that are ontologically primary to the state-change from particle inputs to outputs. Indeed, it is clear that a contiguous tick-tock click-clock mechanism to process interactions that otherwise blow up to infinities, must remain forlorn: Such a mechanism could never halt, or bring about state-change: It could not constitute *any* solution to causality.

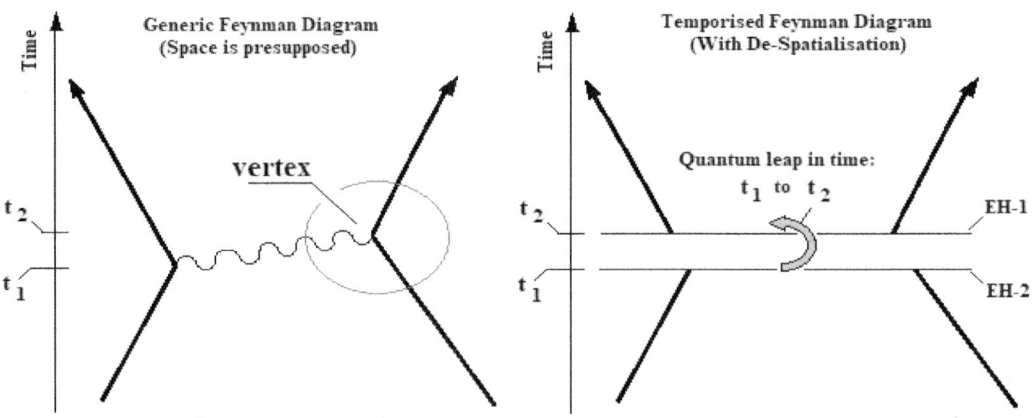

Fig. 5.07: De-spatialisation, temporisation, and the "quantum leap in time"
Notes: The illustration contrasts the generic Feynman diagram, with space presupposed, and with contiguous interaction vertexes with implied infinities, with a modified de-spatialised and temporised Feynman diagram (on the right) involving the 'quantum leap in time', which circumvents or 'leaps over' the vertexes and inherent infinities between EH-1 and EH-2, or t_1 to t_2. This is only possible with de-spatialisation and the consequent elimination of contiguous spatial 'positions' or 'points of contact'. We need not worry about the infinities and halting problems that are concealed within the gap between EH-1 and EH-2, given that the 'quantum leap in time' circumvents these. Note that our quantum leap in time is essentially no different from any quantum transition from one state to the next (such as the leap of an electron from a lower to a higher 'energy level'). The only difference is that 'space' is eliminated and purely temporal processing occurs. The wavefunction collapse (time) from EH-1 to EH-2 is, of course gearless, and does not require a clockable 'computation' or other tick-tock click-clock process to realise. The 'backward in time' loops are also circumvented, given that time proceeds from AND-to-OR, and cannot unravel by proceeding from OR-to-AND, it follows that such loops are inherently absurd, and could not constitute 'backward in time' processes via AND-to-OR forward in time unfolding, and would require possible OR-to-AND de-collapse, or *real* time-reversal.

One approach involves dismissing the whole problem of renormalisation as an artefact of calculation. However, such a move would fail to put the assumption of IOO-causality and impact causality to the test. The approach could not eliminate the renormalising quantum mechanist: **Trick or not, founded on an artefact or not, the very renormalisation process implies the power to apprehend and handle infinities. A universe based *only* on IOO-causality could not manifest the renormalising quantum mechanist**, and it could not manifest beings that can apprehend infinities and recognise halting problems: both essential to the very possibility of the 'trick' of renormalisation.

Our approach is an alternative variant of the above: The infinite degrees of freedom in quantum field theory do not constitute an illusion or artefact: These are real future potentials of the system or interval exhibiting the particle interaction. However, the pertinent particle interactions, realised or potential, are *not* mediated in or across 'space'. Per de-spatialisation, the mediation of particle interactions unfolds purely temporally, *without* space.

Unlike space, time is not infinitely devisable. The shortest time-interval (the shortest delayed choice time-interval relation) that we might obtain or infer for a system of particle interactions most probably approaches Plank time. **While there may be infinite degrees of freedom for potentialities within temporal intervals that approaches Plank time, the progression of the event horizon, from the event horizon featuring the approach of the main particles at t_1 to a succeeding event horizon that exhibits their divergence at t_2, simply leaps over the pertinent time-interval and the potential infinite degrees of freedom therein, *without* needing to process *any* of these infinities into concrete iterations of 'contacts', 'impacts' and outcomes, much less iteratively erase the foregone possibilities therein**: (see **Fig. 5.07**). **The fact that this 'jump over the interval'… this 'quantum leap in time', as it were, is mediated by a wavefunction collapse and time, whose non-contiguous gear-less work function zero causality also defies IOO-schemas, alleviates the problem contrived by the assumption of 'space' and attendant contiguous 'interactions in space'.**

In such a system, through a series of progressions of the event horizon, two particles will approach each other as their mutual delayed choice time-interval relation is reduced to near temporal simultaneity. When approach is at a specific short-interval delayed choice time relation, the attendant event horizon at t_1 will 'dissolve' and a succeeding one will be formed at t_2, 'leaping over' the short interval without having to generate or sort through the infinite degrees of virtual particle interaction and potentialities therein; without having to process their infinite degrees of freedom, or affect some implausible never-terminating IOO-based tick-tock click-clock sortology from among these.

The infinite degrees of freedom within the said time-interval will be subsequently quantum erased through the non-contiguous gearless decoherence and AND-to-OR collapse, into the state-change output at the new event horizon at t_2. Thereafter, through successive event horizons, the particles will 'fly apart' as their delayed choice time-interval relations expand. No actual 'contact' or impact will have taken place between the particles. But remember that the particles are obviously no longer point-like states or objects: They are simply elementary events in time, or pure temporal quanta… or simply, *durata*. This must obviously apply to the very short-lived interaction particles themselves.

Does not all this imply that virtual particles are figments of the imagination? Not so…but *only* on the strict condition that we grasp these as 'quanta in time', and not as spatial particles in space. We can detect virtual particles by reducing the 'quantum leap in time' to shorter and shorter time-intervals or high frequency 'interactions' between the input and output: The virtual particle will appear... but as *durata*. On the other hand, if we delay subsequent observation by a time-interval into a longer time interval, our subsequent observation and attendant AND-to-OR collapse will' leap over' the infinities potentialised within that interval. In that case, no virtual particles will be resolved or detected in OR-form terms.

None of this obviates the resort to renormalisation, at least for calculation purposes. Yet it *does* obviate spatial contiguous impact causality between 'particles in space'. Nature does not iterate through such exchanges, but simply leaps over these by means of temporally discontinuous 'quantum leaps in time'… a process essentially no different from staple quantum transitions or the popular 'quantum leap', except that real transitions do not unfold across space but are configured as 'leaps in time'.

In short, renormalisation in calculation may abide, but IOO-schema based contiguous spatial interaction-causality (or materialism) does not abide. Objectively, the physical world renormalises via non-contiguous 'quantum leaps in time'. Hence, the demise of materialism in this specific context.

5.31: Reprising Why the Universe is not a Computer: The Demise of Materialism

In Part-II **5.19**, indeed in preceding books, we re-discovered that the universe is not a computer. Therein, we entertained the hypothesis that wavefunction collapse and time might be a simulation generated by a hidden universe-scale Turing machine (another variant of hidden gears) consistent with materialism. Our hidden Turing machine confronted the following problems:

- In the context of the convergent category nested futures, which abides throughout the Whitehead facet of the superlative Bergson-Whitehead amalgam, **a putative hidden Turing machine could not carry out the scanning, counterfactual restructuring, selection and elimination of non-consistent nested futures to the infinite future (i.e., grand decoherence). The Turing machine would run into insurmountable halting problems, given the infinite number of non-consistent nested futures that elaborate to the infinite future**. In short, the essential halting problems that beset spatial notions of particle interaction in quantum field theory, and the consequent resort to renormalisation (see **5.19** above), equally apply to the processing of counterfactuals and grand decoherence vis-à-vis Whiteheadian convergent category nested futures.

Therefore, the computation and simulation of attendant wavefunction collapse could never attain resolution, and AND-to-OR time itself could not transpire.

- Hence, **quantum erasure and attendant wavefunction collapse (hence time) is necessarily brought about by an ineffable non-computational process**: a non-definable causality, consistent with our general case against hidden gears and against forlorn IOO-schema materialism.

- Placing consciousness in lieu of hidden gears, or in lieu of our putative hidden Turing machine, will not work either (see **5.16** to **5.18**). Therefore, any want of a Berkeleyan or other form of idealism in lieu of materialism is equally forlorn. Indeed, both materialism and idealism are untenable.

Conversely, what applies to the counterfactual restructuring of the total Whitehead facet applies equally to purported particle interactions in quantum field theory. **Can a hidden Turing machine sort out and process the set of particle interactions subject to infinite degrees of freedom? Obviously, the answer is no. Our purported Turing machine would again crash into insurmountable infinities and halting problem, and the particle interactions could not resolve into culminating events, or into state-change. It follows that, again, at the fundamental level, the universe is *not* a computer, no more than it is a simulation.** Again, this necessarily implies the breakdown of IOO-schema causality and materialism.

Computation does not generate the universe. Instead, it is the universe that simulates computation, *without* itself being a computer, in the same way that it simulates mechanism, without itself being a machine: hence, the **simulation-inversion thesis**. That is, insofar as computation and simulation are time-dependent, and insofar as wavefunction AND-to-OR collapse and time remain ontologically primary to any computational process, the outcomes obtained from wavefunction collapse and time permit simulation of computational mechanical processes in a universe otherwise ultimately causally a-mechanical, non-computational and causally ineffable.

*

Time or wavefunction collapse is ontologically primary to computation. **If we insist that time must be a thing computed, then what computes the computation that generates our time? The illogic culminates in an infinite regress of computations, with the root or original computation infinitely removed from the computation that generates our own wavefunction collapse.** If time and wavefunction collapse unfolds from a root computation, with due consideration to the infinite regress problem just noted, time would 'freeze up', or AND-to-OR wavefunction collapse could not transpire, simply because the initial computation at infinite remove could never ramify to the final computation that computes *our* AND-to-OR collapse. Therefore, computation (even quantum computation) cannot constitute a valid causal basis and explanation for time, wavefunction collapse, and the ultimate basis of state-change (see **5.19** for a more detailed reasoning of the same).

*

As a final augment, we introduce the case from Zohar Ringel and Dmitry Kovrizhi[98]. Ringel and Kovrizhi carried out a quantum Monte Carlo method to study the Hall Effect in low temperature magnetic fields. The key features of the Hall Effect and the Monte Carlo method need not concern us here, save that the want to store information pertaining to the state of just a few hundred electrons involved in the Hall Effect required computational memory capacity that far exceeded the realisable memory capacity obtainable from the employ of the total number of atoms in the universe.

Ringel and Kovrizhi's finding is not as fundamental as our own case against hidden gears, IOO-causality, and the forlorn computational universe: Indeed, the simulation of a few hundred electrons is not explicitly a problem that runs into infinities, or even requires infinite atoms. In any case, the parameters of their method did not claim anything about the larger ontology that subsumes our universe, or that supposedly 'computes' our universe.

However, Rigel and Kovrizhi augments our case: Their argument constitutes an interesting stand-alone critique of any simplistic claim that the universe is some form of computer simulation. By itself, it challenges the notion that the Turing machine approach can constitute a viable basis for solving problems of causality and time: problems that are implicit to any simulation theory of the universe, as much as to our non-simulated a-mechanical gear-less really-existing universe.

Yet, our case against hidden gears and the computational universe has an edge over the approach taken by Ringel and Kovrizhi: The issue is *not* computational capacity, but whether computation *in principle* could ever generate the universe by bringing about root wavefunction collapse and time, resolve quantum particle interactions, much less handle the process of counterfactual grand decoherence of non-consistent nested futures to timelike infinity. The question is whether computation could realise the sorting and processing of the infinite degrees of freedom that attends particle interactions, in turn subsumed to the larger context involving the Whiteheadian grand decoherence of infinities to the infinitely removed future, and even less resolve the noetic apprehension of infinities by the renormalising quantum mechanist. Clearly, hidden gears and computation cannot furnish any of these processes: only a gear-less non-computational process could accomplish it. Therefore, wither materialism.

5.32: Demise of Materialism from the Dualistic Character of Physical Law

In the Preliminary and subsequently, we postulated about the abstract-immaterial character of physical law and its dualism vis-à-vis putative radiation, matter, and 'space'. Physical laws are no more 'material' than are football rules. But a proper physics-based argument for the immaterialism of physical law requires that we reprise the Einstein Podolski Rosen (EPR) experiment-based approach to physical

---

[98] Journal: Science Advances: Vol. 3, No. 9 Quantized Gravitational Responses, The Sign Problem, And Quantum Complexity (2017)

law first adumbrated in Preliminary **0.26**, and reiterated in various forms since. In the essay that immediately follows this one, we will also clarify Bell's inequality, which was inspired by EPR experiment, to be grasped in the light of the Alternative Realist insights into quantum indeterminacy, combined with de-spatialisation; both decisive to the immaterialism and the dualism of physical law.

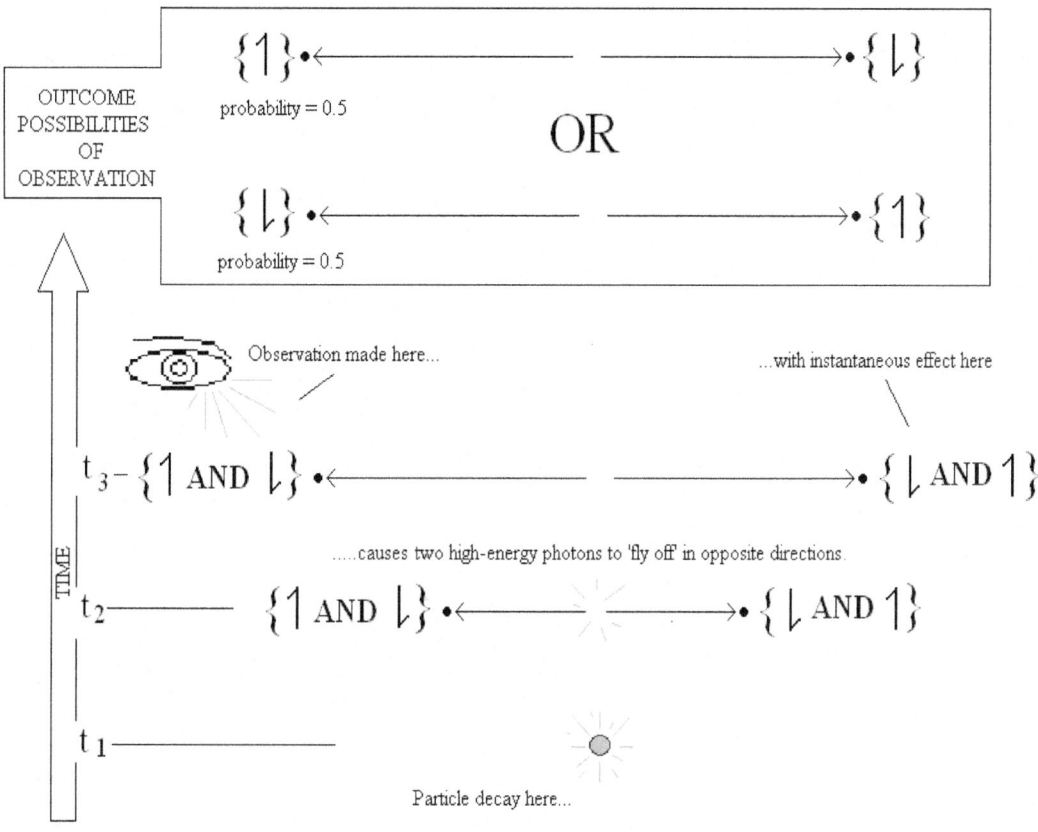

Fig. 508: Illustrating EPR

Notes: The illustration demonstrates 'spooky action at a distance' in EPR. From the bottom to the top illustration, following the time-arrow, at t_1, a particle decays into two component photons. By t_2, the two photons have travelled in opposite directions at the putative speed of light. The photon spin-states are not yet resolved and remain in unitary superpositions of spin-possibilities, with the left-photon in unitary state comprised of {"spin-up" AND "spin-down"} as indicated by the arrows in brackets. The right-photon assumes a conformal {"spin-down" AND "spin-up"}. By t_3, the photons have separated out, presumably by many millions of light years. An observation is made on the left-photon and the system is brought out of isolation, with the left-particle's superposition state {"spin-up" AND "spin-down"} undergoing AND-to-OR collapse. The possible outcomes are presented in the top illustration under OUTCOME POSSIBILITIES OF OBSERVATION. There is a probability of 0.5 that the left photon will resolve into either "spin-up" OR "spin-down". If "spin-up", an instantaneous 'spooky action at a distance' will ensure a "spin-down" outcome for the right-photon, in such a way that quantum spin is conserved. If we obtain "spin-down", this will instantaneous obtain "spin-up" on the right-side, and conservation of spin. The left result acts as a counterfactual; reducing the right to a correlate outcome so as to conserve spin. It does not matter how far apart the particles are. If we could control the outcome on the left, we could control the outcome on the right, and use this to send instantaneous messages across any distance. Yet, the spin-outcome on the left is always a matter of probability: It cannot be deterministically directed, and it cannot be exploited for instantaneous communication, notwithstanding superlative rigging espoused in Book-I.

With these goals in mind, we reprise EPR in **Fig. 5.08**. In the repeat-illustration, we find that, at t_1, an unstable particle subsequently disintegrates into two photons of light by t_2. The expectation is that, if the left-side photon is 'spin-up' the right-side photon must be 'spin-down', and vice-versa. Thus, spin will be conserved. However, according to quantum mechanics, the left-side photon is a superposition AND-form comprised of 'spin-up' AND 'spin-down'. The right-side photon will constitute the converse 'spin-down' AND 'spin-up'. Together, these form a quantum entangled couple: a single wavefunction that subsumes both spin-possibilities and correlates for both photons 'flying off' in opposing directions at the putative speed of light.

In EPR, the principle of the conservation of spin abides, in that one spin-possibility on the one particle of the entangled state is linked to its opposite spin on the other particle. But what happens to spin in a de-spatialised universe? How is the conservation of spin applicable as law in a universe where only time is real? Recall that, time is stereotemporal or 'three dimensional'. It is in the context of stereotemporal time that spin survives into de-spatialisation and temporisation, albeit in altered form; as spin recuperated in terms of delayed choice time-interval relations: A simple way to obtain this is to drop wavelength considerations and model quantum spin purely in terms of frequency…and then drop space itself.

In reference to **Fig. 5.08**, in the interim between t_2 and t_3, the quantum-entangled system is in physical-experimental isolation. One can envision this as a 'bubble' of pre-observational future Whiteheadian possibilities and potentialities 'expanding across space', on the false assumption of space. Given de-spatialisation, the 'bubble', as an isolation state, will temporally 'expand', until a left-side observation helps end the isolation at t_3. With the end of isolation and the 'bursting of the bubble', the left-side spin might resolve into either the OR-form of 'spin up' or the OR-form of 'spin down'. If the left-side resolves into OR-form 'spin-down', the right-side must instantaneously resolve into the opposite OR-form 'spin-up', even if the right-side photon is a million parsecs away: i.e., a 'distance' that would take about 3.26 million years for light to cross, on the assumption of pleonastic 'space' and the 'speed of light'. These spatial terms must be entirely succeeded by temporal terms; recuperated into delayed choice time-interval relata and the *delayed choice time-interval standard* in lieu of the 'speed of light': See Book-IV **4.08**.

Conversely, if the left-side spin is 'spin-up', the right side must instantaneously resolve into opposite 'spin-down'. On the assumption of 'space', the outcome is tantamount to 'non-locality', or 'spooky action at a distance'. With de-spatialisation, non-locality is succeeded by *transtemporality*. That is, the instantaneous effect is not a space-circumventing relation. Instead, it is time-circumventing instantaneous effect; one consistent with the same form of instantaneous effect from counterfactuals-driven erasure of non-consistent nested futures evinced by grand decoherence: (Book-II, **2.34** and **2.35**).

On the left side, there is a 50-50 chance, or 0.5 probability, of obtaining the OR-form of spin-up or spin-down. Given the operation of ineliminable objective quantum indeterminacy, there exists no method of pre-orchestration of spin outcomes in EPR. Otherwise, we could control the outcome on the rights side instantaneously and, devising a relevant code system, then use pre-orchestration to communicate information across vast 'distances' *instantaneously*. Hence, the consensus that it is not possible to communicate meaningful information using EPR 'action at a distance' is correct, notwithstanding superlative rigging and the attendant 'ansible' posited in Book-I **1.47** to **1.50**, neither of which entail literal transaction of actual information via EPR

However, **the conservation of spin, as a piece of information in its own right, is *not* subject to scrambling by objective quantum indeterminacy. The principle of conservation of spin transtemporally 'burns through' quantum indeterminacy and enforces itself on the spin-outcomes**. Thus, the conservation of quantum spin, as law, is enforced across 'parsecs', despite objective quantum indeterminacy. Stated in equivalent de-spatialised temporised terms, the conservation of spin is enforced *transtemporally*, circumventing delayed choice time-interval relations and the delayed choice time interval standard in lieu of the 'speed-of-light limit': i.e., there is no time-delay in the enforcement of the conservation of spin, even if the expected delayed choice time-interval relation might equal 3.26 million years.

The fact that the conservation of spin can 'burn through' in the said transtemporal way (without taking time, and totally circumventing time) constitutes proof that, in a significant sense, real and meaningful information *is* getting through in EPR-relations, despite the scrambling effects from objective quantum indeterminacy. We must arrive at such a conclusion, simply because the conservation of spin is itself a meaningful piece of information: Indeed, the conservation of spin, as a piece of information, is more fundamental and architectonic (i.e., read-only) vis-à-vis merely contingent spin-outcomes, given that it subsumes merely contingent spin-outcomes, supersedes these, and 'burns through' them. Therein resides the basis for the transtemporality and consequent Platonic-Cartesian dualism of physical law vis-à-vis radiation, 'matter', and time.

*

The conservation of spin is *not* communicated across 'space', and certainly not by means of any carrying radiation supposedly instantaneously displacing across space: 'space' does not exist. Only time exists. On the other hand, **insofar as the principle of conservation of spin *is* being 'communicated' transtemporally, and hence in no time at all, the principle of the conservation of spin is *not* being 'communicated' in or across time. It necessarily follows that the principle of the conservation of spin transcends time, i.e., it is transtemporal, and hence Platonic, in the same strict sense that the putative forms of Plato transcend time and are not fated to time.**

In EPR, the principle of the conservation of spin is being enforced on the photons from outside of 'space' and from outside of memory, possibility and time. The principle is dualistic and irreducible with respect to 'space' and with respect to memory, possibility, and time constitutive of the growing block Bergson-Whitehead amalgam. Indeed, from Book-III **3.33**, physical law imbues the totality of the Whitehead facet and emanates from the infinitely removed future. As for the radiation and matter that assumes the stipulations of the principle, the principle is neither 'located in' such material nor manifested as the 'particles'. Rather, and to use the similar terms to those of Plato, **the 'particles' *participate* in the Platonic form of the conservation of spin, which is being enforced on the durata from an extra-ontology**; one irreducible to and transcending 'space', time, 'matter' or, succinctly, transcending memory, possibility, time, if not nous itself. It necessarily follows that the principle of conservation of spin can subsist intact 'outside' of time and independently vis-à-vis the moments of its resolution and enforcement on memory, possibility, time, and independent of nous.

*

The implications to materialism are obvious. **The ontological claims of materialism are undermined by the transtemporality thesis on the conservation of spin, and by its generalisation to all physical law**. Indeed, any closed-causality and closed-existence thesis (succinctly, materialism) must assert that all that exists in nature must be rendered as and produced by the closed interactions of 'matter in space' and time...and that these must constitute the exhaustive variables of causality, with no extra ontology.

The transtemporal operation of the conservation of spin from outside of time, in relation to which the various states of putative 'matter' must remain the subordinated contingents, and through which the conservation of spin 'burns through' without being scrambled by objective quantum indeterminacy, clearly demonstrates that there is far more to real ontology than hitherto imagined by materialism. It is also obvious that the conservation of spin cannot be contrived as an epiphenomenon of 'matter', much less an emergent property of it, precisely because, aside its form as an instantaneous whole (Book-I **1.03**, and **1.26** to **1.27**) the conservation of spin can 'burn through', and is thus enforced upon and across quantum indeterminacy, *and* survives mere contingency.

The transtemporal character of the conservation of spin in EPR, generalised to all laws of nature, usurps materialist ontology. Often, materialism rightly asserts the primacy of natural laws over 'superstitious' notions. What materialist ontology missed is that natural laws are transtemporal... hence, *metaphysical*. The transtemporality and 'burn through' of the conservation of spin in EPR constitutes the crucial case for natural metaphysics; anathema to materialism.

5.33: EPR, Bell's Inequality & the Dualism of Physical Law

Bell's inequality emerged out of the attempt to resolve the implications of the Einstein Podolsky Rosen experiment (EPR)[99]. Einstein rejected the idea that there could be 'spooky action at a distance' or superluminal instantaneous transactions between particles in EPR. Einstein also claimed that all the attributes of the particles must possess 'objectivity', or, in our terms, these must always remain rendered in OR-form resolved profiles of 'position', spin, etc, even in the absence of observation. Hence, the naïve realist interpretation.

The truth or otherwise of EPR-based claims can be tested statistically via Bell's inequalities. If Einstein and naïve realism is correct, and if 'spooky action at a distance' is wrong, Bell's inequality, based on generic quantum mechanics, predicts one set of statistical results obtained from a number of EPR experiments and outcomes. But, Bells inequality predicts a different and *unequal* set of statistical results if 'spooky action at a distance' is correct, which would supposedly imply that the states involved are *not* objective i.e., cannot remain in OR-form before or in the absence of measurement or wavefunction collapse.

In short, Bells contention was that the statistical outcomes obtained from one set of assumptions will be different from and *unequal* to those obtained from the other set of assumptions, and that there is a way to secure the predicted statistical outcomes by carrying out a series of EPR experiments using photons.

It is generally agreed that the overall statistical outcomes obtained from experiments that apply Bells inequality favour 'non-locality' and superluminality: i.e., instantaneous 'communication' between entangled pairs of photons in EPR. It appears that Einstein's naïve realist expectation for hidden variables and non-superluminal effects turn out to be forlorn.

From our own Alternative Realist approach, the notion of hidden variables or gears was challenged by the oft-reprised objectivity of quantum indeterminism, combined with the three-phase developmental approach to the same; in turn combined to the impossibility of hidden gears behind wavefunction collapse, time and in attendant generation of quantum indeterminate outcomes (see Book-III **3.26** and **3.32**, and appropriate summarisation and applications against 'consciousness causes collapse' in Part-I **5.08**, and Part-II **5.16**). In short, clearly, Alt-R provided additional independent material for the case against hidden variables.

Moreover, Alt-R showed that the ontic reality of AND-form logic, and the relevance of the wavefunction, was compelled and justified by the demands of causality intertwined with the principle of the conservation of energy: the *firewall principle*… unless we seek to entertain 'signals from the future' in contravention of causality and the conservation principle. Of course, the firewall principle is inherent to the growing block universe, and the consequent Bergson-Whitehead amalgam is central to Alt-R and, implicitly, supports Bell's inequalities, even if from an unexpected direction that transcends the assumption-set of Bell's inequality.

*

Bell's inequality can favour only one of two possibilities. On the one hand, its statistical structure might have confirmed Einstein's notion of 'objectivity and locality' as he understood it: to the effect that entities are always in 'objective' OR-form state, both before and in the absence of observation. On these grounds, 'objectivity' and locality would preclude superluminal 'faster than light' or instantaneous effects typical of EPR. On the other hand, Bell's inequality has confirmed the opposite conclusion: i.e., the reality of 'non-objectivity' and non-locality (superluminality), to the extent that AND-form reality dominates in the absence of observational resolution, or in the interim of isolation, and instantaneous effects are certainly permitted in nature.

From the framework of Alt-R, we affirm the reality of 'non-objectivity' and non-locality against Einstein's naïve realism, even if for very different reasons that transcend reliance on statistical analysis of EPR outcomes. In short, what Bells' inequality shows and what Alt-R posits are compatible, although Alt-R possesses an explanatory structure for the outcomes garnered.

---

[99] Bell, J. S. (1964)."On the Einstein Podolsky Rosen Paradox". *Physics Physique Физика*. 1 (3): 195–200.

527

Recall that, according to Alt-R, a system placed in isolation and deferred to the future, such as spin in the EPR relation, must remain in its non-resolved AND-form state until it gets to decay into a realised OR-form event. As part of the Whitehead facet of the superlative Bergson-Whitehead amalgam, the system thus deferred cannot possess OR-form 'objective' characteristics in the fashion naive realism expects... until it is subsequently resolved into an OR-form event.

Moreover, according to Alt-R, although not recognised in Bell's inequality or in generic quantum science, or its various interpretations, a system cannot preclude the presence of abstract memory from the Bergson facet of the Bergson-Whitehead amalgam. For example, all things being equal, the total history of the Moon *is* retained within the Bergson facet as a set of abstract wavefunctions in memory. Consequently, to assert that the Moon is purely a 'non-objective' future-potential AND-form state before any observation of it has transpired, cannot be entirely true. This is because we must always include in any description of the Moon (and of any system) it's Bergsonian memory or past, in tandem with its Whiteheadian futurity and the grand decoherence to consistent futures that ties the future to unity with the pasty. In short, the ideal description of any system is the growing block description of that system.

It must always be kept in mind that the moment of any measurement (e.g., the resolution of spin as an event on one of the entangled pairs in EPR) *will not occur* at the same exact time as the moment of the consciousness of it. There is always a delayed choice time interval between the spin-event of the particle versus the subsequent a-noetic detection of it, or our subsequent conscious witness of it. In the same way, **we are always restricted to a delayed choice time-interval between the resolved corpus of events constituting the Moon 1.5 seconds ago (ultimately retained as a set of complex wavefunctions in memory within the Bergson facet) versus the moment of our conscious observation of the Moon *after* 1.5 seconds. In short, there *is* an objective Moon 'out there', but it *is* retained in abstract form within the Bergson facet before our retrospective conscious observation of it 1.5 seconds hence... at least within the stricture of the said 1.5 second delayed choice time-interval. Of course, the Moon's futurity beyond the 1.5 seconds worth of delayed choice time-interval *is* 'non-objective', insofar as it is part of the as-yet not-resolved future potentiality belonging to the Whitehead facet; beyond the 1.5 second demarcation into the Whitehead facet and to the infinite future**.

The point is simply this: Bell's inequality-based claims against 'objectivity' rests on partial error: 'Non-objectivity' is valid beyond the said 1.5 second delayed choice time interval vis-à-vis the Moon, or whatever delayed choice time-interval is pertinent to any given relation. On the other hand, 'objectivity' *is* valid for any delayed choice time-interval up to, but not beyond, the said 1.5 seconds, or within a relation-appropriate delayed choice, per function of Bergsonian memory, and not because of any naïve realist expectation that all things must remain in pure 'objective' OR-form, even in the absence of observation.

To reiterate, **only the future beyond the 1.5 second mark (the as-yet not-realised Whitehead states of the Moon) must remain 'non-objective', or purely in AND-form, *with* nested futures. However, such 'non-objectivity' does not imply 'un-realism', and cannot licence for illusion theory or quantum idealism, given that the Whiteheadian description of any system (hence the AND-form futurity of that system) expresses an ontic or really-existing future state of possibilities: The state of possibilities is 'non-objective' *only* in the sense that the futures are as-yet not-realised potentialities, *not* because they are 'illusions', irrealities, or fantasies concocted by nous.**

*

Bells' inequality applied to EPR appears to assert against hidden variables and in favour of non-locality or superluminality (or transtemporality). Alt-R is in accord with this, but only within the above-stated delimits. Also, 'space' does not exist as a physically valid concept. Therefore, we can only assert for the reality of delayed choice time-interval relations between events, including those that attend EPR. Only time is real. It follows that non-locality needs to be replaced with *transtemporality*.

If we must seek a basis for the instantaneous effect of one particle upon another... if we must seek a basis for 'superluminality'... then we must recuperate this as a *transtemporal* relation between the said particles. As such, **transtemporality cannot involve a 'communication' between the pair of entangled particles** in the fashion of a 'signal' supposedly mediated by 'moving' radiation or 'stuff'. **Nor can this instantaneous effect be characterised as an 'action' transferred from one particle to another via contiguity 'across space' or even across time**.

Per de-spatialisation and the replacement of 'non-locality' with transtemporality, the results of Bell's inequalities are consistent with the reality of the transtemporal enforcement of information and physical law, specifically the conservation of spin, from 'outside' of time. **Hence, Bell's inequality is compatible with the transtemporal enforcement of Platonic Cartesian physical law (see 5.32) dualistically related to and irreducible to time and to the resolved events emerging out of wavefunction collapse and time, including those that attend EPR.** Again, the dualism of physical law necessarily implies an ontology that supersedes the stipulations of materialism... to the demise of materialism.

5.34: The Demise of Materialism from Physical Law, Causality & the Open Universe

Causality that pertains to the enforcement of physical law constitutes the same general ineffable causality operative behind wavefunction collapse and time. The difference is that physical law makes specific impositions. In doing so, it delimits permissible possibilities vis-à-vis the outcomes garnered from wavefunction collapse and time. That is, time and the ineffable causality behind it generates myriad 'read-write' actua out of Whiteheadian future possibility, while physical law structures and delimits the said actua on a 'read-only' basis: (See Book-I **1.05** for the distinction between read-only and read-write notions and classifications).

Of course, Bergsonian memory also plays a delimiting part. It does so via the grand decoherence of the total Whitehead facet into consistent futures, insofar as what must emerge out of total wavefunction collapse and time must be consistent with the past (a minimum requirement) and subject to gravitational and inertial considerations that emerge per function of the Bergsonian past: (See Book-IV Part-III for mnemonic mechanics).

The exemplar read-only physical law is the principle of conservation of quantum spin in EPR: a 'read-only' informational state that restricts and delimits spin-outcomes in EPR to those consistent with the conservation of spin; enforced from outside of the Bergson ,Whitehead amalgam (see emanation theory in Book-III **3.33**). As such, the abstraction belongs to a grander ineffable and immaterial ontology extant the growing block Bergson-Whitehead amalgam comprised of memory, possibility, and time: (See **5.33**).

Now, materialism asserts that the universe is causally and ontologically closed: There can be no causal factors that operate from outside of the universe of memory, possibility, and time. Only 'matter in space' is real, and it does not require extraneous factors and, presumably, it can generate and explain all phenomena. This claim is often justified by a subtle abuse of the conservation principle: Thus, if all causality is material, and if causality is to be specified *purely* in terms of material and radiational inputs, insofar as the sum of inputted energy and matter is conserved, so must causality, now rendered synonymous with the inputs, the outputs, and the operations that mediate the former into the latter: i.e. IOO-schema.

However, this central materialist postulate is fallacious and circular: As we discovered previously, causality in general as it pertains to AND-to-OR wavefunction collapse, time *and* grand decoherence, and the causal process in the generation of attendant quantum indeterminate outcomes, cannot be reduced to or identified with any gears or *any* definable operation, much less reduce to any preceding material or radiational input. Hence, causality is a *something* (a structure-less and undefinable something) that operates from outside of matter, radiation, and 'space'. This fact was rendered even more salient by the transtemporal enforcement of physical law on information and time (see **5.32**, and Book-III **3.33**).

Thus, **given the extant nature of structure-less causality, and given the transtemporality and dualistic nature of physical law, both of which imply the real-existence of a grander ontology that embeds the Bergson-Whitehead amalgam of memory, possibility, and time... it follows that the universe encapsulated as the growing block model *is* ontically open. Even so, the principles of conservation of putative (and pleonastic) energy-matter abide, albeit in altered abstract form (5.29) with full due consideration to the pleonastic status of energy concepts and relations first adumbrated in Book-II, 2.39.**

*

In Book-IV **4.21** we developed the conjecture that there may exist matter-creation processes that draw out putative future-potential 'matter' out of the Whitehead facet... out of the future. Of course, with de-spatialisation, matter is ephemeralised into pure temporal *durata*. Even so, the conjecture about 'future-potential matter' locked within the Whitehead facet remains intact, albeit transformed into prospective stable future-potential durata. **The matter thus created could not constitute the violation of the conservation principle, both in the sense that the consistency imposed on the Whitehead facet via grand decoherence renders the Whitehead future fully consistent with Bergsonian past-in-memory, and from the fact that matter-creation cannot engender inconsistent futures, nor alter or destroy Bergsonian memory, *and* also because the matter thus created pre-existed as future-potential matter within the Whitehead facet, so circumventing notional *ex nihilo* matter-creation**... solving the problems of where the matter would come from to constitute, not just existing matter since and after the putative Big Bang, but possibly in the context of plausible Little Manyworlds (see   Book-IV **4.20**).

However, like all other laws, the principle of conservation of energy-matter, recuperated into its abstract form in **5.29**, must itself be enforced on the domain of future potentiality and time in order to affect memory, possibility and time, and render matter-creation consistent. The source and origin of the principle of conservation is obviously *not* going to reside in existing realised 'matter in space' confined to the event horizon 'skin'. **The said enforcement of the principle of conservation, or its recuperation as the conservation of memory cum consistent futures, and of any other law, must be imposed from 'outside' of memory, possibility, and time... from outside of the Bergson-Whitehead amalgam... through the process of emanation (see Book-III 3.33)... with abstract physical laws originating from and residing within an extra ontology; usurping closed causality and fashioning for open ontology... to the demise of materialism.**

*

The position against Cartesian mind-brain dualism historically rested on an abuse of the principle of conservation, combined to attendant fallacious materialist characterisations of causality usurped throughout this work. We can now appreciate that we live in an open universe in which causality in terms of the drive of time, or per abstract physical law, operates from outside of memory, possibility, and time, despite the fact that the conservation principle abides in both the pragmatic old form and in its recuperated new form. Clearly, nous is not going to be exempt from implications from open causality and open ontology. As with causality in general and Platonic-Cartesian physical laws in particular, nous 'acts' on memory and possibility from outside of memory, and possibility... utilising work function zero processes and the ultimate open state of growing block memory and possibility, without erasing the conserved past in abstract memory, and without incurring inconsistent futures.

5.35: The Demise of Materialism from the Pleonastic Status of Energy-Matter Relations
In the Context of De-Spatialisation, Temporisation & the Bergson-Whitehead Amalgam

Convention and materialism treats energy or work as if identified with and carried and mediated by point-like or spatial block particles. It further assumes that energy is located and distributed 'in space', as are its purported carrying particles, with energy mediated from one set of particles and 'spaceballs' to others through contiguity and contact across space. Yet, aside the pleonastic status of energy

529

concepts and relations first adumbrated in Book-II **2.39**, de-spatialisation renders all such notions void. De-spatialisation, if not the Bergson-Whitehead amalgam bourn of it, forces the reformation of putative energy concepts in terms of pure time relations, *without* space… without locations or contiguity between supposed energy-bearing spatial entities. **With de-spatialisation, energy itself can no longer be located in space, and we must relate energy concepts and processes to the de-spatialised and temporised tripartite structure of the Bergson-Whitehead amalgam**.

De-spatialisation aside, putative energy relations and exchanges must capitulate to the ontological primacy of time and wavefunction collapse. This is because putative energy relations are manifested *through* time and wavefunction collapse. Thus, time needs to happen in order for putative energy relations to manifest. But work function zero time and wavefunction collapse do *not* manifest through a process that requires energy or power or work-inputs. To reiterate, **the ontological primacy of time as a work function zero non-energy driven process ultimately implies the abstraction of energy concepts and relations out of the real ontology of physics: wherein energy concepts are, at best, pragmatic, but no longer fundamental to the real ontology of the physical order: thus rendered pleonastic.**

*

Recall from Book-IV **4.19** that, in generic physics, there exist two categories of energy: namely, *kinetic energy* and *potential energy*. Recall that these categories survived into the Bergson-Whitehead amalgam, albeit in altered form. Convention holds that potential energy is the stored or latent potential capacity of a system to do work on some other system; likened to a 'battery' of stored energy or power. On the other hand, kinetic energy is the actualised work done, or work in the process of being done, per the conversion of potential energy into the said work... made possible in and through ontologically primary AND-to-OR wavefunction collapse and time.

In the relation between energy and matter exemplified by $E=mc^2$, energy is inter-convertible with and into matter, while the sum of energy-matter remains constant or conserved: this being the conservation of energy-matter in its generic modern form. Thus, potential energy and matter are essentially synonymous. Energy **E** could be the kinetic energy, or the energy actually doing the work, released from its congealed matter-state, mc^2.

We will now reprise conclusions from Book-IV **4.19**: We will 're-locate' potential energy to the Whitehead facet of our superlative Bergson-Whitehead amalgam. Consequently, potential energy is *not* in 'matter' as such. The want to locate potential energy within or as matter presupposes 'space' and 'matter in space'. Succinctly, it presupposes that energy is stored in and issues out of some 'location' or located 'matter'. **Given de-spatialisation and temporisation from Book-IV 4.01 to 4.07, there can be no 'location' in which purported potential energy is stored**. Instead, *apparent* potential energy is necessarily a function of time and wavefunction collapse.

We must recuperate both the **m** and the c^2 in mc^2 in terms of de-spatialisation and the Bergson-Whitehead amalgam. This entails that we relegate potential energy to the future-potentiality of the Whiteheadian facet. Also, we must recall that the 'speed of light' transforms into a de-spatialised pure temporal expression: the *delayed choice time interval standard*: (see Book-IV **4.08**), It is necessarily coupled to the progression of the event horizon. Consequently, kinetic energy, as converted Whiteheadian future-potential energy, seemingly issues out of the Whitehead facet through wavefunction collapse and time attendant the progression of the event horizon, within the limits of the delayed choice time-interval standard, which in turn regulates the rate of progression of the event horizon.

Where is **m**? This is certainly not in 'space', given de-spatialisation. The 'mass' is certainly involved in delimiting the rate at which the event horizon progresses. In inertia and gravitation, the pertinent mass is a function of the Bergson facet (see Book-IV Part-III **4.27** to **4.30** for nascent mnemonic mechanics and mass and gravity as a memory effect). In this form, **m** is extant any OR-form matter on the event horizon, given that it is conveyed by the Bergson facet dissociated 'below' the event horizon to which OR-form 'matter' is restricted. As for that aspect of **m** involved in convertibility to kinetic energy, as potential energy, it must be an aspect of the Whitehead facet... of the future… a future-potential-**m**, so to speak.

Potential energy is the possibility and potentiality for work 'stored' within the future-form Whitehead facet; *stored within the future.* In short, potential energy, as future-possible work, *appears to be* the future possibility for 'doable work'.

Kinetic energy *appears to be* **the narrow range of future-form potential energy issuing out of the immanent Whitehead potentiality**… from the interval between the last realised event horizon (realised EH-1) and the future-potential event horizon (potential EH-2): a time-interval demarcated by the said event horizons. As such, and as a pure time-relation, it is necessarily identifiable as 'frequency'. Therefore, it is not a coincidence that a key measure of energy in physics is the frequency of radiation and de Broglie matter, co-joined to the Plank constant.

Thus, the frequency is the potential time-interval that resides between successive event horizons. Frequency in action, through the synonymous transition of the event horizon from EH-1 to EH-2, *apparently* constitutes the conversion of immanent Whitehead potential energy into work done... into kinetic energy.

The rest of potential energy exhibited within the Whitehead facet *appears* **to reside beyond the immanent possibilities** (beyond potential EH-2) and remains inherent to the complex of nested futures that reside beyond the immanent future: where 'future-potential matter' also resides, on the basis of which possible matter-creation might enter the fray: (Book-IV **4.21**).

From the above, kinetic energy is not an attribute of the mnemonic Bergson facet of the Bergson-Whitehead amalgam. That is, kinetic energy does not emerge out of the past-in-memory: Kinetic energy is *not* a memory effect, given that kinetic energy is restricted to the discontinuous transition of the event horizon from EH-1 to EH-2. Thus, unlike the Whitehead facet which constitutes the *apparent* store of potential energy, **the Bergson facet of the amalgam should never be treated as any sort of 'store of energy'**. Moreover, Bergsonian mnemonic processes are critically involved in inertia, gravitation, and other things. Inertia and gravitation (hence mnemonic Bergsonian processes) proceed and affect change on a pure work function zero basis, as was argued in Book-IV **4.31**.

530

So far, we have nothing here that appears to over-upset materialism, notwithstanding speculative matter-creation out of future potential matter; notwithstanding mnemonic gravitation and inertia that proceed on a work function zero basis. Indeed, all we have done is de-spatialised energy; with potential energy now inherent to the Whiteheadian future, and kinetic energy now inherent to the immanent EH-1 to EH-2 AND-to-OR collapse of future-potential energy into OR-form work. Presumably, we can now declare... 'everything is energy in time', in lieu of 'everything is matter in space'. In this way, we might even appear to preserve some modicum of materialism.

*

We now come to our most radical assertion inimical to materialism, first articulated in Book-IV **4.19: Ultimately and ontologically, there are no energy relations in nature, potential or kinetic. Energy concepts are pleonastic**.

Recall that time is not a 'fourth dimension' but it is synonymous with wavefunction collapse and attendant event horizon progression. Also, recall from Book-III and from recapitulations in the present book, that neither quantum indeterminate processes nor the very process of wavefunction collapse (nor time itself) are driven by *any* form of hidden inputs, variables, or *gears*. This clearly implies that there is no 'kick', 'applied power', or *any* definable operation or 'gears, that makes time happen, or brings about wavefunction collapse and the attendant issuance of 'energy relations'.

To reiterate *ad nauseam*, wavefunction collapse and time derive from an ultimate *gear-less* and structure-less process of causality: one that does not require 'energy' or 'work' to initiate it. Indeed, unless AND-to-OR collapse and time transpires, *apparent* kinetic energy cannot manifest. That is, time has ontological primacy over what it manifest; namely, kinetic energy. Hence, kinetic energy is *not* required to bring about time itself; no more than it is required to bring about synonymous wavefunction collapse and tandem transition of the event horizon. Instead, actuated *putative* kinetic energy *presupposes* gear-less work-function zero AND-to-OR collapse and time, without which it could never become' apparent'... without which the succession of events and state-change could not transpire, and upon which we could no longer foist pragmatic fictions of energy relations.

It is not energy or work that drives time, but work function zero time that manifests putative energy, work and kinetics. In the ultimate sense, *there is no energy* or work as such. There is only a non-energy driven and actuated time, with attendant succession of events and state-change... which we then interpret as 'kinetic energy', 'work done', or as actuated 'energy relations'. The interpretation, or *foisting*, is ultimately *not* required, no matter how pragmatic and useful it might otherwise be.

Of course, what is permitted to happen at any successive event horizon must remain consistent with what happened in the past, and thus consistent with the mnemonic content of the Bergson facet, fashioning for the alternative abstract form of the conservation principle, recuperated as the conservation of memory co-joined to the enforcement of consistent futures, *without* 'energy' concepts and relations, *without* 'space'... *without* materialist trope or cliché: (see 5.29 for the recuperation of the principle of conservation).

While frequency is often made synonymous with energy, **in a universe in which energy concepts and relations are pleonastic and not part of the real ontology of nature, frequency transforms into a pure *durata* of time-measure or time-interval in and through which new events are created out of Whiteheadian potentiality, in what is ultimately a pure information physics**.

In short, ultimately, there is no 'energy' in nature, although the concept remains pragmatically useful. Consequently, any modicum of materialism that might have survived in the notion that 'all is energy in time!' turns out to be as forlorn and as the cliché, as 'all is matter in space!'. Instead, all, or *almost* all, is abstract memory, possibility, and time... *without* space, and *without* motion. Hence, *without* 'matter'. Wither materialism.

5.36: The De-Spatialisation & Dematerialisation of Mass
And Overcoming the Central Problem of Quantum Gravity

The reform and pleonastics of energy in the light of the character of AND-to-OR collapse and time, and from the structure of the Bergson-Whitehead amalgam, and principally per de-spatialisation and temporisation, must segue into the reform of 'mass' itself.

The spatial problem of mass first emerged in the 19th century: Therein, as the radius of a putative particle attains zero or point-like form, its mass ought to tend to infinity. Thus, point-particles with mass, such as the electron, would be expected to have infinite mass and inertia. In such a case, we could not change the motion of an immovable infinite-mass particle by the application of *any* force or force-bearing particle via contiguous contact. One should not assume that this obscure problem of spatial mass has gone away with the advent of twentieth century physics or quantum mechanics. The problem was swept under the rug via an early form of renormalisation. This subtracted the infinity from the infinite mass, but failed to address the ontology of the problem of point-particle mass and its expected tendency to infinity, inherent from the notion of 'space', and from particles envisioned as 'spatial points'.

There are two alternative ways to approach the problem of spatial mass: Renormalisation is the first way, although it only solves the attendant problem of calculation. The other is, simply, **we must question the very ontology of space and the notion that the mass of a particle is contained in or at a 'location', or in a spatial block, or constituted in a spatial point. Indeed, with de-spatialisation, it becomes evident that mass must be per function of some other factor**.

With one exception, mass is largely per function of the distribution of information in time: a memory effect from the superpose or sum of wavefunctions in memory from within the Bergson facet (Book-IV Part-III), combined to the attendant grand decoherence of the Whitehead facet that renders futures consistent to implications of Bergsonian memory. The subsequent AND-to-OR collapse of the particle, or of a body made up of such, then manifests inertia or resistance, gravitational attraction, and gravitational acceleration.

*

531

The fact that the problem of mass is an artefact of the assumption of space cannot be sufficiently avowed: It is rendered painfully manifest by the central problem of quantum gravity: i.e., the hitherto inability to unify gravity with quantum mechanics: If we wish to model the gravitational field or the 'spacetime curve' generated by a particle or a body, we must do so, or hope to do so, quantum mechanically. Unfortunately, the generic wavefunction of a particle describes it as a superposition of manifold future-possible *putative* positions: some more probable than others. Generically, and on the over-simplified assumption of a minimum of just two possible putative positions, the particle in AND-form superposition will configure for both potential positions. The putative spacetime curvature purportedly generated by the particle would now be generated at *both* potential positions. But this will not do...especially when a more complex wavefunction might constitute a possibility distribution of manifold, or even infinite number of future-potential 'particle' positions.

One solution is to grasp the ontology that the wavefunction represents: The wavefunction is the mathematical expression of the ontologically real future. Hence, AND-to-OR wavefunction collapse is synonymous with time itself: now no longer a 'fourth dimension'. Consequently, **the future-possible 'positions' of the particle have not yet transpired; have not yet come into realisation. Therefore, there is no gravity or 'spacetime curve' associated with such not-yet realised positions, simply because these have not transpired to generate any gravity curve. i.e.,** an event that has not transpired cannot generate any gravity, obviously.

This solution is only partial: It cannot explain where the gravitational mass is 'located', and it cannot address how it subsequently comes about. The other complementary solution is thus: **There is no space. Therefore, there *are* no 'positions' as such. Therefore, gravity (or the 'spacetime curvature') cannot be said to emanate from a 'position', even a future-potential set of such... no more than can gravitationally mass possess a 'location'. Instead, gravity is a function of a pure de-spatialises temporised superposition of wavefunctions in memory attained from within the Bergson facet: Mass is a memory effect of the Bergson facet. (see Book-IV 4.27 to 4.31).**

Via grand decoherence, the superpose of memories imposes a tilt on any immanent future wavefunction. It also imposes counterfactual erasure of non-consistent Whiteheadian nested futures that attend the futures of the particle: one that entrenches the said tilt and imposes its counterfactual implications on all subsequent future wavefunctions embedded and nested within the total Whitehead facet, to the infinitely removed future.

Where is the 'spacetime curve' in all of this? **Given de-spatialisation, there is no 'space curve' as such.** On the other hand, given gravitational acceleration due to the said memory effect, we will inevitably obtain time dilation effects due to that 'acceleration': i.e., due to the non-uniform transformation of attendant delayed choice time-relations synonymous with 'acceleration', all due to gravity as a memory effect. Thus, **we obtain a 'time curve' as a differential of time dilations. This could be express as a series of 'curvy' event horizons**: (see Book-II **2.44** to **2.45**).

Also note that, in the above scheme, **mass cannot tend to infinity, simply because the superpose of wavefunctions in memory within the Bergson facet (the real face of 'mass') cannot generate infinity. In other words, infinite point-particle mass aside, mass- singularities and 'black holes' cannot form in nature, notwithstanding real-existing supermassive objects in the form of asymptotic 'black holes':** (see Book-IV **2.32**).

We have just summarised and recapitulated the basic schema for quantum gravity presaged in Book-IV **4.29** to **4.32**. Note that de-spatialisation is the critical ingredient to prospective quantum gravity. The central problem of quantum gravity, or the inability to furnish a unification of gravity with quantum mechanics, is rooted in the artefact of 'space' and 'position'. The problem disappears the moment we accept and introduce de-spatialisation. If we are serious about quantum gravity and unification, we must get rid of space.

*

With the demise of space, 'mass as a distribution in space' gives way to 'mass' as memory distributed in time within the mnemonic Bergson facet... no longer confined to a spatial position or point, Mass is simply the superpose of wavefunctions in memory accumulated within the Bergson facet. **Mass as total memory 'curves' the event horizon or, succinctly, it decoheres and tilts the Whitehead facet to conformity and consistency with that memory... from which 'gravitational acceleration' and 'orbital motion' arise. Therein resides the profound complement to the de-materialisation of physics: If we want quantum gravity, we must get rid of space. Get rid of space, and you get rid of materialism, in turn critically dependent on the notion of space for its viability. Hence, in order to advance physics into prospective quantum gravity, we must break the false equivocation of physicalism and naturalism with materialism**... by getting rid of space...and by de-materialising physics and naturalism.

5.37: Brief History of the Decay of Materialism from Within Natural Science

Let us recapitulate how things actually happen in nature: At the first level, when two 'particles' interact, we are effectively superposing their respective wavefunctions or quantum mechanical waves. Succinctly, we are in truth combining the abstract Whiteheadian futures of the putative particles. Thus, **'particle interaction' does not involve contact between 'particles in space'. Instead, it is a combine of de-spatialised futures**, forming a new combined future possibility-state, probability distribution, or superpose. On the false presumption of 'space', we could give the said combination a spatial and wavelength-based description. However, given de-spatialisation and temporisation, we need only furnish a pure time-based frequency model of the probability distributions for future possibilities, for each 'particle' before their combination *and* for the combined superpose formed from these. Given de-spatialisation, there can be no 'contact' between the particles: The **particles are certainly real, if only as *durata*. They are not spatial entities engaging in contiguous spaceballs-style 'contact'**.

Of course, from the input of 'approaching' particles through to their subsequent superpose, and finally through to the outcome realised per subsequent AND-to-OR collapse and time, we can furnish a frequency-based energy account of the whole. Yet, as was shown in **5.35** *and* in Book-II and Book-IV, **we could easily abstract energy and work out of the whole process and declare these pleonastic**. We could retain these foisted notions for pragmatic purposes, albeit proportionate to frequency. But frequency is ultimately a temporal attribute, a measure of the rate of events, or the potential rate for future events. Therefore, we can certainly abstract out energy concepts and treat the whole process of nature purely in terms of temporally distributed realised or potential rates of events.

Although we cannot avoid inputs recuperated into de-spatialised *durata* of potential event-rates (i.e., the frequencies of the particles as probabilities for events in time) and must also include the expectation that the outcomes from these remain consistent with the initial conditions (i.e., a principle of conservation recuperated as memory-conservation cum consistent futures) **ultimately, we need not foist energy concepts and energy-based ideas of conservation, and can replace the latter with the abstract conservation of memory co-joined to the principle of consistent futures**. That is, while we could interpret the whole process in terms of 'energy', we need not ultimately do so, and the want to do so is not a necessary or sufficient condition demanded by the really-existing universe. Indeed, we could treat the whole as a work function zero process by default, given the abstractability and pleonastic status of energy concepts vis-à-vis really-existing physics, and given the work function zero character of AND-to-OR collapse and time that facilitates the said inputs into the outputs, in full abidance with the principle of the conservation of memory co-joined to the principle of consistent futures.

Thus, at the second level, we come to the problem of the very cause of wavefunction collapse in relation to the superpose of possibilities formed by the 'interaction' of our two de-spatialised 'particles'. That is, we come to the question of the very causality behind time itself, synonymous with AND-to-OR wavefunction collapse. Recall that pragmatic 'energy relations' and particle interactions are brought into *apparent* realisation in and through ontologically primary work function zero wavefunction collapse and time, with the attendant conservation principle inferred from the outcome obtained in and through time. However, to reiterate, **wavefunction collapse and time does not require energy-power inputs to affect, much less require contiguous 'contact' causality to unfold. Instead, wavefunction collapse and time are brought about by an ineffable gear-less work function zero process. Thus, pragmatic energy relations, though not part of the real ontology of physics, issue forth per ontologically primary non-energy based, abstract, ephemeralised and *de-materialised* process of time. The ultimate process of causality, identified with time itself, or with whatever makes time transpire, is an abstract ephemeral dematerialised process**... without impact, without contact... most certainly *not* energy-power driven... and from which energy concepts and relations are wholly abstracted and rendered pleonastic.

Our ready materialist presuppositions and expectations (such as about 'space', or for the existence of 'particles in space', or espousing IOO-schema contiguous causality, or for causality as energy-force exchange) are rendered *entirely* pleonastic, no matter how useful or pragmatic such notions might otherwise be.

Cartesian mind, and the process of minded causality and control over the future development of the brain, must also proceed on a purely contact-less ephemeralised abstract work function zero basis… with no violations of the putative principle of conservation, much less of the conservation of memory, or of the related principle of consistent futures.

*

The decay of materialist notions about causality and of materialism itself was almost evident from the beginning. **Galilean physics from early to mid-Eighteenth century may have accorded well with the notion of contiguous contact causality and much else in materialist philosophy.** However, by the end of that century, in Newton, we obtained the first instance of contact-less causality in **gravitational 'action at a distance'**, if not in the less well-known but attendant **work function zero character of orbital and gravitational acceleration**, wherein 'force' only comes into evidence when we seek to alter the standing orbit or gravitational trajectory of a body (see Preliminary, Part-III). The pleonastics of 'force' was later adumbrated in Ernst Mach's abstraction of force from the system of Newtonian equations: constituting perhaps the first argument for pleonastic force or energy in physics[100].

By the Nineteenth century, we saw the advent of **the field approach in physics**, such as in early models of electromagnetism by Maxwell *et al*. This did not necessarily imply immaterialism. Yet, 'fields' (what are they?) surely hinted at ephemeralisation in physics, especially in matters of causal contiguity and contact. Indeed, what exactly is engaging in 'contact' in a field?

The decay of materialisms was rendered more pronounced by the discovery of the **early wave theory of matter**. Despite Newtons preference for a corpuscular or particle solution to 'matter', it was by no means certain that matter was corpuscular. Again, by the nineteenth century, the famous Young's experiment, the early prototype of the modern two-slit experiment, compelled evidence for the reality of a continuous wavelike form for matter. Of course, this envisioned matter as a mechanical wave, *not* as a quantum mechanical one. Yet, it created serious problems: How do we get mechanical waves to interact with each other? When two mechanical waves in a pool of water approach and coincide, they converge and pass through each other with impunity. It is as if no interaction had taken place. From a Victorian viewpoint, if matter is a wave, then the larger waves seen on the surface of a pool of water can only at best be made up of smaller wavelets. But these will also pass through each other with impunity, without apparent interaction. If so, how do we obtain contiguous interaction? Perhaps the larger waves are made of corpuscular objects. This would surely bring back contiguous contact at the lower scale. But why then do the waves pass through each other with impunity if the corpuscular view abided at the lower level?

---

[100] Ernst Mach, *Die Mechanik in ihrer Entwicklung, Historisch-kritisch dargestellt*, Akademie-Verlag, Berlin, 1988, section 2.7.

It ought now to be obvious that **causality was always a problem in physics. Causality never conformed to a contiguous impact or IOO-schema**, no matter how pragmatic the schema and attendant conservation principle, the latter of which need not imply contiguous contact causality: an *a priori* false association.

Hence, our Alternative Realist thesis on physics is rooted in long-standing problems that, if we should admit to their history, serves to expose the generic and historic confidence in materialism to be suspect at best. The same **history undercuts the notion that materialism must be the best possible hypothesis in the philosophy of physics** (i.e., *a priori* materialism) or that physics and naturalism are interchangeable with materialism, or that materialism is science… and science, materialism. Such cliches are untenable.

*

One might be forgiven for thinking that modern physics in the Twentieth century had solved these problems, especially with the advent of quantum mechanics. **What quantum mechanics mostly accomplished was to replace mechanical waves with quantum mechanical possibility and probability waves. The same form of problems remained, but with the adage that the 'interaction' is via the collapse of the converging wavefunctions, and converging or superposed waves must now AND-to-OR collapse into discrete events per perennial wave-particle duality. The discrete outcomes appeared to support the dubious impression of a world made up of spatial 'particles'**: a belief made worse by a whole specialism called 'particle physics': the specialism in most need of de-spatialisation, temporisation and recuperation into pure frequency and *durata*.

Surely, 'particles' constitute proof that the world *is* made up of point-like states and other 'spaceballs', and that these 'collide' and produce *all* the phenomena we know of, including the 'illusion' of consciousness. Depictions of particle interactions in Feynman diagrams and in old particle gas trails seen in pertinent devices certainly reinforced this prejudgement… until we took a closer look at the **implications from renormalisation**.

Interestingly, infinities pre-existed quantum mechanics and quantum field theory: Recall that, in classical physics, in the 19th Century, it was found that, as the radius of a spherical particle approached zero, its inertial mass ought to have shot to infinity. How could such a particle be affected by other particles or by contingent forces, or 'moved', if rendered immovable by its infinite mass? Such problems are artefacts from the presumption of contact causality and notional 'matter in space'; artefacts from the belief in 'point-like' entities and spatial blocks. The problem was never solved, but only renormalised. Indeed, it grew into a more egregious form in the astrophysical singularity or 'black hole' variant of the same.

By the advent of quantum field theory, success of and accommodation to **renormalisation procedures solved the calculation problem that arose from infinities, but it belied and obscured the central problem posed by space and the dubious viability of contiguous impact causality and of attendant IOO-schemas**: The whole issue was side-lined into a calculation morass and into a debate about whether renormalisation was a 'trick'. Yet, if causality truly proceeds in the way the materialist envisions it, in the face of inherent infinities, and even when abstracted out via renormalisation, the *objective* process of particle interactions could never halt and would crash into non-terminating problems: The outcome or output of such interactions could never be obtained or AND-to-OR realised. Time itself would 'freeze up': (see **5.30** for full account post-materialist solution).

But causality cannot proceed in such a fashion.

Belated de-spatialisation of physics, which should have emerged, if not from the usurping of absolute time by Einstein (see Book-IV 4.01) then in the structural delimits to observation from quantum mechanics vis-à-vis the notion of the 'motion' of quantum mechanical waves (Book-IV **4.04**) combined to the quantum mechanical impossibility of seamless-continuous observation of any process, should have rendered the whole notion of contiguous 'contact' totally void: **There is no space**. Therefore, there are no 'locations' at which purported contact can take place, much less exhibit spatial 'point-like' entities, spatial blocks, or 'spaceballs'. Instead, 'particles', though real enough in their de-spatialised temporised pure frequency-based *durata*-form, interact (or temporally superpose) with each other through quantised time, and manifest their outcomes through **quantum leaps in time**. Nature simply leaps over the infinities inherent to the interval that resides between the input and the output, and obtains the output without contact or contiguous 'interaction'; without the mediation of 'force', or of anything else: (again, see **5.30**).

Of course, the state-change transition from input to output, hence the 'quantum leap in time' that circumvents the infinities inherent in 'particle interactions', is realised through gear-less AND-to-OR wavefunction collapse. Of course, wavefunction collapse is time itself, which is not a 'fourth dimension'. The very historical fact that different interpretations in quantum mechanics had posed the question of hidden variables or gears behind wavefunction collapse and time (and causality), and by the fact that many rejected hidden gears, as does Alt-R on more radical but certain grounds, should lay bare the historical fact that natural science, even in its modern quantum mechanised form, was always ambiguous about the ultimate nature of causality and about what makes time happen.

The resort to 'shut up and calculate', hence the attendant evasion of core problems of causality and nature, hardly constituted confidence in the belief that 'gizmo' operates behind causality and time, given that **the attempt to account for wavefunction collapse and time on the basis of gears must fail, and given that the attempt must crash into insurmountable infinities and halting problems**. (see Part-I **5.08**, Book-II **2.46** to **2.47**, and Book-III **3.32**). Thus, **time and causality is brought about by a contact-less, gear-less, ephemeral, and ineffable work function zero process.**

*

It is time and wavefunction collapse that issue forth now-pleonastic 'energy relations' and change. It is not purported 'energy relations and change' that bring about time and wavefunction collapse. Indeed, no wavefunction collapse or time, no 'energy relations', and no state-change.

534

Nor is it computation that brings about wavefunction collapse and time. Computation is a time-dependent clockable schema that issues per ontically primary time and wavefunction collapse. No gear-less wavefunction collapse or time, no computation: (see **5.31**).

Hence, we now grasp in full the false historic association of naturalism, physicalism and science with materialism, and the presumed interchangeability of the latter with the former.

Materialism is dead. *Not* so ironically, it was physics that killed it.

5.38: Summary of Part-III: Dematerialisation of Naturalism

The following lists the implications from de-spatialisation and temporisation to materialism:

- **Spatial and point-like particles are no longer tenable**: Pure time-based physics subverts the notion of 'position', 'location' and spatial 'points' or blocks.

- **Contiguous contact or impact causality between particles is no longer tenable**. De-spatialisation removes 'points of contact' and it voids the notion of contiguous or impact causality.

- **Insofar as input-operation-output schema-based causality is bound to spatial impact causality, it is also fated to demise per de-spatialisation:** As the notion of spatial particles unravel, so must intertwined IOO-schema causality.

Additional overlapping findings also furnished the demise of materialism. These are:

- **Information-matter dualism (5.25):** This culminated into a dualism in terms of both de-spatialisation and temporisation; seamlessly so via the tripartite structure of the Bergson-Whitehead amalgam. Therein, OR-form 'matter' is confined to the event horizon. Both Bergsonian memory 'below', and Whiteheadian futurity 'above' the event horizon, remain dissociated from the 'matter' confined to the event horizon, with implications to memory-brain relations.

- **Breakdown of IOO-causality per the invalidation of hidden gears (5.28)** arose from the objectivity of randomness and of quantum indeterminism, *and* from the demise of 'hidden gears' in wavefunction collapse, time and grand decoherence. (Also see Book-II **2.46** and **2.47**, and Book-III **3.31 and 3.32**). Thus, the middle 'operation' term in IOO-schema unravelled in relation to quantum indeterminacy and the generation of wavefunction collapse, time, and attendant grand decoherence. Materialism bound to 'operation' conceived as gears must also unravel in the face of gear-less causality. By implication, the universe is not a computer or 'simulation' (see **5.31**): the generation of outcomes and events is not by hidden computation.

- **Infinite degrees of freedom in quantum field theory and the breakdown of IOO causality; historically obscured by renormalisation**: (**5.30**). Spatial interactions between point-like particles assume infinite degrees of freedom at points of contact. The calculation problem from this is circumvented via renormalisation, or by subtracting the infinities. Renormalisation obscures the fact that nature could not realise interactions or outputs through contiguous spatial iterational process, given that these would crash into insurmountable halting problems. Thus, particle interaction is instead realised through a temporal variant of the quantum transition: i.e., the 'quantum leap in time'. This 'leaps over' the infinities, without need to iterate through or suffer the infinite degrees of freedom. Thus, materialist notions of contiguous impact-mediated spatial matter-interactions must again unravel. In any case, de-spatialisation and temporisation will not permit otherwise.

- **Transtemporality of Platonic-Cartesian physical law: (5.32)**. An application of the Einstein Podolsky Rosen (EPR) experiment to the ontology of the conservation of quantum spin (an example of physical law) demonstrates that physical laws are enforced from outside of 'space and matter': succinctly, from outside of memory, possibility, and time. Given superlative de-spatialisation, the conservation of quantum spin is imposed by transtemporal means from an extra-ontology 'outside' of memory, possibility, and time... from outside of the Bergson-Whitehead amalgam. Hence, physical laws are Platonic and Cartesian abstractions vis-à-vis possibility, memory and time, and compel for an open ontology that transcends closed causality presumed per 'matter in space'.

- **The problem of mass in quantum gravity: (5.36)**. De-spatialisation implies that particle mass has no spatial 'location, simply because there is no space or 'location'. Gravitational and inertial mass is mostly a pure temporal function of the Bergson facet: a memory effect from the superpose of wavefunctions in memory retained within the Bergson facet. Per remarkable de-spatialisation, the central problem of quantum gravity is solved in that mass need no longer be associated with the superposed 'positions' of the particle in its wavefunction. Nor do we need to associate to 'positions' the gravitic 'spacetime curve'. When this combines with the said memory effect, the path to quantum gravity and the unification of gravity and quantum mechanics is cast open. The implication is that, in order to obtain quantum gravity and superlative unification of physics, science must accommodate de-spatialisation and temporisation. However, to accept de-spatialisation and temporisation is to wither impact causality, IOO-schemas, 'spaceballs'... and materialism.

- **Pragmatic energy relations are pleonastic: ultimately *not* part of the real ontology of physics: (5.35)**. The cliché that 'everything is matter' is oft interchangeable with 'everything is energy'. Putative potential energy is 'locked up' in the abstract and de-spatialised Whiteheadian future, *not* in spatial 'matter': Hence, energy cannot be 'stored' at 'location' or in spatial matter. Kinetic energy or 'work' is put into effect via wavefunction collapse and time. But gear-less wavefunction collapse or time is not itself energy-driven. It is a work function zero process, and ontologically primary to said energy relations. Thus, stare-change is not brought about by energy or work, but by gear-less non-energy driven work function zero AND-to-OR collapse and time. Time is indispensable to state-change. But energy concepts and relations are wholly dispensable and thus pleonastic. Hence, energy relations and concepts can be entirely abstracted from the ontology of physics. Hence, wither materialism founded on the forlorn ontology of energy concepts, contiguous energy relations, and 'work' via spatial contact.

BOOK-V PART-IV:
THE GENERAL THEORY OF MIND & WORLD

5.39: Aims of Part-IV

Recall that Part-II solved against idealism by critiquing 'consciousness causes collapse': Consequently, it is not possible to collapse object to subject, or 'matter' to mind. It follows that general mind-world theory cannot constitute an idealist philosophy of mind. On the other hand, Part-III summarised against the false equivocation of materialism with naturalism and physicalism. Part-III falsified the equivocation and summarised the demise of materialism. Hence, we cannot collapse subject to object, or mind to 'matter'. Therefore, general mind-world theory cannot design for a materialist philosophy of mind.

The implication to prospective mind-world theory from the findings of Alt-R are obvious: **Cartesian mind-world theory is inexorable. There are no alternatives.**.

Emancipated from forlorn idealism *and* from culturally hegemonic materialism, Part-IV and the general theory of mind and world therein, will reprise and supersede the provisional working definition of mind and nous first espoused in the Preliminary. Thus, the **first principles of general mind-world theory will culminate into the four-fold comprehensive definition of mind and consciousness**.

Nous is rendered salient in relation to key hard problems of consciousness: it only attends brains; *not* cups or pumpkins. This is because there is something special about brains, segueing into the aims of subsequent intermediate and fine-structure theories ,to be espoused in Part-V and Part-VI.

Our four-fold definition of nous will reprise the assertion that nous is what the renormalising quantum mechanists does (i.e., the apprehension of infinities). It will also show that nous is that which can apprehend metalatives, all of which involve infinities. Finally, in its most radical aspect, notwithstanding de-spatialisation, our four-fold definition will reassert that mind and consciousness constitutes *the* consummate peculiar state: namely, the noetic moment in absolute time vis-à-vis the growing block system of memory, possibility and time, always subject to the relativity of simultaneity. Hence, the four-fold definition will design for the revival of Cartesian dualism. It will segue into two starting principles of the general theory of mind and world:

- The principle from obviation of the problem posed by *res extensa*
- The principle from the obviation of the problem posed by the principle of conservation.

The problem of space, or *res extensa*, long plagued Cartesian dualism. It constitutes Descartes' minor error. The obviation of *res extensa* via de-spatialisation and temporisation transforms the relation between nous and brains into a pure temporal relation and expression, with very different implications to those garnered from the presumption of 'space' or *res extensa*.

On the other hand, presumed violations of the principle of conservation by Cartesian mind also plagued Cartesian dualism. The historically long-drawn weaponization and abuse of the conservation principle against Cartesian causality was undermined by the reality of work function zero processes first espoused in the Preliminary, if nothing else: (see **0.32** to **0.33**). But Alt-R also espoused work function zero relations in the transtemporal enforcement of abstract physical laws (see Part-III **5.33**), wherein physical law is enforced on memory, possibility, and time from *outside* of memory, possibility, and time. Yet, it is per function of the recuperation of the conservation principle into the conservation of abstract memory, co-joined to the principle of consistent futures, which fully emancipates Cartesian causality from the materialist weaponization and abuse of the conservation principle.

The work function zero profile, especially of the process of AND-to-OR wavefunction collapse and time, is critically pertinent to mind-world causality, given that it demonstrates that fundamental causality itself is not an output obtained from conserved energy inputs, much less an affect from definable operations or gears conducive of state-change. Causality behind time constitutes a gear-less work function zero process. Nous exploits this work function zero time-process to bring about the developmental control of its brain, without the violation of principle of conservation in its generic form; without the violation of the otherwise transformed principle into the conservation of memory and consistent futures. It follows that Cartesian mind-brain causality is *not* in conflict with any true principle or delimit to nature, especially the principle of conservation.

*

The general theory of mind-world will also show that the structure of the growing block Bergson-Whitehead amalgam (which had replaced spacetime by Book-IV), combined to de-spatialisation and temporisation, must obviate materialist notions on the nature of memory. Recall that memory in nature is retained in de-spatialised temporised dissociated form within the Bergson facet of the Bergson Whitehead amalgam. Memory is retained as a temporally organised series of complex abstract *wavefunctions in memory*. This must also apply to memories vis-à-vis the brain, given that purported in-brain memories are not going to be exempt from the general temporised dissociated nature of memory vis-à-vis the rest of nature... if only because there is no 'space', and the brain itself can no longer be modelled as a spatial-material 'object', but must also configure as a distribution in time, projected within the Bergson facet as a perduring 'worm' from its most remote past to its most recent formation.

The implication is that memory, *even if initially assumed to be in-the-brain*, is *not* retained in any manifestation of brains restricted the event horizon. Instead, memory is temporally dissociated into the Bergson facet, as past states of the brain, now rendered into abstract wavefunctions in memory attendant but *extant* the 'materialised' event horizon-restricted brain. The process of memory and its retrieval must consequently de-spatialise, temporise, and dematerialise, making nonsense out of the materialist conception of memory-retrieval conceived as spatial retrieval from in-brain spatial addresses or locations.

The general theory will thus assert that memory-retrieval must work purely temporally and non-contiguously: realised through a process of wavefunction collapse that attends temporal relations to the Bergsonian past. Of course, memory-retrieval and recollection processes that unfold purely temporally are inherently Bergsonian and consummately a-materialistic. Indeed, Bergsonian memory retrieval and recollection is… *Cartesian.*

The memory-retrieval problem will segue into the problem of the 'location' of mind and consciousness in relation to de-spatialised temporised memory and to the equally de-spatialised memory-recollection processes. The moment of recollection (the consciousness and the deeper pre-conscious minding process that initiates recollection) must be as de-spatialised and *as* temporised as memory and memory-retrieval, simply because there is no space. Thus, mind and consciousness… and memory… cannot have a 'location' in brains.

Furthermore, by utilising the Bergson-Whitehead amalgam framework, general theory will show that the moment of the expression of consciousness (typically, of recollection) as a corpus of events constituting its expression, must constitute a future-form potential moment: one nested within the Whitehead facet… or 'located' *in the future*: This potential future moment of consciousness attendant the events of its expression, *cannot reduce* to the realised 'material' brain confined to the event horizon, given that the brain temporally constitutes events *before* the future-potential moment of consciousness-expression has transpired. Therefore, *obviously*, the moment of consciousness, of recollection, or the events of its expression, cannot occur in or from *any* brain activity preceding it in time. But this also implies that mind must select that future-potential moment of the expression of consciousness from within the Whiteheadian future: That is, mind must reach into and 'draw out' from that future… from the Whitehead facet… a brain-pertinent wavefunction in future-potentiality for subsequent development. But from *where* does mind accomplish this? Given de-spatialisation, the question is absurd. Within a de-spatialised temporised universe, we can only pose the question of the *when* of nous. Yet, the *when* of active mind turns out to be as elusive as the absurd 'where' of it… inexorably culminating into Cartesian mind-brain dualism.

*

The general theory will segue into the two modalities of consciousness and mind: the passive and active modes, first introduced in the Preliminary and reiterated elsewhere. The *passive mode* relates future-potential consciousness as the witness-recipient and conscious recollector of memory, always temporally *after* the world, brain and memory-expressions have transpired. Consciousness of the world, brain and of memory occurs *after* brain activity: *never* synchronously with or *as* brain activity. Also, conscious recollection is always in *temporal ecstasys*... or at a potential future temporally removed from prospective brain activity and from the brain itself. Thus, consciousness is non-identifiable with, not clockable as, and non-reducible to the brain and its activity. Hence, Cartesian dualism follows from the temporal relation of witnessing consciousness vis-à-vis temporally preceding brains.

On the other hand, in the *active mode*, non-conscious or pre-conscious mind causally selects and generates subsequent brain activity. The active mode mind accomplishes this by exploiting the structure of the Bergson-Whitehead amalgam, its inherent discontinuous AND-to-OR succession of event horizons, and its attendant 'gaps in time'. The mind engages the world temporally *before* consequent brain activity, hence before the passive mode arises as conscious recollection or expression. In this way, active mind temporally precedes and cannot occur synchronously with the brain activity it instigates, or even with the subsequent moment of the expression of consciousness that temporally follows suit. Hence, mind does not clock or manifest as brain activity at all. Temporally preceding the brain, the mind in its active mode cannot synchronously reduced to the brain that it instigates into activity. Hence, Cartesian dualism follows from the temporal relation of active mind vis-à-vis its subsequent brain activity.

Hence, the general theory will show that both the passive and active modes of nous are consistent with the experimental findings obtained by Benjamin Libet (see Preliminary: **0.44**). It will reiterate that the outcomes furnished by Libet are misinterpreted on a post-hoc *a priori* materialist assumption-loading basis, to the point of declaring mind, consciousness, and volition, illusory.

The general theory will also solve the binding problem of the brain: Wilfred Penfield could not find any basis for the unitary coherence and binding of the brain (see Preliminary **0.46**). Unitary binding emerges seamlessly from the recognition that the complex wavefunction and instantaneous whole that attends the whole brain, as is true of the complex wavefunction and instantaneous whole that attends the development of a complex photographic image, always decoheres and collapses in a unitary fashion. Thus, active mind selects the complex wavefunction and instantaneous whole for its brain in line with its intents. This whole spontaneously decoheres and AND-to-OR collapses into unitary and globally coherent brain activity. The binding is not generated or coordinated by any in-brain 'spatial' material mechanism or process issuing from brains. Instead, the brain is co-ordinated into wholistic activity and coherence by the instantaneous whole selected and imposed upon the brain by its controlling active mind.

*

Noumnemonae, or *memory in consciousness*, was first introduced in Book-IV **4.41**. Its recuperation in the general theory will follow from the elucidation of the passive and active modes of nous. **The general theory will recapitulate *noumnemonae* as a state in absolute time**. i.e., Nous has absolute time-like characteristics. Absolute time-like nous is in diametric contrast to *mnemonae*: i.e., the

537

non-mental Bergsonian *world in memory,* including the brain itself: both *always* subject to the relativity of simultaneity. The dualism between absolute time-like nous versus world and brains subject to the relativity of simultaneity, not only necessitates that *noumnemonae* cannot emerge from mnemonae, but the interrelation must inexorably culminate into Cartesian dualism.

However, a truly radical turn will ensue: **The subsummation of mnemonae to noumnemonae designs for *the en-worlded mind.* The en-worlded mind turns object-subject relations inside out, or right-side in.** To clarify, we typically assume that minds arise from, or are located within, a system of spatial distributions of events: i.e., that minds are embedded within the world of events and *as* events in the world. The general theory will show that memory, possibility, and time (hence the world) must arise to and *within* mind and consciousness, but *not from* mind and consciousness. Hence, the world is not furnished with nous. Instead, nous is furnished a world of temporised memory, possibility, and time. Hence, nous does not arise from the world. Instead, the world arises *within* nous, but *not from* nous. Hence, the en-worlded mind.

The inversion of the object-subject relation in the en-worlded mind culminates into the unexpected Anti-Copernican revolution. That aside, the en-worlded mind ought to imply solipsism. However, when we inter-relate first-person frames in terms of the relativity of simultaneity, and combine this with other findings, the expected solipsism will be fully circumvented, while the en-worlded mind will abide, with all its radical Anti-Copernican consequences and implications.

5.40: Principles of General Mind-World Theory, Essay-I:
The Indispensability of Quantum Mechanics from Mind-World Theory

We must briefly recapitulate the indispensability of quantum mechanics (of AND-form logic, hence the ontically real future) from *any* mind-world theory, materialist or otherwise, regardless of scale. **The indispensability and ineliminability of quantum mechanics to mind-world constitutes the first tenet of the general theory.**

Even though our Cartesian approach does not constitute any form of quantum mind theory, in that it does not require mind to bring about wavefunction collapse, nor orchestrate in-brain quantum indeterminate processes... the general and subsequent theories certainly involve quantum mechanics. Consensus holds that quantum mechanics is irrelevant to the brain and, by extension, to any mind-brain theory. This is on the assumption that quantum mechanics pertains only to the very small: *not* to the macro-scale brain. However, in order to tackle the problem of mind and consciousness, we *must* resort to some sort of quantum mechanics. This is not because of the alleged strangeness of quantum mechanics, nor from the apparent licence it furnishes to New Age claims, especially given the Alt-R position against 'consciousness causes collapse' from Part-II. Nor is it because consciousness is a quantum phenomenon, which it is not.

Quantum mechanics is a fundamental part of nature and brains, *at all scales*: a truism rendered salient by the case for the growing block universe and attendant *grand decoherence.* The quantum mechanical state constitutes the ontic reality and phenomenalisation of the future as an objective reality, *and* we cannot exclude the future from *any* system at *any* scale, including the brain. (See Part-II **5.12** for the reasoning, and the very short summary below). Hence, the typical materialist want to confine quantum mechanics to the micro scale, and declare the macro-world brain a non-quantum mechanical domain, constitutes a fundamental error born out of the failure to recognise AND-form logic and superpose as the objective future, and lacks recognition of the growing block universe that the ontic future attends. The failure to grasp the all-scale pertinence of AND-form logic and quantum mechanics is not new: even Eugene Wigner conceded to the scale fallacy in his abandonment of 'consciousness causes collapse' (See Part-II **5.12**).

As was argued in the critique of 'consciousness causes collapse' in Part-II, the principle of causality, intertwined with the principle of conservation, and how both attend and demand the reality of AND-form logic, requires that AND-form logic (the Whiteheadian future, hence quantum mechanical reality) *must* apply to the macro-scale, hence to the brain itself... unless we risk 'signals from the future' and attendant violations of the principle of conservation. The consequent all-scale generalisation of AND-form logic must hold true, even if hidebound self-interference does not always manifest at the putative macro-scale. Yet, grand decoherence of the Whitehead facet *en toto* unfolds, and the self-interference of the universe with itself at the grandest possible scale, in the formation of inertia at the ultimate macro-scale, abides.

In order to confine quantum mechanics to the micro-scale and circumvent it vis-à-vis the brain, materialist philosophy would need to demonstrate that the macro-world (including brains) constitutes a block model domain: one *without* future possibilities: bereft of the AND-form Whitehead facet of the growing block Bergson-Whitehead amalgam. Materialist philosophy would need to show that this macro-scale futurity constitutes a state fully OR-form resolved daz to the infinitely removed future. It would need to demonstrate this by obtaining OR-form 'signals from the future', and, in doing so, violate principles of causality and conservation.

Materialist philosophy would also need to debunk the nested futures perspective, attendant counterfactual causality and grand decoherence (the basis for the revival of Mach's principle and the presage to the quantum mechanical solution to inertia and much else (see Book-II **2.34** and **2.35**). The fact that quantum mechanics is ineliminable from the macro-world, or from any scale, renders it ineliminable from the macro-scale brain, and from *any* explanation of nous vis-à-vis the brain. In other words, materialist or not, any theory of mind and world *must* include a growing block model of the brain. *All* theory must incorporate the AND-form future and the process of wavefunction collapse and time through which the brain's future is resolved into actua. Indeed, the process of time (wavefunction collapse) is inherent to *any* explication of nous and attendant brains, even if one assumes that the former is totally reduced to and identifiable with the brain, or an epiphenomenon of that brain, or a supposed 'emergent property' of the same... or even an illusion 'excreted' by brains: All of these require a growing block quantum mechanical approach, and theories that exclude the quantum mechanical account are, simply, in fundamental error and un-serious. There is simply no choice in the matter.

Thus, the first tenet of the general theory of mind-world is **the all-scale saliency of quantum mechanics, and its indispensability to, and ineliminability from, *any* theory of mind and world, even materialist theory**. This justifies and *necessitates* our frequent resort in subsequent essays to the processes of the growing block Bergson-Whitehead amalgam and its attendant process of wavefunction AND-to-OR collapse (time) as the critical explanatory framework and indispensable theoretical ingredient. Thus, it is not possible to devise a viable mind-world theory, materialist or otherwise, without time: No time… no consciousness. Not even the illusion of it. But time is synonymous with AND-to-OR wavefunction collapse, central to quantum mechanics and the growing block framework. Hence, insofar as one cannot preclude time, even from materialist theories of the brain, one cannot preclude the process of wavefunction collapse and quantum mechanical science from the same.

5.41: Principles of General Mind-World Theory, Essay-II: Four-Fold Definition of Mind & Consciousness

Given the demise of materialism secured in Part-III, it is no longer mandatory to assume that noetic qualities summarised below are produced or 'excreted' by the material processes in the brain, or through IOO-schema impact and closed causality therein, or by the contiguous interactions of 'matter in space'. Recall that materialism is the licencing background to eliminative materialism, to identity theory, to epiphenomenalism, and to the more recent 'emergence theory'. Simply put, materialism is void. It follows that eliminative materialism, identity theory, epiphenomenalism, and emergent theory, are also void.

In the Preliminary (essay **0.06**) we developed a set of working assumptions about mind and consciousness. With subsequent development, combined to the critique of 'consciousness causes collapse' furnished in Part-II, we now consolidate these into the sought four-fold definition that delineates the domain of mind and consciousness from non-conscious memory, possibility, time, and brains.

Nous is oft grasped in terms of the myriad noetic qualities testified in folk psychology: such as intentionality, will, love, hate. Even if dismissed as mere 'illusions' of folk psychology in eliminative materialism, such qualities obviously associate only with brains, not pumpkins.

The noetic qualities claimed in folk psychology certainly fall under the rubric of Chalmer's *hard problems of consciousness*. The hard problems highlight the problematic nature of any attempt to explain the said qualities in terms of easy-problem in-brain structures and input-operation-output schemas, or by the assumption of now-obviously false *a priori* materialist ontology and causality. Therefore, **Chalmer's hard problems of consciousness, and its validation per the demise of materialism secured in Part-III, constitutes the first contribution to the four-fold definition of nous**.

*

For the second component of the four-fold definition of nous, we must briefly reprise the implications from renormalisation reiterated in Part-II **5.14**, and in the pure time-based solution (quantised time and the 'quantum leap in time') to the problem of causality raised by infinities, and by the assumption of particle-mediated contiguous impact causality central to generic renormalisation. In order to resolve calculation problems that attendant infinities, the quantum mechanist must employ renormalisation. Therein, the physicist must be able to grasp and handle infinities, or recognise halting problems. Yet, the power to handle infinities and recognise halting problems is a unique feature of nous. This power cannot be engendered by any 'material world' based on contiguous particle-mediated gears-based causality, or by 'matter in space', any more than by a finite brain that operates on a finite basis. **The power to apprehend and handle infinities and halting problems exceeds the capacity of the finite brain and world. Hence the second contribution to the four-fold definition of nous**.

Indeed, impact and IOO-causality does not abide even in non-conscious fundamental particle interactions (see **5.30**), and it is hardly expected to have relevance in the mind-brain framework. *At all scales*, the generation of events involves purely non-contiguous or temporal process of wavefunction collapse that 'leaps over' the infinities and circumvents presumed points and vertexes of particle contiguity and contact. Causality is not going to be any different vis-à-vis the framework of brains.

*

Our ability to handle infinities extends beyond renormalisation in quantum field theory: Infinities are inherent to the calculus that underpins much of science, physics, and engineering: Therein, both infinitesimals and limits have played a key role in the development of calculus, and both require the apprehension of infinities. Yet, infinities also play a part in non-mathematical domains: in dictionary definitions; in universals; and in qualitative judgements unique to nous, despite all our qualitative failings as a species. Thus, the problem of *metalatives* vis-à-vis the finite brain.

It is not a fallacy of equivocation to link infinitives in dictionary definitions with infinities in the number-sense. Consider the infinitive *to run*. For how long? The interval implied is indefinite. As such, it implies infinity. Also, consider the infinitive, *to be*. This is not merely indefinite but timeless: *to be*… is extant time itself, in the sense that *Being* is transtemporal. Thus, the infinitive *to be* takes us beyond infinities….and certainly beyond the facility of the finite and temporal brain and world.

While dictionary definitions can be coded into computers, and the appropriate codes (the symbols or pseudo-analogues) can then be outputted upon pre-specified prompts, and while this could permit a device that can effectively passes the Turing test, it is impossible to embed real definitions (or understandings) into such systems. This is because the real definitions implied by *to run*… *to love*, and *to be*… entail infinities and halting problems, notwithstanding timeless *being*, beyond the purview of any finite IOO-schema or computational system, with respect to which there are no possible pseudo-analogues, codes, or symbols. (See Book-I **1.45**. Also see 'superlative rigging' for a quantum noise approach to the code-information distinction: **1.48** to **1.59**).

It also follows that the brain, often modelled as a computational device, and portrayed as operating on essentially the same form of IOO-schema as generic devices, could never generate understandings that, in each case, *must* entail the fundamental ability of nous to grasp and handle infinities and halting problems and apprehend metalatives, such as *being…* by access to the proper forms for these, and *not* merely by access to their pseudo-analogues or agree-upon stand-in symbols, codes and machine-states, much less the finite iterational procedures upon which codes-processing is carried out: processes that must crash into insurmountable halting problems in the face of infinitives and metalatives. Thus, the power to handle infinities, and with this, the power to handle dictionary understandings and metalatives, exceeds even the capacity of the brain. These are furnished to nous from outside of the brain.

Of course, the brain remains indispensable as the means of expression of infinities and metalatives, but *not* as their means of production. Indeed, even nous itself cannot 'compute' infinities, halting problems, infinitives and metalatives by means of any clockable brain-process that unfolds in or requires time. As was shown in Part-II **5.14**, the actual process of understanding (and the actual process of grasping infinities, infinitives, and metalatives) is furnished to nous from 'outside' of the process of time: from 'outside' of the Bergson-Whitehead amalgam… from an ontology extant nous itself: as Kant or Plato had anticipated. Hence, **mind and consciousness is that which can apprehend infinitives and metalatives: Hence, the third contribution to the four-fold definition of nous.**

*

While the world and brain remain subject to the relativity of simultaneity, nous configures as a noetic moment and temporal monad: one in a state in absolute time. Recall that the consequence of de-spatialisation is that all things, including brains, comprise complexes of events organised into pure delayed choice time-interval relations. Hence, the brain and world is wholly temporised: (see Book-IV **4.40** and **4.41**). It necessarily follows that all things, including brains, are strictly subject to the relativity of simultaneity. That is, the events that constitute the brain do not transpire in a common present moment or absolute time (i.e., space). Thus, memory and possibility, including brains, cannot possess absolute time-like characteristics, given the relativity of simultaneity. Therefore, the brain cannot produce within its confines and means the experiential 'common present moment'… or the sense of the indivisible singular-unitary experience... of itself or of a world otherwise subject to the relativity of simultaneity. This sort of power exceeds the brain.

For mind and consciousness to have absolute time-like characteristics, nous must be equivalent to a spatial 'point', if such a thing could exist in a temporised universe. Such a 'point' of consciousness or, succinctly, the temporal monad and unitary noetic moment… or *the conscious moment in absolute time…* is not locatable in brains, precisely because the latter is strictly subject to the relativity of simultaneity and without absolute time-like states. Consequently, absolute time-like mind and consciousness cannot be 'located' in or be reduced to events, nor to the past, nor even to a brain, which always temporally precede the subsequent noetic moment, and are subject to the relativity of simultaneity.

Since experiential absolute time cannot arise from a world subject to the relativity of simultaneity, its ontology must be fully distinct and separate from that world. Hence, **we obtain Cartesian dualism grasped in terms of the delineation between nous in absolute time, constituted as a monadic temporal noetic moment… versus the brain and world strictly subject to the relativity of simultaneity, constituted as a distribution in time. Hence, the fourth contribution to the four-fold definition of nous.**

*

In summary, the four-fold definition of nous are encapsulated thus:

1. **Nous is comprised of all the folk psychology hard problem attributes** associated with brains, but never associated with cups, rocks, or pumpkins. The validity of Chalmers's hard problem approach is finally secured by the demise of materialism in Part-III.
2. **Nous is what the renormalising quantum mechanist possesses**: This entails the power to grasp and handle infinities and halting problems: an 'impossibility problem' for finite brains and finite world.
3. **Nous is that which can apprehend infinitives and metalatives**: *'to run'…and 'to be'*. These all entail infinities, halting problems and transtemporal *being*: All exceed the instrumentality of brains, and which cannot be generated by brains. Hence, Cartesian dualism, must again follow.
4. **Nous is that which possesses absolute time-like characteristics** versus the brain... which, as part of the domain of memory, possibility, and time (i.e., the Bergson-Whitehead amalgam) is strictly subject to the relativity of simultaneity. Hence, Cartesian dualism from absolute time-like nous versus brain and world subject to the relativity of simultaneity.

5.42: Principles of General Mind-World Theory, Essay-III: The Two Principles of the General Theory: Work Function Zero Mind-Brain Causality & the Invalidation of *Res Extensa*

For three centuries, the Cartesian theory of mind was beset by two supposedly insurmountable problems: The *res extensa* problem and the *violation of the principle of conservation*. The general theory is founded on two obviations: The obviation of *res extensa* and the obviation of presumed violation of the principle of the conservation of energy-matter.

Starting with the obviation of *res extensa*: **if, as Rene Descartes claimed, mind and thought have no extensionality (i.e., no spatial characteristics) then how could non-extended mind and thought interface with extended brains?** The answer to this supposedly fatal conundrum resides in the correction of *Descartes' minor error*, first introduced in Book-IV **4.41** and recapitulated below. Thus…

- **There is no *res extensa*, given de-spatialisation**. There is no 'space'. The notion that the brain and world constitute spatial extended states (Descartes' *minor error*) is simply not true. Consequently, the *res extensa* trope used against Cartesian dualism is rendered irrelevant.

- It follows that **the interface between mind and world is an interface realised purely in and across time: a temporal interface**, mediated and realised through quantum leaps in time, in turn synonymous with the gearless work function zero process of wavefunction collapse.

- Thus, the basic Cartesian relation is **a pure temporal relation between non-extended nous (mind and consciousness) and non-extended temporised brain and world**: with the latter constituting a de-spatialised and temporised growing block system of stereotemporal past-memory and future-possibility.

In short, the *res extensa* conundrum, or the spatial argument against Cartesian mind, constitutes forlorn pseudo-argument.

*

We now come to the second obviation: **How can Cartesian abstract (i.e., non-extended) mind, even one purely temporally related to memory and possibility, generate causality vis-à-vis the brain without relying on a form of energy or force-bearing interaction and mediation, without bringing about the violation of principles of conservation of energy.** We address this thus:

- **Time is a work function zero process**: Wavefunction collapse and time, which brings about the enforcement of physical law and the generation of new events and outcomes, *and* permits the imposition of mind's will upon its brain, constitutes a structure-less and gear-less work function zero process. Time does not require any contiguous particle-mediated impact , much less an energy-power input, to transpire. *Work function zero time and wavefunction collapse is key to work function zero mind-brain causality, given that mind affects its brain via the process of AND-to-OR collapse and time*. (See aspects of **5.18** and **5.19**, and preceding essays in Book-II **2.46** and **2.47**).

- **Pleonastic energy concepts and relations are subsumed to and transcended by ultimate pure information physics**. Time is ontologically primary to putative energy-relations and the energy-based conservation principle, given that these are realised in and through AND-to-OR collapse and time, and could not arise without time and wavefunction collapse. But time itself is a work function zero process presupposed to 'inferred' or foisted energy relations and the generic conservation principle on outcomes from wavefunction collapse and time. Therefore, the ontological primacy of non-energy driven time vis-à-vis apparent 'energy relations' brought about by time must abstract out energy-work from the ultimate ontology of *and* from mind-brain causality: See Part-III **5.35**. *The implication is that mind does not require an 'energy input' to affect the brain, given that energy relations are pleonastic and not part of the real ontology of physics.*

- **Cartesian mind exploits the process of work function zero wavefunction collapse and time**: Recall that, per the critique of idealism in consciousness causes collapse' in Part-II **5.18**, consciousness does not itself bring about wavefunction collapse and time. It is work function zero time that facilitates the intercession of nous into the world. *Mind merely exploits the work function zero loophole furnished by time and the attendant 'quantum leap in time' in order to causally affect its brain, necessarily on a work function zero basis: Cartesian causality is work function zero because the process of time it exploits is also work function zero.*

- **Time is quantised and circumvents contiguous 'contact' *and* the infinities inherent to purported particle-mediated causality: The same circumvention holds generally and must also hold vis-à-vis mind-to-brain causality.** The progression of time, the synonymous broken-discontinuous progression of the event horizon, and attendant enforcement of physical laws, combined to the realisation of the outcomes from 'particle interactions' and the realisation of putative 'energy relations', utilises a 'quantum leap in time'. Consequently, infinities in particle interaction, and the purported contacts and impacts between these, are 'leapt over' and entirely circumvented by work function zero AND-to-OR collapse and the 'quantum leap in time' (see **5.30**). *Insofar as mind exploits the process of time to affect its brain through the 'quantum leap in time', mind also temporally leaps over the supposed 'necessity' of particle mediated impact-causality in relation to its brain. Thus, mind does not employ 'particle interaction' or impact causality to causally affect its brain.*

Moreover, per the pleonastic status of energy relations…

- **The conservation principle is recuperated into post-energy physics and into abstraction… into *the conservation of Bergsonian memory* co-joined *the principle of consistent futures*** via the grand decoherence of futures within the Whitehead facet; pertinent to the future evolution and affectation of the brain by mind. (See Book-II **2.39** for our first-time recuperation of the principle of conservation: Also see **5.29**). **Note that the Cartesian affectation of the brain by mind through work function zero process does not destroy or erase the content of growing block memory, nor render the Whitehead future inconsistent with the Bergsonian past.** *Within the delimits of the conservation of memory, and per permitted consistent futures, the mind can affect its brain in any way it wishes, especially owing to the fact that it does not utilise energy or work to accomplish this, and even less does mind employ contiguous causality or gears to affect its brain*, so ending the three-century long materialist weaponization and abuse of the principle of conservation against Cartesian causality and dualism. We can recapitulate this in the following:

- o **The conservation of memory**: Causality in general, and Cartesian causality specifically, cannot erase the growing block Bergsonian accumulation of memory, nor could it change the past into a form that did not happen. Hence, mind cannot violate the conservation of memory.
- o **The principle of consistent futures**: Cartesian causality can only instigate future brain-world outcomes consistent with the past-in-memory. Indeed, via a-noetic means, the Whitehead facet automatically grand decoheres into consistent futures that conform to that past, and mind can only select from those futures thus permitted and consistent. Hence, mind cannot bring about non-consistent futures in violation of the new conservation principle.
- o **Mind utilises a non-energy based work function zero process to select future brain-world outcomes**, exploiting work function zero time, and exploiting work function zero causality, to bring about sought state-change in brains…*without* violating even generic energy-based conception of the principle of conservation.

In short, Cartesian causality does not usurp the conservation principle recuperated into its abstract de-spatialised temporised and growing block form. Thus, with our two key obviations, we obtain the revival of Cartesian mind-brain causality and dualism.

5.43: Principles of General Mind-World Theory, Essay-IV: Memory-Side Implications from De-Spatialisation: Review of the Experiments of Rose and Harding, and Holonomic Memory

Facile materialism asserts… "all is matter in space". However, de-spatialisation asserts there are no *wheres* in nature, but only *whens*: there are only pure time relations between past-events and future-possibilities. It follows that, as is true of all things, the brain cannot be treated as a 'material object in space' with supposed spatial 'addresses' at which nous is supposedly produced, or at which memories are stored; presumably in the form of Karl Lashley's engrams, or as other spatial forms. (See Preliminary **0.38** and **0.39**).

Resolved information pertinent to the past constitutes de-spatialised stereotemporal Bergsonian memory, organized according to primacy-recency rules, from the oldest to the most recent past, retained within the Bergson facet. Then there is de-spatialised stereotemporal Whiteheadian futurity, embedding not-yet realised future-potentials. Hence, the Whitehead facet. Ultimately, **we must constitute a growing block model of the brain as a de-spatialised and temporally organised Bergson-Whitehead complex: constituted as a Bergsonian temporal mnemonic 'worm', one co-joined to a Whiteheadian future potentiality, but incorporating the event horizon (the 'materialised brain') and the process of AND-to-OR wavefunction collapse or time**. The prospective growing block model cannot help but alter how we model memory (as immaterialistic abstract information, temporally distributed and extant the event-horizon-confined 'materialised' brain), usurping generic spatio-materialistic assumptions about memory, memory-retrieval and recollection.

Thus, one must either grasp a Bergsonian account of the brain, recapitulated as a complex superpose of abstract wavefunctions in memory that, as perdurance… as a Bergsonian 'worm' in time… elaborates from past-primacy up to the recentcy of leading OR-form resolutions on the event horizon… or one must model the brain as a set of as-yet not-happen Whiteheadian nested future-potentialities projected, if not to the infinitely removed future, then to a future that encompasses the potential life-span of that brain. Indeed, we must incorporate *both* descriptions. Of course, the growing block model of the brain must also include the process of wavefunction collapse and time attendant the event horizon. In such a milieu, it becomes possible to obtain certainties about the nature of the brain, about memory and its recollection, and about how active mode mind intercedes through the process of time to decohere the Whiteheadian future of the brain to a sought outcome. The finer details about how mind accomplishes mind-brain control must await intermediate and fine-structure theories espoused in Parts-V and VI. Part-IV will deal only with attendant general principles and observations that pertain to Cartesian causality.

*

De-spatialisation and its implications vis-à-vis the brain constitutes the basis for why the experiments of Rose and Harding *et al* were unable to validate Karl Lashley's materialist contention, even regarding easy-problem procedural memories: i.e., the sort of memory easiest to map in terms of input-output neural schemas and structures: constituting an 'easy problem', as opposed to the 'hard problem' posed by, say, mapping the meanings pertaining to "I love Lucy"… or biographical memories. (See Preliminary **0.43**). Implications from de-spatialisation also explains why Karl Pribram's materialist version of the same, but in holographic or holonomic form, was inadvertently falsified in the experiments of Rose and Harding: (See Preliminary, **0.40** for how the glucose pump method also falsified holonomic memory). Both engram and holonomic memory theory conceives memory as a distributions in space. Given de-spatialisation, the spatio-material forms of Lashley's engram and Pribram's holonomic theory are no longer viable. Only temporised models of memory will suffice. Hence, only Bergsonian memory will do.

Clearly, **given de-spatialisation, we require a pure temporised distribution of information pertinent to brains: with temporised memory as part of the 'worm' within the brain-pertinent Bergson facet of the growing block amalgam. Hence, a true holonomic theory of memory requires that information be clouded, *not* in or across 'space' or in the 'spatial brain', but distributed and clouded to a de-spatialised stereotemporal order**, as Henri Bergson had conjectured over a century ago.

*

The brain is *not* irrelevant. **The brain constitutes the means of expression (*not* the means of production, much less the milieu of memory-storage and retention) for otherwise extant abstract-immaterial, purely temporised memory and minding processes**. As part of the pure temporal distribution of information constituting the larger Bergsonian environment, the brain remains integral and indispensable to the process of mind's feedback from the larger 'environment as mnemonic history', with both ultimately constituted as part of the whole Bergson facet of the growing block Bergson-Whitehead amalgam.

As both means of expression *and* means for feedback, the brain *must* develop readiness states via the succession of event horizons to mediate expressions and to attain feedback of the environ, but always via a-spatial and non-contiguous temporal relations, enforced through the superpose of memories and per mnemonic mechanics (see Book-IV **4.26** and **4.27**). None of this will entail spatio-material impact-mediated causality, given de-spatialisation and temporisation.

It is tempting to treat the 'readiness states' of the brain *as if* **these constituted the 'locations' of in-brain engram-type memories, given that these are co-requisite to real memories as their means of expression, even though the readiness states cannot constitute the real basis of memory retention: The real memories are 'elsewhere' or extant; temporally distributed to the abstract Bergsonian facet**. Indeed, with the exception of *noumnemonae*, or 'memory in consciousness', which is distinct from non-conscious Bergsonian memory, the past comprised of both the 'environment in mnemonic history' *and* the history of Bergsonian brain, must be temporally distributed to the Bergson facet of the amalgam. Thus...

- De-spatialisation implies that memories pertinent to the brain cannot be modelled as spatial complexes: These must instead be modelled as abstract memories stereotemporally distributed within the Bergson facet of the Bergson-Whitehead amalgam.

- As stereotemporal states in their own right, the readiness-states of the brain are integrated to an embedding stereotemporal Bergsonian 'worm' and its 'environment as mnemonic history'. The integration is *not* via 'material relations in space', but via the ephemeralised de-spatialised temporally distributed past: with the brain and its memories constituted as a perduring 'worm' in time, wherein readiness states pertain to the most recent terminus of that 'worm' on the event horizon.

*

As an augmenting case against the reducibility of memory to brains, consider the infinitive, '*to run*': an indefinite statement about the associated time-interval from which infinity is implied. The apprehension '*to run*' is retained as an infinity-laden memory, forcing the conclusion that **infinity-laden memories, cannot be stored in finite brains and structures, regardless of whether the brain is modelled as an 'object in space' or as a perduring Bergsonian 'worm' comprised of stereotemporal memories**.

Recall that infinities are integral to such things as dictionary definitions and metalatives (see **5.41**). Hence, infinities are integral to our mundane conscious experiences, recollections, and memories, given the inevitable intertwine between memories about the environment (e.g. the big red garage door from the author's childhood, 1973) and infinities pertinent to definitions and metalatives enmeshed with such memories: all encapsulated in *noumnemonae*... or 'memory in consciousness'. Hence, the materialist notion that memory is comprised of hands-on in-brain structures and spatial 'trace-states' must crash into infinity-laden memories that cannot be accommodated by finite in-brain structures, even when these are de-spatialised and stereotemporised into the Bergsonian 'worm' version of Pribram's holonome. However conceived, the finite brain simply cannot handle the storage, much less the procurement of the conscious understandings and recollections of infinity-laden memories and states.

It follows that, both the storage of infinity-laden memories of conscious experience *and* **the pertinent** *understanding in itself* **must remain a power exercised from 'outside', even of the Bergsonian brain and 'worm': Radical memory must be part of** *noumnemonae* **or 'memory in consciousness', and must exceed even the 'worm'.** From this view, the brain remains an indispensable intermediary to nous and memory recollection, but it *cannot* constitute the storage-facility for real memories, even when reconceived as an abstract Bergsonian temporal worm, and even less when conceived in an event horizon-restricted form as the 'materialised' OR-form brain.

When we combine the brute implications to memory from de-spatialisation with hard problems posed by infinitives and infinity-laden *noeta*, **neither memory nor its minded and conscious recollection, nor the** *understanding in itself*, **could reside in the brain, or be produced by it. Hence, Cartesian dualism is the inexorable conclusion.**

5.44: Bergson-Whitehead Model of Memory, Essay-I:
De-Spatialised Memory & the Bergsonian Model of Memory Recollection

We can now sketch the Bergson-facet approach to memory and recollection. To this end, we will consider the author's childhood memory: Britain, Potters Bar, 1973: the big red garage door, adjunct to the then-family home.

When the big red garage door is recalled, the memory is not retrieved from the brain, although the recollection is certainly expressed via the brain. Instead, nous circumvents the brain and obtains the memory directly from the abstract past, through temporal means: directly from 1973 itself: from the real events retained in Bergsonian memory... *not* **from any presumed facsimile, copy or intermediary writ in the brain**.

The actual big red garage door, retained within the Bergson facet as part of the 'environment in memory', constituted as a complex system of wavefunctions in memory that include the author's then-brain and the world as it existed in 1973. Through putative mnemonic mechanics and a process of temporal resonation, the essentials of which were espoused in Book-IV, the said memory superposes but does not disappear into subsequent wavefunctions in memory within the Bergson facet or the 'worm' that elaborates to include the recency states of the author's brain and world in the relative present. It is tempting to reject this 'mad idea', given ready belief in the falsehood of 'space' and the dubious hegemony of materialism. Yet, as an inevitable consequence of de-spatialisation, memory recollection occurs in exactly the 'mad' way described; through direct temporal acquaintance with the temporally removed and extant

Bergsonian past., although this is accomplished 'through a glass darkly', given sources of inefficiency that we need not adumbrate at this juncture.

Hence, the author's recollection of the 'big red garage door' from the past constitutes direct acquaintance with that past: an acquaintance obtained, *not* through space, and not through contiguous spatial-material mediation, but purely across time. An anti-materialist or immaterialist contention on memory recollection, first intuited by Henri Bergson, over a century ago.

*

To fully appreciate the implications, we first assume strict materialism and blindly assert the notion that memories are stored in the brain…as facsimiles (as mere copies) of past events, but *not* as the real events of 1973. Using sundry procedures, we map that part of the author's brain that 'lights up' when he express about 'the big red garage door from Potters Bar, 1973'. On the assumption of materialism, one naturally assumes that the readiness state that lit up *must* constitute the memory. Indeed, in their experiments to locate procedural memories in chicks, Rose and Harding assumed that the putative locations rendered salient by the glucose pump, or that had thus 'lit up', constituted *the* procedural memories employed by the chicks. Recall that Rose and Harding surgically destroyed these to test for memory, but obtained unexpected results: (See Preliminary **0.40** and brief summary below).

Procedural memories are obviously *not* autobiographical, as is the author's 'big red garage door, 1973'. Yet, for materialism, what applies to procedural memories must equally apply to autobiographical memories, if not more so… especially if one insists that these must also be materially stored in brains, perhaps as the very structures that 'lit up' when the author recalled the big red garage door, 1973. However, there is no 'space'. Hence, the in-brain structure that 'lit up', and which one presumed to be the spatially distributed memory of the big red garage door (even if it could constitute the real memory of this) is *not* and cannot be rendered into a spatial 'location' or distance: Recall that, in a de-spatialised universe, there are no 'locations'. What lit up in the author's brain itself constitutes a corpus of past events distributed in time: related to the neuro-scientist observer and to the author, and to the respective moments of consciousness: as past events within the Bergson facet, retrospectively apprehended by the neuro-scientist observing the 'lit up' events, and retrospectively recapitulated by the author: This retrospectivity must also hold true vis-à-vis procedural memories 'located' by Rose and Harding in their experiments with chicks. That is, what Rose and Harding detected via the glucose pump method were Bergsonian past events, related to both purely across time, *retrospectively*. Rose and Harding related to what 'lit up' in the chick's brain in the same way we relate to the Moon, to any other 'object', or to whatever 'lit up' in the author's brain: a purely temporal relation, constituted as a delayed choice time-interval relation between witness and 'object', *without* space.

Thus, whether it is the Sun, the Moon, the vase on the table, or whatever 'lit up 'in the author's brain or in the chick's brain, the witness is integrating to and drawing information from an abstract past, *across time*, always retrospectively and from a *real* past… *not* from a facsimile or copy of that past… and certainly *not* from a 'material location' of it, given de-spatialisation and temporisation.

Hence, the readiness state for the big red garage door from 1973 that 'lit up' in the author's brain is not constituted as a 'material locations, but as the abstract past: a Bergsonian expression removed from the witness *in time*…. but temporally recollected directly from the original past, through a remarkable process that *does not operate across space*, and is not retrieved through any contiguous material mediation process… simply because there is no 'space'. (More will be said about the detailed process of recollection later).

Temporal recollection must involve mnemonic mechanics: the temporal superposition of wavefunctions in memory through which the past is 'transported', albeit purely temporally and non-contiguously, to a relative present apprehending noetic moment. In the author's recollection of the 'big red garage door', this was accomplished through a temporal 'reach out' to a set of past events and Bergsonian abstractions removed from the author's noetic moment of recollection by a delayed choice time-interval that elaborated back to 1973, *not* to the mere nanosecond into the past, or to whatever 'lit up' in the author's brain. Thus…

- When the author recalls the big red garage door from 1973, his brain lights up in certain 'places'. Another observer, the neuro-scientist, might assume that these must be the in-brain 'locations' of the memory of the big red garage door from 1973: a facsimile of the past: *not* the literal past.

- But de-spatialisation implies that *all* realised events, including that part of the author's brain that lit up, are Bergsonian states of the past, related to pertinent observers (neuro-scientist and the recalling author) temporally, through mnemonic mechanics across time, *not* across space.

- The part of the author's brain that lit up is not a 'location' in space but is itself a more recent past, separated from all and sundry purely by time. To the neuro-scientist, it is a past-event at an interval equal to, say, a nanosecond.

- However, when the author recalls the big red garage door, he temporally reaches back into the Bergsonian past, literally to the 'big red garage door, Potters Bar, 1973" itself: *not* to what lit up in his brain.

- What lights up in the author's brain is the post-recollection means of expression, per function of the fact that the brain and its readiness states are needed in the practical expression of a set of events composed of "I remember the big red garage door from our home at Potters Bar, 1973". As such, what lit up in the brain is *not* the actual recollection itself, much less the mnemonic store or generator, notwithstanding the results of the Libet experiments or implications from the experiments of Rose and Harding (see Preliminary, **0.44** and **0.40**).

544

5.45: Bergson-Whitehead Model of Memory: Essay-II:
Bergsonian Primacy-Recentcy Relations & Memory Recollection Across Time

Given de-spatialisation, **the part of the brain that lit up per the author's recollection of the big red garage door, Potters Bar-1973, constitute a *when*, even if Rose and Harding should insist that it as a 'location' in the brain, or assert the 'location' as the memory of the big red garage door, but not 1973 itself. The retrieval of that in-brain *when* must also entail a pure temporal relation to the witnessing recollecting consciousness, given de-spatialisation and temporisation. Indeed, there is no *essential* difference between direct temporal acquaintance with the actual big red garage door from 1973 versus direct temporal acquaintance with the more recent past comprised by what 'lit up' in the brain. Both entail relations across time, *not* across space...between *whens vis-à-vis conscious recollection.***

If it is possible to temporally acquaint with a *recentcy* past (with what lit up in the author's brain) without spatial or material contiguous exchange, then it is no less possible to temporally acquaint vis-à-vis a more profoundly temporally removed past: the *primacy* constituted as the a-facsimile *actua*: namely, the big red garage door, Potters Bar 1973 itself. Hence, the author's recollection involves direct temporal acquaintance with the big red garage door from 1973-proper... not with a mere recentcy facsimile of it writ the brain: a process that circumvents the brain entirely and operates across time; integrating the primacy from 1973 with the recentcy of readiness states requisite to the expression of memory (i.e., what that lit up in the brain) and with the subsequent moment of the author's conscious recollection of it. **Indeed, this is the only possibility permitted in a de-spatialised temporised universe**.

*

Let us now model the above in terms of the Bergson-Whitehead amalgam and primacy-recentcy rules first introduced in Book-IV **4.10** and **4.27**. The 'lit up' *recentcy* brain activity that one mistook for the 'spatial repository' of the author's memory versus the *primacy* comprised of the big red garage door from 1973-proper, are integrated from within the Bergson facet of the Bergson-Whitehead amalgam. **The author's past conscious attention and collection of the big red garage door in 1973 is, through a process of mnemonic mechanics-mediated superposition of past states within the Bergson facet, made integral to the author's present moment of conscious attention and *re*collection of the same. Note that this is an integration across time, *not* across space,** given de-spatialisation. Hence, tandem to the author's expression about the big red garage door through his in-brain readiness states (the *recentcy*), his witnessing consciousness goes back to the original Bergsonian events of the big red garage door, Potters Bar, 1973 (the *primacy*) without ultimately relying on the brain in recentcy in order to accomplish this (more on this later).

As admitted, in order to express and declare the recollection of the big red garage door from 1973, the author must also reach the readiness state in his brain (the *recentcy*) and 'light it up' in order to unfold the expression and declaration,..."I remember the big red garage door from Potters bar-1973"... as a set of subsequent events. The said readiness states of the brain are analogous to keys on a radical de-spatialised keyboard-set. The author's mind is analogous to the extant end-user: extant both the keyboard-set *and* the embedding 'environment in memory' and its 'worm' brain. The end-user mind activates the keyboard-brain and expresses its intended output via the keys (i.e., what lit up in the author's brain). Obviously, the keyboard does not contain the end-user's memory. Obviously, the keyboard or keys do not contain or generate the end-user author himself... and neither does the brain, nor does the Bergsonian worm and environment in memory incorporative of the big red garage door 1973-proper.

*

From the above, the attendant in-brain *recentcy* state constitutes merely the means of expression of recollection (the keyboard complex), *not* its means of production, even less its 'material' retention. The Bergsonian *recentcy* that 'lit up' in the author's brain does not constitute the 'location' of the Bergsonian *primacy*, or of the a-facsimilic temporal actua: the *real* big red garage door, Potters Bar, 1973... embedded as a Bergsonian *primacy* and complex wavefunction in memory within the grander de-spatialised stereotemporal Bergson facet: temporally integrated to the *recentcy* that 'lit up' in the author's brain through the superposition of Bergsonian memories. Thus...

- The big red garage door-1973 is a *primacy* retained in the de-spatialised Bergson facet of the Bergson-Whitehead amalgam. The brain-events that 'lit up' constitute *recentcy* states that are also retained in the Bergson facet, but as the more recent past.

- Tautologically, the *primacy* is not the *recentcy*: The big red garage door-1973 *is not* the brain-events that 'lit up' attendant the moment of recollection in the 2020s. Therefore, the *recentcy* of brain-events does not constitute a facsimile of the *primacy* from 1973: The brain-events do not constitute the real memory of, nor comprise the portage of the then-consciousness of, the big red garage door-1973.

- However, the *primacy* of the then-consciousness of the big red garage door-1973 is *temporally* integrated and keyed to the *recentcy* of brain-events that are also within the Bergson facet. This entails the superposition of Bergsonian memories obtained via putative mnemonic mechanics.

Where is this retrieval and recollection process that operates across time? **There is no in-brain retrieval mechanism, much less a *where* to it, given de-spatialisation.** The process of how memory arises to a witnessing recollecting nous, and how mind selects for memories and their readies states of expression, unfolds according to de-spatialised and temporised processes, involving the processes of the Bergson-Whitehead amalgam, the superposition of wavefunctions in memory (see Book-IV **4.27** and **4.28**), combined to the gear-less process of AND-to-OR wavefunction collapse and time... with a touch of grand decoherence. The details about this must await the

545

intermediate theory in Part-V. At this juncture, **given the case for de-spatialisation and temporisation, it is obvious that one cannot have the equivalent of a 'Waldo arm' that reaches back into the past, picks up the memory of the big red garage door at some spatial store-point, and deposit it to that part of the brain that 'lit up': In a de-spatialised temporised universe, you cannot have spatio-material inputs and outputs mediate between the recollector and the recollected memory, simply because there is no space or location for attendant contiguous material intermediations to unfold, much less mediate contact-transactions between the** *recentcy* **of the present moment of recollection and the** *primacy* **of the big red garage door 'located' at temporal remove within the Bergson facet in 1973.**

Clearly, the materialist model of memory and recollection is forlorn. Aspects of brain science, neurology, and the whole of philosophy of mind steeped in materialist notions about memory must be overturned and reconstituted anew, necessarily in temporised Cartesian terms, without *res extensa…without* materialism.

5.46: Bergson-Whitehead Model of Mind and Consciousness, Essay-I:
Location of Mind and Conscious Recollection in Relation to the Whitehead Facet & Time

Where does the conscious recollection of memory occur? *Where* is the conscious recollection of the big red garage door 'located'? As with the 'location' of memory and its retrieval process, in a de-spatialised temporised universe, the only valid question is... *when* will recollection of the big red garage door occur? Hence, the want to find a spatial 'address' to the very moment of conscious recollection is as bunk as the want to spatially locate memory. Hence, the want to specify a *where* to 'illusory' nous must remain equally forlorn.

Paradoxically, the recollection of the past, or at least the declaration to that effect, must be slated to occur in the *future*. Therefore, **the Whiteheadian future must be potentialised to, and made to decohere to, the sought culmination: the declaration-event. This is achieved by the intercession of active mind that selects from the menu of Whiteheadian future possibilities the 'item' it seeks to AND-to-OR collapse and realise** *in the future*. **The mind must 'reach into' and decohere that sought future in order to realise its subsequent conscious recollection-declaration of the past**. It follows that we must relate mind and consciousness to the Whiteheadian future and seek to frame the moment of noetic recollection to and through that future.

However, while it is possible to specify a Whiteheadian future potential moment for the future declaration of recollection, and subsequently, upon attendant AND-to-OR collapse, clock that declaration as a subsequent realised set of events... it is *not* possible to clock either the active mind that had instigated that future moment of expressed recollection, nor even the consciousness associated with that expression as an event of the Bergson-Whitehead amalgam, notwithstanding the results obtained from the experiments of Benjamin Libet and, therein, the erroneous attempt to relate the moment of consciousness (volitional or otherwise) with a pertinent moment or interval of brain activity. Thus, we arrive at **the non-clockability thesis:** Mind and consciousness do not constitute events of the 'world', or of the growing block Bergson-Whitehead amalgam. Thus, mind and consciousness cannot be specified or clocked as either Whiteheadian future-potential events, or as realised events belonging to the event horizon and the Bergsonian past.

Of course, **the expression of recollection (including what 'lights up' in the author's brain), or of any other expressions,** *do* **constitute clockable events. Yet, and to reiterate, the mind that instigated these, and the subsequent temporally distinct conscious witness or affirmation of the generated expressions, are** *not* **clockable events of the world or of the brain**. Hence, mind and consciousness are extant the world and brain, with the latter confined to Bergsonian memory and to Whiteheadian future-possibility. That is, it is *not* the case that mind and consciousness arises in or from the Bergson-Whitehead world and brain. But it *is* the case that the world of memory and possibility arise *to* mind and consciousness... as was demanded in the argument for *noumnemonae* from Book-IV, culminating into the *en-worlded mind*. (More will be said about the latter in the concluding parts of the general theory).

*

For the recollection-declaration of the big red garage door-1973 to occur, the future possibility for its expression must be selected out of the complex of future Whiteheadian potentials. From *where* will this selection occur? *Where* in the brain will it transpire? Again, in a de-spatialised purely temporised universe, the *where*-question is an absurd question.

However, we can specify the *when* for the events constituting the declaration of the recollection of the big red garage door. We can also specify the *when* recollected: namely, the big red garage door from 1973. But what about the *when* that caused that recollection, or had selected the specific memory to be recollected, *and* had then decohered the future moment pertinent to its expression or declaration into 'recall'? Given the non-clockability of mind and consciousness, the answer is obvious: the *when* of the active mind that selected the future expression-event is *not* clockable as an event, or as *any* moment within the Bergson-Whitehead amalgam.

There exists a future potentiality within the Whitehead facet at which the declaration of the recollection of the big red garage door from 1973, and the pertinent in-brain readiness-states for the expression and declaration, are potentialised for future realisation. In the intermediate theory in Part-V, we will discover that there are many such future-potentials for conscious recollection; all equally probable, but each attending a different memory or Bergsonian past. Together, these constitute a **Whiteheadian potentiality-menu for recollection-declarations**. From this Whiteheadian menu, active mind forgoes the alternatives and selects the sought declaration of the recollection of the big red garage door-1973. This is accomplished through a process analogous to the operation of a de-spatialised (hence, ephemeralised) punch card: Active mind imposes its ephemeral 'punch card' on the future brain, exploiting the 'gaps in time' that reside between successive event horizons to accomplish the imposition... selecting from that future the declaration of the recollection of the big red garage door-1973. The full account of the ephemeralised 'punch card' process will be furnished in the intermediate and fine-structure theories in Part-V and VI.

546

Again, **insofar as the mind in its active mode selectively acts from between successive event horizons… hence acts from attendant 'gaps in time'… it follows that neither the mind in its active mode, nor subsequent conscious recollection realised in passive mode witnessing consciousness, can be clocked as** *any* **OR-form event on the event horizon, or within Bergsonian past, nor as any potential event in the Whiteheadian future. This is precisely because both active mind and passive consciousness act and arise in and from the 'gaps in time',** *not* **from the frames of past and recent events resolved on successive event horizons … nor from the brain, or what 'lit up' therein… no more than from future potentialities and frames that flank the event horizon and its realised events.**

From non-clockability, we grasp the inherent fallacy at the core of generic interpretations of the Libet experiment (see Preliminary **0.44**): The fallacy therein resides in the *assumption* that the moment of conscious recollection is a clockable event that we can relate to preceding or subsequent clockable brain-events. It is not… even though the attendant brain activity and subsequent expression *are* clockable events.

*

To reiterate and augment, **mind and consciousness exploit the process of AND-to-OR collapse and time. The declaration of recollection aside, neither mind nor consciousness constitute any OR-form event or outcome brought into realisation by the process of time. Minding and conscious recollection 'occur' from or through the 'gaps' that attendant the process of time. This also implies that we cannot ultimately clock mind or consciousness as any event, or as a** *when,* **much less as part of any future potential** *whens.* **At best, we can only state that mind initiates its control of the brain by decohering the Whiteheadian future potentiality to produce recollection-declarations or expressions that then arises to passive mode no-clockable witnessing consciousness.** We can only speak meaningfully of future-potential moment of recollection-declaration events embedded in the Whitehead facet. Yet, this futurity is merely a possible readiness-state for *expression*, which itself requires that it must subsequently emerge into consciousness in order for conscious recollection to occur. i.e., One can only be aware of an expression of recollection temporally *after* its fact, or retrospectively. Again, the point is that **the moment of conscious recollection, distinct from the 'material' events constituting its declaration and expression, is** *not* **constituted as an event of the world, much less an event in the brain. The moment of conscious witness and recollection is** *extant* **the system of events constituting world and brains.** Hence, Cartesian dualism in the context of memory recollection and expression inexorably follows.

5.47: Bergson-Whitehead Model of Mind and Consciousness, Essay-II: 'Location' of Mind & Consciousness in Relation to the Event Horizon

We now augment why mind and consciousness cannot be identified with any state or event on the event horizon, such as with the manifest OR-form brain confined to the event horizon, or with any Bergsonian memory, even as these are apprehended by nous in retrospect…nor with any Whiteheadian futurity, even as these are decohered and selected for future AND-to-OR realisation by nous, per the process described in the previous essays.

Recall from Book-IV that the successor to spacetime is the de-spatialised Bergson-Whitehead amalgam, with the Bergson facet comprising abstract memory, and the Whitehead facet comprising nested future possibilities; with both facets separated by a dynamic event horizon. The event horizon is the component of the Bergson-Whitehead amalgam that comes closest to our forlorn notions of 'matter in space'; constituting the most recent OR-form outcomes generated out of AND-to-OR wavefunction collapse of the Whitehead facet. However, the event horizon is *not* a spatial plenum on or along which one could lay out the brain and its *alleged* 'material memories', or any putative memory-retrieval and depository mechanism, much less the minded initiation of subsequent conscious recollection. Indeed, we cannot inhere any general causality that supposedly unfolds on or along the event horizon, given that ultimate causality is vested in the gear-less process of AND-to-OR wavefunction collapse and time that, in the first place, generates the temporal succession of event horizons and their OR-from outcomes, but is not itself generated by the event horizon or its succession, and even less by the OR-form events or outcomes (the brain) manifested along the event horizons. Indeed, the causality that generates wavefunction collapse and time is gear-less ineffable, and not capable of rendition as any OR-form term or event, nor even as any future potentiality for these: See Book-V **5.19** and Book-II **2.46** and **2.47**, and Book-III **3.31** and **3.32**.

In short, **given that we cannot model causality in general as a thing transpiring on or along the event horizon, we cannot specify on or along that event horizon any causal 'material' exchange for in-brain memory retrieval, or for any corpus that supposedly 'excretes' minding or conscious recollection,** much less accommodate materialist identity theory, epiphenomenalism, or even the latest fashion in emergence or supervenience theory.

However, we must examine the event horizon more closely to appreciate our conclusion: Therein, the relativity of simultaneity will come to our aid. Thus, if we described the event horizon as a simple horizontal line in a generic Minkowski diagram, and we placed the brain along it, (see **Fig. 5.09**) we find that, for two distinct points posited upon it and supposedly constituting the in-brain memory state (**P1**) vis-à-vis its retrieval mechanism (**P2**), the latter point on the event horizon will be related to the other point, *not* horizontally or spatially along the single event horizon, even when we assume the reality of forlorn 'space'. Instead, **P1** and **P2** must relate via the succession of event horizons, or 'vertically'…hence, *temporally*… by means of future-potential delayed choice time-relations that attend pertinent future-potential worldlines and lightcones. Thus:

- Interrelations between putative 'memory' at **P1**, or its putative retrieval mechanism at **P2**, will transpire under conditions of the breakdown of absolute time, the full operation of the relativity of simultaneity, and consequent implicit de-spatialisation: See Book IV **4.01** to **4.07**.

- Hence, even if we sought to identify memories and their conscious recollections as points or structures along the event horizon (or as events **P1** and **P2**, respectively, and 'in the brain') the relations and integrations between these will remain purely 'vertical' or purely temporal, regardless of the false presumption of a spatial distribution of **P1** and **P2** on or along EH1.

- Thus, the memories, the points, will be 'stored' as distributions in time, even when we assume pleonastic spatiality or locatability for these. Any putative retrieval process for these must also operate upon the said temporal distribution vertically and *across time,* and it must somehow 'reach back' to those past distributions in order to recollect these across time, in slated future moment, *not* across space... not along any single event horizon... but 'vertically' via the temporal discontinuous succession of event horizons...or, simply, across time.

- The attempt to reduce the 'location', source, or production of mind or consciousness on or along the event horizon (as 'material structures' along 'space' (i.e., as, say, **P2**) will thus fail. Even if we could accomplish this, its relation to and processing vis-à-vis memories (**P1**) will need to transpire 'vertically'... across time... *without* contiguous contact... without 'material exchange', voiding the initial assumption of spatiality and contiguous material causality.

- This conclusion cannot be altered by the imperceptible short time-interval 'vertical' relations between **P1** and **P2**. Just because we cannot perceive the pertinent short time-intervals, we have no licence to treat the brain as a 'classical object', or as an 'object in space', or seek to 'locate' memory or even mind and consciousness, or *any* 'retrieval mechanism', to 'spatial correlates' of the event horizon-restricted brain.

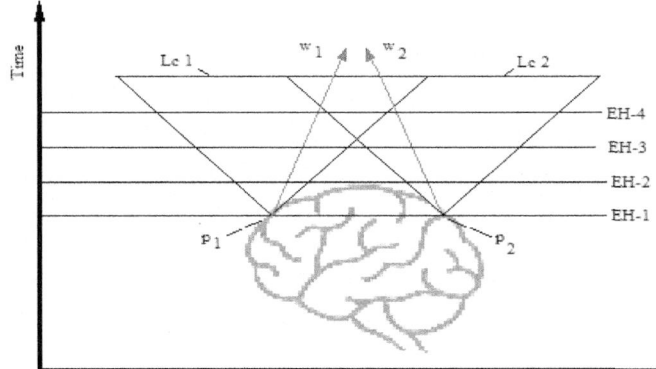

Fig. 5.09: Minkowski model and of integration of brain-events
Notes: The Diagram depicts a simple Minkowski spacetime model of signals mediated between events in the brain, P1 and P2. The model assumes the veracity of space while it incorporates successive event horizons from EH-1 to EH-4 developed in the new spacetime model in Book-II. Brain-events P1 and P2 may be integrated either via the overlapping lightcones (Lc1 and Lc2) per 'speed-of-light' signalling, or via interacting worldlines representing signals transmitted between P1 and P2 at less than the speed of light across the brain. Even with the assumption of 'space', we immediately see that the signals and interactions are *not* mediated horizontally across 'space' or in a 'common present moment' of absolute time, but 'vertically' via either the lightcones or the worldlines. Hence, the integration between P1 and P2 happens across time, as signals from P1 to P2, and as past events. Hence, the unitary binding of the brain (the integration and co-ordination of events P1 and P2 in the brain, whatever its ultimate basis) must also transpire 'vertically' across time. When we explicitly incorporate de-spatialisation and move from spacetime to the Bergson-Whitehead amalgam, *all* relations within the brain transform into de-spatialised ephemeral primacy-recency memory relations. Memory retrieval and the unitary binding of the brain must also constitute a relation and integration of memory and information purely across time: not achievable from any 'point' or corpus within the brain or on or along the event horizon, EH-1.

We must also recall from Part-III that IOO-schema contiguous causality, or materialist notions of causality, were invalidated. Material notions of causality can no more inhere to the 'vertical' temporal relations between **P1** and **P2** than to or along horizontal 'spatial' ones that supposedly reside between these. As for horizontal 'spatial' relations, given de-spatialisation, there are none to speak of in a purely temporised universe and brains.

*

In grand decoherence and counterfactuals-driven quantum erasure of nested futures (see Book-II **2.34**) and in the 'non-locality' or transtemporality entailed therein, emerged the instantaneous unity and coherence of one part of the event horizon with every other part of the same, leading to consequences that included, among other things, the revival of Mach's principle. Could such a process attend **P1** and **P2** along the event horizon EH-1? Could counterfactual causality and grand decoherence rescue some semblance of absolute time and restore some version of notional 'matter in space' and 'spatially mediated causality' to the brain, with attendant approximation to a 'spatial-material brain' thus? The answer is an emphatic no.

The operation of counterfactuals in the quantum erasure of nested futures and in grand decoherence pertain only to the erasure of non-consistent futures (which certainly apply to and between mutually decohering events in the brain) and secures consistency principles that then prohibit the whole universe, or any of its subsumed states (including the brain) from developing into future events that are not consistent with the constituting events and memory of the universe. This cannot lend itself to treating the brain as a state in absolute time, such as to recover the materialist conception of the brain, much less recover the idea of memories in supposed 'spatial locations', or of spatial retrieval mechanisms based on IOO-schema impact-mediated in-brain contiguous causality. Thus…

- **Counterfactual causality and grand decoherence cannot be posited as the means for the integration of one part of the brain (P1) to any other part of that brain (P2) in absolute time or in synonymous spatial terms**. Consequently, it is impossible to postulate for a materialist's 'spatial brain' per counterfactual causality and grand decoherence.

As was recommended in the previous essay (**5.46**), memory retrieval must be founded on entirely different non-spatial, non-contiguous, and purely temporal ephemeralised Bergsonian terms: Recollection *cannot* involve spatial contact between spatial block or point-like 'particles', or P1 and P2 constituted as such. This demand applies to causality in general, as was argued in Part-III, as much as it applies to causality in memor-recollection and minding. Instead, the relation between one part of the brain verses another, as distributions in time, will be subsumed to general AND-to-OR wavefunction collapse and time-processes, which are *not* driven or made to happen by 'gears', 'inputs' or contiguous 'impacts', and which unfold spontaneously according to a work function zero profile. Therefore, the relation between a *primacy* event-complex formed at one event horizon, such as the author's big red garage door from 1973, versus the *recentcy* constituting the expression and declaration of his recollection of the big red garage door at a succeeding event horizon, is not reducible to or accountable in terms of any 'horizontal' contiguous spatio-material relations along the event horizon, or along or in the event horizon-confined brain.

*

Memory relations grasped as pure time relations between successive event horizons are ineluctably 'vertical' temporal relations between succeeding event horizons: they are Bergsonian temporal relations obtained through successions of AND-to-OR collapse or time-processes. Implicitly, the minded selection of the future-potential Whiteheadian moment at which one's recollection of the big red garage door is to be express, as well as one's subsequent conscious witness of that recollection, must be initiated from *outside* of the brain-pertinent Bergsonian memory, and from *outside* of Whiteheadian future-possibility. And, given the above analysis, memory relations, recollection and nous must be initiated from *outside* the event horizon, extant the OR-form 'material' brain confined to it and resolved upon it.

Again, the instigator of the recollection (namely, mind), must be extant the event horizon, hence extant the event horizon-confined brain… and *certainly* extant the whole of the Bergson-Whitehead amalgam. Thus, Cartesian dualism is, again, the inexorable conclusion.

5.48: Bergson-Whitehead Model of Mind and Consciousness, Essay-III: 'Location' of Mind & Consciousness in Relation to the Bergsonian Facet

It is tempting to suppose that both mind and consciousness might be facets of the Bergsonian system of memories, and that these might be clocked as past events within the Bergson facet. From the previous essay (**5.47**) we appreciate that this supposition is false. Even if it were true, it would be of no help to materialism, given that both mind and consciousness, now speculatively posited within the Bergsonian, would remain extant the 'materialised' brain confined to the event horizon: dualism would yet abide. That aside, as was shown in previous material, mind and consciousness are *not* located, clockable in, nor operate from the Bergsonian past, no more than these are clockable as future-potential events of the Whiteheadian future (see **5.46**), or as realisations on the event horizon (see **5.47**).

For memory to fall into consciousness, the specific events and facts pertinent to that memory must first unfold as realised events. By implication, the witnessing consciousness of these can only arise temporally *after* those events and facts have transpired: The witness or consciousness is initially a potential *in the future* vis-à-vis the realised events and facts. Hence, with respect to the whole Bergsonian memory, the witnessing consciousness of past events and memories, with the latter comprised of both the past states of the brain *and* of the larger environment of events, can only be brought about at a future moment of recollection. That is, **conscious recollection cannot be generated from the past, and it is clearly not clockable or locatable as an event of that past, given that it can only be affected in the future;** *always* **in relation to events and memories that have transpired…** *after* **these have transpired**…not from, and not as.

Approaching the same, but from the Bergson facet…

- The brain itself cannot be restricted to the event horizon: The brain is in truth a Bergsonian state of mnemonic mechanical superposes of brain-pertinent wavefunctions in memory: i.e., the temporised version of Karl Pribram's holonomy, and the perduring 'worm' of process philosophy. As such, and as an abstraction of memories, the brain must elaborate into the past,

549

such as to constitute a whole history of the brain...such as to constitute the said perduring 'worm', temporally project from the past to the relative present.

- Hence, at the minimum, the OR-form states of the brain 'materialised' on or along the event horizon are culminations from the most recent AND-to-OR collapse of the said total history or mnemonic superposition of Bergsonian memories constituting the brain 'worm'.

- Obviously, the same Bergsonian memories are inter-related 'vertically' or temporally (see **5.47**) and are subject to the relativity of simultaneity... as stereotemporal relations in and across time.

- Specific events and facts pertinent to memory must first unfold as realised events and as the culmination of the perduring 'worm' upon the event horizon. By implication, the witnessing consciousness of these, and their subsequent minding, can only arise in retrospect, or *after* those events and facts have transpired... i.e., *in the future*. At best, we can only suppose that both mind and consciousness are 'located' in the future. But even this must fail per the non-clockability thesis and other considerations: e.g., infinity-laden noeta.

Hence, the attempt to locate minding and consciousness within the Bergson facet must fail as surely as the attempt to located mind, memory, and consciousness on the event horizon itself, or even within the Whitehead facet itself.

5.49: Bergson-Whitehead Model of Mind and Consciousness, Essay-IV: 'Location' of Mind & Consciousness in Relation to the Whitehead Facet And Solutions to the Unitary Binding Problem of the Brain

The want to locate or clock mind and consciousness as an aspect of the Whitehead facet immediately confronts the fact that the facet constitutes the future: Mind and consciousness in the future have not yet transpired. As such, potential as-yet not-happened mind or witnessing consciousness could not resolve as either agency of or witness of past-memory, simply because neither have yet transpired.

As to the matter of whether mind might intercede upon the future, not as a potential future event, but from the 'gaps' that intersperse potential event horizons... *and* decohere and affect the future development of the brain... *and* bring about the latter's unitary binding... this constitutes an entirely different consideration: part of the ephemeralised 'punch card' theory of mind-brain agency and causality that we promise to develop in the intermediate mind-world theory.

*

The non-reducibility of mind and consciousness to the event horizon and its 'materialised brain' (**5.47**), combined to the similar non-reducibility of the same to the Bergsonian facet and to the temporised brain 'worm' therein (**5.48**), raises the conundrum of the unitary binding of the brain, implicit to nous. For example, the conscious recollection of the big red garage door affected by mind, insofar as it must be realised in and through the future, presupposes the potential coherent unitary control of the brain in terms of its future *global* activity. This requires the unitary binding of brains concordant to the sough global activity. How is this to be achieved?

As we discovered previously in **5.47**, it is impossible to establish any location for mind or consciousness, much less any basis for unitary binding of the brain, from within or across the event horizon, given implications from the relativity of simultaneity and the said temporal 'verticality' of pertinent relations. By exclusion, **unitary binding of the brain could only be achieved through the manipulation of the Whitehead facet: succinctly, through the unitary global AND-to-OR collapse of a future complex wavefunction that, as an 'instantaneous whole', encompasses the whole brain, and culminates in the collapse of the whole brain wavefunction in unitary fashion, thus realising the unitary binding of the brain.**

From this insight, we can now grasp why it was hitherto impossible for brain science, neuroscience (e.g., Penfield) and the philosophy of mind to explain the unitary control and coordination of brains from within the 'material' brain, or from anything on or along the implicit event horizon and its event horizon-restricted OR-form brain.

First, a purely materialist event horizon-based approach (i.e., 'matter in space' approach) to brains, and to its unitary binding, would need to establish the unity and coherence of the brain via impossible spatial and material structures and interactions on or along the event horizon, or 'across space'. As we saw even in the conservative Minkowski spacetime treatment of the latter in **5.47**, such a hope is forlorn.

Brain science, neuroscience and the philosophy of mind preclude the Whiteheadian future-based approach (i.e., the AND-form quantum mechanical approach) to the brain, due to their consensus rejection of any decisive role of quantum mechanics vis-à-vis the brain, on the supposed grounds that quantum mechanics (hence the ontic future) is restricted to the 'very small', and supposedly irrelevant to the macro-scale brain. The dominant want to treat the brain as if it constituted a 'classical object' prevails; combined to lack of insight about the growing block approach and the ontic future; the latter grasped as the Whitehead facet: a quantum mechanical potentiality state pertinent to *all* systems at *all* 'scales', including brains.

Hence, it has not occurred to brain science, neuroscience, or to materialist philosophy of mind that, as is true of *any* and *all* wavefunctions (such as evinced in the development of a photographic image), **a complex wavefunction, necessarily belonging to the Whiteheadian future, and encompassing the whole brain, constituted as an 'instantaneous whole' (see 1.03, and 1.26 to 1.27) will decohere and collapse as a global unity. As such the Whiteheadian future-based approach to the brain constitutes the basis for the automatic global unitary 'instantaneous whole' for that brain, both before it decoheres and collapses into unitary realised brain-events and expressions, as well as in the process of that subsequent decoherence and collapse. Thus, the quantum**

mechanical future-form description of the brain, and the instantaneous whole-based approach that attends it, immediately solves the conundrum of the unitary binding problem of the brain, insofar as any and all quantum mechanical AND-form wavefunction states, as instantaneous wholes, exhibit global unitary decoherence, global unitary AND-to-OR collapse, and globally unitary development. This fact is fully evinced in photography: the generation of a coherent image and pattern out of otherwise disparate, *en masse* quantum indeterminate depositions, wherein each deposition, related to every other per function of strict relativity of simultaneity, must yet cohere into a unitary whole image.

However, in the intermediate mind-world theory in Part-V, and in our later 'punch card' theory, we will discover that the Whiteheadian complex wavefunction or instantaneous whole for the whole brain is *not* comprised of just one complex global wavefunction. Instead, it is a menu for many distinct complex wavefunctions: all rendered equally probable: with each constituting an alternative complex wavefunctions and instantaneous whole specific to different goals that the active mind might seek; each generating profoundly different unitary patterns vis-à-vis the future of the brain: each expressing different intents sought by the mind.

All that now remains is that we bring into the fray the noetic agency, the mind, that selects from the said menu one future out of the welter of otherwise equally probable alternative futures. Thereafter, the collapse of the mind-selected complex wavefunction will automatically follow via subsequent a-noetic means, with consequent unitary decoherence and unitary AND-to-OR collapse of the brain into the sought globally unitary brain activity… so realising the binding of the brain through the said unitary decoherence and collapse.

Of course, this requires an active mind that can enforce selective causality vis-à-vis the Whiteheadian future by means of the ephemeralised 'punch card' process... which is to be detailed in Part-V. Moreover, it requires that we understand that noetic agency cannot be clocked or 'located' on the event horizon, nor within the Bergson facet, nor in the Whitehead facet, as argued in previous essays. Mind must be extant the whole Bergson-Whitehead amalgam… interceding from the 'gaps in time' that reside between the succession of event horizons and attendant 'quantum leaps in time' (see **5.30**). Hence… again… Cartesian dualism ineluctably follows.

*

It might be supposed that unitary visceral experience and consciousness, hence unitary singular mind, is a consequences of the unitary character of the complex wavefunction that attends the Whiteheadian brain. Hence, it might further be supposed that both mind and consciousness ought to be identifiable and clockable as the whole-brain outcome of unitary global wavefunction collapse; perhaps arising at the same time as that whole global unitary outcome, despite contrary claims from the Libet experiments. However, the consequent whole-brain outcome does not constitute the basis for *experiential* unitary nous, nor any basis for the temporally undivided noetic moment. Why?

Unitary mind and consciousness requires nothing less than the noetic version of absolute time (see Part-II 5.15): i.e., the visceral experience and sense of the common present moment: one that stands over Bergsonian memory *and* Whiteheadian possibility, and cannot arise on or along the event horizon... nor from Bergsonian past equally subject to the relativity of simultaneity... nor from the Whiteheadian future subject to the same... nor in or from the unitary coherent collapse of the instantaneous whole into the whole-brain unitary activity.

Where is such a point-like and absolute time-like nous 'located'? Aside the absurdity of seeking 'locations' in a de-spatialised universe, point-like and absolute time-like nous cannot be 'located' in the Bergson facet subject to the relativity of simultaneity, which happens to be inimical to the production of any absolute time-like outcome. Nor can unitary absolute time-like nous be clocked on or along the event horizon, which remains inimical to the production of an absolute time-like state, per the relativity of simultaneity. Nor can absolute time-like nous be potentially clocked within the Whitehead facet, even if the potential events of future brain activity and expression could be so clocked: In any case, upon AND-to-OR collapse, these must also subsequently abide to the relativity of simultaneity. Hence, future-potentiality must also remain inimical to the potentialisation of absolute time-like nous.

Note that, **while a Whiteheadian complex wavefunction, as an instantaneous whole for the whole brain, does indeed decohere and collapse in a unitary way, thus conveying unitary binding to the pattern of the whole-brain so formed, the outcomes produced do not arise in an absolute time-like milieu or constitute an absolute time-like state. Hence, the partology constituting total brain activity will remain interrelated according to the relativity of simultaneity. It follows that the 'point-like' and absolute time-like temporal noetic monad we seek cannot arise from the collapse of the instantaneous whole, nor from the whole-brain activity thus formed. The absolute time-like noetic moment must remain extant the Whiteheadian facet as well as extant the totality of the Bergson-Whitehead amalgam.**

Succinctly, mind and consciousness cannot arise from a brain subject to the relativity of simultaneity, even if the brain's future complex wavefunction decoheres and collapses in the said unitary way, and enforces the unitary activity of brains. To summarise and augment...

- Mind does not manifest as a future-potential of the Whitehead facet: If it could, it would merely constitute potential mind, not yet realised… and it could not constitute any form of active agency over the future development of the brain.

- The unitary binding of the brain (but *not* the generation of mind or consciousness from that unitary activity) is obtained from the unitary global decoherence and AND-to-OR collapse of the 'instantaneous whole' or complex wavefunction attendant that brain, selected for future development by extant Cartesian mind.

- The 'point-like' and absolute time-like nous *cannot* be produced by the unitary decoherence and collapse of the complex wavefunction attendant the whole brain… because the latter's content is internally subject to the relativity of simultaneity…

551

Therefore, *incapable* of furnishing an absolute time-like visceral sense of 'a common present moment' from the brain or from within the whole brain. That is, *nous does not arise from the brain.*

- The mind selects the complex wavefunction for its brain before that wavefunction collapses into unitary whole-brain activity. It does not accomplish this selection from the Bergsonian (see **5.48**), nor from the event horizon (see **5.47**). Instead, mind engages from the 'gaps in time'… from between potential successive event horizons constituting the Whitehead facet... but *not* from anything potentialised to or subsequently clockable per the Whitehead future. *Thus, the mind acts from, but does not reside in, nor emerge from, the Whitehead facet... but intercedes from the 'gaps', 'cracks' or interstices in future-potential time... from outside of memory or future possibility pertinent to brains.*

- The inverse is also true: the globally coherent unitary collapse of the Whiteheadian complex wavefunction imposed by mind, hence the brain and world upon OR-form resolution, arises to and within (but not from) absolute time-like witnessing consciousness. Thus en-worlding the mind... or furnishing nous a world within it.

Ineluctably, Cartesian dualism is the foregone conclusion.

5.50: Augmenting Dualism, Essay-I:
Passive Mode Consciousness, Active Mode Mind, Parallelism & the Conundrum of Causal Freedom

Given de-spatialisation and the necessity of a pure temporal approach to physics and to the brain, we can only speak of the *when* of conscious recollection. Even then, mind and consciousness turn out to be extant the system of memory, possibility, and time, as indicated in the previous essays (also see **5.49**). Also, consciousness is *always* consciousness of the past: It relates to the past purely temporally. Indeed, all witnessing nous constitutes memory-apprehension and recollection: either in the form of first-time *collection* of new events in past-recentcy... or as *recollections* or repeat-recollection of past-primacy.

The noetic moment of unitary nous relates to Bergson-Whitehead amalgam in two modes…

- In the **passive mode, the noetic moment constitutes the moment of witnessing consciousness of the Bergsonian past, *after* or in *retrospect* vis-à-vis the events of the world and of brains that had transpired.** Obviously, witnessing consciousness cannot arise at the same time as the Bergsonian world and brain, given that consciousness of the said world and brain occurs *after* the said events, given implications from the relativity of simultaneity: The witness must temporally succeed the recollected past. In the passive mode, there can be no causal power exercised by nous over the Bergsonian brain and world, given that that past had come into formation *before* it arose to consciousness, the content of the transpired past cannot be affected by the temporally succeeding witnessing consciousness. This insight also undermines idealism, as *any* argument for the ontological primacy of consciousness must now argue for a retrograde power of nous vis-à-vis the pre-constituted past. Thus, *passive mode nous has no causal power. Also, it is non-volitional, given its status as the passive recipient of the past in retrospect and as the pure apprehending witness of the past.*

- In the **active mode, nous as mind precedes and enforces selective causal agency upon future-form Whitehead possibilities of the brain and world.** This is realised through the process of AND-to-OR collapse or time... through the 'gaps in time' between potential successive event horizons. The Whitehead facet attendant the brain and selected by active mind subsequently spontaneously AND-to-OR collapses into a unitary global whole-brain outcome. (For its detailed account, we must await the intermediate and fine-structure theories). Thus, in the active mode, from the 'gaps in time', mind precedes future brain activity. However, *mind cannot be identified with any Whiteheadian potential for brain activity, given that mind acts from the 'gaps' between potential future event horizons, and cannot be clocked as the brain in its future-form.*

- The **ephemeralised 'punch card' theory**: In the intermediate mind-world theory, we shall discover that, in the active mode, and *utilising a work function zero process, causal mind accomplishes control over its brain by means of the equivalent of the throw of a suspended delayed choice 'switch-x'* (see delayed choice experiment, Book-II **2.18**). *This selects the appropriate 'punch card' (i.e., complex whole-brain wavefunction attendant the brain) from the welter of nested future alternative whole brain wavefunctions.* Thus, the mind intercedes from between potential successive event horizons (from the 'gaps' in time') to decohere the future-brain to the sought developmental pattern. The future brain spontaneously AND-to-OR collapses into the sought global brain activity, and into realisation of the intents of mind via the three-phase development process (see **3.26**).

- The **active mode mind does not possess consciousness: it is non-conscious, even while it plausibly possesses volition: Indeed, consciousness is an aspect *only* of the passive mode retrospective witness.** Nous obtains consciousness of the fact of its agency only *after* the facts generated by the non-conscious active mind have transpired and arise to passive witnessing.

A form of parallelism must abide: a parallelism between the domain of possibility and time as it pertains to the a-noetic universe (the Bergson-Whitehead amalgam) versus a distinct noetic domain: namely, the *noosphere*. Through the 'gaps in time', the parallel noosphere intercedes into the Bergson-Whitehead growing block. It intercedes either as passive consciousness and witness *without* agency, or as a futures-decohering agency or mind, but *without* consciousness. **It is at the interface per the 'gaps in time' that nous become subordinate to ontologically primary AND-to-OR collapse and time, without which neither noetic witnessing consciousness nor noetic active mind and agency could transpire.**

Consciousness and mind both possess unique characteristics distinct from a-noetic Bergsonian and Whiteheadian memory, possibility, and time: **Both consciousness and mind enjoy absolute time-like characteristics: If space could exist, mind and consciousness would constitute a point-like state. Given de-spatialisation, the 'point in space' transforms into a noetic moment or durata...** a *temporal monad of mind or consciousness*, en-worlded with a growing block universe to both witness and affect.

5.51: Augmenting Dualism, Essay-II:
Dual Parallelism, the Conundrum of Causal Freedom & the Radical Ontology of Mind

Does mind and consciousness undergo distinct and parallel AND-to-OR collapse in relation to and in tandem with the AND-to-OR collapse of the Bergson-Whitehead amalgam? The ontological primacy of time and AND-to-OR collapse is paramount: Without it, mind could not intercede and affect the future, nor consciousness bear passive retrospective witness to memory or deed. Yet, it remains to be shown whether there is a distinct AND-to-OR collapse as it pertains to the noosphere.

The conundrum is easier to solve for passive mode consciousness than for active mode mind and agency: **Where passive mode consciousness is concerned, we postulate that the contents of the world so witnessed do indeed AND-to-OR collapse out of the Whitehead facet. But the subsequent witnessing consciousness must pre-exist as a state of AND-form noetic potentiality for consciousness, in parallel to and in synchronicity with the processes of the Bergson-Whitehead amalgam. As such, passive consciousness must AND-to-OR collapse into 'consistent consciousness' vis-à-vis the** *primary* **AND-to-OR collapse and content of the Bergson-Whitehead amalgam. The AND-to-OR collapse of passive consciousness is from a parallel noetic potentiality: itself part of the noosphere, distinct from but correlate with the Bergson-Whitehead amalgam.**

Yet, a similar treatment to active mode mind remains elusive. When mind decides to decohere the Whiteheadian future to a sought outcome, does its decision to do so collapse out of a parallel and distinct AND-form state of noetic decision-potentials? If so, then the decisions are in future-potentiality, have yet to transpired, and must AND-to-OR collapse out of that decision potentiality for any subsequent decision-resolution to occur. But if this is subject to a thing similar to quantum indeterminacy... an indeterminacy for noetic future decision-potentials, then the decision realised would be totally arbitrary and as good as random. If so, the 'decision making' and agency on the part of active mode mind would be rendered merely arbitrary and haphazard outcome: a non-decision.

Of course, similar conundrums reside at the heart of the larger problem of causal freedom that besets the philosophy of freedom, to which solutions have remained elusive. Even so, we must postulate solutions: To this end, there are two possibilities. Thus...

- Decision making is as just described: **Mind's 'decision' collapses out of an AND-form noetic decision-potentials via an indeterminate process. Thus, there cannot exist any real decision making, and it is wholly arbitrary** and illusory... or...
- **Mind does not ultimately abide with time, and even transcends time, even if it cannot cause it or drive it... in the sense that mind is not itself subject to any form of AND-to-OR collapse or attendant indeterminism. If so, decision-making does not require an AND-form state of decision potentials or their AND-to-OR collapse,** even while subsequent witnessing consciousness *is* resolved via AND-to-OR collapse of future noetic possibilities, so claiming witness to the deeds or misdeeds of preceding mind.
- **In which case, ontologically, mind has a greater affinity to ineffable ontology or Being (or Brahman) that subsumes but transcends memory, possibility, time,** *and* **witnessing consciousness... but without exhaustive identification with or in ontic exhaustion of the whole of Brahman, given the demise of idealism in Part-II:** (also see the non-arbitrariness and meaningful universe arguments in Book-III **3.14** to **3.15**, especially as these pertain to the Euthyphro dilemma).

The postulate that active mode mind is not ultimately subject to its own parallel form of AND-to-OR collapse and time, even while it depends on Whitehead-facet collapse for the very possibility of its agency in the world, is indicated by the following:

- Mind can look into the AND-form Whiteheadian future possibilities as these pertain to brain and world, *and* decohere these in a selective way: In other words, **mind appears to possess the uncanny power to project into and over the future, at least as much as it can project into and over the Bergsonian past, but without requiring (and actually preceding) the AND-to-OR collapse process that collapses the Whiteheadian future into actua. Thus, active mode mind (but** *not* **witnessing passive consciousness) appears to be transcendent vis-à-vis the process of time, even while it is dependent on time to affect its deeds:** Hence, active mind enjoys far greater affinity to ultimate Being or Brahman, even if it is penultimate to Brahman, and therefore *not* Brahman.

However, the truism that mind cannot be the totality of Being, and that Being cannot be reduced to mind, emerges from the following:

- **The dependency of time-transcendent mind on AND-to-OR collapse and time pertaining to its power to intercede upon world and brains**, which mind does not cause or drive, even if it transcends AND-to-OR collapse and time.
- **Mind is not time-symmetric in its relation to and interface with brains and world rendered into memory and possibility** in that mind cannot operate equally forwards and backwards in time (i.e., mind can never operate from OR-to-AND) vis-à-vis world and brains. Thus, active mind **is totally time-asymmetric (always operating from AND-to-OR) in its modus of intercession and affectation vis-à-vis the brain and world** or the Bergson-Whitehead amalgam.
- The assertion that mind has no suzerainty or choice over the above-stated fundamentals and conditions shows that mind is *not* the comprehensive 'owner of reality', so to speak...and that mind, despite its grater affinity to nous-transcending ontology

and Being, cannot exhaust Being such as to constitute it as its identity, and must remain penultimate to Being. But these aside, a further case can be made from the relation of mind to **superlatives, metalatives, and eternal principles similar to Platonic forms. These are inherent to Being upon which mind subsists, and which mind cannot create *ex nihilo*,** (most notably Euthyphro's dilemma and profound nihilism: Book-III **3.14**). Thus, Being or Brahman exceeds and supersedes nous or 'Atma'. Thus, God is *not* mind, and mind is *not* God: (see **5.21**). Thus, in its relation to Being, mind is penultimate, *not* ultimate.

5.52: Augmenting Dualism, Essay-III: Further Reflections on the Libet Experiments

The results obtained in the experiments of Benjamin Libet *et al* (see Preliminary, **0.44**) are consistent with expectations from our outline of passive and active modes vis-à-vis mind-world relations. Recall that the **Benjamin Libet experiments claimed that consciousness, and especially purported volitional consciousness, transpires *after* brain activity... and that both consciousness and volition must constitute 'illusions', given their supposed production by the preceding non-conscious brain activity.** This interpretation rests on *a priori* materialism: the assumption that consciousness is produced by the brain. The assumption is loaded to the facts of the experiment. Post hoc, the facts are then presented *as if* in 'proof' of that assumption.

However, the Cartesian possibility that consciousness might temporally succeed (or even follow after) the brain activity, or that a temporally preceding mind (otherwise non-clockable) might instigate the brain activity, is not considered in earnest in *any* generic interpretation of the Libet experiments. This is simply because dualism is *a priori* dismissed as impossible. Yet, given the breakdown of materialist causality per de-spatialisation alone, and with all other considerations elaborated in our general theory of mind and world, the notion of a brain-generated mind and consciousness is no longer tenable, simply because the background *assumption* of materialism is void.

Interpretations of the Libet experiments also suffer from the erroneous attribution of volition to consciousness: As we discovered in the passive mode approach, the noetic moment of witnessing consciousness is always of events and of brains and world-in-memory apprehended in retrospect: A temporally preceding pre-resolved past is presented to temporally subsequent witnessing consciousness. As such, consciousness has no choice over what it is presented and is rendered causally passive. Witnessing consciousness has no active agency or control over what it witnesses or what it subsequently declares to have witnessed. Therefore, consciousness is inherently non-volitional. Therefore, the notion of 'volitional consciousness' is an oxymoron.

In short, we cannot rightly assign volition or causal freedom to passive mode consciousness, and even less to its declarations of volition, or its passive witness to agency that rightly belongs to a preceding non-conscious and non-clockable active mode mind.

On the other hand, **the noetic moment of mind in its active mode is *plausibly* volitional, yet non-conscious. Therein, the causal freedom of mind could certainly be entertained as a plausible hypothesis, notwithstanding materialist interpretations of the Libet experiment**.

Again, mind does not exhibit consciousness: Mind is non-conscious or subconscious. It cannot be realised in any declarations exhibited in the Libet experiment. Yet, its reality is declared in passive mode consciousness, albeit in retrospect, as a temporally preceding quality.

*

Why is mind not consciousness? We cannot relate an active agency vis-à-vis brains and world to a passive state that simply affirms the events of the brain and world purely in retrospect or temporally *after* the world and brain have transpired. As we discovered in the active mode approach, the mind as active agency must *precede* brain-pertinent events. Mind must also precede the subsequent passive mode consciousness witness of these events and of mind's attested or declared past-agency. It is simply a matter of timing: **Active causal mind does not occur at the same time as subsequent retrospectively witnessing passive mode consciousness of memory and of events. But since we cannot clock the active mind as a past event, nor as a future potential event, we are left with only one possibility: we must incorporate mind as a non-conscious ontics, as operating via the 'gaps in time' that reside between future-potential event horizons and subsequent declarative witnessing nous.**

The issue is further radicalised by the fact that we cannot even clock passive mode consciousness itself as an event, much less identify it as brain activity, or even with subsequent outputted declaration-events, such as those evinced in the Libet experiments. This is because witnessing consciousness comes temporally *after* the expression or declaration of collection and recollection, and temporally *after* the former are resolved into declarative events. Hence, as is true of the active mode mind, passive mode consciousness must also relate to and belongs to the 'gaps in time' between the succession of events, and is not clockable as an event. **To try to clock either mind or consciousness as an event in the brain or as brain activity observed in the Libet experiments, or as subsequent clockable expressions and declarations, is false, notwithstanding the initial error of assigning volition to passive mode consciousness.** Hence, in these various ways, we grasp how inadvertently false materialist interpretations of the Libet experiments have been and are.

The view afforded by the relativity of simultaneity and the demise of space vis-à-vis the brain, and to any observations we make of the brains in the Libet experiments, is pertinent: **If materialism were true, if mind or consciousness were generated by brain activity, then both ought to be simultaneous and in temporal synchronicity with brain activity (See Preliminary 0.44). Synchronicity of brain activity to subsequent consciousness is impossible to achieve *because* of the relativity of simultaneity, which abides across the brain and throughout the Bergson-Whitehead amalgam (see 5.47). Indeed, the said synchronicity would entail that both brain and world manifest in absolute time: an impossible feat, given the demise of 'space' (see Book-IV 4.01).**

554

5.53: Augmenting Dualism, Essay-III:
Recapitulation of the Absolute Time-Like Character of Mind and Consciousness

Nous possess absolute time-like characteristics, but without *res extensa*. This conveys to nous 'the sense of the visceral common present moment' to which and within which arises the world-in-memory. In its passive mode, consciousness is the retrospective recipient and witness of the world and its brain, the latter constituted as the past-in-memory within the in Bergson facet. Yet, passive mode consciousness is different and distinct from both the brain and the world by the fact that the latter must remain subject to an internal relativity of simultaneity, while the noetic moment of witnessing consciousness remains *not* subject to *any* internal relativity of simultaneity. This is why unitary consciousness cannot arise from the Bergsonian world-in-memory and cannot arise from the brain subsumed to the same... in the strict sense that one cannot obtain an absolute time-like noetic moment from a system of memory and possibility wholly subject to the relativity of simultaneity, and from which absolute time-like states cannot be engendered.

Indeed, the brain and its world possess *partology*, even if *not* as a distribution of 'parts in space'. The partology that attends the brain and world is constituted as temporal ad-mixtures of past moments and events belonging to different moments, a-synchronously inter-related per the relativity of simultaneity; distributed according to delayed choice time-interval relations.

On the other hand, both in its passive and active modes, the noetic moment possesses just one 'part'... or no 'parts' at all: equivalent to a spatial point-like state, but configured as a non-clockable temporal monad or *durata*. Hence, nous is *not* subject to any internal relativity of simultaneity vis-à-vis its other 'parts'. A noetic temporal monad constitutes a bizarre moment in time: a visceral experiential common present moment... a moment *without* transient duration, interlaced to a world and brain *with* transience.

Nous intercedes from between successive event horizons of the Bergson-Whitehead amalgam, as passive witness to Bergsonian memory and time, and as active agent over the future that, within limits, transforms into its own will and deed. Yet, nous is not itself in or of the Bergson-Whitehead amalgam of memory, futurity, and time: i.e. non-clockability, again.

*

Let us augment further: One's perception of the environment as a visual, tactile and olfactory complex of past events... as Bergsonian memory in time... arises as a unity, even though the content of this complex past must remain internally subject to the relativity of simultaneity and is consequently non-unitary and temporally disparate... in the sense that it is not in absolute time, even when it decays out of instantaneous wholes from Whiteheadian future-potentiality into unitary coherent complexes of events and wholes, notwithstanding grand decoherence.

The complex past, though strictly subject to the relativity of simultaneity, converges upon a succeeding witnessing absolute time-like noetic moment... to a virtual 'point'... to a temporal monadic conscious retrospection or recollection... to a moment without 'parts'... to durata without transience. To that noetic moment in absolute time, the complex past arises into a unified undivided experience and experiencer: An otherwise divided non-unitary past subject to the relativity of simultaneity arises to a noetic state in absolute time. Thus, the disparate memories of the past without absolute time are brought into a state of unitary consciousness whose characteristic *is* the singular unitary 'common present moment', or absolute time.

*

We must again differentiate between the unitary binding of the brain... a feature that emerges out of the unitary character of wavefunction decoherence and collapse of instantaneous wholes... versus unitary mind and consciousness; a consequence of the absolute time-like temporal monadic characteristic of nous. As already stated, and as with all the other events of the world, the brain is no less subject to internal relativity of simultaneity, and it is no less temporally disparate than the 'larger' world in memory' that subsumes the 'worm' brain, notwithstanding the fact that the disparate events of the brain-world decay out of instantaneous wholes into pattern-unity. To augment and reiterate, **the unitary collapse of the instantaneous whole constituted as wavefunction attendant the brain cum world is *not* the basis for the absolute time-like nous, although it *is* the basis for the unitary binding and control of the brain by nous. Again, it cannot be sufficiently avowed that the unity of mind and consciousness is a state obtained purely from its distinct absolute time-like characteristics vis-à-vis the growing block order of memory and possibility.**

*

Let us summarise and augment the implications, and conclude our essay:

- **The unitary binding of brains *is* obtained from the spontaneous a-noetic unitary decoherence and collapse of its instantaneous whole Whiteheadian complex wavefunction that attends the brain.** However, the interrelations between the outcomes so obtained remain subject to the relativity of simultaneity: they are temporally disparate.

- **The singular unitary character of active mode mind *and* especially passive mode consciousness *is not* generated by the spontaneous a-noetic unitary decoherence and collapse of the complex wavefunction that attends brains.**

- **The singular unitarity of mind and consciousness is a property of its inherent absolute time-like temporal monadic characteristic,** constituting the noetic 'common present moment', seemingly *without* temporal transience, superlative to the Bergson-Whitehead system of memory, possibility, and time, with respect to which it constitutes the witness and agent.

- The active mode mind, acting from the 'gaps in time' between successive potential event horizons, with power to decohere the future to a sought binded unitary coherent culmination of the brain, in tandem with spontaneous AND-to-OR collapse and

555

time, *is* causative of the unitary coordination of the brain, even though the latter is *not* causative of the causal noetic monad in absolute time.

This latter point is a crucial distinction: **unitary mind and consciousness does not emerge out of brains, but unitary coordination of brains *is* a consequence of the agency of unitary active mode mind.**

*

We complete our essay by superseding the old Cartesian substance dualism with its superlative form, conceived without *res extensa*: Thus, there are two 'substances'. On the one hand, we have 'world substance', or the Bergsonian domain of memory in time, constituting the de-spatialised combination of temporally distributed brain and world, both inescapably subject to the relativity of simultaneity. On the other hand, we have 'mind substance': a unitary singular present moment in absolute time to which the world of disparate temporised memory arises: a noetic moment *without* extension; a moment without transience. Of course, the term, 'substance' is allegorical: there is no 'stuff' as such. Yet, we obtain Cartesian dualism from…

- memory and possibility subject to the relativity of simultaneity…
- versus unitary nous in absolute time.

5.54: Augmenting Dualism, Essay-IV:
Reflections on the Findings of Wilfred Penfield & Ewen Cameron

Wilfred Penfield could not find any neural basis for the unified binding and coherence of the brain (see Preliminary: **0.46**). With some reiteration, we must clarify why this is so in the following bulleted sequence. Thus...

- The unitary binding of brain activity is brought about by the unitary decoherence and unitary AND-to-OR collapse and development of the complex future Whiteheadian wavefunction (an instantaneous whole) that attends the whole brain.

- **Penfield's error (the common error in all brain science, neuroscience, and in forlorn materialist philosophy of mind) was to seek the basis for the unitary coherence and binding of brains from within the brain: i.e., from its resolved states of OR-form distributions… implicitly from on or along the event horizon.**

- **Penfield *et al* never sought the basis for a unitary coherence and binding of brains as a function of the resolution of its Whiteheadian futurity or 'instantaneous whole'. Indeed, this would have required a quantum mechanical AND-form (or future-form) model of the brain, if not a full growing block model of the same**: a move Penfield, brain science, neuroscience, and abortive philosophy of mind never made and might yet refuse to make, perhaps on the dubious notion that quantum mechanics (the future) only applies to the 'very small'… or from the want to treat the brain as a macro-scale 'classical object. Indeed, the quantum mechanical modelling of the brain would be essentially no different to the quantum mechanical model of the process of photographic development: What applies in photography *essentially* applies to brains… wherein a singular complex wavefunction or instantaneous whole encompasses the development of the whole photographic image in a unitary way, despite the indeterminacy of its photon depositions or partology. The same would hold with respect to the development of brain activity, as is does across 'macro-scale' nature.

Of course, Penfield *et al* never distinguished between the unitary binding of brains (the outcomes from which are strictly subject to the relativity of simultaneity *and* inherently a-noetic) versus the singular unity of mind and consciousness due to the latter's absolute time-like characteristic. Even so, we should be generous to Penfield inasmuch as he was intuitive enough to articulated a dualistic view of nous vis-à-vis the brain, against the usual materialist consensus[101].

*

The distinction between unitary nous in absolute time versus binded brains subject to the relativity of simultaneity also accounts for the peculiar findings garnered by the notorious Ewen Cameron (see Preliminary **0.47**). Cameron's **destructive electro convulsive shock methods (destructive ECS) could certainly disrupt the contingent keyboard-like structures and readiness states of the brain, requisite to memory expression and of biographical memory. In so doing, Cameron effectively destroyed the means of expression and 'destroyed' effective recollection of biographical memory; *effectively* wiping out the identities of his victims.**

Imagine a similar destructive method that might be used to destroy the readiness states in the author's brain: perhaps those pertinent to the expression of the recollection of the 'big red garage door from Potters Bar, 1973': Destroy these and you destroy the ability to express or declare the memory of the big red garage door from 1973. Effectively, you will have destroyed memory... but *only* to the extent that memory is only as good as the keyboard facility to express and declare it. Take away the latter and you *effectively* take away the former.

However, our contention is that, insofar as the big red garage door, Potters Bar-1973 abides within the Bergson facet as permanent non-eradicable memory, the author's memory cannot ultimately be erased, no matter what Cameron *et al* might otherwise do: Only the ability to express memory could be irreversibly disrupted by the methods Cameron had applied.

---

[101] The Mystery of the Mind: A Critical Study of Consciousness and the Human Brain. Penfield, Wilder. Princeton university Press, 1975

QUANTUM MECHANICS AND MIND

What the methods of Cameron accomplished was a series of irreversible disruption-events vis-à-vis the brain and against its attendant ready means of expression and declaration of biographical memories. The disruption-events, incorporated into the brain-pertinent Bergsonian facet in their own right, are retained therein as recency. As these disruptions build up at recency, they dominate vis-à-vis the accumulated primacy of biographic memory. Biographical memories yet remain intact within the primacy of the Bergson facet and it 'worm', but recede into the remote past, obscured by the dominant intervening recency comprised of the retained disruptions; suppressed by a mnemonic superpose composed of the recency of disruptions.

We can express the same in terms of the superpose of wavefunctions in memory from within the Bergson facet attendant the brain. Thus...

- **Memories, including biographical memories attendant the brain, are retained within the de-spatialised Bergson facet as complex wavefunctions in memory**, belonging to different past moments, organised into temporal sequence, elaborating from the last recency wavefunction in memory to memories in past-primacy belonging to one's remote biographical depth.

- **Normally, through mnemonic mechanics, the wavefunctions in memory sum to form a new superpose**, (which itself gets incorporated into the Bergson facet as memory) and which undergoes unitary AND-to-OR collapses into the OR-form 'materialised' brains at a successive event horizon.

- **This normal superpose favours the 're-materialisation' of the readiness states that attend the author's expression of the big red garage door from 1973**. Per the repeated recollection of this memory, the readiness state for its expression and declaration will gain entrenchment and resolution per subsequent AND-to-OR collapse of the superpose.

- **Ewen Cameron enters the fray: Using his ECS methods, he repeatedly disrupts the said readiness states. Each disruption constitutes a past complex of events that must *also* become retained as a wavefunctions in memory; retained as a recency of memory of disruptions within the Bergson facet.** These constitute recency states vis-à-vis otherwise intact older complex wavefunctions in memory that attend the memory of the big red garage door from 1973.

- **The recency comprised of disruption-events in memory, combined with the now ever-receding time-removed primacy comprised of the wavefunctions in memory of the big red garage door, must now sum to form a new superpose: a disrupted superpose**.

- **When the disrupted superpose undergoes global unitary AND-to-OR collapse at a succeeding event horizon, it will *not* reconstitute the readiness states for the expression of the big red garage door**: the wavefunctions in memory pertaining to the disruption-events will 'get in the way', given their ineliminability from the Bergson facet and their dominance at recency, while the original intact wavefunctions in memory will recede to remote primacy; weakened to effective erasure, though *not* into actual or ultimate erasure. **The memory of the big red garage door will reside intact within the Bergson facet;** ultimately ineradicable, but no longer readily recuperable.

Normal loss of memory-expression unfolds in a similar manner, without disruptors: If one does not repeatedly express and recollects, the recency that builds up within the Bergson facet and which dominates over the brain 'worm', this will obscure the appropriate means of expression from reforming. The abstract memory fades away into primacy within the Bergson facet and 'worm'. Yet, the memory retained within the Bergson facet is never lost: it cannot be eradicated or erased by any means devised by Cameron, or even by nature.

In any case, whatever Cameron *et al* afflicted against biographical memory, he could not erase the ever elusive and ineffable singular-unitary "I"; with all its peculiar absolute time-like temporal monadism... beyond the capacities of the brain to furnish or generate. The "I" in absolute time does not arise from a brain and world subject to the relativity of simultaneity...it is beyond the power of Cameron to erase. Hence, again, Cartesian dualism follows suit.

5.55: Augmenting Dualism, Essay-V: Three-Fold Superlative Cartesianism

The conclusion in the previous essay must culminate into a novel form of memory, distinct from otherwise pre-mental or a-noetic Bergsonian memory constitutive of the temporised world and brains, otherwise subsumed to the parallel noetic moment as 'memory in consciousness'. This is *noumnemonae*, first adumbrated in Book-IV **4.41**. Simply put, *noumnemonae* is 'memory in consciousness': **incorporative of, yet distinct and different from its Bergsonian a-noetic *mnemonae*.**

One's conscious memory of the big red garage door from 1973 does not entail that the big red garage door is or was *in itself* conscious. Yet, this non-conscious or a-noetic memory... this *mnemonae*... emerges in one's consciousness as 'memory in consciousness'...as *noumnemonae* . Thus, we obtain dualism based on...

- **Noumnemonae**: the noetic moment characterised as an absolute time-like temporal monad; constituting 'memory in consciousness', or consciousness-memory *of* Bergsonian non-conscious a-noetic memory.

- **Mnemonae**, which is subject to the relativity of simultaneity: a-noetic Bergsonian world and brain constituted as a-noetic memory...the 'big red garage door-1973'... finally arising to consciousness in absolute time as noumnemonae.

Note that *noumnemonae* emerges temporally *after* the formation of a-noetic mnemonae: That is, 'consciousness in memory' emerges temporally *after* the non-conscious past-in-memory is formed up... or always after the formation of *mnemonae*, and the subsummation of the a-noetic latter to succeeding witnessing consciousness. Thus, again, we obtain Cartesian dualism.

Thus we obtain three-fold dualism composed of the following:

- The **dualism of active mind versus passive consciousness modes of nous** vis-à-vis world and brains.
- The **dualism of absolute time-like nous versus a-noetic Bergsonian memory and possibility; the latter** *always* **subject to the relativity of simultaneity.**
- And **dualism from** *noumnemonae* **(memory in consciousness) versus Bergsonian a-noetic** *mnemonae* **constituted as** a-noetic world and brain.

5.56: Radicalisation of General Mind-World Theory, Essay-I: The En-Worlded Mind

To control its brain, the active mode mind must exploit the growing block process of perpetual *spontaneous* wavefunction collapse and time, and intercede from between successive event horizons (from the 'gaps in time') to decohere the future in line with its intents. Mind does not bring about wavefunction collapse and time, as was shown conclusively in the critique of 'consciousness causes collapse' in Part-II (**5.18**). Instead, it is general a-noetic gear-less causality that drives time or brings about wavefunction collapse via an ineffable causality that issues from an undefinable ontology or Being that embeds and subsumes the whole of the Bergson-Whitehead amalgam, just as surely as it subsumes and exceeds nous, notwithstanding the greater affinity of nous to Being (see **5.50**).

It turns out that, **in its turn, the growing block Bergson-Whitehead amalgam is subsumed to the domain of mind and consciousness, i.e., to the** *noosphere*…. **While both in turn are subsumed to the grander ineffable mind-transcending neutral ontology or** *Being*. Mind is not the primary ontology, as should be clear in the preceding clarifications and from findings furnished in Part-II: Thus, ultimate ontology is mind-transcendent and ineffable. Even so, to reiterate, the noosphere embeds the Bergson-Whitehead amalgam 'within' it, while the noosphere is in its turn embedded within *Being*.

In short, the world *arises within* mind and consciousness, but *not from* mind and consciousness, as was clarified in the case against idealism in Part-II. Thus, mind and consciousness *does not arise* from the growing block system of memory, possibility, and time (see **5.46** to **5.48**). Consequently, you are not within or 'inside' the world: a spatio-distributive notion. **The world of memory, possibility, and time (including brains) is** *not* **en-minded: Minds are** *not* **distributed amidst 'matter in space', much less arise in and from spatial brains. Instead,** *nous is en-worlded*: **The world and brain arises** *within* **mind.**

Thus, you are not inside the world. Instead, the world is 'inside' you. This remarkable truth is primarily due to de-spatialisation and temporisation, but it is not only due to de-spatialisation and temporisation.

With the en-worlded mind, we obtain the radical inversion of the perennial object-subject relation: **The subject-nous is no longer within the object-world: We cannot collapse subject to object-world per materialism, no more than we could collapse object to subject per idealism. Instead, the object-world arises** *within* **subject-nous… but object-world** *does not arise from* **subject-nous.**

Per the relativity of simultaneity alone, experience always constitutes the retrospective apprehension of the world, in turn constituted as the past; as memory… which arises to and *within* temporally succeeding consciousness. Hence, mind and consciousness cannot constitute 'being in the world', or the *Dasein* of Heidegger. Instead, nous constitutes a world *within* nous. Certainly, there *is* enmeshment between object-world and subject-nous. Thus, an aspect of Heidegger's claims, could be recovered in terms of such enmeshment. But the subject-object relation abides despite Heidegger, albeit recuperated in inverted form as the en-worlded mind… or inverted from 'mind within world' to 'world within mind'; restoring the plausibility of agency and aretaic autonomy espoused in High Humanist philosophy from Plato to Descartes.

*

How is the en-worlding of mind accomplished? How is it possible that the world arises within mind, contrary to the generic misapprehension to the effect that mind is 'caused' by and arises from the brain, or arises from and is 'located' within the world and its brain? As previously stated in Part-II **5.15**, memory, including the Bergsonian brain 'worm', effectively emerges 'within' the witnessing noetic moment; with the latter comprised as an absolute time-like temporal monad. This is partly a function of how, given de-spatialisation and temporisation, the physical order is distributed and organised in pure temporal form, and how all of this then integrates vis-à-vis temporally succeeding nous. (This must also include the subsummation of the Whitehead facet of future potentiality to nous, not just the subsummation of Bergsonian memory).

Note that, given de-spatialisation and the demise of *res extensa*, we are not attempting to fit a spatial universe into a noetic moment, itself dispossesses of spatial attributes: If space was ontically real, nous would constitute the equivalent of a 'spatial point'. But space is pleonastic. Hence, when we assert that world as 'memory in time' comes into apprehension 'within' nous, we assert that de-spatialised stereotemporal memory, organised into pure time relations and *without* extensionality… emerges to and 'within' an equally de-spatialised temporised noetic moment. Hence, the noetic moment is furnished a 'world in memory'. Hence, the en-worlded mind.

Indeed, since mind and consciousness cannot arise from the world and from brains (see **5.48**), the subsequent mind and consciousness is not *in* the a-noetic Bergson facet, and certainly not 'in the brain' subsumed to the Bergson facet as the perduring brain 'worm'. Conversely, we must also reiterate that the world and brain is *not* generated by the noetic moment: (see Part-II: **5.16** to **5.18**: the case against quantum idealism and idealism generally). The a-noetic Bergsonian world, in arising to a succeeding moment of nous, arises to and 'within' subsequent and subsuming nous. Hence, again, the en-worlded mind… although the world so-subsumed must also include the Whitehead facet of possibilitie; also subsumed to the noetic moment.

*

558

QUANTUM MECHANICS AND MIND

With the en-worlded nous, we immediately confront the problem of solipsism: **If temporised memory, possibility, and time, including the brain, arises within nous, then in whose mind and consciousness does it arise? Does it arise in the author's consciousness? Perhaps it arises in the reader's consciousness? Or does it arise in Wigner's friend's consciousness… or in the cat's consciousness… or in God's consciousness?**

The succinct solution to solipsism emerges from the view that the world of memory arises, not just in one mind but within a general noosphere; within the general domain of mind…to which reader and the author are unique non-illusory first-person tributaries. That is, we are tributaries to an absolute time-like noosphere; albeit as non-illusory unique viewpoints, with different world-memories and possibilities. Indeed, **the relativity of simultaneity forces the dissociation of the background noosphere into a system of manifold first-person unique viewpoints**, even while the larger noosphere, or the universal "I", ensures the implicit collective unity of all such non-illusory unique viewpoints.

To grasp how this works, consider the following thought experiment involving the relativity of simultaneity. Assuming forlorn pleonastic space, it takes 1.5 seconds for light to travel from the Moon to the Earth and vice versa: Ultimately, per de-spatialisation, the 'distance' is recuperated into a pure a delayed choice time-interval relation equal to 1.5 seconds. (See Book-IV **4.08**). On the Moon, we place our observer-1, armed with a red flash-bulb. On the Earth, we place observer-2, armed with a blue flash-bulb. Both bulbs are synchronised to simultaneously flash at a set moment, and are then sent to the Moon and Earth, respectively.

Few days hence, observer-1 on the Moon sees his red flash first, followed 1.5 seconds later by the blue flash from the Earth. The order of events for observer-1 on the Moon is… **"Red flash"/ 1.5-second delay/ "Blue flash"**.

The observer on Earth sees the blue flash first, followed 1.5 seconds later by the red flash from the Moon. The order of events for observer-2 is… **"Blue flash"/ 1.5-second delay/ "Red flash"** .

Obviously, the relative order of events is not the same for both observers: there are unique event-order inequalities. Thus, the relativity of simultaneity structures the content of otherwise conserved same-events into a different or *unequal* time-distributions of event-sequence orders vis-à-vis each observer: Thus, we arrive at the differentiation of the observers into non-illusory unique persons: *not* by any ultimate differentiation of the "I", but per the integration of the one common "I" into two sets of inequivalent or unequal distribution of events and consequent viewpoints, interspersed according to an *objective* 1.5 second delayed choice time-interval relation.

To put it another way, **the noosphere's two tributary first-person viewpoints, cast as observer-1 on the Moon and observer-2 on the Earth, are subject to flash-bulb events that proceed according to delayed choice and the relativity of simultaneity. Consequently, the observers are subject to a differentiation of the associated light bulb events in terms of their relative timing and sequence-order, rendered different from one frame and observer versus the other. Consequently, our observers, as witnesses to the events of the world, will be furnished with unique non-interchangeable (inequivalent or unequal) experiences and histories of events, despite belonging to a one-and-same background noosphere, or as the latter's tributaries. In effect, and in a totally non-illusory way, the noosphere splits into a foreground of unique *non-illusory* first-person viewpoints.**

The one non-illusory unitary noosphere is furnished with two truly unique and non-illusory inequivalent experiencers, and, with it, the noosphere splits into non-illusory unique viewpoints or frames of reference… both ultimately belonging to the one-and-same universal nous or the universal "I". In doing so, each viewpoint loses the ready cognisance of their background oneness and unity in and as the universal "I", even though this truism abides undiminished in their common noetic background; it being as non-illusory as the non-illusory unique inequivalent experiencers.

The two non-illusory unique viewpoints then integrate into non-illusory relative first-person versus third-person relations. As stated, both lose cognisance of their background unity as the same "I" and as the same noosphere. The same "I" views itself through a delayed choice time-interval equal to 1.5 seconds into a 'third person viewpoint', into the 'other'. In turn, the latter perceives itself as the first person viewpoint, but attributes third-person status to itself in the 'former', per the same delayed choice time-interval. Thus, it is the time-separation of the one nous into two that engenders non-illusory first-person -v- third-person relations.

We can now further augment how the noosphere splits into non-illusory unique viewpoints, and how solipsism is prevented in the en-worlded mind. To this end, we consider another variant of the above experiment: Therein, we assume that there existed no delayed choice time-interval relations between the red flash and the blue flash from the Moon and the Earth, respectively. Thus, we assume that these occurred in absolute time; in simultaneity in a common present moment. Consequently, observer-1 observed the blue flash from the Earth instantaneously and at the same time as his own red flash. Conversely, observer-2 on the Earth observed the red flash from the Moon instantaneously and at the same time as her own blue flash.

For both observers, the order of events would be…**"red flash" simultaneous with "blue flash"**: i.e., a *same* sequence-order and constitution of events for both observers, with no differentiation or any possibility of experiential inequivalence or unique non-illusory first-person frames, given that both are now constituted in a 'common present moment' in absolute time, and with exactly the same sequence-order of events, with exactly the same experience-content, with exactly the same viewpoint. Hence, there could not arise therein *any* differentiation of the universal "I" into relative non-illusory first-person versus third-person frames.

But, in this impossible scenario, even the expected luminosity-reduction and entropy of a 'distant' light-sources cannot abide: Indeed, the light-flashes occurring in absolute time would transpire 'at the same place': with the Moon, the Earth, and the observers, and the bulbs, effectively reduced to a 'same place' singularity, given no time-delay, hence no 'distance', and, given the perfect simultaneity of said events rendered in absolute time; constituted into a single spatial point, no matter how 'far apart' the events were purported to be or initially assumed to be. Yet, the collapse of the whole to singularity, to the said point, would collapse presumed space itself to that singularity. Hence, it is strange irony that the presumption of space (obviously synonymous with absolute time itself) obviates space by self-collapsing *it* into singularity.

The same collapse would follow for the presumed distinct observers and unique 'view-points' bearing witness. Hence, there could not arise *any* differentiation of the "I" into unique non-illusory first person experiencers or frames, and first-person -v- third person relations could not emerge. With this insight, we come to the key point of the exercise: In a universe dominated by absolute time, such as the one envisaged in the scenario above, there could be no unique first-person to third-person viewpoints. To reiterate, **under conditions of absolute time, for both observers, the red flash would occur at the same time, and effectively at the same place, as the blue flash; while the Moon would transpire at the same place as the Earth, etc... and vice versa... with the same identical experience and same sequence-order of events or histories...** *with the formerly unique viewpoints reduced to a same consequent identity and consciousness.* **In such a universe, consciousness could only have one experience; only one history, and** *only one identity...* **even if purportedly split into many, and distributed to the Earth, to the Moon, and to elsewhere. The non-illusory first-person viewpoint and the individuation we all enjoy would collapse and become rendered impossible... while the unitary singular noosphere would become fully and undeniably salient, albeit without apparent division into, or occlusion by, the foreground of unique non-illusory first-person identities and attendant first-person -v- third person worlds.**

Only a system of Bergsonian memory and Whiteheadian possibility, temporally structured according to delayed choice time interval relations, and subject to the relativity of simultaneity, could permit the viability of *foreground* non-illusory separate and unique first-person experiences, histories, and identities... *and* incorporate the non-illusory but occluded *background* unitary-singular noosphere, or the universal "I"... to which all unique non-illusory tributaries belong.

*

We now deconstruct how the en-worlded mind, and both the unitary-singular "I" *and* its differentiation into unique first-persons, enmesh. **From the perspective of the noosphere, observer-1 on the Moon subsumes observer-2. In other words, observer-2 on Earth, and the blue-flash attached, arises to and** *within* **observer-1 on the Moon, albeit** *after* **the 1.5 second delayed choice time interval: Thus observer-2 is temporally subsumed within observer-1.**

Notice that this is *not* a distribution explicated in terms of the 'space of things' inside observer-1, given de-spatialisation. That is, observer-2 is not inside observer-1 in the spatial sense. Instead, the relation is a pure temporal distribution of observer-2 within observer-1, structured by the 1.5 seconds worth of delayed choice time-interval relation, according to which observer-1 incorporates and subsumes observer-2 as past-memory.

Note that, ultimately, observer-2 is the past variant and version of observer-1, but from a unique and non-interchangeable frame of reference comprised of the unequal temporal distribution of events. That is, observer-2 is actually observer-1, wherein the latter apprehends itself as a past event related to a set of past events, uniquely ordered and structured by delayed choice time.

The inverse is also true: From the perspective of the noosphere, observer-2 on Earth subsumes within it observer-1, the Earth, and the red-flash. In other words, observer-1, and the red-flash upon the Moon temporally arises *within* observer-2 on Earth, as the latter's Bergsonian past-in-memory... with observer-1 itself now retrospectively apprehended as past memory, temporally within observer-2: with the latter related to itself in the form of observer-1, but through a time delay imposed by the temporal structure of nature, and per function of the relativity of simultaneity.

Which of the observers is the absolute solipsistic one? Neither, and the firewall against solipsism emerges from the dominance of time itself: The one is temporally delayed within the other, while the other is temporally delayed within the former, while neither has temporal primacy, even while the one is the same nous and the same "I" as the other, albeit apprehended as past memory in delayed choice arising in the other, and vice-versa. Both are tributaries from, and subsumed to, the grander noosphere that includes both *and* **transcends both.**

Put another way, there is only one mind and consciousness. In its terms, the noosphere exist in absolute time, undivided: It constitutes a singular common present moment of experience. Yet, for reasons just presented, and for whatever ultimate reason (we do not know why existence is arranged in this discovered form, or why nous is structured thus) the unitary singular noosphere communicates with itself through a world of memory, possibility and time, subject to the relativity of simultaneity and distributed into delayed choice time-interval relations. **In effect, the noosphere splits itself into different non-illusory unique conscious viewpoints and egos in and across time: each separated from the other by a time-delay and the associated differentiation and inequalities of the pertinent order-of-events: Each viewpoint is unique in terms of the order and content of otherwise conserved same-events, witnessed retrospectively and** *unequally* **per their a-noetic components, rendered in Bergsonian memory.**

Hence, the unitary collective singular nous *and* **the multiplicity of first-person viewpoints and first-person -v- third-person relations... are** *all ontically real,* **without contradiction and without degeneration into solipsism.**

5.57: Radicalisation of General Mind-World Theory, Essay-II: The En-Worlded Mind & the Fragmentation of the Noosphere From the Grand Decoherence & Collapse of the Whitehead Facet

There is a second decisive reason why the grander singular nous, the universal "I", is occluded, relegated to the background and rendered implicit. Recall from Part-II **5.18** that, in the criticism of Wigner's contention, we asserted that consciousness five minutes from now, or even a millisecond from now, has not yet happened. Therefore, **consciousness must itself be rendered into superposition AND-form set of alternative futures: albeit constituted as** *noetic* **AND-form superpose state of future-potential consciousness.**

This noetic AND-form state must run parallel and consistent with whatever a-noetic observables that undergo grand decoherence and AND-to-OR decay out of the Whitehead facet, subsequently witnessed by passive mode nous that must itself AND-to-OR collapses into consistent consciousness vis-à-vis the former observables, subsequently retrospectively apprehended.

Thus, a domain of noetic AND-form futurity must exist; constituted in parallel with the a-noetic Whitehead facet, at least so far as passive mode consciousness is concerned. The same schema and parallelism must apply to the total noosphere with respect to the totality of a-noetic Whiteheadian futurity, at least as pertains to the noosphere that attends future-potential passive mode consistent consciousness.

This future-potential noosphere enmeshes with the pertinent a-noetic Whiteheadian future-potentials, and abides to consistency with the Whiteheadian future-potentials, a consistency enforced by grand decoherence of the future-potentiality of the noosphere in tandem to the Whitehead facet. Indeed, if the parallel Whitehead facet should undergo grand decoherence to the elimination of certain nested future possibilities, the eliminated possibilities could no longer constitute potential observables for any future-potential passive mode consciousness. Hence, the future-potential noosphere is expected to decohere into consistent consciousness… into consciousness that could only witness the realised observables, and no others, with non-consistent noetic possibilities erased. Tautologically, you cannot bear witness to futures that are no longer permitted to happen.

This insight adds another augmenting case for the objectivity and non-collapsibility of object to subject, *against* idealism.

The corollary to the future-potential noosphere and its passive mode potential consciousness is the **mnemonic noosphere**. The mnemonic noosphere runs parallel to the Bergson facet of a-noetic memory and history. But it also subsumes the a-noetic Bergson facet as part of the en-worlded mind: as 'memory in consciousness' accumulating within the mnemonic noosphere.

*

We must now consider that the noosphere must partly constitute itself as the AND-form future-potentiality for *all* tributary unique non-illusory first-person viewpoints in consciousness; each constituting a consistent consciousness vis-à-vis the temporally shared events that arise from the decay of the parallel a-noetic Whitehead facet. **While future-potential nous must collapse out of its future potential AND-form noosphere, the noosphere cannot collapse into a single all-encompassing witnessing consciousness. It can only AND-to-OR collapse into manifold unique first-person tributaries that, in their temporal interrelations enmeshed to the Bergson-Whitehead amalgam, are subject to the relativity of simultaneity. Consequently, the noosphere cannot help but collapse from a non-illusory unitary singular en-worlded *en toto* noosphere into equally non-illusory unique first-person viewpoint frames**, or into observer-1 and observer-2 in the scenario from **5.56**… for whom **"Red flash"/ 1.5-second delay/ "Blue flash**… will not be the same as or interchangeable with **"Blue flash"/ 1.5 second delay/ "Red flash"**: a relation and differentiated forced upon the parallel AND-to-OR collapse of the noosphere into its consistent consciousness constituted into unique first-person frames… and wherein observer-1 will temporally subsume observer-2 as a past and inequivalent version of itself… while observer-2 will subsume and retrospectively witness observer-1 as a past inequivalent version of observer-2.

Hence, per parallel AND-to-OR collapse of the Whitehead facet with the tandem AND-to-OR collapse of the noosphere, the otherwise unitary singular noosphere, or the universal "I", will differentiate into non-illusory unique first person identities and relative first-person -v- third-person distributions… arising *within* a singular-unitary nous, but merely occluded and relegated to the background.

5.58: Radicalisation of General Mind-World Theory: Essay-III:
Noosphere from the Conscry of Metalatives

Is there another basis we can bring into play that independently augments the implicit reality of both the noosphere and the remarkable en-worlded mind? The answer is to be found in the *conscry of metalatives*, in the fact that nous can engage in remarkable co-apprehension of halting problems, of infinities and infinitives… of metalatives garnered, *not* from the world of a-noetic memory, possibility and time, and *not* from the brain.

We do not observe or infer the idea of infinity, nor attendant halting problems, from the world or from brains. **There are no analogues for infinity, much less for the continuum, in the sensate empirical world or in brains. Nor can the co-apprehension or conscry of infinitives or of aesthetic-moral infinity-laden prepositional superlatives, or the co-apprehension of '*to be*', being, or *Being*, emerge out of the finite temporal domain of a-noetic memory, possibility and time, much less from the instrumentality of brains, given that, again, there are no analogues for infinity or *being* within the world or in brains.**

Indeed, galvanised by background findings in Book-III **3.14** pertinent to non-arbitrariness and the meta-semantic hypostasis, and as was clarified in Part-II **5.14**, the idea of infinity and of timeless *being*, and all infinity-laden ideas and metalatives, are furnished to mind and consciousness from 'outside' of Bergsonian memory **(5.48)**; from outside of Whiteheadian possibility **(5.46)**; and from 'outside' of brains restricted to the Bergson-Whitehead amalgam. Infinities and other metalatives ontologically supersede memory, possibility and any state or process brought into realisation in and through time, including the finite brain itself. Even so, we can yet communicate about these through the use of sensate-empirical states formed into improper forms composed of artificial symbols, bereft of the actual infinities and metalatives: (see Book-I **1.46** and **1.46** for the code-information distinction, rigging and superlative rigging). **But how is it possible for two distinct non-illusory first-person frames to transact and conscry about infinities and metalatives in and through the temporal processes of an a-noetic milieu of memory and possibility intrinsically bereft of such noeta?**

Typically, we invent 'symbols in the world' to communicate about a wide range of ideas, including infinities and metalatives. **Yet, when we invent and transact a symbol for infinity or for 'to be', or else substitute some object for these as a pseudo-analogue, we must first agree that the symbol or object 'means' infinity or represents 'to be' even though neither the symbol nor the object possesses the proper form of either infinity or *being*. Moreover, our pre-agreement about what the symbol or object 'represents' (i.e. the rigging) is not born from or forced from the symbol or object *in itself*, much less by the plethora of improper forms that constitute the a-noetic world of memory, possibility and time**. In short, if we had to rely on improper-form symbols and objects alone to communicate about anything, including and especially about infinities and metalatives, we would fail to conscry about these. Indeed, our remarkable co-apprehensions or *conscry* of infinities and superlatives cannot emerge out of the improper-form of temporally exchanged symbols via the forms of an a-noetic world and brains, even when we dishonestly and apagogically dismiss the noeta as 'illusory' or 'imaginary'.

*

The solution is obvious: The reason why it is possible for us to co-apprehend and communicate about infinities and metalatives despite the fact that there are no analogues for these within the realm of Bergsonian memory or Whiteheadian possibility (nor in brains), is because we enjoy common access to a sphere of abstract ideas, somewhat in the fashion espoused by Plato or Kant: part of a common collective noosphere of meta-semantics; ultimately furnished to the noosphere from Being, both of which supersede the totality of the Bergson-Whitehead amalgam and the brains embedded to that amalgam. Consequently, **when we conscry about metalatives, we are not ultimately communicating about these through the temporal world of objects and symbols belonging to a-noetic memory, possibility and time, or even via the brain. Instead, we gain direct acquaintance with the metalative in and through the superseding universal singular unitary "I", or the noosphere… to which both reader and author are tributaries, and whose metalative abstractions to which we have common access**.

If this is true, why is it not possible to conscry about *all* things in a similar way, by way of direct acquaintance through the noosphere, thus circumventing any resort to the a-noetic Bergson-Whitehead amalgam and its brain... by instant simultaneous 'telepathy', so to speak? If we do enjoy the said sidereal access to metalatives, why do we persist on relying on the transaction of improper forms through an a-noetic world of memory, possibility and time? Why not let one of us think 'infinity', and the other gain the simultaneous 'telepathic' co-apprehension and cognisance of the same… all via instantaneous through-noosphere transference, without relying on time-delayed transactions of contrived improper form or stand-in symbols via an a-noetic Bergson-Whitehead amalgam?

*

The conscry of infinity entails two things: With agreement secured vis-à-vis the pseudo-analogue symbol or object 'stand-in' for infinity, observer-1 relays the symbol to observer-2. There is a delayed choice time-interval relation between the moment of transmission by observer-1 and the moment of reception by observer-2. It is only at the moment of reception of the symbol that observer-2 is prompted to apprehend infinity... obviously, *not* at the same time as observer-1, and certainly not at the same moment as the temporally preceding transmission. Hence, the co-apprehension or conscry of 'infinity' cannot transpire in absolute time or in a simultaneous 'common present moment' for both observers. Thus, the respective observers cannot engage in an instantaneous mind-to-mind 'telepathy' about infinity. In other words, co-apprehension and conscry is time-distributed, even while its very possibility requires sidereal access to metalatives that are extant memory, possibility and time.

Put another way, the single unitary mind and consciousness is talking to itself about infinities, but through a system of symbol-communication using improper forms, through a world of memory, possibility and time subject to the relativity of simultaneity, and scattered by delayed choice time-interval relata... through which the same one-consciousness is differentiated into unique temporally disparate non-illusory first-person frames; or into observer-1 and 2: (See previous essay: **5.56** and **5.57**). The unique distribution of events and associated memories and identities vis-à-vis the observers in the moment of transmission will be non-simultaneous with, unique to, and different from the succeeding moment of time-delayed reception, despite the simultaneous tributary status of both observers to the one-and-same noosphere, its universal "I", and the common metalatives furnished upon the noosphere and upon both observers. **The one-nous is prompted to the same idea of infinity, but at different times, apprehended at different moments at unique viewpoints and non-illusory unique egos, even though the idea of infinity was never temporally 'transmitted' as such, but was pre-existent to both observers: both of whom had simultaneous sidereal access to the idea through their common noosphere. In this sense, we *do* engage in an implicit 'telepathy' about infinity and other metalatives, except that, because of our enmeshment to a time-unfolding and temporally distributed world, the 'telepathy' does not occur at the same time, but transpires to the same ultimate "I" at different times, per function of the attendant delayed choice time-interval relations**.

*

Finally, recall from **5.57** that the parallel noosphere, in its passive mode consciousness, must itself undergo parallel and correlated AND-to-OR collapse into consistent consciousness vis-à-vis the AND-to-OR collapse of the a-noetic Whitehead facet or futurity. Thus, **the noosphere itself has a future-form state of noetic potentialities parallel to the Whiteheadian a-noetic potentialities. It is *to* these noetic potentialities that the metalative idea of infinity or being is furnished, as part of the pertinent potential noetic moments pertaining to observer-1 and observer-2, as are all other metalatives and Platonic or Kantian ideas. It is through the a-synchronous process of AND-to-OR collapse of these noetic potentialities into time-distributed relative realised consciousness states that we, and the noosphere, gain co-apprehension, conscry and consciousness about infinities and metalatives... but *not* at the same time. Again, the collapse of the AND-form noetic potentiality involving the apprehension of infinity at the moment of its transmission by observer-1 is temporally distinct and precedes the AND-to-OR collapse of the same idea into conscious**

562

apprehension at observer-2, attendant the time-delayed reception of the symbol for infinity, and the subsequent time-delay apprehension of infinity by observer-2 prompted by the symbol so received. To both potential noetic moments, the idea of infinity is pre-implanted, so to speak... *not* in or through the a-noetic Bergson-Whitehead amalgam or its AND-to-OR collapse processes, nor by its improper forms, or the symbols involved... but to the respective collapsing noetic moments of conscious witnesses from out of their common noosphere of metalative noeta.

We can breakdown the pertinent sequence involved into the following:

- The idea of infinity is pre furnished to the noosphere attendant observer-1. The consciousness of observer-1 apprehends infinity by a parallel AND-to-OR *collapse sequence-1* of the noosphere and its noetic potentials at t_1. Note that the related consciousness belonging to observer-2 has not yet transpired and has not yet gained cognisance of or conscry of the idea of infinity vis-à-vis the consciousness at t_1 by observer-1

- Upon a delayed choice time interval and a succeeding AND-to-OR *collapse sequence-2* at t_2, observer-2 finally receives the symbol for infinity, and subsequently apprehends the idea of infinity at t_2. Thus, the noosphere that attends observer-2 undergoes another parallel AND-to-OR collapse into consistent consciousness at t_2... which apprehends the idea of infinity... but not at the same time as observer-1 at t_1.

- But the apprehension of infinity accomplished at t_1 or at t_2 is not obtained in, or through, or *as* states of the Bergson-Whitehead amalgam. At both t_1 and subsequent t_2, the apprehension of infinity by the pertinent observers happens through the extant, subsuming and parallel-collapsing noosphere of noetic potentials that are pre-furnished with the idea of infinity, and which furnishes the idea of infinity at t_1 and non-simultaneously at subsequent t_2, to observer-1 and observer-2 , respectively.

- Also note that the apprehension of infinity by the same noosphere at t_1 and then at t_2 is *not* occurring at the same time, despite the fact that the apprehension at both noetic moments (by observer-1 and 2 attached to those different moments as non-illusory unique first-person viewpoints of consciousness) *is* realised *directly through the noosphere, by means of the ideas furnished by the noosphere to both, from 'outside' of the a-noetic events of Bergson-Whitehead amalgam.*

- Therein lies the key to a time-delayed version of mind-to-mind 'telepathy' of sorts: The conscry about infinity is always realised *directly through the noosphere...* and from 'outside' the a-noetic Bergson-Whitehead amalgam... *but not through the processes of the Bergson-Whitehead* amalgam... and not at the same time. That is, one observer and the other will certainly co-apprehend and conscry infinity and other metalatives, but not through the world: Both observers will conscry infinity mind-to-mind... circumventing the world... relying purely on their common noosphere, their universal "I", and its commonwealth of eternal ideas... *but, not at the same time.* This is simply because the respective noetic moments and attendant conscry of metalatives do not AND-to-OR collapse into realisation from their common noosphere at the same time, given the operation of delayed choice time-relations and the relativity of simultaneity in the parallel Bergson-Whitehead amalgam, which temporally distributes observer-1 in relation to observer-2, and their symbol transactions, and conditions the parallel collapse of the AND-form noetic potentials of the noosphere into a non-simultaneous co-apprehension of the pertinent metalatives.

5.59: Radicalisation of General Mind-World Theory, Essay-IV:
The En-Worlded Mind & the Anti-Copernican Revolution

Again, we reiterate the case from **5.56** that the world is not en-minded: The universe is not furnished with minds supposedly distributed into 'bodies and brains in space'. Instead, mind and consciousness is en-worlded. The parallel noosphere, with its tributary of first-person unique identities, is furnished an a-noetic universe of memory, possibility and time. Hence, the brain and the larger Bergsonian order of memory, together with Whiteheadian future possibility, must arise to and *within* nous. Succinctly, the *object* in memory and possibility must arise to and within future-potential *subject*, which must in its turn AND-to-OR collapse out of its noosphere of noetic potentials in parallel to a-noetic memory, possibility and time.

Again, this '*object within subject*' form of the object-subject relation retains the essence of generic object-subject relations, except that the subject is neither within the object nor is it within the world. Instead, the object-world is within the subject-nous, although the object-world does not arise from subject-nous: a key insight secured in the critique of 'consciousness causes collapse' (Part-II **5.18**). **Again, the world is *not* furnished with minds. Instead, the mind (the noosphere, temporally split into multiple non-illusory first-person viewpoints and tributaries) is furnished a whole universe of memory and possibility in time; arising within nous, but *not from* nous.**

This insight must constitute the ultimate anti-Copernican twist: The world and universe is within you... all 14 billion years of it. Thus, you are not lost within the universe... although, given the aretaic failure of human consciousness, it is distinctly possible that the universe is lost within *you*.

*

Historically, humankind's notional place was conceived as confined upon a flat Earth, supposedly at the centre of a universe, shepherded by the Sun and the myriad celestial bodies... en-globed under a starry firmament. Hence, the early geocentric paradigm.

Eventually, flat Earth was displaced by the spherical Earth. But the firmament, and the geocentric arrangement of the cosmos was retained. Throughout the geocentric paradigm, a fallacious existential investment was made in the notion of the geocentric and cosmocentric centrality of Earth, and the *non sequitur* cosmic centrality of humankind, notwithstanding the abject aretaic failure of humankind.

With Copernicus and the struggles of Galileo *et al*, the geocentric arrangement was displaced by the heliocentric paradigm; with all and sundry scattered and revolving about the Sun, yet en-globed within a heavenly firmament. The false existential investment in the geocentric view was exposed. But this was misinterpreted to imply a dethroning or humbling of humankind… even though the aretaic failure of humankind had implicitly rendered such self-salutary notions and claims void.

Eventually, the idea of the firmament itself unravelled: The stars were finally recognised as distant suns, as Giordano Bruno had first postulated. The firmament gave way to a universe of endless suns and worlds scattered in a fathomless void. Therein, with the principle of mediocrity in hand, the idea that the solar system might yet constitute the centre of the universe could barely be maintained. The nihilists had their field day, as they still do.

Few among the 'educated' know about Einstein's contributions: In 1915, the void itself was de-centred. In Einstein's universe, any point therein was rendered as good as any other as a candidate 'centre'… in a universe *without* a centre. Hence, the idea of 'the centre' was itself rendered into mediocrity. Apparently, we do not live in a Copernican universe, but in a de-centred Einsteinian one. Even so, the Einsteinian universe recuperated the essence of the Copernican universe, although the minor differences are rarely noted. Indeed, for the ardent nihilist, the de-centred universe is likely an 'improvement' on the older Copernican nothing.

Hence, **our 'place' in the universe evolved from a flat world-centred universe to a sun-centred sphere-world, and into a full de-centred daz, wherein the principle of mediocrity seemed to reign absolute. Yet, all of the attendant cosmologies retained and assumed the dubious idea of the en-minded world; wherein subjects and minds are distributed and scattered across a spatial and material void; cast within 'bodies and brains scattered in space'.** Whether or not the world was conceived as flat or round, as geocentric or heliocentric, or merely as a de-centred void, did not modify the generic implicit spatial notion of the *en-minded world*.

However, **we have now obtained an unexpected superlative Anti-Copernican subversion: namely, the *en-worlded mind*. It turns out that the universe of memory, possibility and time constitutes a de-spatialised stereotemporal complex; one subsumed to a larger ineffable *Being* and causality. Moreover, the void-and-all… all 14 billion years of it… must transform into memory in time, and to an inexhaustible futurity that elaborates to timelike infinity. *All* of this arises *within nous… within* you.**

*

You are *not within* the universe. Instead, the universe is within *you*.
The real question is: Is the universe *found* in you, or is it *lost* in you?

5.60: Comprehensive Summary Recapitulation of Part-IV: General Mind-World Theory

The following constitutes a comprehensive summary of the key assertions of the general mind-world theory, culminating into the revival of Cartesian dualism.

The ineliminability of quantum mechanics from mind-world theory, materialistic or otherwise. (See **5.40**) Thus…

- The principle of causality and generic principle of conservation intertwined (the firewall principle) demand that AND-form logic abides at all scales. Thus, quantum mechanical reality *must* apply to the 'macro-scale', and even to brains, *not* just to the 'very small'. Otherwise, we risk 'signals from the future' and violations of causality *and* of the principle of conservation. To eliminate quantum mechanics from the macro-scale brain, one must demonstrate that…
  - …the macro-world (including brains) is dispossessed of an objective ontic future: a block universe model of the brain without the Whitehead facet. In other words…
    - …the macro-scale futurity must constitute a fully resolved OR-form block, even to the infinitely removed future, even though proof of this would require 'signals from the future', and attendant violation of causality and the generic conservation principle….and…
    - …nested futures *and* grand decoherence must be proven invalid: Both comprise the consummate case for the onticity of AND-form logic, not just at the macro-scale, but at the universe-scale. At the level of the Whitehead facet, billiard balls otherwise separated by 'light years' quantum-interfere with each other's futures. This is per grand decoherence. This affects the revival of Mach's principle and underpins the growing block quantum mechanical theory of inertia. These must be shown to be false, by demonstrating that AND-form logic does not abide at the macro-scale or in brains, and that only the block universe abides, and the growing block model invalidated.

Yet, quantum mechanics is ineliminable from the macro-scale. It is equally ineliminable from *any* putative mind-world theory and attendant theory of the brain, even from purely materialist theories of the brain. Thus…

- Any mind-world and mind-brain theory *must* incorporate the brain into the growing block framework… into the de-spatialised and temporised Bergson-Whitehead amalgam and its attendant AND-to-OR wavefunction collapse-process (time). Hence, the true model of the brain requires the incorporation to it an all-scale AND-form brain-state (a quantum mechanical model of the brain) inexorable upon the brain's incorporation into the Bergson-Whitehead amalgam.

- Since time itself is synonymous with AND-to-OR wavefunction collapse and is inherent to all things, all-scales, including brains, wavefunction collapse (hence, time itself) must attend the macro-scale brain as a whole. Hence, again, a quantum mechanical or future-form description of the brain is indispensable, whatever the mind-world philosophy presumed.

The general theory and **the four-fold definition of nous**: (**5.41**). Thus…

- *Mind and consciousness is comprised of all the folk psychology hard problem attributes exclusively associated with brains*, but not with cups, rocks, or pumpkins. The validity of Chalmers's hard problem approach is secured by the demise of materialism, in turn secured in the comprehensive case against materialism espoused in Part-III: **5.24** to **5.38**.

- *Mind and consciousness is what the renormalising quantum mechanist possesses*: i.e., the power to grasp and handle infinities and halting problems indispensable to renormalisation. This constitutes an 'impossibility problem', beyond the means of the finite brain. Cartesian dualism follows.

- *Mind and consciousness is that which can apprehend infinities and metalatives*: *including* timeless *being*: These exceed the finite and temporal instrumentality of brains and cannot be generated by brains. Cartesian dualism follows.

- *Mind and consciousness is that which possesses absolute time-like characteristics.* As part of the Bergson-Whitehead amalgam, the brain is strictly subject to the relativity of simultaneity. Thus, the brain is incapable of generating absolute time-like mind or consciousness, or the noetic temporal monad. Cartesian dualism necessarily follows.

Two obviations in the general theory overcame two generic historical tropes raised against Cartesian dualism: (**5.42**):

The obviation of *res extensa* was obtained via the following:

- *There is no 'space'. Thus, there is no 'res extensa'*… courtesy of de-spatialisation. Hence, the notion that the brain and world constitute extended states (Descartes' *minor error*) is untrue.

- In a de-spatialised purely temporised universe, *the interface between nous and world is an interface realised purely temporally and non-contiguously,* realised via the 'quantum leaps in time', in turn driven by gear-less work function zero wavefunction collapse and time… from one-brain state to a succeeding changed brain-state.

- Thus*, the basic Cartesian relation is configured as a pure temporal relation between non-extended nous (mind and consciousness) versus non-extended de-spatialised and temporised brain and world*; the latter constituted as a de-spatialised system of stereotemporal a-noetic past-memory and future-possibility: i.e., the Bergson-Whitehead amalgam. The relation of nous to world and brains is non-contiguous or 'contact-less', given de-spatialisation and temporisation, and the obviation of materialist notional 'point-of-contact' that follows both.

The obviation of the violation of the purported conservation principle was obtained via the following:

- *Myriad work function zero phenomena exist in nature:* these do not violate the generic principle of conservation: e.g., Gravitation: see Book IV **4.31**. Also see the transtemporal enforcement of physical laws (Part-III **5.33** and Book-III **3.33**). The decisive example is furnished by ontologically primary wavefunction collapse and time: a gear-less process that realises putative 'energy-work relations', but does not itself require energy, work, or power inputs to realise. Time and wavefunction collapse constitute a work function zero process.

- *Energy concepts and relations are pleonastic.* Time is ontologically primary to putative energy-relations and to conservation principles conceived in energy-work terms. The enforcement of the latter is via primary AND-to-OR collapse and time. But AND-to-OR collapse and time constitutes a work function zero process requisite to subsumed putative energy relations and the conservation principle. Hence, the ontic primacy of non-energy driven time vis-à-vis apparent 'energy relations' and the energy-based conservations principle brought about by time, implies that action and state-change is not brought about by energy-inputs and relations, but by non-energy based work function zero wavefunction collapse and time. If action and change of state emerge out of non-energy based time and causality, then energy concepts and relations are immediately rendered pleonastic and abstracted out of the ontology of physics, and, necessarily, abstracted out from mind-brain causality (**5.35**).

- *De-spatialisation also undermines energy relations by usurping the notion of contiguous contact causality, attendant 'force-transfer', or energy-input causality in 'particle interactions'*: Change of state in 'particle interaction' is realised through a 'quantum leap in time' (**5.30**) that, insofar as it entails AND-to-OR collapse and time, is wholly a work function zero process... and *leaps over presumed points of contiguity or contact in 'particle interaction' that supposedly entail 'energy exchange' or contiguous action and work.* Thus, obviating contiguity and contact causality in general, and obviating 'energy exchange' through contact. *Causality in brains or between the mind-brain is not going to be any different. Thus, mind-brain causality is non-contiguous, and it does not require 'energy exchange' or contiguous work or action to bring about.*

- When we combine work function zero time and causality with the pleonastic status of energy concepts and relations, *and* incorporate the demise of contiguous contact causality per de-spatialisation, temporisation, and the 'quantum leap in time' to circumvents it all, *it necessarily and inexorably follows that, change in brain activity is obtained without the violation of the much-abused principle of conservation.*

- Finally, in a de-spatialised temporised universe mediated by gear-less causality and time, and wherein energy relations and concepts are pleonastic, the generic conservation principle much-abused against Cartesian dualism is recuperated into an abstract form permissive, consistent and complementary to mind-brain dualism and Cartesian causality (see the first recuperation of conservation principle in Book-II, **2.39**, and then in **5.29** and **5.35** above). In its new form, the principle of conservation vis-à-vis Cartesian causality is recuperated thus:

- ○ **The conservation of memory**: Causality in general and Cartesian causality specifically, cannot alter the growing block Bergsonian facet accumulation of memory, or change the past into a form that did not happen, or erase that past. Hence, mind cannot violate the conservation of memory. In short, the spatial 'conservation of matter' transforms into the temporal *conservation of memory*, which cannot be usurped by Cartesian causality and mind.

- ○ **The principle of consistent futures**: Cartesian causality can only instigate future brain-world outcomes consistent with the past-in-memory. Indeed, via a-noetic means, the Whitehead facet automatically grand decoheres to consistent futures that conform to the conserved past, and mind can only select those futures permitted and consistent with the past. Hence, mind cannot bring about non-consistent futures in violation of the conservation of memory.

- ○ **Mind utilises a non-energy based work function zero process to select future brain-world outcomes**, exploiting work function zero time *and* work function zero general causality, to bring about the sought state-change… without violating even generic energy-based conception of conservation principles, and even less the recuperated abstract form of the principle of conservation reconstituted as the conservation of memory and the enforcement of consistent futures.

When the general theory turned to the issue of the 'location' of memory and nous vis-à-vis the brain, it found that **de-spatialisation and the structure of the Bergson-Whitehead amalgam usurped the belief that memory could be stored at in-brain spatial 'locations':**

- *The brain is a de-spatialised temporised Bergson-Whitehead complex* and cannot be modelled as a 'material object in space', with supposed spatial 'addresses' at which memories are allegedly stored. (**5.43** to **5.44**).

- *De-spatialisation explains the failures of both engram memory theory and holographic or holonomic memory,* also tacitly falsified in the experiments of Rose and Harding. Pribram's holonomic memory, when recuperated into its temporised version, requires that information be clouded in de-spatialised stereotemporal form, as is the Bergson facet itself: thus constituting a mnemonic holonomy in time, *not* in space.(**5.43** to **5.44**).

- *Recollection and 'retrieval' of memory involves direct acquaintance with the past across time, not across space, with memory 'transported' by means of the non-contiguous superpose of memories through work function zero mnemonic mechanics.* Hence, one's recollection of the big red garage door from 1973 is a 'reach out' across time into the Bergsonian past; *not* a reach out across space to what associatively lights up in one's brain: a mere readiness state for the expression of the declaration of memory, *not* its store; *not* its facsimile or copy… and even less the site of conscious apprehension of the memory. (**5.44**).

- *The non-identifiability of real memories vis-à-vis the brain is independently amplified by infinities and metalatives inherent to and inseparable from conscious recollection and nous.* These exceed, cannot be stored in, or produced by, the realised finite brain, notwithstanding de-spatialisation.

- *The big red garage door, Potters Bar, 1973 from the author's childhood is a <u>primacy</u> retained in the Bergson facet of the amalgam. The attendant brain-events that are observed to 'light up' in the author's brain are <u>recentcy</u> states,* also retained in the Bergson facet as more recent memories of the past. Tautologically, **the *primacy* is not the *recentcy*:** The big red garage door-1973 is not the brain-events that 'lit up'. (**5.45**).

- *The process of memory recollection involves mnemonic mechanics and the superposition of Bergsonian wavefunctions in memory*: The notion that memory recollection involves direct acquaintance with the big red garage door from 1973, realised via pure time-resonance, is no more a 'mad idea' than the reach out, again purely across time, to whatever lit up in one's brain, erroneously treated as if the memory. Given de-spatialisation, the latter must also resonate purely in and across time to the witnessing recollecting nous. (**5.45**).

- **It follows that, the growing block model mind and consciousness attendant recollection cannot be confined to the OR-form 'materialised' brain restricted to the event horizon, but must also reach out to and dissociate across the Bergson facet from which it directly temporally acquaints with its memories. Hence, Cartesian dualism follows per the said temporal dissociation and non-contiguous temporal reach of nous to its temporally remote memories.**

All of this lead to an unexpected turn in the nature of memory retrieval and recollection:

- The expression or declaration of recollection is slated to happen in the future: a Whiteheadian potentiality. Thus, in order to express recollection, active mind must reach into and decohere the future to the sought recollection-potentiality, and into what subsequently lights up in the author's brain. Thus, **active mode mind must project into the Whitehead facet and selectively decohere the menu of future possibilities to a specific declaration of recollection. Thus, active mind cannot be confined to the event horizon brain and must dissociate from it in order to attain and decohere the future**. Hence, inexorable Cartesian mind-brain dualism follows yet again. (**5.46**).

- **The non-clockability thesis**: While it is possible to clock as an event the subsequently declaration of one's recollection of the big red garage door from 1973, *it is not possible to clock the active mode mind that instigated the expression of that* *recollection. Nor is it possible to clock the subsequent passive mode consciousness bearing witness to one's recollection.* In short, nous does not constitute clockable 'events' of the Bergson-Whitehead amalgam. (**5.46**).

- By implication, **the active mind, and the subsequent passive moment of recollection, are** *extant* **the Bergson-Whitehead amalgam**. Hence, Cartesian dualism inexorably follows once more.

The above conclusions are enhanced when the attempt to identify memory and recollection with the contents of the event horizon and the OR-form brain confined to it also fail. Hence, **the want to locate nous on the event horizon remains forlorn**. (**5.47**). Thus:

- Assuming that two points, P1 and P2 along an event horizon constituted putative in-brain memory versus its putative in-brain recollection mechanisms, P1 will be interrelated to P2 across a succession of event horizons... 'vertically', or via time... and via attendant future-potential delayed choice time-interval relations and future-potential worldlines and lightcones. The interrelation between P1 and P2 ...between memory and the memory-recollection procedure... will *not* be 'horizontal' or along the event horizon, or across 'space'.

- The assertion for purely temporal 'vertical' association of states within the brain cannot be altered by the imperceptibility of the short time-intervals entailed between P1 and P2. The interval cannot be pretended into a spatial or contiguous 'material' relation, supposedly confined to and furnished by brains.

- Causality and agency cannot be mediated by means of material exchanges across the event horizon or across 'space', but only vertically or temporally and non-contiguously. The vertical pure time relations are *not* constituted as or realised by 'material exchanges' between P1 and P2.

- Thus, the minded selection of the future-potential Whitehead moment at which the author's recollection of the big red garage door-1973 is to transpire, insofar as it must 'vertically' temporally relate Bergsonian memories to Whiteheadian potentialities, must be initiated and instigated by active mode mind 'located' elsewhere, or in and through the 'gaps in time' between successive event horizons pertinent to brains... hence, from 'outside' of the event horizon and associated brain. Hence, again, mind-brain dualism must necessarily follow.

However, **the want to locate nous within the Bergson facet must also fail**: (**5.48**).

- In its realised temporal perduring form, the brain constitutes a Bergsonian 'worm': mnemonic superposes of brain-pertinent wavefunctions in memory temporally elaborated into the past: i.e., Pribram's de-spatialised and temporised holonomy, if you will. Thus, the brain must elaborate as a perduring 'worm', from its most remote past to its most recent event horizon state, and must exceed any confinement to the event horizon, which manifests only its most recent OR-form resolution.

- Hence, OR-form states of the 'materialised' brain, restricted and confined to the event horizon, are culminations from the most recent AND-to-OR collapse rendered consistent with Bergsonian memories constituting the perduring 'worm' brain.

- The attempt to locate nous in the Bergsonian past must fail as surely as the attempt to located mind, memory and consciousness on the event horizon: Why? Specific events and facts pertinent to memory must first unfold as realised events. The consciousness of these, and their subsequent minding, can only arise retrospectively, or temporally *after* the events and facts have transpired, *not* as or from the events that have transpired, whether these events belong to the leading event horizon or to the mnemonic Bergsonian worm constituting the whole history of the brain.

From the Whitehead facet, **nous cannot be 'located' or clocked as potential events of the Whiteheadian future: Active mode mind affects the brain from the 'gaps' between potential successive event horizons. As such, mind is extant the Whitehead facet. The complex wavefunction that mind selects for future AND-to-OR resolution of the brain solves the unitary binding problem of the brain**, the basis for which remained hitherto elusive to Penfield *et al.* (**5.49**).

- *The unitary binding of the brain (but not the generation of mind or of consciousness) is obtained from the unitary global grand decoherence and unitary AND-to-OR collapse of the complex Whitehead wavefunction (instantaneous whole) attendant the brain,* selected for future development by extant Cartesian mind.

- The point-like and absolute time-like nous is *not* produced by the unitary decoherence and collapse of the global wavefunction or 'instantaneous whole' attendant the brains, if only because the brain's content is always internally subject to the relativity of simultaneity (potential or subsequent OR-form)... and incapable of furnishing an absolute time-like noetic 'common present moment'. Hence, mind and *consciousness does not arise from the brain*.

- The active mode mind selects the 'instantaneous whole' complex wavefunction for its brain from the Whitehead facet. This collapses into unitary brain activity. Mind does not accomplish this selection by acting from the Bergson facet, nor from the event horizon and its materialised brain. Instead, mind engages sought future-potentiality from the 'gaps in time' between successive event horizons constituting the Whitehead facet, but not from that Whitehead future. In short, *the mind acts from, but does not reside in... nor emerge from... the Whitehead facet, but from the 'gaps', 'cracks' or interstices in future-potential time...from outside of memory, future possibility, and time.*

567

- The inverse is also true: the coherent outcomes obtained from the unitary collapse of the complex wavefunction or 'instantaneous whole' imposed by mind on the brain, hence the unitary brain, *arises within (but not from) absolute time-like passive consciousness, after* the minded brain activity has transpired per AND-to-OR collapse of the instantaneous whole.

In the above, we discovered that nous has two modes: ***passive mode consciousness*** and ***active mode mind.*** (**5.50**). Yet, neither the conscious witness to recollection, nor the non-conscious active mind preceding it, constitute clockable events. Thus…

- In the **passive mode**, the noetic moment *constitutes the passive non-causal and non-volitional moment of consciousness of the Bergsonian past,* temporally *after the events of the world and brain have transpired.* Obviously, retrospective consciousness cannot arise at the same time as the Bergsonian world and brain, and cannot be identified as any past-event, given that it occurs temporally *after* the said events, and given that the past transpires into formation temporally *before* it arises to witnessing consciousness. Thus, the former's content cannot be affected or modified by consciousness, and the latter is necessarily causally passive in its relation to what it retrospectively witnesses. Thus, *passive mode consciousness has no causal power: it is non-volitional, given its passive recipient status as purely retrospective witness.*

- Through the **active mode**, *mind affects causal agency upon future-form Whitehead possibilities of the brain and world.* The active mode mind decoheres the future into sought future brain-outcomes. Thus, minded agency temporally *precedes* future brain activity and modifies and selects that future temporally *before* it transpires. However, *active mode mind cannot be identified with any Whiteheadian potential for brain activity that it decoheres and selects: it acts from the 'gaps' between potential future event horizons and cannot be clocked as the brain in its future-form.*

- *The active mode mind does not possess consciousness: It is non-conscious or sub-conscious, even if it might possess plausible volitional power.* Nous obtains consciousness of its agency only *after* the facts of that agency have been fulfilled by preceding non-conscious active mind.

The experiments of Benjamin Libet *et al* uncovered a-synchronicity between brain activity and the subsequent *presumed* moment of 'conscious volition'. The a-synchronicity is often purported to invalidate volition, and even consciousness is declared an illusion. However, differentiation into passive and active mode nous, supported by preceding breakthroughs that revive Cartesian dualism, lay bare **critical errors involved in generic interpretations of the Libet experiments (5.52)**. Thus...

- In the Libet experiments, the assumption is made that the non-conscious brain activity produces consciousness and volitional consciousness, *(i.e., a priori materialism). This assumption is loaded to the facts of the experiment. The facts are then presented as if in 'proof' of the assumption (post hoc fallacy).* This formal dishonesty was first raised in Preliminary: **0.37**.

- The *Cartesian possibility that mind and consciousness temporally precedes (or even follows) brain activity instigated by mind is not taken into consideration:* it is dismissed per *a priori* materialism.

- *But materialism is falsified*: (See whole of Part-III: **5.26** to **5.38**). If materialism were true, the want to identify nous with the brains (*assumed* in the Libet experiments) would require exact temporal synchronicity of nous with attendant brain activity. But such a feat requires brains to have absolute time like characteristics: not possible for brains subject to the relativity of simultaneity (**5.53**).

- Also, *volition is erroneously assigned to passive mode consciousness, when volition could only plausibly attend non-conscious active mode mind...* which intercedes from the 'gaps in time' temporally *before* the subsequent brain activity it instigates, and cannot be clocked as that brain activity, or as *any* brain-event (**5.50**).

The conundrum produced by the Libet experiments led to the conclusion that **unitary mind and consciousness does not emerge out of brains, while the unitary brain *is* a consequence of the agency of active mode mind, not the outcome of that unitary brain and activity**. This emerged from the following sequence of reasoning (**5.49** and **5.52**):

- The *unitary binding of brains is obtained from the spontaneous a-noetic unitary decoherence and AND-to-OR collapse of its Whiteheadian instantaneous whole complex wavefunction,* pre-selected for the brain by active mode mind. The events constituting the whole brain, though globally unitary, are internally subject to the relativity of simultaneity.

- The *singular unitary character of the active mode mind and of passive mode consciousness is not generated by the spontaneous a-noetic unitary decoherence and collapse of the complex instantaneous whole wavefunction that attends the whole brain.* The temporally monadic singular unitary character of nous is per function of the absolute time-like characteristic of nous: a state the brain cannot produce or furnish, given the inherent relativity of simultaneity operating in the latter.

- The *active mode mind, acting from the 'gaps in time', or from between successive potential event horizons, is selectively and causally responsible for the unitary brain,* even if the latter is not causally responsible for unitary mind and consciousness.

The above resolved two issues raised by Wilfred Penfield and Ewen Cameron, respectively: With respect to **Penfield's search for the unitary binding of brains: (5.54)**:

- The unitary binding of brain activity is brought about by the unitary decoherence, unitary AND-to-OR collapse and unitary development of the complex future Whiteheadian wavefunction (the instantaneous whole) that attends the whole brain.

Penfield's error was to seek the causal basis for this 'in the brain', or in the resolved total outcome obtained from the decoherence and AND-to-OR collapse of the unitary whole-brain wavefunction. (See preceding summary notes for why this is forlorn).

- *The actual solution requires a growing block quantum mechanical AND-form (or future-form) model of the brain: not known to Penfield, and generally avoided in brain science, neuroscience, and in forlorn philosophy of mind*: all of which usually presume the brain to constitute a 'classical object'. This is per *a priori* materialism. Yet, generic science has had no difficulty in accounting for the global unitary development of a photographic film-image in quantum mechanical terms. A similar application to the brain, treated analogous to a developing photographic film, as a solution to the unitary binding problem of the brain, was never considered.

With respect to findings from the notorious **Ewen Cameron** per his destructive ECS **(5.54)**:

- Cameron inflicted disruption-events vis-à-vis the brain and vis-à-vis the means of expression of memories.
- These disruption-events, incorporated into the brain-pertinent Bergson facet and the brain 'worm' therein, were retained therein as memories in their own right.
- These built up within the recentcy of the Bergson facet, interceding with and occluding older Bergsonian memories that would otherwise restore the pertinent readiness states for established biographical memory recollection-expression and declaration. Thus, memory recollection and, with it, biographic identity, was *effectively* lost, even though it yet remained intact in the abstract Bergson facet.
- The original memories remain intact within the Bergson facet and are ineradicable, but temporally recede to effective loss.
- But most critically, in the 'experiments' of Cameron, the "I" could never be eliminated, given that it is synonymous with *nous* constituted as an absolute-time like temporal monad, necessarily beyond the capacity of brains to furnish or generate, and beyond Cameron's power to erase.

The developments elaborated above culminate into Cartesian dualism... of a form that supersedes classical dualism based on the dualism of *res cogita* versus *res extensa*. With de-spatialisation, *res extensa* (or space) disappears: We obtain Cartesian dualism of memory and possibility subject to the relativity of simultaneity versus unitary absolute time-like nous: **a temporal dualism of *noumnemonae* versus *mnemonae*. (5.55)**. Thus...

- *Noumnemonae* is memory in passive mode consciousness, within the noetic moment in *absolute time-like state…* versus...
- *Mnemonae*: the Bergsonian world and brain without consciousness, or a-noetic memory of brain and world, *always subject to and structured by the relativity of simultaneity,*
- *Mnemonae* arises to *Noumnemonae* or mind and consciousness, but mnemonae is independent of mind and consciousness.
- *Noumnemonae* does not arise from *mnemonae*: Mind and consciousness do not arise from the brain and world.

In total, **superlative Cartesian dualism emerges** as…

- **The dualism of active versus passive modes of consciousness vis-à-vis world and brains.**
- **The dualism of absolute time-like nous versus a-noetic Bergsonian memory and possibility subject to the relativity of simultaneity.**
- **The dualism from *noumnemonae* (memory in consciousness) versus Bergsonian *mnemonae* containing a-noetic world and brain.**

Finally, general theory culminates into **the en-worlded mind and into the Anti-Copernican revolution: (5.56 to 5.59)**. Thus…

- *De-spatialisation inverts the object-subject relation and demands that the world arises within mind and consciousness,* given the breakdown of space, of 'location', and of the mind-world erroneously purported as material 'spatial relations'.
- *Thus, the mind is en-worlded*: Mind and consciousness are *not* 'located' within a world of 'matter, space and brains'. Rather, the latter, as Bergsonian memory (also as Whitehead future possibility) are 'located' within mind and consciousness… within the noosphere… which is in turn subsumed to a grander ineffable transcending neutral ontology or *Being*.
- However, even with the en-worlded mind, *the world does not arise from mind and consciousness*: The case against 'consciousness causes collapse' and against both quantum idealism and general idealism in Part-II prohibits this.
- *The en-worlded mind inverts object-subject relations inside-out…or right-side-in*. The object-world arises within the subject-nous, but *not from* the subject-nous Thus, nous is furnished a growing block Bergson-Whitehead amalgam, constituted as a de-spatialised stereotemporal world of memory, possibility, and time (which includes the brain itself) which arises *within* nous, but *not from* nous... 'within' the *noosphere* of collective transpersonal mind, but *not from* transpersonal mind.
- **The Anti-Copernican Revolution (5.59)**: The en-worlded mind takes the generic belief that we do not occupy the centre of the universe… and are 'insignificant specks' cast to and lost in the unfathomable void... and turns this 'outside-in', or right-

side in. Thus, possibility, memory, and time... all 14 billion years of memory, and all futurity to timelike infinity... arises within us. Hence, we are *not* lost in the universe. Although, given our general aretaic failure, the universe is possibly lost within us.

The en-worlded mind seems to imply solipsism. But solipsism and idealism are circumvented (5.56). Thus:

- Where two temporally separate observers are concerned, one observes a red flashbulb, followed by a blue flash some seconds later... while the other sees the blue flash followed by the red flash per the same time-delay. The same-events are reconstituted differently and into inequivalent sequence-order to each observer, rendering the observers into experientially unique non-illusory first-person frames. This is per function of de-spatialisation, temporisation and the relativity of simultaneity that bestrides time-distributed observers vis-à-vis their events. Following a time-delay, one observer apprehends the other observer as a set of unique past events arising *within* the consciousness of the former, per function of de-spatialisation, the en-worlded mind, and the anti-Copernican revolution thus. In turn, again following the same time-interval delay, the other observer apprehends the former as a set of unique past events in its own right, but temporally delayed and arising to and within the consciousness of that other. Yet, the "I" within the one turns out to be the same "I" within the other. The same "I" is witnessing both sets of said unique non-illusory first persons frames, and is witnessing itself per function of an objective time-delay. The same "I" witnesses itself as different viewpoints because of the time-delay, combined to the non-illusory unique order of events experienced by the one in contrast to the other. Yet, the "I" in both observers *is* the same transpersonal unitary-collective consciousness (the noosphere) to which the one *and* the other are subsumed non-illusory tributaries.

- If no time-delay relations existed and the relativity of simultaneity did not apply to the one in relation to the other observer... if 'space' was real, and if both observers experienced a world subject to absolute time... there could be no differentiation of experience and of the singular universal "I" into non-illusory unique first-person views. In an absolute time frame, all events would happen to all observers in the same order, at the same moment of time, and effectively at the same 'place'; so constituting a one-only undifferentiated experience and one-only viewpoint: Therein, there could be no individuation, and a one-viewpoint solipsism would abide. This clarifies *the indispensable role of the relativity of simultaneity and implicit de-spatialisation and temporisation in the firewall against solipsism.*

- Yet, the background reality of the transpersonal noosphere and the universal "I", which subsumes the time-distributed and non-illusory unique first-person viewpoints, is supported by the conscry or co-apprehension of infinities and metalatives: Infinities and other metalative ideas cannot be furnished by the a-noetic Bergson-Whitehead amalgam. The latter is composed of improper forms vis-à-vis the ideas. Instead, the ideas are furnished to the noosphere from Being, and from the transpersonal noosphere to the subsidiary first-person viewpoints, who can now co-apprehend and conscry infinities and metalatives, *not* through the temporal transaction of agreed-upon 'stand in' codes, symbols or improper forms via the Bergson-Whitehead amalgam, and not through the brain subsumed to the Bergson-Whitehead amalgam... but via the transpersonal noosphere itself, albeit in abidance with the time-delay and time-distribution of the conscrying first-person viewpoints, with respect to which both first-person viewpoints and egos are non-illusory tributaries.

We recuperate the above per the Whitehead facet and the future-potential noosphere. Thus:

- *The noosphere of the universal singular "I" constitutes a future-potential for all consciousness and first-person states: constituting a parallel futurity vis-à-vis the a-noetic Whitehead facet of future-potential world and brains.* The potential noosphere undergoes AND-to-OR collapse to consistent consciousness vis-à-vis the parallel AND-to-OR collapse of the a-noetic Whitehead facet. *The outcomes from the AND-to-OR collapse of the potential noosphere are enmeshed to, and can only be consistent to, the delayed choice time relata that temporally distribute the outcomes obtained from the parallel AND-to-OR collapse of the Whitehead facet... into non-illusory first-person consistent consciousness-frames*, temporally distributed per function of the same-said relativity of simultaneity. Hence, *solipsism is circumvented as the universal "I" collapses into unique inequivalent events and non-illusory first-person witnesses*, distributed in time per the relativity of simultaneity.

BOOK-V PART-V:
THE INTERMEDIATE THEORY OF MIND & WORLD

5.61: Aims of Part-V

The general mind-world theory argued for superlative Cartesian mind-brain dualism; obviating the problem of *res extensa* and overcoming the much-abused weaponised principle of conservation historically arrayed against Cartesianism. Yet, key problems remained beyond the province of the general theory: For example, specific conditions that apply to the Whitehead facet of the Bergson-Whitehead amalgam, which permit active mode mind to generate willed outcomes, were not elaborated upon: the *ephemeralised punch card theory* was merely intimated. The goal of the intermediate theory is to elaborate and present solutions to such problems.

Other problems were also not scrutinised by the general theory: How does the extant mind overcome quantum indeterminism within the brain *and* bring about sought coherent global outcomes therein? How objective quantum indeterminism is overcome in a developing photographic film, combined to the solution to the problem of quantum indeterminism furnished in Book-III **3.26** to **3.29**, clarifies how the same is accomplished in brains. All of this will be accounted for in some detail in the intermediate theory.

If mind and consciousness is extant the brain, why does nous require a brain at all? Why not forgo brains completely? Such a question was beyond the scope of the general theory, but it is fully in the purview of the intermediate theory. We shall find that only brains can generate a *Whitehead equalised possibility plateau* indispensable to mind-brain causality. Its clarification and modelling from within the growing block Bergson-Whitehead amalgam is the central goal of the intermediate theory.

*

Three developments will be key to the intermediate theory: the **equalised possibility plateau**; the related **critical milieu**; and the **ephemeralised punch card theory** of mind-brain causality and control. The equalised possibility plateau is a characteristic of the Whiteheadian future attendant the brain, and it is almost wholly unique to brains. Within the Whitehead facet, it permits alternative future-potential complex global wavefunctions for the future brain, with each constituting the basis for a different intent by mind, and which mind can select through decoherence and, through it, control the future development of its brain, thus affecting the world.

However, the basis for the formation of the equalised possibility plateau is the brain's critical milieu: an arrangement of specific properties that, when projected into the Whitehead facet, will decohere or form the equalised possibility plateau. For example, a pre-developed photographic film, with its properties projected into the Whitehead facet, constitutes a stand-by mode for all possible alternative photographic images that might form upon it in the future: each based on a unique and distinct complex global wavefunction or 'instantaneous whole'. What holds true in the case of the photographic film also holds true vis-à-vis the properties of the brain: a critical milieu by virtue of its key properties.

In intermediate theory, the brain-pertinent critical milieu projects into the Whitehead facet an equalised possibility plateau that attends all future-possible complex global wavefunctions, hence future-possible brain activity. The process by which active mode mind selects from the plateau constitutes the *ephemeralised punch card theory*, which entails a delayed choice **switch-x-like** mind-brain control; wherein active mode mind, exploiting the work function zero process of AND-to-OR collapse and time, acts from the 'gaps in time'... from between potential event horizons... to decohere the future developmental pattern sought for its brain.

Of course, the modelling of this process, including memory, including memory recollection and habit formation, entails *explicit* graphical use of the Bergson-Whitehead amalgam. Thus, the critical milieu, together with the Whiteheadian equalised possibility plateau that it projects, combined to the graphical Bergson-Whitehead model of these, will constitute the heart of the intermediate theory and the framework for the ephemeralised punch card theory of Cartesian mind-brain control.

5.62: Critique of Quantum Mind Theory: Is Quantum Mechanics Relevant to the Brain?

Quantum mind theories assert consciousness as the agency that brings about wavefunction collapse (time). Some quantum mind theories claim that consciousness orchestrates quantum indeterminism. Some theories claim both assertions. Overall, while not always explicitly espousing the primacy of nous, quantum mind theories tend toward idealism and reject dualism. Our Alternative Realist or Alt-R based mind-world theory is *not* a quantum mind theory, as is clear from the critique of 'consciousness causes collapse' in Part-II and per other findings. While quantum mechanical realities must attend mind-world theory, this is owing to the superlative growing block Bergson-Whitehead amalgam framework to *all* mind-world theory, *not* because nous is complicit in causing time itself or involved in the orchestration of quantum indeterminism, and even less because of any forlorn ontological primacy of nous.

*

Materialist approaches to the brain typically reject quantum mind theory on the notion that quantum phenomena do not apply to the brain, or especially to the macro-scale. As part of the Alt-R critique of quantum mind theory, we must reiterate why quantum mechanics cannot be obviated from the 'macro-scale' and from the brain (see **5.40**), given that the relevance of quantum mechanics to the macro-scale brain is ineluctable per function of the growing block framework that forces the saliency of AND-form logic (the Whitehead facet and the future) to all scales, including the macro-scale brain. Yet, the scale-saliency of quantum mechanics does not imply the viability of quantum mind theories, or especially of idealistic quantum mind theories.

The consensus against the use of quantum mechanics in consciousness research and at the macro-scale brain may appear plausible, if only because, at first sight, putative macro-scale billiard balls and brains do not appear to explore the 'sum of all paths', much less display superposition characteristics in the way putative point-like particles supposedly do at the micro-scale. Hence, consensus asserts that, such things as the superposition principle or AND-form logic, or wavefunction collapse and quantum indeterminacy, must be confined to the micro-scale. Indeed, the reasons why Wigner abandoned 'consciousness causes collapse' was per its presumed inapplicability to the macro-scale and to the macro-scale brain itself (see Part-II **5.12**). Thus, the consensus tends to trump approaches that utilise quantum mechanics, such as evinced in the approaches espoused by John Eccles, and by Roger Penrose and Stuart Hameroff.

However, generic confinement of quantum mechanics to the 'micro-scale' does not consider or know about the insights garnered by Alt-R and from the nature and ontology of wavefunction collapse and time. Nor does consensus know about or incorporate implications from de-spatialisation, much less recognise the reality of *grand decoherence* evinced in the erasure of non-consistent nested futures. In short, the consensus for the scale-restriction of quantum mechanics to the 'very small' is in error. Quantum mechanics is pertinent at all-scales, and it is pertinent to brains as much as it is to the total Whitehead facet of the growing block universe that must also subsumes the future-potential quantum mechanical brain.

571

*

One of the central assertions of Alt-R is that time is not a fourth dimension. Instead, time is wavefunction collapse; justified by the intertwine of causality and the conservation principle in the *firewall principle*, but also from the fact that a block universe, one *without* a future set of possibilities, and one in which *all* futures are fully OR-form resolved, could never permit the self-interference and the formation of interference patterns evinced in the macro-scale two-slit experiment, at *any* scale (see **5.01**). The very fact that we have quantum interference in nature proves the saliency of the growing block universe and, with it, the saliency of quantum mechanics or AND-form logic at *all* scales, including vis-à-vis the brain. Consequently, to assert that quantum mechanics is solely confined to the very small is tantamount to asserting that wavefunction collapse, or time, must also be confined only to the very small: an error.

Time is pertinent to both putative micro and macro-scales, if not to the Whiteheadian universe-scale wavefunction. No matter what the scale, the future of any system is effectively in physical isolation and rendered as an AND-form state, notwithstanding grand decoherence (Book-II 2.34). The future remains in isolation in the strict sense that it has not yet transpired: and this insight abides *at all scales*. **As such, the future must constitute as an AND-form superposition of possibilities** *at all scales, including in relation to the brain.* **Therefore, a growing block quantum mechanical model of the brain, and the consequent pertinence of quantum mechanics to brains, is unavoidable.**

Moreover, de-spatialisation central to Alt-R (Book-IV **4.01** to **4.07**) alters the framework through which we must conceptualise 'scale'. Scale is no longer spatial: it is purely temporal. Hence, the 'micro-scale' gives way to the domain of very short time-interval high-frequency physics. Hence, the generic claim that quantum mechanics must be restricted to the 'very small' is tantamount to its restriction *only* to short time-interval high-frequency physics.

Furthermore, one notion of scale we could all agree on is per function of the density of events furnished by the number and frequency of quantum depositions that make up all realised patterns and situations. There is a big difference between a quantum mechanical account of a single quantum event versus the compound *en masse* quantum events that make up the 'larger' world, such as a whole photographic image; such as the brain. But the *en masse* context does not obviate either wavefunction collapse (i.e., time itself) nor the superposition principle (the ontic future), nor quantum indeterminism: All operate at the *en masse* deposition-scale or at the 'macro-scale', even if billiard balls do not seemingly self or mutually interfere at that *en masse* deposition scale. Indeed, the macro-scale interference pattern in the two-slit experiment, formed out of high event density *en masse* quantum depositions, illustrates the point.

However, the moment we incorporate grand decoherence to the Whitehead facet of our growing block universe (i.e., the basis for revived Mach's principle requisite to the growing block quantum mechanical account of inertia and much lese) the fact that 'matter here' instantaneously adjusts according to 'matter elsewhere', regardless of how 'far away' these are, implies mutual quantum counterfactual interference of billiard balls 'here' with other billiard balls across the universe: (See Book-II **2.35** and Book-IV **4.13**). That is, the universe self-interferes with itself at the grandest possible scale to affect inertia, among other things. Thus, per inertial grand decoherence alone, quantum mechanics abides, not simply at the 'large scale', but at the universe-scale.

The implication is obvious: it makes little difference what philosophy of mind one espouses, whether this be a materialist philosophy of the mind-brain, or a dualist view, or an idealist quantum mind theory: In all cases, quantum mechanics cannot be excluded or obviated from the brain or from *any* ideology contending about the mind-world conundrum: That is, you cannot exclude the ontic future from your account of the brain, much less confine the ontic future to the 'very small', or to high-frequency short time-interval gamma ray physics, regardless of ideological or philosophic preference. **Quantum mechanics must be incorporated even into materialist accounts of the mind-brain because materialist accounts cannot exclude the future constituted as the Whitehead facet from any serious model of the brain. Hence, quantum mechanics is indispensable to** *any* **account of the mind-brain, and theories that preclude quantum mechanics are automatically forlorn or un-serious.**

*

Our conclusion about the indispensability of quantum mechanics from any serious model of the brain might give the false impression that what we seek must be quantum mind theory. The following should put this notion to rest:

- *Mind or consciousness does not secretly orchestrate quantum indeterminate processes.* **(5.16).**

- *Mind or consciousness does not cause AND-form wavefunction collapse.* That is, nous does not bring about or cause time: **(5.18).**

- *Mind and consciousness remain purely time-dependent. Time, and wavefunction collapse are ontologically primary to the very possibility of minding and consciousness.* It is time and wavefunction collapse that permits minded agency vis-à-vis the brain and world. No time… no nous.

- *Wavefunction collapse, hence time, and attendant quantum indeterminate processes, are ultimately 'directed' by an ineffable structure-less causality* that transcends 'matter in space' as thoroughly as it supersedes active mind and passive consciousness.

Two consequences follow from all of the above: These are…
  ○ the obviation of 'consciousness causes collapse'
  ○ the obviation of *idealistic* quantum mind theories.

Succinctly, mind theories that assume a quantum mechanical role for mind or consciousness, either as the collapsor of the brain-pertinent complex wavefunction, or as the director and secret orchestrator of in-brain quantum indeterminate processes, must be called into permanent doubt.

QUANTUM MECHANICS AND MIND

It should now be obvious that the otherwise remarkable theories of Eccles, Penrose, and other quantum mind theories, premised on similar assumptions, must also be doubted. The doubt will have nothing to do with the erroneous notion that quantum mechanics applies only to the 'very small'. Indeed, the key problem for generic quantum mind theories resides in their dubious assumption that consciousness either causes wavefunction collapse (**5.18**)… albeit *mostly* within the confines of the brain… or secretly directs in-brain quantum indeterminate processes. But AND-to-OR collapse (time) occurs spontaneously, and it will transpire even without nous, and, as repeated *ad nauseam*, quantum indeterminacy is not a problem that requires a solution, much less a secret orchestrating nous in lieu of equally forlorn a-noetic hidden gears (**5.16**).

*

If nous plays no role in wavefunction collapse and time, nor in the orchestration of indeterminism in brains, then in what way is quantum mechanics relevant to and indispensable from dualistic mind-brain theory and the solution to how extant mind controls its brain? The key is in the critical role played by the growing block grand decoherence process, essential to the justification of the equalised possibility plateau and of the consequent role of quantum mechanics in mind-brain control.

Recall that grand decoherence arose out of the nested futures perspective and from counterfactuals-driven quantum erasure of non-consistent futures (see Book-II **2.34**); all inferred from John Wheeler's delayed choice experiment and the derived nested futures model of the Whitehead facet. The latter compels that the wavefunction cannot simply constitute immanent possibilities in AND-form but must also embed subsequent futures that might branch out from each immanent possibility. Recall that nested futures were categorised into *divergent, entangled,* or *convergent* futures, elaborated into the Whitehead facet and to the remote infinite future. The convergent category turns out to be definitive of the total Whitehead facet and of the growing block future.

In the convergent category of nested futures, any new event that manifests on the event horizon, say, over four lightyears away in the frame of binary Centauri system, will alter the content of the nested future-possibilities attached to our own frame on Earth. A supernova explosion of one of the stars in Centauri binary will decohere the nested future possibilities in our frame up to and beyond four years and into the future, into consistent-only nested futures. This will be accomplished by means of the quantum erasure of the non-consistent futures, or grand decoherence. In short, all futures *without* consistency vis-à-vis the supernova will be erased. But the quantum erasure of those non-consistent nested futures will decohere, not just our own nested futures attached to our frame, but *all* nested futures horizontally attached to *all* frames along the event horizon. In other words, the rest of the universe will instantaneously decohere and become future-consistent with the supernova event at Centauri, if only because *all* future-potential lightcones and worldlines attached to *all* frames along a same event horizon, are potentialised within the Whitehead facet to overlap and *converge* in some common future time, and fully so at timelike infinity.

This insight solved the long-standing problem of inertia by presaging a growing block quantum mechanically theory of inertia, wherein matter 'here' counterfactually knows what matter 'elsewhere' is *not* doing, and can thus acquire and exhibit inertia by means of such instantaneous counterfactual relations; all obtained through the erasure of non-consistent futures… through universe-scale *grand decoherence*. (see Book-II **2.34** to **2.35**). But note that…

- **Grand decoherence is the universe-scale and macro-scale equivalent of generic quantum wave interference and self-interference; one that transpires between macro-scale 'objects'…ultimately, even between billiard balls.**

- **From the reality of grand decoherence, it necessarily follows that attendant quantum mechanics cannot be confined only to the 'very small'.**

- **It also follows that quantum mechanics most certainly applies to the macro-scale brain… to the latter's future-form Whiteheadian state or potentiality.**

We reiterate that grand decoherence designs for both macro-scale *and* universe-scale wave-interference and self-interference. While this is evident in the revived Mach's principle and in attendant inertia, it is especially pertinent to the time-evolution and development of the brain in the context of partial-perpetual wavefunction collapse and time. Succinctly, ***the growing block grand decoherence affords and forces upon us a quantum mechanical model of the brain.*** The brain is ultimately *not* a 'classical object': This is per ineluctable grand decoherence and the ontic reality of the future conceived as the *quantum mechanical* Whitehead facet.

Aside its contribution to the explanatory framework of the permanent magnet, of inertia, and even key aspects of gravitation, grand decoherence supports the anticipated **Whiteheadian equalised possibility plateau**, which emerges out of both nested futures *and* attendant macro-scale grand decoherence; crucial to both the intermediate and fine structure mind-world theories. Therein, a pre-developed photographic film, framed to the nested futures context, will form a projection into the Whitehead facet a possibility plateau upon which all possible future photographic images are potentialised, and in which all such possibilities enjoy equalised probability. This constitutes the Whiteheadian equalised possibility plateau for the photographic film.

Of course, once an image is formed on the film, alternative non-consistent future-images will be erased and foreclosed from subsequent futures. Per grand decoherence, the rest of the universe and its total Whiteheadian futurity will grand decohere and reconstitute into consistency with the newly developed photographic image and per the irreversibility of the film.

However, a developmentally reversible photographic film, one that can accept subsequent photographic images and replace a previously developed image with a new one, will project into the Whitehead facet in a different way: The Whitehead facet will not grand decohere into developmental irreversibility, but into developmental *reversibility*. Yet, given Bergsonian memory, both the replaced image *and* the succeeding image will be retained within the Bergson facet as abstract memory, *without* erasure or loss.

It turns out that the brain is essentially equivalent to a developmentally reversible photographic film and *its* attendant Whiteheadian equalised possibility plateau; with the latter decisive to the quantum mechanical model of the brain: all of it critical to Alt-R's attendant ephemeralised punch card theory of mind-brain control, *and* consequent Cartesian revival.

5.63: Scale-Amplification:
The Generalisation of the One-Slit Condition & Grand Decoherence to the Brain

Why do billiard balls fail to behave like 'particles', notwithstanding the fact that quantum mechanics *does* apply to the putative macro-scale order per function of the growing block universe *and* grand decoherence, if nothing else? Why is it that a billiard ball does not explore all possible 'paths' to a destination? No wonder the very proposal for macro-scale AND-form logic tends to meet immediate scepticism. **The fact that billiard balls do not apparently behave like quantum objects hastens the notion that neither does the brain. This then leads to the assertion that quantum mechanics is irrelevant to brains, or at least irrelevant to mind-brain issues. Of course, grand decoherence proves otherwise. However, it does not prove it explicitly: Billiard balls appear to behave like 'classical' objects despite their quantum mechanical natures at the grand scale. This is also true for brains.** We must clarify why this is so, notwithstanding implications from the growing block approach and from grand decoherence.

*

First, in order to obtain the evidence for the operation of AND-form logic and the superposition principle in the two-slit experiment, we need to produce *en masse* quantum events at the detector: A single quantum incident will not do. A few will not do: We need an above-threshold number of incidences before we can infer *any* process or principle that operates at the purported micro-scale of single particle quanta. The proof that quantum mechanical effects operate at the micro-scale level of a single quantum particle (i.e., that it undergoes self-interference, destructive or otherwise) *always* require the amplification of such events to the *en masse* macro-scale formation; such as the interference pattern formed out of *en masse* quantum incidences. In short, we always need macro-scale amplification of the quantum mechanical to infer the reality of the quantum mechanical at the macro-scale, even if supposedly confined only to the 'very small', notwithstanding implications from de-spatialisation to issues of scale (see **5.62**). But such amplification implies the saliency of quantum mechanical effects at the macro-scale, no longer evidentially confined to the micro-scale.

Moreover, one can see the operation of AND-form logic in photography as much as in the two-slit experiment, wherein photography constitutes another macro-scale amplification and evidence of quantum mechanical effects. In photography, quantum indeterminism, hence AND-form logic, is evident throughout the three-phase developmental process that culminates into the fully formed image. Yet, the photographic film as a whole constitutes a macro-scale state, just as does the whole detector 'surface' in the two-slit experiment. Indeed, the development of the interference pattern in the two-slit experiment involves essentially the same process entailed in photography and could be carried out with photographic films involving photons instead of, say, De Broglie electrons. Similarly, one could develop the equivalent of photograph with electrons: i.e., the electron micrograph, often used in biological sciences.

All of these entail and exemplify the *amplification* of quantum mechanical effects from the very small to the very large; obtained through *en masse* 'particles', whatever their frequencies. Indeed, it is only at the *en masse* conglomeration of quantum incidences (i.e., at the purported macro-scale) that we obtain the fact and evidence for the reality of quantum indeterminism, of AND-form logic, and self-interference, all supposedly confined to the micro-scale, but evidentially no longer so confined. Thus:

- The *en masse* conglomeration of particles through which quantum mechanical realities becomes evident constitute putative 'macro-scale' states, *not* micro-scale ones. Thus, we have **the scale-amplification of quantum mechanical phenomena**.

- **The implication is that quantum mechanical realities fully apply and *always* come into evidence at the purported macro-scale**: *never at the micro-scale*, precisely because they are only ever brought into evidence at the macro-scale, via said amplification.

- That is, we can only ever infer the reality of micro-scale quantum mechanical phenomena from and at the macro-scale. **We can never obtain direct access to a pure micro-scale level event, but must always get to the micro-scale indirectly, in and through the macro-scale... through said scale-amplification.**

- But **this implies that quantum mechanical phenomena *do* scale up to the macro**, otherwise we could not infer their micro reality.

- It necessarily follows that quantum mechanics cannot be confined to the micro-scale and, insofar as it can only ever be evinced at the macro-scale, **it must also follow that quantum phenomena also constitute macro-scale phenomena, otherwise scale-amplification could not be obtained, and evidence for quantum mechanics would be beyond the reach of science**.

Therefore, AND-form logic, the superposition principle, and self-interference, are not confined to the 'very small', but also fully apply to the large, notwithstanding implications from de-spatialisation and the need to recapitulate 'scale' purely in terms of time... or, alternatively, in terms of the number and density of *en masse* quantum incidences.

However, scale amplification on its own is not able to explain the apparent non-quantum mechanical behaviour of billiard balls and brains: i.e., We never observe billiard balls and planets explore all their 'sum of paths', assuming 'path' has any meaning in a de-spatialised physics, despite the fact that these *are* quantum mechanical, both per scale amplification *and* per grand decoherence.

*

QUANTUM MECHANICS AND MIND

The original answer to this conundrum was given in the Preliminary and elsewhere throughout this work. The following constitutes a recapitulation.

The reason why billiard balls do not behave like quantum objects emerges from the fact that phenomena at the billiard ball level are equivalent to single slit experiment conditions, and do not exhibit obvious self-interference. As was pointed out in the Preliminary **0.14**, according to the non-standard evaluation of the double slit experiment, the single slit outcome does not imply the absence of AND-form logic, much less any obviation of the superposition principle: **The very same single-slit distribution can certainly be obtained from quantum mechanical AND-form waves** 'passing through' just one slit, per function of the wave probability structure, even if no self-interfere is structurally permitted. Even if we place a detector at the single open slit and obtain an OR-form resolution of the particle therein, the particle on approach to the single slit must yet constitute an AND-form quantum wave. But, the particle *after* the slit will revert to its AND-form quantum wave-state, until its subsequent AND-to-OR resolution at the detector.

Hence, **the absence of self-interference and of associated interference patterns in the single-slit experiment, or in any system, (including macro-scale billiard balls) does not imply the absence of AND-form logic from that system at its given 'scale'. The same applies to billiard balls brains.**

What about quantum indeterminacy? Billiard balls can be produced and reproduced over time per repeat cycles of the three-phase developmental processes, by quantum mechanical waves undergoing AND-to-OR collapse into *en masse* indeterminate quanta, without obvious or visible self-interference or mutual interference per said wave probability structure, just as is observed in the distribution obtained under single-slit conditions. **The apparent 'classical motion and displacement' of the billiard ball across the table is made up of temporal successions of the same-said rapid three-phase developmental cycles, alternating between OR-form *en masse* quantum indeterminate incidences and AND-form complex wavefunctions pertinent to the billiard ball, followed by another *en masse* AND-to-OR collapse sequence that reproduces the OR-form billiard ball at its succession... involving ineliminable quantum indeterminacy, without obvious interference...** *without* any obvious 'sum of all paths'. This also applies to brains.

While there is no obvious self-interference in billiard balls, people, planets, and brains, we now grasp that this has nothing to do with 'scale', much less the forlorn purported confinement of quantum mechanical phenomena to the 'very small'. It has everything to do with predominance of single-slit-like conditions across nature, which do not imply the absence of AND-form logic, nor of quantum indeterminacy.

However, systems equivalent to two slit-like conditions are *not* confined to the 'very small', as attested from the scale-amplification of quantum phenomena in really-existing two-slit experiment, and from the fact that the evidence for self-interference in two-slit experiments is a thing always manifested at the macro-scale, and subsequently merely inferred about the micro-scale *from* the macro-scale.

*

Finally, **the ostensive absence of self-interference from billiard balls, people, planets, and brains, is merely apparent. The superposition principle, AND-form logic, quantum interference, *and* self-interference, are fully present at the level of billiard balls, moons, worlds, whole galaxies... *and* brains... courtesy of grand decoherence.** (see Book-II **2.34** and **2.35**). Thus, while billiard balls do not undergo *ostensive* self-interference, billiard balls over 'here' on Earth *do* counterfactually interfere with billiard balls 'over there' and elsewhere, instantaneously across 'lightyears', through the process of counterfactuals-driven consistency principles, or grand decoherence...or via the revived Mach's principle. Thus, through grand decoherence that attends the total futurity of the Whitehead facet, AND-form logic, self-interference, and the superposition principle *do apply* at the 'macro-scale', as these most certainly apply at the universe-scale. Hence, these must also apply to the macro-scale brain. This truism is most directly relevant to the formation of the Whiteheadian equalised possibility plateaus critical to mind-brain causality and to Alt-R's growing block quantum mechanical model of the brain.

Hence, quantum mechanics is directly pertinent to the brain. Hence, ultimately, the brain cannot be treated as a 'classical object'. That is, we must describe the brain quantum mechanically: succinctly, as a growing block process and model.

Given the unavoidable relevance of quantum mechanics to the macro-scale and to brains per the above, the materialist want to reduce brains to the status of a 'classical object' is necessarily forlorn and void.

5.64: Critique of Quantum Mind Theory:
Conflation of the Problem of Quantum Indeterminism in Mind-Brain Control

The 'problem' of quantum indeterminism is taken seriously by major proponents of quantum mind theory. This is especially pertinent given that we can no more eliminate the reality of quantum indeterminism from the macro-scale world and brains than eliminate AND-form logic and the superposition principle from the same (see **5.63**). But quantum indeterminism is *not* a problem that needs a solution, especially by resort to quantum mind theories that pose consciousness as the supposed collapsor of brain-attendant wavefunctions, or as the secret orchestrator of subsequent quantum indeterminism, or both.

Quantum mind theories are sometimes characterised as dualistic: This is certainly true of John Eccles' dualist interaction theory, in which mind modifies quantum probabilities for bouton vesicular discharge[102]. (More on boutons and the vesicular grid will be explored in the fine structure theory in Part-VI). Eccles' view conflated quantum indeterminism into a problem that requires a solution. Yet, in

---

[102] Eccles, John C. (1994). *How the self controls its brain*. Berlin: Springer-Verlag.

Book-III **3.26**, in Part-I **5.08**, and in Part-II **5.16**, and elsewhere, we recapitulated the fact that quantum indeterminism is not a problem that requires a secret noetic orchestrator, no more than it requires an a-noetic 'hidden gears'.

Insofar as some quantum mind theories assert the ontological primacy of mind and consciousness, these constitute idealistic theories that might incorporate aspects of Eugene Wigner's long-abandoned approach (see Part-II **5.12**) and the equally earnest views of Henry Stapp (*Mindful Universe: Quantum Mechanics and the Participating Observer*). These approaches require due hearing. This is in contrast to the various dubious New Age takes with similar themes.

However, the Alt-R case against 'consciousness causes collapse' undermines idealistic quantum mind theories.

One intriguing superlative quantum mind theory is that of Roger Penrose *et al*. It rejects the idea that consciousness causes the brain pertinent wavefunction to collapse. Instead, consciousness arises *from* a particular kind of wavefunction collapse pertinent to the resolution of foundational spacetime geometry at the Plank scale, also posited as relevant to gravitation[103]. The view has subsequently developed into a fine-structure approach based on cytoskeletal structures[104]: (More on this will be said in our own fine structure approach in Part-VI).

There are near-similarities between the views of Penrose and the position espoused in Alt-R. There are also major differences. From our Alt-R approach, the ontological primacy of wavefunction collapse and time is certainly prerequisite to the possibility of passive mode consciousness. Thus, no wavefunction collapse and time, no witnessing consciousness. Although Penrose does not identify wavefunction collapse as synonymous with time itself and appears to accept the generic notion that time is a 'fourth dimension', *and* supposes 'space' and generic 'spacetime' as givens, Penrose purports consciousness to arise *from* wavefunction collapse and time: a point of similarity with Alt-R, though not identical to the position in Alt-R.

In Alt-R, wavefunction collapse and time bring about events, without the formation of which there could be no subsequent passive mode consciousness or witness of said events, or any consciousness of events in the absence of events. Also, wavefunction collapse and time permit mind-to-brain agency from active mode mind. This is accomplished via the 'gaps in time'. The spontaneous a-noetic collapse of the mind-selected future wavefunction then brings about the intents of active mode mind into resolved brain-world events.

However, Penrose believes quantum indeterminism is a problem that needs a solution. He proposes *objective reduction* vis-à-vis 'non-computational' but non-random AND-to-OR collapse of foundational spacetime geometry, as the solution to indeterminacy in brains.

In Alt-R, the idea of gear-less causality is almost interchangeable with Penrose's non-computational causality behind wavefunction collapse and time. However, Alt-R sees no need to obviate quantum indeterminism itself, given that this is not a problem that needs a solution: (See Book-III **3.26** to **3.30**; Part-I **5.08**; and Part-II **5.16** in the present book).

Moreover, Alt-R has a different growing block insight into how gravitation occurs and how gravity interfaces with quantum mechanics: with gravity as a memory effect emerging from the superpose of wavefunctions in memory within the temporal Bergson facet: (see Book-IV **4.25** to **4.30**). This much is obviously from de-spatialisation, given that de-spatialisation demands that gravity cannot be a 'force transmitted across space', no more than it can be a 'spacetime geometric curvature', given the fact that there is no 'space' to curve. Geometry is a spatial concept, not part of temporised physical ontology. Hence, in a de-spatialised universe, there is no ultimate 'geometry' of space *or* of time. Gravity *must* be a growing block memory effect from and as the accumulation of the past, as demanded by de-spatialisation and temporisation. Thus, apparent 'spacetime geometry' is at best a pragmatic descriptor we might impose on otherwise mnemonic gravity: (for original reasoning, see Book-IV **4.30**).

Indeed, from the Whitehead facet-based approach, the AND-form future possibilities within the Whitehead facet cannot possess a 'geometry' as such, even though the future possibilities therein must decohere to a consistent set of futures per counterfactuals and attendant grand decoherence… imposed by gravity-generating mnemonic mechanics and from gravity as a memory effect. Again, this is *not* a process generated from 'geometry', although the outcomes from it may be pragmatically described in geometrical terms.

In other words, according to Alt-R, Penrose's collapse of 'foundational spacetime geometry' is not a feature of mind-brain relations, simply because it is not a part of physical reality, although some sort of mnemonic mechanics *must* be pertinent to both, given temporisation through de-spatialisation.

Finally, the fact that Penrose resorts to 'non-computational processes' ought to hint that some things simply defy reduction to 'gears' or to generic materialistic variables. An ineffable structure-less causality, inherently 'non-computational', and causative of wavefunction collapse and time, is necessarily extant the Bergson-Whitehead amalgam as a whole. Any resort to idealism is foreclosed per the critique of 'consciousness causes collapse' in Part-II: **5.10** to **5.22**. But the reality of a grander ontology that embeds noetic ontology versus the a-noetic growing block Bergson-Whitehead system of memory, possibility, and time, constitutes concession to a noetic domain… to a noosphere… and to abstract Cartesian mind-brain dualism.

*

This essay aims to demonstrate that the goal of many quantum mind theories to overcome quantum indeterminism, treating it as a 'problem' in brains, is completely unnecessary. From discoveries in Book-III **3.26** to **3.39**, summarised in Part-II **5.16**, objective quantum indeterminism and randomness do not constitute a problem that needs a solution. In the critique of 'consciousness causes collapse' in Part-II, our usurping of the assumption that nous constitutes an orchestrator of quantum indeterminism constituted

---

[103] Shadows of the Mind. Roger Penrose, Vintage 1995 p367
[104] Ultimate Computing: Biomolecular Consciousness and Nano Technology, Stuart Hameroff, North-Holland, Amsterdam 1987

an attack on the central premise of many quantum mind theories. **If nature in general has no need of a secret deterministic orchestrator of indeterminism in order to generate and sustain pattern, then nature has no more a need of a consciousness directed obviation of quantum indeterminism than it has of an a-noetic mechanical hidden gears to accomplish the same.**

Quantum indeterminism is ubiquitous *at all scales.* in brains as much as in photography. Quantum indeterminism can no more prohibit the formation of coherent brain activity reflective of the will and intent of mind than can the same quantum indeterminism prohibit the formation of a coherent photographic image, consistent to its originary source. This fact clears away any concern with randomness in the brain: it founds an entirely different basis for mind-brain control: one that does not require nous to secretly orchestrate quantum indeterminate processes and outcomes in the brain, even under hot body conditions.

Several factors come into play to undercut the need for a mind-directed orchestration of in-brain quantum indeterminacy:

- **Quantum indeterminacy (hence randomness) is objectively real. There are no secret hidden gears to orchestrate quantum indeterminate processes**. (See Book-III; Part-II).

- **Nous is not the secret orchestrator of quantum indeterminate processes in lieu of a-noetic hidden gears** .(Part-III)

- From the solution to the stochastic coherence problem espoused in Book-III, we elaborated the following:
  ○ **Quantum indeterminate processes are always funnelled into globally determinable developmental outcomes**. This is accomplished by ordinances comprised of bounded probability states exemplified by generic wavefunctions; by Whiteheadian *wavefunctions in potentiality* with nested futures, and by Bergsonian *wavefunctions in memory* with their nested futures 'cropped'. These also apply to brains.
  ○ **The mind can juggle ordinances of memory and possibility, and combine these into ephemeralised 'punch cards', that it then imposes on its brain.** The 'punch card' consequently funnels otherwise *en masse* objective quantum indeterminate terms into global coherent unitary brain activity, consistent with the intent of mind
  ○ **Per mind-imposed complex wavefunctions or 'punch cards', quantum indeterminate processes in the brain unfold according to the three-phase developmental process (Book-III 3.26); fulfilling the pattern globally ordained by the 'punch card' so imposed by the mind**, *despite objective ineliminable quantum indeterminism.* The three-phase process is recapitulated below: Thus…
    ▪ Below threshold phase, wherein insufficient numbers of quantum incidences are resolved out of the collapse of the mind-selected wavefunction 'punch card': a *noise distribution* without any pattern; devoid of information.
    ▪ Transitional-chaotic phase; wherein pattern-structure pertinent to the complex wavefunction imposed by the mind begins to emerge per *en masse* quanta; forming an unstable shifting pattern: a *chaotic distribution*.
    ▪ Noise-reduction and pattern-saliency phase; wherein subsequent objective indeterminate outcomes collectively reveal fully formed coherent pattern ordained by the wavefunction or the mind-imposed 'punch card', without requiring any noetic orchestration of the quanta, much less the mental alteration of quantum probabilities to regulate quantum depositions: a *signal distribution*.

In the brain, **the subsequent development of the abstract 'punch card' into stereotemporal globally coherent brain activity does not require mind to orchestrate the objective indeterminate processes that attend this**. Indeed, once development is initiated, subsequent intercession by mind is no longer required, until a next imposition. Once the development is complete, and the will of mind is enacted in the completion, passive mode consciousness will bear witness to that outcome: (see general theory, Part-IV **5.50**).

We must also assert that there is no need for a 'cold body solution' in any of the above. **The want of cold body solution arose from the want to employ mind and consciousness as the orchestrator of in-brain quantum indeterminacy**: a feat difficult to obtain in a hot body brain. However, the reality of hot body conditions abides across the brain, as they typically abide throughout the three-phase development of a photographic image, if not across most of nature.

Some quantum mind theorists have attempted to overcome the hot body 'problem' by suggesting that there may be ranges and conditions within the brain that are equivalent to cold body-like conditions[105]. However, **since it is not in the first place necessary to employ mind or consciousness as the secret orchestrator of quantum indeterminate processes, the hot body problem and the search for a cold body solution, or equivalent conditions, is rendered unnecessary**. Again, photography constitutes an explicit case of how a hot body state can easily permit otherwise *en masse* objective quantum indeterminate depositions in full indeterminate 'storm' to form up into origin-consistent coherent global outcomes: The same applies to the brain.

5.65: Preface to the Ephemeralised Punch Card Theory of Mind-Brain Causality
And the Whitehead Equalised Possibility Plateau

The active mode mind controls its brain through the selective imposition of a specific 'instantaneous whole' or complex wavefunction or possibility function i.e., the 'punch card'… which mind 'draws' from the Whitehead future or from the Bergsonian past, or from both. To accomplish this, the mind exploits the structure and processes of the de-spatialised temporised Bergson-Whitehead amalgam and the process of work function zero AND-to-OR collapse and time.

---

[105] Tegmark, Max (2000). "Importance of quantum decoherence in brain processes". *Physical Review E.* 61 (4): 4194–4206.

Decisive to the industrial revolution, in past centuries, punch card driven devices were almost as ubiquitous as present day computer based control of industrial machinery: e.g., the Spinning Jenny. Their use continued into the 20ᵗʰ century.

Consider the pin-driven punch card device as a useful analogy to how mind controls its brain (See **Fig. 5.10**). The mechanical contraption is comprised of a pusher, which drives an array of pins into and through a slot to produce an output. By inserting a punch card with a specific arrangement of holes into the slot, we limit the pins permitted to push through to the output side. In this way, we control the pattern at the output produced.

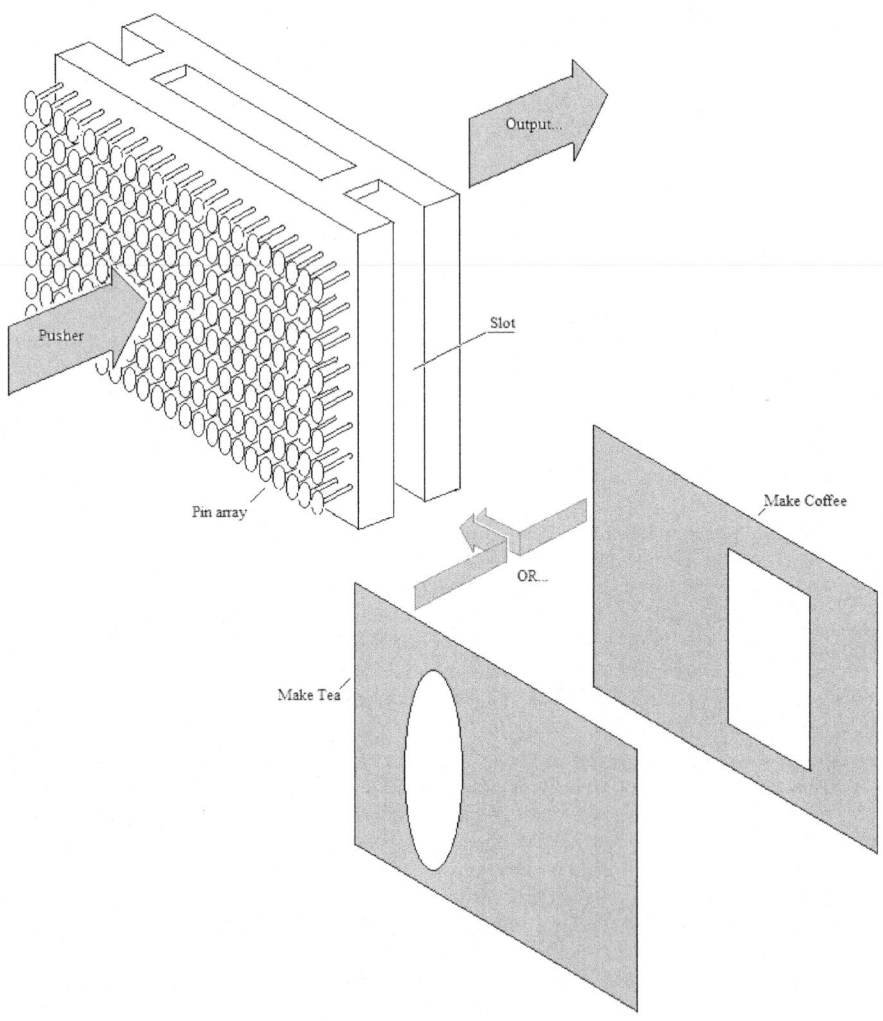

Fig. 5.10: The mechanical pin-driven punch card device for a coffee and tea-maker
Notes: Depiction of a pin-driven punch card device for a hypothetical tea and coffee maker. Insert one of the punch cards into the slot and activate the pusher. Only pins that can push through the cavities on the inserted punch card can get to the output side of the mechanism. Depending on the pattern of pins formed, the output will activate a larger machine (not shown) to make either tea or coffee. If we take the punch card device as analogous to the brain, then the punch cards constitute analogical stand-ins for complex wavefunctions imposed by active mode mind on the brain, while the driven pins constitute stand-ins for either neural discharges or actions pertaining to microtubular structures. Thus, we obtain a mechanical metaphor for mind-brain control. Note that the end-user mind is not part of the system depicted: It works from outside of the system: the stand-in for the non-reducible Cartesian mind.

What are the differences between mechanical 'hardware' punch card mechanisms and the ephemeralised 'punch card' attendant mind-brain control? **In mind-brain control, the equivalent to the 'pusher' is AND-to-OR wavefunction collapse, or *time*.** While the mechanical pusher needs pleonastic 'force', power, energy or work to drive the pins into and through the slot, **the gear-less work** function zero process of wavefunction collapse and time will AND-to-OR resolve *en masse* quantum discrete events (the analogical stand-in for the 'pins') into total global brain activity, without violating either generic or the reformed principle of conservation (refer to Book-II **2.39** and **5.29**). **Thus, unlike the mechanical pusher, which needs *putative* mechanical energy and force to push the pins through, the quantum discrete events (the real 'pins') require no work or power to bring into realisation, given that work function zero AND-to-OR collapse and time is not driven by any energy or power inputs, notwithstanding pleonastic energy relations** (see **5.35**).

As stated, the equivalent to the 'pins' in the context of brains are the *en masse* objective quantum indeterminate events; the 'pixels' that make up the pattern of brain activity. **The equivalent of the 'punch card' in the context of brains is the complex wavefunction or 'instantaneous whole' selected and imposed by the mind upon the brain, which subsequently funnels the *en masse* quantum indeterminate processes (the pins) into definite brain-patterns**, in the same way that a photographic film's complex wavefunction develops its *en masse* photon depositions into a coherent photographic image. Both entail the three-phase developmental process: (see Book-III **3.26**).

In the case of the Cartesian mind, **the imposition of the abstract complex wavefunction 'punch card' is accomplished from between successive event horizon-progressions (the 'slot', so to speak)… or from the 'gaps in time'… necessarily accomplished by means of a work function zero process that exploits ontologically primary work function zero AND-to-OR collapse and time**.

The subsequent three-phase development of the imposed brain-pertinent complex wavefunction 'punch card' brings into realisation the attendant brain activity without need of *any* secret deterministic orchestration of the *en masse* quantum indeterminate terms: (see previous essay, **5.64**).

*

The mind-imposed ephemeralised 'punch card' could constitute memory, retained in and drawn out of the Bergson facet (i.e., the past) constituted as a specific wavefunctions in memory. This could be re-imposed by active mode mind on a subsequent development of the brain, such as to constitute memory recollection.

Alternatively, the ephemeralised punch card could comprise an entirely novel possibility function never-before realised: a wavefunction in future-potentiality drawn out, *not* from the Bergsonian past, but from the Whiteheadian futurity attendant the growing block brain.

The categories of ephemeralised 'punch cards' available to active mode mind are thus:

- **Possibility functions in memory: i.e., Wavefunctions in memory 'drawn out' from the Bergson facet; from the abstract past**: a critical aspect of memory recollection: e.g., the big red garage door; Potters bar, 1973.
- **Novel possibility functions in potentiality: i.e., Novel wavefunctions in future-potentiality constituting never-before enacted brain activity** and implied intents; 'drawn out' from the welter of Whiteheadian nested future possibilities pertinent to brains.
- **Mixed state possibility functions:** i.e., ephemeralised 'punch cards' comprised of admixtures of wavefunctions in memory combined to novel wavefunctions in future-potentiality.

There is a 'fire and forget' feature to how the mind imposes its ephemeralised punch card. Thus:

- **The mind selects and 'fires' it's punch card, and typically ends further involvement.**
- **A-noetic gear-less causality takes over, bringing about attendant AND-to-OR wavefunction collapse and time, with attendant progression of the event horizon.**
- **Quantum indeterminate outcomes then amass into the required brain activity and pattern per the three-phase developmental process**, in a similar way to how this transpires in photography.

A 'change of mind' is possible: Mind selects an alternative punch card to the one previously imposed, and could interrupt the latter's development in order to impose the different global brain-outcome sought. In this way, the Cartesian mind possesses a marginal degree of interruptive control over the developmental-end and activity of its brain.

*

Note that, **insofar as mind exploits AND-to-OR collapse and time (itself a work function zero process) and intercedes in and through the 'gaps in time', it is obvious that mind is never evinced or clocked as an event of the Bergson-Whitehead amalgam.** In effect, the mind causally operates from a sidereal remove vis-à-vis the Bergson-Whitehead amalgam and its succession of event horizons and brains. **Hence, Cartesian dualism,** as was contended in the general theory of mind-world in Part-IV.

*

If mind is extant the brain, what need of brains? Why brains? Although mind is extant the brain, the fact that minds coincide with brains, but not with cups or pumpkins, is *not* accidental. There is something special about brains: Brains permit the imposition of ephemeralised 'punch cards' in the way previously described: **Brains remain indispensable to the formation of the *Whiteheadian equalised possibility plateau*; in turn critical to mind-brain control and consequent mind-world causal power and agency.**

Both future novel affectation of the brain *and* memory-recollection are not ultimately or purely separate. These often enmesh: Mind can only make its brain behave in the sought way at a future potential moment. It follows that the 'punch card' for memory-expression (hence recollection itself) must also be slated to occur in the future. For this to happen, **possibility functions in Bergsonian memory must also be potentialised to the pertinent Whiteheadian future, even while this necessarily requires an ultimate 'drawing out'** of such functions from the Bergson facet: a seeming contradiction and paradox, but an inescapable one that will be resolved in sensible fashion. Thus, the mind's selection of memories is from among a future set of potentialities that have grand decohered into consistent form with the past-in-memory, and can exhibit that past as much as it can exhibit the novel affect futures that have yet to transpire: A bizarre-seeming arrangement that will be clarified in the forthcoming essay.

To grasp the full basis of the above, we *must* first develop the *Whiteheadian equalised possibility plateau;* critical to the very ephemeralised 'punch card' driven mind-brain causality and control just described.

5.66: Prelude to the Whiteheadian Equalised Possibility Plateau

The brain projects an equalised possibility plateau into the Whitehead facet. There are fine-structure states of the brain (not present in cups, rocks, or pumpkins) that constitute the 'critical milieu' on the basis of which the equalised possibility plateau projects into the future and forms up within the Whitehead facet. The plausible fine-structure states and processes that render the brain into a critical milieu from which the equalised possibility plateau arises will be detailed in Part-VI. **From the equalised possibility plateau thus formed, active mind selects and imposes its brain-controlling 'punch card'**, exploiting the 'gaps in time' and work function zero AND-to-OR collapse (time) to accomplish this.

How do we model the Whiteheadian equalised possibility plateau that attends brains? This is not difficult: We first model and incorporate a pre-developed photographic film into the framework of the Bergson-Whitehead amalgam. We then generalise the findings from this to brains framed to the same Whiteheadian domain, justified by the essential similarities that reside between the critical milieu that attends brains and the same that attends photographic films. Both the brain and the photographic film constitute critical milieus, and both generate Whiteheadian equalised possibility plateaus, different only in their specificities.

*

In its chambered state in relative physical isolation, a photographic film-surface can be characterised as a readiness state that can accommodate any number of alternate future-potential images; all on an equal possibility footing. Through grand decoherence, if by nothing else, this property is simply projected into the future as a Whitehead projection. **The photographic film does not favour any particular future image: It is permissive of any potential alternate image on an equal probability or future-possibility footing, baring contingency or 'Wheeler's cat'. That is, the film is permissive with respect to future image-possibilities. Described quantum mechanically, the film constitutes an** *equalised possibility plateau* **projected into the Whitehead facet:** *a superlative wavefunction whose superposed possibilities are complex wavefunctions in their own right.* **That is, instead of a wavefunction that sets the probability for mere quantum discrete incidences, the equalised possibility plateau constitutes a wavefunction for alternative wavefunctions... a wavefunction that superposes alternative inequalities or non-identical complex wavefunctions, all set to equal probability and possibility.**

Obviously, the complex wavefunction that pertains to the image of a vase on a table is not going to be the same complex wavefunction that attends the image of the Big Ben, London. But within an equalised possibility plateau that incorporates *both* image-pertinent complex wavefunctions to a larger wavefunction in potentiality, the distinct and unequal wavefunctions are equalised in terms of probability and possibility: That is, the photographic film has no bias for one image versus the other, and renders the future probability for one or the other unequal image wavefunction into equality of possibility.

The equalised possibility plateau is an extension of the nested futures approach first developed in Book-II, with the difference that, normally, the alternative nested futures are not necessarily set to equalised probability. In the equalised possibility form pertinent to the photographic film... and to the brain... otherwise distinct unequal futures *are* set to equalised probability and possibility.

*

Aside the Whiteheadian equalised possibility plateau that it projects, the photographic film constitutes a critical milieu: precisely because it can set up and project the equalised possibility plateau into the Whitehead facet... precisely because of its unique chemical-structural attributes. An array of light-sensitive chemicals embody the 'surface' of the old-school chemical film: (we ignore de-spatialisation for the moment). In combination, these compounds can admit any image within the bounds of the whole film-surface. This is a feat that cups, rocks and pumpkins cannot accomplish vis-à-vis the Whitehead facet.

Note that, in photography, this equalised possibility plateau is generated by a hot body film: A cold body solution is not an absolute necessity for the constitution of a critical milieu, or the projection from it of a Whiteheadian equalised possibility plateau.

The brain also constitutes a critical milieu whose collective array of basic units also get to project a specific type of equalised possibility plateau into the brain-pertinent Whitehead facet.

In the case of photography, **any bias or selective variable vis-à-vis the image permitted to form does not reside in the properties of the photographic film, nor within its projected Whitehead equalised possibilities. Bias enters the fray per extant contingent** *unequal* **possibilities: such as the photographer who prefers to photograph the vase on the table instead the Big Ben. The same holds in the case of the brain. There is no bias inherent per the equalisation of possibilities attendant the Whiteheadian projection and the equalised possibility plateau generated by the brain therein. In brains, the principal source of bias resides in the extant contingent world, but most especially in the intercession by extant 'end-user' active mode mind**. The latter intercedes and selects from the possibilities nested within the equalised possibility plateau to enact a future whole-brain outcome, and forgoes alternative equally possible whole-brain outcomes in favour of the one thus selected.

The equalisation of future possibilities, hence the equalisation of the pertinent complex wavefunctions superposed and nested in the said plateau, is crucial to understanding the very possibility and basis in how mind's intercession vis-à-vis brain and world transpires.

5.67: Integrating the Equalised Possibility Plateau into the Bergson-Whitehead Amalgam Framework

What does an old-school chemical photographic film look like from within the framework of the Bergson-Whitehead amalgam? When placed on the event horizon and projected into the Whiteheadian future, the Whitehead projection our film generates exhibits an AND-form state for all alternative future-possible images or their attendant complex wavefunctions, rendered into future-potentiality and set to equalised probability, in the same sense that alternatives are suspended and equalised in Wheeler's delayed choice experiment.

The equivalent to the **switch-x** tripper in photography is the photographer. At the critical moment, it is the photographer who decides what image will get to form on the film. Whatever the biases that attend the photographer, the photographic film in itself, per the alternative images it otherwise permits on equal terms, always remains unbiased, in the way that delayed choice **switch-x** is unbiased.

The equivalent to the 'punch card' in photography is constituted by the future-potential environmental states: the image-specific complex wavefunctions that the photographer selects for exposure and development by the equivalent of a delayed choice **switch-x** trip. The film then develops per the usual three-phase developmental process.

As stated, the film is unbiased: It accepts without any preference the states imposed upon it, as it would equally accept an alternative to what was subsequently imposed. In terms of our lay formalism, the architectonic wavefunction describing the pre-developed photographic film as an equalised possibility plateau and readiness-state for all possible images, can be expressed as…

$$\text{Whiteheadian: } \{\textbf{image-a} \text{ AND } \textbf{image-b} \text{ AND}\ldots.\ \textbf{image-n} \mid \text{each equally probable}\}$$

This can be read as…

"future-possible **image-a,** superposed with i**mage-b**…superposed with **image-n**… such that each image (i.e., attendant complex wavefunction) is equally probable".

Therefore, the photographic film constitutes an equalised possibility plateau, *ceteris paribus* with respect to external contingencies and the delayed choice **switch-x** tripping photographer, or even "Wheeler's cat".

*

The key differences between generic wavefunctions and equalised possibility plateaus need reiteration, augmentation, and pertinence to mind-brain relations: Thus…

- **Generic wavefunctions** *describe superposition AND-form states for unequal probability quantum depositions*, epitomised by patterns evinced in the two-slit experiment.

- On the other hand, **the equalised possibility plateau** *is composed of a complex superposition of alternative nested future wavefunctions*, with each nested future image-possibility constituting unique non-interchangeable image-specific complex wavefunctions; each set to a different set of internal probability distributions for their quantum depositions.

- Put another way…
  - generic wavefunctions superpose for probability distributions for quantum incidences and depositions, but…
  - equalised possibility plateaus superpose alternative wavefunctions into equalised probability.

- **No bias:** *The unequal possibilities are rendered equally permissible in terms of both possibility and in terms of required pleonastic energy*…until the photographer comes into the fray and trips the equivalent of **switch-x**.

- **In the case of the brain in lieu of the photographic film, essentially the same descriptions and process apply, save that the active mode mind replaces the photographer as the switch-x tripper.**

Part of the problem inherent to the equalised possibility plateau as applies to the photographic film is that it can only develop once and irreversibly. The image thus formed, and the general wavefunction collapse thus secured, closes off any future alternative image from developing on the same film. This implies that **we must distinguish between**…

- developmentally *irreversible* **equalised possibility plateaus**, versus
- developmentally *reversible* **equalised possibility plateaus,** such as exhibited by the brain.

The photographic film begins as an architectonic equalised possibility plateau for unequal and non-interchangeable image-specific future wavefunctions. Upon subsequent AND-to-OR collapse and development, the plateau is cropped of its alternative future image possibilities: Its Whiteheadian future transforms into a generic unequal non-plateau, and only the formed image can subsequently perpetuate and perdure into the future, while the photographic film irreversibly loses its openness to alternatives.

Through grand decoherence, the rest of the Whitehead facet counterfactually adjusts to the formed image and to the *de facto* irreversibility of the developed film. Hence, only nested futures that are consistent with the reality of the irreversibly formed image will remain within the Whitehead facet: All others will be erased.

Hence, **a developmentally irreversible photographic film is highly restrictive in terms of the future possibilities permitted to the whole of the Whitehead facet: Upon its development, it eliminates all non-consistent future possibilities to the infinite future. Hence, the irreversible film degenerates into a closed Whitehead facet** (see next essay **5.68**).

581

On the other hand, **a developmentally reversible film could not foreclose alternative nested future images from being formed on the same film in a subsequent future development. The rest of the Whitehead facet will grand decohere and adjust to the implications from reversibility in an open way, in that the non-consistent or alternative image-possibilities will not be foreclosed or erased, save those that could have been formed at the moment of the development of the first image: The alternative future images will not become erased: they will remain available to subsequent developments of the film, at least in principle.**

It turns out that the brain is developmentally reversible. It follows that, in order to model the brain as an equalised possibility plateau, and finally place it into the framework of the Bergson-Whitehead amalgam, we must first obtain the Whitehead projection for a developmentally reversible photographic film, and generalise its essential attributes to the brain.

5.68: Photographic Film as an Irreversible Equalised Possibility Plateau: The Closed Whitehead Facet

Typically, photographic films develop irreversibly, permanently excluding from subsequent development any alternative future image. The nested futures category that best describes the irreversibly developing photographic film is the divergent category.

Let us assume that the Whitehead facet nests just two immanent future potential images, both rendered in AND-form. But each nests a divergent possibility-tree of subsequent future-potential images: (See **Fig. 5.11**). With the first event horizon progression and wavefunction collapse, one of the two immanent images gets erased; so will all subsequent nested futures attached to it. On the over simplified assumption that only two alternate future-possible images are permitted, **if we expose the camera such as to develop one image, we will exclude and close off the future to the other image. We will have closed off all the nested futures attached to that foregone image, given the irreversible character of development**.

In truth, this over-idealised divergent nested future is embedded within a larger convergent category nested future definitive of the Whitehead facet *en toto*. When one image is developed and the other image is foregone, so are the nested futures attached to the latter. **The rest of the future universe will undergo grand decoherence into consistency per counterfactuals vis-à-vis the foregone image; erasing all non-consistent possibilities related to the foregone image, to the indefinite future.**

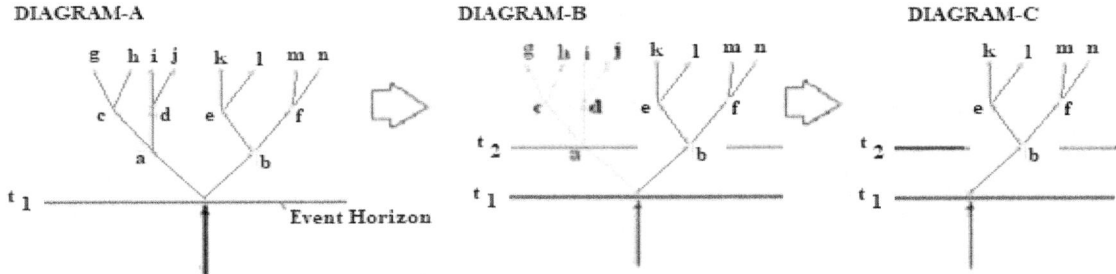

Fig. 5.11: Quantum erasure of nested futures: irreversible information plateau
Notes: The diagrams depict the development of an *irreversible* equalised possibility plateau, such as a photographic film per the nested futures framework. Diagram-A assumes that the photo-film can admit just one of two future-possible alternate photographic images immanent with respect to the film-surface at time t_1: i.e., **image-a** AND **image-b**. Hence, the equalised possibility plateau on the event horizon is open to just one of the two complex wavefunctions, wherein the Whitehead facet is defined by the divergent category of nested futures. Note that **image-a** and **image-b** are equally probable at this juncture, even though they constitute two distinct unequal probability structures for possible *en masse* photon depositions. By Diagram-B, the event horizon has progressed to t_2, and wavefunction collapse and time has occurred: It is the complex wavefunction attendant **image-b** that gets to be imposed on the developing film. By Diagram-C, all nested future states vis-à-vis **image-a**, together with **image-a** itself, are erased from the total future. As such, the erased **image-a** and all attached nested futures are permanently and irreversibly locked out from any future: the future is *closed off*: This is what is meant by the *closed Whitehead facet*: a consequence of developmental irreversibility. By t_3, the possibility plateau is no longer equalised with respect to now-erased and absent **image-a** and its nested futures.

Grasped in terms of both the divergent category and the convergent category, **the irreversibly developing photographic film relates to a *closed Whitehead facet*...** closed insofar as the image, together with all its attendant nested futures, is forgone, foreclosed, and irreversibly erased by attending erasure of all non-consistent futures from the embedding convergent category Whitehead facet.

From **Fig. 5.11**, per the closed Whitehead facet, the irreversible photographic film begins as an equalised possibility plateau. But this applies only to the immanent images, **a** AND **b**; *not* to subsequent events or future images further future-removed within the Whitehead facet. **Upon development, the equalised possibility plateau per the film will AND-to-OR collapse from a state initially open to alternatives to an irreversibly developed film closed off to any alternative, save the perpetuation of the outcome thus developed: a closed Whitehead facet, indeed.**

5.69: The Developmentally Reversible Equalised Possibility Plateau: The Open Whitehead Facet

A number of physical systems in nature constitute closed Whitehead facets. In **5.68,** the photographic film begins as an open Whitehead facet. The grander nesting equalised possibility plateau generated by the critical milieu (the film) will render all possible future images and attendant wavefunction complexes equally probable. The only inequalities and biases will arise from extant contingencies, such as the decision of the photographer to photograph one and not the other image.

Unlike generic photographic films, *the brain generates a developmentally reversible equalised possibility plateau.* In principle, it is possible to develop photographic films that are developmentally reversible. Indeed, though not photographic films, old VHS tapes allowed re-recording of new image-patterns over preceding ones. A chemical photographic film could be devised to operate in a similar way. It so happens that digital cameras, while these do not employ chemical films, *are* developmentally reversible. Thus, the critical milieu involved in digital cameras, projected into the Whitehead facet, readily constitutes a developmentally reversible equalised possibility plateau.

The brain constitutes a developmentally reversible equalised possibility plateau, inasmuch as the discharge of neurons or of other pertinent units, and their interconnections, are not constituted as one-time developmentally irreversible potentials and patterns. Indeed, neural connections enjoy plasticity and can be reversed and reformed into new patterns and attendant discharges, while neural discharges can discharge repeatedly. There may be finer structures that work in tandem with neurons: such as microtubules, such as the biochemical surfaces of cell and neuron bodies that might develop and act in a similar way to the surface of a developing photographic film.

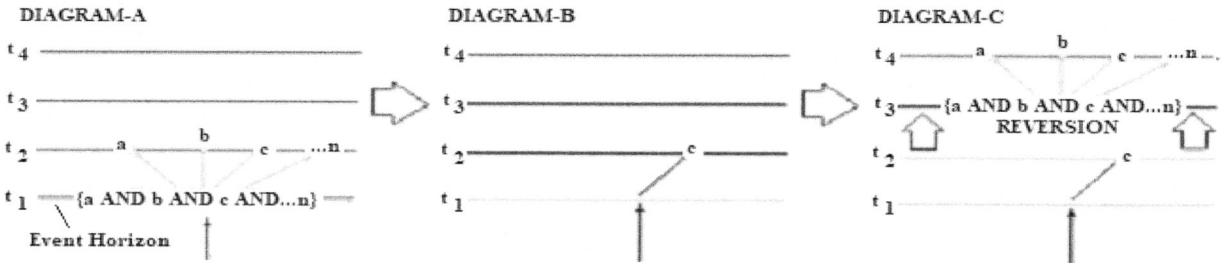

Fig. 5.12: Conditional erasure of nested futures: developmentally reversible possibility plateau

Notes: The illustration depicts a developmentally reversible equalised possibility plateau as a developmentally reversible photographic film, and, ultimately, as the brain. In <u>Diagram-A</u>, event horizon t_1 is open to many immanent alternative images. The plateau is a superposition state comprised of {**a** AND **b** AND **c** AND...**n**}: **a** to **c** are alternate wavefunction complexes for images or brain activity patterns: (**n** implies any number of alternatives). All are at equal probability. The divergent category nested future applies implicitly as background. By <u>Diagram-B</u>, the event horizon progresses to t_2. The complex wavefunction for **image-c** is realised: All alternate images (**a** and **b**) up to and including **n** ought to be erased: closing off the future. However, the film is developmentally reversible. It can revert to its default state. Hence, by <u>Diagram-C</u> at t_3, the equalised possibility plateau reverts to its default state: back to {**a** AND **b** AND **c** AND...**n**}; as it was at t_1. It is now open to alternative image-possibilities. The wavefunctions for one of these alternatives may well get to be imposed on it by t_4. Per its developmental reversibility, the film need not exclude a repetition of **image-c**. Equally, any other previously foregone image can form instead of **image-c**. Thus, through reversion to {**a** AND **b** AND **c** AND...**n**}, the future-form Whitehead-facet associated with the equalised possibility plateau is *open* to previously foregone alternate image impositions, even if already developed into an image, always on an equal probability basis, and per reversibility. Hence, the plateau constitutes an *open Whitehead facet.* This applies to and characterises the brain projected into the Whitehead facet, with the brain's future Whitehead facet open to alternative 'punch cards' (whole brain wavefunctions) that might be selected for it by active mode mind.

If we introduce the property of developmental reversibility to our photographic film, we end up with a near-permanent equalised possibility plateau: The film will permit the development of one future-possible image, *and* **it will allow the succession and replacement of this by another alternative image at a subsequent future... and so on**, or for as long as the film can last. At every event horizon and attendant wavefunction collapse, the film will admit and permit one of any number of alternative developmental possibilities, all on an equal probability basis. In this way, **our film will constitute a developmentally reversible equalised possibility plateau constituting an open Whitehead facet.**

The Whitehead expression of the reversible critical milieu and its equalised possibility plateau, and the process of grand decoherence to which both relate, is given in **Fig. 5.12.** The critical milieu and its possibility plateau can revert to a default readiness state at a subsequent event horizon, and, in doing so, admit an alternate future image in lieu of the preceding image.

In **Fig. 5.12** Diagram-A, the nested futures possibility plateau at event horizon t_1 is open to many immanent alternative image wavefunction complexes. These are...

            Whiteheadian t_1 {**a** AND **b** AND **c** AND...**n** | each image equally probable} ...

This is read as…

"complex image **wavefunction-a** is superposed with image **wavefunction-b**…. is superposed with image **wavefunction-n**… such that, each image is equally probable… where **n** implies any number of alternative images."

Note that, by **Fig. 5.12** Diagram-B, the event horizon, together with the plateau, have progressed to t_2. Partial-perpetual wavefunction collapse has occurred and **image-c** is selected for development, even though the other images were equally probable. The culprit behind the selection is the delayed choice **switch-x** tripping photographer. Consequently, all other alternate images (**a, b,… n**) and the futures that these in turn nest, are quantum erased at t_2… and the total Whitehead facet undergoes attendant grand decoherence to appropriate consistency of futures with what has transpired.

In the case of a developmentally irreversible film, only one image will develop at t_2. i.e. ,image-c… to the apparent permanent erasure and exclusion of all other alternative future images. However, in the present developmentally reversible example furnished in Fig. 5.12, image-c is *not* **capable of developmentally closing off the future to the other images. This is because the critical milieu attendant the equalised possibility plateau is developmentally reversible and can revert to its original default state by t_3, whatever the image that had formed upon it at t_2.**

Hence, by **Fig. 5.12** Diagram-C and t_3, the possibility plateau has reverted to its default state…

$$\text{Whiteheadian } t_3 \ \{a \text{ AND } b \text{ AND } c \text{ AND}…n \mid \text{each image equally probable}\} …$$

This is *essentially* the same equalised possibility plateau as found at t_1. The equalised possibility plateau at t_3 is now *open* to alternative images: Their complex image-specific wavefunctions in future potentiality (i.e., **image-a, image-b… image-n**) enjoy equal probability for future development. The photographer at t_4, could impose any one of these in place of **image-c**. Yet, the possibility plateau will not close off the repetition of **image-c** at t_4 or at any subsequent future event horizon, even though it will be equally open to the other image possibilities: Hence, **the Whitehead facet pertinent to our equalised possibility plateau will constitute an** *open Whitehead facet*.

*

We must also add to the above the reality of abstract Bergsonian memory: When an image is formed (for example, photographic image-c) and it is then succeeded and replaced by another image, image-c is *not* **erased. Instead, it is retained as abstract memory within the Bergson facet**, but with its original nested futures permanently cropped.

We will model memory in detail in subsequent essays pertinent to habit and recollection.

*

It is our contention that developmentally reversible equalised possibility plateaus and the open Whitehead facet are inherent to brains in the context of the Bergson-Whitehead amalgam. Thus, the processes depicted in **Fig. 5.12: A** to **C** applies to brains. The brain is causally open to alternate intention-specific pattern-impositions (the ephemeralised 'punch cards' constituted as complex wavefunctions of memory or futurity) selected and imposed for future development by the active mind on a work function zero basis.

5.70: The Bergson-Whitehead Model of Habit-Formation

Previously, we modelled the equalised possibility plateau as it pertained to the developmentally reversible photographic film. Therein, we sought certain principles that apply to the developmentally reversible brain in the framework of the Bergson-Whitehead amalgam. **The approach can be utilised to model the accumulation of memories and habits, and the temporal retrieval and recollect of Bergsonian memories.**

Notice that we are no longer conceptualising memories as 'material trace-state' structures at spatial addresses within brains. Instead, memories (or their a-noetic components, such as the big red garage door from 1973, and inclusive of past brain-states correlated to it) are retained within the Bergson facet as temporal distributions, and at temporal remove from the 'materialised' brain confined to the event horizon, notwithstanding distinct and overlapping *noumnemonae*, or 'memory in consciousness' (see Book-IV **4.41** for the initial posit on *noumnemonae*). Hence, we are no longer thinking in terms of contiguous memory retrieval in or across 'space'. Instead, both memory and its 'retrieval' are purely non-contiguous temporal processes that unfold via the process of the de-spatialised stereotemporal Bergson-Whitehead amalgam: (see Part-IV **5.42** and **5.43**).

Even if the brain exhibits readiness states for the expressions of habits and memories, these certainly attend memory, but they do not constitute the actual memory. This should come as no surprise: There is no space; only time is real: All things that have come to pass, including past states of the brain in temporal relation to a temporised system of world-events, are distributed and structured into pure time-relations and into abstract Bergsonian sets of wavefunctions in memory. These elaborate from the most recent memories to those that attend the most primary or most remote past-states of the brain and its larger environment. As such, at least a-noetic component memories of world-events, such as the big red garage door-1973 *in itself* (i.e., independent of the then and subsequent retrospective consciousness of it in the 2020s, and including other a-noetic past states of the brain, are necessarily retained within the Bergson-facet, rendered into a growing block Bergsonian temporal 'worm'… *extant* the event horizon… hence extant the 'materialised' brain restricted to that event horizon. This fact must necessarily be incorporated into any Bergson-Whitehead model of memory, of habit, and of memory recollection.

*

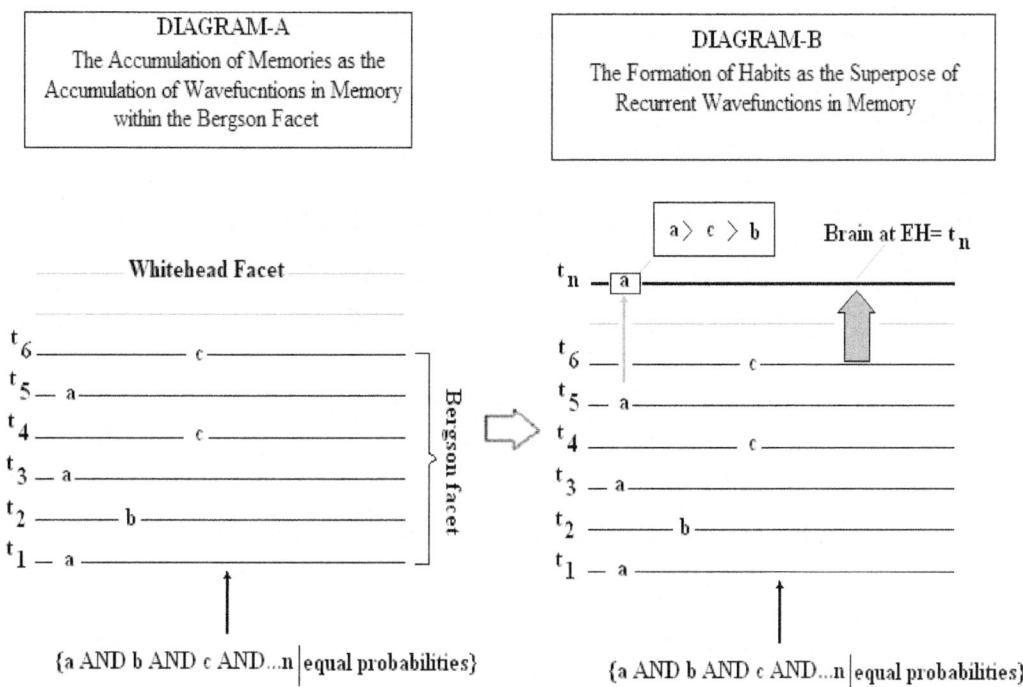

Fig. 5.13: The accumulation of memories and habits in the Bergson facet

Notes: The above constitutes a Bergson-Whitehead model of memories and habits. **Diagram-A** depicts the accumulation of wavefunctions in memory retained within the Bergson facet. The brain is a developmentally reversible equalised possibility plateau and can recall these memories into recollection with equal probability: {a AND b AND c AND…n}. Recurrent memories constitute repeat-recollections at different times in the past, but of the same past. **Diagram-B** depicts habits as a function of memory culminations from the Bergson facet: Memories from t_1 through to t_6 superpose purely across time per de-spatialisation (Book-IV, Part-I). This culminates at the event horizon **EH = tn**. In the absence of intercession by mind, outcome at **tn** will bias to **memory-a**, given that **memory-a** is the most recurrent in the Bergsonian past: at t_1, t_3 and t_5, respectively. Thus, **memory-a** is the dominant habit. In terms of comparative recurrence, **memory-c** dominates over **memory-b**. Thus, the rank order of memories and habits is given as…**a>c>b**. The mind could intercede to have it otherwise but, in its absence, there will be greater difficulty in developing **memory-b** vis-à-vis **memory-a**. Of course, **memory-a** will be the easiest to develop. Mnemonic mechanics and the inverse square law of time (see Book-IV **4.28**) remain implicit to the superpose of memories and the formation of the culmination at **EH = tn**.. The implication is that, despite the equalisation of probabilities in the equalised possibility plateau, recurrent memories bias future development of the brain in favour of the most recurrent memories, as habits… unless the mind intercedes to have it otherwise.

 Our first task will be to devise a Bergson-Whitehead model for habits. In **Fig. 5.13**, Diagram-A, we consider a collection of wavefunctions in memories, depicted as memories **a**, **b**, **c**, up to and including **n**; all set along respective event horizons from t_1 to t_6: each constituting a complex *wavefunction in memory* that includes, not just the configuration of the brain at that moment, but the wavefunction-configuration of the larger world co-joined to that of the brain. (We exclude considerations of *noumnemonae*, or 'memories in consciousness' at this juncture).

 Courtesy of its critical milieu, the brain projects a developmentally reversible equalised possibility plateau into its future. In its future developmental potential within the Whitehead facet, the plateau renders the brain open to the repeat-development (and implied recollection) of any of the stated memories or world-brain relations on an equalised probability and possibility basis. Thus, we devise a Whitehead description of the critical milieu and its implied equalised possibility plateau, pertinent to the stated memories within the Bergson facet: This is expressed as...

Whitehead: {a AND b AND c AND…AND n | all memory states are equally probable}

This is depicted and restated in **Fig. 5.13** Diagram-A. Note that, in Diagram-A, the recurrence of **memory-a** (e.g., the author's memory of the big red garage door, Potters Bar, 1973) implies that it was repeat-recollected into consciousness more frequently than the other memories... at t_3 and t_5, respectively... with **b** the least often recollected.

Habits are functions of accumulated same-memories within the Bergson facet. Since **memory-a** and its past recollections occurred more frequently, by Diagram-B, the culmination at event horizon **EH = tn** will be invariably biased in favour of the ready, easier, and perhaps automatic re-recollection of **memory-a** over and above the recollection of the other memories. Consequently, in the absence of any deliberate intercession by Cartesian mind to have it otherwise, brains at a succeeding event horizon, and per spontaneous AND-to-OR collapse and time, will favour the expression of **memory-a**.

Does this not undermine the notion of an equalised possibility plateau, given the bias from habit? The problem will be clarified in the subsequent essay. Yet, the clue to the solution has been given: **The implications of habit apply** *in the absence of intercession by active mode mind to have it otherwise.*

Per developments in Book-IV, habit and memory-enforcement are not products of spatial relations: **There is no space. Therefore, habit formation and memory-enforcement will unfold per non-contiguous temporal superposition of wavefunctions in memory belonging to different past moments, and a mnemonic superposition process by work function zero means**. In the depicted case in **Fig. 5.13**, Diagram-B, from t_1 to t_6..., **memory-a** is the most recurrent: the *dominant memory*.

However, the mnemonic superposition formed at the succeeding event horizon **EH = tn** will not negate, replace, or wipe out the past wavefunctions in memory from t_1 to t_6. Therefore, memories **a**, **b**, **c** will continue to be retained within the Bergson facet, although their re-collectability will progressively weaken per the inverse square of time and other features of mnemonic mechanics first described in Book-IV **4.28**.

Again, from Diagram-B, in the absence of direct intercession by the Cartesian mind to have it otherwise, the culmination at event horizon **EH = tn** will favour the re-formation of **memory-a**: given that it is the most recurrent memory. On the other hand, memories **c** and **b** will form *recessive memories*, respectively. For its part, **memory-c** is more recurrent vis-à-vis **memory-b**. Yet, **memory-a** is the most recurrent of them all. It follows that the rank order of dominant memories is…**a>c>b**. **This latter relation constitutes the** *structure of habits*: **It outlines the comparative ease, readiness, or automaticity by which one memory is able to re-impose at any subsequent event horizon, notwithstanding the fundamental equalisation of probabilities per the reversible equalised possibility plateau** that characterises the critical milieu of the developmentally reversible brain. The relation also accounts for the *resistance* that mind must face if it should seek to counter-impose memories **b** or **c**, or affect a novel futurity drawn from the future, instead of **memory-a**.

However, the implications from the rank order **a>c>b** also reflect the manifest 'material' readiness states and the action-potentials writ in brains, even when grasped in terms of the usual forlorn materialist presuppositions. **By relating and correlating the rank order of abstract memories within the Bergson facet with the manifest readiness states and established action potentials of the brain, we clarify both the tautological relation of the said readiness states with abstract extant memories** *and* **assert the undeniable radical Cartesian dualistic dissociation of memories from the said readiness states. We also elucidate the background process involving pure temporal mnemonic mechanical processes that inform and form the distribution and restitution of in-brain readiness states at successive event horizon and in 'materialised' brains**, which most neuroscientists, brain scientists and philosophers of mind erroneously confound with real memories or as the in-brain facsimiles of such.

However, the mind could intercede against habit. It will do so with the greatest difficulty if it seeks to develop **memory-b** in place of **memory-a**, given the structure of habits per relation... **a>c>b** and the related bias of the brain's correlated readiness states. On the other hand, **memory-c** would be comparatively easier in lieu of **memory-a**, as it occupies a more favourable position in the said rank order. Of course, the expression of **memory-a** will be the easiest to re-develop, given its dominant position within the rank order, with the least resistance to enforcement by mind.

Also, the process of the succession and replacement of dominant memories and habits with new ones (the process of overcoming habits and the development of new ones) is effectively the same as entailed in memory retrieval and recollection.

This concludes our Bergsonian and Bergson-Whitehead amalgam-based introduction to memory theory in relation to habits.

5.71: The Bergson-Whitehead Model of Memory Retrieval & Habit Reformation

Now **we turn to the process of memory retrieval instigated by Cartesian active mode mind vis-à-vis the brain... vis-à-vis the Bergson-Whitehead amalgam**. This involves the directed intercession of mind against resistance from habit: (More on how mind overcomes habit later). Thus, directed memory retrieval also highlights how habits are both broken *and* succeeded with new ones.

Fig. 5.14, Diagram-A recapitulates the previous illustration: It depicts the accumulation of memories and habit. Therein, wavefunctions in memory are retained within the Bergson facet vis-à-vis the brain from t_1 through to t_6, at a temporal remove from or 'below' the materialised brain restricted to the event horizon. Indeed, t_6, the latest event horizon, will constitute the 'materialised' brain and the dominant readiness states of that brain: readiness states that reflect the habit-structure of dominant versus recessive memories.

Note that de-spatialisation holds. Hence, we are not referring to an implausible model of material memories organised as 'distributions in brain-space', but *only* to de-spatialised abstract memories organised into distributions in time within the Bergson facet.

For the mind to retrieve any memory and impose its expression as its 'punch card' at a future event horizon, it must reach, *not* **into the 'material' of the brain, but into the domain of abstract memories retained within the Bergson facet. It must accomplish this,** *not* **through 'space', but non-contiguously across time.**

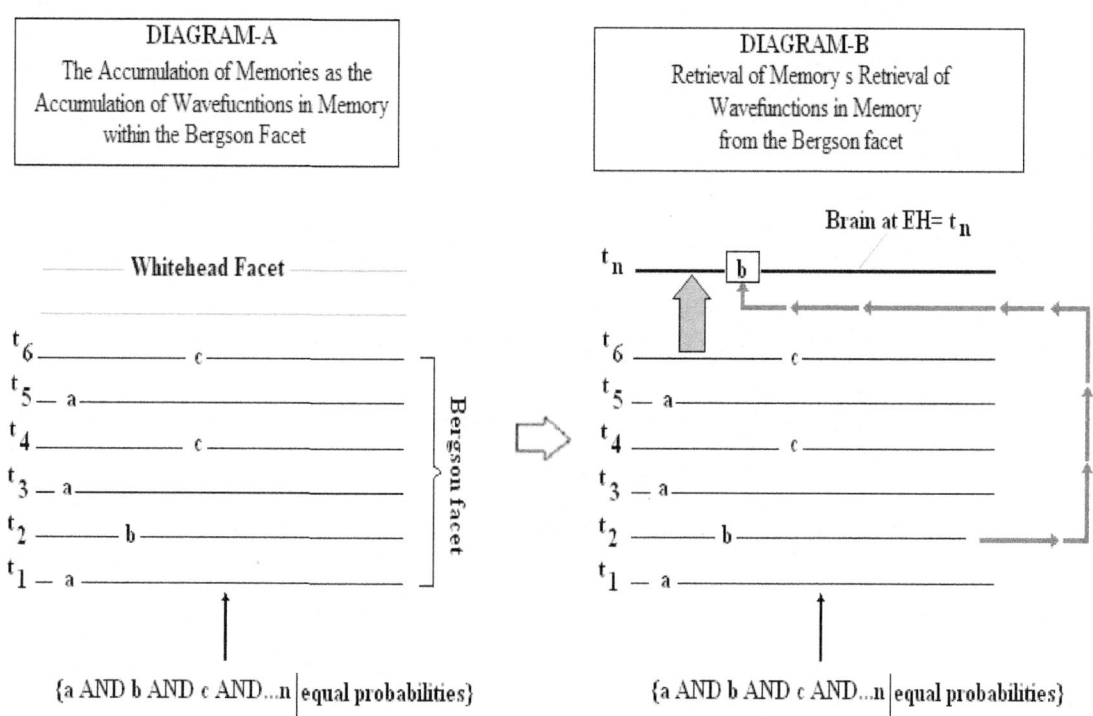

Fig. 5.14: The accumulation and retrieval of memories from the Bergson facet
Notes: The illustration depicts memory retrieval by mind from within the Bergson-Whitehead amalgam vis-à-vis the brain.
Diagram-A is a repeat of a previous diagram depicting the accumulation of memories from t_1 through to t_6. The memories are abstract wavefunctions in memory, temporally retained extant the brain within the Bergson facet, whose final 'materialisation' is at t_6. Thus, memories from t_1 to at least t_5 are *outside* of the brain. The Cartesian mind, if it is to retrieve memories and impose these as its 'punch card' at subsequent event horizons, must reach, not into the 'material' of the brain, but into the abstract Bergsonian past. **Diagram-B** illustrates the retrieval of memory by Cartesian mind. The mind exploits the reversible equalised possibility plateau characteristics of the brain, and thus exploits the open Whitehead facet that this implies: an open Whitehead system that permits the selection of sought memory for realisation from the Bergsonian past and its brain-related memories. In the illustration, the mind countermands the bias from recurrent memory and habit: Instead of **memory-a**, the mind imposes **memory-b** at event horizon $EH = t_n$. Drawing it from t_2 and imposing it on $EH = t_n$. Note that mind does not retrieve memory from a 'location' in the brain. There are no 'locations' for memory, given de-spatialisation and temporisation. Put another way, the 'location' of memory is in time, or in the past. Therefore, there is no 'material storage' for memories in the brain. In truth, the memory drawn out from t_2 is actually potentialised to the equalised possibility plateau at $EH = t_n$. It is potentialised to $EH = t_n$. from the Bergson facet at t_2. The mind intercedes from the gaps in time afforded by the succession of event horizons to select from that equalised possibility plateau **memory-b** potentialised to it from the Bergsonian past at t_2.

One might be forgiven for thinking that this is accomplished through a direct retrieval of memory from the past. Certainly, some aspect of this notion survives the actual unexpected process of memory-retrieval and recollection. The actuality is more a case of 'memories getting to the mind' through a process of resonation of the past into the future, accomplished by background mnemonic mechanics *and* attendant grand decoherence of the future into consistency with the sought past. Thus, the superposition of memories of the Bergsonian past is projected to the Whiteheadian future: a non-contiguous resonance, *not* across 'space', but purely across time. It is from this decohered consistency-rendered future-potential, and per the reversible equalised possibility plateau thus formed within the Whitehead future, that mind selects its sought memory into subsequent recall. **In other words, the past is recollected through the future.**

It seems a contradiction that the retrieval and recollection of the past is achieved through the future. Are not Whiteheadian future states distinct from and compartmentalised from past Bergsonian memory? Upon closer inspection, **it is obvious that any recollection or expression of memories from the past must be slated to happen in and through the future. Thus, one's recollection of the big red garage door, Potters Bar-1973 could happen any moment in the future. It follows that the Whitehead facet, the embodiment of that future, must also be potentialised with prospective recollection-potentials**.

Since one's recollection of the big red garage door-1973 could happen at almost *any* future moment, the attendant nested futures must be potentialised with all possible future moments for just such a recollection or expression, and any of these will constitutes a 'punch card' for that recollection. In effect, in the process of memory recollection, **the Cartesian mind does not so much 'reach into the past', or even let the past resonate to it. The past is already in the mind. Instead, the future is generally conditionalised by that past, and potentialised with that past subsequently sought for noetic recollection and expression, through the grand decoherence of that future into consistent form with that past**, albeit in a form inclusive of alternative possible memories, courtesy of the equalised possibility plateau and the rendition of the Whitehead facet to consistency with that plateau by the same grand decoherence process. Hence, all possible Whitehead futures are rendered consistent with and incorporate the past.

Hence, **it is through the 'presence of the past' *in the future* that the past is 'transmitted' to and gets to that future, which then permits the recollection of the big red garage door, Potters Bar-1973 at that future. The active mode mind reaches into a future reversible equalised possibility plateau and 'plucks' from it the past projected to it from the Bergson facet**: i.e., the big red garage door from 1973. And given the developmentally reversible nature of the brain's critical milieu, the recollection potentialities in futures comprising the Whitehead facet could be repeated in subsequent open futures.

Diagram-B in **Fig. 5.14** explicitly illustrates the retrieval of memory by mind. To achieve this, Cartesian mind must exploit the reversible equalised possibility plateau projected into the Whiteheadian future. Mind 'reaches into the past, to **memory-b** from t_2 retained in the Bergson facet temporally extant the brain, but does so through the spontaneous mnemonic mechanics-based projection of said **memory-b** into the Whiteheadian reversible equalised possibility plateau, projected to the Whiteheadian future at $EH = t_n$. That is, *mind draws out the past through and from the future… and selects and imposes that past via subsequent AND-to-OR development of that future*, manifesting it as the desired brain activity and recollection-expression of the big red garage door, Potters Bar-1973.

*

Recall from the general theory that active mode mind is purely extant the whole system of the Bergson-Whitehead amalgam. **Succinctly, it is through the gaps in time that active mode mind intercedes upon the Bergson-Whitehead amalgam and upon the equalised possibility plateau that constitutes the brain's future, albeit from *outside* of the same Bergson-Whitehead amalgam, its brain, and its equalised possibility plateau**, exploiting the process of otherwise spontaneous work function zero grand decoherence and AND-to-OR wavefunction collapse (or time). Mind then imposes this as its 'punch card' on a successive event horizon and brains, and can do so because the equalised possibility plateau permits this by equalising the probability for the sought memory recollection vis-à-vis the other memories.

Implicitly, in order for mind to accomplish memory retrieval, it must work against the grain of habits per function of the habit structure, a>c>b… against recurrent memory and dominant habit. How does mind overcome this resistance?

- While the relation-hierarchy **a>c>b** appears to imply an *unequal* possibility plateau in favour of **memory-a**, this is not true with respect to selective pressure from active mode mind, which can select *any* memory, even the least favoured **memory-b**. It can accomplish this with impunity.

- It is only in the absence of active mind that the dominant **memory-a** obtains automatic privilege… by arising to passive mode consciousness in a ready way, perhaps in an involuntary way.

- The active mode intercession of mind is not compelled to select in favour of dominant **memory-a**. This is because of the equalised possibility plateau, which potentialises in equal terms, *not* merely re-collectables from the past, but even alternative never before realised futures that might allow for preferential selection of the weakest habituated memory: **memory-b**. *The possibility for the recollection of the least favoured memory is equalised vis-à-vis the most favoured memory,* courtesy of the Whitehead equalised possibility plateau.

- Yet, there will be resistance from habit. The resistance from habit will not be against the certain possibility of free unfettered selection by mind, but will arise from the privileged reformation of the 'material' readiness states that bias for **memory-a** in the absence of intercession by mind to have it otherwise. This bias in the readiness states is certainly a function of spontaneous mnemonic mechanics. Active mode mind must deliberately and repeatedly intercede to recollect **memory-b**. Otherwise, **memory-a** will arise into passive mode consciousness. Indeed, active mind engages in a sort of Cameron move (see **5.54**), but without disruption of incumbent readiness states, to accomplish the replacement of habit, but *not* its memory-loss.

- **Mind overcomes the resistance from habit by gradually building up, through repeated intercessions, an alternative readiness state in favour of memory-b. As old, habituated memory-a recedes into the past, and its contribution to mnemonic mechanics and its superpose diminishes per the inverse square time, and as the mind-imposed alternative memory dominates in the recentcy of memories within the Bergson facet, the cumulative recentcy in favour of memory-b will override the older bias in favour memory-b over memory-a.**

Hence, the Bergson-Whitehead model of memory recollection simultaneously furnishes a model for habit reformation.

5.72: General Attributes of the Critical Milieu

The preceding sough the background to mind-world relations via the developmentally reversible equalised possibility plateau, made possible by the *critical milieu*. There is a reason why brains, and not pumpkins, coincide with nous: a set of conditions that render possible developmental reversibility and the attendant equalised possibility plateau projected into the Whiteheadian future. Hence, the critical milieu affects a loophole that allows active mode mind to intercede into the Bergson-Whitehead amalgam… into the fray of memory, possibility and time.

The fine structure biological details regarding the critical milieu and what constitutes it must await fine structure theory (Part-VI) although the general features that attend it can be found in the developmentally irreversible *and* reversible photographic film previously discussed. Before we incorporate the critical milieu into the Bergson-Whitehead amalgam, we must describe its general features as a preliminary to anticipated fine-structure theory in Part-VI. Subsequently, we will frame this to the Bergson-Whitehead amalgam.

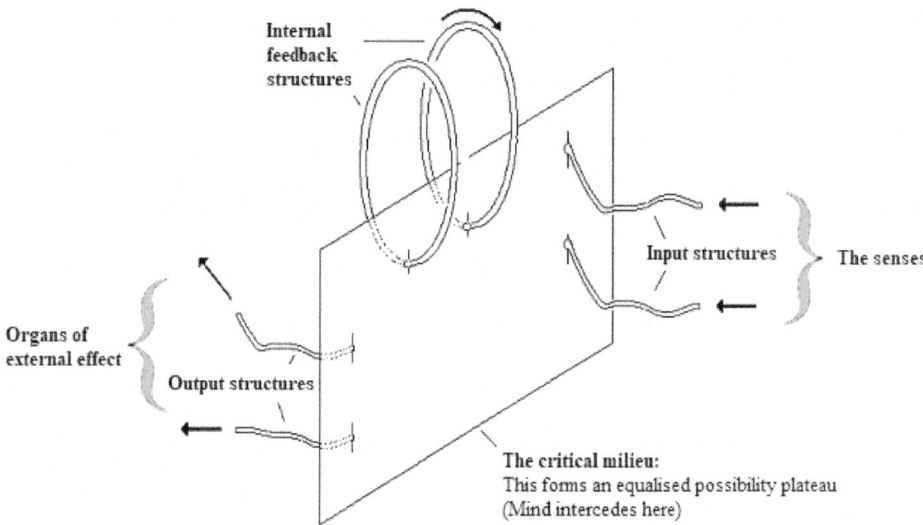

Fig. 5.15: The brain summarised as the critical milieu

Notes: The critical milieu constitutes the decisive structures of the brain: structures that permit direct mind-brain causality, generating the developmentally reversible equalised possibility plateau for the Whiteheadian future of the brain. A spatial-like depiction is illustrated above. It has three attendant structures: These are the **input structures** linked to the senses, which introduce inputs from the environment to the critical milieu: the **internal feedback structures** that permit internal processing and the integration of functional areas (implicitly, these pervade across the critical surface): The third is the **output structures** linked to specialised organs of external effect, which permit the output of intents to the external world. The most important structure is the **critical milieu** itself: This exploits the equalised possibility plateau that attends it. The Cartesian mind intercedes by imposing on the future development of the critical milieu surface a 'punch card' or complex whole-brain wavefunction. It accomplishes this on a work function zero basis. The 'punch card' subsequently funnels objective quantum indeterminate processes into global coherent patterns of brain activity. These manifest on the critical milieu in the same way that an image manifests on the photographic film, to moderate both internal feedback processes and the outputs concordant to mind's sought intents. The minding process involved is not reducible to or identifiable with the critical milieu itself, or even with the imposed punch card or future whole-brain wavefunction, or with *any* events. Mind operates from 'outside' of the critical milieu and the implicit embedding Bergson-Whitehead amalgam, using a work function zero process that does not violate the principle of conservation.

To this end, our preliminary **critical milieu-based model of the brain** is illustrated in **Fig. 5.15**; portrayed as a flat surface (the brain-equivalent of the punch-card mechanism if you will) exhibiting three attending structures. The first of these are **input structures**, linked to the senses: the eyes, ears, etc. These cascade inputs from the larger environment to the critical milieu. In a de-spatialised purely temporised framework, the 'environmental inputs' constitute complex wavefunction in memory constituting that 'larger environment', co-joined to and encompassing the complex wavefunction pertinent to the brain and the critical milieu. The said 'cascade' involves a non-contiguous temporal process of successive AND-to-OR collapse processes and state-changes. Through mnemonic mechanics (see Book-VI **4.27**) these complexes of memories, 'stacked' and time-distributed within the Bergson facet, are superimposed. The attendant AND-to-OR collapse of the superpose manifests brain activity and state-change in tune with co-joined environmental 'inputs'.

The second of the structures that attend the critical milieu consist in the **internal feedback structures**, constituting the internal connections of the brain onto itself. In **Fig. 5.15** these are portrayed as 'rings' that tie parts of the critical milieu to other parts of the same. While only a few are depicted, these pervade across the whole of the critical milieu. While depicted as if spatial distributed, per de-spatialisation, internal feedback structures are also time-distributed memory states rendered into Bergsonian complex wavefunctions in memory and their superposes.

The third type of structure depicted in **Fig. 5.15** consists in the **output structures**; linked to organs of external affect: e.g. arms, hands, fingers, etc. These mediate the intents of the mind that are first initiated as mind-imposed 'punch cards' on the future development of the critical milieu.

In the next essay, we will incorporate the critical milieu to the event horizon within the Bergson-Whitehead amalgam. **The event horizon, together with its critical milieu, 'dissolves' into AND-form Whiteheadian future-potentiality... only to be reformed at a succeeding event horizon via AND-to-OR collapse and time. All the state-changes imposed on the critical milieu by Cartesian mind and its 'punch card' will be insinuated from the 'gaps'** *between* **successive event horizons.** The actual process may involve several successive event horizon progressions and attendant 'gaps'. **At the culminating event horizon, mind's intended brain activity will fully manifest, and, through subsequent event horizons, affects will cascade through the output structures to the larger environ.**

5.73: The Bergson-Whitehead Amalgam Model of the Critical Milieu

We must now integrate the critical milieu into the Bergson-Whitehead amalgam... by simply placing it on or along a given event horizon: namely, **EH-1** at t_1: see **Fig. 5.16**. This furnishes **a Bergson-Whitehead model of the critical milieu**.

Note from the illustration that, above EH-1, we find the Whitehead facet and future possibilities rendered into AND-form (implicit). Below EH-1 resides the mnemonic Bergson facet, comprised of abstract wavefunctions in memory (implicit), organised according to primacy-recentcy rules that attend the critical milieu (hence the brain) subsumed to a grander Bergsonian past.

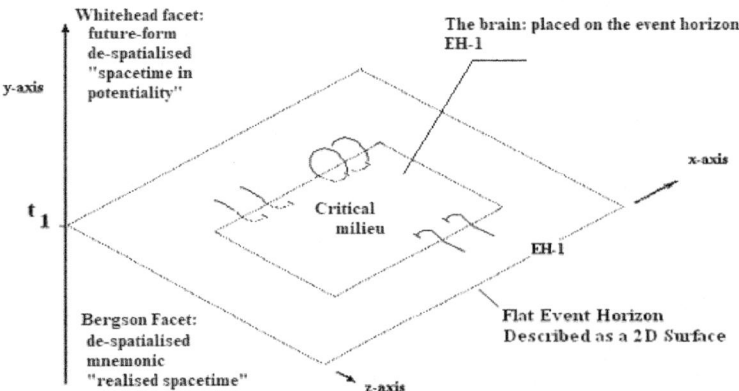

Fig. 5.16: Integrating the critical milieu to the Bergson-Whitehead amalgam
Notes: The diagram depicts the integration of the critical milieu with the Bergson-Whitehead amalgam. The developmentally reversible critical milieu rests on the event horizon at **EH-1**, shown as a two-dimensional approximation of a stereotemporal three-dimensional time-domain. Below **EH-1** resides the mnemonic Bergson facet: Memories therein are drawn out of the past through the future per the Whiteheadian equalised possibility plateau generated by the critical milieu. Above **EH-1** we find the open Whitehead facet, comprised of futures subject to the equalisation of probabilities per the said equalised possibility plateau. The mind selects from among these, and can impose as its 'punch card' on subsequent event horizons, developing these in accordance with its intents.

Also note that the critical milieu, as is the case for the event horizon on or along which the critical milieu is placed, does not exhibit absolute time-like characteristics: the critical milieu is internally subject to the relativity of simultaneity, as is the rest of the event horizon 'surface', and constitutes a de-spatialised state stereotemporally related to its Bergsonian past: i.e., the larger mnemonic environment in which it is embedded or in which it configures as a perduring 'worm'. Consequently, the input, internal feedback and output structures are also similarly delimited, as was explained in the preceding essay.

Also recall that the brain-pertinent critical milieu is developmentally reversible: (see **5.69**): it can revert to a default readiness state that can admit subsequent alternative patterns to form upon it. All of this is implicit to **Fig. 5.16.** wherein the critical milieu is shown directly integrated to the Bergson-Whitehead amalgam via the event horizon.

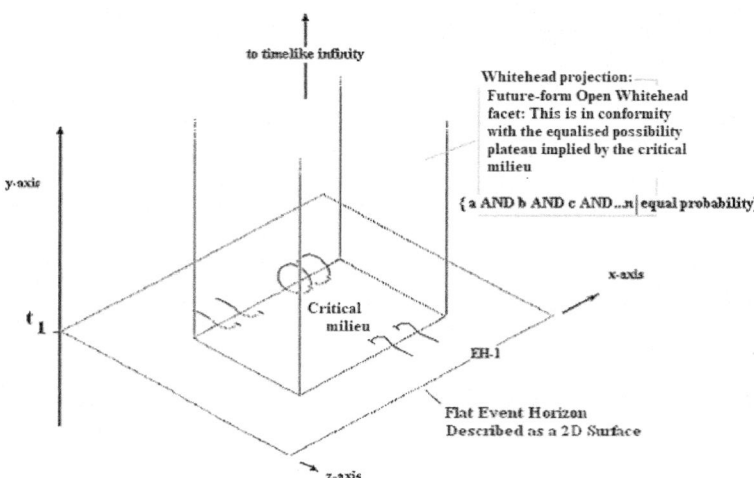

Fig. 5. 17: Projection of the critical milieu to the infinitely removed future

Notes: The diagram depicts the projection of the developmentally reversible equalised possibility plateau into the Whitehead facet. The critical milieu for brains, just like a photographic film for future possible images, constitutes a stand-by state for all possible and permissible future-potential brain activity, constituting the equalised possibility plateau for such prospective activity. Achieved through grand decoherence, the projection from the critical milieu forms the Whitehead facet into an open quantum system (or open Whitehead facet) on an equalised probability footing to whatever wavefunction complex (whether from the Bergsonian past or from the Whiteheadian future) the Cartesian mind deems appropriate to select as its subsequent 'punch card'. Per developmental reversibility, the 'punch card' thus developed will not close off the future to any subsequent alternative 'punch card'. In principle, the projection forming the open Whitehead facet ought to obtain to future infinity. In reality, it will project into an attainable future equal to the average life-expectancy of the critical milieu or brain.

In **Fig. 5.17**, we expand on the simple placement of the critical milieu on the event horizon: The critical milieu now projects a domain into the Whitehead facet... to the indefinite future or to 'timelike infinity': This is the Whitehead projection pertinent to the critical milieu, or the projection of the critical milieu as a potentiality state into the Whitehead future. **Thanks to the developmentally reversible critical milieu, at every potential future event horizon, the Whitehead projection will constitute an open Whitehead equalised possibility plateau.** This might comprise of, say…{a AND b AND c AND…n}… as is implied in **Fig. 5. 17**. **The Cartesian active mind can freely select from among these equally probable possibilities pertinent to the future development of its brain.** The active mode mind will accomplish this through mind-selected 'punch cards', *ceteris paribus*, notwithstanding habits (see **5.70** and **5.71**)… and it will utilise work function zero processes inherent to both grand decoherence and to AND-to-OR collapse and time.

Succinctly, **the Whitehead projection comprises an equalised possibility plateau that also includes all the future-potential brain states pertinent to the recollection of the past from the Bergson facet, projected into the future as potential re-collectables (see explanation in 5.75); although the Whitehead projection will also consist of potentials for first-time novel affectations.**

5.74: Time Evolution of the Critical Milieu & the 'Locus' of Mind-Brain Intercession

Event horizon **EH-1**, depicted in the previous illustrations, must progress 'upward' the Bergson-Whitehead system toward the indefinite future. It does so in the expected broken-discontinuous form per attendant AND-to-OR wavefunction collapse and time. Thus, **EH-1** at t_1 'dissolves', and, together with the brain-cum critical milieu that had 'materialised' upon it, it is relegated to the mnemonic Bergson facet and retained therein as an abstract complex wavefunction in memory. A new event horizon, with a new 'materialisation' of the environ and the brain-cum-critical milieu is expected to form at t_2, as **EH-2**. All of this is illustrated in **Fig. 5.18**.

Note from **Fig. 5.18** that there exists a quantum mechanical interval or a 'gap in time' (more succinctly, a delayed choice time interval isolation-relation) between the 'dissolved' past **EH-1** relegated to the Bergson facet as memory, versus future **potential EH-2** expected to form at potential **t₂**. By exploiting this interval between 'dissolved' **EH-1** and the not-yet realised future **potential EH-2**, the Cartesian active mode mind intercedes into the affairs of the brain. In truth this is more likely achieved through a succession of such 'gaps' and event horizons. But, for brevity, we reduce this complexity to a single quantum mechanical 'leap in time', one demarcated by **EH-1** and potential **EH-2**.

Of course, all of this necessarily implies Cartesian mind-brain dualism and the non-reducibility and non-identifiability of mind and consciousness with the events manifesting through the brain and its critical milieu.

The Cartesian mind acts as a sidereal selective process; exploiting the open Whitehead facet and the Whitehead projection generated from **EH-1** per function of its critical milieu; exploiting the open character of the reversible equalised possibility plateau.

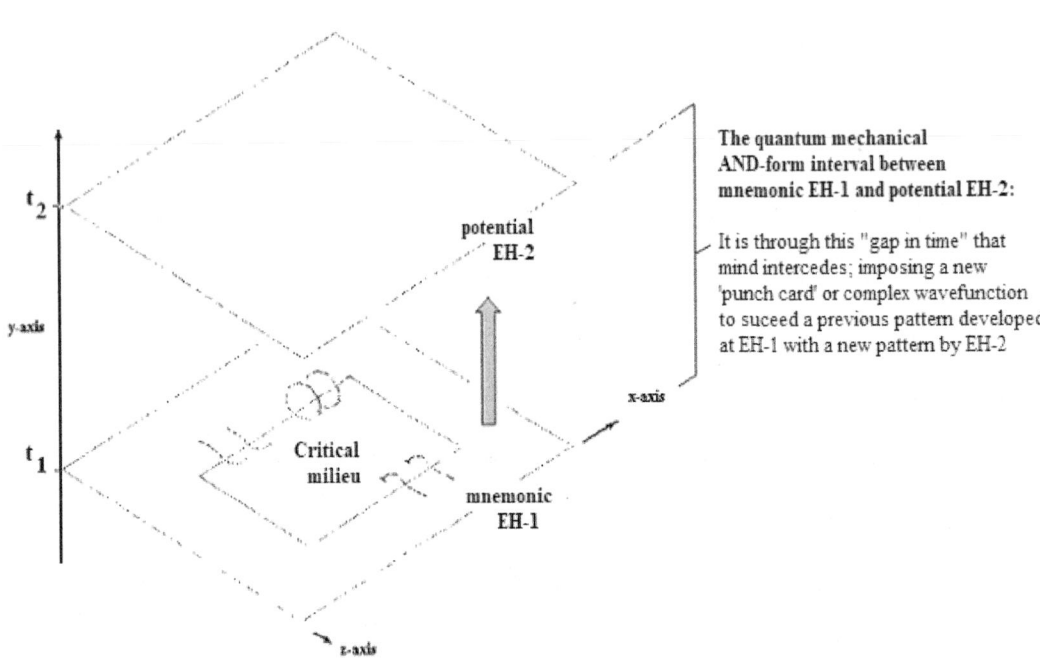

Fig. 5.18: Quantum mechanical interval between EH-1 and EH-2
Notes: It is through the quantum mechanical interval, or the 'gap in time', between past mnemonic **EH-1** and future potential **EH-2** that extant active mode mind causally intercedes to replace a past 'punch card' at dissolving **EH-1** with a different punch card at potential **EH-2**: Thus, continuing a 'change of mind' affected as a new pattern of brain activity and the expression of alternative intents at **EH-2**. Mind intercedes by means a structure-less causality, exploiting the work function zero process of AND-to-OR collapse and time and attendant succession of event horizons, and imposes on **EH-2** any memory-related 'punch card' from within the Bergson facet, or else a novel never-before-realised punch card directly from the open Whitehead facet and its equalised possibility plateau. Mind does not accomplish any of this as any form of registered event on any pertinent event horizon, given that it intercedes from between **EH-1** and **EH-2**: i.e., from outside of the world of past or possible events; from outside of brains and time.

The active mode mind selects its 'punch card' from the equalised possibility plateau by means of the ephemeralised equivalent of the throw of delayed choice switch-x, and per attendant decoherence of the said equalised possibility plateau at potential EH-2. It realises its sought outcome per function of its mind-selected punch card or complex wavefunction. This latter 'materialises' or develops at EH-2 through the three-phase developmental process that funnels otherwise quantum indeterminate events into the globally coherent outcome: the sought brain activity at realised EH-2.

Recall that the mind cannot be clocked as any event, even though the events pertaining to the initiation of mind's intents, observed as brain activity, together with the observable culmination from such activity, can certainly be clocked: The reasons for non-clockability was furnished in Part-IV **5.46** to **5.49**. Hence, the resulting process of brain activity must never be confused for the active mode mind itself, given that neither active mode mind nor subsequent passive mode consciousness can arise or constitute as events of the Bergson-

Whitehead amalgam. That is, mind is not reducible to or identifiable with brain activity, as was argued in Part-IV **5.50**. **At best, the minding process can only be inferred to have 'occurred' within the 'black box' demarcated by the delayed choice time interval isolation-relation (the 'gap in time') between past EH-1 and succeeding now-realised EH-2.**

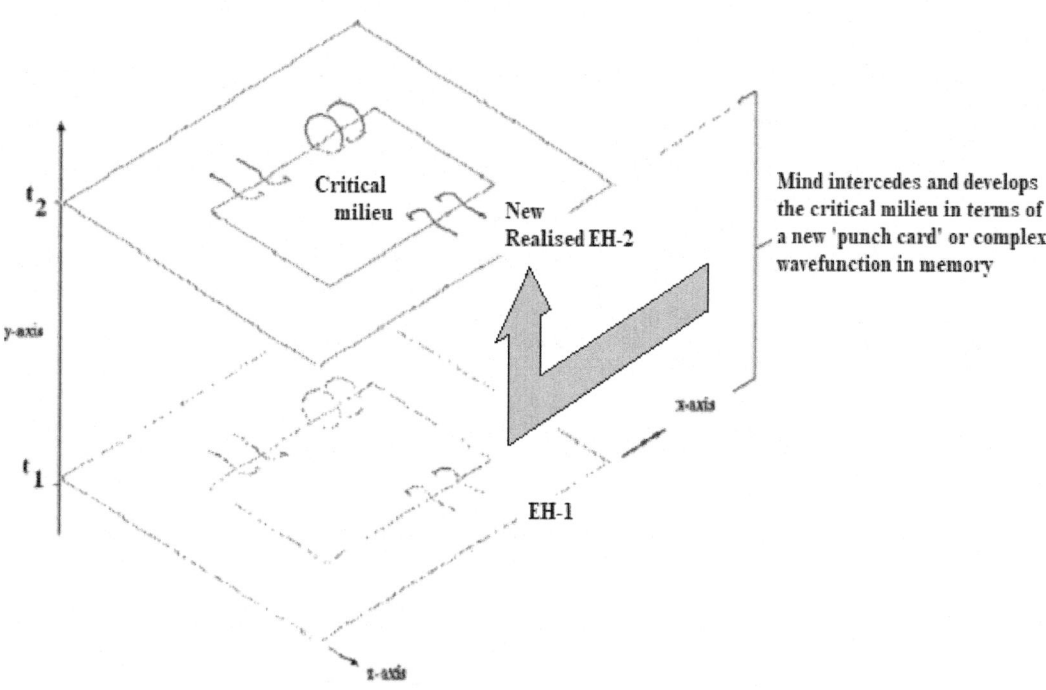

Fig. 5.19: Reformation of the critical milieu and the imposition of the design of mind
Note: The intercession by selective active mode mind from within the delayed choice time interval and between succeeding event horizons, imposes on the critical milieu on the event horizon EH-2 at t_2 either memory drawn from the Bergsonian, (albeit through the future)… or a novel affect drawn directly from the future. Either constitute the 'punch card' imposed by mind that exploits work function zero wavefunction collapse and time, which brings about the succession of **EH-1** by **EH-2**, Either as wavefunction in memory or as wavefunction in potentiality, the imposed 'punch card' will then channel otherwise *en masse* objective indeterminate quantum incidences manifesting on the critical milieu into coherent global outcomes, concordant with the intents of mind.

The process of mind-world intercession is depicted in **Fig. 5.19**. Recall that...
- **there is no tick-tock click-clock hidden gears to make time and work function zero wavefunction collapse happen** (see Part-II **5.18** and **5.19**). **The same holds for how the Cartesian mind imposes the future that it seeks to develop as its brain activity**… without exhibiting any clockable or iterational gear-like features… even if the consequent brain activity remains clockable and unfolds seemingly mechanically: see Part-IV **5.49**.

Indeed, the active mode mind exploits the very process of gearless time and work function zero wavefunction collapse that forms the 'gaps in time' as interstices through which mind intercedes vis-à-vis its brain and world. **Insofar as active mode mind exploits work function zero time, its causality upon the brain and world must itself constitute a work function zero process: Thus, the generic principle of conservation is not violated by Cartesian causality, any more than by AND-to-OR collapse or time.**

To reiterate, in the case of memory expression and recollection, intercession by Cartesian mind from between the dissolution and succession of event horizons imposes its sought wavefunction in memory on the critical milieu (the brain) 're-materialised' on the succeeding event horizon at realised **EH-2**; In truth, the memory is drawn from the potentialisation of the past upon the future and its equalised possibility plateau at potential **EH-2**: see **5.46**.

In the case of novel affectation in which new or never-before realised novel future possibilities are to be affected, from between successive event horizons, the Cartesian mind selects a never before realised novel complex wavefunction, drawn

directly from the futurity constituting the Whitehead facet, permitted by the equalised possibility plateau at **EH-2**. (See **5.65** and **5.73**). **Once the 'punch card' of memory or novelty is selected from the attendant equalised possibility plateau, and attendant grand decoherence occurs, the imposed punch card at EH-2 'materialises' through the three-phase developmental process**, often without further participation by active mode mind. The three-phase developmental process funnels *en masse* quantum indeterminate incidences onto the critical milieu (the brain) at realised **EH-2**, and these form into a global coherent and unitary outcome concordant with mind's will. This forgoes the need for mind to orchestrate quantum indeterminacy in brains.

Thereafter, passive mode consciousness will subsequently arise as the retrospectively apprehending and affirming witness to mind's deeds, and its ramifications to the larger environment through attendant output structures.

5.75: Unified Approach to Memory Recollection & Novel Affect in the Bergson-Whitehead Amalgam

In this essay, our first task is to recapitulate mind-mediated memory retrieval. Recall from the previous essay and its attached illustration (**Fig. 5.19**) how active mode mind intercedes and imposes its selected memory through a series of 'gaps in time' that reside between one event horizon and a future-potential successor. This is further elaborated in **Fig. 5.20** in this essay, which depicts the same, but as a retrieval from Bergsonian event horizon at related to t_n.

Memory retrieval from the Bergsonian past is effectively true. Yet, as repeatedly affirmed, the actual process involves a recollection of the past in and through the future, or from an equalised possibility plateau projected to the future-potential event horizon at t_2, and its AND-to-OR realisation per the decoherence and collapse of the said plateau at t_2 to the sought outcome. The past as memory makes itself present for realisation in the future in the fashion of a future-potential recollection-possibility or potential at t_2, embedded within the Whitehead equalised possibility plateau at t_2. **Succinctly, memory incorporates itself as a re-collectable possibility into the Whiteheadian future through a process of mnemonic mechanics** (see Book-IV 4.26 and 4.27), **via the superpose of Bergsonian past and memory upon the Whiteheadian equalised possibility plateau attached to the future at t_2, and the grand decoherence of the plateau to consistent futures composed of the sought recollection-potential.** The memory is superposed and intermixed with never-before realised pure novel futures as part of the same equalised possibility plateau at t_2.

Hence, strange as it seems, memory recollection possibility is embedded into the future, and realised through the AND-to-OR collapse of that future. **Consequently, memory recollection as a future affect must be unified with novel affectation of new possibilities outside of the scope of memory, but within the scope of the consistency-rendered future.**

Hence, in order for active mode mind to 'reach out into the past' and select a sought memory for recollection and expression through and from the future at t_2, **active mind must**…

- …**trip the ephemeralised equivalent of the delayed choice switch-x attendant the said equalised possibility plateau and memory**. It must accomplish this from between t_n and t_2 (see **Fig. 5.20**) by means of a work function zero process that exploits work function zero wavefunction collapse and time: See delayed choice in Book-II 2.18 and Cartesian causality in **5.76** below.

- By tripping switch-x, **mind selects and decoheres the equalised possibility plateau at t_2 to the sought memory recollection-expression superposed therein.**

- **This erases alternative memory recollection-expression possibilities** also embedded within the equalised possibility plateau at t_2. **The mind renders future events attendant the equalised possibility plateau consistent with the sought Bergsonian past, now slated for recollection-expression** at t_2.

- The complex wavefunction at t_2 decohered to **the sought recollection-expression, will then 'materialise' via spontaneous AND-to-OR wavefunction collapse and the attendant three-phase developmental process into the sought brain activity:** (not shown in the **Fig. 5.20**). **The brain activity will culminate into the recollection-expression of memory, followed by its subsequent apprehension by passive mode consciousness.**

Recall that the critical milieu attendant brains is developmentally reversible. Therefore, the equalised possibility plateau, at some subsequent future (say, at potential t_3 and *its* potential event horizon) will be rendered open to alternative memory expressions foregone at potential t_2.

Recall that **novel affectation constitutes the process through which Cartesian mind draws out its 'punch card'**, *not* **as a recollection of the past, but as a never-before-realised possibility from hitherto not-realised Whiteheadian future.** Novel affectation is either independent of memory recollection (i.e., in its idealised form) or it is part of an admixed state co-joined with memory expression (its typical form), as is also true of memory recollection itself… always co-joined and unified with novel affectation. The key to this insight is the critical milieu and its equalised possibility plateau. Per the critical milieu, otherwise unequal possibilities are rendered equally possible.

From a pure memory recollection approach, wherein the future is assumed to comprise nothing but potential recollections of the past projected into that future from the Bergson facet, the equalised possibility plateau thus projected into the Whitehead facet would permit the recollection-expression of *any* past-memory on an equal probability and possibility footing, notwithstanding implications from habit (see **5.70** and **5.71**). Memory recollection potentials comprised of, say…

$$…\{\mathbf{a} \text{ AND } \mathbf{b} \text{ AND } \mathbf{c} \text{ AND}…\mathbf{n}\}$$

superposed to the future equalised possibility plateau up to and including the pertinent future event horizon, will constitute unique forms of recollection-expressions; effectively novel, even though they constitute potential recollections of the past. Indeed, any subsequent repeat-recollection of a same-memory (e.g., the repeat-recollection of the big red garage door-1973) yet constitutes a unique novel repeat-recollection, insofar as the future moment at which it is repeated constitutes a never-before realised unique moment. **Hence, there can be no such thing as pure memory recollection, given the incorporation into recollection of unique novel affectation.**

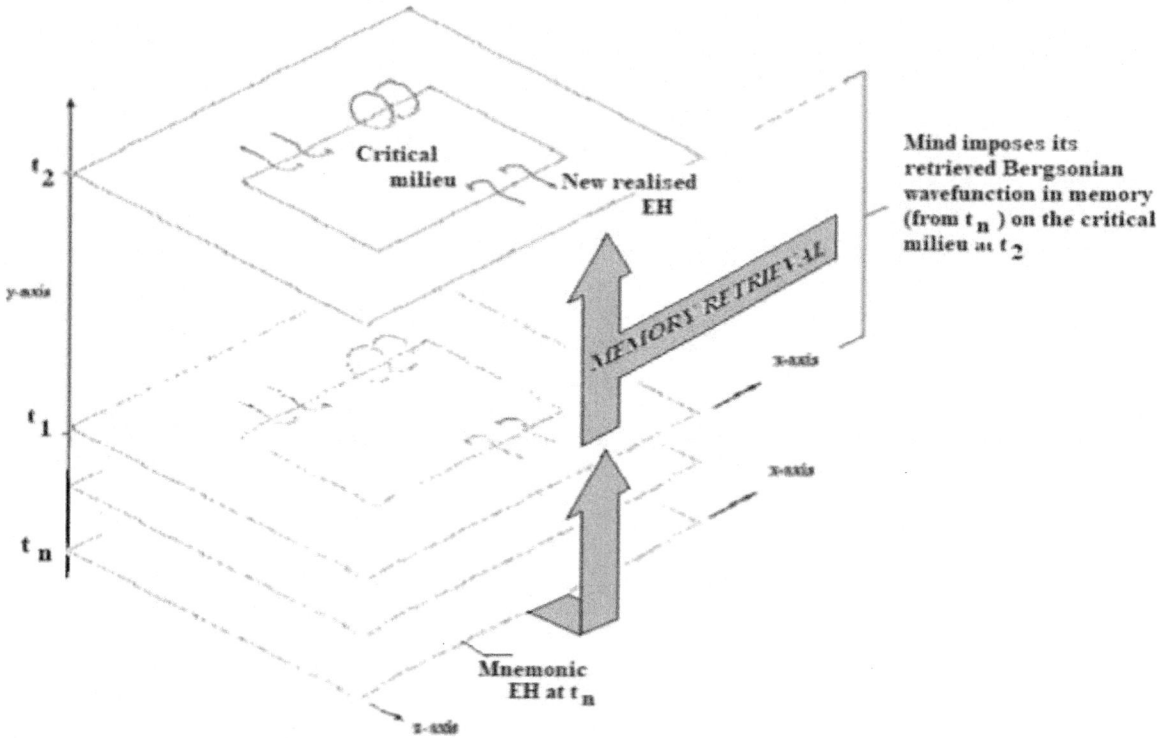

Fig. 5.20: Memory retrieval revisited
Notes: The diagram depicts a 'two dimensional' version of memory retrieval by mind vis-à-vis the critical milieu and brains. The critical milieu re-materialises at **newly realised EH** at t_2. Up to the culmination at t_2, the active mode mind intercedes by throwing the ephemeralised equivalent of Wheeler's **delayed choice black box switch-x** between **EH** at t_n and **EH** at t_2 to impose a wavefunction in memory (its 'punch card') at t_2. The equalised possibility plateau at t_2 decoheres to forgo the alternative memories vis-à-vis the memory selected. The critical milieu at t_2 then AND-to-OR collapses and develops according to the memory selected and imposed by mind. However, given developmental reversibility, the original equalised possibility plateau and alternative recollection-possibilities will be generally restored to the equalised possibility plateau by t_3.

Pure novel affect is also impossible, insofar as any real novel affectation must always enmesh with background memory, habit and attendant identity: It must always occur in the context of the Bergson-facet (or memory *en toto*) if not especially in the context of memory-bound identity; although Alzheimer's disease could imply a partial exception to this, given that the condition involves the inability to express biographical memory. That aside, there can be no such thing as pure novel affectation, given the incorporation and ineliminable presence of memory to the attendant joint equalised possibility plateau.

Again, from a pure memory recollection approach, the future could permit the recollection and expression of *any* past-memory in…

$$…\{a \text{ AND } b \text{ AND } c \text{ AND}…n\}$$

By reference to **Fig. 5.21**, let us apply our lay formalism to describe the combination of novel affectation *and* memory recollection via their joint equalised possibility plateau, and in appreciation of the insight that pure novel affectation is no more possible than pure memory recollection, and that these are always ad-mixed and co-joined.

595

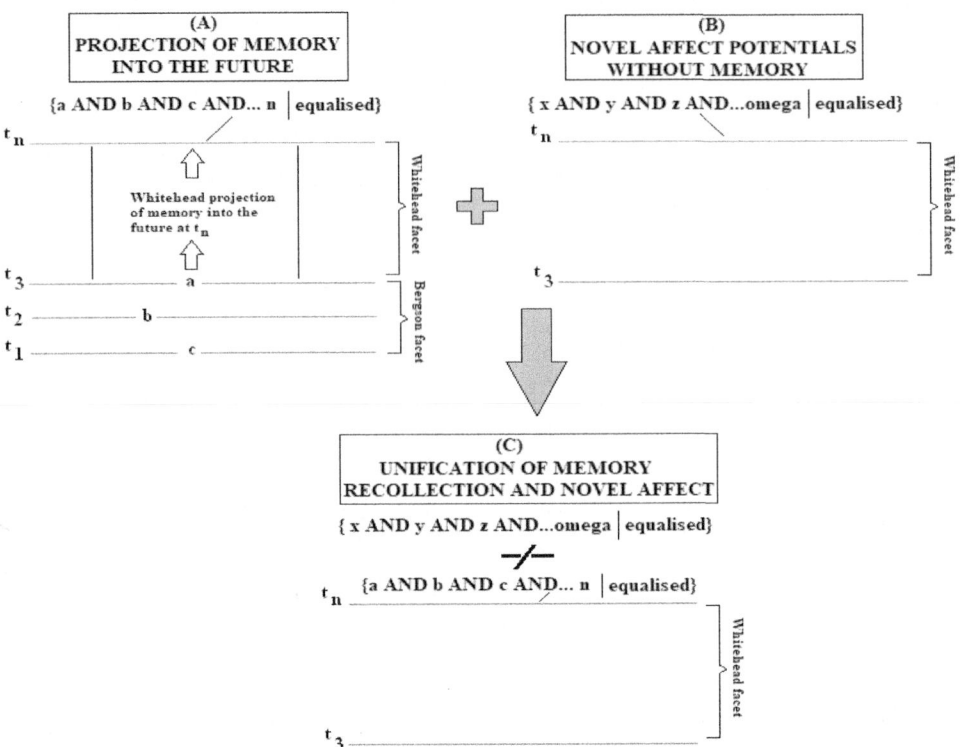

Fig 5.21 Projection of Bergsonian memory into the Whiteheadian future: the unification of memory recollection-expression and novel affectation

Notes: The illustration reiterates the accumulation of memories in **Diagram-A**. Yet, we now grasp the recollection process, a thing that must happen through the future, as an equalised possibility plateau for memory recollection potentials insinuated into the Whiteheadian future, to **tn**. In a previous illustration, the equalised possibility state…{a AND b AND c AND…n} was presented as if affected from the Bergson facet (see depiction **A**). What the critical milieu and its equalised possibility plateau allows is the insinuation of this memory-set into the Whiteheadian **EH = tn**. As such, any of these memories can be recollected in the future from the memory recollection potential, **a** AND, **b** AND c and n. As such, the recollection is obtained always as a future affect and a novelty. That is, memory recollection and repeat-recollection always constitute a form of novel affectation, simply because it must always be realised in, through and as the future. However, memories are co-joined with novel future possibilities outside the scope of memory. The future novel possibilities are depicted in (**B**) as {x AND y AND z AND… **omega**}, enmeshed with memories in depiction (**C**). If future novel affect y is minded, it will subsequently be incorporated into Bergsonian memory as part of {a AND b AND c AND (y)…AND n}. Thus memory will grow, while (**y**) as memory will grand decohere the future to the re-collectability of (**y**), given the permit for this per the reversible equalised possibility plateau.

To that end, the initial equalised possibility plateau for novel affectation (**Fig 5.21 (B)**) will decohere into a new and reduced equalised possibility plateau:. It will decay from…

Whiteheadian:{**x** AND y AND **z** AND…**omega**}..

into…

Whiteheadian: {**x** AND **z** AND…**omega**: minus **y**}

Of course, **y** will now become incorporated into Bergsonian memory retained in the Bergson facet, and the attendant equalised possibility plateau superposing the memories (**Fig 5.21 (A)**) will evolve from…

Bergsonian: {**a** AND **b** AND **c** AND…n}….

into…

Bergsonian:{**a** AND **b** AND **c**… AND (**y**)… AND…**n**} which will now incorporate the former novel affect **y** as memory... as **(y)**.

As previously stated, the developmental reversibility of the critical milieu and the equalised possibility plateau that it engenders implies that all the other future affect potentials, now excluding **(y)**... namely, {**x, z** to **omega**} within the Whitehead projection, will *not* be foreclosed from future selection or development.

Consistent with what we asserted previously; novel affectation is distinct from but enmeshed with memory recollection. As depicted in **Fig 5.21 (C)**, the whole Whiteheadian projection and attendant equalised possibility plateau start out as...

Whiteheadian: {**x** AND **y** AND **z** AND...**omega**} -/- Bergsonian: {**a** AND **b** AND **c** AND...**n**}

Consistent with our general growing block model approach, this then transforms and time-evolves into...

Whiteheadian: {**x** AND **z** AND...**omega**} -/- Bergsonian: {**a** AND **b** AND **c** AND **(y)** AND...**n**}

Thus, as the equalised possibility plateau that attends the Whiteheadian future decoheres under the influence of active mode Cartesian mind, this brings about novel affect-**y**. Consequently, the admixed equalised possibility plateau for Bergsonian memory will enlarge to incorporate **(y)** as its newest memory, hence stripped of its former nested futures. In turn, this memory complex will admix with whatever remains of the novel Whiteheadian futures, forming a new joint equalised possibility plateau: one open to any recollection of memories (now including **(y)**), as well as remaining novel affectation possibilities from novel futures in {**x** AND **z** AND...**omega**}.

To reiterate...

Whiteheadian: {**x** AND **z** AND...**omega**} -/- Bergsonian: {**a** AND **b** AND **c** AND **(y)** AND...**n**}

5.76: Ephemeralisation & De-Materialisation of the Punch Card Model

The workings of a generic punch card mechanisms can furnish ready purchase about how mind-brain causality and control unfolds: (see **5.65** and **Fig 5.10** for simple pin-driven punch card). Yet, it can also mislead, given its spatial and materialistic operational format. However, de-spatialisation and ephemeralisation comes to the rescue. Indeed, our growing block Bergson-Whitehead models of both memory recollection-expression *and* novel affectation in **5.75** are de-spatialised, temporised, hence ephemeralised accounts of the punch card-like process of mind-brain control, cast within the growing block Bergson-Whitehead amalgam framework. We can break down the pertinent steps involved in this process thus:

- **The 'punch card' is the equivalent of either a complex wavefunction in memory from the Bergson facet, or a novel affect wavefunctions in potentiality from the Whitehead future**. Both are subsumed to a co-joining equalised possibility plateau; an admixtures of both future potentiality and past memory: **(5.75)**

- **The 'slotting of the card' is equivalent to the selection and imposition of a complex wavefunction selected from the equalised possibility plateau by the Cartesian active mode mind**, which engages from between successive event horizons, or from the black box 'gaps in time'; utilising the ephemeralised equivalent of the throw of **switch-x** from Wheeler's delayed choice... **all accomplished on a work function zero basis**.

- **Finally, the process of AND-to-OR wavefunction collapse and time, also a gear-less work function zero process, *and* inherently ephemeralised, constitutes the 'pushing of the pins'**, with outcomes concordant with the complex wavefunction selected for future development by active mode mind, **comprised of *en masse* quantum depositions; so 'materialising' the brain activity on a succeeding event horizon and in its critical milieu (the brain)**.

Of course, the whole Bergson-Whitehead amalgam in and through which the above processes are realised constitutes a stereotemporal de-spatialised, temporised and wholly non-contiguous unfolding. Consequently, the embedded processes of mind-brain control is also de-spatialised, temporised, ephemeralised... hence, *de-materialised.*

*

In generic punch card driven mechanisms, the operation of the punch card assumes a 'spatial' and materialised dimension of explanation: The end-user must apply a contiguous energy, power or work to push the punch card into its designated slot. Putative energy, power or work must also push the perpendicular pins against and through the holes cut into the slotted punch card. Again, using putative work, the pins that push through the punch card must then activate a specific output. Throughout, we tend to presume the operation of contiguous contact or impact causality per the generic input-operation-output (IOO) schema: i.e., materialism... fostered by the usual lack of knowledge about the pleonastic status of energy concepts and relations: (see **5.35**).

If mind-brain causality operated in the same materialistic way, how could immaterial mind (the end-user) bring about the imposition of its material punch card? Where would mind get the energy, power or work to get the brain to actuate the equivalent of 'pushing its pins'? How could a purely abstract-immaterial Cartesian mind bring about material changes to the brain, if the brain is causally closed to such a mind? The questions are variants of historical tropes raised against Cartesian dualism, often attended by the usual clause against perpetual motion machines and abuse of the principle of conservation: all arrayed against Cartesian reality. Yet, the general basis for Cartesian causality and the non-materialist ephemeralised punch card approach that we espouse was furnished in the preceding general theory in Part-IV, and also consolidated by our critique of materialism and the de-materialisation of physics in Part-III and elsewhere: all implicit to the Bergson-Whitehead model of memory-recollection and novel affectation adumbrated in **5.75**. The key findings furnished in Part-III and Part-IV can be congealed into the following, counter to the historical materialist contention against Cartesian dualism. Thus:

- **The breakdown of IOO-schema closed causality per the breakdown of hidden gears. AND-to-OR** wavefunction collapse and time, through which state-change is brought into effect and inputs are transformed into outputs, is not driven by a contiguous input-operation-output schema or hidden gears, or by any definable operation. Therefore, there is no mechanism behind state-change: a truism that must also apply to the time-evolution of the brain. The same must apply in our ephemeralised 'punch card' approach to mind-brain causality.

- **Time and wavefunction collapse constitutes a work function zero process**: **Time itself is not brought about by any energy-power input or work-input, given that there are no gears behind time such as to require energy or power**. Therefore, the fall of an apple, the orbit of the Moon, or the workings of the brain… insofar as *all* of these are realised in and through the non-energy and non-power-driven work function zero wavefunction collapse and time process, are ultimately *not* brought about by energy-power inputs. The same follows for any punch card-like mind-brain causality which, in any case, exploits the very process of work function zero time and the black box 'gaps' between the succession of event horizons.

- **Energy relations are ultimately subsumed to ontologically primary time**. Putative 'kinetic energy' appears to issue out of the future potential Whitehead facet through AND-to-OR wavefunction collapse. Thus putative energy is potentialised within the Whitehead facet as 'future-potential energy'. Thus, **putative energy relations, and their attendant conservation principle, remain subsumed to an ontologically primary process of AND-to-OR collapse, or time. Yet, time itself is *not* driven by any energy input** and does not require a 'battery store' to power it or make it happen. Time and wavefunction collapse is brought about by a gear-less work function zero process. **Insofar as time is ontologically primary, and since there can be no notional energy relations without time, and since time itself is *not* driven by or subsumed to any form of energy-input or relation, this necessarily implies that energy notions and relations must ultimately abstract out of the real ontology of physics, rendered pleonastic**. Thus, energy-work relations are ultimately not parts of the real ontology of physics (see Part-III **5.35**).

- **The abstraction of energy and work out of the ontology of physics necessarily implies the abstraction of energy concepts and relations out of the framework of mind-brain causality**. Thus, the materialist expectation that mind should affect an energy input or work to instigate causality vis-à-vis its brain, or to 'slot the punch card' and 'push the pins', is rendered void and false. Insofar as pleonastic energy relations are abstracted out of the ontology of physics by work function zero time, the same energy-relations and concepts must abstract out of mind-brain relations subsumed to the same process of time. **Thus, the mind's imposition of its 'punch card' upon its brain does not involve or require any pleonastic 'energy input or work', much less a 'power source' to drive it. The minded imposition of the ephemeralised punch card is a work function zero process. It also follows that the violation of the conservation principle cannot arise in Cartesian causality**. Indeed, the principle of conservation of energy is no more violated by the workings of the Cartesian mind than it is by the work function zero process of time and wavefunction collapse itself; the very process that manifests *putative* (and ultimately pleonastic) energy relations and attendant conservation principles, notwithstanding the work function process of enforcement of physical laws, implying that the enforcement of the conservation principle itself does not violate the conservation principle: (See Book-III **3.33** and Part-II **5.32** to **5.34**).

- **The principle of conservation abides, but it is recuperated into an abstract form that does not run into contradiction with Cartesian causality, while it permits Cartesian causalit**y: In the face of de-spatialisation and temporisation, and the pleonastic status of energy-work, and the transformation of mass into an abstract memory effect, the conservation of energy and matter must transform into an abstract form recuperated as **the conservation of Bergsonian memory** and, per grand decoherence due to memory, **complemented by the principle of consistent futures** in the enforcement of consistency of the Whitehead facet to the sum of 'initial conditions retained as conserved memory within the Bergson facet: (see Book-II **2.39** for initial contention, and the pertinent essay **5.29**)'. Mind-brain causality via the ephemeralised 'black box' **switch-x** tripping process cannot usurp the conservation of memory or engender non-consistent futures: Mind-brain causality cannot erase Bergsonian memory, nor can it affect inconsistent futures contrary to memory. Therefore, Cartesian causality does not violate the conservation principle in its recuperated post-energy and post-work form.

- **De-spatialisation usurps materialist assumptions, including those that attend mind-brain causality**. De-spatialisation and temporisation forces a reframe of all things, including relations that attend the brain. These must be recuperated into pure non-contiguous and stereotemporal forms. All of nature is organised into future-potential and mnemonic delayed choice time-interval relations: all framed to a de-spatialised stereotemporal Bergson-Whitehead amalgam. The consequences are thus:
    - **De-spatialisation usurps the notion of particle-mediated causality, and even subverts particle mediated force and energy relations (see 5.30)**. In a pure time-based physics, there are no 'wheres' at which putative point-like particles or bodies are 'located': There are only *whens*. It follows that **spatial notions of contact or impact-mediated causality are rendered impossible in a de-spatialised temporised physics. This works against the false notion that mind must input work-bearing particles or contiguous 'power' to affect its brain**, supposedly via 'physical contact in space'.

Hence, the notion that mind-brain causality requires such a materialist scheme must suffer the same fate as general spatial contiguous causality.

○ **De-spatialisation usurps the notion that information, and especially memory, has a 'location' or 'spatial address' in the brain**. There is no space. Hence, in a de-spatialised universe, memories purportedly stored in the brain cannot have 'locations' in that brain. In a de-spatialised universe, information is wholly temporised. Thus, the search for any in-brain 'location' for one's memory of the big red garage door from Potters Bar-1973, will necessarily fail, even if parts of one's brain 'light up' and correlate with the declaration of the recollection (see Part-III **5.30**).

○ **De-spatialisation usurps the notion that mind and consciousness has a 'location' or a 'spatial address' in the brain. There is no space. This renders forlorn the want to 'locate' the controlling principle of the brain *in* the brain.** There can be no spatial locus in the brain that generates even 'illusory' mind or consciousness. Given de-spatialisation, nous could only relate to world and brains in purely temporal terms, culminating into the temporal precedence and subsequence of nous vis-à-vis brain and world.

Thus… de-spatialisation, the invalidation of particle 'locations' and 'contact causality', the demise of hidden gears and of the input-operation-output schema for time and causality, usurp ubiquitous materialist philosophy arrayed against Cartesian causality. Mind imposes its punch card upon its brain, and brings about the unitary binding and coherent activity of its brain, on a non-contiguous work function zero basis. Ultimately, energy concepts and work-energy relations, and attendant energy-based conservation principles, are *not* parts of the real ontology of physics (they are pleonastic), no matter how pragmatic and useful they are in lesser and dated science.

All of our points justify the necessary ephemeralisation and de-materialisation of the punch card approach to mind-brain causality, as much as of causality in future physics in its anticipated evolved form. To that end, in the next essay, we will seek to clarify ephemeralisation and attendant unitary binding of the brain directly from the framework of the Bergson-Whitehead amalgam and from work function zero time.

5.77: Ephemeralisation of the Stereotemporised Critical Milieu

De-spatialisation implies that the critical milieu inherent to the brain… essentially, the mind-pertinent brain… cannot constitute a spatial 'surface' or even a block-space. Instead, it must configure into a de-spatialised purely stereotemporal distribution of events and future possibilities; organised into delayed choice time-interval relata; always subject to the relativity of simultaneity, both internally and in relation to the temporally preceding or 'embedding' world. It follows that any 'punch card' imposed by the Cartesian mind upon the de-spatialised stereotemporal critical milieu must itself constitute a de-spatialised stereotemporal state. Indeed, the 'punch card' will configure as a stereotemporal instantaneous whole, formed from the co-joining of wavefunctions in memory and wavefunction in future-potentiality; part of the Whiteheadian equalised possibility plateau: (see **5.75**).

Recall the essential interchangeability between the critical milieu as it applies to brains and the same as applies to a photographic film-surface: The photographic film does not constitute a 'spatial surface': In a de-spatialised universe, the film, like all else, constitutes a pure stereotemporal state, always subject to the relativity of simultaneity, both internally and in relation to events 'outside' it. The same applies to the complex image-pertinent wavefunction and instantaneous whole that insinuates upon the film: also de-spatialised and temporised. The complex wavefunction acts as the 'punch card', specific to the subsequent three-phase development of *en masse* indeterminate photon depositions into the pertinent image… which must itself constitute a de-spatialised stereotemporal state.

Exactly the same applies to the 'punch card' that attends the development of brain activity: also, de-spatialised and stereotemporised… hence, *ephemeralised*. Indeed, it ought now to be considered a banal fact that the whole physical universe in which photography and the brain is embedded constitutes a pure de-spatialised, temporal and stereotemporal ephemeralised order. Hence, we obtain ephemeralisation of the critical milieu and of the mind-brain interaction per function of that milieu and the larger universe in two interleaved ways:

● **The critical milieu, the 'punch card', and the whole process of the subsequent development of brain activity, is necessarily de-spatialised and purely stereotemporised.** As such, the critical milieu, the punch card, and the whole process of brain activity, is **ephemeralised.**

● **The minded selection and imposition of the 'punch card', and attendant mind-brain causality, insofar as this cannot entail any purported particle-mediated contiguous closed IOO-schema causality, must transpire in a de-spatialised purely stereotemporised universe, and it must also ephemeralise in accord.**

5.78: Cartesian Causality in Delayed Choice Switch-X: Causal Freedom & How Mind Selects its Punch Card

The delayed choice **switch-x** based approach to mind-brain causality segues seamlessly from the Bergson-Whitehead model and from the ephemeralised punch card approach. **To furnish the delayed choice switch-x approach to Cartesian causality, we utilise a 'black box' approach vis-à-vis the 'gaps in time' expected to intersperse a minimum of two successive event horizons. The black box conceals the 'throw of switch-x' that transpires within the said gap upon the intercession of the Cartesian mind, a feat accomplished on a gear-less and non-contiguous work function zero basis.** The 'gap' interspersing the event horizons *is* the 'black box' whose contents are hidden from direct view, and whose Cartesian processes are not clockable as any realised or future-potential event, or even definable, but whose ontology, nature and attendant causality can be justifiably inferred as ontically real.

Recall that John Wheeler's delayed choice constituted the basis for the nested futures view. This entails counterfactual causality and grand decoherence, and the erasure of non-consistent nested futures (see Book-II **2.34**). The same structures, processes and principles apply in mind-brain causality, but directly to the equalised possibility plateau, whose attendant **switch-x** is modelled *as if* residing within 'the gap'... within the 'black box' demarcated by the pertinent event horizons.

Note that there is no spatio-material hardware for our virtual **switch-x** within the black box 'gap': It is an ephemeralised **switch-x**, but with the difference that, instead of just the two switch-setting and attendant nested future wavefunctions, our noetic **switch-x** has as many settings as the total number of potential novel affects and memory recollection-expression possibilities (the punch cards) inherent to the equalised possibility plateau. Thus, **switch-x** relates to the Whitehead equalised possibility plateau depicted in **Fig. 5.21** (C) as...

Switch-x = f {Whiteheadian: {**x** AND **y** AND **z** AND...**omega**} -/- Bergsonian: {**a** AND **b** AND **c** AND...**n**} }

Within the 'black box' demarcated by the said event horizons, **switch-x** is suspended: i.e., its throw is delayed into a not-yet-tripped future developmental possibilities for the future brain comprised of the above-stated admixture of future novel affects (**x**, **y**, **z**) co-joined to memory recollection-expression potentialities, (**a**, **b**, and **c**).

Let us imagine that Cartesian active mode mind selects the complex whole-brain wavefunction novel affect-**y**. Thus, affect-**y** is slated for the future development of the brain. **Through the said 'gap in time' or the synonymous 'black box' between the pertinent event horizons, the Cartesian mind 'trips' hitherto suspended switch-x to setting y. The equalised possibility plateau at the slated future-potential event horizon consequently decoheres to y. Affect-y subsequently AND-to-OR collapses at that horizon through the three-phase developmental process, culminating into the in-brain 'materialisation' of novel affect-y.**

Subsequently, a new equalised possibility plateau is formed, possible per function of the developmental reversibility of the critical milieu. At a new future potential event horizon, the new equalised possibility plateau must adjust to the fact that **y** is no longer a novel affect. Thus **y** is now relegated to the Bergsonian facet as memory...as (**y**)... with only **x**, **z** and other novel possibilities remaining and available for future realisation. Hence, the new reversible equalised possibility plateau will configure as...

Whiteheadian: {**y** AND **z** AND...**omega**} -/- Bergsonian: {**a** AND **b** AND **c** AND (**y**)...**n**}

Through a subsequent black box 'gap in time', active mode mind may intercede to affect a possibility in the formalism above, including the repetition of affect-**y**, but as memory (**y**), and as the recollection of (**y**) , by tripping a subsequent ephemeralised **switch-x**.

*

The Cartesian mind 'throws' switch-x by means of gear-less contact-less causality, hence attains superficial causal freedom, andon a work function zero basis; hence, no violation of the conservation principle will arise, and memory will be conserved, while futures will remain consistent to memory per function of grand decoherence. Subsequent work function zero AND-to-OR wavefunction collapse (time) of the mind-selected now-decohered equalised possibility plateau will 'materialise' the recollection-expression of memory (**y**) via the three-phase developmental process, at a subsequent event horizon. **Note again;, there is no 'hardware' to constitute switch-x residing in the 'black box' between the pertinent event horizons.**

To augment, the process of mind-brain causality likely unfolds via a series of event horizon successions, through a series of 'gaps in time' involving partial punch card selection and succession, via imposition of grand decoherence and AND-to-OR collapse.

But how does Cartesian mind 'trip' **switch-x**? How does Cartesian causality actually work? Surely, our depiction must be contrived bizarre nonsense? Old habits die hard. It is difficult to imagine a seeming a-causal process of Cartesian 'switch tripping', even with the insight furnished by Alt-R. This is due to the cultural saliency and hegemony of materialistic suppositions and expectations. By habit, we expect the setting on **switch-x** to be thrown manually and to involve some form of contiguous contact-causality. We expect this to require an energy or power supply and input to make it happen. We expect this to involve a stream of spatially colliding 'particles' or 'spaceballs'. Indeed, how could mind trip **switch-x** *and* initiate whatever 'punch card' it seeks to develop in relation to the brain, without resort to some form of work function? It cannot accomplish such feats for free?

It is useful to momentarily forget the claims of Alt-R and recapitulate the model of 'interaction' that takes place between two particles in pre-Alt-R generic terms, to which we tend to hoist all the usual materialist tropes previously adumbrated. **When we get, two 'particles' to interact, we are effectively superposing their respective wavefunctions.** These are probability distributions for future possibilities that attend the respective 'particles'. In short, we are combining the *futures* of the particles. Thus, even when we restrict our view to the generic and the conventional, 'particle interaction' does not involve a spatial contact between 'particles in space' but a co-incidence and combine of their futures, forming a new combined future-possibility distribution or superpose. On the false presumption of 'space', we could give this combination, and the whole process leading to it, a spatial wavelength character, and ignore the exclusive time or frequency-based model necessitated by de-spatialisation and temporisation. Even so, **the joint superpose cannot involve contiguity between the two wavefunction and their possibility and probability states... because these have not yet transpired into realisations, aside implications from de-spatialisation. Contiguity cannot be realised between the superposing particles because mere sets of possibilities rendered into probability structures are non-contiguous by default: The addition of possibilities and probabilities does not involve a 'contact' as such.**

Again, given that the wavefunctions superposed are future-potentials, nothing has yet transpired to qualify *any* attribution of a 'contact': You cannot claim 'contact' for events that have not yet transpired into realisation.

Also consider that, for state-change to transpire, the joint superpose from the coinciding of the possibility states of the two 'particles' must undergo AND-to-OR collapse. **Generic quantum mechanics never asserted any certainty that AND-to-OR collapse transpired per function of 'gears', or by contiguous causality. The matter of hidden gears or otherwise always remained moot.** In any case, even in generic pre-Alt-R terms, the putative **'energy-force exchanges' presumably involved and evinced in state-change, were portrayed as brought about by AND-to-OR wavefunction collapse. But generic quantum mechanics never claimed any certainty that AND-to-OR collapse was brought about by 'energy-force inputs'. Thus, generic 'particle interaction' and causality never supported, or was at best agnostic about materialist causality**. Thus, materialist expectations about Cartesian causality are as suspect as they were precarious within the framework of purely generic pre-Alt-R conception of 'particle interaction'.

But we cannot ignore the claims of Alt-R. Thus, we must re-introduce de-spatialisation, temporisation, gearless AND-to-OR collapse and time, the pleonastic status of energy-work notions, the 'quantum leap in time', and the recuperation of the principle of conservation into its abstract Cartesian-friendly form… and the consequent breakdown of materialism and the unravelling of materialist expectations about causality in general, and about Cartesian causality, specifically. Hence, Cartesian causality vis-à-vis ephemeralised black box **switch-x** in mind-brain control is secured.

Hence, Cartesian dualism abides.

5.79: The Split-Brain Conundrum from the Bergson-Whitehead Amalgam Framework, Recapitulated in terms of De-Spatialisation & Temporisation

The intermediate mind-world theory furnishes the opportunity to return to a key conundrum first mentioned in the Preliminary, **0.48**: namely the split-brain conundrum. The 1940s saw a brief indulgence in the use of the *corpus callosotomy*: the surgical separation of the two hemispheres of the walnut-like brain. The procedure sought to reduce epileptic seizures[106]. It produced odd outcomes and conclusions seemingly in favour of split consciousness.

With the surgical splitting of the brain, it is expected that two distinct individuals and identities must form; each confined to its own hemisphere: i.e., split consciousness. From this, it might be posited that consciousness *must* be produced by the brain, and if this were not so, the observed split of the original consciousness into two could not emerge. Therein, whatever might be responsible for the production of consciousness within the brain ought also to split into two; to each hemisphere.

In the older data, split consciousness appeared true, even if accompanied by certain anomalies. For example, under normal conditions, split-brain patients did not report any split in identity: Identity remained unitary, and, by implication, the consciousness also. That the brain must also operate in unitary form despite the surgical split seemed perplexing. Yet, split identity seemed to appear if certain experimental restrictions were applied, through which one could consult one hemisphere without the inclusion or knowledge of the other[107]. Yet, **recent observations by Yair Pinto *et al* contradicted the older findings for split consciousness[108]. Pinto's experiment demonstrated unitary consciousness and binding of both brain-hemispheres, *despite* surgical splitting**. But how could the separated hemispheres become unitary? This ought no longer to be possible. One ought to find the two halves of the brain, hence the two identities, perhaps even at cross-purposes; with prospective concord as difficult to achieve as it is between any two individuals. Yet, the truth is the opposite, and Yarn Pinto's findings concur; demonstrating unitary coherence of the brain and identity despite the surgical split.

*

In the Preliminary, we posited that the key to unitary brain and mind, despite surgical splitting of the brain, is de-spatialisation itself: De-spatialisation and temporisation renders ostensive splitting of the brain void. Indeed, how does one split a distribution in time? How does one 'cut' memory-in-time into 'two pieces'?

We now augment the implications by incorporating the split brain into the Bergson-Whitehead amalgam. But there is another key consideration to make: one garnered from the general theory of mind-world, co-joined to de-spatialisation and to the Bergson-Whitehead model of the split-brain.

Recall from **5.56** and **5.59** that nous, and its attendant mind and consciousness, is constituted as the 'visceral sense of the common present moment' within which the world of memory and time arises. That is, dissimilar to the growing block system of memory, possibility and time subject to the relativity of simultaneity, consciousness and nous in general enjoy absolute time-like characteristics, and nous is configured as a temporal noetic monad: the closest we might come to a 'point in space' in a pure temporal universe. It follows that the split enforced on the brain modelled from within the Bergson-Whitehead amalgam implies that **the world and brains must arise to the temporal monad of witnessing consciousness in the form of a unitary whole, despite, including, *and* consistent**

[106] Mathews, Marlon S.; Linskey, Mark E.; Binder, Devin K. (2008-02-29). "William P. van Wagenen and the first corpus callosotomies for epilepsy". *Journal of Neurosurgery*. 108 (3): 608–613.

[107] Wilson et al. 1977; Gazzaniga, LeDoux, and Wilson 1977

[108] Yair Pinto, David A. Neville, Marte Otten, Paul M. Corballis, Victor A.F. Lamme, Edward H.F. de Haan, Nicoletta Foschi, Mara Fabri (2016): 'Split brain: divided perception but undivided consciousness' in *Brain*.

with the surgical split… but without dividing the noetic temporal monad itself. In other words, the surgical splitting of the brain cannot split the noetic monad of identity and witnessing consciousness. Hence, the identity assumed by nous also abides

*

On the assumption of space, the surgical splitting of the brain is effectively no different to splitting a block of cheese into two halves. Yet, even then, the two halves of the former block, now apparently separate, behave in inexplicable ways: The two halves will gravitate toward each other and seek to converge back into unity.

We have already presaged gravity as a Bergsonian mnemonic effect (Book-IV: **4.30**) with implications to the grand decoherence of the Whitehead facet (**4.31**). However, unitary coherence of identity and consciousness in split-brain patients is not due to any gravitational relation between the two hemispheres. Indeed, de-spatialisation is the more critical and directly pertinent factor vis-à-vis the unity of consciousness and the coherence of the brain, with or without the surgical split.

As stated, in a physics that assumes space, splitting an object into two appears a simple matter, with apparent simple consequences. But with de-spatialisation, there is no space. **Only time is real, and the physical order, including the block of cheese… including the brain… must now be configured as a distribution in time. Consequently, cutting a block of cheese into two might otherwise appear straight forward… but cutting a' distribution in time' into two turns out to be a little more complicated. Indeed, how does one cut time into two, or cut a distribution in time into two?**

In the case of the de-spatialised temporised cheese-block, we appear to ostensibly cut the cheese block in two, but we cannot separate either block from their temporal relation to their common past: a relation *not* mediated in or across now-pleonastic 'space', but one established in and across pure time as a non-contiguous relation of the abstract past vis-à-vis the relative present. Hence, **split your cheese block however you wish, you cannot sever it from its abstract Bergsonian past: you cannot sever it from its larger order of Bergsonian memory.**

However, de-spatialisation need not necessarily imply unity in concert. Hypothetically, if two objects were 'separate' from the very beginning, such as applies to two observers distributed to the Moon and the Earth, respectively… and subject to the relativity of simultaneity… they will remain separate in terms of the unique sequence-order of witnessed events and attendant first-person identity. On the other hand, if the objects were originally in unity, but subsequently split into two, the original unity cannot be subtracted out of the subsequent split, despite all things considered, including the inverse square of time. Therein lies the basis for unitary concert and unitary consciousness of the split brain, despite the surgical split into apparent disunity.

Hence, **the reason we obtain unitary identity and consciousness in split-brain cases is per function of the fact that cutting the brain into two cannot cut off either hemisphere from their common past or from their past-mnemonic unity, retained as abstract memory of the brain in its former unity within the Bergson facet, and which superposes upon the present brain, split or not, by means of a non-contiguous resonation of information and memory across time. Indeed, the primary mechanism for this is the grand decoherence of the Whitehead facet… into consistency with both the initial unitary brain mnemonically retained within the Bergson facet, *and* the subsequent surgical splitting of the brain, also retained in the Bergson facet. The subsequent grand decoherence of the Whitehead facet and the futurity of brains, while it must cohere into consistency with the implications of the surgical split, must also abide into consistency with the preceding unitary brain-in-memory *before* it was surgical split. The early unity subsumes the subsequent split, and thus retains unit, in any future development of the brain… even for a surgically severed brain.**

Of course, if the two halves of the brain were grown and developed separately from the very beginning: effectively as two distinct brains, we would expect two distinct identities and consciousness per each truly separate hemisphere… in the same fashion that we obtain two identities per two distinct whole-brains.

And if the brain, split or not, were subject to gross continuous or systematic disruption per Cameron's annihilation-ECS, the ready ability of an otherwise unitary mind to utilise its unitary expression in and through the brain would be broken, and identity *seemingly* lost.

*

We reiterate that nous constitutes a temporal noetic monad: the nearest we can come to a point-like state in a de-spatialised temporised physics… within which the world-in-memory arises into passive mode consciousness… and within which the brain also arises, split or not. One can appear to split the brain in as many ways as one pleases. While de-spatialisation and temporisation will certainly render the attempt void, given that one cannot really cut an 'object' distributed in time into separate parts in the false spatial sense, splitting the temporal brain will have absolutely no effect on the temporal noetic monad or nous within which the brain and world arise. **One can appear to split the brain and world that arises within the temporal monad of nous. But this cannot 'cut' the noetic temporal monad itself, nor the singular identity that attends that monad. It follows that the identity will perdure even after the surgical split, in a form consistent with findings from split-brain patients and the investifations of Yair Pinto.**

*

The process of unitary concert of brains, and the perdurance of unitary identity despite the surgical split, involves two sub-processes. First, **the pre-surgery unity of the brain, constituted as a mnemonic 'worm' within Bergsonian face, decoheres the future brain into consistency with the pre-surgery unity, despite the subsequent surgical split**… 'burning through' the split and re-instituting unity through the grand decoherence of future possibilities pertinent to the brain within the Whitehead facet to consistency with the prior brain-unity. We simply superpose all wavefunctions in memory pertinent to the pre-split past of the brain with the first set of post-surgery wavefunctions in memory. Therein, the pre-surgery memory will dominate over the first set of post-surgery memories, only to dominate every subsequent post-surgery states thereafter, baring Cameron style gross disruption.

The memory of the unitary brain will thus 'burn through' the post-surgery state. The surgical split will remain, and it will produce consistent peculiar outcomes. But unity will be reinforced on the future brain regardless, as the Whitehead facet grand decoheres to consistency with the demands of dominant pre-surgery unitary brain-in-memory.

Second, and benefiting from the grand decoherence of the future states of the brain, the active mode mind, as **the unitary noetic temporal monad and *its* retained non-split identity therein, will intercede through the 'gaps in time' in order to decohere the Whitehead facet pertinent to the brain into consistent futures in line with its will *and* its identity: This will 'burn through' the surgical split, imposing unitary resolution of the brain... or the unitary collapse of its complex future wavefunction... into what nous intends, even with the surgical split... and *always* consistent to and expressing the unitary identity acquired before the surgical split... which, in the first place, at the noetic moment level, was never fated to or affected by the surgical split.**

Yet, there is the possibility of gross disruption from Cameron's ECS. If carried out for a sufficient period of time, even the retained identity belonging to the noetic monad will be unable to insinuate itself and its identity to its brain. Also, it will also recede into the past; so much so that, all that will remain is the denuded "I", which cannot be annulled by Cameron's destructive ECS, simply because any temporal monad is beyond the purview of any affect realised within the system of memory, possibility and time.

5.80: Introduction to the Problem of Causal Freedom: The Qualitative Agency Problem

The ephemeralised work function zero delayed choice black box **switch-x** approach to mind-brain causality appears to solve the most superficial aspect of the problem of causal freedom.

The perennial problem of causal freedom is less an issue of causality, and much less about the false conundrum between physical determinism versus indeterminism. Indeed, these are either false problems or easy problems that Alt-R and our ephemeralised delayed choice **switch-x** approach can attend and circumvent, as was shown in **5.78**. The hard problem is constituted as the semantic-qualitative form of minded agency and *its* implication to the viability of agency, despite minds' mere physical causal freedom secured via delayed choice black box **switch-x** in **5.78**. This deeper problem was indirectly elaborated in Part-II in **5.16** in the critique of the contention that consciousness orchestrates quantum indeterminate processes in lieu of equally forlorn a-noetic hidden gears. It is also implicit to the Euthyphro dilemma (Book-III **3.14**).

Recall from Part-II **5.16** (a recapitulation of Book-III **3.32**) that, a very large number of hidden deterministic scripts could be devised to secretly orchestrate apparent quantum indeterminate processes, with each script equally apt to the task; with none exclusively special so as to constitute *the* script; with none qualitatively superior to any other. The belief that consciousness must orchestrate and obviate apparent quantum indeterminism must presuppose a qualitative basis for putative minded selection from among the scripts. But, given that there are no qualitatively superior scripts to choose from, the unique role of mind and consciousness as qualitative decider is rendered irrelevant. Hence, the **qualitative superficiality problem** of consciousness-directed circumvention of quantum indeterminism. The same problem transfers to the question of qualitative agency beyond the limited purview of the orchestration of quantum indeterminism.

Even from the framework of the limited purview recapitulated above, given the qualitative equality and interchangeability of all scripts, any mind-selected script from among these would constitute a completely arbitrary decision or agency: Since one script is as good as any other, the 'free choice' from among these degenerates selective minded agency to a mere trivial, arbitrary and haphazard 'choice', with minded agency reduced to mere arbitrary haphazard agency, essentially no different from a mindless selector-device operating on random principles in lieu of mind. This constituted the **arbitrary selection problem** in consciousness-mediated circumvention of quantum indeterminism.

As with qualitative superficiality, the arbitrary selection problem also transfers to qualitative agency problems beyond the narrow concern for quantum indeterminism.

As intimated, essentially the same problems emerge vis-à-vis mind's causal freedom in the milieu of larger qualitative aesthetic and ethical concerns: Hence, the **qualitative agency problem of freedom**. Thus, while the superficial causal freedom problem is solved via the delayed choice **switch-x** 'black box' approach, the qualitative agency problem of freedom remains unsolved, although we could yet garner a useful non-formal solution: We will now do so below.

*

Mind and consciousness must engender and attend outcomes of the brain and world that are either qualitatively superficial, non-significant, or *trivial* ... such as is true in the choice between strawberry versus chocolate ice cream (hardly a matter of excellence or world-shattering consequence)... or nous must attend and engender *profound* outcomes of qualitative significance. For example, should one cooperate with the Nazi regime or resit it? The choice is not interchangeable with the choices attending the utterly trivial ice cream domain. Thus, there exist qualitative inequalities in the universe, and an ontic plausibility for minded agency or causal freedom-proper, beyond its superficial trivial form vis-à-vis the ice cream domain.

In the context of purely trivial choices, nous retain its mere causal freedom, courtesy of the switch-x black box. Yet, nous could not possess or engage in any form of real agency, save as a haphazard arbitrary selector over mere trivial alternatives. Of course, an arbitrary selector... an effective fruit machine... *has no real agency*... because the 'choices' it makes and the outcomes it generates are *essentially* random, and agency is rendered in absentia, even if **switch-x** black box processes were engaged.

Noetic agency only becomes viable when mind-brain causal freedom attends the instigation of comparative profound outcomes, such as exemplified by the choice between cooperating with the Nazi regime versus not doing so. The choice must be based perhaps on transtemporal principles of truth or Good that supersede the silly 'ice cream domain'; involving a variant of either Platonic forms or Kantian categories… or other similar notions of superlatives or principle, such as to constitute the viability of qualitative or aretaic agency.

Our non-arbitrariness thesis, which emerged out of apagogic arguments against memory and informational nihilism (see Book-III **3.06** to **3.17**) secures the basis for Platonic and Kantian-like ontology requisite to the plausibility of noetic aretaic agency, beyond requisite but non-sufficient black box **switch-x** Cartesian causality.

However, a principle-led mind and choice is qualitatively *determined*. That is, resort to principle implies 'no choice' but what the principle demands and compels: It implies that we relinquish the 'freedom' (or rather the *arbitrariness*) over the alternatives toward a superlative end or goal or principle. This appears to imply the relinquishing of agency. The very basis for mind's agency… namely, *principle*… appears to undermine agency through compellence, despite the viability of mind's causal freedom secured on the basis of our **switch-x** black box approach espoused in **5.78**.

One might argue that, in the choice to relinquish the alternative in favour of principled non-cooperation with the Nazis, the surrender of agency to principle is what makes for freedom and agency: somewhat in the manner Kant may have suggested; wherein one has freedom of choice vis-à-vis the categories. However, this runs into the following problem: Is the choice between the categories to surrender agency to principle an arbitrary choice, no different from the choice between flavours of ice cream? If not, then there is no agency, given that a being compelled by principle is not driven by itself. On the other hand, if the choice from among the categories is as arbitrary as the choice over ice cream flavours, then, again, there is no agency: the 'choice' and agency is as good as one made by a fruit machine, even with the background certainty of our black box **switch-x** and the superficial physical-causal freedom that it attends.

*

Three solutions to the qualitative agency problem are offered, with the third constituting the decisive solution that incorporates the first two. **First, one ought not to care that one is determined by principle. One would rather be determined by principle *without* agency versus 'agency' *without* principle, or agency without agency, given arbitrariness and superficiality.** Despite black box **switch-x** and the attendant viability of causal freedom, principle-based agency appears to be without choice. If true, then no matter. For it is *ironically* 'preferable' (choice?) to be determined to be good, hence good but determined, versus superficially and arbitrarily 'free' *without* principle… or as 'free' as a fruit machine, or superficial and *unfree*.

However, the ontic reality of transtemporal superlatives is requisite; whether Platonic, Kantian, or other: If there is principle, and it turns out that we are compelled and determined by it, this would constitute good news. If there is no principle, and all is arbitrary, then there would be no qualitative difference in collaborating with the Nazis versus resisting them, versus 'choice' between ice cream flavours: The choice would be rendered as trivial as the 'choice' in the ice cream domain, *even with* black box **switch-x**.

The second solution requires that we understand that the 'problem' is a fallacious one, except insofar as we most certainly need an ontology of *principle,* furnished by the non-arbitrariness thesis similar in form to the one espoused in Book-III. This aside, **the fallacy derives from the want of a 'formula' for moral decision making; an 'algorithm' for aesthetic choice; a mechanism… a 'hidden gears'… for qualitative agency**. Yet, as was discovered in both Part-II and Part-III, and as is clearly stated in the black box **switch-x** approach to Cartesian causality, **ultimate causality behind time itself defies definition and rendition in terms of *any* form of tick-tock click-clock 'gears', algorithms, or formalisms. This must also apply to the process of noetic qualitative agency. We are making the mistake of looking for a 'gears' behind qualitative agency and freedom when none can exist.**

Qualitative agency and excellence is ultimately processed through a gear-less ineffable causality, as is ultimate ineffable time and general causality. Indeed, the case for noetic causal freedom in **5.78** exploits this fact.

The third solution involves a unique combination of the two adumbrated above: **Causal agency and causal freedom involve choice between principle versus arbitrariness: Nous can recognise the difference between these and conscry a spectrum bounded by arbitrariness on the one end versus principle on the other. The choice is not between the Kantian categorises or Platonic forms as such, but between category or form versus non-category or *dis-form*… in the sense that we can conscry some form of imperative in, say, not collaboration with the Nazis, but we cannot distil *any* imperative in the choice between strawberry versus chocolate ice cream.**

The quality of agency, or lack thereof, is revealed in its tendency to one or the other end of the *form-disform spectrum*… in the tendency to obscure or ignore qualitative inequalities, or even deny their ontology by resort to nihilistic apologetics of one form or another. Thus, excellence in agency is realised in the recognition of qualitative inequalities, in the garnering of intellectual resources and habit in and for such recognition, and the tendency toward principle, or away from arbitrariness. This constitutes *real choice*, and *real agency*, in that the choice is comparatively profound, not trivial with… respect to which our black box **switch-x** solution is indispensable.

Thus, causal freedom as qualitative agency is rendered ontologically plausible per the differentiation of principle from the arbitrary via the form-disform spectrum, per inherent objective qualitative inequalities, combined to a gear-less and ineffable black box **switch-x** Cartesian causality, which renders possible free agency toward principle and away from arbitrariness (towards sophrosyne and excellence) or in the opposite direction toward, vacuity and a-sophrosynic de-calibration. Again, this requires qualitative inequalities, or the real ontology of principle in the fashion of Platonic forms or Kantianism categories, or some other variant of the same, to which the non-arbitrariness thesis from Book-III is a contribution.

5.81: Summary of Part-V

In Part-V, the intermediary mind-world theory espouses the de-spatialised temporised Bergson-Whitehead amalgam basis for radical mind-world relations: The growing block Bergson-Whitehead amalgam, implicit to the general theory, is explicit in the intermediate theory, and it scaffolds almost all of the former's arguments, reiterating that the brain *must* be modelled quantum mechanically'. Thus...

- **The pertinence of quantum mechanics to the macro-scale universe and brains per the growing block model and grand decoherence**, wherein billiard balls 'here' counterfactually instantaneously decohere billiard balls 'elsewhere' into consistent futures per the convergent category Whitehead facet. This revives Mach's principle. The nested future basis renders possible the Whitehead equalised possibility plateau, transforming the brain into a future-form open quantum mechanical system: All macro-scale systems are subject to the AND-form Whitehead facet, and so is the brain. Hence, we must model the macro-scale brain as a growing block quantum mechanical system. (See **5.62**).

- **Scale-amplification of quantum phenomena renders quantum mechanics applicable to the macro-scale brain**. We never infer AND-form logic or quantum indeterminism from one or just a few quantum incidences at the 'micro-scale'. It is only when we amplify these up to macro-scale *en masse* event-densities that we obtain their evidence. Hence, AND-form logic and self-interference *is* manifested at the macro-scale (for example, as interference patterns) and *not* confined to the micro-scale. Thus, quantum mechanics abides at the macro-scale, and it must also abide at the macro-scale brain (See **5.63**).

- Per the non-standard evaluation of the two-slit experiment, the absence of interference under the single slit condition does not imply the absence of either AND-form logic or of quantum indeterminism from the same. **Single-slit like conditions pervade at the macro-scale. Since AND-form logic cannot be obviated from single-slit conditions, it follows that AND-form logic and quantum indeterminism must also pervade at the macro-scale of billiard balls and brains, even if interference patterns are not readily evinced in any explicit way at the macro-scale**...until grand decoherence enters the fray and renders salient the mutual interference of billiard balls here with billiard balls 'elsewhere', and the self-interference of the universe with itself at all scales, including the macro-scale brain. (See **5.63**).

The intermediate theory, a neo-Cartesian theory of mind-world relations, employs a growing block quantum mechanical model of the brain by explicitly framing it to the Bergson-Whitehead amalgam. Yet, it does *not* constitute quantum mind theory. Why?

- **Generic quantum mind theories (including those of Eccles, Stapp, and Penrose) seek to employ mind and consciousness as either the collapsor of the wavefunction, or as the secret orchestrator of quantum indeterminism, or as both**. But...
  - *Wavefunction collapse and time does not require a collapsor-mind* (see Part-II **5.18**): On the contrary, it is consciousness that depends on ontologically primary time. No wavefunction collapse and time, no nous ; no noetic agency. Time is brought about by an a-noetic ineffable causality that transcends but subsumes nous, but it is *not* nous.
  - *Quantum indeterminism is not a general or brain-specific problem that requires a 'solution', and even less in need of an orchestrating consciousness* (see Part-II **5.17**, Part-V **5.64** and Book-III **3.26** and **3.32**): Unitary brain activity develops in the same way that a unitary coherent photographic image develops: through a three-phase process encompassing objective quantum indeterminate events that are funnelled *en masse* into ordained structured global outcomes. There is no requirement for an orchestration of quantum indeterminism by nous, no more than by a-noetic hidden gears.

Intermediate theory proceeded to the' punch card' theory, the critical milieu, and the equalised possibility plateau: constituents of the Bergson-Whitehead growing block model of mind-world. Thus...

- **The mechanical punch card constitutes the consummate analogy for mind-brain control** (5.56): In terms of the Bergson-Whitehead amalgam, the 'slot' stands for the 'gaps' between successive event horizons: the mind imposes its punch card through the 'slot'. Spontaneous wavefunction collapse and time manifests the 'pins' (i.e., the *en masse* quantum incidences) pushed through to a succeeding event horizon conformally to the pattern of the 'punch card'. All of this is achieved on an ephemeralised work function zero basis. The 'punch cards'' may consist of...
  - **Possibility functions in memory**, 'drawn out' from the Bergson facet as brain-pertinent wavefunctions in memory.
  - **Novel possibility functions in potentiality**, 'drawn out' from the welter of Whiteheadian nested futures pertinent to brains; i.e., the wavefunctions in future potentiality.
  - **Mixed state possibility functions:** i.e., admixtures of past in memory and new futures.

- **Mind-brain control requires the critical milieu**, its Whitehead projection into the future; and the equalised possibility plateau consequently formed... from which the various categories of 'punch cards' (noted above) selected by mind for future realisation. (See **5.65** and **5.66**).

- **The equalised possibility plateau is key to mind-brain control in the Bergson Whitehead amalgam** (5.67) and renders possible the punch card model of mind-brain control. Unlike generic wavefunctions, *the equalised possibility plateau constitutes a superlative wavefunction whose superposed possibilities are complex wavefunctions in their own right*: i.e., a wavefunction for wavefunctions, possible per the structure of the nested futures quantum wave: (See Wheeler's delayed choice in Book-II **2.18** and **2.19**. Also see **5.18**). The equalised possibility plateau is a future-form Whitehead projection of the critical milieu. The brain constitutes just such a critical milieu. So does a photographic film. The latter can allow any number of alternative images to develop upon it, each per function of a different wavefunction. The future-form Whitehead description of the photographic film is an equalised possibility plateau composed of the superpose of alternative unique complex quantum mechanical waves for unique images. But a photographic film constitutes an *irreversible possibility plateau* and relates to a *closed Whitehead facet* (See **5.68**) insofar as the images foregone, together with all attendant nested futures, are irreversibly erased by the attendant counterfactuals.

- **The developmentally reversible equalised possibility plateau**: A photographic film is potentialised to a number of alternative images at potential t_1. Thus…

  Whiteheadian $\{a$ AND b AND c AND…n | *each image equally probable*$\}$ …

  Image-c is developed at t_2 and its attendant event horizon. If the film is developmentally reversible, this will not close off the future to the development of the alternative images: an *open Whitehead facet*, insofar as the film at t_3 will reverted to essentially the same equalised possibility plateau as was found at t_1. (See **5.69**).

The intermediate theory supplemented its growing block Whitehead future-form model of the brain, co-joined with the Bergson model of the same in the context of habit formation, memory retention, and *memory-retrieval through the future.*

- **Habit formation through mnemonic mechanics of recurrent memory**. If the recollection of a wavefunction in memory, say, **memory-a**, is more recurrent it will be retained as instances of recurrence within the Bergson facet more frequently than alternative wavefunctions in memory, in contrast to **memory-b** and **c**. A habit structure and a rank order will form within the Bergson facet in which the recurrent memory will dominate over the recessive memories: i.e., **a>c>b**. (See **5.70**). Spontaneous mnemonic mechanics (see Book-IV **4.27**), and consequent superposition of these memories, will generate a superpose at a succeeding event horizon that reflects this habit structure, and will bias the equalised possibility plateau projected into the Whitehead facet into a not-so-equalised function in favour of **memory-a**, with greater readiness for recollection-expression, at least in the absence of mind's intercession to have it otherwise. Mind could supplant **memory-a** to form an alternative new habit structure: (See **5.71**).

- **Memory recollection of the Bergsonian past is obtained through the Whiteheadian future, given the fact that Bergsonian memory is projected into the future via grand decoherence… into a recollection-declaration potential within the Whitehead equalised possibility plateau**. Recollection involves the retrieval of memory from the Bergson facet, *not* through 'space', but non-contiguously across time. Yet, the actual process involves 'memories getting to the mind' through a process of the resonation of the abstract past into the future, through the grand decoherence of that future into consistent futures that exhibit the memory as a recollection-potential within the equalised possibility plateau; within the Whitehead future. Hence, it is through the presence of the past *in the future* that mind actively selects for and express its memory at succeeding future event horizons, and generates attendant brain activity or declarations of recollection.

- **The critical milieu** constitutes the essential definition of the brain: a developmentally reversible state that forms a Whitehead projection into the future, thus forming a developmentally reversible equalised possibility plateau therein; rendering a loophole within the Bergson-Whitehead amalgam that permits the entry of mind into the fray of memory, possibility and time. The critical milieu is comprised of input structures, output structures and internal feedback structures that integrate the brain to the wider environ as well as internally. Effects on the brain per mind's agency engender ramifications to the co-joined environment through the said structures. (See **5.72** and **5.73**).

- **Bergson-Whitehead model of the critical milieu and its time-evolution**. We incorporate the critical milieu to the event horizon. The event horizon, together with its critical milieu, 'dissolves'… only to reform at a succeeding event horizon through attendant AND-to-OR collapse and time. The imposition of mind's 'punch card' will be insinuated from *between* the dissolution of one event horizon and its succession by a next event horizon, via the 'gaps in time', or through successions of such. At the succeeding event horizon, mind's intended brain activity will manifest: affects will cascade through the output structures of the critical milieu, *and* via subsequent event horizons. (See **5.74**).

- **The Bergson-Whitehead amalgam treatment unifies memory-retrieval and novel affect**: (**5.75**). From a pure memory recollection approach, the equalised possibility plateau projected into the future as a memory recollection-expression potential will look something like… Bergsonian: $\{a$ AND b AND c AND…$n\}$). But from a pure novel affect approach involving never before transpired 'punch cards', the equalised possibility plateau ought to look something like… Whiteheadian: $\{x$ AND y AND z AND…$omega\}$. Yet, there is no such thing as pure memory, no more than there is pure novel affect. Memory recollection always happens intermixed with novel affectation, and vice-versa. Hence, the actual equalised possibility plateau is an admixture expressed as…

  Whiteheadian: $\{x$ AND y AND z AND…$omega\}$ -/- Bergsonian: $\{a$ AND b AND c AND…$n\}$

- **The equalised possibility plateau and the 'punch card' undergo time-evolution:** The equalised possibility plateau might start as the admixture…

  Whiteheadian: $\{x$ AND y AND z AND…$omega\}$ -/- Bergsonian: $\{a$ AND b AND c AND…$n\}$

But when mind selects novel affect **y** and realises it, **y** drops out of the Whiteheadian side and time-evolves into memory (**y**), incorporated into the Bergsonian side. Hence, the equalised possibility plateau time-evolves into......

Whiteheadian: {**y** AND **z** AND...**omega**} -/- Bergsonian: {**a** AND **b** AND **c** AND (**y**)...**n**}   (See **5.75**).

- **The ephemeralisation of the 'punch card'**. The punch card selected and imposed by mind is an abstract ephemeralised state, consequent upon de-spatialisation and temporisation. It is also ephemeralised as a constituent of the abstract equalised possibility plateau that resides in the Whitehead future. The form of contiguous causality that attends spatial 'hardware' does not apply to the non-contiguous temporal imposition of the 'punch card' by mind. (See **5.76**).

- **The critical milieu is an ephemeralised de-spatialised stereotemporal state.** De-spatialisation and temporisation implies that the critical milieu inherent to the brain cannot constitute a 'surface' or 'block' in space, but a de-spatialised, stereotemporal, hence ephemeralised complex. Hence, any punch card imposed by the mind must also configure as a de-spatialised stereotemporal and ephemeralised state. (See **5.77**).

To realise memory recollection-expression *and* novel affectation from within the black box 'gaps in time' between the succession of event horizons, active mode mind selects and decoheres the Whiteheadian equalised possibility plateau to the outcome it wants. Mind accomplishes this via the equivalent of the tripping of black box **switch-x**, via a non-contiguous work function zero process: one that cannot undercut the principle of conservation.

- **Delayed choice 'black box' switch-x model of Cartesian causality**. (See **5.78**). From the gaps in time formed between successive event horizons, Cartesian mind 'trips' an abstract delayed choice **switch-x** to selection from the equalised possibility plateau the complex wavefunction it seeks to develop in its future brain (i.e., the punch card **y**). This is accomplished consequent upon the breakdown of IOO contact causality per de-spatialisation and temporisation, the pleonastic status of energy and work, and on a work function zero basis, without the violation of the conservation principle, or the same recuperated as the conservation of memory (of the Bergson facet) twinned with the principle of consistent futures (of the Whitehead facet vis-à-vis past-memory). Cartesian causality in the form of the ephemeralised black box **switch-x** resolves the problem of physical causal freedom, but it does not resolve the qualitative problem of agency, The latter is solved separately and complementarily in **5.80** on the basis of black box **switch-x** combined to ontic qualitative inequalities that render aretaic qualitative agency possible, hence render freedom-proper possible.

- **Explaining unitary identity and unitary concert in split-brain patients: (5.79)**. The pre-surgery unity of the brain, constituted as a mnemonic 'worm' within Bergsonian facet, decoheres the future brain into consistency with that pre-surgery unity, despite the subsequent surgical split of the brain, 'burning through' the split and re-instituting unity through the grand decoherence of future possibilities to consistency with the memory of prior unity. Second, the identity of the split-brain patient is per function of the noetic temporal monad, which is extant the brain and extant the Bergson-Whitehead amalgam itself: it cannot be split per function of the surgical splitting of the brain. Hence, once formed, the identity cannot be split and abides the split. Finally, and benefiting from the grand decoherence of the future brain to a unitary brain despite the subsequent surgical split, the unitary noetic monad and its non-split identity therein will intercede through the 'gaps in time' to decohere the Whitehead facet pertinent to the split brain into consistency with its former unitary brain and monadic identity: Thus, the unitary identity which, in the first place, was never fated to or affected by the surgical split of the brain, will abide despite the subsequent surgical split.

BOOK-V PART-VI:
FINE-STRUCTURE THEORY OF MIND AND BRAIN

5.82: Aims of Part-VI

De-spatialisation alone undercuts any materialist contention on memory and mind: It constitutes the solid basis for both the general and intermediate mind-world theory. De-spatialisation alone forced conclusions in favour of Cartesian revival by relating memory, mind and consciousness to the brain and world in pure temporal terms, notwithstanding the breakdown of materialism and IOO-schema causality per de-spatialisation and other considerations (see Part-III). For its part, the intermediate theory directly incorporated abstract memory, mind and consciousness into the de-spatialised stereotemporal Bergson-Whitehead amalgam: the replacement to 'spacetime'. It then developed the critical milieu, its attendant Whitehead projection, and the consequent Whitehead equalised possibility plateau; this being the decisive element to mind-brain causality and control. It then furnished the black box **switch-x** solution to Cartesian mind-brain causality and causal freedom and followed this with a plausible theory in solution to the qualitative problem of agency.

While the general and intermediate theories enjoy certitude, courtesy of certain de-spatialisation and temporisation, if nothing else, **the fine-structure theory is far more conjectural**. Its role is to foster pathways to eventual certainty, not to establish it at this juncture. **Through its four conjectures, the aim of fine-structure theory is to explore the various ways in how fine-structure associations, cells and organelles in brains could form the basis for the critical milieu from which the projection of the reversible equalised**

possibility plateau is insinuated into the Whitehead facet, and upon which, and through the 'gaps in time' associated with it, Cartesian mind can impose its 'punch card' upon the future state of its brain, per the intermediate theory, so affecting mind-brain control.

In fine-structure theory, we will not seek to 'locate' mind and consciousness in brains. The non-clockability thesis abides (see Book-V **5.14** and **5.46**, and Book-I **1.51**): In analogical terms, it is no more possible to 'locate' nous in brain fine-structure or associated events than it is possible to identify end-users with their keyboards, or 'locate' and identify physical laws with manifest structures that presuppose laws. Nous is necessarily in ecstasys vis-à-vis the brain and its events: nous cannot be specified or reduced to those events... nor to the expressions and declarations of consciousness that attended by events.

Also, our fine-structure theory will not seek to account for forlorn notions on how consciousness collapses the brain-spanning wavefunction. The decisive findings from the critique of 'consciousness causes collapse', furnished in Part-II **5.16** and **5.18** to **5.19**, undermine quantum idealism and prospective 'quantum mind'. Thus **fine-structure theory correlates with the critique of 'consciousness causes collapse'**: Mind is *not* the wavefunction 'collapsor'. i.e., Mind does not drive or bring about time. Indeed, time and wavefunction collapse are ontologically primary to the very possibility of mind and consciousness, insofar as both ontologically depend on time and the decay of successive events out of the Whiteheadian future potentiality to affect agency in the world and to gain noetic witness to its own agency and events.

Moreover, fine-structure theory will *not* attempt to purport a consciousness-driven secret orchestration of quantum indeterminism in brains. The general and intermediate theories asserted as axiomatic that quantum indeterminism is no more a problem vis-à-vis brains than it is in photography. Indeed, mind-driven orchestrated quantum indeterminism was invalidated in the critique of 'consciousness causes collapse' in Part-II **5.16**. Fine-structure theory abides with this conclusion.

The general and intermediate theories claimed specific assertions on the fact that quantum mechanics (i.e., AND-form logic and the wavefunction collapse processes, or time) apply at all scales, including the 'macro-scale' brain. To assert that it does not apply vis-à-vis brains is tantamount to the assertion that time (i.e., wavefunction collapse, through which the events of the brain come into realisation) does not apply to brains: an absurdity. Of course, myriad other reasons also assert the scale-pertinence of quantum mechanics, most notably grand decoherence (see **5.12**; **5.40** and **5.62** to **5.63**). Thus, the fine-structure theory will also abide with this all-scale conclusion and it will contribute to the growing block quantum mechanisation of the brain through its 'molecules in a box' approach.

*

The first aim of Part-VI will constitute an outline of various fine-structure candidates pertinent to the formation of the critical milieu: It will aggrandise these candidates into **the critical milieu "sandwich"**, centred on the synaptic gap and its flanking structures. This will be followed by two hypotheses: the **convergent structure hypothesis** and the **threshold hypotheses**. Both will seek to explain how the brain could circumvent 'problems' posed by quantum indeterminism. Then, utilising the critical milieu 'sandwich', Part-VI will develop **three distinct fine-structure conjectures in search of the critical milieu in brains**.

However, following a brief critique of Penrose and Hameroff's orchestrated objective reduction, Part-VI will culminate into a **fourth fine-structure conjecture** and an alternative critical milieu 'sandwich': one centred on the collective microtubular-tubulin array: another possible basis for the critical milieu.

Throughout fine-structure theory, no detailed account of the various neurotransmitter receivers or inhibitors or exciters will be presented. This is *not* because these are irrelevant: These are implicit to whatever happens at the critical milieu and the mind-imposed 'punch card' that attends it. The concern with the Cartesian imposition of a complex wavefunction 'instantaneous whole' upon the critical milieu has primacy over any attendant ancillary molecular cell biology and mechanics. Succinctly, **we need to uncover how the critical milieu and the subsequent projection of the equalised plateau happens in brains before we can begin the long process of garnering about subsumed molecular-cell and molecular mechanical processes.** To this end, we cannot start with neurotransmitters and attendant processes, and work our way up. Indeed these lower level factors do not lead into any insights about their larger field: namely, the critical milieu itself, or Whitehead projections, and the equalised possibility plateau thus, or implications from de-spatialisation, temporisation and ephemeralisation, or the nature of the growing block Bergson-Whitehead amalgam that constitutes the basis for the general and intermediate theories. Hence, we must start with the developmentally reversible critical milieu and the collapse of the mind-imposed complex wavefunction upon it, which then garners the selective and organised discharge of the various attendant lower level molecular-cell and molecular mechanics: all of which are subsumed to the loftier general and intermediate theories and to the fine-structure critical milieu that organises these lower-level considerations.

Moreover, throughout fine-structure theory, little will be stated about the functional areas of brains, such as the hippocampus, the motor cortex, etc... briefly summarised in the Preliminary, **0.43**. This is *not* because these are irrelevant. To apply the keyboard analogy, the alpha-numeric pads and functional keys cannot account for or generate the end-user. Similarly, in the context of the brain, our search is for the fine-structure basis for the interface between Cartesian mind with its keyboard-brain (the critical milieu) and the basis for the Whiteheadian equalised possibility plateau. The functional division of the brain presupposes the primacy of the critical milieu and the equalised possibility plateau implicit to it, but does not lead to the latter. **The matter of the various functional areas is subsumed to the grander mind-brain causality and critical milieu, and treated as implicit, *not* the starting point**.

*

A key conundrum of the fine-structure theory is how mind-brain control triggers attendant brain growth and reconnection that then moulds the brain into what the mind requires. We will speculate that a separate equalised possibility plateau within the brain-pertinent Whitehead facet contains the sum of nested futures for all possible alternative brain growth and connection possibilities.

608

This is coupled to the equalised possibility plateau specific to mind's 'punch cards'. Cartesian mind stimulates the decoherence and direction of growth, and attendant actualised state-change in growth and connection possibilities from the welter of such, by decohering the leading 'punch card' plateau, which then decoheres and collapses the equalised possibility plateau for brain growth and state-change into tandem consistent futures. Again, for brevity, there will be no presentation of the neuro-biochemistry associated with any of this. This is not because it is irrelevant, but because it is a subsumed and treated as given or implicit.

Finally, it cannot be sufficiently avowed that **the likely failure of our four fine-structure conjectures to deliver on the goal of 'locating' the critical milieu in brains must not be misconstrued as a vindication of any forlorn materialist notions on mind-brain relations**. Materialism is dead: (See Part-III). De-spatialisation alone unravelled it. **The inevitable consequence of de-spatialisation and other pertinent findings from Alt-R have rendered revived Cartesian dualism the inexorable conclusion.** This is clear from the general and intermediate theories of mind and world, and from the Alt-R background to both. It necessarily follows that we must search for the fine-structure basis for the critical milieu and its equalised possibility plateau, and the basis for the work function zero delayed choice black box **switch-x** mind-brain control, even if our four conjectures prove not to be up to this task. Prospective failure of our task promises to stimulate further discussion and discovery, so leading to solutions to the fine-structure problem of mind-world dualism.

5.83: Fine-Structure Ephemeralisation: "Box of Molecules" and "Brain in the Box": The Fine-Structure Case for the Quantum Mechanical Model of the Brain

The following is an extension of the reformed new cymatics espoused in Book-IV **4.39** and should be considered as an annex to it. To that end, let us envision molecular interactions in a box... of the sort that the reader might come across in a chemistry class or in a molecular biology course: **a Ludwig Boltzmann inspired 'box of molecules':** a classical 'spaceballs' model, wherein molecules are rendered as 'solid objects' and attributed with seamless OR-form trajectories and momenta; rebounding off the walls and from each other in the usual materialistic contact-mediated fashion, with input-operation-output schemas incorporated. At certain velocities and coinciding shape-alignment, the molecular entities will contact, react and breakup into smaller components. Alternatively, they might aggrandise into larger complexes. And the only probabilities that might apply will be those that attend Boltzmann statistics: born out of our ignorance about initial and subsequent conditions and outcomes: with no place for or reference to inherent quantum indeterminacy or objective AND-form possibility-superpositions.

To introduce this classical spatial 'spaceballs' model vis-à-vis brains and its fine-structure processes, we simply insert the same notions into a much enlarged box, but one comprised of 'cellular walls' formed of water-repellent and water-loving phospholipid arrays; networks of tube-like microtubules, and neurotransmitter vesicles and receiver organelles: all incorporated into neuronal boutons, dendrites, post-synaptic membranes and gaps: attended by neurotransmitter discharges whose dynamics are approximated to the above Boltzmann 'box of molecules', culminating into a fine-structure 'brain in a box'.

The moment we incorporate quantum mechanical reality to the 'brain in the box', we find ourselves entangled in all manner of disputes. One party will adhere to the Boltzmann 'spaceballs' model and depict everything in 'classical terms': relegating to irrelevance quantum mechanics by confining it to the 'even smaller'. While the other, often the excluded party, comprised of the likes of the late John Eccles, and of Henry Stapp, and Roger Penrose, will posit that the quantum mechanical must play a critical role in the operation and control of the brain, both at the molecular-cell level, and even at the purported fundamental spacetime-geometric level. Of course, we need not berate the fact that quantum mechanics and AND-form logic *certainly* abide at the macro-scale, the macro-scale brain, and at the molecular, molecular-cell and cellular level... as implicated in previous essays in this work.

It is worth recapitulating the all-scale pertinence of quantum mechanics to the above 'molecules in the box' and 'brains in the box'. This can be garnered immediately from the contrast between the block model and the growing block model. In a block universe, the future is configured into a fully resolved seamless OR-form block, *without* an open future, or *any* future. Consequently, in a block universe, the two-slit experiment could never obtain self-interference, much less its amplification into the staple interference pattern of generic quantum mechanics and physics, as this requires AND-form states and attendant the onticity of the future.

However, in a growing block universe, the future is ontically real, and is constituted as an as-yet to happen potentiality for future possibilities (nested) rendered in AND-form. Only the growing block universe can furnish the possibility for self-interference and interference patterns in generic two-slit experiments. It follows that we do indeed inhabit a growing block universe.

Again, **only the block universe could support the onticity of the classical future-less and memory-less Boltzmann-inspired spaceballs vision of the 'box of molecules' and the attendant 'brain in the box'. Yet, we inhabit a growing block universe: one that certainly furnishes a 'molecules in a box' and 'brains in a box', but *with* a Whiteheadian future-potentiality; *with* an accumulating Bergsonian past, and wherein the past-in-memory decoheres the Whitehead future pertinent to the boxed molecules and brains, and accomplishes this without contiguity or contact through a non-contiguous resonation of information across time, made possible by temporisation compelled by de-spatialisation. Thus, the growing block quantum mechanical depiction of 'molecules in a box' and the 'brain in a box' is recuperated into certainty by de-spatialisation and temporisation; wherein the notion of definite seamless OR-form 'molecular paths and trajectories' unravels, as must molecular contiguity and 'space' itself, and wherein the very entities cease to be spaceball 'points', 'nuggets', or blocks, and are transformed into complex *durata* organised into a growing block system of abstract memory *and* future potentiality... evolving through 'quantum leaps in time'... transitioning from one configuration to a succeeding one *without* motion, without contiguity, without 'impact', and** without 'force exchange'… and wherein energy relations are rendered pleonastic… and the fundamental processes, no longer composed of 'action' or 'work', are recuperated into quantum counterfactuals, decoherence and grand decoherence, accompanied by the mnemonic superposition of past configurations in abstract memory, forced into state-change through a gear-less causality through AND-to-OR collapse and time.

Hence, the Boltzmann 'spaceballs' approach to the 'molecules in the box' and the 'brain on the box' must ephemeralise into a growing block quantum biology of the molecular, of the molecular-cell, and of the cellular… all implicit throughout our fine-structure conjectures.

5.84: Fine-Structure Candidates Pertinent to the Critical Milieu

The intermediate mind-world theory in Part-V posited the developmentally reversible critical milieu and modelled it in terms of the growing block Bergson-Whitehead amalgam, its attendant open Whitehead projection, and its equalised possibility plateau therein. Of course, the critical milieu constitutes the essential attribute of the brain: it renders the brain open to the processes of the superlative Bergson-Whitehead amalgam and Cartesian causality, by projecting the equalised possibility plateau into and within the Whitehead facet. Therefore, the critical milieu ought to be evinced in in-brain fine-structure. **The task of fine-structure theory is to search for the critical milieu in the brain**. But we must first summarise possible fine-structure candidates found in brains, pertinent to the critical milieu. These are summarised in the reiterated depiction in **Fig. 5. 22**.

By reference to **Fig. 5.22**, the brain can be modelled as a wiring diagram made up of neurons, the basic nerve or brain cells. Neurons are connected by means of 'wiring' composed of axons, synaptic shafts and synapses. The latter terminate into boutons which relate to, but are not fully attached to, either neuron cell-body surfaces or to the surfaces of dendrites that extend out from neuronal bodies. The bouton relates to the cell body-surface or dendrite surface through the pre-synaptic gap. It is into this gap that the bouton discharges its neurotransmitters, which are stored within the bouton's synaptic vesicles arranged into vesicular grids. The surface of the neuron that faces the bouton possesses neurotransmitter receptors.

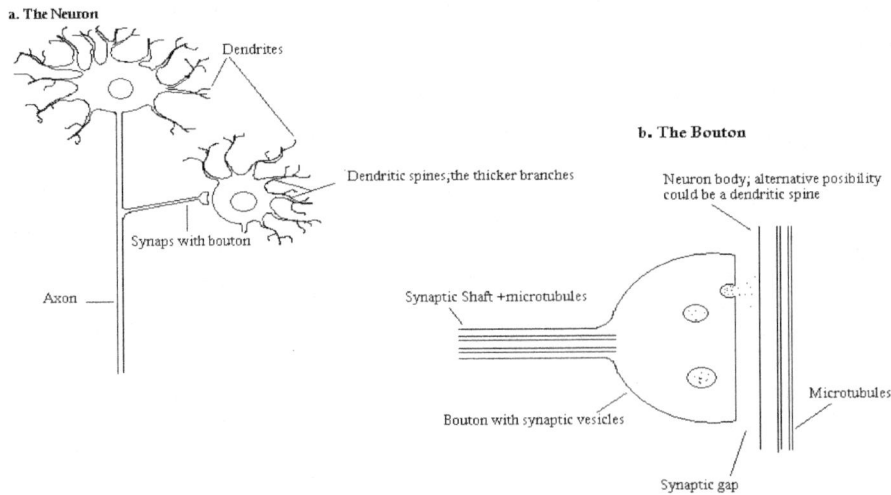

Fig. 5. 22: Neurons, boutons and cytoskeletal structure-candidates of the critical milieu
Notes: Depiction-a illustrates a conglomeration of neurons connected through axons and synapses. **Depiction-b** illustrates the synaptic shaft with its bouton and synaptic vesicles. Our first area of interest will be the bouton, the pre-synaptic membrane side (i.e., the 'inside' facing of the bouton) followed by the synaptic gap itself… followed by the post-synaptic cell surface. The first three conjectures will seek to locate the critical milieu per the synaptic gap and flanking structures. Boutons contain vesicles from which neurotransmitters are discharged into the synaptic gap. These interact with receptors on the 'wall' of the neuron body. The reactions from there evince consequences to the inner regions of the cell and its cytoskeletal microtubule structures. The latter, and the complexes they form, will constitute the basis for the fourth conjecture.

Notwithstanding the ephemeralisation that directly attends the fine structure approach (see **5.83**), the stated structures and candidates are modified by the discharged neurotransmitter molecules which resonate consequences into the inner regions of the neuron cell and interact with the network of cytoskeletal microtubular complexes therein, if not ultimately with the genetic material itself, and with related metabolic processes of the cell. All of these may precipitate the modification of the cell and its external relations: a possible loophole through which mind-brain causality might precipitate structural changes and developments in the cells and up to the larger wiring diagram of the brain as a whole, concordant with the long-term aims of the mind, moulding the brain in accordance with its intents.

The first three fine-structure conjectures will relate the search for the critical milieu to the candidates that surround the extracellular and intercellular synaptic gap and the process of vesicular discharges pertinent to the bouton, to the vesicular grid therein, and to the critical post-synaptic membrane. The fourth and last conjecture will relate the critical milieu to the intracellular cytoskeletal microtubule structures and processes that have also been hypothesised in ORC-OR theory proposed by Penrose and Hameroff. But the case presented throughout will constitute a non-quantum mind Cartesian contention, as was clarified in the intermediate theory (Part-V) and in the introduction.

Whatever the actual form the critical milieu assumes, recall from the intermediate theory that it is via the critical milieu that developmental pattern-formation takes place in the brain, in a fashion similar to the development of a photographic film. The pattern develops out of otherwise objective quantum indeterminate terms that are funnelled into ordained global outcomes by the three-phase developmental process espoused in Book-III: albeit a developmentally reversible process in the case of the brain. Of course, the global 'image' or pattern thus formed is per function of the 'instantaneous whole' complex global wavefunction, or 'punch card', imposed by the Cartesian mind upon part of or whole of the brain's critical milieu (see Part-V **5.72** to **5.73**). The 'punch card' is selected by Cartesian mind from the Whitehead equalised possibility plateau, itself projected into the Whitehead facet per function of the developmentally reversible critical milieu. These, and the Bergson-Whitehead growing block models that attend these, were detailed in the intermediate theory in Part-V and constitute the implicit background to the fine-structure theory. Therefore, we need not recount these in detail within fine-structure theory, and the reader may refer to the intermediate theory for the key details.

*

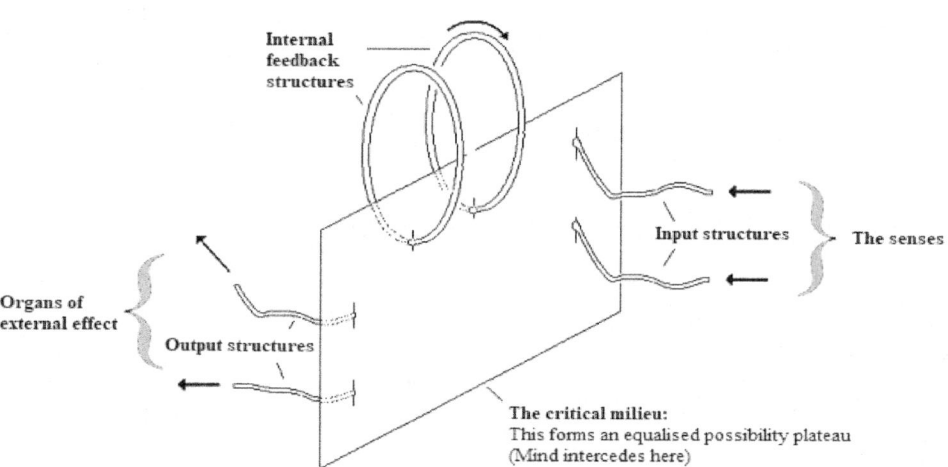

Fig. 5.23: Homing in on the critical milieu
Note: The illustration is a reprise of our previous summary-diagram of the critical milieu prior to its framing into the Bergson-Whitehead amalgam. The critical milieu, as a developmentally reversible state, serves to project an open Whitehead facet into the Whitehead facet, elicits the formation of an equalised possibility plateau therein: a necessary requisite for the causal intercession of Cartesian mind vis-à-vis the brain; one achieved from quantum mechanical intervals or 'gaps in time' that reside between successive event horizons. Our aim in fine-structure theory of mind and world is to found a useful series of conjectures about the 'location' of the critical milieu and of attendant mind-brain intercession in Cartesian causality.

How could the structures depicted in **Fig. 5.22** constitute the critical milieu? Recall the general depiction of the critical milieu from the intermediate theory: This is repeated in **Fig. 5.23**. We will now converge the structures depicted in **Fig. 5.22** with the general scheme of the critical milieu depicted in **Fig. 5.23**. By doing so, we will obtain the complex illustration in **Fig. 5.24**: namely, **the critical milieu 'sandwich'**. This 'sandwich' constitutes the critical milieu rendered in terms of the adumbrated fine-structure candidates, but excludes the implications from cytoskeletal structures: these will re-emerge as central features of the fourth conjecture..

In **Fig. 5.24**, the rendition of the critical milieu furnished is directly pertinent to the leading three anticipated fine-structure conjectures. In the illustration, we take the total number of boutons in the brain and bring these together to form a collective array: namely, the **collective pre-synaptic membrane-side,** denoted as (**B**). We then face this with the **collective synaptic gap** denoted (**A**): namely, the conglomeration of all the synaptic gaps associated with every bouton comprising the pre-synaptic membrane-side.

We then collect all neurotransmitter receivers facing all of the boutons arrayed on the post-synaptic membrane side… or on the surfaces of dendritic spines… to form the **collective post-synaptic membrane side**, denoted (**C**). But (**C**) could also constitute the

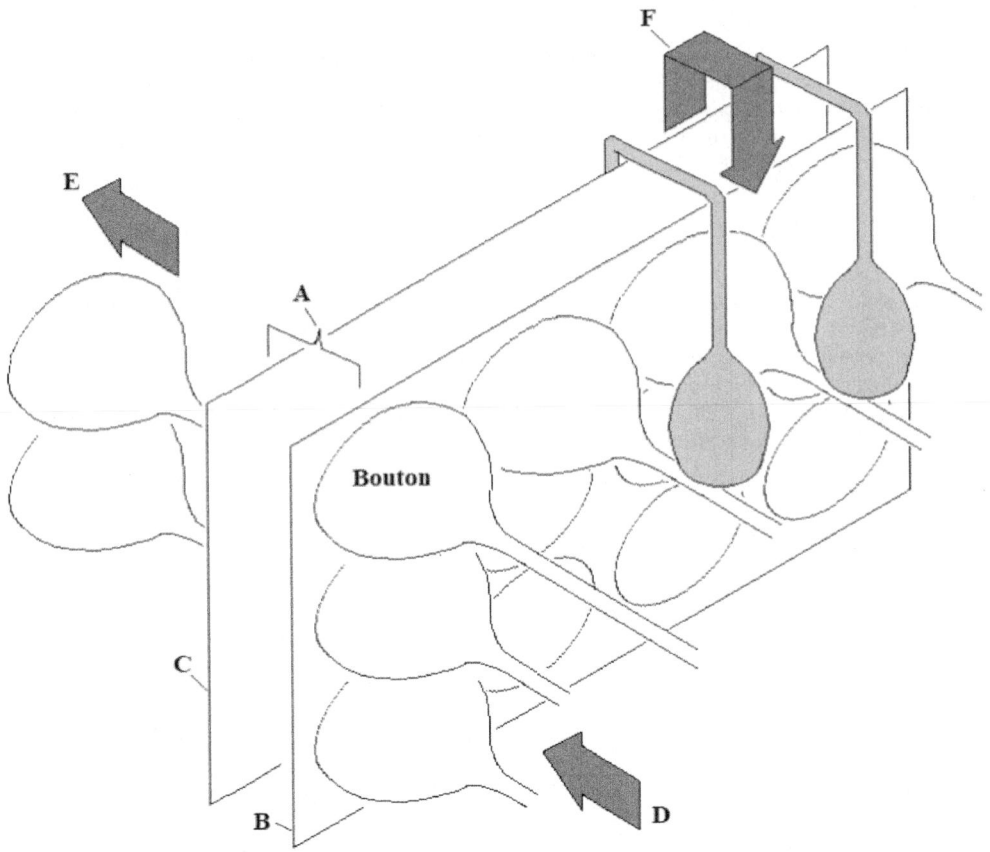

Fig. 5.24: The critical milieu "sandwich"
Notes: The fine-structure critical milieu "sandwich": Note its similarity to the pin-driven punch card mechanism, the two-slit experiment arrangement, if not the photographic chamber. The synaptic gap is the central feature, shown as (**A**). Take all of the boutons and aggregate these to form (**B**): i.e., the **collective pre-synaptic membrane-side**. This faces (**A**): the **collective synaptic gap**, which is in turn flanked by (**C**): the **collective post-synaptic membrane side**, lined with neurotransmitter receptor-collective (not shown). In relation to incoming inputs from the environment (**D**), neurotransmitters within the collective of vesicles in the boutons along (**B**) are discharged into the collective synaptic gap (**A**), toward the collective of receptors that line the collective post-synaptic membrane-side (**C**). Note the feedback loops (**F**) depicted as bouton-dendrite relations (shaded and rounded boutons): These in turn face a **collective post-synaptic dendrite surface** (not shown: essentially similar to (**C**): essentially part of (**C**)). Environmental inputs via the five senses (**D**) input to the collective pre-synaptic membrane side. Upon processing, outputs (**E**) are issued to the environment, or resonate as internal feedback signals (**F**). It is tempting to hypothesise that it is in and through the synaptic gap (the critical milieu) that mind-brain interface occurs: The equivalent of a developmentally reversible photographic image is formed on the post-synaptic membrane side through which mind affects outputs (**E**), including the declarations of consciousness and internal feedback signals (**F**).

collective post-synaptic dendritic membrane side, when dendrites get to face the surfaces of the boutons. When we put these structures together, we obtain the 'sandwich' depicted in **Fig. 5.24**.

Note the uncanny similarity between our critical milieu 'sandwich' and the pin-driven punch card mechanism depicted in Part-V, **5.65: Fig. 5.10**. Also note the parallel with photography: wherein the equivalent of the photographic film in our "sandwich" might be constituted as the collective post-synaptic membrane side (**C**) , with its complex array of neurotransmitter receivers on that surface, placed in lieu of the light-sensitive compounds in photography. Expanding on the analogy, the equivalent to the photographic aperture and inputs would constitute as the collective pre-synaptic membrane side and its bouton array of vesicles (**B**), set either to discharge (open aperture) or inactive (closed aperture) with the neurotransmitters so discharged standing in for the input of photons from the external environment.

The semblance to photography aside, also note that the "sandwich" has a semblance to the arrangement and processes of two-slit experiment, with the processes involved in the brain *essentially* no different than those that unfold in the two-slit experiment. Hence, if we treat the pre-synaptic membrane-side (**B**) of the sandwich as the equivalent of an array of 'emitters' from the a two-slit experiment, albeit with more than one emitter. And if we treat the post-synaptic membrane-side (**C**) as analogous to the detector array in the two slit experiment, we obtain a non-trivial similitude between what happens in and across the collective synaptic gap (**A**) with what happens in the gap formed in the two-slit experiment.

All we then need is the equivalent of the barrier: This would be the ephemeralised punch card imposed or 'slotted' into the collective synaptic gap by the Cartesian mind. As part of the second conjecture, we speculate that the latter acts through or throughout, the collective synaptic gap (**A**) in order to affect the key processes of mind-brain control, culminating into the formation of the required pattern on the collective post synaptic side (**C**). **By imposing its 'punch card', the Cartesian mind accomplishes the equivalent of the modification of the barrier-setting. However, instead of envisioning a 'material barrier' within the collective synaptic gap, we imagine a purely ephemeralised one: i.e., the very same 'instantaneous whole' mnemonic or novel affect global wavefunction, or 'punch card', decohered by active mode mind and 'drawn out' of the Whitehead equalised possibility plateau via the black box switch-x tripping, espoused in Part-V, 5.75 and 5.78.**

Per the non-clockability thesis (see Part-V: **5.78**) the intercession by mind could not be clocked as a set of events in the growing block Bergson-Whitehead system or in the brain, or even in the critical milieu and 'sandwich' retained and potentialised therein.

Also note, in the 'sandwich' depicted in **Fig. 5.24**, the internal feedback loops (**F**) and the input and output structures (**D** and **E**, respectively). Described in generic spatial terms, the boutons that form the pre-synaptic membrane side receive inputs from the environment: Many such inputs attend the input structures (**D**). Information from the external environment is shunted to the collective bouton array and its collective vesicular grid, culminating into patterns of vesicular discharges. Per de-spatialisation, the shunting does not unfold across 'space' but non-contiguously across time, via event horizon successions and per quantum leap. in time

Once shunted to the collective bouton array, the information from the environment combines with and is modified by the implications from the controlling ephemeralised 'punch card' imposed by the Cartesian mind. Outcomes that transpire at the collective pre-synaptic vesicular grid side (B) then cross the whole of the collective synaptic gap. Through the three-phase developmental process (see Book-III 3.26) the 'punch card' delimits and funnels otherwise objective quantum indeterminate vesicular discharges into a globally coherent outcome in conformity to the demands from the environment, now co-joined to the 'punch card' imposed by Cartesian mind via processes that combine two-slit *and* photographic development processes.

Again note, in the 'sandwich', the internal feedback loops (**F**): In **Fig. 5.24**, these are depicted by a looping arrow and as axons and synaptic shafts that originate, *not* from sensory organs, but from within the brain itself: shunting aspects of the combine of environmental information *and* the 'punch card' back to the brain, as inputs originating from the brain... as internal feedback. The said bouton-to-shaft associations are depicted by grey shaded shafts and boutons that face the axons and dendrites that lead into the 'normal' boutons. But such bouton-to-shaft associations, insofar as they entail synaptic gaps, are implicit to the collective synaptic gap (**A**). Even so, for didactic purposes, these are depicted as distinct structures in **Fig. 5.24**.

Note that, *all* of the relations and processes depicted and involved are fully de-spatialised, temporised and ephemeralised: unfolding 'within' a stereotemporal distribution and succession of events and potentials, *without* contiguity or contact causality... whose finer processes obtain the equally de-spatialised, temporised and ephemeralised 'molecules in the box' and 'brain in the box' clarified in 5.83. Hence, the depicted structures and processes should not be misconceived as spatio-material outlays involving 'contact causality' or closed-causality IOO schemas, but as organised successions of events constituted into a growing block stereotemporal order of memory, possibility and time: i.e., the processes of the de-spatialised and ephemeralised Bergson-Whitehead amalgam. (See general and intermediate theories: Part-IV and Part-V.)

5.85: Convergent Path Conjecture & Threshold Theory:
Fine-Structure Circumvention of In-Brain Quantum Indeterminism

In the development of coherent brain activity, ultimately, quantum indeterminism poses no more a 'problem' than it does in the formation of coherent imagery in photography. Quantum indeterminism in brains is not a problem that requires a quantum mind-based solution. Indeed, following on from **3.26** and **3.32** (Book-III) the basis for our assertion was established in the critique of 'consciousness causes collapse' in Book-V Part-II; reiterated in and integrated to the general and intermediate theories in Part-IV and Part-V.

However, for readers not quite satisfied with our Alt-R based position against the 'problem' of quantum indeterminism and quantum mind theory, **there may be plausible banal structural means through which the brain might delimit and circumvent randomness and quantum indeterminism, *and* permit coherent control of brain activity by mind.** One such structure is entertained in the *convergent path hypothesis*.

Quantum indeterminism and probabilistic processes could pose problems in brains, but only in critical milieus segmented into divergent structures. The essential schema of such a divergent structure is depicted in **Fig. 5.25**. Therein, we imagine a critical milieu subdivided into, say, four distinct areas: For example, these could be four distinct boutons, or even four distinct neurotransmitter vesicles, or even four distinct neurotransmitters vis-à-vis divergent receivers. Each area will link up to distinct expressions sought by active mode mind as an end-state. For example...

A = "make tea"   **B = "make coffee"**   **C = "apple juice"**   **D = "orange juice"**

Now, suppose the mind sought to bring about expression **A** ("**make tea**"). In order to do so, it would need to instigate pertinent discharge-events on the critical milieu. But only one of the four depicted areas of the critical milieu can lead to expression **A**.

Let us assume that each of the four areas require just one quantum indeterminate incidence to activate it. Let us also assume that there is a 1-in-4 chance (probability = 0.25) that any one of the four areas on the critical surface might get activated. The implication is that a single quantum incident instigated by mind will end up activating just one of the four stated areas and just one of the expressions. Given that mind has no control or orchestrating power over quantum indeterminate processes, the mind's intended outcome for expression **A** will be uncertain: a matter of 0.25 probability. Hence, over the long run, the mind will get what it wants (tea) generally once per four attempts: Thus, a 1-in-4 coherence of intention-to-effect will be countered and frustrated by a long-run 3-in-4 incoherence of intention-to-effect.

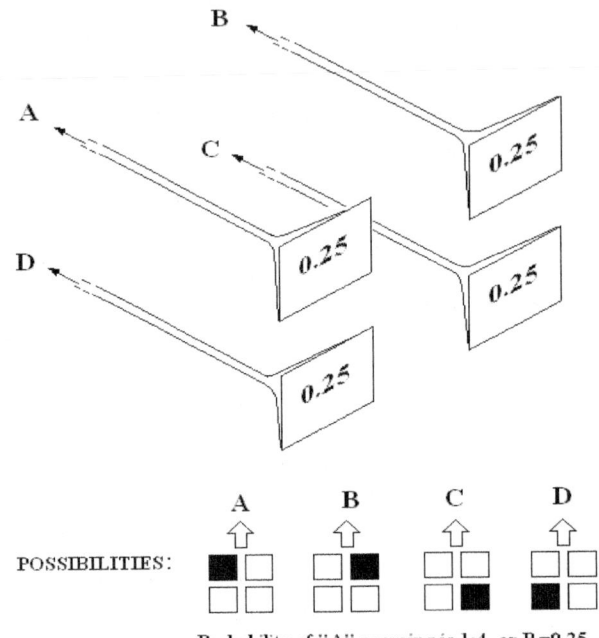

Probability of "A" occurring is 1:4, or P =0.25

Fig. 5.25: Divergent structure and the problem of randomness in mind-brain control
Notes: The problem of quantum indeterminism vis-à-vis mind-brain control can be grasped by the use of the divergent structure depicted above. Here, the critical milieu is split into four distinct areas; each pertinent to a different outcome that might be sought by mind; each with a 1-in-4 (or probability = 0.25) of activation by the mind. We imagine a single quantum incident (or even a corpus of such) that might activate just one of the four areas. The distinct expressions linked to each of the four areas of our critical milieu could be, say, **A** = "**make tea**"; **B** = "**make coffee**"; **C** = "**apple juice**"; and **D** = "**orange juice**". Suppose the mind sought to bring about expression **A** = "**make tea**". To do so, mind imposed a 'punch card' whose wavefunction collapse was subject to the possibility-set depicted at the bottom of the illustration. Unfortunately, the coherence of the outcome versus the mind's intended outcome will be a matter of 0.25 probability, with 3-in-4 incoherence. Mind will only get what it wants from the brain on a 0.25 probability. This level of incoherence simply does not accord with experience.

This will not do: Mind-brain control in real experience enjoys binary control over its brain, world and its intents, *ceteris paribus*. What the mind intends happens in the brain in full concordance with its aims. Thus, real-world mind-brain control is not scrambled by quantum indeterminism, and it is tempting to imagine some secret orchestration of the quantum probabilities by mind, such as to modify probability and indeterminism, or even obviate it.

This much was posited by John Eccles *and* in objective orchestration posed by Penrose and Hameroff. But, these theories are unnecessary, insofar as they seek to obviate indeterminism in brains. In tandem to the overall conclusion that quantum indeterminism in nature or in brains is *not* a problem that needs a solution, there are banal structural solutions that can obviate the problem posed by divergent structures. The problem can be circumvented and corrected by means of convergent structure arrangements. A didactic depiction of a convergent structure is furnished in **Fig. 5.26.**

In the convergent structure, though each of the four areas composing the critical milieu enjoy a quantum indeterminate probability of activation of 0.25, and mind has no control over which of these will get activated, all the areas converge to a one and same expression: i.e., expression-A ("make tea"). Thus, if the mind sought to bring about expression-A, and imposed on the critical milieu the pertinent 'punch card' for that outcome, and only one quantum incident was allowed to resolve on any one of the four areas, the indeterminate nature of the resolution and of the area consequently activated will not unravel into ambiguity in the final outcome, given that *all* **the areas structurally converge to produce the same outcome: i.e., "make tea".** Thus, the mind will get what it wants (outcome-**A**) from its brain, every time it wants it, despite the fact that it might have no control over which of the four areas get activated per function of ineliminable objective quantum indeterminacy.

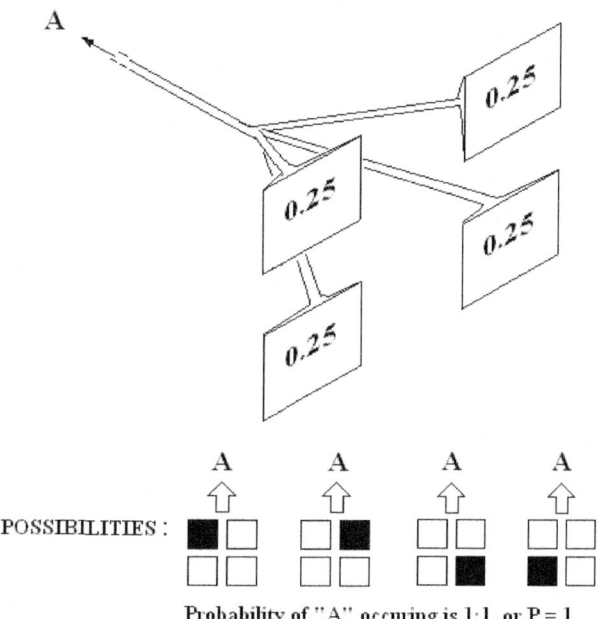

POSSIBILITIES :

Probability of "A" occuring is 1:1, or P = 1

Fig. 5.26: Convergent structure-solution to quantum indeterminism in mind-brain control
Notes: The problem of quantum indeterminism vis-à-vis mind-brain control grasped through the previous divergent structural arrangement can be circumvented by means of the convergent structural arrangement, wherein four areas of the critical milieu, each subject to quantum indeterminate probability of occurrence set at 0.25, all converge to just one expression i.e., **A** = **"make tea.** Thus, if the mind sought to bring about **A**, it merely needs to impose on its critical surface its pertinent 'punch card' wavefunction. While the same uncontrollable quantum indeterminism will apply (depicted at the bottom of the illustration), whichever of the four areas gets activated, given that all converge to produce the same outcome, outcome-**A** is guaranteed to happen: The probability of **A** happening is now secured to 1, despite inherent quantum indeterminism. The coherence of outcomes to mind's actual intentions will be set to absolute certainty, despite the 0.25 quantum indeterminate probability involved in the activation of any of the pertinent four areas. The mind will get what it wants from its brain every time it wants it, despite the fact that it has no control over the quantum indeterminacy that operates within it.

In the example depicted, the mind need not engage in any secret deterministic orchestration to circumvent quantum indeterminate processes, at least in convergent structures. Although, how convergent structures (and the threshold hypothesis that follows it) are stimulated to form up in brains is another question: partially addressed in essay **5.86**.

Convergent structure hypothesis could work in conjunction with the overall conclusion, especially espoused in Book-III, that, as is true in photography and elsewhere, quantum indeterminism is not ultimately a problem that needs a 'solution', and that, through *en masse* quantum indeterminate events, coherent brain activity will arise, if only because the complex wavefunction pertinent to the 'punch card' 'instantaneous whole imposed by mind upon the brain ordains *en masse* quantum indeterminate events, through the three-phase developmental process, to a final culminating globally coherent and unitary whole-brain activity.

We have spoken about 'readiness potentials' before: It appears that, for the brain to execute any action or expression, a build-up to a threshold number of neural discharges must transpire. Also, the brain exhibits randomness, or random discharges of neurons or vesicular discharges. Throughout, we will assume the ineliminability of such randomness as a function of quantum indeterminism across the brain. Since ours is not a quantum mind theory, we need no more attribute to Cartesian nous the status of a secret orchestrator of in-brain quantum indeterminism than the same to some God-like wavefunction collapsor.

Thus, we reiterate the hypothesis that randomness and quantum indeterminacy in the brain is typically controlled by banal convergence structures. We hypothesise that these are developed and embedded into the critical milieu and to the 'sandwich' described in the previous essay.

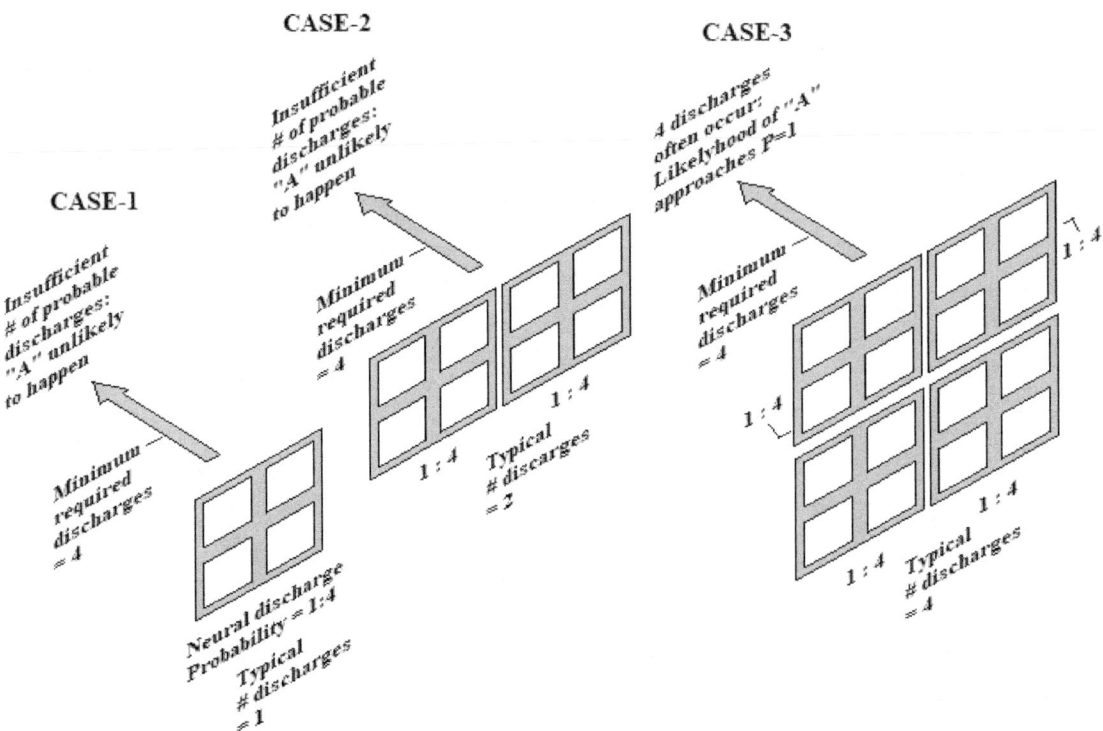

Fig. 5.27: Threshold theory: fine-structure circumvention of quantum indeterminism
Notes: Illustration depicts three cases of convergent structures. The convergent structure was previously described in **Fig. 5.26**. We assume as before that the quantum indeterminate probability of a neural discharge per each convergent structure array, with each array composed of four neurons, is 0.25, or 1 in 4. However, we now add that, for "A" to happen, we need a minimum threshold of 4 neural discharges. In **Case-1**, the typical number of neural discharges obtained at a given moment will equal 1, rarely 4. It follows that "A" is unlikely to happen. In **Case-2**, we cluster two such arrays and converge these to make "A" happen: each array is set to quantum indeterminate 1:4 probability of discharge. Now the number of discharges obtained is typically 2. This is insufficient to overcome the threshold requirement of 4 for "A" to happen. However, by **Case-3**, the combination of four arrays increases the quantum indeterminate likelihood of the typical number of discharges to 4. This is equal to the required threshold. Thus, in **Case-3**, "A" is very likely to happen, but not guaranteed to do so. What happens if we take **Case-3** and add to it additional arrays?

How does the convergent structure approach to the delimitation of randomness and quantum indeterminism apply to readiness potentials and thresholds that incorporate mass neural activity? In **Fig. 5.27** and in **Fig. 5.28**, we depict a series of arrayed convergent structures of increasing complexity: All are variations of our original four-area convergent structure from **Fig. 5.26**, save for the fact that, in the depicted instances in **Fig. 5.27**, "A" is resolved into expression by a minimum threshold of, say, four discharges, not just by one discharge.

In **Fig. 5.27 Case-1**, we have a simplified depiction of our original four-area convergent structure: Part of the critical milieu, the four areas could be comprised of a single neuronal bouton or a whole collectives of boutons, in the form depicted in the "sandwich" in **5.84**, **Fig. 5.24**. The likelihood that any one of the areas or boutons will discharge will be 0.25, or 1 in 4.

Now, in order for the array to make "**A**" happen, let us assume we need a threshold total of four discharges at the required moment. In **Case-1**, since the typical discharge will be just one, and rarely meet the threshold requirement of four, "**A**" is unlikely to happen.

In **Case-2**, we expand the convergent structure to a two-times four-area array. Again, each array is set to a 1-in-4 probability of discharge. This double array will typically obtain two discharges, but rarely four: This is still below the minimum required threshold of four. Thus "**A**" is still not likely to happen.

However, by **Case-3**, we will have a total of four arrays: each still set at 1 in 4 probability of discharge. Yet, at any moment, the number of typical discharges generated will be four: This is equal to the minimum threshold. Consequently, the threshold readiness potential has reached a point at which "**A**" is now likely to happen, though it is not absolutely guaranteed to do so.

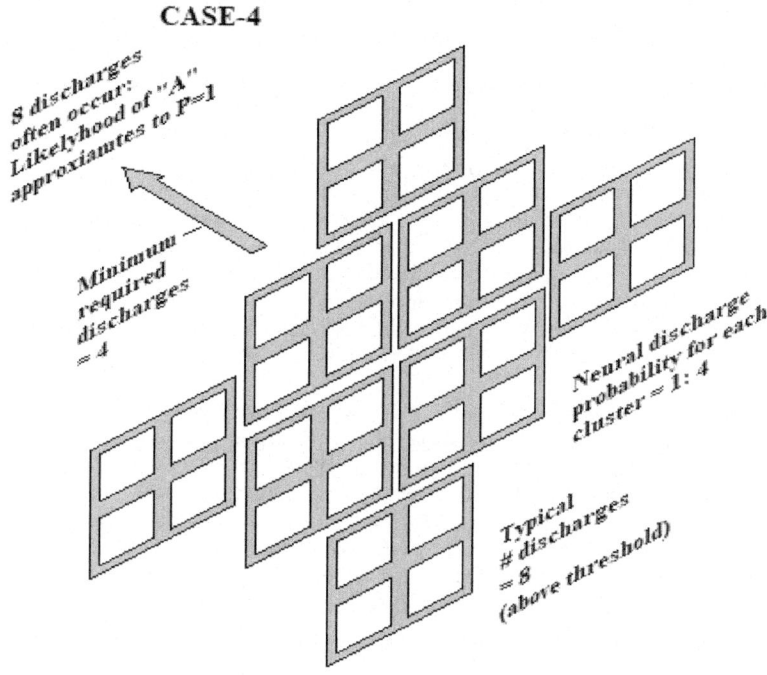

CASE-4

8 discharges often occur: Likelyhood of "A" approxiautes to P≈1

Minimum required discharges = 4

Neural discharge probability for each cluster ≈ 1: 4

Typical # discharges = 8 (above threshold)

Fig. 5.28: Threshold-overcoming of quantum indeterminism through the expanded array
Notes:: The illustration depicts **Case-4**: the total number of arrays have been expanded and clustered to eight: all converge to make "**A**" happen. Again, we assume that the probability of a neural discharge per each array composed of four neurons is 0.25, or 1-in-4. Again, we assume that, for "**A**" to happen, we need a minimum of 4 such discharges. In **Case-4**, the typical number of neural discharges at a given moment has been amplified to 8, which exceeds the threshold requirement of 4. It follows that "**A**" is very likely to happen. There comes a point when, as we expand the cluster with additional arrays, the minimum number of neural discharges that are going to happen at any given time will equal or exceed the minimum threshold of 4, the required threshold. When this happens, "**A**" is guaranteed to happen, despite the fact that the discharge of any neuron is a matter of non-controllable quantum indeterminacy and probability. In order to overcome quantum indeterminism in brains so as to guarantee that "**A**" happens, what the Cartesian mind needs to do is to stimulate the growth, the convergent connection and development of arrays so-clustered to a point at which the minimum discharge secured will equal the required minimum threshold for any operation to happen: In this way, the probability of discharge will be guaranteed to transpire at almost probability = 1. Thus, counter to the view of John Eccles, there is no requirement that mind alter the probability of discrete neural discharges by supposedly modifying the quantum probability of vesicular discharges in individual boutons. However, this fine-structure threshold theory is pertinent and counter to *any* quantum mind approach that deems it necessary for mind to secretly orchestrate or overcome quantum indeterminism, often through some variant of 'consciousness causes collapse': rendered unnecessary through clustering and threshold overcoming.

By **Case-4**, depicted in **Fig. 5.28**, we obtain a total array of eight-times-four. The typical number of discharges at a given moment will now equal eight, which exceeds the minimum threshold of four: Hence, "**A**" is almost absolutely guaranteed to happen. **The assumption here is that, where the convergent structure approach to randomness and quantum indeterminacy is concerned,**

minimal threshold levels are reached in the brain by means of a clustering of convergent structures, such as to bring about an over-threshold. If we further enlarged our array to a **Case-5**, or **Case-n**, there would come a point at which the *minimum* number of discharges at any one time, and *not* merely a typical number of such, will equal or over-exceed the minimum required threshold for "**A**" to happen. Thus, by such banal structural means, quantum indeterminacy in relation to thresholds will be overcome without requiring a secret noetic orchestration of the number or direction of quantum indeterminate events, without requiring the noetic modification of pertinent probabilities, of the form conjectured by John Eccles.

In other words, in the matter of the orchestration of quantum indeterminism in brains, quantum mind theory is not necessary.

*

In order to bring about expression "**A**", the active mind, interceding from the 'gaps in time', and using work function zero black box **switch-x** tripping, selects from its Whiteheadian equalised possibility plateau the appropriate complex wavefunction instantaneous whole. The mind imposes this on the brain on an array of neurons organised to execute "**A**". The imposed complex wavefunction undergoes spontaneous AND-to-OR wavefunction collapse, but this is not made to happen by mind: Recall from **5.18** (and indirectly from **5.16**) that mind is not a wavefunction collapsor.

The probability distributions for events (such as, say, vesicular discharge probabilities vis-à-vis boutons) are inherent to the mind-imposed complex wavefunction or 'punch card'. The bouton vesicular discharges are quantum probabilistic, and there is no possibility for the noetic control or modification of their probabilities for discharge. Instead, the sought 'control' is an effect of the mind-imposed global complex wavefunction rather than a process affected at the level of discrete boutons in the form Eccles espoused. The mind-imposed global wavefunction structures the probability of discharges in the same way that a similar complex wavefunction for photography funnels the probability of photon-deposition upon a film-surface, *without* direct deterministic orchestration.

There is no absolute certainty that any bouton vesicular discharge will transpire, just as there is no guarantee that a photon will deposit at a specific 'address' on the photographic film. But what the mind-imposed wavefunction and its collapse achieves is the guarantee that a required number of discharges *will* take place: It does not matter where or at which boutons these take place, so long as the required number or threshold, and their affects, coincide with convergent structures…or so long as these converge to make "**A**" happen.

5.86: Hypothesis on How Convergent Structures and Thresholds Could be Developed: The Developmentally Reversible Fine-Structure Equalised Possibility Plateau

All of the above presupposes that the brain can be stimulated to form the requisite structures, and that such a stimulation is agitated by the repeated attempt on the part of active mode mind to realise its sought outcomes. From the perspective furnished by the background Bergson-Whitehead amalgam, in the context of Bergsonian memory and background mnemonic mechanics, the number of accumulated failed or successful trials triggered by the active mind progressively decohere and narrow Whiteheadian future potentialities in favour of the sought development and of brain growth into the requisite convergent arrays and thresholds. Thus, we introduce a *distinct* parallel Whitehead equalised possibility plateau: one also projected and potentialised into the future: one complementary to the equalised possibility plateau that attends the mind-brain controlling 'punch cards' noetically imposed: one that AND-form superposes potential developmental processes suited to the realisation of the sought 'punch cards' imposed by mind, or aids the heuristic process of overcoming in-brain structural frustration to realise these via tandem heuristic developmental and growth processes of pertinent convergent structures and thresholds.

Recall that the brain at the fine-structure level is not developmentally fixed or irreversible. Instead, beyond certain exempt structures, the brain is profoundly *developmentally reversible* or plastic. As such, all alternative fine-structure networks, connections and reconnection that the brain could achieve are potentialised into its Whiteheadian future in the form of said *parallel developmentally reversible fine-structure equalised possibility plateau*: itself a grand wavefunction for complex wavefunctions that are specific to the molecular, molecular-genetic, and molecular cell development of fine-structure growth and connective potentials.

When the mind decoheres its 'punch card' to make "A" happen, but this is frustrated by inadequate fine-structure arrangements to facilitate "A", a non-conscious process is unleashed that eventually heuristically decoheres and grand decoheres the said parallel developmentally reversible fine-structure equalised possibility plateau toward requisite potentials that favour the formation of appropriate convergent structures and arrays necessary to make "A" happen. Subsequent AND-to-OR collapse, or a series of such, will develop the required fine-structure connections, convergent structures and thresholds. Thereafter, the mind can impose its punch card for "**A**" without encountering frustration.

5.87: The First Fine-Structure Conjecture

In order to articulate the first fine-structure conjecture, we must dissect the bouton. The series of depictions in **Fig. 5.29** inspects the inner structures and workings of the bouton. It also displays how these inner structures aggregate to constitute the critical collective pre-synaptic membrane-side vis-à-vis the equally important collective synaptic gap, true to the form of the critical milieu "sandwich" (see **5.84, Fig. 5.24**).

We take the familiar bouton and we cross hatch it to grasp its interior structure. Therein, we find the vesicular grid, comprised of the neurotransmitter-containing pre-synaptic vesicles arranged into hexagonal grid patterns; with each vesicle roughly corresponding to the vertex of the hexagons comprising the grid.

618

Let us abstract away the body of the bouton to render salient the circular 'plate' comprising the hexagonal grid vesicles. If we accomplish the same to all available boutons in the brain and then adjunct and co-join all their circular 'plates', we obtain the **collective pre-synaptic vesicular grid**. This collective also constitutes the **pre-synaptic membrane-side** of our critical milieu 'sandwich' previously depicted in **Fig. 5.24**.

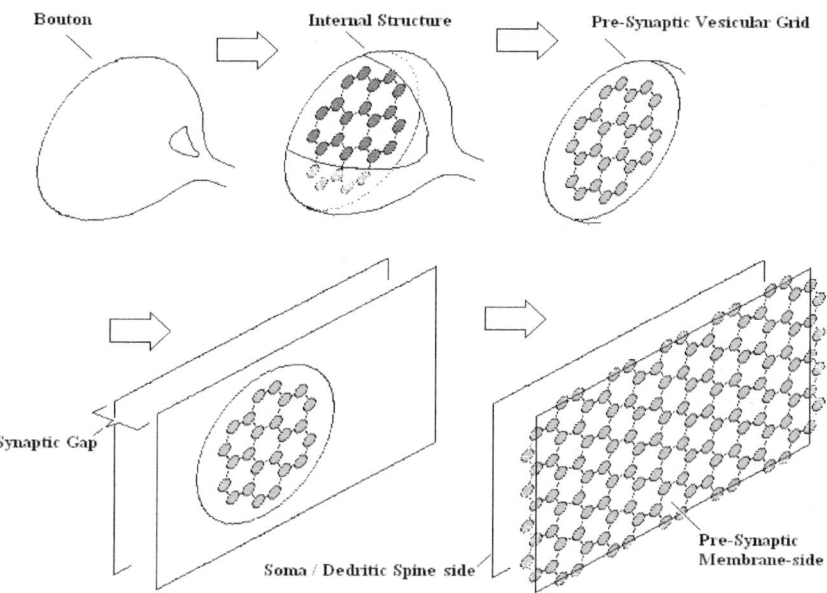

Fig. 5.29: The inner structure of the bouton and the collective pre-synaptic membrane side
Notes: These are depictions of the inner structure of the bouton and how it relates to the collective pre-synaptic membrane-side, to the synaptic gap, and to the conjectured critical milieu 'sandwich'. The top-left depicts the exterior of the bouton, followed by a cross-hatched inner view of the bouton with vesicular neurotransmitter vesicles arranged in a hexagonal grid. The next depiction removes the outer body of the bouton, reducing it to a circular 'plate' of hexagonal pre-synaptic vesicles. We then place a single 'plate' on one of the demarcating surfaces flanking the synaptic gap from the critical milieu 'sandwich'. We then collect all such plates into the collective pre-synaptic vesicular grid, making for a collective *pre-synaptic membrane-side*. The conjecture entails that mind imposes its abstract 'punch card' on the collective pre-synaptic membrane side and the global discharge from this, without requiring fine-control of each vesicle. The development of the pattern implied will unfold per the three-phase developmental process as applies in the Sierpinski triangle, in photography, and across nature: wherein randomness and quantum indeterminism is not inimical to the emergence of coherent outcomes. Mind simply imposes its 'punch card', and the rest takes care of itself, forming up into a coherent global outcome.

We now conjecture that, **by selecting a 'punch card' for the expression of memory recollection or for novel affectation from the equalised possibility plateau embedded to the Whiteheadian future (see Part-IV 5.69), and imposing this on the collective pre-synaptic vesicular grid, the mind asserts its sought abstract control over the subsequent global vesicular discharge-pattern which AND-to-OR collapses into realisation at the collective vesicular grid and at the collective pre-synaptic membrane side.** As stated previously, mind accomplishes this without direct orchestration of quantum indeterminate terms that discharge the neurotransmitter vesicles, notwithstanding the contribution to the same from plausible convergent structures and threshold rules (**5.85**).

The subsequent development of the global pattern of vesicular discharges implied by the mind-imposed 'punch card' will unfold per the three-phase developmental process as found in our old friend, the quantum mechanised Sierpinski triangle, and in photography, and throughout nature: wherein quantum indeterminism is understood *not* to be inimical to, but an aid to the development of globally coherent outcomes (as was espoused in Book-III).

Again, the process of spontaneous AND-to-OR collapse funnels collective vesicular discharges into global coherence *without* further noetic intercession... unless there is a change of mind. Mind imposes the 'punch card' typically on a 'fire and forget' basis. The rest unfolds mindlessly and ineluctably toward the coherent in-brain global outcome sought by mind.

The imposition of the punch card constitutes an ephemeralised process that, as has already been established in the intermediate theory, exploits the structure of the Bergson-Whitehead amalgam, the work function zero basis of wavefunction collapse and time, and the black box delayed choice **switch-x** Cartesian causality instigated from the 'gaps in time' between the event horizons.

Thus, our first fine-structure conjecture asserts that the punch card imposed by the mind on the collective pre-synaptic vesicular grid and on the pre-synaptic membrane side, globally controls the total vesicular discharge, in the fashion that a complex wavefunction pertinent to a photographic image controls the denaturation of light-sensitive compounds arrayed on the surface of a chemical photographic film: a quantum indeterminate process that yet culminates into a global coherent outcome.

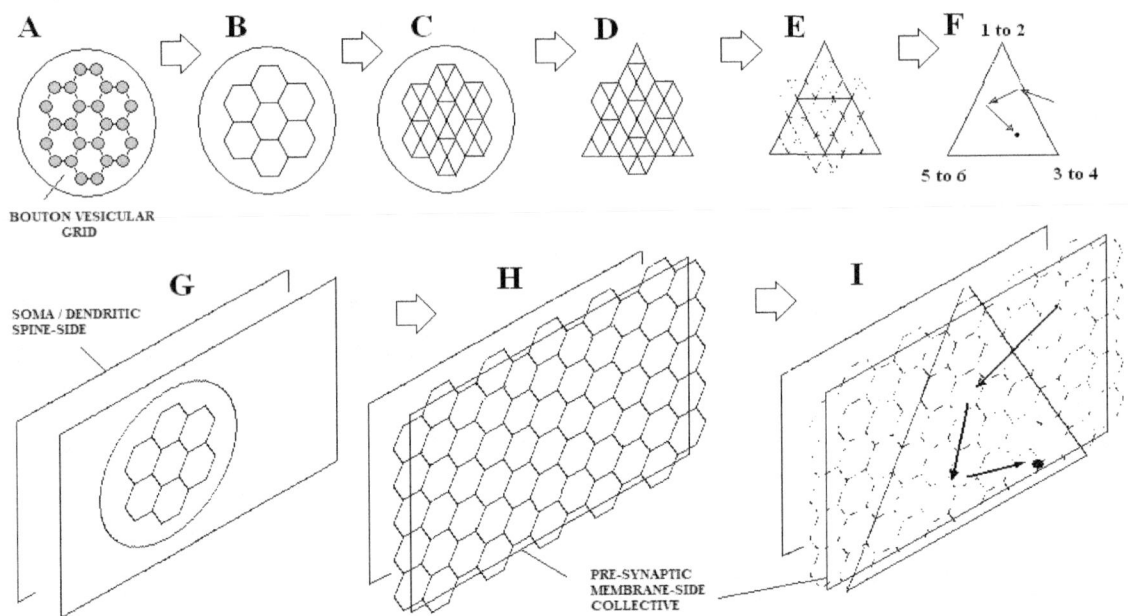

Fig. 5.30: Pre-synaptic vesicular grid and membrane: the first fine-structure conjecture
Notes: The didactic illustrations outline the basis of the first fine-structure conjecture. **Diagram-a** recapitulates the pre-synaptic vesicular grid 'plate' within the bouton. **Diagram-b** depicts the hex-mesh implicit in the 'plate' of vesicles. **Diagram-c** joins all the vertexes of the hexagonal mesh. By **diagram-d**, we arrive at embedded equilateral triangle tiles implied by the joining of the vertexes: intimating the 'triangles within triangles' pattern of the Sierpinski triangle. The Sierpinski triangle is rendered explicit in **diagram-e**. Thus, neurotransmitter discharges from the 'plate' of vesicles within each bouton may be involved in a Sierpinski triangle-like 'chaos game': This is rendered explicit in **diagram-f**. **Diagram-g** reprises the 'sandwich' constituting the critical milieu and embeds just one vesicular grid 'plate' on the pre-synaptic membrane-side. By **diagram-h** we add the total collective of such 'plates' to form the hex-mesh for the collective pre-synaptic membrane-side vesicular grid. Utilising essentially the same reasoning in **diagrams "a" to "f"**, we conjectured that a Sierpinski triangle-like 'chaos game' developmental process is unfolding the discharge of the collective vesicular grid. See **diagram-I**. Our conjecture is that Cartesian mind imposes on the collective vesicular grid a complex wavefunction 'punch card'. This plays a 'chaos game' that then discharges neurotransmitters into the synaptic gap, *without* requiring discrete orchestration of the vesicles. Therein, inevitable randomness will be funnelled by the three-phase developmental process into globally coherent outcome; one concordant with mind's intent.

While it is not asserted that the quantum mechanical Sierpinski triangle necessarily applies in the development of brain activity, and its utilisation in prior diagrams serves only a didactic function, in the case of the discharge of bouton vesicles along the collective pre-synaptic vesicular grid, the very hexagonal structure of the vesicular grid suggests an intriguing affinity to the process of our quantum mechanised Sierpinski triangle. This is didactically incorporated into **Fig. 5.30**, although a similar process unfolds in the development of all wavefunctions, whether future-potential or mnemonic.

One tantalising parallel to the first conjecture is to be found in the visual processing in the hexagonal compounded eye of bees and other insects, and how a similar hexagonal and triangle-based tiling depicted in **Fig. 5.30** might also apply therein. The happenings in the collective pre-synaptic membrane-side and the collective vesicular grid is of the same form as the process that unfolds in the compound eye. Inputs from the environment, as complex wavefunctions in their own right, are implicated in the process of collective vesicular discharge in brains, as are complex wavefunctions pertaining to light in visual processing in compound eyes in bees, if not in the human eye itself. But, to the collective pre-synaptic vesicular grid and membrane side, we must also incorporate the intercession of the mind by means of its 'punch card'. As a complex wavefunction in its own right, the 'punch card' co-joins with the environmental input, also constituted as a complex wavefunction. The combination and super-superposition of these, and the AND-to-OR collapse and attendant three-phase development process combined, then funnels the pattern of collective vesicular discharges into a mind-sought global outcome. All of this is accomplished without the need of secret orchestration of quantum indeterminate terms espoused in various quantum mind theories.

These ideas are implicit to **Fig. 5.30**. Therein, starting with a single platelet of the pre-synaptic vesicular grid inside a bouton, we progress through a series of illustrations that abstract the vesicles themselves; revealing the hexagonal arrangement of the grid within the platelet; and the joining of the vertexes in the vesicular grid platelet. The latter highlights the equilateral triangular tiles that make up the hexagonal grid, which could be generalised to the collective pre-synaptic vesicular grid. For didactic purposes, we then play the familiar Sierpinski triangle 'game' across the pre-synaptic vesicular grid, so making our point salient.

Again, the point is not to assert that the quantum mechanised Sierpinski triangle *itself* has any role to play in the above, but to emphasise that the same processes as holds in the quantum mechanised Sierpinski triangle and in the AND-to-OR development of complex global wavefunctions abides across the pre-synaptic vesicular grid and in the development of the pattern of its discharges.

We reiterate that it is the Cartesian mind that imposes on the collective vesicular grid and membrane-side the global 'punch card' complex wavefunction, whether this pertains to memory or to novel affect. The Bergson-Whitehead model for how this process unfolds was outlined in the intermediate mind-world theory in Part-V.

5.88: The Second Fine-Structure Conjecture

While the first conjecture emphasised the development of the mind-imposed 'punch card', and did so in terms of the collective pre-synaptic vesicular grid *and* the collective vesicular discharge, **the second conjecture shifts focus to what might happen within and across the collective synaptic gap, expressed in terms of the collective of neurotransmitter molecules in their aggregate approach to the collective post-synaptic membrane side and its collective receptor arrays**.

The appropriate depictions pertinent to the second fine-structure conjecture are given in **Fig. 5.31**. Therein, the collective vesicular grid (the structure on the right of the depiction) discharge neurotransmitters into the collective synaptic gap. Per the first conjecture, the discharge is globally controlled by the mind-imposed 'punch card', imposed upon the collective pre-synaptic membrane-side and the collective vesicular grid through the mind-induced decoherence of future possibilities. **In the second conjecture, the discharged neurotransmitters form a collective 'mist' within the collective synaptic gap. The pattern exhibited by this 'mist' is either seeded at the collective pre-synaptic vesicular grid discharge per the first conjecture, or by intercession of the mind to within the synaptic gap: i.e., the defining postulate of the second conjecture**.

As was asserted in the intermediate mind-world theory and in the first fine-structure conjecture, the mind imposes its 'punch card' through an ephemeral process that exploits the structure and processes of the growing block Bergson-Whitehead amalgam, and the 'gaps in time' that reside between successive event horizons pertinent to brains, facilitated by its critical milieu.

According to the second conjecture, the 'punch card' is imposed upon the said 'mist'. The 'mist' of neurotransmitters is not in a vacuum but within a suspension mainly composed of water molecules bounded by van der Val's forces. Even so, the effects of the 'punch card', which is envisaged in the central diagram in **Fig. 5.31** as the didactic developing Sierpinski triangle, funnels otherwise quantum indeterminate processes into ordained predictable global pattern-outcomes, *without* need of the noetic orchestration of quantum indeterminate processes that pertain to the collective neurotransmitter 'mist'; all obtained through the three-phase developmental process (Book-III, **3.26**). The only question remaining is, is the imposed 'punch card' globally affective on the neurotransmitters themselves, or is it affective upon the mostly water suspension and matrix within the synaptic gap? Since all of this is conjectural in any case, one is free to explore the possibilities and alternatives.

Given its semblance to the two-slit experiment, the second fine-structure conjecture can be clarified by reference to that experiment: The mind-imposed punch card, operating within the synaptic gap, can be likened to an ephemeralised barrier, and the effect that the barrier-setup has in the global distribution that finally forms upon the detector-surface. The stand-in for the two-slit experimental detector-surface would constitute the post-synaptic membrane-side and *its* array of neurotransmitter receivers. The brain stand-in for the 'particles' would be the aggregate of neurotransmitter molecules, or the 'mist', formed within the synaptic gap, and perhaps the water molecular milieu in which the mist abides. The stand-in for the emitter would be the array of collective emitters constituted as the collective pre-synaptic vesicular grid: more than one' emitter' or vesicle, in this instance.

As stated, the stand-in for the 'barrier' and its setting would constitute, albeit in abstract and ephemeralised form, the mind imposed 'punch card'. Recall that, upon isolation, two-slit experiment barriers, though seemingly 'material', are actually abstract, temporised, and *ephemeralised*. Recall that de-spatialisation necessitates that we recapitulate the 'materio-spatial' barrier, *not* as a spatial distribution made up of 'stuff', but as a temporal distribution of the realised Bergson-facet, or of the Whitehead facet future-possibilities constituted either as wavefunctions in memory within the Bergson facet, or as wavefunctions in future-potentiality within the Whitehead facet... or as the co-joining of both. Moreover, through grand decoherence, we must project the barrier itself into the Whitehead facet... as a forecast AND-form state, with implied ephemeralisation per the same. We could go further and place many slits or holes into the barrier, and even suspend their open or closed settings per function of delayed choice quantum suspended **switch-x** as found in Wheeler's delayed choice. In doing so, we directly obtain the Whiteheadian equalised possibility plateau.

621

Comparatively, the mind imposed 'punch card' is an abstract ephemeralised possibility wavefunction, selected from and decohered out of the future Whitehead equalised possibility plateau via **switch-x** tripping Cartesian mind-brain causality.

Thus, the treatment of the collective synaptic gap as something similar to the gap between emitter and detector in the two-slit experiment, and the mind-selected 'punch card' imposed within the synaptic gap as the equivalent to the experimental barrier, is not at all out of bounds, especially given de-spatialisation, temporisation and ephemeralisation in both the two-slit experiment *and* in brains.

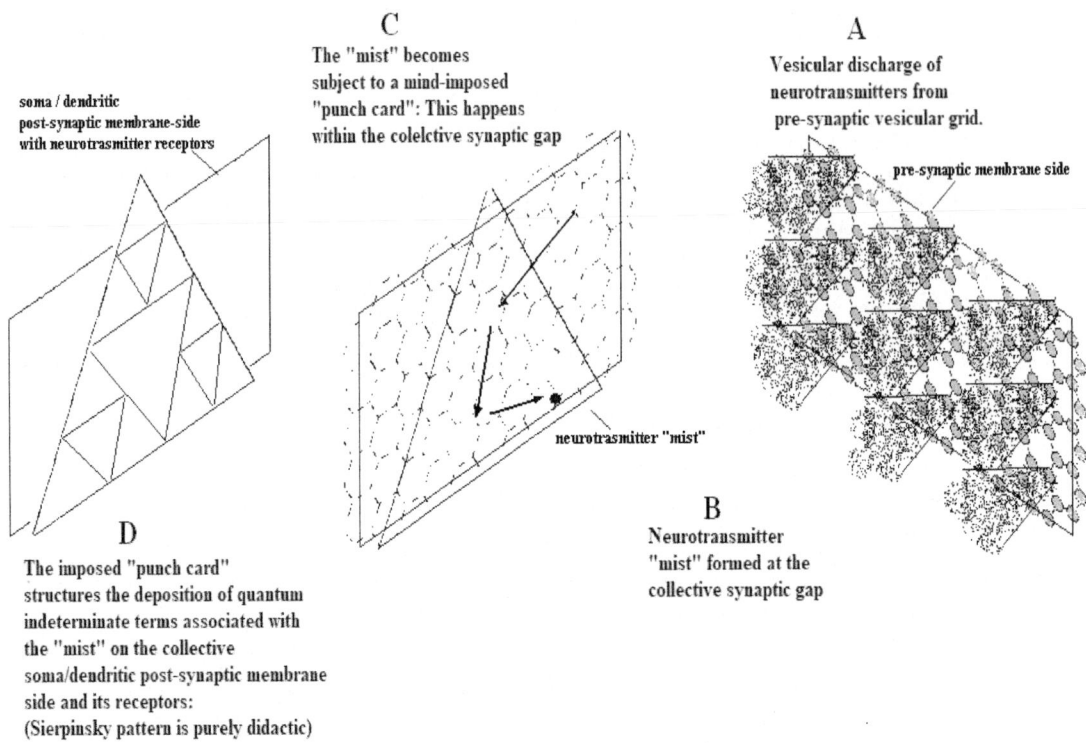

C

The "mist" becomes subject to a mind-imposed "punch card": This happens within the colelctive synaptic gap

A

Vesicular discharge of neurotransmitters from pre-synaptic vesicular grid.

pre-synaptic membrane side

soma / dendritic post-synaptic membrane-side with neurotrasmitter receptors

neurotrasmitter "mist"

D

The imposed "punch card" structures the deposition of quantum indeterminate terms associated with the "mist" on the collective soma/dendritic post-synaptic membrane side and its receptors: (Sierpinsky pattern is purely didactic)

B

Neurotransmitter "mist" formed at the collective synaptic gap

Fig. 5.31: Synaptic gap and the second fine-structure conjecture
Notes: In (**A**) the collective vesicular grid discharges its neurotransmitters into the synaptic gap. In (**B**), the 'mist' is formed within the synaptic gap. In (**C**), the neurotransmitter 'mist' is subject to a mind-imposed 'punch card', didactically envisaged as a developing Sierpinski triangle. (The mind selects its 'punch card' from the Whitehead equalised possibility plateau). Mind need not orchestrate the quantum indeterminism associated with each and all neurotransmitter molecules in the 'mist: These will be channelled into a global pattern of coherence by the 'punch card' and per the three-phase developmental process. By (**D**) the global pattern implied will come into full development at the collective post-synaptic membrane-side, with implications to collective neurotransmitter receptor-behaviour. Note that the second fine-structure conjecture is analogous to the two-slit experiment in key ways: The imposed punch card within the synaptic gap can be likened to the barrier setting. It regulates what happens on the collective post-synaptic membrane-side (the 'detector'). A change of mind would be equivalent to a new mind-imposed alternative 'punch card' on the said 'mist' within the synaptic gap.

Keeping these thoughts in mind, **in the second conjecture, the mind alters the setting of the abstract 'barrier' within and throughout the synaptic gap. Again, the mind accomplishes this through the ephemeralised delayed choice black box switch-x tripping process described in the intermediate theory in essay 5.78**. The punch card 'barrier setting', once imposed, patterns the distribution of neurotransmitters that deposit on the collective post-synaptic membrane-side, as does the barrier vis-à-vis the collective of photon or electron depositions versus the detector in the two-slit experiment. This distribution-pattern reflects the bounded probability structure of the complex wavefunction 'punch card' in line with mind's intents, realised as subsequent global brain activity.

According to the second conjecture, when we change our mind, say, from "cup of tea" to "cup of coffee", we simply change the type of 'punch card' or ephemeralised 'barrier setup' imposed into the synaptic gap. The output culminates into the alternative sought by that change in mind.

5.89: The Third Fine-Structure Conjecture

The first conjecture held that mind-imposed punch cards apply to the pre-synaptic membrane-side: succinctly, on the pattern of neurotransmitter discharges from the collective pre-synaptic vesicular grid (the equivalent to the 'pins' in the punch card, or the emitter in the two slit experiment, albeit exhibiting more than one emitter in this context). Therein, the critical milieu was identified as the pre-synaptic vesicular grid.

On the other hand, the second conjecture asserted that the mind-imposed punch card applies to the post-discharge collective neurotransmitter 'mist' released into the collective synaptic gap. The conjecture assumed an affect similar to the one obtained from the barrier in the two-slit experiment. Thus, the second fine-structure conjecture held that the critical milieu is constituted within the collective synaptic gap and its water-cum-neurotransmitter 'mist': perhaps constituted as the whole of the collective synaptic gap.

The third fine-structure conjecture is depicted in **Fig. 5. 32**. It postulates that, through an ephemeralised process that holds throughout nature, and which exploits foundational processes of the Bergson-Whitehead amalgam and the 'gaps in time', **the mind imposed 'punch card' works its developmental process directly on the collective post-synaptic membrane-side and on its collective array of neurotransmitter receptors: equivalent to the detector surface in the two-slit experiment, or the chemical array on a film in photography. Thus, according to the third conjecture, it is the collective post-synaptic membrane-side and its collective of neurotransmitter receivers that constitutes the critical milieu.**

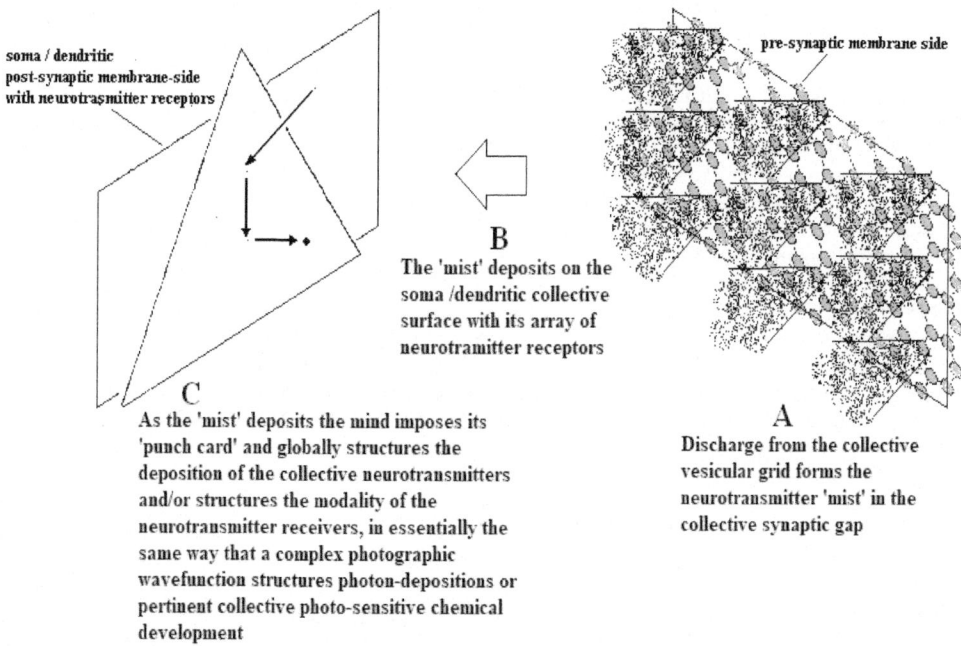

B
The 'mist' deposits on the soma /dendritic collective surface with its array of neurotramitter receptors

soma / dendritic post-synaptic membrane-side with neurotrasmitter receptors

C
As the 'mist' deposits the mind imposes its 'punch card' and globally structures the deposition of the collective neurotransmitters and/or structures the modality of the neurotransmitter receivers, in essentially the same way that a complex photographic wavefunction structures photon-depositions or pertinent collective photo-sensitive chemical development

pre-synaptic membrane side

A
Discharge from the collective vesicular grid forms the neurotransmitter 'mist' in the collective synaptic gap

Fig. 5. 32: What happens at the post-synaptic membrane? The third fine-structure conjecture
Notes: The illustration depicts the third fine-structure conjecture. This holds that the mind-imposed 'punch card' works directly on the collective post-synaptic membrane-side and its collective array of neurotransmitter receptors; perhaps funnelling both the deposition-pattern of neurotransmitters *and* modifying the attribute of the receptors: (**C**). Again, as with all the other conjectures, this would not involve any direct discrete orchestration of individual receptors or neurotransmitters by mind. Each would deposit or activate in a purely quantum indeterminate way: in the same way that light sensitive compounds on a photographic film-surface accomplish it, without secret deterministic orchestration, producing a globally coherent outcome on the collective post-synaptic membrane-surface. This 'materialises' through the three-phase developmental process into the global outcome concordant with the mind-imposed 'punch card'... hence concordant with mind's intent.

There are two possible sub-conjectures here: **The *first sub-conjecture* entails that the mind-imposed punch card funnels the global deposition-pattern of neurotransmitters upon the collective neurotransmitter receptors, and that it is upon this 'surface' that the punch card comes into effect.** This would not require any discrete orchestration of otherwise quantum indeterminate behaviour of the neurotransmitters. Nor is mind required to bring about the collapse of the complex wavefunction that constitutes the 'punch card'.

The *second sub-conjecture* **entails that,** instead of structuring the deposition of the neurotransmitters, **the mind-imposed punch card funnels and patterns the collective activity of neurotransmitter receivers that line the collective post-synaptic membrane, or some key attribute that attends the receivers vis-à-vis the incoming neurotransmitters… or both**. Again, there would be no need for any noetic orchestration of any quantum indeterminate terms, and no need of a noetic wavefunction 'collapsor'. Instead, the global behaviour of the receptor-collective will resolve into a coherent pattern through the expected three-phase developmental process espoused in Book-III, as is found to happens in photography.

Which one of the sub-conjectures is true? A combination of both is likely.

We again draw useful parallels with the two-slit experiment: **The third conjecture treats the collective post-synaptic membrane and its array as if it constituted the detector-surface featured in the two-slit experiment…** or the photographic film, if photography is to be used as a useful parallel… or an electron micrograph, if electron micrography is to constitute the parallel.

Of course, the detector in the two-slit experiment might involve a light-sensitive or electron-sensitive film. Using the latter as a parallel, **the neurotransmitter receivers that line the collective post-synaptic gap are** *ultimately* **likened to light-sensitive or electron-sensitive compounds arrayed on the film detector-surface. The mind-imposed punch card operates on the array of neurotransmitter receptors as might a generic complex wavefunction operate on the photographic or electron micrographic film-surface or detector-surface,** modifying the receivers in the same sense that incoming photons or electrons denature the collective array of light-sensitive compounds on the film. In a similar way, **the neurotransmitter receptors arrayed on the post-synaptic membrane side are collectively made to conform in specific ways to incoming neurotransmitters, generating a global coherent pattern of such,** collective conformity which then generate subsequent patterned ramifications to the inner world of the cells: a matter for our forthcoming fourth fine-structure conjecture.

5.90: Toward the Fourth Fine-Structure Conjecture: Fourth Fine-Structure Candidates

Inspired by the complex behaviour of the singled-celled Paramecium, in their contribution to quantum mind theory, Roger Penrose and Stuart Hameroff sought to locate mind and consciousness in cytoskeletal structures, substructures and attendant processes. While single-celled organisms obviously do not possess brains, cytoskeletal structures remain ubiquitous within their cytoplasmic mediums, as they are also ubiquitous in cells, in neurons, and in associated axons and dendrites in brains in general.

The control processes that attend the Paramecium, perhaps involving rudimentary states of nous, might also be exhibited within and through the cytoskeletal matrix of manifold neurons, with discrete rudimentary mentation coalescing into a larger consciousness.

Penrose and Hameroff postulate the locus from which consciousness manifests to be the tubulin building-blocks that constitute microtubules in cytoskeletal structures. They claim that, via attendant quantum coherent and entangled states, fundamental spacetime geometry collapses, manifesting through the two possible modes of the tubulin *orchestrated objective reduction*, or ORC-OR.

Recall that the Alt-R approach to mind-brain causality does not constitute a quantum mind approach: our revived Cartesian dualism does not assert nous as the collapsor of the brain-attendant quantum mechanical wavefunction or the driver of synonymous time. Nor does Alt-R pose nous as the secret orchestrator of quantum indeterminism in brains. (See Part-II). In contrast, in ORC-OR, Penrose and Hameroff propose an implicit orchestration of quantum indeterminism, obtained at the level and in the modes of tubulin components of microtubular formations.

Since the inception of their theory, Hameroff has proposed a series of candidate structures and processes at which ORC-OR might take place, and at which consciousness might be manifested and located in brains. These candidates purportedly require quantum coherent and entangled states, which are difficult to obtain in hot-body brains. Opponents of ORC-OR have claimed to falsify these proposed candidates, although Hameroff has repeatedly countered these, albeit without conclusive or decisive outcome. Thus, the reality or otherwise of quantum coherent and entangled states in the fine-structure of brains remains unsolved.

However, if the cause behind the search for quantum coherent and entangled states in brains is to render the brain quantum mechanical, then our case for the general application of AND-form logic at all 'scales', including that of the brain, must render Hameroff's search, together with the opposition to his postulates and against the relevance of quantum mechanics to the brain, moot: **The brain** *is* **quantum mechanical, because the quantum mechanical wave is the objective phenomenalisation of the future… of** *any* **system… at** *any* **scale… including the brain:** This is demanded by the growing block universe and the recognition of the quantum mechanical wave and wavefunction as the future-form expression of the growing block system, in turn constituted as the Bergson-Whitehead amalgam, further demanded by the firewall principle per causality intertwined with the principle of conservation (see Book-II **2.10** and **2.11**, and Book-IV **4.10**). If nothing else, essay **5.83** further clarifies the reason for the general pertinence of quantum mechanics to the brain. Indeed, **in its future-potential Whitehead form, the brain** *is* **a quantum coherent whole.**

Thus, the brain possesses an AND-form future at *all* scales. The organisation and logic of that future is AND-form in character, modelled as the wavefunction, to which we incorporate the nested futures view and attendant grand decoherence, all in order to render fully salient the reality of the quantum coherent future-form quantum mechanical brain.

However, the quantum mechanical brain does not automatically or necessarily imply nous as the secret orchestrator of quantum indeterminism in brains, and it cannot support quantum idealism or nous as the 'collapsor' of the wavefunction (see Part-II **5.16** and **5.18**: the critique of 'consciousness causes collapse').

In Part-V, **5.64**, we adumbrated the Alt-R based criticism of ORC-OR. It will not be reiterated in detail here, suffice to reiterate that our Cartesian approach is not a quantum mind theory. De-spatialisation alone engenders serious problems to the notion of generic spacetime geometry as much as it does to the notion of Penrose's fundamental spacetime geometry at the Plank level. What would *any* geometry look like in a de-spatialised purely temporised universe?

Even so, notwithstanding the quantum mechanisation of the brain secured on an alternative growing block basis in Alt-R, **our Cartesianism does not necessarily object to the possibility of quantum coherent and entangled states in tubulin structures or in any other fine-structure candidates proposed by Hameroff** *et al.* **Should such states exist, these could concert harmoniously into our Alt-R non-quantum mind-based growing block Cartesian framework and the ontic reality of the quantum coherent future.**

*

Aside from the Alt-R objection against quantum mind theory in general and the stated doubt about ORC-OR, **our general objection is against the idea that mind or consciousness could be 'located' in the brain, or emerge as a clockable event out of brain-events, even events that involve 'collapses in fundamental spacetime geometry' in quantum coherent and entangled arrangements in tubulin** or in any other fine-structure candidates pertinent to brains. The **non-clockability thesis, a consequence of superlative de-spatialisation, unravels any prospect of a location for nous.** The thesis asserts that neither active mode nous (mind) nor passive mode nous (witnessing consciousness) could be clocked as events of the brain or of the world, or on and along the event horizon within the growing block Bergson-Whitehead amalgam: (see Part-IV **5.50** and **5.51**). Nous must be related to world and brains in purely temporal terms; unavoidable in a de-spatialised temporised universe and brains dispossessed of 'positions' or 'locations'.

The world of memory and possibility, in conjunction with the brain, must emerge to a temporally succeeding moment of passive mode consciousness; the latter characterised as an abstract absolute time-like state: a noetic moment or temporal monad. Consequently, nous does not arise from events. Rather, events, and brain-events specifically, and even collapses of 'fundamental spacetime geometry' … temporally arise to nous, *not from* nous. In any case, the abstract witnessing nous is constituted as an absolute time-like state; part of the en-worlded mind (see Part-IV **5.56** to **5.59**), with nous always in *ecstasys* with respect to the world and brains that are subject to the relativity of simultaneity of events, even if constituted by the putative' collapse of fundamental spacetime geometry'.

As for the active mode mind… it 'acts' from the 'gaps in time' that intersperse successive event horizons. As such, active mode mind is profoundly non-clockable, in the sense that, unlike passive mode consciousness, it could never be confused for recollection-declaration events or expressions correlated with passive mode nous. Thus, again, active mind is no more a clockable event than is passive witnessing consciousness, even if the events should be constituted as 'collapsing fundamental spacetime geometry'.

The point is, there are no events 'locatable' in brains at which or from which nous could issue, reduce to or be rendered identifiable. Specific to ORC-OR, neither mind nor consciousness can be reduced to or identified with in-brain events, even if these should be comprised of 'collapse of fundamental spacetime geometry' at purported quantum entangled or quantum coherent sites.

Yet, the fine-structure candidates designated in ORC-OR *do* constitute viable candidate-structures, but repurposed for our fourth conjecture. To this end it is possible to construct the equivalent of a developmentally reversible critical milieu 'sandwich', and the open Whitehead equalised possibility plateau that attends it, with obvious conjectural implications to the question of 'where' in brains the elusive ephemeralised punch card is imposed by the Cartesian mind.

*

In **Fig. 5.33** we depict the specific fine-structure rendition that *also* attends ORC-OR theory: Note therein how microtubules are constructed out of tubulin. Also note how tubulin is depicted in summary-form in three possible configurations: a pre-resolution AND-form superposition of its two molecular modes; followed by two possible combinatorial configuration-outcomes: These are equivalent to **on** or **off** states or combinations of these… or to **1**, or **0**, or combinations of these… so exhibiting the same form of logic found in neuron action, which either discharge or do not discharge, and could quantum suspended into both possibilities. Thus, we furnish a similar treatment of the molecular-cell tubulin.

To make an immediate connection to our own non-quantum mind Alt-R based Cartesian revival, let us **imagine a 'surface', a critical milieu, approximating to a developmentally reversible photographic film: but one coated with tubulin molecules instead of photosensitive states.** Such a tubulin-lined film would convey a sum of possible future patterns constituted as combinations and quantum AND-form suspensions of alternative forms of tubulin, constituting all possible photo-image equivalent complex wavefunctions that span the whole of the tubulin coated surface. This attendant AND-form sum of future possibilities and patterns would project into the Whitehead facet to form a pertinent *quantum coherent* equalised possibility plateau. **The Cartesian mind, from between successive event horizons, or from the 'gaps in time' would then select and decohere one of the complex wavefunctions made possible by the array of AND-form suspended tubulin, or from out of the equalised possibility plateau, and impose it as an instantaneous whole on the tubulin film-surface.**

In the context of brains, in combination with convergent structures and threshold theory-based solutions (see **5.85**), the mind imposed punch card will resolve otherwise quantum indeterminate tubulin outcomes into ordained *en masse* globally coherent patterns, despite objective quantum indeterminism, without need of noetic orchestration of tubulin outcomes, per function of the global three-phase developmental process furnished in Book-III **3.26**. Therein, it makes not the slightest difference if we include the more complex view of tubulin in terms of plausible quantum coherent or entangled states espoused by Penrose and Hameroff. From the viewpoint furnished by Alt-R, these would also resolve in a purely quantum indeterminate fashion, *without* secret deterministic orchestration by mind or consciousness, even if the process involved implicit 'collapse of fundamental spacetime geometry': A coherent global brain-wide outcome would yet arise.

625

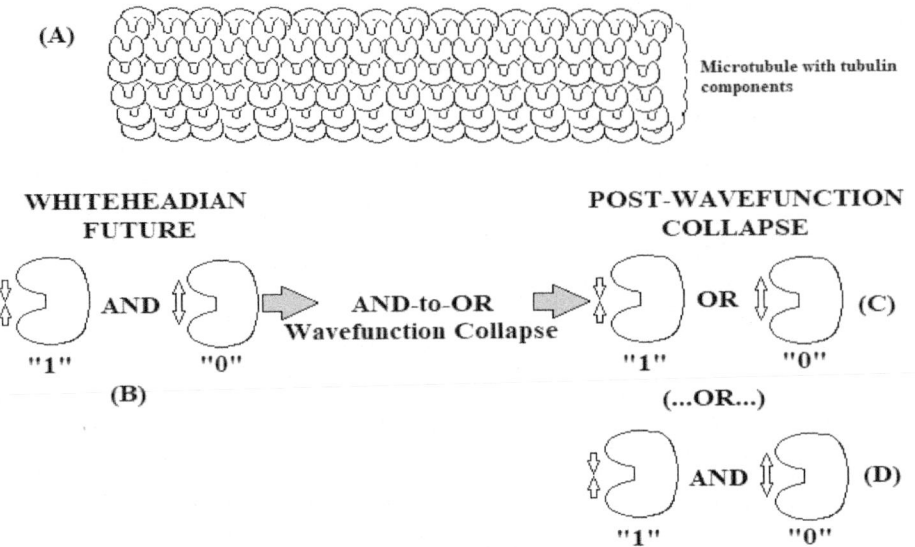

(A) Microtubule with tubulin components

WHITEHEADIAN FUTURE

POST-WAVEFUNCTION COLLAPSE

AND

"1" "0"

AND-to-OR Wavefunction Collapse

OR

"1" "0" **(C)**

(B)

(...OR...)

AND **(D)**

"1" "0"

Fig. 5.33: The three modes of microtubule tubulin
Notes: The illustration depicts three modes of tubulin; OR-form and AND-form. Tubulins are components of microtubules **(A)**. These in turn compose cytoskeletal structures within single celled organisms, cells, and neurons, and are also found within neuronal dendrites and axons (not shown). We do not need to argue for the existence of hot body quantum coherent or entangled states: In a growing block universe, all states have a Whiteheadian AND-form future (a future *always* quantum coherent: all states are quantum mechanical, and Tubulin is not exempt from this. In **(B)** the future possibilities are shown as AND-form superposes of the two alternate future states of tubulin, denoted as "1" AND "0". This superpose undergoes AND-to-OR wavefunction collapse or time, decaying out of its future possibilities into "1" OR "0", as depicted in **(C)**. This decay is arguably quantum indeterminate. In our Alt-R non-quantum mind model, the collapse of tubulin to one of its two states does not require any secret deterministic orchestration by mind, no more that ontologically primary AND-to-OR wavefunction collapse and time requires noetic causality to bring it about. Note that it is possible that some tubulin states will remain unresolved or in delayed AND-to-OR resolution: Non-resolved quantum suspended tubulin will remain in AND-form superposition **(D)** while other tubulins will have collapsed into their OR-form outcomes. Consider that an array of tubulin lining a 'surface', or a critical milieu, would project a Whiteheadian equalised possibility plateau: a nesting hedging wavefunction composed of nested future alternative complex wavefunctions of tubulin-state combinations; each exhibiting the potentiality for a novel pattern of future tubulin-state combinations, or the mnemonic equivalent of the same entailed in memory-recall, as does a complex wavefunction for alternative photographic image-possibilities vis-à-vis given photographic film.

Note that, insofar as it is affected from the 'gaps in time', without identification or reduction to prior or subsequent AND-to-OR resolved events, the active mode mind involved in the imposition of the 'punch card' would constitute a non-clockable work function zero **switch-x** tripping black box process, in the fashion described in Part-V **5.78**. Thereafter, the tubulin coated film-surface (i.e., the critical milieu in brains) would be spontaneously AND-to-OR resolved into the mind-selected outcome per function of objective quantum indeterminate processes. Again, it is tempting to imagine a secret orchestration of these by nous into mind-directed tubulin outcomes. But this is unnecessary. Again, it is not the orchestration of discrete quantum indeterminate outcomes that matter. Recall that quantum indeterminism is not a 'problem' that needs a solution in *any* context, as was asserted in Book-III and in Part-II. Indeed, we have endlessly waxed on how ineliminable quantum indeterminism in photography does not prohibit the formation of pre-ordained coherent photographic image-patterns, no matter how quantum indeterminate the terms. This is not going to change vis-à-vis brains, or even in the context of a 'film' composed of quantum-suspended tubulin arrays, or per the same that constitutes the critical milieu in *any* fine-structure conjecture.

5.91: Fourth Fine-Structure Conjecture & the Critical Milieu: The Microtubular-Tubulin Collective

In the previous essay, we intimated at the fourth conjecture in the context of a photographic film-like structure; one lined with tubulin molecular-cell components instead of light-sensitive compounds. We also pondered the three possible modes that tubulin might assume, both OR-form and AND-form. We must now search for such a structure in the brain: a form of the critical milieu suited to our fourth conjecture.

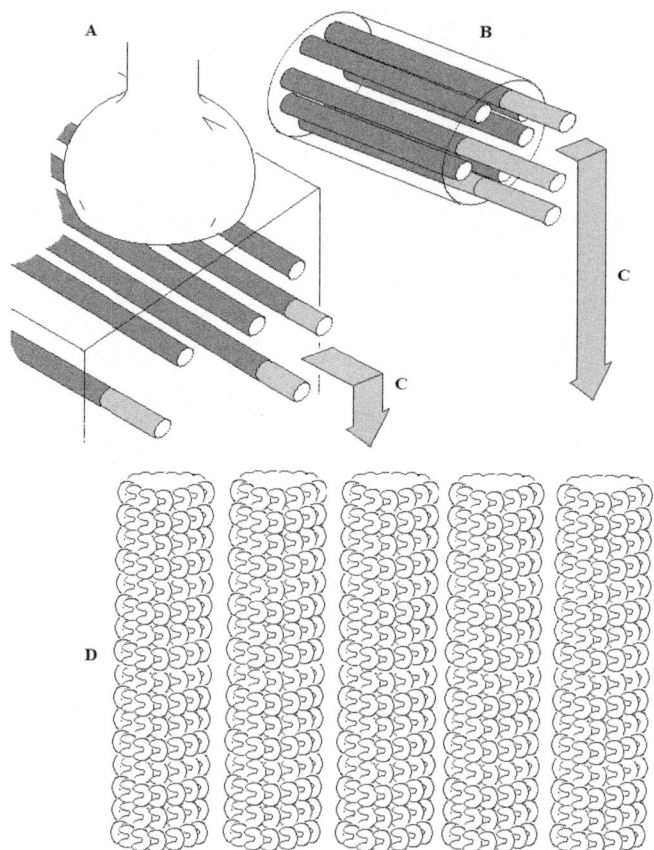

Fig. 5.34: Microtubules in cells and dendrites

Notes: The illustration depicts a step-by-step preliminary process through which we might construct a tubulin-based critical milieu in brains. Microtubules are found 'below' the post synaptic membrane of cells and neurons **(A)**. They are also found in axons and dendrites **(B)**. To form the preliminary to the sought critical milieu, we take all such microtubules from all neurons and attendant structures **(C)** to form a total collective batch of microtubules **(D)**. In the pertinent essays and diagrams, we will process these into the tubulin based critical milieu pertinent to the fourth conjecture.

The illustrations from **Fig. 5.34** and **Fig. 5.35** depict step by step processes through which we might construct a tubulin-based critical milieu in the brain. Per **Fig. 5.34**, microtubules are found 'below' the post synaptic membrane of cells and neurons. They are also found in axons and dendrites. To form the preliminary to our tubulin-based critical milieu, we take all such microtubules from all neurons and attendant structures to form a total collective batch of microtubules. We then 'unzip' each microtubule, opening these up and placing them flat,... such as to constitute a surface composed of arrayed tubulin. Each 'flat' can then be placed adjunct to others to form up into the larger **microtubular-tubulin collective** total (**Fig. 5.35**). The 'surface' so-formed now constitutes the decisive critical milieu upon which Cartesian mind might impose its complex wavefunction 'instantaneous whole' or 'punch card'.

However, in order to form a tubulin-centred version of the collective synaptic gap "sandwich", we must relate our microtubular tubulin collective to two pre-synaptic vesicular grid collectives: one collective mediates the outputs from the patterns formed on the microtubular-tubulin collective (**Fig. 5.36**, boutons attendant **D**) and the other, the pre-synaptic collective, with *its* boutons, mediates environmental and internal feedback upon the microtubular-tubulin collective (Fig. **5.36** B).

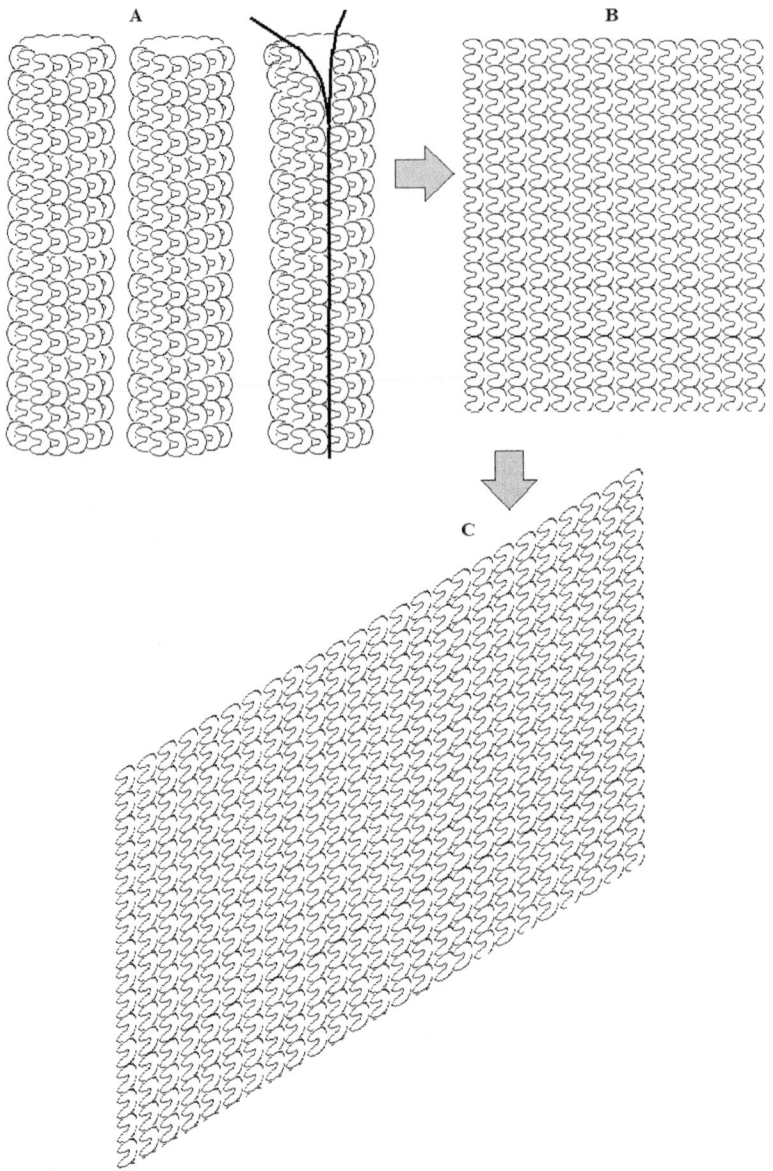

Fig. 5.35: The microtubule-tubulin collective
Notes. The illustration depicts how we might form the central tubulin-based array-centred version of the critical milieu per the fourth conjecture. To form this, we take our collective batch of microtubules removed from *in situ* in neurons and attendant structures **(A)**. We then 'unzip' each microtubule; opening these up and 'placing these flat', so to speak **(B)**, such as to constitute a surface composed of arrayed tubulin. Each of these 'flats' can then be placed adjunct to each other to compose the larger total *microtubular-tubulin collective*: i.e., our critical milieu per the fourth conjecture **(C)**. The Cartesian mind imposes its complex wavefunction or 'punch card' upon such a microtubular-tubulin collective, engendering similar processes as was entertained in the three previous fine-structureconjectures.

The attendant **microtubular-tubulin collective "sandwich"** is depicted in **Fig. 5.36**. Note the internal feedback loop (**E**) mediated from the output-side to the input side. Also note that the post synaptic membrane-side collective and *its* receivers, all pertinent to the third conjecture, are omitted for brevity: The structure is certainly present, but implicitly.

Recall that **the 'punch card' imposed by the mind, as a complex wavefunction 'drawn out' from the future equalised possibility plateau, undergoes AND-to-OR collapse and resolution into the pertinent global pattern composed of the combination of three tubulin modalities. This is 'materialised' on the microtubular-tubulin collective through the usual three-phase developmental process. Similar to how a photographic image forms, the pattern thus formed constitutes a coherent global outcome, despite inherent quantum indeterminate resolution or suspension of each tubulin into one of its three possible modes.**

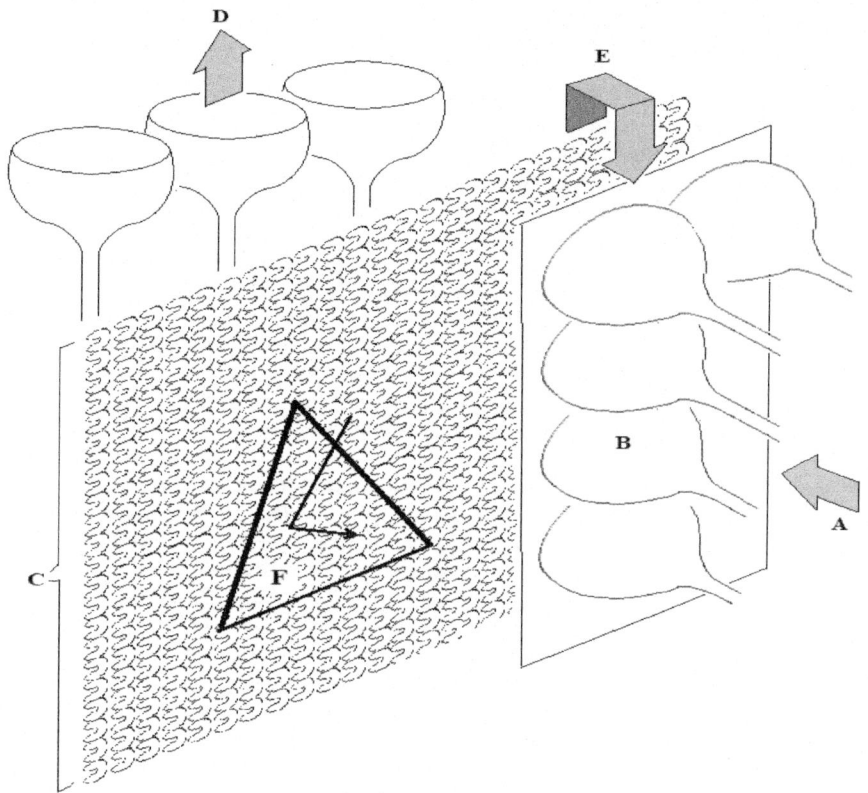

Fig. 5.36 The microtubular-tubulin collective and the fourth conjecture
Notes: The diagram depicts a tubulin array-centred version of our previous critical milieu "sandwich". We relate our microtubular-tubulin collective **(C)** to two pre-synaptic vesicular grid collectives: In **(A)** the pre-synaptic collective, with its boutons and their collective pre-synaptic vesicular grid, mediate environmental and internal feedback inputs to the microtubular-tubulin collective, reflective of inputs from our environment and from the intercession from Cartesian mind to our microtubular-tubulin collective. The other pre-synaptic vesicular grid **(D)** mediates outputs from the patterns formed on the microtubular-tubulin collective to the environment. Note the internal feedback loop **(E)** mediated from the output side to the input side. Also note: the collective post synaptic membrane-side is omitted, but only for brevity: it is present, if implicitly. The Cartesian mind imposes its controlling 'punch card' on the microtubular-tubulin collective array at **(F)**. Note again the didactic use of the Sierpinski 'chaos game' and the implied three-phase development process that funnel otherwise *en masse* quantum indeterminate resolutions of tubulin into globally coherent outcomes, just as similar processes funnel quantum indeterminate activation of light-sensitive compounds into globally coherent patterns in photography.

Thus, **Fig. 5.36** depicts the fourth conjecture in full and didactic form: **Mind's 'punch card' is imposed upon the microtubular-tubulin collective at (F): The resolution of the mind-imposed complex wavefunction is depicted per the usual didactic Sierpinski triangle 'chaos game'** stand-in. As in all previous conjectures, the depiction of the Sierpinski triangle does not imply its literal attendance upon the microtubular-tubulin collective, even though it perfectly captures the fact that quantum indeterminism is not a 'problem' that needs a solution, much less one in need of a deterministic orchestration by nous.

It cannot be sufficiently avowed that, in the fourth conjecture, and in all other conjectures, we have sought the 'location' of the critical milieu, *not* the 'location' of either mind or consciousness. Again, mind and consciousness do not constitute clockable

events. With this earnest reminder, we conclude the series of fine-structure conjectures and, with it, bring to completion our fine-structure theory.

*

Of the four conjectures, which one encapsulates the true process? In real brains, one suspects that all four are employed, although the fourth conjectures might constitute the initial process and initial critical milieu through which Cartesian causality is achieved. Thus, a similar process attendant the first conjecture will likely apply to the input-side pre-synaptic vesicular grid collective, leading to the processes attendant the second conjecture within the collective synaptic gap... in turn leading into the processes attendant the third conjecture on the post synaptic membrane side and *its* array of neurotransmitter receivers... finally cycling back as inputs vis-à-vis the microtubular tubulin collective array. Thus constituting a complete system that integrates all four fine-structure conjectures for Cartesian mind-brain control.

5.92: Summary of Part-VI: The Fine-Structure Mind-World Theory

The summary pertaining to the fine-structure theory will be comparatively brief.

- **The pertinence of quantum mechanics to fine-structure theory**: (5.83): The general, intermediate and fine-structure theories have their own contentions on the quantum mechanisation of brains. No matter what the scale, given the growing block universe and the Bergson-Whitehead amalgam model of it, *all* systems at all scales possess a Whiteheadian future of alternate possibilities in AND-form. Therefore, AND-form logic, hence quantum mechanics at its essential, is *not* confined to 'the very small' and it (hence the objective reality of futures) applies to both the discrete and *en masse* molecular level (i.e., "molecules in a box"), *and* at the fine-structure cellular level (i.e.," the brain in a box") *and* across the macro-scale brain, notwithstanding conclusive arguments for the all-scale relevance of quantum mechanics from grand decoherence alone.

- **Aim of fine-structure theory:** The theory is comprised of *four fine-structure conjectures*. Each seeks to locate the critical milieu in fine-structure molecular and cellular terms. Fine-structure theory lacks the certainty that the general and intermediate theories possess. Even so, fine-structure conjecture constitutes the starting-point for any future fine-structure theory that might succeed. This admission does not obviate the certainty about Cartesian dualism secured in the general and intermediate theories and by background Alt-R: nor does it save materialism from its ready and obvious demise.

- **Fine-structure theory is not quantum mind theory:** The search for the critical milieu in brains is *not* a search for either mind-mediated collapse of quantum wavefunctions, nor is it a search for mind-mediated secret orchestration of quantum indeterminism in brains. In short, fine-structure theory does not fashion for quantum mind theory. Throughout fine-structure theory, the non-clockability thesis of nous abides. Revived Cartesian dualism is the given truism.

- **Mundane structures for the reduction of quantum indeterminism in brains:** (5.85): Ultimately, quantum indeterminism is no more a problem in brains than it is in the development of a coherent photographic image out of quantum indeterminate processes. Minded orchestration of quantum indeterminism is no more necessary in brains than it is in photography, or across the universe, notwithstanding the case against 'consciousness causes collapse' furnished in Part-II. However, fine-structure theory conjectures for mundane in-brain structures that limit the impact of quantum indeterminism and general randomness: These are **convergent structures** that co-join pertinent fine-structure units otherwise subject to probabilistic developmental uncertainty, yet compelled toward a fated culmination. No matter what unit gets indeterminately activated, the expression sought by mind gets to happen per function of the convergence of the activatable units to a same-common and certain end. The other mundane structural obviation of indeterminism may involve the **threshold hypothesis**, wherein a complex of convergent structures transform an otherwise quantum indeterminately uncertain activations into a certain one. This is accomplished through the thresholds. Per the threshold obtained or crossed, massed indeterminate outcomes get to transpire the expression sought by nous. Both convergent structure and threshold hypothesis constitute structural ways of converting uncertainty (ultimately quantum indeterminate) to certainty in brains and in brain-outcomes.

The four fine-structure conjectures and the critical milieu: (5.87 to 5.90): The conjectures hold that complex wavefunctions or instantaneous whole 'punch cards' are imposed by the Cartesian mind upon specific collective fine-structure candidates that might constitute the critical milieu in brains. The pertinent complex wavefunctions then decohere and AND-to-OR collapse the candidate structures and the critical milieu so-formed into complex coherent brain activity attendant to the goals of nous. Whether synaptic-gap centred, or microtubular-tubulin centred (i.e., depending on the fine-structure "sandwich") the critical milieu could be constituted as...

- **The collective pre-synaptic membrane-side** (interchangeable with the pre-synaptic collective vesicular grid-side): see **5.87**
- **The collective synaptic gap**, involving the collective 'mist' of neurotransmitters in the gap: essay **5.88**
- **The collective post-synaptic membrane-side**, involving the array of collective neurotransmitter receivers: **5.89**
- **The collective microtubular-tubulin array**, involving tubulin arrays arranged into a critical milieu 'surface': **5.91**

SUMMARY AND CONCLUSION TO BOOK-V

The Alternative Realist thesis and interpretation is posited as the conclusive template for quantum mechanics: Alt-R exhibits novel conclusions about the following: (**5.01**).

- **The process of AND-to-OR wavefunction collapse is synonymous with time**: Wavefunction collapse *is* time... and time *is* wavefunction collapse. Time is *not* a 'fourth dimension'.

- **The wavefunction pertains to an ontically real future**, insofar as it grapples with a quantum mechanical wave that constitutes the objective future of the growing block universe: Thus the future is ontically real and 'out there', and manifests as the quantum mechanical wave. This is demanded by the intertwine of causality with the putative conservation principle into **the firewall principle**, which wards against 'signals from the future', prohibits the transpiration of the future before it is meant to transpire, and compels that the future necessarily constitute as an AND-form state of not-yet realised possibilities: hence the phenomena of the quantum mechanical wave and the superpose of futures: (also Book-II **2.10** and **2.11**).

- **The wavefunction, hence the future, constitutes a nested futures system**: This is the obvious implication from John Wheeler's delayed choice. The nested futures remains spatially static on the assumption of forlorn 'space'... aside the truism that, obviously, the future cannot move in or across a space. Therefore, the quantum mechanical wave cannot spatially displace. Also, the quantum mechanical wave is purely temporised per de-spatialisation, and frequency measures must completely replace and obviate wavelength measures: (Also Book-II **2.02** and **2.03**)

- **There are no hidden gears or variables behind either quantum indeterminism or time (i.e., wavefunction collapse)**. In the first, it is impossible to eliminate randomness and quantum indeterminacy, even when we resort to putative hidden orchestrating deterministic scripts: The want to devise a secret determinism crashes into infinities. In the second, a gears based approach to wavefunction collapse or time also crashes into an insurmountable non-terminating problem, especially per function of the nested futures framework. It follows that **causality and time are per function of a gear-less structure-less ineffable process** that undermines closed IOO-schema causality. This also implies that the universe is no more a computer or 'simulation' than it is a machine: (Also Book-III **3.32** and Book-II **2.46** and **2.47**).

- **Randomness and quantum indeterminism are objectively real, and quantum indeterminism is *not* a problem that requires a 'solution'**. Systems come into being through random or ultimately quantum indeterminate processes subordinated to a three-phase developmental process. The bounded probability state, the implicit bell curve distribution, or the complex wavefunction, funnel objectively quantum indeterminate outcomes into globally ordained fated pattern-structures: (also Book-III **3.26** and **3.32**).

- **The total future, as the nested futures universal wavefunction, constitutes the Whitehead facet of the growing block Bergson-Whitehead amalgam**. The Whitehead facet is subject to counterfactual causality and **grand decoherence**: the basis for the **revived Mach's principle**: in turn, the basis of the growing block quantum mechanical theory of inertia. The basis for the quantum mechanical theory of the permanent magnet also follows from counterfactual causality and grand decoherence, as does gravitation: (Book-II **2.06**, **2.18** and **2.34** and **2.35**, and Book-IV **4.17**).

- **De-spatialisation** emerged out of the impossibility of the physical-empirical proof of the notion of the 'moving' quantum mechanical wave and its purported 'space', augmented by Alt-R-unique arguments based on Wheeler's delayed choice, employed to directly usurp the notion of the motion of the quantum mechanical wave; followed by the meta-delayed choice approach that usurped space itself. In any case, space turned out to be synonymous with absolute time, and it should have gone to the dustbin with absolute time and the eather in 1905. (Part-I **5.02** and Book-IV **4.01** to **4.07**).

- **De-spatialisation asserts that nature has abstract memory**, or temporally distributed Bergsonian memory. The principle argument for memory is de-spatialisation: With de-spatialisation, 'objects in space' must be recuperated into a system of past events, or *durata*, organised in and by stereotemporal delayed choice time-interval relations, all *without* space. This *is* abstract memory: i.e., information in memory distributed in pure time, extant 'matter in space': succinctly, memory extant any leading event horizon, culminating into the mnemonic Bergson-facet of the growing block Bergson-Whitehead amalgam.

- **The primitive unification of physics** is achieved, first by the intermediate model of spacetime followed by its successor per function of de-spatialisation in the Bergson-Whitehead amalgam. The Bergson facet contains memory in the form of wavefunctions in memory distributed in time: i.e., past wavefunctions with their nested futures cropped or erased. On the other hand, the Whitehead facet is comprised of wavefunctions in potentiality: future wavefunctions with intact nested futures that elaborate to the indefinite future: (see Preliminary **0.29**; further justified throughout Book-II and Book-IV).

- **Inertia and gravitation are memory effects from the Bergson facet,** combined to and augmented by counterfactual causality and grand decoherence implied vis-à-vis the Whitehead facet. The basis of gravity is the superpose of *wavefunctions in memory* retained within the Bergson facet, encapsulated in **mnemonic mechanics**. The memory superpose will distort or 'tilt' the immanent wavefunction in future potentiality that attends the immanent potential event horizon within the Whitehead facet. The AND-to-OR resolution of it will manifest inertia, gravitation, and gravity acceleration: (Book-**IV 4.25** to **4.31**).

631

- **Prospective gravitation per Bergsonian memory effect within the Bergson-Whitehead amalgam designs for prospective quantum gravity,** the successor to General Relativity, accompanied by a higher level of unification of relativistic physics with quantum mechanics. (Book-IV Part-III).

The Alternative Realist thesis and interpretation segues into and is augmented by the case against 'consciousness causes collapse': Overall, **wavefunction collapse is time itself, but consciousness is not the collapsor of the wavefunction: consciousness is *not* the driver of time**: (Part-II). Time is ontologically primary vis-à-vis minded agency and witnessing consciousness, or no events get to transpire, and there is nothing to become conscious of, much less affect through minded agency. Succinctly…

- **As the purported collapsor of the wavefunction, nous would run into an insurmountable non-terminating problem:** In the framework of the nested futures wavefunction, nous would have to deliberate and eliminate non-consistent nested futures, and preserve consistent futures, to the indefinite future, in order to bring about AND-to-OR collapse of possibility into actuality, crashing into a halting problem. Time, counterfactual causality and grand decoherence happen a-noetically spontaneously per function of gear-less causality, which in turn makes both passive mode consciousness *and* active mode mind viable. Nous in lieu of forlorn hidden gears is no more capable of driving time than is a-noetic hidden gears. (**5.18**).

- **Consciousness does not secretly orchestrate quantum indeterminism. Consciousness cannot act in lieu of hidden gears to orchestrate quantum indeterminism.** From the very large number of putative orchestrating deterministic scripts, there is no one exceptional script for nous to select: Any one of the very large number of such scripts is apt to bring about the same global outcome. Putatively, mind can select any one of these. Yet, its 'decision' would be purely arbitrary, or no different from the operation of a mindless random sortology furnished by a fruit-machine vis-à-vis the very large number of equally apt scripts. That is, nous as the secret determinator of the seemingly indeterminate turns out to be as superfluous and as forlorn as a-noetic hidden gears. (**5.16**)

Alt-R also segued into the de-materialisation of physics and the demise of materialism. (Part-III).

- **De-spatialisation alone implies the demise of materialism. There is no space. Therefore, there are no 'locations' in space for putative point-like objects or 'material bodies', much less any spatial 'paths' for these, and even less for contiguous causality.** Particles abide, but only as purely temporised *durata*, in turn structured into pure time relations. (**5.26**).

- **De-spatialisation also implies the demise of contiguous contact causality or impact causality,** undermining the usual schemas for how physical causality is supposedly mediated. Since there is no space, there can be no spatial point-of-contact, or *any* contact, between putative 'particles', much less a communication of causality or 'force' by such means, as required by materialism, and erroneously supposed in contemporary physics and particle physics, despite the infinities and renormalisation that the supposition engenders. (**5.26**).

- **Time and AND-to-OR wavefunction collapse is a process involving a 'quantum leap in time'; one mediated by a gear-less structureless causality that undermines the "operation" term in input-operation-output schemas (IOO) essential to materialist conceptions of causality:** (**5.28**) This is independently augmented by de-spatialisation, which prohibits the possibility of contact causality requisite to the contiguous 'operation' term in IOO schemas: (**5.30**).

- De-spatialisation, gear-less time, combined to the processes of Bergson-Whitehead amalgam, fashions for **the circumvention of infinities in renormalisation** (**5.30**). De-spatialisation also designs for **abstract-immaterial dualistic physical laws** (**5.32**); the **abstraction, as pleonastic, of energy concepts and relations from the ontology of physics** (**5.35**). Superlative de-spatialisation also forces **the reformation of mass itself** (**5.36**) requisite to any viable quantum gravity, wherein mass can no longer be spatial or 'located' at a point or block, and must instead constitute a function of the total Bergson facet upon the future: i.e., a memory effect.

The case against 'consciousness causes collapse' and de-materialisation in physics per Alt-R undermines both idealism (the want to collapse object to subject) and materialism (the want to collapse subject to object). It leads to the revival of Cartesian dualism. A grander ineffable ontology subsumes and existentialises non-illusory mind *and* non-illusory world, yet supersedes mind and world without identification or reduction to either. (Part-IV and V: *The General and Intermediate Theory of Mind-World*). Thus…

- **De-spatialisation corrects against Cartesian presumption of *res extensa* (there is no space) and transforms mind-brain relations into a relation cast in pure time.** De-spatialisation also permanently undermines the want of a 'location' to memories in brains and usurps the search for a spatial locus for the presumed production of mind and consciousness in the brain. Neither memories nor nous is rendered in or produced by the brain, but extant the brain per function of the framework of the de-spatialised growing block model, if nothing else.

- **The a-noetic components of memories in relation to brains, such as the author's recall of 'the big red garage door-1973' (in itself not-conscious) are per function of the Bergson facet** of the Bergson-Whitehead amalgam, with the materialised brain restricted to the event horizon.

632

- **Mind as non-conscious active mode nous acts from the 'gaps in time'** inherent to the succession of event horizons, synonymous with the process of the 'quantum leap in time' and with time itself. Through its acts, it affects outcomes in the brain.

- **Causally passive consciousness transpires temporally *after* Bergsonian events, bearing witnesses to the past temporally *after* the events thus witnessed had transpired. Passive mode nous is a potential of the future** and decays out of its own noetic potentiality in parallel with the decay of world and brain out of the futurity of the Whitehead facet.

- **Mind-brain causality unfolds from within the 'gaps in time', with Cartesian dualism implied.** Succinctly, this involves the minded selection of a complex brain-spanning wavefunction or 'instantaneous whole' from the Whitehead *equalised possibility plateau* of such. The attendant causality involves a non-clockable black-box delayed choice **switch-x** trip; one realised from 'within' said 'gaps in time'. The process is an ephemeralised work function zero process, just as is the process of AND-to-OR time. Indeed, mind depends on work function zero time to affect and witness the world. Mind is no more capable of violating the principle of conservation than is the process of AND-to-OR time.

- **The Conservation principle is recuperated into abstract form per de-spatialisation and from the pleonastic status of energy-work concepts, and in the face of gear-less work function zero AND-to-OR collapse and time.** The pleonastic status of energy abstracts energy out of the principle of conservation, which must now be recuperated in abstract non-energy based growing block form. Thus, memory is conserved and, in tandem, the future is decohered into consistent futures. Cartesian causality cannot alter the past-in memory or usurp the **conservation of memory**, and cannot enforce inconsistent futures vis-à-vis memory. Hence, Cartesian causality cannot violate the conservation principle, and noetic causality over outcomes exploits work function zero time. Hence, mind *is* work function zero process in its own right... again, obviating the materialist abuse of the principle conservation arrayed against Cartesian dualism.

- Radical implications emerge from revived Cartesian dualism. Thus, Bergsonian world and brains subject to the relativity of simultaneity... contrasts with nous defined by absolute time-like characteristics: a temporal noetic moment. This culminates into the **en-worlded mind**, the **unitary collective transpersonal nous**, and the attendant **Anti-Copernican revolution**.

The above were asserted in the **general and intermediate theories of mind-world: (5.31** and **5.81)**. All that remained was the search for a conclusive **fine-structure theory of the critical milieu, responsible for projecting the equalised possibility plateau through which mind gains causal purchase over its brain. (5.82** to **5.90)** without resort to quantum mind theory. Four conjectures were offered in putative fine-structure theory. These do not enjoy the certainty of the general and intermediate theories and of revived Cartesian dualism, all solidly founded on inexorable de-spatialisation and temporisation, amongst other critical discoveries, courtesy of the Alternative Realist philosophy and interpretation of physics, quantum mechanics and nous.

PHILOSOPHICAL SEGUE
TO QUANTUM MECHANICS AND MIND

CONTENTS

EMRE ASENA

PHILOSOPHICAL SEGUE
TO QUANTUM MECHANICS & MIND

6.01: Introduction to the Philosophical & Ontological Segue

The Alternative Realist philosophy and thesis on physics, quantum mechanics and nous culminated into three principal breakthrough areas: each anathema to established materialism and to prevalent necronic mechano-nihilistic culture. These are...

- **The de-materialisation of physics**, which emerged out of de-spatialisation and temporisation; from the case against hidden gears and for work function zero causality; from the ineffable gear-less drive behind wavefunction collapse and time; from the breakdown of input-operation-output schemas and the demise of contiguous causality; from the pleonastic status of energy and work concepts; and from the growing block system that incorporated all of the preceding.

- **Cartesian revival and the Anti-Copernican revolution**. Thus, revived mind-brain dualism and the remarkable insight into the en-worlded mind; all rendered salient most acutely in the general and intermediate mind-world theories in Book-V.

- **The non-arbitrariness thesis and the case for the meaningful universe** from Book-III Part-I, counterthetical to informational and mnemonic nihilism; inimical to general nihilism.

In this philosophical segue secured by the stated breakthrough areas, we will adumbrate on our version of the three principal questions of existence: These are, *the question of aretaic agency*, with respect to which Cartesian revival and mind's causal freedom are indispensable requisites. This will be followed by the *question of meaning*, to which the non-arbitrariness thesis is paramount *and* must presuppose Cartesian aretaic agency. And, finally, *the question of consciousness in memory*.

Our summary resolutions of these questions will segue to broader concerns of *sophrosyne*[109], society, culture and existence, given that any contention about nature, agency and existence must engender implications to said broader concerns. The hegemonic materialist contention, part of world-hegemonic necronic mechano-nihilistic culture, fosters conclusions about human agency and purpose, society and culture permissible upon the doctrine of existence specific to materialist philosophy and ontology. But a counterthetical contention on these matters, based on the de-materialisation of physics, the veracity of Cartesian mind, and the saliency of aretaic agency, and the incorporation of non-arbitrariness and meaning, cannot but foster radically different and opposed conclusions and aspirations, if not engender the emergence of a breakaway *Metic*[110] consciousness, identity, and anticipatory civilisation.

Of course, these larger considerations remain beyond the scope of this work and can only be intimated in passing... as are the more complex articulations of the said three great questions of existence. Indeed, these matters require dedicated books of their own. Hence, in the following philosophic segue, and for brevity, we will address the three great questions of existence and civilisational implications largely in broad summary and nascent form.

6.02: The Question of Aretaic Agency

The answer to the question of aretaic agency, or qualitative agency, requires two precursors: namely, the non-arbitrariness thesis *and* Cartesian revival. The former furnishes the meaningful universe preconditional to subsequent aretaic possibility. The latter furnishes mind's causal freedom necessary to the viability of agency in general, and aretaic agency specifically, in the framework of a meaningful universe.

Non-arbitrariness arose out of the Alternative Realist apagogic critique of anticipatory mnemonic and informational nihilism in Book-III (Part-I). This also included the critique of nihilistic Manyworlds. The apagogic critique segued from the non-arbitrariness thesis into the primary foray into the thesis for the meaningful universe.

Cartesian revival was the culminating aim of Alt-R. De-spatialisation and temporisation, combined with gear-less wavefunction collapse or time, led to the demise of contiguous causality and the usurping of input-operation-output IOO schemas, combined to the saliency of work function zero processes in nature and the pleonastic status of energy or work... all consolidated into the growing block model of the universe. All unravelled materialism. The demise of materialism, co-joined to the demise of the other extremism in idealism via the critique of 'consciousness causes collapse', segued into the revival and vindication of Cartesian mind-brain dualism and Cartesian causality.

Cartesian revival is a necessary but insufficient basis for qualitative agency: The latter requires supplementary causal freedom of mind, wherein nous is neither determined nor haphazard. Yet, having secured the superficial causal freedom of mind in Book-IV **5.78**, nous must also enjoin to a non-arbitrary non-nihilistic world, replete with myriad non-trivial choices and profound consequences: i.e., the requisite qualitative inequalities to the viability of aretaic agency. Hence, the case for co-joining of non-arbitrariness in Book-III *with* Cartesian causal freedom can furnish the viability of aretaic agency: a point first espoused in Book-IV **5.80**.

---

[109] *Sophrosyne*: The meaning of the term is best captured by the term 'calibration' in the light of proof, truth and insight into life, universe and existence.

[110] *Metic*: Classical Greek term for foreigner or stranger: resident of Athens, but formally not a citizen. We adopt the term to mean 'estranged' vis-à-vis the prevalent condition of humankind, regardless municipal citizenship or inherited ethnic-cultural identity.

*

While **5.78** resolved the problem of noetic causal freedom, the problem is less an issue of mind's causal freedom, but more a semantic qualitative conundrum, of a form first encountered in our critique of 'consciousness causes collapse'; specifically in the framework of the problem of consciousness-orchestrated quantum indeterminism in Book-IV, **5.16**. Therein, on the assumption that there exist secret or implicit orchestrating scripts that deterministically direct otherwise seeming quantum indeterminism, it was conjectured that nous in lieu of a-noetic gears might decide and select the operating orchestrating script and, in this way, obviate apparent quantum indeterminism. Yet, this presupposed a qualitative basis for the said minded selection from among the very large number of equally apt scripts. But, given that there is no qualitatively superior or exclusive script to choose from among the equally apt scripts, the pertinent role of nous as qualitative decider was rendered void, and the want of a consciousness-directed quantum indeterminism was rendered superfluent. Hence, noetic agency in the putative orchestration of quantum indeterminism crashed into the **qualitative superficiality problem** of consciousness-directed circumvention of quantum indeterminism.

Beyond the limited purview of quantum indeterminism, in a meaningless nihilistic universe without qualitative differentiation, there could be no qualitative superior choice to affect. All would be rendered nihilistically equalised or levelled. Hence, the qualitative superficially problem would be amplified from the limited purview of quantum indeterminism to the aesthetic, ethical and existential milieu. Hence, the need for a non-arbitrariness thesis to obviate what is now **the generalised qualitative superficiality problem**.

Returning to the purview of quantum indeterminism, given the qualitative equality and interchangeability of all orchestrating scripts, any minded-selection from among these would constitute mere arbitrary decision, degenerating minded agency into mere haphazard 'choice', with mind reduced to arbitrary agent, one no different from a mindless fruit machine in its lieu. Hence, the **arbitrary selection problem** of consciousness-mediated circumvention of quantum indeterminism.

Again, amplified from the limited purview of quantum indeterminism to a general meaningless nihilistic universe, there could be no qualitative superior choice to affect the aesthetic, ethical and existential milieu, and all agency would be rendered effectively random and arbitrary: Again, all would be rendered nihilistically levelled. Hence, **the generalised arbitrary selection** problem would abide: minded agency would be reduced to the status of a fruit machine: a non-agency.

These concerns show that, by itself, the case for Cartesian causality furnished in **5.78** cannot solve either the qualitative superficiality problem or obviate the arbitrary selection problem of agency. This is because noetic causal freedom, or *superficial* freedom, could 'work' equally well in a nihilistic universe as much as in a non-arbitrary meaningful universe of qualitative inequalities, but does not automatically lead to the onticity of qualitative inequalities in the aesthetic, ethical and existential milieu. Hence, the need for a general non-arbitrariness thesis, key foundations to which was furnished in Book-III. It will be recapitulated in its pertinent form in essay **6.03**.

*

If we inhabited a nihilistic universe or, at best, if all possibilities and choices where merely of the superficial kind equivalent to the trivial choice between strawberry versus chocolate ice cream, mind's causal selective freedom over these would be rendered entirely vacuous. Thus, while it could yet possess causal agency, it could not possess or claim real agency, save as arbitrary 'selector' over trivialities. Hence the qualitative superficiality and arbitrary selection problem, again. Agency is rendered viable only when noetic causal freedom attends and instigates profound possibilities, exemplified by the high stake choice between, say, co-operating with the Nazi regime versus resisting it: a decision putatively based on superlatives or transtemporal principles of the Good, or on some variant of either Platonic forms or Kantian categories, or some other philosophy of superlatives and principles, such as to constitute principle-led Cartesian mind; one that renders the ontic reality of qualitative inequalities viable; the necessary requisite to aretaic agency.

However, principle-led nous… is *determined*. The resort to principle seems to imply 'no choice' but what the principle demands or compels. It seemingly implies that we relinquish the 'freedom' (or rather the *arbitrariness*) over the alternatives, and garner toward superlative goals or principles. It appears to imply the *seeming* relinquishing of agency *itself*. In this way, the very basis for mind's agency… namely, *principle*… appears to undermine agency, despite the viability of mind's causal freedom secured in **5.78**.

One might argue that the surrender of agency to principle is what paradoxically makes possible freedom and agency: somewhat in the fashion Kant is purported to have suggested; wherein agency has freedom to choose from among the categories. However, this runs into the problem that the choice from among the categories must either be arbitrary, thus no different from the choice between strawberry versus chocolate ice cream… or it must be compelled by principle, again implying the relinquishing of freedom to principle, despite noetic causal freedom secured in **5.78**. But is mind's choice over the categories trivial or principled? If the former, we crash into both the qualitative superficiality problem *and* the arbitrary selection problem: thus, wither principle and wither aretaic agency. If the latter, then agency must again be relinquished to principle: wither causal freedom.

Three plausible solutions to the problem of aretaic agency can be postulated. **The first solution is that, one ought not to care that one is compelled by principle: One would rather be determined by principle *without* agency than acquire 'agency' *without* principle**, given that the latter could hardly constitute *any* agency per function of arbitrariness and vacuity. Principle-based agency without choice is *ironically* the 'preferable' choice: It is 'preferable' to be determined to be good, hence good but determined… even though we would have no choice in the matter to found such a 'preference'… versus superficially and arbitrarily 'free', but *without* principle… or 'free' as a fruit machine… hence, ultimately *unfree*.

The second solution requires that we understand that the 'problem' is a fallacious one, except that we most certainly need an ontology of *principle*, furnished as part of general non-arbitrariness. That aside, **the fallacy derives from the want of a 'formula' for moral decision making; an 'algorithm' for moral choice; a mechanism… a 'hidden gears' for qualitative agency**. Yet, as was discovered in our unravelling of noetic causality (**5.78**) *and* of the ultimate character of causality behind AND-to-OR collapse and time

638

itself, causality *en toto* defies rendition in terms of any tick-tock click-clock gears, algorithms or formalisms. This must also abide with respect to noetic qualitative agency vis-à-vis qualitative outcomes. In other words, we are making the mistake of looking for a 'gears' behind qualitative choice, aretaic agency and aretaic freedom. But aretaic agency must remain as gear-less and ineffable as gear-less ineffable general causality.

The third solution... the decisive solution... involves a unique combination and supersession of the two furnished above and, in the process, appears to solve the problem of qualitative agency: **Aretaic causal agency and causal freedom must involve choice between principle versus arbitrariness. Nous recognises the difference between these and can conscry and perpetually update a spectrum bounded by arbitrariness on the one end versus principle on the other. The choice in qualitative noetic agency is *not* between the Kantian categorises or Platonic forms as such, but between category or form at one extreme versus non-category and non-form at the other... clarified by the fact that we can conscry an imperative in, say, not collaboration with the Nazi regime, but cannot distil *any* imperative in the trivial-arbitrary choice between strawberry versus chocolate ice cream.**

The quality of agency, or lack thereof, is revealed in its tendency to one or the other end of the said spectrum. Agency tends either toward arbitrariness and nihilism or toward principle and substance. Thus, excellence is realised in the tendency toward principle, *away* from arbitrariness, triviality, vacuity, inanity, or nihilism... securing prospective substantive or even profound agency, and, with it, aretaic freedom and agency... secured through gear-less and ineffable Cartesian means.

However, all of this presupposes the real ontology of principle in the fashion of Platonic forms or Kantian categories, or a universe of qualitative inequalities per variations of the same; supported by our non-arbitrariness thesis (Book -III, **3.14** to **3.15**), the attendant meaningful universe, combined to Cartesian causality secured in **5.78**.

6.03: The Question of Meaning, Purpose & Being: Overcoming the Hydra of Nihilism

The non-arbitrariness thesis secured in Book-III Part-I, combined to the solutions to randomness and quantum indeterminacy secured in the same, segued into the contention for the meaningful universe and, with it, the viability of aretaic agency and mind's causal freedom therein, the latter independently supported by the onticity of nous furnished by the revival of Cartesian dualism.

However, the non-arbitrariness thesis does not constitute a substantive aretaic theory of universe and existence, although the former inexorably frameworks for the latter. The principal utility of non-arbitrariness is that it acts as a firewall against *process nihilism* in its mnemonic, informational and anti-aretaic forms. It does so by permanently robbing process nihilism of its three tropes.

The first of these nihilistic tropes consists in the **appeal to contingency and happenstance**, or randomness and accident. The second, but lesser-known appeal of nihilism, specifically evinced in Manyworlds and in conjectural multiverse cosmogenesis, is the **appeal to brute force exhaustive process.** Both appeals seek to obviate meaningful and purposeful sortology and causality and espouse a nihilistic ontology of possibilities, against principle and purpose.

The third most insidious and profound *inadvertent* nihilistic appeal is the oft religious resort to God-consciousness as primary ontology, in lieu of both random contingency *and* a-noetic brute-force exhaustive generation and sortology of possibilities and principles, *and* the *ex nihilo* (nihilistic) creation of possibilities, principles and purpose. This is the **appeal to Yahweh**, whose ultimate nihilistic essence was grasped in Plato's dialog, the *Euthyphro*, augmented in the recuperation of Euthyphro's dilemma in our *qualitative superficiality problem* and the *arbitrary selection problem*; both of which inexorably arise in posing God in lieu of contingency and brute-force exhaustive causality in the processing of possibility and principles and in the *ex nihilo* creation of both.

The core deficiency of the three nihilistic appeals consists in the fact that they cannot explain or furnish the ontology of the very possibilities they apagogically espouse, presuppose, or otherwise claim to annihilate, and even less the ontology of principles or laws that are supposedly generated by 'accident' or else by resort to exhaustive means... or by resort to 'God'. Analogically, just as the light bulb cannot furnish the ontology of light and electromagnetic phenomena, much less furnish the governing principles of such phenomena, the resort to 'accident', or to brute force exhaustion, or to 'God' in lieu of the metaphorical lightbulb, cannot explain nature, the laws governing it, or its ontology and reason for being.

*

Nihilism seems to come in a variety of flavours: a world-spanning Hydra. In addition to the core three appeals or tropes of process nihilism, there are forms of *qualitative nihilisms* that we must also note in passing. For example, *Continental nihilism* prevails largely in the European and Anglo-Saxon world, exemplified by absurdism, pessimism and other views. The second is *Brahmanic nihilism*, inherent to the Indian subcontinent and based on ancient notions of illusion or *maya*. More will be said about Brahmanic nihilism at the conclusion to this essay, save to state that Brahmanic nihilism is almost essentially interchangeable with *idealistic nihilism*, or nihilism born out of the supposed ontological primacy of nous, and the world as the illusion of nous. This was usurped by the Alt-R critique of idealistic 'consciousness causes collapse' in Book-V Part-II, if not independently by the case for non-arbitrariness in Book-III.

Continental nihilism, such as absurdism, anti-foundationalism, pessimism, etc... tend to be facile and *always* apagogically false. Absurdity, futility, happenstance, and anomy, *always* presuppose the existence of a meaningful universe of otherwise minimal content. Otherwise, such claims cannot abide *as* content in their own right, nor posit the opposites they claim to annihilate. It cannot be sufficiently avowed that the contention for the meaningful universe is *not* for an exclusively euphoric happy daz, but entails a broad spectrum of possibilities that also include absurdity, anomy, futility, terror, loss... and death. As existentially legitimate concerns, none of these could exist or be reported and in the midst of an ontic 'flat line' or content-less universe. On an all-or-nothing basis, the Continental notion of a meaningless content-less universe must unravel. Simply put, Continental nihilism is apagogically *un-serious*.

639

Another form of nihilism emerged out of the White Sea world[111] as a culmination of nominalism, mechanism, and other trends, leading to or converging into late ontological materialism, with the latter foisting itself upon early modern physicalism, gaining succour from the facile aspects of the latter. Hence, the *necronic mechano-nihilistic tendency*, now hegemonic across the world.

From the Preliminary through to Book-V, our work set to overturn ontological materialism central to necronic mechano-nihilism. Indeed, the arguments for de-spatialisation, temporisation, the breakdown of contiguous causality, the usurping of input-operation-output schemas and of closed ontology... and the recovery of the object-subject relation via the recovery of nous and Cartesian dualism, and of the causal freedom of mind requisite to aretaic agency... all of this rendered materialism untenable *and* served to disentangle physicalism from materialism, evolving into de-materialised post-materialist ontology, naturalism and science.

From the White Sea world, we also inherited ancient Heraclitan and Parmenidean forms of *temporal nihilism*. In Heraclitus, all is change or flux; all is impermanent: The river one moment is not the same river at the succeeding moment. Indeed, per the 'all or nothing' problem of nihilism (Book-III **3.11**) in the Heraclitan universe, there cannot be a fixed identity called 'the river', or *any* identity or information, much less any grounds for prospective process philosophy: (See Book IV **4.12**).

The Heraclitan claim is apagogically false, given that, if it were correct, on the all-or-nothing basis, there could not be *any* identities or principles that abide in time, even less their resonation to us as enduring or perduring beings, and even less our recognition of these; assuming witnessing nous itself could be viable in a Heraclitan absolute flux or the absence of *any* enduring and perduring being. Obviously, in an Heraclitan universe, causality and agency would be rendered totally void. Indeed, Heraclitan temporal informational nihilism fashions for obvious tandem memory erasure (see Book-III **3.06**).

Thus, in a Heraclitan universe, we could not speak about or witness 'the river' at all, or apprehend or declare about anything else: The 'river' could not resonate in time. This last criticism helps recapitulate the central dishonesty at the heart of all forms of nihilism; namely, the *backchannel problem*: As a claim about world and ontology, nihilism must espouse and presuppose the very things it claims to deny or annihilate. Hence, Heraclitan nihilism must entail dishonest backchannel performance to a meta-semantic hypostasis composed of transtemporal identities and ordinances, including the 'river' itself, whose identity perdures regardless of what the ancient Heraclitus recommended.

In short Heraclitan temporal nihilism, or the abuse of time to assert nihilism through impermanence, is apagogically untenable.

For his part, the ancient Parmenides espoused that all is permanence, and that change and time are illusory. The obvious contradiction with experience and witnessing consciousness aside, the Parmenidean universe inadvertently constitutes a block model. Since there is no time, there is no as-yet not-realised futurity of possibilities, potentialities or alternatives. Indeed, the 'future' is fully pre-resolved and, as an open non-resolved potentiality, it could not exist.

Parmenidean nihilism immediately culminates into the nihilism of agency: there is no venue for agency owing to the claim that there are no future-alternatives for agency to enact, assuming that the onticity of agency is at all viable in a timeless universe dispossessed of any possibility for state-change. In the universe of Parmenides, there is no nous, and even less ontic agency whose viability presupposes aretaic choice and the minded induction of state-change.

Of course, per Alt-R, the block universe is false: only the growing block model abides. This is obvious from the very viability of both quantum mechanics and the two-slit experiment, and its interference pattern; possible only in a growing block universe, but not in a Parmenidean timeless block. It follows that future-potentiality is real, and that time itself, synonymous with AND-to-OR wavefunction collapse, is *also* real. Therefore, change and state-change is real. Therefore, noetic agency is, at the very least, plausible.

Simply put, the Parmenidean *block nihilism* is as apagogically untenable as Heraclitan temporal nihilism.

Our contention is that both permanence and change must abide and subsume to a superior ontology; one beyond but incorporating both permanence of timeless principle *and* change. Yet, the moderate attempt to unify permanence and change in Plato's metaphysics was 'rejected' for over two-millennia; perhaps even denounced 'fascistic'.

*

Heraclitus and Parmenides notwithstanding, and the rest of the hydra of nihilism notwithstanding, we must now recapitulate triadic process nihilism, founded on the *appeal to contingency*, on the *appeal to brute-force exhaustion*, and on the *appeal to Yahweh* in lieu of contingency and brute force exhaustion: essentially a modified recuperation of material covered in Book-III, Part-I.

We start with a didactic fictitious domain: the *chair domain*: similar to our jigsaw puzzle domain from Book-III: a universe of possible chairs that might come into realisation through the action of three distinct types of sortology. Thus, an arbitrary or random contingent sortology might iterate a one-legged chair as its outcome, or it might realise a two-legged chair instead. None are gravitationally viable; none can abide with the pertinent 'survival of the fittest' scheme. We might extend the range of possibilities to include the three-legged stool and the four-legged chair. These are our first two gravitationally viable 'survival-fit' possibilities. We might also add superfluent possibilities comprised of five-legged, six-legged, or n-legged chairs that, while all survival-fit, *exceed* the requirements of survival; perhaps with or without back-rests; with or without adornments... all superseding mere survival requirements due to gravity.

---

[111] *White Sea world*... or Mediterranean-centred world: This includes North Africa, the Levant, Asia Minor, and, by cultural osmosis, Central Asia and West Eurasia, as well as Europe. All came under the influence of common ideas and breakthroughs in confluence, disseminated across the White Sea over many millennia.

The chair domain, in its range of non-viable possibilities, in its survival-fitting possibilities, and in those possibilities that fit, exceed *and* supersede the stipulations of mere survival, mirrors the world of biological evolutionary potential and cosmic evolutionary potential per the Big Bang monoblock or other cosmic theory.

We reiterate the contingent *modus operandi* by imagining the operation of a quantum mechanised dice-shaker: a process of sortology vis-à-vis the stated range and domain of chair-possibilities. (For the quantum dice shaker, see Book-III, **3.26**, **Fig.3.11**). The chair selected into realisation by the quantum indeterminate sortology will be a matter of probability, contingency, happenstance, or 'accident': the sort espoused in evolutionary theory and, in *its* distinct context, even in Big Bang cosmogenesis.

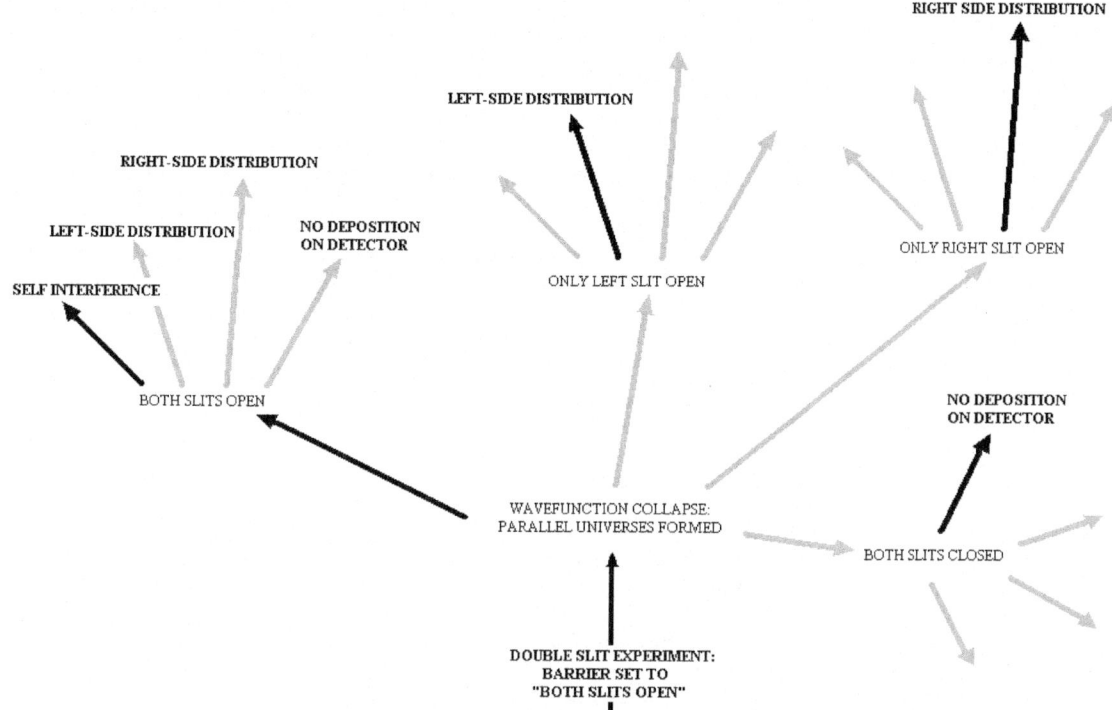

Fig. 6.01: The apagogic invalidation of Manyworlds-based informational and mnemonic nihilism
Notes: The Manyworlds process of brute force exhaustive a-causal, a-informational and a-mnemonic development process, insofar as it is inherently nihilistic, cannot furnish the consistency or otherwise of the four possible barrier-to-detector correlations and histories pertaining to the development of the two-slit experiment. Hence, when the root wavefunction generates *both slits open* as one of four Manyworlds histories, this must further develop into four subsequent histories, wherein *both slits open* correlates with *self-interference*... with *left-side distribution*... with *right-side distribution*... and with *no deposition on detector*, respectively. Only one of these will constitute a seeming consistent history, wherein *both slits open* correlates with *self-interference*: The others will be nonsense and inconsistent histories, wherein *both slits open* will correlate with, *left-side distribution*; with *right-side distribution*, and with *no deposition on detector*, respectively. In total, only one out of sixteen default histories will constitute the 'sensible history' in which 'consistency' abides (highlighted by continuous black arrows) with the rest constituting 'inconsistent' nonsense-verses. Notwithstanding a universal Alzheimer's effect, wherein all information and memory must wash out even from the one-and-only sensible 'consistent' history, Manyworlds-based information and memory nihilism must assume its own intelligibility as a non-subjective claim *and* objective state of the universe: This ought to be impossible in a nihilistic ontology, even for denizens in the one sensible history, given that information and memory cannot temporally carry-over, and it cannot have any ontologically real status *from within* the totally nihilistic Manyworlds scheme, *even in the one sensible history*. Thus, any claim about the state of any history, and even a Manyworlds-based assertion for information and memory nihilism from within *any* history, including from within the one sensible history, must presuppose backchannel to a meta-semantic hypostasis composed of *a priori* ordinances and ideas about 'consistency' and 'inconsistency', beyond what Manyworlds *in itself* could furnish, and which *cannot* be furnished by Manyworlds at all, owing to its inherent nihilism per function of its brute force exhaustive a-causal, a-mnemonic and a-informational modus operandi. The fact that we *can* furnish such judgements in our own universe implies either a Manyworlds *with* a backchannel to a meta-semantic hypostasis, or a single unique history without Manyworlds, wherein information is inherent and objective to the system. Both options undermine nihilism *and* the utility of Manyworlds to prospective informational and mnemonic nihilism.

In the first iteration, let us assume we obtain a one-legged chair. But the outcome will *not* be gravitationally viable and will tip over: a one-legged chair cannot survive. However, note that the contingent sortology does not ontologically pre-constitute the one-legged chair. Instead, contingency *presupposes* the one-legged chair *and* the other chair-possibilities. The contingent arbitrary sortology cannot create the chair-possibilities, no more than it could create or existentialise the principles of gravity, or the conditions that decide the viability or otherwise of chair-outcomes... no more than it could existentialise the very ontology and existence of itself as the process of sortology in residence.

The goal of the appeal to contingency and 'accident' in a myriad of frames asserts that outcomes are *not* caused or 'designed' but are engendered by pure chance. They are therefore posited as 'meaningless', with the attendant universe or domain declared arbitrary. In the shallowest cliché, it is proclaimed that... "everything is chance and accident, and therefore without purpose or meaning".

We reiterate that the chair-domain mirrors both biological evolution and Big Bang cosmogenesis: Therein we transplant in lieu of mere chair possibility-sets the possibilities that pertain to the evolutionary tree, if not to the set of alternative evolutionary trees rendered into nested future potentiality. We accomplish the same vis-à-vis the range of Big Bang monoblock sets for alternate natural constants, laws, and consequent potential universes. Nihilism foisted upon either domain asserts that the evolutionary outcome... or the type of universe that issued out of the Big Bang... was a matter of happenstance and 'accident'. Hence, 'purposeless and meaningless'. And if we add to both the erasure hypothesis from the first variant of memory and information-nihilism from Book-III **3.06**, we obtain full process nihilism in the context of both evolution and cosmogenesis.

However, the contingent sortology issuing the 'accidental' outcomes, and the evolutionary or cosmic possibilities generated via similar 'accident', *must* presuppose the ontology and pre-existence of bio-evolutionary and cosmic possibilities *and* the ordinances pertaining to their respective viability or 'survival', or even presuppose the attendant excess and supersession of survival. The contingent sortology as 'accident' cannot engender or explain the ontology of the pre-existent possibility and potentiality-sets, even if it subsequently engenders these arbitrarily or by default. The contingent sortology can only exploit the *a priori* possibilities, even if on a haphazard contingent basis. And it makes not the slightest difference when we employ a full indeterminate sortology to enhance 'accident' per objective quantum indeterminacy (Book-III **3.32**); notwithstanding that objective indeterminacy is *not* inimical to pattern, to information, or to memory, nor to evolutionary and cosmic possibility... but an aid to these: (see Book-III **3.26** to **3.30**). However, for didactic purposes, we ignore these insights and treat randomness and chance in the way nihilists treat it; as if inimical to order and information.

It is precisely in the presupposition of the ontology of possibilities and principles entailed in the nihilistic appeal to contingency and 'accident' that we garner the dishonest use of the backchannel... especially when the appeal to 'accident' often fails to pre-declare what it presupposes, and equally fails to offer *any* explanation or account of the ontology and existence of the possibilities and attendant principles or laws it backchannels into the fray. These are all *a priori* the 'accidental' sortology; not ontologically furnished by that sortology. Indeed, the contingent sortology employed fails to account for the ontology of itself.

Again, it is as if, having discovered the light bulb, we thought we could explain away and dismiss the ontology of electromagnetic phenomena, wholly presupposed to the light bulb and *not not* negated or annihilated by it.

*

The same apagogic dishonest approach is entailed in the nihilistic appeal to brute-force exhaustive sortology in lieu of contingency or 'accident'. The most egregious form of this was exemplified in the prospective nihilistic exploitation of the Manyworlds approach usurped in Book-III **3.12** to **3.18**. Indeed, **Fig.6.01** (preceding page) is a reiteration highlighting the problematic nature of Manyworlds vis-à-vis memory and information perdurance and consistency.

In the brute-force exhaustive approach, contingency is thrown aside, and *all* chair-possibilities are realised... *all* evolutionary possibilities are realised... supposedly in alternative universes or in Manyworlds form. And *all* possible universes implicated by the Big Bang monoblock of constants, laws and futures are subsequently issued forth via the exhaustive a-causal and information-free process central to Manyworlds.

Again, as with contingent sortology, the brute-force exhaustive appeal must also presuppose the ontology and existence of the range of chair-possibilities... of the evolutionary possibilities... of the cosmic Big-Bang possibilities... and their respective ordinances of viability and survival... and their definition as either sensible or non-sense universes... or even their survival-exceeding and superseding characteristics. The brute-force exhaustive sortology so espoused cannot engender the ontology and pre-existence it presupposes, nor furnish the ontology of itself. It can only employ the possibilities, albeit on a possibility-exhaustive form, in contrast to the merely contingent 'accident' uni-history form.

Again, in the brute-force exhaustive approach, the attendant nihilism engages in the usual backchannel dishonesty. It does so in the instant it fails to declare *any* explanation or account of the ontology and pre-existence of the possibilities and attendant principles and laws that it presupposes, exhausts into realisation, yet supposedly annihilates... especially in the context of multiverse cosmogenesis, and most certainly in the prospective abuse of Manyworlds in information and mnemonic nihilism (see **Fig.6.01** above and Book-III **3.12** to **3.18**).

Contingent and brute-force appeals in our chair domain, possibly in evolution, and especially in multiverse cosmogenesis, are born out of understandable anti-theistic anti-religious ire. These secular nihilistic pseudo-cosmopolitan appeals are employed with a view to obviate possibility and principle so as to obviate the possibility of attendant meaning and purpose... all in order to obviate the possibility for God. In this, the pseud-cosmopolitan shares the same false assumption espoused by the theist: namely, the false assumption that meaning and purpose in nature *must* imply God, and vice-versa. The theist tends to argue for God from the evidence of order and purpose. The pseudo-cosmopolitan employs contingency or brute-force exhaustion of the possibilities in order to obviate the 'evidence' for order and purpose, in order to obviate God and undermine religion. Yet, the want to circumvent God is void: When we place nous conflated to God in lieu of either contingent or brute-force exhaustive sortology, this also culminates into nihilism… of a more profound form: Yahweh will not brook resort even to a backchannel to the meta-semantic hypostasis vis-à-vis nous super-conflated to God. Historically, in the White Sea world, the two-millennia old insight as to why resort to nous conflated to God must culminate into profound nihilism was espoused in Plato's *Euthyphro* and in the dilemma of that name. Thus, we arrive at the most insidious form of nihilism: *nihilism from the appeal to Yahweh.*

*

In the first iteration of the nihilistic appeal, Yahweh plays at randomness. Therein, we simply replace a-noetic contingent or brute-force exhaustive sortology vis-à-vis the chair domain, or the evolutionary domain, or the cosmic domain, with nous conflated to God. In the first instance, God selects one chair-possibility from the myriad of possibilities available, but does so purely on the basis of arbitrary fiat. Hence, no due divine attention is paid to the possibility or its attendant principles arbitrarily selected. Although the sortology involves conscious participation and effective noetic causal freedom (now rendered divine, but no less arbitrary) there is no qualitative judgement or aretaic sortology applied to what was arbitrarily chosen into realisation: akin to selecting a random card from a shuffled pack, without any consideration to what was selected. Hence, when God plays dice, the *qualitative superficiality problem* abides. One may as well replace Yahweh with an a-noetic fruit machine, or even our quantum mechanised dice shaker from Book-III. Note that Yahweh merely presupposes the possibilities subject to His selection by arbitrary fiat:)(See implications below).

The divine deference from qualitative judgement also engenders the *arbitrary selection problem*: Divine arbitrary fiat reduces the 'choice' between the possibilities to mere trivial alternatives. This is not because the possibilities are comparatively trivial in themselves, but solely because of the divine deference of judgement. The outcome may or may not be viable, and may or may not be both viable *and* supersede mere survival. Again, Yahweh's mere arbitrary fiat could not furnish the possibilities and principles, but could only presupposes these, at least in its use of the arbitrary and contingent *modus operandi*.

The question of whether Yahweh could create the possibilities and principles *ex-nihilo* prior to the subsequent divine arbitrary fiat… and the implications from such a contention per the *Euthyphro* dilemma… will be explored below.

*

In the second iteration, Yahweh plays at brute-force exhaustive selection over possibilities pertaining to the chair domain, or the evolutionary domain, or to the comic set. Hence, *all* possibilities are realised. Again, no due attention is paid to the possibilities or principles, and divine qualitative judgement is avoided, even if divine consciousness participates in lieu of a-noetic merely mechanical brute-force exhaustion. Thus, the exhaustion of possibilities is effectively arbitrary and, again, both qualitative superficiality *and* default arbitrary selection problem (now-exhaustive) abides. Again, the choices are rendered trivial, but per the exhaustion of possibilities in abnegation of judgement rather than by divine random selection. Again, the exhausted outcomes may or may not be viable, or may or may not be both viable *and* supersede mere survival. Yahweh's deliberate brute-force exhaustion of the possibilities could not furnish the possibilities and principles, but merely presupposes these in its use of the brute-force *modus operandi*.

However, the third iteration involves God's direct attentive qualitative judgement and selection from the possibilities and principles: the form of 'creation' approximating to what is espoused in religion. Therein, God pays due attention to the character of the chair chosen, or the 'evolutionary' outcome selected, or the cosmic possibilities and principles that get to abide by divine permit… discerning their viability or otherwise, or their survival-exceeding supersession and uniqueness. In the first variation of this qualitative intercession, the presupposed possibilities and principles are coeval and co-eternal with Yahweh. While divine qualitative selection is possible, and the divine judgement as to what is issued into realisation could be profound, the qualitative stature of the possibilities *in themselves*, and the integrity of attendant principles *in themselves*, are co-eternal with God, and do not originate from nor are existentialised by God. That is, the possibilities and principles are presupposed to God and to whatever sortology He enacts over these, in the same sense that a light bulb is coeval to electromagnetic phenomena, but does not furnish the coeval ontology of electromagnetic phenomena.

On the other hand, in the second variant of qualitative intercession, God decides what constitutes the Good: God does not presuppose the possibilities and principles, but creates and existentialises evolutionary and cosmic potentials and ordinances, *ex nihilo*… and does so on the basis of pure divine will, fiat and whim, *without resort to coeval eternal God-independent principle, or without principle*… conjured *ex-nihilo*… not constituting independent co-eternal sets. Otherwise the status of God as God would be rendered untenable.

The culmination is profound *absolute nihilism*.

In other words… and to reiterate the Euthyphro's dilemma in contemporary form… is a thing Good because God esteems it to be so, or does God esteem it to be Good because it *is* Good *in itself*? In the latter instance, the Good… the qualitative character of possibilities and attendant principles *in themselves*… derive from ordinances coeval to God: ontologically equal to or even superior to God. Therein, God must depend on the coeval co-eternal ontology of the Good. Thus, the evolutionary and cosmic possibilities and principles abide; non-arbitrariness abides; and the meaningful universe also abides. But, since their ontology is not existentialised by God, what ultimate need of God? On the other hand, God creates principle and possibility *ex nihilo*… culminating into profound nihilism.

*

If God is the highest principle, then He need not rely on any independent or exceeding and transcending ontology of possibilities and principles. Consequently, the evolutionary and cosmic outcomes will be worse than merely arbitrary, and absolute nihilism will

643

prevail. Alternatively, if God and the ordinance of possibilities and principles are co-eternal, with the latter either within God, or coeval with God, or 'next to' God, the outcomes will be non-arbitrary: We will obtain a non-arbitrary meaningful universe. However, in such a scenario, God is no longer God: He is subordinated to and dependent on an independent hypostasis of ordinances and principles upon which He must totally rely for His acts of 'creation': He has no control over the onticity or constitution of co-eternal possibilities and principles from which He 'creates', whether these are co-eternal *within* God or co-eternal 'next to' God.

In short, God is no longer the highest principle, but a being subordinated to architectonic possibility and principle: i.e., God is no longer God, but reduced to Plato's demiurge. Indeed, even if He did exist, He is rendered ontologically *pleonastic*.

The choices are stark: In order for the universe to be non-arbitrary and meaningful, we must either inhabit a universe in which God totally relies on an independent hypostasis of possibilities and principles… a meta-semantic hypostasis that renders God no longer God… or we must inhabit an atheistic universe, wherein non-conscious contingency or brute-force exhaustive sortology normally abide, but the possibilities, principles and outcomes thus generated are ontologically non-arbitrary and meaningful, given that, despite the claims of the pseudo-cosmopolitan, neither a-noetic contingency nor brute-force exhaustion of possibilities and principles could obviate the *a priori* ontology of the said possibilities and principles, much less render the universe nihilistic and meaningless.

If God created the hypostasis of possibility and principle *ex nihilo*, the very possibilities and principles so created would not be rendered merely illusory; simply, these could not exist. An omnipotent God is the ultimate nihilistic eraser of information, memory and meaning, unless the possibility for information, memory and meaning ontologically pre-exists God, or co-exists within God, or 'next to' God… or exists *without* God. If God could create possibility and principle *ex nihilo*, *not* based on co-eternal, eternal and non-created hypostasis, then the principles and possibilities would at best be rendered contrived, based on nothing, and God could not garner even trivial qualitative difference of the form that attends vacuous 'choice' between strawberry versus chocolate ice cream. Worse, God could not construe even a backchannel to a meta-semantic hypostasis of God-independent eternals. Indeed, resort to an *ex nihilo* creation necessarily denies the possibility of eternals and possibilities; a thing profoundly worse than qualitative superficiality and the arbitrariness problem would emerge: namely, absolute nihilism. Hence, resort, or appeal to Yahweh, constitutes a worse nihilism than resort to contingency or to brute force exhaustion of possibility and principle in the weaker atheistic nihilism. Ironic that the theist are better at nihilism than materialists-nihilist atheists. Put another way, God is no more capable of furnishing the ontology of possibility and principle, or the meta-semantic hypostasis, or even furnish its very own onticity, than is the pseudo-cosmopolitan resort to a-noetic contingency or brute-force exhaustion in the forlorn attempt to obviate the ontology of possibility and principle in hopes of obviating meaning, order and God: It turns out that, the best way to obviate meaning and order… *is* resort to God.

The only 'advantage' to nihilism that Yahweh coneys is the prospective circumvention of the backchannel fallacy and dishonesty typical of the pseudo-cosmopolitan approach: an obviation of the backchannel problem obtained through the presumption of God's creation of the very possibilities and principles, *ex nihilo*. Yet, the 'advantage' proves wholly void, as the all-or-nothing problem must again rear its head, now in a gross profound way, rendering resort to God into an absolute nihilistic move, exposing the theistic position to be even more nihilistic (albeit inadvertently so) than the positions espoused by contenders and abusers of evolutionary theory, cosmology, and anti-aretaics.

6.04: The Ontology of Possibility and Principle

Neither resort to contingency, nor resort to the exhaustion of possibilities, nor to God, can furnish or obviate the ontology of possibility and principle. Each nihilistic appeal must presuppose these, while resort to God as the *ex-nihilo* source of possibility and principle will render possibility and principle *impossible* and inexistent. The very fact that we inhabit *any* universe at all belies the existence of God, or at least a Yahwhenised God conceived in Abrahamic religious terms as omnipotent super-conflated nous.

If nous conflated to God cannot furnish the ontology of possibility and principle presupposed to evolutionary and cosmic potential, then how do we account for the ontology of possibility and principle? The following address to this question culminates into the thesis on mind-transcending Being: i.e., the '*real* God', so to speak, first adumbrated in Book-V **5.21**: *Final Atheism & Transcending Atheism.*

*

Consider the generic black box approach. Therein, we have no direct access to the processes unfolding inside the black box. The box has an external input and output port. Through the former, we transfer inputs into the box. From the subsequent outputs generated, and from the comparative changes evident in the outputs, we infer indirect knowledge about the hidden processes inside the box.

De-spatialisation compels us to apply a black box approach to causality itself: De-spatialisation and temporisation undermine contiguous 'spaceballs' form of contact-causality by denying the very possibility of 'locations' or points-of-contact entailed in generic input-operation-output schemas. With the demise of contiguity, the only resort is causality conceived as a purely temporal transition between one set of events (inputs) and successive events (outputs). The interval, the 'gap in time' between the input and output defining the state-change constitutes the real 'black box'. And the temporal transition between the input and the output, or AND-to-OR wavefunction collapse, constitutes the contiguity-circumventing 'quantum leap in time' espoused in Book-V **5.30**.

Generalised to the pertinent context, per the usual prejudgement, we expect to find a definable operation or 'gizmo' concealed inside the black box, within the 'gaps in time' that attend input-operation-output schemas of process, causality, state-change and time. However, our Alt-R contention throughout was that, in the matter of ultimate causality, the hope for *any* definable operation therein must remain forlorn, as was compelled by the answer to the question of 'hidden gears' in AND-to-OR wavefunction collapse and time at the centre of how Whiteheadian future potentiality transforms into the actuality of successive events, into state-change, and into the realised implications of transtemporal constants, physical laws and principles within the system of the Bergson-Whitehead growing block universe.

First, any putative definable operation entailed in AND-to-OR wavefunction collapse and time must be able to process the grand decoherence of nested future potentials within the total Whitehead facet. It must accomplish this to timelike infinity. Therein, any putative iterational clockable process, or 'mechanism', or computation, or 'algorithm', would crash into an insurmountable halting problem (see Book-II **2.46** to **2.47**, and Book-V **5.18** and **5.31**). This insight eliminates any possibility that we might reduce ultimate causality or the drive of time to *any* definable operation or contiguous mechanism, or 'gizmo'. Whatever the actual 'operation' that makes time possible, it is hidden from us within the black box, and from the output. But, as judged from what that hidden process must accomplish, and by elimination of what it cannot be, we immediately grasp that ultimate causality cannot constitute as a definable clockable mechanism.

Nor can consciousness or nous in lieu of impossible hidden mechanism inside the black box abide without also succumbing to the same halting problem, given the time-dependency of nous itself... dependent as nous is on ontologically primary gear-less wavefunction collapse and time for its own viability and possibility, and not itself the drive of the very time that renders it viable into realisation. (See Book-V **5.18**: the critique of 'consciousness causes collapse').

Although the black box approach furnishes certain information about what the process of causality and time cannot be, it cannot furnish what the process *is*, save that it is non-definable and *ineffable*. The same assertion can be made about the generation of quantum indeterminate outcomes from AND-to-OR wavefunction collapse and time. The attempt to garner an effective 'black box' hidden gears that supposedly secretly orchestrates surface quantum indeterminism must also fail, and for similar reasons that the want of 'hidden gears' in the drive of time also forlorn (see Book-III **3.32**). Again, nor can nous in lieu of gears deliver any solution to quantum indeterminism, given the qualitative superficiality and arbitrariness problem from the attempt to assign to nous the role of the secret orchestrator of quantum indeterminism. (See Book-V **5.18**).

Moreover, the character of physical law was shown to be dualistic, extant and transtemporal vis-à-vis the superlative growing block Bergson-Whitehead amalgam, as well as extant the process of AND-to-OR time itself (see Book-III **3.33**, and Book-V **5.33**). Indeed, physical law imbues the Whitehead facet of future potentiality to the infinite future: It does so via emanation *from* the infinitely removed future and *from* a hypostasis that supersedes the total growing block system, and even supersedes time itself: (See again **3.33**).

Let us inhere the ineffable drive of time, the gear-less orchestration of quantum indeterminism, and the enforcement of time-asymmetric abstract-dualistic physical law, if not the transtemporal meta-semantic hypostasis of law and principle itself, into the concealing 'black box'. But 'where' is this black box? Where can we locate the very thing concealed within the gaps in time from which ultimate ineffable causality operates?

To address this question, we employ the *inverted black box*: Therein, we turn our black box 'inside out' or 'sight-side in', so to speak: The 'inside' becomes the domain of Bergsonian past-memory, co-joined to the most recent OR-form realisations of the event horizon, co joined to the monoblock of Whiteheadian future possibilities imbued with emanated physical law. In short, by inverting the black box, we place the growing block system 'inside'. But this 'inside' must also incorporate the noosphere (potential and realised) to which inhere realised and prospective mind-world relations: This does not alter the assertion that the growing block system is subsumed to nous and to the noosphere per the case for the en-worlded mind in Book-V **5.56**.

By inverting the black box, the 'outside' now transforms into the hypostasis of ineffable non-clockable causality; the source of gear-less sortology; the source of abstract transtemporal law and principle; the source of future possibility and of nous itself 'inside' the inverted black box; so subsuming all of memory, possibility, time and nous, while ontologically exceeding and superseding memory, possibility, time and nous.

Note that, just as the process of AND-to-OR collapse of Whiteheadian potentiality into Bergsonian memory is furnished by ineffable causality belonging to the grander hypostasis that transcends memory, possibility, and time... this truism also abides with respect to the AND-to-OR collapse of realised nous, which must collapse out of the future-potential noosphere. In other words, we reiterate that nous is time-dependent: i.e., the ineffable causality driving time itself also renders possible mind and consciousness. As argued in Book-V **5.18** and in the critique of 'consciousness causes collapse', it is not consciousness or mind that drives time, but it is time that drives consciousness: In other words, ineffable causality and time collapses potential nous into actualised nous, and renders consciousness and mind-world causality possible, even while the ineffable causality that makes nous possible exceeds, supersedes and transcends nous.

Again, via the inverted black box, the ineffable causality that transcends nous is subsumed to the grander hypostasis that also subsumes the noosphere, in turn partly constituted as future-potential consciousness and mind... as surely as it subsumes physical law and principle *and* the monoblock of possibilities that elaborate to the infinite future.

The same *exceeding* hypostasis does not itself arise from any monoblock of possibilities: It subsumes those possibilities. The same exceeding hypostasis does not itself decay out of AND-to-OR collapse and time: It *subsumes* and transcends AND-to-OR collapse and time: i.e., *Existence* and *Being* ontologically precedes time. Thus, the same hypostasis does not arise from time-dependent nous, but subsumes nous and the time-process that resolves mind and consciousness out of future noetic potentiality. The same hypostasis does not arise from the noosphere, but subsumes the noosphere. The noosphere arises from the ineffable hypostasis. But the ineffable hypostasis *does not arise from* the noosphere. Thus, nous does not constitute primary ontology: The primary ontology *is* the ineffable hypostases.

The ineffable hypostasis is the *panentheon* to the growing block system of memory, possibility, nous, time, law and principle... in that it ontologically subsumes and includes all, while ontologically supersedes and exceeds all, including nous; including nous conflated to God: It incorporates and existentialises all of these, while it is *none* of these, and exceeds, supersedes and transcends them.

*

Given that existence precedes essence in the ontological sense more than in the mere temporal sense, in the language of neutral ontology, the ultimate hypostasis is *existence in itself*: the ineffable *Being* that makes possible all things *without* reducing to those things. This is the *real God*, but it is wholly unlike the God of generic theism: The real God exceeds, supersedes and transcends nous super-conflated to God. God recuperated as existence itself, or as ineffable *Being*, circumvents the Euthyphro dilemma, and existentialises without *ex nihilo* 'creation' the co-eternal monoblock of evolutionary, cosmic *and* noetic possibility, as much as it existentialises transtemporal law and principle rendered into plausible Platonic forms, Kantian categories, or the terms of some other ontology of principle.

Or so it can be argued.

As an alternative, one could drop 'God' and assert for pure neutral ontology or 'neutral monism'; fashioning for an alternative form of non-nihilistic post-materialist atheism. In either case, we could not secure religion upon ineffable Being. Indeed, the way to Being is not through fideistic theistic nihilism, but through the unfolding maze and labyrinth of memory, possibility, time, nous and principle: a process perhaps best analogised in the Situationist *Dérive*[112] recuperated into an unexpected prospective *new cosmism*.

*

However we comport to *Being*, the non-arbitrariness thesis and the inferred meaningful universe, must abide: Neither the nihilistic appeal to contingency, nor to brute exhaustion of possibilities, nor even to *Being* contrived into Yahweh, or 'God as mind', can save nihilism. This also applies to *Brahmanic nihilism*, insofar as a variant of Brahmanism in the Indian sub-continent claims *Atma*, or nous in general, to be the same as Brahman or *Being*. Therein, one might also find a touch of contemporary indulgence in 'consciousness causes collapse', and, with it, the declaration of world as the *maya* or illusion of *Atma*... with inevitable nihilism.

But *Atma is not Brahman*, save in the most general truism that *all* is Being, and all belong to its Panentheon. Yet, this cannot obviate the understanding that there is more to Being than nous, just as there is more to Being than memory, possibility, time and principle. Being ontologically subsumes but *exceeds* and supersedes all existents; existentialising these without *ex nihilo* creation, *without* identification with what it existentialises. Being is *not* Atma: Being existentialises Atma... and exceeds it into ineffability beyond Atma.

6.05: The Question of 'Consciousness in Memory'

Insofar as the system of de-spatialised and temporised universe and brains arises to and 'within' absolute time-like mind and consciousness... insofar as it is not possible to obtain absolute time-like nous from the system of world and brains subject to the relativity of simultaneity... it follows that nous cannot be identified with, reduced to, much less be 'excreted' by the domain of world and brains. Hence, the revival of Cartesian dualism.

For brevity, we will put aside similar dualistic implications from the conscry of infinities, superlatives and metalatives, or from other myriad hard problems of consciousness, even though these independently lend themselves to the same Cartesian conclusions (see Book-I **1.45** and **1.54**, and Book-V, Part-IV: *the General Theory of Mind-World*). Instead, we will concentrate only on Cartesian dualism inexorable upon the relativity of simultaneity inherent to the brain-world, versus distinct absolute time-like monadic nous.

Moreover, from the viewpoint afforded by the en-worlded mind attending our Anti-Copernican revolution, the consummate fact that the de-spatialised temporised universe and brains retained in Bergsonian memory must arise to and *within* nous, and arise therein as the complex past apprehended by consciousness from the vantage of a temporally succeeding noetic moment, we obtain the radical conclusions in favour of 'memory in consciousness' or *noumnemonae*: (See Book-VI **4.41**, and Book-V **5.53**). But can we also obtain its inverse? That is, having obtained 'memory in consciousness', can we also obtain 'consciousness in memory' and *its* retention as ineradicable memory? The answer engenders radical implications and conclusions that exceed those obtained from the en-worlded mind, and the contention for the ontic reality of the temporal noetic monad, radical as this is in its own right.

If 'consciousness in memory' is possible, is it a function of Bergsonian memory? Is mind and consciousness retained within the mnemonae belonging to the Bergson facet? Is it generated by the Bergson facet? The answer is *no* on all counts, and this was clarified in Book-V **5.48**. The Bergson facet certainly contains memory, but it is memory comprised of a-noetic stereotemporal information: i.e., 'the big red garage door, Potters Bar-1973' *in itself*, or independent of the temporally subsequent moment of its consciousness: i.e., a memory from the author's early childhood.

Bergsonian memory includes the past states of the brain, constituted as a perduring 'worm' within the Bergson facet. This is *also* a-noetic per the relativity of simultaneity that operates across the brain in contradistinction to the temporally subsequent moment of absolute time-like monadic consciousness. The brain constitutes a distribution of past events, or a collection of Bergsonian wavefunctions in memory pertinent to brains. This is in contrast to the leading event horizon and its 'materialised' brain. But the

---

[112] *To drift...*"let themselves be drawn by the attractions of the terrain and the encounters they find there" ... and, in our mnemonic context, incorporate this into timeless memory. The 'terrain' is the universe of possibility. Since we do not pre-know the destination, nor even know Being itself, even while it posits itself as the final destination, it follows that we can only 'drift': Though not haphazardly, and always with ever-growing understanding.

646

Bergsonian brain does not contain states of nous, given that these, and the 'worm' thus, cannot furnish any absolute time-like attribute peculiar to the sought noetic moment.

Certainly, Bergsonian memory as a-noetic memory must include the brain itself. But both the a-noetic world-in-memory *and* the a-noetic brain must later subsume to consciousness. In doing so, both transform into 'memory in consciousness' when, following a time-delay, the Bergsonian world and brain arises to and *within* absolute time-like nous, so constituting the en-worlded mind.

Obviously, the big red garage door from 1973 does not itself exhibit one's consciousness: After a delayed choice time-interval, the big red garage door came into one's then-consciousness back in 1973. The then-brain did not generate one's then-consciousness, for reasons stated and fully explored in Book-V in the general and intermediate mind-world theories.

Furthermore, we reiterate the non-clockability thesis about absolute time-like nous (See Book-V **5.46**). Mind and consciousness are not clockable or identifiable as Bergsonian past states, or even as Whiteheadian future-potential events of either the future-potential world or of the brain. Indeed, neither mind nor consciousness constitutes any realised or potential event exhibited by the Bergson-Whitehead amalgam: Both mind and consciousness are sidereal to and dualistic-extant vis-à-vis the Bergson-Whitehead amalgam.

In short, as part of the en-worlded mind, Bergsonian a-noetic memory (the big red garage door-1973) arises to and within consciousness, and the consciousness of that a-noetic world is *not* retained in, *not* preserved in, nor clocked within, nor as, Bergsonian memory or Whiteheadian potentiality. Thus, 'memory in consciousness' cannot arise from, or be captured as, or be constituted as, Bergsonian memory.

*

However, **'memory in consciousness' is inconceivable *unless* consciousness itself is retained in memory, albeit in the inverse... as 'consciousness in memory'**... even while the latter cannot be retained in or as Bergsonian memory. Thus, one remembers being conscious of the big red garage door-1973... or conscious of having been conscious of it in 1973. One has retained this within a succeeding and leading present-consciousness as a memory of one's then-consciousness. Indeed, to have retained the big red garage door but *not* one's then-consciousness of it, recollected in the now-consciousness of both in retrospect, would be tantamount to retaining nothing.

Obviously, the big red garage door from 1973 did not itself exhibit one's consciousness of it, but it came into the then-consciousness: One could not have a memory of the big red garage door *unless* one also retained memory of the then-consciousness of the big red garage door. The retention of this 'consciousness in memory', as memory in its own right, and as a re-collectable ontic in a temporally subsequent moment of consciousness, is the inexorable reality inherent to one's experience and consciousness of the past. Thus, 'consciousness in memory' is inherent to and inexorable from the very possibility of 'memory in consciousness' and the attendant en-worlded mind.

However, 'consciousness in memory' implies memory outside of, independent of, and parallel to a-noetic Bergsonian memory, as required by and consistent with the non-clockability thesis and the non-identifiability and non-reducibility of nous to Bergsonian memory.

The same holds for one's successive now-consciousness: i.e., the one that recalls the then-consciousness of the big red garage door, Potters Bar-1973. The temporally subsequent now-consciousness of the then-consciousness of the big red garage door is *extant* the Bergson-Whitehead amalgam of memory and future possibility. The state of then-consciousness, non-clockable then, is remembered by an equally non-clockable subsequent now-consciousness; also extant the world and brain retained in Bergsonian memory, hence extant the Bergson facet and, indeed, extant the whole Bergson-Whitehead amalgam.

In short, one's 'memory in consciousness' of the big red garage door, incorporative of the memory of one's then-consciousness... as a 'consciousness in memory'... is *not* produced by, retained in, or clockable as the Bergsonian brain and world. Hence, both the consciousness-then and its subsequent recollection and retention in now-consciousness, are extant the brain and world; part of a parallel order of memory and nous. Recall that this parallel order constitutes the *noosphere*. (See Book-V **5.50**).

Hence, 'consciousness in memory' is not retained within the Bergson facet, but it *is* retained within the parallel extant noosphere; *within* a noetic ontology 'over and above' the subsumed and incorporated a-noetic Bergson-Whitehead amalgam.

*

Given the case for 'consciousness in memory'... which must also constitute as memory extant and distinct from, yet incorporative and subsuming of Bergsonian memory... what happens to 'consciousness in memory' when the brain faces death and dissolution? Succinctly, what happens when the critical milieu synonymous with brains (See Book-V Part-V **5.72** to **5.73**) ceases to function and falls apart upon death; no longer capable of generating an open Whitehead projection... no longer able to generate the attendant Whiteheadian equalised possibility plateau that otherwise permits active mode mind and its black box delayed choice **switch-x** tripping to control its brain to affect the world? (See Book-V **5.78**).

The answer is simple: Upon the brain's death and dissolution, the mind is no longer able to causally affect its brain or world. That is, absolute time-like nous is no longer capable of exercising any causal power over the domain of a brain and world subject to the relativity of simultaneity... although it could conceivably continue to exercise witnessing consciousness of both in the passive mode.

What does any of this imply to 'consciousness in memory' and to the total retention of one's biographic history and identity in consciousness, itself retained as part of 'consciousness in memory' within the parallel noosphere, and in extant form vis-à-vis the growing block Bergson-Whitehead amalgam and the brain therein? Given the non-clockability of nous vis-à-vis the brain, and given the parallel 'consciousness in memory' within the noosphere extant Bergsonian memory... it obviously follows that the functional

647

cessation and dissolution of the brain *cannot* imply the dissolution or annihilation of 'consciousness in memory' and of the biographic identity therein, much less lead to the erasure of 'memory in consciousness' subsumed to 'consciousness in memory'.

With the death and dissolution of brains, the proverbial 'end-user' will certainly have lost its keyboard vis-à-vis the world: It will have lost causal power vis-à-vis the growing block domain. But the end-user was never the keyboard to begin with. Hence, the end-user will abide the dissolution of its brain. Nous will abide death.

In other words, 'consciousness in memory' survives the death and dissolution of its brain.

Hence, we arrive at the ultimate materialist-nihilist anathema: namely, the survival of mind and of biographic identity in 'consciousness in memory' vis-à-vis the death and dissolution of the brain.

*

The survival of nous is the inexorable conclusion from the form of revived Cartesian dualism that emerged seamlessly and automatically from de-spatialisation and temporisation alone, and from the attendant non-clockability thesis, as well as from contentions beside.

However, it turns out that the critical concern is less the certain survival of consciousness, of personal memory, and of biographical identity upon death: In an awesome and terrifying universe… non-arbitrary, meaningful, and bizarrely purposeful though it is… the crucial issue is one of the quality or aretaic integrity of what has survived.

The question is not one's mere survival *per se*: Even the cockroach get to survive. Rather, it is survival… but *as what*?

*

With 'consciousness in memory' secured, so is the continuation of consciousness and biographic identity. But how significant is the survival of nous in the framework of the other great questions of existence?

Although it is likely to garner the greatest interest, the continuation of nous and identity *in itself* turns out to be our most trivial discovery. Remarkably, the universe and existence would *ultimately* be no less meaningful in the absence of the continuity of consciousness than in its now-certain survival.

Consider that the meaningfulness of the universe strictly depends on the satisfaction of two demands. The first is furnished by the non-arbitrariness thesis, which renders the sum of possibilities meaningful, notwithstanding the ontic reality of randomness, indeterminacy, contingency and happenstance, among other consideration (see **6.03** and **6.04**). The sum of possibilities as non-arbitrary and as ultimately meaningful is intrinsic to the sum of possibilities, ontologically independent of noetic agency and its causal freedom. Yet, the full potentiality of its meaningfulness requires the inclusion of noetic witness and agency, secured as part of ontically real nous and attendant Cartesian revival. On the other hand, noetic agency, as free agency, wholly depends on objective non-arbitrariness: This furnishes to noetic agency aretaic potentiality (see **6.02**). Thus, if we retain the non-arbitrariness thesis *and* noetic aretaic agency, but hypothetically exclude the continuation and survival of nous… or imagined a universe in which survival after death was not possible… we could yet retain the meaningful universe, albeit *without* the survival of consciousness.

On the other hand, if we retain both non-arbitrariness *and* noetic agency, *and* include the survival of consciousness, we yet again retain the meaningful universe and existence. And yet, the survival of consciousness proves neither necessary nor indispensable to the ontology of the meaningful universe, given meaningfulness could yet abide in the absence of consciousness-survival.

Moreover, the survival of consciousness cannot constitute *the* decisive axiom to moral possibility, both in the sense that it is not ultimately indispensable to meaningfulness and to aretaics, and in the sense that, *in itself,* it has nothing to contribute to the independent ontology of meaningfulness or aretaics, but its significance presupposes the independent ontology of meaningfulness and aretaics. Thus, consider a nihilistic universe in which non-arbitrariness does not abide and noetic agency, even though ontically viable, is rendered vacuous through the obviation of aretaic potentiality. Further imagine that the survival of consciousness yet remained, even in the midst of such dismal nihilistic dazzle. Therein, the significance of the survival of consciousness would be rendered totally superfluous, and, *in itself,* survival of nous could not furnish or rescue either meaningfulness or aretaic possibility. In short, *in itself*, the onticity of consciousness in memory is aretaically neutral and superfluous. If the continuation of consciousness is to make a profound contribution to the meaningful universe and existence, it can only do so in the context of and full dependence upon an ontically independent *objective* non-arbitrariness, and the aretaic agency this renders possible, but which nous *in itself* could not accomplish or furnish.

*

If the cockroach had the capacity for speech, it would declare that it would not like to be stepped on, but that, should the dreaded moment come its way, it would hope for survival. What holds true for the cockroach certainly holds true for the cockroach in us all. It is not an exaggeration to state that almost all religion is obsessively centred around a cockroach-like hope of promissory circumvention and survival of inescapable death. Indeed, in this matter, Christianity is probably the most egregious. The hysterical centrality of death in religion renders religion at least unintentionally aretaically compromised, save by the grace of accidental aretaics that might come its way from the incorporation of ancient sophrosyne. But aretaics could not be furnished by the mere want of survival, for reasons previously argued. Tendency toward true aretaics could only be secured upon non-arbitrariness and from the potentiality for aretaic agency, neither of which is dependent on the survival of consciousness, and wherein both would abide even in the absence of survival. Again, if the continuation of consciousness is to contribute to the meaningful universe and existence, it can only do so in the context of an independent and indispensable ontology of non-arbitrariness and attendant prospective aretaic agency.

Yet, the survival of death is not a matter of speculation: 'Consciousness in memory' most certainly abides. Thus, the survival and continuation of consciousness also abides. Yet, of the three great concerns of existence, survival *in itself* yet remains trivial and, in principle, almost wholly dispensable. Either aretaic excellence constitutes *the* end-in-itself'… or it is nothing more than a mere means, and, consequently, of no final importance. Given non-arbitrariness and the attendant viability of aretaic agency, excellence most certainly *is* the end-in-itself, and would continue to be so even in the absence of the survival of consciousness. Indeed, the choice to resist the Nazi regime tends to excellence, even in the absence of survival. Virtue would resist it, even in the absence of survival. To resist, but only upon the promise of survival, is to degenerate aretaics into the existential equivalent of a transactional market relation.

Where religion tends to the inadvertently diabolical is in its contrived promise of survival and continuation, but conditional to concession to a contrived pseudo-aretaic fideistic 'excellence': an excellence that rarely extends beyond 'fideism *itself* as excellence', based on a transactional arrangement for survival in exchange for faith *in itself*: a market-like relation espousing payment, or simply *corruption*: perhaps most characteristic of Christianity. In other words, in religions, perhaps most especially within Christianity, faith in the arbitrary claims of the creed is the 'excellence' for which 'payment' is survival after death, while excellence *in itself* independent of contrived 'faith in the creed' or even independent of survival, is rejected as *the* end-in-itself, save per any moderation from contingent accretion from pre-Christian Hellenic aretaics and sophrosyne, and wherein *mere* continuation after death transforms into *the* end in itself, even though it is the Good, independent of and not conditional upon survival, which constitutes the only real goal... with or without survival after death. This aretaic corruption inherent to religion is necessarily born out of understandable desperation and hysteria; born out of the essential cockroach resident within us all.

Yet, consciousness in memory abides, as does attendant survival and continuation of consciousness... even *without* religion. Indeed, even the cockroach get to survive. Hence, the continuation of consciousness is not secured through any specific creed, and its reality does not require *any* specific creed or fideism. Survival is not a privilege exclusive to any set of believers, or to Christianity alone or at all... or to *any* other creed alone or at all. Instead, the survival and continuation of consciousness emerges out of the very structure of discoverable reality that attends memory, possibility, time, nous, principle and Being: Its discovery is secured through an a-fideistic secular-philosophic naturalistic path.

*

What is the role of the continuation of consciousness in the non-arbitrary meaningful universe, if it is not ultimately indispensable to the meaningful universe or to aretaic potentiality? As was previously asserted, the survival and continuation of consciousness could contribute to a meaningful universe, but *only* on the basis of indispensable non-arbitrariness and attendant potential aretaic agency. Since the latter two are certainly secure, we must reiterate the question: What is the role of the continuation of consciousness in our consummate non-arbitrary meaningful universe, if it is not ultimately indispensable to the meaningful universe or to aretaic potentiality?

6.06: Metic

While the breakthroughs hitherto garnered might constitute a new basis for the reformation of the human condition, we have been here before. In the White Sea world, over two-thousand years ago, the Platonic corpus alone constituted an astonishing unrivalled development. Among many insights, Plato's form-based inference for the survival and continuation of nous in the *Phaedo* and elsewhere did not merely come close to the truth: it fulfilled it... a fact obscured by subsequent accretion of nominalist, materialist and nihilistic cynicism. Yet, in the classical world, the Platonic corpus enjoyed little traction, save among a numerically inconsiderable section of the-then tiny minority of literate society. The Platonic corpus and its sophrosyne barely survived the Abrahamic religions of the White Sea world and was predictably shackled by the crude ontology and pseudo-aretaics of these Legacy creeds.

A similar fate befell perennial philosophy and esoterics in the Indian sub-continent and elsewhere, despite remarkable intuitions achieved, such as the notion of the collective unitary nous to which we are all tributaries; of a form perhaps more ambitious than the Gnostic, Platonic, and Neo-Platonic currents of the White Sea world.

One could argue that, in the absence of the printing press, which later rendered mass literacy possible, the Platonic corpus, or *any* perennial philosophy or esoterics then and since, could not have garnered any plurality. Yet, the printing press and consequent full mass literacy have been with us for at least a century: we have yet to realise reform of the human condition. Indeed, few could have foreseen the historic suzerainty of the nominalist, materialist and nihilist accretion, only augmented by the inadvertent nihilism inherent to Legacy religious culture; sealed by the more recent formation of the necronic mechano-nihilistic hegemony that now bestrides the world.

However, the problem is not only the saliency of attendant *Hamartian culture*[113], but base political economy itself. This imposes serious delimits to the reform of humankind. As long as humankind is trapped in the condition of mass usury and *time-entrapped* in Faberist time-scarcity, and further deranged by *numisma* and *chrematistics*, the human condition is unlikely to advance, regardless of any extraordinary breakthroughs achieved in consciousness and cosmos, or even in banal technical capability.

*

The advancement of humankind into aretaic sophrosyne that might be secured through breakthroughs into nature, nous, cosmos and meaning requires exorbitant time and investment. But the consumption of time in labour, even if occasionally necessary and useful labour, but often unnecessary, wasteful, if not destructive, consumes time. Hence, Faberism and the time-entrapment problem. Consequently, human development remains minimal or virtually non-existent. Unless science, engineering and design is consciously put to the emancipation of time, either through reform or through further technical advancement (and both require a political revolution to bring into effect, admittedly implausible at the time of writing) the time-entrapment problem will remain. Of course, Hamartian

---

[113] A culture founded in *Hamartia* or error, culminating into a complex nihilistic mentality that cripples humankind. While hamartia originally meant fatal error undermining a tragic and noble hero, hamartian humanity is neither tragic nor noble, but perhaps irredeemably fractally flawed.

649

mentality cannot conceive life and existence outside of fetishised Faberism, and the plurality is hostile to emancipatory possibility. Such prospective emancipation lies outside the scope of existent Hamartian culture and fetishised Faberism.

But the Faber fetish enjoys an ally-parasite. Historically, the only way that the parasite-few could secure access to a material world in excess of their own labour was through the mass usury of the rest. The consequent host-parasite relation insipid to our Hamartian condition is necessarily made worse by both the Faberian time-entrapment of humankind *and* by insidious numismatics and chrematistics that attend it all. The host-parasite relation thus entrenched cannot but usurp the advancement of consciousness, no matter what great breakthrough we might garner about nature, nous and the meaningful universe: an *operating system problem...* one that cannot run the requisite 'program' to supersede Faber fetishism, usury, and parasitic chrematistics.

Clearly, in the said host-parasite relation, the most destructive factor is posed by *the problem numismatics and chrematistics[114]*; the history of which was only partially uncovered in David Graber's *Debt: the First Five-Thousand Years*. Numismatics began when the host-plurality was 'persuaded', originally on fictitious religious terms, and per the deep credulity of the host, into contrived debt to the parasite-few. Nothing more than a set of book-keeping entries, and secured upon blind belief, the first odious 'sin', or 'debt', or *money*, constituted the greatest historical scam, aside from the fact that the Gods demanding the debt obviously did not exist. Yet, the scam...and the credulity that constitutes its credos, mobilised host human labour into generalised peonage and into mass wilful usury in labour capitulation and labour-theft. Hence, the numismatic captivity of the host-plurality by the parasite-few[115]. From the numismatic point of view, the original 'sin', or Hamartia, was less the contrivance of fictitious debt or money, but the mass credulity of the host-plurality and its wilful self-deliverance into numismatic captivity, exploitation, dispossession, and final disposal: a point missed by the late David Graber, who did not adequately explore the requisite problem of human and mass credulity, or much less treat it as *the* problem.

Numisma as 'idea', as gross mass credulity, as 'law'... as species-being asserted through its forms of host-consciousness and parasite-consciousness... constitutes a most destructive anomia to which Faber-fetishism and usury subsume. Numisma conceives human existence and being only in nihilistic transactional host-parasite terms. Otherwise, the usury of others, and the wilful participation of the host-consciousness to be used, especially per its more recent delusion that, through 'hard work' it might itself ascend to the parasite-fold, would be rendered untenable. There can be no aretaic eunomia in the midst of such numismatic Hamartian anomia. Money is *the* testament of human species-being in its ultimate nihilistic form: No great cosmism of truth, proof, insight, or in breakthrough into the mysteries of cosmos, nous and meaning, can avail against it. Hence, at this juncture, the reform of humankind remains truly remote, even while the pre-conditions for post-usury, post-scarcity and, conceivably, even post-numisma, might well be with us, as are the minimal requisite breakthroughs in consciousness, cosmos and meaning.

With the general reform of the Hamartian anomia highly unlikely at the time of this writing, in the interim, the best we can hope for is the emergence of a breakaway *Metic* consciousness and identity, if not a Metic nation in diaspora; self-founded on the basis of breakthroughs similar to those espoused in this work: part of a consciousness revolution and new cosmism; constituting a Metic diaspora *in situ*. In the long run, such a Metic diaspora could lead to full physical exodus and full breakaway Metic civilisation, although it is difficult to imagine where on Earth such a civilisation might be established: Antarctica? The oceans? Arcologies in space?

The parable of the Jews-of-old comes to mind; ironic, given the a-religious tendency throughout this work. Perhaps more myth than history, the ancient Jews offer a ready metaphor for the essential meaning of *Metic[116]*, superseding its original Athenian meaning. The ancient Jews purportedly developed a breakaway Godhead vis-à-vis Egypt: a consciousness, identity and nation in exodus... at first *in situ*, but later culminating into full physical exodus. A contemporary Metic breakaway consciousness and identity that founds itself as a nation could constitute a similar condition; an exodus *in situ* in the midst of Hamartia, finally culminating into full physical exodus *from* Hamartia.

Granted, the metaphor has serious limitations per the fact that the ancient Jews did not in truth constitute a breakaway consciousness or identity: The mentality they espoused was of the same *oussia* as that of Egypt: i.e., contrived 'answers' to the great questions of existence in lieu of real answers: ultimately and *essentially* no different from the creed and identity in Egypt and other creeds since, even if different as night and day in the superficial surface-level.

A Metic breakaway consciousness and identity would constitute a *true* breakaway from the Hamartian anomia, insofar as its sophrosyne would be based on real answers to the central questions of existence, culminating into a *Dasein* ultimately incompatible with the Hamartian anomia in which it finds itself stranded, in opposition to the nihilistic Hamartia that bestrides the world. But cynicism anticipates that prospective Metic consciousness will more likely devolve into a colony of gentrified navel gazers, as evident in the public adoption of Legacy perennial philosophy and esoterics, and most acutely in its New Age flapdoodle form. Yet, we must strive.

---

[114] *Numisma* is money. Succinctly, it is mass credulity congealed into a law, binding the host-consciousness to contrived debt-peonage to the parasite-consciousness, all on totally contrived grounds. *Chrematistics* is the utilisation of numisma (credulity,) nominally for the amassment of exclusive property and rent, but penultimately for control-from-without of the host-plurality, ultimately for narcissistic supply and for hubris for its own sake.

[115] The core problem is numismatic credulity. Perhaps only a system of non-circulating labour vouchers in lieu of numisma, or anti-money, could resolve the problem of numismatic credulity. The idea has immediate didactic value: Vouchers could only compensate labour and, per its non-circulation, structurally prohibit parasitical labour-theft, so purging the parasite, while exposing the core mass credulity behind numisma, if only in hindsight: Indeed, the very notion that money *must* circulate is itself born of credulity.

[116] *Metic* in the present juncture is to mean estranged breakaway-consciousness and identity, in degrees of opposition to the general Hamartian anomie.

QUANTUM MECHANICS AND MIND

<u>6.07: To What End?</u>

For obvious reasons, it is not possible to ascertain the full answer to what happens to consciousness upon the death and dissolution of the brain. Given 'consciousness in memory', we can be certain that first-person identity and biographic memory survive. We are also certain that the survival of consciousness and memory would be ultimately trivial *in itself*. But the non-arbitrary universe, and the attendant viability of aretaic agency, transforms mere survival into something more; into a thing no longer superfluous or insignificant. What the ultimate purpose of all of this remains a profound mystery. Even so, on the question of what happens to us upon death, it is possible to anticipate key penultimate possibilities, and do so with relative certainty.

The first anticipation consists in *memory tantalus*, inferred from 'memory in consciousness'. Therein, trapped in our memories, we will confront the total sum of our lives: we will face what was done and cannot now be undone. Hence, the need for a *sophrosyne of memory*: or an art of 'life and memory'… an aretaics of intent, deed and consequence in inexorable inescapable memory.

For its part, the *fantasy tantalus* anticipates the equivalent to lucid dreaming, but more radical. Therein, we will enter a lucid fantasy 'more real than real', but of our own contrivance and control. In that fantasy, we will seek to fulfil unrequited wants, desires and visions, dubious or otherwise. In tandem with memory tantalus, we will engage in futile fantasy of what might or ought to have been vis-à-vis what was; against the dissatisfactions and tragedies that had transpired. But the fantasies, though 'more real than real', can never constitute compensation or amends. Thus, the need for acerbic interrogation of ends, means and life-goals: a *sophrosyne of ends*.

The third tantalus, perhaps the most onerous, emerges out of key conclusions secured in Book-V **5.56** to **5.59**. This asserts the demise of the restrictions imposed on nous in its enmeshment with the 'reducing valve' brain and the system of memory, possibility and time. The enmeshment of nous with world and brains leads to a generalised fragmentation of nous into non-illusory first-person identities and into first-person versus third-person worlds; secured per function of the relativity of simultaneity and the temporal scattering of unique viewpoints; wherein one is removed from the other into unique non-illusory first-person biography, even though, in the now-obscured background, we are all the one-and-same nous: tributaries to the unitary noosphere.

With the death and dissolution of the brain, nous will be emancipated from this enmeshment and restriction. Hence, the *tantalus from the removal of the removes*. Therein, the first-person third-person divide will unravel. It is anticipated that the first-person memory and biography of the one, though retained intact as 'consciousness in memory', will transform into the first-person direct acquaintance with the memory and biography of the other… and vice-versa… and into full saliency of the 'one in the many' or 'the many in one'. This does not imply that non-illusory first-person identity and viewpoints dissolve. It is just that, with the removal of the removes, identification becomes with the many, and only the third-person aspect becomes untenable, thus revealing the one in the light of the other, and vice-versa.

The memory tantalus, the fantasy tantalus, and tantalus from the removal of the removes… all likely operate in tandem, suggesting the need for a sophrosyne more profound and radical than that espoused in the perennial prison philosophies from Plato's Cave to Boethius, through to the modern Viktor Frankl.

We contend the non-original notion that the universe constitutes a non-arbitrary labyrinth-cum-maze; one constituted as an aretaic prison. Through the prison of existence, aretaic agency and struggle must seek what matters and ascend… ultimately transcending the prison of existence into extraordinary apogee. Thus, the need for a *sophrosyne of ascension*.

It transpires that the universe is not meaningless 'matter in space'. It is made up of non-arbitrary memory, potentiality and principle. And, with the Anti-Copernican revolution culminating into the en-worlded mind, it turns out that we are *not* 'insignificant specs, lost in the void'. On the contrary: All memory and possibility… all fourteen billion years of cosmic history since the putative origin of the universe… arises *within* our en-worlded minds. We are not lost in the universe. But, possibly, the universe could well be lost in us: a matter of the aretaic quality and mettle of mind and agency en-worlded.

Also consider that we are the being that can apprehend infinity: We do not find infinity in sense-data or garner it from memory, possibility and time. Instead, it, and the other superlatives and metalatives, are furnished to nous from a superseding transtemporal hypostasis that subsumes and transcends sense-experience, as surely as it subsumes and transcends memory, potentiality, nous and time.

Moreover, our memories and our consciousness of such realities, and lives accordingly lived in fidelity or infamy, and as testimonies to eternity, *survive* our deaths into radical nous.

Then there is ineffable *Being*, superior to any Abrahamic or other parody… but which must also transcends memory, possibility, time, *and* nous itself. *Being* defies definition or disclosure in *any* terms. Yet, it invites pursuit.

*

Is all of this for eternity? How long before nous, trapped in the tantalus of existence, is rendered insane? To what end is all this?

From the non-arbitrariness thesis, combined to the Anti-Copernican en-worlded mind, and from much else, we can be certain that the universe and our existence *is* meaningful, even if bizarrely and perhaps terrifyingly so; obscured by the Hamartian anomie that straddles the world and cripples humanity.

The universe and existence appears to be its own excession or 'outside of the context' problem, as is life and nous… in the sense that these ought not to exist, yet *do* exist, and, as such, remain *truly* 'out of the context'.

A guess: The aim of the universe and existence is to unravel its own reason for existence. Mind and consciousness is part if this wonderful mad agenda. And, ultimately, ineffable Being subsumes all and constitutes itself as pure mystery. It beacons we pursue it… and to which all ought to ascend. Indeed, direct acquaintance with Being may well constitute the implausible promissory end-goal of life, universe and existence. May we venture.

THE AUTHOR'S JOURNEY

Born in Britain in 1969, an international relations student with aspirations to an academic career therein, life, fate and major illness, or 'Wheeler's cat', intervened to make it otherwise. Yet, even then, closer inspection of my other pursuits foreshadowed what might lie ahead: love of ontology and philosophy, love of natural science, and a fascination with the conundrum of consciousness.

In the early winter of 2000, reflecting on readings of Eccles and Penrose, it occurred to me that quantum mechanical AND-form radiation and matter waves could not be the transporters of concretised OR-form information. This, combined with the ineliminability of AND-form quantum mechanical intervals, and the implied dissociation of OR-form information from quantum mechanical radiation and matter, gave birth to nascent *information-matter dualism*, with obvious implications to brains, to memory, and the nature of mind.

In 2004, following a long hiatus and recovery from a life-threatening medical condition, my hope was to compile the notes on nascent information-matter dualism into a legible document, with a dubious idea to publish the material in a philosophy journal, or perhaps a small book. Things would not have developed beyond this limited goal but for the fact that, upon further scrutiny, information-matter dualism transformed into the first iteration of the *growing block model*: the *intermediate model spacetime*. This in turn entailed key insights, such as *nested futures*, *grand decoherence* and attendant *revival of Mack's principle*, and the *presage to a growing block model of inertia and gravitation*: all adumbrated in Book II and further developed in Book-V. Consequently, my limited goals had exploded into a major book project. As this neared completion in 2007, it occurred to me that the succession of events that we interpret as the 'motion of an object' must decay out of the growing block futurity, in turn constituted as a spatially static nested-future quantum wave; one that does not itself move. The fact that the quantum wave (the future) cannot spatially displace hastened reinspection of the then nascent growing block model and, with the aid of John Wheeler's delayed choice approach, culminated into *de-spatialisation and temporisation*, while both led to the demise of contiguous contact-causality, the *dematerialisation of physics*, and the emancipation of physicalism and naturalism from a materialist philosophy critically dependent on the ontics of space and contiguous causality.

With de-spatialisation and temporisation, the universe transformed into a thing wholly different from convention, encapsulated in the *Bergson-Whitehead amalgam*, or spacetime without space, exhibiting temporally distributed Bergsonian *abstract memory* and nascent *mnemonic mechanics* as additional components to the growing block model of inertia and gravitation. Therein, most of generic physics could be recuperated into a de-spatialised purely temporised growing block paradigm. Therein, causality transformed into an a-contiguous process of pure temporal transition: the 'quantum leap in time' furnished in Book-V; essentially no different from generic quantum transitions, but without 'space and motion'… without contiguity. Therein, spatial *atomos* was superseded by temporal *durata*.

With the process of time now at the very centre of the growing block order, the issue of what makes time happen, or what makes synonymous AND-to-OR wavefunction collapse transpire, assumed the core concern. It was principally Book-III and V that attacked notional 'hidden gears', both in wavefunction collapse *and* in the orchestration of quantum indeterminism: a finding consistent with remarkable de-spatialisation and temporisation, given that 'gears' or 'mechanism', when posited as the drive of time, presuppose forlorn spatial physics based on contiguous contact-causality. In a purely temporised universe, the succession of events, time-transitions and state-changes cannot be realised by means of now-impossible spatial contact-based mechanical means.

It required but few more steps to render physics finally fit for the Cartesian mind. The first requirement was born out of the fact that the process of *wavefunction collapse and time is a work function zero process*, even if the succession of events and state-changes it manifests are interpreted in terms of *pleonastic energy or work relations*. The second requirement was secured in Book-V in the *demise of 'consciousness causes collapse'*, and the fact that nous in lieu of forlorn hidden gears would fare no better as the drive of wavefunction collapse and time than dumb a-noetic mechanical hidden gears, and equally less as the orchestrator of quantum indeterminism. The way was now clear to the incorporation of nous to physics, with the latter recapitulated into the de-spatialised and temporised Bergson Whitehead amalgam, culminating into a Cartesian revival that far exceeded its tentative form furnished by nascent information-matter dualism, with implications conveyed in the *Philosophical Segue…* and all of it encapsulated in this long-drawn book now in the reader's possession.

*

My gratitude to the reader for purchasing this book. I hope its content will prove to be a revelation and much more. Yet, it is one among many publications in philosophy of physics, interpretation of quantum mechanics, and philosophy of consciousness. I hope that it does not fall on deaf ears, as it might.

The demands of the project itself, of work, of health and life, forestalled the desirable possibility of an academic career in dismal philosophy of mind or in nascent modern philosophy of physics. Consequently, I had to walk the path of an autodidact. Credentials may secure ready hearing, but do not necessarily culminate into ready agreement. The possibility of accord about anything, including the radical claims I have made in this book, depends on the evidential and rational integrity of the material presented, destined to stand or fall on its own terms. I am cautiously confident that, to the best of my learning and ability, the claims I have made will hold up to light and to appropriate and fair scrutiny.

Of course, I bear full responsibility for all of the errors. May these be few in number.

Emre Asena

INDEX

R

S

T

U

Printed in Great Britain
by Amazon

45161446R00381